European Commission
and the
Belarus, Russian and Ul
on Chernobyl Affairs, Er

The radiological consequences of the Chernobyl accident

**Proceedings of the first international conference
Minsk, Belarus
18 to 22 March 1996**

Editors
A. Karaoglou, G. Desmet, G. N. Kelly and H. G. Menzel

EUR 16544 EN

Published by the
EUROPEAN COMMISSION

Directorate-General XII
Science, Research and Development

B-1049 Brussels

Cataloguing data can be found at the end of this publication

Luxembourg: Office for Official Publications of the European Communities, 1996

ISBN 92-827-5248-8

Printed in Luxembourg

FIRST INTERNATIONAL CONFERENCE

The Radiological Consequences of the Chernobyl Accident

ACKNOWLEDGEMENTS

The Scientific Secretariat and the Editors wish to express their thanks to the Steering Committee, to all members of the Programme Committee, to the special guests and to all those who prepared papers and posters included in these Proceedings. Very special thanks are due to all the 32 coordinators for the splendid job they have done, for their willingness and for their ability to remain patient at all times.

This Conference could not have taken place without the help and cooperation of many people. It is impossible to mention all of them but we are aware of and grateful for their willing help. The local organisers Drs Petriaev and Sokolik have played a tremendous role in the preparation of this Conference and we cannot thank them enough.

Our thanks go also to the whole team of SCIC who were responsible for the local organisation, a task not always easy and to Mrs Scholz for her professionalism and attention to detail. We acknowledge the work of Lucien Cécille who took the first steps to organise the Conference.

The whole Radiation Protection Research Action team has been very "active" and cooperative all along the preparations. Special thanks go to all members of the administrative secretariat, to Irina Sombré for her help in dealing with all CIS participants, to Christine Mottet who developed the large mailing list and retyped many of the manuscripts and to Horst Stöbel for arranging all the official documentation. Finally, we would like to express special thanks to Danielle Larivière for her untiring, enthusiastic and efficient involvement in the whole planning and execution of the Conference.

............., Brussels-Luxembourg, 1995.

Publication no. EUR 16544 EN of the Commission of the European Communities,
Dissemination of Scientific and Technical Knowledge Unit,
Directorate-General Telecommunications,
Information Market and Exploitation of Research, Luxembourg

Contents

II.B   Exposure of the population

PART 1:   Main Papers

PART 2: Posters

II.C   Management of contaminated territories

PART 1:   Main Papers

III.   HEALTH EFFECTS FOLLOWING THE ACCIDENT

III.A   Treatment of accident victims

PART 1:   Main Papers

III.B   *Thyroid cancer in children living near Chernobyl*

PART 1:   **Main Papers**

III.C   Epidemiology of exposed populations

PART 1:   Main Papers

IV. OFF-SITE MANAGEMENT OF FUTURE ACCIDENTS

Decision support systems for the off-site management of future acccidents

THE RADIOLOGICAL CONSEQUENCES OF THE CHERNOBYL ACCIDENT

Proceedings Editors

A. Karaoglou, G. Desmet, N. Kelly & H. Menzel

Steering Committee

I.A. Kenik	Minister of EMERCOM of Belarus
V.I. Kholosha	Acting Minister of MinChernobyl of Ukraine
V.Ya. Vozniak	First Deputy Minister of EMERCOM of Russia
E. Andreta	Director of DGXII "Energy", European Commission, Brussels
J. Sinnaeve	Head of the Radiation Protection Unit of DGXII, European Commission, Brussels
K.H. Chadwick	Head of Sector "Radiation Biology and Health Effects", DGXII, European Commission, Brussels

Scientific Programme Committee

G. Desmet	DGXII, European Commission, Brussels
A. Karaoglou	DGXII, European Commission, Brussels
G.N. Kelly	DGXII, European Commission, Brussels
H. Menzel	DGXII, European Commission, Brussels
L.I. Anissimova	Ministry EMERCOM, Russia
V.G. Bariakhtar	National Academy of Sciences, Ukraine
S.T. Beliaev	Research Centre "Kurchatov Institute", Russian Federation
N.A. Kryssenko	Ministry of Health Protection, Belarus
E.P. Petriaev	Belarus State University
V.M. Ponomarenko	Ministry of Health Protection, Ukraine
B.S. Prister	Ukrainian Institute of Agricultural Radiology
A.F. Tsyb	Medical Radiological Research Centre, Russian Federation

Scientific Secretariat

A. Karaoglou	DGXII, European Commission, Brussels

Organiser

J.P. Scheins	"Conference Techniques and Organization", Joint Service Interpretation - Conferences (SCIC), European Commission, Brussels

Administrative Secretariat

H.G. Stöbel	Liaison officer with the CIS partners participating in the Chernobyl Projects, FZR-Rossendorf (Germany)
I. Sombré	DGXII, European Commission, Brussels
D. Larivière	DGXII, European Commission, Brussels
Ch. Mottet	DGXII, European Commission, Brussels

Preface

On April 26, 1986, at 01.23 Moscow time, the reactor operators lost control of the number four unit of the Chernobyl nuclear power station; the reactor went from a very low power level into a power surge, exploded and caught fire. The explosion ejected large amounts of radioactive debris into the atmosphere and the fire melted the fuel elements in the core of the reactor, releasing the more volatile radioactive fission products and the radioactive fission product noble gases such as krypton and xenon. The accident resulted in widespread radioactive contamination over large areas of Belarus, the Russian Federation and the Ukraine and seriously affected the local population.

Following the accident the European Radiation Protection Research Programme defined additional research requirements, re-oriented some existing research contracts and started some new contracts. A revision of the Radiation Protection Research Programme 1985-1989 (COM(87)332 Final) was proposed with the specific aim of launching 10 coordinated multinational projects for assessing and mitigating the short-term consequences within the Community. The studies of the 10 specific lines of research yielded a comprehensive overview of the short-term consequences in the Community and were the first step towards the evaluation of the medium-term consequences. The results from these ten research investigations have been published as a series of EUR Reports and an overview is presented in the EUR Report published in 1990 (EUR 13199).

In 1991, the Communication of the Commission to the Council (SEC(91)220 DEF, 12.02.91) outlined, as part of the continued concern with the consequences of the Chernobyl accident, the background for a collaboration between the Commission's Radiation Protection Research Programme and the "Chernobyl Centre for International Research" (CHECIR). This centre was created in the frame of an agreement for international collaboration between the All-Union Ministry for Atomic Power and Industry and the International Atomic Energy Agency. In the frame of the ""Activités complémentaires de préparation, d'accompagnement et de suivi - Collaboration with the former Soviet Union in Radiation Protection" (APAS-COSU) programme, a preparatory phase of 7 projects, dealing with radioecology and nuclear emergency management, was launched.

In order to formalise the research cooperation an "Agreement for International Collaboration on the Consequences of the Chernobyl Accident"- was signed in June 1992 between the EC by Vice-President of the Commission Mr Filippo Maria Pandolfi and the relevant ministries of the three Republics, namely Belarus by H.E. Mr Ivan A. Kenik, The Russian Federation by H.E. Dr Yuri S. Tsaturov and the Ukraine by H.E. the late Mr. Georgy A. Gotovchits.

Within the terms of this agreement a Co-ordination Board has been established to monitor the implementation with representation from the EC and each of the three republics. The Coordination Board meets twice a year and has a number of roles including oversight of the programme, confirmation of projects, coordinators and participating institutes. The projects are implemented by formation of groups from the CIS and the CEC. They are realised as partnerships between Eastern and Western research institutions and hospitals. For each project approved by the Coordination Board, scientific coordinators are appointed in the EC and the CIS. Their roles are to coordinate the work of the various institutes in the EC and CIS, respectively, and jointly to organise the work of the projects to a successful completion and to report on their results.

The aims of this scientific collaboration have been to increase the possibilities for training of CIS scientists; provide financial support to CIS institutes for allocating staff to the joint projects; introduce new technology and train medical specialists; improve the local infrastructure and create a regional research facility in each of the three republics Belarus, Russia and Ukraine. Thus, the budget of this programme includes local assistance, equipment and training for the CIS. The financial resources provided by the Commission for the whole programme (1991-1995) amounted to 20 million ECU and the total budget together with the contribution of those EC institutes participating in the programme is of the order of 30 million ECU. This budget has been used to support 16 research projects, involving some 80 research groups in Western Europe and around 120 research groups within the three Republics.

The collaborative programme is composed of three different areas:

- radioactive contamination of the environment, its assessment and mitigation (8 projects);
- health effects, evaluation and treatment (6 projects);
- off-site emergency management (2 projects).

A great deal of experience has been gained from this scientific collaboration with the CIS. Numerous difficulties were encountered and a lot of effort and initiative was needed when the programme was launched. These difficulties, however, have been largely overcome in the course of the programme certainly also thanks to cooperation with colleagues from the "Bundesamt für Strahlenschutz" in Berlin and the "Forschungszentrum Rossendorf" in Dresden. Personal relationships between scientists and effective working relationships between the institutes in the EC and CIS have developed creating the necessary infrastructure for successful collaboration. The overall objectives of the joint programme were to enable the scientists to improve their understanding of knowledge of the health and environmental impact of radioactive contamination and of how it can be reduced; to assist those in the three republics responsible for evaluating and mitigating the consequences of the accident; to help the governments of the affected republics to form a rational policy in response to the accident and to alleviate the suffering of all the people affected by the accident.

E. Andreta
Director DG XII.F
Energy

I.A. Kenik
Minister of EMERCOM
Belarus

J. Sinnaeve
Head of Unit DG XII.F.6
Radiation Protection research

V.I. Kholosha
Acting Minister of
MinChernobyl
Ukraine

V.A. Vladimirov
Vice-Minister of EMERCOM
Russia

Introduction

Five main objectives were assigned to the EC/CIS scientific collaborative programme:

- improvement of the knowledge of the relationship between doses and radiation-induced health effects;

- updating of the arrangements for off-site emergency management response (short- and medium term) in the event of a future nuclear accident;

- assisting the relevant CIS Ministries to alleviate the consequences of the Chernobyl accident, in particular in the field of restoration of contaminated territories;

- elaboration of a scientific basis to define the content of Community assistance programmes;

- updating of the local technical infrastructure, and implementation of a large programme of exchange of scientists between both Communities.

The topics addressed during the Conference mainly reflect the content of the joint collaborative programme:
1. Environmental transfer and decontamination;
2. Risk assessment and management;
3. Health related issues including dosimetry.

Conference objective

The main aims of the Conference are to present the major achievements of the joint EC/CIS collaborative research programme (1992-1995) on the consequences of the Chernobyl accident, and to promote an objective evaluation of them by the international scientific community. The Conference is taking place close to the 10th anniversary of the accident and we hope it will contribute to more objective communication of the health and environmental consequences of the Chernobyl accident, and how these may be mitigated in future.

The Conference is expected to be an important milestone in the series of meetings which will take place internationally around the 10th anniversary of the nuclear accident.

It also provides a major opportunity for all participants to become acquainted with software developed within the framework of the collaborative programme, namely:

- Geographical Information Systems displaying contamination levels and dose-commitments;
- Decision Support Systems for the management of contaminated territories;
- Decision Support Systems for off-site emergency management (RODOS), etc.

Contributors

The predominant part of the conference is made up of papers from scientists involved in the collaborative programme which is part A of each session in the Proceedings. Poster sessions enabling contributors - both internal and external - to report in more detail on their work is also organized. Full papers of the posters are in part B of each session in the publication.

I. THE RADIOACTIVE CONTAMINATION
AND
REMEDIAL MEASURES

The Atlas of Caesium-137 Contamination of Europe after the Chernobyl Accident

Yu A Izrael
Institute of Global Climate and Ecology, Moscow, Russia

M De Cort, A R Jones
Environment Institute, CEC Joint Research Center, Ispra, Italy

I M Nazarov, Sh D Fridman, E V Kvasnikova, E D Stukin
Institute of Global Climate and Ecology, Moscow, Russia

G N Kelly
CEC, DG XII.F.6, Brussels, Belgium

I I Matveenko, Yu M Pokumeiko
Republic Centre of Radiation and Environment Monitoring, Minsk, Belarus

L Ya Tabatchnyi
Minchernobyl, Kiev, Ukraine

Yu Tsaturov
Roshidromet, Moscow, Russia

Abstract. The Atlas, which was compiled under the Joint Study Project (JSP6) of the CEC/CIS Collaborative Programme on the Consequences of the Chernobyl Accident, implemented into the European Commission's Radiation Protection Research Action, summarizes the results of numerous investigations undertaken throughout Europe to assess the ground contamination by caesium-137 following the Chernobyl accident. The Atlas incorporates about 100 color maps at a range of scales (1:200k - 1:10M) which characterize the contamination in Europe as a whole, within state boundaries and for zones where the contamination levels are above 40 kBq/m^2 ($\approx$ 2.0% of the European territory) and above 1480 kBq/m^2 ($\approx$ 0.03% of the European territory). Investigations have shown that around 6% of the European territory has been contaminated for more than 20 kBq/m^2 after the Chernobyl accident. The total amount of deposited caesium-137 in Europe is 8× 10^{16} Bq and distributed in the following manner: Belarus 33.5%, Russia 24%, Ukraine 20%, Sweden 4.4%, Finland 4.3%, Bulgaria 2.8%, Austria 2.7%, Norway 2.3%, Romania 2.0%, Germany 1.1%.

1. Introduction

The Chernobyl nuclear power plant (CNPP) accident of 26 April 1986 was followed by a partial destruction of reactor IV which resulted in a significant release of radioactive material into the natural environment. The release from the Chernobyl incident was much greater than either the Windscale (United Kingdom) or the Three Mile Island (USA) reactor accidents.

As a result of the complicated meteorological situation which persisted after the accident and the relatively long exposure of the reactor to the atmosphere, radioactive materials were deposited over a wide area. In the vicinity of the CNPP, graphite and particles from the destroyed reactor were deposited while finer particles were found at substantial distances from the site. Depending on the prevailing wind direction and precipitation events during the weeks immediately following the accident, volatile products such as iodine-131 (with a half-life of 8 days), tellurium-132 (3.2 days) and long-lived caesium-137 (about 30 years) were spread over thousands of kilometers. In the days immediately following the accident, contaminated air masses moved west, then north-west then north-east. As a result territories, in the Ukraine, Belarus, the European part of the Russian Federation and, to a lesser extent, Scandinavia were heavily contaminated. Subsequently, the wind direction switched to the south then swung to the south-west, bringing the radioactive cloud over the Balkans and the Alps. Several days after the accident, the air masses carrying the radioactive particles had traversed almost all European countries.

Based on a series of radioactivity measurements carried out after the accident, it was determined that volatile fission products (Te-132 and I-131) deposited in close proximity to the accident (i.e. up to a distance of 40 km) amounted to about 5% of those in the reactor. Similar studies of refractory products amounted to about 1% while Cs-137 was approximately 2%.

The total radioactive release which was deposited over the European territory amounted to 4% of the total radioactivity accumulated in the reactor, of which Cs-137 was about 15% or $\approx 7.8 \times 10^{16}$ Bq ($\approx 4 \times 10^{16}$ Bq of this amount was deposited over the former USSR territory).

During the initial period of the accident the largest doses resulted from I-131. After the initial period, especially for the area outside of the evacuation zone, caesium-134 and caesium-137 were the major contributors to the exposure of the population. The external dose from caesium-137 between one and fifty years on differently directed patterns was 75-90% of the dose of the total sum of radionuclides deposited over the terrain.

After the Chernobyl accident various compilations have been made of the contamination of particular countries or regions in Europe. These compilations have been made for different purposes and consequently there are significant differences in their resolution and quality. To date no attempt had been made to compile a comprehensive presentation of the contamination over the whole territory of Europe, the continent on which by far the majority of released material was deposited. In many cases improved data have since been, and continue to be, obtained through more refined and extensive monitoring, in particular in those areas where greater contamination occurred.

Therefore it is opportune to prepare a comprehensive atlas of the radioactive deposition of the whole of the European territory consequent upon the Chernobyl accident. The publication of such an atlas by the tenth anniversary of the accident would have wide public and scientific interest. In addition to the more obvious interest in and use of the factual content of the atlas, it would provide most useful and needed perspective, especially in the former Soviet Union, for judging the significance of the contamination.

2. The Objectives

Since caesium-137 presents a long-term threat for the population of Europe and given its wide dispersion across the continent, the European Commission accepted a proposal on a joint study to compile "The Atlas of caesium contamination of Europe after the Chernobyl accident". The goals of the Atlas are:

- to provide generalized and detailed information on the distribution of caesium-137 in soil over the whole European territory, and separately by countries.
- to provide an estimate of the total amount of caesium-137 deposited across Europe and separately by country as a result of the Chernobyl NPP accident.
- to assess the external gamma dose from the Chernobyl caesium-137 and compare it with that from natural radionuclides in soils and rocks, as well as that from cosmic radiation.
- to familiarize the general public, governmental and municipal bodies with a comprehensive view of the pattern of caesium-137 across the whole of the European continent.

In the future, based on the data analysis made for the Atlas, the problem of the harmonization of the different sampling and measurement methods between different countries can be analyzed in order to improve the quality of the information that could be exchanged. Also, because the measurements on which the Atlas is based have been made sometimes at identical locations but at different times, it should be possible to investigate the behavior of the caesium-137 in a wide variety of European soils and under different climates. Finally, the spatial analysis of these data should bring us new information about the way to improve the sampling structures and the analysis of these data.

3. The Content and the Structure of the Atlas

The Atlas contains about 100 color maps, mostly of A2 format, with accompanying texts. Although other artificial radionuclides were released, it was decided to present only caesium-137 levels, because of the availability of the many measurements performed throughout the European countries and because it is by far the major contributor to dose other than in the very short term following the accident. Since this radionuclide was already deposited due to the atomic bomb tests, the situation before and after the Chernobyl accident can be compared.

The following chapters describe the five sections of the Atlas.

3.1. The Introductory Section

This section deals with:

- an overview of the phenomena of radioactivity for the layman (natural and artificial radioactivity, scientific units) as an explanation of external dose. Additionally this is illustrated with small scale maps of Europe containing information about natural and artificial radioactivity:
  - natural radionuclides including external gamma doses;
  - dose rate from cosmic radiation;
- a brief description of the history of the Chernobyl accident, the temporal dynamics of radionuclide fallout together with their volumes, as well as an assessment of the scale of the catastrophe;

3.2. The Data Section

This is by far out the largest part of the Atlas. Because the radioactive material was deposited in a highly inhomogeneous way, and because the various European countries adopted different sampling strategies resulting in maps with varying sampling densities, the Atlas presents the caesium-137 deposition at various scales. All deposition levels are normalised to 10 May 1986, the day at which the radioactive release from the reactor stopped. The scale with isoline values (see Table 1) is based on scientific and administrative considerations: since deposition is purely a physical phenomena, it is normal practice to present it by a consistent logarithmic scale. On the other hand, political and administrative deposition levels which were adopted in the former USSR, i.e. 185, 555 and 1480 kBq/m^2 (resp. 5, 15 and 40 Ci/km^2) have to be considered.

The values chosen for the caesium deposition isolines depend on the scale of map. A summary is presented in Table 1. In order not to overload the European overview map with information, alternative values on this scale are presented. The levels of the highest contamination are only shown on the local maps with deposition values > 1480 kBq/m^2: these areas are relatively small and require a separate scale to show appropriate details.

Table 1: *Isoline values of the caesium-137 contamination density by map type.*

Contamination levels [*]		Map type			
		European	Country	Local	
kBq/m^2	Ci/km^2			> 40 kBq/m^2	> 1480 kBq/m^2
0.4	0.01	+	+		
1	0.027		+		
2	0.054	+	+		
4	0.1		+		
10	0.27	+	+		
20	0.54		+	+	
40	1.08	+	+	+	
100	2.7		+	+	
185	5	+	+	+	
555	15		+	+	
1480	40	+	+	+	+
4000	100				+
10000	270				+

(*)   The values shown are preliminary and may be subject to changes with respect to those presented in the atlas

3.2.1. The Overview Section

This subsection contains radiological information at a European scale presented in the following maps:
- pre Chernobyl caesium-137 deposition (normalised to 10 May 1986) at scale 1:20M (1:20,000,000);
- post Chernobyl caesium-137 deposition (normalised to 10 May 1986) at scale 1:10M;
- caesium-137 external gamma dose.

Figure 1 shows an initial attempt to map caesium-137 deposition in Europe. It is possible to show from the map in Fig. 1 that levels > 20, > 40 and > 1480 kBq/m^2 were deposited on respec-

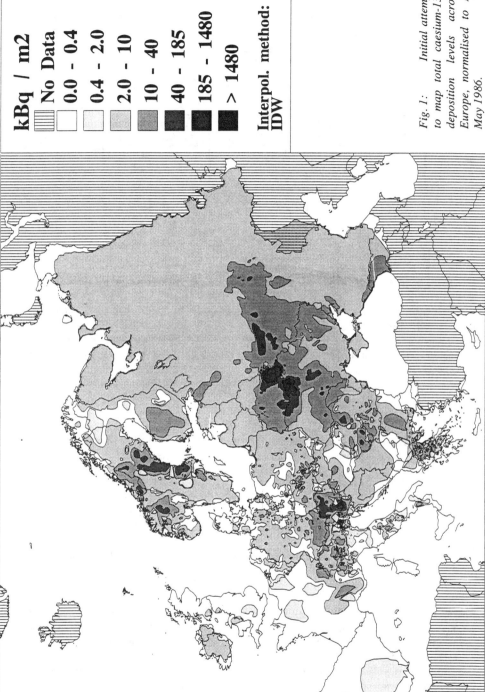

kBq / m2

▦	No Data
☐	0.0 - 0.4
☐	0.4 - 2.0
▤	2.0 - 10
▥	10 - 40
■	40 - 185
■	185 - 1480
■	> 1480

**Interpol. method:
IDW**

Fig. 1: Initial attempt to map total caesium-137 deposition levels across Europe, normalised to 10 May 1986.

tively ≈6%, ≈2% and ≈0.03% of the European territory. Table 2 shows the areal extent of contamination by caesium-137, for the various deposition intervals, for Europe as a whole. The total amount of caesium-137 deposited in Europe is about 8×10^{16} Bq.

*Table 2: The areal extent of total caesium-137 (bomb fallout +
Chernobyl) deposition in Europe*

Cs-137 deposition interval (kBq/m^2)	Area (×1000 km^2)	% of the European territory
>1480	3.1	0.03
555-1480	7.2	0.03
185-555	19	0.2
40-185	211	1.7
20-40	432	3.6
10-20	871	11.6

The map in Fig. 1 also indicates the direction of caesium-137 deposition patterns. The eastern pattern, passing from the Chernobyl NPP across the Russian territory to the Urals and further to Siberia, is clearly seen. In the Ukraine, several southern patterns, interrupted by the Black Sea, are observed with their onward contamination being recorded in Bulgaria, Turkey and Greece. The south-western patterns leave noticeable spots in the Ukrainian Carpathians, later on appearing in the Balkan mountains. The western patterns, passing across the territory between the Ukraine and Belarus show a series of northward branches, then turning eastward. This leads to the deposition patterns for Belarus, Poland, Germany, Lithuania, Sweden, Norway, Finland and the Leningrad oblast of Russia. In the Alps, some anomalies with levels above 40 kBq/m^2 are observed.

3.2.2. The Country Map Section

In order to give more geographical and radiological details to reflect the national or regional situation, the country map section includes maps showing the caesium-137 deposition in almost each European country at a medium scale (1:1M - 1:2.5M), together with the sampling/measuring locations .

Table 3 shows the areas of Chernobyl contamination by caesium-137 in European countries as calculated from the map shown in Fig. 1. The results were obtained by multiplying the average deposition value with its corresponding area. These areas were calculated by means of Autocad. Special attention was given to the region around Chernobyl with deposition levels > 1480 kBq/m^2, where the calculation of the corresponding areas was performed on 1:200k maps.

3.2.3. The Section on High Contaminated Zones

This section contains maps that present deposition information for local zones:
- maps with levels above 40 kBq/m^2: zones of enhanced contamination (i.e. parts of Scandinavia, the Alps, Greece, Rumania, Russia, Belarus and Ukraine) are highlighted by means of large scale maps (1:500k);
- maps with levels above 1480 kBq/m^2: the highest contaminated zones, i.e. certain areas of Briansk-Mogilev and Chernobyl-Pripiti, are shown on very large scale maps (1:200k). An example for the 60 km zone around Chernobyl can be found in Fig. 2.

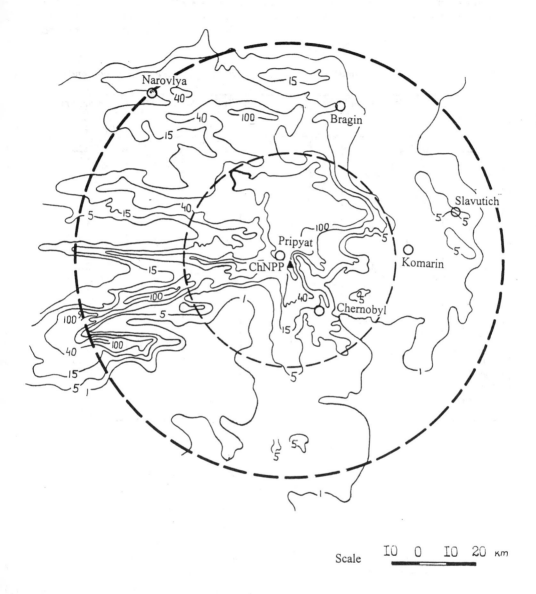

Scale ‖ 10 0 10 20 km

Fig. 2.   *sixty km zone map around Chernobyl. Deposition levels of caesium-137 (1989) for 185, 555, 1480 and 3700 kBq/m² (on the map indicated as 5, 15, 40 and 100 Ci/km²)*

The quality of the mapping is determined largely by the density of sampling and measurement points. Hundreds of thousands of measurements were performed in the Ukraine, Belarus, Russia and Sweden by aerogamma surveys conducted at scales of 1:200k and 1:1M at flight altitudes of 50-150 m. About ten thousand soil samples were taken in Central and Western European countries. The territories of Norway, Finland, UK, Greece, Germany, the Netherlands, Austria, and Switzerland are most completely investigated.

Table 3:   caesium-137 (total) contaminated areas in European countries in thousand km^2

Countries	Area (in 1000 km^2) contaminated above specified levels (kBq/m^2)						% of contamination deposited in Europe
	10-20	20-37	37-185	185-555	555-1480	>1480	(%)
Belarus	60	30	29.9	10.2	4.2	2.2	33.5
Russia	300	100	48.8	5.7	2.1	0.3	23.9
Ukraine	150	65	37.2	3.2	0.9	0.6	20
Sweden	37.4	42.6	12.0	-	-	-	4.4
Finland	48.8	37.4	11.5	-	-	-	4.3
Bulgaria	27.5	40.4	4.8	-	-	-	2.8
Austria	27.6	24.7	8.6	-	-	-	2.7
Norway	51.8	13.0	5.2	-	-	-	2.3
Romania	14.2	43.0	-	-	-	-	2.0
Germany	28.2	12.0	-	-	-	-	1.1
Greece	16.6	6.4	1.2	-	-	-	0.8
Slovenia	8.6	8.0	0.3	-	-	-	0.5
Italy	10.9	5.6	0.3	-	-	-	0.5
Moldova	20	0.10	0.06	-	-	-	0.45
Switzerland	5.9	1.9	1.3	-	-	-	0.35
Poland	8.6	1.0	-	-	-	-	0.23
Estonia	4.3	-	-	-	-	-	0.08
Czech Rep.	3.4	0.36	-	-	-	-	0.09
Slovak Rep.	2.1	-	-	-	-	-	0.05
Lithuania	1.2	-	-	-	-	-	0.02

3.3. The reference section

The reference and information section of the Atlas includes supporting maps on population density, soil type, elevation and vegetation for Europe, at scales of 1:15-20M.

3.4. The meteorological section

Deposition patterns depend largely on the wind fields and precipitation patterns. Meteorological data, (daily precipitation and twelve hourly wind fields) during, and for two weeks after, the initial release are presented on 1:40M scale maps.

3.5. Technical appendices

Technical appendices to the Atlas consist of a description of the methods used for soil sampling, remote and laboratory measurements of caesium-137 contamination density together with the procedures used to process the data and compile the maps.

4. The map compilation procedure

One of the important elements of this project is the use of a Geographic Information System (GIS) for the preparation and production of the maps showing the density of caesium-137 deposition across Europe. A GIS is a set of software tools designed to efficiently capture, store, update, manipulate, analyze and display all forms of geographically (or spatially) referenced information. Certain complex spatial operations that would be very difficult, time consuming or impracticable in traditional database or computerized drawing packages are possible only with a GIS [1]. Individual datasets can be stored as separate layers which can then be combined with each other as required allowing relationships, trends and patterns to be visualized. The GIS being used in this project is ARC/INFO, version 6.1., a powerful software developed by ESRI Inc. of California. ARC/INFO includes a relational database interface for integration with commercial database management systems (DBMSs) and a fourth generation macro language for developing customized applications.

The cartographic detail for the European and Country scale maps is provided by information contained in the Digital Chart of the World (DCW). The DCW, produced by the US Defense Mapping Agency [2], is an established dataset of assorted digital cartographic features for the world at a scale of 1:1M. This provided a common base from which all the maps within the Atlas could be produced. Where necessary, the DCW data have been supplemented by additional information from the Lovell Johns 1:5M European Digital Database and the European Commission's Eurostat GISCO Database. Some digitizing was undertaken in order to add further geographical detail to the larger scale maps (e.g. to add localities in order to improve the visualization of the large scale maps). Substantial editing of the DCW was necessary prior to its effective use in the Atlas.

The information on radioactive deposition from the collaborating laboratories came in the form of point data, geographically located by a latitude and longitude coordinate. This information is stored in the GIS which creates a 'point' coverage (or theme) of the sampling locations for the area of interest whereby each point is tagged with a unique identification code. Additional information, such as caesium level and any other attribute information, can then be attached directly to the location through the point's identification code. The deposition sampling points are then transformed to a suitable equal area map projection (in this case the Lambert Azimuth Projection). Once the data have reached this level, cartographic data (e.g. coastline) can be overlaid for checking the locational accuracy of the sampling point coordinates. This primary analysis of the data includes also the analysis of the relation between different layers: e.g. the meteorological parameters and the elevations have been compared to the spatial distribution of the contamination; the display of the cities could explain in certain cases the lower deposition levels in those areas as the heat generated by the cities can be an obstacle to the radioactive deposition.

The next stage in the project requires the generation of maps that display isolines of deposition. This task, depending on the density of the points and the requested degree of resolution of the map, requires a degree of interpolation and generalization of the radiation data. More details on the methods used can be found in the Atlas.

In case of densely distributed points, the inverse distance weighted interpolation method has generally been used. In other cases, deeper and more complex investigations were necessary and have required external software like GEOEAS [3], GSLIB [4], VARIOWIN [5] and basic statistical packages. The very general steps of such analysis were finding the populations which were presenting different spatial distribution of the contamination, finding models which would describe these distributions and finally interpolate these data on the base of these models. The result of these interpolations are new point coverages with regular structures and with data generated at the unsampled places. This data can then be contoured and represented with isolines.

5. Conclusions

By collecting more than 500,000 data related to the spatial distribution of caesium-137 in Europe after the Chernobyl NPP accident, the Atlas has clearly shown the importance of such a dataset. For the first time it is possible to provide a comprehensive map of European contamination after the Chernobyl accident, useful for scientific community and also enable layman to better appreciate the extent of the contamination and its relative impact. Since the Atlas was fully electronically prepared the data could be made available on CDROM, useful for further scientific study. Taking into account the radioactive decay for caesium-137, the user of the Atlas can estimate the radioactive levels in the future over all Europe. Further onwards, the data generated by interpolation during the preparation of the Atlas can be used as a reference to which scientists can compare new measurements in order to analyze the contamination in time, and this for different regions and for different conditions. Further to these conclusions, it is hoped that this study can be expanded to other long-lived (e.g. strontium-90, plutonium-238, -239, -240, and americium-241) and short-lived (e.g. I-131) radionuclides.

Acknowledgments

The Atlas was compiled under the Joint Study Project (JSP6) of the CEC/CIS Collaborative Programme on the Consequences of the Chernobyl Accident, implemented into the European Commission's Radiation Protection Research Action.

The authors gratefully acknowledge the invaluable contribution to this project by all the collaborating and participating institutes in terms of the supply of data and in the constructive comments that they have provided, as well as Gregoire Dubois for his important efforts in introducing and applying spatial analysis techniques for the deposition interpolation procedures.

References
[1] Tomlin, Dana, 1990, Geographic Information Systems and Cartographic Modelling. Prentice Hall.
[2] Danko, D., 1992, The Digital Chart of the World Project. Photogrametric Engineering and Remote Sensing, 58,8,1125-1128
[3] Geoeas - Geostatistical Environmental Assessment Software - Computer Science Corporation, Public domain software.
[4] Deutsch C:, A.G. Journal (1992), 'GSLIB. Geostatistical Software Library and User's Guide. Oxford Univ. Press, 340 p
[5] Pannatier Y., (1993), 'VARIOWIN: Ms-Windows Programs for Exploratory Variography and Variogram Modelling in 2D', Statistics of Spatial Processes: Theory and Applications. Bari, Italy, Ed Capasso V., Girone G & Posa D., pp. 165-170

Outline of main bilateral collaborative agreements dealing with the migration of Chernobyl consequences in Ukraine

V.G. Bariakhtar, V.M. Ponomarenko

1. Introduction

The liquidation of consequences of accidents respecting to human technical activity of such scale, as the disaster on Chernobyl nuclear power plant, can be transformed into fruitless waste huge human, political, material, financial and other resources, if the manipulation with these resources will not be coordinated with the last scientific achievement (facts of large losses in connection with liquidation of consequences by failure, scientific data stipulated by insufficiency and representations).

The scientific achievement recognized at an international level promote, as a rule, growth of mutual trust and effective interaction of interested countries in work on liquidation of consequences of failure (examples of scientific results, use of which has allowed to international community to achieve scale effects at liquidation of consequences of Chernobyl accident).

The agreement on scientific cooperation between CEC and USSR (Belorussia, Russia and Ukraine) was signed in 1991 the Purpose of cooperation consisted of valuation of the main factors of a radiating conditions and development of the recommendations on extenuation of consequences of Chernobyl accident.

In 1992 in Bruxelles of Ministries of Ukraine, Russian federation and the Republics Belorussia on problems of Chernobyl accident together with CEC have signed the bilateral Agreement on organization of joint international scientific researches.

Fulfillment of 5 joint projects in ecological area (experimental projects ECP 1,2,3,4,5) and 2 in the field of development of radiological criteria for acceptance of the decision (research projects JSP 1,2) was originally planned.

In 1993 the number of the projects was finished up to 10. The projects ECP 6,7 and JSP 3, devoted to medical questions were added: methods biological dosimetry (ECP6), treatment of beam illness (JSP3), study malignant tumors at the liquidators (ECP7).

In 1993 under the initiative of countries SIC the agreement on increase since 1994 quantities of the projects up to 16 is achieved. The projects on problems of public health services · molecular gears and treatment thyroid cancer among the children (ECP8 and JSP4), methods of retrospective dosimetry (ECP10), role of the pasturable animals in receipting of radionuclides into human organism (ECP9), drawing up of the atlas of pollution of territory of Europe (JSP6), way of formation of a doze облучения of the person (JSP5) are again begun.

Thus, within the framework of the Agreement an extensive network of joint work with connection more than 200 laboratories and organizations of appropriate specialization in countries the EC and CIS was created.

2. Condition of cooperation

The management joint by the Chernobyl research projects pursuant to the Agreement is carried out COORDINATION Council.

From the moment of signing of the Agreement was held of 7 meetings of Coordination Council.

Within the framework of each project joint meetings were systematically organized. Effectively passed training the scientist in leading laboratories of Europe and countries CIS. Thanking the help the EC has managed essentially supplying Ukrainian research organization with new-type equipment.

From the party CEC the financing is allocated only to European countries through organization, with which the Commission concludes the contract.

The budget of the contracts CEC with European institutes is not considered and not supervised by Coordination Council. However the budget of each project includes 3 articles of rendering assistance to countries CIS: purchase and delivery of the equipment, exchange by the experts and local help (grant to the scientists). The applications for financing under these articles are made by the national coordinators, transfer to their coordinators from CIS, which coordinate them with the European coordinator. The distribution of means under all three articles is discussed and taken on a meeting of the representatives of all institutes - participants of the projects in the beginning of each year.

The summary plan of financing of countries CIS under all projects is approved by a meeting of Coordination Council in March.

Delivery of the equipment, registration of training and exchange by the scientists and allocation of the local help in 1992-1993 rr. The European coordinators executed, that did not permit to supervise completeness of fulfillment of the budget and registration of the customs documents etc. For the improvement of coordination of work under the projects required from each recipient of the equipment and management of the appropriate account the Coordination Council has taken the decision to entrust execution of the considered articles of the budget to research centre Rozendorf (Germany, Dresden), having created there special Secretary. Accordingly, and in countries CIS national Secretaries from 5-7 persons are created.

The purpose of Secretaries · to organize distribution of the information between the participants, to ensure duly drawing up of the uniform plan of work for a year and preparation of the reports under the projects, reception of the equipment on customs house and transfer to its participants of the projects, control at use of the equipment, registration of the necessary documents in bank "Vidrodzhennia" and rendering assistance in reception of the grants.

One of major problems of Secretaries: the analysis of results of work under the projects and their integration and issue of the recommendations on improvement of work and introduction of results.

On a meeting of Coordination Council in October 1994 Kiev the decision on creation of group of management in a structure 4 scientific representatives (on one from countries CIS and Europe) and chief · of the chapter of department of radiating protection XII of management CEC is accepted.

The problem of group of management · to execute connection between CEC and countries of CIS in period between meetings of Council and preparation of materials for consideration on Coordination Council.

The efforts within the framework of the Agreement were directed on development of the approaches to effective struggle of international community with harmful consequences on polluted territories on the basis of new knowledge as about behavior радионуклидов in an environment, as about effect of radiation on the person.

The main purposes of the Agreement were defined by the following practical directions:

- Influence of radiation to occurrence of oncological diseases and predisposement to cancer on molecular level. Epidemiology of the exposed population. Determination of a doze облучения and medical help to the exposed population and etc.

- Criterion of acceptance of the decisions about decontamination of specific strongly polluted territories. Perspective of restoration technology. Revealing of dumps of radioactive waste, leading to potential polluting of underground waters and various ways of storage waste, including their processing, as well as measure on storage received waste.

- the Help of mass media in understanding of a nature and sizes of risk of residing on polluted territories and ways of decrease of this risk with the help special ration and style of life. An establishment of trust to measures, taking steps for achievement of safe conditions of life in polluted regions;

- a System of accident precaution.

In the plan of progress to these purposes achieve significant progress. At an international level new scientific results, enabling are received and agreed to give multidimensional valuations of consequences of Chernobyl accident and to produce the important practical recommendations for overcoming of these consequences.

The first stage of fulfillment of the Agreement has suited to the end. On the summons of a day a problem of effective use of results of researches and acceptance them to a management in state, political, public and scientific circles of the participants of the Agreement as a whole.

3. The sums of collaboration

- Scientific data used in CIS countries for the estimation of radioactive situation (contamination of environment objects, food products, intensity of radionuclides migration in soil and in biological chains) are verified by the leading institutes of European countries.

- There was considered correctness and coincidence of systematic approaches to the study of radioactive situation and the dynamics of dose rating formation on population.

- Ukrainian scientists have run in basically new methods of the absorbed doses determination, determination of forms of existence of radionuclides in water and soil solutions, the estimation of reliability of radionuclides absorption by soils, cancer disease forecast. Computer simulations of radionuclides behavior in ecological chains, estimation of countermeasures efficiency, etc.

- There were delivered by CEC Institutes and run in the work station and software for the creation of information system for the forecast and estimation of current radioactive situation in case of large-scale nuclear accidents, using RODOS model system.

- Chernobyl Central Laboratory was created to provide the researches on joint projects. The Laboratory is completely equipped with modern alpha-, beta-, gamma-spectrometer devices and furniture at the expenses of CEC (about 450 KECU). The Laboratory is run by RTC RIA "Pripyat". During the period of 1994 the Laboratory had carried out more than 4000 analyses.

• The implementation of cooperation program of 1991-1995 allows to unify the criteria and standards of nuclear security in case of accident, that makes the reasons for the integration of Ukraine into the common security system of Europe.

• As a result of the agreement there was provided the information exchange between Ukraine from one side and Russia, Byelorus and CEC countries from the other side, that promotes the full use of information, necessary for the overcoming of the consequences of Chernobyl accident. The program realization extended the understanding of the situation in CIS countries and of the necessity to take the number of countermeasures at the suffered territories.

• During the collaboration period the Ukrainian institutes - participants of the Program have received for the carrying out of project researches the scientific equipment and hardware that allowed to increase the efficiency and reliability of the results.

• This Collaboration made it possible to support the trained specialists of Ukraine drawn intosolution of problems of Nuclear security and also to promote the increase of professional level of Ukrainian and western specialists as well, at the expenses of scientists exchange funds.

• The results of Collaboration, that will be presented in scientific reports, articles, monographs of Conferences , will work for the future progress of scientific community of Europe and the strengthening of the stability in Europe in general.

4. Problem of the future cooperation

The close interaction with other programs of European Union, which concern to the technical help (to countries CIS) and humanitarian help (ECHO-2), should amplify and to be developed in the same direction, as the large number of auxiliary activity in sphere дезактивации or medical help victims of radiation, such interaction corresponds main results of the research program. The areas of development of a joint network of work, including action on management of data and general information, can become main objects pursuant to the above mentioned programs of the help, from the point of view of extenuation of a extremal situation after accident.

The general agreed efforts, mentioned here, are intended in order force practical application of the certain appropriate approaches to extended struggle with consequences on polluted territories and to improve wide international struggle with possible failures. In the special cases limited quantity of more fundamental researches can be necessary to ensure the contribution to teleological research.

The main purposes are, first, to promote improvements in long-term management of territories polluted by a radioactive material in former Soviet Union, as well as improvements in sphere of public health services and material maintenance of the population, subject danger заражения, and, secondly, to use available experience as the contribution to development of more effective ways of struggle with consequences of radiation of possible nuclear failure, in general, to improve radiological protection by the best study of behavior radionuclides in an environment and effect of radiation on the person.

The research work will be continuation of that work, which was begun in 1991 under aegis of the Agreement on international cooperation on overcoming of consequences of Chernobyl accident between a Commission and Ministry on Chernobyl problems in Belorussia, Russia and Ukraine.

INTERNATIONAL CO-OPERATION OF BELARUS ON OVERCOMING THE CHERNOBYL CATASTROPHE CONSEQUENCES

I. V. Rolevich and A. V. Semeshko

Ministry for Emergencies and Population Protection
from the Chernobyl NPP Catastrophe Consequences
14 Lenin St., 220030 Minsk, Belarus

Experience of the last years shows, that the effective counteraction to the consequences of the Chernobyl catastrophe requires constructive international co-operation, attraction of intelligent and material resources of international organizations, different countries, research centers. Taking into account complex economic situation in the country, the help of foreign countries in minimization of the Chernobyl catastrophe consequences in the framework of international and bilateral relations is important and urgent.

While discussing Chernobyl questions with foreign organizations and for elaborating of international projects and programs for rendering assistance to the Republic of Belarus, the greatest attention is given to the following priorities:

REDUCTION OF THE GENERAL RISK OF LOSS OF A HEALTH OF THE AFFECTED POPULATION AND LIQUIDATORS OF THE CHERNOBYL NPP CATASTROPHE CONSEQUENCES:

Health of the liquidators of the Chernobyl NPP catastrophe consequences; general oncology; cytogenetic effects; reconstruction of individual and collective doses; thyroid cancer diseases; general state of health of the population; creation of national Chernobyl registers;

REDUCTION OF THE ADVERSE ECOLOGICAL, ECONOMIC AND SOCIO-PSYCHOLOGICAL EFFECTS OF THE CHERNOBYL NPP CATASTROPHE CONSEQUENCES:

Establishment of the socio-rehabilitation centers for the affected population; development of optimal technological chains of decontamination of the territories, contaminated by radionuclides; development of optimal technological chains of use of wood in the zones with a various levels of radioactive pollution; development of the enterprises for manufacturing medicines and medical equipment; construction of factories for food products processing and manufacturing of children food

RESTORATION OF NORMAL CONDITIONS OF LIFE AND ECONOMIC ACTIVITY ON THE POLLUTED TERRITORIES:

Construction of settlements for the affected population, with social, cultural and economic infrastructure; perfection of a system of radiation control of agricultural products; development and application of counter-measures in agricultural forestry sector on the polluted territories

In accordance with the agreement on joint activities on minimization and overcoming the Chernobyl accident consequences, signed by the Governments of Republic of Belarus and Governments of Russia and Ukraine, it is planned to continue the work on maintenance of the information exchange system between the three affected states, including creation of joint data bank on the radiation conditions, and also on the population migration from these regions; co-ordination and material provision of the working group activities on fulfillment of the Belarussian-Russian protocol on the Chernobyl accident consequences overcoming co-operation.

Belarus actively interacts with the Government of Russia on direction into the republic material, technical and financial means for joint programs realization. These means are used for rehabilitation of population and liquidators of the Chernobyl NPP accident, restoration of economic potential of the contaminated territories, production of the ecologically clean products.

According to the Memorandum of co-operation with Japan on the Chernobyl NPP accident consequences minimization for the population health, signed in Tokyo on April 18, 1991, scientific researches are carried out, as well as exchange by the experts dealing with evaluation and reconstruction of doses of the thyroid gland and the whole body, exposure, epidemiological researches, preventive measures, diagnostic and treatment of thyroid gland diseases, pre-leukemia condition and leukemia itself.

The project Chernobyl -Sasakawa is carried out within the framework of the long-term programme on examination and study of the state of health of children living on the contaminated territories. As to the information available on June 1995, the Japanese party delivered medical equipment and supply materials to it, vehicles to the Gomel oblast totalling about 5 million dollars, to Mogilyov oblast - 3216483 dollars. 16 experts have been trained abroad.

The Memorandum of mutual understanding with Germany has created a political platform for maintenance of the numerous private German initiatives, already rendering the significant help in the salvation of the Chernobyl problems; priority projects of the republic have also been included there.

Germany is one of countries, which were the first to offer help to the affected by the Chernobyl NPP accident population of Belarus. German humanitarian help makes about 50% from amounts, received by Belarus from abroad during the post-accident years, that includes: shipment of humanitarian goods to Belarus (medicines, medical equipment, food products, clothes); organization of health improvement and rest for children in Germany; financing of construction of rehabilitation facilities in Belarus and help in equipping and maintenance of such institutions; training of the doctors from the contaminated regions in German clinics; treatment of seriously ill children.

One of main spheres of co-operation is organization of children rehabilitation. For example, during 1994 spring-summer period 14000 children from Belarus during 1994 spring-summer period had rest and improved their health in Germany. Main part of the children improved their health in German families was invited by public organizations, established with the purpose of rendering assistance to people affected by the Chernobyl NPP accident.

Children rehabilitation center "Hope" ("Nadezda", Minsk oblast) has been established with the assistance of the Committee "Leben nach Chernobyl" and organization Mennerarbeit of Kassel (Germany) Evangelic Church. The German party allocated 988 579 DM as on 1.01.95, there is 1 079 455 DM allocated as building materials, equipment and other materials to equip the center. Totally in 1992-1994 the assistance received was equal to 2 068 034 DM .

Contacts with Holland in realization of the so-called "Gomel project" are examples of fruitful international co-operation. Under this project a joint Belarussian-Dutch consultative-diagnostic polyclinic and health informational center has been created.
The assistance, rendered by Holland is estimated as more than US$3 million.

A project on optimization of the radiation control system of the agricultural production on the example of Kalinkovichi meat factory was carried out together with the British firm Moushel in 1993-1994.
The Government of Great Britain acted as a donor for this project through the "Know-How" fund. In the result of the project a complete radiological laboratory evaluated in US$75000 was delivered to the Kalinkovichi meat factory. Five employees of laboratory have been trained in Great Britain.

Realization of the pilot project on construction of a mini-milk factory with a line for milk processing and its cleaning from radionuclides began in 1994. In 1995 the British Government through the "Know-How" fund allocated 50000 pounds sterling for the development of the project feasibility report, the development of the business - plan will be completed by firms Moushel and Bradtek (Great Britain).

The agreement for co-operation between the Scientific and Research Institute of Radiation Medicine of the Republic Belarus and National Health Institute (Betedza, USA) on studying cancer and other thyroid gland diseases of the population, affected by radiation as a result of the Chernobyl NPP accident, was signed. The delivery of the equipment, training of the experts will be financed by the American party.

In 1993 the joint Belarussian-American project on rapeseed cultivation and its re-processing on the contaminated territories was launched. During its realization devices and equipment estimated in US$286000 were delivered by the American party. 8 types of summer and winter rapeseeds were given for tests. Five Belarussian experts have been trained in the USA on technology of rapeseed cultivation and processing. For the expiring period biannual experiments on rapeseed cultivation and processing on the contaminated territories have been conducted. Their results permit to make a conclusion, that oil cultures cultivation on the contaminated territories with their further processing on the technical purposes reduces a collective doze exposure of the population of republic and permits to receive competitive and highly profitable production on affected lands. Now the project reached the level of commercial introduction.

The co-operation with the French companies "Electricite de France" and "Kozema" has been conducted, basically, in the field of evaluation of radiological consequences of the Chernobyl accident on the population health and on environment of regions, highly polluted by radioactive fall outs.

Pursuant to the Protocol on intentions signed in June 1992, "Electricite de France" delivered medical and radiological measuring equipment estimated in 10 000 000 francs to Belarus. It also organized Belarussian doctors and experts training in France.

During recent years many assistance projects were launched and are in realization now on mitigating the Chernobyl consequences . This creates significant preconditions for effective international co-operation on minimization of accident consequences, ioining up the intelligent and material resources of different countries for overcoming the Chernobyl aftermath. Thank to the international co-operation on Chernobyl problems the Republic receives additional means for protection of the population health and creation of normal living conditions for people in the affected regions.

Characteristics of the Development of the Radiological Situation Resulting from the Accident, Intervention Levels and Countermeasures

S.T. BELYAEV, V.F. DEMIN, V.A. KUTKOV

Russian Research Center "Kurchatov Institute", Moscow, Russia

V.G. BARIAKHTAR

Ukranian Academy of Sciences, Kiev, Ukraine

E.P. PETRIAEV

Chernobyl State Committee, Minsk, Belarus

Abstract

Great efforts have been made in the frame of the national and international research programs to get complete data on the radioactive releases, environmental contamination and radiological situation resulted from the Chernobyl accident. Beginning from the first publication (IAEA meeting, August 1986) these data have been considerably improved and added. The most important change of them with their influence on the decision making in the mitigation activity and the current situation is described and analyzed.

The national and international regulatory documents at the moment of the accident were neither complete nor perfect in some necessary aspects especially in respect to the countermeasures at the intermediate and long-term phases. New documents have been worked out during the intervention activity. From 1986 series of documents were developed on the national and international levels. These documents are considered and analyzed in the context of their practical implementation and by the modern experience and research results. The history of countermeasures adopted on the different intervention phases are described.

These documents mainly establish intervention levels in terms of averted doses and regulate only radiation protection. They don't content any intervention levels in terms of residual doses and risk, which are necessary for regulation of social and health protection of population suffered from the accident. Other restriction for the optimal regulation comes from use of the effective dose for establishing intervention levels. These and other respective aspects are discussed.

1 Introduction

"Classical" principles of radiological protection are based on radiation doses, intervention levels and effective countermeasures. Clear and logical in principle, these basic parts need a specific clarification on each after-accident period. Being unavoidable and useful on the first early stage, "the classical" principles and criteria meet specific obstacles in introduction and practical application on the next, long-term stage.

First, one needs a reliable prediction on future dynamics of radiological situation and effectiveness of specific counter-measures. Most of the dynamical data critical for decision making are case and area dependent and the time needed for providing measurement and analysis. Natural "self cleaning" processes and their time scale, which are crucial for dynamical dose prediction is not the only example of the kind.

The "classical" pattern of radiological protection considers mostly the radiation factor. The choice of protective measures is governed by effective doses, both received and projected, also established and adopted intervention levels, respectively. The effectiveness of the countermeasures is measured by the value of an averted dose.

The lessons learned from Chernobyl show that the above single-factor pattern of radiological protection is appropriate only at an acute post-accident phase (days and weeks after an accident) when the radiation factor prevails and basic countermeasures are proceeded from pre-arranged intervention levels.

At the next long-term phase (months, years after the accident) there is enough time for a human factor to come fully into force. This factor implies the psychological and social acceptance, by the public, of the countermeasures to be implemented and the response of the public to their implementation, the reflection of the situation by mass media, the reaction of Legislative and Administrative Bodies. A non-optimal, or wrong, strategy at the long-term stage can aggravate essentially the situation as a whole. In this context it is instructive to consider the chain of decision making and corresponding feedback events after the Chernobyl accident.

2 The measures implemented on the early stage of the Chernobyl accident

In 1986, at the early stage of the accident, the regulation was based on the Criteria for the Decision Making on Measures to Protect the Public in the Event of a Reactor Accident, adopted by the USSR Ministry of Public Health in 1983 [1, 2, 12].

On 12 May 1986 the USSR Ministry of Public Health set annual dose limit for the public exposure for the first post-accident year of 10 rem/a (100 mSv/a). On 23 April 1987 the same ministry set new annual dose limit for 1987 of 3 rem/a (30 mSv/a) and on 18 July 1988 - the annual dose limit for 1988 - 1989 of 2.5 rem/a (25 mSv/a).

By the end of 1986 the situation had seemed fully under the control:

1. *The necessary relocations were made according to the "Population protection and evacuation plan for 30-km radius zone of the Chernobyl Nuclear Power Plant"* [17]:

- 27 April 1986 – relocation of 49,360 residents from the town of Pripyat.
- 28 April 1986 – evacuation of 5,000 members of the Chernobyl NPP staff from the town of Pripyat to the resorts within 30-km zone.
- 2 May – beginning of partial temporal evacuation of children and pregnant women from some settlements of Belarus.
- 2 May – 6 May – relocation of 28,242 residents from the Belarussian sector of the 30-km zone.
- 2 May – 3 May – relocation of 10,090 residents from the Ukranian sector of the 10-km zone.
- 4 May – 5 May – relocation of 28,242 residents from the Ukranian sector of the 30-km zone.
- 6 May – 31 May – relocation of 35,700 residents from the settlements situated near 30-km zone on the western trace (Ukraine).
- 3 June – 10 June – relocation of 6,017 residents from the settlements situated near 30-km zone on the northern trace (Belarus).
- 30 September – the finish of the relocation of the residents from the settlements situated near 30-km zone.
- In 1986 relocated were in all 116,000 people from 188 settlements (including the towns of Pripyat and Chernobyl).

2. *The territories affected by the accident were zoned:*

- 30 April – drawing out a map of gamma-fields near the Chernobyl NPP and the affected regions of the Ukraine and Belarus with isolines from 5 mR/h and up.
- 5 May – the first map of ground contamination.
- 10 May – the more detailed map of gamma-field with isolines from 0.5 mR/h and up.
- May – based on the map of gamma-field and the established dose limit of 10 *rem* for the first post-accident year, the following zone scheme was suggested by the USSR Ministry of Health and the USSR State Committee for Hydrometeorology:
  - an exclusion zone (above 20 mR/h; about 400 km^2) where even temporal stay of people is prohibited;
  - a relocation zone (above 5 mR/h; 1100 km^2), where the stay of shift personnel engaged in eliminating the consequences of the accident is only authorized;
  - a controlled zone (3 to 5 mR/h; about 3000 km^2) with temporal relocation of children and expectant mothers, strict radiological survey and exclusion practically the consumption of local food products.
- July – August – the USSR State Committee for Hydrometeorology submitted detailed maps of contamination by ^{137}Cs, ^{90}Sr and $^{239,240}Pu$. These maps became the basis for decision on additional relocation of residents from some settlements (from 29 settlements in Belarus, 4 in Russia and one in the Ukraine).

3. *The continuous monitoring of radiation situation and the control of food supply were arranged:*

- 30 April – the derived intervention levels for ^{131}I in milk.
- 30 April – beginning of partial iodine prophylaxis in some settlements of Belarus [16].
- 6 May – the temporal derived intervention levels for ^{131}I in foodstuffs.
- 12 May – the dose limit 10 *rem* for the first year established by the USSR Ministry of Health.
- At the beginning of May the isotopic composition of radioactive fallout was determined for three directions of air-mass transport. This allowed to estimate the first-year and 50-year external doses for the population of the main contaminated regions.
- 16 May – the temporal derived intervention levels for beta-radioactivity levels in foodstuffs.
- 30 May – the temporal derived intervention levels for beta-radioactivity levels in foodstuffs instead of DILs from 6 May and 16 May.

4. *The delivery of clean foodstuff was organized to the regions where consumption of local food products was restricted.*

5. *Mass decontamination of the Chernobyl NPP site and settlements was under way:*

- October – the putting into operation the Unit I of the Chernobyl NPP.
- November – the construction of the "Sarcophagus" – (the cover for the damaged Unit IV of the Chernobyl NPP) was completed.
- November – the putting into operation the Unit II of the Chernobyl NPP.

6. *Measures on improvement of the medical servicing of the population have been taken.*

In 1986 – 1991 all decision making and measures implementation were promptly coordinated by the Governmental Commission with an Operational Group situated on April 26-27 in Pripyat and from April 28 – in Chernobyl.

3 Evolution of regulation related to the Chernobyl accident in the Former USSR

It has been often noted that the management of post-accident countermeasures was implemented under conditions when in some necessary aspects the regulatory documents were neither complete nor perfect, especially in respect to measures at the intermediate and long-term phases. The same may be said about the scientific basis (guides, recommendations, procedures, etc.).

Limited time, incompleteness and inperfectness of the relevant scientific data and recommendations, the changing social-political situation impeded development and adoption of adequate regulatory documents. Many of them adopted and acting in the period 1988-1990 were of temporary character.

3.1 Chernobyl 1988 Concept

In 1988, the USSR National Commission on Radiation Protection, headed by L.I. Il'in developed the "Concept of safe residence in populated areas contaminated after the Chernobyl accident" [1, 2]. That Concept was adopted by the USSR Ministry of Public Health on 22 November 1988. The discussions of this Concept (also known as the "35 rem Concept") aroused, coincided with the period of social agitation before the elections, when Chernobyl Affair just taken out from the "secret vaults" was exploited repeatedly. The harsh criticism concerned all the events, which led to the accident, also the protective measures put into practice soon after. The debates were led not always on a professional level. Moreover, the opinions of renowned specialists more often were ignored as "delegates of the parties concerned", and nonspecialists became "recognized experts". The mass media quickly took advantage of the situation by disclosing untrue or misinterpreted non-confirmed data. In this quarrelsome atmosphere, the "35 rem Concept" became the major point of the controversy (all the criticism being purely emotional and poorly argued), and was finally labeled as inhuman. On 25 April 1990 as a result of that discussion Chernobyl 1988 Concept was rejected by the Supreme Soviet of the USSR.

This Concept was based on using a lifetime dose as a measure of the lifetime radiological hazard. The main principles and criteria of Chernobyl 1988 Concept are the following [1]:

1. To establish limit of the individual lifetime dose 35 rem applied to the summary doses resulting from external and internal exposures.

2. The observation of represented dose limit is regulated by the mean individual dose equivalent in the critical group of each populated area.

3. The fixed standard includes the doses, which the population had been exposed to since April 26, 1986.

4. The fixed standard does not include doses incurred from natural background radiation.

The lifetime dose intervention level was established to limit late health effects caused by radiation exposure. On the territories where the predicted lifetime dose would not exceed that value, the limitation withdrawal was suggested from 1990. In areas where it was envisaged that lifetime dose intervention level would be exceeded, relocation should be made. A critical review of the given Concept and other possible ones was made in the International Chernobyl Project during the conferences on decision making [7].

For various reasons this Concept was not adopted by authorities. Main reason is connected with incompetent interference of social and scientific organizations, press, and representatives of local and central organs of power during discussion and adoption of the "35 rem Concept". Also it should be noted that the Concept itself was a simplified reflection of the scientific and practical experience of that time and the situation established in 1988-1989 in the regions affected by the Chernobyl accident. The one level system of making decisions on protective measures (below the specified level of 35 rem for lifetime no actions are required, above that level the relocation is obligated) would not allow any optimal use of the whole complex of possible protective measures.

Concerning the revision of the coefficients of radiological risk made in the late 1980s - early 1990s and accumulation of the post-Chernobyl experience, the International Regulatory Bodies (ICRP and IAEA) drawn up and published new Recommendations on Radiation Protection [11] and Safety Standards [6]. The USSR NCRP also planned to issue new Notinal Safety Standards in 1990. This work was not completed as the Commission's powers were cancelled about the disintegration of the USSR. That is why the USSR Radiation Safety Standards approved in 1976 and slightly amended in 1987 is formally in force in the territory of Russia now. The roots of Chernobyl 1988 Concept are originated from that Safety Standards, worked out on the base of the early ICRP Publications [8, 9]. It should be noted that they are markedly different from the ICRP 1990 Recommendations [11] for the same purpose because of the USSR NCRP's own opinion on radiation protection standardization and management was based on the concept of critical organ. Practically this position corresponds to the use of the threshold approach in assessing the radiological hazard. The first appearance of the term "effective dose" in an official regulatory document occurred in developing the Chernobyl 1991 Concept [3]. That is why in Chernobyl 1988 Concept was not used:

- the concepts and indices of radiological risk,

- the concept of effective dose, and

- the principle of optimization.

3.2 Radiological situation in the affected regions on the long-term post-accident phase

For the period of 1990–1992 relatively fast process of natural "selfcleaning" has continued. Half-time of decreasing of Cs content in agriculture production and respective internal local public doses is equal to 1.5–2 years and was in a good agreement with the ICRP model predictions [10]. Gamma-dose rate in open air and respective external public doses decreased by 10–20 % per year.

The annual radiation doses for population decreased in 1.5–2 times for this period due to influence of this process only. It was not expected earlier. More low decreasing process was predicted after 1990.

Previously adopted and currently applied countermeasures continue effectively to work.

"Soft" countermeasures (without strong limitations for the lifestyle and life and working conditions) such as food controls and corresponding agricultural measures make it possible to decrease internal doses in 10 times in those places where these doses might have essential values.

Everywhere in the regions with high original radioactive contamination ($> 15\,Ci/km^2$) at present:

- internal doses D_{int} are small (often very small). In 1992 they were equal to 0.02 - 2 mSv, the process of their decrease is continuing;

- ratio D_{int}/D_{ext} as a rule does not exceed 0.2 (without countermeasures this ratio in nonchernozem regions is equal or more than 1.);

- sum of doses $D = D_{int} + D_{ext}$ in 1992 did not exceed 5 mSv including population settlements designated for relocation by the recent decision of Bryansk region administration.

Exceptional is the situation in two villages where in 1992 the doses slightly exceeded established control level (5 mSv/a)

- doses received due to recreation in natural environment and due to consumption of natural foodstuffs (fish, mushrooms, berries, meet of wild animals and birds, etc.) are relatively small and could not essentially contribute to the cumulative local public doses.

In the regions with the initial contamination in the range of $5 - 15\,Ci/km^2$ the natural "selfcleaning" process and "soft" countermeasures can now guarantee the annual doses, which are much less than the control level of 5 mSv and in the most of settlements less than the non-action level of 1 mSv. In the regions with the ^{137}Cs fallout $1 - 5\,Ci/km^2$ where no countermeasures were adopted public doses in 1991 and 1992 were everywhere below the "non-action" level of 1 mSv per year.

3.3 Chernobyl 1991 Concept

The Soviet Government recognizing the vital importance for the Chernobyl Concept to be acceptable by the public, entrusted the Academy of Sciences (In which Coordinating Council for Chernobyl Scientific Problems was working since November 1986) with the examination of the "35 rem Concept" and all its possible alternatives. The goal of that work was to elaborate a synthetic position, which could be acceptable for the concerned Republics and institutions. The long-lasting and heated debates, which took place during the Council meetings led to nothing but precarious compromise. An agreement already, signed by the specialists from all official institutions, was changed many times after separate debates in each Republic.

After rejection of the "35 rem Concept" the USSR Supreme Soviet and Government assigned the USSR Academy of Sciences to draw up a concept of long-term protective measures in cooperation with representatives of other organizations. To elaborate this concept an ad hoc working group of about 60 specialists headed by academician S.T. Belyaev was formed and confirmed. The working group consisted of representatives of some All–Union and republican scientific organizations including also those who subjected to criticism the recommendations of national and international competent organizations on post-accident protective measures. The leadership of the USSR NCRP refused to send its representatives in this working group.

The Chernobyl 1991 Concept of protective post-accident measures was drafted [3] by the end of 1990 and after discussion and correction was approved on the 8th of April, 1991 by the USSR Government. Authorized Bodies of Belarus, Russia and Ukraine worked out and adopted also their version of the Concept, which were in main principle points very similar to the All-Union Chernobyl 1991 Concept [18].

Main points of the Chernobyl 1991 Concept are the following.

1. Initial premises

- protective measures made it possible to reduce essentially the public exposure in 1986 - 1990;

- socio-psychological factor (stress, state of fear or anxiety) has taken on great importance alongside the radiation exposure. This is a post-accident syndrome typical in any extreme situation, but intensified by incomplete or misrepresented public information about the real situation, inadequate decisions on protective measures and their untimely execution;

- implementation of radiation protection measures must simultaneously be aimed at relaxation of socio-psychological tension and stresses;

- as far as possible, to avoid mandatory mass relocations, taking into account of the radiation protection criteria recommended by International Bodies during the implementation of the State Union-Republic program of urgent measures (1990–1992) for mitigation of the accident consequences.

2. Principles and criteria

- The irradiation dose due to the Chernobyl accident shall be the basic index for making decisions on protective measures, their character and scale, as well as compensating for damages.
- The excess (over the natural and technogenic radiation background for given locality) of the public exposure from the Chernobyl fallout is permissible and doesn't demand any intervention if an average annual effective dose is lower than 1 mSv for 1991 and following years.

  At the level of 1 mSv and lower, the conditions of living and working activity of the population do not require any restrictions.
- At a higher level than 1 mSv (0.1 rem) per year (over the natural and technogenic background), protective actions (countermeasures) should be taken.

  Complex of protective measures should be aimed at the continuous reduction of radiation exposure and of the level of contamination of foodstuffs, while simultaneously weakening those restrictions which upset the usual way of the life and vital functions. Achievement of these goals should be optimized with the condition that an average individual effective dose equivalent does not exceed 5 mSv in 1991, with a maximum possible decrease of this limiting level up to 1 mSv in future.
- Voluntary relocation can by reckoned among the countermeasures. Each person living in a contaminated territory shall have the right to make own decision about continuing to live in the given territory or going to another place of residence, based on unbiased information about the radiation situation, socio-economic and other aspects of life. Any decision adopted should not give a direct economic advantage.

Then basing on these Concepts the Chernobyl All-Union [13] and republican laws [18] were worked out and adopted by Supreme Soviets. These laws turned out to be considerably different by some their principle points from the relevant Concepts and each other. Below the differences and contradictions between the Concept [3] and Law [13, 14] in Russia are considered.

4 Post-accident management in Russia

4.1 Governmental structure

The Russian State Committee on mitigation of consequences of the Chernobyl accident was set up according to the Decree of the Russia Supreme Soviet Council by 19 September 1990, N^0 172-1. Powers and responsibilities of the Committee were additionally specified by the Decree of the Russian Council of Ministers by 15 March 1991, N^0 151. It was declared by the last Decree that the Committee should control all activity related to mitigation of consequences of the Chernobyl accident and other abnormal radioactive contaminations on the territory of Russia including - coordination of the activity of other Ministries, Committees and Governmental Departments related to the Chernobyl affairs, - preparation and fulfillment of the State Program on mitigation of consequences of the Chernobyl accident, - organization of scientific research, - international cooperation related to the Chernobyl affairs.

By the Russian Federation Government Decrees by 16 February 1992, N^0 91 and 14 May 1992, N^0 316 the Russian Scientific Commission on Radiation Protection (Russian NCRP) headed by A.F. Tsyb was set up. Main objectives of the Commission are working out conceptual regulation documents on radiation protection of the public in normal and abnormal conditions.

By the Decrees of RF President and Council of Ministers (24.08.92, N^0 445-pn and 25.03.93, N^0 243) the Chernobyl State Committee was transformed into the State Committee of the Russian Federation for the social protection of the public and rehabilitation of regions affected by Chernobyl and other radiation accidents. The responsibilities of the Committee were expanded to other territories abnormally radioactive contaminated.

The modern state governmental structure reflects the recent RF President decision (Decree N^0 66 by 10 January 1994) on the reorganization of the RF Council of Ministers. The main element of this structure is the Ministry of the Russian Federation for Civil Defense Emergencies and Elimination the Consequences of Natural Disasters (MCDE). It was combined from the former RF State Committee of Emergency management and the former RF State Committee for the social protection of the public and rehabilitation of regions affected by Chernobyl and other radiation accidents.

4.2 Current regulation documents

Since mid-1991 the practical activity on the elimination of consequences of the Chernobyl accident has been regulated by the Law [13] and the Concept [3]. However it became clear already in 1992 that the regulatory documents connected with both the elimination of consequences of the Chernobyl accident and other applications needed to be further improved and developed. It was caused by several reasons. Main of them are the following.

1. The Chernobyl 1991 Concept [3] and the Law [13] have limited application: only to the situation, turned out after 1990 (1991 and following years) in the regions suffered from the Chernobyl accident.

2. In the Law there are serious contradictions and nonjustified principles that prevents from optimal implementation of long-term protection and restoration measures. Besides implementation of the Law created additional social problems.

3. In the period from 1990 - 1993 the fast reduction of the contamination levels from natural "self cleaning" levels has continued. The annual public exposure have decreased by a factor of 1.5 - 2 by this process only. A more slowly decreasing process was expected after 1990. Previously adopted and currently applied countermeasures continue to work effectively.

4. In Russia there are several contaminated regions besides the regions suffered from the Chernobyl accident (Ural region, territories near nuclear weapons test sites, etc.). Since 1991 the issues of radiation protection, social rehabilitation and economic compensation in these territories have been under consideration by scientists and local and state authorities. The experience from these areas was used to reconsider the past recommendations on intervention strategy and intervention levels.

In the first part of the Law ("General Provisions") the public dose is taken (agree with the Concept) as a main index for decision making. It established that average annual dose equal to 1 mSv is acceptable and does not require any intervention (non-action level). Nevertheless inspire of these justified provisions another parts of the Law are based on another index: a level of soil contamination (Ci/km^2). This index is used for decision making in the territorial zoning, population relocation and other countermeasures allowances and compensations. Between this index and the dose there is no direct relationship. Moreover due to the process of fast natural "selfcleaning" and due to countermeasures adopted, the difference between them is increasing.

Now, even on territories of the Bryansk region with high level of initial contamination (more than 15 and even more than 40 Ci/km^2) the reliably measured doses are small and continue to decrease further quickly. So called "soft" countermeasures, i.e., countermeasures without strong limitations for the lifestyle and working conditions such as foodstuff control and corresponding agricultural countermeasures, can reduce the internal doses from intake of contaminated foodstuffs by a factor of up to 10 in those areas where ingestion doses are significant.

So it is important to note that in the end of 1992 the levels of radiation exposure and socio-psychological conditions have changed considerably. Rehabilitation of these areas is therefore possible, at least in the Russian territories that suffered from the accident. This rather quick changing of the situation in the Chernobyl region that was not fully recognized previously should be considered in decision making on an optimal intervention strategy.

Considering these demands on improved regulation documents it was planned by some responsible organizations (Russian NCRP; the Chernobyl State Committee and others) to develop in 1992 - 1994 new improved recommendations and guides on carrying out protection and restoration measures after nuclear accidents considering those occurred in the past and probable future one's. It was understand that in these documents one should

- consider as interacted all post-accident phases: early, long-term and a final restoration (rehabilitation) one's,

- develop in more details not only radiation but also social protection aspects.

Here we note three documents now being finished in development:

1. Recommendations on practical realization of the current Concept of social protection of the public in the regions suffered from the Chernobyl accident in the conditions of rehabilitation and restoration phase begun (Chernobyl 1993 Recommendations);

2. A Concept of rehabilitation of the public and normalization of ecological, sanitary and socio-economical situation in settlements of Altai region, located in the zone affected by nuclear weapon tests on the Semipalatinsk proving ground (Altai 1993 Concept);

3. A Concept of radiation, medical, social protection and rehabilitation of the public of the Russian Federation affected by accidental exposure (Post-accident 1995 Concept).

Below a short description of these documents are given.

4.3 Chernobyl 1993 Recommendations

The main goal of the "Recommendations on practical realization of the current Concept of social protection of the public in the regions suffered from the Chernobyl accident in the conditions of rehabilitation and restoration phase begun" (Chernobyl 1993 Recommendations) [15] was to make proposals on somewhat urgent changing radiation protection strategy in the Russian regions suffered from the Chernobyl accident

- considering quick improvement of the situation in these regions,

- not going far from the frames of the current Law [13, 14].

It is proposed:

1. To state as a worthwhile the transfer in the public social protection:

- from strict measures of radiation protection (mandatory relocation, rigid limitations to lifestyle and to economic activities, etc.) to the measures for rehabilitation of territories and to restoration of normal life and economic activity;
- from providing compensations and allowances to individuals to the allowances for improving social and economic conditions in the public sector.

2. While continuing radiation controls to soften limitations on recreation in natural environment and on consumption of natural foodstuffs such as fish, mushrooms and meet of wild animals and birds.

3. For the settlements left after relocation to prepare the programme for rehabilitation with the aim to stop the process of degradation of agricultural lands and the process of degradation of residential, industrial and social infrastructure.

For this purpose it is proposed by creation preferential economic conditions:

- to stimulate return of formerly relocated settlers if they wish to return;
- to attract there socially and economically active people by creating legislative and tax preferences and advantages for developing industry and agriculture and for restoration of economic infrastructure and social and cultural life.

4. Relevant Competent Bodies to display legislation initiative to set paragraphs of the current Law and new Laws that are under consideration in the Russia Supreme Soviet and concerned to radiation protection policy in Russia into accordance with new conditions on the contaminated territories and experience in the Chernobyl accident consequences mitigation.

4.4 Altai 1993 Concept

This "Concept of rehabilitation of the public and normalization of ecological, sanitary and socio-economical situation in settlements of Altai region, located in the zone affected by nuclear weapon tests on the Semipalatinsk proving ground" (Altai 1993 Concept) [4]; was developed following the Russian Government commission (1992) decision to work out a program of rehabilitation of the public in settlements of Altai region, located in the zone affected by nuclear weapon tests on the Semipalatinsk proving ground.

As it was recognized after open publication data about these nuclear weapon tests (in 1949 - 1962 and 1965 years) and the recent scientific research considerable part of the Altai region territory was seriously suffered from them. Peculiarities of the modern situation are the following:

- exposure of the population took place in the past, in the period of the nuclear weapon tests, now the relevant current public doses are insignificant;
- due to specific features of the radiation source considered the public exposure should be mainly ascribed to so called "acute" irradiation. For such exposure radiological risk coefficients are higher then for chronic exposure;
- there are some problems with health of the local population.

In the Concept two dose levels were established for decision making only on social protection:

$$
\begin{aligned}
D_{S1} &= 0.05\,Sv \\
D_{S2} &= 0.25\,Sv
\end{aligned}
\tag{1}
$$

The low level D_{S1} is practically a non-action level: any social protection measures are introduced if only total individual doses D from the nuclear weapon tests are more then D_{S1}. In dependence on value of D population under social protection activity is subdivided into two categories:

- the first - persons with doses $D > D_{S2}$ and their children and grandchildren;
- the second - persons with doses D in the interval $D_{S1} < D \leq D_{S2}$ and their children and grandchildren.

The collective and individual (only collective) social protection measures are envisaged for the first (second) category. They include:

- economic compensation,
- improvements in social sphere (medical and other services, culture, etc.),
- better environment protection.

4.5 Post-accident 1995 Concept

The main goal of the "Concept of radiation, medical, social protection and rehabilitation of the public of the Russian Federation affected by accidental exposure " (Post-accident 1995 Concept) [5] was to make proposals on changing radiation, medical, social protection and rehabilitation strategy in the Russian regions suffered from the radiation accidents and nuclear weapons test many years ago:

- the Chernobyl accident (Central and Western Russia);

- the Kyshtym accident and other Ural radiation accidents (Ural region);

- the nuclear weapons test sites (Altai and other regions);

- other radiation accidents.

Taking into account the demands of new international recommendations [6, 11] it was planned to develop in 1992 – 1994 new improved recommendations and guides on carrying out protection and restoration measures after nuclear accidents considering those occurred in the past. Current Post-accident 1995 Concept was worked out by the Russian NCRP, headed by A.F. Tsyb. The Concept was adopted on 17 August 1995 by the Russian NCRP and the Ministry of Russian Federation for Civil Defense Emergencies and Elimination the Consequences of Natural Disasters.

That Concept is the extension of the Chernobyl 1991 Concept [3], Chernobyl 1993 Recommendations [15] and Altai 1993 Concept [4]. Main topics of the Post-accident 1995 Concept are the following.

1. Radiation protection of the public, including

- set of recommended countermeasures;
- procedure for zoning the affected territories; and
- procedure for forming the register of exposed persons and persons suffered from the accident.

2. Medical protection and rehabilitation of the public.

3. Psychological protection and rehabilitation of the public.

4. Social and economical protection and rehabilitation of the public.

5. Juridical protection of the public.

Three annual effective dose levels (E_Z) are established in the Concept for decision making on radiation protection including zoning:

$$
\begin{aligned}
E_{Z1} &= 1\,mSv/a \\
E_{Z2} &= 5\,mSv/a \\
E_{Z3} &= 20\,mSv/a,
\end{aligned}
\tag{2}
$$

where E_{Z1} is the non-action level. That levels have to be compared with E_A, an annual effective dose, averaged over inhabitants of the settlement. The value of E_A for real territory one mast find under the condition, when the main remedial actions are stopped. Two zones were suggested:

- zone of radiation control ($E_{Z1} < E_A < E_{Z2}$);

- zone of limited living ($E_{Z2} < E_A < E_{Z3}$).

Those who want to leave the last zone can obtain the help from Governmental Bodies. It is not desirable for families with children to arrive here for living. It is necessary to explain possible health risk for those who plan to arrive here for living. Zoning should be reconsidered once every three years.

Considering the situation in 1995 one should note that there are practically no settlements where the annual dose exceeds 5 mSv. Really the action levels 5 and 20 mSv/a regulate the arrival of people for living and restoration of life on the territories suffered from the accidental radioactive contamination. They are not levels that regulate relocation. Relocation should not be called on the late phase of a post-accident activity. Decisions on relocations should be done only in the early phase. It means that in Russia regulation level 20 mSv/a is not the action level for relocation.

Medical and social protection and rehabilitation of the public in the Concept is based on the definition the cohort of *exposed persons*, cohort of *persons, suffered from the accident* and *high risk cohort*. The organization base for all action is the National Radiation and Epidemiologic Register.

By the Concept an *exposed person* is the person with effective dose more than 50 mSv of acute accidental exposure or the person with effective dose more than 70 mSv of chronic accidental exposure.

By the Concept a *person, suffered from the accident* is the person with disease, which connection with radiation accident was confirmed in response to special procedure, established by the Ministry of Public Health and Medical Industry of the Russian Federation.

Following the Concept the National Radiation and Epidemiologic Register includes:

- persons, suffered from the accident;

- exposed persons and their kids were born after the accident;

- exposed *in utero* persons with average body equivalent dose more than 10 mSv;

- adults with thyroid dose more than 0.5 Gy;

- kids with thyroid dose more than 0.2 Gy; and

- persons, relocated from the contaminated territories.

The *high risk cohort* includes the persons, suffered from the accident and exposed persons with high doses:

- adults with effective dose more than 250 mSv of acute accidental exposure or the person with effective dose more than 350 mSv of chronic accidental exposure;

- exposed *in utero* persons with average body equivalent dose more than 50 mSv;

- adults with thyroid dose more than 2.5 Gy; and

- kids with thyroid dose more than 1.0 Gy.

The collective and individual protection measures are specially envisaged for each cohort.

5 Conclusion

In this paper we considered needs and grounds for developing new generation of the post-Chernobyl regulation documents. To meet urgent needs some special regulation documents [3] – [5], [13] – [15] (see [18] for more details) were worked out and adopted in Russia, Ukraine and Belarus until 1995.

More general and principle regulation documents are in the stage of preparation and adoption. The most important among them is the Safety Standards (Norms) for Protection Against Ionizing Radiation – "NRB–95", prepared by joint working group of Belarussian and Russian scientists, headed by the member of ICRP, professor P.V.Ramzaev. That working group worked in a strong contact with NCRP of Russia and Belarus. The draft of that document includes chapter appeared as a result of analysis of our Chernobyl experience: "Requirements to restriction the public exposure in emergency situations", inclusive the guidelines for intervention levels in emergency exposure situations.

Now there is a real hope to have in coming future considerably developed and improved regulation documents of a new generation. One of the main task during their preparation is

- to learn all necessary lessons from the Chernobyl and other accidents, post-accident activity and results of relevant scientific research,

- to use effectively the results of the international cooperation in the post-Chernobyl scientific research, primary the results of seven CEC - CIS projects (ESPs and JSPs).

References

[1] Avetisov G. M. Policy of the National Commission on Radiation Protection after the Chernobyl accident. Derived interventional levels for food // Historical perspective of the countermeasures taken following the Chernobyl accident. Reflections on the Concepts and Regulations adopted in the CIS for Post-Accidental Management.– Centre d'etude sur l'Evaluation de la Protection dans le domanie Nucleare (CEPN).– Fontenay-aux-Roses, France.– CEPN Report N^0 225.– 1994.– P. 113 – 136.

[2] Avetisov G.M., Buldakov L.A., Gordeev K.I., Il'in L.A., The policy of the USSR NCRP on the substantiation of temporary annual dose limits for public exposure due to the Chernobyl accident // Medical Radiology, N 8, P. 3.– 1989 (in Russian).

[3] A Concept of living conditions for people in the regions affected by the Chernobyl accident (prepared by the USSR AS working group (chairman - Belyaev S.T.), adopted by USSR Government on 08.04.91, Resolution N^0 164. The English translation: Concept of Safe Living Conditions for People in the Regions Affected by the Chernobyl Accident.– Ibidem [1], P. 331 – 338.

[4] A Concept of rehabilitation of the public and normalization of ecological, sanitary and socio- eco-nomical situation in settlements of Altai region, located in the zone affected by nuclear weapon tests on the Semipalatinsk proving ground, prepared by the Chernobyl State Committee working group (chairman Gordeev K.I.) and the Russian NCRP working group (chairman Demin V.F.), adopted by the Russian NCRP on 07.05.93.

[5] A Concept of radiation, medical, social protection and rehabilitation of the public of the Russian Federation affected by accidental exposure, prepared by the Russian NCRP working group (chairman Tsyb A.F.), adopted by the Russian NCRP on 17.07.95.

[6] International basic safety standards for protection against ionizing radiation and for the safety of radiation sources.– Vienna.– IAEA.– (Safety series, 115).– 1994.

[7] The International Chernobyl Project. Technical Report. Report by an International Advisory Committee.– IAEA.– Vienna.– 1991.

[8] ICRP Publication 1.– Recommendations of the ICRP Adopted 9 Sept. 1958., Oxford, Pergamon Press, 1959.

[9] ICRP Publication 9.– Recommendations of the ICRP Adopted 17 Sept. 1965., Oxford, Pergamon Press, 1966.

[10] ICRP Publication 29. Radionuclide release into the environment: Assessment of doses to man // Ann. ICRP.– 1979.– V. 2.– N 2.

[11] ICRP Publication 60. Recommendations of the Commission – 1990.– Ann. ICRP.– V. 21.– N. 1 – 3, 1991.

[12] Konstantinov Y.O. Interventional levels in the USSR before the Chernobyl accident and early countermeasures.– Ibidem [1].– P. 29 – 58.

[13] The Law of Russian Federation "On social protection of the public suffered from radiation exposure due to the Chernobyl accident", 15 May 1991.

[14] The Law of Russian Federation "On amendments to the RF Law "On social protection of the public suffered from radiation exposure due to the Chernobyl accident", 18 June 1992, N 3061-1.

[15] Recommendations on practical realization of the current Concept of social protection of the public in the regions suffered from the Chernobyl accident in the conditions of rehabilitation and restoration phase begun, Russian NCRP working group (chairman S. Belyaev), adopted by the Russian NCRP on 13.01.93.

[16] Savkin M., Niggiyan A. Iodine prophylaxis countermeasures.– Ibidem [1].– P. 59 – 86.

[17] Savkin M. History of zoning processes after the accident at the Chernobyl nuclear power plant.– Ibidem [1].– P. 157 – 186.

[18] Sivintsev Y. New concepts of population radiation protection in radioactive contaminated areas adopted in the CIS.– Ibidem [1].– P. 137 – 156.

II. ENVIRONMENTAL ASPECTS
OF THE ACCIDENT
A. Behaviour of radionuclides in contaminated territories

EC Contribution to The Evolution of The Objectives of Radioecological Research in Relation To The Radioactive Deposition and Its Impact on Land Use and Environmental Management After The Nuclear Accident at Chernobyl

Gilbert DESMET

EC-DGXII-F-6

Rue de la Loi, 200, 1049 Brussels

Abstract. The uncontrolled release of radionuclides coming up after the Chernobyl accident has led to a large number of scientific and political activities to assessing the contamination of the environment and the consequences for the population. A large scale of measures were deployed attempting to mitigate the consequences and initiatives were launched to follow the fate of the radionuclides in and around the Chernobyl area. Some of these efforts are described in this paper. It summarizes which way radioecologists had chosen to evaluate the problem, to compare the scientific culture existing in East and West, to sharpen their views on the fundamentals of radioecology and to test their knowledge in the real field.

1. INTRODUCTION

In April 1986 Chernobyl 4 exploded and large areas around the nuclear power station were contaminated. The cloud of contamination spread further on largely over Europe and elsewhere. The track of this cloud is well known. Activities were deployed to comprehend the amplitude and intensity of the contamination of the environment. A few million people were potentially affected by the deposition of radioactivity from this accident. People were affected by this accident as there were potential health risks due to external radiation or through inhalation of the volatile, but short-lived Iodine-139. The environment was influenced by the short-lived radionuclides, equally though longer lived radionuclides were deposited such as Caesium-137, Strontium-90 and some even longer lived such Plutonium, Ruthenium and others.

Agricultural land, water catchments with their tributaries and rivers, as expected, became contaminated; the deposition of radionuclides had no preferences though, and forests as well as extensively used pasture and semi-natural land were touched by the deposits. Around the reactor zone, later called "exclusion zone", radionuclides were deposited as simple chemical ionic entities as well as in particulate form with a more complicated chemistry. The more distant from the nearby deposition area, the more the radionuclides deposited were in the normal ionic form.

Radioecology which is the scientific discipline studying problems of "interaction between the ecosphere and radioactivity and how to deal with it" [2] was confronted with a real problem!

2. DESCRIPTION OF THE RADIOECOLOGICAL CONCEPT

2.A The fundamentals of radioecology

The use the population makes of its food and industrial products and of their general interaction with the same environments, may lead to external and internal radiation doses.

Different stands could have been taken to assess the size of the environmental impact of the Chernobyl event. Such an event could have been traditionally quantified by a simple monitoring of contamination levels in the environment and changes against time. This would signify lumping together sometimes very complex events in simple numbers and expressions such as 'e.g.' Concentration Ratios, being the ratios of concentrations in successive compartments of the environment (soil, plant, animal, water, fish etc.). By straightforward multiplication of Concentration Ratios or Transfer Factors a global and quick view on the temporary contamination in target compartments such as food and water can be obtained.

The objective, however, of all the experimental works in the radioecology area is not only to be able to determine the dose to man from different ecosystems. It is also and even to a larger extent its aim to understand the parameters which determine the fate of radionuclides at the **medium and long term in the environment** and therefore the **medium and long term** contribution to the dose. This knowledge is also essential when envisaging adequate measures which would sufficiently reduce the dose without though a drastic and unacceptable change of the ecological quality and thus of the economic value of the treated ecosystems.

There are some ecological rules that govern the fate of radionuclides in the environment and questions to be solved therefore. It is the task of a radioecologist to unravel the dominant mechanisms that govern this interaction. They are e.g. the fate of the radionuclides after deposition; how relevant are deposition rates for long-term dose calculations; what are the mechanisms that determine the long-term fate of radionuclides. When any xenobiotic substance comes into contact with the biosphere a number of dynamic processes changes the chemical properties of these substances. They are thence transferred as **chemical** entities from one ecosystem to an other. Gradually ratios of initial deposited amounts are appearing in neighbouring ecosystems. Is it correct to consider the initial values after e.g. deposition as "real" data expressed as they are usually as "amount of contamination per kg, or amount of contamination per square meter?" Phenomenologically it may look so, mechanistically however these data are only lumping a number of events together giving the semblance of a single number, whereas in reality it is a composed one.

Also, nuclides are assumed to reach equilibrium with the environment, and transfer coefficients are defined as steady-state concentration ratios between one physical situation and another. Straightforward chemical equilibrium is, however, totally strange to a dynamic biological nature. In its composition every living being is very different from its surroundings. It is also the concrete materialisation at a time t of a throughflow of matter and energy against time (**flux**). The thermodynamics applicable to living systems (including ecosystems, thus also comprehending non-living elements interacting with the living element) is the thermodynamics of **open** systems instead, being crossed by a flux of matter and energy, that keeps these systems **far from the equilibrium.**[3] It means "in concreto" that radionuclides are constantly flowing from one ecosystem to another, that the ecosystems' components in the mean time grow and decline, and that nothing is constant at all. It means changes in concentration can embody as well changes in total inventories of the radioactive material as changes in the size of the area or volume of contamination. These fluxes of matter are in the mean time paralleled by fluxes of energy consumed for maintenance of the environmental stability and thence very much also for the stability of the human consumption.

This rules out the use of simple straightforward multiplication of concentrations and concentration changes at modelling of such events or when planning any measure to mitigate the accidental situation.

The principles outlined above apply therefore; the deposition on agricultural land can be transferred from the contaminated soil to its subsoil or to the crops grown on that land; the deposited radionuclides are captured by forest canopies, littered to the soil and recycled in the typical vegetation such as trees, shrubs and mushrooms, all potentially of use for industrial or human consumption; rivers and water of lakes can be directly contaminated or the radioactive material which was caught by plants and soils (catchments) can run off to these water reservoirs or reach the very same reservoirs through seepage to the underground water table.

2.B   Application of the fundamentals of radioecology in the mitigation of the consequences of an accident and environmental management

Areas highly contaminated by a radioactive release may require measures that render them again ecologically accessible and economically usable. Several approaches can be envisaged to reach this objective and the measures may vary in their intensity of action!

In the management of contaminated areas and application of a countermeasure strategy for reduction of both the external and internal doses to the population, there are several factors to be considered. They are such as the potential radiation dose to the public to be averted by application of the countermeasure or the dose to the workers carrying it out; the direct economic cost of the countermeasure and the existence of dose and activity concentration limits imposed by authorities. The selection of the most appropriate countermeasure depends on the specific circumstances, such as availability of resources and the extent of the affected area, as well as its ecological and agricultural characteristics. Countermeasures can have significant economics effects.

3.   IMPLEMENTATION OF RADIOECOLOGICAL RESEARCH AFTER THE CHERNOBYL ACCIDENT. WHERE HAVE WE WORKED and WHY?

During the Chernobyl accident about 5% of the estimated tons of Uranium dioxide fuel and fission products were ejected into the atmosphere. The radioactivity was then very irregularly dispersed into mainly the Northern hemisphere. Roughly 65% of the contamination was deposited over Belarus, 20% over Russia and 10% over Ukraine. Belarus contains most of the highly contaminated areas (>40 Ci km^{-2} = >1480 Kbq m^{-2}). The inhomogeneous distribution can be demonstrated by the fact activities inside the 30-km zone are sometimes as low as 1 Ci km^{-2}, whereas at 50 km southwest of Chernobyl and 300 km northeast soil activities of more than 40 Ci km^{-2} have been measured.

In the environmental area, problems of contamination of land used for intensive agriculture, of zones of extensive agriculture, of zones of consumption of natural foodstuffs and problems of contamination of surface water and their food products were investigated through four directed Experimental Collaborative Projects (ECP). In the area of radioecology four ECP's were aiming at a basic understanding of environmental mechanisms, governing the fate of radionuclides.

- ECP 2: The Transfer of Radionuclides through the Terrestrial Environment to Agricultural Products, Including the Evaluation of Agro-Chemical Practices for Countermeasures.
- ECP 3: The Modelling and Study of the Mechanisms of Transfer of Radioactive Material from Terrestrial Ecosystems to an in Water Bodies.
- ECP 5: The Behaviour of Radionuclides in Natural and Semi-Natural Ecosystems (Forests, Marches, Heather, etc.)
- ECP 9: Transfer of Radionuclides to Animals, their Comparative Importance under Different Agricultural Systems and Appropriate Countermeasures.

In relation to this inhomogeneity, the work of the ECP's 2, 3, 5 and 9 have not been restricted to the Chernobyl exclusion zone. Investigations have been carried out in areas where there was little change in the occupation of the land by the population, and were the traditional living habits were maintained.

The contamination was dispersed over various sorts of landscapes of economic use. Research was thence conducted by four teams looking at the effect of contamination on agricultural land and on pasture land (meadows) (ECP2 and 5), on forested areas (ECP5) and on drainage areas of rivers and lakes (ECP3). A global study on the impact of land use and food consumption habits was carried out by ECP9. Such included the comparison between the effect of the consumption of food products from private (small-scale) farming and those of industrial (collective) farms.

Research activities were conducted in the exclusion zone and in the large periphery where either the activity was high or where mainly the population makes intensive use of the local products. The necessaries of life consist of provisions for food, water, clothing, housing etc. They come mainly from farming and general husbandry, water reservoirs and deducted water supplies, manufacturing and so forth.

Another criterion for the selection of research location is the soil type. Soil type indeed does play a dominant role in the behaviour and fate of radionuclides. Their (bio)-availability to a large extent depends on the way and intensity the soil binds and releases the deposited radionuclides for further integration in the foodchain or for deeper migration.

The places of research are located
- in Ukraine in the 30-km zone at Chistogalovka and Kopachi, outside the zone at Poleskoe, the Rovno region (Sarny), etc.
- In Belarus at Bragin, Vietka, Mogilev, Gomel, etc.
- In Russia in the Bryansk, Kaluga and Tula region, etc.
- The drainage area of the Pripyat-Dnieper river-reservoir system

4. EVALUATION OF THE RESULTS OF THE RADIOECOLOGICAL RESEARCH PROJECTS FOR A LONG-TERM ASSESSMENT OF THE PERSISTENCE OF ENVIRONMENTAL CONSEQUENCES OF THE ACCIDENT

4.A The fundamentals of radioecology

Important phenomena came to the attention through the radioecological research after the Chernobyl accident.

Research on plainly distribution and redistribution as well as on the physico-chemistry and bioavailability of the deposits gave the main following results:

- The deposition of the radionuclides after the accident was very heterogeneous, not only from a geographical point of view but also from a physico-chemical point of view. In the exclusion zone around the exploded reactor a considerable amount of the radioactive material was deposited as "hot particle", it means complex amorphous particles containing variable amounts of elements. The further away from the location of the accident the more though the form of the deposits became simpler and more ionic.

- "Hot particles" are such that they have a low bio-availability and are less swiftly assimilated by the vegetation or animals. In the course of time though "hot particles" are beginning to disintegrate and the availability of the composing elements such as Sr increases, and effects its radio-ecological half-life.

- Soil types and their concurring chemistry are paramount for the long-term behaviour (bio-availability and migration) of radionuclides through its control of their absorption/desorption features. The availability of Cs in arable soils, with a high clay content is rather low when at least perturbations are under control or envisageable. The problem of availability of Cs in soils with high organic content is however still considerable such as in meadow pastures with low quality soil. The availability remains considerable as these soils hardly can benefit of the strong binding properties of clay particles, if present. The high proportion of organic matter confuses the clay capacities, and the radioactive material shows a persistent bioavailability. Moreover, there always remains the problem of problem of some "reversibility" of the bonded radioactive nuclear material, especially Caesium; it is clear from the research done that a great deal of attention still should be paid to kinetic aspects of radionuclides fixation in soils!

- Seminatural ecosystems, especially coniferous forests intercepted considerable amounts of radioactive material, and litter fall takes it to the soil surface. The cycling of matter in semi-natural environments is a well-known phenomenon. The matter flows as well through migration as through native soil processes of soil horizon production below the upper surface. From there on the radionuclidesbecome available for further migration to deeper layers if the horizon profile is fit for it and could from there on seep through to the underground water table; they become also available for tree root uptake and further cycling in the tree; it takes up to four years before the RN's deposited on forest canopies fully enters the biogeochemical cycling of a dynamic ecosystem. "Root" uptake thence prevails. The radionuclides are mainly delayed in Oh (Organic humic layer zone) horizon. The depth depends on the soil type, and microbial life can perturb profiles. Mushrooms also are part of the cycle as their mycelium (fungal roots) are "grazing" their nutrients at different soil depths and thence accumulate the radionuclides in a species dependent way.

- Wild animals are a potentially important source of radioactivity; the transfer is high but very variable, so predictions of mean contamination levels are difficult. Wild animal contamination is considerable but being though connected mainly to the "exclusion zone" also a revival of the wildlife is noticed.

- In zones with a complicated hydrological system of vague river tributaries and canalization such as in northwest Ukraine, flooding of river foreland may result in increases in radionuclide concentration especially Caesium; this leads to important redistribution over the land.

- Two main sources of long-term contamination of water prevail:
  - annual flooding of the Pripyat flood plain
  - catchment transport, meaning that water reservoirs can become a long-lasting sink for contaminants through releases of radioactivity from peat bogs; the water pathway is thus a very important pathway for eventual transfer of contamination in the long-term to uncontaminated areas.

- The sediments of fresh water systems (rivers and lakes) can act as sinks but are subject to the surrounding chemical conditions for the further fate of the radionuclides; clays are important compounds of these sediments and the absorption/desorption dynamics depends on competitive ions such as potassium and ammonia. Models including straightforward "distribution factors" (Kd) ought to be considered with caution.

- Phenomena of very high concentration in fish in comparison to the level in the surrounding water were observed where fish species and habitat are key parameters.

4.B **Application of the fundamentals of radioecology in the mitigation of the consequences of an accident and environmental management.**
Beyond basic radioecological research, there was striven simultaneously after the application of the basic findings, and equally after the obtention of insights and a grip on the influence of features of general ecological nature including agricultural practices and living habits of the concerned population.

The research on the effect of agricultural practices included all sorts of soil amendments, such as the use of fertilisers or soil ameliorants. Some major results can be summarized as follows:

- The interaction between competitive ions (fertilisation!) for the reduction of the dose depends very much on the soil type and soil condition. Potassium (K) can for instance be successfully applied where no K saturation condition prevails, i.e. on poor or poorly fertilised land. Otherwise, the effect would be marginal! For the assessment of the effect of other soil ameliorants, processes and systems such as application of some zeolites or mulching of soils, that work well under laboratory or pilot conditions are still in a stage of circumstantial knowledge when applied in the field. From a scientific-technical point of view, a better "STRUCTURE" of the available and new knowledge has been attained. Loose statements about the use of fertilizers or manure, etc., can be tightened now, and all sorts of amendments to soils and animals such as mulching or of a chemical nature (Prussian Blue) can be better assessed and quantified. It is still true though that more data are available than there have been properly evaluated. Too many "random" measures have been tried and performed, which have not been profoundly (statistically) tested against the insights on mechanisms of behaviour of radionuclides in the environment, recently acquired! Efficiency could be enhanced if a consistent correlation exercise would be carried out further on, based on the wealth of data obtained. This is definite not only for soil and vegetation but true for all compartments of the environment.

- The economic structure of the ex-Soviet Union is special in the sense that the main provision of food stuffs is still provided by so-called Collective Farms, controlled by governmental regulations. Therefore arable land in the "exclusion zone" as well as outside on Collective Farm lands was studied in the three Republics, and the mechanisms exerting an influence on the dose to man assessed.

Consumption though of seasonally collected or privately grown food products is becoming an important common practice on large rural areas. The most salient result of these studies is the revelation food products from seminatural zones are playing a major role in determining intake to several particular groups and possibly also to the entire population. The zones of extensive agriculture production are private farms, semi-natural ecosystems and forests. This means herding cattle grazing on semi-permanent pastures and forested land, picking of mushrooms in the season, or producing food on the own garden. Rural populations including town habitants rely on subsistence farming and the use of nearby forest products, mainly mushrooms and are exposed to higher doses than in "collective farm" systems. Assessments of doses have to be comprehensive and include all pathways, since general knowledge of important sources may not be sufficient to describe radionuclide intake.

-       During the years a number of somewhat fortuitous measures were taken regarding hydrological measures to reduce or to avert the risk that could be caused by the flooding of the Chernobyl area. The years have taught to treat these floodings with caution in order to avoid undesirable side-effects and unexpected contamination redistributions. Fluvial planning and flood averting dike construction has thence been considered carefully and eventually became based on computer-simulated hydrological management.

GENERAL CONCLUSIONS

Conclusions regarding progress in methodology, some examples:

-       Questions such as "how deep should one muster in order to take a representative soil sample" are false ones if one does not consider horizon formation in natural soils!

-       The lessons we learned especially display how essential it is to remain constantly aware that random sampling is to be carefully interpreted in the light of the realization of the complexity of the habitat of animal and man; it means test samples are to be representative for the total habitat. They are being a society of life, living in and feeding from simple private farming and collection of foodstuffs in small villages, over industrially organised and exploited farming up to collecting provisions in urban areas.

Conclusions regarding the progress of radioecological science:

The complexity of the post-Chernobyl situation is immense and the need for an integrated scientific approach has been clearly shown. All systems are tightly connected .It is pointless to do just some isolated measurements or monitoring of contamination levels. These data are to put in a frame where urban zones, use of agricultural land and forest and finally water supply are looked upon as complex dynamically interacting environment. It is realised now that this complexity of contamination necessary leads to investigations of total inventory changes which are best defined as fluxes of matter and energy between different ecosystems.

Despite the progress made we still know not enough about *long-term effects of the contamination nor of the mitigative measures*, or in other words what will be the ultimate ecological quality and socio-economic usefulness of the treated zones? To which extent and for how long is the population of especially rural areas going to be affected by the fact that their land has a reduced nutritional quality and an economic value, and will be the health impact on people living constantly under the pressure of a low exposure?

Conclusions regarding the progress in environmental management:

It is necessary to do the research in a thrust of attempting to comprehend the complexity of the problem in an integrated way which lead at least to a "wake up" and to the awareness of the need for integrated environmental management .Not only plainly features of chemical and biological nature are to be investigated, but also features of human habitats and human industrial and agricultural activities have to be taken into account. The consideration of the latter features and activities have now become an integral part of the knowledge structuring of radioecology!

Acknowledgments : the author of this paper could not have written this paper without the dedication of the coordinators and all the scientists, both from the EU and Associated Countries and from the CIS, involved in the work of the Experimental Collaborative Projects, ECP2, 3, 5, 9.

References

Specific:

[1] Desmet, G.M., Van Loon, L.R. and Howard, B.J. 1991. Chemical speciation and bioavailability of elements in the environment and their relevance to radioecology. *Sci.Total Environ.,***100,** 105-124

[2] Desmet, G.M. 1992. Position of radioecology in the context of radiation protection. *I.U.R NEWSLETTER* **n° 7** 7-9

[3] S. Frontier, D. Pichod-Viale, *Ecosystèmes:structure-fonctionnement-évolution.*Masson, Paris, Milan, Barcelona, Bonn, 1991

General:

[4] O'Riordan, T. *Environmental Science for Environmental Management.* Longman Scientific & Technical, Harlow, 1994

[5] Allen, T.H. and Hoekstra, T.W. *Toward a Unified Ecology. (Complexity in Ecological Systems Series)* Columbia University Press, New York, 1992

Fluxes of Radionuclides in Agricultural Environments:
Main Results and still unsolved Problems

Rudolph ALEXAKHIN[1], Slava FIRSAKOVA[2], Gemma RAURET I DALMAU[3],
Nicolay ARKHIPOV[4], Christian M. VANDECASTEELE[5], Yuri IVANOV[6],
Serguei FESENKO[1] and Natalya SANZHAROVA[1]

[1] *Russian Institute of Agricultural Radiology and Agroecology (RIAERAE), 249020, Kaluga Region, Obninsk, Kievskoe str., Russia*
[2] *Byelorussian Institute of Agricultural Radiology (BIAR), Fediuninsky str., 16, 246007, Gomel, Belarus*
[3] *Facultad de Quimica, Universidad de Barcelona, Avenida Diagonal 647, E-08028 Barcelona T_c^1*
[4] *RIA "Pripyat", Libknecht str., 10, Chernobyl, Ukraine*
[5] *SCK•CEN, Boeretang 200, B-2400, Mol*
[6] *Ukrainian Institute of Agricultural Radiology (UIAR), Mashinostroiteley str., 7, 255205, Chabany7, Kiev Region. Ukraine*

Abstract Agricultural products originating from the areas subjected to high radioactive deposit after the Chernobyl accident are a main contributor to the radiological dose to local populations. The transfer fluxes of radionuclides through agricultural food chains, to food products consumed by humans, depend on the characteristics of the ecosystems considered (first of all, on the soil type), on the type of agricultural product of concern, as well as on the physico-chemical properties of the radioactive element and its speciation in the released. The parameters describing the fluxes through the compartments of the agricultural ecosystem are dynamic; they change with the time after the accident an are strongly influenced by agricultural practice, including the application of countermeasures. The influence of these factors on the ^{137}Cs fluxes in the main agricultural ecosystems of the Chernobyl accident zone are quantitatively determined and the main topics, where further investigation is needed, are identified.

1. Introduction

Agricultural products originating from those regions in Belarus, Ukraine and Russia heavily contaminated after the Chernobyl accident in 1986 are an important source of additional radiological exposure of populations. On soils with light mechanical composition - soddy podzols and peaty soils - which are the most representative in the affected area, foodstuffs (principally agricultural produces) and water can contribute up to 90% of the total irradiation dose [1]. The remedial actions applied to control the fluxes in the various agricultural systems, as well as in natural and semi-natural biocenoses, allowed to limit the radionuclides transfer to humans, and hence the internal radiological dose delivered to individuals. These actions aiming at reducing the radionuclide fluxes trough food chains constitute the most effective way to reduce the total radiological dose to exposed populations; methods based on the reduction of the external dose exposure are substantially more difficult to apply and very expensive.

The intensity of the radionuclide fluxes trough the food chains depends on different factors. Among those, the most important ones are:

1. the type of (agricultural, semi-natural or natural) ecosystem considered,
2. the physico-chemical properties of the radionuclides, their speciation in the fallout, and
3. the change of these characteristics with time due to environmental factors (weathering, aging) and human interventions (agricultural practices and application of countermeasures) [2].

2. Influence of the type of ecosystem

The intensity of the radionuclide transfer through food chains, which governs the rate of radionuclides incorporation into human body, is mainly influenced by the agricultural land use pattern (meadows, pastures, arable soils etc.), the nature of the plant and/or animal productions on these lands (plant products -cereals, leafy vegetables, tubers -, fodder, milk, meat, ...) and specific biogeochemical parameters of the ecosystem considered.

The intensity of radionuclide fluxes in agroecosystems depends on whether or not, mechanical treatments (ploughing, discing, ...) are performed immediately after the fallout; in other words, the fluxes intensity on arable soils differs substantially from those on soils not disturbed (meadows, pastures) after the deposit. The mat present in meadows and pastures, where accumulated radionuclides remain weakly fixed and hence available for root uptake, is an important factor explaining the higher content of ^{137}Cs in meadows and pastures grass than in temporary pasture grass installed on arable soils. Moreover, the decrease of ^{137}Cs availability for plants (binding to the clay fraction) as a result of aging proceeds more slowly in meadows and permanent pastures than in arable soils, with the consequence that the ecological half-time in pasture and meadow plants is longer than that in plants grown on arable lands. A quasi-equilibrium situation considering the ^{137}Cs transfer from soil to plant in meadows and pastures in the Chernobyl zone was established approximately 5 to 6 years after the deposit whereas, for arable soils, a relative constant accumulation rate for ^{137}Cs in plants had already be reached earlier. These respective delays correspond to the time required to come to an equilibrium between the different water-soluble, exchangeable, acid soluble fractions of ^{137}Cs in the soil [3].

The type of soil and their physico-chemical characteristics are other important parameters ruling the radionuclide fluxes in agrocenoses. For instance, when ^{137}Cs TF's to different types of agricultural products grown on 4 different soil types representative of the Chernobyl zone (i.e. peat, soddy podzolic sandy soils, soddy podzolic loamy soils and chernozem) are compared, the highest values are obtained on peaty soil and lowest, on chernozem. The mean ratios of the transfer values for ^{137}Cs between these two extremes soil types amount to 10.0 and 26.6 for hay (respectively on amended and non-amended meadows), 8.3 for potato tubers, 32.5 for cereal grains. These ratios calculated for animal products are equal to l 5.5 and 15.0 for beef and pork, and 10.0 and 26.6 for milk from collective farms and the private sector, respectively.

In the late phase after the accident, when production of foodstuffs with a radionuclide concentration exceeding DIL is practically excluded special attention should be paid to the restriction of the collective doses to the population, which are consuming contaminated food. In that case it is desirable to estimate the radionuclide fluxes in agricultural land per unit area, i.e. taking into account these products that are representative for main diet. These parameters may be used for planning and implementation of methods of agricultural practice in such a way that proposes the minimal collective doses and minimal risks of irradiation [4, 5].

The relative intensity of the radionuclide annual fluxes from the soil to the main foodstuffs produced in the areas of Ukraine and Belarus affected by the accident at the ChNPP are presented in Table 1. The ^{137}Cs fluxes to plant products (grain and potato) are, in general, higher than those to animal products (milk and meat). On both amended and non-amended lands (natural pastures and meadows), the ^{137}Cs fluxes to milk and meat are lower than the fluxes to potato, by a factor ranging form 2 to 24 and from 6 to 42, respectively; compared to the fluxes to cereal grains, respective differences by a factor up to 7 and 12, are observed. On another hand, the radiocaesium fluxes to meat and milk are similar on amended pastures and meadows (e.g. for lands where countermeasures were implemented to decrease the ^{137}Cs transfer to agricultural products), and non-amended lands, in the likeness of the fluxes to fodder (hay): a higher transfer on non-amended lands being compensated by a proportionally lower overall productivity.

Table 1. Annual ^{137}Cs fluxes in different agroecosystems in the Chernobyl accident area in 1994 (relative to a ^{137}Cs contamination density equal to 1 Bq/m^2)

Agricultural product	BELARUS (Data from the Byelorussian Institute of Agricultural Radiology)			UKRAINE (Data from the Ukrainian Institute of Agricultural Radiology)		
	productivity (kg.m^{-2}.y^{-1})	TF (10^{-3} m^2.kg^{-1})	flux (y^{-1})	productivity (kg.m^{-2}.y^{-1})	TF (10^{-3} m^2.kg^{-1})	flux (y^{-1})
Grains of cereals	0. 35	0.1	3.5 10^{-5}	0.45	0.3	1.3 10^{-4}
Potato	2.0	0.05	1.0 10^{-4}	2.5	0.18	4.6 10^{-4}
(non-amended lands) Hay	0.1	6.4	6.4 10^{-4}	0.1	2.2	2.2 10^{-5}
Milk	0.083	0.64	5.1 10^{-5}	0.083	0.22	1.9 10^{-5}
Meat	0.006	2.6	1.6 10^{-5}	0.0125	0.88	1.1 10^{-5}
(amended lands) Hay	0.3	2.2	6.6 10^{-4}	0.3	0.74	2.2 10^{-4}
Milk	0.25	0.22	5.5 10^{-5}	0.25	0.074	1.9 10^{-5}
Meat	0.018	0.88	1.6 10^{-5}	0.0375	0.29	1.1 10^{-5}

Table 2. Annual ^{137}Cs fluxes in different agroecosystems of the Bryansk region averaged over the period 1992-1994 (relative to a ^{137}Cs contamination density equal to 1 Bq/m²)

Agricultural product	RUSSIA (Data from the Russian Institute of Agricultural Radiology and Radioecology)		
	productivity (kg.m^{-2}.y^{-1})	TF (10^{-3} m^2.kg^{-1})	flux (y^{-1})
Grains of cereals	0.3	0.2-1.3	0.6-3.9 10^{-4}
Potato	2.0	0.03-0.05	0.6-1.0 10^{-4}
Hay (non-amended lands)	nd	1.2-6.4	na
	nd	0.47-1.4	na
Hay (amended lands)	0.04-0.06	0.047-0.14	1.9-8.4 10^{-6}
Milk (collective farms)	0.006-0.01	0.33	2.0-3.3 10^{-6}
Meat (collective farms)			

Comparable data (annual productivity, transfer factors and estimated radionuclide fluxes) gathered in one of the most heavily contaminated area in Russia (Bryansk region, Novozybkovsky district) are reported in table 2. In this case, the ^{137}Cs relative fluxes to plant products are substantially higher (up to a factor 10 to 20 in average) than those to meat and milk. The ^{137}Cs fluxes to plant foodstuffs (potatoes and cereal grains) are in the same order of magnitude; those to animal products (milk and meat) are also similar. The differences in ^{137}Cs flux intensities observed for the same agricultural products (plants as well as animal products) between the three republics (tables 1 and 2) are essentially explained by differences in soil type, crop yield and animal productivity.

This approach, considering fluxes instead of transfer factors, has the main advantage that it reflects the most important characteristics of the contaminated areas, including the influence of soil type and the specific productivity of the foodstuffs. However, it has the weakness that it does not take into account the real land use nor the agricultural management structure on the contaminated territories. Therefore, in order to overcome this drawback and to obtain a more realistic evaluation of the ^{137}Cs fluxes to agricultural products and of their change with time, we suggest to calculate the fluxes on actual foodstuffs production in the region considered, weighted by the radionuclide concentrations in each specific food product (table 3). The results provided by this method allow to draw two main conclusions. First of all, they show that in real situation the milk and meat pathways contribute the most to the radionuclide ingestion by the populations living in contaminated areas. Secondly, they demonstrate that the relative contribution of the different pathways varies depending on time evolved since the accident. This can be explained by the considerable changes of the ^{137}Cs concentration in different crops induced by the implementation of countermeasures. Moreover, the yields in the contaminated areas can also vary with time, due to restrictions imposed to the production of certain crops, as well as, to economical and social reasons.

Table 3. Annual ^{137}Cs fluxes in agriculture of Novozybkovsky district of Bryansk region in 1987 and 1992

Agricultural products	Total quantity produced in the district (10³ t/y)	Average ^{137}Cs concentration in the product (Bq.kg⁻¹)	Total flux (Bq.y⁻¹)	Flux per unit of area (Bq. y⁻¹.m⁻²)	Flux relative to the flux in cereal grains
1987					
Grain of cereals	65.0	187	12.2 10⁹	4.1 10⁻⁵	1.00
Potato	140.0	108	15.1 10⁹	5.1 10⁻⁵	1.25
Vegetables	13.0	112	1.5 10⁹	5.0 10⁻⁶	0.25
Milk	515	1820	1.5 10¹¹	5.8 10⁻⁴	15.0
Meat	11.7	6300	7.4 10¹⁰	2.9 10⁻⁴	7.50
1992					
Grain of cereals	53.0	29	1.5 10⁹	5.3 10⁻⁶	1.0
Potato	129.0	26	3.4 10⁹	1.2 10⁻⁵	2.3
Vegetables	11.2	22	0.3 10⁹	0.9 10⁻⁶	0.1
Milk	38.0	220	8.3 10⁹	4.4 10⁻⁵	7.6
Meat	8.1	590	4.8 10⁹	1.9 10⁻⁵	3.3

The differences in intensity between the ^{137}Cs fluxes for different foodstuffs make clear which one is critical in terms of a source of irradiation of population. In the Chernobyl region, milk (and meat to a lesser extent) is undoubtedly the main contributor to the internal dose; the relative contribution of other foodstuffs like cereals (bread) and potatoes is less important. For this reason, restriction on milk (and possibly meat) production in zones where the radionuclide concentration is expected to exceed the DIL appears as the first action to be taken in the immediate aftermath of an accident. But, to decrease the radiological exposure (collective doses) in the longer term, the fact that, in some agricultural area, the ^{137}Cs fluxes to grain and potato can be higher than those to milk and meat must be kept in mind.

3. Influence of the physico-chemical properties of radionuclides, including their forms in fallout

Two main forms of ^{90}Sr and ^{137}Cs were released by the Chernobyl accident: fuel particles from the destroyed reactor core (mainly deposited in near zone , in a radius of 30 km from the ChNPP) and condensed forms (which traveled longer distances). This resulted in the fact that the fluxes of these two radionuclides in agroecosystems were dependent from the ChNPP distance and also exhibited a different dynamic pattern in course of time after the deposit. Indeed, the intensity of the radionuclide fluxes through food chains depends not only on their physico-chemical properties, but also on their speciation in the deposit. If the influence of the physico-chemical characteristics of the radioelement has been largely addressed in the past in agricultural radioecology, that of their speciation in the fallout is much less documented and constitutes a rather novel item. Most of the information regarding the behaviour and accumulation in plants of ^{90}Sr and ^{137}Cs deposited as fuel particles released from an exploded reactor core were obtained from the numerous investigations carried out in the near zone of the ChNPP (until 30 km from reactor). These studies have demonstrated that the ^{90}Sr and ^{137}Cs mobility in soil and availability for plant uptake depend on the interaction of two opposite processes:

1.   on one hand, the weathering of the fuel particles leading to the subsequent release of the radionuclides from their matrices, making them available for root uptake and,

2.   on the other hand, their concomitant fixation on the soil solid phase (exchange complex), which reduces their availability for plant assimilation.

In the areas where condensed form dominated, ^{90}Sr and ^{137}Cs were mobile and, subsequently, highly available for plant uptake. The condensed form of the Chernobyl ^{137}Cs was even more mobile than that from the global fallout. On the contrary, ^{90}Sr and ^{137}Cs released into the environment as fuel particles was less available for root uptake until the radioelements were leached from their matrix.

In the Bryansk region where the contamination was due to condensed forms, ^{90}Sr was present in soil, during the period 1986-1987, mainly in exchangeable forms, easily available forms for plant assimilation. On the contrary, in the near zone of the ChNPP, the ^{90}Sr bio-availability was very low, so that their soil-to-plant concentration ratios were, in the first years after the accident, 2 to 4 times lower than those calculated for ^{137}Cs. A lower ^{90}Sr availability compared to that of ^{137}Cs does not correspond to common knowledge regarding these radionuclides as learned from their behaviour in the global fallout, the data from the Kyshtym accident (1957) or research carried out on experimental systems artificially contaminated with ionic forms. In areas contaminated by fuel particles, the root uptake of ^{137}Cs decreased with time, due to irreversible fixation of this radionuclide on clay minerals (aging), whereas ^{90}Sr accumulation, on the contrary, progressively increased as a result of fuel particles weathering and ^{90}Sr redistribution within the soil profile. The exchangeable ^{90}Sr fraction increased rapidly from 1987 until 1990 and remained practically constant thereafter [6, 7]. ^{137}Cs aging in the soil, as the result of its gradual fixation on poorly reversible binding sites on clay minerals (namely illite type clays), is one of the most important factors governing its availability for plant uptake.

4. Role of countermeasures in agricultural practice

The experience gathered with the mitigation of the consequences of the Kyshtym accident in the South Urals (1957) and the Chernobyl accident (1986) has taught that the introduction of countermeasures in the agricultural practice is an effective way to decrease of the intensity of radionuclide fluxes through the soil-plant-animal-human food chains. The main attention is usually paid to soil-plant steps of the food chains, taking into account that soils represents the most important reservoir of radionuclides in a contaminated terrestrial environment.

On meadows, among the possible agrotechnical treatments, rotating ploughing and disking decrease the ^{137}Cs transfer to plants by a factor 1.2 to 1.8, ploughing results in reduction factor ranging from 1.8 to 3.3 times, ploughing with deep-placement of upper contaminated layer (to the depth up 0.5 m) reduces the plant contamination by a factor 8 to 16. A surface amendment of meadows promotes a Cs transfer decrease from 1.3 to 3.1 times, and a radical amendment, from 2.3 to 11.2 times.

On arable soils the most important countermeasures are liming and fertilization. Lime application leads to a 1.6 to 2.3 times decrease of the ^{137}Cs transfer to plants, addition of phosphorus and potassium fertilizers induces a 1.2 to 2.2 times reduction and manure, a 1.3 to 1.6 times decrease. Combined application of lime, manure and high P and K doses can be even more effective, with reduction factors ranging from 2.5 to 3.5.

For ^{90}Sr, agrotechnical and agrochemical countermeasures have similar efficiencies: deep-ploughing (up to 50 cm depth) decreases of the ^{90}Sr concentration in plants by 1.3 to 2.3 times, deep-placement of the upper layer allows a 1.6 times reduction, liming lowers the transfer by 1.1 to 2.7 times and application of mineral fertilizers induces a 1.1 to 1.9 times decrease.

On soils with light mechanical composition, the application of alumino-silicates can be a useful way to decrease of radionuclide uptake by plants The application of zeolites leads however to contradictory results, but, in the most favorable situation, this countermeasure can decrease the ^{137}Cs concentration in plants up to 15 times. The application of high N-fertilizer doses must be avoided as it leads to an increase the ^{90}Sr and ^{137}Cs concentration in plants.

The efficiency of agricultural countermeasures depends on the time of their implementation: the earliest they are applied, the most effective they are.

In the period 1993-1995, the difficult economical situation in Russia caused a reduction of the restoration effort in the Chernobyl zone. As a consequence, the addition of fertilizers (mineral as well as manure) were diminished and, subsequently, the ^{137}Cs concentration in agricultural products increased. According to data from the Ministry of Agriculture and Food of Russia, the ^{137}Cs concentration in cereal grains and potatoes in the Bryansk region was increased, in 1993, in comparison with the 1991 situation by a factor about 3 (from 70 Bq/kg in 1991 to 26 Bq/kg in 1993 for grains and from 107 to 37 Bq/kg for potatoes). Between 1991 and 1993 the K_2O fertilization was decreased from 81 to 18 kg/ha.

5. Dynamics of the radionuclide fluxes in agroecosystems

The intensity of the radionuclide flux tends to decrease with time after the deposit. Two different ecological half-times (T_c), corresponding to periods of time, can be identified: the decrease is more rapid in the first years after the accident and slows down thereafter until a steady state is reached (Tables 4 and 5). The analysis of the results presented in table 4 shows that the ecological half-time for ^{137}Cs in milk, averaged over the whole observation period (1987-1992) amounts to 1.6 to 2.3 years, and 2.3 to 4.8 years, for Bryansk and Kaluga regions, respectively. The different half-times between the two regions can be principally attributed to differences in the extent of applied countermeasures. When the three first year only are taken into account, the T_c^1 for ^{137}Cs in milk in the districts of Bryansk and Kaluga are rather similar (from 0.8 to 1.8 y); lower values are obtained in areas like the Krasnogorsky (1.0 y) and Novozybkovsky (0.8 y) districts, in the Bryansk region, where countermeasures were started earlier and carried out to a greater degree. Considering the later period (1990-1992), T_c^2 for ^{137}Cs concentration in milk estimated for different districts in the Bryansk region vary from 2.6 to 4.7 y, and in the Kaluga region, between 5.7 and 11.7 y. This discrepancy between the two region is again in agreement with the scale of countermeasures carried out in each respective region. Moreover, in areas where early countermeasures were taken, and completed before

1990, the ratio between the half-times of the first and second periods is larger (3.5 at Krasnogorsk to 4.5 at Novozybkov) than at other locations in the same region, because the countermeasures influence on the changes in ^{137}Cs levels in milk during the second period is still appreciable.

Table 4. Half-time periods of ^{137}Cs concentration decrease in milk, T_c (y) [8]

Types of production	T_c (y) Whole period (1987-1992)	T_c^1 (y) First period (1987-1989)	T_c^2 (y) Second period (1990-1992)
Bryansk region			
Gordeyevsky	1.7	1.2	2.6
Klimovsky	1.8	nd	4.7
Klintsovsky	2.3	1.3	2.6
Krasnorsky	1.8	1.0	3.5
Novozybkovsky	1.6	0.8	4.2
Kaluga region			
Khvastovichsky	2.3	1.5	6.3
Ulyanovsky	4.8	1.8	11.7
Zhizdrinsky	2.9	1.1	5.7

Table 5. Half-time periods of ^{137}Cs concentration decrease in plant products, T_c (y) [8]

Types of production	T_c (y) Whole period (1987-1992)	T_c^1 (y) First period (1987-1989)	T_c^2 (y) Second period (1990-1992)
Bryansk region			
Grain	1.4 - 1.9	1.2 - 2.1	1.7 - 4.3
Potato & root crops	1.5 - 2.8	1.4 - 1.9	2.2 - 7.7
Hay	1.5 - 1.9	0.9 - 1.4	1.7 - 7.3
Kaluga region			
Grain	3.1 - 4.9	nd	nd
Potato & root crops	3.2 - 5.6	2.9	5.8 - 14.1
Hay	2.0 - 3.0	1.4 - 2.3	3.3 - 4.8

Similar conclusions are also valid concerning the decrease of the ^{137}Cs concentrations in plant products. In most cases, the T_c calculated for the main plant products are in the same order of magnitude than those for milk. In the Bryansk region, the ecological half-times range from 0.9 to 2.1 y and from 1.7 to 7.7 y for T_c^1 and T_c^2, respectively. The corresponding values estimated for the districts of Kaluga region vary from 1.4 to 2.9 and from 3.3 to 14 y. Thus, the overall T_c, estimated over the whole period 1987-1992, for all types of plant products in the districts of the Bryansk region are considerably less (1.4 to 2.8 y), than in the districts of the Kaluga region (2.0 to 5.6 y), in agreement with the pattern of countermeasures implementation in both regions. Half-times (from 0.7 to 1.5 y) quite similar to those reported here regarding the decontamination of agricultural products have been mentioned earlier [9].

As a general conclusion, the results presented here above demonstrate that the changes with time of the intensities of the radionuclide fluxes in agroecosystems is ruled by three main factors:
1. radioactive decay,
2. changes in the biological availability of radionuclides under the influence of biogeochemical processes and
3. application of countermeasures.

For contaminated areas in Russia, in regions where countermeasures were intensively applied, the relative contribution of these three processes to the decrease of ^{137}Cs content in agricultural products were estimated to 0.06-0.07, 0.33-0.36 and 0.57-0.61, respectively.

6. Some actual problems

The intensity of radionuclides fluxes in different types of agricultural ecosystem, which determine the internal irradiation dose, can be changed in the time course after the accidental contamination. Therefore the study of the main factors governing the radionuclides transfer processes is extremely important. The investigation of the long-term consequences of countermeasures implementation on the fluxes of radionuclides is also essential. Such information must be used for planning the agricultural practice and the countermeasure strategy on contaminated lands, in order to mitigate the negative consequences of a radioactive contamination.

References

[1] Guide-book on radiation Situation and Radiation Doses received in 1991 by population of Areas of Russian Federation Subjected to Radioactive contamination Caused by the Accident at Chernobyl NPP. M.J. Balonov (ed.) (Saint-Petersburg: "Ariadna" - "Arcadiya" Publishers) 1993 (in Russian)

[2] Agricullural Radioecology. R.M. Alexakhin and N A. Kornoyev. (eds) (Moscow: Ecologiya Publishers) (1992). (In Russian)

[3] N.I. Sanzharova, S.V. Fesenko, R.M. Alexakhin, V.S. Anisimov, V.K. Kuznetsov and L.G. Chernyayeva, Changes in the Forms of ^{137}Cs and its Availability in Plants as Dependence On Properties of Fallout after Chernobyl Nuclear Power Plant Accident. The Science of the Total Environment 154 (1994), 9-22.

[4] B.S. Prister, N.K. Novikova, N.V. Trachenko, L,.I. Nagovizina, I.I. Berezhnaya, N.D. Semenyk and V.M. Rudoi, The Ways of Collective Dose Exposure Decrease at the Agricultural Land Contamination, Gigiena and Sanilariya 8 (1990) 5-75.

[5] R.M.Alexakhin. Countermeasures in Agricultural Production as an Effective Means of Mitigating the Radiological Consequences of the Chernobyl Accident. The Science of the Total Environment 137 (1993) 9-20.

[6] S.V Krouglov, A.D. Kurinov and R.M Alexakhin. Chemical Fractionation of ^{90}Sr, ^{106}Ru, ^{137}Cs and ^{144}Ce in Chernobyl Contaminated Soils - An Evolution in a Course of Time. (Submitted to The Journal of Environmental Radioactivity. 199S).

[7] S.V. Krouglov, A S. Filipas, R.M. Alexakhin and N. P.Arkhipov. Long-term study on transfer of ^{137}Cs and ^{90}Sr from Chernobyl-Contaminated soils to grain crops. (Submitted to The Journal of Environmental Radioactivity. 1995).

[8] S.V Fesenko, R.M..Alexakhin, S.I.Spiridonov, N.l.Sanzharova. Dynamics of ^{137}Cs concentration in Agricultural Products in Areas of Russia Contaminated as a Result of the Accident at the Chernobyl Nuclear power Plant. Radiation Protection Dosimetry 60 (1995) 155-166.

[9] V.N.Shulov, GY.Bruk, M.I.Balonov, V.I. Parkhomenko and I.Y. Pavlov. Cesium and Strontium Radionuclide Migration in the Agricultural Ecosysltem and Estimation of Internal Doses to the Population. The Chernobyl papers, vol.I. Research enterprises. Washington, 1993, pp.167-218.

Distribution of Radionuclides in Urban Areas and their Removal

J. ROED[*], K.G. ANDERSSON[*], E. SOBOTOVITCH[+], E. GARGER[#], I.I. MATVEENKO[□]

* Risø National Laboratory, P.O. Box 49, DK-4000 Roskilde, Denmark
+ IGMR AS Ukraine, Palladin st. 34, Kiev 252142, Ukraine
# Ukraine
□ Belarus

Abstract. The major contamination processes in the urban environment are wet and dry deposition with the former leading to much greater deposition per unit of time. Typical deposition patterns for radiocaesium in urban areas have been identified for these processes and recent in situ measurements have been used to verify these relations and to investigate the urban weathering effect over long periods. The results of a recent series of field trials of decontamination methods in urban or suburban Russian areas are reported, and this experience has been incorporated in an example of formation of strategies for clean-up in an urban contamination scenario.

1. Introduction

The accident at the Chernobyl nuclear power plant in April 1986 has lead to high levels of outdoor surface contamination in parts of Ukraine, Byelorussia and Russia. Residential areas within the 30 km zone around the nuclear power plant are still unoccupied due to unacceptably high levels of radiation (almost entirely ^{137}Cs) deposited on the ground and on various man-made surfaces in the urban environment. Ever since the accident, research has been carried out by the Contamination Physics Group of the Risø National Laboratory in collaboration with CIS institutes with the ultimate goal of developing strategies for decontamination of these areas.

Nine years after the release of some 3.7 EBq of radioactivity to the atmosphere from the Chernobyl unit 4 reactor, the external radiation dose in the contaminated areas is essentially determined by the aount of ^{137}Cs (half-life 30 years) still present.

2. The urban contamination pattern following deposition

2.1. Initial urban deposition phase

Airborne gases and particulate matter arrive at the ground surface in the form of either wet or dry deposition. Wet deposition occurs in precipitation (generally rain) and leads to a very different distribution pattern from dry deposition. Again, there can be variations within a particular mode of deposition. For example, the deposition pattern resulting from a light shower giving very little surface run-off will be different from that produced by prolonged heavy rain.

The dry deposition pattern of caesium and ruthenium aerosols in an urban complex is difficult to describe in simple terms. Some of the important factors influencing dry deposition are:

*   the size of the aerosol particles. This will largely determine the deposition mode (Brownian diffusion, atmospheric diffusion, impaction or sedimentation).

*   the surface roughness of the receptor surface.

*   turbulence of the air.

A series of radiocaesium dry deposition measurements made by Roed in urban areas shortly after the Chernobyl release in 1986 revealed that the deposition velocities (ratio of the air concentration at 1 m above the surface to the dry deposition flux to the surface) were very small compared with those found for rural areas.

The measured deposition velocities on different types of urban surface are listed in Table 1.

Table 1. Deposition velocities measured in Denmark after the Chernobyl accident (in 10^{-4} m/s).

Paved areas	Walls	Windows	Grass (clipped)	Trees	Roofs
0.7	0.1	0.05	4.3	7	2.8

Estimates by Roed [1] based on measurements of Chernobyl fallout in areas which received solely dry deposition and in areas in which deposition took place with heavy rain, revealed the typical relationships shown in Table 2 between caesium contamination levels per unit area of different types of surface. These findings are consistent with measurements made in various parts of Europe ([2], [3], [4], [5]).

Table 2. Relative source strengths for various surfaces shortly after deposition of Chernobyl fallout caesium, relative to that on a cut lawn, where there is no penetration in the soil.

SURFACE:	Dry deposition	Wet deposition
Gardens, parks	1.00	0.80
Roofs	1.00	0.40
Walls	0.10	0.01
Streets, pavements	0.40	0.50
trees with leaves	3.00	0.10

The factor of 0.8 rather than 1 for wet deposition in gardens and parks is explained by the slight penetration of the isotopes into the soil giving some attenuation. This gave a shielding factor of 0.8. On roofs, only about 40 % of the deposited radiocaesium is retained after wet deposition : the remaining 60 % remains in the run-off water. The figure of 3.0 for trees indicates that the leaves of the trees constitute a very efficient 'aerosol filter'. The relative deposition figures for trees relate to a plan-projected area of ground covered by the trees.

2.2. Long-term changes of the urban contamination relationship

The surface contamination relationships will change with time, due to the weathering of the surfaces on which deposition took place. The decrease with time of the contamination level has been carefully followed in heavily contaminated areas in different parts of Europe and the time-dependence modelled [6]. After nearly one decade, the dose-rate from road pavements has generally decreased by at least an order of magnitude. Meanwhile, the dose-rate from areas of soil and roofs fell by about 50 %, while that from walls has decreased by only 10-20 %.

Weathering effects have been investigated in 1993 and 1994 in situ in areas of the former Soviet Union. In the town of Pripyat, less than 4 km away from the Chernobyl power plant, high levels of radioactive contamination were recorded compared with other investigated areas, although this area received a dry deposited contamination. In other, more remote Ukrainian towns and villages where measurements were made, such as Poleskoie, Vladimirovka and Varovice as well as the Russian towns Yalovka and Novozybkov, more than 100 km away from the power plant, deposition of Chernobyl debris occurred with rain.

Gamma-ray measurements on a sandstone wall in Pripyat in 1993 showed the contamination level to be in the range 199-350 kBq/m^2, compared to 0.9-28 kBq/m^2 found on similar surfaces in the more remote towns and villages in the Novozybkov area. Also in Pripyat, it was found that the contamination level on a grass covered surface (2.9 MBq/m^2) was approximately ten times that on walls, while in the areas where wet deposition occurred, levels on grassed surfaces were about two orders of magnitude greater than on walls. These findings were in line with the previously recorded results in Table 2. On impervious horizontal surfaces, such as roads and pavements, 'natural' decontamination (weathering) due to traffic and surface run-off water had reduced the contamination level from presumably 1.5 MBq/m^2 (estimated from the levels on grass and walls) to 30-350 kBq/m^2 since 1986. The highest residual levels were found on concrete pavements and the lowest on asphalt roads. Since by far the greatest contribution to the external dose at this point comes from the open (grassed) areas, the study of soil radiocaesium penetration in the heavily contaminated areas is an important task. The average of a series of soil radiocaesium profiles in the Halch area in Russia in 1994 is shown in Figure 1. As can be seen, the profiles are now rather deep, and for instance removal of a thin top soil layer might not have a great effect on the dose rate.

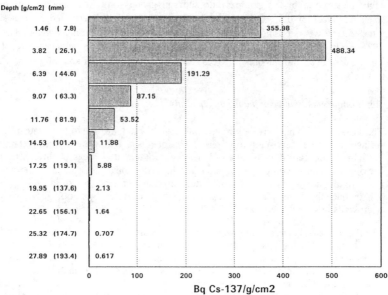

Depth [g/cm2] (mm)

	Bq Cs-137/g/cm2
1.46 (7.8)	355.98
3.82 (26.1)	488.34
6.39 (44.6)	191.29
9.07 (63.3)	87.15
11.76 (81.9)	53.52
14.53 (101.4)	11.88
17.25 (119.1)	5.88
19.95 (137.6)	2.13
22.65 (156.1)	1.64
25.32 (174.7)	0.707
27.89 (193.4)	0.617

(mm) Mean depth of the sample slices.

Figure 1. Average soil radiocaesium profile in Halch, Russia in the summer of 1994.

3. Decontamination in the urban environment

Although the fixation of residual radiocaesium on any urban surface certainly becomes stronger with time, a series of experiments made 7-9 years after the Chernobyl accident took place has shown that it is still possible to substantially reduce the external dose in the urban environment.

3.1. Decontamination of walls

A series of decontamination trials was made on walls in the Halch area of Russia in 1994. The objective was to test in the heavily contaminated areas those methods which have in previous experiments been found to have the greatest effect.

3.1.1. High pressure water treatment

In an experiment on a limed wall (an area of 21 m²) water hosing through a high pressure jet at 120 bar was applied. The amount of water used in the test was 20 l/m². The time consumption was 1.6 minutes per m². The method generated 0.4 kg/m² of solid waste.

The treatment was found to have removed 26.5 % of the contamination. The corresponding decontamination factor is 1.36 with a 95 % confidence interval from 1.24 to 1.48, which is less than what has previously been obtained on similar walls. The reason for this is that the migration of the radionuclides into the wall has reached such an extent that little can be removed by surface-cleaning methods on this type of wall.

3.1.2. Sandblasting

On a wall of the same type (test area : 5.0 m^2) an experiment with the more abrasive sandblasting method was conducted, and the effect detected with the beta counter. As sandblasting goes deeper into the material, a greater effect was recorded.

The water requirement for the test was 54.5 l/m^2, and the amount of sand required was 2.25 kg/m^2. Treatment of each square meter took 2 minutes. The amount of waste generated by the method was found to amount to the applied materials plus 1 to 5 % (loosened material).

The effect was found to be a decrease in the radiation level by about 80 %, corresponding to a decontamination factor of 5.02 with a 95 % confidence interval from 4.31 to 5.73.

3.1.3. Cation exchange

In an experiment in 1993, about 100 litres of ammonium nitrate solution (0.1 M) was sprayed onto areas of about 16 m^2 of similar sandstone walls in Pripyat and in Vladimirovka. The treated area soon became saturated with the solution and was then left for about half an hour before being thoroughly rinsed with 100 litres of clean water.

Beta measurements showed that in Pripyat, about 67 % of the contamination was removed but in Vladimirovka the figure was only 21 %.

The influence of the deposition mode seems evident: the fallout over Pripyat (very close to the Chernobyl plant) was almost entirely in the form of small particles of irradiated uranium oxide fuel (up to 100 mm diameter). Fallout over Vladimirovka was essentially in the form of condensed radiocaesium: i.e. material which had been discharged from the reactor core in the gas phase, condensed in the atmosphere and scavenged by falling rain.

3.2. Decontamination of roofs

Experiments were conducted in the heavily contaminated Halch area to investigate the effect of roof decontamination by two different methods: high pressure water treatment and a specially constructed roof-washer device.

3.2.1. High pressure water treatment

The decontamination factors in this experiment were identified on an asbestos roof by direct beta measurements, but similar results have been recorded in the laboratory by gamma measurements on removed roof pavings.

The high pressure water hosing was performed through a turbo nozzle at a pressure of 150 bar. The treatment was found to have removed almost 40 % of the caesium contamination.

3.2.2. Special roof washer

In order to meet the demand for improved countermeasures for roofs a special roof-cleaning method has been developed at Risø and tested in situ in the town of Halch.

This newly developed specially constructed device consists of a rotating brush with a shape that follows the curves of the roof and which is driven by pressurized air at 700 l/min. (7-8 bar) and uses water at ordinary tap water pressure. The fact that the cleaning is carried out at low pressure and in a wet medium makes the method specially suitable for work in contaminated areas, as the resuspension hazard is minimized. Further, this relation satisfies the existing legal demands in Europe regarding treatment of asbestos surfaces. The brushing pressure and cleaning function has been specially adjusted to ensure that only coatings such as moss and algae are removed while the abrasive effect on the roof material is negligible. A test on an asbestos roof gave less than 25 mg of loosened asbestos fibres per liter of applied water. This is much lower than any existing limits. An earlier experiment has shown that very little caesium contamination will be in the liquid phase of the washed off material from such operations (< 1%), while the rest will be associated with the solid fraction (moss, algae, etc.). There will thus only be a very limited amount of waste for disposal.

The roof-cleaner can be applied by an operator standing on the top of the roof or - in case of single-storey buildings - from ground level, and is easily driven back and forth by the brush requiring very little action from the operator.

In the experiment, which was conducted on an eternite roof in Halch, 120 l of water were used (recycled 3 times). This corresponds to 12.8 l/m^2 of the applied roof surface. The time requirement was found to be 3.3 min./m^2. The cost of the method would be much lower than other suggested methods such as for instance sandblasting, since it requires only water.

The result was found to be a reduction of the radiation level by 56 %, corresponding to a decontamination factor of 2.29, with a 95 % confidence interval from 2.20 to 2.38, which is a remarkably high number compared with previously reported results of other procedures for removal of aged contamination on a roof.

3.3. Reduction of dose rate contribution from areas of soil

This section describes recent tests in the former USSR of methods to reduce the dose rate contribution from areas of soil and grassed areas in the urban environment, such as gardens and parks.

3.3.1. Triple digging

Triple digging (Fig.2) has shown to be an excellent method to reduce the dose to people, both where the uptake to plants is considered, and for external dose reduction. This method can be used in gardens and other places where it is impossible or expensive to use e.g. skim and burial ploughing. For external dose the dose reduction factor, R, is defined as:

$$R = (D'_a - D'_b) / (D'_i - D'_b),$$

where D'_i and D'_a are the dose rates 1m above the surface, respectively before and after the tripple digging. D'_b is the dose rate contribution from the background radiation. R is dependent on a variety of parameters, the most important being:
1. The isotope(s) considered.
2. The initial distribution of the isotopes in the soil.
3. The depth of the digging.

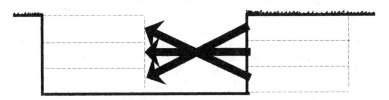

Figure 2. Diagrammatic representation of how triple digging is carried out and how the order of the 3 layers of soil is changed. The total digging depth in the experiment was 30 cm

An experiment was performed in 1995 in Novo Saboriche, where a 10 by 10 m plot was dug. The isotope of greatest concern was ^{137}Cs. From the result of this experiment a model was created to calculate the dose reduction factor R as a function of the size of the plot and the initial distribution (see Fig.3). In this case the depth of the triple digging was 30 cm. It can be seen that if the initial contamination is in the uppermost 10 cm of soil then the dose reduction factor will range from 0.08 to 0.5, depending on the size of the plot and the initial distribution.

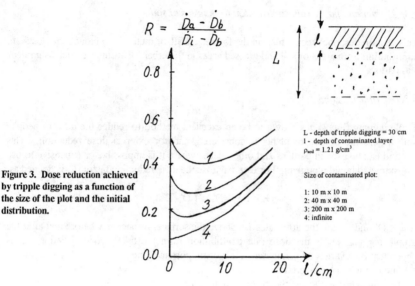

Figure 3. Dose reduction achieved by tripple digging as a function of the size of the plot and the initial distribution.

$$R = \frac{\dot{D}_a - \dot{D}_b}{\dot{D}_i - \dot{D}_b}$$

L - depth of tripple digging = 30 cm
l - depth of contaminated layer
ρ_{soil} = 1.21 g/cm³

Size of contaminated plot:

1: 10 m x 10 m
2: 40 m x 40 m
3: 200 m x 200 m
4: infinite

3.3.2. Skim and burial ploughing

In the urban environment, the application of skim and burial ploughing would be restricted to large areas, such as parks.

This new type of plough has been developed jointly by Risø National Laboratory and Bovlund Agricultural Engineers Denmark. The plough skims off the topmost layer of soil (about 5 cm) and buries it at a depth of some 40-50 cm without inverting the intermediate layer. Hence the name 'skim and burial plough'. The removal of only about a 5 cm layer of topsoil rarely affects the fertility of the land and poorer quality subsoil is not brought to the surface.

Overall, the skim and burial plough greatly reduces radiation levels at the ground surface, the resuspension hazard is eliminated, most of the contamination is made inaccessible to plant roots and soil quality is unaffected.

The effect of the procedure, which has been tested in the former USSR, has been found to be a reduction of the dose-rate by some 94 %, but in very sandy soils it may be difficult to achieve the objective with this method..

4. Development of a strategy for decontamination

Based on in situ measurements of urban contamination and decontamination Risø National Laboratory has specialized in development of decontamination strategies for the urban environment. The concept of the strategy formation is essentially the same as that introduced more than five years ago [7], although more detail is now available.

In the following a short demonstration will be given of how to apply the obtained information in the development of a strategy for clean-up of an urban area. Many considerations must be taken into account, although certain simplifications must obviously be made.

Consider an environment consisting of a single-storey house surrounded by a garden, as they typically occur in the suburban areas.

If it is assumed that the deposition of radiocaesium to this area takes place in the early autumn, in the absence of precipitation, and that the relative distribution of caesium on different urban surfaces is as shown in Table 2 (dry deposition), the average dose rate and accumulated dose to people living in the particular type of environment can be calculated using a model such as URGENT [6]. This model takes into account the loss of radioactive matter on each surface by weathering. The weathering model is based on the results of extensive field measurements in different parts of Europe.

In this case it has been assumed that the average person spends 85 % of the time indoors and 15 % of the time outdoors. The fact that we are dealing with a single-storey building simplifies the case somewhat, as for instance the dose contribution from the roof of a multistorey building may be large to the people living on the top floor, but negligible to people living at ground level in the same house. Also, in urban centres the dose contribution from contaminated roads and other pavements will become significant. In some dry deposition scenarios indoor decontamination may also be important.

Many different methods might be considered, but the most advantageous in the particular case might be to use the described roof-washer system for roofs, high pressure water treatment for walls, cutting of trees, and triple-digging for grassed areas and soil areas.

Table 3 shows the calculated % dose reduction achievable by these methods if they are initiated 6 months or 10 years after the deposition took place. Figures are given for both the reduction of the total accumulated life-time dose over 70 years by application of a method and for the reduction in dose-rate at the particular time when the method is applied. Also given is an estimate of the costs of the procedures.

Table 3. Costs and benefits of application of different clean-up methods in a suburban environment. Percent reduction of accumulated doses over 70 years and immediate dose rate reduction are given, assuming that clean-up is initiated 6 months or 10 years after deposition.

Surface:	Roofs	Walls	Trees	Soil
% 70-y dose red. (6 months after)	2.9	1.8	5.9	65.2
% dose rate red. (6 months after)	6.4	1.2	30.2	45.5
% 70-y dose red. (10 years after)	0.3	0.5	0.3	23.1
% dose rate red. (10 years after)	1.2	1.3	3.9	69.4
Costs (ECU/m^2)	2	1.7	7	0.5

From Table 3 it is clear that although the trees will play a very important role in the beginning, due to the very high deposition velocity, a decontamination of trees will be less important regarding the accumulated dose over 70 years, due to the rapid weathering. The costs of cutting trees are also rather high.

The dose contribution from roofs will be comparatively large in the beginning, but due to weathering the roofs will soon become less important to decontaminate.

It is clear that the grassed areas will play the major role, and as the cost of tripple digging is relatively small, clean-up of grassed areas would be given first priority in a clean-up strategy for this scenario. Even if 10 years go by before the garden is dug, it is still possible to reduce the total accumulated dose by one-fourth by this simple method. However, it is clear that in urban areas with smaller garden areas, the other surfaces will be much more dominant. It is therefore important to tailor a strategy for use in a specific type of area.

5. Conclusions

Typical initial contamination patterns have been observed after the Chernobyl accident. Some of these relations were found to be in-line with results of recent in situ measurements in the former USSR. The weathering effects in urban areas of Russia were investigated in 1994 and it was found that the caesium contamination level on roads and pavements had decreased by a factor of 4-50, depending on the traffic. In the Halch area, the highest soil radiocaesium concentration was found at a depth of about 3 cm. As much of the radiocaesium has now reached considerable depths, decontamination by removal of a thin top soil layer would no longer be possible. Various methods for decontamination of urban surfaces have been investigated in situ, and the currently most effective have been described in detail. An example of strategical decontamination planning has been given.

References

[1] J. Roed, Deposition and removal of radioactive substances in an urban area, Nordic Liaison Committee for Atomic Energy, 1990.

[2] O. Karlberg and B. Sundblad, A Study of Weathering Effects on Deposited Activity in the Studsvik and Gävle Area, Studsvik Technical note NP-86/78, 1986.

[3] F.J. Sandalls and S.L. Gaudern, Radiocaesium on Urban Surfaces in West Cumbria Five Months after Chernobyl, Draft report, Environmental and Medical Sci. Div., Harwell Laboratory, 1986.

[4] J. Roed, Run-off and Weathering of Roof Material following the Chernobyl Accident, Rad. Prot. Dos., vol.21, No. 1/3, pp. 59-63, 1987.

[5] P. Jacob, R. Meckbach and H.M. Müller, Reduction of External Exposure from Deposited Chernobyl Activity by Run-off, Weathering, Street Cleaning and Migration in the Soil, Rad. Prot. Dos., vol. 21, No.1/3, pp. 51-57, 1987.

[6] K.G. Andersson, URGENT - a model for prediction of exposure from radiocaesium deposited in urban areas, presented at the GSF/CEC/IAEA/GAST workshop on dose reconstruction in Bad Honnef, June 6-9, 1994.

[7] J. Roed, K.G. Andersson, J. Sandalls, Reclamation of Nuclear Contaminated Urban Areas, in proc. of the BIOMOVS Symposium and Workshop in Stockholm, ISBN 91-630-0437-2, 1990.

Behaviour of Radionuclides in Meadows including countermeasures application

B.S.PRISTER[1], M.BELLI[2], N.I.SANZHAROVA[3], S.V.FESENKO[3], K.BUNZL[4],
E.P.PETRIAEV[5], G.A.SOKOLIK[5], R.M.ALEXAKHIN[3], Yu.A.IVANOV[1],
G.P.PEREPELYATNIKOV[1], M.I.IL'YN[1]

[1] Ukrainian Institute of Agricultural Radiology (UIAR), 255205, Chabany, Kiev region, Ukraine.
[2] National Environmental Protection Agency (ANPA-DISP), I-00144, Via Vitaliano Brancati,48, Rome, Italy
[3] Russian Institute of Agricultural Radiology and Agroecology (RIARAE), 240020, Obninsk, Kaluga region, Russia,
[4] GSF-Forschhungzentrum fur umwelt und gesundheit GmbH, D-8042 Neuherberg, Germany
[5] Belorussian State University (BUS), 220080, Fr.Skorini av.,4, Minsk, Belarus

Summary. Main regularities of the behaviour of ChNPP release radionuclides in components of meadow ecosystems are considered. Developed mathematical model of radionuclide migration in components of meadow ecosystems is discussed. Radioecological classification of meadow ecosystems is proposed. Effectiveness of countermeasures application in meadow ecosystems is estimated.

1. INTRODUCTION

Natural and semi-natural meadow ecosystems represent critical systems determining the major contribution to the exposure dose of population after a radiation accident. One of the critical chains of radionuclide uptake by animal products is the consumption of fodder when the animals are grazed on meadows.

As a result of the ChNNP accident the different meadows types such as dry, flood plain, low land (wet) and peatland were contaminated. The soil cover of these meadows is presented by a wide set of soils. These soils are characterised by significant variety which defines the differences in the intensity of radionuclide migration.

2. MATERIALS AND METHODS

In 1991-1994 the research was carried out on the territory of Russia, Ukraine and Belarus, contaminated with radionuclides after the accident on ChNPP.

Samples of soil and meadow plants were taken in 40 sampling points inside 30-km ChNPP zone and outside it: 12 sampling sites in Russia, 9 - in Belarus and 19 - in Ukraine. These sites include the most typical types of meadows of radioactive contamination zone, and, respectively, the most typical meadow soils - soddy-podzolic (of various mechanical composition) and peaty. Sampling of soil and plants and measurement of radionuclides content in the samples was carried out by the standard techniques. Extracted radionucliudes species were identifier by the techniques of Tesser at al (1) and F.Pavlotskaya (2).

3. DYNAMICS OF RADIONUCLIDE IN MEADOW ECOSYSTEMS

3.1. RADIONUCLIDES MOBILITY IN SOILS

Mobility of Chernobyl radionuclides in soils is predetermined by some groups of factors, including: i) physical-chemical forms of fallout; ii) physical-chemical properties as well as granulometric and mineralogical composition of soils; iii) water regime of soils etc..The content of some extractable forms of radionuclides in soils is used usually as a criteria of radionuclides mobility in soils.

Results of sequential extraction procedures by Tessier et al method have shown the following distribution of radiocaesium between the extractable fraction (% from the total content) - *mineral soils*: F1 (readily exchangeable): 0.5-17.3; F4 (persistently bound): 12.6-60.0; F5 (residual): 10.0-78.7; *organic soils*: F1: 0.3-2.2; F2: 0.4-1.7; F3): 4.4-16.6; F4: 16.5-33.7; F5: 50.7-70.5. Thus, the dominant activity of radiocaesium is represented by persistently bound and fixed fraction in soils. It general, one should note the significantly low level of radiocaesium exchangeable forms in organic soils in comparison with the mineral soils.

Results of sequential extraction procedures by F.Pavlotskaya method have been shown the following distribution of radiocaesium between the extractable fraction (% from the total content) - *mineral soils*: P1 (soluble in distilled water): 0.07-2.8; P2 (exchangeable): 0.5-15.7; P3 (soluble in 1N HCl): 0.55-10.2; *organic soils*: P1: 0.05-3.5; P2: 0.22-8.5.

Dynamics of the content of radiocaesium exchangeable forms in soils depends on above-mentioned factors (Fig.1). In general, content of ^{137}Cs exchangeable forms in soils is decreasing with time, different decrease rate is noted for conditions, characterised by different soils conditions as well as various initial forms of fallout (ratio of fuel component of fallout to condensed one). Significantly lower content of radiocaesium exchangeable forms as well as higher intensity of its decrease is noted for groups of hydromorphous soils (Fig.1, groups of soils III and IV), characterised by higher value of CEC and higher content of clay minerals. However, this phenomena is not an absolute one and has some exceptions. Increase of the content of ^{137}Cs exchangeable forms in soils of some meadows in 30-km since 1990-1991 and its further decrease have been observed (Fig.1, soil VI). It should be noted also that less content of radiocaesium in soils of ChNPP immediate zone is connected not with high content of fuel component in fallout only, but with high sorption by solid phase of hydromorphous soils also.

The association processes between dissolved soil organic matter and radionuclides will have effects on the behaviour of radionuclides in soils. For a better understanding of the radionuclides mobility in soils and for reliable predictions, gel filtration of the soil solution could be applied increasingly in the future. Data, obtained in the framework of the project, show that not only Pu, Am and Sr, but radiocaesium also is associated to a considerable extent with dissolved soil organic matter.

Migration ability of Chernobyl radionuclides in soil profiles is governed by mentioned above groups of factors also (i.e. physical-chemical forms of fallout; soils properties etc.): most intensive vertical migration of Caesium radioisotopes is noted for hydromorphous organic soils, the least intensive vertical transfer - for automorphous mineral soils, characterised by heavy granulo-

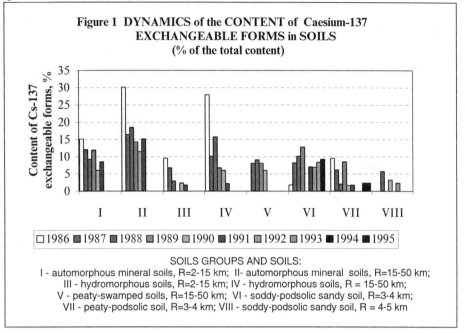

Figure 1 DYNAMICS of the CONTENT of Caesium-137 EXCHANGEABLE FORMS in SOILS (% of the total content)

Content of Cs-137 exchangeable forms, %

☐1986 ■1987 ■1988 ■1989 ▦1990 ■1991 ☐1992 ▨1993 ■1994 ■1995

SOILS GROUPS AND SOILS:
I - automorphous mineral soils, R=2-15 km; II- automorphous mineral soils, R=15-50 km;
III - hydromorphous soils, R=2-15 km; IV - hydromorphous soils, R = 15-50 km;
V - peaty-swamped soils, R=15-50 km; VI - soddy-podsolic sandy soil, R=3-4 km;
VII - peaty-podsolic soil, R=3-4 km; VIII - soddy-podsolic sandy soil, R = 4-5 km

metric composition. In general, ^{90}Sr is characterised by more intensive vertical migration in soils in comparison with Caesium radioisotopes. The highest migration rate of ^{90}Sr is noted for mineral soils with ligth granulometric composition (weakly-humous sand and sandy soils). Migration ability of Plutonium isitopes in soils profile is more closed to ability of Caesium radioisotopes. It should be also noted that migartion rate of ^{90}Sr and Plutonium isitopes in soils profile depends on fraction of fuel component in fallout as well as on the destruction intensity of fuel particles in soils. The quantitative characteristics of ^{137}Cs vertical transfer in soils of different meadows types are presented below (Chapter **4. MODEL OF RADIONUCLIDE MIGRATION ...**).

3.2. RADIONUCLIDES TRANSFER TO MEADOWS PLANTS

Investigation of ^{137}Cs transfer from soils to the vegetation of natural meadows in 1988-1994 demonstrated that the dynamics of TF change is specific and is determined by the type of meadow, which takes into account the soil type and its water regime. The role of soil type as one of the main factors which determines the mobility of ^{137}Cs in soil-plant system is examined in the previous chapter of the report. The role of meadow water regime is of similar importance, especially in the regions with dominating lowland meadows on organogenic soils.

In the conditions of water regime, optimal for meadow plants, the values of TF have decreased by 20-25 times within 5 years. Nevertheless, considerable decrease of TF in the conditions of deficient of redundant moisturising of meadow soils was not found in these years (Fig. 2)

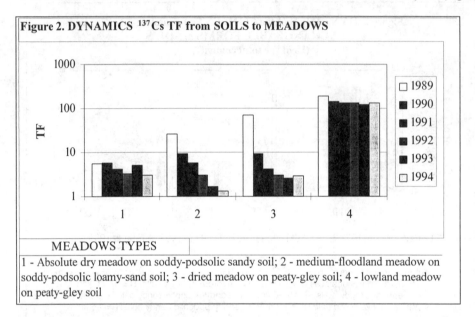

Figure 2. DYNAMICS ^{137}Cs TF from SOILS to MEADOWS

MEADOWS TYPES

1 - Absolute dry meadow on soddy-podsolic sandy soil; 2 - medium-floodland meadow on soddy-podsolic loamy-sand soil; 3 - dried meadow on peaty-gley soil; 4 - lowland meadow on peaty-gley soil

4. MODEL OF RADIONUCLIDE MIGRATION IN COMPONENTS OF MEADOW ECOSYSTEMS

Model of radionuclide migration in the components of meadow ecosystems has been developed by RIARAE. The schematic representation of the model is shown in **Fig.3**. The following processes were considered in the model:

- ^{137}Cs transfer in the forms of easily soluble aerosols and coarse dispersed particles and fuel ones to sod and soil;
- destruction of fuel particles in soil and sod followed by the change of ^{137}Cs state to the form available for plant uptake;
- decomposition of sod and ^{137}Cs leaching from it into root-inhabited soil layer;
- ^{137}Cs redistribution between different soil fractions;
- plant uptake of ^{137}Cs from soil and sod;
- ^{137}Cs removal from root-inhabited soil layer.

The processes which determine radionuclide content in the components of meadow ecosystems in the first period after the fallout (initial interception, weathering and others) were not considered in the model because the objective was to predict ^{137}Cs bioavailability for long-term period after the deposition. Each land unit is represented in the model by 10 state variables, which indicate the quantity of radionuclide (Bq/m^2) in specific compartments.

Values of ^{137}Cs residence half-time (T_e - ecological half-time, T_{eff} - effective half-time) in root-habitated soil layers of different meadow types are presented in Table 1. These data are shown obviously the principal dif-ferences on the intensity of radionuclide vertical trans-fer in meadows soils according to landscape-geo-chemicaql conditions.

Analysis of the results, obtained with use of described model, allows to make the following conclusions:

- *Half-lives of ^{137}Cs in vegetation as well as its residence times in soils in exchangeable and mobile forms are rather close for peatlands of various types;*

Dynamics of ^{137}Cs content decrease in the root layer of soil for peatlands of various types differs both in the quantity and in the quality, so, for the some types of peatlands the first residence time exceeds the second one which is connected with the regularities of its for-mation;

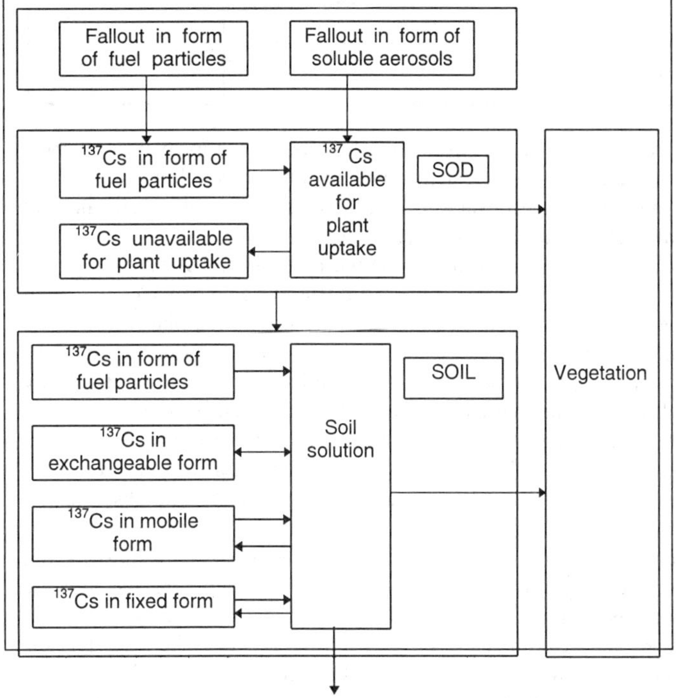

- *The average half-lives of ^{137}Cs in plants dur-ing the period for which the calculations were carried out (1-50 years after the fallout) depend on the decrease of both the total radio-caesium in the root layer of soil and con-tent of exchangeable and mobile forms of ^{137}Cs in soil.*

Thus, the main conclusion can be drawn, based on the analysis of results presented in this chapter, that the change of radionuclide bioavailabil-ity in meadow ecosystems is governed by numerous factors. The role of these factors and its contribution to the decrease of ^{137}Cs

Figure3 CONCEPTUAL SCHEME OF THE MODEL OF RADIONUCLIDE MIGRATION IN THE COMPONENTS OF MEADOW ECOSYSTEMS

content in meadow plants depend on the time lapsed after the deposition and is greatly variable.

The results of the comparison of experimental data with the model predicted are shown that the model based on the assumptions which were made above, fits the experimental data quite well.

RADIOECOLOGICAL CLASSIFICATION OF MEADOW ECOSYSTEMS

The meadows of Polessye zone of Ukraine, Belarus and Russia, subjected to radioactive contamination after the accident on Chernobyl NPP, are the secondary formations, as they are formed on after-forest lands. The meadows occupy about 10% of the territory and are divided into

Table 1. Caesium-137 residence half-time in root-habitated soil layers of different meadow types

Meadow type	Soil type	T_e, years	T_{eff}, years
Dry meadow	Soddy-podzolic loamy sand	110	23,5
Dry meadow	Leached chernozem	306	27,3
Dry meadow	Dark grey forest	163	25,3
Wet meadow	Soddy-podzolic low gleyed, loamy sand	48	18,5
Wet meadow, low peatland	Peaty	27	14,2
Wet meadow, transient peatland	Peaty	9,2	7,1

two types: mainland and floodland. The main part of the meadows is located in river floodlands. Mainland meadows include waterless meadows, located on upland parts of the relief, and lowland meadows. Existing waterless and lowland meadows were preserved only on small areas. They are located on the plots, unfit both for agriculture and forestry.

Waterless meadows occupy up to 25% of the whole meadows area. They are character-ised by unhomogenious water regime, lack of moister, as subsoil waters lie on considerable depth (more than 3 m). Rain water and water from melted snow are fully use by the plants due to the quick flow. The dominant soil types here are soddy-podzolic sandy and sandy-loamy; the defi-ciency of organic matter is evident, which can be explained by the intensive leaching of soils by atmospheric precipitation. If these lands are not constantly used for farming, brushwood of trees and shrubs are formed on them, and later - forest caenoses. Considerable areas of waterless meadows are ploughed at present, but partially they are again used as grasslands. Therefore on them sown meadows are formed regularly. Waterless meadows are subdivided into absolute wa-terless and waterless of redundant (temporarily redundant) moisturising.

Absolute waterless meadows are located on upland plots of the relief (hills, hillocks, up-per parts of slopes). Rainwater and waters from melted snow are not retained here, the soils are poor, depleted in nutrition elements, and, as a rule, acid.. Vegetation is presented by sparse low herbage , where the following species are dominant: *Festuca ovina L., Nardus stricta L., Agrostis sp.* In summer the grass is often burned by the sun and after cutting and feeding to animals grows very badly. The yield of dry mass is 200-400 kg/ha. On more fertile (e.g. carbonate) soils the yield is higher (400-700 kg/ha), hay is of better quality, as the dominant grass species are *Agrostis, Poa pratensis L., Anthoxanthum odoratum L.* and *Achillea millefolium L.* As a rule, absolute dry meadows are used as pastures. Radical improvement is recommended on them.

Normal waterless meadows are located in watershed valleys and in the middle part of the slopes. Soils are of medium moisture content, they better retain precipitation, content of organic matter is not very high. Cereals grow on these meadows - *Agrostis vulgaris With., Anthoxanthum odoratum L., Nardus stricta L.,* on more fertile loamy soils dominant species are: *Phleum pratense L., Poa pratensis L., Dactylus glomerata L., Bromopsis inermis Leyss.,* as well as *leguminous, Trifolium pratense L., Lathyrus pratensis L.,* and grass mixtures - *Centaurea jacea L., Potentilla anserina L.* The yield of hay on normal waterless meadows is 600-1500 kg/ha, they are used as hayfields and pastures. The most important measures on their improvement are optimum application of mineral fertilisers and overhead irrigation during summer grazing period.

Waterless meadow of redundant (temporary redundant) moisturising are located in the valleys and slightly lowered parts of watersheds with seasonal stagnation of surface waters, especially in spring and autumn. The soils here are soddy-podzolic, swampy in some places. On these meadows the dominant species are: *Poa trivialis L., Poa pratensis L.,Deschampsia caespitosa L.,Agrostis alba L., Agrostis canina L., Carex, Potentilla anserina L.* The yield of hay of medium quality is 1000-2000 kg/ha. Such meadows, if not very intensively used for farming, are overgrown with grasses of low value (*Deschampsia caespitosa L. Holcus lanatus L.*), radical improvement is recommended for them.

Lowland meadows are very widely spread in Polessye. They are located on the terraces, low valleys, flat depressions with stagnant waters, slight flowless depressions on watersheds. They are characterised by the supply with atmospheric, flowing and ground moisturising, which becomes excessive sometimes. These meadows are flooded in spring due to accumulation of flowing waters and rise of subsoil waters. Soil cover is presented by soddy soils with various podzol and gley content, meadow and silty-swampy soils, as well as peaty-gley and peaty soils. The vegetation is presented by water-loving cereals , carex , Juncus, Eriophorum, grass mixtures. Lowland meadows are fertile and after draining can be used for sowing new grasses or growing various agricultural crops.

Floodland meadows are located in the valleys (floodlands), which are annually flooded with waters from melted snow in spring and sometimes with rainwater in summer. After flooding alluviums or silt, containing many nutritious elements, remain on the surface, which, together with regular moisturising, create favourable conditions for the development of grass. The soils are mainly of alluvial origin (silty-swampy, soddy-gley and peaty), their mechanical composition depends on the particles brought with flooding. Soils with great amount of brought sand are less fertile and get very dry in summer. Floodland meadows provide higher and more stable yields as compared to mainland meadows, hay and pasture grass are of higher quality, as there are many valuable cereals in the herbage (*Alopecurus pratensis L., Festuca pratensis Huds., Phleum pratense L., Poa pratensis L.*) with admixture of *Trifolium,* and grass mixtures. However, not all grasses can survive durable flooding, and in this case water-loving plants (carex with admixture of *Poa palustris, Phalaroides, Glyceria fluitanus L.* etc.) appear instead of valuable cereals and leguminous. By the duration of flooding or depending on the height of location above the river-bed floodland meadows are subdivided into shortly flooded or dry (on high level), medium-flooded or humid (on medium level) and long flooded or swampy (on low level).

Shortly flooded meadows are flooded during up to 15 days a year and not annually. They are located in the floodlands of Polessye small rivers and upper elements of relief in the flood-lands of big rivers (the Dnieper, the Pripyat, etc.). On these meadows soddy and podzolled soils with sandy-loamy and sandy mechanical composition prevail.. They vary greatly in floral composition and character. Main areas of shortly flooded meadows are hayfields and pastures of a good quality. The yield on these meadows is 800-1500 kg/ha.

Medium flooded meadows are flooded during 15-25 days a year, they are located mainly in the floodlands of big rivers. As a rule the soils are soddy sandy-loamy and loamy. These meadows are very good hayfields and pastures. Valuable cereals and leguminous are dominant in the herbage. On such meadows 2 cuts of grass is done, the yield is 2000-3000 kg/ha.

Long flooded meadows are flooded during not less than 25 days a year and are located in the floodlands of big rivers. Most of them are used as hayfields and pastures of good quality, but durable flooding causes the formation of hydrophilic carex groups with small amount of cereals and grass mixtures (leguminous are rare). The yield is 1000-2000 kg/ha.

In this connection the accumulation of radionuclides in meadow grass differ up to 2 orders of magnitude. The investigation of radioecological characteristics of natural meadows allows to range the natural ecosystems. The main factors, determining the radioecological evaluation at meadows are the following : soil properties (especially mechanical composition), moisture regime and geobotanical composition. The problem of differentiation approach when determining intervention levels for different types of meadow ecosystems is not solved.

The main aim of this classification consists in the separation of meadow ecosystems with different rates of radiation safety, the radioecological meadow classification should be based on the typical practice classification using in agriculture. Specific peculiarities of meadow ecosystems should be taken into account for prognostication at radiation situation after the accident as well.

Three radioecological parameters were used for the elaboration of radioecological meadow classification (Table 2). Content of exchangeable forms of ^{137}Cs and transfer factors are

Table 2. Radioecological classification of meadow ecosystem

Meadow type	Soils groups	Content of ^{137}Cs exchangeable form in soils, %	T_e ,years	TF of ^{137}Cs
Waterless	Sandy, loamy sand	5 - 20	50 - 100	1 - 20
	Light loam, middle loam	1 - 15	100	0,2 - 15
	Heavy loam, clay	0,5 - 5	110 - 140	0,1 - 0,7
Floodland	Sandy, loamy sand	1 - 15	40 - 90	1 - 40
	Light loam, middle loam	0,5 - 10	110	0,5 - 15
	Peaty	0,3 - 10	80	17 -45
Lowland	Sandy, loamy sand	0,6 -10	40 - 90	3 - 20
	Light loam, middle loam	0,3 - 8	60	2 - 10
	Peaty	1,3 - 10	40 - 90	10 - 50
	Transient peat	5 - 12	16 - 21	15-30
	Top peat	8 - 20	18	50 - 90*

* - to 200 for Ukraine

important parameters for biological availability. Ecological half-life of ^{137}Cs in root zone (0-10 cm) is an integral parameter of migration intensity. This parameter depends on model, used for calculation, but regularities of ^{137}Cs migration on meadow type do not change. These parameters of ^{137}Cs migration are given in previous chapters. We would like to underline, that preceding classification cannot be consider as absolute one, since it is practically impossible to take into account the numerous factors and conditions that influence the TF, change dynamic of which could not be forecasting. Nevertheless our classification is enough universal for practical use on the area of middle part of Europe.

6. APPLICATION OF COUNTERMEASURES IN MEADOW ECOSYSTEMS

Use of natural fodder on contaminated areas should be intensified by conduction of agroameliorative measures. Measures providing decrease of radionuclides transfer in grass stand of meadows include agrotechnical treatment of vegetative layer and tillage, selection of grass species with minimum rates of radionuclide accumulation, application of fertilisers and agroame-liorants. Preliminary deaqation and area levelling is carried out. Two methods of meadow regress-ing - surface and radical are recognised. On flooded meadows as well as in the chernozem zone where erosion processes are developed the surface resting is used which include damage of vegetative layer, seedling growth of perennial grasses and application of fertilisers and agroame-liorants. In meadow management two methods of radical resting are distinguished: the common method with preliminary cultivation (1-3 years) of annual grasses and rapid one with seedling of perennial grasses immediately after the cultivation of vegetative layer.

A large volume of research before and after the ChNPP accident showed that by grassing of meadows on contaminated areas the use of methods of the type such as liming, application of

Table 3. Countermeasures effectiveness on the natural meadows

Countermeasure	Reduction factor, time	
	Mineral soils (sandy, loamy sand)	Organic soils (peaty)
Deaquation	-	2 - 4
Disking or rotary cultivation	1,2 - 1,5	1,8 - 3,5
Ploughing	1,8 - 2,5	2,0 - 3,2
Ploughing with turn-over upper layer on depth 35-40 cm	8 - 12	10 - 16
Liming	1,3 - 1,8	1,5 - 2,0
Mineral fertilisers:		
N	-(1,1 -3,0) **(Increased factor)**	-
K60-K240	1,5 - 3,0	1,5 - 3,0
NPK (1:1,5:2)	1,2 - 2,0	1,5 - 2,0
Surface improvement	1,6 - 2,9	1,8 - 14
Radical improvement	3,0 - 12	4,0 - 16

increased doses of phosphoric and potassium fertilisers and organic fertilisers, zeolites and clay minerals is effective for the decrease of radionuclide transfer into grass stand.

Based on the common regularities of radionuclide migration in soil-plant system the effectiveness of ameliorants is to be defined by their effect on soil properties. According to this fact, the maximum effectiveness from application of ameliorants on low-productive acid soils of light mechanical composition is observed. Application of lime is the usual practice for the improvement of meadows on acid soils. Long-term investigations demonstrated that the decrease of radionuclides transfer to plants becomes more resistant by limiting in the dose of 1.5-2 norms calculated by hydrolytic acidity. Application of lime causes reduction in ^{137}Cs transfer to grass stand by a factor of 1.2-2.4, while the transfer coefficients of ^{137}Cs into grass stand decrease under the influence of increased doses of phosphoric and potassium fertilisers within the range from 1.1 to 2.0 times. Application of zeolites as ameliorants affects ambiguously ^{137}Cs transfer to grass stand - at times no effect is noted and the other times - redite acts as lime. As a rule, the most effective practice is combined application of ameliorant and fertilisers.

Considerable variability of countermeasures efficiency is determined by the variability of soil characteristics, climate conditions and technological amelioration. The countermeasures efficiency (Table 3) should be taken into account for the determining of intervention levels after the accident.

The table illustrates the efficiency of countermeasures implemented on the contaminated areas in Belarus, Russia and Ukraine.

REFERENCES
1. Tessier A., Campbell P.G.C. and BissonM. (1979). Sequential extraction procedure for the speciation of particulate trace metals. Anal.Chem., Vol.51, No 7. P.844-851.
2. Pavlotskaya F.I. (1974). Migration of the radioactive product of global fallout in soils. Moscow: Atomizdat.- 216 p. (Rus.)

Dynamics of Radionuclides
in Forest Environments

Maria BELLI
National Environmental Protection Agency, via V. Brancati, 48, 0144 Rome, Italy
Fjodor A.TIKHOMIROV, Alexei KLIASHTORIN & Alexey SHCHEGLOV
Moscow State University, Soil Science Faculty, Leninskie gory,119899, Moscow, Russia
Barbara RAFFERTY
Radiological Protection Institute of Ireland
3, Clonskeagh Square, Clonskeagh road, Dublin, Ireland
George SHAW
Imperial College of Science Technology and Medicine
CARE, Buckhurst road, Silwood Park, Ascot, SL5 7TE, Berkshire, United Kingdom
Erich WIRTH, Lothar KAMMERER, Werner RUEHM & Martin STEINER
Bundesamt fuer Strahlenschutz, Institut fuer Strahlenhygiene
Ingolstaedter Landstrasse 1, 85764 Neuherberg, Germany
Bruno DELVAUX, E. MAES & Nathalie KRUYTS
Université Catholique de Louvain, Unité des Sciences du Sol
Place Croix Du Sud, 2-Bte 10, 1348 Louvain-La Neuve, Belgium
Kurt BUNZL
Forschungszentrum fuer Umwelt und Gesundheit
Ingolstaedter Landstrasse 1, 85764 Neuherberg, Germany
Alexander M.DVORNIK
Forest Institute, Gomel, Belarus
Nicolay KUCHMA
RIA "Pripyat", Radiology and Land Restoration Department
Libknecht Str.10, Chernobyl, Ukraine

Abstract. In the CIS countries, during the Chernobyl accident, more than 30000 km^2 of forested areas received a ^{137}Cs deposition higher than 37 kBq m^{-2} and about 1000 km^2 a deposition of radiocaesium higher than 1.5 MBq m^{-2}. Before the accident only few data were available on the behaviour of radionuclides in forests and during last eight years, the understanding of the fate of radionuclides in these ecosystems has been improved significantly.

This paper reports the results achieved in the frame of 1991-1996 EU/CIS collaborative project on the consequences of the Chernobyl accident. The ECP-5 project deals with the impact of radioactive contamination on natural and semi-natural environment. The investigations were carried out in different forest ecosystems, located in the near field (within the 30-km zone around the Chernobyl nuclear power plant) as well as in the far field in the CIS and in the western Europe countries. The results achieved have been used to develop a simplified model representation of the behaviour of radiocaesium within forest ecosystems.

1. Introduction

After Chernobyl accident, it was difficult to define the contribution of forest ecosystems to the dose to man and to adopt well justified countermeasures in these environments, because there was very little information relating to the impact to man of radioactive fallout on forests. After the fallout from atmospheric nuclear weapons testing, in the 1960's, considerable attention was paid to the effect of fallout on agricultural products, drinking water, etc. but only the lichen-caribou-man and lichen-reindeer-man food chains were studied for natural and semi-natural environments [1,2]. The few observations in forest environments in the 1960's and the studies carried out after the Kystym accident (in which mainly ^{90}Sr was released) showed that in these environments, radionuclides remain available for a longer time than in agricultural systems [3]. In the wake of the Chernobyl accident it became apparent that these ecosystems are very important sources of dose to man which demand careful management. Nine years after the Chernobyl event, the ^{137}Cs concentrations in plants grown in forest, did not significantly decline. Meat and milk from animals grazing on forest clearings as well as mushrooms, wild berries and game, might contribute a significant dose to man. Restrictions in the use of food products coming from semi-natural ecosystems are still necessary in some areas heavily contaminated of Ukraine, Belarus and Russia. Additionally external dose may be received by forestry workers and groups of population using timber for furniture or building material. Wood industries, like pulp mills, consuming large amount of wood, concentrate radionuclides in their waste products. In highly contaminated areas these wastes can be a source of external dose to workers in wood industries. Furthermore in the heavily contaminated areas of CIS countries, forests are a potential reservoir of secondary contamination and fires represent a resuspension risk. In conclusion, in the long-term, the contribution to the dose to man from forest environments might be, for some groups of population, more important than that from agricultural and urban areas. Due to the lack of understanding of radionuclides behaviour in these systems, it was essential to implement a programme of intensive research.

The studies performed in the frame of EU/CIS collaborative ECP-5 project "Behaviour of Radionuclides in Natural and Semi-Natural Environments" [4] had as a central aim the understanding of the migration and transfer mechanisms, the quantification of the main parameters playing a role in these processes and their dependence on time; reliable predictions of the long-term dose to man from forests and meadows cannot be done without this knowledge. Therefore within this project, field studies were carried out in near field (within the 30km zone around Chernobyl) and in far field forest scenarios (in other areas of CIS and western Europe countries). Forest systems can be partitioned into four major components: overstorey, understorey, organic and mineral soil horizons. This paper gives a short overview of the behaviour of radiocaesium in the different components of forest systems and discusses the main processes that play a role on the dynamics of nutrients and trace elements throughout the forest system. Furthermore this paper briefly reports the results of a model developed on the basis of the ECP-5 experimental data and shortly gives some suggestions for soil-based countermeasures.

2. Results and Discussions

2.1. Radiocaesium Behaviour in Trees

Immediately after the Chernobyl accident, a programme of field monitoring started in the 30-km zone around the Chernobyl nuclear power plant. These investigations showed the role of the tree canopies in the interception of dry deposition; in the 30-km zone 60-90% of ^{137}Cs deposition was intercepted by tree leaves/needles. The variability of the fraction intercepted by the vegetation was mainly attributable to the type of forest; deciduous trees, leafless at the time of deposition collected the lowest fraction, while coniferous stands, characterised by high densities of trees, intercepted about 80-90% of the deposited material. The effectiveness of tree canopies in the interception of dry deposition was also observed in Belarus, 50-100 km away from Chernobyl, where forests showed a higher deposition than open field areas [5]. The initial capture and retention of radioactive material by forest canopies is highly dynamic and the duration of radioactive deposit in tree canopies is considerably shorter than in other compartments of the forest ecosystems. Immediately after the deposition event, wind action can resuspend and redistribute the deposited material to adjacent agricultural areas. This was observed after the Kyshtym accident when a broadening of an initially narrow band of contamination near a forested area was seen within 1 month following the accident [5]. Season of deposition seems to influence the half-life of radionuclides on tree canopy. Experiments in controlled conditions, with ^{89}Sr in aqueous solution [6], show that in spring and in summer, when tree growth is active the half-life of loss is about 30 days, irrespective of rainfall events, while during autumn and winter period, when the trees are physiologically dormant, the half-life is about 120 days.

Table 1.
TF (m^2 kg^{-1} 10^{-3}) of ^{137}Cs to the various structural parts of woody plants. Data collected in 1987.[137]Cs depositions at the sampling sites are:D1 - 155 , D3 - 160 kBq m^{-2} and biomass is 0.001 m^2 kg^{-1} dry mass.

Tree	Site	external bark	internal bark	wood	branches	young needles/leaves	old needles
Pine (Pinus sylvestris)	D1	29.7	4	0.83	7.7	7.2	36.2
	D3	-	-	-	-	-	-
Birch (Betula pendula)	D1	28.2	3.2	1.2	10	11.2	-
	D3	37.8	5.5	2.3	17.8	32.3	-
Oak (Quercus robur)	D1	21.3	2	0.66	8.8	5.2	-
	D3	-	-	-	-	-	-
Alder (Alnus nigra)	D1	-	-	-	-	-	-
	D3	30	3	0.83	7.8	7.3	-

In the Chernobyl area the initial contamination of the forest stands was due to fallout interception by the aerial structure. Thus the main contamination pathway was external, and in this case, the main influence on the behaviour of radionuclides was the physico-chemical properties of the fallout. In Table 1 are reported the data collected in two forests located in the 30-km zone (D1, 26 and D3, 28 km from Chernobyl nuclear power plant). Data are expressed as TF; ratio of radiocaesium concentrations in dry plant material (Bq kg^{-1}) to radiocaesium deposition (Bq m^{-2})

Concentrations of ^{137}Cs in various above ground organs and tissues of woody plants decrease in the following series: external bark (cork)>branch wood>leaves and needles>internal bark (bast)>trunk wood. In the first year after the accident, radiocaesium found in the internal part of the tree demonstrates translocation of external contamination to the internal parts.

Following the initial contamination, radionuclides deposited on the canopies are transported to the forest floor by weathering processes (rainfall and wind action) and by shedding of leaves and other parts of the trees. Weathering processes and leaf/needle fall transfer most of the radionuclides, from the canopy to the forest floor, where they enter in the bio-geochemical cycle and the main pathway of contamination of forest vegetation is via root uptake from the soils. Data collected in the 30-km zone around Chernobyl nuclear power plant show that in autumn 1987 about 90% of the total deposition was in the first layers of the forest soil and about 85% was found in the Ol layer (mainly formed by needles and leaves fallen down after the Chernobyl accident). Eight years after the accident, the radiocaesium distribution found along the soil profile at D3 and D1, show that about 60, 70% of deposition is still in the organic layers of forest soils (Ol, Of, Oh), with a maximum value in the Oh horizon. When root uptake became the dominant pathway, soil properties have a major influence on the extent of radionuclide transfer to wood. Figure 1 shows results achieved in two sites with different soil characteristics; site D1 is located on soddy podzolic sandy soil, while D3 is located on peaty gley soil. The initial TF is quite low and its decline in the years immediately after the accident is similar at both sites. Subsequently, in peaty site, there is an increase in TF as the contamination migrates down the soil profile to the rooting depth. In contrast, at the podzolic site, sorption of radiocaesium in the clays of this soil type results in a continuous decline in TF.

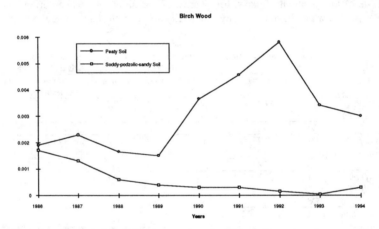

Figure 1. TF (m^2 kg^{-1}) of ^{137}Cs to birch wood on two soil types

Tree samples collected in Ukraine (near-field, Kopachi) and in Ireland, in 1993/1994, show that in far field site, 26% of ^{137}Cs deposition was found in the tree biomass (18% in the overground biomass), while in Kopachi this value was 1.7% (1.3% in the overground biomass). The factors that can explain the low uptake of ^{137}Cs at the Ukrainian site are: the age difference between the trees at the two sites, (the Ukrainian trees are 60 years, while the Irish trees are only 30 years) and the low availability of radiocaesium for root uptake at Kopachi, where it is mainly incorporated into fuel and condensed particles. The influence of tree age on radiocaesium concentrations was investigated in Belorussian forests. The results show that 60 to 80 year old trees have a 6 times lower ^{137}Cs concentrations compared with 20 to 40 year old trees [4].

2.2. Radionuclide Fluxes from Canopy to Forest Floor

To predict the long-term behaviour of radionuclides in forest ecosystems it is necessary to determine the major fluxes involved in the transfer of elements from the canopy to the forest floor. The transfer rates of elements were assessed by measuring their concentrations in the fluxes (throughfall, stemflow, litterfall etc.) and then multiplying concentrations by the flow rate of the respective fluxes. Two coniferous forests, the first one located in the 30-km zone and the second one in Ireland, were equipped with litter traps and water collectors. The fluxes of radiocaesium associated with litterfall (material derived by shedding of dead leaves and other components of the trees), with throughfall (water derived by the interaction of precipitation with the canopy) and with stemflow (water derived by the interaction of precipitation with the stems of the trees) were regularly determined. The selected areas are representative of the near field and far field scenarios. Table 2 summarises the radiocaesium fluxes measured in 1992/1993 at Kopachi site (near field) and at Waterford (Shanagarry Wood) in Ireland (far field). Data are expressed as percentage of total deposition on the sites.

Table 2.
Annual fluxes of ^{137}Cs at Ukrainian and Irish sites in 1992/1993

Fluxes	Kopachy (Ukraine) (%)	Waterford (Ireland) (%)
Throughfall	0.044	0.60
Stemflow	0.007	0.01
Litterfall	0.14	0.50

2.3. Radionuclide Migration in Forest Soils

Eight years after the accident, the radiocaesium distribution along the soil profile of forests located in the 30km zone shows that the percentage of deposition in the organic layers of forest soils (Ol, Of, Oh), ranges from 60 to 80%. The variability between the different forests is mainly attributable to the different physico-chemical characteristics of fallout. In the areas nearer to the point of release (Shepelichi and Kopachi), the highest percentage of deposition was found in the superficial layers of forest soils, because the radionuclide deposited were mainly associated with fuel and condensed particles with larger dimensions.

To evaluate the migration rates between forest soil horizons, residence half times (τ) for the horizons Ol, Of, Oh, Ah and A1 were derived considering each horizons as a compartment.

Table 3.
Caesium (^{137}Cs and ^{134}Cs) residence half times, in years and migration rates in cm y^{-1} for individual soil horizons at each of the three study sites. Mean half times are given ± 1 standard deviation.

Soil Horizon	Shepelichi Half Time (y)	Shepelichi Migration Rate (cm y^{-1})	Kopachi Half Time (y)	Kopachi Migration Rate (cm y^{-1})	Dityatki 1 Half Time (y)	Dityatki 1 Migration Rate (cm y^{-1})
Ol	0.99 ± 0.01	1.01	1.1 ± 0.4	0.90	1.01 ± 0.03	1.19
Of	1.76 ± 0.04	1.42	3.0 ± 0.7	0.82	3.60 ± 0.07	0.42
Oh	200 ± 17	0.01	24 ± 3	0.06	10± 2	0.10
Ah	1.29 ± 0.01	0.78	6 ± 1	0.17	3 ± 2	0.30
A1	6.61 ± 0.06	0.61	17± 13	0.24	28 ± 19	0.14
	n = 2		n = 4		n = 4	

The vertical migration rate coefficients between the compartments were assessed by fitting the experimental data on vertical radionuclide profiles at known times after deposition. For the sampling site Hochstadt, Bavaria, the compartment model was fitted to an extensive time series of soil profiles.

The residence half-times and the migration rates were assessed for three sites within the Chernobyl 30-km zone (Table 3) and for sites at Waterford in Ireland and Hochstadt in Bavaria (Table 4).

The derived migration rates for radiocaesium within the Chernobyl 30-km zone, demonstrate the tendency for the Oh horizon in particular to retard the vertical migration of caesium isotopes, at least within the first few years following deposition. Examining the trend of migration rates within this horizon, there is a clear increase from Shepelichi to Kopachi to Dityatki 1. This is most likely a reflection of the decreasing importance of 'hot particles' at each of these sites. The very large Oh residence half time and the low migration rate observed at Shepelichi is a good indication, that when radiocaesium is initially introduced to a forest ecosystem in 'hot particle' form, its migration through the soil is likely to be considerably slower than if it is present as a fine aerosol, as would be expected in the far field situation.

Table 4.
Summary of caesium (^{137}Cs and ^{134}Cs) residence half times, in years, and rates, in cm y^{-1}, for individual soil horizons at each of the two study sites.

Soil Horizon	Waterford Half Time (y)	Waterford Migration Rate (cm y^{-1})	Hochstadt Half Time (y)	Hochstadt Migration Rate (cm y^{-1})
Ol	0.9	1.12	2.9	0.34
Of	1.5	1.95	4.6	0.44
Oh	3.3	0.90	6.3	0.32
A1	3.5	1.43	8.7	0.17
A2	14.2	0.35	1.8	3.33

The migration rate estimates at the far field sites (Waterford and Hochstadt) indicate a generally lower retention of radiocaesium within the Oh horizon than at the sites in the 30-km zone.

One of the major causes of variability in observed residence half times between different forest sites and between different radionuclides is likely to be the physico-chemical speciation of radionuclides. Two aspects of this phenomenon must be considered: the initial speciation (ie. speciation of the depositing radionuclide) and post-deposition alterations to radionuclide speciation within the soil during the migration processes.

Table 5.
Average concentration and annual leaching of ^{137}Cs by intrasoil flow in Ukraine and Ireland. Annual leaching data are presented also as a % (per unit area) of the ^{137}Cs of the soil horizons being leached.

Site	Soil Horizon	Depth (cm)	^{137}Cs (Bq l^{-1})	^{137}Cs (Bq m^{-2})	(%)
Ukraine	Ol+Of	0-3	13.3	3900	0.5
	Ol+Of+Oh	0-4	17.7	3800	0.2
	O+Ah/A1	0-5	8.5	815	0.07
	O+A1	0-10	7.0	372	0.03
	O+A1+B1	0-20	5.4	99	0.007
Ireland	Ol+Of	0-3	0.4	129	13
	Ol+Of+Oh	0-6	0.3	102	2

The role of organic layers in retarding the vertical migration of radiocaesium in forest soil is also confirmed by the results of the lysimetrical studies, summarised in Table 5 which show that the amount of ^{137}Cs leached diminishes with increasing depth down the soil profile. The ^{137}Cs content of leachate was subject to seasonal variability and tended to be positively related to rainfall, but other seasonal factors such as microfloral and microfaunal activity could have a co-incident influence. However, overall the total amount of ^{137}Cs present in these leachates was low - of the order of 1-2% of total deposition.

2.4. *Radiocaesium Fluxes in Decomposing Litter*

Migration of radiocaesium within the forest soil profile is slow and most of the Chernobyl deposition of radiocaesium is still in the organic material of the forest litter horizons. In this layers occur the most important processes in nutrients and pollutants recycling within the forest. Here processes of decomposition remobilise elements which are contained within dead forest products and make them available for reuse by the organisms of the forest. The breakdown of litter by decomposition is a key process in the mobilisation of elements which are incorporated into plant material.

To determine the rate of radiocaesium loss from decomposing litter, litter bags were located in the Ol and Of layers of the two coniferous forests above reported (Kopachi, Ukraine and Waterford, Ireland.

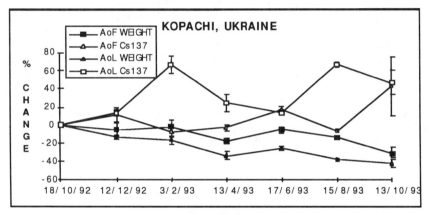

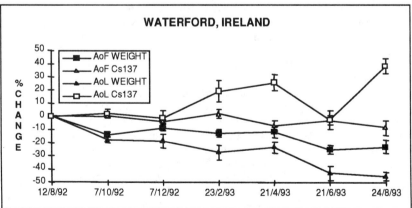

Figures 2.a,b: Mean percentage change in weight and mean percentage change in ^{137}Cs content of Ol and Of litter bags, following incubation for periods from 2-12 months in Ol and Of horizons of pine forest soil in Ireland and Ukraine. Error bars represent standard deviation of 5 replicates.

Figure 2 reports the percentage of change of weight and radiocaesium content as to the original content versus time. A negative % change indicates a loss of elements from the litter bags, while a positive value indicates importation of elements to the bags. In both sites weight loss was greater in the Ol layer, because Ol material is less resistent to the decomposition than Of material. The pattern of weight loss is similar to that described by Hunt [7], in which weight loss occurs in the early stages when easily broken down material is degraded, and thereafter the rate declines as more recalcitrant material accumulates. At both sites the ^{137}Cs content of the Ol samples increased periodically indicating an importation of ^{137}Cs throughout the experimental period despite a simultaneous decrease of the sample mass. There was almost no change in the Of samples, however, the maintainance of an approximately constant ^{137}Cs content in the Of litter bags in spite of up to a 20-30% weight loss, shows that some compensatory importation of ^{137}Cs took place. Two pathways can explain this importation from other compartements of the forest to the litter layer: transfer from the canopies to the forest floor via water fluxes and transport from Oh and Of (with higher content of radiocaesium) to Ol layer through fungi mycelia. The first source is water washing from the forest canopy. The annual input to forest floor as throughfall, stemflow and litter fall amounts to approximately 100 Bq.m^{-2} (Ireland) and 13000 Bq m^{-2} in the Ukrainian site. The area occupied by five litter bags is equivalent to 0.11 of a square meter. Even if all of the ^{137}Cs associated with these inputs was retained in the litter bags the amount added would not account for the observed increment in ^{137}Cs concentration. The second possible source of radiocaesium is the underlying litter layers; fungi in forest soils are generally more abundant in the Of and Oh horizons and migrate from these horizons to attain nutrients from the more easily decomposed Ol litter. The Of and Oh horizons contain 20-35 times more ^{137}Cs compared to the Ol horizons. Therefore, in the two deeper horizons there is great potential for accumulation of high levels of ^{137}Cs by fungal mycelia. Invasion of fresh Ol litter by fungi which have accumulated ^{137}Cs from the Of and Oh horizons, may significantly enhance the ^{137}Cs content of the Ol material. Fungal invasion of Of litter is less likely to significantly enhance the already relatively high ^{137}Cs content of the Of.

2.5. Uptake of Radionuclides by Understorey Vegetation and Mushrooms

The samplings carried out in the 30-km zone show that radiocaesium transfer factors in understorey vegetation vary by a factor 90 depending on species and site. Comparison of TF_{org} (defined as the ratio of the concentrations in plants as Bq/kg dry weight to the concentrations in the organic horizons of soil as Bq/kg dry weight) calculated for radiocaesium averaged in time from 1992 to 1993, reveals that at D1 and D3 the values for TF_{org} increased from 1992 to 1993, at K2, however, they decreased with time. This effect has been observed for nearly all kinds of plants. Great sesonal variations on radiocaesium uptake by plants were observed by several authors [8, 9]. To rule out this effect, all plant samples were taken end of June and beginning of July, respectively. The averaged values for TF_{org} also appear to be different at the three sampling sites. *Fragaria vesca* is growing at every plot. Its TF_{org} is similar at D1 (TF_{org} = 0.07) and K2 (TF_{org} = 0.05) but significantly higher at D3 (TF_{org} = 0.27). *Rubus saxatilus* takes up about 2 times more radiocaesium at D3 (TF_{org} = 0.23) than at D1 (TF_{org} = 0.10). In 1993, the fern *Athyrium filix femina* took up 2.9 times more ^{137}Cs at D3 compared with K2. This finding is in good agreement with the soil characteristics: D1 and K2 are a podzol and D3 is a peaty-gley soil type.

Uptake of radionuclides from soil is the main contamination pathway for mushrooms. Soil properties, as well as mushroom species, appear to be the main factors influencing contamination rate of mushrooms. An important species feature is, in this respect, mycelium distribution in the soil profile.

Generally, fruitbodies of some mushroom species are characterised by the highest ^{137}Cs transfer factors to above-ground phytomass in comparison with any other forest vegetation. Data collected in 1992/1993 on peaty-gley soil (D3 forest) show that the transfer factors values range from 80 to 130 for berries, while they range from 20 to 1700 $(m^2 \ kg^{-1} \ 10^{-3})$ for mushrooms. Thus, edible mushrooms can significantly contribute to the ingestion dose of the local population. A fundamental factor is the incidence of the mycelium in the most contaminated litter layers (Of Oh). Interspecies variation of ^{137}Cs transfer to mushrooms can be more than one order of magnitude. Some of this variation may be accounted for by some mushrooms which accumulate potassium taking up more ^{137}Cs, since potassium is chemically analogous to caesium. Furthermore, mycelium distribution in the different soil layers varies depending on mushroom species. The living habit of different fungal species influences ^{137}Cs uptake [10]; Saprophytic mushrooms show accumulation rates which are comparable with those of green plants. The representatives of species group called Xilophytes (Xilogenous) are more effective ^{137}Cs accumulators while highest transfer factors are characteristic of mycorrhizal species, however significant species-dependent variations in transfer factor were observed within this group. These variations are possibly due to differences in mycelium distribution in the soil profile.

Soil properties influence the magnitude of the radiocaesium transfer factor. The transfer factors can differ up to 200 times in magnitude depending on the soil type. At site D3 a very high two-year averaged Tf_{org} of 14 was determined for *Paxillus involutus*. This result agrees very well with measurements in Bavaria, where this species has also one of the highest uptake rates for radiocaesium [11]. *Paxillus involutus* belongs to the group of symbionts, which accumulate significantly more caesium than saprophytes or parasites [10]. Therefore, the low averaged TF_{org} of 0.9 for *Paxillus involutus* at D1 is surprising. In general, mushrooms at podzol sites exhibit lower transfer factors than those from peaty-gley site. Similar behaviour has been found for green plants and trees.

Broad variability of the transfer factors was noticed for the same mushroom species over the years since 1987. The data obtained revealed no apparent trend in radiocaesium transfer factors with time, neither for fungi as a whole, nor for any single mushroom species.

2.6. *Modelling*

Three forest models have been developed by groups working in the ECP-5 Project. These models are described in a paper [12] presented at this Conference. RIFE I model has been specifically developed using the data collected in the ECP-5 project. The output of RIFE I is a series of radiocaesium inventories and/or concentrations within the principal forest components.

Table 6.
Rate coefficients, half times and migration rates derived for the Kopachi field site within the Chernobyl 30-km zone.

Transfer from	Transfer to	Rate Coefficient (y^{-1})	Transfer Half Time (y)	Migration Rate $(cm \ y^{-1})$
Tree	0-5 cm soil	0.143	4.852	-
0-5 cm soil	5-10 cm soil	0.019	36.34	0.138
5-10 cm soil	10-20 cm	0.009	79.71	0.063
10-20 cm soil	deeper soil	0.015	45.05	0.222
Soil (total)	Tree	0.002	311.0	-

This model takes into account the major flux pathways involved in the recycling of elements between the principal components of the ecosystem. As before reported fluxes were determined in field monitoring carried out by ECP-5 at forest sites in the Chernobyl 30-km zone and in Ireland. In addition to the major compartments for which fluxes are calculated dynamically, RIFE I has two compartments representing herbaceous understorey species and mushrooms. Radionuclide concentrations in these compartments are calculated using ranges of soil-plant (or mushroom) transfer factors derived both by ECP-5 and from other literature sources. The rate coefficients, transfer half times and (in the case of soil layers) migration rates derived from field measurements for the Kopachi site are shown in Table 6.

Bearing in mind the discrepancies between radiocaesium distributions at the Waterford and Kopachi sites it is interesting to make two principal observations from the table. First, loss of radiocaesium from the top 10 cm of the soil profile at Kopachi is slow with a half time greater than the physical decay half time of ^{137}Cs. Secondly, the apparent uptake rate of radiocaesium by trees at Kopachi is extremely low, accounting for the very small fraction of the total radiocaesium inventory within the tree biomass as compared with that at Waterford.

Rate coefficients derived from field data from Waterford and Kopachi have been input into the RIFE I model as calibration or 'benchmark' parameter groups which can be recalled as default values. Figure 3 shows comparisons between field data and RIFE predictions for the Kopachi site for the year 1993. A time simulation (25years) of radiocaesium distribution in the forest components show that the longevity of contamination of organic soil layers is considerable in the near field, whereas in the far field contamination of tree tissues and deeper soil horizons is likely to be a longer lived problem [12].

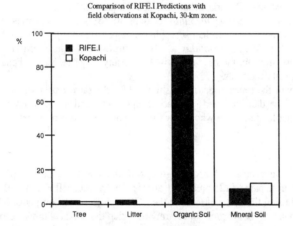

Figure 3 Measured and modelled ^{137}Cs distributions

2.7. Countermeasures: Approach for Application in Semi-Natural Systems

Soil based countermeauseres are used in common agricultural practice and are therefore relatively inexpensive, available in large quantities and easy to apply to large areas of land. The application of these remediation actions to forest stands requires further investigations on the effectiveness of these treatments on multi-layered soils and on the secondary possible negative effects. Chemicals used in agriculture may disturb the ecological balance of forest systems and literature reports many examples of the negative effects of fertilisers on natural environments (watertable contamination and reduction in plant species diversity [13]). In addition, the inacessability of many semi-natural ecosystems and the difficulty on the application of these measures in forest stands, due to the problems of adequately incorporating these additives with the forest soil, may prevent the general application of soil-based countermeasures in these environments. Nevertheless, these treatments may have a number of potential advantages and may reduce the transfer of radiocaesium in several ways: some fertilisers compete with radiocaesium for plant root uptake; other soil treatments increase the radionuclide fixation in soils, reducing consequently its mobility and transfer to plants [14]. In conclusion for semi-natural environments, the actual stage of knowledge allow only to identify the potential effectiveness of some soil based countermeasures on different types of soils. Further investigations are required to assess their effectiveness and the possible negative effects of these amendments on semi-natural systems.

On the basis of experimental and field data achieved in the frame of ECP-5 project [4], a preliminary approach is suggested to identify the most suitable countermeasures for different soil categories.

Soil characteristics in semi-natural environments located under latitudes 50-60°N, can be classified in three categories:

- Category 1: soils rich in clay minerals having a high fixing potential for radiocaesium; these soils have very thin organic layers (humus type is a mull) and are often called brown earths (brown soils, acid brown soils) or black earths (chernozems, brunizems);
- Category 2: acid soils poor in clay minerals; these soils have thick organic horizons (humus type is a moder or a mor) and belong to the group of podzolic soils as well as some acid brown earths (acid brown soils , soddy podzolic soils, podzols);
- Category 3: peaty soils very poor in or devoid of clay minerals able to fix radiocaesium; peats and peaty soils have very thick (>30 cm) organic horizons (in acid peaty soils, radiocaesium is very mobile and it can be associated with organic compounds).

Generally these soil categories show an increase in radiocaesium uptake by vegetation in the following order: Category 1<Category 2 < Category 3 [4].

The most suitable countermeasures for each soil category are outlined below.

- *Category 1.* on a long term basis (as long as the fixing minerals are stable), these soils fix large quantities of radiocaesium. Application of K fertilisers could be an adequate soil based countermeasure for soils of this category.
- *Category 2.* these soils have a lower fixing potential for radiocaesium. Improving the fixation process may require one or more of the following measures: the application of fixing clay materials, the mixing of O and Ah horizons by ploughing, K fertilisation, liming.
- *Category 3.* these soils do not fix radiocaesium on a long term basis. Soil based countermeasures must consider here, both the application of fixing clay materials and the correction of soil physico-chemical conditions (liming, K fertilisation), particularly if the peats are acid.

3. Conclusions

In the initial period immediately after the Chernobyl accident the main influence on the behaviour of radionuclides is the physico-chemical properties of the fallout. After 1-2 years weathering processes and leaf/needle fall transfer most of the contamination to the ground and within 2-4 years the radionuclides entered in the bio-geochemical cycle. In this second phase, in which the main pathway of contamination of vegetation is via root uptake from soil, soil properties influence radionuclide uptake by vegetation and mushrooms.

The results achieved in ECP-5 project show that the availability of radiocaesium to the transfer between forest compartments is different in the near and in far field scenarios. In fact in forests located in the near-field (in the range of about 5 km from the Chernobyl nuclear power plant), the derived migration rates demonstrate the tendency for Oh layer in particular to retard the vertical migration of caesium, at least until the degradation of the 'hot particles'. Furthermore the bioavailability of radiocaesium is lower in the near field than in the far field. The low mobility of radiocaesium in the near field is mainly attributable to the large amount of radiocaesium attached to fuel and condensed particles deposited in the vicinity of the source of release.

The experiments on litter decomposition showed the importance of soil micro-organisms in the binding of radiocaesium in the superficial layers of the forest soil (Ol, Of, Oh). The partitioning of radiocaesium between soil micro-organisms and clay minerals, where these are present, is still not well understood. Furthermore was clearly focused the role of fungal mycelium on the upward migration of radiocaesium in the forest organic horizons.

On the basis of experimental data collected in the frame of the ECP-5 project three compartmental models were derived. These models describe with good accuracy the radiocaesium distributions observed in the different sites investigated, but they are still site specific. Then there is a need to expand the data base of parameter values for radionuclide transport within forest. In addition, while processes such as soil migration have been studied in great detail other processes as the behaviour of radionuclides in the organic horizons of forest soils, tree uptake and distribution are yet to be understood at a level which will allow confident predictions of wood contamination and associated doses over the decades to came.

References

[1]   Bird, P.M.; Studies of Fallout ^{137}Cs in the Canadian North; Archives of Environmental Health; 1968, 17, 631-638.
[2]   Bird, P.M.; Radionuclides in Foods; The Canadian Medical Association Journal; 1966, 94, 590-597.
[3]   Alexakhin, R.M., Ginsburg, L.R., Mednik, I.G, Prokhorov V.M.; Model of ^{90}Sr Cycling in a Forest Biogeocenosis; The Science of the Total Environment; 1994, 157, 83-91.
[4]   Belli, M., Tikhomirov, F.A.; Behaviour of Radionuclides in Natural and Semi-Natural Environments; 1996 Final Report of ECP-5 Project (1991-1996).
[5]   Tikhomirov, F.A., Shcheglov, A.I.; Main Investigation Results on the Forest Radioecology in the Kyshtym and Chernobyl Accident Zones; The Science of Total Environment; 1994, 157, 45-57.
[6]   Tikhomirov, F.A., Shcheglov, A.I. & Sidorov, V.P.; Forests and Forestry: Radiation Protection measures with Special Reference to the Chernobyl Accident Zone; The Science of Total Environment; 1993, 137, 289-305;
[7]   Hunt, H.W.;. A simulation model for decomposition in grasslands; 1977, Ecology, 58: 469-484.
[8]   Sandalls, J, Bennett, L.; Radiocaesium in Upland Herbage in Cumbria, UK: a Three Year Field Study; Journal of Environmental Radioactivity; 1992, 16, 147-165.
[9]   Colgan, P.A., McGee, E.J., Pearce, J., Cruickshank, J.G., Mulvany, N.E., McAdam, J.H., & Moss, B.W.; Behaviour of Radiocaesium in Organic Soils: Some Preliminary Results on Soil-Plant Transfer from a Semi-Natural Ecosystem in Ireland; in: Desmet, G.& al. (eds): Transfer of Radionuclides in Natural and Semi-Natural Environments; 1990, Elsevier, London & New York, 341-354.
[10] Römmelt, R., et al.; Untersuchungen über den Transfer von Caesium 137 und Strontium 90 in ausgewählten Belastungspfaden; 1991, ISH-Heft 155/91. Neuherberg.
[11] Kammerer, L., et al.; Uptake of radiocaesium by different species of mushrooms; J. Environ. Radioactivity; 1994 23, 135-150.
[12] Shaw, G., Kliashtorin, A., Mamikhin, S., Shcheglov, A., Rafferty, B., Dvornik, A., Zuchenko, T.& Kuchma, N.; Modelling Radiocaesium Fluxes in Forest Ecosystems; 1996 This Proceedings.
[13] Schumacker, R., Fraiture, A., Loneux, M. & Marchal, A.; Conséquences des Fumures sur l'Ecosystème Forestier et la Qualité des Eaux. Supp.à la revue Environnement; April 1989.
[14] Lembrechts, J.; A Review of Literature on the Effectiveness of Chemical Amendments in Reducing the Soil to Plant Transfer of Radiocaesium and Radiostrontium; The Science of Total Environment; 1993, 137, 81-95.

Physical and Chemical Factors Influencing Radionuclide Behaviour in Arable Soils

G. RAURET[1], R.M. ALEXAKHIN[2], S.V. KRUGLOV[2], A. CREMERS[3],
J. WAUTERS[3], E. VALCKE[3], Y. IVANOV[4], M. VIDAL[1].
1 - University of Barcelona (Spain)
2 - Russian Institute of Agricultural Radiology and Agroecology, Obninsk (Russia)
3 - Catholic University of Leuven (Belgium)
4 - Ukrainian Institute of Agricultural Radiology, Kiev (Ukraine)

Abstract. Soil-to-plant transfer of radionuclides integrates plant physiological and soil chemical aspects. Therefore, it is necessary to study the factors affecting the equilibrium of the radionuclides between solid and soil solution phases.

Desorption and adsorption studies were applied to the podzolic and peat soils considered in the ECP-2 project. In the desorption approach, both sequential extraction and "infinite bath" techniques were used. In the adsorption approach, efforts were directed at predicting Cs and Sr-K_D on the basis of soil properties and soil solution composition.

Desorption approach predicts time-dynamics of transfer with time but it is unsufficient for comparatively predicting transfer. Adsorption studies informs about which are the key factors affecting radionuclide transfer. For Sr, availability depends on the CEC and on the concentration of the Ca+Mg in the soil solution. For Cs, availability is mainly dependent on the partitioning between FES -frayed edge sites-, which are highly specific and REC -regular exchange complex-, with low selectivity for Cs. Moreover, availability depends on the K and NH_4 levels in the soil solution and fixation properties of the soil. Considering these factors, the calculation of the in situ K_D values helps to predict the relative transfer of radionuclides.

The calculation of the K_D of the materials that could be used as countermeasures could permit the prediction of its suitability to decrease transfer and therefore to help in producing cleaner agricultural products.

1. Introduction

After the Chernobyl Nuclear Power Plant accident, some CIS and Western Europe institutes carried out a broad range of studies in contaminated areas in the former Soviet Union. These studies dealt with radionuclide behaviour in soil-plant systems: radionuclide distribution in soils, time-dynamics of radionuclide behaviour, soil-to-plant transfer and strategies to decrease the plant uptake (countermeasure actions). This joint work was carried out in the frame of the ECP-2 project, entitled "Transfer of radionuclides through the terrestrial environment to agricultural products, including the evaluation of agrochemical practices".

One of the important steps in the radiocontamination of the food chain, following a nuclear accident, is the soil-to-plant transfer of radionuclides. Part of the ECP-2 project has been focused on studying in depth the basics of the processes governing such transfer, which integrates the action of both plant physiological and soil chemical aspects, ECP-2 being concentrated mainly on this last one. Besides, the overall effect of countermeasures also integrates these two processes. Therefore, if we wish to rationalize field observations and to design actions to reduce transfer, it is necessary to get a quantitative insight in the basics of the various processes governing radionuclide transfer.

Plant uptake appears to be directly related to radionuclide level in the liquid phase plus a concentration factor that is plant specific and dependent on the soil solution composition. In general, the fraction of a radionuclide present in the soil solution is believed to be a very small fraction of the total amount. Therefore, transfer depends on the capacity of the solid phase to replenish radionuclide levels in the soil solution, subsequent to uptake or elution. Furthermore, this capacity depends on the fraction of the available radionuclide (pool of total radionuclide in the soil that participates in solid-liquid partitioning) and also on the distribution coefficient (soil solution-solid phase). The first magnitude, which quantifies the potential of soil to fix radionuclides is time dependent (ageing processes) and the second one is related to the sorption selectivity of solid phase and composition of soil solution, especially of sorption competitive species.

The combined influence of these parameters makes the correlation of the experimental transfer factors with only any one of them to be incorrect. The fraction of radionuclide available is often estimated by desorption techniques, whereas the K_D can be calculated from soil sorption potential and soil solution composition. However, if the other parameters remain constant, for a given soil the time dependence of the available fraction can be related to changes of transfer factors with time.

With respect to radiocaesium behaviour, soil-chemical availability is essentially a function of radionuclide partitioning between REC (regular ion exchange complex) and FES (specific sites in the frayed edges of clay layers) in the solid phase. The K and NH_4 levels in the solid and liquid phases are of paramount importance since they are Cs-competitive ions. On the other hand, the key factors of radiostrontium retention in soils are the cation exchange capacity and the composition of the REC, especially Ca and Mg levels, which are Sr-competitive species, since the sorption-desorption of radiostrontium is essentially controlled by ion-exchange processes. From this knowledge a mechanistic approach can be applied to quantify radionuclide speciation and ascertain which are the key parameters responsible from their behaviour.

To reduce the radionuclide transfer to plants, the design of countermeasures is of paramount importance. The emphasis in countermeasure strategy has been adressed recently at the soil chemical level, trying to reduce the level of radionuclides in the soil solution by applying adsorbents characterized by high radionuclide adsorption properties or, in some cases, increasing the concentration in soil solution of competitive species, as K and Ca. Zeolites and clay minerals have been usually considered as potential soil amendments to reduce plant uptake. Similarly as for soils, these amendments can be characterized in terms of radiocaesium and radiostrontium adsorption potentials. These quantitative insights are useful to estimate the effect of the addition of an adsorbent to contaminated soils, by comparing radionuclide adsorption potentials before and after adsorbent addition. Based on this approach, a methodology was designed to predict the amendment effectivenes for a given scenario at the soil chemical level.

2. Experimental

2.1. Sampling

Sampling areas were distributed in three different republics (Russia, Belarus and Ukraine), as shown in figure 1. Soils were collected in 1991 and 1992.

Figure 1. Mape of the sampling areas.

The field plots are described as follows:

· Chistogalovka -CHI- (Ukraine). This field plot is less than 5 km from the Chernobyl reactor, with sandy-podzolic soil. This plot contained hot-particles due to fuel deposition.

· Polesskoye -POL- (Ukraine). This field plot is at 60 km west of the nuclear plant, with loam-sandy, podzolic soil.

· Bragin -BRA- and Vetka -VET- (Belarus). These field plots are about 150 km northeast of Chernobyl: the soils are peaty and sandy-podzolic respectively.

· Novozybkov -NOV- and Komsomoletz -KOM- (Russia). These field plots are at about 200 km northeast of the reactor: the soils are sandy-podzolic and peaty respectively.

2.2. Soil characterization

The methods used in soil characterization are available in literature. Total cationic exchange capacity (CEC) values were measured by the AgTU method, whereas K, Ca, and Mg were estimated after NH_4Cl 1 mol·l^{-1} extraction, and NH_4 after KCl 1 mol·l^{-1} extraction [1]. Radiocaesium Specific Interception Potentials in K scenario (SIP_K) values were obtained as described in recent publications [2]. SIP values in NH_4 scenarios (SIP_{NH4}) were deduced considering SIP_K/SIP_{NH4} ratios equal to 5 (5 is the mean value of the NH_4-to-K selectivity constant in specific sites).

2.3. Experimental design

Two approaches, based on desorption protocols and calculation of distribution coefficients respectively, were used to study the physical and chemical factors that influence radionuclide behaviour in the soils described above.

In order to estimate the available fraction and to study the time dynamics of radiostrontium and radiocaesium, a sequential extraction scheme was applied to samples recently contaminated with soluble ^{137}Cs and ^{85}Sr, which allowed us to define an initial stage in the ageing process (4-day ageing samples). ^{85}Sr was chosen to estimate the behaviour of ^{90}Sr in soils contaminated by the Chernobyl accident.

The distributions obtained were compared to those coming from the application of the same scheme to samples directly contaminated by the Chernobyl accident, taken 6 years after.

These distributions defined the final stage for ^{137}Cs and ^{90}Sr (6-year ageing samples). Moreover, another desorption technique ("infinite bath" technique) was also applied to calculate the maximum pool of available radiocaesium in both types of sample (4-day and 6-year ageing soils).

The mechanistic approach consisted mainly in the calculation of the in situ K_D values for radiocaesium and radiostrontium, as well as the study of the response of the radiocaesium K_D to different levels of NH_4, which allowed the speciation study between REC and FES sites.

Finally, the calculation of distribution coefficients was also used in the laboratory back-up for countermeasure strategy.

3. Results and discussion

3.1. Desorption studies

In order to estimate the radionuclide available fraction, an operational approach was applied, based on sequential extractions. Such procedure is based on subjecting a soil sample to a sequence of increasingly aggressive extractant reagents, setting up a serie of different available fractions.

In the ECP-2 project, a scheme was designed and used in common by all the groups, which uses water (Fraction 1), NH_4OAc 1 mol·l^{-1} (Fraction 2), HCl 6 mol·l^{-1} (Fraction 3) and HNO_3 8 mol·l^{-1} (plus some drops of H_2O_2) (Fraction 4) as extractant reagents [3]. The desorbed radionuclide in Fractions 1 and 2 can be related to exchangeable fraction, whereas Fraction 4 and residue may be associated with radionuclides irreversibly fixed in the mineral phase [4,5]. Because of the significant relation between desorption yields and experimental conditions, the main potential of such method lies in the study of the time dynamics of radionuclide partitioning and explanation of long-term (mobility) behaviour.

During the ECP-2 project a new method was developed to desorb radionuclides from a solid phase. In this new procedure, a desorption flux from the soil is generated by an "infinite-bath" technique, using an ion-exchange technique: the soil is dispersed in a solution of an extractant (NH_4Cl 10^{-3} mol·l^{-1}) and brought into dialytic contact with a high capacity ion exchanger (Giese granulate -ammonium-copper-hexacyanoferrate-), which breaks up the equilibrium of the soil solution and generates a radionuclide desorption flux from soil into the liquid phase, characterized by "near-zero" levels of radionuclide [6].

3.1.1. Application of the sequential extraction scheme to obtain radiocaesium distributions

Figures 2a and 2b show ^{137}Cs distributions obtained in podzolic and peaty soils, respectively, using the sequential extraction scheme, for samples recently contaminated with ^{137}Cs in soluble form (4-day ageing) and for soil samples contaminated by Chernobyl accident from ploughed field plots (6-year ageing).

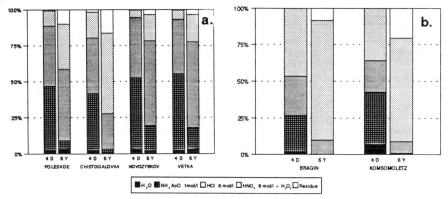

Figure 2. Time dynamics of radiocaesium distributions.

The distributions obtained 4 days after laboratory contamination showed that there seemed to be two different distribution patterns, corresponding to podzolic and peaty soils respectively.

It can be deduced that radiocaesium seemed to be rapidly fixed by the soil, although more than 40% remained in exchangeable form (Fractions 1 and 2, using water and ammonium acetate as extractant reagents), with the exception of Bragin soil, in which this fraction was only around 25%: in general, in both peaty soils, available fractions (F1+F2) seemed to be lower and mineral fractions (mainly desorption with HNO_3) clearly higher.

With respect to the distributions obtained 6 years after the nuclear accident, a similar trend was observed. Lower desorption yields in the fractions related to exchangeable radionuclide (Fractions 1 and 2) were obtained in the peaty soils (Bragin and Komsomoletz), suggesting that in spite of the high organic matter content in these samples (70% and 84% for Bragin and Komsomoletz respectively), mineral phase plays a key role in radiocaesium retention. In these two soils, radiocaesium was mainly extracted in Fraction 4 (HNO_3), which may be associated with the mineral-organic matter phase.

With the exception of the soil from Chistogalovka, which had a significant contribution of fuel deposition [7], the pattern distribution was again quite similar in podzolic soils, the desorption yield in Fraction 3 (extraction with HCl) being the greatest. Taking into account the different behaviour observed in Chistogalovka, in which radiocaesium is less soluble, and considering the results obtained with laboratory contaminated samples, it may be concluded that the lower radiocaesium solubility observed in Chistogalovka soil, 6 years after the nuclear accident, was related to fuel particulates.

The conclusions of ageing can only be drawn when condensed deposition is the main contribution of radionuclide contamination, since condensed deposition behaves as soluble contamination. For Chistogalovka soil, for instance, no ageing studies can be performed since two different types of contamination are present in 4-day and 6-year samples. For the rest of the soils, time dynamics of the radiocaesium distribution revealed a significant decrease in the available fraction (defined as the sum of Fractions 1 and 2) and a clear increase in

Fraction 4 and residue, which are related to radiocaesium fixed to the mineral fraction, this behaviour being especially marked in both peaty soils. Further comments on ageing are discussed in the next section.

3.1.2. Comparison between radiocaesium desorption yields obtained applying "infinite bath" and sequential extraction techniques.

Table 1 shows the desorption yields obtained after applying the "infinite bath" technique and the available fraction defined from the application of the sequential extraction scheme (sum of desorption yields in Fractions 1 and 2).

Table 1. Radiocaesium available fraction (expressed as percentage of total activity).

	4 Days			6 Years				
	IB	ΣFi	IB/ΣFi	IB	ΣFi	IB/ΣFi	IB4D/IB6Y	ΣFi4D/ΣFi6Y
CHI	77	41.8	1.8	-	3.4	-	-	12.3
POL	60	46.8	1.3	20.3	9.3	2.2	3.0	5.0
BRA	62	26.8	2.3	5.8	n.q.	-	10.7	-
VET	86	55.1	1.6	32.1	17.9	1.8	2.7	3.1
NOV	85	52.7	1.6	36.7	19.2	1.9	2.2	2.7
KOM	78	42.4	1.8	11.6	1.0	11.6	6.7	7.8

IB: Desorption yields obtained using 'infinite bath' technique.
ΣFi: Sum of desorption yields obtained in Fractions 1 and 2 of the sequential extraction scheme.
IB4D/IB6Y: Ratio between desorption yields obtained using 'infinite bath' technique in 4-day and 6-year samples.
ΣFi4D/ΣFi6Y: Ratio between the sum of desorption yields obtained in Fractions 1 and 2 of the sequential extraction scheme, in 4-day and 6-year samples.
nq: less than quantification limit.

The desorption yields from the "infinite bath" technique are always higher than those deduced from the ordinary shaking procedures. Different factors may explain this behaviour. On the one hand, the different experimental conditions used in the two procedures, especially the longer extraction times in the "infinite-bath" technique, should be taken into account. Other factor to be considered is the possibility of collapsing clay layers by the higher concentrations of NH_4 when using common extraction procedures. Therefore, it can be concluded that "infinite bath" technique better defined the total pool of available radiocaesium, although both techniques give similar information.

The differences in the behaviour of Chistogalovka soil, because of the presence of fuel particulates, were highlighted when the "infinite-bath" desorption was studied in detail, as shown in Figures 3a and 3b, which represent the desorption yields obtained over time in 6-year samples. For podzolic soils there were two zones in desorption, one relatively fast at the start of the extraction, and a subsequent "pseudo-plateau". For peaty soils, desorption process was slower and lower. On the other hand, Chistogalovka replicates had quite low reproducibility: it appears that during the 90-day extraction, some fuel particulates were dissolved.

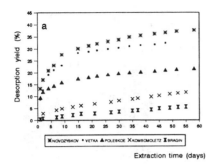 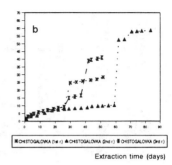

Extraction time (days)   Extraction time (days)

Figure 3. Desorption yields obtained from the application of "infinite bath" technique.

Both techniques - "infinite-bath" and sequential extraction- gave the same qualitative information about the time dynamics of radiocaesium availability. In both cases, and for the period studied, it was observed a decrease in the available fraction of around 2-3 times, for podzols, and about 7-11 fold for peaty soils. These values should be in agreement with the decrease of transfer factors observed in experimental fields. Few studies have been reported to date with reliable data about changes of transfer over time, but Alexkhakin et al. reported a decrease with time in transfer factors of around 2-3 times in podzolic soils coming from the Bryansk region [8].

Considering the values of soil-to-plant transfer obtained for radiocaesium in field experiments, it can be concluded that these desorption studies may account for the global decrease in transfer factors with time, but fail to explain in which soil the transfer factor should be higher: the experimental results show that in peaty soils the transfer factors are higher, in spite of what could be deduced from "infinite bath" techniques or sequential extractions. Therefore, besides the pool of available radionuclide, as quantified by these desorptions, other factors at soil level appear to rule soil-to-plant transfer.

Similar conclusions with respect to the lack of correlation between the changes of soil-to-plant transfer and radionuclide available fraction in soils are drawn from a parallel experiment, carried out by the UIAR, in which it was studied the influence of different fertilizers and manure on transfer to oats. In this study, the sequential extraction scheme was applied to control soil and to soils where fertilizers and manure had been added. These treatments, which led to an eventual decrease of the transfer in respect of the transfer in reference soil, did not have any significant influence on the radiocaesium distribution obtained [9].

3.1.3. Application of the sequential extraction scheme to obtain radiostrontium distributions.

Figures 4a and 4b show radiostrontium distributions obtained in podzolic and peaty soils respectively, using the sequential extraction scheme, for 4-day and 6-year ageing samples.

The distributions obtained in 4-day ageing samples are quite similar among the podzolic soils, with more than 80% of radiostrontium being easily desorbed (Fractions 1 and 2). In the peaty soils this percentage is slightly lower, what can be related to the higher percentage of organic matter. In any case, the radiostrontium available fraction is clearly higher than that for radiocaesium.

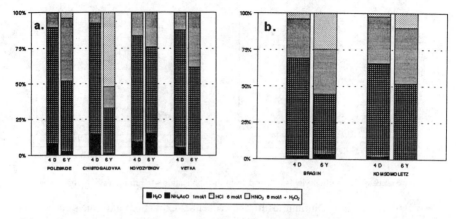

Figure 4. Time dynamics of radiocaesium distributions.

When studying the distributions obtained with 6-year samples, the different behaviour pattern of Chistogalovka soil because of the presence of fuel particulates is again highlighted comparing the distributions obtained with the other podzolic soils. In Chistogalovka soil, the radiostrontium fraction desorbed in the fourth step is the most significant, the lowest being the exchangeable fraction.

When comparing soils with mainly condensed deposition, there was a slight decrease over time in the exchangeable fraction and an increase in the fraction desorbed with HNO_3, related to fixed radiostrontium. However, this ageing process was clearly lower to that observed for radiocaesium. No conclusions can be drawn for Chistogalovka soil since the change of type of radionuclide contamination overcame the ageing process.

As stressed for radiocaesium, the application of the sequential extraction scheme is not a good method to ascertain in which soil the radiostrontium transfer may be higher. Actually, this scheme levels out the different soils. For instance, comparing with the transfer factor data available, the difference found in the exchangeable radiostrontium fractions in Vetka and Bragin soils (slightly lower for Bragin) can not explain the differences found in the transfer factors obtained in field experiments, which were of around one order of magnitude [10].

3.2. Mechanistic approach

3.2.1. Radiocaesium partitioning between REC and FES.

Table 2 shows some of the main soil parameters (pH, CEC, SIP_K) that have to be considered with studying radionuclide sorption.

CEC values, which are representative for radiostrontium adsorption potential, varies between very low values for the podzolic soils to very high values for the peat soils. Consequently, although depending on the Ca-Mg levels in the soil solution, it can be foreseen that with respect to soil vulnerability to radiostrontium contamination, differences can be expected to amount to one order of magnitude.

Table 2. Soil parameters as well as radiocaesium partitioning between REC and FES.

SOIL	pH	CEC[a]	K_{exch}	$NH_{4\,exch}$	SIP_K[b]	SIP_{NH4}[c]	RIP[d]	% REC[e]
POL	6.58	36	0.94	0.94	460	90	9.4	2 - 9
CHI	7.73	69	2.61	1.52	290	60	20.6	7 - 25
VET	6.08	24	3.38	1.57	710	140	24.8	3.5 - 15
NOV	7.00	50	2.49	1.77	470	95	21.3	⁴ - 18
BRA	5.98	1030	4.43	11.60	520	100	16.0	3 - 15
KOM	6.25	1140	6.23	8.64	260	50	14.9	5 - 25

a: CEC, K_{exch} and $NH_{4\,exch}$ concentrations are expressed in $\mu eq \cdot g^{-1}$.
b: Specific Interception Potential in K scenario, SIP_K (in $\mu eq \cdot g^{-1}$).
c: Specific Interception Potential in NH_4 scenario, SIP_{NH4} (in $\mu eq \cdot g^{-1}$).
d: Regular Interception Potential, RIP (in $\mu eq \cdot g^{-1}$).
e: Percentage of radiocaesium in the Regular Exchange Complex, % REC. The lowest values have been calculated from $RIP \cdot 100/(SIP_K + RIP)$, the highest values being $RIP \cdot 100/(SIP_{NH4} + RIP)$.

With respect to radiocaesium, SIP_K values, which are representative of radiocaesium interception potential, vary in a relatively narrow range.

It is easy to predict the partitioning of radiocaesium between the 'frayed edge sites' (FES) and the regular exchange complex (REC). The calculation is made from the values of the SIP_K, i.e. the product of $[FES] \cdot K_c(Cs/K)$ (Cs-to-K trace selectivity coefficient multiplied by the capacity of FES), the Specific Interception Potential in NH_4-scenario (SIP_{NH4}), i.e. the product of $[FES] \cdot K_c(Cs/NH_4)$ (about 5 times lower than SIP_K). The REC pool is evaluated from the Interception Potential of the Regular Exchange Complex (RIP), i.e. $(NH_4 + K)_{exch} \cdot K_c(Cs/K)$. For peaty soils, $K_c(Cs/K) = 1$, and around 5 for podzolic soils.

From the interception potential values obtained, it can be observed that the lower limit of the radiocaesium ratio RIP/SIP seems to correspond to a K-scenario in the FES, the upper limit corresponding to a NH_4-scenario, as is known due to the higher competitivity of this cation in the FES. Furthermore, and considering that for all systems relatively high NH_4 saturations can be predicted in the FES, the RIP/SIP_{NH4} ratio can be expected to be a good estimate of the field situation. On this basis, it is found that radiocaesium percentages in the regular exchange complex are in the range of 10-25%. These values are significantly lower than the desorption yields shown in table 1, even considering the values obtained using the "infinite bath" procedure 4 days after laboratory contamination (60-85%), clearly showing that significant fractions of radiocaesium are also desorbed from the FES.

It is clear that the effective radiocaesium levels in soil solution are controlled by the FES pool. Furthermore, it is highlighted the significant role of NH_4 in the radiocaesium partitioning: in situ K_D values for radiocaesium will be ruled predominantly by NH_4.

3.2.2. Response of the $K_D(Cs)$ in a mixed scenario

In the previous section, the calculation of the radiocaesium partitioning between REC and FES showed that radiocaesium was mainly intercepted by the FES in the micaceous clay minerals in soils. This conclusion was actually indirect, so a procedure that gives direct evidence is needed.

This procedure is based on measuring the K_D response of radiocaesium in a mixed-scenario containing NH_4, K and Ca. Distribution coefficients of radiocaesium ($K_D(Cs)$) were obtained in ten scenarios, all of them with the same concentrations of Ca and K (0.1 mol·l^{-1} and 10 mmol·l^{-1} respectively) and increasing NH_4 concentrations (up to 5 mmol·l^{-1}), in order to study the response of the system to the increasing concentrations of NH_4^+. Further details can be found in [11].

In this experiment, there is no masking of sites in the regular exchange complex. If the Cs were associated with the FES, then K_D should be quite sensitive to NH_4 since NH_4 is much more competitive than K in these sites. On the other hand, if all Cs were in the regular exchange sites, K_D should not change since NH_4 and K in these sites are equally competitive. In short, this is shown from the following theoretical explanation:

$$K_D = \frac{(K_D \cdot m_K)}{m_K + K_C(NH_4/K) \cdot m_{NH_4}}$$

or

$$\frac{(K_D \cdot m_K)}{K_D \cdot m_K} = 1 + K_C(NH_4/K) \cdot m_{NH_4}/m_K$$

where:

m_K, m_{NH4}: concentrations of K and NH^4, in mmol.l^{-1}.

$(K_D \cdot m_K)$: $K_D \cdot m_K$ product a zero loading of NH_4.

$K_D \cdot m_K$: $K_D \cdot m_K$ product in a K-NH_4 scenario.

$K_C(NH_4/K)$: NH_4-to-K selectivity coefficient in FES.

Therefore, if the ratio $(K_D \cdot m_K)/K_D \cdot m_K$ is represented versus m_{NH4}/m_K, the slope should be the selectivity coefficient of NH_4-to-K in FES ($K_C(NH_4/K)$).

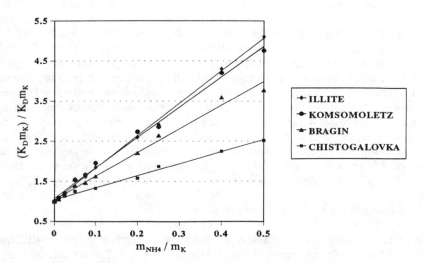

Figure 5. $K_D(Cs)$ response to increasing NH_4^+ concentrations in a mixed scenario.

Figure 5 represents the results obtained after the application of the mixed scenario to some of the soils studied. The main result is that Komsomoletz and Bragin (both peaty soils) exhibit a behaviour which is similar to illite. Therefore, it can be deduced that the specific sites in clays govern radiocaesium behaviour, more than the total CEC capacity does, even in soils with a high organic matter content.

3.2.3. Calculation of in situ K_D

In order to make predictions on in situ K_D values for radiocaesium and radiostrontium, quantitative data on the soil solution composition is needed. Such data were obtained from the values of K and Ca+Mg concentrations in water extracts, using a solid/liquid ratio of 1/2 $g \cdot ml^{-1}$. NH_4 values were obtained on the basis of experimentally measured NH_4/K ratios in the exchange complex and the K concentrations in the liquid phase extracts. These values are shown in table 3. Cation concentrations in field capacity conditions were also estimated considering a field capacity of 30% in podzolic soils and of 70% in peaty soils. The differences of NH_4 concentration in soil solution make expect a nearly one order of magnitude effect for the radiocaesium in situ K_D.

Table 3. Results of water extractions as well as estimation of the in situ K_D from field capacity values.

SOIL	RATIO 1/2		FIELD CAPACITY (FC)			IN SITU K_D	
	K	Ca+Mg	NH$_4$	K	Ca+Mg	K_D-Cs	K_D-Sr
POL	0.11	0.68	0.31	0.31	5.38	250	6.7
CHI	0.39	3.56	0.63	1.07	25.3	70	2.7
VET	0.50	0.32	0.56	1.19	3.75	180	6.4
NOV	0.34	2.91	0.69	0.96	21.69	107	2.3
BRA	0.50	22.81	2.31	0.89	70	42	14.7
KOM	0.90	12.34	2.19	1.58	40	20	28.7

These concentrations allow us to calculate radiocaesium in-situ K_D values, obtained on the basis of the following equation:

$$\text{For Cs} \qquad \text{in situ } K_D = SIP_K/(m_K + 5 \cdot m_{NH4})_{FC}$$

For radiostrontium, in situ K_D values are obtained as the ratio of CEC and the Ca+Mg concentration in the soil solution.

$$\text{For Sr} \qquad \text{in situ } K_D = CEC/(m_{Ca} + m_{Mg})_{FC}$$

The results of these calculations are also given in table 3. For radiocaesium, the differences in the in situ K_D values, are mainly due to the levels of NH_4 in soil solution. The predicted high levels of K and, especially, of NH_4 in the soil solution of the peat soils lead to the lower in situ K_D values.

The opposite situation is observed for radiostrontium. In situ K_D values are all quite low for podzolic soils, and the differences are essentially due to the different Ca+Mg concentrations in soil solution, the higher values being for the peat soils.

3.2.4. Prediction of a relative scale of radionuclide transfer

A radiocaesium TF scale can be set up in terms of the product of the available fraction (defined from the extraction using the "infinite bath" technique) and the reciprocal sum of K+NH$_4$ in soil solution, divided by the in situ K$_D$. In this relative scale the same plant should be considered, to allow the comparison, as well as a passive role of this in radionuclide uptake. Due to the time dependence of the available fraction, this sequence can be defined for fresh contaminated samples and for six-year aged samples. The two series, normalized against fresh contaminated Vetka, are:

· Fresh contaminated soils:
    Vet (1) < Pol (1.5) < Bra (1.6) < Nov (1.7) < Chi (2.7) < Kom (3.4)

· Six-year aged soils (not considered Chistogalovka):
    Bra (0.2) < Vet (0.4) < Pol=Kom (0.5) < Nov (0.7)

In spite of being closer to experimental results than a scale only derived from desorption values, this sequence does not match perfectly with experimental data, especially for six-year aged soils. However, this sequence assumes a comparable fertilizer state for the soils.

For radiostrontium it is possible to define a similar sequence. In this case the restriction to comparable soil chemical scenario in examining aging effects is less severe since the Ca+Mg concentration term cancel each other in the soil chemical and plant factor, the sequence being ruled by the available fraction divided by the CEC:

· Fresh contaminated soils:
    Kom (0.8) < Bra (1.4) < Chi (27) < Nov (33) < Pol (50) < Vet (73)

· Six-year aged soils:
    Kom (0.8) < Brag (0.9) < Chi (8.8) < Pol (29) < Nov (31) < Vet (53)

This sequence fits nicely with transfer factors derived from experimental fields.

3.3. Laboratory back-up for countermeasure strategy

3.3.1. Laboratory tests

A set of adsorbents were studied and characterized, as shown in table 4. They have all high adsorption potentials for Cs (SIP$_K$) or Sr (CEC). For zeolites, the SIP$_K$ values were dependent on the PAR value of the soil solution [12]. Zeolites show extremely high adsorption potentials for Cs or Sr but unfortunately, the adsorption of these radionuclides is completely reversible.

Table 4. Chemical characterization of the adsorbents studied.

AMENDMENT	TYPE	CEC (eq/kg)	SIP$_K$ (eq/kg)
Sapropel	organic (73 % o.m.)	1.05	0.154
Donegal	peat (97 % o.m.)	0.53	0.002
Bentonite	clay	1.21	-
Clinoptilolite	natural zeolite	20317	35-55
Natural Mordenite	natural zeolite	4973	60-125
Na-Mordenite	synthetic zeolite	-	60-125
5A	synthetic zeolite	27586	-

To test the effect of the amendment known amounts of soil samples with or without the adsorbents (at a given dose) were weighted in dyalisis membranes and dialytically equilibrated with a solution representative for the soil solution. After presaturation, dialysis tubings were equilibrated with the same solutions labeled with ^{137}Cs or ^{85}Sr. Further details can be found in [13].

Some results of the laboratory experiments are presented in tables 5 and 6. Table 5 shows the changes of radiocaesium K_D values obtained for a set of soils, when applying (K_D^{ads}) or not (K_D^{ref}) natural mordenite, at a dose of 1%. To allow the comparison, results are also available of the application of both clinoptilolite and mordenite to a podzol coming from Mol, Belgium (CEC: 0.017 eq/kg; SIP$_K$: 0.21 eq/kg).

Table 5. Results of laboratory tests on the effect of adsorbent amendments for radiocaesium.

SOIL	K_D^{ref} (ml/g)	K_D^{ads} (ml/g)	K_D^{ref}/K_D^{ads}
MOL (CLINOPT)	311	689	2.2
MOL	291	1315	4.5
POL	807	2569	3.2
CHI	362	1435	4.0
VET	896	1832	2.0
NOV	562	1383	2.5
BRA	638	1751	2.7
KOM	151	498	3.3

Table 6 represents radiostrontium K_D obtained for Vetka soil, after a number of treatments with different amendments.

Table 6. Results of laboratory tests on the effect of the adsorbent amendments for radiostrontium.

TREATMENT	K_D^{ref} (ml/g)	K_D^{ads} (ml/g)	K_D^{ads}/K_D^{ref}	pH
5A 1%	3.02	31.1	10.3	7.05
5A 2%	3.02	51.3	17.0	7.27
5A 4%	3.02	85.0	28.2	7.53
Bentonite 1%	3.02	4.17	1.4	6.07
Bentonite 2%	3.02	5.13	1.7	6.22
Bentonite 4%	3.02	7.39	2.4	6.26
Bragin 1%	3.02	4.43	1.5	5.91
Bragin 2%	3.02	5.71	1.9	6.19
Bragin 4%	3.02	8.30	2.8	6.60
Donegal 1%	3.02	3.69	1.2	5.64
Donegal 2%	3.02	4.33	1.4	5.29
Donegal 4%	3.02	4.95	1.6	5.06

It is seen that the addition of adsorbents characterized by high radionuclide adsorption potentials to soils with low radionuclide adsorption potentials effectively increase the K_D. For radiocaesium, a 1% of dose of the natural mordenite results in an effect of a factor 2 to 5. Clinoptilolite proves to be less effective, but as it is available at much larger quantities and lower prices than mordenite, the application of larger doses of this material may be economically more feasible than the addition of mordenite.

For radiostrontium it is clear that distribution coefficient is governed by the CEC, which is demonstrated by the observed values that agree with those deduced for the individual CEC values. For radiostrontium the changes of pH have to be considered. CEC in podzolic soils mainly depends on organic matter sites, which are pH dependent. Therefore, a decrease in pH leads to a decrease in CEC and thus in the distribution coefficient. This fact allows us to explain the differences observed between Bragin and Donegal soils (both peat soils with similar CEC at a given pH, but actually higher for Bragin since Donegal is an acidic soil). Furthermore is one of the factors that explain the excellent results obtained for the synthetic zeolite 5A.

3.3.2. Plant growth experiments

Some plant growth experiments derived from these studies, were also performed. The soil chosen was the Mol podzol, and 1%-clinoptilolite and 1%-Na$^+$-mordenite (for Cs) and 1%-5A, 2%-sapropel and 2%-Bragin soil (for Sr) were used as amendments. The plant chosen was spinach.

The main results are shown in table 7. In this podzol, characterized by low radiocaesium and radiostrontium adsorption potentials, the addition of mineral and organic adsorbents considerably reduces the transfer of the two radionuclides.

Table 7. Results of the plant growth experiment.

TREATMENT	TF control/TF treatment	PREDICTED EFFECT	pH
	Radiocaesium		
Na$^+$-Mor 1%	4.57	5.23	5.11
Clinop. 1%	2.99	2.88	6.19
	Radiostrontium		
5A 1%	24.95	8.97	7.68
Bragin 2%	2.75	2.20	5.27
Sapropel 2%	2.06	1.93	4.86

The observed effect coincides reasonably well with the predicted effect, except for 5A zeolite. In this case, an additional dilution effect should be considered due to the huge amount of Ca ions that are brought into the solution.

4. Conclusions

Sequential extractions are useful to explain changes in radiocaesium and radiostrontium behaviour over the time, since a decrease in the available fraction as well as an increase in the radionuclide fixed to mineral fraction were observed.

The speciation capacity of the desorption studies is limited, since only with the data coming either from the sequential extraction scheme or from "infinite-bath" technique it not possible to estimate the fraction of radiocaesium associated with the REC or with the FES sites.

Furthermore, the "prediction" capacity of the available fraction of radiocaesium and radiostrontium defined from this desorption approach, with respect to radionuclide mobility and changes of it over time, can be only used for the same soil, assuming minimum changes of the soil solution composition over time. It is not possible the prediction of in which soil the transfer (root uptake) is going to be higher only on the basis of the pool of available radionuclide, since other factors (adsorption properties, composition of the soil solution) are of most importance. Furthermore, the value of the total pool is not clearly related to soil properties.

The quantification of the partitioning between FES and REC shows that only 10-25 % of radiocaesium is associated with REC.

The key role of illitic clays (especially FES sites) in radiocaesium retention is clearly shown by studying the changes of K_D(Cs) in mixed K-Ca scenarios, with increasing NH$_4$ concentrations. In both podzolic and peaty soils, the behaviour observed is similar to that seen for illite.

The calculation of in situ K_D values, which considers NH$_4$, K and Ca+Mg status in the soil solution, seems to be a key parameter to predict in which soil the soil-to-plant transfer can be higher.

The use fo Cs and/or Sr specific adsorbents as countermeasures will be effective if the radionuclide adsorption potential exceeds that of the contaminated soil by a factor of 100. In practice, only soils characterized by low Cs or Sr adsorption potentials are to be considered. Based on a soil chemical characterization of soil and adsorbent, the solid/liquid distribution behaviour of Cs and Sr in soils and adsorbents can reasonably well be predicted and the effect of the addition of an adsorbent on the Cs or Sr solution levels can be estimated optimally. The laboratory test presented may be a first step to assess whether the addition of an adsorbent will be affective and, if so, to what dose it must be applied to obtain the desired effect.

References

[1]. R. Chhabra, J. Pleyser and A. Cremers, 'The measurement of the cation exchange capacity and exchangeable cations in soils: a new method', Proceedings of International Clay Conference, Mexico, Appl. Publ. Ltd., 1975, pp. 439-449.

[2]. A. Cremers, A. Elsen, P. De Preter and A. Maes, 'Quantitative analysis of radiocaesium retention in soils', Nature, 335 (1988) 247-249.

[3]. M. Vidal, M. Roig, A. Rigol, M. Llauradó, G. Rauret, J. Wauters, A. Elsen and A. Cremers, Two approaches to the study of radiocaesium partitioning', Analyst, 120 (1995) 1785-1791.

[4]. D.H. Oughton, B. Salbu, G. Riise, H. Lien, G. Ostby and A. Noren, 'Radionuclide mobility and bioavailability in Norwegian and Soviet soils', Analyst, 117 (1992) 481-486.

[5]. M. Vidal, J. Tent, M. Llauradó and G. Rauret, 'Study of the evolution of radionuclide distribution in soils using sequential extraction schemes', J. Radioecology, 1 (1993) 49-55.

[6]. J. Wauters, L. Sweeck, E. Valcke, A. Elsen and A. Cremers, 'Availability of radiocaesium in soils: a new methodology', Sci. Total Environ., 157 (1994) 239.

[7]. N.P. Arkhipov, A.N. Arkhipov, L.S. Loginova, and G.S. Meshalkin, Report for the project 'Transfer of radionuclides through the terrestrial environment to agricultural products and livestock, including the evaluation of agrochemical practices - ECP 2', Pripyat Research and Industrial Association, Ukraine, 1992.

[8]. N.I. Sanzharova, S.V. Fesenko, R.M. Alexakhin, V.S. Anisimov, V.K. Kuznetsov and L.G. Chernyayeva, 'Changes in the forms of ^{137}Cs and its availability for plants as dependent on properties fo fallout after the Chernobyl nuclear power plant accident', Sci. Total Environ., 154 (1994) 9-22.

[9]. Y. Ivanov, B. Prister, P. Bondar, G. Perepelyatnikov, N. Omelyanenko, L. Perepelyatnikova, L. Oreshich and S. Zvarich, Report for the project 'Transfer of radionuclides through the terrestrial environment to agricultural products and livestock, including the evaluation of agrochemical practices - ECP 2', Ukrainian Institute of Agricultural Radiology, Ukraine, 1994.

[10]. N.V. Grebenschikova and S.K. Firsakova, Report for the project 'Transfer of radionuclides through the terrestrial environment to agricultural products and livestock, including the evaluation of agrochemical practices - ECP 2', Byelorussian Institute of Agricultural Radiology, Byelorussia, 1993.

[11]. J. Wauters, M. Vidal, A. Elsen and A. Cremers, 'Prediction of solid-liquid distribution coefficients of radiocaesium in soils and sediments. Part II: a new procedure for solid phase speciation of radiocaesium', Applied Geochemistry, in press.

[12]. E. Valcke, B. Engels and A. Cremers, 'The use of zeolites in radiocaesium and radiostrontium contaminated soils: a soil chemical approach. Part I: Cs/K exchange in clinoptilolite and mordenite and the influence of Ca'. Submitted to Zeolites.

[13]. E. Valcke, C. Vandecasteele, M. Vidal and A. Cremers, 'The use of mineral and organic adsorbents as countermeasures in contaminated soils: a soil chemical approach'. Submitted to Zeolites.

ELEMENTS OF A UNIFIED PROGNOSTIC MODEL FOR SECONDARY AIR CONTAMINATION BY RESUSPENSION

F. Besnus[1], E. Garger[2], S. Gordeev[3], W. Holländer[4], V. Kashparov[5], J. Martinez-Serrano[6], V. Mironov[7], K. Nicholson[8], J. Tschiersch[9], I. Vintersved[10]

Abstract. Based on results of several joint experimental campaigns and an extensive literature survey, a prognostic model was constructed capable of predicting airborne activity concentrations and size distributions as well as soil surface activity concentrations as a function of time and meteorological conditions. Example scenario calculations show that agricultural practices are of lesser importance to secondary air contamination than dust storms immediately after primary deposition and forest fires.

1 INTRODUCTION

The Chernobyl accident has affected and still affects many people. Mitigation of these effects is one priority. Another priority must be to analyze and understand the behaviour of radionuclides in the environment so that in the event of a future accident, predictions can be made and remedial actions can be taken based on rational strategies.

1.1 THE IMPORTANCE OF RESUSPENSION

Once soil particles have become airborne, they will be picked up by the wind. The turbulent motion of the wind disperses the particles so that their overall concentration decreases while the dimension of the cloud increases. In addition to this process, which conserves the total airborne mass, particle deposition may take place: cloud or rain processes may scavenge particles and bring them down to the surface by this so-called *wet deposition*. In addition, particles may have a relative motion with respect to the air parcel to which they belong. Eventually, they may settle due to gravity, or impinge on flow obstacles like plants, rocks etc due to their inertia or by diffusive motion which is prevalent for small particles. All the latter processes contribute to *dry deposition*.

From the surface, they may migrate into the soil, become dissolved and carried away by groundwater, and taken up by plants. Yet there is still the possibility that nuclides already deposited on the surface may become airborne again by a process called resuspension: It is important since it provides a potential vector for secondary contamination of already de-contaminated areas, and an occupational risk for people working in agriculture.

In the Experimental Collaborative Project ECP1, natural and anthropogenic nuclide resuspension was studied in the neighbourhood of Chernobyl. The general objective of ECP1 was to enhance understanding of the various mechanisms leading to resuspension and secondary contamination, and to start building up a predictive capability for airborne

[1] CEA, France; [2] UAAS, Ukraine; [3] RSPEAC, Russia; [4] FhITA, Germany; [5] UIAR, Ukraine; [6] CIEMAT, Spain; [7] IRBAS, Belarus; [8] AEA Technology, U.K.; [9] GSF, Germany; [10] FOA, Sweden

concentration (and its radiological consequences) caused by resuspension by using all available literature data.

1.2 ELEMENTS OF RESUSPENSION, ATMOSPHERIC TRANSPORT, AND DRY DEPOSITION

The following information is required for understanding and predicting resuspension, atmospheric transport, and dry deposition:

- The soil nuclide specific activity and its depth distribution as a function of time. The most important parameter is of course the specific activity at the very surface from where resuspension actually takes place.
- The airborne mass concentration (in mg/m³) which also depends on wind speed, soil humidity, and anthropogenic activities of different kinds
- The soil-air transfer factor which relates the specific activity of the soil surface with the specific activity of the airborne dust for a variety of soil types, and as a function of wind speed, soil humidity, and anthropogenic activities of different kinds.
- The atmospheric transport factor which takes into account atmospheric dilution as well as dry deposition. Wet deposition is not considered here since it was found that during wet weather resuspension is greatly reduced.
- Based on these data, the atmospheric nuclide concentration and the inhalation exposure of risk groups can be assessed as well as the potential speed of secondary contamination of already decontaminated areas, and the possibility of land reclamation. This requires, however, knowledge of the primary airborne particle size distribution and its temporal evolution.

2 MODELLING

The ECP1 experiments were performed in order to generalize the description of resuspension as far as possible with the goal of building up a predictive capability for secondary airborne radioactivity. Questions to be answered (preferably on the basis of data collected during ECP1) concern

- the migration of hot spots, and the recontamination of already decontaminated areas by natural as well as anthropogenic resuspension, and
- the health impact on occupationally exposed agricultural workers

Normally, there are two main steps in resuspension. First of all, through hydrodynamic (wind) stress or mechanical disturbance (like agricultural activity), particles carrying activity will be separated from the soil matrix and become airborne in a layer close to the ground. Therefore, for modelling we need source-specific activities (of soil, firewood etc), and gravimetric source strengths which depend on soil properties and the types of resuspension. Subsequent atmospheric transport has to be covered in the next step. All our own data, and many from the literature are compiled in the present chapter and parameterized so that easily airborne concentrations for a wide variety of situations can be predicted in terms of size distribution and concentration as a function of source distance. To achieve this, we need a

- source characterization module, a
- transfer module, and a
- resuspension/transport/deposition module.

Using these modules we can perform comprehensive modelling and discuss the results of selected example scenarios.

2.1 SOURCE CHARACTERIZATION MODULE

There is a variety of potential sources like wood, pollens, sea spray etc which cannot be discussed here but are dealt with in the ECP1 Final Report [1]. Focusing on soil resuspension, the single most important aspect is without a doubt the soil surface activity. Initially, the deposited nuclides rest in a very thin layer at the surface. Later on, the contamination will migrate into the soil by leaching, dissolution, and capillary effects which can be described mathematically by convective diffusion. The specific soil activity (A_s in Bq/kg) is a solution of the convection-diffusion equation fulfilling the boundary condition $\frac{\partial A_s}{\partial z}\big|_{z=0} = 0$ which is initially δ-distributed and is given by (with the soil density ρ)

$$A_s(z,t) = \frac{\sigma}{\sqrt{2\pi}\cdot\rho\cdot s(t)}\cdot\left\{\exp\left[-\frac{(z-u\cdot t)^2}{2s(t)^2}\right] + \exp\left[-\frac{(z+u\cdot t)^2}{2s(t)^2}\right]\right\} \tag{1}$$

The above boundary condition is approximately fulfilled if the mass transfer rate into the soil is much faster than into the air which is the case unless there is a dust storm. The dispersion s(t) is given by $s(t) = \sqrt{2Dt}$ with the diffusion coefficient D and the convective velocity u. Then, the relation between the surface contamination density σ and the specific soil activity $A_s(z,t)$ is given by $\sigma = \rho\int_0^{-\infty} A_s(z,t)dz$. Note that σ does not depend on time if radioactive decay is disregarded. If it is taken into account, $\sigma(t) = \sigma_0\cdot\exp(-\lambda_N t)$. λ_N is the effective half life of nuclide N including radioactive decay and terrestrial transport by e.g. solution etc. Both, the convective velocity and the diffusion coefficient can be determined from experimental nuclide profiles at a certain time. Typical values for D are in the range 10^{-8} cm²/s and for u in the range 10^{-10} cm/s. Examples for depth distribution profiles for different times are shown in Fig. 1. It is important to note that the the temporal evolution of the surface activity concentration $A_s(z=0,t)$ follows a $t^{-0.5}$-dependence.

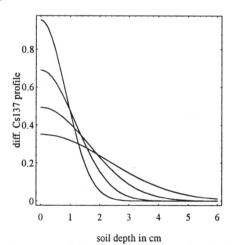

soil depth in cm

Fig. 1: *The convective diffusion model based on fits to experimental profiles allows the activity concentration profile for other times to be calculated: shown are curves for 2, 4, 6, and 8 years after deposit in Zapolye. Distributions spread out in time, and surface concentration decays with the inverse square root of time.*

2.2 TRANSFER MODULE

The purpose of this module is to describe depletion and enrichment during the resuspension process and to link the easily available soil-specific activity information with the specific activity of airborne dust. This is not a trivial question as radionuclides may have their own particle size distribution which can be different from the soil distribution; therefore, dispersion and fractionation may occur. The soil-air transfer function $T_{s \to a}$ relates the specific activities of air and soil via $A_a = T_{s \to a} A_s(z=0)$. It is expected to have a complex dependence on soil type and humidity, wind speed and mechanical activity. Indeed, this was observed as Figure 2 shows, but so far, we have no explanation for this behaviour: It may be due to the dispersion process or simply reflect the size and transport distance dependence of the specific activity since the median diameter changes from approx. 4 μm for low concentrations to about 50 μm for concentrations measured by means of the SSRT [2].

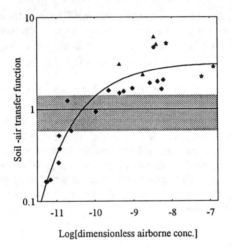

Figure 2: *Soil-air transfer function* $T_{s \to a}$ *as a function of the decadic logarithm of the dimensionless airborne Cs137 concentration (which is given by the ratio of airborne activity concentration in Bq/m³ to soil surface concentration also in Bq/m³). All data from natural and anthropogenic resuspension in Zapolye are included (diamonds). Anthropogenic experiments from Bragin are shown as triangles, and SSRT experiments from Bragin as stars. The shaded area is the typical experimental scatter of soil samples. It can be clearly seen that the transfer factor is positively correlated with airborne concentration, and becomes saturated for high concentrations at* $T_{s \to a} \approx 3.4$.

As of now, not enough data are available for a scientifically sound parameterization procedure. It is therefore suggested to assume neither depletion nor enrichment during the resuspension process, i.e. $T_{s \to a} = 1$. For conservative estimates, $T_{s \to a} = 3.4$ could be used for all nuclides.

2.3 RESUSPENSION/TRANSPORT/DEPOSITION MODULE

From the two preceding modules, specific activities of airborne dust can be obtained. If we now provide a scheme to calculate gravimetric airborne particle size distributions and concentrations as a function of source type and transport distance, the prognostic problem is essentially solved.

Wind is the only factor stirring up particles in natural resuspension. In contrast, anthropogenic resuspension is characterized by mechanical soil disturbance with the wind acting as vector only. Because of this different nature both processes are treated separately.

2.3.1 NATURAL RESUSPENSION

Depending on whether an actual or an average forecast is necessary, the input data required will be different. The two following subsections describe possible approaches.

2.3.1.1 Gravimetric area source strength

Natural resuspension and wind erosion are extremely complicated and not well understood processes, partly because measurements are extremely difficult [3]. Due to the lack of a general consensus concerning the numerical dependence of vertical flux density on wind speed, we (more or less arbitrarily) choose the relation of Gillette [4]

$$j_\uparrow = \psi \cdot u_*^5 \text{ where the empirical constant is } \psi = 3 \cdot 10^{-6} \frac{kg \cdot s^4}{m^7} \tag{2}$$

The power exponent is well in the range of values reported in the literature. According to Gillette, this relation approximately holds for many soils but the error margin may well be plus/minus two orders of magnitude.

As an applicability check, the above formula was applied to our measurements in Zapolye. Assuming resuspension - deposition equilibrium, for each particle size $j_\uparrow = v_\downarrow \cdot c$ (3)

Since the soil particle size distribution was found to be close to log-normal with a mass median aerodynamic diameter of $d_0 \approx 100 \mu m$ and a geometric standard deviation $\sigma_g \approx 3.55$, the airborne equilibrium particle size distribution of the concentration c is proportional to

$$c \propto j_\uparrow / v_\downarrow \propto j_\uparrow / d_{ae}^2 \tag{4}$$

It is therefore given by the -2nd moment of the soil particle size distribution (for details of the method see ECP1 Final Report). This is due to the fact that the deposition velocity is closely approximated by the sedimentation velocity which is $\propto d_{ae}^2$. According to the Hatch-Choate conversion formula [5], the median of the q-th moment of a log-normal distribution with median d_0 is given by

$$d_q = d_0 \cdot Exp\left[q \cdot \ln^2 \sigma_g\right] \tag{5}$$

Then, with q = - 2, we obtain a median diameter of 4 μm and a $\sigma_g \approx 3.55$ for the airborne soil resuspension-deposition equilibrium size distribution from the above considerations. This is in remarkable agreement with the experimentally determined distribution of the natural resuspension event shown in Fig. 3.

Using the total airborne mass concentration of 57 μg/m³, we arrive at a measured resuspension flux density of $j_\uparrow$=3.7*10[-7] kg/(m²s) which implies an ensemble average deposition velocity of 6.5 mm/s corresponding to a particle size of 14 μm which is reasonable for the distribution of Fig. 3. These values also compare favourably to the flux density of 9*10[-8] kg/(m²s) suggested by Gillette's formula assuming Berlin wind statistics and u*=0.1*u. We therefore consider it a reasonable basis for assessing the resuspension flux density under natural conditions in the Chernobyl area.

Equation (2) should be applied for (nonequilibrium) area source scenarios in conjunction with the specific soil activity and the soil-air transfer function yielding the activity area source strength in Bq/(m²s). The resulting airborne activity concentrations can then be calculated from a suitable theory [6], not always easily, however.

2.3.1.2 Long-term resuspension behaviour

A more direct and frequently used approach to immediately estimate airborne activity concentrations is via the resuspension factor R. It is defined by the ratio of airborne contamination concentration c (in Bq/m^3) at a certain reference height (which is usually not well defined) and surface contamination density σ (in Bq/m^2) i.e.

$$R = \frac{c}{\sigma} \tag{6}$$

From this definition it follows that it can be strictly applied under equilibrium conditions only since for instance contamination tranported by advection into an uncontaminated area where σ is zero would lead to an infinite resuspension factor, and we still have the problem with σ. On the other hand, under equilibrium conditions, the resuspension rate $\beta_\uparrow = j_\uparrow / \sigma$ is proportional to the resuspension factor i.e. $\beta_\uparrow^{eq} = R \cdot v_\downarrow$ (7)

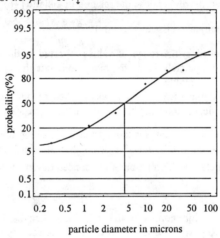

Fig. 3: *Typical airborne cumulative mass size distribution during natural resuspension. Mass median aerodynamic diameter was 4.3 μm, and geometric standard deviation was $\sigma_g = 4.0$ at a total mass concentration of 57 μg/m^3.*

Because of its practical importance and well-established empirical data base, we consider the temporal behaviour of the resuspension factor. Resuspension data for a short time after Chernobyl are not abundant. In Hannover, about 15 days after primary deposition resuspension factors in the range $2*10^{-6}$ to $2*10^{-5}$ m^{-1} were found depending on wind speed [7]. These data are shown together with UAAS experimental data from Chernobyl [8] in Fig. 4. The curve is a regression to the Chernobyl data with a best fit representation to the time t in days

$$R(t) = 2.09 \cdot 10^{-4} \, m^{-1} \cdot \left(\frac{t}{d}\right)^{-1.67} \tag{8}$$

The best fit exponent of - 1.67 is very reasonable considering the surface activity concentration decay proportional to $t^{0.5}$ which would have to be added to the exponent of -1.07 expected from theory (see Reeks et al. [9]).

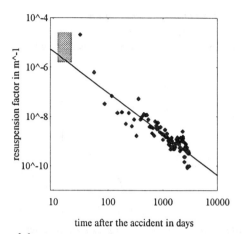

Fig. 4: *Experimental data of the resuspension factor measured in Chernobyl (corrected for radioactive decay of Cs-137). First data point is from June 1986. The straight line shows the fit to the data with a power exponent of -1.67. Short-time fluctuations caused mostly by wind speed and soil humidity fluctuations do not exceed ± half an order of magnitude. The rectangle in the upper left corner are data from Hannover 12 - 22 days after primary deposit.*

Adopting the above average equilibrium deposition velocity of 6.5 mm/s, we can integrate the equilibrium resuspension rate using equations (7) and (8) with respect to time, and obtain the integrated resuspended fraction (IRF) during the first 10 years (Fig. 5). It is important to realize that this IRF becomes effective only at a contamination edge since inside the contaminated zone there is a resuspension - deposition equilibrium. The consequences of this transport into the clean area is discussed in the next section.

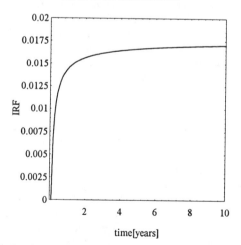

Fig. 5: *Integrated resuspended fraction (IRF) of Cs-137 calculated from equilibrium natural resuspension at Chernobyl as a function of time (years) starting one month after the accident.*

2.3.1.3 Assessment of contamination migration

Another important questions is how quickly de-contaminated zones become re-contaminated again under natural resuspension conditions without storm events. The situation is illustrated in Fig. 6.

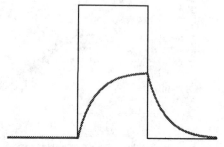

Fig. 6: *Diagram of a contamination strip. For explanations see text.*

Wind coming from the left enters a contaminated strip (thin line) and the air concentration (thick line) increases as a consequence of resuspension. Past the end of the contaminated zone, the concentration decreases again due to deposition, and at the same time soil contamination builds up at such a rate that the total amount of contamination is conserved. If the contaminated strip is very wide, resuspension-deposition equilibrium is established. The situation of a clean area embedded into a contaminated zone is the same, mirrored at the horizontal axis (i.e. upside down): past the clean zone edge, the air concentration decreases at the expense of the increasing surface contamination. Although rigorous treatment of the coupled soil-air transport equation (see e. g. [10]) is beyond the scope of this study, the order of magnitude can be easily assessed using a few simplifying assumptions [1]. The result of a simplified mathematical model is shown in Fig 7. It can be clearly seen that even in the close neighbourhood of the contaminated zone, the natural resuspension does not suffice to cause more than a relative contamination density of approx. 0.5. At a distance of 100 km, values of 0.1 can be reached after several years. It has to be emphasized that the above model exaggerates the actual contamination transport since constant wind speed and direction were assumed while in reality the wind changes; therefore, the resulting stochastic transport is considerably smaller than calculated above.

2.3.2 ANTHROPOGENIC RESUSPENSION

Probably the hardest experiments to explain quantitatively are the ones on anthropogenic resuspension. The reason is threefold: first of all, we have to sample close to the source. This means that we have an extremely non-homogeneous situation where size-dependent dilution and deposition of particles have to be taken into account simultaneously. Secondly, the particle size distribution changes extremely rapidly, and instrumental insufficiencies may distort the data. Finally, these data have to be evaluated using adequate three-dimensional size-dependent particle transport models in spite of the fact that physically correct models are virtually intractable mathematically. Nevertheless, a simple model for a line source derived on a semi-empirical basis works reasonably well, as demonstrated in the following. The 2D concentration distribution is assumed to be

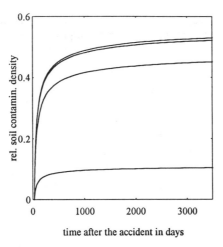

time after the accident in days

Fig. 7: *Temporal evolution of relative soil contamination density which is given by the ratio of actual contamination density $\sigma(t)$ to upstream contamination density σ_{up} of an infinitely wide zone. The curves refer to distances of 0.1, 1, 10, and 100 km downstream of the contamination edge. Calculation was started 30 days after the primary deposition event with actual values of the natural power-law Chernobyl resuspension factor.*

$$c(x,z,v_{sed}) = \frac{Q_l(0)}{h_0 u + (\kappa u_* - v_{sed}) \cdot x} \cdot Exp\left(-\frac{z}{h_0 + (\kappa u_* - v_{sed}) \cdot \frac{x}{u}}\right) \cdot \left[1 + (\kappa u_* - v_{sed}) \cdot \frac{x}{h_0 u}\right]^{-\frac{v_{sed}}{\kappa u_* - v_{sed}}}$$

(9)

where x is the horizontal and z the vertical coordinate, u is the (constant) wind speed, u_* is the friction velocity, κ is von Karman's constant, $Q_l(0)$ is the line source strength, and h_0 is the mixing depth at the location of the anthropogenic disturbance. The concentration profile will also depend on the particle sedimentation velocity v_{sed}.

The model validation strategy was as follows: Starting from the soil particle size distribution the airborne particle size distribution evolution with distance and height was calculated and compared with the vertical and horizontal flux densities such as a cloud would produce and the 'equivalent line source strength' for the activity determined in this way. As an example, the calculated vertical flux density vs transport distance is shown in Fig. 8 in comparison with experimental data. It has to be emphasized that the absolute height of the curve is not fitted but rather calculated from the model using the source strength (73 mBq/(ms)) obtained by the airborne concentration measured by the WRAC [11]. The complete discussion of the model and the experimental results is given in a separate publication [12].

As demonstrated, the transport/deposition model works reasonably well provided the source strength is given. While for agricultural activities the source strength was determined experimentally using the method described above, it would be interesting to generalize the method for other situations. Unfortunately, the understanding of the detailed physical mechanisms of traffic resuspension on unpaved roads is not good at present. As a consequence, empirical formulae were derived [13] which describe the emission strength q in grams per metre travelled as functions of the soil characteristics and the truck properties. If the number of wheels on the truck is w, its mass m in tons, its speed v in km/hr then

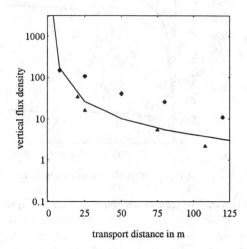

transport distance in m

Fig. 8: *Vertical flux density in mBq/(m²s) as determined experimentally (diamonds: UIAR; triangles: UAAS) and theoretically from the above model (curve) with the same source strength as above. The data point closest to th source has no well-defined distance, but the others have. The experimental data of UIAR are probably too high since their comparison with the horizontal flux density would result in an extremely high deposition velocity representing particles of > 100 microns.*

$$q = 0.61 \cdot \left(\frac{s}{12}\right) \cdot \left(1 - \frac{p}{365}\right) \cdot \left(\frac{v}{48}\right) \cdot \left(\frac{m}{2.7}\right)^{0.7} \cdot \left(\frac{w}{4}\right)^{0.5} \qquad (10)$$

The soil properties are taken into account by the silt content s (in %) of the road surface material, and the soil humidity influence parameterized by the number of days p with ≥ 0.254 mm of precipitation. It is claimed that the 95% confidence interval for the above equation is a factor of 1.46. The above equation applies for days with less than 0.254 mm of precipitation, and zero emission is assumed for rainy days. The point source strength Q_p in g/s of a moving truck is then proportional to the vehicle speed. Instead of the moving point source, a quasi-instantaneous line source described by the preceding formula can be assumed if the distance is large enough so that the truck travel time is much smaller than the plume travel time. We have only two experiments on resuspension with a truck on an unpaved road. They were performed in Zapolye, and yielded gravimetric mass concentrations (as determined by the WRAC) of 2-3 mg/m³ connected with cesium activities of 24-39 mBq/m³ which is the same order of magnitude as during agricultural activities, and in reasonable agreement with formula (10).

2.4 COMPREHENSIVE MODELLING AND EXAMPLE SCENARIOS

By comprehensive modelling we can understand the application of the whole chain of parameterized formulas discussed above. In this way, example calculations for radiologically important situations and prognoses are easily possible, and the results obtained in our field experiments can be generalized and become applicable to other locations and contamination patterns. A major problem is that published data on resuspension vary vastly [14]. Therefore, faced with the task of comprehensive modelling, a choice between the different publications had to be made based on an estimate of the reliability of the data. Without doubt, the choices are not entirely free from some arbitrariness and personal convictions but we hope to have made clear the rationale for the choices we have made in the previous sections.

2.4.1 NATURAL RESUSPENSION

The annual average value for the natural resuspension gravimetric flux density for a certain wind speed distribution is important as it allows the relevance of the omnipresent atmospheric circulation on the nuclide transport to be assessed. Using the soil - air transfer function, the annual average activity resuspension rate can be estimated, and compared with the values obtained by our spot measurements. Using the wind speed distribution of Berlin (which should be quite similar to the situation near Chernobyl), and Gillette's resuspension flux formula, we find an annual average flux density value of $4.1*10^{-8}$ kg/(m^{2}s) if we assume the friction velocity as 10% of the wind speed. If we further assume a specific activity of 10 Bq/g and an activity surface contamination density of 10^5 Bq/m^2, we obtain an average resuspension rate of $4.1*10^{-9}$ 1/s. This value is certainly plausible and compatible with our experimental results. An alternative, more classical approach would be via the resuspension factor formula (8) taking into account the square root decay of surface activity concentration due to migration. The advantage of this approach is that it actually describes average behaviour at certain locations for certain soils. Independently of the actual procedure, one has to conclude that natural resuspension long after an accident occurs at rather low speed, and therefore does not pose an immediate threat.

Of course, this is by far not a worst case scenario which might look as follows: Resuspension by wind shear is a highly nonlinear event which means that considerable fractions of the annual resuspension yield are produced within short time periods by gusts, dust devils and dust storms [15]. Gillette and Dobrowolski [16] measured deposition flux densities of 290 - 490 g/(m² yr), and similar values ranging between 20 - 500 g/(m² yr) are reported for Karakalpakia [17]. Wind erosion rates around Lake Aral averaged 2 - 3 mm per year over the last 30 years [18] which is equivalent to 3200 - 4800 g/(m² yr). *It is absolutely clear that such high erosion rates immediately after a primary deposition could effectively resuspend the whole contamination, which would be still at the very top of the soil at that time.* Even for a less dramatic scenario with a surface contamination density of 10^5 Bq/m² and a contamination depth of 5 mm, a soil surface-specific activity of $1.25*10^4$ bq/kg as in Zapolye, for semi-arid and windy conditions one would have contamination rates of 3600 - 6100 Bq/(m² yr) under the above assumptions.

2.4.2 ANTHROPOGENIC RESUSPENSION

The anthropogenic measurements were used to derive line source strengths. The conclusion we can draw is that anthropogenic resuspension is negligible compared to natural resuspension as far as transport is concerned. However, from the occupational point of view it is far more important than natural resuspension for risk groups like tractor drivers.

A sporadic but potentially very important means of contamination relocation are forest fires which have not been discussed here. The reason is that atmospheric transport calculations of buoyant plumes are very complicated. Forest fires can mobilize about 4% of the total inventory per event as compared to about 1.7 % for natural resuspension immediately at the border between a contaminated and a clean zone.

3 CONCLUSIONS

The prognostic model introduced above is based on physical principles, structurally complete and consists firstly of a source characterization module which describes the temporal evolution of the activity concentration of various sources. Secondly, the transfer module describes potential enrichment/depletion processes occurring during the resuspension process itself facilitating prognosis of activity data. Finally, the resuspension / transport / deposition module then describes source strengths, airborne transport (including change in concentration and particle size distribution) and deposition on a mass basis which links our radio-ecological approach with data on wind erosion. The model was validated by field experiments. Combining these modules, average and worst case assessments of activity can be made for any time after the accident. However, the uncertainty range for an actual forecast is rather large as for any meteorology-related event: It is estimated to ± one to two orders of magnitude while for monthly averages the uncertainty is probably only ± half an order of magnitude. This is (at least partly) due to the fact that not enough data were available for all dependencies on soil type, soil humidity, source type etc to allow reliable parameterization. Therefore, future research is necessary if more detailed prognostic power is desired.

References
[1]   ECP1 - Contamination of surfaces by resuspended material. Final Report, November 1995
[2]   The Standardized Soil Resuspension Test (SSRT) consists of a strong jet directed towards the soil surrounded by a circular sheath flow which takes up the resuspended particles. More information in: W. Holländer et al: A Standardized Soil Resuspension Test (SSRT) for quick ranking of soil erodibility; Atmospheric Environment; in preparation.
[3]   see e.g. Special Issue on Field intercomparison of dry deposition monitoring and measurement; J. Geophys. Res. D1 1990
[4]   Gillette,Fine particulate emissions due to wind erosion, Trans. ASAE 1977, Fig. 4, p 894
[5]   see e.g. Hinds, Aerosol Technology, Wiley 1982
[6]   P. I. Onikul and L. G. Khurshudyan, K voprosu o rasprostranenii pyli ot ee nazemnykh ploshchadnykh istochnikov; Trudy Glavnoi Geofizicheskoi Observatorii 1983, No. 467, 27 - 36
[7]   Holländer, Resuspension factors of Cs137 in Hannover after the Chernobyl accident, J. Aerosol Science 25 (1994):789 - 792)
[8]   Garger E. K. and F. O. Hoffman; Uncertainty of the long-term resuspension factors; subm.: Atm Sciences
[9]   Reeks, M. W., J. Reed and D. Hall (1988) On the resuspension of small particles by a turbulent flow; J. Phys. D: Appl. Phys. 21, 574 - 589
[10]   Buikov, M. V. (1994) The pollution exchange between soil and the near-surface air layer through turbulent transfer, resuspension and dry deposition, J. Aerosol Sci. 25: 859 - 866
[11]   Holländer, W., G. Dunkhorst, G. Pohlmann; A sampler for total suspended particulates with size resolution and high sampling efficiency for large particles, Part. Part. Syst. Charact. 6 (1989): 74-80
[12]   Holländer W. et. al. ; Particle transport and deposition past an agricultural line source; Atm. Environment; in preparation
[13]   U.S.EPA 1985/1986/1988/1990 Compilation of Air Pollution Emission Factors, AP-42, 4th edn Research Triangle Park
[14]   see e.g. Nicholson, K. W.; A review of particle resuspension; Atm. Environment 22 (1988): 2639 - 2651
[15]   see e.g. The Soviet-American experiment on the arid aerosol, Atm. Env. 27a(1993), issue 16
[16]   pp 2519 - 2525, above
[17]   p 58 in: Kuksa, Southern Seas (Aral, Caspian, Azov, and Black) under anthropogenic stress (in Russian), St. Petersburg, Gidrometeoizdat 1994
[18]   Hydrometeorological problems of Priaralye, ed. Chichasov, Leningrad, Gidrometeoizdat, 1990

Resuspension and Deposition of Radionuclides Under Various Conditions

E. Garger[1] , S. Gordeev[2] , W. Holländer[3] , V. Kashparov[4] ,V. Kashpur[1] ,J. Martinez-Serrano[5] ,V. Mironov[6] , J. Peres[7] , J. Tschiersch[8] , I. Vintersved9, J. Watterson[10]

Abstract: The resuspension of Cs-137 and Pu-239+240 has been assessed at sites within and outside the 30 km exclusion zone around Chernobyl. Measurements were made during periods of wind-derived resuspension and during simulated and real agricultural activity. From these data, resuspension rates (fraction of deposit removed in unit time) or emission rates (fraction of deposit removed in unit time or unit area) have been calculated. Resuspension rates of Cs-137 have declined by at least an order of magnitude 7 years after the accident and were found to be of the order of 10^{-10} s^{-1}. During agricultural activity, the resuspension rate may exceed background levels by four orders of magnitude.

Introduction

During experimental investigations of " Experimental Collaboration Project 1(ECP 1)" in 1992-1994 detailed information about the airborne concentration and size distribution of the radioactive atmospheric aerosol was obtained for the resuspension by wind and during anthropogenic activity. Measurements were made at sites with different surface and soil characteristics within and outside the 30 km exclusion zone around Chernobyl. These measurements have allowed a number of important parameters to be calculated which will help to predict the spread of contamination and the inhalation dose.

Resuspension as a source of Cs-137 in the atmosphere

The data presented in Figs.1a and 1b demonstrate how the atmospheric concentrations of Cs-137 and Ce-144 have changed with time since the accident and how resuspension of deposited material has affected the airborne concentrations of these two radionuclides. The measurements were made at two sites, Chernobyl, and Pripyat [1]. The two straight lines show the expected atmospheric concentrations of Cs-137 and Ce-144, allowing for radioactive decay, standardised to the activity concentration in June 1989. After 1989, the majority of the decontamination work in the 30 km zone was complete, and resuspension by mechanical activity would have become less important. The figures show the monthly mean atmospheric concentrations of Cs-137 and Ce-144 have declined by one or two orders during the four year measurement period. During 1987 and 1988, there was a large effort made to decontaminate the area surrounding the Chernobyl NPP, and there is some evidence that this work enhanced the atmospheric concentrations of Cs-137 and Ce-144. The decrease in atmospheric concentration of Ce-144 and Cs-137 is higher than would be expected from radioactive decay alone and this is particularly well illustrated in Figure1a. This feature demonstrates that there are processes which are responsible for the reduction of the activity concentration in the air,

----

[1] UAAS ,Ukraine; [2] RSPEAC, Russia ; [3] FhITA, Germany ; [4] UIAR , Ukraine ; [5] CIEMAT, Spain ; [6] IRBAS , Belarus ; [7] CEA, France ; [8] GSF , Germany ; [9] FOA , Sweden ; [10] AEA , Tecnology, UK

for example, vertical migration in soil, and run-off of radionuclides with rain or melting water from snow cover. The atmospheric concentrations are apparently influenced by the levels of anthropogenic activity with atmospheric concentrations attributable to: wind-driven resuspension, ~ 10 to 100 μBq m^{-3}; light agricultural activity, ~ 200 to 400 μBq m^{-3}; and a forest fire 17 km from the measurement site, ~ 1000 to 2000 μBq m^{-3}.

Studies in Sweden have illustrated the potential for long range transport of Cs-137 from the heavily contaminated region around Chernobyl. There are events in Sweden where Cs concentrations increased simultaneously over a large region, and on these occasions, the weather situation in northern Europe has been dominated by an anticyclone over Russia. The receptor orientated trajectories calculated for these periods indicate that the Chernobyl area is a possible source for the extra atmospheric activity.

Range of measurements made

European and CIS collaborators made measurements of atmospheric concentration and deposition using a wide variety of equipment. The institutes represented by their initials in the text and tables are shown in the footnote at the bottom of the first page of this article. The devices used to measure atmospheric concentrations included: passively aspirated samplers (fabric 'cone' sampler , UAAS; gauze screen; IRBAS) and actively aspirated size selective samplers (impaction surfaces in a wind tunnel, AEAT; Andersen PM10, CIEMAT; 'GRAD', high volume, and 'PK' impactors, UAAS; rotating arm impactor (RAI),GSF and Aerodynamic Particle Sizer (APS), GSF).

The teams made measurements of the atmospheric concentration of Cs-137 at six sites; three within the 30 km zone and three outside the zone. In some cases, these measurements have been combined with parameters from a meteorological station (atmospheric stability, wind direction) to allow the calculation of specific resuspension parameters, for example, wind-driven resuspension rates and emission rates from agricultural activity.

Intersite variability in the resuspension factor.

A simple, easily measured estimate of the level of resuspension is given by the resuspension factor [2]:

$$R = \frac{c_{am}}{\sigma_a} \qquad\qquad \text{(Equation 1)}$$

where: R =   resuspension factor   (m^{-1})
         c_{am} =   atmospheric concentration   (Bq m^{-3})
         σ_a =   soil activity to a specified depth   (Bq m^{-2})

The data in Table 1 summarises the mean resuspension factors for Cs-137 and Pu-239+240 at the measurement sites. The Cs-137 resuspension factors are mean values calculated from a range of atmospheric samplers. The Pu-239+240 resuspension factors are for particles < 10 μm in diameter (data from the PM10 sampler).

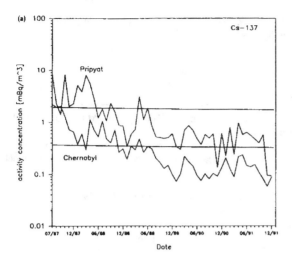

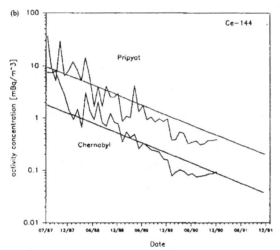

Figure. 1: *Air concentrations of Cs-137 and Ce-144 at Pripyat and Chernobyl (July 1987 to December 1991)*

The table shows that the highest resuspension factor was recorded at Novozybkov and this may be because the surface at this site was bare soil which would have been relatively erodable. Despite the range of environmental conditions during the work and the variability of the sites examined, there is only a four-fold difference between the lowest and highest measured mean values of the resuspension factor. A relatively small value was recorded at the Beach-Pripyat, although this site has the highest surface contamination. This is probably because the surface particles on the beach were larger than at the other sites, and the radioactivity had penetrated to greater depths. Localised activity at the Beach site, for example people moving around, enhanced the resuspension rate by a factor of ten.

Table 1: *Mean values of the resuspension factor of Cs-137 and Pu-239+240 measured between 1992 and 1994*

	Beach-Pripyat	Zapolye (1993)	Kopachi	Novozybkov	Mikulichi	Kovali
Surface characteristics	> 90 % sand	grassland	grassland	bare soil	rye field	barley field
Direction [distance (km)] from Chernobyl NPP	W [4]	S [14]	S [2]	NE [150]	N [45]	N [45]
Cs-137 contamination (Bq m^{-2})	3.3x10^6	5.9x10^5	2.3x10^6	1.1x10^6	4.2x10^5	5.3x10^5
Mean resuspension factor of Cs-137 ± SD (x10^{-10} m^{-1})	2.2 ± 1.9	4.4 ± 2.3	2.0 ± 1.4	7.7 ± 3.5	3.1 ± 3.1	6.3 ± 7.2
Pu-239+240 contamination (Bq m^{-2})	5x10^4	5.7x10^3	3.9x10^4			3.4x10^2
Mean resuspension factor of Pu-239+240 (x10^{-10} m^{-1})	1.1x10^{-10}	1.8x10^{-10}	2.4x10^{-11}			1.1x10^{-9}

The resuspension factors of Pu-239+240 are comparable to those measured for Cs-137. The highest mean value was recorded at Kovali, although it is difficult to ascribe reasons for this.

1 Resuspension factor in relation to particle size

Measurements of the atmospheric concentration of Cs-137 according to particle size were made by several samplers. Figure. 2 shows the resuspension factor of Cs-137 according to these two particle size fractions at all the field measurement sites. The figure clearly shows the large contribution of particles greater than 15 μm in diameter to the total resuspension factor at the heavilycultivated sites of Novozybkov, Mikulichi and Kovali.   At Zapolye, during periods of wind driven resuspension, there was a good linear correlation between the resuspension factor of Cs-137 and particles below 20 μm in diameter. This is shown in Figure 3. In general, the resuspension factor increases with particle diameter.

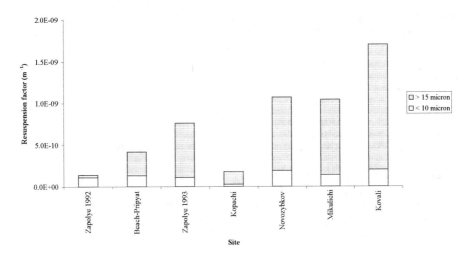

Figure 2: *Resuspension factors for Cs-137 associated with airborne particles < 10 μm and > 15 μm at all the measurement sites*

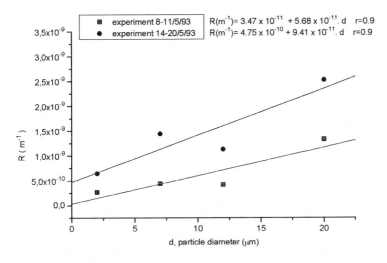

Figure 3: *Resuspension factor as a function of particle diameter during wind resuspension experiments. Zapolye field site. Atmospheric concentration data from PK impactor.*

Intersite variability in the resuspension rate

The resuspension rate, potentially, is a more useful measurement of resuspension than the resuspension factor, since it enables downwind inhalation doses and deposition levels to be predicted. However, it is difficult to measure in practice. It is defined as [2]:

$$\beta_{\uparrow} = \frac{j_{\downarrow m}}{\sigma_a}$$ (Equation 2)

where: $\beta_{\uparrow}$ = resuspension rate (s^{-1})
$j_{\downarrow m}$ = resuspension flux $(Bq\ m^{-2}\ s^{-1})$
s = soil activity to a specified depth $(Bq\ m^{-2})$

Shortly after the Chernobyl accident in 1986, it were made several measurements of the resuspension rate within 30 km of the Chernobyl plant in [3]. These measurements were made at sites with very different surface characteristics. All the measurements relate to periods when advection of material from upwind sources was considered to be small, the ground surface dry and with a moderate wind ($\sim$2 m s^{-1} at 1.0 m). Table. 2 presents these data.

Table. 2: *Resuspension rate according to surface type in 1986*

Site	Resuspension rate $\pm$ SD x 10^{-9} (s^{-1})		
	Ce-144	Cs-137	Zr-95+Nb-95
Zapolye	0.3 ± 0.1	1.0 ± 0.7	0.4 ± 0.2
Forest	2.1 ± 0.9	2.1 ± 0.8	3.7 ± 0.9
Beach-Pripyat	2.2	3.7	2.4

The work suggests that resuspension rates of between 1 and 4×10^{-9} s^{-1} were appropriate for a fresh deposit of Cs-137 at all the sites, but, the resuspension rates of Ce-144 and Zr-95+Nb-95 were approximately three to four times lower at Zapolie. Table. 3 summarises the resuspension rates of Cs-137 measured at sites during the field campaigns of the ECP1 programme.

The resuspension rates determined are highly variable, both with respect to the magnitude of estimates at individual sites using the same measurement technique and at the same site using different measurement techniques. It must be noted that the determination of the resuspension rate in this way is prone to high levels of uncertainty. Factors such as systematic measurement differences, including different measurement heights and integration periods could account for some of the differences between techniques also. The resuspension rate would be expected to vary with time, anyway. The technique used by IRBAS provided the highest estimates of resuspension rate at Novozybkov and Mikulichi .The values reported are of the same order as those recorded shortly after the accident, although the resuspension rate would have been expected to decline sharply in the first year or two after the accident (see Cs-

Table. 3: *Resuspension rate of Cs-137*

Site	Date	Cs-137 mean $\beta_\uparrow$ x 10^{-10} ± SD (s^{-1})			
		cone	GRAD	impactor+PM10	gauze screen
Zapolye	13/5/92 to 11/8/92	4.20± 0.16		0.0011	
Beach-Pripyat	14/7/92 to 11/8/92	0.08 ± 0.07		0.11	
Zapolye	06/5/93 to 01/6/92	1.60 ± 1.20	0.8 ± 0.7	0.017	
Kopachi	28/7/93 to 03/8/93				0.44
Novozybkov	17/5/94 to 24/5/94			0.0044	4.6
Mikulichi	13/7/94 to 29/7/94			0.13	0.09

137 data in Figure. 1). Although these measurements of resuspension have been made in the absence of agricultural activity, personnel moving whilst preparing equipment has enhanced the resuspension rate by up to an order of magnitude.

Resuspension during agricultural activity

At Zapolie and Kopachi in 1993 approximately half of the experiments were designed to simulate agricultural activity (harrowing) while the others consisted of vehicles being driven along a dirt track. Line sources were prepared (raked bare soil) around the measurement site to cater for different wind directions. Prior to carrying out the experiments, both of these sites were undisturbed grassland, sparsely vegetated, with generally dry conditions prevailing throughout. Vehicles used included two different sizes of tractor (pulling a spiked harrow) and a large, six-wheeled army truck. In 1994 the emphasis of the fieldwork was on real agricultural practice and measurements were undertaken for a variety of operations at three different sites. At Novozybkov fertilisation, cultivation and planting were carried out on working agricultural land, i.e. the soil was bare and well mixed. Whole field areas were worked as opposed to a single strip as at Zapolie. At Mikulichi a rye field was harvested and ploughed. Harvesting (of barley) was also carried out at Kovali. To maintain a strict quality control over the results, only data from experiments satisfying certain criteria have been interpreted. The basic criteria were as follows. Experiments were only selected when:
(1) The wind direction was consistently blowing across the line (or area) source towards the measurement site;
(2) The wind speed was sufficiently strong to provide a steady, well-mixed plume. Experiments conducted during variable, thermally generated winds were rejected;
(3) The type of vehicle, it's speed and operation must have been constant throughout the duration of the experiment;
(4) The meteorological conditions should be the same for the whole experiment.
The Aerodynamic Particle Sizer provided a detailed picture of the dust concentration of the different particles sizes during the experiments (see Figure 4). The data obtained demonstrated that the agricultural activity increased the atmospheric concentrations by a factor of several thousand in comparison to the background concentrations at distances of 20 to 30 m from the dust sources and 10 to 100 times at 100 m or more,

depending on conditions. The increase in particle concentration was not uniform for the whole particle size range. The increase due to agricultural activity was the highest in the giant particle fraction. This finding makes the assessment of the activity connected to the giant particles important.

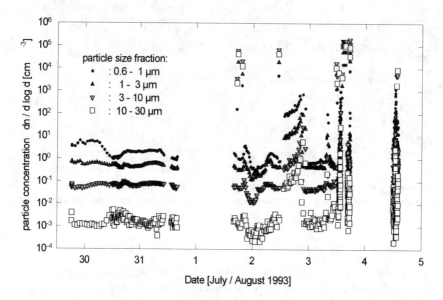

Figure 4: *Normalised particle number concentration in four different size ranges as measured by an Aerodynamic Particle Sizer*

The measurement of the number concentrations of hot particles showed an increase of three orders of magnitude, reaching 0.7 to 1.0 hot particles per m^3 with a maximum activity of 1.5 to 2.0 Bq per particle. The atmospheric concentration of resuspended Cs-137 was found to depend strongly on soil humidity r_{sh}. The relationship is roughly exponential i.e. $R = R_0 \exp(\lambda r_{sh})$ with $R_0 \approx 3 \times 10^{-7} m^{-1}$ and $\lambda \approx 60$. This means that a soil humidity of 10% reduces R by a factor 300 as compared to dry soil.

1 Emission rates during agricultural activity

The analitical solution of the semi-empirical diffusion equation for a stationary infinite crosswind strip source was used for the estimation of the emission rates during agricultural activity [4]. These experiments have been considered as approximating to a finite dust strip. In reality, the source was neither instantaneous nor stationary as the agricultural equipment moved many times during an experimental period. The emission rates reached very high values during the work (see Table 4). These rates exceed background resuspension rates by four orders of magnitude for the whole particle size spectrum, and by three orders of magnitude for the respirable size range.

Table 4: *Emission rates for various types of agricultural vehicles and agricultural activity derived from PK impactor measurements and Equation (4.9)*

Date (Zapolye)	Kind of vehicle	Emission rate, 10^{-6} s^{-1}	Emission rate for d$\leq$12 μm, 10^{-6} s^{-1}
12.05.93 morning	MTZ-82	0.027	0.004
13.05.93 morning	T-150	1.5	0.16
13.05.93 afternoon	T-150	2.0	0.20
22.05.93 morning	ZIL-131	1.0	0.09
22.05.93 afternoon	ZIL-131	1.4	0.13
24.05.93 afternoon	ZIL-131 and ZIL-130	0.48	0.09

Results calculated for the agricultural work at Novozybkov, Mikulichi and Kovali using a box model are shown in Table 5. Here, the results are expressed in terms of an emission rate per unit area. The mean value for all the sites was 9.2 mBq s^{-1} m^{-2} (s.d. = 7.8) for >15μm particles. For an area source the emission rate can be considered equivalent to a resuspension flux and this enables calculation of the resuspension rates presented in Table 5. Soil contamination densities of 1.2, 0.42 and 0.56 MBq m^{-2} have been used for Novozybkov, Mikulichi and Kovali respectively . Overall, the mean resuspension rate (for >15 μm particles) was 1.4 x 10^{-8} s^{-1} (s.d. = 1.5 x 10^{-8}). This is several orders of magnitude greater than the background values for these areas.

6.2 Deposition velocity

The values of deposition velocity measured during agricultural work are systematically higher (by nearly a factor of three) compared with natural conditions. Close to the source (x=22m) the mean deposition velocity is (0.144± 0.136) m s^{-1} while at greater distances (x $\geq$130m), a value of (0.113±0.087) m s^{-1} has been calculated. The large scatter in values of $v_\downarrow$ does not permit detailed discussion of the dependence of deposition velocity on distance. If Stoke's formula is used to estimate the mean radius of particles with observed $v_\downarrow$ = 0.115 m s^{-1}, then a value of r $\approx$ 20 μm is obtained. Using the equation $v_\downarrow = bu_* + v_{sed}$ [5], where b=0.01-0.08 for particles of diameter greater than 5 μm depositing to a dry grass surface and with a soil density of 2.3 g cm^{-3}, gives a value for the mean radius of 18 μm. From impactor measurements it was found that particles of this order of size had the second highest activity.

Table. 5: *Summary of results from area source equation (Novozybkov and Bragin 1994)*

Experiment (reference)	c (mBq m^{-3})	Q (<10 μm) (mBq s^{-1} m^{-2})	Q (>15 μm) (mBq s^{-1} m^{-2})	Resuspension rate (total) (s^{-1})
Novozybkov[a]				
13.5 (Agric 2/Nov 94)	19.3	-	8.1	6.8×10^{-9}
14.5 (Agric 3/Nov 94)	43.5	-	20.4	1.7×10^{-8}
14.5 (Agric 4/Nov 94)	39.9	-	16.3	1.4×10^{-8}
17.5 (Agric 7/Nov 94)	11.7	-	5.3	4.4×10^{-9}
Mikulichi				
30.7 (Agric 1/Mik 94)	25.6	0.2	5.5	1.4×10^{-8}
31.7 (Agric 2/Mik 94)	8.7	0.4	1.2	3.8×10^{-9}
01.8 (Agric 3/Mik 94)	30.7	1.6	4.7	1.5×10^{-8}
02.8 (Agric 4/Mik 94)	127.0	0.9	20.6	5.1×10^{-8}
Kovali				
06.8 (Agric 1/Kov 94)	5.6	0.3	0.8	2.6×10^{-9}

Note: [a] Novozybkov data are for particles >15 μm only

7. Contamination of the atmospheric surface layer and yield for agricultural work

Measurements of the temporal change of the mean concentration of Cs-137 showed a sharp increase in concentration at the start of harvesting. Concentrations for agricultural work were higher by a factor of ten over background concentrations. The mean background deposition flux density was 0.003 mBq m^{-2}s^{-1} at Mikulichi and 0.009 mBq m^{-2}s^{-1} at Kovali. The maximum deposition flux densities were 1.4 mBq m^{-2}s^{-1} and 0.53 mBq m^{-2}s^{-1} for rye harvesting and barley harvesting respectively. Thus, for anthropogenic activities the deposition flux density was 60-400 times higher than natural background levels.

Summary

The atmospheric concentrations of Cs-137 have declined by over two orders of magnitude since the accident. However, long term measurements extending to some years after the accident show clear evidence that resuspension influences the atmospheric concentrations. The data highlight the potential for events such as forest fires to raise the atmospheric concentrations by up to 100 times.

Team members of have made extensive measurements of atmospheric concentrations several sites in the CIS and have combined these with meteorological measurements to provide estimates of the resuspension in these areas. This work has included calculating resuspension factors and also resuspension rates.

Atmospheric concentrations of Cs-137 were assessed in all the field campaigns, but only a few measurements of Pu were made. Resuspension factors were found to be in the range 2.0 to 8×10^{-10} m^{-1} for ^{137}Cs for the particle size range covering a few microns up to tens of microns in diameter. Resuspension factors were similar for Pu, $\sim 10^{-9}$ to 10^{-10} m^{-1}, but these measurements only relate to particle sizes less than 10 μm in diameter.

Despite the wide range of soil types and environmental conditions at the sites where resuspension was assessed, there was only a four-fold difference between the lowest and highest estimates of the mean resuspension factor of Cs-137. The inter-equipment variability in the resuspension factor was around an order of magnitude, and this needs to taken into account when comparing data from a range of types of equipment. It is not an easy task to relate the magnitude of the resuspension factor to environmental parameters such as soil moisture content and wind, since the effect of individual parameters cannot be controlled. However, large scale human movement, for example, preparing the field sites, increased the resuspension factor of Cs-137 by up to an order of magnitude.

Resuspension rates of Cs-137 have declined by at least one order of magnitude 7 years after the accident. For example, at a grassland site, the resuspension rate has fallen from 1.0×10^{-9} s^{-1} in 1986 to 1.6×10^{-10} s^{-1} in 1993 for the particle size range covering a few microns up to tens of microns in diameter. The variability in the resuspension rates was much greater than in the resuspension factors. Values of the resuspension rate, calculated from the same instrument, varied by up to two orders of magnitude at individual sites and a similar variability was observed in the values of the mean resuspension rate calculated from different instruments.

The magnitude of the measured resuspension rates was of the same order as those found during a review of the literature, taking into account the age of the Chernobyl deposit. From the limited data available, the resuspension rate for Pu appears comparable to that of Cs-137 but these Pu measurements only relate to particle sizes less than 10 μm in diameter.

The data from field based measurement campaigns of agricultural work have been used in conjunction with models to predict the airborne radioactive contamination in the atmosphere at ground level. The fraction of material removed from the surface in either unit time or per unit area have been calculated. Values varied from 2.7×10^{-8} to 2.0×10^{-6} for all particle sizes and from 4.0×10^{-9} to 2.0×10^{-7} for particles of 12 μm in diameter.

Harvesting, harrowing, cultivation, planting, etc., are operations which give an increase in secondary contamination of the underlying ground surface at distances of between 50-200 m from the agricultural work being undertaken. On the ground the contamination is increased by a factor of between 60-200 while in the surface layer of the atmosphere levels are increased by 10-20 times.

The creation of a buffer zone (an untreated agricultural area) of 200 m width between a settlement and nearby farmland would allow a decrease of approximately two orders of magnitude in the transfer of secondary radioactive substances into the settlement during agricultural work on the bordering fields.

References

[1] Garger, E.K., Kashpur, V.A., Gurgula, B.I., Paretzke, H.G. and Tschirsch, J. (1994) Statistical characteristics of the activity concentration in the surface layer of the atmosphere in the 30 km zone of Chernobyl. *J. Aerosol Sci.*, 25(5), 767-777.

[2] Nicholson, K.W. (1988) A review of particle resuspension. *Atmos. Environ.*, 22(12), 2639.

[3] Garger, E.K., Zhukov, G.P. and Sedunov, Yu.S. (1990) Estimating parameters of wind lift of radionuclides in the zone of the Chernobyl Nuclear Power Plant. *Soviet Meteorology and Hydrology*, N 1, 5-10 (in Russian).
[4] Onikul, R.I. and Kchurshudyan, L.G. (1983) K voprosu o rasprostranenii pyli ot ee nasemnych plochshadnych istochnikov. Trudy Glavnoy Geophysicheskoy Observatorii, N 467, pp. 27-36 (in Russia).

[5]Chamberlain A.C. (1967) Transport of lycopodium spores and other particles to rough surfaces. Proc. Roy. Soc. Ser. A, Vol. 226, N 1444, pp. 63-70.

Physico-Chemical and Hydraulic Mechanisms of Radionuclide Mobilization in Aquatic Systems

A.V.KONOPLEV[1], R.N.J.COMANS[2], J.HILTON[3], M.J.MADRUGA[4], A.A.BULGAKOV[1]
O.V.VOITSEKHOVICH[5], U.SANSONE[6], J.SMITH[3], A.V.KUDELSKY[7]

1) Institute of Experimental Meteorology, Obninsk, Russia;
2) ECN, Petten, The Netherlands; 3) Institute of Freshwater Ecology, UK
4) DGA-DPSR, Lisbon, Portugal; 5) UHMI, Kiev, Ukraine; 6) ANPA, Rome, Italy
7) Institute of Geological Science, Minsk, Belarus

Abstract. This paper presents main results of joint studies carried out in frame of EC-coordinated ECP-3 Project "Modelling and study of the mechanisms of the transfer of radioactive material from the terrestrial ecosystem to and in water bodies around Chernobyl" in part of geochemical pathways. Physico-chemical models of specific migration processes are developed and recommended for application as sub models for inclusion in the decision support system (JSP-1). Main parameters, determining the behaviour of radionuclides in aquatic ecosystems are identified and methods for their estimation in emergency situations are proposed.

1. Introduction

Following the Chernobyl accident significant quantities of radioactivity were released into the reservoir system on the Dneper river, potentially creating a major radiation hazard to several million people. As a result, in the lower reaches of the Dneper basin between 10-20 % of the dose to man has been calculated to be introduced via water (mainly irrigation). In the city of Kiev the importance of this pathway is much less (a few per cent) but recent studies have shown that, in public perception, water is equal in importance to food as a vector. This is mainly because the Kiev reservoir forms the only direct transport pathway between the reactor and the city. Hence, it is important to understand the processes which are likely to remobilize radioactivity and transport it via aquatic pathways, so that the detailed basis of all decisions can be given to, and accepted by the local population.

Although in Kiev the aquatic pathway is not of major importance, in some country districts the cumulative dose is much higher than average level calculations suggest. This is because some small lakes, which are heavily used by the local population, retain very high levels of radionuclides in the water column. This increases the relative importance of the water pathway for these populations. If effective counter-measures are to be carried out in these local areas it is important to determine the major mechanism in each system which causes high concentrations to be maintained in the water column.

Table 1. Basic characteristics of water bodies under studies in ECP-3

Water body	Location	K^+, mg/l	NH_4^+, mg/l	Sediment type	Deposition type
Kiev reservoir	Ukraine	3 - 4	0.5 - 1	mixed	mixed
Devoke Water	UK	0.4	0.4	organic	condensation
Kozhanovskoe	Russia	10-15	0-5	organic	mixed
Svyatoe	Russia	18-30	15-30	organic	mixed
Glubokoe	Ukraine	10-15	2-12	clay	hot particles

The EC-Coordinated ECP-3 Project "Modelling and study of the mechanisms of the transfer of radioactive material from the terrestrial ecosystem to and in water bodies around Chernobyl" had two main objectives in part of geochemical pathways studies:

1) to identify the fundamental factors controlling radionuclide transport in aquatic systems. To incorporate this understanding into better sub-models of radionuclide behaviour and to extensively test these models before recommending to JSP-1 for inclusion in the decision support system;

2) to identify appropriate methods of study for individual, highly contaminated water bodies in order to understand the processes involved and propose appropriate countermeasures.

Five water bodies, presented in Table 1, were being studied with a view to a) comparing methodologies; b) identifying the main causes of residual high concentrations of radiocesium in the water column and c) verifying models of radiocesium remobilization.

2. Hot particles as a specific feature of the Chernobyl accident

During the explosion and the fire at the fourth unit of the Chernobyl Nuclear Power Plant in April-May 1986 a great amount of dispersed nuclear fuel (fuel hot particles), structural materials, and substances dumped into the reactor and formed in it (condensation hot particles) was released into the atmosphere. The fuel particles were of dense or loose structure and consisted of uranium oxides. Their sizes ranged from hundreds of microns to fractions of a micron. The radionuclide composition of the fuel particles was similar to the fuel make-up in the damaged unit with some depletion of volatile nuclides ($^{134,\ 137}Cs$, ^{106}Ru etc.). Fuel particles account for more than 90% of the total amount of hot particles. The condensation particles were generally characterized by smaller size and regular form. They can include either a wide spectrum of radionuclides or 1 to 2 radionuclides (for example $^{134,137}Cs$; ^{106}Ru; $^{144}Ce + ^{95}Zr$; $^{144}Ce + ^{106}Ru$ etc.).

Release of these fuel and condensation particles into the environment was the main distinguishing feature of the accidental contamination following the Chernobyl accident.

The nuclear weapon testing fallout had more than 90% of ^{90}Sr and ^{137}Cs in water soluble and exchangeable form (i.e., extractable by neutral salt solutions). The fraction of water soluble and exchangeable forms in the Chernobyl fallout was much lower due to the presence of water insoluble hot particles, and depended on the distance from the damaged unit. For example, the fraction of non exchangeable ^{137}Cs in the fallout near Chernobyl was about 75% [1], in the Bryansk region-50-60% and in Cumbria (UK) - about 10% [2]. As a result, the proportion of exchangeable radionuclides in soils in the near zone of the plant during the first years following the accident was much lower than that after the nuclear weapon testing [3]. For this reason, the Chernobyl radionuclides had higher values of distribution coefficient in the "soil-water" system and, hence more slower migration. Wash-off of dissolved ^{90}Sr with surface runoff in the 30-km zone in 1986-1987 was much lower than those obtained at the sites on which ^{90}Sr was applied as a soluble salt [1]. Apart from the differences in the radionuclides speciation in the Chernobyl and nuclear testing fallout, they also differ by the direction of transformation in soil. The mobility of nuclear testing radionuclides decreases with time because of fixation by soil components, while in the near zone of the NPP in the first years following the accident, the predominant process was leaching of radionuclides from hot particles which led to increased migration ability [4].

Thus, the peculiar features of the Chernobyl fallout did not make it possible to use directly the results of the studies of nuclear testing radionuclides and that is why the behavior of hot particles in environment needs in-depth investigation.

3. Kinetics of radionuclide leaching from hot particles

The physically based modelling of leaching processes is not easy because of non uniformity of sizes, forms, and chemical nature of particles. Besides, to use the results of such modelling for practical purposes, experimental data are needed about distribution of particles of different composition and sizes over a significant area as well as about characteristics of each type of particles. In view of this, it seems reasonable to use for forecasting integral parameters characterizing the rate of radionuclide leaching in different parts of the contaminated zone. The first order rate constant k_l (leaching) could serve as such a parameter, i.e. $dP/dt = -k_lP$, where P is concentration of radionuclide incorporated in fuel particles, t is time and k_l is a function of surface area of the particle.

As the hot particles dissolved their sizes decrease and the proportion of radionuclides in the particles of worst solubility increase. The first process results in the increase of k_l, and the second in its decrease. The assumption of the balance between the above processes makes it possible to use the first order equations to describe the kinetics of radionuclide leaching from the hot particles. The values k_l may be obtained from data on the transformation of radionuclide forms in soils. The proportion of a radionuclide in fuel particles is assumed to be equal to the fraction of a radionuclide in the non exchangeable form minus that fixed by the soil. In that calculation, it is better to take data on ^{90}Sr which is weakly fixed by soils. The results of calculating of the leaching rate constants are given in Table 2.

The leaching rate constants given in Table 2 agree with the values of k_l calculated from the dissolution rate of fuel particles [5]. The latter constants ranged from 10^{-3} to 10^{-4} day^{-1} for the particles of 50 to 500 μm (1-km zone) and from 10^{-2} to 10^{-3} day^{-1} for the particles of 5 to 50 μm (the region of Chernobyl).

Table 2. Rate constants for ^{90}Sr leaching from the fuel hot particles

Distance from the NPP, km	Location	Rate Constant 10^{-3} day^{-1}
4 - 10	Flood plain of R. Pripyat	0.15 - 0.87
	Kopachi	0.40 - 0.48
10 - 20	Benevka	1.4 - 1.7
	Chernobyl	1.5
	Korogod	1.0 - 3.6

The calculation and measurements of the proportion of the water soluble and exchangeable forms of ^{90}Sr indicate that at present the amount of radionuclides occurring as a part of the fuel particles is considerable only in the 10-km zone of the NPP, whereas at farther distances, the hot particles have mostly disintegrated.

4. Basic transformation processes of chemical speciations in soil and bottom sediments

The most expedient way to classify radionuclide speciations is to divide them into water soluble, exchangeable, and non exchangeable (fixed) chemical forms in soil [3, 4]. Different chemical forms can be separated by the technique of sequential extraction with solutions of different compositions. The equilibrium between water soluble and exchangeable forms is established quickly and it is reasonable in many cases to consider them as a single mobile form. In terms of chemical forms, fixation of radionuclides is transfer from their mobile form to fixed form. For mathematical description of this process, models were proposed in which fixation was treated as an irreversible process [4, 6]. The data available in literature suggest that at the initial stage of fixation and for processes of relatively short time-scale this assumption is quite warranted. However, the data for long-term transformation of radionuclide chemical forms in soil indicate the existence of a remobilization process opposing to fixation. After ^{137}Cs and ^{90}Sr solutions are applied to soil, the fraction of their exchangeable forms decreases not to zero, as is supposed to happen during irreversible fixation, but to a certain stationary level independent of the amount of radionuclide applied and not changing significantly, at least for several years [7]. With allowance for fixation reversibility and two mechanisms of fixation-selective adsorption and diffusion into the solid phase of soil, transformation of radionuclides chemical forms in soil can be summarized in scheme presented in Figure 1.

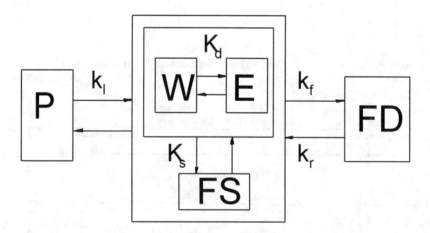

Fig.1 Scheme of Transformation Processes of Chemical Forms in Soil-Water System.
W, E, F are water soluble, exchangeable, mobile and fixed forms, respectively; FS, FD are radionuclides, fixed as a result of selective sorption and diffusion into solid phase respectively; P is radionuclides in fuel particles; k_l, k_f, k_r are first order rate constants of radionuclide leaching from fuel particles, fixation and remobilization respectively; K_d is exchangeable distribution coefficient; K_s is equilibrium constant of selective sorption.

Analytical solution of differential equations corresponding to the scheme of Figure 1 gives following time dependence of mobile radionuclide form after the Chernobyl deposition on soil:

$$m = (1 - K_s)\left[\left(\frac{k_f m_o + k_r - k_l}{k_f + k_r - k_l} - \frac{k_r}{k_f + k_r}\right) e^{-(k_f + k_r)t} + (1 - m_0)\frac{K_e - k_r}{k_f + k_r - k_l} e^{-k_l t} + \frac{k_r}{k_f + k_r}\right] \quad (1)$$

Where m and m_0 are current and initial fraction of mobile forms in soil, respectively; K_s - equilibrium constant of selective sorption; k_l, k_f, k_r are first order rate constants of radionuclide leaching from fuel particles, fixation and remobilization, respectively.

Experimentally determined transformation parameters of ^{137}Cs for different soil types of 30-km zone are presented in Table 3.

It is expected that fixation rate must be dependent so on the liquid phase chemical composition, for example on the concentration of ammonium in the pore waters of bottom sediments [8]. Experiments have shown that the fixation process proceeds at a faster rate in the interlayers of illite at low levels of competing cations (NH_4 and K), which causes an increase of cesium K_D^{tot} values. Although not enough data have been collected for a statistically significant evaluation, an apparent correspondence of lower contents of exchangeable potassium with higher fractions of fixed ^{137}Cs is observed. Batch experiments carried out with the organic soil (peat bog) from Devoke Water shows this dependence of cesium fixation rate and potassium concentration in solution even for soil aged for a short time (1 hour) (Table 4).

To study in more detail the kinetics of basic ^{137}Cs transformation processes in soils, a long-term laboratory investigation was carried out. It was shown that the interaction of cesium with soils is characterized by three phases with distinctly different time scales: initially rapid uptake (characteristic time about several minutes), relatively slow intermediate uptake (characteristic time about several days) and the third phase with characteristic time about several months. The initial phase is interpreted as the rapid cation-exchange sorption on adsorption sites located on the surface of soil particles, the intermediate phase is a result of cesium redistribution from non-selective to high selective adsorption sites and the third phase seems to be associated with cesium diffusion in mineral particles.

Table 3. Parameters of Chemical Forms Transformation Processes of ^{137}Cs in Typical Soils of 30-km Zone of the Chernobyl NPP

Soil Type	$k_f \cdot 10^2$, day^{-1}	$k_r \cdot 10^3$, day^{-1}	K_s
Sod-podzol, clay	1.1 ± 0.3	2.9 ± 1.1	0.41 ± 0.04
Sod-podzol, sand	1.7 ± 0.2	4.4 ± 0.5	0.54 ± 0.02
Peat bog soil	3.9 ± 0.4	5.2 ± 0.8	0.70 ± 0.02

Table 4 - Fraction of extracted and fixed radiocesium in organic soil (Devoke Water) at different potassium concentrations. Aging time 1 hour.

KCl concentration	Fraction of ^{137}Cs (%)	
M	NH$_4$Ac-extracted	Fixed
0.01	20.2	40.7
0.025	29.5	31.9
0.05	35.3	22.9
0.1	40.7	17.7

An increase of total distribution coefficients with time is observed. For example, the total distribution coefficient for the soddy-podzolic (mineral) soil MS increases from about 600 l.kg^{-1} after 1 hour to about 20000 l.kg^{-1} after 210 days, implying that the kinetic process is very slow. The time period of ^{137}Cs equilibration in this soil is about one year. In contrast, fixation in the peat soil occured very rapidly so that within two weeks the radiocaesium concentration in the water phase and in the first ammonium extraction were lower than the limit of detection [7].

This implies that, strictly speaking, the total cesium distribution coefficient cannot be used for mathematical modelling of migration processes with characteristic times less than several months. Therefore, it appears that the exchangeable distribution coefficient, which increases slightly during the first few days and then remains constant [7], is a more relevant parameter for modelling short-term (2days - 6 months) transportation processes. It can be calculated as follows:

$$K_d^{ex} = \frac{A_{ex}}{A_W} * \frac{V}{M} \qquad (2)$$

where A_w and A_{ex} represent the radiocaesium activities in the water phase and in the first ammonium acetate extraction respectively, V is a volume of liquid phase and M is a mass of solid phase.

The exchangeable distribution coefficient can also be used to predict radiocaesium K_d^{tot} in bottom sediments and soils under the assumption of pseudo-equilibrium conditions.

5. Prediction of Radionuclide Distribution Between Solid and Liquid phases of Soils and Bottom Sediments

The need for a specific distribution coefficient for modelling particular systems stems from the fact that, on the one hand, this parameter is governing the migration rate of radionuclides in natural water systems, their build-up in fish and vegetation, and, on the other hand, there are no all-purpose K_d values to be used for modelling the behavior of radionuclides in any natural water ecosystem. For example, the quoted K_d values for ^{137}Cs in surface water range over 4 orders of magnitude. Moreover the difference in K_d at the same site for different layers of lake bottom sediment can be as great as two orders of magnitude [8].

The K_d value is also dependent on the time elapsed since after radionuclide contamination of the soil or bottom sediment (ageing effect). Therefore, a reliable prediction of the behaviour of radionuclides is possible only if distribution coefficients for a given soil-water system at a given time are assessed based on the knowledge of mechanisms and kinetics of sorption-desorption processes.

Prediction of the radionuclide distribution in soil-water systems should, generally, be performed in two stages. The first step is prediction of transformation of species and determination of the part of radionuclides in the exchangeable form. The goal of the second step is to make assessment of K_d for the exchangeable form of a radionuclide.

For estimating the distribution coefficient of the exchangeable form of the radionuclide two approaches can be made. Both approaches assume that ^{137}Cs and ^{90}Sr in solution occur in the cationic form and that their exchangeable sorption is described by the equation of the cation exchange. The major difference between them is how non-uniformity of adsorption centres of soils and sediments is accounted for. Under the first approach, external non-homogeneities of adsorption sites are minimized by using a cation-analogue of the radionuclide as a competing cation M which has the closest possible sorption properties [4] to the radionuclide. For determination of required parameters a standard procedure of sequential extractions is used [9,10]. Concentrations of radionuclides and competing cations are determined in a water extraction and the proportion of exchangeable forms in the solid phase is calculated from their concentrations in the acetate extraction. The data obtained are used for calculation of the selectivity coefficient:

$$K_c = \frac{[R]_{ex} * [M]_w}{[R]_w * [M]_{ex}} \qquad (3)$$

$$M = K, NH_4 \text{ for } Cs - 137$$
$$M = Ca, Mg \text{ for } Sr - 90$$

The calculated selectivity coefficient is then used to predict the distribution coefficient of the exchangeable form of the radionuclide:

$$K_d^{ex} = \frac{K_c [M]_{ex}}{[M]_w} \qquad (4)$$

Among the advantages of the procedure is the simplicity of experimental determination of parameters. In addition, parameters can also be assessed from literature data. For example, data concerning the proportion of exchangeable forms of radionuclides and competing cations in soils and sediments determined by the standard method are available for many natural soil-water systems.

The method has been successfully used for predicting distribution of ^{90}Sr and ^{137}Cs in lake bottom sediments and their concentrations in surface runoff [4]. A limitation of the method is that the K_c value can, strictly speaking, depend on the composition of the liquid phase and in those cases when it differs significantly from the composition at which K_c was found, the value should be recalculated. This adds uncertainty to the prediction.

In the approach proposed by A.Cremers and co-workers [11] the non-uniformity of adsorption sites is taken into account through dividing them into selective: frayed edge sites (FES) and regular sites (RES). The distribution coefficient is then calculated as a sum of K_d for FES and RES. Most substrates have radiocesium sorption dominated by the frayed edge sites, highly organic soils being the only relevant exception [12]

The characterization of a system in terms of FES capacity and trace selectivity coefficient K_c (Cs/K) allows predictions of radiocesium distribution coefficients in scenarios for which the FES are homoionically potassium saturated. If the product of the selectivity coefficient K_c (Cs/K) and the capacity of the FES can be treated as a constant defining the radiocesium interception potential (RIP) for the FES, the distribution coefficient for radiocesium can be expressed as:

$$K_d^{FES} = \frac{K_c (Cs/K)[FES]}{m_K}$$ (5)

The radiocesium interception potential [$K_d^{FES} m_K$] can be extended to an ammonium scenario [$K_d^{FES} m_{NH4}$] allowing the measurement of ammonium to potassium selectivity coefficients and an assessment of the relative competitive effect of these cations in the hyper selective sites. Table 5 summarizes the [$K_d^{FES} m_K$] and [$K_d^{FES} m_{NH4}$] values calculated from the mean of the plateau values for a set of soils and sediments. These data are in good agreement with results obtained in other soil and sediment systems [12-17].

The $K_c(NH_4/K)$ which reflects the higher sorption competitivity of ammonium in the FES pool covers a range of 3 to 7. These values provide a rationale for the remobilization of radiocesium from reducing sediments [8,17].

Reasonable predictions of K_d^T values in different ionic scenarios can be made on the basis of eq. (5), taking into account the action of both potassium and ammonium ions. The equation required is:

$$K_d^T = \frac{\left[K_d^{FES} m_K\right]}{K_c (NH_4/K) m_{NH4} + m_K}$$ (6)

where the denominator expresses the total competitive effect of potassium and ammonium ions.

The predicted K_ds values calculated from eq.(6) taking into account $K_d m_K$, $K_c(NH_4/K)$ (Table 5) and potassium and ammonium water composition values are in good agreement with the "in situ" K_ds obtained from the ^{137}Cs contents of sediment and water (column or pore water) [18]. So, the solid/liquid partitioning of radiocesium in field scenarios can reliably be predicted on the basis of sediment laboratory characterizations and water composition. Moreover, it was shown [19] that in situations following a nuclear accident, a first estimate of the radiocesium distribution coefficient in sediments can be deduced solely from the pore water ammonium concentration. This approach, although useful in rationalizing sorption behaviour on the basis of readily measurable properties of the system is, however, of limited interest for the behaviour of radiocesium in the field. The time scale for the model scenario is quite short and the relevance of these characterizations for the long-term behaviour of radiocesium is questionable because of fixation.

Table 5 - Relevant parameters for sediments and soils from different ecosystems.

Sampling places	Sample	Sampling points	CEC meq.100g^{-1}	K_{dm_K} meq.g^{-1}	$K_{dm_{NH4}}$ meq.g^{-1}	$K_c(NH_4/K)$
Kiev reservoir, Ukraine	bottom sed.	KRS2(1)*	45-70	4.40	0.67	6.6
		KRS2(2)*		4.67	0.67	7.0
		KRS1(1)*	40-44	1.77	0.36	4.9
		KRS1(2)**		1.63	0.35	4.7
Kiev reser. Dnieper river Pripyat river	suspended matter	6	-	8.64	-	-
		7	-	1.74	-	-
		9	-	0.88	-	-
Kajanovskoe lake, Russia	peat bog soil	807	126	0.24	-	-
Glubokoye lake, Ukraine	bottom sed.	GL3	-	1.04	0.18	5.8
Svjatoye lake, Russia	bottom sed.	1-14cm	37.5	1.03	-	-
		14-28cm	39.5	1.34		
Devoke water, England	bottom sed.	0-2cm	24.3	0.74	0.22	3.4
		2-5cm	16.1	0.60	0.16	3.8
		8-11cm	17.7	0.71	0.21	3.4
		11-15cm	18.0	0.76	0.23	3.3
	peat bog soil	0-5cm	76.7	0.026	-	-
		5-10cm	70.0	0.033	-	-
		10-15cm	75.7	0.019	-	-
		15-20cm	81.5	0.010	-	-
		20-25cm	68.0	0.036	-	-
Chernobyl 10 km zone	alluvial sod soil	201	28.0	0.44	-	-

*- upper layer; **- bottom layer

When considering the long-term behaviour of radiocesium the desorption potential appears to be a key property of the system, particularly when the environmental conditions change (as a result of washout from the water column, or a change in redox conditions in sediments leading to a change in interstitial fluid composition). The behaviour dynamics of radiocesium will be controlled by the potential of the sediment to release the radionuclide, i.e. the size of exchangeable fraction. The virtue of this method is constant values of K_c(FES) for different soils.

It can be concluded that application of both procedures will make it possible to have satisfactory estimates of the distribution coefficient of the exchangeable form of the radionuclide and the choice of method in each particular case should be dictated by the availability of necessary parameters.

6. Study of the effect of organic matter of soils and bottom sediments on the physical-chemical state of radiocesium.

The commonly accepted mechanism of radiocesium fixation is by diffusion into interplanar spaces of clay minerals [20]. It is believed that natural organic matter adsorbs cesium reversibly and non-selectively and clay minerals alone are responsible for cesium fixation, even in soils very rich in organic matter[12]. According to this hypothesis, the correlation between cesium exchangeability and organic matter content would be positive. It was found, however, that the radiocesium exchangeable fraction is very low in some organic soils. The ammonium acetate extractable fraction of Chernobyl ^{137}Cs in the peaty soil from the Kojanovskoe lake watershed was about 2% [8], while the typical value for mineral soils of the same region was 10% [3,4]. Very high radiocesium retention values were also found for peaty soils from the Devoke Water watershed after ^{137}Cs labelling. In these soils, after 3 days aging time, the ammonium chloride extractable fraction was about 2% and it decreased to 0.1% after 2 years.

Qualitative determination of mineral components suggests that the clay mineral content in this soil is very low (below the detection limit). On the basis of these observations, a study has been carried out in order to evaluate how organic matter influences the physico-chemical state of ^{137}Cs in soil and bottom sediments. In this study the standard method of sequential extractions [3,9,10] was used to determine water soluble, exchangeable and acid soluble (HCl 1:1) forms and insoluble residue.

Soil and bottom sediment samples were divided into two groups: for the first group, the samples were analysed without incineration prior to acid extraction. For the second group, the samples were subjected to incineration at 500^0 C, then extracted with ammonium acetate. In this way in the first group of samples, high molecular weight organic matter was preserved during the whole procedure, whereas in the second group of samples the organic matter was removed before the acid extraction. For the bottom sediments no marked difference was noted in radionuclide species determined with and without incineration of the samples. The fraction of ^{137}Cs fixed in the insoluble residue after water, ammonium acetate and acid extractions for these samples is 5 to 20% close to that which is normally detected in soils and bottom sediments. Thus, the organic matter in bottom sediments either does not fix radionuclides, or fixes them in such a way that they are extractable with hydrochloric acid. A similar picture was observed with the sod podzol gley soil.

For the peaty soil the trends are the opposite: the ^{137}Cs fraction in the insoluble residue is 52.8-63.9% of that of the untreated samples and an order of magnitude higher than the fraction of insoluble radionuclide in the burned samples (5.6-8.2%). Similar results were obtained for peaty soil from Devoke Water watershed 2 years after spiking with [137]Cs. The ^{137}Cs fraction fixed in the insoluble residue was 16.4% in the burned sample and 59.0% in the unburned. This suggests that 40-50% of ^{137}Cs in these soils is probably fixed by the organic matter and not extracted by concentrated solutions of ammonium acetate and hydrochloric acid. The other possible explanation for these data is a modification of the mineral structure, resulting in an increase in radiocaesium extractability. To check this possibility, two additional experiments were performed. In the first the dependence of the Chernobyl radiocaesium extractability on the temperature pretreament of the samples was studied. The

Table 6 - Exchangeable radiocesium in peaty soil from Kojanovskoe lake catchment as a function of the incineration temperature.

Sample number	Temperature	Lost weight %	Exchangeable ^{137}Cs %
1	room	0	1.2
2	105 ^{o}C	7.7	1.3
3	150 ^{o}C	9.8	1.3
4	200 ^{o}C	74	8.0
5	300 ^{o}C	79	9.5
6	500 ^{o}C	81	11.1

results of this experiment are summarised in Table 6. The data indicate that the main increase in radiocaesium extractability occurs at 200° C when the organic matter starts to burn.

In the second experiment, organic matter from the Kojanovskoe lake peaty soil samples was destroyed by hydrogen peroxide at 60° C over a period of 1 month and radiocesium was extracted from the mineral residue by 1M ammonium acetate solution. Other samples were incinerated at 500°C before ammonium acetate extraction to study how the high temperature treatment alters the extractability of cesium associated with the mineral components of the soil. Table 7 shows the results of these experiments. It can be seen that there is no difference in the radiocaesium exchangeable fraction for the two sample treatments. It seems that caesium released from organic matter was subsequently fixed by mineral components. This process is quite likely because of the wet conditions (H_2O_2 treatment), and high temperature and duration of the experiment. Incineration of the mineral residue at 500° C did not significantly change the caesium extractability. This indicates that this treatment has no effect on sorption properties of the mineral components and therefore the increase of exchangeable caesium content in the original soil is a result of removal of organic matter.

These results can not be explained by any current concept of caesium fixation in soils and bottom sediments. Therefore, the mechanism of radiocaesium interaction with the organic matter of peaty soils calls for further investigation. A probable hypothesis is that radiocaesium fixation by organic matter of peaty soils occurs due to its diffusion into the volume of organic particles. This hypothesis is in good agreement with the result of radiocaesium fixation kinetics by organic and mineral soils [7]. Furthermore, the organic peaty soil from Devoke Water fixed radiocaesium faster than mineral sod podzolic soil from

Table 7 - Exchangeable radiocesium in peaty soil from Kojanovskoe lake catchment as a function of different treatments.

Sample number	Treatment	Lost weight %	Exchangeable ^{137}Cs %
1	H_2O_2	75	2.9
2	$H_2O_2+500^{o}$C	81	3.4

the Chernobyl area. This phenomenon can be attributed to faster caesium diffusion in amorphous organic particles than in crystalline mineral particles.

7. Radionuclide wash-off with surface run-off from catchment

Surface run-off from catchments is an important source of radionuclides in rivers and lakes in contaminated areas. Normalized wash-off coefficients, defined as the ratio of radionuclide concentration in run-off to surface density of catchment contamination are used to quantify this process [1,4,21,22]. In previous studies [21,22] it was found that ^{137}Cs and Sr-90 concentrations in run-off are independent of plot size and rainfall intensity. This indicates that the radionuclides in the overland flow are in equilibrium with the upper layer of soil and that the rate of attaining sorption-desorption equilibrium is quite high.

The principal mechanism of ^{137}Cs and ^{90}Sr transfer from soil to surface run-off is ion exchange. At ion-exchange equilibrium, the concentrations of identically charged cations are related by the ratio:

$$A_w = A_{ex}B_w/K_c(A-B)B_{ex} \qquad (7)$$

where A_w, B_w are concentrations of cations A and B in the liquid phase, respectively; A_{ex}, B_{ex} are concentrations of cations A and B in the solid phase, respectively and $K_c(A-B)$ is the selectivity coefficient for cation A exchange for cation B.

The concentration of radionuclide, as follows from Eq. (7), is directly proportional to the concentration of competing macro-cations in run-off. Evidence for this is given by simulated rainfall experiments on run-off plots with different cation concentrations in the rainfall water. The cation composition of the run-off is determined by the mineralization of the rainfall water and release of cations from soil solution into run-off. Given free infiltration of rainfall water into the soil, soluble compounds are washed out from the surface layer at the beginning of the rainfall and at some time mineralization of the run-off becomes identical to that of the rainfall. In all experiments in 30-km zone [21,22], this occurred prior to the run-off event. As a result, the equilibrium concentration of radionuclides in the run-off changes only as a result of depletion of exchangeable species in the upper soil layer.

In addition to the cation concentration in run-off, Eq. (7) includes the concentration of exchangeably sorbed cations and radionuclides in the upper soil layer. This can be determined by the standard technique of 1N ammonium acetate extraction [9,10]. However, given non-uniform distribution of radionuclides and macro-cations in soil profile, the result will depend on the depth of the sampled soil layer. One of the ways to solve the problem is using a concept of a complete mixing layer (CML) or in other words, a surface soil layer in which the concentrations of dissolved compounds in soil solution are equal to the run-off concentration. Values of CML in our experiments were determined from the time dependence of ^{90}Sr concentration in run-off in special long-term experiments and were about 5 mm or less at the all plots. According to our data, exchangeably sorbed forms of ^{137}Cs and ^{90}Sr are distributed in the upper soil layer fairly uniformly and the concentration of exchangeable forms of radionuclides and macrocations in the 0-5 mm layer can be used as a measure for their concentration in the CML.

The selectivity coefficients in the "run-off - upper soil layer" system which are required for calculations were obtained from the results of plot experiments

Ca^{2+} and K^+ were used as a competing ions for ^{90}Sr and ^{137}Cs respectively. The mean value of selectivity coefficient $K_c(Sr-Ca)$ was 2.1 ± 0.9 and $K_c(Cs-K)$ was 20 ± 4.

The values of selectivity coefficients and concentrations of radionuclide exchangeable forms and macrocations in the upper half-centimeter were used to make predictive estimates of the normalized wash-off coefficients of radionuclides in the rainfall run-off from the catchments of the rivers Pripyat and Ilya. The concentration of cations in the runoff was assumed to be identical to that in the rainfall water. The results of this calculations are given in Table 8. Estimated normalized wash-off coefficients are in good agreement with the experimentally measured values.

The approach described for wash-off coefficient estimation was applied to Devoke water catchment. Calculations were carried out using extraction data for the 0-4 cm layer for four soil cores representing the main soil types of the catchment and potassium concentration in rain-water or lake water [23]. Unlike the 30-km zone around Chernobyl, the dominant macrocation in Cumbria rainfall water is sodium. Predicted dissolved radiocesium concentrations ranged from 4-17 mBq/l and 11-54 mBq/l for rain-water and surface water (lake water) respectively. The values based on the potassium concentration in the lake were all within a factor of 2 of the measured lake water radiocesium activity (21 ± 4 mBq/l)

Comparison of Devoke water soil profiles with results for soils in the 30-km zone show that radiocesium from the Chernobyl deposition has been much more mobile in soil and sediments of Devoke water than in 30-km zone. This is unexpected as the soils in the 30-km zone mainly consist of a layer of sand between 0.5 and 2.0 m thick, overlying a podzol. The sands would be expected to have high ^{137}Cs mobilities. However, most of the activity remained near the soil surface. The reason for this apparent inconsistency is that, immediately after the accident, the mobility of ^{137}Cs was governed by the different forms of radiocesium in the deposition. In western Europe (United Kingdom) the deposition was dominated by condensation particles containing at least 85% of the radiocesium in mobile forms [2]. However, in the town of Chernobyl, at the time of the accident, only 26% of the ^{137}Cs was in mobile forms [1].

This fact was exacerbated by at least two other factors:

1) the large difference in annual precipitation: 1840 mm/year for Devoke compared to 550 mm/year around Chernobyl which creates a difference of about a factor of four in run-off coefficients, i.e. 60-70% of the precipitation in the Devoke water catchment is lost through outflowing streams compared to 15-29% in the Chernobyl catchment. Hence much more rain water passes through the surface soil of Devoke water compared to the Chernobyl soils; 2) the very low clay mineral content of upland Cumbrian soils compared to soils in the 30-km zone so that mobile forms represent up to 30% in Cumbria and 5-10% in the 30-km zone causing higher mobility in Cumbrian soils.

Table 8. Predicted and measured normalized radionuclide wash-off coefficients

Location	Normalized Wash-off Coefficients (10^{-6} mm^{-1})			
	Predicted		Measured	
	Sr-90	Cs-137	Sr-90	Cs-137
Pripyat catchment				
Chernobyl	22±10	1.6±0.8	24±13	2.5±1.2
Kopachi	12±7	-	18±8	-
Ilya catchment				
Staraya Rudnya				
Kliviny	34±18	3.7±1.4	36.9±12.2	5.0±2.2
Rudnya Ilyinetskaya				

As a result ^{137}Cs wash-off coefficients range up to a few percent per annum in the Cumbrian Lake District, whereas coefficients of 0.01-0.1% are found in the 30-km zone [1].

8. Conclusions

The behaviour of radionuclides in aquatic systems is largely dependent on:
- the chemical forms of radionuclides present in atmospheric fallout;
- the rates of transformation processes in soils and bottom sediments;
- environmental parameters, influencing the distribution of exchangeable forms between solid and liquid phases.
The main transformation processes are:
- disintegration of fuel particles, resulting in additional radionuclides passing into the solution;
- fixation of the exchangeable form;
- remobilization, i.e. transfer of the non exchangeable form into exchangeable form.
Transformation of chemical species of radionuclides in soils/sediments can be described by first-order kinetic equations with rate constants determined in field and laboratory experiments.
For predicting vertical and horizontal migration of radionuclides in soils and sediments the values for the following parameters are needed: the proportions of chemical forms in the fallout; the rate constants of transformation processes; adsorption capacities (CEC, FES) and selectivity as a function of depth and cationic composition of the solution in contact with soil or sediments.

References

[1] A.V.Konoplev and Ts.I.Bobovnikova, Comparative analysis of chemical forms of long-lived radionuclides and their migration and transformation in the environment following the Kyshtym and Chernobyl accidents, In.: Proceedings of the Seminar on Comparative Assessment of the Environmental Impact of Radionuclides Released During Three Major Nuclear Accidents: Kyshtym, Windscale, Chernobyl. CEC, Luxembourg (1990), 371-396

[2] J.Hilton et al., Fractionation of radioactive cesium in airborne particles containing bomb fallout, Chernobyl fallout and atmospheric material from the Sellafield site, J. Environ. Radioactivity. 15 (1992) 103 -108.

[3] A.V.Konoplev et al., "Distribution of radionuclides in the soil-water system fallen out in consequence of the Chernobyl failure", Soviet Meteorology and Hydrology, 12 (1988) 65-74

[4] A.V.Konoplev et al.,. Behaviour of long-lived radionuclides in a soil-water system, Analyst, 117 (1992) 1041-1047

[5] S.A.Bogatov et al., On stability of radiologically hazardous radionuclides in different forms of fuel release after the Chernobyl accident, In.: Abstracts of the Conference "Geochemical Pathways of Radionuclide Migration in the Biosphere, USSR, Gomel (1990)

[6] R.N.J.Comans et al., Kinetics of cesium sorption on illite", Geochim. Cosmochim. Acta, 56 (1992), 1157 - 1164

[7] A.V. Konoplev et al., Long-term investigation of Cs-137 fixation by soil, Paper presented on International Seminar on Freshwater and Estuarine Radioecology, Lisbon, 1994

[8] Modelling and study of the mechanisms of the transfer of radioactive material from the terrestrial ecosystem to and in water bodies around Chernobyl. EC-Coordinated ECP-3 Project (COSU-CT93-0041), Final Report 1993-94.

[9] A. Tessier et al., Sequential extraction procedures for the speciation of particle trace metal, Anal. Chem., 51 (1979) 844-851.

[10] F.I Pavlotskaya, Migration of Radioactive Products from the Global Fallout in Soils, Moscow (1974) 270 pp, in Russian.

[11] A.A. Cremers et al., Quantitative analysis of radiocesium retention in soils", Nature, 335 (1988) 247-249.

[12] E. Valcke, The Behaviour Dynamics of Radiocesium and Radiostrontium in Soils Rich in Organic Matter, PhD Thesis, Katholieke Universiteit Leuven, Leuven (1993).

[13] P. De Preter, Radiocaesium retention in aquatic, terrestrial and urban environment: a quantitative and unifying analysis, PhD Thesis, K.U. Leuven, Belgium, April 1990, p. 93.

[14] M.J. Madruga, Adsorption-desorption behaviour of radiocaesium and radiostrontium in sediments, PhD Thesis, K.U. Leuven, Belgium, October 1993, p. 121.

[15] J. Wauters, Radiocaesium in aquatic sediments: sorption, remobilization and fixation, PhD Thesis, K.U. Leuven, Belgium, February 1994, p. 109.

[16] D. Evans et al., Reversible ion-exchange fixation of caesium-137 leading to mobilization from reservoir sediments. Geochimica et Cosmochimica Acta, 47 (1983) 1041-1049.

[17] R.N.J. Comans et al., Predicting radiocaesium ion-exchange behaviour in freshwater sediments. ECN-RX-93-108 (1993).

[18] M.J.Madruga et al.,Radiocesium sorption-desorption behaviour in soils and sediments. This conference.

[19] R.N.J. Comans et al., Mobilization of radiocesium in pore water of lake sediments. Nature, 339 (1989) 367-369

[20] D.G. Jacobs and T. Tamura, The mechanism of ion fixation using radio-isotope techniques, 7th Intern. Congr. of Soil Sci., Madison USA (1960), 206-214.

[21] V.A.Borzilov et al., An experimental study of the washout of radionuclides fallen on soil in consequence of the Chernobyl failure, Soviet Meteorology and Hydrology,11 (1988), 43-53.

[22] A.A.Bulgakov et al., Removal of long-lived radionuclides from the soil by Surface runoff near the Chernobyl nuclear power station, Soviet Soil Sci.,23 (1991) 124-131.

[23] J.Hilton et al., Retention of radioactive cesium by different soils in the catchment of small lake. Sci. Tot. Env.,129(1993), 253-256

Processes and parameters governing accumulation and elimination of radiocaesium by fish.

Rolf H. HADDERINGH

KEMA, Environmental Services, P.O. Box 9035, 6800 ET Arnhem, the Netherlands

Oleg NASVIT

Institute of Hydrobiology, Geroev Stalingrada prospect 12, Kiev 254210, Ukraine

M. Carolina V. CARREIRO

DGA/DSPR, Estrada Nacional 10, 2686 Sacavem Codex, Portugal

Igor N. RYABOV

*Institute of Evolutionary Morphology and Ecology of Animals,
Leninsky prospect 33, Moscow 117071, Russia*

Victor D. ROMANENKO

Institute of Hydrobiology, Geroev Stalingrada prospect 12, Kiev 254210, Ukraine

Abstract. After the Chernobyl accident in April 1986, high levels of fish contamination with radiocaesium were observed in different water-bodies in countries outside the former Sowjet Union, particularly in some lakes in Scandinavia. Therefore, due to the high levels of radionuclide deposition in Ukraine, Belarus and Russia, high contamination in freshwater fish could be expected in these countries. It was considered that the contribution of the fish pathway to the individual and collective doses to the Ukrainian population was of lower significance in comparison with other pathways. However for local populations inhabiting contaminated areas, fish consumption could represent a higher risk. Large differences in contamination levels in fish were found in different water-bodies, and even in the same water-body between species. One of the important findings was the wide range of the levels of contamination within the same species by the "size effect". The European Commission was interested in reducing uncertainties in population doses estimates. Therefore the following studies, in collaboration between EU and CIS countries were started in 1992:
* laboratory experiments concerning the dependence of radiocaesium biological half-life in fish upon the potassium concentration in water and food, water temperature and fish size
* field studies in different water-bodies in Ukraine and Russia to understand the "size-effect" and reveal the critical groups of population.

1. Introduction

After the Chernobyl NPP accident high contamination levels in fish with ^{137}Cs were observed in different water-bodies in countries outside the former Sowjet Union, particularly in some lakes in Scandinavia and England [1,2,3]. Therefore due to high levels of radionuclide deposition in Ukraine, Belarus and Russia, high contamination in freshwater fish could there be expected. It was considered that the contribution of the fish pathway to the individual and collective doses to the Ukrainian population was of lower significance compared to other pathways. However for local populations inhabiting contaminated areas, fish consumption could represent a much higher risk.

The level of ^{137}Cs contamination in fish show large differences in relation to the water-body. This was already found in the period 1960-1970 after the nuclear weapon tests [4,5]. After the Chernobyl NPP accident these differences can be recognised again in Europe [2,6] and in Russia [7]. This can be explained only partly by the level of deposition and water chemistry such as the potassium concentration. Differences in ^{137}Cs levels can also be found between fish species; these might be due to metabolic and diet differences [3]. Differences within the same species in relation to fish size ("size effect") are also well known [3,8,9,10]. Feeding rates, food type and metabolic activity may explain the large variations of ^{137}Cs concentrations between different size classes and species [11]. This can result in different biological half-lives between species and between size classes within one species as reported by some authors [12,13].

In aquatic systems fish are top predators in the food chain. Local inhabitants can receive a protracted dose. The contamination level of inhabitants depends on the position of the fish catch in the prey predator series. Therefore nutritional habits and fish habitats and connectedly the fish size are important. The contribution of these parameters to the eventual contamination of the fish catch is a matter of importance.

To study the above aspects, field surveys were carried out in different water-bodies in Ukraine, the northern part of Kiev Reservoir, the Cooling Pond of NPP Chernobyl and in Russia, Kajanovskoje Lake located in Bryansk Region. Laboratory experiments concerning the dependence of radiocaesium biological half-life in fish upon the potassium concentration in water and food, water temperature and fish size were carried out. Recommendations are given for fish consumption.

2. Material and Methods

2.1. Field studies

For the field studies samples were collected in the years 1992-1994 at four locations (Figure 1). The first location is near the village Stracholesye in the northern part of Kiev Reservoir. Within this area, sampling was carried out at distances of 0.3, 0.7 and 1.2 km from the shore. The average water depth is about 1 m. This area has a dense vegetation of water plants and reeds. The second area is an old arm of the Pripyat river just upstream from the inflow into Kiev Reservoir. The maximum depth is about 9 m. The shores of this area have a dense vegetation of submerged water plants and reeds. Sampling was carried out in the river arm itself and also in a shallow area at about 500 m from the old river arm. The third location is the Cooling Pond of the Chernobyl NPP where samples were collected in the cold and in the heated part. The fourth location is Kajanovskoje Lake, situated in the Bryansk area in Russia with a surface area of about 6 km^2 and an average depth of 1.5 m. Sampling was carried out at three locations in the open water and in the vegetation on the eastern shore.

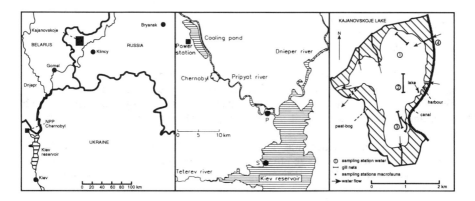

Figure 1.   Area maps showing the sampling areas in Ukraine and Russia.
left map : overview map; middle map : Northern part of Kiev Reservoir
with three sampling stations, Chernobyl NPP Cooling Pond, old arm
Pripyat river (**P**) and Stracholesye (**S**); right map , Kajanovskoje Lake

In total 12 fish species were investigated. The English and latin names of these
species and the families are as follows: Esocidae: pike (*Esox lucius*); Cyprinidae: roach
(*Rutilus*), rudd (*Scardinius erythrophthalmus*), asp (*Aspius*), tench (*Tinca*), silver
bream (*Blicca bjoerkna*), bream (*Abramis brama*), sabre carp (*Pelecus cultratus*),
goldfish (*Carassius auratus gibelio*); Ictaluridae: channel catfish (*Ictalurus punctatus*),
Percidae: perch (*Perca fluviatilis*), pike perch (*Stizostedion lucioperca*).
        Fish was collected with gill nets, trawl nets, rod and line and with dipnets. The
length of all individuals was measured to the nearest mm and the weight to the nearest
g. For identification of the consumed food, stomach contents of were preserved. Samples were taken from individual fish or from pooled fish. Pooled fish consisted of
individuals of about the same size. Very small fish like young of the year were always
pooled. Samples were taken from total body or from muscles. Each sample was dried
at 105 °C and ground.
        Macroinvertebrates were collected with dip nets or by hand from waterplants.
Water was collected by filling buckets at the water surface. From each station 3-4 l
water was concentrated by evaporation to about 100 ml. ^{137}Cs and other gamma
radionuclides were measured by hyperpure Ge detectors and Ge(Li) and NaI(Tl)
detectors.
        To assess the size effect, regression analysis was carried out on all sets of fish
data. The type of size effect is determined by the values of the slope. The size effect
was judged as positive ("+") if the slope b ± 2 S.E had a positive value, negative
("-") if the slope b ± 2 S.E was negative and as neutral ("0") if the slope b ± 2 S.E
had values between positive and negative. To compare the level of ^{137}Cs contamination
in fish at the four sampling locations, the concentration for a "standard" fish of 0.5 kg
was calculated from the regression equations. For pike a standard weight of 2.5 kg was
used because the adults of pike reach a high weight.

2.2. Laboratory experiments

2.2.1 Fish size effect

Carp (*Cyprinus carpio*) from the Chernobyl NPP Cooling Pond, where this species showed a size effect, were brought to the laboratory to study radiocaesium excretion, for about 160 days. Fish weighed approximately 0.3 to 1.2 kg and were fed with uncontaminated food, about 2 % of their weight. Measurements were made on live fish.

Radiocaesium excretion was also studied in goldfish (*C. auratus gibelio*), a species that does not show a size effect. Fish weight ranged from 35 to 110 g and each fish was orally given ^{137}Cs in gelatine solution. Afterwards the procedure was the same.

2.2.2. Changing on environmental parameters

The effect on the biological half-life of different K^+ concentrations in water (0.35, 3.5 and 35 ppm) and watertemperature (20, 12 and 5 °C) was studied for the cyprinid species *Chondrostoma polulepis polylepis*. Fish weighing between 1 to 4 g, were held in aquaria in artificially prepared water with the required K^+ concentrations. They were fed 5 times a week in separated aquaria, with uncontaminated bivalves, each meal representing 5 % of the total fish weight. After the radiocaesium (^{134}Cs$^+$) uptake phase during 30 days, the elimination in the same artificial water was measured, on live fish, during periods of 35 up to 250 days. The biological half-lives were evaluated.

The effect on the radiocaesium biological half-life of different K^+ concentrations in food was studied in carp. Carp of 61 to 211 g were contaminated by injection of ^{137}Cs in a 10 % gelatine solution. Food with different amounts of K^+ (1.6-7 mg) was injected to individual fish. Radiocaesium excretion was measured during periods up to 80 days.

3. Results

3.1. Contamination level in fish at the four sampling locations

Concentrations of ^{137}Cs in fish with the standardized weight of 0.5 and 2.5 kg are presented in Figure 2. Differences in concentration levels between the locations are large. In 1994 perch reaches a ^{137}Cs concentration of about 40.000 Bq/kg in Kajanovskoje Lake, whereas in Kiev Reservoir (Stracholesye) only 600 Bq/kg was found in 1993. These differences between the locations are expressed in the ratios of the average concentration levels: Stracholesye 1, Pripyat 1.1, Cooling Pond 36 and Kajanovskoje Lake 58.

3.2. Contamination levels in different fish species

Figure 2 shows clearly the differences in ^{137}Cs concentration between the sampled fish species. At almost all locations silver bream, bream, roach and goldfish (group I), all Cyprinidae, have the lowest concentrations. Highest concentrations were present in perch, pikeperch (Percidae) and pike (Esocidae) (group III). A group with intermediate

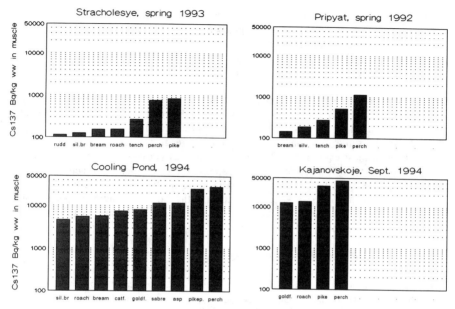

Figure 2.   Calculated concentrations of ^{137}Cs (Bq/kg ww) in fish with a standardized body weight of 0.5 kg (pike 2.5 kg) collected from the four sampling stations.

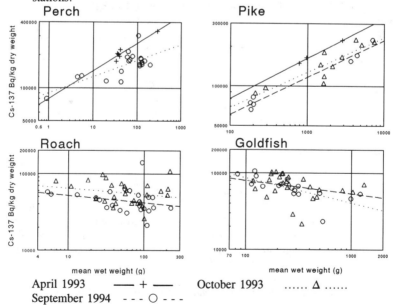

April 1993   —— + ——   October 1993   ...... Δ ......
September 1994   - - - O - - -

Figure 3.   Relationship between ^{137}Cs concentration (Bq/kg dw in muscle) and the body weight (ww) of perch (*Perca fluviatilis*), pike (*Esox lucius*), roach (*Rutilus*) and goldfish (*Carassius auratus gibelio*) collected in 1994 in Kajanovskoje Lake, Bryansk region.

levels are sabre carp and asp (group II); these species belong to Cyprinidae. The contamination level in perch, pike and pikeperch is approximately four times higher than in silver bream, bream, roach and goldfish.

3.3. Relation between ^{137}Cs and fish weight ("size effect")

Regression analysis was carried out on the 49 sets of fish data collected at the four sampling locations. The regression lines of four fish species from Kajanovskoje lake are presented in Figure 3. Positive size effects were found for perch, pikeperch, pike, channel catfish, sabre carp and asp. Tench and bream are species without a clear relationship: in most cases the ^{137}Cs concentration does not change with the weight of the fish (size effect "0"). Among the remaining species, goldfish, roach, rudd and silver bream, the majority has a neutral or a negative relationship (size effect "0" or "-"). The strongest positive size effect was found in the Cooling Pond; the species were perch, pikeperch, sabre carp, asp, bream and goldfish. For these species the values of the slope ("b") were above 0.44.

3.4. Effect fish size on the biological half-life

Experiments with carp from the Cooling Pond, i.e., chronically contaminated, and goldfish, contaminated in the laboratory (see Figure 4), revealed that the biological half-life increases with the fish weight, but only to a certain weight (0.5-0.6 kg in carp). For both species the observed tendencies were well described by power functions, with good correlation:

carp (up to 0.6 kg) $T_b=533\ M^{0.52}$ $r^2=0.97$ (M in kg)
goldfish $T_b=59.4\ m^{0.11}$ $r^2=0.90$ (m in g)

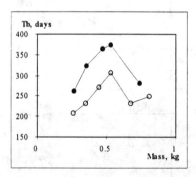 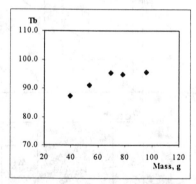

Figure 4. ^{137}Cs biological half-life in fish of different mass class. left: carp (*Cyprinus carpio*) calculated at 76th (O) and 166th (●) day of the experiment right: goldfish (*Carassius auratus gibelio*) at the 71th day of the experiment.

3.5. Effect of potassium concentration and temperature on biological half-life

Laboratory experiments with *C. polylepis polylepis* showed that different potassium concentrations in water influence the uptake and release of radiocaesium by fish. At K^+ concentrations of 0.35 mg/l the uptake of radiocaesium was higher than at concentrations of 3.5 and 35 mg/l, mainly at 20°C and 12°C. At 5°C no difference was found. Retention of radiocaesium was highest at the lowest potassium concentration of 0.35 mg/l. This effect was clearly demonstrated at 12 °C (see Figure 5). At K⁺ concentrations of 0.35, 3.5 and 35 mg/l the biological half-lives of the long-term radiocaesium component were indeed different, respectively 305, 167 and 105 days.

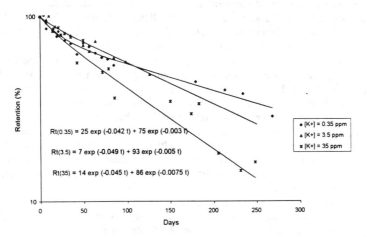

Figure 5. ^{134}Cs retention of the cyprinid species *Chondrostoma polylepis polylepis* at different K⁺ concentrations in water and at temperature of 12°C. Correlation coefficients are respectively: $r^2=0.98$, $r^2=0.97$, $r^2=0.95$

4. Discussion and conclusions

4.1. Variations in contamination level in relation to location, species and size

The great differences between the concentration levels of ^{137}Cs in fish at the four locations are caused by differences in contamination level in the water. The measured concentrations in water (Bq/l soluble ^{137}Cs) at the sampling sites were respectively: 0.085-0.10 in Stracholesye (1993), 0.085 in the old arm of Pripyat river (1992), 4 in the NPP Cooling Pond and 6 in Kajanovskoje Lake. The high concentrations of the last two locations are strongly related to the high deposition of ^{137}Cs and low turnover time. Large differences in the ^{137}Cs concentration in fish were found in biotopes situated at short distances, as the Cooling Pond and the Northern part of Kiev Reservoir. This is also the case at great distances, more than 200 km from Chernobyl: fish collected from the Iput river, revealed ^{137}Cs concentrations 10-100 times lower than measured in Kajanovskoje Lake, which is situated at a distance of only 20 km northwest from the Iput river, due to differences in the deposition level [7]. Thus the contaminating conditions in water-bodies situated in the same area, can differ very strongly. The relation between the contamination in fish and in water was assessed by calculating bioaccumulation factors for each species. The variation of these factors among the locations was smaller than a factor 2 for each species. This means that the rate of accumulation of ^{137}Cs from the aquatic environment (food and water) is of the same magnitude in the four locations. The highest factors were found for the predatory species. An important parameter which might influence the contamination level in fish, is the potassium concentration in water. At K^+ concentrations below a value of about 2 mg/l, accumulation of ^{137}Cs will increase exponentially in freshwater fish [5] and was confirmed by data concerning elimination experiments with different K^+ concentrations in water, carried out in the framework of EU and CIS countries collaboration. This phenomenon cannot be expected at the four investigated locations where the potassium concentrations vary between 2 and 4 mg/l. Almost all ratios found in this study lay within the range of bioaccumulation factors, 200-8000, mentioned by Blaylock [14] for different fish species in water with a potassium content of 2 mg/l.

Large differences in ^{137}Cs level were found between predatory and non-predatory fish species. An important reason for this phenomenon is probably the ^{137}Cs content in the food. In Kajanovskoje Lake for example the ^{137}Cs concentration in roach and goldfish, being the food of the predators perch and pike,is four times higher than in molluscs, the food of non-predators. Also high levels have been found in two cyprinid species, asp and sabre carp. Adults of these species are predatory [15] and occupy a special place within the cyprinids, which are mainly non-predatorous. However, the food factor cannot be the only reason. In small perch for example we found a much higher ^{137}Cs concentration than in roach of the same or greater size. These small perch eat predominantly macroinvertebrates as do the cyprinids. So the type of food does not seem the only reason for the difference in Cs concentration.

A second factor which might influence the release of Cs from the food is the pH level in the stomach or intestine. A low Ph might facilitate the release of Cs from the food and subsequently the radiocaesium uptake in the tissue. Production of acid gastric fluids (HCl) occurs in most fish species with a stomach as percids [16,17] but not in species with a intestinal duct as the cyprinid genera *Cyprinus*, *Rutilus* and *Gobio* [18]. This phenomenon was confirmed in our samples. In cyprinids (roach, tench, gibel carp) an average Ph value of 7.3 +/- 0.3 was measured. In pike, perch, ruffe we measured pH's values 3.9 and 5.6. Experimental work is necessary to investigate the possible role of the Ph. Also the residence time of food in the stomach or intestine might play a role. In perch and pike this time seems to be longer than for some cyprinids [16]. Another factor might be the longer biological half life for predators [19]. For perch a half life of 200 days for radiocaesium was found and 57-150 days for roach at 15 °C [20].

One of the main factors for a positive size effect might be the high radiocaesium short after the deposition. Therefore fish living at that time show the highest accumulation. After the deposition, in most cases, the radiocaesium concentration will decrease by turnover of the water and also the accumulation rate in fish will decrease with time. So fish born years after deposition will have lower concentration than older fish. The strongest positive size effect was found in the Cooling Pond, for almost all species including some non-predatorous cyprinids as bream and goldfish. Probably, the reason is the fast decrease of the ^{137}Cs concentration in the water. In Kajanovskoje Lake on the contrary positive relation between the ^{137}Cs concentration in fish and the fish weight is much weaker. Even a clear negative relationship for the cyprinids roach and goldfish was found. This phenomenon might be explained by the very slow decline of the ^{137}Cs in water, not more than 10% in 1 year. In that situation the difference between young and old fish is smaller. Also the change of food composition with length/weight of the fish might be a reason for the size effect. In this study it was found that perch changes from plankton (juveniles) and macroinvertebrates (smaller individuals) to fish (greater individuals) while cyprinids take more or less the same food at different life stages. Exceptions within the cyprinids are asp and sabre carp which take fish as food as adults (see above). Theory assumes that the ^{137}Cs concentration in preyfish is higher than in other organisms. This was found for Kajanovskoje Lake but not in Kiev Reservoir. An other possible reason might be the increase of the biological halflife with weight as found in our experiments with the two cyprinid species.

4.2. Human fish consumption aspects

Freshwater reservoirs in Russia, Belarus and Ukraine are actively used for commercial fishing and fish-breeding. For instance, in the 6 Dnieper reservoirs the total catch for the period 1980-1990 was about 250 000 tons. In the five years following the accident (1986-1990) the average year catch was 21.080 tons of fish [21]. So the catch after the accident did not change significantly. The ^{137}Cs level in fish from the most contaminated Kiev Reservoir is now below the EU-standard of 600 Bq/kg ww.

Before the Chernobyl accident 17% of the Bryansk population consumes fish from local rivers and lakes [22]. From our own observations semi-commercial fishing with gill nets and rod and line are still practice in Kajanovskoje Lake, despite of the very high contamination level of the fish, up to about 40 kBq/kg ww. Model predictions indicate still levels around 5 kBq/kg ww in muscle of pike and perch after 30 years [23]. Adding potassium as a countermeasure to reduce the ^{137}Cs content in fish is not useful as the potassium concentration in the water is above the critical level of 2 mg/l. Only for water-bodies with lower K^+ concentrations (namely the nordic lakes, where it is below 1 mg/l), to add potassium would increase fish excretion rate. The only possible countermeasure for Kajanovskoje Lake should be a complete foodban for fish and interdiction of any fishery activity.

From the present study it can be concluded that predatory fish species, like perch and pike, contain the highest ^{137}Cs concentrations, especially the large specimen due to the positive size effect. Highest contamination can be expected from water-bodies with low potassium concentrations and with long turnover time. For monitoring the useability of fish for human consumption it is advised to measure the ^{137}Cs concentration in muscles of 0.5 kg perch and 2.5 kg pike. If the level in these species is below the official standard concentration this will be also the case for all other fish species.

References

[1] Hammar J., Notter M., Neumann G.,(1991). Northern reservoirs as sinks for Chernobyl cesium: sustained accumulation via introduced *Mysis relicta* in arctic char and brown trout. The Chernobyl fallout in Sweden, Ed. L. Moberg, The Sweden Radiation protection Institute, pp. 183-205.

[2] Elliot, J.M., Hilton, J., Rigg, E., Tullet, P.A., Swift, D.J. and Leonard, D.R.P., 1992. Sources of variation in post-Chernobyl radiocaesium in fish from two Cumbrian lakes (north-west England). Journal of Applied Ecology, 29:108-119.

[3] Elliot, J.M., Elliot, J.A. and Hilton,J., 1993. Sources of variation in post-Chernobyl radiocaesium in brown trout, *Salmo trutta* L., and arctic charr, *Salvelinus alpinus* (L.), from six Cumbrian lakes (northwest England). Annls. Limnol. 29:79-98.

[4] Kolehmainen, S., Häsänen, E., and Miettinen, J.K., 1966. [137]Cs levels in fish of different lymnological types of lakes in Finland during 1963. Health Phys. 12:917-922.

[5] Fleishman,D.G. 1973. Accumulation of Artificial Radionuclides in Freshwater Fish, in D. Greenberg (Ed.), Radioecology, John Wiley & Sons, New York, pp. 347-370.

[6] Hadderingh, R.H., 1989. Distribution of ^{137}Cs in the aquatic foodchain of the IJsselmeer after the Chernobyl accident. In: The radioecology of natural and artificial radionuclides (W.Feldt, Ed.), Proceedings of the XVth Regional Congress of IRPA, Visby, Gotland, Sweden, 10-14 September, 1989, pp. 325-330.

[7] Fleishman, D.G., Nikiforov, V.A. and Saulus, A.A., 1994. [137]Cs in fish of some lakes and rivers the of the Bryansk Region and North-West Russia in 1990-1992. J. Environ. Radioactivity, 24:145-158.

[8] Andersson, E., 1989. Incorporation of ^{137}Cs into fishes and other organisms. In: The radioecology of natural and artificial radionuclides (W. Feldt, Ed.), Proceedings of the Xvth Regional Congress of IRPA, Visby, Gotland, Sweden, 10-14 September, 1989, pp. 312-317.

[9] Koulikov, A.O., Ryabov, I.N. (1992). Specific Cesium Activity in Freshwater Fish and the Size Effect. The Science of the Total Environment, 112:125-142.

[10] Lindner, G., Pfeiffer, W., Robbins, J.A. and Recknagel, E., 1989. Long-lived Chernobyl radionuclides in lake Constance: Speciation, sedimentation and biological transfer. In: The radioecology of natural and artificial radionuclides (W.Feldt, Ed.), Proceedings of the Xvth Regional Congress of IRPA, Visby, Gotland, Sweden, 10-14 September, 1989, pp. 295-300.

[11] Meili, M. (1991). The Importance of Feeding Rate for the Accumulation of Radioactive Caesium in Fish after the Chernobyl Accident. The Chernobyl Fallout in Sweden. Ed. L. Moberg. The Swedish Radiation Protection Institute, pp. 177-182.

[12] Ugedal, O., Jonsson, B., Njastad, O., Neumann, R., 1992. Effects of temperature and body Size on Radiocaesium retention in brown trout, (*Salmo trutta*). Freshwater Biology, 28:165-171.

[13] Foulquier, L., 1979. Étude bibliographique sur la capacité et les modalités de la fixation du radiocésium par les poissons, CEA-BIB-231 (2), (1979), pp. 360.

[14] Blaylock, B.G., 1982. Radionuclide data bases available for bioaccumulation factors for freshwater biota. Nuclear safety, 23:427-438.

[15] Ladiges, W. and Vogt, D.,1965. Die Süsswasserfische Europas. Paul Parey, Hamburg und Berlin, pp. 250.

[16] Fänge, R. and Grove, D., 1979. Digestion. In: Fish physiology (Eds. Hoar, W.S., Randall, D.J. and Brett, J.R.), pp. 161-260. Academic Press, London.

[17] Nikolski, G.V., 1963. The ecology of fishes. Academic Press, London, pp. 352.

[18] Western, J.R.H. and Jennings J.B., 1970. Histochemical demonstration of hydrochloric acid in the gastrictubules of teleosts using an in vivo prussian blue technique. Comp. Biochem. Physiol., 35:879-884.

[19] Foulquier, L. and Baudin-Jaulent, Y. 1990. The impact of the Chernobyl Accident in Continental Aquatic Ecosystems. A literature review, Proceedings of the Seminar on Comparative Assessment of the Environmental Impact of Radionuclides released during three major nuclear Accidents: Kyshthym, Winscale, Chernobyl. Luxemburg, 1-5 Oct. 1990, pp. 239-704.

[20] Häsänen, E., Kolehmainen and Miettinen, J.K., 1968. Biological half-times of Cs^{137} and Na^{22} in different fish species and their temperature dependence. In: Radiation protection, Part I, (Eds. W.S.Snyder et al.) pp. 401-406. Pergamon Press, New York.

[21] Ryabov I.N., 1991. Analysis of countermeasures to prevent intake of radionuclides via consumption of fish from the region affected by Chernobyl accident. International Seminar on Intervention Levels and Countermeasures for Nuclear Accidents, Cadarache, 1991, pp. 379-396.

[22] Balonov, M.I., Travnikova, I.G.(1990). The role of agriculture and natural ecosystems in forming of the internal irradiation of citizens from the contaminated zone. First international working group on severe accidents and its consequences, 30 October-3 November 1989, Dagomys, Sochi, Moscow, "Nauka", pp. 153-160.

[23] Heling, R., Nasvit, O., Ryabov, I.N. and Hadderingh, R.H., 1995. In: Final Report ECP-3.

Evolution of the contamination rate in game

O. Eriksson[1], V. Gaichenko[2], S. Goshchak[3] , B. Jones[4], W. Jungskär[1] , I Chizevsky[3], A. Kurman[3], G. Panov[2], I. Ryabtsev[5], A. Shcherbatchenko[6], V. Davydchuk[7], M. Petrov[7], V. Averin[8], V. Mikhalusyov[8] and V. Sokolov[5].

[1] Department of Ecological Botany, Uppsala University, Uppsala, Sweden, [2] Schmalhausen Institute of Zoology, Academy of Sciences of Ukraine, Kiev, Ukraine, [3] Restoration Department, RIA Pripyat, Chernobyl, Ukraine, [4] Department of Clinical Chemistry, Faculty of Veterinary Medicine, Swedish University of Agricultural Sciences, Uppsala, Sweden, [5] Severtsov Institute of Evolutionary Morphology and Ecology of Animals, Russian Academy of Sciences, Moscow, Russia, [6] Institute of Nuclear Research, Academy of Sciences of Ukraine, Kiev, Ukraine, [7] Institute of Geography, Academy of Sciences of Ukraine, Kiev, Ukraine, [8]Byelorussian Research Institute of Agricultural Radiology, Gomel, Byelarus.

Abstract
The Chernobyl accident caused considerable contamination of natural environments in large parts of Europe. In the most heavily contaminated areas close to the reactor site tissues from two wild animal species (wild boar, *Sus scrofa* and roe deer, *Capreolus capreolus* [L]) have been sampled for determination of radioactive contamination. The level of ^{137}Cs has been determined in a large number of samples and other radionuclides like ^{90}Sr in some samples. Systematic samplings during different seasons of the year have been made of these two species within the framework of the ECP9 project. The results show that there is a considerable individual variation within each season for both these species and also a seasonal variation most pronounced in wild boar. For the wild boar the minimum levels of ^{137}Cs are seen during summer (end of August) and autumn (end of October) and maximum levels in winter (end of February). In the roe deer the maximum levels are seen in spring (end of May) and minimum in summer. These variations reflects the feed selection during different seasons of the year. The level of ^{137}Cs contamination in muscular tissue has not decreased noticeably during the study period from summer 1992 to winter 1995.

Introduction

Environmental contamintion with radioactivity will cause intake of radioactivity by wild animals living in the contaminated area. Such contamintion of game was detected already in the first years of atmospheric nuclear bomb testing [for a review see 1]. Studies performed in that period also showed a seasonal variation caused by the differences in forage selection at different seasons of the year. Systematic studies of game in a heavily contaminated environment seems, however, not to have been done earlier.

The objectives of the present study were to study contamination levels in muscular tissue of roe deer (*Capreolus capreolus* [L.]) and wild boars (*Sus scrofa*) and to evaluate the importance of meat from wild animals as source of radioactive contaminants to humans. The results can be used to calculate radiation doses to humans consuming this type of food products. Data was also obtained on the radioecology of these animals and the importance of different forage plants used by the animals during different periods of the year.

Material and methods

Sampling of roe deer and wild boars were performed in three areas heavily contaminated by the Chernobyl acident in Ukraine, Byelarus and Russia. The study was part of a larger project (ECP5 and ECP9) where transfer of radionuclides were studied in rural communities where people are living to a great extent on products produced in natural and semi-natural ecosystems.

Details of the three sampling areas are given below as well as the actual sampling periods.

The research area in Ukraine

The samples were all obtained from within the evacueted zone around the damaged nuclear power plant in Chernobyl. The area is charachterized by a mainly low flat relief with river flood-plains, terraces, end moraine ridges, moraine fluvioglacial and limnoglacial plains. Rather infertile sandy soils dominate the upper levels, where fallow arable lands occupies the major part. Grassland with *Elytrigia repens* (L.) Nevski, *Festuca ovina* L., *F. rubra* L., and *Oenothera biennis* L. is dominant. Planted forests with Scots pine (*Pinus sylvestris* L.) of various ages, sometimes with oak (*Quercus ruber* L.) in the understory occur in the upland sandy areas. *Alnus glutinosa* (L.) Baertn. dominates low-laing swamp forests. The lower level of the zone is occupied by the flood plains of the river Pripyat and its tributaries. The vegetation in these areas is dominated by *Salix acutifolia* Wells., xerophytic shrubs and graminids.

During this study totally eleven sampling were made in the Chernobyl area. The first sampling was in early June 1992 and after that samplings have been made in mid-August (summer), late October (autumn) and end of February beginning of March (winter).

Samples have been obtained in two areas with different contamination levels. The surface contamination ^{137}Cs in the high contamination area is 1100 to 1500 kBq per m^2 and in the low contaminated area 180 to 700 kBq per m^2.

Stomach content from the wild boars and rumen content from the roe deer, faeces, forelimb flexor muscles, liver and one metacarpal bone, but also other tissues, were sampled and deep frozen for future analyses. The stomach/rumen samples are preserved by mixing with an equal volume of 95 % ethanol. After a rapid field check of the stomach/rumen content about 100 g (wet weight) of plant species occuring in noticable quantities are collected around the point where the animal is sacrifized.

Animal sampling

Most animals were collected during early mornings or during afternoons and nights. Efforts were made to obtain undisturbed specimens. Only during the winter period with snow, when animals were hiding in thick cover, a certain amount of beating was executed.

Animal tissue samples were measured for ^{137}Cs in a γ-spectrometer system (Ortec) with a GeLi detector. The system has been intercalibrated and corresponds to established requirements. The results are expressed in Bq per kg fresh weight .

For quantitative botanical analysis of stomach/rumen samples the technique described by Eriksson et al. [2] was used. Briefly the methods utilises ethanol preserved material which is stirred in a bucket and 400 to 500 ml is washed through a set of sieves with mesh sizes ranging from 4000 to 500 μm. Particles smaller than 500 μm are discarded. Each fraction is put on a transparent tray and the plant fragments to be examined are indicated by an underlaing 100 point grid. The area frequency of plants/plant groups is converted to frequency by weight using weight constants specific for the plant groups and particle sizes found.

A subsample of the preserved stomach/rumen samples was freeze dried and the ^{137}Cs content was determined. The dried material was then ashed at 600°C over night in a muffle furnace to determine the ash content of the stomach/rumen content.

Plant sampling

About 100 g (wet weight) of plant material was collected in the immediate vicinity of the spot where the animal was sacrifized using the initial field check to identify the plant species last grazed by the animal. The plant material is dried and ground to a powder wich is measured in the γ-spectrometer. The radioactivity was also determined in 5 cm soil samples taken on the location of animal sampling.

The research area in Belarus

The sampling was performed near the villages Savichi and Dvor-Savichi in the evacuted zone just north of Chernobyl. The geographical charachteristics are similar to the Ukrainian sampling site. The level of ^{137}Cs contamination is 555 to 1480 kBq ^{137}Cs per m^2.

Samplings were performed in the selected area during February, winter sampling, and during July, summer sampling, 1994 and 1995. In February roe deer and wild boars were obtained as well as moose (*Alces alces* [L.]) and during the July sampling only wild boars. Different tissues from the animals and contents from the gastro-intestinal tract were taken for determination of ^{137}Cs and [90]Sr. Rumen/stomach content was also sampled in 1994 for determination of botanical composition using the same method as in Ukraine [2]. The plant species identified were measured for radioactivity content and the results were used to calculate the daily intake of radionuclides by the animals.

The research area in Russia

During August 1995 a wild animal sampling expedition was performed in the Novozybkov area within the ECP9 programme. Wild animal, roe deer and wild boar, samplings had earlier been performed during the period October to December in 1992, 1993 and 1994.

The samplings were done in an abandoned area with a ^{137}Cs surface contamination of 1.25 to 10 x 10[5]Bq/m[2].

Results and discussion

UKRAINE

The number of animals and the ^{137}Cs levels in muscular tissue are shown in Figure 1 (wild boar) and Figure 2 (roe deer) for the entire study period. As can be seen from thesa figures the results show a considerable seasonal variation for both species. For the wild boar the lowest values are seen during summer and autumn, while roe deer have the lowest levels in winter. There is a considerable variation in radiocesium levels in animals of the same species obtained from the same area. But roe deer obtained from a specific area generally have lower contamintion levels than wild boars obtained from the same area. This variation is also seen in the contamination of the dried and ashed rumen/stomach samples due to the differences in individual forage selection seen in free ranging wild animals. This seasonal variability in ^{137}Cs contamination has earlier been reported in free ranging ruminants (e.g. [3, 4]) from other parts of Europe after the Chernobyl accident.

The mean ash content in stomach samples from wild boars varies from 32 % d.m. (dry matter) content in the winter to 6 % of d.m. in the summer and with intermediate values, mean 14 %, in the spring and autumn. But also here the individual variation is considerable. In the roe deer the seasonal differences are smaller with ash contents of 9 to 15 % d.m. for spring and winter, respectively.

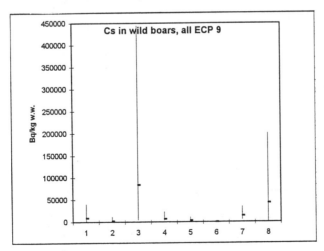

Figure 1. The contamination of muscular tissue (range and median) from wild boars (*Sus scrofa*) sampled from 1992 to 1994 in the Chernobyl zone. The figures on the abscissa denotes the sampling periods and number of animals sampled (n). 1 summer 1992 (n=5), 2 autumn 1992 (n=7), 3 winter 1993 (n=11), 4 spring 1993 (n=8), 5 summer 1993 (n=6), 6 autumn 1993 (n=4), 7 winter 1994 (n=6) and 8 spring 1994 (n=7).

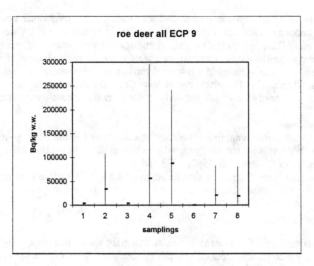

Figure 2. The contamination of muscular tissue (range and median) from roe deer (*Capreolus capreolus* [L.]) sampled from 1992 to 1995 in the Chernobyl zone. The figures on the abscissa denotes the sampling periods and number of animals sampled (n). 1 summer 1992 (n=5), 2 autumn 1992 (n=6), 3 winter 1993 (n=6), 4 spring 1993 (n=6), 5 summer 1993 (n=6), 6 autumn 1993 (n=6), 7 winter 1994 (n=6) and 8 spring 1994 (n=6).

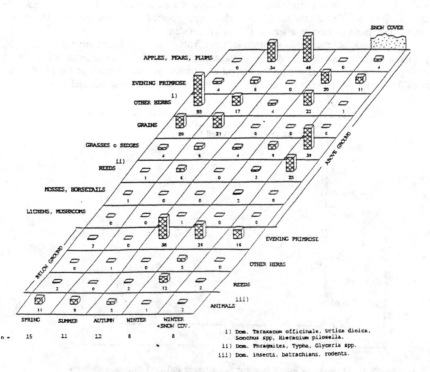

Figure 3. Mean results of quantitative botanical analysis of stomach content from wild boars obtained during the different seasons in the Chernobyl zone.

- 150 -

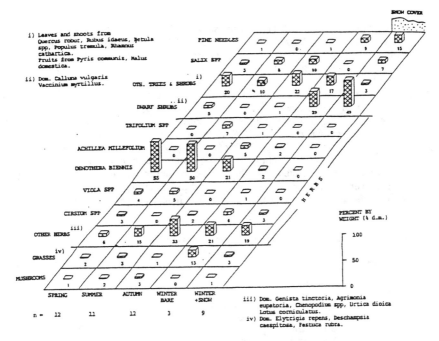

Figure 4. Mean results of quantitative botanical analysis of rumen content from roe deer obtained during the different seasons in the Chernobyl zone.

Although the intake of ^{137}Cs with soil is considerable the uptake from the gastro-intestinal tract is expected to be limited in from soils high in organic matter [5]. A certain part of the measured contamination of the muscular tissue must, however, come from ingested soil both in the most contaminated wild boars and roe deer as it is diffcult to account for these levels (> 0.2 MBq/kg tissue w.w) with the contamination levels found in forage plants eaten by the animals.

Table 1. Concentration of ^{137}Cs and ^{90}Sr in tissue samples from ungulates sampled in the Dvor Savichi area in Byelarus during February, 1994 and 1995.

Animal species	Sampling year	Animal age, years	^{137}Cs in muscle kBq/kg wet weight	^{90}Sr in muscle kBq/kg dry matter	^{90}Sr in bone
Moose	1994	4	7.03	0.004	29.63
(*Alces alces*)	1995	1	0.19	--	11.42
	-"	6	11.7	0.029	35.99
	-"	5	4.35	0.095	95.10
Roe deer	1994	1	6.29	0.023	61.30
(*Capreolus*	-"	2	10.73	0.012	5.13
capreolus)	1995	5	4.84	0.004	4.57
	-"	5	3.67	--	4.25
	-"	2	4.21	0.020	5.63
Wild boar	1994	5	3.70	0.008	4.56
(*Sus scrofa*	-"	3	40.70	0.005	9.15
[L])	-"	5	4.44	0.010	4.21
	1995	3	17.3	0.005	30.88
	-"	2	8.40	0.003	9.64
	-"	0.8	61.90	--	11.31

BELARUS

The measured concentrations of ^{137}Cs and ^{90}Sr are shown in Table 1 for the winter sampling of moose (*Alces alces*), roe deer and wild boar. Both in July 1994 and 1995 three wild boars were obtained in this study area. In 1994 the ^{137}Cs contamination of muscular tissue was from 0.3 to 15.5 kBq/kg w.w. and in 1995 from 8.4 to 19.8 kBq/kg w.w. Both in winter and in summer the contamination levels vary considerably for the wild boars although the levels appear lower in summer than in the winter like in Ukraine. The roe deer sampled in winter showed more comparable levels of contamination due to feeding on plants with a more even level of ^{137}Cs contamination.

The quantitative botanical analysis of rumen/stomach samples showed that young top sprouts of aspen and willow made up 60 to 75 % of the feed intake of these animals. Other tree species (e.g. birch, Scots pine, oak and apple) and shrubs constituted another 20 to 25 % of the ration (see Table 2). During the winter the diet of wild boars was dominated by lower and under ground parts of a few plant species while the diet of the ruminats contained components from several plant species. During the summer the three wild boars sampled had an even more one-sided diet with a mean content of 95 % oats grain, maize and potatoes. The remainder was other parts of oat straw and grasses.

Table 2. Botanical composition of rumen/stomach content from ungulates sampled in the Dvor-Savichi area in February, 1994.

Animal species	Amount %	Plant species
Moose	40	Young sprouts and bark of aspen
	35	Young sprouts and bark of willow
	15	Young sprouts and bark of birch
	10	Young bark of Scots pine
Roe deer	30	Bark of aspen
	30	Young sprouts of willow
	21	Young sprouts of different trees (e.g. birch, apple, oak) and shrubs
	12	Young sprouts of bilberry
	3.5	Young sprouts of raspberry
	3.5	Young sprouts of cowberry
Wild boar	90	Roots, rhizomes, lower parts of marsh plants
	5	Tissues of animals (moose, wild boar, roe deer, hare)
	2	Rodents (mice, voles, shrews)
	1	Earthe worms, larvae
	2	Soil

Table 3. ^{137}Cs contamination (kBq/kg w. w.) of forage plants selected by wild animals in the Dvor-Savichi area.

Plant parts and species		^{137}Cs, kBq/kg fresh weight
Young sprouts of aspen (*Populus tremula*)	1.54	
Bark of aspen	7.19	
Young sprouts and bark of willow (*Salix* spp.)		1.06
Young sprouts of birch (*Betula* spp.)		2.15
Apple tree (*Malus dom.*)		2.32
Young sprouts and bark of pear (*Pyrus comm.*)		2.57
Young sprouts of billberry (*Vaccinium myrtillus*)		2.03
Young sprouts of raspeberry (*Rubus ideus*)	4.30	
Raspberries		3.01
Bark from Scots pine (*Pinus sylvestris*)	4.14	
Cough-grass (*Agropyron repens*)		0.37
Oenothera biennis		0.99
Phragmites communis		0.74
Genista tinctoria		1.49
Melilotus alba	0.09	
Potentilla erecta		5.48
Urtica dioica	0.32	
Chamaenerium	0.73	
Artemisa vulgaris		0.07
Animal tissues	4.71	
Rodents		0.97
Earth worms, larvae		2.73
Soil		3.07

The ^{137}Cs contamination of forage plants sampled in the Byelarussian study area where the wild animals were sampled in February is shown in Table 2.

The grain consumed by the wild boars during summer contained on average 13 Bq/kg fresh weight and the other feeds 25 Bq/kg fresh weight.

There is an obvious discrepancy between observed intake of contaminated plants and the measured levels of contamination in muscular tissue of the sampled ungulates. This is partly due to the fact that animals are moving between areas with different levels of contamination so the animal body is not in balance with the intake.

The ^{90}Sr contamination of muscular tissue is low compared to the levels found in bone (results not shown) which generally are more than two orders of magnitude higher. There is no dirct correlation between the measured levels in these two tissues most certainly due to the much faster turno over of Sr in soft tissue than in bone. There was not any correlation between the surface area contamination with ^{90}Sr and the levels in muscular tissue.

RUSSIA

A number of wild boars and roe deer have been sampled in the Novozybkov area in Russia from 1992 to 1994. The results are presented in Table 4. All these samples were obtained from October to December so no obvious seasonal change is seen. In these animals the forage intake was not studied but the aggregated transfer ($t_{(ag)}$, m^2/kg muscular tissue w.w.) have been calculated both for the wild boars and the roe deer. The results from this area also show a considerable variation between different individuals as in the two other study areas. In all these cases this is an effect of the considerable variability in surface area contamination of radioactivity.

Table 4. Concentration of ^{137}Cs in muscular tissue samples from wild boars and roe deer sampled in the Novozybkov area mainly during the period October to December 1992, 1993 and 1994.

Species	Sampling year	n	$t_{(ag)}$, $\cdot 10^{-3}$, range	^{137}Cs kBq/kg in muscle, range	
Wild boar	1992	7	1.37-7.89	1.24-7.10	high contaminated area,
	1993	12	0.39-6.50	0.25-3.40	8 - 10 $\cdot 10^5$ Bq per m^2
	1994	4	0.13-68.97	1.19-62.15	-"
	1992	-	—	—	low contaminated area
	1993	4	0.85-3.42	0.17-0.69	1.25-2.5 $\cdot 10^5$ Bq per m^2
	1994	-		—	
Roe deer	1992	5	6.56-9.03	5.90-8.13	high contaminated area,
	1993	9	1.27-18.45	1.14-10.6	8 - 10 $\cdot 10^5$ Bq per m^2
	1994	1	0.86	0.80	-"
	1992	2	6.73, 7.45	1.35, 1.49	low contaminated area
	1993	2	2.05, 2.60	0.41, 0.52	1.25-2.5 $\cdot 10^5$ Bq per m^2

Conclusions

The results of both this study show a considerable seasonal variation in contamination levels of muscular tissue with ^{137}Cs with lower levels in the summer and autumn for the wild boar and lower levels in the winter for the roe deer. All Russian samples except one has been obtained during late autumn or early winter so possible seasonal changes can not be seen in these samples.

Generally the contamination level is lower in roe deer than in wild boars obtained from the same sampling area. But the individual variations within each species is considerable during all seasons especially in the Chernobyl material reflecting the great variability in surface area contamination.

There is no obvious decrease in the level of contamination of animal tissue samples obtained from the different study sites during the three years studied.

In the wild boarr the lowest contamination is seen during summer and autumn. These periods would be the best hunting periods to avoid intake of radiocesium from game by hunters and their families. In the roe deer the lowest contamination is seen during winter which hence would be the preferable hunting time to avoid human contamination with radiocesium.

Samples from the Byelorussian site have been analysed for ^{90}Sr. The levels in muscular tissue are from 5 to 25 Bq/kg fresh weight and not correlated to the levels in bone or the level of surface contamination with ^{90}Sr. The levels in bone are three orders of magnitude higher than in the soft tissue.

References
[1] Whicker, F.W. & Schultz, V. 1982. Radioecology: Nuclear energy and the environment. Chapter 5, Vol. I. CRC Press, Boca Raton, FL, USA.
[2] Eriksson, O., Palo, T. & Söderström, L. 1981. Reindeer grazing in winter time. Uppsala. Växtbiologiska studier, 13. ISBN 91-7210-813-4. (In Swedish).
[3] Howard, B.J. & Beresford, N.A. 1992. Transfer of radiocesium to ruminants in natural and semi-natural ecosystems and appropriate countermeasures. *Health Phys*, 61, 6, 715-725.
[4] Karlén, G., Johanson, K.J. & Bergström, R. 1991. Seasonal variation in the activity concentration of ^{137}Cs in Swedish roe-deer and in their daily intake. *J. Environ. radioactivity*, 14, 91-103.
[5] Belli, M., Blasi, M., Capra, E., Drigo, A., Menegon., S., Piasentier, E. & Sansone, U. 1993. Ingested soil as a source of ^{137}Cs to ruminants. *Sci. Total Environ.* 135, 217-222.

II. ENVIRONMENTAL ASPECTS
OF THE ACCIDENT
A. Behaviour of radionuclides in contaminated territories

Posters

Resuspended Hot Particles in the Atmosphere of the Chernobyl Area

F. Wagenpfeil and J. Tschiersch
GSF Forschungszentrum für Umwelt und Gesundheit GmbH
Institut für Strahlenschutz, Ingolstädter Landstr. 1,
D-85758 Oberschleissheim, Germany

Abstract. In this paper the results of the collected resuspended hot particles in the Chernobyl area are presented and discussed. The measurements were carried out during the resuspension experiments in the project ECP1 from 1992 until 1994. Aerosol samples were taken with new designed rotating arm impactors to collect simultaneously during the same experiment three samples of giant particles with different particle size ranges. After γ-spectroscopy and digital autoradiography the nuclide ratios and number concentrations of airborne hot particles could be derived. For wind resuspension a maximal concentration of 2.6 particles per 1000 m³ and during agricultural practice of 36 particles per 1000 m³ was measured.

1. Introduction

After the Chernobyl reactor accident a big amount of nuclear fuel was released into the atmosphere. The particle formation was first caused by two mechanical explosions. In the next days volatile fission products like I, Cs, Sr and particles consisting of pure nuclear fuel were released due to high temperatures in the reactor core. The emitted particles are called *hot particles* and can be classified in the following way [1]:

    a) A-typ particles: pure ^{103}Ru- and ^{106}Ru-particles originated after oxidation, evaporation and finally condensation,

    b) B-typ particles: Fuel Hot Particles (FHP, consisting of pure nuclear fuel).

The study of hot particles is of importance because of their different behaviour in respect to several radioecological aspects compared to particles with attached condensation products. In general the activity per particle is much higher and the radionuclides are fixed in a different matrix. This might mean a different inhalation risk because of the not uniform activity distribution and the different solubility. The different bounding of the radionuclides may cause a different migration of activity in the soil and as a result of this, a different availability of the deposited activity for resuspension.

In this paper airborne measurements of hot particles are presented which were carried out during resuspension experiments of ECP1 from 1992 until 1994 at different locations of the contaminated Chernobyl area.

2. Methods of sampling and analysis

A new rotating arm impactor (first single stage version see Jaenicke and Junge, 1967 [2]) was designed to collect simultaneously during the same experiment three samples of giant particles with different particle size ranges. The particles are collected on three sticky sampling slides (flat geometry) mounted on a rotating rod (Fig. 1). The distance from the rotating axis and the width of the slides are different. With a choosen fixed geometry three different lower cut-off diameter depending on the rotating frequency can be realized. The rotating frequency is stabilized and can be adjusted by a static frequency changer controlling a three phase current motor. An IR reflection light barrier is mounted close to the rotating arm to monitor the rotating frequency. The size range of the particle diameter at a rotating frequency of 5 Hz is from 10 μm, 20 μm and respectively 28 μm up to all larger particles [3].

The rotating arm impactors were operated simultaneously in up to three different heights above ground (1.7 or 2.2 m, 3.8 m and 6.4 m) during the resuspension experiments. Samples were taken during wind resuspension and during different simulated agricultural activities.

To determine resuspended γ-activity all the slides were first measured with Ge-detectors (Canberra Inc.). An inhomogenious activity distribution on the surface of the slides (hot particles are point sources) could be measured with a digital autoradiograph (Berthold). This instrument measures the position, intensity and distribution of ionizing radiation (β⁻, α-emitters) of a sample with a flat geometry. By that means particles collected on the slides of the rotating arm impactor can be measured without a special preparation of the samples.

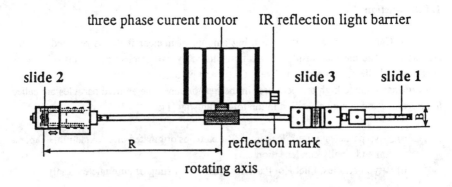

Fig. 1: New designed rotating arm impactor with three sampling slides; at a rotating frequency of 5 Hz the size range of the collected particles is larger 10 μm, larger 20 μm and larger 28 μm aerodynamic diameter [3].

3. Results and discussion

A good tracer for hot particles is the low mobile fission product ¹⁴⁴Ce, which can be measured with γ-spectroscopy. The ¹⁴⁴Ce-activities of all collected hot particles in 1992 until 1994 at the sites Pripyat Beach, Zapolie and Kopachy calculated back to the moment of the accident were in the range between 30 and 300 Bq. In addition to ^{137}Cs also [134]Cs, [154]Eu and [155]Eu were found. The average nuclide ratios are in good agreement with the theoretical calculations of the radionuclide-composition of the fourth block at the moment of the accident and the measured hot particles in soil [4,5] (see Tab. 1).

From the measured [144]Ce-activity of one hot particle and the activity of [144]Ce per g nuclear fuel, the minimum particle size can be derived (assumed, that the particles are spherical and consist only of fuel components). The calculated diameter was between (6 ± 4) µm and (12 ± 4) µm. The number of hot particles per m³ air was found to be very high in the Pripyat Beach during wind resuspension: 20 FHP's in 1.7 m height (1 FHP per 380 m³) and 6 FHP's in 3.8 m height (1 FHP per 1270 m³). In Zapolie and Kopachy only during technogenic activities hot particles could be found: 13 FHP's (1 FHP per 290 m³) in Zapolie and 16 FHP's (1 FHP per 38 m³) in Kopachy (see Tab. 2 and Fig. 2). In Novozybkov no resuspended hot particle in the giant particle range could be found.

From slides without hot particles the activity of one aerosol particle can be estimated. With the assumption of equipartition of ^{137}Cs on the particles and the determination of the particle number on the sampling slides a mean particle activity in the range 1 - 10 µBq was determined [6]. Hot particles were measured to have 1 - 10 Bq ^{137}Cs, which is about a factor of 10^6 more in ^{137}Cs-activity.

Tab. 1: Average nuclide ratios obtained by the rotating arm impactor [6], after measuring of 1200 hot particles in soil and by theory.

results obtained by	average nuclide ratios			
	$\dfrac{A(^{137}Cs)}{A(^{144}Ce)}$	$\dfrac{A(^{134}Cs)}{A(^{144}Ce)}$	$\dfrac{A(^{154}Eu)}{A(^{144}Ce)}$	$\dfrac{A(^{155}Eu)}{A(^{144}Ce)}$
airborne sampling [this paper]	0.06 ± 0.02	0.03 ± 0.01	0.0011 ± 0.0003	0.0033 ± 0.0016
hot particles in soil [4]	0.041	0.020	0.0015	0.0017
theory [4]	0.062	0.036	0.0030	0.0043

Tab. 2: Number concentration of hot particles in air depending on the sampling height during wind resuspension (site Pripyat Beach) and during agricultural activities (sites Zapolie and Kopachy).

Site	soil contamination [Bq/m²]	sampling height [m]	airborne number concentration $[10^{-3} \cdot m^{-3}]$
Pripyat-Beach (wind resuspension)	$1.0 \cdot 10^7$	1.7	2.6
		3.8	0.79
Zapolie (agricultural work)	$5.4 \cdot 10^5$	1.7	2.5
		3.8	7.3
		6.4	1.7
Kopachy (agricultural work)	$2.2 \cdot 10^6$	1.6	28
		2.2	36

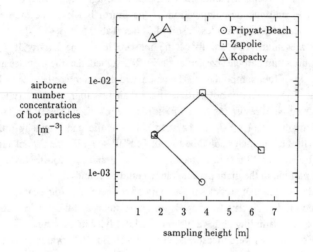

Fig. 2: Number concentration of hot particles in air depending on the sampling height during wind resuspension (site Pripyat Beach) and during agricultural activities (sites Zapolie and Kopachy).

References

[1] F. Steinhäusler, Summary of the Present Understanding of the Significance of Hot Particles from the Chernobyl Fallout, Division of Biophysics, University of Salzburg, Hellbrunnerstr. 34, A-5020 Salzburg, Austria, from 'HOT PARTICLES FROM THE CHERNOBYL FALLOUT', Bergbau- und Industriemuseum, Band 16, 1988, pp 143-144
[2] R. Jaenicke, C. Junge, Beiträge zur Physik der Atmosphäre, Vol. 40, 1967, pp. 129-143
[3] F. Wagenpfeil, T. Härtl and J. Tschiersch, Size-fractionating sampler for giant particles, Journal of Aerosol Science, Vol 25, 1994, pp. 111-112
[4] V. D. Kuriny, Yu. A. Ivanov, V. A. Kashparov, N. A. Loshchilov, V. P. Protsak, Ye. B. Yudin, M. A. Zhurba and A. E. Parshakov, Particle-associated Chernobyl Fallout in the local and intermediate zones, Ukrainian Institute of Agricultural Radiology, Ann. Nucl. Energy, Vol. 20, No. 6, 1993, pp. 415-420
[5] N. A. Loshchilov, V. A. Kashparov, Ye. B. Yudin, V. P. Protsak, M. A. Zhurba and A. E. Parshakov, Experimental Assessment of Radioactive Fallout from the Chernobyl Accident, Ukrainian Institute of Agricultural Radiology, Kiev, Ukraine, Sicurezza e Protezione, N. 25-26, pp. 46-49, Gennaio Agosto 1991
[6] F. Wagenpfeil, Messung der Resuspension von Riesen-Aerosolpartikeln in der Region von Tschernobyl, Dissertation, Ludwig-Maximilians-Universität München, to be published

Scientific Provision and Control in Plant - Breeding on Contaminated Territory of Ukraine

P.F. BONDAR, Yu.A. IVANOV, B.S. PRISTER, G.P. PEREPELYATNIKOV
Ukrainian Institute of Agricultural Radiology (UIAR)
7, Mashinostroïtelei st., 255205 Chabany (Kiev Region), UKRAINE

Abstract. Complex of tasks and methods of its solution during the assessment of radiological situation, as well as some ways of its improvement on farmlands, subjected to radioactive contamination after the accident on ChNPP, are demonstrated.

The assessment of radiological situation on farmlands, subjected to radioactive contamination, after the accident on ChNPP, is a complex problem, which requires an operative solution for a range of tasks.

One of the first and complicated tasks is a detailed detection of soil contamination levels on each field, hayfield, pasture. Having vast areas of contaminated land and great spottiness of farmlands contamination density within one farm, the solution of this task by sampling and measuring soil samples is unrealisable due to the enormous volume of work. In Ukraine this problem was solved with the help of an express-method of soil contamination levels assessment by th data of ground gamma-survey in-situ, with the account of fallout radionuclide composition and radionuclides distribution in soil profile. (For gamma-survey SRP-68-01 radiometer was used). Soil contamination density was calculated by the formulas:

for arable fields: σ, $kBq/m^2 = d\,3700\,(P - P_0)$, mR/hour;

for meadows: σ, $kBq/m^2 = d\,1665\,(P - P_0)$, mR/hour;

where: σ = soil contamination;

P = gamma-background on contaminated territory;

P_0 = natural gamma-background on this territory;

d = correction, which takes into account radionuclide composition of the fallout.

The second urgent task is the assessment of biological availability of Caesium-137 and Strontium-90 in the fallout on the territory, subjected to radioactive contamination. Biological availability of radiocaesium in the fallout was investigated by the comparison of radionuclide transfer from the same soil, contaminated with the fallout of ChNPP release, and at application of radiocaesium to soil in water-soluble form.

$RBA = CR_0/CR_1 = OR_0/OR_1$

where: RBA =   relative biological availability of radionuclide in fallout

CR_0 and OR_0 =   radionuclide concentration ratio and observed ratio (in soil-plant system), calculated for radiocaesium, fallen out to soil with ChNPP fallout;

CR_1 and OR_1 =   the same indices, calculated for radiocaesium introduced to soil in water-soluble form.

Table 1.   Accumulation of radiocaesium in maize (dry weight) and relative biological availability of radiocaesium in the fallout.

N	Location of plots in relation to ChNPP		CR$_0$		CR$_1$		RBA	
	Direction	Distance km	1989	1992	1989	1992	1989	1992
1	West	2	0.54	1.02	1.05	1.42	0.53	0.72
2	NW	4	0.37	0.21	0.58	0.27	0.57	0.78
3	NE	6	2.04	2.10	4.28	2.66	0.48	0.79
4	NE	7	1.45	0.97	2.03	0.96	0.71	1.01
5	NE	8	1.43	0.93	2.23	1.59	0.64	0.59
6	West	10	0.94	0.42	1.56	0.45	0.60	0.93
7	North	17	0.57	0.49	0.63	0.66	0.90	0.74
8	SW	23	0.51	0.30	0.77	0.30	0.65	1.00
9	SW	35	0.59	0.66	0.77	0.74	0.77	0.89
10	SW	42	0.56	0.80	-	0.75	-	1.06
11	SW	55	0.44	0.49	0.54	0.66	0.80	0.74
12	SW	64	0.48	0.97	0.79	0.96	0.60	1.01
13	West	65	0.65	0.80	0.68	0.75	0.96	1.06
14	SW	75	0.41	0.66	0.56	0.74	0.69	0.89
15	SW	75	0.32	0.77	0.60	0.83	0.53	0.93
Average							0.67	0.87
Average for 30-km zone							0.63	0.82
Average outside 30-km zone							0.73	0.94

The research carried out demonstrated that in 1988-89 the biological availability of radiocaesium in the fallout inside and outside 30-km zone was, on the average, 59 and 74 % respectively, in 1992 - 82 and 94 % respectively. Thus, biological availability of radiocaesium in the fallout outside 30-km zone was comparable to the availability of the nuclide, introduced to soil in water-soluble form (Table 1).

For the assessment of the radiobiological situation on farmlands and development of measures on its improvement by correct location of the crops, data on the levels of yield contamination depending on plants biological peculiarities are needed. In similar soil and climate conditions on the contaminated territory quantitative characteristics of the yield contamination were investigated for more than fifty crops. The differences in radiocaesium contamination levels for cereals and leguminous crops go up to a factor 72 (Table 2), for the main part of vegetables yield to a factor 40. In this connection the correct location of the crops favours a decrease of products contamination levels. At similar soil contamination density levels, contamination on private plots (on highly cultivated soils) are considerably lower than on commercial fields.

Table 2.   Transfer factors (TF) of ^{137}Cs to plants and its content in the yield at contamination density of soddy-podzolic sandy-loamy soil 0.56 Mbq/m^2.

Crop	TF, (Bq/kg) / (kBq/sq.m)		Content in the yield, Bq/kg	
	grain	straw	grain	straw
Maize	0.09	0.50	50	278
Winter wheat	0.11	0.43	61	239
Barley	0.13	0.22	72	122
Triticale	0.18	0.29	100	161
Spring wheat	0.12	0.33	67	183
Millet	0.21	0.93	117	516
Rye	0.24	0.41	133	228
Oats	0.62	0.96	344	533
Beans	0.79	1.26	205	700
Buckwheat	1.05	1.20	582	666
Soya beans	0.88	1.20	484	677
Peas	1.12	1.45	622	805
Vetch	1.29	1.48	716	821
Yellow lupin	6.56	3.90	3641	2165
Differences of TF for grain is 72.9 times				

A certain improvement of the radiological situation on farmlands can be achieved by application of fertilizers and ameliorants. Liming of acid soils allows to reduce the yield contamination by 2-3 times. Additional application of various ameliorants, including zeolites, to limed soil, did not provide further decrease of yield contamination. On low fertile soddy-podzol soils with exchangeable potassium contact up to 10 mg/100 g, potassium fertilizing is the effective means for the decrease of yield contamination.

Maximum practically significant effect of potassium fertilizing is achieved at doses of potassium 300-400 kg/ha (Table 3). On soils with a content of exchangeable potassium above 10 mg/100 g the effect of potassium fertilizing decreases.

Table 3.   Effect of various doses of Potassium fertilizers for yield and accumulation of radiocaesium in it

Potassium dose kg/h	Expected efficiency %	Real efficiency %			Yield % to control*	
		oats	peas	potatoes	oats	peas
60	29	19	17	14	103	119
120	45	48	20	23	106	119
180	56	57	27	60	104	127
240	63	70	40	63	109	117
300	68	78	57	63	107	124
360	71	76	68	60	108	120
480	77	79	60	60	109	129
600	81	81	66	63	112	123

* Control yields: oats (grain) - 12500 kg/ha, peas (grain) - 10300 kg/ha, potatoes (tubers) - 135800 kg/ha

Nitric fertilizers provide an increase of radiocaesium accumulation in plants, and when applied as a part of complete fertilizer, they reduce the efficiency of potassium fertilizers. For a decrease of the product contamination, the potassium dose in the complete fertilizer should exceed nitrogen dose by 2-3 times. Similar doses of fertilizers have a different influence on radiocaesium accumulation in different crops. In this connection the identification of optimum fertilizers doses for various crops in specific conditions is a task of great practical importance. The investigations demonstrated that application of mineral and organic fertilizers, as well as ameliorants, do not provide statistically reliable change of mobile radiocaesium in soil.

It was found experimentally that a radical improvement of the soil is one of the most effective methods of increasing the productivity and reducing the levels of herbage contamination on natural meadows and pastures.

Conclusion

A scientifically based solution of the above-listed tasks allows to carry out a radiological survey of the fields, hayfields and pastures with a high economy of labour and expenses within the shortest period of time. It allows also to make maps of farmlands contamination for all farms on the contaminated territory. The fact of a high biological availability of radiocaesium outside 30-km zone allowed to use to full extent the information accumulated before the accident on radiocaesium behaviour in biological systems for the prediction of possible yield contamination in the first period after the accident. The presence of the maps of farmlands contamination density and quantitative characteristics of the yield contamination for various crops, obtained in real radiation and natural conditions, allowed to develop a recommendation on the rational use of fertilization for a maximum yield with the least content of radionuclides on arable fields and pastures.

Mechanisms of Radionuclides ^{137}Cs and ^{90}Sr Uptake by Plant Root System

Anatoliy SOKOLIK, Galina DEMKO, Vadim DEMIDCHIK

Belarus State University, F. Skarina Avenue, 4, 22080 Minsk, Belarus

Abstract. Using the model object (single plant cell of freshwater alga) it had been shown that radionuclides enter a cell by selective (mainly potassium channels) and unselective (membrane ionic leakage) pathways. The apoplast (cation-exchangeable cell wall) influences on this process. The same selective and unselective ways were shown to be involved in the radionuclide accumulation by seedling roots of the barley from solution. The relative share of both pathways was determined under different environment conditions. It is concluded that radionuclide accumulation could be reduced without disturbing the mineral nutrition of a plant by means increasing the share of the selective ways on the basis of elaborating of agromeliorative measures or choice of appropriate species and sorts of agricultural plants.

1. Introduction

The presence of the radionuclides from global fallout, from Chernobyl fallout and from similar (possible) accidents in soils of large territories call forth the task of searching for possibilities to decrease radionuclides accumulation in the plants. This problem can be solved only successfully and with minimal expenditures on the basis of knowledge of the mechanisms of radionuclides (RN) transfer between plants and the environment, as well as inside the organism along its organs and tissues. The knowledge of these mechanisms can give a possibility to evaluate the capability of plants to decrease RN accumulation without disturbing their growth and development, and to choose the ways of elaborating appropriate agromeliorative measures.

Previously we studied basic mechanisms of ionic transport through plant cells membrane [1,2]. Selective and unselective ways were shown to be present as ionic (mainly potassium) channels and as unselective ionic leakage of plasmalemma. The characteristics of these mechanisms were studied [1,2]. Our approach its to use model objects and model experiments. Radionuclides, accumulated in soil, enter a plant in ionic form together with elements of mineral nutrition. That is why we firstly have studied the mechanisms of RN accumulation by single plant cell of a freshwater Characeae alga, which is a convenient model object in studying ion transfer into a plant cell. Then we have modelled the soil-solution/root RN transfer to evaluate the role and relative share a mechanism of ion transport in the process.

2. Objects and Methods

The cells of freshwater Charophytes Nitella flexilis have been used [1]. Due to their large size (up to 0.5 mm in diameter and 5 cm in length) these are very convenient for the electrophysiological experiments and direct studies of ion fluxes using a radioisotope.

In the course of the experiments the curves of radionuclides accumulation and washing-out from an isolated cell against time have been obtained by periodical withdrawal of cells from the solutions and by their activity determination . The value of the specific activity in the solutions amounted to 10-20 kBq/ml for ^{90}Sr and 50-100 - for ^{137}Cs. The nuclides have been used as CsCl and SrCl$_2$, their concentration being less than 10^{-6}M.

Cs: The cells, withdrawn from the solution for the activity calculations, after rinsing in distillate were subsequently held for 5 min in two solutions of artificial pond water (APW) for the radionuclide removal from the cell wall (apoplast). For the estimation of its cation-exchanging capacity no soaking has been used.

Sr: Strontium and yttrium penetrate into cells differently due to their different ion charges. The measurements were conducted using an Al filter to chop of a fraction of 0.99 of Sr radiation and one of 0.50 of Y, and without the filter - to obtain an adequate estimation of Sr content. The cell activity has been determined immediately upon rinsing, the apoplast capacity has been estimated by counting the rate after the first 10 min. of exposure in a labelled solution. The basic APW solution contained 10^{-4}M KCl, 10^{-3}M NaCl and 10^{-4}M CaCl$_2$; pH value was 8.2 and were obtained with the help of TRIS and MES buffers.

3. Results and Discussions

3.1 ^{137}Cs and ^{90}Sr accumulation by single plant cell

Curves of a time course of accumulation and washout of radionuclides in a single cell were registered. Curves reveal three exponential phases according to three cell compartments. The first one, the fastest phase (about ten min) represents the exchange between solution and apoplast (cell wall); the second one, about ten hours, is the main and reflects ion transport through cell membrane to cytoplasm; the more slow phase presents accumulation in vacuole. The main results of a data analysis are shown in Table 1.

Table 1.   *Parameters of the ^{137}Cs (upper, left) and ^{90}Sr (bottom, right) accumulation by single plant cell (alga Nitella) under varying environmental conditions. T-time constant of the first stage of the processes of nuclides accumulation (in), or washing (out) by the cell, when the cell wall stage was rejected. N-cell activity, divided on the cell surface area. K_d - distribution coefficient as the ratio of specific cell activity and one of the solution. Each of the data is a mean at least 5 cells, pH was 8.2 D, H and N - dark, high and normal illumination.*

Solutions	APW	APW	APW + 5mMK$^+$	APW + 5mMCa^{2+}	APW + 5mMk$^+$ + 5mMCa^{2+}	APW + 5mMCs$^+$
Light conditions parameters	H	D, N	N	N	N	N
T in, min.	550 -	950 1140	1350 2120	940 2160	1540 2040	2050 -
T out, min.	580 1610	1100 1300	850 830	- 380	1510 -	1560 -
N$_{st}$, s^{-1}	880±190 14360±3520	510±200 10100±2100	360±80 8300±1650	400±110 7300±1430	120±40 7690±1250	340±80 -
N$_{ap}$, s^{-1}	- 1520±180	1390±240 1770±290	1100±250 1710±320	- 880±170	960±210 800±190	- -
K$_d$	14.4±3.1 320±75	8.2±1.4 210±42	6.0±2.1 160±32	6.7±1.3 156±30	2.1±0.7 166±30	5.6±1.5 -

The relative share of both the selective and the unselective pathways of radionuclides in dependence of main environmental factors have been determined by the analysis of above data [2]. In particular, it has been found, that in dilute solutions (10^{-4}M potassium) half of the radiocesium pass plasmalemma trough potassium channels, (selective way) and another half - through unselective ionic leakage [3]. The influx of radiostrontium is mainly through leakage: only 0.3 of whole flux pass through channels. When solution's concentration is risen ($5*10^{-3}$MK), the influx rate of radiocesium and radiostrontium was significantly lower due to depolarization of the plasmalemma. Calcium lowers the radiostrontium influx owing to the effect on channels and decrease cation-exchange capacity of cell wall (apoplast) [2]. As the illumination increases the nuclides uptake rate through the plasmalemma and also the value of accumulation coefficient are growing appreciably.

3.2 Uptake of radionuclides by seedling roots

The curves of radionuclide accumulation against time from a water solution by 4-5 days old barley seedlings, grown in a water culture, were registered. They were about linear for 4-6 hours. It was shown that direct influence of conditions of mineral nutrition (ionic and concentration composition of the accumulation medium) on the rate of radiocesium entering into roots is not high. Varying the composition of the growing medium is considerably more effective: for example, potassium deficit accelerates more than four times radiocesium accumulation under the circumstances (Table 2).

The effect of varying calcium and magnesium concentrations in growing medium was studied also and great influence of these treatments on radionuclides accumulation was revealed. It was also proved that radiostrontium entrance into a root proceeded with the participation of the apoplast and non-selective ionic leakage of the plasmalemma, the share of each of them was determined in the growing medium. For example, Ca and Mg deficit causes and increase of a share of the apoplast. A considerable share of the membrane pathway (ionic leakage) in general radiostrontium entrance was shown as well as a direct influence of the composition of the accumulation medium on the rate of nuclide entrance.

Table 2.   Parameters of ^{137}Cs accumulation by barley root seedlings, growing in different nutrient solutions. Kn-potassium, calcium, calcium and magnesium are the normal Knope's solution with the low levels (10^{-4} mole/liter) corresponding element. N_{ap}, N_{root}, $N_{green part}$ are the activity of apoplaste, root and green part of a seedlings, after accumulation for 4 hours, divided to the root surface area.

Parameter	Growing Medium			
	Knope's medium	Kn-potassium	Kn-calcium	Kn-calcium and magnesium
N_{ap}	2.08± 0.70	2.32± 0.65	2.65 ±0.64	2.86±0.30
N_{root}	0.90± 0.26	4.13± 0.49	2.20± 0.60	1.1±0.34
$N_{green part}$	0.11± 0.05	0.057± 0.025	0.082± 0.025	0.35±0.013

4. Conclusions

The results show that RN accumulation by a seedling root is realized both by selective and unselective pathways. Their relative shares vary and reveal a small dependence on the mineral conditions of the accumulation media. Conditions of mineral nutrition influence the accumulation of ^{137}Cs by roots, mainly through the process of formation of components of the ionic transport system through ontogenesis. [90]Sr entrance in roots of barley seedlings occurs through the apoplast and non-selective ionic leakage of plasmalemma; the share of them is determined by the composition of growing medium. For example, potassium and magnesium deficit induced an increase of the apoplast share.

The obtained results make possible to propose a direction of reducing radionuclides accumulation by the plants without disturbing their mineral nutrition; it is increasing the share of the selective pathway by means of both using special agrochemical measures and choosing of the corresponding species of the agricultural plants.

References

[1]   A.I. Sokolik, V.M. Yurin, Potassium channels in plasmalemma of Nitella cells at rest. J. Membrane Biol., 1986, V.89, pp. 09-22.

[2]   V.M. Yurin, A.I. Sokolik, A.P. Kudryashov, Regulation of ion transport through plant cell membrane. Minsk, 1991, 217 pp. (in Russian).

[3]   A.I. Sokolik, V.M. Yurin, Ionic leakage of the plant cell plasmalemma-real ion pathway into the cell. Biologia plantarum, 1994, v. 36 (suppl.) p 246.

^{90}Sr Levels in Water Samples of Adriatic Sea

L. QAFMOLLA, M. SINOIMERI, B. GRILLO & K. DOLLANI
Institute of Nuclear Physics, Tirana, Albania

Abstract. The Institute of Nuclear Physics, Tirana, performs systematic measurements of the radioactivity in different samples for the national network. Hundreds of samples of air, fallout, surface water food and vegetation collected in various regions of the country were measured. Water samples of the Adriatic Sea are measured for the determination of ^{90}Sr, ^{137}Cs, since 1992 as presented in this paper.

The radiochemical separation of ^{90}Sr from Barium and Calcium are described in this paper. The oxalic acid $C_2H_2O_4 \times 2H_2O$ (as oxalate, carbonate and oxide) method for the determination of the chemical separation ratio was used. The activity concentration of ^{90}Sr in water sea is evaluated by its daughter ^{90}Y in equilibrium with it. The collective dose assessment by ^{90}Sr is done.

1. Introduction

Parallel with the widening of the nuclear power and radioactive isotopes use, environmental radioactive contamination measurements are developed for radiation protection scope. Since 1975, some environmental measurement methods for beta and beta-gamma low level measurement were developed. Global beta counting with proportional gas detectors, liquid scintillation counters and low level gamma spectrometry are some of them [1[.

Beginning from April 29, 1986 the works for identification of the eventual radioactive contamination of Albania territory caused by Chernobyl accident, for radioisotopes with long half-life periods, as ^{90}Sr, ^{137}Cs etc. are carried out [1]. The presence of ^{90}Sr in biosphere is a long-term risk, due to its chemical similarity to Calcium and its long biological half-life [2]. Concerning the ^{90}Sr, taking into account the level and the fact that both ^{90}Sr and its daughter ^{90}Y are pure beta emitters a general measurement procedure is based on chemical separation and low-level beta counting. Since 1992 sea water samples are collected every 3 months.

2. Method

Sea water samples of 10-60 liters depending on strontium concentration are taken for ^{90}Sr concentration determination.

The separation method used was as follows: the separation of alkaline-earth elements as phosphate, the nitric acid fuming method to remove the calcium ions, and the dregs method to remove the baryium ions as chromates is applied [3]. The tracing by non radioactive strontium salt of the strontium nitrate waterfree $Sr(NO_3)_2$ was used to determine the chemical separation ratio. The chemical separation ratio of calcium Ca^{+2}, Sr^{+2}, ions was determinated by the X-ray fluorescence measurement method. A value higher than 99 % was obtained. The overall chemical yield of ^{90}Sr was 70 - 80 %, while that of ^{90}Y was 96.8 - 99.7 %. The strontium activity was determined by the ^{90}Y activity in equilibrium measurements, which source were produced in oxalate form. The measurements were performed mainly in a low-level beta counting system with plastic scintillation detector and anti-coincidence protection. The lower limit of detection of system is 12 mBq/l, while the registration effectivity is 0.41. The uncertainty of ^{90}Sr activity concentration in water sea is estimated to be about 10 %, at the 68 % confidence level.

3. Results and conclusions

The ^{90}Sr activity concentration level of the samples collected (located in beach of Durres) are presented in Table 1.

Table 1 ^{90}Sr activity on water Adriatic Sea (10^{-1} Bq/m^3)

Place	1992			1993				1994		
Trimester	Jul	Sep	Dec	Mar	Jul	Sep	Dec	Mar	Jul	Sep
Durres	5,4	5,9	4,9	1,2	1,5	1,7	8,4	2,5	1,1	3,0
Aritmet. mean	5,4			3,2				2,2		

The permitted maximal concentration (PML) for ^{90}Sr in water reservoirs and water sea for our country is 4.10^2 Bq/m^3.

For the Danube river the ^{90}Sr concentration for 1992 is 10 Bq/m^3 and for the Black Sea is 15.4 Bq/m^3 [4,5].

From results of the Table 1 the concentration of ^{90}Sr for Adriatic Sea water is in average 3.34 Bq/m^3. In comparison with PMC, the mentioned value is around 100 time lower, therefore contribution of ^{90}Sr to population dose with bathing or with use the water sea is unconsiderable. For the population the consumption rate id 2 kg/y of Sardele fish 1 kg/y of Barbunj fish and 2 kg/y of Merluc fish, e.g. in total 5 kg/y per person. The activity concentration for each kind of the fish is: Sardele 0,051 Bq/kg, Barbunj 0,056 Bq/kg and merluc 0,049 Bq/kg. The arithmetic mean for three species of fish is 0,052 Bq/kg. The annual intake of ^{90}Sr is 0,052 Bq as result of multiplying concentration with consumption per person adding a factor of 2 as recommended by IAEA: 0,052 Bq/kg x 5 kg x 2 = 0,52 Bq.

While the dose conversion factor for ^{90}Sr is $3,3.10^{-9}$ Sv/Bq, the dose commitment for Albanian population is 0,05 man.Sv. Considering that collective dose from natural and artificial source of radiation is nearly 10.000 man.Sv, it is clear that the dose contribution from ^{90}Sr (0,05 man.Sv) is negligible. From literature [5] it is shown that for France the contribution of ^{90}Sr in the collective dose is 0,08 man.Sv, e.g. is nearly the same with the value presented by us.

References

[1]   P. Skende, Xh. Myteberi. Bull. Nat. Sc., 2 (1989).
[2]   M. Eisenbud, Environmental Radioactivity, Acc. Pres. 1978
[3]   HASL-300 Health and Safety Lab. Procedures 1972.
[4]   Rules on ionising radiation works, 1973.
[5]   Radiation Protection 70, Report EUR 15564 EN, 1994

Fuel Component of ChNPP Release Fallout: Properties and Behaviour in the Environment

Yu. IVANOV[1], V. KASHPAROV[1], J. SANDALLS[2], G. LAPTEV[3], N. VICTOROVA[3],
S. KRUGLOV[4], B. SALBU[5], D. OUGHTON[5], N. ARKHIPOV[6]

[1] *Ukrainian Institute of Agricultural Radiology (UIAR), 255205 Chabany
(Kiev region), Ukraine*
[2] *AEA Technology, Culham Laboratory, Abingdon, Oxfordshire, OX12 3DB, UK*
[3] *Institute of Hydrometeorology (IHM), Nauki av. 37, 252028 Kiev, Ukraine*
[4] *Russian Institute of Agricultural Radiology and Agroecology (RIARAE), 240020
Obninsk (Kaluga region), Russia*
[5] *Research and Industrial Association (RIA) "Pripyat", 255620 Chernobyl, Ukraine*

Abstract. Main characteristics of Chernobyl fuel particles and particles spatial distribution around ChNPP are considered. Main regularities of the behaviour of fuel particles in the components of terrestrial and aquatic ecosystems (migration, transformation, etc.) are discussed.

1. Introduction

During the accident on Chernobyl NPP a significant part of the radionuclides, including Strontium, Actinides and other refractory chemical elements, has been released to the atmosphere in the form of small particles of nuclear fuel. The majority of these particles fell to the ground within 30 km zone of the reactor, although some hot particles were found in at least 11 countries outside the boundary of the ex-Soviet Union [1]. Peculiarities of the introduction of radionuclides, initially contained in the matrix of fuel particles, in to migration chains and their transfer in components of the environment are considered.

2 Fuel particles: properties and spatial distribution around Chernobyl NPP

Chernobyl fuel particles were essentially high burn-up uranium oxide fuel with a composition similar to that of the fuel in the reactor core but with some depletion of the radioisotopes of the more volatile chemical elements Iodine, Ruthenium and Caesium. Some particles were spherical, others were angular shards [2-4]. Apart from Uranium oxide and fission products, many hot particles also contained Zirconium and traces of Iron, Molybdenum, Nickel, Copper, Zinc, Silica, Aluminum and Lead [5,7]. Within the ChNPP 30 km zone fuel particles were estimated to account for more than 75 % of the total radioactive contamination on the ground.

Essentially all the radiostrontium and Plutonium were associated with particles at the time of deposition [2,3]. The relationship between i-th radionuclide's activity (Ai) and ^{144}Ce activity in fuel particles for the moment of accident is shown in the Table 1 [2,3].

Table 1. Experimental relationship between i-th radionuclide's activity (Ai) and ^{144}Ce activity in the fuel particles on the moment of accident

Radionuclide	$A_i/A(^{144}Ce)_{FHP}$
^{90}Sr-	0.05
^{95}Zr	2.3
^{103}Ru	1.08
^{106}Ru	0.26
^{125}Sb	0.006
^{134}Cs	0.02
^{137}Cs	0.04
^{141}Ce	1.34
^{144}Ce	1.0
^{154}Eu	0.0015
^{155}Eu	0.0017
^{239,240}Pu	0.0004

The fraction of radionuclides, initially contained in fuel particles (FP) matrix, in total activity of these ones in soils at the moment of accident, using the data on soil contamination with refractory radionuclides (^{144}Ce etc., taking into account the radioactive decay) could be estimated by equations (1,2) [3].

$$q(^{137}Cs) = (^{137}Cs/^{144}Ce)_{FP} / (^{137}Cs/^{144}Ce)_{soil} = 0.04 / (^{137}Cs/^{144}Ce)_{soil} \quad (1)$$
$$q(^{90}Sr) = (^{90}Sr/^{144}Ce)_{FP} / (^{90}Sr/^{144}Ce)_{soil} = 0.05 / (^{90}Sr/^{144}Ce)_{soil} \quad (2)$$

Within the 10 km zone the particles ranges in size from some micron up to about 150 μm [2,4]. Few particles with a diameter in excess of 20 μm were found outside the 10 km zone. In 1987 median radius of fuel particles in soil was 1.5 - 2.5 μm, and in 1989 - 2.0 - 3.5 μm [8,9]. Decrease of fine particles fraction with time is connected with its lower chemical stability in soil, as fine particles have a higher degree of oxidation in comparison to large ones. Particle size distribution in flood-plain soil has been described by a lognormal distribution law with MAD, equal to 1-2 μm, whereas in bottom sediments of Glubokoye Lake - up to 5 μm.

The contribution from fuel particles was largest in the 30 km zone. Southwards and southeastwards from the reactor, more than 50 % of the radiocaesium was in the form of fuel particles, the fraction of particles in the fallout decreases with increasing distance from the NPP [2]. Fuel particles were dispersed in aquatic systems not only due to atmospheric fallout, but also as a result of riverborn suspended transport. Thus, hot particles have been identified in deep bottom sediment of the Kiev Reservoir (middle part, old river channel) in the layer of peak activity 25-35 cm dated as 1986-88.

3. Behaviour fuel particles in components of the environment

3.1   Migration ability of fuel particles in soils, waters and sediments

Results of *in situ* observations testify that in the first 2-4 years after the release the intensity of migration in the soil profile for radionuclides, both for fuel and condensed fallout components (Caesium, Cerium, partially Strontium radioisotopes) was practically the same, i.E. no spatial differentiation of the content of radionuclides with different properties by the horizons of soil profile was observed [10,11]. Comparison of these data with results of model (column) experiment allow to show the significant role of the mechanical transfer of radionuclides in soils as well as the one of fuel particles through pores in the soil profile, especially in dry sandy soils [11].

Particle penetration in bottom sediments of lakes and river is clear; the registered depth is 5cm in soils and up to 10 cm in bottom sediments. The vertical transfer in the soil profile depends on soils properties as well as the radionuclides' initial forms (fuel or condensed component of fallout) [10,11].

3.2   Transformation of fuel particles in soils, water and sediments

Latest data obtained in 1993-1995 demonstrate that in some soils there is high quantity of undestructed fuel particles (up to 60-80 % in accordance to the data of autoradiography and assessments of the content of ^{90}Sr ion-exchangeable forms in soil). These results are observed mainly for soddy-podzolic soils with a low moisture content in the immediate zone of the ChNPP. The data need some refinement of the previous estimations of the fuel particles half-life in soils (0.8-4 years) [11].

The content of radionuclides, contained in fuel particles matrix, in Chernobyl Lakes and Pripyat old arms bottom sediment is still very high and causes a continuous secondary contamination of the water column. Part of activity, contained in fuel particles matrix in per cents of the whole bulk activity of bottom sediments is reached in the upper layers, up to 40-50 %. In general, particle destruction processes in soils are more intensive compared to bottom sediments.

A significant dependence of the dissolution velocity of fuel particles on the level of physico-chemical transformation of the particles matrix (incinerated or non-incinerated fuel, particles, subjected to leaching in soil in natural conditions, etc.) and on characteristics of the environment (pH, Redox-potential, etc.) is shown in model experiments [11].

The proportion of exchangeable radionuclides in the near zone of the plant during the first years after the accident was much lower than that after the nuclear weapon testing [12]. For this reason, the distribution coefficients of Chernobyl radionuclides in the "soil-water" system were quite high, causing a reduction of migration. The predominant process was leaching of radionuclides from hot particles which led to increased migration ability [13]. Radiological monitoring in 1986-1994 of the Pripyat River showed an increase of ^{90}Sr concentration in water only in the section of the river in the Chernobyl exclusion zone as a factor of two and even more during the ice-jam in 1991. In fact, only breakdown of the fuel particles could explain this observation [14,15].

3.3 Role of fuel particles on formation of the dynamics of the radionuclides bioavailability

Bioavailability of Caesium radioisotopes, presented by condensed component of ChNPP fallouts, is comparable with biological availability for ^{137}Cs of global fallouts and is primarily determined by physical-chemical properties of soils.

During the first 3-4 years after the accident, the biological availability of "Chernobyl" Caesium-137 at the territory, contaminated mainly by fuel component of fallout, was 1.5-2 times lower than this one of Caesium-137, introduced into the soil in initial water-soluble form and/or at the territory, contaminated by condensed component of fallout.

Bioavailability of ^{90}Sr, represented mainly by the fuel component of fallout, is increasing till now. This conclusion has been supported by experimental data of RIA "Pripyat", presented in Fig. 1 [16]. Thus, the presence of the fuel particles in the fallout of Chernobyl accidental release has modified the intensity of introduction of the radionuclides from fuel component to migration chains in terrestrial ecosystems. The modification depends on both the soil-climatic conditions and on the fuel particles properties.

Fig. 1   Dynamics of ^{137}Cs TF, (Bk/kg) / (kBk/m^2) to the grasses of natural meadows at the fuel tracks of ChNPP release fallout

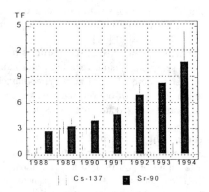

3.4 Behaviour of radionuclides presented by fuel particles in animals

According to the experimental data a lower availability of ^{137}Cs, contained in matrix of fuel particles ingested by cows in comparison with water-soluble one, causes both a lower transfer of radionuclide to milk and a slower rate of this process [11]. Results of a joint AAS-NLH and UIAR experiment with goats showed that the retention of fuel particles in the Gi-tract may be significantly longer than expected. The release of radionuclides from the fuel particles may occur during digestion in the GI-tract followed by a subsequent uptake of Caesium radioisotopes [17].

References

[1] Chernobyl accident. (1995). Kiev, "Naukova dumka" Publisher.-560 p. (Rusk).

[2] N.A. Loschilov, V.A. Kashparov, Ye.B. Yudin et al. (1991). Experimental assessment of radioactive fallout from the Chernobyl accident. Sicurezza e Protezione. No. 25-26.-P.46-50.

[3] V.D. Kuriny, Yu.A. Ivanov, V.A. Kashparov et al. (1993). Particle-associated Chernobyl fallout in the local and intermediate zone. Annals of Nuclear Energy (1993). Vol.20 No.6-P 415-420.

[4] F.J. Sandalls, M.G. Segal, N.V. Victorova (1993). Hot particles from Chernobyl: a review. J. Environ. Radioactivity, 18, pp.5-22.

[5] Borovoj et al. (1990). New products of Fuel Interaction with Constructing Materials of the 4th Block of the Chernobyl NPP. Radiochimiya. N.6, p.103-113. (Rus.).

[6] B. Salbu, D. Oughton, A. Ratnikov et al. (1994). The mobility of ^{137}Cs and ^{90}Sr in Agricultural Soils in the Ukraine, Belarus and Russia. Health Phys. Vol.67 No.5.

[7] F.J. Sandalls (1994). Physico-chemical properties of hot particles and plant uptake. UK AEA Report AEAT/43705001/REMA-133.

[8] V.A. Kashparov, Yu.A. Ivanov, S.I. Zvarish et al. (1994). Formation of hot particles during the accident on Chernobyl NPP. Radiochimiya No1, p.87 (Rus.).

[9] V.V. Demchuk, O.V. Voytsekhovitch, V.A. Kashparov et al (1991). Analysis of Chernobyl Fuel Particles and their Migration Characteristics in Water and Soil. Proc. of CEC Semin. on Comparative Assessment of the Environmental Impact of Radionuclides Released During Three Major Nuclear accidents: Kyshtym, Windscale, Chernobyl. Luxembourg, 1-5 October 1990. EUR 13574.

[10] Yu. Ivanov, S. Levchuk, B. Prister et al. Migration parameters of ^{137}Cs and ^{90}Sr from Chernobyl fallouts in the soils of Ukraine, Belarus and Russia. J. Environ. Radioactivity (Submitted to publication).

[11] Yu.A. Ivanov, V.A. Kashparov, N.M. Lazarev et al (1994) Physico-chemical forms of ChNPP release fallout and long-term dynamics of released radionuclides in components of agroecosystems. Mater. of IV Intern. Conf. "Results of 8 years works on the liquidation of ChNPP accident consequences" (Zeleny Mys, Dec. 1994). (submitted to publication).

[12] A.V. Konoplev, V.A. Borzilov, Ts.I. Bobovnikova et al (1988). Distribution of radionuclides in the soil-water system fallen out in consequence of the Chernobyl failure, Soviet Meteorology and Hydrology. 12. P.65-74 (Rus.).

[13] A.V. Konoplev, A.A. Bulgakov, V.E. Popov et al (1992), Behaviour of long-lived radionuclides in a soil-water system. Analyst, 117. P. 1041-1047.

[14] O. Voitsekhovitch, V. Demchuk, G. Laptev (1990). Analysis of secondary contamination sources for Pripyat river. UHMI Proc, Moscow. (Rus.).

[15] G. Laptev, O. Voitsekhovitch (1993). Study of radionuclide wash-off from Pripyat River catchment soils under flood. UHMI Proc. N269, Moscow. (Rus.).

[16] A.N. Arkhipov, N.P. Arkhipov, D.V. Gorodetsky et al (1994). Development of the radioecological situation on agricultural lands within 30 km zone of Chernobyl NPP. Preprint. Chernobyl. 44 p.

[17] B. Salbu, T. Krecling, K. Hove et al. Biological relevance of hot particles ingested by domestic animals (submitted to publication).

Peculiarities of Particulate ^{137}Cs Transport and Sedimentation in Kiev Reservoir

O. VOITSEKHOVITCH, V. KANIVETS, I. BILIY, G. LAPTEV
Ukrainian Hydrometeorological Institute
Nauka Av. 37, 252028 Kiev, Ukraine
U. SANSONE, M. RICCARDI
ANPA, 48 via Vitaliano Brancati, 00144 Rome, Italy

Abstract The paper presents the data on Chernobyl radiocaesium bound to suspended matter and bottom sediments at different locations along the sampling rout from Rivers of Chernobyl zone to upper Reservoirs of Dnieper River. These data were collected as a result of joint Ukrainian-Italian field exercises in the frame of ECP-3 project. It was found out that total ^{137}Cs concentration in the water column decreases downstream the Chernobyl zone while K_D *in situ* values substantially increase with approach to the Kiev HPS dam. Taking account of uniform hydro-chemical conditions in investigated area one can explain this phenomenon only by gradual elimination of coarse sandy component with low sorption capacity from the river flow by sedimentation. In contrary, radiocaesium which is selectively sorbed and fixed on fine clay particles travels much longer distances and ensures observed higher K_D *in situ* values. This conclusion is supported by the analyses of three sediment cores taken in upper, middle and lower parts of Kiev reservoir.

1. Introduction

An exclusive role of Kiev reservoir as a huge natural trap for Chernobyl radionuclides is defined by its geographical position and hydrological regime. Actually it is an about hundred kilometres long watercourse from the Pripyat river (which is the main carrier of radionuclides from the Chernobyl zone) and the Dnieper river confluence to the HPS dam in the Northern suburbs of Kiev City. According to previous observations in 1986-1993, up to 60% of particulate cesium (i.e. 18-25% of total activity) coming into the Kiev reservoir was trapped and buried in its bottom sediments [1].

In order to better understand the transport mechanisms and to validate models it was extremely important to trace out the particulate ^{137}Cs fate for different size fractions of suspended particles through the pathway from the Chernobyl zone to the Dnieper River reservoirs. This is done by focusing on the key transport parameter K_D distribution coefficient *in situ* which is defined as the ratio between concentrations of radiocesium in particulate [Bq/kg] and soluble [Bq/l] phases. In this particular study it was done it two ways. Firstly, water sampling survey downstream of the Chernobyl zone presents "an instant picture" of particulate cesium transport, and secondly, bottom sediment cores taken at three characteristic point give an idea of temporal variations in suspended cesium concentration in different parts of the Kiev reservoir.

2. Materials and Methods

Samples of water and different fractions of suspended materials were collected in 1992, 1994 and 1995 in two reservoirs (Kanev and Kiev) and four rivers (Dnieper, Pripyat, Uzh and Ilya) which are different in morphological scale and hydrological regime caused in different level of suspended particles contamination. Subordination sequence of the hydrological relationship between above rivers and reservoirs are as follows :

Ilya → Uzh → Pripyat → Dnieper (Kiev res.) → Dnieper → Dnieper (Kanev res.)

Sampling exercise was carried out using two different sampling devices.

The first system was capable to perform both size fractionation of suspended materials and concentration of dissolved caesium directly at time of taking sample. The different fractions of suspended materials were separated on sequentially arranged cartridges with nylon filters (PALL filters, HDC II, 1000, diameter 60 mm, filtration area 0.49 m^2). Filters were used with nominal pore sizes 40 μm, 10 μm and 0.45 μm. The suspended matter mass determined as difference of dried filters weight before and after filtration.

Dissolved ^{137}Cs was concentrated on columns filled in ammonium hexocyanocobaltferrate (NCFN) ion-exchange resins. In order to assure the resin efficiency to concentrate radiocaesium, series of two ion-exchange columns were used (diameter 20 mm, height 160 mm and 80 mm, respectively). The volume of filtered water ranged from 500 to 2000 litres, and samples in each sampling point were taken in two replicates.

The second device was based on a "Millipore" Tangential filtration and allowed to collect the suspended material using cartridge of filters with 0.1 μm pore size. Dissolved radiocaesium was also concentrated on ion-exchange columns described above. Volume of filtered water ranged from 500 to 800 litres.

All filters and resins were subjected for gamma spectrometry using HPGe detectors. The time of sample counting 20 h provided a standard deviation <10%.

An automatically operated logger was continuously used to record changes of pH and turbidity (suspended material content) of water during the sampling time 8 hrs. Some components (e.g. Ca, Mg, Na, K and NH_4) were measured in the laboratory.

One sediment core was taken using 1 meter long mini-Mackereth pneumatic corer in the upper part of Kiev reservoir at the Pripyat mouth (Strakholesye village, 56 km upstream the Kiev HPS dam) and two other ones - by mechanical ADT (30 cm long) corer near Sukholutchye village (37 km upstream the Kiev HPS dam) and in old river channel (18 km upstream the Kiev HPS dam). The location of sampling points is fitted to the old river channel and characterised by very high sedimentation rates (about 4, 3 and 2.3 cm per annum, respectively). The cores were sliced in intervals of 1 or 2 cm, dried and γ-counted in a standard geometry.

3. Results and Discussion

It is noted worthy that hydrochemical conditions along the sampling route turned out to be fairly uniform. Typical values for Ca^{+2} were reported as 50 mg/l, for Mg^{+2} - 10 mg/l, for Na^+ - 15 mg/l and for K^+ - 3 mg/l; pH values ranged from 8 to 8.5 except for the Ilya river (sampling point 11) where it was registered as low as 6.9. Nevertheless the difference in hydrological conditions in rivers and reservoirs stipulated drastic variations in the results related to suspended radiocaesium.

Table 1 shows the ^{137}Cs concentrations in water and ^{137}Cs distribution coefficient (K_D) calculated "*in situ*" at each sampling site for the different grain sizes of suspended material. At all sampling sites ^{137}Cs seems to be more strongly associated with the largest size fractions of suspended matter (40 and 10 μm). These results appear surprising and could be attributable to the high bloom of algae observed during the sampling periods (June-July). To explain the higher values of ^{137}Cs found in the largest size fractions of suspended matter (40 and 10 μm) the following hypothesis has been formulated : algae surfaces could present natural traps for the finest suspended particles. In this way the material collected in the filters of 40 and 10 μm pore sizes could be affected by the presence of mineral fractions and this could justify the higher values of caesium K_D found. To clarify the role of phytoplankton the uptake of radionuclides and role of suspended material adhesion onto surfaces of algae and aquatic plants, a more detailed experiment has been planned.

Table 1. ^{137}Cs Dissolved in Water and K_D *in situ* Values (1993) in the Different Fractions of Suspended Material

Sampling Point	#	Susp. Part. (mg/l)	^{137}Cs in water (mBq/l)	"In situ" K_D *10^{-3} (l/kg)			
				Total	40 μm	10 μm	0.45 μm
Illya River (2.5 km upstr. Uzh)	11	7	420±10	8±1	23±1	9±1	3.2±0.2
Uzh River (15 km upstr. Pripyat)	10	10	73±2	50±6	70±8	38±6	20±1
Pripyat River (Chernobyl)	9	22	70±10	69±33	74±36	44±13	17±5
Dnieper River (10 km upstr.Kiev res.)	7	18	16±1	108±2	106±5	181±59	22±5
Kiev Reservoir (Pripyat&Dnieper confl.)	6	8	90±10	79±4	90±8	73±16	28±1
Kiev Reservoir (45 km upstr. Kiev dam)	8	4	106	106	124	100	47
Kiev Reservoir (25 km upstr. Kiev dam)	4	3	94±2	113±1	143±7	103±1	50±6
Kiev Reservoir (3.5 km upstr. Kiev dam)	2	3	74±1	87±3	109±1	69±15	64±14
Dnieper River (Kiev City)	3	9	50±3	69±2	77±4	64±3	39±1
Kanev Reservoir (68 km dnstr. Kiev dam)	5	6	45±1	79±5	91±5	79±4	53±3

Since the 40 and 10 μm filters may have been affected by the presence of mineral fractions, only the data relating to the fraction retained on the 0.45 μm filters are discussed.

It can be clearly seen from the data that when downstream the Pripyat river and Kiev reservoir the water velocity in the main channel slows down, suspended matter content decreases and radiocesium K_D *in situ* increases to the highest values in the lower part of the reservoir (p.2). Further down the Kiev HPS dam in the Dnieper river channel (p.3) the K_D value substantially decreases and then increases again in the lower part of Kanev reservoir (p.5).

The most appropriate explanation of this phenomenon seems to be a drastic change in the mineral composition within the finest fraction along the sampling route. It appears that heavier sandy inclusions with low sorption capacity are mostly deposited near the Pripyat river mouth in upper part of Kiev reservoir and further according to formation of sedimentation "cone". One the contrary finer clay fractions travel down the main channel and ensure higher K_D values in lower parts of the reservoir.

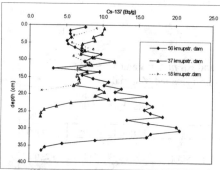 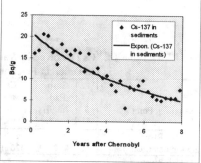

Fig.1. ^{137}C profiles in bottom sediment at different parts of Kiev Reservoir

Fig.2. Exponential time trend for dated core from upper part of Kiev Reservoir

This hypothesis is also confirmed by the results of measurements of bottom sediment cores presented in Fig. 1. It shows much higher sedimentation rate near the Pripyat river mouth and substantial decrease of radiocesium concentration in sediment in recent years (probably because of river channel flushing). When plotted versus time of deposition (Fig.2) "post-Chernobyl" part of this profile shows the trend $exp(0.18t)$ with $R^2=0.76$, where t is time measured in years after the accident. On the other hand similar regression for total ^{137}Cs concentration in water in the Pripyat river near Chernobyl for the same period (2500 records corrected to 1986) gives $exp(-0.21t)$ with $R^2=0.81$ [2]. So, it seems that in the upper part of the reservoir near the Pripyat river mouth, the concentration of particulate radiocesium is determined by the concentration in river water. On the contrary looking at the profiles from the middle and upper parts of the reservoir one can conclude that here particulate concentration is more or less stable and defined by an irreversible contamination of fine clay material eroded from the Pripyat catchment [3].

In conclusion, a very important practical implication from the analysis of the presented data could be formulated. Modelling of radiocesium transport in the river-reservoir system with suspended matter should be based on detailed knowledge about changes in mineral content of each size fraction and mechanisms of their contamination (i.e. selective and non-selective sorption, fixation, etc.) and consequent K_d variation along the water course. In this regard a standard approach which implies use of constant K_d values for some different size fractions [4] seems to be very simplified and somewhat misleading.

References

[1]   O. Voitsekhovitch, V. Kanivets, G. Laptev, I. Biliy, Hydrological Processes and their Influence on Radionuclide Behavior and Transport by Surface Water Pathways as Applied to Water Protection after Chernobyl Accident, In: Proc. UNESCO, Hydrological Impact of NPP, 1992, Paris (1993) 85-105

[2]   I. Biliy (Unpublished data 1995)

[3]   A.V. Konoplev et al., Behavior of Long Lived Chernobyl Radionuclides in a Soil-Water-System, Analyst **117** (1992) 1041-1047

[4]   M.J. Zheleznyak et al., Mathematical Modelling of Radionuclides Dispersion in the Pripyat-Dnieper Aquatic System after the Chernobyl Accident, Science of the Total Environment, **112** (1992) 89-114

Interstitial soil solutions -- medium and factor of active redistribution of radionuclides in the system "soil substratum - groundwater"

Anatoly KUDELSKY[1] , Vasily PASHKEVICH[1] , Alexander PETROVICH[1]
Evgeny PETRYAYEV[2] , Galina SOKOLIK[2], Svetlana OVSYANNIKOVA[2]
[1] *Institute of Geological Sciences of the Academy of Sciences of Belarus. Zhodinskaya str.,*
7, Minsk 220141, Belarus
[2] *Belorussian State University. Leningradskaya str., 14, Minsk 220080, Belarus*

Abstract. Pore solutions of bog soils have been studied in high bog ecosystems situated in the zones contaminated by Chernobyl-born radionuclides (Belarus, UK, Russia). High ^{137}Cs concentrations in soil solutions (up to 48 Bq/l), low Kd (average 418±50), as well as close correlations between ^{137}Cs and K^+ from interstitial water of highly aerated vegetation-bog soils have been established. Radiation effects of the mobile ^{137}Cs migration in pore solutions to the groundwater surface have been revealed.

1. Introduction

Concentration of radioactive caesium in bog water (pore solutions, bog water itself, drainage water, etc.) is generally well above its concentration in other natural water types (river, lake, groundwater, etc.). In this context it is interesting to analyze data on the ^{137}Cs behaviour in pore solutions of bog soils obtained by the authors during research carried out within the ECP-3 project for the territory of Belarus.

2. Objects of investigations

Investigations carried out in the Republic of Belarus under a programme of the ECP-3 project have concerned problems of ^{137}Cs distribution and migration in high bog systems, as well as radiocaesium transfer from bog to river systems.

High bog systems located within the Devoke (UK) lake catchment area, in the Pripyat river basin of the Belarus territory (Peat bog Opromokh), as well as bogged territories found within the catchment basins of the Svyatoye and Kozhanovskoye lakes (Russia, Bryansk region) have been studied thoroughly and extensively. The tasks of the project have been fulfilled successfully due to a comprehensive study of the problem of ^{137}Cs distribution in pore solutions.

Interstitial water (or solution) is meant water infilling pore space of soil and fine capillaries and retained in soils by both the closed cavity system itself, and the intermolecular coalescence force. Like the firmly bound types (crystallization, zeolitic, film water, etc.), interstitial water is geneticly linked to soil and experiences changes together with soil during its evolution. However, unlike the first ones the interstitial water is highly involved in geochemical processes that take place in the system: mineral soil medium $\Leftrightarrow$ water $\Leftrightarrow$

organic matter ⟺ gas, as on the one hand, it is a good solvent, and on the other hand, constitutes a medium, where chemical elements and compounds are initially accumulated and redistributed. Interstitial water interact with the mineral and organic soil fractions for the longer time, than groundwaters do, and so represents a peculiar buffering and binding member in the system: subsurface water ⟺ mineral-organic soil complex ⟺ plants.

Considering the interrelations existed between pore solutions, soil solid complex and free groundwater, it seems possible to suggest a preliminary three-membered classification of pore solutions: I - water of open pores flowing or squeezed out by manual squeezing (at 0.8 MPa); II - water of rather closed large pores partially, or completely squeezed by manual squeezing at pressure ranging from 0.8 to 9.8 MPa; III - water of fine isolated pores squeezed at pressure about 9.8 MPa and higher; IV - free water.

Hence, relationships between pore water, solids and free water is as follows:

$$S(soil) \leftrightarrow \underset{\underset{s}{\updownarrow}}{\overset{\overset{s}{\updownarrow}}{III}} \leftrightarrow \underset{\underset{s}{\updownarrow}}{\overset{\overset{s}{\updownarrow}}{II}} \leftrightarrow \underset{\underset{s}{\updownarrow}}{\overset{\overset{s}{\updownarrow}}{I}} \leftrightarrow IV$$

3. Materials and Methods

In September 1993 the flooded peat bogs in the catchment of Dewoke Water (Cumbria, UK) were sampled in order to investigate the strong retention but high availability of radiocaesium in peat soils. Studies included measurement of in-situ K_D's and assessment of the contribution of fine dispersed suspended fraction in to "dissolved" radiocaesium. In July 1994 a further sampling exercise in Belarus, Ukraine and Russia was carried out. Peat cores were taken using 22 cm diameter stainless steel peat cores fabricated in the UK. The cores were sliced in the field at 5 or 6 cm intervals to a depth of 30-60 cm, the samples being returned to the laboratory in sealed plastic bags. Samples were also taken of surface standing water and from the outflow of the Opromokh bog at various dates during 1993 and 1994.

Pore waters from peat cores were removed by squeezing the samples using special plastic press moulds at a presure of 0.8 and 9.8 MPa. The remaining solids were removed by filtration through 3 μm then 0.2 μm cellulose acetate filters. A sub-sample of the total solids > 20 μm was taken and analysed for ^{137}Cs on a low-background Ge-Li coaxial γ-detector. Solids within the range 0.2 μm - 3.0 μm were counted separately. A sub-sample of the pore water was removed for measurement of the major cations (Ca^{2+}, Mg^{2+}, K^+, Na^+), ammonium. The pore water from each core was also analysed for anions (SO_4^{2-}, NO_3^-, Cl^-), pH, DOC and alkalinity. The pore water was then preconcentrated by evaporation and analysed for ^{137}Cs. For the purpose of intercomparison of results, the above experimental analyses were also carried out on a core from the Devoke Water site (#7).

4. Results

4.1. Pore solutions of the bog systems

Pore water content of bog soil amounts to as much as 0.22-0.75 of the total soil samples volume (average 0.44). ^{137}Cs concentration in pore water varies from 1.6 to 48.0 Bq/l, the highest nuclide concentrations being typical of the upper parts of hydrochemical profiles (Table 1). Relative to the total ^{137}Cs in soil substratum, the radionuclide content of pore water varies from 0.1 to 6.0%, averaging 2.20±0.37%. Data on the bogland Opromokh (Belarus) and peat bog Devoke Water (core #7) were taken into consideration, when the average value was calculated.

Table 1. Liquid phase content and ^{137}Cs distribution in vertical profile of bog soil

| Layer, depth, cm | Pressure, MPa | Porewater volume, ml | Porewater portion in total sample volume | Filtrate (d<0.2 μm) | | | Soil residue, Bq/kg |
				Bq/sample	% of ^{137}Cs content of layer	Bq/l	
Peat bog Opromokh, core #8. 19.03.1994							
0-5	0.8	553	0.37	24.2	2.4	43.7	5536.9
-"-	9.8	35		2.9	0.3	82.5	5520.6
5-10	0.8	412	0.28	6.0	0.7	14.4	6743.3
-"-	9.8	41		1.1	0.1	27.7	6694.2
10-15	0.8	484	0.32	1.8	0.3	3.7	2889.4
-"-	9.8	26		0.4	0.1	15.4	2887.7
15-20	0.8	476	0.32	1.4	0.4	2.9	670.4
-"-	9.8	36		0.4	0.1	10.5	669.7
Peat bog Opromokh, core #10. 29.05.1994							
0-5	0.8	901	0.57	23.34	1.6	25.90	11180
5-10	0.8	607	0.39	9.61	0.8	15.83	10690
10-15	0.8	650	0.41	3.62	0.6	5.57	4030
15-20	0.8	766	0.49	0.94	1.0	1.23	320
20-25	0.8	843	0.54	1.56	3.2	1.85	110

It was found that ^{137}Cs concentration in soil water squeezed from the organic and mineral substratum at a pressure of 9.8 MPa is two or three times that in water squeezed at 0.8 MPa. That is water layers closely associated with solid soil substances and bog vegetation typically show the higher isotope concentrations than those relatively remote from solid surfaces. In other words, solid material of soil and plant associations is a source of ^{137}Cs, which is desorbed and delivered to bog water by pore water.

Summing up major data obtained as a result of investigation of pore solution of bog soils it is necessary to note that:

a. Chernobyl ^{137}Cs fallout is redistributed between components of bog system: vegetation cover (A); soil pore solutions (B); bog plants impregnating water (C); open surface bog water (D); drainage water (E). According to ^{137}Cs content, these components may be arranged in the following order:

$$A > B > C > D > E$$

b. Radiocaesium activities in the peat bog are high both in the solids and in the aqueous phase. In general, the in-situ Kd-values varies in a system "bog soil -- pore solution" from 127 to 1037 l/kg (average 418 ± 50). The Kd maxima are noted in the layers of 5 to 10 and 10 to 15 cm with subsequent monotone decrease of Kd down the soil profiles. For mobile ^{137}Cs forms the Kd value varies from 8 to 364, averaging 17.79±2.16 % of the Kd value for total ^{137}Cs of bog soil within the Opromokh bog land.

c. The relationship between $\log_{10}($ aqueous ^{137}Cs) and $\log_{10}[K^+]$ in pore and drainage waters is well represented by the equation:

$$\log_{10}[^{137}Cs] = 0.83 \log_{10}[K^+] + 0.21$$

with correlation coefficient $R^2 = 0.83$. Some relationship was also observed between $\log_{10}$ (aqueous ^{137}Cs) and $\log_{10}[NH_4^+]$, however the correlation coefficient was much less significant ($R^2 = 0.45$). Correlation between a logarithm of the aqueous activity of ^{137}Cs and that of the sum of cations in pore water solution show a correlation coefficient $R^2 = 0.64$.

d. ^{137}Cs removal from bog to river systems has been assessed. This assessments suggest that the total removal of radiocaesium comprises 0.30 % of the catchment inventory per year.

4.2. Pore solution and groundwater

Pore solutions constitute a medium where mobile radionuclide forms are redistributed along the profile "contaminated soil - groundwater". Such a redistribution (migration) of radionuclides results in the radioactive contamination of groundwater, which is an important source of drinking water supply within the contaminated territories of Belarus.

The ^{137}Cs content of groundwater ranges from 0.04 to 0.47 Bq/l, and that of ^{90}Sr - from background values (below the instrument detection limit) to 2.1 Bq/l. There is an intimate relation between the radionuclides inventory of the soil aeration zone and their contents of groundwater.

5. Discussion and conclusions

Pore solutions that the significant part of inventory of radionuclides able to migrate is associated with are important for the radionuclides redistribution in natural objects and natural decontamination of contaminated lands.

High concentrations of ^{137}Cs in pore solutions are responsible for high intensity of radiocaesium migration in peat-bog system and for respectivly high water removal of the radionuclide outside the contaminated territory.

It is sufficient to mention that, according to our estimates, about $130.5 \cdot 10^6$ MBq of ^{137}Cs activity in a layer of 0-25 cm is associated with pore solutions of peat-bog soils in contaminated regions of Belarus (6700 km^2), and annual ^{137}Cs removal from bog ecosystems to rivers is as high as $168 \cdot 10^4$ MBq (1.29 % of the total inventory of dissolved mobile forms). Annual water migration of ^{137}Cs is 0.3-0.5 % of the total ^{137}Cs inventory of bog ecosystems, which is at least ten times ^{137}Cs removal from catchments composed of mineral soils (0.05 - 0.005 %) with high content of clay minerals. These are recent and rather surprising data, since the idea of low rates of ^{137}Cs remobilization in a section of highly organic bog soils was traditionally advanced.

The information obtained by the study of pore solutions is used as the basis for forecasting estimates of the radiochemical conditions of natural water. This is very important in the context of elaborated projects for social and economic rehabilitation of vast territories in southeastern Belarus contaminated as a result of the accident at the Chernobyl nuclear power plant.

The Removal of Radionuclides from Foodstuffs During Technological Treatment and Culinary Processing

L.V. PEREPELYATNIKOVA[1], N.M. LAZAREV[1], T.N. IVANOVA[1], F.A. FEDIN[1],
S. LONG[2], D. POLLARD[2]

[1]*Ukrainian Institute of Agricultural Radiology (UIAR),*
7, Mashinostroitelei street, 255205 CHABANY (Kiev region), Ukraine
[2]*Radiological Protection Institute of Ireland (RPII),*
3, Clonskeagh Square, Clonskeagh Road, Dublin 14, Ireland

Abstract. Comparative data on the efficiency of industrial and domestic processing techniques of plant and animal products, which reduce radionuclides content foodstuffs, are summarized. Methods, which provide the greatest decrease of radionuclides content in the final products with minimum losses of nutritional value, were identified.

Agricultural production in the Ukraine is conducted on the territories, where soils contamination does not exceed 555 kBq/m^2. However, in the western regions of Ukraine, the Rovno region in particular, where up to 40-50 % of the soil cover are peaty soils with high transfer of Cs^{137} to agricultural crops (10-100 times higher than on mineral soils), these limitations are up to 1-5 Ci/m^2 = 3.700 - 17.500 kBq/m^2. Besides, a lot of cattle in private sector (2-3 cows per family) are determined by the lack of good pastures, and forest and swampy pastures are used, where transfer factors are the highest (100-200 times higher). In these regions the greatest contribution of foodstuffs to internal irradiation dose is observed.

At low doses of external irradiation, reduction of internal dose by agricultural products processing (reduction of Cs specific activity in the products, consumed by population) can significantly improve the radiological situation in the region and reduce social tension.

Analysis of the contribution of separate human diet components to the total intake of radiocaesium by the human organism, demonstrates that 70-85 % of radiocaesium is transferred with animal products (milk, meat). Therefore it is necessary to apply measures for the reduction of radiocaesium intake with these products.

Radioprotective measures on all stages of food chains allow to reduce considerably the radionuclides content in the final products. In the experimental conditions one can achieve decontamination of foodstuffs by a factor ten, but in industrial conditions countermeasures efficiency is much lower to a factor 2-4..

For the reduction of radiocaesium content in foodstuffs, produced from milk and meat, there are enough simple and practicable methods, thus, with milk processing to fresh cheese the specific radioactivity decreases 1.3-1.4 times, at producing sour cream with 20, 30 and 40 % fat content - by 1.2, 1.4 and 1.6 times respectively, at cheese-making-up to 8 times, at processing to butter - 5 and more times.

Presently in the collective sector, due to the complex of radioprotective measures, one managed to reduce radiocaesium content in animal products below intervention levels. In the private sector the situation is different. In two villages of Dubrovitsa district, Rovno region, we analyzed per example the diets of inhabitants and radiocaesium intake with foodstuffs. The diet of village inhabitants consists mainly of foodstuffs, produced on private plots. Each inhabitant on his private plot is producing a limited range of foodstuffs, most of which is consumed by the family.

Specific radioactivity of cheese samples and samples of sour cream with 30 % fat content, produced in domestic conditions, was, on the average, about 1.4 times lower than the specific radioactivity in the milk. This reduction is in proportion to the reduction of the milk water phase in dairy products, where radiocaesium is concentrated. In order to reduce radiocaesium transfer with dairy products, produced and consumed by local population, we recommend to process milk to such foodstuffs, the technology of which provides a significant reduction of Cs^{137} transfer from milk to ready product.

The specific activity of Cs^{137} in cheeses prepared by standard methods was found to be between 1.3 and 5 times lower than the specific activity of the milk from which they were prepared. Under all conditions the conversion of milk to cheese results in the cheese retaining less than 8 % of the total Cs^{137} activity of the original milk. The specific activity of Sr^{90} in cheeses whose coagulation pH was greater than 5.5, was found to be between 5 and 10 times greater than the specific activity of the milk from which they were prepared. However, the specific activity of Sr^{90} in quark, an acid cheese (coagulation pH approx. 2), was 1.5 times lower than the specific activity of the initial milk. This is consistent with the binding characteristics of the Sr^{90} protein complex. Acidification of the milk to pH of 5.1 (the isoelectric point of this protein) or less results in a liberation of most alkaline ions from the proteins into the aqueous phase. At coagulation pH values greater than 5.1 between 50 and 70 % of the total Sr^{90} activity of the original milk remains in the cheese, whereas at a coagulation pH of 4.39 this amount is reduced to 8 %.

Nutritional analysis of the cheeses showed that the nutritional content of cheeses produced by modified techniques did not differ significantly from those produced by standard techniques.

Within the project framework, the effect of technological processing and culinary preparation of meat for the reduction of radionuclides content was investigated.

Optimum conditions for the removal of radiocaesium from contaminated meat were determined, which include the effect of the following parameters: the size of meat pieces treated, the meat type, the treatment type, the brine-to-meat ratio, the NaCl concentration, the treatment temperature, the effect of the pH of the treatment solution.

The experiments demonstrated that for salting of meat an additional one hour treatment will result in approximately 60 % decontamination. These treatment conditions will achieve maximum decontamination while minimizing losses of nutrients. If greater decontamination is required, the treatment time may be increased or the meat size decreased. Such modifications will, however, result in increased losses of water soluble nutrients.

Among different domestic meat cooking methods boiling, stewing and frying was investigated. The work was done with meat of bull, fed in 10 km ChNPP zone. Meat radioactivity was about 30000 Bq/kg.

The results obtained testify that boiling is the most effective method for radiocaesium removal from meat (Fr = 0.13 - 0.14). Besides, additional salting accelerates the process of radiocaesium transfer to broth. Less than a half of the initial activity was found in stewed meat (Fr = 0.4). However, extracted juice at this method of culinary preparation is usually consumed with meat, while in the first case it is not used. At frying Fr value was 0.77-0.80, as both meat and fat are used for food. Deep freezing of meat does not influence radiocaesium migration in the course of meat culinary preparation.

Together with milk human diets include plant products - potatoes, vegetables, which are not themselves dangerous from the point of view of Cs accumulation, but taking into account that in the diet of rural inhabitants potatoes consumption is 200-300 kg per year, and that of cabbage 50-100 kg, it is necessary to consider the contribution of these products to the dose. Due to the peculiarities of living and eating habits in many contaminated districts, the people are picking up and eating traditional forest products - mushrooms and berries, which accumulate a lot of caesium, and, therefore, should be taken into account as well.

Complex agrochemical decontamination measures for the reduction of radionuclides transfer to animal products is not always practicable and effective enough..

A considerable reduction of radionuclides transfer to foodstuffs is provided by technological processing and culinary preparation. It is reasonable to start the processing of raw foodstuffs from mechanical cleaning of their surface from soil.

The effects of various culinary and preservation techniques, such as mechanical processing, cooking, canning, marinating and brining, on the radiocaesium content of these food types are evaluated. Those techniques resulting in the greatest reduction of the radiocaesium content of the final product were identified.

At salting of vegetables and mushrooms, the amount of radiocaesium consumed with these products will be 1.5 - 2 times less than in the initial product, if the brine is not going to be used for food (sauerkraut, salted cucumbers, mushrooms).

The most effective method of culinary plant products preparation is thermal processing, which allows to reduce radiocaesium content (from 2 to 10 times) (boiling of potatoes, mushrooms).

An essential reduction can be obtained also by soaking dry and fresh mushrooms and berries in water (1.5 - 3 times) for 30 - 35 hours.

Consequently, the above-mentioned technologies allow to considerably reduce the caesium content in foodstuffs consumed by people, living on contaminated territories, thus decreasing internal irradiation dose.

Mechanisms Controlling Radionuclide Mobility in Forest soils

B. DELVAUX[1], G.I. AGAPKINA[2], K. BUNZL[3], B. RAFFERTY[4], A. KLIASHTORIN[2],
N. KRUYTS[1] & E. MAES[1]

[1]*UCL-Sciences du Sol, Place Croix du Sud, 2/10, 1348 Louvain-la-Neuve, Belgium*
[2]*Moscow State University, Leninskie gory, 119899 Moscow, Russian Federation*
[3]*GSF-Forschungszentrum für Umwelt und Gesundheit, Institut für Strahlenschutz,
85764 Neuherberg, Germany*
[4]*RPII, Clonskeagh Road, Dublin 14, Ireland*

1. Introduction

After a radioactive pollution of forest ecosystems, hot particles and aerosols are first intercepted by the tree canopy. Rainfall and throughfall transport the contaminants onto the soil surface. Runoff being limited, forest soils act as sinks for the radiopollutants. The fate of radionuclides in soils therefore influences the radioecological hazards for the entire food web through biological uptake. It depends upon the dissolution of the particles and the mobility of the radioactive soluble forms.

Soil processes strongly influence the radionuclide mobility in soils. The underlying mechanisms involve both abiotic and biotic factors. Three major processes are illustrated herebelow: sorption by soils colloids, complexation and/or association with organic matter, biological uptake and transport. A large attention is paid to radiocaesium.

2. Materials

The above-mentioned processes are characterized in some soils belonging to the soil collection illustrated hereafter.

Soil type/horizon		Soil net retention (%)[1]	% organic matter	% clay	% K[2] (total)	pH (water)	CEC[3]
I *Eutric*	Ah	82	5.5	46	2.05	6.8	20
cambisol	Bw	96	1.6	55	2.31	8.1	7
IIa *Dystric*	OAh	62	30.2	--	1.45	3.7	38
cambisol	Ah	81	24.3	26	1.68	3.7	32
	AB	64	10.7	22	1.99	3.9	19
IIb	Ah	57	23.5	8.5	0.96	3.9	32
Podzoluvisol	E	86	6.3	13	1.00	4.0	14
III *Podzol*	Ah	44	29.9	8	0.75	4.1	55
	E	50	1.6	3	--	4.9	3
IV *Histosol*	H	3	98	0	0.09	3.0	126

[1]Percentage of the initial loading. [2]Total K content is directly related to the content in micaceous clay minerals.
[3]Cation Exchange Capacity pH7 (cmol(+)/kg).

The figure illustrates a broad morphological schematic view of some major forest soil types observed under northern latitudes 50-60° N. the most distinctive property is the thickness of both the organic (O and H) and hemiorganic (Ah) horizons. Histosols (IV) are characterized by a large accumulation of organic matter (peats). An intense biological activity leads to the rapid decomposition of litter in Eutric Cambisols, Chernozems and Luvisols (I). Thicker O horizons are related to a poorer biological activity in acid soils such as Dystric cambisols, Podzoluvisols (II) and Podzols (III).

3.   Results and discussions

Radiocaesium net retention in soils
Soil components adsorb the soluble forms of radionuclides. Under reversible conditions, desorption occurs: radionuclides can be leached or uptaken by biota. If adsorption is irreversible, radionuclides are fixed and become hardly available. Adsorption/desorption phenomena thus strongly influence the fate of radionuclides in soil environments. Sorption-desorption data for radiocaesium were obtained for five soil materials representative of the collection illustrated above (I→IV), using a recent methodology [1].

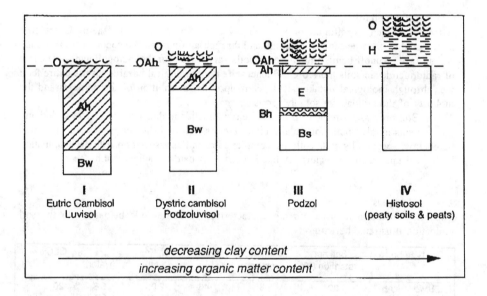

The selected soils differ in their content in clay, organic matter and micaceous clay minerals. The net retention of radiocaesium strongly discriminates the tested materials: in the Ah and H horizons, it largely decreases from 82 to 3 % (I>IIa>IIb>III>IV), with decreasing clay and total K content and increasing organic matter content.
Radiocaesium net retention is, from a *physico-chemical point of view*, under the full dependance of micaceous clay minerals and ionic conditions prevailing at the soil-solution interface [2,3].

In acid forest soils, major cations such as NH_4^+ and H^+ strongly influence radiocaesium desorption through respectively the trapping effect (interlayer collapse) and K depletion. Despite of its high CEC, organic matter does not directly contribute to irreversibly sorb the radiopollutant, but it may influence the reactivity of mineral surfaces for radiocaesium fixation through Al complexation. In the soil IIa, the net retention is higher in Ah than that in AB: aluminic cations block interlayer specific sites for Cs fixation in the AB horizon, while they are complexed by organic acids in the Ah. Organic acids may also contribute to K depletion, hence thereby Cs desorption.

Contamination hazards to above standing vegetation and other living organisms are therefore expected to be high on peaty soils and forest acid soils with thick organic layers. Greenhouse experiments using *Lolium multiflorum L.* indeed show that radiocaesium bioavailability largely increases with increasing organic matter content and decreasing content in clay and in micaceous clay minerals (IV>>III>IIbIIa>>I).

Association of radionuclides with mobile organic compounds
In some acid forest soils (II and III), the slow decomposition of organic materials accumulating in the upper layers produces soluble organic acids. These acids bear deprotonated groups and thus behave as anions in the soil solution. These anions can be associated with cationic radionuclides through weak electrostatic attraction (Cs), ionic bounding (Sr), or complexation (Ru, Ce, Am, Pu). Such associations have been characterized in the soil solution of the Of, Oh and (Oh + Ah/E) horizons of a heavily contaminated soddy podzolic soil (podzoluvisol: II), situated 6 km west of the nuclear reactor near Chernobyl (Novoshepelitchi, Ukraine).

Gel filtration isolates five "organic" fractions with decreasing molecular weight. $^{239+240}Pu$, ^{238}Pu and ^{241}Am are mostly associated with the highest molecular weight fractions. ^{90}Sr is concentrated in the lowest one (inorganic). ^{137}Cs is mostly associated with the intermediate fractions in the Of horizon and is more uniformly distributed with all the fractions in the deeper layers [4].

The different distributions of plutonium and americium, but particularly of strontium and caesium between the five fractions obtained by gel filtration of the soil solution from the three horizons clearly show that:
- the nature of soil organic matter present in the three horizons is different;
- these differences largely affect the association between organics and radionuclides; radiocaesium is most sensitive while plutonium is complexed very strongly.

In the soil solution of acid forest podzolic soils, organic compounds of various molecular weights are thus able to associate with cationic radionuclides. Such associations should undoubtedly have effects on their mobility and biological availability. In addition, the extent of this association is different for soil solutions sampled from distinct horizons: similar Kd values observed for a given radionuclide in various horizons do not necessarily imply a similar radioecological behaviour of this radiopollutant.

Radiocaesium biological uptake and upwards transport
Nutrient cycling is essential in forests ecosystems. An important process is the decomposition of contaminated litter material: radionuclides incorporated in dead plant remains are mobilized by other organism living in the surface organic horizons.

The decomposition of contaminated litter material in sandy podzolic soils (type II in the figure) had been studied *in situ* in two sites: in the Ukraine at Kopachi, a 60 years old stand of *Pinus sylvestris* situated 6 km south east of the Chernobyl NPP, in Waterford, Ireland, a 35 year old stand of *Pinus contorta*.

Litter material from Ol and Of organic horizons was used to fill nylon mesh litter bags that were further installed in their respective soil layers. Changes in weigh and radiocaesium content of the bags were recorded during one year.

The soil profile analysis reveals that the radiocaesium pool of the Of horizon is 35 (Ireland) and 70 (Ukraine) times greater than that of the Ol. However, large increases in radiocaesium content are observed in the litter bags introduced in the Ol.

The net importation of the radiopollutant in the litter Ol is not due to infiltration by canopy washings and litter leachate. A proposed explanation is that the decomposer fungi, which invade the Ol litter from the more contaminated Of, introduce radiocaesium to the litter: some fungi are indeed know as radiocaesium accumulators. Similar observations have been carried out under laboratory conditions [5].

By this mechanism radiocaesium is constantly transported upwards to the fresh litter layers: transport may partially explain the low downard migration of radiocaesium in the organic horizons of the forest soils.

Conclusions

The mobility of radionuclides in forest soils is governed by several processes involving both abiotic and biotic factors.

The sorption-desorption process chiefly governs the activity of radionuclides in the soil solution, hence thereby their mobility and biological availability. Radiocaesium exhibits a very low mobility in mineral soils. Both mobility and bioavailability however increase as the thickness of organic layers and their content in organic matter increases. Clay minerals of micaceous origin strongly act as slinks for radiocaesium in forest soils. The magnitude of Cs mineral fixation in topsoils is expected to be the highest in mineral soils of type I, and, to a lesser extent, of type II.

A low mobility of radiocaesium in the surface horizons of forest soils may also be partially explained by a biological mobilization: fungi absorb radiocaesium and transport it to upper layers, thereby contributing to constantly recycle the radioelement in the organic horizons. This mechanism is probably important in soils with thick organic layers (III, IV, and, to a lesser extent, II).

Radionuclides can be associated with soluble organic anions in the soil solution of forest acid soils. Such associations are highly mobile: they are stable in conditions of poor biological activity (low temperatures, acid soil infertility, water excess, etc.). Their magnitude is expected to be the highest in thick acid organic layers (soils of type III and IV).

References

[1]   J. Wauters, L. Sweeck, E. Valcke, A. Elsen and A. Cremers (1992) Availability of radiocaesium in soils: a new methodology. SPRI-CEC Meeting *The dynamic behaviour of radionuclides in forests*, May 18-22 1992, Stockholm, Sweden.

[2]   D.D. Eberl (1980) Alkali cation selectivity and fixation by clay minerals. Clays and Clay Minerals 28(3), 161-172.

[3]   Maes and Cremers A. (1986).

[4]   G.I. Agapkina, F.A. Tikhomirov, A.I. Shcheglov, W. Kracke and K. Bunzl (1995) Association of Chernobyl-derived [239+240] Pu, [241]Am, ^{90}Sr and ^{137}Cs with organic matter in the soil solution. *Journal of Environment Radioactivity* (In Press).

[5]   J.S. Olsen and D.A. Crossley, Jr. (1966) Tracer studies of the breakdown of forest litter. In: *Radioecology*, V. Schultz & A.W. Klement (eds.). Chapman and Hall Ltd., London, pp 411-416.

Mathematical modelling of radionuclide migration in components of meadow ecosystems

S. FESENKO [1], K. BUNZL [2], M. BELLI [3], Yu. IVANOV [4], S. SPIRIDONOV [1], H. VELASCO [5] and S. LEVCHUK [4]

[1] RIARAE, Obninsk, Russia, [2] GSF, Neuherberg, Germany,
[3] ANPA-DISP, Rome, Italy, [4] UIAR, Kiev, Ukraine
[5] Universidad Nacional de San Luis Conicet, Argentina

Abstract. The models of radionuclide behaviour in meadow ecosystems developed in the framework of the ECP 5 project to predict the long-term impact of nuclear accidents are described. The parameters of the models as well as significance of various processes and factors governing radionuclide mobility in different types of meadows are discussed. Evaluations for the ecological half-lives of radionuclides in meadow plants and residence half-times for different soil layers are presented.

1. Introduction

The results of investigations carried out after the accident on the Chernobyl NPP have demonstrated that semi-natural meadow ecosystems are one of the most important contributors to both external and internal irradiation of population living on radioactive contaminated territories [1,2]. That shows a need for the development of mathematical models to predict the long term-behaviour of radionuclide in meadow ecosystems. Two groups of models have been developed to predict behaviour of radionuclides on the base of the experimental data achieved within the frame of the ECP 5. The first group includes models of vertical migration. The development of these models was aimed at predicting the vertical distribution of radionuclides for the calculation of external irradiation doses and evaluation of the role of this process for decreasing the radionuclide transfer in food chains. The second one includes a model describing the dynamics of radionuclide behaviour in soil-plant system to predict the contribution of semi-natural meadow ecosystem to the long-term internal dose to the population. The purpose of this paper is to present the modelling exercises in the frame of the Project and to evaluate the significance of the processes governing radionuclide bioavailability in meadow ecosystems after a single accidental release of radioactive substances into the environment.

2. Modelling of ^{137}Cs availability in 'soil-plant' system

The data obtained from the network of experimental sites within the 50 km zone of the ChNPP have been used for the development of models describing quantitative regularities of ^{137}Cs behaviour in soil-plant system. The presence of particles, containing ^{137}Cs in the soils results in the development of simultaneous opposite processes, including both an increase of the plant 'available' amount, due to destruction of fuel particles and a decrease of its 'mobility' attributable to sorption of ^{137}Cs by soil [3].

Then the dynamics of ^{137}Cs root uptake is influenced by the characteristics of fallout and by ^{137}Cs sorption processes that are driven by soil properties and meadow type [1-3]. Therefore, taking into account the conclusions from the analysis of experimental data obtained within the project [3] and applicable data from the literature [1], the following processes were considered in the model: ^{137}Cs transfer in easily soluble aerosol forms and coarse dispersed particles and fuel ones to mat and soil; destruction of fuel particles in soil and mat followed by change of ^{137}Cs state to forms available for plant uptake; decomposition of mat and ^{137}Cs leaching from it into root-containing soil layer; ^{137}Cs redistribution between different soil fractions; plant uptake of ^{137}Cs from soil and mat; ^{137}Cs removal from root zone of soil profile. A full description of the model (including adaptation and validation exercises) is given elsewhere [5]. One of the tasks of the development of models of radionuclide migration was an explanation of the regularities of ^{137}Cs behaviour in meadow ecosystems observed experimentally. A series of calculations was carried out on the basis of the model of radionuclide migration in 'soil-plant' system presented earlier to clarify the dynamics of changing ^{137}Cs content in meadow vegetation and to evaluate the role and significance of factors controlling the variety of biological availability of radionuclides in different time periods after the fallout. A detailed analysis of these results has been published [4], and only a few selected examples can be given here.

2.1. Fallout properties and bioavailability ^{137}Cs in soil-plant system

The results of investigations carried out in zones of the Chernobyl NPP with different compositions of fallout [2,3] allowed to conclude that the fallout type has a great deal to do with the dynamics of radionuclide availability in soil-plant system. Results of the calculations presented in Fig. 1 confirm this conclusion quantitatively and show that the effect of fuel particles on radionuclide plant uptake is of complicated character. So, in 1986 ^{137}Cs plant uptake decreases with the increase of the fuel component in fallout because during this period a high fraction of fallout is bound in fuel particles and not available for plant uptake. In the following, as a results of destruction of fuel particles and ^{137}Cs leaching therefrom the character of this dependence varies and in the 3rd year after the fallout the radionuclide plant uptake depends in inverse proposrtion on the share of the fuel component, i.e. an increase of ^{137}Cs bioavailability in the zones with high content of fuel particles in soil. In 5-7 years after the deposition the ^{137}Cs availability in the zones with different fallout characteristics levels off due to the decreasing of radionuclide amount in the form of fuel particles and radiocaesium sorption in soil. It has been also shown that the influence of fallout properties on the dynamics of ^{137}Cs TF's to meadow vegetation depends also on soil characteristics.

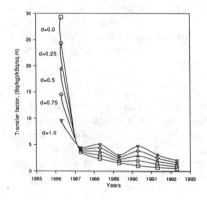

Fig. 1. Dynamics of ^{137}Cs TFs to wet meadow vegetation for different shares of fuel component.

2.2. Meadow soil and mat characteristics

Fuel particles were present (while in varying quantity) in all soils of the experimental sites of ECP5. Therefore, to assess the effect of meadow soil properties on the dynamics of ^{137}Cs bioavailability the decrease in soil-plant system for the case of fallout in the form of easily soluble aerosols was calculated. These calculations were performed for hydromorphous and automorphous soils with contrasting characteristics. The results indicate that in the first year after the fallout the differences in the dynamics of decreasing ^{137}Cs content in plants are not significant. Starting in the 4th year after the fallout the decrease of radiocaesium in the vegetation on hydromorphous soils occurred more rapidly and beginning with the 2nd- 4th years after the deposition the effect of soil properties on the change of bioavailability in meadow ecosystems becomes important and the decrease rate of ^{137}Cs content in plants dependent on soil characteristics can significantly (up to 5-7 times) differ.

3. Modelling of vertical migration and evaluations of the radionuclide residence half-times in meadow soils.

3.1. Residence half-times of in different soil layers (dependence on time and type of fallout)

To assess significance of the factor influencing the residence half-times in different soil layers the compartment model adapted by GSF from the work of Boone [5] was used. The advantage of this model is that no information on actual radionuclide transport processes in soil (e.g. sorption properties, speciacion, water infiltration, etc.) is needed. Moreover, in this model the variety of soil properties can be taken into account, which effects vertical migration of radionuclide. The values of the parameters for this model were estimated for two sites located nearby the ChNPP (fuel type of fallout) and in Bavaria (Germany) (gas-aerosol type of fallout). The results have shown that the character of changing of the residence half-times as a function of time is considerably variable for the sites with different type of the fallout. For the site in Bavaria, these residence half-times increase significantly with time for all soil layers. Most likely, it is due to the ^{137}Cs sorption by soil. At the near-field site of the ChNPP the time dependence of the residence half-times in the top soil layer (0-1 cm) decreases significantly with time. It can be explained by the fact that the physico-chemical form of the deposit was quite different in these areas. In the nondistributed soil of the ChNPP site a considerable fraction of ^{137}Cs fallout is incorporated into the fuel matrix, these particles remain in the 0-1 cm layer for a long time, until they are dissolved by weathering processes. As a result, the mobile fraction of ^{137}Cs can there increase and the residence time of total radioceasium in this layer will be decreasing. These results are quite well agreed with those of the modelling bioavailability of ^{137}Cs in soil-plant system presented in this paper and indicate necessity to use for realistic evaluation of residence time the models which take into account types of fallout and processes governing the decrease of radionuclide mobility in soil.

3.2. Dynamic of the decrease of ^{137}Cs content in various types of meadow soil

Results presented earlier show that simple compartment and diffusion models could be not used for realistic long-term prediction of radionuclide transfer in soil as they do not take into consideration the change of the mobility of radionuclides with time. Therefore, two models were developed in frame of the Project to describe the variety of mobile ^{137}Cs fraction in soil as a result of sorption of radionuclides by soil. The first (Rabes) model is based on analytical representation of radionuclide distribution in undisturbed soil profile. Radionuclide sorption in this model described as a exponential function of the time. [6]. The second model considers the soil profile as a sum total of layers, between which the radionuclide exchange takes place. Each separate layer, from the point of view of ^{137}Cs content, is a sum of compartments, each being characterised by a definite potential sorption capacity for radionuclides. Since for evaluation of parameters of this model in parallel with the distribution of radionuclides in soil the data on radionuclide content in exchangeable and mobile forms for several years are needed these calculation were made for meadow ecosystems on which a comprehensive information from experimental studies in frame of ECP5 was available [2]. The result achieved show that the half-times of radionuclides in the root zone of soil (0-10 cm) can vary from 17 - 60 years for peatlands up to 2000 years for dry meadow on loamy and heavy loam mineral soils as depend on type of meadow and granulometric composition of soils (Table 1). The ecological half-times decrease in the different type of meadows in the following order:

Dry meadow (loamy sand, heavy loam soils) > Dry meadow (Sandy soils)> Lowland (wet) meadow (Light loam)> Flood plain meadow (Loamy sand) >Lowland (wet) meadow (Peaty)> Flood plain meadow (Peaty)> Low peatland (Peaty)>Transient peatland (Peaty)

Table 1. Half-lives of radionuclides in various types of meadow soil

| Type of meadow | Granulometric composition | T_{ec}, year | | | T_{eff}, year |
| | | mode | 95 % confidence interval | | |
			lower bound	upper bound	
Dry meadow	Sand	1870	1380	2515	29,5
Dry meadow	Loamy sand	2030	1690	2400	29,6
Dry meadow	Heavy loam	1970	1260	3070	29,5
Lowland (wet) meadow	Light loam	530	403	692	28,4
Lowland (wet) meadow	Peaty	280	200	387	27,1
Flood plain meadow	Loamy sand	440	220	890	28,1
Flood plain meadow	Peaty	60	49	73	20,0
Transient peatland	Peaty	17,4	14,2	21,3	11,0
Low peatland	Peaty	26	19	35	13,9

This indicates that contribution of vertical migration to the decrease of ^{137}Cs amount in the root zone on mineral soils is a negligible. On contrary, on wet meadow and on peatland it can be an important factor which influences the decrease of the transfer of ^{137}Cs into foodchains.

4. Conclusion

The results obtained show that fallout properties, type of meadow and soil characteristics are main factors which affect significantly the ^{137}Cs transfer in meadow ecosystem. The main conclusion to be drawn, based on an analysis of the results presented in this paper, is that the radionuclide in component of meadow ecosystems is governed by numerous factors. The role of these factors and their contribution to a decrease of ^{137}Cs bioavailability in soil-plant systems depend on the time elapsed after the deposition and are considerably variable.

References

[1.] Agricultural Radioecology. Alexakhin, R.M. & Korneyev, N.A. (eds.), Ecology, Moscow, 1992 (In Russian).
[2.] The Behaviour of Radionuclide in Natural and Semi-Natural Environments. Final report for the period 1993-1994. EC-Co-ordinated ECP5-Project. COSU-CT93-0043, ANPA, Rome, 1994
[3.] Sanzharova, N.I., Fesenko, S.V., Alexakhin, R.M., Anisimov, V.S., Kuznetsov, V.K., & Chernyayeva, L.G.. Changes in the forms of ^{137}Cs and its availability in plants as dependent on properties of fallout after the Chernobyl Nuclear Power Plant accident. The Science of the Total Environment. 154 (1994) 9-22
[4.] Fesenko, S.V., Spiridonov, S.I., Sanzharova, N.I. &. Alexakhin, R.M. Dynamics of ^{137}Cs Bioavailability in Soil-Plant System in Areas of the Chernobyl Nuclear Power Plant Accident Zone with Different Physico-Chemical Composition of Radioactive Fallout. The Journal of the Environmental Radioactivity. 1996 (In press)
[5.] Boone, K., Kantello, M.,V., Mayer, P.,G. & Palms, J.M. Residence half-times of ^{129}I in undisturbed surface soils based on measured concentration profiles. Health Physics. 48 (1985) 401-413.
[6.] Velasco, R.H., Belly, M., Sansone, U. and Menegon, S. Vertical Transport Of Radioceasium in Surface Soil: Model Implementation And Dose-Rate Computation. Health Physics. 64 (1993) 37-44

Long-Term Study on the Behaviour of Chernobyl Fallout Radionuclides in Soil

S. KROUGLOV, R. ALEXAKHIN[1] and N. ARKHIPOV[2]

[1]*Russian Institute of Agricultural Radiology and Agroecology*
Obninsk, Kaluga Region, Russia
[2]*Research and Industrial Association "Pripyat"*
Chernobyl, Ukraine

Abstract. The evolution with time chemical species of ^{90}Sr, ^{106}Ru, ^{134}Cs, ^{137}Cs, ^{144}Ce in soil, and the data on variation of ^{90}Sr, ^{134}Cs, ^{137}Cs transfer to grain and straw of four cereal crops has been used to estimate the rate of radionuclide release from fuel particles and caesium fixation in the soil. Field experiments were carried out in the 30km restricted zone around Chernobyl NPP.

1. Introduction

Different physical and chemical characteristics of the radioactive deposits after the Chernobyl accident affected the pattern distribution of radionuclides in the environment and their particular behaviour at local sites. Chernobyl fallout contained particles of dispersed nuclear fuel in addition to more homogeneously distributed radioactive material. These particles were found in the fallout at large distances from the site of the accident, but the major part of them was deposited on the nearest heavily contaminated zone [1, 2].

The fate of fuel fragments and their role in the ecosystems depended on the amount deposited, the rate of leaching and on the bioavailability of associated radionuclides in soil conditions. It is possible to assume that release from fuel matrix into the soil solution and following interaction with other soil constituents would change the physico-chemical form of the radionuclides deposited and thus would affect their mobility and availability for root uptake. However, the influence of the environmental parameters on fuel particles and on the rate of radionuclide release from them is not fully investigated. The Chernobyl accident occurred about ten years ago. The follow-up of the radioactive contamination of some specific areas could give us some clues on the long-term behaviour of the radionuclides deposited.

The purpose of this investigation were :

- to evaluate the chemical species of selected radionuclides held in soil contaminated after the deposition of fuel fragments and to study the evolution of physico-chemical forms of radionuclides in the course of time;
- to present data on the long-term variation of ^{90}Sr, ^{134}Cs and ^{137}Cs transfer to grain crops that were grown under field conditions on soil contaminated with particles of irradiated nuclear fuel;
- to estimate the life-time of fuel particles and the rate of caesium fixation in soils under natural conditions using the results of field experiments.

2. Materials and Methods

Studies were carried out in 1987-1994 at the experimental sites located inside the 30km restricted zone around Chernobyl Nuclear Power Plant (ChNPP). The speciation of ^{90}Sr, ^{106}Ru, ^{134}Cs, ^{137}Cs and ^{144}Ce in soils and the evolution of radionuclide physico-chemical forms with time was investigated through the analysis of soil samples that were collected annually. A slightly modified version of the method introduced early by Pavlotskaya [3] has been used to evaluate species of radionuclide in the soils. The procedure consists in subjecting a soil to increasingly powerful reagents in order to separate the following fractions: free ion and water-soluble (water); cation exchangeable (ammonium acetate); extractable by diluted mineral acid (1M HCl at ambient temperature); strongly fixed (boiling 6M HCl) and the unextracted forms corresponds to the residue in soil after extractions.

To study the long-term variation of transfer factors (TF) the ^{90}Sr, ^{134}Cs and ^{137}Cs concentrations measured in grain and straw of four cereal crops grown under field conditions have been used. The selected crops were winter rye and wheat, spring barley and oat, which are the most common grain crops in this region. Field experiments were carried out on soddy podzolic and peaty gley soil under conventional agricultural conditions.

Gamma-emitting nuclides were determined in whole soil, soil extracts and plant material by high resolution gamma spectrometry using hyper pure germanium detectors with a relative efficiency of 20%. Activity of ^{90}Sr in the soil and plant samples was measured by separating radiostrontium from other nuclides and bulk soil matrix elements as oxalate and carbonate, then separating and counting its daughter product ^{90}Y after it has grown in at or near radioactive equilibrium.

3.   Results and Discussion

3.1   Speciation of Radionuclides

The first set of soil samples was taken in May 1987. Chemical extractions revealed a clear pattern for radionuclides speciation in the soil. Water-soluble fraction did not exceed 1% for all radionuclides examined whereas the value of exchangeable fraction was increased from 0.1% to 15% in the sequence Ru $\approx$ Ce << Cs < Sr. In the case of ^{134}Cs and ^{137}Cs, the unextractable fraction represented 20-25% of the total caesium activity in soil, more than 60% existed in the form extractable by boiling 6M HCl, and small amounts being distributed in the other fractions. Pattern of ^{134}Cs and ^{137}Cs disposition in soil correspond closely. Both ^{106}Ru and ^{144}Ce were also completely immobilised in the soils when leaching with water or when treated with ammonium acetate solution. However, when followed by exposure to the cold dilute HCl, more effective extraction of ^{144}Ce resulted. Although Ru can exist in anionic or cationic forms, more than 90% of ^{106}Ru remained in soil after all treatments [4].

At the same time, chemical fractionation did not reveal clear differences between soil types. Data obtained show a large variation from place to place but this variation could not explained by spatial variability by soil characteristics only. However, with all radionuclides considered except ^{106}Ru the actual speciation in soil was influenced by variables such as the deposition level and time elapsed since the accident. The reason is that a certain number of years after soil contamination a significant part of radionuclides was included into physically and chemically persistent fuel fragments. Thereafter the fuel particles containing the fission products have been leached and weathered under natural soil conditions. Due to the alteration effect of the environment on the fuel fragments, levels of exchangeable ^{90}Sr and ^{144}Ce fraction extractable by diluted acid increased rapidly, while the importance of insoluble or slightly soluble fractions was decreased for both nuclides. The amount of ^{90}Sr fraction available to the root plant uptake starts to increase in 1987 and remains approximately on the same level since 1990.

The pattern of ^{137}Cs behaviour was not defined as for ^{90}Sr and [144]Ce because ageing is more noticeable for caesium nuclides. A closer look at the experimental data reveals, that there was a decrease of the exchangeable fraction and a slight increase of caesium activity in acid-soluble fractions during 1987-1989. Considering the levels of residual ^{137}Cs in soil, one finds a pronounced minimum at the same period. The levels of mobile fractions decreased and the relative importance of the residual fraction increased. This clearly indicates that there must have been radiocaesium release fuel fragments, but caesium released readily interacts with other soil constituents and this becoming more strongly fixed.

It is possible to use changes in ^{90}Sr and [144]Ce speciation in soil for the calculation of the relative rate of radionuclide release from fuel particles under natural conditions. On the contrary, the evolution of ^{137}Cs chemical forms with time can be applied for estimation the ageing of the caesium deposited. A least squares fit has been applied to the experimental data, and the most important ^{90}Sr, ^{137}Cs and [144]Ce fractions in soils have been taken into consideration. This gives the result that radionuclides have been released from fuel particles with a half-time varying among 25 and 45 months for ^{90}Sr, and among 25 and 60 months for [144]Ce, with a mean of 36 and 45 months respectively. At the same time, the result shows a strong fixation of radiocaesium with half-time ranging from 35 to 55 months, with a mean of 45 months.

3.2   Dynamics of ^{90}Sr and [137]Cs Transfer from Soil to Grain Crops

The investigation started in 1987, when soil-to-plant transfer became the most important pathway for plant contamination. In 1988 levels of ^{90}Sr, [134]Cs and ^{137}Cs TF have varied widely according to the soil types and, within the same type of soil, according to the levels of deposition and crop species cultivated.

The variations among the different sites in the ^{137}Cs transfer to grain crops were very large and can be described by a factor 100. On the peaty gley soil, TF of ^{137}Cs was higher by a factor 15 to 40 relatively to that on the soddy podzolic. Due to a higher mineral content in the straw, the caesium transfer to the straw was about twice as high as to the grain. Among those considered, oat was the most contaminated crop. The behaviour of [134]Cs and ^{137}Cs was similar in all cases [5].

The range of ^{90}Sr TF values was lower compared to that of ^{137}Cs TF values, while the difference between transfer to the straw and grain was greater. The activity ratio ^{90}Sr in straw/^{90}Sr in grain has varied from 5 to 12. In contrast to radiocaesium, data show a decreased radiostrontium root uptake from the heavily contaminated soils of the same type. In addition, it must stressed that the rate of ^{90}Sr accumulation by crops was comparable or even slower than that of ^{137}Cs, which is not in accordance with the usual findings. Another evidence of marked but unusual difference in bioavailability of ^{90}Sr and ^{137}Cs on both soils came from the fact that the average nuclide ratio in crops was sometimes smaller than the ratio in soil. The possible explanation of these observations may be that almost all of ^{90}Sr deposited at this area has been incorporated into fuel particles, and thus it was less available for root uptake by plants. In the following years this ration notably increased due to transforming processes taking place in the soil environment, and because of the different binding capacities of soils regarding Cs and Sr nuclides.

For both ^{90}Sr and ^{137}Cs nuclides the TF values were depending on the time elapsed since the accident. A small increase and then a sharp decline of ^{137}Cs uptake by plants has been noticed in 1988-1990. In the following years the TF of ^{137}Cs reduced only by a factor 2 to about 4. This is the net result of the two competing processes that were taking place in the soil : leaching of radionuclides from fuel particles into the soil solution and radionuclide fixation due to interaction with soil constituents. The rate of TF reduction differed mainly between soddy podzolic loamy sand and peaty gley sand soil. As a whole, during the period 1988-1994 ^{137}Cs transfer to the straw and grain of crops was reduced by a factor of about 50.

Contrary to the radiocaesium, TF values of ^{90}Sr increased in 1988-1990 and remain on the approximately same level since 1991. In total, for straw and grain of crops the TF values increased by a factor of 3 to 8. Today, about 10 years after Chernobyl accident, we have not seen any decrease in the ^{90}Sr activity concentrations in crops. These results indicate that during the time considered the major part of ^{90}Sr has been leached from fuel fragments and redistributed in soil.

A least squares fit has been applied to the experimental data from 1988 to 1991 assuming that the changing TF values for crop products can be represented by a single decaying exponential function. This gives the result that the ^{90}Sr concentration in crops increased with a half-time varying among 7 to 44 months, with a mean of 20 months. On the contrary, ^{137}Cs concentration in crops decreased with a half-time varying among 8 to 25 months, with a mean of 15 months.

4. Conclusions

The type of fallout after Chernobyl accident and fallout level at the different sites was an important factor affecting the radionuclides speciation in soil and their availability to plant uptake. As the leaching rate of fuel particles in soils is a time dependent process, the behaviour of radionuclides changed due to alterating effect of the environment on these particles.

References

[1]   Yu. V. Dubasov, A.S. Krivohatsky, V.G. Savonenkov and E.A. Smirnova, The varieties of fuel particles in the fallout of the near zone of Chernobyl NPP (In Russian), Radiochimiya **33**, n5 (1992) 102-103

[2]   F.I. Sandals, M.J. Segal and N. Viktorova, Hot particles from Chernobyl : A review, J. Environ. Radioactivity **18** (1993) 5-22

[3]   F.I. Pavlotskaya, Migration of radioactive products from global fallout in soils (In Russian), Atomizdat, Moscow 1974 215 p.

[4]   S.V. Krouglov, A.D. Kurinov and R.M. Alexakhin, Chemical fractionisation of ^{90}Sr, [106]Ru, ^{137}Cs and [144]Ce in the Chernobyl-contaminated soils - An evolution in a course of time, Submitted to the Journal of Environmental Radioactivity (1995).

[5]   S.V. Krouglov, A.S. Filipas, R.M. Alexakhin and N.P. Arkhipov, Long-Term Study on Transfer of ^{137}Cs and ^{90}Sr from Chernobyl-Contaminated Soils to Grain Crops, Submitted to the Journal of Environmental Radioactivity (1995).

Contamination characteristics of podzols affected by the Chernobyl Accident.

F. Besnus[1], J.M. Peres[1], P. Guillou[1], V. Kashparov[2], S. Gordeev[3], V. Mironov[4], A. Espinoza[5], A. Aragon[5]

[1]Institut de Protection et de Sûreté nucléaire - France ; [2]Ukrainian Institute of Agricultural Radiology - Ukraine ; [3]Russian Scientific Practical and Expert Analytical Centre - Russian federation ; [4]Academy of Sciences of Belarus, Institute of Radiobiology - Belarus ; [5]Centro de Investigaciones Energeticas, Medioambientales y Technologicas - Spain

Abstract

In the framework of ECP1 project, the soils from 6 experimental sites contaminated after the Chernobyl NPP accident have been studied in order to characterize source terms for resuspension effects in rural or agricultural areas. Except for one sand deposit located within a few km from the nuclear plant, the selected sites were podzols which had undergone important contamination during cloud transfer above districts of Ukraine, Belarus and Russia. The soils have been sampled during 5 field campaigns carried out in 1992 to 1994. Radionuclides of major importance for dose delivery, i.e. Cs-137, Sr-90, isotopes of Pu and Am-241 were measured by organisations involved. The specific activity distributions in the first 30 cm of top soil and the surface contamination densities representative of each site were determined for the radionuclides above. Complementary experiments and studies such as size specific activity distribution in soil fractions, fuel hot particles numbering and selective extraction, were carried out in order to identify contamination mechanisms and try to predict their evolution. Finally, nuclide ratios were estimated for each site and compared to those representative of fuel composition at the time of accident. Interpretations of the results obtained are given in present paper. It appears that despite the fact that weak retention properties are expected from investigated podzols, the migration of studied nuclides has been rather slow during the past 9 years, allowing 70 to 90% of initially deposited activity to remain within the first 5 cm of soil in almost all cases. Nevertheless, there are some evidence of differences in the nature of deposited radionuclides (condensed forms or fuel particles), increasing with the remoteness of studied sites from accident location. Some attempts have also been made to simulate the evolution of the distribution trends. Results from these attempts are given in present paper.

1 Introduction

The present work was performed in the framework of ECP1 project, aiming at assessing the processes of contamination by resuspension of material from rural areas strongly affected by the Chernobyl Nuclear Power Plant accident. Soil characterization is a necessary step for the determination of sources of resuspended material. A compilation of all results concerning the characterization of 6 contaminated sites and acquired during 5 experimental campaigns from 1992 to 1994, has been achieved. Systematic measurements have been carried out in order to determine surface contamination densities, activity distribution according to depth and particle size, and the distribution of fuel particles in soils. Though a rather wide spectrum of radionuclides has been measured, the present paper focuses on results from Cs-137, Sr-90, Pu-239 and Am-241 measurements, for reasons of their key importance for radiological impact and for the understanding of soil contamination mechanisms. Finally, a few attempts to predict the contamination evolution of studied sites were made. More details can be found in EUR report n° 16527 [1].

2 Material and methods

For soil sampling, top soil cutters were used on 20, 25 and 30 cm depth, 4 cm inside diameter, made from 0,3 cm thick cold-rolled steel. Generally about 20-25 cores were taken. The separation of granulometric classes from soil samples was obtained by wet sieving using polyester meshes disposed on a shaker table. For fractions of lower diameter than 20 µm, a centrifugation and sedimentation of the remaining was necessary. The grain sizes were measured by Laser Beam Diffraction (LBD) using a CILAS HR-850 granulometer. The size distribution of samples are determined by means of a computerized system for particles ranging from 0.1 to 600 µm.

Activity measurements were performed by -γ-spectrometry on hyperpure Germanium detector for Cs-137 and Am-241. Isotopes of Pu were extracted by dissolution in a mixture of acids (HNO3 and HF) and separated from Am by standard methods. They were measured by α-spectrometry. Contents of Sr-90 were determined on daughter radionuclide Y-90. Measurements for Y-90 and Pu-241 were carried out with windowless gas-flow counter "TESLA".

Finally, selective extraction was performed by leaching samples with NH_4^+ cations (exchangeable fractions), with a reducing solution (dithionite for determination of oxyde bound fractions), and with oxidizing solutions ($H_2O_2 + HNO_3$ for determination of organic and sulphide bound fractions). The strongly fixed fractions have been determined by acid leaching of soil samples (1N and 6 N HCl or 7N HNO_3).

3 Results and Discussion

The six selected sites are located over three independent states on areas strongly affected by the Chernobyl NPP accident (3 sites in Ukraine : the "Kopachi", "Zapolye" and "Beach" sites, which can be considered as virgin soils situated within 10 km from the Chernobyl NPP, two sites in Belarus: the "Mikulichi" and "Kovali" sites located at 60 km from the NPP in the Bragin district, which were disturbed by agricultural activities, and one site in Russia : the "Novozybkov" site located at about 150 km from the NPP, for which data related to both virgin and disturbed soils were available. All sites are of similar podzolic nature, except for the Beach site which is an artificial sand deposit. According to direct measurement of grain size by LBD method, the studied podzols textures are mainly constituted of silts and fine sands for about 70% of their volume. The fine volume fraction of soil (<2µm) is about 10%. The main mineral component of former soils is quartz, whatever fine or coarse fraction is considered. Mineralogy studies carried out on the fine fraction showed, nevertheless, some content in clay minerals. For the Beach site, the frequency distribution was found to be very narrow, with more than 75% of coarse sands (>200 µm) and only 1% of particles under 50 µm.

The specific activities of Cs-137, Sr-90 and Pu-239+240 have been measured on samples taken out from the first 30 and 50 cm of soil. For virgin soils, despite a rather high variability of results coming from the same locations, due probably to local perturbations induced by bioactivity or/and heterogeneity of radionuclide initial deposition, the mean vertical distributions show similar trends from one site to another. In nearly all cases, 70 to 90% of Cs, Pu and Sr activity are still present in the first 5 cm of soil 9 years after the accident. Only Sr90 distribution in Novozybkov was found to be significantly different, with a deeper penetration of Sr in soil. Some typical distributions are given in Fig. 1.

On cultivated fields, a good homogenisation of activity was observed as a result of the implementation of standard technogenic activities such as ploughing or harrowing. The amount of total activity remaining in the first 5 cm of soil drops down to 20% in such cases, whatever radionuclide is considered.

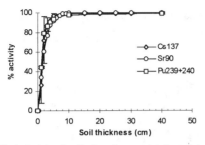 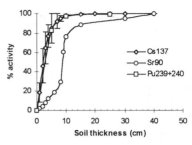

Fig.1 :Activity distributions (as cumulated percentages) in podzols from Kopachi (left) located within 10 km from NPP, and Novozybkov (right) located at 150 km from NPP

Am-241 activity build-up from Pu-241 radioactive decay has been estimated from field data and by use of results on initial fuel composition at the time of accident, taken out from various theoretical and experimental studies [2], [3]. It appears that Am-241 activity will rise until 73 years after the accident and is likely to reach at this date 43 times its initial value, which would lead to activity levels of about 9×10^4 to 4×10^5 Bq/m² (350 to 1500 Bq/kg) in soils near NPP, and 4×10^3 to 5×10^3 Bq/m² (15 to 20 Bq/kg) in soils from the Bragin district. Am-241 may therefore become of prime importance for future dose delivery.

Surface contamination densities (σa) and nuclide activity ratios were reconstructed from activity measurements and are given in table1.

Tab.1:   Orders of magnitudes of surface contamination densities (σa in Bq/m²), specific activity values (As in Bq/g) in 0-1 cm layers, and radionuclide ratios related to field or fuel values. Site location is given as the average distance from NPP.

	Cs-137		Sr-90		Pu-239-240		Pu/Sr		Pu/Cs	
Site location	σa	As	σa	As	σa	As	field	fuel	field	fuel
10 km	1E+06	1E+01	1E+06	1E+01	1E+04	1	1E-02		1E-02	
60 km	5E+05	1E+01	2E+04	1E-01	5E+02	5E-03	1E-02	1E-02	1E-03	1E-02
150 km	5E+05	1E+01	2E+04	1E-02	5E+01	5E-04	5E-03		1E-05	

It appears that the levels of Cs-137 activity stay roughly the same in soils located at distances ranging from 4 to 150 km of nuclear plant, but σa for Pu-239+240 and Sr show a steep decrease of 1 to 3 orders of magnitude depending on site location. Comparative studies of Pu/Cs activity ratios measured in the various sites and proposed values of ratios representative of fuel particles [4], [5], show that Cs is probably mainly contained in fuel particles in sites located within a 10 km zone around the Chernobyl NPP. But with increasing distance, the part of Cs activity in soils issued from initial deposition of volatile forms becomes progressively dominant (90% at 60 km and around 100% at 150 km). The distribution of Sr activity observed in Novozybkov, along with the information given by Pu/Sr ratio values as well as some results on exchangeable fractions obtained by selective extraction, which showed increasing values with the distance [6], [7], would indicate a similar trend for Sr90.

Finally, Cs, Ce and Eu distributions according to soil particle sizes have been realized on samples from Zapolye site, located within the 10 km zone. For Ce and Eu, identified as good tracers of fuel particles, the size distribution showed a maximum of activity for particles of 2μm mean diameter, which is consistent with size distributions typical of fuel particle emission during the burning phase [8]. Size distribution of Cs activity realized on the same sample showed a different pattern with a sharp increase of specific activity for the lowest particle sizes, which may be due to the sorption of a small part of « condensed » Cs on clay minerals, though of weak proportion in investigated soils. Such hypothesis needs however to be confirmed.

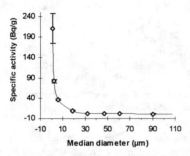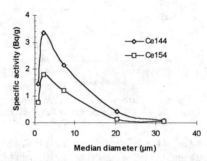

Fig.2 :Size specific activity distributions for Cs-137 (left) and Ce-144, Eu-155 (right) in Zapolye soil

Attempts were made to simulate the basic migration trends observed by means of models based on diffusive and convective transfer of radionuclide in soils or transfer in layers of soils described as discrete compartments. Simulations were performed in order to make a prognosis of the Cs-137 distribution evolution from 1995 to 2005. Results would indicate a decrease of about 20% of Cs-137 activity in the 0-5 cm layer of soil, within 10 years from present date. A decrease of about 40% is expected for Sr-90 under the same conditions. No major differences in migration speeds representative of close and remote sites can be derived from simulations. These results may be consistent with already observed evolution since the time of accident accounting for the rather large uncertainties showed by the activity distributions determined experimentally in the first 10 cm of soil. Additional verification in future would nevertheless be worth to carry out.

References

[1] EUR report n°16527-EN, « Contamination of surfaces by resuspended material », 1996
[2] Kashparov V. et al. :« Research of the physical-chemical and dosimetric properties of the Ukraine territory radioactive contamination as the result of the ChNPP accident » - Report UIAR, N.35, V.1, 1990, Kiev (in Russian)
[3] Kuriny V.D. et al. : "Particle-associated Chernobyl fall-out in the local and intermediate zones", Annals of Nuclear Energy, Vol.20 N°6 PP.415-420, Pergamon press Ltd, 1993
[4] Buzulukov Y. P. et al. : « Release of radionuclides during the Chernobyl accident » - The Chernobyl Papers, *Doses to the Soviet Population and Early Health Effects Studies*, Volume 1, p.3-21, S.E. Merwin and M.I. Balonov eds., Research Enterprises Inc., 1993, Richland, Washington
[5] Kashparov V. et al. : « Formation of hot particles during the accident on Chernobyl NPP » - Radiochemistry, N.1, p.87, 1994, Kiev (in Russian)
[6] Le Cocguen A., Besnainou B. : "Mobilité des radioéléments dans des sols contaminés" - CEA/DCC report, NT/SEP n°362, 1995, Cadarache (*in French*)
[7] Mironov V. et al. : « Contamination of surfaces by resuspended material » - ECP1 Progress report, Academy of Sciences of Belarus, Minsk, 1994
[8] Ter-Saakov A.A. et al. : "Radiation and ecological investigations performed by RSPEAC in 1986-1993 period. Survey" - Russian Scientific and Expert Analytical Centre, 1993, Moscow

Radiocaesium sorption-desorption behaviour in soils and sediments

M.J. MADRUGA[1], A. CREMERS[2], A. BULGAKOV[3], V. ZHIRNOV[3],
Ts. BOBOVNIKOVA[3], G. LAPTEV[4], J.T. SMITH[5]

[1]*DGA/DPSR, E.N. 10, 2685 Sacavém, Portugal*
[2]*Laboratory of Colloid Chemistry, K.U. Leuven, Belgium*
[3]*SPA 'TYPHOON', Obninsk, Russia*
[4]*Institute of Hydrometeorology, Kiev, Ukraine*
[5]*IFE, Wareham, Great Britain*

Abstract. The main goal of this investigation is to gain a deeper understanding of the processes governing the radiocaesium fate in the environment. This study has been conducted in the following directions: radiocaesium specific adsorption and kinetics of radiocaesium fixation. Methods of quantification the specific sites of radiocaesium adsorption (frayed edges sites - FES) were used to predict the radiocaesium distribution coefficients (K_d) in environmental conditions. It is shown that the predicted K_d values give a good indication of the 'in situ' K_ds for different aquatic systems and that the concentration of competitive cations (potassium and ammonium) in the overlying water could influence the 'in situ' K_d values. Kinetic studies of radiocaesium fixation show that the retention process is very slow and that the radiocaesium exchangeable K_d seems to be more adequate than total radiocaesium K_d for modeling of short-term radiocaesium migration processes..

1. Introduction

The radiocaesium mobility in the environment is controlled by its distribution between solid and liquid phases. The adsorption-desorption of this radionuclide in aquatic ecosystems is complicated by two phenomena: specific adsorption and fixation. It is generally accepted that the specific adsorption of radiocaesium in soils and sediments takes place at the frayed edge sites (FES) of the illitic clay particles rather than at the regular exchange sites (RES). The product of the FES pool capacity and the value of the trace selectivity coefficient of caesium with respect to potassium and ammonium in the FES could be quantified [1] and is denominated radiocaesium interception potential (RIP). This parameter allows from laboratory sediment characterizations, the prediction of radiocaesium distribution coefficients. After the radiocaesium adsorption on the solid phase surface it can migrate and be fixed in the interlayer sites of the clay particles [2]. The fixation process which governs the decrease over time of radiocaesium concentration in solution is commonly interpreted as a transformation of the exchangeable fraction of radiocaesium into the fixed form. As the fixation process occurs over timescales greater than, of the same order as, many environmental processes, the kinetics of radiocaesium sorption-desorption are very important to evaluate, at long-term, its uptake and migration in the environment.

2. Objects and Methods

The radiocaesium interception potential (RIP) values with respect to potassium $[K_d^{FES}m_K]$ and ammonium $[K_d^{FES}m_{NH4}]$ ions and the radiocaesium sorption-desorption kinetics were determined experimentally using the protocols described in [3] and [4] respectively.

3. Results and discussion

Table 1 summaries the predicted K_ds values, for bottom sediments, calculated using RIP theory [3] and potassium and ammonium water composition data, for three different aquatic ecosystems. These values give a good indication of the 'in situ' K_ds obtained from the ^{137}Cs contents of sediment and water (column/pore water).

Figure 1 shows the total and exchangeable radiocaesium distribution coefficients values versus contact time for the soddy podzolic soil (MS). It is verified an increase with time in both distribution coefficients. The total distribution coefficient increases from about 6×10^2 $1.kg^{-1}$ after 1 hour to about 2×10^4 $1.kg^{-1}$ after 210 days, implying that the kinetic process is very slow. The time period of ^{137}Cs equilibrium in this soil is about one year. So, the total ^{137}Cs distribution coefficient cannot be used for mathematical modelling of migration procedures with characteristic times less than several months.

Table 1. - Predicted and 'in situ' radiocaesium distribution coefficients for different aquatic systems

Sampling site	water composition meq. 1^{-1}		$K_d m_K$ meq.g^{-1}	K_C (NH_4/K)	pred. K_d $1.g^{-1}$	'in situ' K_d $1.g^{-1}$
	K^+	NH_4^+				
Kiev reservoir	0.26^*-$0.38^\#$	0.05^*-$0.18^\#$	4.4-1.8	6.6-4.9	7.5-1.4	3.3
Devoke water	$0.014^@$	$0.016^@$	0.74-0.76	3.4-3.3	10.8-11.4	10
Glubokoye lake	0.077^*-$0.18^\#$	0.028^*-$1.1^\#$	1.04	5.8	4.3-0.16	3.5-0.15

* = pore water (upper layer); $^\#$ = pore water (bottom layer); $^@$ = water column

It appears that the exchangeable distribution coefficient which increases slightly during the first few days and then remains constant is more relevant parameter for modelling short-term transportation procedures.

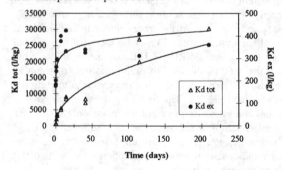

Fig. 1 - Radiocaesium K_d^{ex} and K_d^{tot} versus time for the soddy podzolic soil (MS).

Figure 2 shows an apparent linear relationship between K_d^{tot} and K_d^{ex} of the Chernobyl radiocaesium for a Glubokoye lake sediment profile. Since there is an uniformity on the sediment physico-chemical properties, only different concentrations of competitive cations (potassium and ammonium) in pore water could influence the radiocaesium distribution coefficients. These results are in agreement with data for predicted and 'in situ' K_d presented in Table 1.

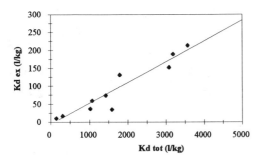

Fig. 2 - Relationship between radiocaesium K_d^{ex} and K_d^{tot} for Glubokoye lake sediment.

4. Conclusions

It is concluded that the total radiocaesium K_d in field scenarios can reliably be predicted on the basis of sediment laboratory characterizations (short contact time periods) and overlying water composition. It is shown that in environmental conditions the concentration of competitive cations (potassium and ammonium) in the overlying water could influence the radiocaesium distribution coefficients. It is seen that exchangeable radiocaesium distribution coefficient seems to be the more relevant parameter for modeling of short-term transportation processes.

References

[1]   A. Cremers, A. Elsen, P. De Preter and A. Maes, Quantitative analysis of radiocaesium retention in soils. Nature, 335, 6223 (1988), pp. 247-249.

[2]   E. Brouwer, A. Baeyens, A. Maes and A. Cremers, Caesium and rubidium ion equilibra in illite clay. J. Phys. Chem., 87 (1983), pp. 1213-1219.

[3]   M.J. Madruga, Adsorption-desorption behaviour of radiocaesium and radiostrontium in sediments, PhD Thesis, K.U. Leuven, Belgium, October 1993, p 121.

[4]   Modelling and study of the mechanisms of the transfer of radioactive material from the terrestrial ecosystem to and in water bodies around Chernobyl. EC-Coordinated ECP 3 Project (COSU-CT-93-0041), Final Report 1993-1994.

Radioecological Phenomena of the Kojanovskoe Lake

I. RYABOV[1], N. BELOVA[1], L. PELGUNOVA[1], N. POLJAKOVA[1] and
R.H. HADDERINGH[2]

[1]*A.N. Severtsov Institute of Ecology and Evolution of RAS, Moscow, Russia*
[2]*KEMA Environmental Services, Arnhem, Netherlands*

Abstract. During post Chernobyl radiation monitoring of water ecosystems made by
the Complex Radioecological Expedition of the Russian Academy of Sciences in
1992, a lake with abnormal high ^{137}Cs contents in fish muscles was discovered in the
Bryansk area. The concentration of ^{137}Cs in fish muscles from this lake is about a
hundred times higher than in fish from other lakes in this area and even higher than
in fish from the Cooling Pond of the Chernobyl NPP. This lake is called
Kojanovskoe. It was suggested to investigate this phenomenon in the frame of the
ECP 3 project. During 1993-1995 the ichthyofauna of the lake, the fish feeding
specificity and the ^{137}Cs contents in fish muscles in comparison with fish from other
water bodies was investigated. Some suggestions about reasons for the phenomenon
are made. An idea to organize an international radioecological reserve on the lake
Kojanovskoe is put forward.

1. Introduction

After the Chernobyl accident in April 1986 a large-scale contamination of aquatic system
located in CIS countries took place [1,2,3]. Because of active using of rivers, lakes and
reservoirs for fishing in CIS, radioecological investigations of the contaminated areas are
of great practical and scientific interest. From 1986 the Complex Radioecological
Expedition started studying the consequences of the accident. From 1993 on the work is
continued in the frame of the ECP 3 project. Fish from Cooling Pond of the Chernobyl
NPP, river Pripyat and the northern part of Kiev reservoir was studied. In 1992 ten fish
farms in the Bryansk area in Russia also were investigated. The results achieved during
1993-1994 on the catchment areas of different water bodies indicated Kojanovskoe lake as
the most highly contaminated.

The Kojanovskoe lake is situated at 400 km northwest from Chernobyl. A swampy
landscape surrounds it. There is a thick layer of sediments on the bottom. The surface area
of the lake is about 6 km^2, the average depth is about 1,5 m, the out-flowing stream is about
0,25 m^3/sec. Vegetation is abundant in the lake. The deposition measured in 1993 is about
100 kBq/m^2, the concentration of K^+ in dependence of seasons is 1,6-2,7 mg/l, concentration
of ^{137}Cs in water 1,2-1,9 Bq/l. It was estimated that there is 3,7 x 10^9 kBq of ^{137}Cs in this
lake.

For the radioecological investigations all fish was collected with gill nets (stretch mesh
0.16 - 12 cm) and separated in species. More than 300 samples were taken. Each sample
consisted of whole fishes and muscle tissue.

Eight species of fish were found in the lake. Among them the main marketable fish species are: golden cart *(Carassius auratus gibelio)*, pike *(Esox lucius)*, perch *(Perca fluviatilis)*, ruff *(Gynocephalus acerinus)* and roach *(Rutilus rutilus)*.

2. ^{137}Cs concentration in fish muscles

During the investigation at 10 fish farms in the Bryansk area in 1992, in samples from 9 of them, the ^{137}Cs concentration in fish muscles was not beyond 100 Bq/kg ww. Only in one farm it was 300 Bq/kg ww. During the radioecological monitoring in the Disna river no specimens with a concentration above 60 Bq/kg ww has been measured. In the Iput river six places were investigated in 1992. Only in one of them one specimen of pike was discovered with a ^{137}Cs concentration of 800 Bq/kg ww.

To compare these data with those in the Cooling Pond of the Chernobyl NPP in 1987, the concentration reached 200 kBq/kg ww among predatory fishes [4]. Till 1990 through it did not exceed 10 kBq/kg ww. At the same time in the Kiev reservoir the average ^{137}Cs concentrations were: 1,6 kBq/kg ww in pike, 1,9 kBq/kg ww in perch. In 1992 the ^{137}Cs concentration decreased approximately 3 times. It became 0,6 kBq/kg ww in pike and 1 kBq/kg ww in perch.

Meanwhile it turned out that the ^{137}Cs contents in fish muscles of some specimens from the Kojanovskoe lake was about a hundred times higher than in fish from other water bodies. On the whole ^{137}Cs contents varied in different species. The highest ^{137}Cs concentration of 30 kBq/kg was noticed among predatory species (pike, perch) while the lowest was in roach and ruffe (5-8 kBq/kg ww). Golden carp was in an intermediate position (14 kBq/kg ww). Figure 1 shows the average of the ^{137}Cs concentration in different fish species from Kojanovskoe lake (1994).

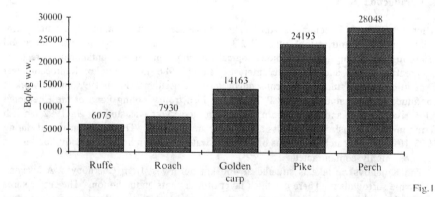

Fig.1

Figure 2 gives a picture comparing the ^{137}Cs concentration in fish muscles from Kojanovskoe lake river Iput and Kiev reservoir.

It is noteworthy that for the three years of investigations the ^{137}Cs concentration among the fish species belonging to different trophic levels has not reduced to a great extent. In the meantime the ^{137}Cs concentration in similar fish species from other water bodies, for instance from the river Iput (Bryansk region) has decreased about twice and from the Kiev reservoir 3-4 times.

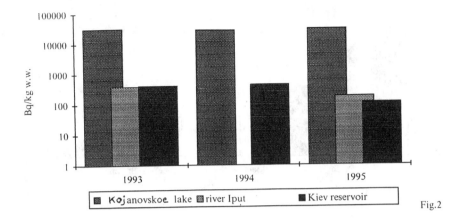

Fig.2

3. Specificity of the fish feeding

During a comparison of food contents in some specimen of the same species taken from different water bodies some peculiarities are found. Thus in spike from Kojanovskoe lake the number of food items was less than in spike from the Kiev reservoir. Besides the latter used much more golden carp in their diet. It is worth to pay a special attention to this fact as in this species the ^{137}Cs contamination is usually higher than in other cyprinid fishes.

Figures 3 and 4 present some results concerning pike feeding from Kojanovskoe lake.

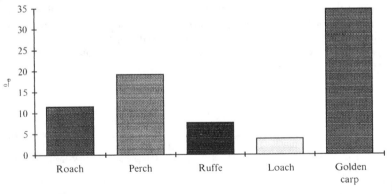

Fig.3

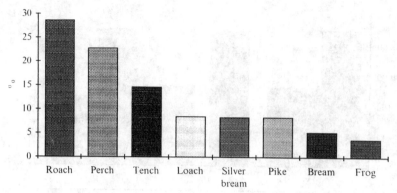

Fig.4

4. Conclusions

The results of the investigations indicated that during 1993-1995 the ^{137}Cs concentration in fish muscles from the Kojanovskoe lake practically has not decreased while the concentration in fish from Kiev reservoir has decreased in 4 times. In pike in average 113 Bq/kg in 1995 while 429 Bq/kg in 1993. In the Iput river the concentration decreased twice since 1995. In 1995 it was only 184 Bq/kg ww. In all water reservoirs except the Kojanovskoe lake the ^{137}Cs concentration was lower than permissible limits (600 Bq/kg ww). It may be supposed that because of specific local conditions of some reservoirs like the Kojanovskoe lake the ^{137}Cs concentration in hydrobionts is supported at high level during long time.

It may be suggested that the following factors serve as a reason of the relatively high contamination level of the Kojanovskoe lake: the high level of the contamination of food items and the water in the lake; the low turnover-time of the water in the lake; the low content of K^+ in water. For predator fish (pike, perch) and increased percentage of bentos fishes in their diet in comparison with other species may serve as an additional factor. Considering that the natural clearing rate of the lake is slow it may persist several decades. No doubt that it is necessary to take measures for banning fishing and other economic uses of the Kojanovskoe lake. At the same time it is reasonable to organize an international radioecological reserve in this place.

References

[1]   R.H. Hadderingh, Distribution of ^{137}Cs in the aquatic foodchain of the Ijsselmeer after the Chernobyl accident. In: W. Feldt (Ed.). The radioecology of natural and artificial radionuclides. Proceedings of the XVth Regional Congress of IRPA. Visby Press, Sweden, 1989, pp. 325-330.

[2]   I.N. Ryabov, Analysis of countermeasures to prevent intake of radionuclides via consumption of fish from the region affected by the Chernobyl accident. In: Proceedings of International Seminar on Intervention Levels and Countermeasures for Nuclear Accident. Commission of the European Communities Press, 1992, pp. 379 -395.

[3]   A.O. Kulikov and I.N. Ryabov, Specific Cesium Activity in Freshwater Fish and the Size Effect. The Science of the Total Environment 122 (1992) 125 - 14.

[4]   I.N. Ryabov, Evaluation of radioactive pollution on impact on hydrobionts in the 30 km control area of the Chernobyl nuclear power station. Radiobiologia 32 (1992) 662 - 667.

Radiation Dose From Chernobyl Forests: Assessment Using The FORESTPATH Model

W.R. Schell, I. Linkov and E. Belinkaia
University of Pittsburgh,
Pittsburgh, PA 15261, USA

V. Rimkevich, Yu. Zmushko and A. Lutsko
International Sakharov Institute on Radioecology,
Minsk, Belarus

F.W. Fifield, A.G. Flowers and G. Wells
Kingston University, UK

Abstract. Contaminated forests can contribute significantly to human radiation dose for a few decades after initial contamination. Exposure occurs through harvesting the trees, manufacture and use of forest products for construction materials and paper production, and the consumption of food harvested from forests. Certain groups of the population, such as wild animal hunters and harvesters of berries, herbs and mushrooms, can have particularly large intakes of radionuclides from natural food products. Forestry workers have been found to receive radiation doses several times higher than other groups in the same area. The generic radionuclide cycling model FORESTPATH is being applied to evaluate the human radiation dose and risks to population groups resulting from living and working near the contaminated forests. The model enables calculations to be made to predict the internal and external radiation doses at specific times following the accident. The model can be easily adjusted for dose calculations from other contamination scenarios (such as radionuclide deposition at a low and constant rate as well as complex deposition patterns). Experimental data collected in the forests of Southern Belarus are presented. These data, together with the results of epidemiological studies, are used for model calibration and validation.

1. Introduction

The Republic of Belarus is one of the areas most significantly affected by radionuclides from the Chernobyl NPP accident. After the accident, immediate attention was given to the prevention of human external irradiation due to the gamma shine from the radionuclide cloud and to the clean-up of agricultural areas contaminated by radionuclide deposition. Natural and semi-natural ecosystems, especially forests, initially were not a high priority but subsequently have been found to be efficient reservoirs for deposited radionuclides. The residence times of stable elements and long-lived radionuclides can approach several thousand years. Radionuclides and nutrient elements incorporated into forest biota, harvested as construction materials and consumed as foodstuff can contribute significantly to human radiation dose. This contribution was found to be as high as 37% of the total radiation dose due to consumption of forest-related foods by several populations in Belarus [1].

Several pathways exist whereby people are exposed to radionuclides from contaminated forest ecosystems. In general, external and internal components of the total radiation dose can be distinguished. The external dose results from direct irradiation due to the radionuclides present in the local environment, while ingested and inhaled radionuclides contribute to the internal dose. In the case of contaminated forest ecosystems, the external dose can be received by the general public via direct gamma shine while walking in the forest or through irradiation from radionuclides incorporated into construction materials and paper. The occupational exposure can be received by working in the contaminated environment and by irradiation from by-product materials and from timber cut for forest industries. The internal dose arises from the ingestion of forest products (berries, mushrooms, birds, game, etc.) and from inhalation of resuspended radionuclides from soil and plants; in addition, radiation dose is received from ash caused by forest fires and from wood burned for heating purposes. These exposure routes can affect populations far removed from the contaminated zones.

Actual doses received by members of the public and forest workers vary widely, depending on the individual characteristics, workplace and living habits as well as on the environmental level of contamination. General dose assessment guidelines are required to set standards for radiation protection. Radiation doses can be measured using individual dosimeters, whole-body counting, hematological tests, etc., but these techniques are expensive and provide only individual-specific data. Modeling can provide not only estimates of the doses but also can predict future trends of dose accumulation.

We have developed a generic model, FORESTPATH [2], which describes the major kinetic processes and pathways of radionuclide movement in forests and natural ecosystems and which can be used to predict future radionuclide concentrations. The FORESTPATH model was successfully applied in a general evaluation of remedial policies for contaminated forests [3] and used to direct a sampling program in the Chernobyl Exclusion Zone [4]. In this paper, the FORESTPATH model is used to evaluate internal radiation doses resulting from the consumption of forest berries and mushrooms as well as external dose due to working in highly contaminated forests. Results of model simulations show that forests can contribute significantly to the human radiation dose and thus need to be considered for the purpose of radiation protection.

2. Exposure Assessment

We consider a rural population such as that inhabiting an area near the Exclusion Zone of Belarus contaminated by 5 Ci/km^2 of ^{137}Cs. The average annual consumption of forest products for this population is about 10 kg of fresh mushrooms and 20 kg berries collected from a pine forest near a village which has 5 Ci/km^2. The biomasses of tree, understory and organic layer are chosen to be 14, 0.2 and 6 kg/m^2 respectively, which is typical of the 30-year-old pine plantations in Belarus [1]. Large segments of this population are involved in forestry and agriculture activities. Work in forests implies an annual occupational exposure of about 1,000 hours in areas characterized by a surface deposition of 20 Ci/km^2.

The two major routes of exposure for this population are external, due to gamma emission from soil, and internal, due to radionuclide ingestion from forest berries and mushrooms (hunting is not considered to be popular for this population). The resulting average external dose for the i-th year following the deposition can be calculated using:

$$E_i = R_i * O * DR, \qquad \text{where} \qquad (1)$$

E   is the annual average external dose (Sv/yr),
R   is the activity which remains in forests at i-th year following the deposition (Bq/m^2),
DR   is the dose rate factor (1.1 nSv/h per Bq/m^2 for ^{137}Cs [5])

The average internal dose can be calculated using

$$I_i = (OL_i * M + U_i * B) * DC * F_r, \qquad \text{where} \qquad (2)$$

I   is the annual average internal dose (Sv/yr),
OL   is the radionuclide concentration in mushrooms (Bq/kg fresh weight),
U   is the radionuclide concentration in berries (Bq/kg fresh weight),
M   is the ingestion rate for mushrooms (kg/yr),
B   is the ingestion rate for berries (kg/yr),

DC   is the ingestion dose coefficient ($1.4 * 10^{-8}$ Sv/Bq for ^{137}Cs [6]), and
Fr   is the food processing retention factor (0.7 for berries and 0.2 for mushrooms [7]).

Additional irradiation dose can arise from radionuclide resuspension by soil or ash (if timber is used for heating purposes), use of contaminated construction materials and paper, forest fires, etc. The resulting dose from these and other processes is a subject for future investigation and are not considered in this paper.

3. Model Description

The FORESTPATH model calculates a time series of inventories for a specific radionuclide distributed within the following six compartments: Understory, Tree, Organic Layer, Labile Soil, Fixed Soil and Deep Soil. Six coupled ordinary differential equations describe the transfer of a radionuclide between the forest compartments. The residence times are the major parameters governing the transfer. The initial data for a given radionuclide is the radioactivity to be distributed within the forest compartments. A complete description of the FORESTPATH model can be found in reference [2].

Berries are assumed to be a part of the Understory compartment and have similar contamination density. Fungi can be important media for radiocesium migration in forests. According to estimates and experimental data [8], they can be responsible for holding up to 40% of the radiocesium present in the Organic Layer. As a first approximation, mushrooms are, therefore, assumed to be a part of the Organic Layer and thus have the same contamination density. A more advance model for the mushroom contamination is currently being developed [9].

4. Results

Figure 1 presents a 50-year FORESTPATH simulation for ^{137}Cs concentrations in mushrooms and berries found in coniferous forests. An initial contamination density of 5 Ci/km^2 and generic FORESTPATH parameters [2] were used, as well as the forest characteristics presented above. We assume that mushrooms and berries are parts of the Organic Layer and Understory compartments and thus have the same contamination density. The current radioactivity of mushrooms is significantly higher because they take up nutrients, such as K, and, therefore, absorb radionuclides from the Organic Layer. On the other hand, the activity in berries, which have a deeper root zone and extract their nutrients from the Labile Soil, is shown to increase with time reaching a maximum at about 10 years following the accident [2]. Our model simulations represent well the concentrations and time trends which have been measured in different species of mushrooms and berries in Belarus [1].

Figure 2 presents the external radiation dose resulting from living in an area with contamination density of 5 Ci/km^2 as well as the additional radiation doses due to occupational exposure in the forest and by the forest product food consumption. Work in a contaminated forest of 20 Ci/km^2 leads to an annual external dose of 4.2 mSv (sum of public and forest worker) for the fifth year following the accident. This is more than twice the average external dose received by the non-forest-worker population in this area. Dose rate measurements for forest workers were conducted at several Belarussian sites [1]. The occupational exposure was found to be 1.75 to 2.9 times higher than the external dose of the public living in the same area, which is in agreement with the model predictions. According to the model, the human radiation dose is slowly decreasing with a half-time of about 30 years due to the physical decay of ^{137}Cs.

5. Conclusions

The complex problem of radionuclide contamination in forest ecosystems requires the use of a model to synthesize and analyze the properties of the entire ecosystem as well as to evaluate the radiation dose resulting from forest usage. The generic forest model developed and used here provides a beginning point for evaluating internal and external radiation doses over long time periods. The model uses dynamics consistent with biological processes and

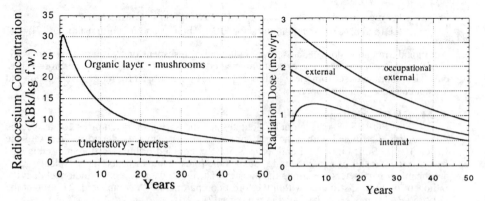

Figure 1. ^{137}Cs concentration in compartments of a coniferous forest following an acute deposition.

Figure 2. Radiation doses received by different populations.

calculates the resulting human radiation doses which would be received over some 50 years. While not yet validated, the initial results on mushroom and berry consumption do provide dose estimates that can be compared with experimental measurements. Contaminated forests constitute a significant hazard to the public over long periods of time depending primarily on the forest food intake. Occupational exposure can be several times greater than that for the general public. The use of a dynamic model can facilitate the decision-making process and help in the design of efficient abatement, remedial and social policies.

Acknowledgments

The authors are very grateful to Dr. J. Vives-Batlle, Dr. B. Morel and Dr. M. Tobin for the fruitful discussion and to P.M. Schell for the editorial and technical assistance. This paper was prepared with the partial support of the US National Academy of Sciences under the CAST program.

References

[1]   V. Ipatyev, ed. Forest and Chernobyl, Forest Institute of the Belarussian Academy of Sciences, Minsk, Belarus 252p; 1994.

[2]   Schell, W.R.; Linkov, I.; Myttenaere, C.; and Morel, B., A dynamic model for evaluating radionuclide distribution in forests from nuclear accidents. Health Physics; 1995 (in press).

[3]   Schell, W.R.; Linkov, I. Radiologically-contaminated forests: A modeling approach to safety evaluation and management, in: R. Lewis, ed. Challenges and Innovation in the Management of Hazardous Waste. Air and Waste Management Association; 1995 (in press).

[4]   Schell, W.R.; Linkov, I.; Rimkevich, V.; Chistic O.; Lutsko, A; Dvornik, A.M.; Zhuchenko, T.A. Model-directed sampling in Chernobyl forests: general methodology and 1994 sampling program. Science of the Total Environment; 1995 (in press).

[5]   U.S. Nuclear Regulatory Commission. Calculation of annual doses to man from routine releases of reactor effluents for the purpose of evaluating compliance with 10 CFR part 50, Appendix I. Regulatory guide 1.109; 1977.

[6]   ICRP. Age-dependent doses to members of the public from intake of radionuclides: Part 2, Ingestion dose coefficients. ICRP Publication 67; 1994.

[7]   IAEA. Handbook of parameter values for the prediction of radionuclide transfer in temperate environments. Vienna: International Atomic Energy Agency; Technical Reports Series No. 364; STI/DOC/010/364; 1994.

[8]   Guillitte, O.; Melin, J.; Wallberg, L. Biological pathways of radionuclides originating from the Chernobyl fallout in a boreal forest ecosystem. Science of the Total Environment 157:207-215; 1994.

[9]   Schell, W.R.; Linkov, I.; and Morel, B. Application of a dynamic model for evaluating radionuclide concentration in fungi. Abstract for 1996 International Congress on Radiation Protection, Vienna, April 14-19; 1996.

Modelling Radiocaesium Fluxes in Forest Ecosystems

Results from the ECP-5 project conducted under the Agreement for International Collaboration on the Consequences of the Chernobyl Accident between the European Commission and the Ministries for Chernobyl Affairs in Belarus, Russia and Ukraine

G. Shaw: Centre for Analytical Research in the Environment, Imperial College at Silwood Park, Ascot, Berkshire, SL5 7TE, United Kingdom
A. Kliashtorin, S. Mamikhin, A. Shcheglov: Radioecology Laboratory, Soil Science Faculty, Moscow State University, Moscow, 119899 Russian Federation
B. Rafferty: Radiological Protection Institute of Ireland, 3 Clonskeagh Square, Clonskeagh Road, Dublin 14, Ireland
A. Dvornik, T. Zhuchenko: Byelorussian Research Institute of Forestry, Gomel, Belarus
N. Kuchma: 'Pripyat' Research and Industrial Association, 1-a B. Khmelnitsky Street, Chernobyl, 255620 Ukraine

Abstract: Monitoring of radiocaesium inventories and fluxes has been carried out in forest ecosystems in Ukraine, Belarus and Ireland to determine distributions and rates of migration. This information has been used to construct and calibrate mathematical models which are being used to predict the likely longevity of contamination of forests and forest products such as timber following the Chernobyl accident.

1. Introduction

Little information on radionuclide migration processes within forest ecosystems existed before the Chernobyl accident. Yet in countries such as Belarus, where approximately 20% of the national forest cover is contaminated to levels in excess of 15 Ci km^{-2} (555 kBq m^{-2}), the post-contamination management of forests is a highly important economic and social problem [1].

During the period 1992 - 1995 forest sites contaminated by the Chernobyl accident in Ukraine, Belarus and Ireland have been monitored by ECP-5 to determine the magnitudes of radionuclide fluxes (principally radiocaesium) between the major components of the forest ecosystems concerned. A range of sites was chosen for this study which represents a variety of forest types receiving a wide range of initial contamination inventories from the Chernobyl pulse.

The aim of this collaborative study has been to provide dynamic data which can be used to develop prototype mathematical models of forest contamination to be employed in accident emergency response and post-accident forest management. This paper describes, first, the philosophy underlying the approach to measurement and evaluation of radionuclide fluxes in contaminated forest ecosystems and, secondly, the development and implementation of three mathematical models by members of ECP-5.

2. Conceptual Framework for Forest Monitoring & Modelling

Analyses of the fluxes and storage of isotopes such as ^{137}Cs in and between discrete ecological compartments can provide useful information for subsequent numerical analysis and modelling of radionuclide fate and persistence [2]. Such analyses are required not just for ecological curiosity but because they provide an unavoidable starting point to the problem of estimating individual and collective effective doses to potentially large groups of people following nuclear contamination events.

The general framework adopted in this study for the compartmental analysis of radiocaesium behaviour in forests is shown in Figure 1. Deposition of radionuclides from the Chernobyl plume was predominantly dry in the immediate vicinity of the ChNPP itself, whereas further afield (particularly in western Europe and Scandinavia) wet deposition resulted in highly localised 'hot spots' of contamination. However, as deposition from Chernobyl can be considered to approximate a single 'pulse' in the long term the most important consideration in flux modelling of forests is the initial total deposit per unit ground area and the relative interception of this by the tree canopy. Losses of contamination from the tree canopy can be rapid, effected by stemflow, throughfall and the loss of leaves or needles. These processes transfer contamination to the litter layer on the forest floor, after which underlying soil horizons become contaminated at a rate which is controlled by soil migration. Ultimately radionuclides may be lost to the drainage waters flowing from forests, but Tikhomirov *et al.* [3] have ascertained that such losses are trivial in forests of the Chernobyl 30 km zone. More important is the return of radionuclides, especially ^{137}Cs, to the standing biomass via biological uptake processes which completes the cycle of radionuclide movement within the ecosystem.

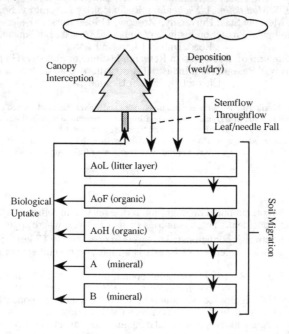

<u>Figure 1</u>: Conceptual framework for measurement and modelling of radiocaesium fluxes in contaminated forest ecosystems.

For the radiologically important isotope ^{137}Cs, with a radioactive decay half life of ~30 years, the primary question of interest is whether the persistence of the contaminant within the forest ecosystem will be controlled by physical decay or by ecological self decontamination processes. If self decontamination of the system as a whole is slow then it is relevant to ask whether certain components of the ecosystem will accumulate a greater proportion of the total ^{137}Cs inventory than others. If so, this has important implications for assessment and aversion of effective doses.

3. Modelling Approaches

Three forest models have been developed by groups working as part of ECP-5. These are ECORAD (Moscow State University), FORESTLIFE (Belorussian Forest Institute) and RIFE (Imperial College, London). Each of the models is based on a compartmental approach which can readily be fitted to the conceptual framework shown in Figure 1. Each model identifies the soil and the tree compartments as being the major components of the forest ecosystem. Both ECORAD and FORESTLIFE use a diffusion based approach to model radiocaesium movement in the soil which Mamikhin [4] has previously described for

ECORAD. In FORESTLIFE a quasi-diffusion equation is used to describe vertical migration to any depth, x, in the forest soil

$$\frac{dq}{dt} = D \cdot \frac{d^2q}{dx^2} \tag{1}$$

where q is the activity and D is the diffusion coefficient for radiocaesium. In order to describe observed distributions of radionuclides within forest soils in Belarus 'fast' and 'slow' diffusion coefficients must be applied which approximate to advection and diffusion components of migration, respectively. Following the prediction of migration in the soil profile ECORAD and FORESTLIFE calculate radiocaesium distributions within the tree component of the forest ecosystem based on transfer coefficients measured after the Chernobyl accident at a series of calibration sites. For ECORAD, four of these sites are in Russia (in the Bryansk and Kaluga regions) and four of are in the Ukraine, situated within the 30km exclusion zone. For FORESTLIFE, 13 monitoring plots in forests in SE Belarus were used to monitor intensively radiocaesium within soils and vegetation between 1992 and 1994. These plots contained pine stands of different ages allowing a relationship between tree age and transfer coefficient to be determined which demonstrated that young trees (10 - 20 years) can accumulate up to 7x more radiocaesium than older trees (70 - 80 years) at the same soil activity concentration.

The RIFE model consists of five compartments which represent the major radiocaesium storages within a forest ecosystem. The general equation describing fluxes within the compartmental system can be written as

$$\frac{dQ_x(t)}{dt} = I_x + \sum_{x \neq y} k_{yx} Q_y - Q_x \left\{ \sum_{x \neq y} k_{xy} + k_x + \lambda \right\} \tag{2}$$

where $Q_x(t)$ is the activity of radiocaesium in compartment x at time t (Bq), I_x is the rate of radiocaesium input into compartment x (Bq t^{-1}), k_{yx} is the transfer coefficient from compartment y to x (t^{-1}), k_x is the loss rate coefficient from compartment x (t^{-1}) and λ is the physical decay rate of the isotope under consideration (t^{-1}). The data on which RIFE is based are derived from 'near-field' sites within the Chernobyl 30km zone as well as from a 'far-field' site in Ireland and the model provides a framework within which experimental field data may be evaluated.

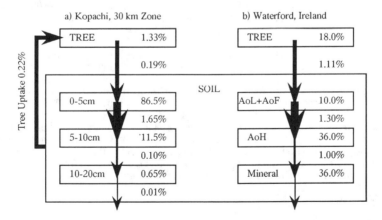

Figure 2: Summary of radiocaesium inventory distributions and annual fluxes at Ukrainian and Irish forest sites in 1992/1993.

4. Measurement & Modelling of Radiocaesium Fluxes in Forest Ecosystems

Field monitoring carried out by ECP-5 at forest sites in the Chernobyl 30km zone (Kopachi) and in Ireland (Waterford) has been aimed at quantifying radiocaesium fluxes along the major pathways shown in Figure 2. These two sites were selected to represent 'near-field' and 'far-field' scenarios, respectively, with the former site receiving a significant number of hot particles. Two major discrepancies are evident between the results for each site. The first is that downwards migration of the radiocaesium inventory into the mineral soil has been much higher at Waterford than within the 30km zone, probably reflecting the influence of hot particles in retarding the leaching of radiocaesium. The second major difference is in the fraction of the total radiocaesium inventories at each site present within the trees. At Waterford this fraction is 0.18, while at Kopachi it is 0.013. Again, this major difference between sites is most probably due to the difference in the physico-chemical form in which the radiocaesium was deposited; at Waterford the bioavailability of Chernobyl-derived radiocaesium is evidently much higher than at Kopachi.

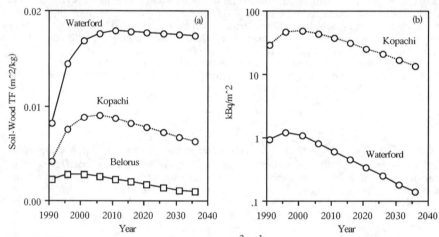

Figure 3: (a) Predicted soil-to-pine wood transfer factors ($m^2 kg^{-1}$) at Kopachi and Waterford (RIFE) and for 20-40 year-old trees in Belarus (FORESTLIFE). (b) RIFE predictions of absolute pine wood contamination ($kBq\ m^{-2}$) at Kopachi and Waterford.

A major goal of ECP-5's forest modelling studies is to predict the likely contamination of wood and the effective decontamination half times of timber following peak contamination. Figure 3(a) shows predictions of soil-to-pine wood transfer factors made using the RIFE and FORESTLIFE models. Both models predict similar time courses for Kopachi and for the Belorussian pine forests, with peak contamination occurring between 1995 & 2005 and similar wood decontamination rates at each site. For Waterford the predicted decline in soil-to-pine wood transfer factor is slower due to the more bioavailable radiocaesium at this site being recycled within the standing biomass. However, Figure 3(b) shows that the absolute contamination of pine wood at Waterford declines more quickly than at Kopachi (effective half times 16 years and 24 years) due to the loss of mobile radiocaesium from the tree rooting zone by leaching. This loss by leaching is retarded by hot particles in the 30km zone.

5. References

1. Crick, M. (1994) Reflections on Forestry in Belarus. Presented at VAMP, IAEA, Vienna. Nov. 1994.
2. Heal, O. W. and A. D. Horrill (1983) The terrestrial ecosystem: an ecological context for radionuclide research. pp. 31 - 46 in: Ecological Aspects of Radionuclide Release, Blackwell Scientific Publications, UK.
3. Tikhomirov, F. A., A. I. Shcheglov and V. P. Sidorov (1993) Forests and forestry: radiation protection measures with special reference to the Chernobyl accident zone. Sci. Tot. Environment, 137, 289 - 306.
4. Mamikhin, S. V. (1995) Mathematical model of ^{137}Cs vertical migration in a forest soil. J. Environ. Radioactivity, 28(2), 161 - 170.

Dynamic of Radionuclides Behaviour in Forest Soils

W. Rühm, M. Steiner, E. Wirth, A. Dvornik, T. Zhuchenko, A. Kliashtorin,
B. Rafferty, G. Shaw, N. Kuchma

Within the research project ECP-5, the dynamics of radionuclides in automorphic
forest soils within the 30-km-zone of Chernobyl and of hydromorphic forest soils in
Belorussia have been investigated. In upland forest soils, the lower layers of the
organic horizons are characterized by the highest residence times for radiocesium and
represent the largest pool for all radionuclides investigated. According to a
preliminary estimate, radiocesium is more mobile compared to ^{125}Sb, which in turn
migrates faster than ^{60}Co, ^{144}Ce, and ^{154}Eu. ^{106}Ru shows the lowest mobility. With
regard to radiocesium, hydromorphic soils exhibit migration rates and transfer factors
from soil to trees, which by far exceed those in automorphic soils. Based on a two-
component quasi–diffusional model the average bias of ^{137}Cs in mesotrophic swamp
soils was predicted. The activity concentrations of U, Pu, and Cs suggest that U and
Pu were originally deposited as hot particles and that U is naturally accumulated in
organic horizons.

1. Materials and Methods

1.1. Sampling sites

Our studies were performed at four forest sites, which characterize more than 80% of the forest
within the 30–km–exclusion zone of Chernobyl:

Ditiatky 1 (D1): Mixed forest (50% oak, 30% pine, 20% birch) with an age of about 55 to 60
years. 28.5 km south of NPP. Podsol on sandy fluvio–glacial deposits. ^{137}Cs deposition 0.21
MBq/m^2. Dry forest.
Ditiatky 3 (D3): Mixed forest (50% alder, 40% birch, 10% pine) with an age of about 60 to 75
years. 26 km south of NPP. Peat–gley on sandy fluvio–glacial deposits. ^{137}Cs deposition 0.16
MBq/m^2. Wet forest.
Kopachy 2 (K2): Pine forest with an age of about 60 years. 7 km southeast of NPP. Podsol on
sandy fluvio–glacial deposits. ^{137}Cs deposition 2.7 MBq/m^2. Dry forest.
Shepelichy 1 (Sh1): Mixed forest (80% pine, 20% birch, small portions of oak and alder) with an
age of about 40 years. 7 km west of NPP. Podsol on sandy fluvio–glacial deposits. ^{137}Cs
deposition 23 MBq/m^2. Dry forest.

Furthermore, Belorussian swamp forests, located about 200 km from the NPP, were investigated.
The ^{137}Cs inventory within the uppermost 60 cm thick layer ranges from 0.55 to 1.6 MBq/m^2.

Oligotrophic swamp: Sphagnous pine forest. Sphagnum peat on wood–sedge–sphagnum peat.
Average ground–water level 10 cm.
Mesotrophic swamp: Sedge–sphagnous pine forest. Sphagnum–wood–sedge peat on reed–wood
peat. Average ground–water level 12 cm.
Eutrophic swamp: Nettle alder forest, mixed stand. Wood peat on light gleyish loam. Average
ground–water level 34 cm.

1.2. Sampling methods

At each plot up to ten complete profiles, randomly distributed within an area of approximately 100 m * 100 m, were sampled. From the upper organic L-, O$_f$-, and O$_h$-layers, and the underlying first mineral horizon, which still contains organic material (A$_h$-layer), squares of 20 cm * 20 cm, resp. 25 cm * 25 cm, were cut by knife. The horizons were gathered separately by hand. From the deeper mineral horizons three layers with a thickness of 3 cm each were sampled with a cylindrical soil-corer covering an area of 54 cm^2. After determining the fresh weight, the material was dried in an oven at 85°C and thereafter milled.

The movement of radiocesium between the different forest compartments, especially between canopy and soil, due to water flow and litter fall was monitored using litter traps, precipitation traps, stemflow traps, and lysimeters.

2. Dynamics of radionuclides in automorphic soils

2.1. Distribution of radionuclides in soil

As can be seen from Fig. 1, in 1994 at K2, the dominant portion of ^{137}Cs activity was still found in the organic horizons. The same is true for all other radionuclides investigated. The O$_h$-horizon represents the largest pool for radionuclides. Apparently, this layer retains radionuclides effectively. Its small contribution to the radiocesium content of leachates further supports the importance of the O$_h$-horizon as a sink of radiocesium. The L-horizon, which is built up by falling needles and leaves, is characterized by low specific activities. The reason is that only 1% of the total ^{137}Cs inventory moved from canopy to soil per year at K2. For other radionuclides this fraction is even smaller, since they are present in fresh litter only in negligible amounts.

Comparison of the vertical distributions of radionuclides in 1993 to those in 1994 revealed a clear movement to deeper soil horizons for all radionuclides investigated. The dominant contributions shifted from the O$_f$- to the O$_h$-horizon.

2.2. Estimate of relative migration velocities

To investigate qualitatively the migration velocities of different radionuclides, a new approach has been developed, which is based on the so-called "cesium ratio" (CR) and the "normalized cesium ratio" (NCR). The CR of radionuclide X in a certain horizon denotes the ratio of the specific activities of X and ^{137}Cs within this soil layer, where both activities are corrected for radioactive decay. The NCR is calculated by dividing the actual CR of radionuclide X by the CR of the initial fallout.

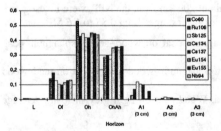 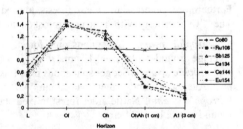

Fig. 1: Distribution of radionuclides in soil at K2 in 1994, based on activity per area.

Fig. 2: Normalized cesium ratio (NCR) for different radionuclides at D3 in 1993.

As an example, Fig. 2 depicts the NCRs for different radionuclides at D3 in 1993. Radionuclides moving downwards more slowly compared to ^{137}Cs are relatively enriched in upper soil layers and, therefore, show NCR > 1. Vice versa, the relative shortage of these radionuclides in deeper soil layers leads to NCR < 1. Altogether the NCRs of radionuclides slower than ^{137}Cs are expected to diminish with increasing soil depth. ^{60}Co, ^{106}Ru, ^{125}Sb, ^{144}Ce, and ^{154}Eu clearly exhibit this behaviour and thus prove to be less mobile compared to ^{137}Cs. The NCR < 1 observed for the L–horizon might be attributed to radiocesium recycling via root uptake by plants and subsequent litter fall.

The normalized antimony ratio (NAR) is defined analogously to the NCR, when the reference nuclide ^{137}Cs is substituted by ^{125}Sb. Using the same arguments as above we conclude that the isotopes ^{60}Co, ^{106}Ru, ^{144}Ce, and ^{154}Eu are less mobile compared to ^{125}Sb. As indicated by the steepest descent of NAR with increasing soil depth, ^{106}Ru is characterized by the lowest mobility. Since ^{125}Sb is present in leaves and needles only in negligible amounts, the influence of recycling via litterfall on the NAR of the L–horizon is not present.

2.3. Quantitative analysis: Compartment model

For a more quantitative approach a simple linear compartment model of vertical radionuclide migration has been developed to evaluate the residence half–times for different soil horizons. The compartments represent the decay–corrected specific activities (Q_n) of radiocesium within the different soil layers. The model is described by a set of first order differential equations, which can be written as $dQ_1/dt = - \lambda_1 Q_1$ for the uppermost litter layer (L–horizon) and $dQ_n/dt = \lambda_{n-1} Q_{n-1} - \lambda_n Q_n$ for the n–th soil layer, respectively.

The residence half–times $T_{1/2,\ n}$ for the sampling site K2, as derived from the experimentally determined coefficients via $T_{1/2,\ n} = \ln 2/\lambda_n$, are summarized in Table 1.

Tab. 1: Residence half–times for radiocesium in different horizons at the sampling sites K2 and Hochstadt.

Horizon		L	O_f	O_h	A_h	A_l/B
K2:	$T_{1/2}$ [y]	1.1 ± 0.4	3.0 ± 0.7	23.9 ± 3.4	6.0 ± 1.1	16.7 ± 13.2
Hochstadt:	$T_{1/2}$ [y]	2.9 ± 0.6	4.6 ± 0.9	6.3 ± 2.3	8.7 ± 6.5	1.8 ± 0.3

The migration half–times demonstrate the tendency for the lower organic horizon (O_h at K2) to retard the downward migration of cesium isotopes very effectively. To a lesser extent this finding also holds for the Bavarian reference site near Hochstadt (see Tab. 1), where the O_h– and A_h–horizons exhibit the highest residence times.

Based on the extensive data set obtained from the Hochstadt site, the reliability of the model itself has been demonstrated by comparing the present distribution of ^{137}Cs in soil due to nuclear weapons fallout with corresponding predictions.

3. Dynamics of radionuclides in hydromorphic soils

Since the type of soil, especially its physico–chemical properties, is expected to crucially determine the dynamics of radionuclides, automorphic and hydromorphic soils have been compared to reveal relevant parameters for the migration of radionuclides.

The rate of ^{137}Cs migration in hydromorphic soils by far exceeds that in automorphic soils, most probably because of the moisture and the aggressive soil medium. ^{137}Cs mobility decreases from oligotrophic to mesotrophic swamps and is lowest for eutrophic swamps. The latter exhibit a clear–cut zone of radiocesium accumulation at the transition from the peaty soil layers to the mineral ones. A qualitatively similar behaviour has been observed in upland forests, where the lower organic horizons represent the largest pool for radiocesium.

Transfer factors for ^{137}Cs uptake by trees turned out to be more than one order of magnitude higher for swamp soils compared to upland soils.

For a quantitative description of the vertical migration of ^{137}Cs in swamp soils, a convective–diffusional model and a two–component ("slow" and "fast" component) quasi–diffusional model have been developed. Within the framework of the latter, the average bias for ^{137}Cs in mesotrophic swamps was predicted. Seven years after the initial deposition, the average bias is 6.0 cm. After 30 years it is 12.6 cm and will have reached the average ground–water level. Compared to sandy soils, these values are nearly 7 times higher.

4. Physico–chemical form of deposition

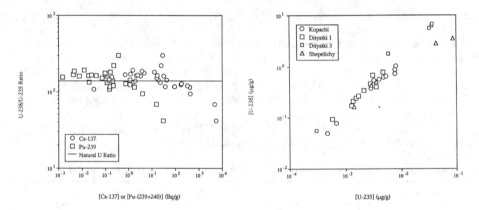

Fig. 3: Isotopic ratio $^{238}U/^{235}U$ vs. activity concentrations of ^{137}Cs and $^{239+240}Pu$.

Fig. 4: Activity concentrations of ^{238}U vs. those of ^{235}U.

Obviously, not only the soil properties but also the physico–chemical form of the deposition might influence the migration behaviour of radionuclides. Therefore, the activity concentrations of U, Pu, and Cs isotopes and their ratios were employed to test the probability that U and Pu were originally deposited as "hot particles".

The uranium isotopic ratio $^{238}U/^{235}U$ hints at ^{235}U enrichment in the O_f– and O_h–horizons, which indicates a non–natural source of U, most likely hot particles from irradiated reactor fuel. Furthermore, at high activity concentrations of $^{239+240}Pu$ and ^{137}Cs, respectively, the uranium isotopic ratio falls below its "natural value" of 138 (see Fig. 3), which supports the idea of reactor–derived uranium fragments. At high U concentrations, the relationship between ^{238}U and ^{235}U reveals two distinct groups (see Fig. 4): Near the NPP, at Sh1, a low isotopic ratio hints at hot particles. At the more distant site D3, the isotopic ratio lies at or above the natural value, which again suggests a natural concentration of U by organic soils. The latter might be caused by relatively long residence times for reactor–derived uranium in organic soil layers.

For plutonium, the isotopic ratios $^{238}Pu/^{239+240}Pu$ range between 0.51 and 0.54. Therefore, Pu is likely to derive exclusively from the Chernobyl accident, as expected because of the high plutonium activity concentrations.

Radioecological Aspects of Radiosorbents Application and their Place in the System of Countermeasures on the Contaminated Territory of Ukraine

G.A. BOGDANOV[1], V.A. PRONEVICH[1], S. SYASKIY[1], B.S. PRISTER[2],
N.M. LAZAREV[2], A.S. SOBOLEV[2], L.M. ROMANOV[2]
[1] *Ukrainian Academy of Agricultural Sciences,*
9, Surovora street, Kiev, Ukraine
[2] *Ukrainian Institute of Agricultural Radiology (UIAR)*
7, Mashinostroïtelei street, 255205 Chabany (Kiev region), Ukraine

Abstract. The work presents comparative characteristics of the efficiency of mineral silicates (zeolites, saponite, palygorskite, vermiculite) and ferrocine preparations. The experiments were performed with various species of agricultural animals and poultry. Depending on the type of sorbents, reduction of Cs^{137} content in animal products was up to 1.5-4 times.

1. Introduction

Agricultural production, especially dairy and meat cattle-breeding in the period of the accident, put forward the problem of sorbents application. The Ministry of Chernobyl, the Ministry of Food and Agriculture of Ukraine, the Ukrainian Academy of Agricultural Sciences with the participation of scientific institutions of the National Academy of Sciences, regional radioecological centres, within a comparatively short period had performed complex research and wide-scale industrial verification. These results were the basis for the application of mineral silicates and synthetic sorbents as a countermeasure for the reduction of radionuclides transfer from the diet to animal organism. Milk and meat in contaminated districts even 10 years after the accident remain the main dose-forming products for rural population, especially in private sector as the farmers have to graze their animals or make hay on the lands with high density of radioactive contamination or characterized by high transfer factors of radionuclides from soil to plants.

Therefore together with countermeasures in the link soil-plant, it was reasonable to investigate the efficiency of radiosorbents, which decrease the transfer of radionuclides from the diet to animal organism and animal products.

2. Main results of the work

In connection with the reserves of mineral silicates in Ukraine and in the necessity to develop strategy directions and principles of ecological and economic background of their use the following sorbents were investigated: zeolite and humolite, saponite, palygorskite, vermiculite, ferrocine and its derivatives - bifezh, ciom, boli.

The biological and radioecological was given both for natural silicates and their modified forms, and for synthetic sorbents (ferrocianides) with high sorption capacity.

1. Problems of sorbents norms in the diets of animals were developed and technological parameters of special sorbing fodder additives are being worked out.

2. Investigation with cows and young cattle (production of milk and beef in private and collective sector), pigs, rabbits, poultry and minks allowed to make a comprehensive biological assessment of natural and synthetic sorbents.

2.1. Concentration of Cs^{137} in milk under the effect of feeding, for example humolite, as compared to control, decreased by 1.5-4.4 times depending on the level of milk contamination, and when ferrocine preparations were applied by 4-10 times. When young cattle was fed in such conditions, Cs^{137} content in meat decreased by 1.6-2.6 times. In all experiments addition of humolite and ferrocines did not affect the nutritional value of milk. Altogether 30 experiments were carried out with cows and 20 experiments with young cattle. In this industrial verification and implementation about 30 thousands heads of cattle were involved.

2.2. Mineral silicates as a source of biogenic mineral substances do not have a detectable digestibility, which characterizes it as products both safe for nutrition and possessing radiosorption abilities all the way through the transport in the gastrointestinal tract of animals and having a favourable effect for metabolism.

3. A decorporating effect of mineral silicates, especially palygorskite, was observed in relation to heavy metals. At similar intake with fodder, the removal of lead in the faeces of animals in the experimental groups increased by 7.3-9.5 times, and that of strontium by 3.5-4.5 times.

4. The addition of natural and modified minerals to forage provides both a strong effect of radionuclides binding during ingestion in gastrointestinal tract and causes a reduction of their assimilation by blocking the transport and accumulation and an increase of radionuclides removal from animal organisms.

5. When ferrocine and its derivatives (ciom, bifezh) are added to forage in the dose of 3-6 g per head, the concentration of radiocaesium in the milk of cows decreases from the third day of the experiment by 3-8 times.

6. When these preparations are fed to sheep, pigs (3 g per head) and to young cattle (6 g per head), concentration of $Cs^{137, 134}$ in muscle tissue in 30 days of the experiments decreases by 3-10 times; in parenchyma organs this index is higher (10-80 times). The positive effect was obtained at the use of ferrocine in the fodder of adult and young geese.

7. In the experiments it was shown that ferrocine can be used in the form of powder, in combination with concentrates, with salt-licks, in granules and in the form of boli. Boli with the sorbent were used for the first time in October 1990 and applied using Norwegian technology and with the participation of Norwegian specialists. Starting from the 3-6 day of the experimental period, radiocaesium concentration in the milk of cows decreased by 3-7 times (at the experimental farm 'Radioecologist') in Rovno region by 10 times. In the experiments with salt-licks, where the concentration of ferrocine was 10 %, by the end of experimental period (21 days), a 3-fold reduction was reached.

8. Performing of special toxicological experiments and veterinary-sanitary expertise of the product allowed to prove the harmlessness of using natural silicates (humolite) and synthetic sorbents (ferrocine).

Comparative efficiency of radiosorbing preparations

Forage additives	Decrease of Cs^{137} concentration, times			
	Milk	Meat of cattle	Rabbits	Chicken
Zeolites: . natural . modified	1.5 3.0	1.4 2.6	7.8 9.0	1.9 38.0
Vermiculite: . natural . modified			4.0 40.0	7.0 40.0
Ferrocines: . salt-licks . powder . boli . bifezh . ciom	2-5 4-10 3-4 2-8 4-18	1.7-2.9		3.5

Efficiency of ferrocine application for the decrease of radiocaesium concentration in animal products

Indices	Type of animal products				
	Milk of cattle	Meat of cattle	Meat of pigs	Meat of sheep	Meat of geese
Dose, g/day per head	6	6	1	1	0.15
Time for achieving maximum effect, days	15	30	30	30	30
Decrease of radionuclides concentration in product	8-10	3-4	3-4	7-8	4-5

II. ENVIRONMENTAL ASPECTS
OF THE ACCIDENT
B. Exposure of the population

Pathways, Levels and Trends of Population Exposure after the Chernobyl Accident

Michael Balonov
Institute of Radiation Hygiene, St. Petersburg, Russian Federation,
Peter Jacob
GSF - Institut fur Strahlenschutz, Oberschleissheim, Germany,
Ilya Likhtarev
Scientific Centre for Radiation Medicine, Kiev, Ukraine,
Victor Minenko
Institute of Radiation Medicine, Minsk, Belarus

Abstract. In this paper main regularities of the long-term exposure of the population of former USSR after the Chernobyl accident are described. Influence of some natural, human and social factors on the forming of external and internal dose in the rural and urban population was studied in the most contaminated regions of Belarus, Russia and Ukraine during 1986-1994. Radioecological processes of I, Cs and Sr nuclides migration in biosphere influencing the processes of population dose formation are considered. The model of their intake in human body was developed and validated by large-scaled measurements of the human body content. The model of external exposure of different population groups was developed and confirmed by the series of individual external dose measurements with thermoluminescent dosemeters. General dosimetric characteristics of the population exposure are given along with some samples of accumulated external and internal effective doses in inhabitants of contaminated areas in 1986-1995. Forecast of the external and internal population effective dose is given for the period of 70 years after the accident.

1. Introduction

After the Chernobyl accident the environment appeared contaminated with complex mixture of radionuclides - products of nuclear fission and neutron activation in the reactor Unit 4 - released in the atmosphere during 10 days, at least. The Chernobyl NPP is located close to common borders of Belarus, Russia and Ukraine, this being the reason why the population in all these republics of the former Soviet Union were affected by external and internal irradiation from many radionuclides with different radiological properties. Ratio of external and internal dose, of activities of different radionuclides incorporated in the human body, significantly depends on radionuclide composition of the cloud and fallout from it in each particular region, on meteorological conditions of fallout (mainly, amount of precipitations) and later on - from dominating soil type, agricultural practices and countermeasures applied.

This paper distinguishes between two groups of people that were exposed to significantly higher doses than the entire population of the former Soviet Union as a result of the Chernobyl accident:

-    urgently evacuated population;
-    population of contaminated area.

The urgently evacuated population consists of forty nine thousand inhabitants in the town of Pripyat, situated 3 km from the Chernobyl NPP, who were evacuated on 27 April

1986 owing to the danger of acute radiation injury, and fifty three thousand inhabitants in the 30-km zone around the Chernobyl NPP evacuated over a period of ten consecutive days. It is known that they were subject to external exposure and incorporation of radioactive substances during a period from 1-11 days. No cases of acute radiation sickness have been found in the population that were evacuated in 1986.

Population of the contaminated area. About 4 million people permanantly live and are subjected to external and internal irradiation in the area with Cs-137 surface activity over 0.04 MBq/m^2 (1 Ci/km^2) with the area over 131 thous. km^2. About 270 thousand of them remained in 1986 in the so called "controlled area" (CA) of 10.3 thousand km^2 with a surface activity of Cs-137 over 0.6 MBq/m^2 (15 Ci/km^2) and higher, including: in Russia - 112 thousand people in 2.4 thousand km^2 in the Bryansk region, in the Ukraine - 52 thousand people in 1.5 thousand km^2 in Kiev and Zhitomir regions, and in Belorussia - 109 thousand people in 6.4 thous. km^2 in the Gomel and Mogilev regions. The package of active countermeasures for radiation protection of the population has been constantly performed in the CA since 1986: delivery of non-contaminated meat and dairy products, decontamination of settlements, measures on decrease of radionuclide content in agricultural products, etc. The population of the CA villages with the greatest level of radioactive contamination was during the period from 1986-1992 gradually resettled to noncontaminated areas.

This paper presents a review of main regularities of exposure of two indicated population groups, both from external and internal (incorporated in the body) sources of radiation during the past ten years after the Chernobyl accident. We also consider prognose of population exposure for the future sixty years and altogether estimate the exposure for seventy years after the event, which is close to the average duration of human life.

2. Radionuclide composition of Chernobyl fallout

The analysis of regularities of population exposure in contaminated areas is naturally based on the data on radionuclide composition and levels of contamination. Radionuclide composition of the release of the Chernobyl accident products into the atmosphere varied considerably as far as the temperature and conditions were concerned [10]. During the same period, the change in weather conditions caused distinctions of radioactive fluxes in different directions, even in their initial nuclide composition [2, 11]. The basic cause of further separation of radioactive mixture in the cloud was the different deposition rate of aerosol particles of different dispersivity and density. The largest of them, mainly of fuel composition, fell out in the so called "near zone", i.e., at the distance of some tens of kilometers from the source. To some extent, the element composition of the radioactive depositions was influenced by deposition mechanism: wet deposition with precipitations or dry one under the action of gravitation, atmospheric mixing and diffusion.

In Table 1 we present the data on composition of radioactive fallouts in the whole spots separately for the near (up to 100 km) and the far zones in the form of the ratio of the surface activity σ_r of the r-th radionuclide on soil to the surface activity σ_{137} of the most radiologically significant nuclide of the Chernobyl accident, cesium-137 [2,11]. The list of radionuclides is separated into three groups: volatile (isotopes and compounds of I, Te, Cs. etc.) refactory nonvolatile (Zr, Ce, Pu, etc.), and intermediate ones as to this attribute (Ru, Ba, Sr). Regarding volatile radionuclides, considerable separation with respect to Cs-137 was not noted. Considerable difference of relative content of refractory radionuclides in the fallouts in the far zone as compared with the composition of the release engages our attention. It thereby, quantitatively characterizes the process of the radioactive cloud

Table 1.

General composition of radionuclide depositions in different regions of the European territory of the former Soviet Union [2, 10, 11] (σ_r/σ_{137} on April 26, 1986) *

Radionuclide	Entire release [10]	Near zone (< 100 km)	Far zone (> 100 km)	
			"Cesium spots" in Belorussia and Russia	South of Kiev region
^{131}I	20	15-30	10-14	(1.0)
^{132}Te	5	13-18	(13)	(1.0)
^{134}Cs	0.5	0.5	0.5	0.5
^{137}Cs	1.0	1.0	1.0	1.0
^{125}Sb	-	0.02-0.1	0.03-0.07	0.1
^{110m}Ag	-	0.005-0.01	0.005-0.014	(0.01)
^{103}Ru	2.0	3-12	1.7-2.0	2.7
^{106}Ru	0.4	1-5	0.5-1.4	1.0
^{140}Ba	2.0	3-20	0.7-1.1	(0.5)
^{89}Sr	1.0	1-12	0.2-0.3	0.3
^{90}Sr	0.10	0.1-1.5	0.01-0.03	0.03
^{91}Y	-	3-8	0.06	0.17
^{95}Zr	2.0	3-10	0.03-0.11	0.3
^{99}Mo	-	3-25	(0.11)	(0.5)
^{141}Ce	2.3	4-10	0.07-0.16	0.5
^{144}Ce	1.6	2-6	0.04-0.15	0.3
^{239}Np	20	7-140	(0.6)	(3)

* values in parentheses were estimated from indirect data.

depletion and of the fallout from the cloud as it moved off the release source, of less volatile particles predominantly of the fuel composition, including refractory products of fission and activation. This also relates within certain limits to barium and strontium isotopes.

The Central, Bryansk-Belorussian, and Kaluga-Tula-Orel spots are separated in the map of radioactive contamination of the European part of the former USSR [2]. The Central radioactive spot was formed around the Chernobyl NPP during about 10 days of the release with predominant direction of contamination to the west and north-west in the territory of Ukraine and Belorussia. The Bryansk-Belorussian spot centered at 200 km to the north-north-east from the Chernobyl NPP was formed on 28-30 April 1986 as a result of rainfalls at the interface of the Bryansk region of Russia and the Gomel and Mogilev regions of Belorussia. The surface activity of Cs-137 (σ_{137}) in the most contaminated soil sites here reaches 3-5 MBq/m². The Kaluga-Tula-Orel spot in Russia with its center at the distance of about 500 km to the north-east from the Chernobyl NPP is also "cesium" one and was formed because of rains on 28-30 April from the same radioactive cloud that the Bryansk-Belorussian spot was formed. Here the levels of depositions from the depleted cloud are lower, and σ_{137} does not reach 0.6 MBq/m² [2,11].

3. Pathways of population exposure

The relation between sources and types of radiation, ways of incorporation of radionuclides into body and the role of separate nuclides turned out to be considerably different for the categories of exposed persons indicated above. This difference is primarily determined by their staying in different zones of radioactive contamination and in different periods after the accident. This caused different isotopic composition of nuclides mixture and the spectral composition of their radiation, and different levels of irradiation.

Taking into account these considerations, in Table 2 we systematized basic exposure pathways of separate categories of people during early and long-term periods after the Chernobyl accident. The squares of the table present basic nuclides or groups of nuclides responsible for the given pathway. In the case of ingestion, only the radionuclides absorbed in the alimentary tract are shown.

The urgently evacuated population of the near zone of the accident were subjected to external gamma exposure from the cloud and to beta and gamma exposure from the radionuclides deposited on the surface during 1-11 days after the reactor explosion. When estimating the committed effective dose, we take into account the role of inhalation of radionuclides from the cloud and of resuspended ones. Food products were subjected to surface contamination. Intake of I, Cs, and Sr radionuclides with local milk could be decreased because population was notified about the accident and prohibited to consume contaminated products. According to the data of our measurements in 1986, the dose in thyroid of some children reached tens of Gray.

The population in contaminated areas has been subjected to exposure during many years. In connection with long-term accumulation of the dose, the contribution into it of the exposure from the radioactive cloud is insignificant in comparison with the external exposure from deposited radionuclides: according to available data of measurements, less than 10 % of the dose during the first year. The contribution of inhalation from the initial cloud and of resuspended radionuclides from wet deposition is also insignificant as compared with ingestion of I, Cs, and Sr isotopes. It is necessary to take into account external exposure of skin by high-energy beta radiation in the first months after the accident. Later on, as the radionuclides decay and deepen into soil, the role of this factor decreases.

Table 2

Main pathways and nuclides of population exposure
after the Chernobyl accident

Category of Exposed Persons	Time after Accident, days	External Exposure		Internal Exposure	
		β	γ	Inhalation	Ingestion
Evacuated Population	1-11	^{106}Ru/Rh ^{144}Ce/Pr ^{132}Te/I	^{132}Te/I ^{131}I IRG*	^{131,133}I ^{132}Te/I TUE**	^{131}I ^{132}Te/I ^{134,137}Cs
Population	< 100	^{106}Ru/Rh ^{132}Te/I	^{132}Te/I ^{131}I ^{134,137}Cs	^{131}I TUE**	^{131}I ^{134,137}Cs ^{89}Sr
of contaminated area	> 100	-	^{134,137}Cs ^{106}Ru/Rh	TUE** ^{106}Ru/Rh ^{144}Ce/Pr	^{134,137}Cs ^{90}Sr/Y

* IRG denotes Inert Radioactive Gases (Kr, Xe).
** TUE denotes radionuclides of TransUranium Elements (Pu, Am, Cm).

The leading factor of internal exposure during the first month after the accident was incorporation of I-131, especially in children with milk. Later on, cesium radioisotopes (Cs-137 and Cs-134) play the leading role in internal exposure of population. Their intake into the body during the first months is determined by surface contamination of the environment, and later it strongly depends on the properties of local soil and the time after contamination.

4. Dosimetry of external exposure

The deterministic model of exposure of different age and social groups of the population has been developed on the basis of experimental investigations in 1986-1994 [5, 6,13]. These studies included hundreds of thousands of measurements of the exposure dose rate in different periods after the accident above the virgin soil, in typical plots of settlements, including residential, industrial, and social buildings; the results of the poll of about one thousand inhabitants of the Bryansk region of Russia [5] and of rural inhabitants of the Ukraine [13] about their mode of behavior during different seasons; the results of 450 analyses of Cs-137 and Cs-134 in profiles of virgin soil taken in 1986-1994 in Russia, Ukraine, Bavaria and Sweden, the results of over 5 thousand measurements of individual doses in inhabitants done with thermoluminescent method [5, 6, 13].

Dr. V. Golikov in Russia [5] and Dr. D. Novak in Ukraine exposed anthropomorphous phantoms of people of different age ranging from one year to adult, containing TL-detectors in many organs showing conditions of exposure to Cs-137 and Cs-134 radionuclides outdoors and indoors, to determine conversion factors from the exposure dose to organs and effective doses.

For reconstruction of the external dose in evacuated population, results of measurements of the exposure dose rate in the town of Pripyat and in many settlements of the 30-km zone, beginning from the 26 April 1986, data of the polling of about 35 thousand inhabitants on the mode of their behavior after the accident and before evacuation together with isotopic composition of environmental samples are used [12].

According to the deterministic model presented on Fig.1 the average annual effective dose E_k in the k-th group of a settlement inhabitants depends on **dose rate in the air** at the height of 1 m above an open plot of virgin soil in this settlement and its vicinity $\dot{D}(t)$; **location factor** LF_i equal to the ratio of dose rate at the i-th typical plot in the settlement to $\dot{D}(t)$; **time factor** TF_{ik} equal to the part of time spent during a year at the i-th plot; **conversion factor** CF_{ik} from the absorbed dose rate in the air to the effective dose.

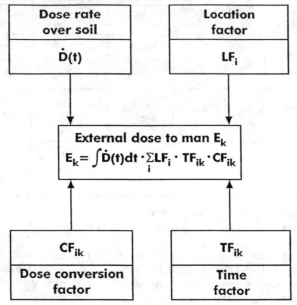

Fig.1. Model of external dose formation in k-th occupational group of population

In April-June 1986, short-lived radionuclides as Te/I-132, I-131, Ba/La-140 etc. made great contribution to the dose rate. This contribution was about 80-90% in May, 40-60% in June and in the whole year 1986. Since 1987, the dose rate was mainly determined by gamma radiation of Cs-134 and Cs-137. Isotopic composition of fallout and initial dynamics of the dose rate are well known [1, 2, 5] and were used in the model. Dose rate in the air $\dot{D}(t)$ is considered as sum of dose rate from short-lived and medium-lived radionuclides and from long-lived Cs radionuclides:

$$\dot{D}(t)=\dot{D}_{sm}(t)+\dot{D}_{Cs}(t), \ \mu Gy \cdot h^{-1} \ (1)$$

$$\dot{D}_{sm}(t)=\sigma_{137}\cdot\sum_r \sigma_r \ /\sigma_{137}\cdot d_r\cdot \exp(-ln2\cdot t/T_r) \ , \ (2)$$

Where: σ_{137}, kBq $\cdot$ m^{-2},   -   average surface activity of Cs-137 on soil in the locality on a day of the accident, 26.04.86;

σ_r , kBq $\cdot$ m^{-2} ,   -   the same quantity for radionuclide r with decay half-period T_r;

d_r , (pGy$\cdot$ m^2)/(Bq$\cdot$ h) -   dose rate coefficient equal to initial dose rate in air 1 m above ground created with the plane thin source of 1 kBq$\cdot$m^{-2} of radionuclide r in soil for dry or wet fallout, respectively.

After decay of short-lived radionuclides and initial intensive migration of Cs-134, Cs-137 in the environment of settlements further dynamics of the dose rate is determined mainly by decay and penetration of cesium radionuclides in soil. Investigation of about 450 soil samples during eight years was the basis for obtaining by P.Jacob and V.Golikov of two-exponential expression for the average dose rate of Cs-134, Cs-137 gamma radiation in the open area at a distance of more than 100 km from Chernobyl NPP:

$$\dot{D}_{cs}(t) = \sigma_{137}(\ d_{137}\ \cdot \exp(-\ln 2\ t/30) + 0.54 \cdot\ d_{134}\ \cdot \exp(-\ln 2 \cdot t/2.1)) \cdot AT(t),\ \mu Gy \cdot h^{-1}\ (3)$$

$$AT(t) = 0.57\ \cdot \exp(-\ln 2 \cdot\ t/1.5) + 0.57\ \cdot \exp(-\ln 2\ \cdot\ t/50),\ rel.\ un.\ (4)$$

where: 0.54, rel..un.,   -   average ratio of Cs-134 to Cs-137 activities in Chernobyl fallout;

30 and 2.1, years,   -   half-periods of Cs-137 and Cs-134 decay, respectively;

AT(t), rel.un.,   -   attenuation function for dose rate in air due to migration of Cs radionuclides in soil. Second term of AT(t) is obtained with the help of data from [15];

d_{134} and d_{137} are taken as for a plane thin source at the soil depth of 0.5 g. cm^{-2}.

Table 3

Reconstruction and prognosis of the average effective dose of external exposure of an adult population in the areas contaminated after the Chernobyl accident.

Country, reference	Population group	E/σ_{137}, μSv per kBq $\cdot$ m^{-2}				
		1 year	1986-1990	1991-1995	1996-2056	1986-2056
Ukraine [13]	Rural	16	33	-	(32)**	(74)**
Russia [5,6]	Rural	13	28	8	28	64
	Urban	8	17	6	17	40

σ_{137} is given for 1986
** Preliminary estimates

As follows from the described model further decrease of dose rate in open air due to continued deepening and decay of Cs-137 is expected with a half-period of about 19 years or 3 to 5 per cent in a year. In Table 3 the generalised estimations are presented both in the external exposure dose accumulated during past 10 years and expected in the future. The average dose for adult inhabitants of rural (villages, settlements) and urban (towns with the population between 10 and 100 thousands persons) localities with socio-professional composition of population and structure of housing resources typical for temporary latitudes of the European part of former USSR in 80-th and 90-th is given . The dose is standardised to average soil surface contamination with Cs-137 in the locality and its vicinity in 1986. For more specific groups of inhabitants, the average dose calculated by means of Table 3 should be multiplied by factor of 0.6 to 1.7. Ratio of the dose in inhabitants of rural and urban localities with equal level of contamination is about 1.5. From the Table 3 one can conclude that the inhabitants of the contaminated zone have obtained during 10 years about 60% of the external dose and about 40% will be received in future.

To validate the model used for dose calculations Fig. 2 presents the time dependence of annual effective dose of external exposure of rural inhabitants of the Bryansk region of Russia estimated according the model described above and from large-scaled individual dose measurements with thermoluminescent dosemeters [6]. Agreement between calculated curve and measurements data corraborates correctness of accumulated and forcasted dose estimates.

According to the data by I. A. Likhtarev and V. K. Chumak [12], the reconstructed average effective dose from external gamma exposure in inhabitants of Pripyat town evacuated on 27 April 1986 was 12 mSv, and in inhabitants of the villages of the 30-km zone evacuated before 4 May - 18 mSv. In some Ukrainian villages of the 30-km zone, the average dose during this period reached 110-130 mSv and individual ones - up to 383 mSv.

After evacuation of inhabitants of the 30-km zone, villages located in the Bryansk-Belorussian spot became the most contaminated with radionuclides inhabited areas. The average effective dose accumulated in inhabitants of the contaminated area during past ten years was from 1.3 mSv for $\sigma137 = 0.04$ MBq/m2 (1 Ci/km2) to 150 mSv for $\sigma137 = 4$ MBq/m2 (v. Zaborje, Bryansk region). The dose in separate groups of population, varies by 0.6-1.7 times, and in individuals - up to the factor of three to both sides.

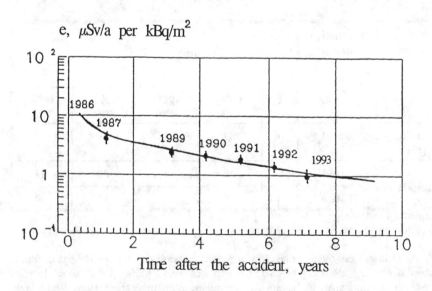

Fig. 2. Time dependence of annual effective dose of external gamma radiation in rural population of Bryansk region [5,6].

5. Dosimetry of internal exposure

The structure of the simple model of internal exposure of a man located in area contaminated with radionuclides is presented on Fig. 3. Main pathways of radionuclide intake in a body of a man of *k-th* age and gender group are considered: inhalation with average inhalation rate IR_k (m³.c⁻¹) of the air with the time-dependent concentration of *r*-th radionuclide AC_r (Bq.m⁻³) and ingestion of the set of *f*-th food products including drinking water with consumption rate CR_{fk} (kg.day⁻¹) with time dependent specific activity SA_{fr} (Bq .kg⁻¹). Data on radionuclide content in the air, drinking water, agricultural and natural food products are obtained during current radiation monitoring and radioecological studies. Rate of air inhalation by persons of different age and gender for different activities is well known from physiological studies. The consumption rate of different food products varies significantly both from age and sex and of local traditions of agricultural production, collection of natural food, dietary habits. For internal dose estimation after the Chernobyl accident these data were obtained by population polls [9,19] and from analysis of statistical data. Dose conversion factors are usually taken from ICRP-56 [16].

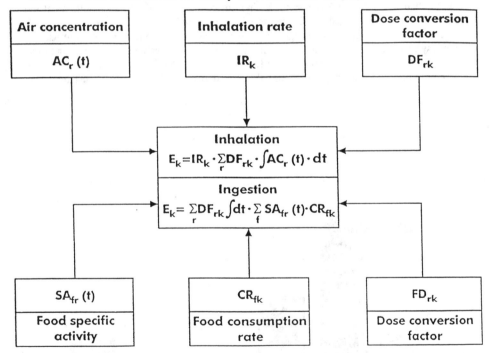

Fig.3. Model of internal dose formation in *k*-th age and gender group of population
   r - radionuclide index;
   f - food product index

Not many reliable experimental data on radionuclide content in air and food products were collected during the most important early period of the Chernobyl accident. This is why for the purposes of internal dose estimation about 350 thousand measurements of I-131 in thyroid gland were done in inhabitants of the three republics, 1 million measurements of Cs-

134,137 content in the body, hundreds of analyses for Sr-90 and tens of analyses for Pu isotopes in autopsy samples of tissues were performed. The results of these measurements are the basis for reconstruction of internal exposure dose in different groups of population of the contaminated area. Both these reconstructions and prognose require taking into account the time dependence of intake in the body.

5.1. Dose of radioiodine radiation in thyroid

As the function of intake of radioiodine in inhabitants of contaminated areas the two-exponential function was used, describing dynamics of iodine-131 content in milk [9], or the sum of this function with the function of inhalatory intake during the first days, or the similar simplified function that represents homogeneous intake during 10-15 days, and consequent decrease of intake with the period of 5 days [8] typical for natural decontamination of milk from I-131. The difference in results of the dose reconstruction by means of different intake models does not exceed 30 %.

By means of the dosimetric models described briefly above and in detail in [8, 9,20], the individual doses from I-131 content in thyroid were calculated. In each settlement the average dose value was usually estimated for six age intervals: under one year, 1-2, 3-6, 7-11, 12-17 years, and adults. In an age group, the frequency distribution in the dose usually has an asymmetric form close to lognormal one. The maximal value of the individual dose sometimes exceeds mean arithmetic value for the settlement by 3-5 times. Figure 4 shows an example of age dependence of average doses in thyroid of the inhabitants in Kiev [21]

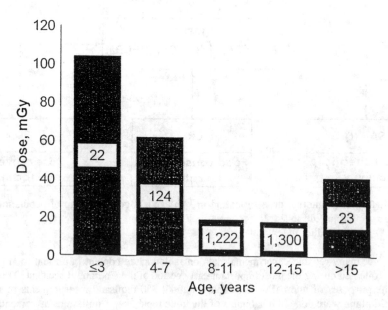

Fig. 4. Age dependence of the average thyroid dose in Kiev residents after the Chernobyl accident [21]

The relation of the average dose in thyroid determined with the measurements of the thyroid in May-June 1986 with different parameters of radioactive contamination of the environment in Russia (surface activity of Cs-137 on soil σ_{137}, dose rate in air on 10-12 May 1986, average concentration of I-131 in milk on 5-12 May 1986, the content of radiocesium in the body of adult inhabitants in the same village in August-September 1986) was analyzed within the technique of linear regression. The correlation coefficients of the parameters in all cases were highly significant: within the limits from 0.86 to 0.95 [8]. According to the data by V.T.Khrush and Yu.A.Gavrilin, this relation is nonlinear in the Southern Belorussia: the dose is relatively higher in the area of lower radioactive contamination [19].

Actual average levels of exposure in adult inhabitants of Pripyat town, of the most contaminated districts of the Kiev and Zhitomir regions of Ukraine, and of the Gomel and Mogilev regions of Belorussia, are estimated about 0.4 Gy, and in the Bryansk region of Russia - 0.1-0.2 Gy. Accordingly, average doses in children of preschool age (1-6 years) are by 3-10 times higher and reach 3-5 Gy in separate settlements of these districts. Individual doses in children of Russia reached 8-10 Gy, and in Ukraine and Belorussia - 30 and even 50 Gy [8,9,21].

5.2. The dose from cesium, strontium and transuranium radionuclides

The process of dose formation from internal exposure to population due to intake of long-lived Cs-137, Cs-134, and Sr-90 through food chains can be separated into the stages of the surface contamination of vegetation in spring and summer 1986 and of the system (root) transfer of them from soil to plants since summer and autumn 1986. The parameters of deposition and migration at the first stage strongly depend on the distance from the source of release, the state of vegetation and weather conditions, and on the second stage - on agrochemical properties of soil and on agricultural practice. During a long period of time, the main amount of radiocesium (60-90 %) was incorporated in the body of local inhabitants with dairy and meat products.

Beginning from the autumn 1986, the "soil - plant" system is the most variable link of the food chain, which determines protracted incorporation of Cs and Sr radionuclides to the body of man with vegetable and animal products. The types of dominating soils in different regions of radioactive contamination differ considerably: black earths and grey forest soils in the Tula and Orel regions of Russia, turf-podzol soils in Belorussia and the Bryansk region, peat soils in the west of Ukraine and Belorussia. Depending on the type of soil, the average value of the cesium transfer factor to grass varies almost by three orders of magnitude. The transfer factor for Sr-90 from soil to natural grasses varies within narrower limits, namely up to 60 times [7, 17].

Besides soil properties, the process of natural fixation of cesium radionuclides in soil structures with time after the radioactive contamination of the area considerably influences their content in agricultural production. Fig. 5 illustrates the dynamics of the aggregated transfer factor for Cs-137 from soil to milk of cows pastured in semi-natural pastures. The specific activity of Cs-137 in grass, milk and meat of cattle decreased since 1987 to 1991-92 by 1-2 orders of magnitude. Similar decrease is observed in the content of Cs-137 in agricultural production from ploughed soils: cereals, potatoes, etc. The period of decrease of Cs-137 content in all these products in districts with turf-podzol and black soil varied within 0.6-2 years, and, was on the average, 1.2 years. In some districts of Rovno region in Western Ukraine where peat soil dominates transfer factors soil-vegetation-milk are significantly higher and their decrease is lower than in other areas. Along with annual decrease, the content of radionuclides in milk and meat varies considerably during a year in seasons depending on the ration of the cattle: it is increased during the pasture period and decreases when feeding cattle by fodder root-crops and mixed fodder during the stalled period . Since 1991-92, there is the tendency to slowing down of the process of natural decontamination of agricultural products, in the first place, in the black earth zone. On the contrary, cesium radionuclide content in many species of natural food products collected from the forest (mushrooms, berries) did not reveal significant decrease since 1986 extra to radioactive decay - see Fig. 5.

In the most contaminated areas of the former USSR, the measures of radiation protection of population were widely performed beginning from May 1986, including prohibition for consumption of local food products, in the first place, of dairy, meat and natural ones, and supply of noncontaminated food products. These actions considerably decreased intake of radionuclides in the body. The degree of decrease of intake in different settlements widely varies and is characterized by the factors from 1.5 to 15 [18]. Therefore, in the most contaminated area, the best method for assessment of the current dose in inhabitants is measurement of Cs-137 and Cs-134 content in the body, and calculation of actual dose during the investigated period, taking into account seasonal variations.

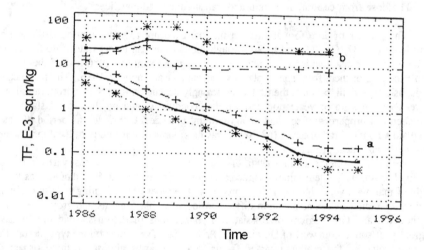

Fig. 5. Cs-137 transfer factor from soil to milk (a) and mushrooms (boletus luteus, b) in Bryansk region as a function of time after the Chernobyl accident [V. Shutov and G. Bruk, in press]

The maximal content of Cs-134,-137 in the body of inhabitants of the Chernobyl accident zone was reached in the first months after the accident. In summer 1986 the average content of the sum of these nuclides in the body of adult inhabitants of some villages in Belorussia and Russia reached 0.4-0.6 MBq, and individual one - up to 4 MBq. In the Novozybkov town of the Bryansk region with population of 46 thousands, the average Cs-134,-137 content in the body in summer 1986 was about 0.06 MBq. Further on, in the settlements, where "radiation-free" food products were supplied, the content of radiocesium decreased at first with the period less than one year, then with a period of 1-1.5 years. This process considerably slowed down in 3-4 years. In the villages, where consumption of local food products continued, the "fast" part of the process was absent, and slowing down of the decrease took place also in 3-5 years.

Table 4 shows in the generalized form the assessments of the average effective dose of internal exposure in adult rural inhabitants of Russia and Ukraine during different periods after the Chernobyl accident. The assessments are based on the above described radioecological data and dosimetric model for conditions without consequent active countermeasures, as applied to the areas with dominating peaty, turf-podzol sandy and black earth soils. According to our data, the dynamics of external exposure practically does not depend on soil type. It is seen with the comparison in Tables 3 and 4 that in the absence of protective measures the contribution of internal exposure to the total dose during the first year is greater than from the external one, for all these soil types. Already during the first year, the higher cesium transfer factor to vegetation in poor soils tells on the value of the dose. Further on, the dose of internal exposure prevails only on peaty and turf-podzol soil, and in black earth cesium-137 is strongly fixed, and weakly migrates to man through food chain. Actual doses in population of the area, where active countermeasures of protection from internal exposure (replacement of local food products, agricultural countermeasures) are considerably lower than those calculated according to Table 4.

Table 4

Reconstruction and prognosis of the average effective dose of external
exposure of an adult rural population in the areas
contaminated after the Chernobyl accident.

Country, reference	Soil type	E/σ^*_{137}, μSv per kBq $\cdot$ m-2				
		1 year	1986-1990	1991-1995	1996-2056	1986-2056
Ukraine [23]	Peaty	20-150	90-570	-	-	-
Russia [7]	Turf-podzol	90	150	20	14	184
	Black	28	29	1	1	31

*   Prognosis does not take into account the future consumption of natural food products
** σ_{137} is given for 1986

Prognose of population internal exposure in particular area   as well as current exposure depends strongly from the agricultural soil type, level of its contamination and from dietary habits of people. In the areas where black and turf-podzol clay soils dominate contribution of internal exposure in the total dose is insignificant since 1987 and in the future. But in the areas with sandy and especially peat soils this contribution remains significant for a long time and needs special attention.
The decrease of internal dose with time is expected different for population groups consuming agricultural food only which contains low Cs-137 activity and for people also consuming local natural food, especially forest mushrooms with high concentration of this nuclide. Typical ratio of Cs-137 concentration in forest mushrooms and cow milk obtained in 1990-th in the same area contaminated due to Chernobyl accident mainly varies in the

range of 10 to 100. Future decrease of Cs-137 content in agricultural food products is expected with the effective half-period of about 10 years, according to previous observations [20] but in particular areas with peat soils - with half-period up to 20 years. The intake of Cs radionuclides with natural food did not change significantly during the last 10 years, except for radioactive decay. This means, relative contribution of natural food in internal dose increases with time. For mushroom-eaters in the contaminated areas it already contributes from 20 to 80% of the annual internal dose. According to modern models of radionuclide migration in forest, content of Cs-137 in "forest gifts" is not expected to decrease significantly faster than with half-period of about 30 years.

Thus, we expect for decrease of internal dose in persons not consuming natural food with half-period of about 10 years in most regions and up to 20 years in some regions with peaty soils. For people intensively consuming forest food, especially mushrooms with the rate of about 10 kilogramme per year and more, we can expect higher level of internal exposure decreasing with the half-period up to 30 years. Prediction of internal exposure dose for inhabitants of intermediate zone not consuming forest products is quantitatively presented in Table 4. For soil types indicated in this table consumption of natural products may increase future internal dose by a factor of 2 to 4.

In our methods of prognose we also take into account ingestion of Sr-90 and inhalation of transuranium isotopes with soil particles. Due to low content of Sr-90 in the Chernobyl release and fallout outside the 30-km zone its contribution in the internal effective dose does not exceed 5 - 10%, according to intake calculation and direct measurements of Sr-90 in human bones (autopsy samples). Similar contribution from the inhalation of Pu-238, -239, -240 and Am-241 originated from Pu-241 will not exceed 1% even for outdoor workers.

Aknowledgements. This work was partially supported by European Commission according the Contract COSU-CT93-53.

References

1. Sources, effects and risks of ionizing radiation. Annex D. Exposures from the Chernobyl accident. UNSCEAR, 1988 Report, United Nations, New York, 1988.

2. Izrael, Yu.A.; Vakulovskii, S.M.; Vetrov, V.A.; Petrov, V.N.; Rovinskii, F.Ya.; Stukin, Ye.D. Chernobyl: Radioactive contamination of the environment. Leningrad: Gidrometeoizdat; 1990 (in Russian).

3. Ilyin L.A., Balonov M.I., Buldakov L.A. and others. Radiocontamination patterns and possible health consequences of the accident at the Chernobyl nuclear power station. J.Radiol.Prot. 10(1) 3-20 (1990).

4. The Chernobyl Papers. V.I: Doses to the Soviet Population and Early Health Effects Studies Ed. by S.E.Merwin and M.I.Balonov, Research Enterprises, Richland, 1993, 440 p.

5. Golikov V.Yu.; Balonov, M.I.; Ponomarev, A.V. Estimation of external gamma radiation doses to the population after the Chernobyl accident. In [4], p.247-288.

6. Erkin V.G.; Lebedev O.V. Thermoluminescent dosemeter measurements of external doses to the population of the Bryansk region after the Chernobyl accident. In [4], p.289-312.

7. Shutov, V.N., Bruk, G.Y., Balonov, M.I., Parkhomenko, V.I., and Pavlov, I.Y. Cesium and strontium radionuclides migration in the agricultural ecosystem and estimation of internal doses to population. In [4], p.167-220.

8. Zvonova, I.A. and Balonov, M.I. Radioiodine dosimetry and prediction of thyroid effects on inhabitants of Russia following the Chernobyl accident. In [4], p.71-126.

9. Likhtarev I.A., Shandala N.K., Gulko G.M., Kairo I.A., Chepurny N.I. Ukrainian

thyroid doses after Chernobyl accident. Health Physics 64, N6, 594-599 (1993).

10. Buzulukov, Yu.P. and Dobrynin, Yu.L. Release of radionuclides during the Chernobyl accident. In [4], p.3-22.

11. Orlov, M.Yu., Snykov, V.P., Khvalensky, Yu.A., Teslenko, V.P., Korenev, A.I. Radioactive contamination of Russian and Belorussian territory after the Chernobyl accident. Atomnaya Energiya, v.72, issue 4, 1992, pp.371-376 (in Russian).

12. Likhtarev T.A., Chumack V.V. and Repin V.S. Retrospective reconstruction of individual and collective external gamma doses of population evacuated after the Chernobyl accident. Health Physics 66, N6, p.p. 643-652 (1994).

13. Likhtarev I., Kovgan L., Novak D., Vavilov S., Jacob P. and Paretzke H. Effective doses due to the Chernobyl external irradiation for different population groups of Ukraine. Submitted to Health Physics Journal (1994).

14. Jacob, P.; Meckbach R. External exposure from deposited radionuclides. Proceedings of the seminar on methods and codes for assessing the off-site consequences of nuclear accidents, May 7-11, 1990. Athens; 1990.

15. Kevin Miller, S.L.Kuper, Irene Helfer. Cs-137 fallout depth distributions in forest versus field sites: implications for external gamma dose-rates. J.Environm. Radioact.12 (1990), 23-47.

16. Age-dependent Doses to Members of the Public from Intake of Radionuclides. ICRP Publication 56. Annals of the ICRP, V.20. 1989.

17. Shutov V.N. Influence of soil properties on Cs-137 and Sr-90 intake to vegetation. In: "Report of the Working Group Meeting". Madrid, Spain, 1992, p.11-15.

18. Balonov M., Travnikova I. Importance of diet and protective actions on internal dose from Cs radionuclides in inhabitants of the Chernobyl region. In [4], p.p. 127-166.

19. Gavrilin Yu. I., Khrusch V.T., Shinkarev S.M. Internal irradiation of the thyroid in the residents of regions contaminated with radionuclides in Byelorussia. Medical Radiology, N6, 15-20 (1993); (In Russian).

20. Ionizing radiation: sources and biological effects. UNSCEAR 1982 Report, United Nations, New York, 1982.

21. Likhtarev T.A., Gulko G.M., Kairo T.A., Los T.P., Henricks K. and Paretzke H.G. Thyroid doses resulting from the Ukraine Chernobyl acccident - Part I: Dose estimates for the population of Kiev. Health Physics 66, N2, p.p. 137-146 (1994).

22. JSP-5. Pathway Analyses and Dose Distributions, Ed. by P. Jacob. GSF, D-85746, Oberschleissheim, Deutschland, 1995.

23. Likhtarev T.A., Kovgan L.N., Vavilov S.E., Gluvchinsky R.G., Perevoznikov O.N., Litvinets L.N., Anspaugh L.R., Kercher T.R., Bouville A. Internal exposure from the ingestion of foods contaminated by Cs-137 after the Chernobyl accident. Submitted to the Health Physics Journal (1995).

Exposures from External Radiation and from Inhalation of Resuspended Material

Peter Jacob and Paul Roth
GSF-Institut für Strahlenschutz, Oberschleißheim, Deutschland,
Vladislav Golikov, Michael Balonov, and V. Erkin
Insitute of Radiation Hygiene, St. Petersburg, Russia,
Ilya Likhtariov
Scientific Center for Radiation Medicine, Kiev, Ukraine,
Eugeni Garger
UAAS-Institute of Radioecology, Kiev, Ukraine, and
Valerj Kashparov
Ukrainian Institute of Agricultural Radiology, Kiev, Ukraine

Abstract. In the modelling of external exposures due to cesium released during the reactor accident of Chernobyl, gamma dose rates in air over open undisturbed sites are considered to be different according to the unsoluble fraction in the deposit. This is taken into account by forming different classes according to the distance from the Chernobyl NPP. The effect of the different migration behaviour in these distance classes on the gamma dose rate in air is found to increase with time. Predictions of gamma dose rates in air are based on measurements of the nuclear weapons tests fallout. Various population groups in the CIS countries are defined according to their place of residence (rural or urban), their occupation or age (indoor resp. outdoor workers, pensioners, school-children, or preschool-children), and their kind of residence (wooden, brick, or multistorey house). Model results for various population groups are compared with the results of TLD-measurements of individual external exposures. For the calculation of inhalation doses, the new ICRP model for the respiratory tract was used. The dose assessments were conducted for measured size resolved activity distributions of resuspended material, obtained at different locations and for several kinds of agricultural operations. Inhalation doses vary considerably with respect to different kinds of work. Tractor drivers receive much higher doses than other agriculturual workers, especially when the cabin window of the tractor is open. Effective doses due to the inhaltion of resuspended plutonium are assessed to be a few µSv per initial deposit of one kBq/m^2. Inhalation doses from ^{137}Cs are usually smaller by an order of magnitude than the doses from Pu, provided a high solubility is assumed for resuspended Cs.

1. Introduction

Several years after the reactor accident of Chernobyl, radiation exposures of the population in contaminated areas still concern the living conditions. Currently and in the next decades, main exposure pathways are due to the deposited cesium. For most of the contaminated areas, the external irradiation contributes most to the population exposures [1,2]. An improved modelling of this exposure pathway is the main scope of the second part of the present paper.

As a general rule, the internal exposure due to incorporation of cesium by ingestion dominates the population exposure in regions with a high transfer of cesium to grass and other foodstuffs. The areas with the highest population doses due to the Chernobyl accident belong to these high transfer regions. Recent developments in improved modelling of ingestion doses are described in [3]. Another pathway of concern was in the recent years the potential hazard due to resuspension of radioactive material. A critical group consists here of agricultural workers on fields contaminated with plutonium. Potential inhalation doses are studied in the third part of the paper.

2. External exposures

The purpose of the present study is to evaluate data and develop a model for the external doses distributions of the population in contaminated areas. The developed model is assumed to be applicable from the year 1990 to the end of the lifetime of people being born before the reactor accident of Chernobyl.

At the time t after a deposition of a cesium isotope, the effective dose rate $\dot{H}_i(t)$ of a member of a population group i may be calculated by

$$\dot{H}_i = A \cdot \dot{g} \cdot \exp(-\lambda \cdot t) \cdot r(t) \cdot \sum_j f_j(t) \cdot p_{ij}(t) \cdot k_{ij}(t), \tag{1}$$

where A is the activity deposited per unit area on a reference site, $\dot{g}$ the gamma dose rate in air per activity per unit area with a reference distribution of the cesium in the ground, λ the decay constant, and r(t) the gamma dose rate in air at the reference site divided by the gamma dose rate in air for the reference distribution. The summation index j indicates types of locations, f_j the gamma dose rate in air at a location j relative to the gamma dose in air at the reference site, $p_{ij}(t)$ the relative frequency of stay for members of population group i at location j and $k_{ij}(t)$ the conversion factor from the gamma dose in air to effective dose.

In this report, reference sites are considered to be open fields with undisturbed soil, normally lawns or meadows. A plane source below a soil slab with a mass per unit area of 0.5 g cm^{-2} has been chosen as a reference distribution to approximate the energy and angular distributions of the radiation field in air over an undisturbed field during the first years after the deposition. For this geometry, a value for $\dot{g}$ of 1.7 nGy h^{-1}per kBq·m^{-2} has been obtained for ^{137}Cs and of 4.7 nGy h^{-1}per kBq·m^{-2} for ^{134}Cs [4].

2.1 Gamma dose rates in air over open fields

A data base on the attenuation of the gamma dose rate in air due to the migration of cesium into the soil has been established. Main sources of information were [5], [6] and new results obtained in the framework of the projects JSP 5* and ECP 5**. The data base has about 450 data sets, each containing besides other information the time of measurement, the value of r(t), and the distance from the Chernobyl nuclear power reactor plant.

* Supported by the EC under contract COSU-CT94-0091
** Supported by the EC under contract COSU-CT94-0081

The data set indicates the factor $r(t)$ to decrease with increasing distance from the release point. Two factors may be responsible for this observation. First, cesium bound to fuel and condensation particles is known to be less available for migration than cesium that has been evaporated and then attached to the background aerosol. Second, in the data base for short distances, observation points with dry deposition dominate, whereas for long distances observation points with wet deposition dominate. Data grouped in three distance groups are given in Table 1.

Prediction of the future behaviour of $r(t)$ may be based on observations of the fallout from nuclear weapons tests. An analytical approximation of the data for measurement sites at large distances from the release point and for the nuclear weapons tests in the form

$$r(t) = a_1 \cdot \exp(-\ln2 \cdot t/T_1) + a_2 \cdot \exp(-\ln2 \cdot t/T_2), \qquad (2)$$

yielded half lifes to be $T_1 = 1.5$ years and $T_2 = 50$ years. Based on the observation that fuel particles tend to dissolve over a period of several years in the environment, the same half lifes were assumed for the other distance categories and obtained values for a_1 and a_2 are indicated in Table 1.

The weapons fallout was deposited over long periods and column experiments have shown that under these conditions the initial penetration of the cesium into the soil is negligeable [7]. However, the sites in the far field from Chernobyl studied here had different meteorological conditions and an initial migration of the cesium during the wet deposition. It may be assumed that this initial difference retains and that the Chernobyl cesium will have migrated more deeply than corresponding profiles for the weapons fallout at the same time after deposition. Nevertheless both data sets are pooled here, since no better information is available for long times after deposition.

2.2 Location factors

In settlements of urban and rural type, the characteristics of radiation field differ from those over an open plot of virgin land due to shielding and to varying source distributions on the surfaces as a result of different deposition velocities, run-off and weathering. Location

Time after deposition	Distance category		
(years)	$D \leq 100$ km	100 km $< D \leq 1000$ km	$D > 1000$ km
	$a_1=0.48; a_2=0.81$	$a_1=0.60; a_2=0.63$	$a_1=0.53; a_2=0.51$
0.75	1.17 (1.15)	0.98 (1.05)	0.87 (0.88)
2.75	0.88 (0.93)	0.84 (0.78)	0.67 (0.64)
4.75	0.87 (0.82)	0.68 (0.66)	0.55 (0.53)
6.75	0.79 (0.77)	0.60 (0.60)	0.46 (0.48)
24	— (0.59)	— (0.45)	0.39 (0.36)
30	— (0.54)	— (0.42)	0.33 (0.33)

Table 1. Average values for $r(t)$ of entries in data base for the first seven years after Chernobyl and derived from measurements of the atomic weapons tests fallout. The values in parentheses give the results of Eq. (2) with the paramters indicated in Fig. 1.

factors have been measured with thermoluminescence detectors in the higher contaminated areas of Ukraine ($A^{Cs137} > 185$ kBq $\cdot$ m^{-2}) in the period 1987-1989 [8]. In Russia, first measurements were performed in summer 1989 [9] in three large villages of the Bryansk region with Cs-137 activities per unit area above 1000 kBq/m^2 (Nikolayevka, Yalovka and Svyatsk). During this campaign, several thousand gamma dose rate measurements were performed in different points in the settlements and in their vicinity. The analysis included only data obtained in villages before decontamination actions were taken. Similar measurements were performed in the period 1992 to 1994 in ten villages in the contaminated area of Russia. The results obtained in the two countries were found to be consistent. Location factors measured after 1987 in rural environments were found to be independent of time. Values are given in Table 2.

The presence of a snow cover during the time period November-March reduces according to experimental measurements performed in the Brjansk region in the average the value of the dose rate in air over virgin plots by a factor of 0.76. Since the average reduction over streets may be assumed to be less, location factors during winter time might be a little higher than during summer time. This is confirmed by TLD-measurements of individual doses performed in the same settlements in winter and in summer time. According to these measurements, the mean reduction of the annual external dose to snow cover of 0.94 was derived. This small effect is not considered in the following.

2.3 Relative frequencies of stay

Information about relative frequencies of stay of the adult rural population at various locations were obtained in Ukraine from 8984 responses of evacuees from the 30 km zone to a questionnaire [8]. In Russia, in 1989, 1992, and 1993 corresponding responses of the rural population to questionnaires were obtained and results are summarized in Table 2.

Location	f_j	p_{ij} (anual)				
		Indoor workers	Outdoor workers	Pensioners	School children	Preschool children
Living areas						
house (wooden)	0.13					
house (brick)	0.07	0.50	0.48	0.68	0.60	0.52
multi storey house	0.02					
outside of houses	0.55	0.21	0.15	0.30	0.24	0.14
Work areas						
buildings	0.07	0.25	0.07	0.0	0.0	0.0
multi storey house	0.02	0.0	0.0	0.0	0.15	0.25
work yard	0.30	0.03	0.07	0.0	0.0	0.09
ploughed field	0.50	0.0	0.18	0.0	0.0	0.0
virgin land	1	0.0	0.04	0.0	0.0	0.0
Rest areas						
forest, meadow	1	0.01	0.02	0.02	0.01	0.0
$\sum p_{ij} \cdot f_j$ *	—	0.22/0.19/ 0.17	0.31/0.28/ 0.26	0.27/0.23/ 0.20	0.22/0.19/ 0.16	0.18/0.14/ 0.12

* The first number is for residents in wooden houses, the second in brick houses and the third in multistorey houses

Table 2. Location factors f_j and relative frequencies of stay p_{ij} for rural population groups.

2.4 Effective dose per gamma dose in air

The migration of the radionuclides into the soil changes the spectral-angular distribution of the photon fluence exposing the human body. In principle, the value of k_{ij} will change as well. To estimate experimentally conversion factors for a real vertical radionuclide distribution in soil, in summer 1991 and 1992 a series of phantom experiments was carried out. In the experiments three antropomorphical phantoms were used, the Alderson Rando phantom presenting adults and two phantoms of children of 5 and 1 year of age (produced by ATOM Ltd, Riga, Latvian Republic). As experimental sites, two open sites .: virgin land, one open ploughed site, and a location inside a wooden house in contaminated areas were chosen.

The experimental results for the three outdoor locations agreed within the limits of only 10% for each of the phantoms. Results of Monte Carlo calculations [10] for the reference distribution (plane source at 0.5 g/cm^2) of radionuclides were found to be intermediate between the experimental results for outdoor and indoor locations. In the present model a value of 0.9 Sv•Gy^{-1} for preschool children (0-7 years), of 0.8 Sv•Gy^{-1} for school children (8-17 years) and of 0.75 Sv•Gy^{-1} for adults is assumed for all locations.

2.5 Model results

Annual effective doses of rural indoor workers living in woodframe houses are shown in Fig. 1. For comparison results of the UNSCEAR model [11] are given approximating the attenuation due to the cesium migration into the soil by a constant factor. Dose estimates for Russian settlements in 1991 [1] agree within 10 % with the current model. In the present model for the period 1990-2056, annual effective doses of different population groups differ by a constant factor. The factors for rural population groups are given in Table 3. Urban population groups have lower external exposures due to the Chernobyl accident [12]. According to Table 3 rural outdoor workers living in woodframe houses are the critical group for external exposures. Among them, forestry workers and herdsmen receive the highest dose [12].

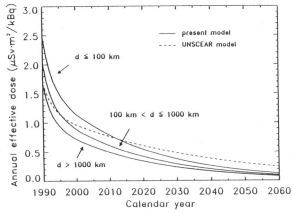

Fig. 1. Annual effective doses due to external exposures of rural indoor workers living in woodframe houses. Results are normalized to the ^{137}Cs activity initially deposited per unit area, a ratio of the ^{137}Cs activity to the ^{134}Cs activity of 1.8 has been assumed. The dotted line represents the UNSCEAR model [11].

Population group	Wooden houses	Brick houses	Multi storey houses
Indoor	1.00	0.86	0.77
Outdoor	1.36	1.23	1.09
Pensioners	1.23	1.05	0.91
Schoolchildren	1.07	0.92	0.78
Preschoolchildren	0.98	0.76	0.65

Table 3. Annual effective doses due to external exposures of rural population groups, normalized to the annual effective dose due to external exposures of rural indoor workers living in woodframe houses.

2.6 Comparison with measurements of individual external doses

Monthly external doses were measured with TLDs during the spring-summer periods of 1990-1993 in 21 settlements of the Bryansk region with average ^{137}Cs activities above 0.4 MBq/m^2 [13]. The conversion factor from readings of an individual dosimeter to the value of the effective dose was determined on the basis of the results of phantom experiments, and it was 0.9 Sv/Gy for adults, 0.95 Sv/Gy for schoolchildren and 1 Sv/Gy for preschoolchildren. Table 4 shows a good agreement between the average doses in the population groups obtained by the model and by measurements. A comparison with measured individual doses showed some of those to be larger than the largest of the average doses in the population groups. Doses to individuals are not the scope of the present model.

3. Inhalation doses

Resuspended material may be inhaled and deposited in the human respiratory system. Radionuclides attached to the inhaled particles lead to radiation doses to the lung and other tissues. Besides the radiation characteristics of a particular radionuclide, the main physico-chemical factors that affect inhalation doses are the size distribution of the inhaled particles and the solubility of the material. Consequently these factors require particular attention.

Year	Monthly effective dose (μSv)					
	Wood houses			Brick houses		
	Model	Measurement	Ratio	Model	Measurement	Ratio
1990	290	285(174)	1.02	271	230(100)	1.18
1991	246	235(414)	1.05	230	205(103)	1.12
1992	216	200 (95)	1.08	202	200 (49)	1.01
1993	195	150(195)	1.30	182	115 (84)	1.58
Average			1.11			1.22

Table 4. Monthly effective doses (μSv) of outdoor workers in the Bryansk region in the periods (April-November) of the years 1990 to 1993. Results of the present model and of TLD measurements [13] are given, the figures in parentheses give the number of measurements.

3.1 Method of inhalation dosimetry

The new ICRP lung model was used for the assessment of inhalation doses due to resuspended radioactivity from the Chernobyl accident. The dose calculations were performed applying the computer program LUDEP 1.0, developed at NRPB [14], which implements the model structure and parameter values approved by ICRP and described in ICRP Publication 66 [15].

Standard values for an adult male worker were assumed for the physiological and activity related parameters. This includes the assumption of a breathing rate of 1.2 m^3/h. Particle density was assumed to be 3.0 g/cm^3 The rate of particle dissolution and subsequent uptake to blood is described by the absorption parameters: type 'F' (fast), type 'M' (moderate) and type 'S' (slow), which are similar to the former lung retention classes D, W and Y, as specified in ICRP Publication 30 [16]. Radionuclide transformation data were taken from ICRP Publication 38 [17], and the biokinetic models were selected as given in ICRP Publication 30 [16]. Doses were calculated for various organs of the body and for the different regions of the respiratory tract, as specified in the new lung model. This includes explicitly the extrathoracic region of the respiratory tract, which was not considered in previous lung models.

3.2 Inhalation doses for agricultural activities

Dose calculations for inhalation of radionuclides were performed for actually measured size resolved aerosol concentrations of radionuclides and surface soil contaminations of the sampling campaigns of ECP1 in 1993 and 1994 [18, 19]. Due to the time-consuming analytical procedures, particularly for α-emitters, the particle size distribution was measured only for several experiments and also the complete radionuclide spectrum was measured only at a few places. As an example, Table 5 shows the airborne radionuclide composition and activity concentrations at one of the experimental sites.

Size resolved measurements of plutonium and ^{137}Cs inside a tractor cabin were made during several kinds of agricultural operation at Novozybkov and Bragin. Doses for the different radionuclides were calculated for each size fraction separately and summed up. The resulting lung doses and effective dose values for Novozybkov (surface soil contamination 1.6 x 10^2 Bq/m^2 for $^{239+240}$Pu and 1.1 x 10^6 Bq/m^2 for ^{137}Cs) are shown in Table 6a, and for Bragin (surface soil contamination 6.4 x 10^2 Bq/m^2 for $^{239+240}$Pu and 5.3 x 10^5 Bq/m^2 for ^{137}Cs) in Table 6b. It is obvious from this table that the doses may vary considerably with respect to different kinds of work, with highest values during cultivation at Novozybkov (with 18 μSv effective dose from $^{238+239+240}$Pu and 5 μSv from ^{137}Cs). However, there is a remarkable increase in ^{137}Cs doses by an order of magnitude, when the

Site	Agricultural operation	Refe-rence			Activity concentration (mBq/m^3)				
			^{137}Cs	^{90}Sr	^{238}Pu	^{239}Pu	^{240}Pu	^{241}Am
Zapolye	harrowing	ECP1 [18]	700	460	2.5	2.2	2.8	4.0

Table 5. Airborne activity concentrations at Zapolye during harrowing on 13 May 1993.

a) Novozybkov 1994

Kind of agricultural	Lung Dose (μSv)		Effective Dose (μSv)	
operation	$238+239+240_{Pu}$	137_{Cs}	$238+239+240_{Pu}$	137_{Cs}
ploughing	38	2.8	8.1	3.1
cultivation	74	4.4	18	4.9
fertilization	-	2.1	-	2.3
potato planting	77	1.8	16	1.9
potato planting (open cabin window)	-	23	-	26

b) Bragin, 1994

Kind of agricultural	Lung Dose (μSv)		Effective Dose (μSv)	
operation	$238+239+240_{Pu}$	137_{Cs}	$238+239+240_{Pu}$	137_{Cs}
ploughing	36	0.6	11	0.6
cultivation	5.1	1.4	1.3	1.5
rye harvesting	10	0.2	3.0	0.2
straw harvesting	6.8	0.1	1.5	0.1

Table 6. Inhalation life time doses per 832 h work (one working year) of tractor drivers for various agricultural activities at Novozybkov (a) and Bragin (b). Airborne activity concentrations and surface soil contaminations from ECP1 campaigns [18].

cabin window of the tractor is open (Table 6a). Unfortunately there are no data available for the plutonium activities under this condition. It is therefore difficult to assess the relevance of the plutonium doses for this situation.

Similar dose calculations for other sites and activities show that the internal doses for tractor drivers are between one and two orders of magnitude higher than for other agricultural workers.

The dose calculations of the above examples were performed using the actually measured size resolved aerosol concentrations. For the solubility of the material, the standard solubility classes (type 'F' for ^{137}Cs, and type 'S' for all plutonium isotopes) were used due to the lack of more precise information. Some experimental work on the solubility of inhaled dust particles was conducted at the UIAR in Kiev. In these investigations, simulated lung fluid (SLF) was used to evaluate the solubility of Cs, Sr, and Pu. In these measurements, all these nuclides exhibited a very low solubility. Although these *in vitro* experiments may be of limited significance for the real situation in the human lung, they strongly emphasize the need for better information of the physico-chemical characteristics of the resuspended material. If resuspended ^{137}Cs would be considered as a 'S'-type material, then the dose coefficients would change significantly. The longer retention of the material in the lung would increase the lung dose by a factor of 60 as compared to 'F'-type material, and also the effective dose would increase by a factor of 8. In this case, the dose contribution from ^{137}Cs would be quite significant.

As stated before, for an estimation of the total inhalation dose from resuspended material from the Chernobyl fallout, the full spectrum of the relevant radionuclides as shown in Table 5 must be taken into account. Table 5 shows that the activity concentrations in air are not very different for the three plutonium isotopes ^{238}Pu, ^{239}Pu and ^{240}Pu for this measuring site. If this finding can be generalized, then the radiation doses from the inhalation of these nuclides can be estimated, even where not all of these isotopes are determined separately, since also the inhalation dose coefficients are very similar for these isotopes. The concentration of ^{241}Pu was not measured here, but since the dose coefficients for this plutonium isotope are by several orders of magnitude lower than for the previous mentioned ones, the resulting doses seem to be not significant. The dose coefficients for ^{241}Am are again similar to those of the plutonium isotopes and dose contributions from this nuclide can be roughly estimated from the above activity ratios in the cases where no experimental data are available. It should be kept in mind, however, that americium is generally considered to be of considerably higher solubility than plutonium. As for ^{137}Cs, also for ^{90}Sr significant dose contributions must be expected if the solubility of the resuspended material is low.

With regard to the these uncertainties, it is obvious that it is difficult to derive an estimate for the total internal dose from the inhalation of resuspended radioactive material. With sufficient care it can be concluded, however, that even at sites inside the 30 km zone, life time doses per year of work generally will hardly exceed 200 µSv for lung dose and 35 µSv effective dose for agricultural workers.

4. Conclusions

A deterministic model has been developed to calculate annual external exposures to population groups in areas of Russia, Ukraine and Belarus that have been contaminated after the Chernobyl accident. Rural outdoor workers living in woodframe houses are the group with the highest average dose. In the period 1990 to 1993 annual external doses were obtained to be a few µSv per initial ^{137}Cs deposit of one kBq per m^2.

Effective doses due to the inhalation of resuspended plutonium were assessed to be a few µSv per initial ^{238}Pu + $^{239/240}$Pu deposit of one kBq per m^2, if 800 hours of agricultural work under dusty conditions are assumed. Therefore, effective doses due to the inhalation of resuspended material are under these conditions of the same order of magnitude as external exposures, if the ^{238}Pu + $^{239/240}$Pu activity initially deposited is of the same order of magnitude as the ^{137}Cs activity. For lung doses this would be the case if the ^{238}Pu + $^{239/240}$Pu activity was one fifth of the ^{137}Cs activity. The probability to find areas contaminated by the Chernobyl accident with such high relative plutonium activities is low even within the 30 km zone.

Acknowledgement: The present work was supported by the European Commission under the contracts COSU-CT93-0039, COSU-CT-93-0053, COSU-CT-94-0077, and COSU-CT94-0091. The authors also would like to thank Dr. Jochen Tschiersch for support in acquiring ECP1 data

References

[1] Reference book on the radiation situation and exposure doses in 1991 for the population of the districts of the Russian Federation affected by the radioactive contamination after the Chernobyl accident. Edited

by I. Balonov, Institute of Radiation Hygiene, State Committee on Sanitary and Epidemiological Control, St. Petersburg, 1993.

[2] Catalog of exposure doses of the population of settlements of the Republic Belarus. Ministry of Public Health, Scientific Research Institute of Radiation Medicine, Minsk, 1992.

[3] I.A. Likhtarev, L. Kovgan, M. Balonov, M. Morrey, P. Jacob, G. Pröhl, Assessing Internal Exposures and the Efficacy of Countermeasures from Whole Body Measurements. First International Conference of the European Commission, Belarus, Russian Federation, and Ukraine on the Radiological Consequences of the Chernobyl Accident, Minsk, 1996. These Proceedings.

[4] P. Jacob, H. Rosenbaum, N. Petoussi, M. Zankl, Calculation of Organ Doses from Environmental Gamma Rays Using Human Phantoms and Monte Carlo Methods. Part II: Radionuclides Distributed in the Air or Deposited on the Ground. GSF-Bericht 12/90. GSF, D-85764 Oberschleißheim, Deutschland, 1995.

[5] P. Jacob, R. Meckbach, H.G. Paretzke, I. Likhtarev, I. Los, L. Kovgan, I. Komarikov, Attenuation Effects on the Gamma-Dose Rates in Air after Cesium Depositions on Grasslands. *Radiat. Environ. Biophys.* **33** (1994) 251-257.

[6] K.M. Miller, J. L. Kuiper, I. K. Helfer, [137]Cs Fallout Depth Distributions in Forest Versus Field Sites: Implication for External Gamma Dose Rates. *J. Environ. Radioactivity* **12** (1990) 23-47.

[7] W. Schimmack, K. Bunzl, F. Dietl, D. Klotz, Infiltration of Radionuclides with Low Mobility (^{137}Cs and ^{60}Co) into a Forest Soil. Effect of the Irrigation Intensity. *J. Environ. Radioactivity* **24** (1994) 53-63.

[8] I. Likhtarev, L. Kovgan, D. Novak, S. Vavilov, P. Jacob, H.G. Paretzke, Effective Doses due to the Chernobyl External Irradiation for Different Population Groups of Ukraine. *Health Phys.* **69** (1995) 1-12.

[9] V. Yu. Golikov, M. I. Balonov, A.V. Ponomarev, Estimation of External Gamma Radiation Doses to the Population after the Chernobyl Accident. In: The Chernobyl Papers, Vol. I. Eds. S.E. Merwin, M.I. Balonov. Research Enterprises Publishing Segment, Richland, Washington 1993, pp. 247-288.

[10] K. Saito, N. Petoussi, M. Zankl, R. Veit, P. Jacob, G. Drexler, Organ Doses as a Function of Body Weight for Environmental Gamma Rays. *J. Nucl. Sci. Techn.* **28** (1991) 627-641.

[11] United Nations Scientific Committee on the Effects of Atomic Radiation, Sources, Effects and Risks of Ionizing Radiation, 1988 Report to the General Assembly. United Nations, New York, 1988.

[12] V. Golikov, M. Balonov, A. Ponomarev, V. Erkin, P. Jacob, I. Likhtarev, External Exposures. In: JSP 5 Pathway Analyses and Dose Distributions. Ed. P. Jacob. GSF, D-85746 Oberschleißheim, Deutschland, 1995.

[13] V.G. Erkin, O.V. Lebedev, Thermoluminescence Dosimeter Measurements of External Doses to the Population of the Bryansk Region after the Cherenobyl Accident. In: The Chernobyl Papers, Vol. I. Eds. S.E. Merwin, M.I. Balonov. Research Enterprises Publishing Segment, Richland, Washington 1993, pp. 289-312.

[14] A. Birchall, M.R. Bailey, A.C. James, LUDEP: A Lung Dose Evaluation Program. *Radiat. Prot. Dosim.* **38**, (1991) 167-174.

[15] Human respiratory tract model. ICRP Publication 66, *Annals of the ICRP*, Vol. **24**, No. 1-4 (1994).

[16] Limits for intakes of radionuclides by workers. ICRP Publication 30, Part 1, *Annals of the ICRP*, Vol. **2**, No. 3-4 (1979), (also Part 2, 1980; Part 3, 1981; Part 4, 1988).

[17] Radionuclide transformations. ICRP Publication 38, *Annals of the ICRP*, Vol. **11-13**, (1983).

[18] ECP1 Contamination of surfaces by resuspended material; CEC/CIS Joint program on the consequences of the Chernobyl accident, EUR-Report, 1995.

[19] E. Garger, V. Kashpur, V. Kashparov, M. Buikov, N. Talerko, I. Vintersved, J. Peres, J. Martinez-Serrano, K. Nicholson, J. Tschiersch, Resuspension and deposition of radionuclides under various conditions. First International Conference of the European Commission, Belarus, Russian Federation, and Ukraine on the Radiological Consequences of the Chernobyl Accident, Minsk, 1996. These Proceedings.

Exposures from consumption of agricultural and semi-natural products

P. Strand[1], M. Balonov[2], L. Skuterud[1], K. Hove[3], B.Howard[4], B.S. Prister[5], I. Travnikova[2], A.Ratnikov[6]

1) Norwegian Radiation Protection Authority, Norway
2) Russian Institute of Radiation Hygiene, Russia
3) Agricultural University of Norway
4) Terrestrial Ecology Insitute, Merlewood, UK
5) Ukrainian Academy and Agricultural Sci, Ukraine
6) Russian Insitute of Agricultural Radiology and Agroecology, Russia

Abstract. The importance of food from different production systems to the internal dose from radiocaesium, was investigated in selected study sites in Ukraine and Russia. Food products from semi-natural ecosystems are major contributors to the individual internal dose to rural population in areas affected by the Chernobyl accident. At the selected study sites it is estimated in 1995 that foods from private farms and forests contribute on average 35% to 60%, to the individual internal dose, variation relating to soil types and implemented countermeasures. The importance of food products from private farms and particularly forest products increases with time since ^{137}Cs concentration in some of the natural food products have longer ecological half life than food products from agricultural systems. A significant relationship was observed between consumption of mushrooms and whole body content of radiocaesium in rural people. The contribution to the collective dose of food products produced in the semi-natural ecosystems is less than the contribution to the individual internal dose for the local rural population.

1. Introduction

Ingestion pathways are important routes leading to radiation doses in man after deposition of radioactive fallout. Several factors will influence the extent of intake of radionuclides. The importance of semi-natural ecosystems, compared to agricultural systems, in determining dose to rural populations has been uncertain. Transfer of radiocaesium to food products in agricultural systems is usually lower than those from semi-natural ecosystems. In particular, some products such as certain mushroom species and game are known to contain relatively high amounts of radiocaesium in comparison with agricultural products. Therefore, it has been suggested that the comparative importance of different farming systems and ecosystems needs to be reassessed with regard to the transport of radionuclides to man.

Village residents have farming and dietary habits which potentially predispose them to higher rates of radiocaesium intake. On the private farms one reason for this is lack of mineral fertilises, which give potential for higher transfer of radiocaesium to both vegetation and animals. Furthermore, village residents have easy access to mushrooms and berries from the forest.

In contrast, the greater total quantity of food produced in agricultural systems needs to be considered when calculating overall collective dose, and compared to that from private farms and forests. Hence, it is important to know what proportion of people's diets which arises from the various ecosystems and food production systems, and how this varies when calculating collective or individual dose. The differences in long term transfer of radiocaesium between food products from agricultural and semi-natural ecosystems may also change their comparative importance with time.

One of the prime objectives of EU funded ECP9-study was to make an initial assessment of the comparative importance of different food producing or collection systems in determining the individual internal dose for rural populations in a small number of selected settlements. In this paper we have extended the study to compare the collective internal dose received from consumption of foods produced in agricultural and semi-natural ecosystems.

Ecosystems where food are produced

The food production systems in the CIS country can be divided into two groups, the agricultural system, which include collective farms, and most of private farming. Whilst collective farms routinely use land rotation combined with ploughing and fertilisation to improve productivity, private farmers seldom apply artificial fertilisers, and tend to use animal manure for improving yields, particularly in their vegetable gardens.

Products from semi-natural ecosystems fall into two categories. Firstly there are the natural food products such as, mushrooms and berries which are used by people. In addition, clearings in forests and non improved pasture, are often used to provide fodder for the winter period, or as additional grazing land for animals.

Farming systems

Collective farms produce food through intensive management of the major soil and animal resources of their area, and provide labour opportunities for the rural population. Typically 2-5 villages are located within the area of a collective farm. Many village families are also allocated a plot of land in which to grow foodstuffs for their own use. Therefore within the village a subsistence farming economy operates, partly based on income from the collective farm and partly on exchange and sale of home grown vegetable and animal products. Traditionally, private farms have one or two cows, and milk is used for personal consumption as well as food for animals. Roughage for cattle is most often harvested from forest or scrubland. The grazing regime of privately owned cattle varies but to some extent relies on utilisation of marginal land that is not used by the collective. This includes the use of river banks, natural pastures and clearings in the forest. Sometimes they graze on fields which are lent to the farmers by the collective farm. The manure produced by the cattle is normally the only additional source of plant nutrients used for the private plots.

As the private farms utilise semi-natural ecosystems (e.g. grazing of domestic animals in natural pasture or in the forest) for milk production, this production has only been described as 'agricultural' when the pastures have been ploughed and/or fertilised. However it is sometimes necessary to differentiate further between private and collective farming systems, so in this study three different product groups are considered: Those from collective farming (intensive agricultural), from private farming (partly less intensive farming) and natural food products (e.g. food gathering). The use of countermeasures after the Chernobyl accident made these groups less well defined since in many areas, countermeasures (ploughing and fertilising) applied for reducing radionuclide transfer, made private farming less dependent on unimproved land.

Countermeasures

Countermeasures for reducing internal doses were applied to most contaminated areas, especially those area used intensively as part of the agricultural food producing system. Countermeasures used on the collective farm system included soil management, feeding regimes, radiocaesium binders, change of crops, and also the abandonment of some land used for food production. In the private farming system the more drastic countermeasures included restriction on the ownership of private cattle in the most heavily contaminated areas. Ploughing of natural pasture was also used widely, and private farmers given advices concerning soil management. Fewer countermeasures were available for reducing transfer of radioactivity to natural food products. However, the population were given information about change of diet and methods of food preparation to reduce radioactive content of foods.

2. Methods

The comparative importance from different sources of radiocaesium was considered for both individual doses for the rural population and collective doses from radiocaesium. Dietary survey and questioners about food production in the private farms were performed in the study sites. For both categories the time dependence and effect of countermeasures has been considered.

Detailed information was collected from selected study sites and combined with information about larger areas. Information on the following topics were collected.

- activity levels in food
- transfer to food products and variation over time
- dietary habits and dietary changes
- use of countermeasures
- quantity of food produced in the different production systems

3. Individual dose assessment from food consumption for rural populations in selected areas

Detailed studies were conducted in four different study sites where the influence of countermeasures, soil type and dietary changes were considered. The study areas selected for the detailed study of assessment of internal dose are shown in Table 1 [1]. The study site at Voronok was selected mainly for comparison with the Russian study site at Novozybkov. (Later "Russian site" indicates the Novozybkov area if Voronok is not spesified).

Table 1 Study areas in Russia and Ukraine

Country	Study areas	Deposition kBq m^{-2}
Russia	Novozybkov	700
	Voronok	50
Ukraine	Dubrovitsa	130

The volume of food products collected in the forest is small compared to that from agriculture (private or collective). However, the transfer of radiocaesium to natural products, such as mushrooms, wild berries, fish, and wild animals is often considerably higher than in the collective farming system. From the dietary survey [1] it was obvious that natural products are consumed by a significant proportion of the population. For all the study sites a considerable part of the population consume mushrooms (1/3 to 1/2). However, only a very few people (<1%) at the study sites consume meat from game. In addition to ranking use of natural foods the actual consumption of mushroom and forest berries (in kg d^{-1}) was estimated. In Table 2 the contribution of different food production systems is shown.

Table 2 Comparative importance (given in %) for the dose received from consumption of food contaminated with radiocaesium, from different ecosystems and food producing systems

Ecosystem	Food producing system	Ukrainian study site	Russian study site
Agricultural	Collective farming	17	17
	Private farming	48	22
Forest	Natural food	35	61

At the Ukrainian study site the agricultural system contributes on average about 65% of the radiocaesium intake, at the Russian site the agricultural system contributes about 40%. All the food products from private farming are, in this overview, attributed to the agricultural system. However, part of the milk production in the private system would been partly attributed to the semi-natural ecosystem because of private cows grazing natural and forest pastures which have not been fertilised or ploughed as a countermeasure. In the Russian study site natural food products are the main contributor to the internal individual dose from radiocaesium. In Ukraine food from the private farming system is a more important source.

For people who do not consume mushrooms we find that private farms in the Russian study site contribute 15% to radiocaesium intake, whilst 83% is due to intake of food produced in the collective farm. Among "non-mushroom-eaters" milk and meat contribute about 60% of the ^{137}Cs intake, while for mushroom eaters milk and meat contribute only about 20%; in this group about 70% of the ^{137}Cs dietary intake is due to mushroom consumption. The contribution of agricultural vegetable products is about 10%. In all cases, locally produced foods contributes more to the intake of ^{137}Cs than food bought from shops.

In Ukraine, among "non-mushroom-eaters" milk and meat contribute 60% of the ^{137}Cs intake. For those eating mushrooms, about 50% of their 137Cs intake come from mushrooms, and milk and meat supplies about 30% of the ^{137}Cs to this group. The contribution of agricultural vegetable products is about 10-15% for the different groups. As at the Russian site, products from private farms contribute more ^{137}Cs to the total diet than imported food sold in state shops. The private farming and natural food products contribute 83 % on average of the intake.

For all the study sites a strong, and consistent relationship exists between the extent of mushroom consumption and whole body radiocaesium content, as shown in Figure 1.

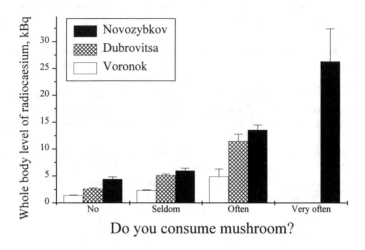

Do you consume mushroom?

Figure 1   Relation between frequency of mushroom consumption and wholebody levels (kBq) of radiocaesium in 1994 and 1995.

Private farming is of little importance at the Russian study site because the private farming system no longer produces significant quantities of milk products. The consumption of milk from private production decreased following the Chernobyl accident to 1-3% of pre-1986 levels [4]. This dramatically reduced the intake of radiocaesium to man from the private farming system compared with the situation prior to 1986. In Russia changes in diet and agricultural practices adopted after the Chernobyl accident have given and still give, significantly reduced intake of radiocaesium to the population. From the information about milk consumption in another study site outside the controlled area (Voronok) where private cattle were not removed, it is possible to estimate the importance of the private farms' contribution to the dose at the Novozybkov site. The comparative importance between the groups for the Russian study site would then be as shown in Table 3.

Table 3 Estimated mean contribution (given in %) for the dose received from consumption of food contaminated with radiocaesium, from different ecosystems and food producing systems. Under traditional private farming practices.

Ecosystem	food producing system	Russia
Agricultural	Collective farming	5
	Private farming	50
Semi-natural/Forest	Natural food	45

This is similar to the results from the Ukrainian study site.

The agricultural system is in 1995 the major contributor to radiocaesium intake, contributing on average about 55%. However, again, the milk production in the private farming system is fully attributed to the agricultural ecosystem, mainly because of the intensive soil treatment countermeasures applied after the Chernobyl accident. Hypothetically, if this had not been the case the semi-natural ecosystem would have had a greater role. This is demonstrated in Table 4 where it is assumed that 50% of private milk production is associated with unimproved or forest pastures. To estimate the radiocaesium levels in private milk at the Novozybkov study site the present levels at the Voronok study site have been used with allowance made for the difference in deposition, between the two sites. For the Ukrainian study sites it was also assumed that 50% of private milk production is associated with unimproved or forest pasture. The result of this scenario is shown in Table 4.

Table 4 Comparative importance (given in %) for the dose received from consumption of food contaminated with radiocaesium, of different ecosystems and food producing systems, if countermeasures had been used in much lesser extent on natural pasture.

Ecosystem	Food producing system	Ukrainian study site	Russian study site
Agricultural	Collective farming	17	7
	Private farming	32	31
Semi-natural/Forest	Private farming	16	17
	Natural food	35	45

The agricultural system under such assumption is estimated to contribute on average 49% and 38% to the radiocaesium intake, respectively for the Ukraine and Russian study site.

It is clear that at present the collective production system is a comparatively unimportant source of radiocaesium for residents of villages in contaminated areas with similar characteristics to those included in this study and with continued traditional private farming. If individuals have a high intake of natural food products the radiocaesium intake is strongly influenced by consumption of these products. By introducing countermeasures in the private farming system the transfer of radiocaesium to man has been considerably reduced. However, that continued use in private production of semi-natural ecosystems, for which few countermeasures are available, maintains the importance of this production system.

Time dependence of the internal individual dose

The results given above describe the present situation concerning the importance of ecosystems and food producing systems for the radiocaesium intake to man. However, a clear difference has been demonstrated between the long term transfer in intensive agricultural systems and in semi-natural systems [2,3,4]. Therefore, the comparative importance of different food production systems will probably change with time.

To illustrate changes with time the intake of radiocaesium in the first year after fallout has been estimated by applying reported transfer factors for natural pasture in the area [1,4], and by using information about ecological half lives for different food products [5]. Fesenko et al [5] consider the reduction in the different food products to be dependent on three factors 1) natural biogeochemical processes, 2) countermeasures and 3) radioactive decay. By taking the relevant radiocaesium levels of food products or the ecological half life into account for the products, the comparative importance for the different ecosystems and food producing systems can also be estimated at an early stage after an accident (e.g. one year after the accident, 1987). In the Bryansk region the effective half life for ^{137}Cs decrease in plant material varied between 1.4 and 1.9 year [5]. The long term behaviour of natural food products e.g. mushrooms, has been described by Shutov et al [3]; for mushrooms they found no significant reduction in radiocaesium levels in the Bryansk region. The comparative importance of the different system is shown in Table 5 for the Russian study site. Relevant information was not available from the Ukrainian study site.

Table 5 Comparative importance (given in %) for the dose received from consumption of food contaminated with radiocaesium, from different ecosystems and food producing system after one and nine year following the deposition.

Ecosystem	Food producing system	Russian study site 1987	Russian study site 1995
Agricultural	Collective farming	5	17
	Private farming	47	22
Semi-natural/Forest	Private farming	39	
	Natural food	8	61

At the Russian study site the intensive agricultural system contributed on average of 52%, to the internal dose one year after the accident compared to 39% at present. However the situation changed drastically concerning the importance of natural food products compared to the two other food producing systems. Natural food products contribute only about 8% in 1987 to the total intake of radiocaesium whilst in 1995 they contributed about 61%. For the Russian study site in 1987 the products from the private farming system would have been the main contributor to the intake of radiocaesium to man for the village residents.

4. Collective (total) internal dose estimation from consumption of total food production at the study sites

To dived the collective internal dose between different production systems at the study sites requests combination of knowledge of the total amount of food produced in each of the three food production systems. This includes products which are relevant for food consumption by either local people or by people outside the area. In Table 6 the amount of food, present radioactivity concentrations, and collective dose from different systems, estimated from productivity and dietary surveys performed in 1994 and 1995 of the local population is shown. The number of people living at the two study sites are 850 and 3000 respectively in the Russian and Ukrainian study sites. The yield of natural food products such as fungi and berries may be considerably underestimated since gathering by people from outside the area who also use the forests is not included.

Table 6 Total activity of radiocaesium in the food products produced at the two study sites.

Food producing systems	Food products	Russian study site			Ukrainian study site		
		Quantity produced	Activity levels	Total activity	Quantity produced	Activity levels	Total activity
		t y⁻¹	Bq/kg	kBq	t y⁻¹	Bq/kg	kBq
Collective	milk	1241	60	74460	1300	50	65000
	meat	155	273	42315	350	100	35000
	potatoes				813	20	16260
	vegetab						
	grain				1673	250	418250
	sub-total			116775			534510
Private	milk				543	140	76020
	meat	1	113	113	98	172	16856
	vegetab	95	11	1045	113	45	5085
	potatoes	217	11	2387	528	27	14256
	sub-total			3545			112217
Natural	mushrooms	4	10000	40000	49	1000	49000
	berries	1	403	403	30	210	6300
	sub-total			40403			55300
Total				160723			702027

The comparative importance for the different food producing systems for the collective dose from intake of radiocaesium by food produced at the study sites is shown in Table 7.

Table 7 Comparative importance (given in %) for the collective dose received in 1995 from consumption of food contaminated with radiocaesium, from different ecosystems and food producing systems at the study sites in Russia and Ukraine.

Ecosystem	Food producing system	Ukraine study site	Russian study site
Agricultural	Collective farming	76	74
	Private farming	16	2
Semi-natural	Natural food	8	25

Again at the Russian site private farming production is considerably affected by the removal of private cattle. If this was not done in 1986 the importance between the system would change from what shown in Table 7 to about 66%, 11% and 22%, respectively for the collective farming, private farming and natural food products at the Russian study site . The semi-natural ecosystem would probably contribute to the internal dose on average with 13% and 26% respectively at the Russian and Ukrainian study sites if natural pasture were not improved. This shows that the semi-natural ecosystem is also important in relation to collective dose. However there is need for better knowledge about the consumption of natural food products both in the rural and specially for the urban population. Since the radioactivity concentration of radiocaesium in some natural food products has not significantly reduced during the years after the Chernobyl accident whilst the agricultural system have declined. The importance of agricultural systems for collective dose is therefore expected to decrease with time after an accident.

Relevance of the result for other affected areas in CIS
The study site in Novozybkov region has the soddy-podzolic soil which occurs in major part of the affected areas in Russia. The collective farms in the affected area are similar to the collective farm in the study sites with major food products being potatoes, milk and barley. The study site in Ukraine is representative for some areas in the Rovno region with specially high transfer of radiocaesium due to the soil type. Other affected areas in Ukraine are more similar to the Russian study site.

References
[1] P.Strand, B.Howard and V.Averin, Fluxes of radionuclides in rural Communities in Russia, Ukraine and Belarus. Post-Chernobyl action report. Commision of the European Communities. 1996.

[2] K.Hove and P.Strand, Prediction for the duration of the Chernobyl radiocaesium problem in non-cultivated areas based on a reassessment of the behaviour of fallout from Nuclear weapons tests. In: Flitton S, Katz EW (Eds), Environmental contamination following a major nuclear accident. IAEA 306. 1:215-223. 1990.

[3] V.Shutov., G.Bruk and M. Balonov, Internal exposure of inhabitants by cesium-137 in villages Dobrodeevka and Bobovichi of the Bryansk region (in press).

[4] M.I.Balonov and I.G.Travnikova, Importance of diet and protective action of internal dose from ^{137}Cs radionuclides in inhabitants of the Chernobyl region. In: The Chernobyl papers, Vol.1.,pp. 127-167 1993.

[5] S.V.Fesenko, R.M. Alexakhin, S.I. Spridonov and N.I Sanzharova, Dynamics of ^{137}Cs concentration in agricultural products in areas of Russia contaminated as a result of the accident at the Chernobyl nuclear power plant. Radiation Protection Dosimetry. Vol. 60 No 2 pp 155-166. 1995.

EXPOSURES FROM CONSUMPTION OF FOREST PRODUCE

J. Kenigsberg [a], M. Belli [b], F. Tikhomirov [c], E. Buglova [a], V. Shevchuk [a],
Ph. Renaud [d], H. Maubert [d], G. Bruk [e], V. Shutov [e]

[a] Research Institute of Radiation Medicine, Masherov ave. 23, 220600 Minsk, Belarus
[b] ANPA Via Vitaliano Brancati 48, 00144 Roma
[c] Moscow State University, Russia
[d] Nuclear Protection and Safety Institute, CE/Cadarache 13108 Saint-Paul-Lez-Durance,
Cedex, France
[e] Institute of Radiation Hygiene, Mira str. 8, St.Petersburg, Russia

Summary

Traditionally, the diet of people from a number of regions of Belarus, Russia and Ukraine includes the foodstuffs from natural environment. After the Chernobyl accident there are some increase of forest gifts consumption for several categories of population in these countries. As these products very often have the relatively high level of radioactivity compared with agricultural foods, they may play an important role in the intake of radioactivity and in the infernal dose formation.

Data about the values of the transfer factor for different forest gifts during the years after the Chernobyl accident were obtained. For mushrooms two types of species according to high and low values of transfer factor were identified. For berries the clearly distinction in the transfer factor of different species were obtained also. The transfer factor for different species, gathered on hydromorphic soil higher than those gathered on automorphic soil. For game one transfer factor for all species was obtained. We examined the modification of the radioactivity contamination of forest gifts during processing and culinary preparation and determined data about the frequencies of usage of different types of culinary practices for forest gifts. On the basis of obtained data the model for the calculation of the radiocaesium intake due to the forest gifts were proposed.

The created model may be useful for the purpose of calculation of radiocaesium intake due to forest gifts consumption in the time of accidental situation.

The object of this study is to provide an appreciation of the role of consumption of forest products in the internal dose formation for people, who are living at the territories, contaminated after the Chernobyl accident.

For this the following questions were investigated:

- the transfer of radiocaesium to the forest gifts from the contaminated forests, taking into account types of forest soil and time after the accident,
- the modification of the activity in forest gifts during processing and culinary preparation and characterisation this transformation with the help of the processing factors values;

• dietary habits of population (determination of the frequency of usage of different types of culinary preparation and the quantities of forest gifts consumed by citizens of Belarus and Russia);

In this study we intend to attribute a transfer factor to the principal edible mushrooms and berries species, and one transfer factor for all game in the dynamics.

More than 4000 samples for 18 species of mushrooms were examined.

On the basis of samples of most consumed species of mushrooms it were obtained main parameters which characterise the data distribution of transfer factor for mushrooms as a whole (Fig.1).

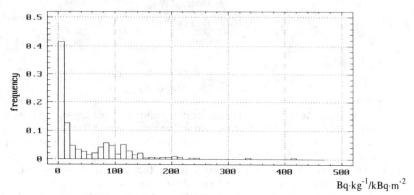

$$Bq \cdot kg^{-1}/kBq \cdot m^{-2}$$

Figure 1: The distribution of the transfer factor for mushrooms.

Analysis of the distribution of transfer factors for all mushrooms shows a lognormal character with bimodal component within the distribution.

It is known that different species of mushrooms have a great distinction in the ability for radionuclides accumulation. Rantavaara (1), Randa et al. (2) and Mascanzoni (3) note a strong difference between some types of species. The first one including *Cantharella cibarius, Boletus edulis* among others, for which the transfer factor is about 5 to 10 Bq.kg⁻¹ / kBq.m⁻² and the second one including *Boletus badius, Lactarius, Suillus variegatus* for which it is about 80 to 400 Bq.kg⁻¹ / kBq.m⁻² . The reason given for this differentiation is often the depth of layer at which mycelium sources its nutrients. For example *Boletus edulis* has deeper mycelium development compared to others and remains less contaminated.

Results of our investigation clearly demonstrates the possibility of identifying types of mushrooms in accordance with their ability for accumulation of radiocaesium. The comparison between the values of transfer factor shows that the border between high and low levels of transfer factor in our study is near 40 Bq.kg⁻¹ / kBq.m⁻². The parameters of transfer factor for *Leccinum scabrum* have an intermediate character. This fact is the next confirmation of the bimodal component in the lognormal distribution of all mushrooms. The first peak is formed by transfer factor of those species of mushrooms without an ability for radionuclides accumulation and the second one represents transfer factors for species with a strong accumulation capacity.

According to our results and literature data first type of mushrooms includes *Agaricus arvensis, Agaricus augustus, Armillaria mellea, Armillaria tubescens, Boletus edulis, Cantharellus cibarius, Hygrophorus species, Leccinum aurantiacum, Lepista nuda,*

Macrolepiota procera, Macrolepiota Rhacodes, others *Boletus* species. The second one is formed by the following species: *Hydnum rependum, Hydnum rufescens, Leccinum scabrum, Russula species, Rosites caperata, Suillus luteus, Suillus variegatus, Suillus Bovinus, Xerocomus Badius* and others *Xerocomus* species.

The investigation of the transfer factors for different species separately shows the following (table 1).

Table 1.

Main statistical parameters of the transfer factor distribution for different species of mushrooms, Bq.kg⁻¹ / kBq.m⁻² .

Species of	1986-1994 in Russia			1989-1990 in Belarus			1994 in Belarus		
mushrooms	A	M	SD	A	M	SD	A	M	SD
Boletus edulis	7.3	4.6	7.2	7.8	6.4	6.0	13.9	9.9	11.0
Cantharellus cibarius	6.2	4.6	5.8	8.6	5.2	10.4	11.7	6.8	14.6
Xerocomus badius	-	-	-	110.6	99.1	50.9	83.6	81.2	74.4
Russula	10.0	6.3	10.0	28.3	17.0	42.7	50.3	25.6	74.8
Tricholoma flavovirens	11.0	5.7	11.0	8.4	5.9	7.3	43.7	44.0	25.0
Suillus luteus	32.0	23.0	26.0	98.3	90.1	68.3	41.7	32.9	33.4
Armillaria mellea	1.6	1.2	1.5	7.4	7.9	4.2	4.8	3.4	4.6
Leccinum scabrum	15.0	10.0	17.0	46.4	40.0	57.2	48.9	35.0	72.9

A - average, M - median, SD - standard deviation, R - range.

We obtained the high values of transfer factor for *Xerocomus badius, Suillus luteus*. The average levels of transfer factor for *Boletus edulis, Cantharellus cibarius, Armillaria mellea, Tricholoma flavovirens* are between 7.4 and 8.6 Bq.kg⁻¹ / kBq.m⁻². The transfer factor for *Russulla* and *Leccinum scabrum* are also at these low levels. According to the data of Randa, the levels of transfer factor for *Boletus edulis* after the Chernobyl accident are lower than for other species of mushrooms (4). The mean values of transfer factor reported by Rantavaara are 0.8-20.0 Bq.kg⁻¹ / kBq.m⁻² for *Boletus edulis*, 6.1-13.0 Bq.kg⁻¹ / kBq.m⁻² for *Cantharellus cibarius* (1). The results of the factor investigation of Mascanzoni are as follows: the average value of transfer factor for *Boletus edulis* is 5.0 Bq.kg⁻¹ / kBq.m⁻², for *Cantharellus cibarius* - 14.0 Bq.kg⁻¹ / kBq.m⁻² (3).

From the data on the transfer factor for different species of mushrooms it is clear that there are no significant differences between values obtained during different years after the Chernobyl accident. Figure 2, completed on the basis of Russian data, confirm this observation.

We did not find statistically significant decrease of radioactive contamination of mushrooms during 8 years after the Chernobyl accident, or their decontamination was very slow. Relatively high rate of caesium-137 specific activity decrease in mushrooms was noted for *Cantarellus cibarius* - the half-period of decontamination (T) is equal to approximately 5 years and for *Suillus luteus* - T = 7-8 years. However, all these decrease are not statistically significant.

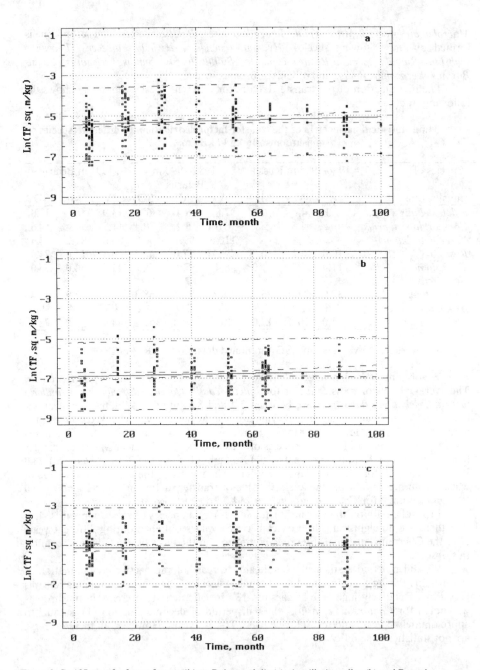

Figure 2. Cs-137 transfer factor from soil into Boletus edulis (a), Armillaria mellea (b) and Russula cyanoxantha (c). The inner pair of dashed line defines the domain of the 95% confidence limits of the mean values; the outer pair of dashed line bounds 95% of individual data points.

Furthermore, for almost all mushrooms species, gathered in Russia, were found an increase of caesium-137 specific activity in the first 2-3 years after the Chernobyl accident. A similar effect was noted by other investigators (5,6). The explanation is, on one hand, the penetration of the radionuclide from the upper forest fall to the layer of mushroom mycelium, and on the other hand, an increase in the Cs-137 content of soil due to its additional ingress with fallen leaves in autumn 1986 and with needles of coniferous trees over several years after the accident. This effect is most obvious in the case of *Boletus edulis*, whose mycelium is relatively deep.

It is important to emphasise, that the transfer factor for different types of mushrooms depends on the type of forest soil. The investigation of the transfer factor for *Boletus edulis, Cantharellus cibarius, Tricholoma flavovirens* and *Xerocomus badius*, gathered at the different types of forest soil Belarus during 1989-1994 years shows the following (table 2).

Table 2.

Descriptive parameters of the transfer factor for mushrooms, depending on types of soil, Bq.kg^{-1} / kBq.m^{-2}.

Species of mushrooms	Automorphic soil				Hydromorphic soil			
	Number of samples	Average	Median	Standard deviation	Number of samples	Average	Median	Standard deviation
Boletus edulis	279	14.8	13.8	5.3	194	41.6	40.4	7.1
Cantharellus cibarius	311	7.7	5.7	6.3	341	10.7	9.3	5.9
Tricholoma flavovirens	208	8.4	7.3	4.7	167	14.9	14.3	4.8
Xerocomus badius	221	56.2	51.6	20.8	183	122.4	116.1	34.8

According to our data, it is possible to conclude, that there is dependence of the transfer factor values on the type of soil. The transfer factor for different species, gathered on hydromorphic soil higher than those, gathered on automorphic soil. This fact is more obvious for *Boletus edulis, Tricholoma flavovirens and Xerocomus badius.* In general, variation of transfer factor for uptake of Cs-137 by all mushrooms in automorphic and hydromorphic soil are 10 - 100 Bq.kg^{-1} / kBq.m^{-2} and 20-1700 Bq.kg^{-1} / kBq.m^{-2} correspondingly.

Investigation of the transfer factor of berries were carried out on the basis of examination of more than 2000 samples for 8 species. The main parameters of the transfer factor for all berries which were obtained are the following (Fig. 3).

The investigation of the transfer factors for species of berries separately shows that there are clearly distinction in the transfer factor of different species (Table 3).

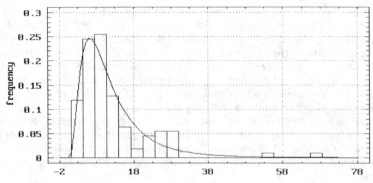

$$Bq \cdot kg^{-1}/kBq \cdot m^{-2}$$

Figure 3: Distribution of the transfer factor for berries.

Table 3.

Main statistical parameters of the transfer factor distribution for different species of berries, Bq.kg⁻¹ / kBq.m⁻².

Species of berries	1989-1994 in Belarus			1986-1994 in Russia		
	A	M	SD	A	M	SD
Vaccinium myrtillus	7.7	5.4	7.9	6.5	5.3	4.4
Vaccinium oxycoccus	8.9	8.5	2.7	13.0	13.2	10.0
Vaccinium vitis-idaea	-	-	-	10.0	-	6.0
Fragaria vesca	2.0	1.6	1.6	3.8	2.6	3.8
Rubus idaeus	-	-	-	2.6	1.9	2.6

A - average,   M - median, SD - standard deviation,   R - range.

By the data of Rantavaara (1) and Balonov (7), the average values of transfer factor for different species of berries are between 0.9-13.4 Bq.kg⁻¹ / kBq.m⁻². For *Vaccinium myrtillus* this values are 2.8 - 5.8 Bq.kg⁻¹ / kBq.m⁻² , 7.9 Bq.kg⁻¹ / kBq.m⁻² .

It is known that there are great differences in transfer factor of berries which connect with the peculiarities of forest soil. We investigated the transfer factor for *Fragaria vesca*, gathered at the hydromorphic and automorphic soils of Belarus during 1989-1994 years. Transfer factor for *Fragaria vesca*, gathered at the hydromorphic soils is 3.9 times higher than those for automorphic soils. Variation of transfer factor for uptake of Cs-137 by all berries in automorphic and hydromorphic soil are 10 - 15 Bq.kg⁻¹ / kBq.m⁻² and 80-130 Bq.kg⁻¹ / kBq.m⁻² correspondingly.

The level of contamination of game varies widely. This is connected with the large home range of wild animals and different levels of radiocaesium density contamination within this territory. In this connection, the examination of the transfer factor for game is an extremely complicated process. We prepared the distribution of the transfer factor for game, without differentiation between types of wild animals on the basis of investigation about 100 samples (Fig. 4).

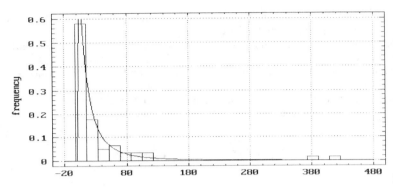

Figure 4: Distribution of the transfer factor for game.

As can be seen, the distribution is lognormal, the average value of the transfer factor is 36.2 Bq.kg^{-1} / kBq.m^{-2} , median - 11.0 Bq.kg^{-1} / kBq.m^{-2}, standard deviation - 72.7 Bq.kg^{-1} / kBq.m^{-2}. According to data, which were obtained in Russia the average values of this parameter are 0,5 - 28,0 Bq.kg^{-1} / kBq.m^{-2} .

It is necessary to emphasise the importance of the investigation of the radioactivity modification during culinary preparation of forest gifts and dietary habits of the rural citizens in consumption of natural products.

The results of such investigation about values of processing factors for different culinary practices and the frequencies of its usage are the same (Table 5).

Table 5.

The values for processing factors for different culinary practices of forest gifts

Type of culinary preparation	Processing factor	Normalised frequency of usage, %
Preparations of mushrooms		
Cleaning and washing	0.8	100
Cooking with pouring out of the first water	0.6	20
Cooking with pouring out of the second water	0.2	13
Cooking with the pouring out of the third water	0.2	9
Drying	10.5	22
Frying	0.3	19
Pickling	0.3	18
Preparations of berries		
Washing	0.9	100
Cooking of jam	0.5	49
Beating up with sugar	0.65	45
Drying	9.0	6
Preparations of game		
Steeping and salting	0.3	100
Cooking	0.25	32
Frying	1.3	56
Smoking	1.3	12

Therefore, the types of processing and culinary preparation provide for a considerable decrease of the initial contamination of a product.

For the purpose of calculation of radiocaesim intake due to forest gifts consumption and on the base of the data obtained the approaches to setting up a deterministic model were investigated. During this study different parameters, which are important to take into account in the principal structure of the model were discussed. According to this, the following parameters are included in the model (Fig.5).

Calculations have been made with the help of the following equations:

$$INTK_{m, b, g} = DC \bullet TF \bullet Q \bullet \sum_{i} F_{li} \bullet PF_i$$

with: $INTK_{m, b, g}$ - daily intake due to mushrooms,
                                   berries or game ingestion $Bq \cdot day^{-1}$
       DC - density of contamination $Bq \cdot m^{-2}$
       TF - transfer factor to mushrooms, berries or game $m^2 \cdot kg^{-1}$
       Q - quantity consumed per day $kg \cdot day^{-1}$
       F_{li} - frequency of culinary practice i   (-)
       PF_i - processing factor of the culinary practice i   (-)

$$INTK_{fg} = INTK_m + INTK_b + INTK_g$$

with: $INTK_{fg}$ - daily intake due to consumption of forest gifts $Bq \cdot day^{-1}$
       $INTK_m$ - daily intake due to consumption of mushroom $Bq \cdot day^{-1}$
       $INTK_b$ - daily intake due to consumption of berries $Bq \cdot day^{-1}$
       $INTK_g$ - daily intake due to consumption of game $Bq \cdot day^{-1}$

$$[INTK_{fg}]_a = \int_0^{365} INTK_{fg} \, dt$$

with: $[INTK_{fg}]_a$ - annual intake due to the consumption of forest gifts $Bq \cdot year^{-1}$

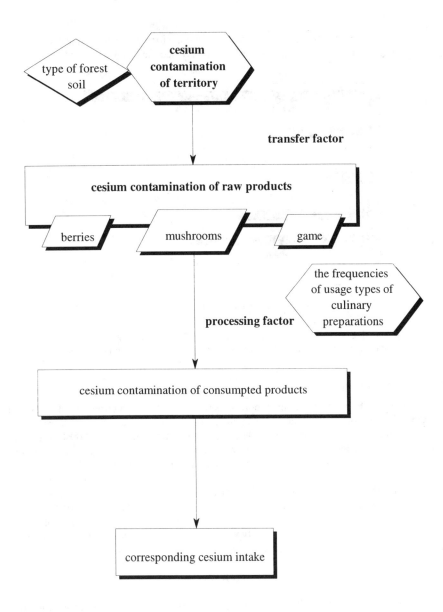

Figure 5. Principal scheme of the model for radiocaesium intake calculation due to forest gifts consumption.

The estimation of different foodstuffs contribution in radiocaesium intake among the rural population of the contaminated area of Belarus revealed the following (Figure 6).

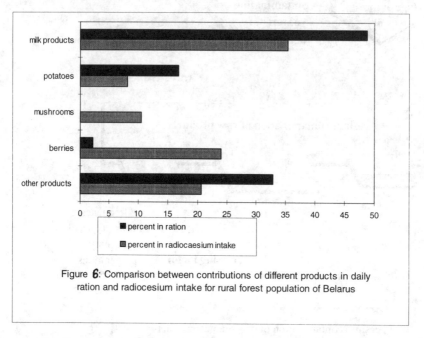

Figure **6**: Comparison between contributions of different products in daily ration and radiocesium intake for rural forest population of Belarus

This figure shows the great disproportion between daily consumption rate of forest gifts and their contribution to the intake due to their high level of contamination in relation to others foodstuffs of the same region. In Belarus, mushrooms which make 0.09% of the total ration volume, contribute 10.5% into the daily radiocaesium intake. Therefore, foodstuffs from forest will make 39% of daily alimentary radiocaesium intake and will form a dose of internal exposure equal to 0.34 mSv/year vs. 0.86 mSv/year, formed by the whole ration. This data confirm the important contribution of forest gifts in the internal exposure.

Data obtained during the investigation about the transfer factors of forest gifts for different types of soil and different species of mushrooms and berries, data about processing factors and frequency of usage of some types of culinary practices are the background for model of calculation the internal exposure due to forest gifts consumption.

References

1. Rantavaara, A.H. (1990). Transfer of radiocaesium through natural ecosystems to foodstuffs of terrestrial origin in Finland. In: Transfer of radionuclides in natural and semi-natural environments. Ed.: G.Desmet, London, 202-209.

2. Randa, Z., Benada, J., Horina, J., Klan, J. (1990). Mushrooms - significant source of internal contamination by radiocaesium. In: Transfer of radionuclides in natural and semi-natural environments. Ed.: G.Desmet., London, 169-178.

3. Mascanzoni, D. (1990). Uptake of Sr-90 and Cs-137 by mushrooms following the Chernobyl accident. In: Transfer of radionuclides in natural and semi-natural environments. Ed.: G.Desmet, London, 459-467.

4. Horina, J., Randa, Z. (1988). Uptake of radiocaesium and alkall metals by mushrooms. J.Radioanal.Nucl.Chem., Letters 127/2,107-120.

5. Shutov, V.N. (1992). Influence of soil properties on Cs-137 and Sr-90 intake to vegetation. 8-th Report of the IUR working Group Soil-to-Plant Transfer, IUR, Balen, Belgium, 11-15.

6. Rantavaara, A.H.(1987). Radioactivity of vegetables and mushrooms in Finland after the Chernobyl accident in 1986. Finish Centre for Radiation anu Nuclear Safety. Report STUK-A59, 88p.

7. Balonov, M.I., Travnikova, I.G., (1990). The role of agricultural and natural ecosystems in the internal dose formation in the inhabitants of a controlled area. In: Transfer of radionuclides in natural and semi-natural environments. Ed.: G.Desmet et al., Elsevier, London, 1990, 419-430.

Exposures from Aquatic Pathways

Vladimir BERKOVSKI
*Ukrainan Scientific Center for Radiation Medicine, Melnikova Str, 53 - 252050 Kiev,
Ukraine*
Oleg VOITSEKHOVITCH
Ukrainian Hydrometeorological Institute, Nauka Avenue, 37 - 252028 Kiev, Ukraine
Oleg NASVIT
Institute of Hydrobiology, G. Stalingrada Av., 12 - 254655 Kiev, Ukraine
Mark ZHELEZNIAK
Inst. of Cybernetics, Ukrainian Acad. of Sciences, Glushkova, 42 - 252187 Kiev, Ukraine
Umberto SANSONE
National Environmental Protection Agency, Via V. Brancati, 48 - 00144 Roma, Italy

Abstract. Methods for estimation aquatic pathways contribution to the total
population exposure are discussed. Aquatic pathways are the major factor for
radionuclides spreading from the Chernobyl Exclusion zone. An annual outflow of
^{90}Sr and ^{137}Cs comprised 10-20 TBq and 2-4 TBq respectively and the population
exposed by this effluence constitutes almost 30 million people. The dynamic of
doses from ^{90}Sr and ^{137}Cs, which Dnieper water have to delivered, is calculated.
The special software has been developed to simulate the process of dose formation
in the of diverse Dnieper regions. Regional peculiarities of municipal tap, fishing
and irrigation are considered. Seventy-year prediction of dose structure and
function of dose forming is performed. The exposure is estimated for 12 regions of
the Dnieper basin and the Crimea. The maximal individual annual committed
effective doses due to the use of water by ordinary members of the population in
Kievska region from ^{90}Sr and ^{137}Cs in 1986 are 1.7×10^{-5} Sv and 2.7×10^{-5} Sv
respectively. A commercial fisherman on Kievske reservoir in 1986 received
4.7×10^{-4} Sv and 5×10^{-3} Sv from ^{90}Sr and ^{137}Cs, respectively. The contributions to
the collective cumulative (over 70 years) committed effective dose (CCCED$_{70}$) of
irrigation, municipal tap water and fish consumption for members of the population
respectively are 18%, 43%, 39% in Kievska region, 8%, 25%, 67% in Poltavska
region, and 50%, 50%, 0% (consumption of Dnieper fish is absent) in the Crimea.
The predicted contribution of the Strontium-90 to CCCED$_{70}$ resulting from the use
of water is 80%. The CCCED$_{70}$ to the population of the Dnieper regions (32.5
million people) is 3000 person-Sv due to the use the Dnieper water.

1. Introduction

After a large radiation accident, as was the case in Chernobyl, the water pathway (both
surface and ground) represents in the long term, the only real way by which radioactive
materials can be transferred from high contaminated areas (as the evacuated zone around
Chernobyl) to uncontaminated areas (as the Black Sea). The Dnieper system is the principal
source of freshwater supply in Ukraine. This system provides drinking water for about 8
million people living in the Dnieper basin as well as irrigation water and fisheries for a
population of up to 30 million. The annual commercial fish catching in the Dnieper cascade
is more than 25000 tons. The radioactive contamination of the Dnieper cascade due to the

Chernobyl accident has been extensively described [1,2]. The annual ^{137}Cs and ^{90}Sr inflowing into the Dnieper cascade are respectively in some years reached 2-4 TBq and 10-20 TBq. These data confirm that there is a long-term input of ^{137}Cs and ^{90}Sr, from the northern heavily contaminated areas, into the Dnieper and Pripyat rivers. From these rivers, most of the ^{137}Cs is distributed along the six artificial reservoirs composing the Dnieper cascade system, while the dissolved form of ^{90}Sr reach more easily the Black sea.

Dose assessment due to the Dnieper water usage, based on measurements of population diet components, is extremely difficult. A very extensive area was affected by the radioactive contamination and no representative data are available on radionuclides content in the diet of population living along the different Ukrainian regions crossed by the Dnieper cascade. Consequently, the long term exposure (70 years) from aquatic pathways has been estimated on the basis of monitoring data of the Dnieper water and using environmental transport models. All available information concerning agricultural production, as well as irrigation water and fisheries were considered.

To estimate the collective dose for population of Ukraine from the consumption of Dnieper water during 70 years after the accident (till 2056) WATOX code was used in the version based on three months averaged input data. The three month averaged discharge of the Dnieper River, Pripyat River and tributaries to the Dnieper Reservoirs since 1895 were used to create hydrological data base for a long term prediction. The scenario of the worst radiological conditions should be based on the sequences of high runoff years since 1994. The constructed set of the hydrological data used 1970-1992 (high runoff period) and then 1912-1950 (low runoff period) as a "hydrological forecast" for 1994-2057. For the prediction of radionuclides concentration in the tributaries to the Kiev Reservoir the regression relations between concentration and water discharge were constructed based on the experimental data of UkrHMI.

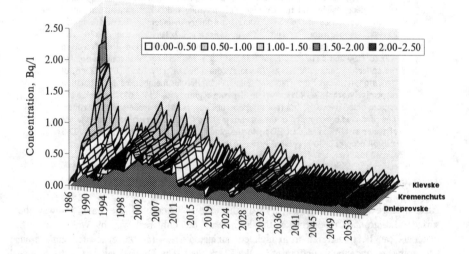

Fig. 1. Measured and predicted ^{90}Sr concentration in the Dnieper reservoirs

These relations were used for simulation of 15 years after the accident (till 2001). It is supposed that the amount of water exchangeable form of ^{90}Sr on the watersheds will remain constant this period, balanced by the leaching from fuel particles and by the decay and the percolation into the soil.

For simulation since 2001 the regressions were used with the attenuation coefficient equals to the double decay rate. The simulated results (Fig. 1) demonstrate that the large southern Kakhovka reservoir, damping the seasonal oscillations, will have after some years practically the same level of ^{90}Sr concenration as the Kiev reservoir, where three month averaged concentration will change from 1 Bq/l in the initial period to the 0.1-0.2 Bq/l in 2056 [3]. These data and data of the better from radiological conditions hydrological scenarious was used to calculate dose for the development of the post-Chernobyl.

2. Internal Radiation Dose

The internal radiation dose has been estimated for the population located in 12 Ukrainian regions and in the Republic of the Crimea. To this end a computer code has been developed [4] for the processing of a large set of site-specific data (age-structure of population, food consumption for different ages of population, agricultural production, irrigated areas, etc.). and the prediction of the radionuclides transport in the food chain as a function of time.

Table 1
Water Usage of the Dnieper Cascade

Regions	Water Bodies (rivers and reservoirs)	Total Population (x 10^6)	Population Consuming Dnieper Water (x 10^6)	Irrigated Areas (ha x 10^3)	Total area of Agricultural Land (ha x 10^3)
Chernigivska	Dnieper and Desna	1.4	-	6	2034
Kievska	Kiev and Kanev	4.5	0.75	116	1670
Cherkaska	Kremenchutske	1.5	0.2	66	1420
Kirovogradska	Kremenchug	1.2	0.4	56	2015
Poltavska	Kremenchug	1.7	0.3	57	2086
Dniepropetrivska	Dniprodzerdzhinsk Zaporozhie Kakhovske	3.8	2.0	254	2387
Zaporizka	Kakhovske	2	1.0	272	2225
Mikolaevska	Kakhovske	1.3	0.4	190	1953
Kharkivska	Dniprodzerdzhinsk	3.2	0.4	-	-
Luganska	Dniprodzerdzhinsk	2.9	0.1	-	-
Donetska	Dniprodzerdzhinske	5.3	2.2	-	-
Khersonska	Kakhovske	1.2	-	464	1932
Crimea	Kakhovske	2.5	0.5	390	1748
Total		32.5	8.1	1871	19470

The mathematical base of the software module, which simulates irrigation and radionuclide transport in food chain derives from ECOSYS-87 model [5]. The simulation of the metabolic processes in human body, under chronic intake, is based on the ICRP Publication 56, 67 (metabolic models of caesium and strontium) [6,7], and the 1990 ICRP Recommendations [8] were used. For dose calculation, it was applied the "Internal

Dosimetry Support System" or IDSS code, adapted for ICRP needs. For this calculation are taken into account the characteristics of the population of the Dnieper regions, presented in Table 1.

The potential exposure of the population, due to the intake of radionuclides from the Dnieper cascade water, was calculated in terms of "Annual Committed Effective Dose" (ACED), expressed in Sv. The resulting dose refers for one year intake, committed to age 70, and the "Collective Cumulative (for 70 years) Committed Effective Dose" (CCCED$_{70}$) as person-Sv - the age-dependent ACED, integrated over 70 years of intake and over age structure of population. One very important characteristic of dose formation is an ACED to the Maximally Exposed Member of General Public (MEMGP). Dose to MEMGP is calculated on the assumption that diet is "average" for the particular region, includes products produced on irrigated land only, contains tap water from the Dnieper. Consumption rate of Dnieper fish is determined by the commercial catch. It should be noted that dose to MEMGP does not reflect the level of doses in a critical group, because such a group was considered separately.

In the long-term prediction, only the intake of ^{137}Cs and ^{90}Sr were considered. The contribution to dose of the other radionuclides was relevant only during 1986. The estimation of the CCCED$_{70}$ from ^{131}I in tap water (Kiev city), based on ^{131}I concentration in the Kiev reservoir [9] was about 100 person-Sv for a population of 750.000

3. Water Usage

3.1. Municipal tap water

The Dnieper cascade represents the main water-supply system for a population of up to 8 million, located in 10 regions and the Crimea (Table 1). The main consumers are located in the Dniepropetrivska and Donetska regions. The following reference levels of tap water consumption were considered.

Population Age	liter/day
3 months	0.20
1 year	0.50
5 years	0.70
10 years	1.00
15 years	1.25
Adults	1.50

The clearance factor on processing of water has been assumed to be equal to 2 for ^{137}C and ^{90}Sr.

3.2. Irrigation Water

More than 1.8 million hectares, composed by productive farming lands, are irrigated by the Dnieper cascade water and 72% of them, directly by the Kakhovske reservoir (Table 1). About 50% of the irrigated lands are used for the production of fodder for farms animals (meat and milker). Vegetables production represents less than 10% of the total irrigated lands.

The activity in irrigated agricultural plants is accumulated because of root uptake of radionuclides from primary Chernobyl soil contamination and earlier worldwide fallout, root and foliar uptake from irrigation water. The routes are similar in the case of contamination of agricultural animal products.

The transport of radionuclides from the soil to the human body is simulated. Additionally to primary fallout on soil, which is considered by ECOSYS, the irrigation, drinking water and fish consumption are also taken into account.

Ukraininan and site-specific data on crop yield, types of irrigated crops and consumption rates were used in the dose prediction [10]. Considering the difficulties to obtain representative site-specific radionuclide transfer factors for foliar and root uptake, for all the regions crossed by the Dnieper cascade, data collected on the most contaminated northern regions of the Ukraine (where intensive investigations were performed after the accident) as well as the ECOSYS data were used.

The short-lived radionuclides were not considered in the calculation and consequently in the model for animals it was assumed that the concentration in milk is in continuous equilibrium with the concentration of radionuclides in the diet.

The food basket considered for human consumption contained: cereals, milk and dairy products, potatoes, beef, leafy vegetables, root vegetables, fruits, poultry and eggs.

Figure 2 presents the dynamic of ^{90}Sr concentration (from irrigation water only) in foodstuffs in Kievska region.

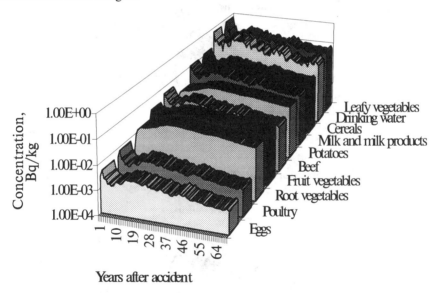

Fig. 2. Model prediction of ^{90}Sr concentration in different foodstuffs of Kievska region

The highest levels of contamination are observed for leafy vegetables, wheat, and milk. The predicted concentrations in leafy vegetables and tap water are practically equal.

Figure 3 reports for the Kievska region, the ^{137}Cs concentrations in different agricultural products contaminated in 1986 and those predicted for 2055.

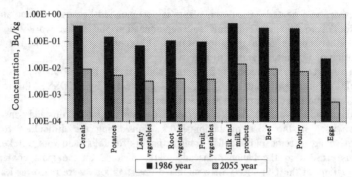

Fig. 3. Model prediction of ^{137}Cs concentration in different foodstuffs of Kievska region for 1986 and 2055

Figures 4 and 5 show the predicted contribution of ^{90}Sr and ^{90}Cs in different foodstuffs to the "Annual Committed Effective Dose" to the "Maximally Exposed Member of General Public", due to the usage of the Dnieper water in Kievska region in 1993.

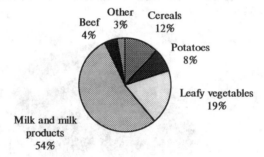

Fig. 4. Contributions of different foodstuffs to ^{90}Sr dose to MEMGP as a result of use of Dnieper water in Kievska region in 1993

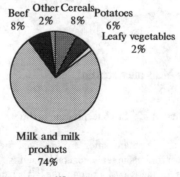

Fig. 5. Contributions of different foodstuffs to ^{137}Cs dose to MEMGP as a result of use of Dnieper water in Kievska region in 1993

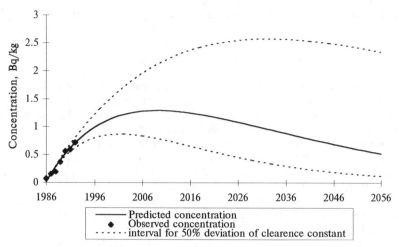

Fig. 6. Predicted ^{90}Sr concentration in rice. (observed data from [11])

Using the data from Perepeliatnikov and Prister [11] and the contamination scenario of the water of Kakhov reservoir, the following function has been derived to estimate the activity of ^{90}Sr in Ukrainian rice (see Fig. 6):

$$C_{rice}(t)=(0.014+0.15t)e^{-0.043t}, \text{ Bq kg}^{-1}$$

where:

$C_{rice}(t)$ is an activity concentration in rice at time t after the Chernobyl accident (Bq/ kg);
t is time after the accident (yr.);

The clearence costant of ^{90}Sr in water is 0.043 yr^{-1} ($T_{1/2}$ =16 year) was derived from data in [11]. Figure 6 presents a comparison of prediction to observations of the concentrations of ^{90}Sr in rice.

3.2. Commercial fishery

The Dnieper cascade is intensively used for commercial fishery (more than 25 thousand tons per year). After the Chernobyl accident, there was no significant reduction in fish catch in all the reservoirs. In this paper site-specific data from 1986 to 1993, concerning the commercial fishery and contamination of fish were used [12 ,13]. The prediction of fish contamination is based on simple bioaccumulation model. The bioaccumulation factors used are respectively 30 and 1000 for ^{90}Sr and ^{137}Cs [13].

Figures 6, 7and 8 show respectively the contribution of fish consumption to the total collective dose for the population of Kievska and Poltavska regions and the Crimea. The highest values of radionuclide concentrations in fish were found between 1987 and 1989. At present time these values are from 2 to 4 times lower.

The exposure of professional fishermen, that consume a considerable amount of fish, has been taken into account. On the basis of a consumption of 360 kg of fish per year, the fishermen of the Kiev reservoir, received in 1993, 4.7x10^{-4} Sv and 5x10^{-3} Sv respectively for ^{90}Sr and ^{137}Cs. During 1993 these values were 1.7×10^{-4} Sv and 1.6×10^{-3} Sv. These data indicate that the fishermen are to be considered as the most important critical group.

4. Assessment of total dose

Fig. 9 presents the prediction of ^{90}Sr accumulation in bone structure of an adult "Maximally Exposed Member of General Public", due to the water usage in the Kievska region. The 50-year accumulated equivalent dose is 1.3 mSv to red bone marrow and 2.6 mSv to bone surface. The committed effective dose is 0.22 mSv.

Figures 10 - 12 give the computed contribution of different water pathways of ^{137}Cs and ^{90}Sr to the "Annual Committed Effective Dose" respectively for Kievska and Poltavska regions and the Crimean Republic. The dotted curves indicate the dynamic of ACED. The structure of irrigated land and tap water consumption were taken into account. In 1986, the collective dose in Kievska region exceeded the present dose level by 5-7 times. An opposite effect is observed in the Crimea Republic. This is attributable to the low level of radioactive deposition occurred in the lowest part of the Dnieper cascade.

The "Annual Committed Effective Dose" in 1986 to the "Maximally Exposed Member of General Public", due to water usage in the Kievska region results respectively 1.7×10^{-5} Sv and 2.7×10^{-5} Sv for ^{90}Sr and ^{137}Cs. The assessable contributions of irrigation, water from the municipal tap water supply and fish consumption to $CCCED_{70}$ respectively are 18%, 43%, 39% in Kievska region, 8%, 25%, 67% in Poltavska region, and 50%, 50%, 0% (consumption of the Dnieper fish is absent) in the Crimea.

The predicted average contribution of strontium-90 to $CCCED_{70}$ to the population of the Dnieper regions resulting from the use of water is 80%. The $CCCED_{70}$ to the population of the Dnieper regions (32.5 million people) is about 3000 person-Sv due to the use of water.

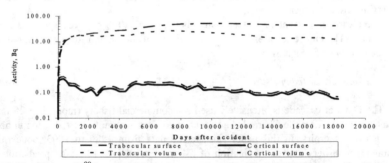

Fig. 9. Prediction of ^{90}Sr activity in bone structure of adult MEMGP due to the Dnieper water usage in the Kievska region

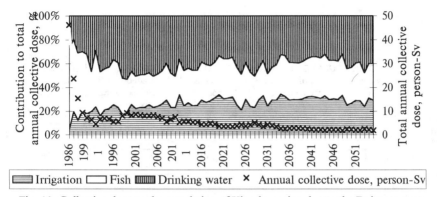

Fig. 10. Collective dose to the population of Kievska region due to the Dnieper water usage

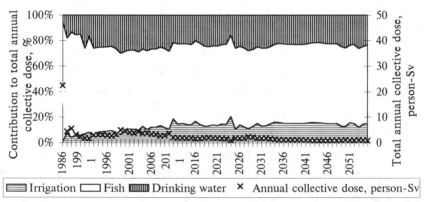

Fig. 11. Collective dose to the population of Poltavska region due to the Dnieper water usage

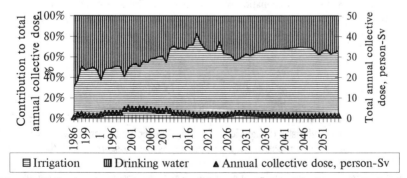

Fig. 12. Collective dose to the population of the Crimea Republic due to the Dnieper water usage

Table 2

CCCED$_{70}$ from ^{90}Sr and ^{137}Cs to population of Dnieper regions

Region	Total population (x 10^6)	^{90}Sr and ^{137}Cs CCCED$_{70}$ as a result of primary soil contamination (person-Sv)	^{90}Sr CCCED$_{70}$ as a result of use of Dnieper water (person-Sv)	^{137}Cs CCCED$_{70}$ as a result of use of Dnieper water (person-Sv)	Contribution of Dnieper water dose to Total CCCED$_{70}$ (%)
Chernigivska	1.4	340	4	2	1.7
Kievska*	4.5	2400	290	190	17
Cherkaska	1.5	620	115	50	21
Kirovogradska	1.2	320	140	40	36
Poltavska	1.7	200	130	60	49
Dniepro-petrivska	3.8	560	610	75	55
Zaporizka	2	170	320	35	68
Mikolaevska	1.3	190**	150	20	47
Kharkivska	3.2	N/A**	60	4	-
Luganska	2.9	N/A**	15	1	-
Donetska	5.3	N/A	330	20	-
Khersonska	1.2	70	100	20	63
Crimea	2.5	160	175	5	53
Total	32.5	5030	2439	522	37

* The northern districts of Kievska region with high level of soil contamination are not considered because Dnieper water is not used in these areas. Due to important differences in water consumption pattern that values is not reflect situation the Kiev city.
** The data on primary soil contamination are not available.

In order to demonstrate the role of water factors in dose formation the estimations of the contribution of the Dnieper cascade to total CCCED$_{70}$ from ^{90}Sr and ^{137}Cs were also performed (Table 2). The assessment of CCCED$_{70}$ which resulted from primary (Chernobyl and non-Chernobyl) contamination of the soil by ^{90}Sr and ^{137}Cs [14] was performed in the same manner as estimation of the dose from water. This assessment is based on the data on primary soil contamination and does not consider short-term processes after the Chernobyl accident. It should be noted that reliability of total dose estimation is not too high because of the data on primary soil contamination was obtained only for preliminary estimation and may not reflect the low-level contamination is present in the majority of regions. Such regions marked are characterized by uneven levels of contamination. In most of them it is practically impossible to identify the post-Chernobyl component.

5. Conclusions

The presented results indicate that each of the three main components of dose formation (tap water, irrigation and fish) is significant in the Dnieper water problem. ^{137}Cs is of primary importance in the fish pathway. That pathway is the most important for Kievska region, which consumed fish from Kievske reservoir and for Middle Dnieper regions, which consumed a significant amount of fish from Kahovske reservoir. The calculation demonstrates that a separate and very important problem is the high level of individual doses (up to 2 mSv per year) to fishermen and members of their families. Further investigation including measurements by WBC and excreta analysis is required to develop countermeasures to address the problem.

^{90}Sr plays an important role in the tap water and irrigation pathway. In Kiev city that radionuclide delivered about half of total ^{90}Sr dose. The role of ^{90}Sr is increasing in southern Dnieper regions. In the Crimea ^{90}Sr formed more than 95% of total "water dose." As mentioned above, the total contribution of ^{90}Sr to $CCCED_{70}$ from Dnieper water is 80%. This fact necessitates a broader application of bioassay methods in dose monitoring.

All presented results were obtained by means of specialized PC software. That permits the simulation of enviromental and biokinetic processes. The authors would be grateful for offers of on collaboration in testing and improving that software.

References

[1]   O. Voitsekhovitch, U. Sansone, M. Zhelesnyak, D. Bugai. Water Quality Management of Contaminated Areas and its Effects on Doses from Aquatic Pathways. First International Conference of the European Commission, Belarus, Russian Federation and Ukraine on the Radiological Consequences of the Chernobyl Accident. Minsk, Belaru, 18-22 March 1996.

[2]   O. Voitsekhovitch, V. Kanivets et.al , Hydrological Processes and their Influence on Radionuclide Behavior and Transport by Surface Water Pathways as Applied to Water Protection after Chernobyl Accident .,In:. Proc. UNESCO,Hydrological Impact of NPP.1992 . Paris.(1993), 85-105.

[3]   M.J.Zheleznyak et al., Mathematical Modeling of Radionuclides Dispersion in the Pripyat-Dnieper Aquatic System after the Chernobyl Accident, Science of the Total Environment, 112 (1992) 89-114.

[4]   V. Berkovski, V.S. Repin, G. Ratia at al. Estimation of the Dnieper Water Contribution to Dose Formation of Ukrainian population. Final Report. Research Contract No. 35-93. Ministry on Chernobyl Affairs. Kiev. 1994, 152 pp.

[5]   H.Muller, G.Prohl. Ecosys-87: A dynamic model for the assessment of the radiological consequences of nuclear accidents.//Health Phys.-1993.-V.64.- P.232-252.

[6]   Age-dependent doses to members of the public from intake of radionuclides. Oxford: Pergamon Press; ICRP Publication 56, Part 1; 1989. 122 p.

[7]   Age-dependent Doses to Members of the Public from Intake of Radionuclides: Part 2 Ingestion Dose Coefficients. ICRP Publication 67. Annals of the ICRP. Vol. 23. No.3/4 1993. 167 p.

[8]   1990 Recomendation of the International Commission on Radiological Protection. Oxford: Pergamon Press; ICRP Publication 60; 1990.

[9]   Vakulovskij, S. M.; Voitsekhovich, O. V.; Katrich, I. Yu.; et al. Radioactive contamination of water system in the area affected by releases from the Chernobyl NPP accident. Proc. of the IAEA Symposium. Vienna, 1990.

[10] The agricultural production in Ukraine. Statistical data. Kiev: Ukrainian Ministry of Statistics. 1992.

[11]   Perepeliatnikov, G. P.; Prister, B. S. A migration of the Chernobyl radionuclides in fields, which irrigated by Dnieper water. Problems of agricultural radiology. Rel. 2. Kiev. 1992.

[12]   Napier, B. A.; Templeton, W. L.; Ryabov, I. N.; Kryshev, I. I.; Sazykina, T. G. Long-term radiation dose and effects from contamination of Dnieper aquatic ecosystem. Report to U.S. DOE Contract DE-ACO6-76RLO 1830. September 1992.

[13]   Tkachenko, N. Private communication. 1993.

[14]   Nosko, B. S.; Prister, B. S. Loboda M. B. Reference book on agrochemical and agroecological state of Ukrinian soil. Kiev: Urojaij. 1994.

Assessing Internal Exposures and the Efficacy of Countermeasures from Whole Body Measurements

Likhtarev, I., Kovgan, L., Gluvchinsky, R., and Perevoznikov, O.[1]
Morrey, M. and Prosser, S.L. [2]
Jacob, P. and Prohl, G.[3]
Kenigsberg, Y. and Skryabin, A.M.[4]
Colgen, P.A.[5]

1. The Research Centre for Radiation Medicine, Kiev, Ukraine.
2. National Radiological Protection Board, Chilton, UK.
3. GSF, Insitut fur Strahlenschultz, Neuherberg, Germany.
4. Institute for Radiation Hygiene, St Petersburg, Russia.
5. CIEMAT, Madrid, Spain.

ABSTRACT. Traditional procedures for modelling the ingestion dose pathway combine environmental transfer models with human metabolic models in order to assess the doses received. In general, these models have been developed for specific ecological and socio-economic circumstances, rather than for globally-averaged conditions. Experiences which occurred following the Chernobyl accident have demonstrated that in the event of a large scale radiation accident it will be virtually impossible to monitor adequately all the radiologically significant components of human diet which may have become contaminated with radionuclides.

This paper describes an internal dosimetry model based on the most widely available measurements following an accident: radiocaesium measurements in soil and milk, and whole body measurements in humans. One application of the model to estimate ingestion doses received by inhabitants of the northern region of the Rovno Oblast in Ukraine is also described. In addition, this model enables the effectiveness of food countermeasures to be estimated.

This study formed part of Joint Study Programme 5 (JSP5) on pathway analysis and dose distributions and was jointly funded by the European Comission (EC) and Commonwealth of Independent States (CIS).

INTRODUCTION

Following the accident at Chernobyl, a very wide area of territory in Russia, Belarus and Ukraine became contaminated with radionuclides. In order to estimate ingestion doses received by the inhabitants of these areas, a large database of measurements is required as input to the environmental and metabolic models utilised. These data must be representative of the wide range of ecological conditions in the contaminated areas and must also adequately reflect the habits of individuals in these regions, if dose assessments are to be realistic. The data commonly utilised are radionuclide concentrations in soil and food, and food consumption rates. It is also necessary to determine the proportion of diet which is obtained from local sources, since studies undertaken as part of Joint Study Programme 5 (JSP5) on pathway analysis and dose distributions, have shown that locally-produced foods contribute to the majority of the radiocaesium intake in the contaminated territories[1].

For post-accident management, it is necessary to be able to estimate the radiation doses received or being received as quickly as possible. It is therefore important to develop alternative approaches to dose estimation which do not place such emphasis on large measurement data sets.

Such approaches usually rely on a range of whole body measurements (WBM) to 'anchor' the dose model results, but otherwise they can be tailored to require as input data only those measurements which exist or can reasonably be provided, given the nature of the post-accident situation. Within Task 4 of JSP5[1] , a model for estimating average regional ingestion doses has been developed based on only the three most commonly recorded types of measurement: soil contamination, privately produced milk concentrations and WBM. This paper will outline the basis of the model, describe its structure and its mechanism for determining the effectiveness of countermeasures. One application of the model will then be described: the region under study being the area encompassing the northern part of the Rovno Oblast.

MODEL OVERVIEW

The schematic structure of the model is outlined in Figure 1. In summary, the model consists of two parts. The first part employs a simple phenomenological approach to estimate doses, but uses the concept of 'milk equivalent intake' to reduce the need for detailed data on consumption rates and radionuclide concentration levels for all foods. It is well established that the consumption of locally produced milk is the major contributor to the average regional dose, in the contaminated territories[2]. In the model the following two assumptions are made: that, in the absence of knowledge of the accident, dietary habits would remain unchanged; and that, for a given region, the time variation of food concentration has been similar in form for all foods. For the last to be true, measurement data should not be aggregated across areas with widely differing soil types, although it may be assumed that relative food concentrations are the same between different soil types. In addition, ideally, the region either should not have been subject to agricultural countermeasures, or such countermeasures should have been applied with equal effect to all the land from which food has been obtained. However, since the consumption of locally produced milk is the dominant exposure pathway, the selective implementation of agricultural countermeasures will only have a small influence on the validity of this assumption. Based on these assumptions, information on consumption rates prior to the accident, and knowledge of the radionuclide concentrations in all the foods of interest for a typical part of the region and a single representative time period, it is possible to determine the relative contribution of all parts of the average diet to the total dietary intake of caesium-137. From this, the equivalent consumption of locally produced milk can be determined that would exactly reproduce the calculated total intake of caesium-137. In the model, this is termed the 'absolute milk equivalent'. This absolute milk equivalent, together with information on the time variation of the concentration in locally produced milk, can then be used to estimate the intake of caesium-137 throughout the region and time period. This is termed the 'reference' intake.

The second part of the model uses WBM to estimate the actual or real intake of caesium-137. This 'real' intake can be compared with the reference intake and, provided the assumptions made in the calculation of the absolute milk equivalent were valid, the difference between the two is a measure of the influence of people's modified dietary behaviour in response to the accident.

In both parts of the model it is necessary to make an allowance for the contribution to average dose from foods obtained from outside the area. This contribution arises from several sources, including the wide official distribution of foods produced in contaminated regions across Ukraine, (to prevent their concentrated consumption by a few individuals), and foods illegally entering the market. This component can be estimated from the caesium body burdens of people living well away from the contaminated regions. This information can be combined with a knowledge of the fraction of diet obtained from 'imported' foods to provide an estimate of the intake of radiocaesium from this component of diet.

Detailed description of model

Among the inhabitants of a given settlement it is possible to single out some i subgroups ($i = 1,...,n$), which are distinguished by their type of diet. Typical subgroups include infants, children younger than seven years, children 7 - 15 years old, teenagers 15 - 18 years old, and several adult groups (*eg.* employed indoors, agricultural workers, pensioners). *The total diet W_i, of the members of i-th group may be represented as:*

$$W_i = \sum_j w_{ij} \tag{1}$$

where w_{ij} is the consumption rate (kg or L d^{-1}) of the j-th food-component in the diet of members of group i. Usually the diet consists of two parts: that which is locally produced and that which comes from other areas. With the locally produced fractions denoted by f_{ij}, eqn (1) may be rewritten as:

$$W_i = \sum_j f_{ij} j_{ij} w_{ij} + \sum_j (1 - f_{ij}) w_{ij} \tag{2}$$

In contaminated territories the first term, representing locally grown food, is more important.

If the ^{137}Cs concentration in the j-th component of locally-produced diet is c_j, then the intake of ^{137}Cs, q_i, in the total diet for members of the i-th group is:

$$q_i = \sum_j c_j f_{ij} w_{ij} \tag{3}$$

The value of c_j may be determined as the product of the parameters σ_{sj} and k_{sj} where σ_{sj} is the level of deposition on the s-th field or pasture where the j-th food or pasturage is produced, and k_{sj} is defined as the "j-th transfer factor", which is determined as:

$$k_{sj} = \frac{c_j}{\sigma_{sj}} \tag{4}$$

Since the value of k_{sj} (and, accordingly, the value of c) changes with time, the intake function q_i is also a function of time. Thus, combining eqns (3) and (4), the intake function for the i-th group is:

$$q_i(t) = \sum_j k_{sj}(t)\sigma_{sj} f_{ij} w_{ij}$$

(5)

Eqn (5) describes the time-dependent ^{137}Cs-intake function for the group of people living on the contaminated territory, if the style of life (diet and agricultural practices) has not changed after the accident. This function may be calculated, if detailed information on all parameters is available.

Reference parameters of the model

In actual situations it is practically impossible to obtain detailed, valid information about thousands of communities, or even to obtain detailed information for one community for the values of σ_{sj}, k_{sj} and f_{ij}. Thus, it is important to introduce a system of generalised, reference indexes of the radiation situation for a single settlement, and to consider the primary dosimetric variables relative to the reference values. The average (for the settlement and its environs) ^{137}Cs-deposition density, σ_0 is used as the primary reference index, where:

$$\sigma_0 = \frac{\sum_{sj} \sigma_{sj}}{n}$$

(6)

and n is number of measurements for σ_g for each settlement. For the parameter σ_g we will use the term *"reference ^{137}Cs-soil deposition"*. It is also useful to modify the previous definition [eqn (4)] of the transfer factor and to introduce the term *"reference transfer factor"* k_j^0 for the j-th locally produced food:

$$k_j^0(t) = \frac{c_j(t)}{\sigma_0}$$

(7)

Parameter $k_j^0(t)$ represents the main transfer between Compartments 1 and 2. Now eqn (5) may be rewritten as:

$$q_i(t) = \sigma_0 \sum_j k_j^0(t) f_{ij} w_{ij}$$

(8)

The ^{137}Cs-intake function, $q_i(t)$, which results from the ingestion of contaminated foods in the case where no limitations and countermeasures had been taken ("people knew nothing about the accident") is noted as the "reference-intake function" q_i^0. Equations (4) and (8) determine the ^{137}Cs flow with the total diet. But in the actual situation of a large-scale accident, usually only one or a few diet components can be measured reliably over large territories. In practice, in Ukraine following the Chernobyl accident, only the ^{137}Cs concentration in milk was well characterised. If the average individual consumption rate of milk by persons in group i is $w_{i,m}$ litres per day (m is used here and below to denote milk), $f_{i,m}$ is the part of milk locally produced, and the ^{137}Cs contamination of locally produced milk is $c_m(t)$, the total intake of ^{137}Cs with milk is:

$$q_{i,m}(t) = c_m(t)_0 \, f_{i,m} \, w_{i,m} \tag{9}$$

Further, the *relative milk equivalent*, $p_{i,m}(t)$ is defined as the ratio of ^{137}Cs intake with the whole diet to that consumed with locally produced milk alone:

$$P_{i,m}^0(t) = \frac{q_i^0(t)}{q_{i,m}(t)} \tag{10}$$

Then,

$$P_{i,m}^0(t) = \frac{\sum_j c_j(t) f_{ij} \, w_{ij}}{c_m(t) f_{i,m} \, w_{i,m}} \tag{11}$$

The parameter $P_{i,m}^0(t)$ is non-dimensional and is greater or equal to one. Now, the term "whole diet absolute milk equivalent", defined in the model overview, $w_{i,m}^0(t)$ may be introduced as:

$$q_i^0(t) = c_m(t) w_{i,m}^0(t) \tag{12}$$

where

$$*w^0_{i,m}(t) = p^0_{i,m}(t)f_{i,m} \, w_{i,m} = \sum_j l_{j,m}(t)f_{ij} \, w_{ij} \tag{13}$$

In Equation 13 the term $l_{j,m}(t) = \dfrac{c_j(t)}{c_m(t)}$ is the function of ratio of ^{137}Cs concentrations in the j-th component of diet to one in milk. Absolute milk equivalent $w^0_{i,m}(t)$ is equal to the amount of locally produced milk that could be consumed to provide the value of $q_i(t)$ with the whole diet.

In reality after the Chernobyl accident food production in the contaminated areas was not stopped. Instead, the produced foods (butter, sour cream, meat and others) were distributed to the other less contaminated territories. For this reason, some low level of food contamination existed throughout the whole country, and resulted in both the so-called "imported ^{137}Cs-food contamination" $c_{j,imp}$ and a permanent (background) level of ^{137}Cs body burden Q_{bac}. The imported ^{137}Cs-food contamination may be expressed as:

$$q_i(t) = q^0_i(t) + q_{i,imp}(t) \tag{14}$$

where $q_{i,imp}(t) = \sum_j c_{j,imp}(t)(1 - f_{ij})w_{ij}$ represents the intake of ^{137}Cs resulting from the consumption of imported food, $c_{j,imp}(t)$ is the ^{137}Cs concentration in the imported fraction of the j-th component of diet. If $q^0_i(t)$ is known, the corresponding committed effective dose, $D^0_T(t)$ ("reference dose") due to ^{137}Cs intake up to time T after the accident for members of group i is derived from $q^0_i(t)$ as:

$$D^0_{i,T} = K_i \int_0^T q^0_i(t)dt \tag{15}$$

where $K_i(Sv \, Bq^{-1}ingested)$ is the effective dose-coefficient for group i.

Countermeasures

The main food countermeasures implemented in Ukraine were the partial or complete replacement of locally produced foods with ones produced in relatively clean areas. These countermeasures were initiated both by the government, which organised deliveries of "clean" foods, and by individuals, who modified their own diets. Food replacement is considered in the model by the introduction of the *food-replacement function*, $h_{ij}(t)$, which is equal to or less than one and which characterises the decreased consumption of locally produced food as a result of countermeasures. Let the function $q_i^0(t)$ modified by $h_{ij}(t)$ be denoted by $q_i^*(t)$ and termed the *"countermeasure intake function"*. Thus,

$$q_i^*(t) = \sum_j c_j(t) \, h_{ij}(t) \, f_{ij} \, w_{ij} \qquad (16)$$

As above, *the countermeasure whole diet absolute milk equivalent* $w_{i,m}^h(t)$ can be introduced as:

$$q_i^*(t) = c_m(t) h_{ij}(t) w_{i,m}^h(t) \qquad (17)$$

where $w_{i,m}^h(t)$ is the daily consumption of locally produced milk that would equal the ^{137}Cs intake with whole diet when the countermeasures take place.

If countermeasures were realised, the *countermeasure committed effective dose,* D_T^*, is calculated as follows:

$$D_{i,T}^* = K_i \int_0^T q_i^*(t) dt \qquad (18)$$

Effectiveness of the countermeasures

There are two important considerations that must be discussed in order to define the effectiveness of countermeasures. First, it must be decided whether the degree of effectiveness is the achieved decrease in the contamination level of the food crops produced in the area under consideration, or whether it is the decrease in the internal dose received by the subpopulation of interest. Second, if reduced dose is the accepted criterion, it is necessary to determine exactly in which subpopulation the dose decreased, *ie.* the inhabitants of the settlement where these foods were produced or, if these foods were distributed via a system of governmental purchasing and redistribution, the inhabitants of the whole area or country. Here the reduction of dose is considered by the authors as the main criterion of the effectiveness of countermeasures. Moreover, the Chernobyl experience has shown that, from the point of view of dose, the exact consumption rates of the locally produced foods are critical; and the critical subpopulation is the inhabitants of the rural settlements where these foods are produced and consumed. So, the *dose effectiveness of countermeasures* for the members of group i up to time T after the accident may be characterised by a coefficient $H_{i,T}$ which is the ratio of the reference and countermeasure committed effective dose up to time T:

$$H_{i,T} = \frac{D_{i,T}^0}{D_{i,T}^*} \qquad (19)$$

In addition, *the function of countermeasure effectiveness,* $H_i(t)$, which characterises the relative decrease with time of the intake of ^{137}Cs due to the countermeasures is defined as:

$$H_i(t) = \frac{q_i^0(t)}{q_i^*(t)} = \frac{\sum_j c_j(t) f_{ij} w_{ij}}{\sum_j c_j(t) h_{ij}(t) f_{ij} w_{ij}} = \frac{w_{i,m}^0(t)}{w_{i,m}^h(t)} \qquad (20)$$

APPLICATION OF MODEL

This ingestion dose model has been applied to one region within Ukraine, to illustrate some of the information that can be derived from it. The region chosen is the area encompassing the northern parts of the Rovno oblast. This is a predominantly rural area of low economic status, in which fallout from Chernobyl was relatively low, but the soil-to-milk transfer coefficient is frequently high. All these factors combine to make the potential dose from ingestion higher than that from external exposure. For application of the model, the region has been divided into three sub-regions, according to the effective soil-milk transfer factor. The sub-regions were defined as follows: soil-milk transfer factors of 1-5 Bq/l per kBq/m^2, 5-10 Bq/l per kBq/m^2 and >10 Bq/l per kBq/m^2.

The absolute milk equivalent consumption rate and other consumption rate data appropriate to the region are available[3,4,5] as shown in Table 1. The dietary data are for a year before the Chernobyl accident, and therefore, before people's dietary behaviour was modified in response to the accident. By combining all these data, it can be shown that the predominant ingestion dose pathway for average dose is privately produced milk. This is indicated in the final column of Table 1. No other single foods make a major contribution; the remaining intake is spread over a wide range of foods. Therefore, for the estimation of average potential dose in this region, it is reasonable to assume that normal variations in the relative concentrations foods and in relative consumption rates can be neglected, so long as the concentration and consumption rate of privately produced milk is well-characterised. As indicated in the final column of Table 1, assuming the statistics on locally produced milk consumption are reliable, the average potential intake of caesium-137 from total diet can be assumed to be equivalent to 1.6 l/day of locally produced milk.

Food	Concentration relative to milk	Consumption rate (kg/day)	Local fraction	Component of caesium-137 intake (kg/day)
Milk	1	0.9	1	0.9
Milk products	0.4	0.2	0.6	0.092
Beef	3.5	0.01	0.01	0.0035
Pork	0.9	0.016	1	0.144
Wild game	9	0.002	1	0.018
Poultry	1.5	0.01	1	0.015
Potatoes	0.08	0.35	1	0.08
Leafy vegetables	0.5	0.09	1	0.045
Grain (bread)	0.001	0.4	0.01	-
Mushrooms	10	0.02	1	0.2
Fish	2	0.038	1	0.076
Milk equivalent				1.6

Table 1 Dietary components of adult caesium-137 intake for rural areas of Rovno oblast

The model also includes the possibility of a correction to the reference intake to allow for contamination in imported foods. Table 2 gives whole body measurements made on individuals in Kiev (ie individuals not obtaining their diet from locally produced foods) over a period of several years. From these whole body measurements total intakes of caesium-137 from imported foods can be inferred. From Table 1 it is clear that less than 20% of the average diet in Rovno oblast came from outside of the local area in the years preceding the Chernobyl accident. Therefore, the total potential intake of caesium-137 can be obtained (conservatively) as the sum of the reference intake determined from the absolute milk equivalent and the intake determined for individuals living in Kiev.

Year	Number of measurements	Geometric mean (Bq)	Geometric standard deviation
1986	220	850	6.3
1987	62	740	5.1
1988	752	560	1.6
1980	726	480	1.6

Table 2 Body burdens of caesium-137 in Kiev residents

The real intake of caesium-137 by adults, as a function of time, can be inferred from whole body measurements. This real intake function is compared with the reference intake function in Fig. 2 for the three sub-regions of the Rovno region studied. It can be seen that at all times the reference intake exceeds the real intake. If all the model assumptions are valid, it follows that the only reason for this difference is the modification of people's dietary habits in response to the accident (in particular, the substitution of locally produced milk with imported milk).

In order to explore this further, the ratio of the reference intake function to the real intake function, $H_{adult,real}(t)$, was computed. This function was fitted by the following formula:

$$H_{adult,real}(t) = L\ e^{\lambda_h t} \qquad (21)$$

where L and λ_h are constants, and t is the time after the accident. These parameters characterise both the level and the change with time of the countermeasures. In addition, the effective half-time (in years) of countermeasure introduction, T_h, can be defined:

$$T_h = \ln 2\ /\ (365\lambda_h) \qquad (22)$$

In Table 3 the values of these parameters are given for each sub-region. It can be seen that, at early times after the accident, measures to modify dietary habits were most successful in areas with an intermediate soil-milk transfer factor (highest value of L). This is most likely to be related to practical aspects, for example, the quality of roads etc for transporting substitute milk supplies to the settlements. However, in those areas where initial measures were less effective, the half-times of countermeasure introduction are relatively shorter. This can be interpreted as rapidly increasing effort being applied to making the countermeasure effective in those areas where initially it was not effective. This is something that warrants further investigation.

Milk-soil factor (Bq/l per kBq/m^2)	L	λ_h (days^{-1})	T_h (years)
1-5	1.9 ± 0.7	(10.2 ± 3.7) 10^{-4}	1.9
5-10	6.9 ± 6.1	(2.6 ± 5.9) 10^{-4}	7.3
>10	3.1 ± 1.6	(4.4 ± 3.9) 10^{-4}	4.3

Table 3 Effectiveness of countermeasures: parameter values from the model

By comparing the reference and real intake functions, it is possible to calculate the average individual ingestion doses that have been averted by changes in dietary behaviours. These are shown in Table 4 for the three defined areas and integrated to various times after the accident. The effectiveness of such countermeasures, particularly in the areas with an intermediate value of soil-milk transfer factor and within the first year of the accident is clear.

Soil-milk transfer factor	Integration Time Relative to Time of Accidents, y		
(Bq/l per kBq/m^2)	0-1	0-4	0-6
1-5	2.4	3.6	4.2
5-10	7.2	8.1	8.5
>10	3.4	4.1	4.5

Table 4   Estimated average individual ingestion doses averted by changes in diet in northern regions of Rovno oblast, mSv

Conclusions

In conclusion, ingestion is potentially a very significant exposure pathway for caesium-137 following an accident. Owing to the rural structure and lifestyle in much of the contaminated territories, the consumption of local produce forms a major part of the diet. The uneven pattern of contamination and the varying availability of some local foods (eg forest produce) means that it is difficult to predict the doses that will be received in settlements based on consumption rates and food concentrations, without detailed knowledge of each settlement. However, regional average consumption rates are presented to enable the order of the doses to be predicted.

Once whole body measurements have been made, it is possible to develop models which estimate doses already received with much improved accuracy. The model described sucessfully compares the 'real' intake function derived from whole body measurements with that predicted from consumption rates appropriate to the years prior to the accident and food concentrations (the reference intake function) to provide good estimates of the internal doses received by a population. This modelling and comparison procedure enables an analysis, among other things, of the effectiveness of countermeasures. It is clear that the substitution of local milk with clean milk imported from uncontaminated areas has resulted in a very substantial reduction in the exposure resulting from ingestion.

References
1. CEC. JSP5 Pathway analyis and dose distributions: Final Report. (To be published).
2. Skryabin et al. Distribution of doses received in rural areas affected by the Chernobyl accident. NRPB-R277. HMSO. (1995).
3. Likhtarev, I.A., Kovgan, L.N., Vavilov, S.E., Gluvchinsky, O.N., Litvinets, L.N., Anspaugh, L.R., Kercher, J.R. and Bouville, A. Internal exposure from the ingestion of foods contaminated by caesium-137 after the Chernobyl accident. Report 1: General model. Ingestion doses and countermeasure effectiveness for the adults of Rovno Oblast of Ukraine. Health Phsics (submitted, 1995).
4. Shutov, V.N., Bruk, G.Y., Basalaeva, L.N., Vasilevitskiy, V.A., Ivanova, N.P., Kaplun, I.S., Koslovskaya, I.S. and Pushonik, S.I. The role of mushrooms and berries in the formation of internal exposure doses to the population of Russia after the Chernobyl accident. Radiation Protection Dosimetry (submitted, 1995).
5. CEC. ECP9 Fluxes of radionuclides in rural communities in Russia, Ukraine and Belarus: Interim report for the period 1993 to 1994. (1994).

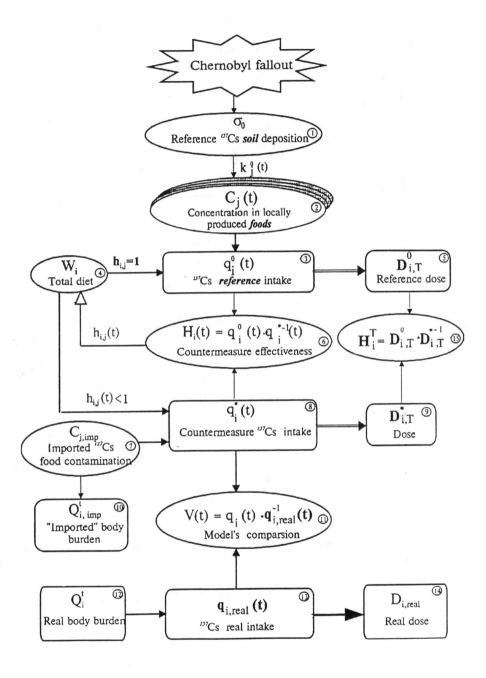

Figure 1: Schematic structure of the model

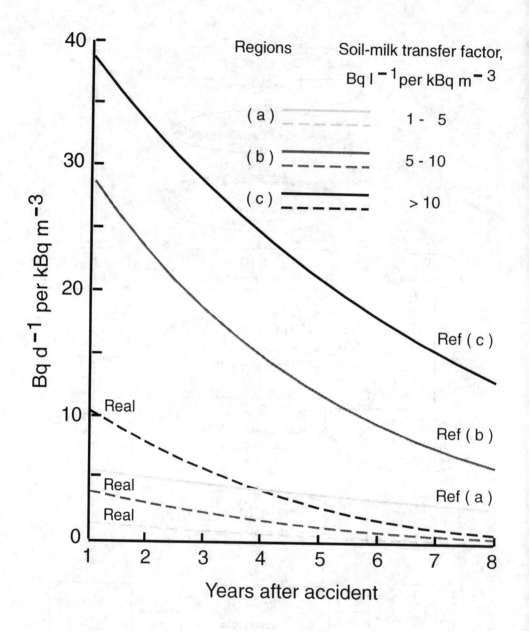

Figure 2: Real and reference intake functions

II. ENVIRONMENTAL ASPECTS
OF THE ACCIDENT
B. Exposure of the population

Posters

Radionuclide Mobility in Soils and its Effect on the External Radiation Exposure

K. Bunzl, U. Hillmann, P. Jacob, R. Kretner, W. Schimmack
GSF-Forschungszentrum, Institut für Strahlenschutz, Neuherberg, Germany

F. Tikhomirov, A. Scheglov
Moscow State University (MSU), Moscow, Russia

N.P. Arkhipov, A.N. Arkhipov
Research and Industrial Association (RIA) "Pripiyat", Chernobyl, Ukraine

R.M. Alexakhin, S.V. Kruglov
Russian Inst. Agricult. Radioecology (RIARE), Obninsk, Russia

N. Loschilov, Y. Ivanov, S. Levchuk. V. Kashparov, L. Oreshich
Ukrainian Inst. Agric. Radiology (UIAR), Kiev, Ukraine

Abstract. In order to predict the gamma-dose rate of ^{137}Cs for a period of about 100 years after the Chernobyl accident, the present vertical distribution of radiocesium in several meadow soils in the Chernobyl area and in Germany was determined and the corresponding residence half-times of this radionuclide in the various soil layers were evaluated with a compartment model. The resulting residence half-times were subsequently used to calculate the vertical distribution of ^{137}Cs in the soil as a function of time and finally to predict the external gamma-dose rates in air for these sites at various times.

The results show that the time dependence of the relative gamma-dose rate in air $D/D_{t=0}$ at both sites can be described by a two-term exponential equation. At a given time, the relative gamma-dose rate at the Chernobyl sites is always higher as compared to the German sites. This difference is the result of the slower vertical migration of ^{137}Cs at the Chernobyl sites.

1. Introduction

The deposition of radiocesium to a meadow will not only contaminate the plants but also cause an external radiation exposure of persons working on the contaminated land. With time, radiocesium migrates into the soil, and part of the gamma radiation of ^{137}Cs is attenuated by the overlaying soil layers. On a flat area, the resulting decrease of the external gamma radiation thus depends on i) the radioactive decay of ^{137}Cs, ii) the time dependence of the vertical distribution of ^{137}Cs in the soil, and iii) on the wet bulk density of the soil as a function of depth. The aim of the present investigation was to estimate the long-term external radiation exposure due to ^{137}Cs of contaminated grassland at various sites in the Chernobyl area and in Germany.

2. Experimental

2.1. Sites

In Bavaria (Germany) 7 undisturbed grassland sites with different soil properties, located mainly south east of Munich were selected. In the CIS countries the following four grassland sites were chosen:
1) Near Zapolje, circa 15 km south of the reactor site. 2) Near Chistogalovka, circa 12 km southwest of the reactor site. 3) Near Kopachi, circa 6 km south of the reactor site. 4) Near Dityatki, circa 25 km off the reactor site.

2.2. Soil sampling and measurements.

At the four corners and in the middle of a 10 x 10 m square 3 soil cores (5 cm in diameter) were taken in the following layers: 0-1, 1-2, 2-4, 4-7, 7-10, 10-15, 15-20, 20 - 30 cm. Immediately after sampling, the moisture content and the bulk density of each soil layer was also determined by standard methods. Subsequently, all samples were air dried and sieved to 2 mm using standard procedures. Radiocesium in each soil layer from each site was determined by direct gamma spectrometry. The experimental error of the ^{137}Cs determination, which depends on the activity present in each soil layer and on the counting time, was always < 10%.

3. Evaluation of the residence half-time of ^{137}Cs in the various soil layers.

The residence half-time τ of Cs in the various soil layers were evaluated from the observed depth profiles of this radionuclide with a compartment model. In the present case, we write for the transfer of activity A_i (Bq m^{-2}) of a radionuclide in the compartment "i" in a small time interval Δt (day)

$$\frac{\Delta A_i}{\Delta t} = K_{i-1} A_{i-1} - K_i A_i - \lambda A_i \qquad (1)$$

where K_i (day $^{-1}$) is the fractional rate of transfer from compartment "i-1" to compartment "i", and λ is the disintegration constant of the radionuclide. To obtain for a given radionuclide in each soil layer the corresponding value of K_i, the system of Eq.1 is integrated numerically. The residence half-time of the radionuclide in layer "i" is given then by $\tau_i = 0.693/K_i$. To compare the values of τ observed in soil horizons of different thickness L, we report for each horizon the value of τ/L.

4. Time dependence of the residence half-times of ^{137}Cs in soils.

For two sites in the Chernobyl area and in Bavaria/Germany sufficient data were available to evaluate the time dependance of the residence half-times of ^{137}Cs in the various soil layers since the fallout event in April 1986. These data are shown in Fig. 1.

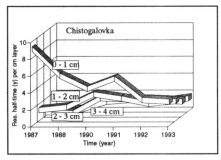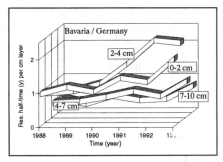

Fig.1. Time dependence of the mean residence half-time of Chernobyl-derived ^{137}Cs in various soil layers of a grassland soil. All values are given with respect to the thickness L of each soil layer (i.e. τ/L). Left: near-field site of the ChNPP at Chistogalovka. Right: site in Bavaria/Germany.

The main difference between these two sites with respect to the migration of radiocesium was apparent in the *top soil layer*, where the residence half-time increased at the German site between 1987 and 1994 from about 0.7 to 2 years/cm, while it decreased at the ChNPP site from about 9 to 3 years. In the deeper soil layers the residence half-times increased at both sites with increasing time in a rather similar way. The opposite behavior of the residence half-time in the top soil layer at the two sites results from the fact that ^{137}Cs was deposited at the ChNPP site predominantly in the form of rather *insoluble fuel particles*, while at the German site this radionuclide was attached in a more available form to natural aerosols. The increase of the ^{137}Cs residence half-times with time at both sites in the deeper layers can be explained by the progressive fixation of radiocesium by clay minerals of the soil.

5. Time dependence of the external gamma-dose rate

The residence half-times of ^{137}Cs as for the various soil layers as evaluated with the compartment model from the depth profiles observed in 1993 are shown in Fig. 2.

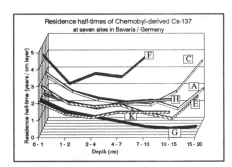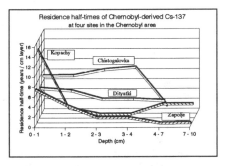

Fig.2. Residence half-times of Chernobyl-derived ^{137}Cs in the soil as a function of depth at 7 sites in Germany (left) and 4 sites in Chernobyl area (right).

These data were subsequently used to obtain (again with help of the compartment model) the vertical distribution of radiocesium in the soil for the coming years. Because Fig.1 indicates that the residence half-times tend to increase with time even after 1994 (exception for the 0 - 1 cm layer at the Chernobyl sites), we doubled the residence half-times in the layers 0 - 4 cm, and quadrupled them in the layers 4 - 30 cm for t > 11 years after deposition. The resulting depth profiles of ^{137}Cs (not shown here) were subsequently used to calculate the resulting gamma-dose rate of ^{137}Cs in air (1m height above the surface) according to Saito and Jacob (1995).

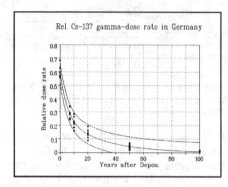 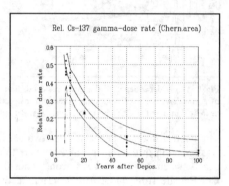

Fig. 3. Predicted relative gamma-dose rate as a function of time for meadow sites in Germany (left) and in the Chernobyl area (right) due to migration in the soil and radioactive decay of ^{137}Cs. Regression lines and 95% prediction limits.

The results are given in Fig.3 for the German sites and for the Chernobyl sites. The time intervals were 7, 10, 20, 50, and 100 years after the Chernobyl accident. All dose rates are given as relative dose rates, i.e. with respect to a dose rate D_0, which was calculated by assuming that all ^{137}Cs deposited at a given site was present at the soil surface only. The decrease of the dose rate with time due to radioactive decay of ^{137}Cs is included in these values.

Fig. 3 reveals that even at t = 0. i.e immediately after the fallout, the relative dose rate in air is < 1. This was caused at the Chernobyl sites by inhomogeneities of the soil surface layer (surface roughness). The 95%-prediction limits for this value are, however, so large (see Fig. 3, right), that a statistically meaningful value cannot be obtained in this way. At the German sites, where observed ^{137}Cs-depth profiles shortly after the deposition are available, the extrapolated relative dose rate in air is obtained more accurately (see Fig. 3, left) as 0.593 (95% prediction limits 0.519 - 0.665). This rather low value is due to surface roughness and partial infiltration of ^{137}Cs in to the soil by several heavy rain showers during the fallout period (about 3 days).

The curves shown in Fig. 3 were obtained by regression analysis. Because for the German sites a simple exponential function did not prove to be satisfactory, we used a two-term exponential function $D(t) = a \cdot exp[-t/b] + c \cdot exp[-t/d]$, where $D(t)$ is the relative gamma-dose rate and t is the time in years after April, 1986. The corresponding correlation coefficients were for the German and the Chernobyl sites 0.987 (p < 0.001). The values of the constants a, b, c, d are for the German sites (and Chernobyl sites): a = 0.34 ± 0.084 (5.81 ± 234); b = 5.05 ± 1.60 (1.27 ± 9.76); c = 0.26 ± 0.083 (0.60 ± 0.10); d = 28.7 ±8.55 (24.3 ± 4.07). The data show that after 50 years the relative gamma-dose rate will have decreased at the German sites to 5% (95% predicition interval: 0 - 12%) and at the Chernobyl sites to 8% (95% prediction limits 0.1 - 15%) of their initial value. The curves given in Fig. 3 should be considered preliminary. Additional research will be required to verify them.

Redistribution of Chernobyl ^{137}Cs in Ukraine wetlands by flooding.

Peter A. BURROUGH
Department of Physical Geography, Utrecht University, Post box 80.115,
3508 TC Utrecht, the Netherlands.

Morna GILLESPIE and Brenda HOWARD
Institute of Terrestrial Ecology, Merlewood Research Station, Windemere Road,
Grange-over-Sands, Cumbria, LA11 6JU, UK.

B. PRISTER
Ukraine Institute of Agricultural Radiology, Machinostroiteley Street 7, Kiev, Ukraine 25505.

Abstract. In northwest Ukraine, some soils in the Rovno region near the Byelorus border 300km west of Chernobyl have unusually high radiocaesium levels with strong evidence of rapid uptake in the food chain. In a study area covering 76.5 km² near Dubrovitsa, radiocaesium levels vary strongly both spatially and temporally from less than 50kBq/m² to more than 1200kBq/m²: at some sites near major streams, 1993 levels are more than three times those of 1988. Geostatistical methods linked to geographic information systems (GIS) demonstrate that the elevated 1993 levels result from transport and concentration by river flooding, a problem which probably affects all areas regularly inundated by the river Pripyat and its tributaries along a 400km stretch of the Ukraine-Byelorus border.

1. Introduction

Much of the Ukraine was affected by the April 1996 explosion at the Chernobyl nuclear plant and the subsequent radioactive plumes, which were carried to the north and west by the prevailing winds. Radioisotopes, including ^{134}Cs and ^{137}Cs, were deposited over a wide area. Although modal deposition levels of radiocaesium in the Rovno region of the Ukraine some 350 km west of Chernobyl were low compared with other much more highly contaminated areas nearer the accident, the ecological conditions of the dominant soil types in the area allow comparatively high plant uptake of radiocaesium, leading to persistent long-term contamination of various food products, particularly milk, meat and mushrooms [1].

Samples of soil collected in 1988, 1993 and 1994 demonstrate that in some areas the ^{137}Cs levels have declined in line with isotopic decay, but there are significant areas where the maximum levels of radiocaesium in 1993 and 1994 exceed the maximum levels measured in 1988 by two to three times. The major difference between areas where levels have declined and areas where levels have increased is proximity to rivers and canals.

The redistribution and concentration of radiocaesium in fluviatile sediments is well understood. Not long after the 1986 accident Walling *et al* demonstrated the increased concentration of Chernobyl radionuclides in channel and floodplain sediments of the River Severn in Wales [2]. Therefore similar processes may cause the local enhancement of ^{137}Cs levels in the Rovno area of the Ukraine.

This paper demonstrates how geographic information systems and methods of geostatistical analysis have been used to identify and confirm the role of hydrological processes such as flooding and deposition of polluted sediments in the severe post-Chernobyl enhancement of radiocaesium levels in a part of the Rovno region.

2. Study area and data collection

Data were collected from an intensively studied area, namely a typical collective farm area within the Rovno area (the Chapayev and Kolos Collective Farms) which was identified in 1988 by the Ukrainian authorities as a site having a large background level of radiocaesium with strong spatial and temporal variability [1]. The topography is extremely flat and the main drainage is by meandering river channels that drain an area of approximately 6500 km² in a northerly direction, uniting north of the Ukraine-Belarus border to flow east as the River Pripyat which joins the Dnieper River at Chernobyl. The whole Belarus-Ukraine border area is characterized by extensive areas of soddy gley and peaty soils which are marked on modern and historical 1:3000000 and 1:500000 maps. The actual study site measures some 10.5 x 12.5 km and covers 76.5 km² of the Dubrovitsa District of the Rovno region [1].

A digital geographical information system (GIS) of sampling locations, natural and artificial drainage, areas affected by annual spring floods, field parcel numbers, soil series and landuse maps was created by digitizing 1:10000 scale paper source maps. A gridded database with a resolution of 50 x 50m was created from the basic map data for quantitative spatial analysis and interpolation. Samples of contaminated soil were collected in 1988, 1993 and 1994 at 72, 87 and 47 sites respectively and were analyzed for ^{134}Cs and ^{137}Cs. The location of each sample was digitized.

2.1 Correction for ^{134}Cs and decay.

To study temporal changes caused by hydrological processes the raw data were first corrected for the decay rates of the two isotopes using an isotopic ratio for 1986 of $^{137}Cs/^{134}Cs = 2:1$ [3]. All analyses refer to radiocaesium data that have been converted to 1986 equivalents (Table 1). The data are strongly positively skewed and were normalized by converting to natural logarithms. All spatial analysis uses transformed data but the results have been back-transformed for display and ease of interpretation.

Table 1 Statistics of the transformed data as 1986 equivalents (kBqm^{-2}).

Year	N	Min	Max	Mean	Median	Mode	Cv
1988	72	9.3	406.0	134.4	97.6	93.0	67.1
1993	87	20.6	1267.8	239.4	164.6	107.3	95.2
1994	47	17.3	531.5	135.1	135.1	multiple	83.0

3. Spatial analysis.

3.1 Risk of flooding

Simple buffering of the drainage map gave a map of site proximity to (or risk of) flooding for each location in the study area with respect to major rivers, lakes and canals (Fig. 1).

3.2 Geostatistical analysis: comparing data between the years.

Plots of ^{137}Cs levels against distance of the site from annual flooding or water showed that in each year the highest levels occur in the zone that is flooded (DFLD = 0 m) or within 1 km from water or annually flooded areas [4].

However, because the soil samples were collected each year at different locations the variations in ^{137}Cs levels over the years could be due to different samples from a highly variable population. Geostatistical methods based on *regionalized variable theory* known as *conditional simulation* [5,6] were used to compare measured and predicted ^{137}Cs levels at the same locations. The aim was to examine the likelihood that measurements in 1988 could have yielded values as large as those found in 1993 if samples had been taken in 1988 at the 1993 sample sites.

Using the 1988 data, 100 simulations were made of the expected value of ^{137}Cs at each of the 87 data points sampled in 1993. The procedure was repeated to predict the values at the 1994 sampling sites from the

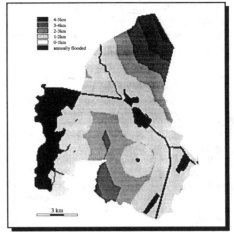

Figure 1 Flood zones in the study area.

1993 data. The 100 simulated values at each grid point were averaged to determine the expected value and its standard deviation.

The simulated predictions and their standard deviations were used to compute a *normalized* index of increase or decrease over the years between the two sampling periods defined as

$$\text{NDIF}_{ij,x} = (M_{i,x} - P_{ij,x,y}) / \text{SDIF}_{ij,x,y} \qquad (1)$$

where: $M_{i,x}$ is the measured value for year i at sites x, $P_{ij,x,y}$ is the mean value of 100 simulations of ^{137}Cs for year i at sites x predicted from measurements made in year j at sites y, and $\text{SDIF}_{ij,x,y}$ is the standard deviation of the 100 simulations of $P_{ij,x,y}$. Note that all normalized differences were computed using the log-transformed data.

Fig. 2 shows the normalized indices plotted against distance from water or annual flooding and demonstrates the close association between proximity to flooding and large (and variable) ^{137}Cs levels. The spatial distributions of the normalized indices were obtained by interpolating to the 50 x 50m grid using Ordinary Point Kriging [5] and back-transforming the values from natural logarithms (Fig.3). Once combined in the GIS with the maps of soil and land use, these maps demonstrated the close association between floodplain soils, landuse types and areas with elevated levels of radiocaesium. Although the match between flood zones and enhanced normalized levels is not exact, the close fit reinforces the hypothesis of a strong association between flooding, flood potential and enhanced levels of [137]Cs. This result clearly supports the initial hypothesis that radiocaesium can be redistributed by water: indeed, some of the elevated levels in the south west of the area may have been transported from further upstream.

4. Discussion.

The combined GIS and geostatistical analysis show clearly how data on flood frequency, soil types, land use and soil samples can be combined to evaluate the hypothesis that the enhanced levels of ^{137}Cs measured after 1988 are due to flooding and deposition of polluted sediments. Further data on the enhancement of ^{137}Cs levels in milk before and after a major flood in July 1993 confirm the direct effect of flooding on uptake into the foodchain, particularly on unimproved pastures on peaty soils [4]. The detailed interactions between flood deposition, retention in the soil and uptake into the food chain have yet to be fully investigated but together with studies to determine the total extent of the problem along the whole Pripyat drainage network they will be major objects of future research.

References

[1] Strand, P., Howard, B., & Arverin, V. (1994) Editors: *Fluxes of radionuclides in rural communities in Russia, Ukraine and Belarus.* ECP9 Annual Report November 1993-December 1994. NRPA.

[2] Walling, D.E., Rowan, J.S. & Bradley, S.B. (1989) Sediment-associated transport and redistribution of Chernobyl fallout radionuclides. *Sediment and the Environment* (Proc. Baltimore Symposium 1989), IAHS Publication no 184, 37-45.

[3] Beresford, N.A., Howard, B.J., Bennet, C.L. & Crout, N.M.J., 1992. The uptake by vegetation of Chernobyl and aged Radiocaesium in Upland West Cumbria. *J. Environmental Radioactivity* 16: 181-195.

[4] Burrough, P.A., Gillespie, M., Howard, B.J., Howard, D.M., Pronevich, V., Prister, B., Strand, P., Skuterud, L. & Desmet, G.M. (1995). *Redistribution of Chernobyl ^{137}Cs in Ukraine wetlands by flooding.* Joint publication Department of Physical Geography, University of Utrecht, NL & Institute of Terrestrial Ecology, UK, 49pp.

[5] Deutsch, C.V. & Journel, A. (1992). *Geostatistical Software Library and User's guide.* Oxford University Press, New York. 340pp.

[6] Gómez-Hernández, J.J. & Journel, A.J. (1993). Joint sequential simulation of multigaussian fields. In: A. Soares (ed), *Geostatistics Troia '92.* Association of Mathematical Geology, 1992, 85-94.

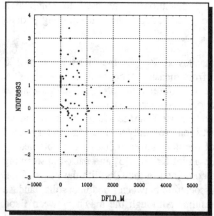

Figure 2 Plot of normalized differences for 1988-1993 against distance to flooding or water

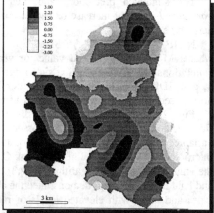

Figure 3 Interpolated levels of normalized differences for 1993.

Estimation of the balance of radiocaesium in the private farms of Chernobyl zone and countermeasures with regard to the reduction of health risk for rural inhabitants

B.S.Prister, A.S.Sobolev
Ukrainian Institute of Agricultural Radiology (UIAR),
7, Mashinostroitelei st., Chabany, Kiev region 255205, Ukraine

G.A.Bogdanov
Ukrainian Agricultural University,
15, Geroev Oborony st., 252041, Kiev, Ukraine

I.P.Los
Ukrainian Research Centre of radiation medicine,
53, Melnikov st., 252050, Kiev, Ukraine

B.J.Howard,
Institute of Terrestrial Ecology (ITE),
Merlewood Research Station, Grange-over-Sands,
Cumbria LA11 6JU, Great Britain

P.Strand
Norwegian Radiation Protection Authority (NRPA)
P.O.Box 55 1345 Osteraas, Grini naringspark 12, Norway

Abstract. The intake of Cs-137 to human organism with various components of the diet was investigated. The highest amount of caesium is consumed with milk, but after the implementation of countermeasures the content of Cs-137 in milk decreases, and in some settlements of Polessye relative intake of radiocaesium with mushrooms increases

The accident on Chernobyl NPP caused the wide-scale radioactive contamination of the environment, which provided the transfer of the radionuclides to foodstuffs by the food chains.

The diet of the rural population of Polessye consists mainly of the foodstuffs, produced on private plots.

The production of foodstuffs on private plots has a number of peculiarities as compared to collective farms. Thus, the cattle is grazed on the pasture, where in some cases agrotechnical measures for the reduction of Cs-137 transfer from soil to agricultural products where not implemented.

MATERIAL AND METHODS.

Caesium-137 concentration in the samples was measured by means of gamma-spectrometer Nokia LPA 4900 with germanium-lithium detector. The samples of foodstuffs were taken from the typical families of Ukrainian Polessye, which consist of 3-4 adults and 2-3 children below 16.

According to the statistical data, land-use structure in Polessye, both on private and collective farms, varies in the ratio of arable lands and natural pasture: for collective farms - 1:2 - 2:1; for private farms - 1:5 - 1:10.

The decrease of Cs-137 content in milk is the result of Cs-137 binding by the soil absorption complex and application of countermeasures. These two processes provided the 10-20-fold decrease of Cs-137 content in milk on collective farms 987 to 1994, while on private farms it was only 4-6-fold decrease. Therefore Cs-137 activity in the flux with milk from private farms is about 80% , and from collective farms - only 20% of Cs-137 activity in all milk, produced in the regions. (Table 1).

The data from Table 2 also testify that the activity of Cs-137 in milk and meat from private farms is several times higher than from collective farms.

Another peculiarity of the diet of Polessye inhabitants is high consumption of mushrooms and berries. The content of Cs-137 in mushrooms in various Polessye regions varies from 800 to 32000 Bq/kg (Table 3). As a result, Cs-137 intake with mushrooms will vary as well. In 1994-1995 detailed analysis of this radionuclide intake with various components to the diet of a family in "Khliborob" collective farm, Dubrovitsa district, Rovno region, was carried out . It was found that the people in the families from the villages at a distance 3-5 km will have great differences in Cs-137 intake with various components of the diet. The population of Milyachy village, which consumed milk (Cs-137 content in milk 80-110 Bq/l). produced on the pastures with countermeasures, received 49% if Cs-137 with milk and 6% with mushrooms; without countermeasures - 63% with milk (230-550 Bq/l in milk) and 11% with mushrooms.

In the villages Velyun and Zagreblya (Cs-137 content in milk 14-30 Bq/l) 13-15% of Cs-137 is received with milk and 21-26% with mushrooms).

The data of sociological survey demonstrated that the population of investigated villages can be divided into 3 groups by the consumption of forest products: pensioners, workers of the collective farm and forestry workers. Forestry workers are the critical group, which consume twice as many mushrooms as the workers of the collective farm. Besides, forestry workers make hay in the forest (Cs-137 activity in hay is 10000-15000 Bq/kg), and in winter the content of Cs-137 in milk during a certain time was 300-450 Bq/l. After the replacement of hay produced in the forest by the hay from the cultivated field of the collective farm (Cs-137 activity in the hay is 800 Bq/kg) content of Cs-137 in milk decrease to 45 Bq/l, and relative intake of Cs-137 with milk to the diet of a family decreased from 80 to 55%.

CONCLUSION

Cs-137 intake to the diet of a family in Polessye depends mainly on the content of Cs-137 in the diet of cows and amount of mushrooms consumed by a family.

In order to reduce the intake of Cs-137 by human organism it is necessary to provide the local population with concentrated fodder and sorbents and to exclude contaminated fodder from the diet of cows. In the settlements, where after the application of countermeasures on the pastures relative intake of Cs-137 with milk is less than with mushrooms, the inhabitants should be informed by the radio and local newspapers about the most contaminated areas of collecting mushrooms and about the species of mushrooms with the highest accumulation coefficients of this radionuclide from soil.

Table 1. Cs-137 fluxes with milk from private and collective farms

Number of cows in the villages where the milk contamination in private farms is >110 Bq/l	Milk production, tons	Cs-137 concentration, Bq/l	Cs-137 flux, kBq/year	Cs-137 flux, %
Volyn region				
Private 7748	186	140	2603328	80
Collective 6500	162	40	662000	20
Zhitomir region				
Private 13389	321	120	3856032	97
Collective11510	228	35	1007125	21
Rovno region				
Private - 21 475	515	175	9109500	82
Collective - 20 320	508	40	2032000	18
Chernigov region				
Private - 360	8.6	110	95040	80
Collective - 485	12.1	20	24250	20
Kiev region				
Private - 224	5.4	140	75264	57
Collective - 385	9.6	60	57750	43

Table 2. Content of Cs-137 in milk and meat of private and collective farms in 1995, Bq/kg

Region	District	Milk		Meat	
		Private farms	Collective farms	Private farms	Collective farms
Zhitomir	Emilchensk	90-240	40-70	30-120	10-70
	Novovolynsk	30-100	30-60	50-180	50-170
	Luginsk	60-500	20-150	100-500	100-300
	Olevsk	140-900	60-300	170-500	100-300
	Korosten	40-500	80-200	40-220	40-901
	Ovruch	30-260	40-200	80-400	100-320
Rovno	Dubrovitsa	14-500	30-70	80-330	40-110
Kiev	Polesskoye	60-190	50-80	70-250	50-130

Table 3. Content of Cs-137 in mushrooms and berries in Zhitomir and Rovno regions in 1995, Bq/kg

Region	District	Forest berries	Fresh mushrooms
Zhitomir	Emilchensk	1000-1800	1800-3500
	Novovolynsk	800-1600	80-1600
	Luginsk	500-8000	1300-3400
	Olevsk	1000-2700	3700-5000
	Korosten	1000-3500	2600-5000
	Ovruch	1000-2500	2000-4000
Rovno	Dubrovitsa	800-4500	5000-32000

II. ENVIRONMENTAL ASPECTS
OF THE ACCIDENT
C. Management of contaminated territories

Management of Contaminated Territories Radiological Principles and Practice

Per HEDEMANN JENSEN
Risø National Laboratory, Roskilde, Denmark

Spartak T. BELYAEV, Vladimir F. DEMIN
Russian Research Centre "Kurchatov Institute", Moscow, Russia

Igor V. ROLEVICH
Chernobyl State Committee, Minsk, Belarus

Ilja A. LIKHTARIOV, Leonila N. KOVGAN
Ukrainian Scientific Centre for Radiation Medicine, Kiev, Ukraine

V. G. BARIAKHTAR
Ukrainian Academy of Sciences, Kiev, Ukraine

Abstract. The current status of internationally agreed principles and guidance for the management of contaminated territories and the international development of intervention guidance since the Chernobyl accident is reviewed. The experience gained after the Chernobyl accident indicates that the international advice on intervention existing at the time of the Chernobyl accident was not fully understood by decision makers neither in Western Europe nor in the former USSR and that the guidance failed to address adequately the difficult social problems which can arise after a serious nuclear accident.

The differences between CIS practice and international guidance, both conceptually and practically, are identified. The general response of the authorities in the former USSR regarding many early actions for protection of the affected population after the Chernobyl accident were broadly reasonable and consistent with internationally established guidelines pertaining at the time of the accident. During the years following the accident, decisions on countermeasures in the former USSR were based on four different criteria: annual dose, lifetime dose, temporary permissible levels in foodstuffs and surface contamination density of ^{137}Cs. Due to socio-psychological and political factors, requirements for radiation protection were made more and more strict.

The CIS criteria of today for different protective actions and strategies are given in terms of annual doses or activity concentrations af ^{137}Cs in different foodstuffs. International guidance is given as *intervention levels* in terms of *avertable doses* by specific countermeasures and as *action levels*. Action levels refer to different protective actions or protection strategies. Action levels are levels above which remedial actions are taken and below which they are not, and they refer to the maximum *residual dose* without any action. The CIS criteria are conceptually a system of *action levels* rather than a set of intervention levels. The numerical values of these action levels are not directly comparable to international numerical guidance, but they seem not to be in direct conflict.

Unresolved issues have been identified to be the interaction between radiological and non-radiological factors in decision-making. Both radiological and non-radiological factors will influence the level of protective actions being introduced. Social-psychological countermeasures are a new category of action, in the sense that social protection philosophy has not yet been developed to fully include their application after a nuclear accident. It has been suggested that the inclusion of such countermeasures into the intervention decision making framework should be as a part of the radiation protection framework. It is argued here that optimization of the overall health protection is *not* a question of developing radiation protection philosophy to fully include socio-psychological factors. It is rather a question of including these factors - in parallel with the radiological protection factors - in cooperation between radiation protection experts and psychological specialists under the responsibility of the decision maker. The overall optimization of the total health protection is thus the responsibility of the decision maker(s) with guidance from radiation proction experts as well as experts in the fields of social and psychological sciences.

1. Introduction

The experience after the Chernobyl accident was that many actions taken led to an unnecessarily large expenditure of national resources, and many instances occurred of contradictory national responses. Therefore, there was a strong need for a simple set of internationally consistent intervention levels and action levels and for clear guidance on application of the principles in planning and preparedness for response to nuclear accidents or radiological emergencies. Therefore, the international radiation protection organizations ICRP and IAEA both have prepared a set of clear and coherent intervention principles and also numerical values for generic intervention levels.

The protective measures taken in the CIS after the Chernobyl accident included long-term countermeasures such as relocation of the population and continuing agricultural countermeasures. The levels at which these measures were introduced were based on different rationales. During the years following the accident, these levels have been adjusted by the competent authorities in the CIS. This work is still in progress in Russia, Belarus and Ukraine.

2. Evolution of international guidance on intervention after the Chernobyl accident

At the time of the Chernobyl accident guidance on protection of the public after a nuclear accident in which radionuclides have been dispersed into the environment existed both internationally and in the CIS. The experience gained after the Chernobyl accident indicates, however, that the international advice on intervention was not fully understood by the decision makers, neither in Western Europe nor in the former USSR. The intervention guidance was mixed up with dose limits for practices and the guidance failed to address adequately the difficult social problems which can arise after a serious nuclear accident.

2.1. International Guidance at time of the Chernobyl Accident
The basic principles given by the ICRP [1] and IAEA [2] for planning intervention in accident situations and for setting intervention levels were the following at the time of the Chernobyl accident:

(a) Serious deterministic effects should be avoided by the introduction of countermeasures to limit individual dose to levels below the thresholds for these effects;

(b) The risk from stochastic effects should be limited by introducing countermeasures which achieve a positive net benefit to the individuals involved;

(c) The overall incidence of stochastic effects should be limited, as far as reasonably practicable, by reducing the collective dose equivalent.

Upper dose levels above which the introduction of the countermeasure was almost certain and lower dose levels, below which introduction of the countermeasure was not warranted were given for irradiation of the whole body and individual organs. Between the recommended upper and lower dose levels, site-specific intervention levels were expected to be set by national authorities. The intervention levels covered both the early and intermediate phase after an accident. For the late phase no values were recommended, since it was considered that the main questions facing the decision maker would be whether and when normal living could be resumed, and that the situations would vary too widely to give any generic numbers for that purpose. The ICRP and IAEA *two-tier-system* on intervention levels is summarised in Table 1.

Table 1. ICRP and IAEA intervention level ranges for introducing countermeasures.

Countermeasures	Whole body	Single organs
Sheltering	5 – 50 mSv	50 – 500 mSv
Iodine prophylaxis	–	50 – 500 mSv
Evacuation	50 – 500 mSv	500 – 5,000 mSv
Relocation[a)	50 – 500 mSv/y	not anticipated
Control of foodstuffs[a)	5 – 50 mSv/y	50 – 500 mSv/y

[a) The projected dose for relocation and foodstuff control were only defined for the first year.

2.2. Main problems in the past recommendations

Regarding the international guidance on intervention levels a number of problems were identified when applying it after the Chernobyl accident, although the basic principles still were considered to be valid. Confusion was created because intervention levels for introducing countermeasures were interpreted as doses *received* and not as doses *averted* which wrongly was interpreted as if these intervention levels were dose limits. In addition, major difficulties in the application were:

- how to apply intervention levels, *eg*, in the case of foodstuffs, did the intervention level refer to the sum of food items or to each of the foodstuffs separately ?
- how to compare the dose with the intervention level; was the projected dose or the avertable dose relevant ?
- how was the principle (c) to be applied ? What was the relationship between principles (b) and (c) ?

Major confusion was also created by the references in the ICRP Publication 40 [1] to the dose limits in justifying the numerical values of the intervention levels.

2.3. Current status on internationally agreed intervention principles

The latest recommendations from the International Commission on Radiological Protection [3] outline the systems of protection for *practices* and *interventions*. Human activities that *add* radiation exposure to that which people normally incur due to background radiation, or that increase the likelihood of their incurring exposure, are termed *practices*. The human activities that seek to reduce the existing exposure, or the existing likelihood of incurring exposure which is not part of a controlled practice, are termed *interventions*.

The dose limits recommended by the ICRP are intended for use in the control of practices. The use of these dose limits, or of any other pre-determined dose limits as the basis for deciding on intervention might involve measures that would be out of all proportion to the benefit obtained by the intervention.

In some situations the sources, the pathways and the exposed individuals are already in place when the decisions on control measures are being considered, and protection can therefore only be achieved by intervention. The avertable dose by the protective action, ΔE, can be found as the difference between the dose *without* any actions and the dose *after implementation* of a protective action. If a protective measure were introduced at time t_1 and lifted at time t_2, the avertable dose, ΔE would be equal to the time-integral of the dose per unit time over this time interval, τ. The concept of an avertable dose is shown in Fig. 1.

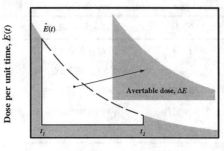

Time after start of accident

Fig. 1. Avertable dose when the protective measure is introduced at time t_1 and lifted again at time t_2. The length of the time interval is $\tau = t_2 - t_1$.

The process of justification and optimization *both* apply to the protective action, so it is necessary to consider them *together* when reaching a decision. *Justification* is the process of deciding that the disadvantages of each component of intervention, i.e. of each protective action or, in the case of accidents, each countermeasure, are more than offset by the reductions in the dose (avertable dose) likely to be achieved.

Optimization is the process of deciding on the method, scale and duration of the action so as to obtain the *maximum net benefit*. In simple terms, the difference between the disadvantages and the benefits, expressed in the same terms, e.g. monetary terms, should be positive for each countermeasure adopted and should be *maximised* by setting the details of that countermeasure.

Intervention levels refer to the dose that is expected to be averted (avertable dose) by a *specific countermeasure* over the period it is in effect. If an intervention level is exceeded, *ie*, if the expected avertable individual dose is greater than the intervention level, then it is indicated that the specific protective action is likely to be appropriate for that situation. The intervention level for a specific countermeasure can be determined from optimization as indicated above. The optimized intervention level would be expressed either as an avertable *individual dose* per unit time or as an avertable *collective dose* per unit mass of a given foodstuff.

Table 2. Summary of recommended intervention levels from ICRP Publication 63 [4] and IAEA Safety Series No. 109 [5] for urgent and longer-term countermeasures.

Protective measure	IAEA generic optimized intervention levels	ICRP range of optimized values
Sheltering (less than 1 day)	10 mSv	5 – 50 mSv
Administration of stable iodine	100 mGy to thyroid	50 – 500 mSv to thyroid
Evacuation (less than 1 week)	50 mSv	50 – 500 mSv to thyroid
Temporary relocation	initiate at 30 mSv in a month suspend at 10 mSv in a month	almost always justified at a dose level of 1 Sv optimized range of: 5 – 15 mSv/month
Permanent resettlement	if lifetime dose would exceed 1 Sv	–

Recommendations from the ICRP [4] and IAEA [5] on intervention levels for urgent and longer term countermeasures are shown in Table 2. They represent the international consensus achieved on intervention levels as recommended by six international organisations [7]. For foodstuff restrictions and agricultural countermeasures, the intervention/action levels for long-lived β-emitters like ^{137}Cs is in the range from a few hundreds to a few tens of thousands of becquerels per kg foodstuff [5].

3. Evolution of intervention guidance in CIS

The protective measures taken in the CIS after the Chernobyl accident included early countermeasures like sheltering, administering stable iodine and evacuating those parts of the population who might be exposed to the plume. Long-term countermeasures, such as relocation and foodstuff restrictions, were taken to mitigate the effects of lower, but still significant levels of radiation from surface and soil contamination. The levels at which these measures were introduced were based on different rationales, and the levels have been changed by the competent authorities during the years following the accident.

3.1. Russia

Since the early years of development of nuclear power in the USSR, serious attention has been given to the planning of measures to protect the population in the event of a release of radioactive materials into the atmosphere from a nuclear reactor. A two level set of criteria, level A and level B, was used for different protective actions [8,9].

Based on measured radionuclide composition in the environment obtained in the first days of the Chernobyl accident and corresponding long-term predictions of external radiation doses and intake of radionuclides of caesium into the body, it was considered appropriate to establish a Temporary Permissible Level (TPL) of dose of 100 mSv to the whole body in the first year. Of these 50 mSv was allocated to internal radiation and 50 mSv to external radiation. For the following years of 1987, 1988 and 1989 the TPLs were 30, 25 and 25 mSv, respectively [10].

Since the accident, surface contamination criteria have been used to delineate affected areas for such matters as the payment of compensation. Strict Control Zones are those areas with a surface contamination density of ^{137}Cs above 15 Ci/km^2 (555 kBq/m^2) and Controlled Zones with a surface contamination density between 5 and 15 Ci/km^2 (185–555 kBq/m^2).

In 1988 the USSR NCRP developed the concept of safe living of the population based on a so-called lifetime dose limit (LDL) [12]. This was adopted to limit the lifetime risk of late health effects. The USSR Ministry of Public Health approved the LDL-concept with its numerical value set at 350 mSv. In areas where it was envisaged that the LDL would be exceeded, protective measures should be implemented, including relocating the population. The LDL concept was critically analyzed in the International Chernobyl Project. The concept was a simplified reflection of the scientific and practical experience of that time reflecting the situation in 1988-1989 in the regions affected by the Chernobyl accident. This one-level system of decision-making on protective measures would, however, not allow any optimization of the whole complex of possible protective measures. The LDL concept was rejected by the USSR Supreme Soviet in April 1990.

In April 1990, the Supreme Soviet of the USSR implemented criteria for relocation in terms of surface contamination density. A surface contamination level for ^{137}Cs of 40 Ci/km^2 (1480 kBq/m^2) was adopted as the criterion for compulsory relocation [11]. For pregnant women and children, the level was 15 Ci/km^2 (555 kBq/m^2).

A Committee of the USSR Academy of Sciences was assigned to develop an alternative concept [13]. According to this concept the main criterion for further implementation of protective measures was the annual effective dose from Chernobyl fallout, starting from 1991. When this dose were less than 1 mSv, no intervention were needed. If the dose would exceed 1 mSv, a complex of protective measures should be carried out. Implementation of radiation protection measures should simultaneously be aimed at relaxing socio-psychological tension and stress. Achievement of these goals should be optimised with the constraint that the average individual effective dose should not exceed 5 mSv in 1991, and a maximum possible reduction to 1 mSv/y in the future.

With the countermeasures already adopted there were no settlements where the actual individual doses in 1991 could exceed the maximum level of 5 mSv/y. This level was established as a control level and not as a level for relocation. In territories where actual doses exceeded 1 mSv/y the population had the right to be relocated. One of the main point of the concept was to avoid mass relocations. Based on this concept the Chernobyl All-Union Laws as well as Republican Laws were prepared and adopted in 1991.

It should be emphasized that some important points of the laws appeared to be internally inconsistent and in contradiction with the concept. In addition to the effective dose the surface contamination density of ^{137}Cs was used in the Laws as an index for protective measures, including mandatory relocation. Radiation and social protection measures should be implemented at a surface contamination density of ^{137}Cs above 37 kBq/m^2. As a result of this very strict criterion the regions officially recognized as affected by the accident increased from a few to seventeen and the population from a few hundred thousands to about three millions. The consequences were that decisons were made on additional obligatory mass relocation, and it became impossible to achieve an optimised protection from a complex of countermeasures. In addition, the implementation of the Laws created additional social problems and consequently negative consequences.

It became clear already in 1992 that the regulatory documents connected with both the elimination of the consequences of the Chernobyl accident and other applications needed to be further improved and developed. The objective of implementing protective and rehabilitative measures in contaminated territories in the Russian Federation was to improve the policy taking into account the new accumulated data and experience. Some concrete recommendations for this were made [14]. The main recommendations were a shift in protection policy from strict measures of protection (mandatory relocation, rigid limitations to lifestyle and economic activities etc.) to rehabilitation of territories and restoration of normal life and economic conditions and a shift from providing compensations and allowances to individuals to allowances for improving social and economical conditions in the public sector.

The new concept was worked out [15] and adopted by the Russian NCRP, mainly applicable to existing post-accidental situations. In accordance with the concept the objective was to provide a high health standard for the population living in these areas. The territories contaminated by radionuclides were to be subdivided into *zones*. This zoning was based on the annual effective individual dose to the population due to the contamination. Two zones of annual doses were suggested:

- *zone of radiation control (1–5 mSv/y)*, and
- *zone of restricted residence and voluntary relocation (5–20 mSv/y)*.

Those who desire to leave the zone of 5–20 mSv/y can obtain help from governmental bodies. It is not desirable for families with children to settle here for living. People who plan to settle in those areas should be explained the possible health risks. The zoning should be reconsidered once every three years.

In 1994 there were no settlements in Russian territory where the annual dose could exceed 20 mSv. In territories where the effective doses were below 1 mSv/y lifestyle and economic activities were free from any restrictions. This dose level defines in practice a border between normal and abnormal conditions [15]. A few levels of residual dose were established for social and health protection.

Considering the situation in 1995 it should be emphasized that there are practically no settlements where the annual individual doses exceed 5 mSv.

In fact, the action levels of 5 and 20 mSv/y regulate the settlement of people for living and restoration of life in the territories affected by the contamination from the Chernobyl accident. The levels are not for regulation of relocation which is a countermeasure not to be implemented in the late phase of a post-accident activity. Decisions on relocation should be taken in the earlier phases of the accident (this is considered in the new Russian regualtions, see para. 4.1). Therefore, the Russian regulation level of 20 mSv/y is *not* an action level for relocation.

3.2. Ukraine

The Ukrainian strategy for setting criteria in terms of intervention levels for countermeasures on the radioactive contaminated territories in Ukraine has been changing during all the years following the Chernobyl accident. The evolution in the criteria is shown in Fig. 2.

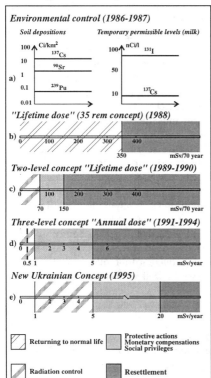

Fig.2. Evolution of Ukrainian criteria for countermeasures during the years following the Chernobyl accident.

At the early stages of the accident (first few years), the level of environmetal radioactive contamination and radiation, *eg*, γ-exposure rate in air, surface contamination density and activity concentration in foodstuffs were used as the main criteria for the introduction of protective actions (Fig. 3 (a)).

In 1988 the total dose from all exposure pathways due to the accident were proposed as as a criterion for decision making on protective actions. In accordance to the recommendations of the National Committee on Radiation Protection of USSR, the first of these dose criteria was the so-called "lifetime dose concept" (Fig. 3(b)). This dose was considered to be a conservatively estimated total dose from external and internal exposure from all exposure pathways due to the accident over a time period of 70 years under normal living conditions in the contaminated territories. A lifetime dose level of 350 mSv was suggested. Because of a number of reasons, mainly due to the reaction of the public, this concept - refered to as the "35-rem concept" - was never accepted.

In 1989-1991 the "two-tier-lifetime-dose concept" for decision making was considered in Ukraine (Fig. 3(c)). Two lifetime dose levels over 70 years of 70 mSv and 150 mSv were suggested as intervention levels. This concept has not been applied in Ukraine as well.

In 1991 the so-called "three-tier-annual dose concept (Fig. 3(d)) for intervention was proposed in Ukraine. In accordance to this concept the annual dose for decision making in certain settlements was to be the sum of the external γ-dose from deposited activity, the internal dose from ingestion of locally grown contaminated foodstuffs containing ^{137}Cs and ^{90}Sr, and the inhalation dose from resuspension of plutonium.

This "three-tier-annual-dose concept" for intervention has been used in Ukraine up until 1995. The intervention levels were 1 mSv/y and 5 mSv/y and since 1994 also 0.5 mSv/y. In accordance to these levels all contaminated territories in Ukraine were divided in zones. Different sets of countermeasures, monetary compensations and privileges were introduced in each zone. The radiological control were provided in the territories where the annual doses were 0.5–1 mSv. The complex of protective actions were to be taken in territories where the annual doses were in the range 1–5 mSv. In territories where the annual doses exceeded 5 mSv relocation (resettlement) were to be considered.

3.3. Belarus

After the accident, surface contamination criteria have been used to delineate areas affected by the Chernobyl accident. Strict Control Zones are those areas with a surface contamination density of ^{137}Cs above 15 Ci/km^2 (555 kBq/m^2) and Controlled Zones with a surface contamination density between 5 and 15 Ci/km^2 (185–555 kBq/m^2).

In 1988 the USSR NCRP developed the concept of safe living of the population based on a so-called lifetime dose limit (LDL) [12]. This was adopted to limit the lifetime risk of late health effects. The USSR Ministry of Public Health approved the LDL-concept with its numerical value set at 350 mSv. In areas where it was envisaged that the LDL would be exceeded, protective measures should be implemented, including relocating the population. The LDL concept was rejected by the USSR Supreme Soviet in April 1990.

In April 1990, the Supreme Soviet of the USSR implemented criteria for relocation in terms of surface contamination density. A surface contamination level for ^{137}Cs of 40 Ci/km^2 (1480 kBq/m^2) was adopted as the criterion for compulsory relocation [11]. In Belarus a level of 15 Ci/km^2 (555 kBq/m^2) was adopted as the criterion for compulsory relocation.

A Committee of the USSR Academy of Sciences was assigned to develop an alternative concept [13]. According to this concept the main criterion for further implementation of protective measures is the annual effective dose from Chernobyl fallout, starting from 1991. When this dose were less than 1 mSv, no intervention were needed. If the dose would exceed 1 mSv, a complex of protective measures should be carried out. Achievement of these goals should be optimised with the constraint that the average individual effective dose should not exceed 5 mSv in 1991, and a maximum possible reduction to 1 mSv/y in the future. Based on this concept the Chernobyl All-Union Laws as well as Republican Laws were prepared and adopted in 1991.

The basic documents that regulate the use of countermeasures in the Republic of Belarus are the following:

(1) *Republican concept of living in the territories contaminated with radionuclides as a result of the Chernobyl Nuclear Power Plant catastrophe, adopted by the Bureau of the Presidium of the Belarussian Academy of Sciences of 19 December, 1990;*

(2) *Concept of residing the population in the regions affected by the Chernobyl Nuclear Power Plant catastrophe, adopted by the decree of 8 April, 1991, N164.*

The essence of the second concept is that two levels of annual individual effective doses of "Chernobyl origin" were established for introduction of countermeasures:

- when annual individual effective doses do not exceed 1 mSv no interventions should be made;
- when annual individual effective doses fall in the range of 1–5 mSv a complex of protective measures should be used aimed at constantly reducing the dose rate;
- when annual individual effective doses exceed 5 mSv resettlement should be implemented.

According the the Republican concept the level of 5 mSv/y from the Chernobyl accident should constantly be reduced. The reduction rate were defined as:

1990	5 mSv/y
1993	3 mSv/y
1995	2 mSv/y
1998	1 mSv/y

The above mentioned concepts have been used as basis of the law of the Republic of Belarus *On Social Protection of the Citizens Affected by the Chernobyl Nuclear Power Plant Catastrophe.*

4. Current status on intervention in CIS

The basic intervention philosophy in the three republics have been under constant evolution. At present, the different concepts being developed include - in addition to radiation protection factors - social protection considerations. The development also includes a suggestion to unify the systems of radiation protection for interventions and practices.

4.1. Russia
New basic recommendations applicable to the existing contaminated territories and possible future accidental situations are being developed to include all experience on liquidation of the consequences of nuclear accidents and nuclear weapons tests. The guidance will be based on both radiation and non-radiation risks and will be expressed in *three* sets of different intervention levels:

(a) *General Intervention Levels* (projected doses) establishing strategies of intervention;

(b) *Specified Intervention Levels* (avertable doses or risks) for radiation protection purposes;

(c) *Specified Intervention Levels* (residual doses or risks) for social protection purposes.

The last set of levels was primary introduced for social protection of the population in the Altai region affected by the nuclear weapons tests at the Semipalatinsk test site. Obviously, this set of levels should have a wider application and should be improved taking into account new data and new experience.

The General Intervention Levels include two principal levels: *an upper dose level* (the dose constraint) above which the introduction of any countermeasures is compulsary in preventing people from receiving doses above this level, and *a lower dose level* having the role of a non-action level. In addition to the general recommendations on intervention strategy and intervention levels, specific recommendations on methodology of risk analysis for post-accidental situations and optimization of strategies of protective and rehabilitation measures are being developed for approval. These developments are using the results from the EU/CIS cooperative research projects (ECPs and JSPs). The new recommendations are expected to be finalized in 1996.

4.2. Ukraine

At present, in the large part of the affected territory of Ukraine the annual individual effective doses do not exceed 1 mSv. This level of dose is identical to the dose limit for exposure of the population in practices used in many countries including Ukraine. However, because the exposure from the Chernobyl accident and the exposure from non-accidental sources were considered separately, the social security of inhabitants receiving the same doses depends on the origin of the source giving rise to the exposure.

Taking into account this curiosity the possibility of a new concept of intervention in territories contaminated by the Chernobyl accident is now being considered. It is proposed to determine the annual dose for intervention as the sum of accidental doses (from all Chernobyl exposure pathways) and industrial exposure (practices).

Annual dose levels of 1, 5 and 20 mSv are considered as the international levels for the whole Ukrainian territory (Fig. 3(e)). The protective actions related to these levels are:

- no special protective actions have to be implemented if the annual dose in settlements is below 1 mSv;

- radiological control has to be provided in territories where the annual doses are in the range of 1–5 mSv; effective protective actions should be introduced for the settlements where the annual doses are in the range of 5–20 mSv;

- the possibility of resettlement must be considered if the annual doses would exceed 20 mSv.

Such type of guidance for interventions is a compromise and, in some sense, a composition of practice and intervention criteria for the late phase of the Chernobyl accident. Therefore, the realization of this guidance is possible only if the additional accidental exposure is comparable to the exposure from normal practices for most of the territories.

This concept has been revised and already accepted not only by single scientists but also by the National Commission of Radiation Protection of Ukraine, the health Ministry, the Ministry of Chernobyl and the Council of Ministers. At present, it is under consideration by the Supreme Soviet of Ukraine.

4.3. Belarus

At present, Belarussian scientists develop a concept of protective measures in a rehabilitation period for the population living in territories of the Republic of Belarus affected by the Chernobyl accident. This concept is at the stage of adoption.

According to the concept, in territories where the annual individual effective doses do not exceed 1 mSv, living conditions and economic activities should not be limited by radiation protection factors. Consequently, additional exposure of the population due to radioactive fall-out resulting in annual individual effective doses lower than 1 mSv is permissible and should not require any limitations (Article 3 of the law *On social protection of citizens affected by the Chernobyl nuclear power plant catastrophe*). Monitoring of objects in the natural environment and of agricultural production should be carried out to calculate and estimate real radiation doses to the population and to implement, if needed, limited and local protection measures.

In territories where annual individual effective doses exceed 1 mSv but are lower than 5 mSv, well-grounded activities aimed at further reduction of individual and collective doses should be implemented. These measure would include, in addition to radiation monitoring of the natural environment and of agricultural production, local decontamination of sites where the external exposure is the dominating exposure pathway.

In territories where the annual individual effective doses exceed 5 mSv, residing would not be recommended and economic activities would be limited.

5. Differences between CIS guidance and international guidance on intervention

According to the international guidance from ICRP and IAEA intervention levels refer to the dose that is expected to be averted (avertable dose) by a *specific countermeasure* over the period it is in effect. If an intervention level is exceeded, *ie*, if the expected avertable individual dose is greater than the intervention level, then it is indicated that the specific protective action is likely to be appropriate for that situation. Intervention levels are specific to accident situations.

Action levels refer to different protective measures or strategies like agricultural countermeasures or radon reducing measures in houses and they relate to the residual dose without any remedial actions taken.

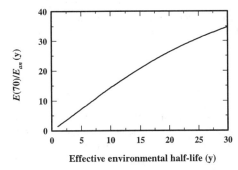

Fig. 3. Ratio of lifetime dose over 70 years, $E(70)$, to annual dose, E_{an}, as a function of the effective environmental half-life of the source of exposure.

The intervention guidance used in the CIS republics all have the character of action levels expressed as annual doses above which different protective actions would be needed. For a given exposure situation there would be a fixed ratio of avertable dose to action level as shown in Fig. 3. An action level for resettlement of, say 20 mSv/y, would be equivalent to an avertable lifetime dose of 150–300 mSv over the following 70 years if the effective environmental half-life is 5–10 years. CIS action levels are thus another way to express avertable doses, and, conceptually, they are in line with international guidance. However, the numerical values differ somewhat from international numerical guidance with a tendency of CIS-levels being lower than international numerical guidance.

6. Unresolved issues

Following a nuclear or radiological emergency, especially in the later phases, many complex human, social and economic considerations will have to be taken into account by the responsible authorities. The decisions and protective actions taken may themselves induce social and psychological impacts. Internationally, the application of different intervention levels in similar circumstances resulting from a single accident would cause much confusion in the public mind. At the national level, taking decisions about lines of demarcation between those areas where protective measures are applied and those where they are not, might create anxiety or even fear by people living on the 'safe' side of the demarcation line. From the experience in CIS following the Chernobyl accident, countermeasures to mitigate social-psychological impacts have obviously been needed.

It has been suggested that the inclusion of such countermeasures into the intervention decision making framework should be as a part of the radiation protection framework. This suggestion seems awkward as radiation protection factors are related to the level of protection achieved including those factors describing the dose distribution averted and costs and other disadvantages incurred in averting doses. The level of socio-psychological impact would depend not only on the presence of radiation but to a large extent on other non-radiological protection factors, such as the attitude of the mass media, the politial climate and the general level of information in the population. Non-radiological protection factors would also be different from country to country, and would probably be highly dependent on the existing political situation. To include socio-psychological factors in the radiation protection framework would thus give very random levels of radiation protection which could result in loss of credibility of the radiation protection community.

Therefore, to achieve an optimized overall health protection, non-radiological protection factors should enter the optimization process in parallel with radiological protection factors to form an optimized countermeasure strategy. The optimization of the overall health protection would thus be the responsibility of the decision maker with guidance from radiation proction experts as well as experts in the fields of social and psychological sciences. It is of great importance that the decision makers present the protection strategy to the public in a transparent way so all the factors and their relative importance in reaching the optimized strategy are revealed.

7. Conclusions

Following the accident at Chernobyl, it became evident that some clarification of the basic principles for intervention was necessary. In particular, it became clear from the experience of the Chernobyl accident recovery that there was a need for a simple set of internally consistent intervention levels that could have some generic application internationally. Such a set of values was considered desirable to increase public confidence in authorities charged with dealing with the aftermath of an accident.

Over the past decade considerable progress has been made in developing internationally recognized principles for decisions on protective measures following accidents involving radioactive material. The development of ICRP Publication No. 63 [4] and IAEA Safety Series No. 109 [5] represents an international understanding on the principles for intervention and numerical values for generic intervention levels. The recommendations in Safety Series No. 109 form the basis for the standards and numerical guidance related to intervention contained in the International Basic Safety Standards for Protection against Ionizing Radiation and for the Safety of Radiation Sources of the FAO, the IAEA, the International Labour Organisation, the Nuclear Energy Agency of the Organisation for Economic Co-operation and Development, the Pan American Health Organization and the WHO [7].

The guidance on protective actions in CIS is still being developed. The philosophy is based on the concept of action levels for specific countermeasures as well as for strategies of countermeasures. For resettlement, an action level of 20 mSv/y is suggested which would correspond to an avertable effective lifetime dose of about 300 mSv if the effective half-life of ^{137}Cs in the environment is about 10 years. The ICRP and IAEA intervention level for permanent resettlement is given as an avertable effective lifetime dose of 1 Sv. The international recommended action levels for foodstuff countermeasures is 1,000 Bg/kg of ^{137}Cs [7], which is identical to the WHO recommendation for foodstuffs moving in international trade. For specific strategies of foodstuff countermeasures, optimized action levels would be of the order of a few hundreds of Bq/kg of ^{137}Cs. CIS action levels for ^{137}Cs in foodstuffs appear to be somewhat lower than the international recommended action levels.

Radiological protection factors have been used in developing international numerical guidance on intervention level for implementing countermeasures to reduce doses after a nuclear or radiological emergency, but explicit guidance is not provided on how psychological and social factors should be included in the optimization of overall health protection. However, the optimization of radiation protection and certain psychological and social protection should probably not be carried out independently as separate and independent entities, as the overall health protection would depend on both radiological and non-radiological protection factors. The overall health protection should thus be based on an optimized countermeasure strategy, which would be the responsibility of the decision maker(s) with guidance from radiation proction experts as well as experts in the fields of social and psychological sciences.

8. References

[1]   INTERNATIONAL COMMISSION ON RADIOLOGICAL PROTECTION, *Protection of the Public in the Event of Major Radiation Accidents: Principles for Planning*, Publication 40, Pergamon Press, Oxford, New York (1984).

[2]   INTERNATIONAL ATOMIC ENERGY AGENCY, *Principles for Establishing Intervention Levels for Protection of the Public in the Event of a Nuclear Accident or Radiological Emergency*, Safety Series No. 72, IAEA, Vienna (1985).

[3]   INTERNATIONAL COMMISSION ON RADIOLOGICAL PROTECTION, *1990 Recommendations of the International Commission on Radiological Protection*, Publication 60, Pergamon Press, Oxford, New York, Frankfurt, Seoul, Sydney, Tokyo (1990).

[4]   INTERNATIONAL COMMISSION ON RADIOLOGICAL PROTECTION, *Principles for Intervention for Protection of the Public in a Radiological Emergency*, Publication 63, Pergamon Press, Oxford, New York, Seoul, Tokyo (1993).

[5]   INTERNATIONAL ATOMIC ENERGY AGENCY, *Intervention Criteria in a nuclear or radiation emergency*, Safety Series No. 109, Vienna (1994).

[6]   COMMISSION OF THE EUROPEAN COMMUNITIES, *Radiation Protection Principles for Relocation and Return of People in the Event of Accidental Releases of Radioactive Material*, Radiation Protection-64, Doc. XI-027/93, (1993).

[7]   INTERNATIONAL ATOMIC ENERGY AGENCY, *International Basic Safety Standards for Protection Against Ionizing Radiation and for the Safety of Radiation*

Sources, Safety Series No. 115-I, Vienna (1994).

[8]   Dibobes I.K., Il'in, L.A., Konstantinov Y.O. et. al., *The adoption of urgent measures to protect the population in the event of an accidental release of radioactivity into the environment, Handling of radiation accidents* (Proc. Symp., Vienna, 1969).

[9]   Konstantinov Y.O., *Criterion for urgent decision making on protective measures in case of an accident at nuclear power plants.* Radiation safety and protection in nuclear power plants. Issue 9. Moscow, Energoatomizdat, p. 148 (1985) (In Russian).

[10]   Avetisov, G.M, Buldakov, L.A, Gordeyev, K.I., Il'in, L.A., *Policy of the USSR National Commission on Radiation Protection on the Substantiation of Temporary Annual Dose Limits for Public Exposure due to the Chernobyl Accident.* Medicinskaya Radiologiya, No. 8, p. 3 (1989) (in Russian).

[11]   SUPREME SOVIET OF THE USSR, *All-Union State and Republican Programme for the Rectification of the Consequences of the Chernobyl Accident, 1990-1992.* Moscow, 1990 (in Russian).

[12]   Il'in, L.A., *Statement on the lifetime dose limit concept.* Institute of Biophysics, Moscow, 21 November 1990.

[13]   Belyaev, S.T., *A Concept of living conditions for people in the regions affected by the Chernobyl accident.* Document prepared by a USSR AS Working Group under the Chairmanship of S.T. Belyaev. Adopted by the USSR Government April 8, 1991. Resolution N 164.

[14]   Belyaev, S.T., *Recommendations on the practical realization of the current Concept of social protection of the population in the regions affected by the Chernobyl accident in the conditions of rehabilitation and restoration phase that have begun.* Document prepared by a Russian NCRP Working Group under the Chairmanship of S.T. Belyaev. Adopted by the Russian NCRP 13 January, 1993.

[15]   Tsyb, A.F., *A Concept of radiation, health and social protection and rehabilitation of population of the Russian Federation suffered from emergency radiation exposure.* Report prepared by the Russian NCRP Working Group under the Chairmanship of A.F. Tsyb. Adopted by the Russian NCRP, 1995.

Management problems of the restricted zone around Chernobyl

Vladimir KHOLOSHA, Emlen SOBOTOVITCH, Nicolay PROSCURA,
Sergey KOZAKOV, Pavel KORCHAGIN

MinChernobyl of Ukraine, 8 Lvovskaya sg. 254655. Kiev. Ukraine

The Problems of Management in the Zone of Exclusion

In this brief report we will try to consider the main problems on minimization of the consequences of the accident and management of actions provided at present in the Chernobyl zone at the territory of Ukraine in decade retrospect

On April,26,1986,at 1.23.58 a.m. the accident occured at the Chernobyl NPP,the scales and consequences of which have not estimated unambiguesly and during the past ten years. The works on researching and understanding of the whole complex of problems related to Chernobyl tragedy most probably will be studied by numerous scientists and practical workers of some nearest generations.

The most important task of the first stage required the urgent managed actions was to reduce the radioactive irradiation of the population, living at the territories around the Chernobyl NPP and personnel,taking an active part in the accident works, up to the most permissible level. Firstly, it was taken a decision to evacuate population from the most contaminated areas and to keep the personal on watch.

As a result of evacuation of more than 92 thousand people from two towns (Pripyat and Chernobyl) and 74 villages it was formed the Chernobyl zone of exclusion (conventionally named 30km-zone)by square of 2044.4 sq.km in Ukraine.The boundaries of the present zone of exclusion finally was approved by Law of Ukraine " About the Law Regime of the territory,affected the radioactive contamination after the Chernobyl catastrophy" adopted on 27 February,1991.

The radioactive contaminated ground of exclusion zone was out of national economy.

The official tasks of elimination of the consequences of the accident at the Chernobyl NPP, formulated in the initial period, were as follows:

- to provide the normal vital activity for population at the territories,affected radioactive contamination;

- to prevent the distribution of radioactive contamination at the adjacent region to Chernobyl NPP;

- to create the conditions for entering into operation the first and the second units.

Now it should be considered the management by works provided in the exclusion zone at the different stages for the last ten years. In the mid - day of April,26,1986 in the frame of the former USSR the Government Commission on the inquest of reasons of the accident at the Chernobyl NPP headed by B.Scherbina,the Deputy Chairman of the Council of Ministers of the USSR, was organized. The main tasks of the Government Commission were:

to determine the accident scales; to develop and implement the measures on localization and elimination of its consequences; health safety and assisstance to population; to study carefully the reasons of the accident and to develop on its basis operative and long-term measures to prevent the similar accident in future.

The Government Commission provided an active work in the difficult conditions and from time to time (due to radiation conditions) changed its staff.Initially the Government Commission was based in Pripyat town and then in Chernobyl. The analysis of the accident scales immediately in-situ, the lack of regulations on the elimination of such out of drafted accident, responsibility of a high level for taken a decision in the limited time determined the special course of activity of the government Commission. Since September 1986 the Commission had been working in unchanged staff.

An example of such important decisions may be the decision of the Government Commission dated 27 April 1986 at 12 o' clock concerning the evacuation of people of Pripyat town when 45 thousands people were evacuated from 10 km-zone for three hours and further it was the staged evacuation of inhabitants from 30 km-zone.It was finished in 1986.

Later the State Commission of the Council of Ministers of the USSR on extraordinary situations headed by V.Doguzhiev,the Deputy Chairman of the Council of Ministers of the USSR, was organized.

Taking into account the great importance of works on elimination of the consequences of the accident at the Chernobyl NPP and the lack of the state system of the actions in such extraordinary situations it was accepted a decision to concentrate the arised problems to be solved in the hands of the government of the country. Since 29 April 1986 the Active group of Political Bureau of the Central Committee of the CPSU initiated the work related to elimination of the consequences of the accident,headed by N.Ryzhkov, the Chairman of the Council of Ministers of the USSR. This Active group had the meetings almost every day.

The most important problems were submitted for consideration of Political Bureau of the Central Committee of the CPSU and the government of the USSR on which it was taken almost 20 resolutions. One of the principal decisions was to fulfill the work concerning elimination of the consequances of the accident at the Chernobyl NPP, including the construction of the encasement under the broken reactor, to the Ministry of the Middle Machine Engineering of the USSR.

The another example may be the meetings of the Active group for considering the radiation situation in the settlements and the predicted irradiation doses of the inhabitants.

It should be noted that one of the main tasks to prevent distribution of the radioactive contamination at the area adjacent to NPP is the scientific and economically justified and realized in the first year after the accident.

After the disintegration of the USSR the work management in the zone of exclusion was provided by subdivisions of the State Committe of Chernobyl of Ukraine, which was later reorganized into the Ministry on affairs of the protection of the people against the consequences of the accident at the Chernobyl NPP (MinChernobyl of Ukraine).

Now it should be discussed the radioecological situations in zone of exclusion. The radioactive contamination of nature environment. Distribution of radioactive contamination on the territory of zone with nonuniform density of radioactive fallouts and radionuclide compound, relations of various forms of radioactive fallouts are present in general by radionuclides of Cs-137, Sr-90 and transuranium elements. Up to 95% of radioactive contamination are concentrated in the upper layer of the ground in thick of 5 cm.

Surface radioactive contamination of this territory (not accounting the sites of localization of the radioactive wastes and industrial square of the Chernobyl NPP) is equal to 110 thousands Ci of Cs-2137, 127 thousands KI of Sr-90 and 800 Ci of Pu-239-240. The territory of the contamination level of up to 15 Ci/km of radiocaesium, 3 Ci/km of radiostrontium and 0.1 Ci/km of plutonium elapsed 1856 sq.km.

The "Shelter" encasement. As to maximum estimation in the "Shelter" encasement there are about 80 tonns of nucleal fuel, contained the radioactive materials with the activity of up to 20 mln.Ci. Besides the fuel-contained materials in the encasement "Shelter" there are a large quantity of radioactive wastes,which consists of the portions of the active zone of the unbroken reactor, reactor graphite, contaminated metal and building consructions of the energy unit. Then at the operative units of the Chernobyl NPP a large quantity of wasre nuclear fuel and wastes after operation are accummulated.

The sites of disposal and temporary localization of radioacive wastes. There are radioactive materials, the total activity of which is equal up to 380 thousands Ci and in volume of 1 mln. cube m in the built-in three points of disposal and the sites of temporary localization of radioactive wastes, making during the exrtaordinary and decontamination works. The temporary points are the concentration of 800 constructionally simple installations . The cooling pond of the Chernobyl NPP is a reservoir in square of 22.9 sq.m with the water volume of 160 mln.cube.m. In 1989 - 1993 the mean annual water contamination was between 140-330 pCi/l of Sr-90. The total activity of radionuclides in the bottom sedimentations is reached to 3.5 thousands Ci of Cs-137, 800 Ci Sr-90 and3 Ci of Pu.The main migration paths of radionuclides out of zone of exclusion are the annual water flow (formed by way of surface flow and outflow of radionuclides by the underground water), air (wind-driven), biogenic and industrial transfer.

Presently industrial activities in the Exclusion Chernobyl zone are being developed in following directions: operating of the Chernobyl NPP, liquidation of the consequences of Chernobyl accident and also management of economical activities in the Chernobyl Exclusion zone. The five major institutions are involved in these activities: Industrial Associates Chernobyl NPP, Research and Industrial Associates (RIA) Pripyat; Building Associated Company of ChNPP; Research and Technical Corporation "Shelter" (Ukrytie) of National Academy of Science and Governmental Specialized Production Associates "Chernobylles". Besides more than

100 different research, design and other institutions have a different research and applied businesses in the Chernobyl close-in zone and their activities require coordination and management.

The Ukrainian Government since the first days following the Chernobyl events is carrying out a huge work aiming to minimize the consequences of the Accident. After the collapse of the Soviet Union the Ukrainian Government took the responsibility for liquidation of consequences of the greatest technological disaster in history. To coordinate all the activities carried out in the Exclusion Chernobyl zone the special Governmental Management Department of Minchernobyl - Administration of the Chernobyl exclusion zone was created. The Administration is to manage and coordinate all the activities carried in the close-zone, deal with funding and provide a security and guard control of the Chernobyl zone and also radiation safety and health management and registration of the personal involved in these activity. The public relations in term of information about social, ecological, economical and research activities are also under responsibility of the Administration. The decisions of Administration are compulsory for all institutions working in the Chernobyl Exclusion zone. The supervision on the institution activities is being carried by the Governmental Supervision Departments of Ukraine.

The funding of activities in the Exclusion zone is carried out via Administration's structure from specially created "Chernobyl foundation" of Minchernobyl Budget. This budged has forming from special taxes income of the any salary payment of all legal persons in Ukraine. Annual expenses on the covering of all activities in the Chernobyl exclusion zone have reach 4-5% of the total amount of the Chernobyl Foundation budget, including funding of the work on providing safety of the destroyed Reactor N-4 and the "Shelter".

The Government implements its policy by development of relevant legislation and documents of central executive institutions.

In 1994-95 a new "Concept of the Chernobyl Zone on the territory of Ukraine" for the period to 2020-25 was developed and adopted. It was created by team of best Ukrainian scientists and specialists. The Concept is based on the laws in force and defines the system of organizational, ecological, medical, scientific and technical principles and priorities of scientific and economical activities in the zone. The main objective of the Concept is minimization of negative ecological, economical and social consequences of Chernobyl accident. The Concept includes: functional district division of the zone territory; the main approaches to building an ecologically secure system around the "Shelter" unit; the issues of radioactive waste management; radiation monitoring of the environment; approaches to building an administrative system in exclusive zone; scientific research priorities; countermeasure priorities; ecological forecast for the zone etc.

According to recent additions to and changes in "Chernobyl" laws of Ukraine all territories of compulsory evacuation (to the West from the exclusive zone) are to be under jurisdiction of Administration of the exclusive zone. Under these circumstances the Administration faces the problem or total evacuation of the population, transfer of lands and forests belonging to different agencies, revision of borders of contaminated land lots, transfer and utilization of infrastructure on these territories and a number of other important issues concerning long-term activities in that zone. In practical aspect there arises the problem of security and management on territories of compulsory evacuation that are not directly adjacent to the exclusive zone.

CONCLUSIONS:

1. At the moment of the accident at the Unit 4 of Chernobyl NPP in USSR there was no developed system of Emergency response.

2. In different phases of post-accidental activities different governmental structures are needed.

3. It is necessary to generalize an experience of Emergency response activities and to develop appropriate law basis.

Information-Analytic Support of the Programs of Eliminating the Consequences of the Chernobyl Accident: Gained Experience and its Future Application

R. V. Arutyunyan, L. A. Bolshov, I. I. Linge, I. L. Abalkina, A. V. Simonov,
and O. A. Pavlovsky

IBRAE — Nuclear Safety Institute of the Russian Academy of Sciences
52, Bolshaya Tulskaya, 113191, Moscow, Russia

On the initial stage of eliminating the consequences of the Chernobyl accident, the role of system-analytic and information support in the decision-making process for protection of the population and rehabilitation of territories was, to a certain extent, underestimated. Starting from 1991, activity in system-analytic support was the part of the USSR (later on, Russian) state programs.

This activity covered three directions:

Development of the Central bank of the generalized data on the consequences of the radiation catastrophes [1, 2]; development, implementation, and maintenance of the control informational systems for the Federal bodies; computer-system integration of the.

To our regret, the financial support of the Federal program is permanently shortened during recent years; the perspectives of a new program for 1996 - 2000 are still unclear. A new concept of radiation, medical, and social protection and rehabilitation of the population of the RF is under consideration, but is not yet adopted.

The formed banks of data and models, developed methods of processing, analysis of information, and their integration by modern computer tools make it possible now to perform system analysis and predict the situation on the contaminated territories at the satisfactory level of reliability and accuracy. In order to make a prognosis of the situation evolution, it is important to account for the previous steps in eliminating the consequences of the Chernobyl NPP accident. At the initial stage, a number of mistakes were made: certain protection measures (evacuation, preventing the thyroid exposure, etc.) were implemented too late; interruption levels were changed hustily; hundreds thousand persons were involved into emergency actions in the region of the Chernobyl NPP, unprepared for the work in the radioactive contaminated territories and objects; wrong informational policy was conducted with respect to the local population and administration. In subsequent periods, the territory, where counter-accident measures were undertaken, was only expanded.

By the end of 1988, there was an unsuccessful attempt to move to the rehabilitation phase on the base of the definition of the additional life doze as 350 mSv [8]. The attempt caused strong negative reaction of the public. As a result, instead of the expected localization of the problem, a substantial increase in its scale was obtained.

Mistakes, made rather by politicians than by experts in radiological protection, and stereotypes, immanent to the previously existing system, affected the strategy of the activity aimed at the elimination of the consequences of the Chernobyl accident. At the same time, the situation concerning the participants of the elimination of the accident consequences (the liquidators) grew more aggravated. Regardless of the exposed doze, the certificate of "the participant of works aimed at the elimination of the consequences of the Chernobyl NPP accident" was delivered to any person who at least one day stood in the region of the Chernobyl NPP or in the resettlement zone. Owing to the existing problems concerning data reliability, inevitable cases of morbidity, disability, and mortality in the whole cohort of liquidators were interpreted as a common and regular manifestation of the harmful effect of radiation.

As a result, the "Law about the social protection of citizens subjected to the effect of radiation owing to the Chernobyl NPP catastrophe" was adopted in 1991. The law took into action several very simple principles for dividing the territory into zones (according to soil contamination by ^{137}Cs) and implementation privileges for the liquidators. Inhabitancy to simplicity and old-fashioned measurement units caused the reception of the contamination level in 1 Ci/km^2 as a bound for the zone separation. It is important that, arbitrary reception of merely simple figures for making decision on such a complicated problem resulted in the multiple increase in the total square of the territories where counter-accident measures were undertaken. The final decision was made even without an exact estimate of the zone sizes. By the summer of 1991, the maximum number of residents of the territories subjected to radioactive contamination was estimated as 1.5 Million. Later on, as a result of the improved estimate of the radiation situation, additional settlements with total number of residents exceeding 2.8 Million persons were included into contaminated zone. It is possible to assume that, if more rational approaches to the strategy of the protective and rehabilitation measures were adopted, they would be restricted by the territory with the total population within 500 000 persons; measures of social protection and medical rehabilitation would be applied to 30 - 50 thousand persons who participated in the works with maximum exposure dozes.

Nevertheless, by the beginning of 1992 there existed (though doubtful) regulating base — the "Law.." and the Federal program that was developed on the base of the "Law...". Realization of the program made it possible to reduce the population exposure dozes and the grating of the agriculture production. In addition to the population resettling, an opposite migration flow appeared. According to certain demographic parameters (for example, the birth-rate), the resettlement zone and the zone with the right for resettlement were even more successful than uncontaminated territories. The results of the sociological inquiries confirmed the compensating effect of the program. These circumstances created rather good perspective for completing work on many territories.

The performed prognosis for the evolution of the radioecological and social-psychological situation on the contaminated territories [4] led to a conclusion that, for the present stage, radiological approaches to the choice of the strategy for the realization of the state program are inadequate to the existing situation. Such possibilities were lost in 1991. There are two possible variants for the further development of the situation.

The first, "safe" variant assumes realization of the program up to 1998 for all territories, including weakly contaminated, where radiation situation is already completely normalized. In this case, in the time period 1998 - 2000, it is possible to localize rehabilitation and compensating measures on the territories of just four regions (Bryanskaya, Kaluzhskaya, Tulskaya, and Orlovskaya) with total population 300 000 - 400 000 residents. The average dozes of additional exposure for the residents of the settlements by the year 2000 will not exceed 1 mSv per year virtually everywhere but several small settlements with total population not exceeding 1000 persons. The specific features of the motivation, psychological precepts, people relations and mood in various zones of contamination, together with objective differences in radiological situation, lead to necessity of the differentiated approach to changes in the zone status in the transient period, and of the forced appeal to the simplest characteristics of the radioactive contamination (like the rate of the external gamma exposure).

The second variant assumes abrupt shortening of the program, which might be stipulated by the new "Concept of the radiation, medical, and social protection and rehabilitation of the population of the Russian Federation, subjected to accidental exposure". This concept suggests zone separation according to the year doze only (zones of radiation control and voluntary residence); it is much more consistent than the actual Law. According to the new concept, even the most conservative dosimetric estimates show that, the zones of radioactive contamination should include territories with the population up to 100 000 residents. However, the measures aimed at the transition to the new concept cannot be implemented quickly. Without a transient period, very unfavorable social-psychological situation would be developed inevitably. Personal and political potentials of the territories covered by the actual program make it quite probable that, new legislative authorities will be again forced to make a decision about the necessity to adopt "the general concept of the safe residence on the contaminated territories, acceptable for the wide strata of the population".

There are a lot of Chernobyl lessons. One of the most important is that, the problem of elimination of the consequences of the large-scale radiation catastrophes is such varied that any "straightforward" approach will not give positive results. Making strategic decisions requires both qualitative and quantitative description of all consequences, including radiological, social-economical, psychological, demographic, medical, and, to the same extent, political.

The gained experience in the elimination of the consequences of the radiation catastrophes can be applied in the tasks of prevention and improvement of the preparedness of the related civil, defense services and local authorities for the actions in emergency situations. In particular, during 1993 - 1995, IBRAE prepared and carried out several practical games devoted to the problems of making decisions for the population protection in course of the radiation accidents [6 - 8]. Special systems of the full-scale modeling of the radiological information in course of the radiation accidents are developed for the training goals. These systems generate arbitrary amounts of all kinds of the radiological information.

References

[1]L. Bolshov, I. Linge, R. Arutyunyan et al., Chernobyl Experience of Emergency Data Management. Proc. of NEA Workshop Emergency Data Management, 12-14 Sept. 1995, Zurich, Switzerland (in press).

[2]L. Bolshov, I.Linge, R. Arutyunyan et al., The Information System "Chernobyl" of EMERCOM of Russia (this issue).

[3]R. Arutyunyan, L. Bolshov, V. Demianov, A. Glushko et al., Environmental Decision Support System on Base of Geoinformational Technologies for the Analysis of Nuclear Accident Consequences (this issue).

[4]I. Linge, R. Arutyunyan, I. Ossipiants, et al., Experience of the Complex Analysis of the Situation on the Territories Contaminated as a Result of the Chernobyl NPP Accident. Proc. Conf. "Radioecological, Medical, and Social-Economical Consequences of the Chernobyl NPP Accident. Rehabilitation of the Territory and Population", Golitsino, 21-25 May 1995.

[5]L. Il'in, Reality and Myths of Chernobyl, ALARA Ltd, Moscow, 1994, pp. 385-410.

[6]R. Arutyunyan, I. Linge, O. Pavlovsky, J. Brenot, P. Ginot, H. Maubert, and D. Robeau. Franco-Russian Role-Play on Decision Making in the Event of Radiological Contamination of Large Areas of Land. PORTSMOUTH-94 Proc., Nuclear Technology Publishing, 1994, pp. 329 - 332.

[7]Command and Headquarter Training, Guideline Materials, 22 - 24 November, 1994, Moscow, NSI-IBRAE.

[8]Command and Headquarter Training "Polyarnye Zori-95", Practical Game Report, 22 - 24 November, 1994, Moscow, NSI-IBRAE.

Fluxes of Radiocaesium to Milk and Appropriate Countermeasures

[1]HOWARD, B.J., [2]HOVE, K., [3]PRISTER, B., [4]RATNIKOV, A., [5]TRAVNIKOVA, I.,
[6]AVERIN, V., [7]PRONEVITCH, V., [8]STRAND, P., [9]BOGDANOV, G.
and [3]SOBOLEV, A.

[1] *Institute of Terrestrial Ecology, Merlewood Research Station, Grange-over-Sands, Cumbria, LA11 6JU, United Kingdom*
[2] *Department of Animal Science, Agricultural University of Norway, P.O. Box 5025, N-1432 Ås-NLH, Norway*
[3] *Ukrainian Institute of Agricultural Radiology, 7 Machinostroiteley St., p.g.t. Chabini, Kiev 255205, Ukraine*
[4] *Russian Institute of Agricultural Radiology and Agroecology, 1 Kievskoe St., Obninsk, Kaluga Region 249020, Russian Federation*
[5] *Russian Institute of Radiation Hygiene, 8 Mira St., St. Petersburg 197101, Russian Federation*
[6] *Belarussian Institute of Agricultural Radiology, 16 Fedunsky St., Gomel 246050, Belarus*
[7] *Sarny Scientific and Experimental Research Station, Sarny, Rovno region, Ukraine*
[8] *Norwegian Radiation Protection Authority, Postboks 55, N-1345 Østerås, Norway*
[9] *Ukrainian State Agricultural University, Geroev Oborony Street 15, Kiev-41, 252041 Ukraine*

Abstract. Radiocaesium contamination of milk persists in some areas of Belarus, Ukraine and Russian Federation which received fallout from the Chernobyl accident. In general, effective countermeasures have been used which ensure that radiocaesium activity concentrations in milk from collective farms does not exceed intervention limits. However, farming practices differ greatly between the large collective farms, and the small, family operated private farms which are responsible for a major part of the food consumed in many rural areas. As a result of comparative low rate of use of fertilizers and utilization of poor quality land ^{137}Cs activity concentrations in milk from family-owned cows continues to exceed intervention limits in some areas. It is therefore important to be able reliably to quantify the rates of transfer of ^{137}Cs to private milk, so that all areas where persistent problems occur are identified, and appropriate countermeasure strategies applied. Where there is considerable variation, within a few km², in both soil type and deposition, the ^{137}Cs content of private milk is highly variable. However, combining information about soil type, transfer rates for each major soil type, deposition, pasture size and grazing strategies can be a useful method of quantifying transfer of radiocaesium to milk. Geographical information systems provide a promising new tool to integrate these factors. Effective countermeasures are available to reduce radiocaesium transfer to private milk. Private farmers are more sceptical of such methods than the scientists, administrators and agriculturalists in their society, particularly those methods involving the use of chemical additives given to their animals. Increased information efforts on the local level appears to be a prerequisite for a successful implementation of necessary countermeasures.

Introduction

This paper briefly summarises some studies in ECP9 which have focused on the transfer of radiocaesium to milk and appropriate countermeasures in small selected study sites [1]. Milk is an important potential source of radiocaesium for humans. It has become evident after the Chernobyl accident that the extent of radiocaesium ingestion via milk by humans depends on a large number of ecological, agricultural and socio-economic factors. To be able to reliably quantify the intake of radiocaesium in milk all these factors need to be considered.

In the short and medium term after the Chernobyl accident the contamination of milk from collective farms was the focus of concern, because of its importance in contributing to collective dose. Much effort has been devoted to ensuring that radiocaesium contamination of collective milk supplies was below intervention limits. Furthermore, in heavily contaminated areas private dairy animals were removed from residents and replaced by supplying uncontaminated milk from central sources. Over the medium and longer term there has been a realization that radiocaesium activity concentrations in milk of private farmers living in villages in some less highly contaminated areas continues to exceed intervention limits.

Assessments need to be carried out on a large scale to give an estimate of the overall importance of milk as a source of radiocaesium in the contaminated areas for decision makers. However, it is helpful to carry out detailed studies in smaller areas to ensure that the focus of the modelling exercises is correct, and that they take into account all the important factors determining radiocaesium intake via milk. ECP9 has been focusing its studies in selected contaminated study sites in Ukraine, Russia and Belarus to try and quantify the contribution of different potential sources of radiocaesium to the diet of village residents. The study sites were centered on:

•Milyach village, Kolos and Kliporop collective farms, Dubrovitsa district, Rovno region, Ukraine
•Shelomi village and Rodina collective farm, Novosybkov district, Bryansk region, Russian Federation
•Savichi village, Bragin region, Belarus

Milk production in the study areas occurred both in large collective farms which were mainly responsible for the supplies to the towns in the area, and on the private farms in the villages where the workers on the collective farms live. Milk production on the collective was mechanized to a considerable extent. Herds were large (usually >200 dairy cattle) and milking, harvest and transportation carried out with the help of machines. Equipment for cooling and storage of milk was available. The private farmers produced milk for consumption in the family, each of which commonly have one cow. Milk yields were between 2000 and 2500 kg y^{-1} in both production systems.

This paper will summarize the information ECP9 has obtained on transfer and fluxes of radiocaesium to milk, and briefly report on a questionnaire which attempted to critically evaluate the attitudes of farmers and other relevant professionals towards potential countermeasures. Aspects of the transfer to milk will be illustrated using data from the Ukrainian study site, which was the most variable from a radioecological perspective, incorporating a wide variation in soil types and including both collective and private farming.

ECP9 evaluated historical data on transfer to milk for the study sites and carried out further directed sampling in 1994 and 1995 of soil, vegetation and milk. Transfer to milk was quantified using classical radioecological methods, but also considering spatial variation using geographical information systems (GIS) and total fluxes to milk by evaluating agricultural production methods.

Factors affecting radiocaesium transfer to milk

Many factors affect the extent to which milk will be contaminated by radiocaesium. The major factors include:

- Deposition rates of radiocaesium
- Soil type and consequent varying rates of transfer of radiocaesium to milk
- Rates of decline in radiocaesium activity concentrations in milk
- Production strategies
- Application of countermeasures

To understand the processes which are important for the transfer of radiocaesium, knowledge of the farming system is also required. In the study sites, the grazing season starts in May, and feeding is again necessary from the beginning of October. During the summer season both the collective and private cows obtain the most of their energy intake from grass, and no (private) or little grain or commercial concentrates are fed. In Table 1 the rations provided for the cattle in the two productions systems are given.

Table 1 Daily rations and ^{137}Cs content for dairy cattle in a collective farm and a typical private farm in the Dubrovitsa study area

Farm type		Grazing period		Feeding period	
Collective farm	^{137}Cs	May - September		October - April	
Feed	Bq kg^{-1} fw	kg d^{-1} fw	Bq d^{-1}	kg d^{-1} fw	Bq d^{-1}
Pasture		45			
Hay	1170	-		1	1 170
Straw (Rye)	230	-		4	920
Silage (grass, maize)	340	-		15	5 100
Concentrate (grain)	250	1	250	2	500
Total					7 690
Private farm	^{137}Cs	May - September		October - April	
Feed	Bq kg^{-1}	kg d^{-1}	Bq d^{-1}	kg d^{-1}	Bq d^{-1}
Pasture		45			
Hay	720	-		10	7 200
Concentrate	250	-		0.3	75
Total					7 275

Concentrated feeds were generally available only in small quantities to the private farmers, while the collective cows received 1-2 kg daily. Grass and roughages used by the collective cows were harvested on fields where extensive countermeasures had been undertaken. Private cattle grazed partly on improved and rough pastures, and also within forest clearings where ^{137}Cs activity concentrations in the grass was high compared with other pastures. Hay for winter feed was collected in forests and rough pasture. As a result, the ^{137}Cs activity concentration of winter hay for private cows was highly variable (range 41 - 1 299 Bq kg^{-1} dw in 1994 and 104 - 7700 Bq kg^{-1} dw in 1995). Some of the hay was exchanged for hay with a lower ^{137}Cs content by the collective farm as a way to reduce the ^{137}Cs content of locally produced milk.

<u>Deposition rate</u>

The deposition rate is commonly quantified on a large scale for planning purposes into broad categories, defined by mean deposition rates. However, it is well known that deposition of radiocaesium was heterogeneous, and can vary over more than one order of magnitude within a few metres. There is often good sample data, at a local level, on the variation in deposition in contaminated areas, and if such information is evaluated in a GIS, such as kriging and interpolation, then the areas of greatest uncertainty in the deposition measurements can be identified [2].This strategy has been used by ECP9 to direct sampling requirements to improve deposition information where it is most needed.

<u>Soil type and consequent varying rates of transfer of radiocaesium to milk</u>

It is well known that uptake rates of radiocaesium vary considerably with soil type. Whilst soil types can be classified broadly in large areas, local variations in soil types can give rise to great variability in transfer of radiocaesium to milk. It is fortunate that for many of the contaminated regions soil type has been classified precisely, and mapped within each collective farm. At the highly variable Ukrainian site, which has a wide variety of different major soil types, the transfer values measured for soil to plant transfer (Table 2) have been used to predict the activity concentrations in vegetation for each pasture used by collective and/or private cows.

Table 2 Aggregated transfer coefficients (T_{ag}) (Bq kg^{-1} dw/Bq m^{-2}) for the dominant soil groups at the Dubrovitsa study site for 1994-1995

Soil type	Soil to grass T_{ag} (m^2 kg^{-1} x 10^{-3})	
	Range	Mean ± SD
Peaty - cms*	9.36 - 120.94	33.06 ± 30.89
Peaty + cms*	0.64 - 42.23	9.77 ± 9.64
Soddy podzols	<0.34 - 8.74	3.21 ± 2.50
Gleys	0.33 - 3.57	1.23 ± 1.01
Alluvium	0.11 - 0.72	0.24 ± 0.18

* - cms = without countermeasures, + cms = with countermeasures

There are obvious differences between transfer rates for pastures with different soil types, with T_{ag} values showing a large standard deviation comparable to the mean for all soil types. The effectiveness of countermeasures can be seen clearly for the peaty soils. Soil type varies within some pastures, and therefore T_{ag} values for only one soil type applied across whole pastures may be too simplistic and inaccurate, particularly for estimating transfer to milk since cows, by their grazing, will integrate over the whole pasture.

Predicted ^{137}Cs contents in grassy vegetation were calculated for each pasture used for collective and/or private dairy cows by combining information on deposition and the proportion of each soil type in each pasture with the respective T_{ag} values, taken as the mean value from Table 1, in a GIS. This provided a mean predicted radiocaesium activity concentration for grassy vegetation for each pasture which took account of the relative proportion of each major soil group in each pasture. Such an approach takes account of the total surface area of the ecosystem being assessed as recommended by [3]. Mean values were then calculated for ^{137}Cs deposition and vegetation activity concentration for all the pastures used by the two collective herds in the study site and these "weighted" values are given in Table 3. Such weighted values take into account the relative sizes of each of the pastures grazed by the collective cows which is an important consideration since pastures sizes can vary by a factor of ten, and averages taken for all pastures used by each herd would be inappropriate. In addition, the ^{137}Cs activity concentration of milk from the summer grazing period has been predicted by applying a concentration ratio for milk/grass of 0.13, based on recommended values for F_m and daily intake [4]. The independently predicted ^{137}Cs activivty concentration of milk is compared with that actually measured in Table 3.

Table 3 Weighted mean values for ^{137}Cs deposition and vegetation contamination for the pastures used by the collective cows, and predicted milk activity concentrations

Collective	Predicted mean ^{137}Cs values			Measured
	Deposition (kBq m^{-2})	Vegetation (Bq kg^{-1} dw)	Milk (Bq l^{-1})	Milk (Bq l^{-1})
Kolos	117	546	71	61
Kliporop	80	611	79	79

Measured ^{137}Cs activity concentrations for collective milk in 1994-95 varied from 50 to 104 Bq l^{-1} for Kolos cows and 42 to 103 Bq l^{-1} for Kliporop cows. The predicted values agree well with the observed and suggest that this approach has some merit. The spatial variation in predicted ^{137}Cs content of milk based on the predicted ^{137}Cs values in vegetation is shown in Fig. 1.

Fig. 1 Predicted ^{137}Cs content of milk over the Dubrovitsa study site

Similar calculations to that above have been carried out to estimate T_{ag} values comparing the weighted mean values for ^{137}Cs deposition and vegetation contamination with the mean measured ^{137}Cs activity concentration in milk over the summer grazing period (including all values from May to September 1995). The resulting T_{ag} values for Kolos and Kliporop are 0.52 and 0.98 m^2 kg^{-1} x 10^{-3} respectively. Such T_{ag} values, whilst giving an overall value for the herd, will be affected by the proportion of each soil type in each herds' pasture area. For the Kolos herd 75% of the pastures used by the herd have a gley type of soil, whilst the Kliporop cows graze pastures with 76% peaty soils.

Weighted mean values for ^{137}Cs deposition and vegetation activity concentration for all pastures used by the private cows in each village have been calculated and the values are given in Table 4, which also gives the predicted milk values, calculated as above, for comparison with measured values for 1994 and 1995. The division of Milyach into -cms and +cms is because some of the cows in this village graze a pasture which has not been ploughed or treated with fertilizers. All other pastures have been treated to reduce radiocaesium uptake by vegetation.

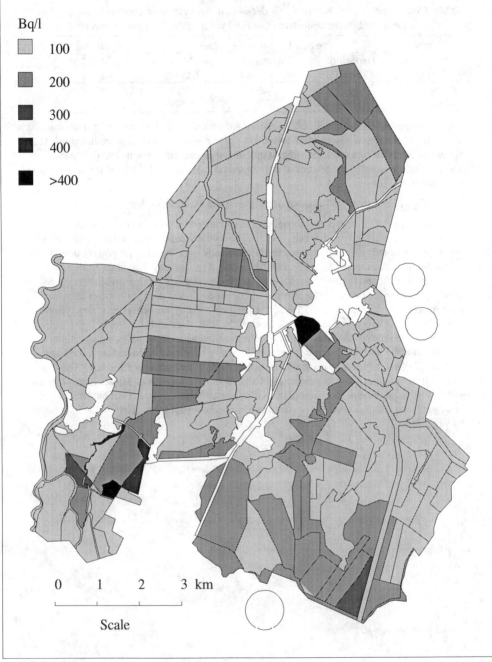

Fig. 1 Predicted Cs-137 activity concentrations in milk calculated by field for the Dubrovitsa study site, Ukraine

Bq/l

100
200
300
400
>400

0   1   2   3 km

Scale

Table 4 Weighted mean values for ^{137}Cs deposition and predicted vegetation content for pastures used by each village and comparison with predicted and measured ^{137}Cs content of milk

| Village | Predicted mean ^{137}Cs values | | | Measured |
	Deposition (kBq m^{-2})	Vegetation (Bq kg^{-1})	Milk (Bq l^{-1})	Milk (Bq l^{-1})
Villages within the Kliporop farm				
Milyach - cms	111	3587	466	414
Milyach + cms	75	576	75	238
Lugovoye	95	305	40	159
Bila	194	1128	147	208
Villages within the Kolos farm				
Vilun	224	545	71	33
Zagreblya	153	1072	140	20

The agreement between predicted and measured milk activity concentrations is again reasonable, but not as good as for the collective herd. The observed discrepancies are probably due to the more variable grazing conditions of private cows and to sampling of milk from sub-herds at a single time, which would not be representative for the total pasture available to each herd over the summer grazing season. This emphasizes the importance of ensuring that samples are collected in a representative manner. For instance, the private cow milk in Zagreblya was sampled from cows which graze an area of the pasture dominated by alluvium soils, which have a comparatively low T_{ag} value. However, the total pasture at Zagreblya has more peaty soil than alluvium and therefore the predicted value is higher than that measured. Nevertheless the general trend where all the villages associated with the Kliporop collective farm have higher contamination levels in milk, even though the pastures which they graze are less contaminated. This is due to the predominance of peaty soils on the Kliporop farm. In contrast the two villages in the Kolos area have higher deposition of ^{137}Cs, but lower milk contamination because of the predominance of gley soils.

In addition, the data in Table 4 has been compared with measured ^{137}Cs levels in milk to calculate T_{ag} values which take account of all pastures grazed by the private cows of each village. These T_{ag} values are given in Table 5, and are compared with the dominant soil types for each villages pastures.

Table 5 T_{ag} values for private milk comparing weighted mean values for all pastures used by the cows with predicted and measured ^{137}Cs in milk

| Transfer parameter | Village | | | | | |
	Milyach -cms	Milyach + cms	Lugovoye	Bila	Vilun	Zagreblya
Predicted milk T_{ag} (m^2 kg^{-1} x 10^{-3})	4.20	1.00	0.42	0.76	0.32	0.92
Measured milk T_{ag} (m^2 kg^{-1} x 10^{-3})	3.73	3.17	1.67	1.07	0.15	0.13
Dominant soil type in pastures	Peat bog -cms	Peat bog + cms	Soddy podzol	Peat bog + cms	Gley	Alluvium & peat bog

Using an approach of calculating T_{ag} values with predicted vegetation contamination for all pastures and measured milk values results in estimated transfer parameters which appear to be unrealistically high (eg. in Milyach + cms, Lugovoye) or low (eg Zagreblya), presumably for the same reasons as those discussed below Table 4.

Rates of decline in radiocaesium activity concentrations in milk
It has proved particularly difficult to quantify changes in radiocaesium levels in milk with time in any generic way, because most contaminated pasture has been treated to minimise radiocaesium transfer to milk. Consequently measured radiocaesium levels reflect both time-dependent changes and responses to countermeasures and calculated half-lives will be site specific.

A comparison has been made of the changes with time in transfer to collective and private cow milk, using milk data for Milyach private cows which graze the peaty-soil pasture which has never received countermeasures (Fig. 2). Radiocaesium activity concentrations in the collective milk declined most rapidly in 1988, when extensive countermeasures were used at the site. For the private milk the decline in ^{137}Cs content of milk has occurred gradually over 1991 to 1995, with the exception of a flooding event in 1993. The calculated ecological half-lives, discounting the 1993 temporary rise, are 3.8 y for the private milk and 2.7 y for the collective milk with R^2 values of 0.96 and 0.77 respectively.

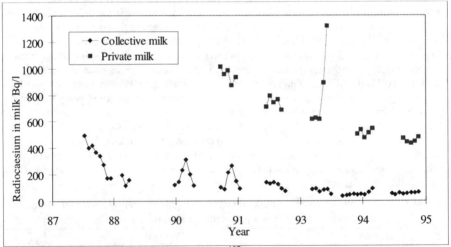

Fig. 2 Comparison of the changes with time in ^{137}Cs content in collective cow milk and private milk from cows grazing untreated pasture.

Production strategies
The extent to which radiocaesium is transferred to milk depends on the agricultural procedures used in the contaminated area. The type of land used, its management, and the consequent transfer of radiocaesium to milk differs between the collective and private farming systems, as previously indicated. Major differences in radiocaesium intake in the two production systems are related to the type of land used for production of feeds, ie. improved pastures and fields for the collective as opposed to a certain amount of unimproved and forest land for the private sector.

Transfer of radiocaesium from soil to hay on the collective fields was typically 2 to 10 times higher than the transfer to rye. Per unit of land used for production, cattle which are fed grains would therefore receive less ^{137}Cs than those which graze or feed on hay. Cattle on the collective farm received about 18% of the dietary energy from grains, while 36% of the energy requirements were covered by pasture and the remaining 46% from roughages given during the indoor feeding season. The private farms had little grain to give their animals; more than 95% of the energy came from pasture and hay. This feeding pattern with a low intake of grains made private cows more likely to produce milk with a high ^{137}Cs content than collective cows, especially when hay is cut on unimproved rough pasture and forest clearings.

Although plant productivity has been maintained at a reasonably high level during the years after the Chernobyl accident, partly as an effect of the deliberate use of fertilizers as a countermeasure, the nearly complete lack of mineral fertilizers which has been experienced by the collective farms during 1994 and 1995 will have a strong effect on plant productivity in the coming years, and may also lead to increased transfer of ^{137}Cs to vegetation. Mineral fertilizers were only available to the private farmers in very limited quantities; instead these farmers depend heavily on the use of cattle manure for their plots. The continued application of comparatively highly contaminated cattle manure may increase the ^{137}Cs content of soil in the private vegetable gardens.

In a study of the fluxes of energy and radiocaesium which has been carried out on the basis of the production figures from the collective farms, it is apparent that a large proportion of the energy (77% of the brutto production in plant crops) is either undigestible or is being used for maintenance of the production animals (Table 6). Likewise, only 12% of the radiocaesium in the plant crops is contained in grains and other products exported from the farm. The remaining 88% is recycled via manure and compost to the collective fields.

Table 6 Estimate of fluxes of energy and ^{137}Cs through the collective farms at the Dubrovitsa study site during a one year production cycle. Energy is calculated as the net energy available from the feeds after digestion and absorption

	Energy (TJ y^{-1})	[137]Cs Flux (GBq y^{-1})
Total plant production	167	8.5
Use of plant production:		
Grains to state	27.4	0.54
Grains/potato to private	5.5	0.3
Animal feeds	131.0	7.7
Animal production:		
Milk and meat sold to state	6.5	0.22
Meat sold to private	0.2	0
Recycled (manure and compost)	124.0	7.5

The average private farm cultivates an area of 0.3 to 0.4 ha to grow a variety of vegetable crops. On each farm there are on average 3.4 adults and 2.3 children below 16 years of age. A variety of domestic animals were kept: with means of 1.3 dairy cows, 0.1 goat, 10 chickens, 2.8 pigs, 2.2 geese, 2.7 rabbits, 1.9 turkeys and 1.2 ducks. The calculated energy content of the products from these animals were 11.5 GJ, from which 9.1 GJ or 79% can be attributed to milk. The total plant production on the farm was 29.5 GJ, mainly from potatoes and grain. A large, but undetermined part of this plant produce was used to feed the animals on the farm. In particular, a significant part would have to be used for the yearly production of about 240 kg of pork (17% of animal produce). However, the major feed energy in the private farming system was hay and pasture for the dairy cows (85 GJ per year). With an estimated mean concentration of ^{137}Cs of 400 Bq kg^{-1} in samples of winter hay for the 1994-1995 season which were measured in 18 private farms in 1994, and an average pasture contamination of 3587 Bq kg^{-1} (Milyach -cms, Table 4) roughages represented the dominating flux (9.3 MBq y^{-1}) of ^{137}Cs in the private farming system. Comparable figures for the families in Milyach where countermeasures were applied was 2.3 MBq y^{-1}. Although uncertainty is involved in the estimation of radiocaesium concentrations in the winter feeds, as demonstrated in samples taken towards the end of the winter feeding season (which had average values from 700 to 2300 Bq kg^{-1}) roughages used to feed dairy cattle were by far the most significant influx of radiocaesium to the private farms. For comparison, plant produce from the private plot contained 313 kBq, and animal produce excluding milk a total of 34 kBq y^{-1} of ^{137}Cs. In the villages of Vilun and Zagreblya, with low concentrations of ^{137}Cs in milk, a total of 100 kBq y^{-1} was produced, while in the farms in Milyach which used pastures where no countermeasures had been performed, the milk produced contained 760 kBq y^{-1}. Berries, fungi and other forest products, which were used for human consumption, contributed 16 kBq y^{-1} on the average farm.

Food energy from vegetables, potatoes, grains and animal products were sufficient to maintain a family of 10 adults. Thus there is a substantial food production potential in the private farms, and the production per unit area in the vegetable gardens are generally higher than in the collective fields. The large fraction of food energy from milk clearly emphasizes the crucial importance of the dairy cow in providing animal products for the rural population in this study site. Since milk is also a major contributor to the flux of ^{137}Cs, countermeasures which can limit the transfer of radiocaesium to milk will be particularly important in reducing radiocaesium intake by the rural farming population.

Application of countermeasures

In the ECP9 study area a variety of different countermeasures had been implemented. Deep ploughing and reseeding of all pastures was carried out during 1987-1988. Fertilizers were used on all fields in the collective farms. Both vegetable- and grass crops received 1.5 times the pre-Chernobyl rate of K during 1988-90. New schemes for crop rotations which were beneficial with respect to reducing ^{137}Cs transfer were devised. A radiology service was established in Milyach since 1992 which monitored milk and meat from the collective farm, and food products from the private farms in the villages. The reduction factors obtained with the available countermeasures are shown in Table 7.

Table 7 Use of countermeasures in agricultural production at Chapayev collective farm during 1988-93, and observed reduction factor

Countermeasure	Area (ha)	Number of animals	Total use	Reduction factor
Deep ploughing	990			1.3-2
Mineral fertilizer	720		360 t	2-2.5
Liming	420		1260 t	1.5-2.5
Fertilizer, pasture	250		75 t	2.5-3
Fertilizer + Sapropel*, pasture	440		13200 t	1.7-1.9
Prussian Blue boli for animals	-	80	250 boli	2.2-2.8
Prussian Blue powder	-	65	3250 g	1.5-1.9

* Sapropel - organic deposits with some clay minerals from reservoir bottom sediments

A questionnaire was completed with 14 of the private farmers involved in the agricultural survey in Milyach to collect information on the actual use of countermeasures. The results showed that only small changes in agricultural practices had taken place in the village after the Chernobyl accident. Partly as a result of international studies, caesium binders had been used for the private cattle on the farms where interviews were carried out. In 1995 mineral fertilizers were used on less than half of the farms, and countermeasures were only needed for the few remaining cows which gave milk with ^{137}Cs activity concentrations above the intervention limits (Table 8).

Table 8 Results from interviews with owners of 14 private farms in Milyach and Vilun on the use of countermeasures to control ^{137}Cs contamination of agricultural produce

Countermeasures	Yes N	No N	Comments
A) Countermeasures used in 1986-94			
Additional soil treatment practices	0	14	
Mineral fertilizers	12	2	Manure used on all farms
Special chemicals in any form	0	14	
Changes in plant production	0	14	Fields changed, but not crops
Changes in animal production	0	14	
Caesium binders for animals	12	2	Hexacyanoferrate powder, humalite
B) Countermeasures in use in 1995			
Additional soil treatment practices	0	14	
Mineral fertilizers	6	8	Manure used on all farms
Liming	5	9	Planned for autumn 1995
Caesium binders for animals	3	11	Limited current need
Apply restrictions in grazing of cattle	0	14	

It is evident that the usefulness of countermeasures is dependent not only on their radiological effectiveness, but also on other factors such as practicality (including availability, cost and effort) and acceptability [5]. ECP9 has used a questionnaire with a range of different types of people to evaluate attitudes to countermeasures which could be used for village residents.

The range of countermeasures incorporated included all methods which ECP9 participants considered to be potentially useful for reducing radiocaesium contamination levels in private produce. The countermeasures were sub-divided into three categories: pasture-based, animal-based and household-based.

Pasture-based countermeasures included two approaches, one of pasture improvement, ranging from radical improvement including deep or surface ploughing, reseeding and fertilization to less intensive procedures such as surface fertilization only. The second approach considered was exchange of pastures. It is possible to allocate land to private farmers which is either less contaminated, or produces less contaminated vegetation due to the soil type.

Animal-based countermeasures considered five approaches, one of "clean feeding" when uncontaminated feed can be provided to animals, and four involving the use of radiocaesium binders to reduce gut uptake of radiocaesium, thereby reducing radiocaesium contamination levels in milk and meat. The binders considered were Prussian blue provided as a powder, in a boli which resides in the rumen and release the binder for a period of 4-8 weeks [6] or within a salt lick. A further binder, clay minerals, which can be given together with fodder or incorporated into concentrate was also included.

Household-based countermeasures considered four approaches. The first two were selling contaminated private produce to state enterprises or the collective (or swapped for less contaminated equivalents) and local radiometric inspection of food. The last two approaches were dietary advice and advice on methods of cooking food to reduces the intake of radiocaesium.

The questionnaire was used for five different groups of five people in each republic:
- Private farmers - people who were resident in the respective study sites
- Collective farm managers (not Russia)
- Agricultural scientists
- Administrators, who deal with the Chernobyl problem at a local level
- Radioecologists - individuals who were not directly involved in ECP9.

For each countermeasure the participant was asked to give a rating from 0-5 (least to most positive) on four different aspects concerning the practical use of these countermeasures. Respondents were asked to comment on the effectiveness, effort required and availability of each of the 11 categories. In addition, the participant was also asked to answer "Would you do it?" on the same scale. Therefore, the maximum possible total score for each countermeasure over all classes was 20, and for all 11 countermeasures, 220.

All questionnaires were completed as intended with the exception of collective farm managers in Russia. Household-based measures were scored very differently in the three republics with Ukrainians giving high scores and Belarussians relatively low scores. Selling foodstuffs scored lowest for most groups, presumably because the people are subsistence farmers and need to keep all the food they produce for themselves. Overall the highest scores were given, in declining order, for:

pasture improvement > provide clean feed > local radiological inspection of food

and the lowest overall scores were given, in declining order, for:

salt licks > boli > selling contaminated produce

It is fundamentally important whether the private farmers, to whom these countermeasures are directed, will consider applying the countermeasures. There was a tendency among them to be more reluctant to use countermeasures involving the direct treatment of their animals, probably because of the overriding importance to the private farmers of their cows. The total scores are compared for each groups in Fig 3. Private farmers in Russia and Belarus scored consistently lower than the other respondent groups, but this was not the case for the Ukrainians, possible because they were familiar with many of the measures because they are living in an area which has been the focus of much research activity, and many of them have personal experience of the countermeasures.

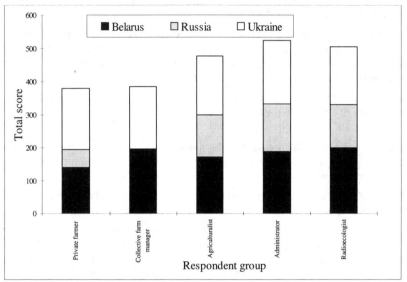

Fig 3 Comparisons between total scores of the different respondent groups.

Such assessments of countermeasures may be a useful tool and these initial results suggest that effort needs to be devoted to involving and informing the private farmers themselves.

Conclusion

There are a wide variety of ecological and production strategy factors which determine the extent of radiocaesium contamination of both collective and private milk. Because there is detailed information available on soil types within most of the contaminated area it should be possible to combine information on deposition rates, soil type and grazing strategies to identify the areas where pasture grass or fodder would have the highest radiocaesium activity concentrations. However, if there is considerable variation in the spatial distribution of major soil groups and deposition within a few km then it is likely that transfer to milk will also vary significantly within a small area and it will be difficult, and probably inappropriate, to make generalizations.

To predict radiocaesium contamination of private milk it is important to have good information on transfer rates between the different soil groups and grass, and to ensure that sampling strategies for soil, vegetation and milk are representative. It is of particular importance to be able to monitor the winter feeds, which may be highly variable in radiocaesium content. The criteria used to amalgamate soil types and quantify appropriate soil to plant transfer is a crucial element in modelling at the small scale and merits critical evaluation.

Many effective countermeasures are available to reduce radiocaesium activity concentrations in milk and such countermeasures could still be of value in the study areas. For local use in contaminated rural villages it is important to inform and involve the private farmers adequately about the methods to maximize their use and effectiveness.

References

[1] Strand, P., Howard, B.J & and Averin, V. (1996) Fluxes of radionuclides in rural Communities in Russia, Ukraine and Belarus. Post-Chernobyl action report. Luxembourg: Commission of the European Communities.

[2] Luuresma, K., Howard, B.J., Howard, D.C. & Averin, V. (1995) The application of GIS and geostatistics to assist in the interpretation of information about the Chernobyl accident. In: Landscape ecology: theory and application. 205-208, Aberdeen: International Association of Landscape Ecology.

[3] Desmet, G.M., Van Loon, L.R. & Howard, B.J. (1991) Chemical speciation and bioavailability of elements in the environment and their relevance to radioecology. Sci. Total Environ., 100, 105-124.

[4] International Atomic Energy Agency (1994) Handbook of parameters values for the prediction of radionuclide transfer in temperate environments Technical report Series No. 364. Vienna: International Atomic Energy Agency.

[5] Nisbet, A.F. (1995) Evaluation of the applicability of agricultural countermeasures for use in the UK. NRPB-M551. Chilton: National Radiological Protection Board.

[6] Hove, K., Strand, P., Salbu, B., Oughton, D., Astasheva, N., Sobolev, A., Vasiliev, A., Ratnikov, A., Aleksakhin, R., Jigareva, T., Averin, V., Firsakova, S., Crick, M. & Richards, J.I. (1995) Use of caesium binders to reduce radiocaesium contamination of milk and meat in Belarus, Russia and Ukraine. In: Environmental Impact of Radionuclide Releases (IAEA-SM-339/153). 539-547. Vienna: International Atomic Energy Agency.

Impact of the Chernobyl accident on a rural population in Belarus

X. Aslanoglou[a], P.A. Assimakopoulos[a], V. Averin[b], B.J. Howard[c], D.C. Howard[c],
D.T. Karamanis[a] and K. Stamoulis[a]

[a]Nuclear Physics Laboratory, The University of Ioannina, 451 10 Ioannina, Greece.
[b]Belarus Institute of Agricultural Radiology, Feduninsky 16, Gomel 246050, Belarus.
[c]Institute of Terestrial Ecology, Merlewood Research Station, Windermere Road,
Grange-over-Sands, Cumbria LA11 6JU, England.

1. Introduction

In a recent research endeavour under programme ECP9[1] three distinct sites in the Republics of Russia, Ukraine and Belarus were selected for detailed radioecological study. The objective of this investigation was to identify the sources of radiocaesium and radiostrontium intake to a specific segment of the population, i.e. subsistence farmers residing in areas where high contamination levels persist after the Chernobyl accident. In what follows, we present the results obtained from the District of Bragin in Belarus.

Contamination levels in foodstuffs produced in the selected site were assessed by means of two approaches (1) using a geographical modelling approach of estimating contamination levels in food products through deposition information and transfer parameters, and (2) via direct measurements of activity levels in foodstuffs from private households. This information was combined with food consumption rates derived from dietary surveys on the population of the area in order to calculate radiocaesium and radiostrontium intake. The results were then compared to data from whole body activity measurements.

2. The study site

The study site consisted of two villages, *Savichi* and *Dvor-Savichi* in the Bragin District of Belarus, 30 km north north-east of the Chernobyl Nuclear Power Plant. These villages received relatively high depositions of both ^{137}Cs and ^{90}Sr after the Chernobyl accident. In Savichi, the population was evacuated in August-September 1986 but gradually returned during 1987 and 1988 without official permission. The population in 1994 was 100 households, totalling 164 persons. In Dvor-Savichi, the population was not evacuated and in 1994 consisted of 59 families, with 180 people. The population in both villages consists of elderly people, mostly pensioners, with a mean age of 62 y in Savichi and 58 y in Dvor-Savichi.

The study site is situated at the edge of the 30 km zone and the villages were previously within a single collective farm area. After the accident the collective farm was closed down and a large amount of the arable land was abandoned. Today, the inhabitants of Savichi and Dvor-Savichi obtain the main food products from small farms. Hunting, picking of mushrooms and berries and bee-keeping is a side and in some cases the main occupation of the villagers. The soils in Bragin are not very fertile and it is difficult to obtain high yields. The private plots are therefore fairly large and average 0.9 ha. The main cultivated crops are potatoes, grain (barley, rye, wheat), cabbage, onions, carrots, mangle and sugar-beet, tomatoes, cucumbers, fruits and berries. Potatoes and cereals occupy the largest arable lands. A two-field crop rotation system with the above mentioned crops is common. Every year the plots of equal area under potato and cereals are interchanged. The yearly treatment of

---

[1] *Transfer of Radionuclides to animals, their comparative importance under different agricultural systems and appropriate countermeasures (ECP9)*. Programme COSU--CT-94 of the Commission of the European Communities.

plots includes ploughing, harrowing, sowing and earthing up is carried out using horses and the most simple agricultural implements (plough, harrow, hillier). Soil is fertilised with the manure obtained from domestic animals. In Byelorussian villages an average consumption of potatoes is 250-300 kg. Potatoes are also very important in the ration for domestic animals, especially pigs. Cereals are used for feeding pigs and poultry. In some farms flour (rye, wheat) is produced, which is used for cooking of flour dishes and bread. Pork and fat are important for the food supply and each family can have from 1 to 7 pigs for slaughter. Milk and milk products (cream, sour-cream, curds and butter) are also produced at the farm. Vegetables, fruit and berries are consumed fresh, dried or marinated for the winter-spring supply. Wool, fells feathers and down is used for clothing and bedding.

Natural products (meat of wild animals and birds, fish, mushrooms, wild berries, nuts) may contribute to a small extent to the food supply. Practically all components of the food supply are produced by the people themselves. Bread, confectionery, tea, salt, industrial goods for everyday necessities (clothes, footwear, TV and radio sets, bicycles, etc.) are bought in state shops.

As noted earlier, the population of the villages are people of middle and old age. Young people and young families live in farm centres and in small and big cities. Their visits to relatives in the village is short-term and usually connected to holiday or help for seasonal works. People in the villages seldom visit regional centres and such visits are usually related to needs for medical care or purchasing of industrial goods.

2.1 LAND USE

A map from the original collective farm was used as a basis to define the study site. Although some of the land use has now changed, the field boundaries generally remain as they were prior to the accident. The land use is now governed by private farming requirements and consists of four main types: pasture, hayfields, abandoned and built up (the two villages). A digital map was produced using land use boundaries from existing information and the land use map of the study site in 1995 was prepared (Fig. 1). Land use information is also summarised in Table 1.

2.2 SOIL TYPE

As with land use, information describing the soil type was available from collective farm records. A digital soil map was produced which contained the various sub-soil types, but for modelling and predictive purposes these were aggregated into four main types: soddy podzol, soddy podzol swampy, soddy swampy and peaty soils. Total areas covered by these soil types are summarised in Table 2, while the corresponding soil map is given in Fig. 2. A cross tabulation of land use and soil type is presented in Table 3.

Table 1 Summary of land cover at the Bragin study site

Land cover	Area (sq km
Pasture	30.42
Hayfield	39.49
Built Up	16.75
Abandoned	53.52
Total	140.18

Table 2 Summary of the soil types at the Bragin study site.

Soil type	Area (sq km)
Soddy podzol swampy	81.03
Soddy podzol	27.22
Soddy swampy	22.63
Peaty	9.30
Total	140.18

2.3 DEPOSITION

Deposition data were obtained initially from 391 samples measured for ^{137}Cs and 43 samples measured for ^{90}Sr activity concentration. To allow interpretation of deposition across the whole study

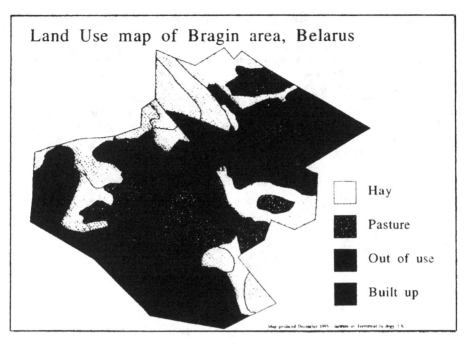

Fig 1. Land Use at the study site

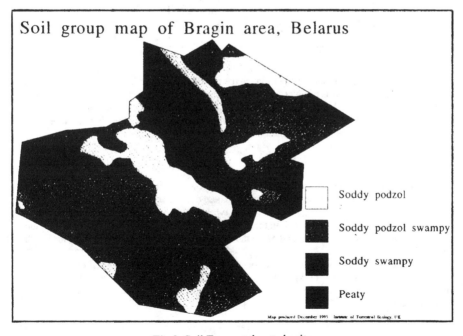

Fig.2. Soil Types at the study site

site the point data were interpolated using Ordinary Kriging [see e.g., P.A. Burrough (1995), *Spatial Aspects of Ecological Data* in: Jongman, R.H.G.; ter Baak, C.J.F. and van Tongeren, O.F.R., Data analysis in community and landscape ecology, Cambridge University Press, p. 239-242). In 1995, extra data for ^{90}Sr deposition were collected to improve the deposition map. The sample site location was selected by combining the standard deviation from the Kriged surface with the land cover map and choosing areas with the highest uncertainty and relevant land use. The corresponding maps for ^{137}Cs and ^{90}Sr deposition in the District of Bragin are shown in Figs. 3 and 4. The uncertainty in deposition values is also mapped in the corresponding figures. As expected, the interpolated surface produced for ^{90}Sr has a higher uncertainty than the estimated deposition for ^{137}Cs largely due to the smaller sample size. The uncertainty of the interpolation varies across the surface, being highest in areas remote from sample points or where neighbouring values differ dramatically.

Table 3 Cross tabulation of land use and soil type in the Bragin study site.

Soil type	Land use	Crop type	% of total area
soddy podzol swampy	hayfield	hay	12
soddy swampy	hayfield	hay	10
soddy podzol swampy	pasture	grass	8
soddy podzol	pasture	grass	8
soddy swampy	pasture	grass	8
soddy podzol	village	vegetable	7
soddy podzol swampy	village	vegetable	7
soddy podzol	hayfield	hay	3
soddy swampy	village	vegetable	2
peaty	hayfield	hay	1
peaty	pasture	grass	0
peaty	village	vegetable	0
Abandoned			33
Total			100

3. Radiocontamination of foodstuffs in the District of Bragin

Samples of foodstuffs produced by the villagers or derived from the natural environment were collected during 1994 and 1995 and measured for radiocaesium and radiostrontium concentrations. In addition to the general picture of food contamination in the district, these data also yielded information on the transfer of radionuclides from soil to grass and milk.

3.1 SOIL-TO-GRASS AND SOIL-TO-MILK AGGREGATE TRANSFER

Several samples of soil and vegetation growing on it were collected during 1994 and 1995 and measured for ^{137}Cs and ^{90}Sr activity concentration. These data were used for the calculation of aggregate transfer coefficients (T_{ag}) for the various types of soil encountered in the region. The results are contained in Tables 4 and 5. In these Tables the T_{ag} values have been calculated on the basis of individual sites and then averaged, so the values are not directly taken from the means shown in the table.

Depending on the type of soil, the soil-to-vegetation aggregate transfer coefficient for ^{137}Cs is in the range of (30-700) x 10^{-6} m^2 kg^{-1}. One exception shown in Table 4 is the value of T_{ag} for undisturbed soddy podzol swampy which is almost 30 times higher than the correspondent value for ploughed soil.

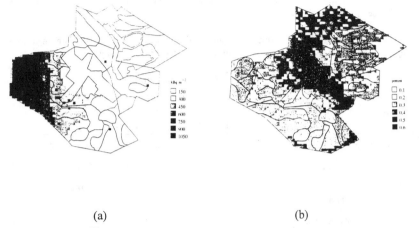

(a)   (b)

Figure 3 Deposition of ^{137}Cs in the Bragin site. (a) Deposition map obtained from interpolation of sampling data. (b) Kriging coefficient of variance in deposition values.

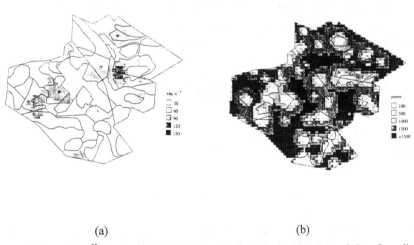

(a)   (b)

Figure 4 Deposition of ^{90}Sr in the Bragin site. (a) Deposition map obtained from interpolation of sampling data. (b) Kriging coefficient of variance in deposition values.

Aggregate transfer coefficients for ^{90}Sr are about two orders of magnitude higher than those for ^{137}Cs. For each of the "soddy" classes of soil the T_{ag} values are consistently similar at 3 - 5 m^2 kg^{-1} x 10^{-3}. Surprisingly, the ^{90}Sr soil-to-vegetation aggregate transfer coefficient was highest for the peaty soil, although it should be noted that this value is based on only two measurements made in 1994 for this soil type.

Table 4 Cs-137 activity concentration in soil and grass samples collected in the Bragin study site during 1994 and 1995. N is the number of samples employed in each calculation.

Year of sampling	Soil type	N	Activity concentration (mean ± SD)		T_{ag} soil-grass (m^2 kg^{-1} x 10^{-3})
			Soil (kBq m^{-2})	Vegetation (Bq kg^{-1} dw)	
1994	Peaty, not ploughed	2	280 ± 57	4320 ± 764	15.4
1994/95	Soddy podzol	7	247 ±247	39 ± 6.9	0.26 ± 0.22
1994/95	Soddy podzol swampy (ploughed)	21	389 ± 233	100 ± 135	0.23 ± 0.15
1994/95	Soddy podzol swampy (undisturbed)	3	147 ± 27	6440 ± 1153	43.7
1994/95	Peaty swampy	3	57 ± 5	1679 ± 54	29.6
1994/95	Soddy swampy	13	137 ± 67	54.6 ± 63	0.67 ± 0.43

Table 5 Sr-90 activity concentration in soil and grass samples collected in the Bragin study site during 1994 and 1995 (mean ± SD). N is the number of samples employed in each calculation.

Year of sampling	Type of soil	N	Mean deposition and activity concentration		T_{ag} (m^2 kg^{-1} x 10^{-3})
			Soil (kBq m^{-2})	Vegetation (Bq kg^{-1} dw)	
1994	Peaty	2	90	655 ± 64	7.28
1994/95	Soddy podzol	5	84 ± 31	291 ± 62	3.7 ± 0.7
1994/95	Soddy podzol swampy	20	65 ± 53	260 ± 145	4.3 ± 2.0
1994/95	Soddy swampy	11	42 ± 15	184 ± 62	3.5 ± 1.5

3.2 PRODUCTION AND CONSUMPTION OF PRODUCTS

During July to September 1994 a total of 10 families, 5 families in Savichi and 5 families in Dvor-Savichi, were interviewed to record the production of crops and animal products. At the same time, samples were taken of soil, pasture grass and food products for measurements of ^{137}Cs and ^{90}Sr content. Interviews and sampling of food products was repeated during the summer of 1995. Results of these interviews with regard to production in private farms and consumption per family are given in Tables 6, 7 and 8. The last columns in Table 6 contains also the calculated yield (in kg m^{-2}) in each village. As seen from these Tables the cultivated area for each crop per family, production and yields are very similar in the two villages. With the exception of potatoes, all produce is consumed by the family. Equally similar are the habits of the inhabitants of the two villages with regard to keeping domestic animals, as seen from Tables 7 and 8.

Table 6 Area, production and consumption of vegetables and fruits per family plot in Savichi and Dvor-Savichi. Mean and range.

Product	Area (m^2)		Production (kg)		Consumption per family (kg y^{-1})		Yield (kg m^{-2})	
	Savichi	Dvor-Savichi	Savichi	Dvor-Savichi	Savichi	Dvor-Savichi	Savichi	Dvor-Savichi
Potatoes	3200 (2500-4000)	3400 (2000-4000)	4800 (4000-7000)	5800 (5000-8000)	750 (550-950)	850 (600-1100)	1.5	1.7
Cabbage	75 (25-100)	91 (10-200)	126 (40-200)	119 (15-200)	all	all	1.7	1.3
Onions	144 (40-500)	58 (10-200)	106 (50-300)	161 (25-500)	all	all	0.74	2.8
Carrots	23 (4-50)	49 (15-80)	28 (5-50)	42 (20-90)	all	all	1.2	0.86
Beets	170 (50-150)	58 (40-100)	334 (50-1000)	132 (100-200)	all	all	2.0	2.3
Beans	36 (4-100)	25 (10-50)	7 (2-15)	7.5 (5-10)	all	all	0.19	0.3
Cucumbers	51 (25-80)	74 (20-150)	48 (35-100)	58 (20-100)	all	all	0.94	0.78
Tomatoes	54 (40-80)	82 (30-200)	56 (30-120)	53 (20-80)	all	all	1.0	0.64
Berries	75 (50-100)		28 (8-50)		all		0.37	

3.3 ACTIVITY LEVELS IN FOODPRODUCTS

Measurements of activity levels of foodstuffs produced in the private plots of Savichi and Dvor-Savichi were performed during June - July, 1994 and in February, May and August, 1995. Samples of milk and vegetables were purchased from five families in each village and measured for ^{137}Cs and ^{90}Sr activity concentration. All gamma-ray measurements were performed with an intrinsic Ge detector and all radiostrontium measurements with a Camberra a, b, ã-spectrometer. An intercalibration exercise was carried out by measuring 12 samples both at the Belarus Institute of Agricultural Radiology in Gomel and at the Nuclear Physics Laboratory in Ioannina. All such duplicate measurements were in satisfactory agreement.

Table 7 Domestic animals in Savichi.

Animal	Main product	Average number per family	Average age (y)	Yield per animal	Consumption by family per year	Sold on market	Consumption by relatives and friends	Feeding rations (kg animal⁻¹ d⁻¹)			
								Grass	Hay	Potatoes	Grains
Cattle	milk	1	5	Milk 2960 kg y^{-1}	Milk 1400 kg	1260 kg (animal feed)		53	3000	3300	225
Pigs	meat, fat	3		Weight at slaughter 160 kg	185 kg	200	95	-	-	2600	262
Hens	eggs, meat	13		Eggs 1240 y^{-1} Meat 11 kg y^{-1}	1100		140	-	-	20	30
Sheep	wool	0.4						2500	1100	-	-
Turkeys, ducks, geese	meat	none									

Table 8 Domestic animals in Dvor-Savichi.

Animal	Main product	Average number per family	Average age (y)	Yield per animal	Consumption by family per year	Sold on market	Consumption by relatives and friends	Feeding rations (kg animal⁻¹ d⁻¹)			
								Grass	Hay	Potatoes	Grains
Cattle	milk	1	5	Milk 2660 kg y^{-1}	Milk 1400 kg	1260 kg (animal feed)	200	50	2700	3000	260
Pigs	meat, fat	5	1	Weight at slaughter 185 kg	225 kg	300	210	-	-	2000	150
Hens	eggs, meat	15	2	Eggs 1410 y^{-1} Meat 10 kg y^{-1}	1200			-	-	25	18
Sheep	wool	3	4	-				2200	900	-	-
Turkeys, ducks, geese	meat	2	2	Eggs 4.5 y^{-1} Meat 10 kg y^{-1}	4.5			-	-	90	60

Average values of activity concentration for both ^{137}Cs and ^{90}Sr in all food items contained in the survey questionnaire as obtained from five replicate samples, are contained in Table 9.

3.4 ACTIVITY LEVELS IN FOODSTUFFS DERIVED FROM THE UNIMPROVED OR SEMI-NATURAL ECOSYSTEM

Replicate samples of mushrooms commonly consumed by the inhabitants of the Bragin area were taken and measured for both ^{137}Cs and ^{90}Sr activity concentration. The data are presented in Table 10. The Table contains also measurements of the soil on which the mushrooms were growing and litter (leaves, mulch, etc.) in the surrounding area. All measurements refer to dry weight of the samples with the exception of ^{137}Cs in mushrooms which refers to wet weight. Radiostrontium concentration was found at relatively low levels, whereas ^{137}Cs was present in high concentrations. The variability in the values obtained, as seen from the ranges in Table 10, was also considerable.

Table 9 Average ^{137}Cs activity concentration in food samples from the villages of Savichi and Dvor-Savichi.

Product	Savichi (Bq kg^{-1})		Dvor-Savichi (Bq kg^{-1})	
	^{137}Cs	^{90}Sr	^{137}Cs	^{90}Sr
Milk	13.5 ± 1..7	10.1 ± 2.2	6.8 ± 2.3	5.5 ± 0.9
Flour	3.2 ± 0.6	3.1 ± 0.2	2.7 ± 0.3	3.0 ± 0.1
Bread	6.4 ± 1.0	2.2 ± 0.4	5.3 ± 1.1	1.9 ± 0.3
Pork	52 ± 12	-	32	-
Poultry	39.6 ± 5.6	-	24.9 ± 3.0	-
Eggs	7.5 ±1.6	5.5 ± 1.0	8.7 ± 2.1	5.2 ± 2.5
Potatoes	7.9 ± 1.2	5.5 ± 1.7	5.3 ± 0.9	2.1 ± 0.4
Tomatoes	5.8 ± 2.8	3.0 ± 0.8	3.4 ± 1.3	2.7 ± 0.4
Beet root	12.8 ± 3.1	22.2 ± 3.7	5.3 ± 1.1	25.4 ± 6.9
Cabbage	11.7 ± 1.8	-	10.9 ± 1.7	-
Onions	12.9 ± 4.9	16.4 ± 3.1	15.9 ± 3.4	21.9 ± 6.9
Carrots	6.1 ± 1.7	15.9 ± 4.9	4.9 ± 1.1	10.9 ± 1.2
Cucumbers	2.9 ± 1.1	13.8 ± 6.0	2.7 ± 1.2	4.4 ± 1.7
Blueberries	6,540 ± 710	166 ± 35	6,540 ± 710	166 ± 35

Table 10 Cs-137 and ^{90}Sr activity concentration measured in mushrooms collected in the Bragin district. Mean and range.

Species	Soil (kBq m^{-2})		Litter (Bq kg^{-1})		Mushrooms (Bq kg^{-1})	
	^{137}Cs	^{90}Sr	^{137}Cs	^{90}Sr	^{137}Cs	^{90}Sr
Cantharellus cibarius	1811 ± 126 1626 - 2160	101 ± 8 91 - 125	70450 ± 507 64500 - 87500	2738	217000 ± 33880 110000 - 315000	592 ± 98 328 - 853
Boletus edulis	2385 ± 300 2085 - 2685	104 ± 6 97 - 110	60600 ± 56000 4170 - 117000	450 ± 230 218 - 680	253000 ± 140000 7420 - 500000	623 ± 149 409 - 911
Russula app.	1560	83	88100	5042	889000	400
Mixed sample	1251	65	47300	4437	215000	563

In addition samples of meet from wild animals hunted in the Bragin district were measured for ^{137}Cs and ^{90}Sr activity concentration. The results are contained in Tables 11 and 12.

Table 11 Ranges of ^{137}Cs activity concentration measured in meet of wild animals hunted in the District of Bragin during 1994 and 1995.

Animal	Activity concentration (Bq kg^{-1})
Elk	200 - 12000
Roe-deer	3000 - 10000
Wild boar (winter)	8000 - 60000
Wild boar (summer)	8000 - 20000
Hare	15 - 250
Wild duck	35 - 110
Partridge	7.9 ± 0.3

Table 12 Ranges of ^{90}Sr activity concentration measured in meet and bone of wild animals hunted in the District of Bragin.

Animal	Activity concentration (Bq kg^{-1})	
	Muscle	Bone
Elk	30 - 100	10000 - 100000
Roe-deer	4 - 20	4000 - 6000
Wild boar	3 - 5	5000 - 30000

4. Intake of radionuclides to man

The data contained in the previous sections, combined with the dietary habits of the inhabitants in the two villages considered here, can lead to an estimate of radionuclide intake.

4.1 DIETARY SURVEY

In 1994 ten individuals from each village were interviewed with regard to their dietary habits and personal precautions against radioactivity contamination from food consumption. During the summer of 1995 a more systematic survey was carried out by means of a more detailed questionnaire Fifty individuals from each village were interviewed.. Table 13 contains the food consumption of the population in the villages of Savichi and Dvor-Savichi.

In order to assess the importance of food products from the semi-natural ecosystem in peoples' diet, Figs. 5 - 7 show the percentage of the surveyed population consuming

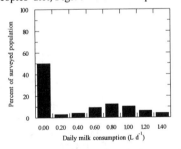

Figure 5 Daily consumption of domestically produced milk by the population in the District of Bragin.

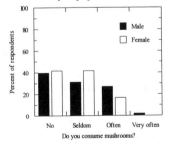

Figure 6 Consumption of mushrooms by the population in the District of Bragin

domestically produced milk, mushrooms and fish from the local lakes. As seen from these Figures about 50% of the population surveyed responded that they do not consume any local milk or milk products, whereas the mean consumption of milk for the rest of the population is approximately 0.8 L d^{-1}. Forty percent of the people surveyed answered that they do not consume mushrooms picked at the nearby forests, whereas very few people stated that they eat mushrooms "very often". Only 20% of the surveyed population answered that they "seldom" eat fish caught in the local lakes. All people interviewed stated that they never consume game animals.

As seen in Figs 5 - 7, there was no difference between the eating habits of the male and female population in the two villages.

4.4 INTAKE OF RADIOCAESIUM AND RADIOSTRONTIUM

The data in Tables 9 and 13 may be used to extract the average daily activity intake for ^{137}Cs and ^{90}Sr of the population in the District of Bragin. This calculation for all products, except for dietary items derived from the natural ecosystem (mushrooms, berries, game, etc.) is contained in Table 14. It is noted that in both villages almost 85% of the average daily activity intake is derived from three items: domestically produced milk, pork and potatoes. It is noted that the total ^{137}Cs daily intake figures in Table 14 do not include contributions from natural dietary items, especially mushrooms, since accurate data for mushroom consumption by the population in the Bragin area are not available.

Table 14 Average daily activity intake from agricultural dietary items of the inhabitants of Savichi and Dvor-Savichi.

Product	^{137}Cs (Bq d^{-1})		^{90}Sr (Bq d^{-1})	
	Savichi	Dvor-Savich	Savichi	Dvor-Savich
Milk	15.1 ± 2.4	11.1 ± 4.8	11.9 ± 2.6	11.6 ± 1.9
Bread	3.6 ± 0.6	1.7 ± 0.4	1.2 ± 0.2	0.6 ± 0.1
Pork	8.8 ± 2.3	6.4 ± 2.4	-	-
Eggs	1.5 ± 0.4	0.9 ± 0.3	0.9 ± 0.2	0.4 ± 0.2
Potatoes	10.3 ± 2.0	5.9 ± 1.4	9.0 ± 2.8	2.6 ± 0.5
Vegetables[a]	1.4 ± 0.3	1.8 ± 0.4	2.3 ± 0.7	3.1 ± 0.9
Total	**41 ± 4**	**28 ± 5**	**25 ± 4**	**18 ± 2**

[a] Mixed vegetables (tomatoes, beets, cabbage, onions, carrots, and cucumbers).

4.5 WHOLE BODY MEASUREMENTS

The 100 individuals interviewed during the summer of 1995 were subjected to whole body measurements in August, 1995. All measurements were performed with a portable NaI detector. Figure 8 shows the frequency of whole body radiocaesium activity measured at the surveyed population in the District of Bragin. Some parameters of the whole body

Table 15 Radiocaesium whole body activity measurements in the District of Bragin.

Village	Sex	Whole body activity (kBq)		
		Mean value	Median	Range
Savichi	Male	22.0 ± 5.0	11.8	3.3 - 92.5
	Female	19.6 ± 4.0	8.5	3.0 - 107.3
Dvor-Savichi	Male	8.4 ± 1.7	6.5	1.5 - 27.8
	Female	7.8 ± 1.7	4.6	1.5 - 20.7

measurements are also contained in Table 15. The mean whole body activity versus age is presented in Figure 9. As seen in this Figure, people in the age range 50 - 70 y are considerably more burdened than younger individuals. This may be the result of conscious efforts made by younger people to avoid heavily contaminated food.

Table 13 Comparison of food consumption of the villages Savichi and Dvor-Savichi, kg d^{-1} (Mean + SE).

Source	Village	Milk and dairy products	Meat and meat products	Bread and bakery products	Potatoes	Vegetables	Fruit	Eggs	Fish
Private Produce	Savichi	0.78 ± 0.03	0.15 ± 0.02	0.53 ± 0.04	0.68 ± 0.03	0.11 ± 0.01	0.04 ± 0.01	0.10 ± 0.02	0.015 ± 0.001
Private Produce	Dvor-Savichi	1.13 ± 0.03	0.16 ± 0.01	0.32 ± 0.03	0.73 ± 0.03	0.19 ± 0.01	0.04 ± 0.01	0.08 ± 0.01	0.022 ± 0.003
Private Market	Savichi	0.40 ± 0.03	0.020 ± 0.003	-	0.95	0.05 ± 0.01	0.05 ± 0.01	0.06 ± 0.01	0.05
Private Market	Dvor-Savichi	0.97 ± 0.05	0.035 ± 0.003	-	0.50	0.05 ± 0.01	0.01	-	0.25
State shop	Savichi	0.24 ± 0.02	0.022 ± 0.003	0.37 ± 0.03	-	-	0.019 ± 0.003	0.02	0.016 ± 0.001
State shop	Dvor-Savichi	0.17 ± 0.03	0.005 ± 0.001	0.37 ± 0.02	-	-	-	-	0.018 ± 0.001
Total	Savichi	1.42 ± 0.05	0.19 ± 0.02	0.90 ± 0.05	1.63 ± 0.03	0.16 ± 0.02	0.108 ± 0.007	0.18 ± 0.01	0.081 ± 0.001
Total	Dvor-Savichi	2.27 ± 0.07	0.20 ± 0.01	0.70 ± 0.04	1.23 ± 0.03	0.24 ± 0.01	0.015 ± 0.010	0.08 ± 0.01	0.290 ± 0.003

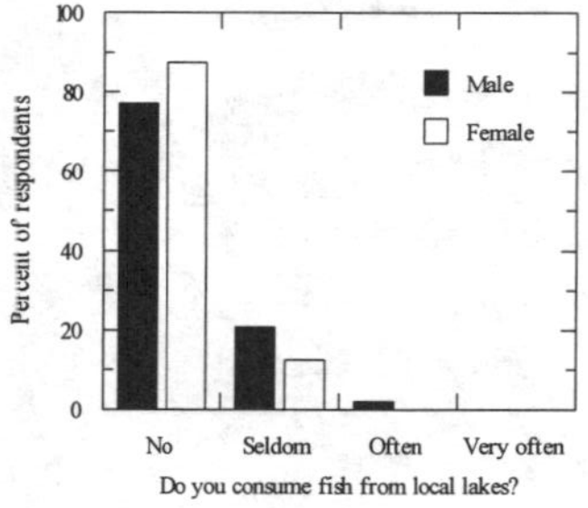

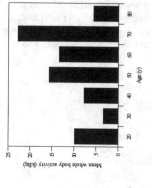

Figure 9 Whole body radiocaesium activity measured in the population (100 individuals) of the District of Bragin versus age.

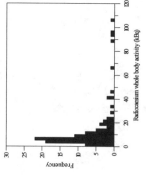

Figure 8 Frequency of whole body radiocaesium activity measured in the population (100 individuals) of the District of Bragin

Figure 7 Consumption of fish from local lakes by the population in the District of Bragin

The correlation between dietary habits of the population in the District of Bragin and whole body activity measurements was also investigated. Figure 10 contains the mean whole body activity versus domestic milk daily consumption, in which no correlation is evident. Similarly, Figures 11 and 12 present the mean whole body burden versus mushroom and local fish consumption, respectively. Again, no relation is seen between the consumption of local lake fish and whole body burden. On the contrary, a very significant correlation emerges for consumption of mushrooms.

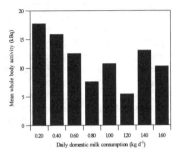

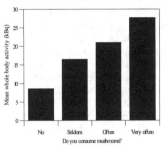

Figure 10 Mean whole body activity measured in the population of the District of Bragin (100 individuals) versus milk consumption.

Figure 11 Mean whole body activity measured in the population of the District of Bragin (100 individuals) versus mushroom consumption.

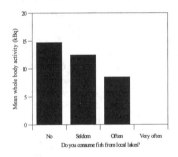

Figure 12 Mean whole body activity measured in the population of the District of Bragin (100 individuals) versus consumption of fish from local lakes.

5. Discussion and conclusions

The data contained in the previous pages may be considered as representative of the radiological situation in a closed rural society, which was heavily affected by the Chernobyl accident, and which since then has been operating under a subsistence farming economy. The information amassed in this research can be employed in the assessment of the radiological impact to the population in the area and to plan for appropriate countermeasures. In particular, the information supplied through GIS modelling with respect to land use and deposition, in conjuction with aggregate transfer parameters contained in Tables 4 and 5, may be employed for the optimum reassignment of land usage. It is, for instance, evident from these data that soddy podzol soils should be used for the growing of animal feed, whereas ploughing (and reseeding) of the fields emerges as an effective countermeasure for both ^{137}Cs and ^{90}Sr. The information contained in the maps of Figs. 2 - 4 should greatly facilitate the strategic reallocation of land assets.

In order to assess intake of radiocontaminants detailed studies were made of the private production upon which the local population is primarily dependent. As mentioned earlier, each village family has a plot of land in which it grows foodstuffs for its own use. Each family also owns one or at most two cows for milk production (used both for private consumption and as food of animals), three to five pigs for the family's meat and fat needs and an average of 15 hens. Private farmers in the study site were interviewed about the level of their production and samples of their crops were measured for radiocaesium and radiostrontium concentration. Potatoes was found to be the major crop, used as food for both humans and animals, while production levels of other vegetables and fruits were substantially lower. Radiocontamination of foodstuffs produced in both villages was measured for both ^{137}Cs and ^{90}Sr with the highest levels obtained for milk (Table 9).

The data obtained from the dietary survey in the two villages, combined with activity levels measured in food samples, lead to an estimate of radionuclide intake of the population through the consumption of locally produced foodstuffs. The data in Table 14 reveal that approximately 1/3 of the total daily activity intake arises from the consumption of milk, with potatoes as the second most important source of radiocontamination.

A further source that was considered potentially important, especially with regard to radiocaesium intake, was the consumption of foodstuffs derived from the unimproved or semi-natural ecosystem. As indicated in Table 9, activity concentration measured in blueberries gathered from nearby forests were found at substantial levels, especially for ^{137}Cs. The same is true for mushrooms and wild animals sampled in the region. However, two important factors preclude the quantification of radionuclide intake by the population from this source. First, it was found very difficult to determine the amount of such products consumed regularly by the population. Thus the dietary survey was only able to classify the population in four broad categories as those who include in their diet such products "Never", "Seldom", "Often" or "Very often". Second, the variability in activity concentration of the samples measured was so wide that in most cases only broad ranges could be determined (Tables 10 - 12).

It should be noted that from the rudimentary dietary survey with regard to consumption of natural products described above, as well as from the correlation with whole body burden of the inhabitants of Bragin discussed later, mushroom consumption emerges as a primary pathway for radiocaesium intake. Other items derived from the natural or semi-natural ecosystem were either consumed in very small quantities (nuts, blueberries) or very seldom (fish from local lakes). Game animals are very rarely consumed. It is therefore evident that mushrooms should be singled out for further investigation as an important source of radiocaesium burden to the population. Such an investigation, in addition to quantifying the yearly consumption, should differentiate among the species of mushrooms, which, from the data of Table 10, are seen to differ substantially with regard to radiocaesium content. It should further take into account the various methods of preparation and storage of mushrooms (cooking, drying, storage in water or brine) which may significantly alter the radiocaesium content of the consumed product.

The importance of mushroom consumption as a pathway for radiocaesium intake is further verified from whole body radiocaesium measurements carried out for 100 inhabitants of the area under study. As seen in Fig. 11, a very strong correlation emerges for mushroom consumption versus whole body radiocaesium content.

It is hoped that the research presented here will lead to a better understanding of the particular production system under consideration in contributing to radionuclide intake by man. A more informed implementation of countermeasures could then be made. This unique combination of radioecological and animal production information should ultimately aid the decision making process.

Countermeasures Implemented in Intensive Agriculture

S. FIRSAKOVA
Belorussian Institute of Agricultural Radiology, Belarus

K. HOVE
Agricultural University of Norway, Norway

R. ALEXAKHIN
Russian Institute of Agricultural Radiology and Agroecology, Russia

B. PRISTER
Ukrainian Institute of Agricultural Radiology, Ukraine

N. ARKHIPOV
RIA "Pripjat", Ukraine

G. BOGDANOV
National Agricultural University, Ukraine

Abstract. The wide application of countermeasures in agriculture at different times after the Chernobyl accident provided an opportunity to estimate the most efficient means of reducing radionuclide transfer through the chain soil-plant-animal-man. The choice of countermeasures against the background of traditional practices was governed by a variety of factors, including the composition of the radioactive fallout, the soil types, land usage, the time elapsed since the accident and the economics. The efficiency of the various countermeasures was assessed in terms of both reduction of individual dose in the contaminated areas and of the collective dose.

The estimation of the various countermeasures comparative efficiency is presented, their impact on individual doses reduction and the contribution reduction of produce produced in the contaminated area into the collective dose of the population is shown.

1. Introduction

During the first weeks after the Chernobyl accident the experts in agricultural radiology elaborated for the Ukraine, Belarus and Russia the guidelines for agricultural farming in the contaminated areas. For the period of 10 years these guidelines were supplemented and made more precise.

The experience of agricultural farming in the area contaminated with long-living radionuclides after the Chernobyl accident shows that the implementation of certain land-reclamation and organizational measures make it possible to produce plant and animal products within temporary permissible levels of Cs-137 and Sr-90.

The technological cycles of agricultural production provided an opportunity to perform countermeasures in the key links of the radionuclides transfer through food chains: - soil - plant, feed - animal, raw materials - ready food products.

The most critical food product on Cs-137 content in agricultural areas in the Ukraine, Belarus and Russia after the releases of the Chernobyl NPP appeared to be the cows milk the contribution of which into the internal dose of the population varies within the limits of 40-80%.

The variety of soil cover in contaminated areas and the radioactive deposition heterogeneity cause the differential approach to the choice of countermeasures in all the three technological links.

2. Countermeasures in the chain soil-plant

2.1 Arable lands

The countermeasures on arable lands are the most available and easily feasible. The applying of mineral and organic fertilizers does not require additional equipment, the change in methods of crops cultivating and improves physical and agrochemical soil properties and also increases the yield of field crops.

For the time period elapsed after the accident the radiocaesium transfer from soil into plants reduced significantly at the expense of the "aging" processes, migration, the radioactive decay of Cs-134 (96%) and Cs-137 (19%) and the realization of countermeasures (Fig. 1).

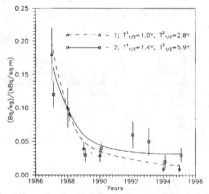

Fig.1. Cs-137 content related to the contamination density in main field crops (1 - cereals, 2 - potato).

The period of the intensive applying of countermeasures is characterized by the rapid decrease of Cs-137 uptake into field crops ($T_{1/2}$ 1.0-1.4 y). During the more remote periods after the accident when the volume of the countermeasures carried out on arable lands reduced (1991-1995) Cs-137 transfer into plants also became slower ($T_{1/2}$ 2.8-5.9 y). According to the estimate of R.M. Alexakhin *et al* [1] the countermeasures contribution into the total rate of Cs-137 transfer reduction on arable land is up to 57%.

In table 1 the comparative estimate of countermeasures efficiency on arable lands in the contaminated areas in Russia, the Ukraine and Belarus is presented [1,2,3,4].

The experiments carried out within ECP-2 project on the lands of the three CIS countries showed the efficiency of the traditional countermeasures (the applying of manure, phosphorous-potassium fertilizers in increased doses).

Table 1.

Effectiveness of agrochemical countermeasures for three settlements in CIS countries

Countermeasure	Soil type	Reduction of root uptake transfer		
		Belarus	Russia	Ukraine
Liming, 1.5 Hg	Mineral	1.3-2.5	1.6-2.3	1.5-2.5
	Organic	-	-	2.0-3.0
Fertilizers $N^{60}P^{90}K^{120}$, 1.0:1.5:2.0	Mineral	1.1-2.8	1.5-2.2	1.5-2.2
	Organic	-	-	1.4-2.4
Manure, 60 ton per hectare	Mineral	1.4-2.8	1.3-2.6	1.2-2.5
	Organic	-	-	-
Liming, Fertilizers $N^{60}P^{90}K^{120}$	Mineral	1.5-2.3	2.0-2.7	1.8-3.5
	Organic	-	-	2.0-4.0
Liming, Manure	Mineral	1.5-3.0	1.7-2.9	1.5-3.0
	Organic	-	-	-
Liming, Fertilizers $N^{60}P^{90}K^{120}$, Manure	Mineral	2.0-3.9	2.3-3.6	2.5-4.7
	Organic	-	-	2.7-5.2
Manure, Fertilizers $N^{60}P^{90}K^{120}$	Mineral	2.8-6.5	2.1-3.7	2.0-3.6
	Organic	-	-	-

The attempt to reduce Cs-137 uptake from soil into plants by means of applying of clay minerals: bentonite, clynoptilolite and zeolite appeared to be ineffective. It is accepted to consider the clay minerals mentioned above as caesium-selective sorbents and that was confirmed by the laboratory experiments while the imitation of fresh caesium fallout. However, during quite a long period (8 y) the process of Cs-137 "aging" have taken place that was confirmed by the reduction of mobile forms, i.e. the fixing of it with native clay minerals. Just because of this the applying of additional clay minerals in soddy-podzolic soils under natural conditions did not make an impact on the reduction of Cs-137 uptake into plants [5].

Thus, during the remote period after the accident the most effective and expedient countermeasures for arable lands are countermeasures aimed at the increase of their fertility. They are available, traditional and are covered by the increase of the yield.

Within the program of the Chernobyl accident consequences liquidation the following volumes of countermeasures are financed (thousand. ha):

	Ukraine 1994	Russia 1994	Belarus 1995
Liming	10.3	14.8	260
Additional applying of P	8.3	-	800.5
Additional applying of K	8.3	56.8	1239

To reduce the production of crops accumulating the radionuclides intensively in the contaminated areas cereal-leguminous and leguminous perennial grasses were excluded from the crop rotation. The part of technical crops, including oil-yielding rape-seed increases.

2.2. Meadows and pastures

As it was mentioned above, the cows milk is one of the main dose-forming food products. That is why it is necessary to use countermeasures reducing radionuclides uptake in feeds to decrease internal irradiation doses. Meadows and pastures used for milk and meat production occupy about 60% of the total balance. Therefore the applying of countermeasures on meadows and pastures (the radical improvement of natural meadows and resowing of grasses on cultivated lands) contributes the most greatly into the reduction of irradiation doses of the population in the contaminated areas.

The results of various methods efficiency of the radical improvement of meadows on mineral and peaty soils [6,7] are presented in Fig. 2 and 3. All the methods of soil processing in combination with various methods of applying of lime and mineral fertilizers give stable reduction (3-5 times) of Cs-137 and Sr-90 accumulation in the herbage of meadows on mineral soils. The radical improvement of meadows on peaty soils gives a sharp decrease of Cs-137 uptake in cultivated grasses. However, these methods are not effective in relation to Sr-90.

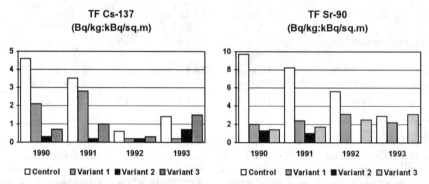

Fig. 2. Efficiency of different methods of the radical improvement of meadows on mineral soils for Cs-137 and Sr-90 reduction in herbage dry weight (Bq/kg : kBq/m^2). Control - natural meadow, Variant 1 - 6 t/ha of lime on the sod, disking, Variant 2 - 6 t/ha of lime on the sod (Sr-90 did not determine in 1992 and 1993), N^{30} P^{90} K^{120}, disking, Variant 3 - K^{250} on the sod, ploughing, N^{30} P^{90} K^{120} + 6 t/ha of lime on the sod.

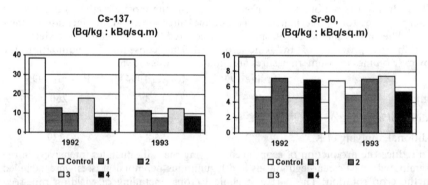

Fig. 3. Efficiency of different methods of soil treatment and fertilization for Cs-137 and Sr-90 reduction in herbage dried weight on peaty meadows (Bq/kg : kBq/m^2). Control - natural meadow, 1 - N^{60} P^{60} K^{120}, disking, 2 - K^{250}, disking, N^{60} P^{60} K^{120}, 3 - ploughing, N^{60} P^{60} K^{120}, 4 - K^{250}, ploughing, N^{60} P^{60} K^{120}.

The investigations carried out by the other authors confirmed the reliability of these countermeasures [8,9,10], the implementation of which made it possible to reduce sharply the production volumes of contaminated milk (table 2).

Table 2.

Milk production in the collective sector with Cs-137 content exceeding TPL, thousand t

Year	Ukraine	Russia	Belarus
1986		110.9*	524.6*
1987		96.8*	308.9*
1988	78*	95.9*	193.3*
1989	61*	74.8*	69.6*
1990	62*	48.8*	7.2*
1991	1*	10.97*	22.1**
1992	0*	6.2*	9.4**
1993	0*	1.6*	14.9***
1994	0*	0.59*	12.4***

Note: * - TPL 370 Bq/l, ** - 185 Bq/l, *** - 111 Bq/l.

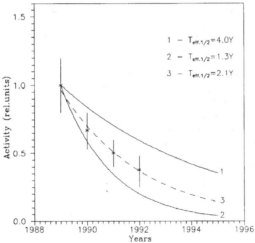

Fig. 4. Dynamics of activity reduction in milk produced on natural and cultivated meadows. 1 - natural meadows, 2 - cultivated meadows, 3 - production data.

Dynamics of activity reduction in milk produced on natural and cultivated meadow lands is presented in Fig. 4. The half-periods of milk calculated for these curves are 4 and 1.3 y, respectively.

Curve 3 reflects the exponential recession of milk activity under real conditions with the half-period of 2.1 y and sums up the relative contribution of 2 types of meadows.

The radical improvement of meadows not only reduced the activity of produced milk but also accelerated the rates of its 'cleaning'. The results presented indicate that the potentialities of countermeasures have not been realized fully.

The means for the radical improvement of meadows and pastures are allocated at the expense of the state in the following volumes (thousand. Ha): Ukraine (1994) - 52.7, Russia (1994) - 6.9, Belarus (1995) - 95.0.

3. Countermeasures in the chain feed - animal

Though the radical improvement of meadows and pastures is a reliable method of the radionuclides uptake reduction into the feeds for animals but not all natural lands can be cultivated. It concerns the forest and boggy pastures where it is very complicated to perform the radical improvement and sometimes even impossible. In this case the selective sorbents preventing from the absorption of radionuclides in the gastro-intestinal tract are introduced into the diet of animals. The most wide-spread and well-known are caesium-selective sorbents of a number of ferric hexacyanoferrates (Prussian Blue, ferrocyn, Giese salt, Nigrovich salt).

The Prussian Blue and ferrocyn have the high selectivity and absorption capacity.

Within the joint project realized by IAEA, FAO, Agricultural University of Norway and Radiation Hygiene Institute of Norway, Belarussian Research Institute of Agricultural Radiology, Ukrainian Research Institute of Agricultural Radiology, All-Union Research Institute of Agricultural Radiology and Agroecology there were carried out the investigations on the applying of ferrocyn in the composition of boli and salt licks in order to reduce Cs-137 content in milk and muscle tissues of cattle.

The investigations in the three CIS countries showed the reliability of such countermeasures and the stability of the effect. Cs-137 content in milk and muscle tissues of cattle decreases 2-5 times [11, 12].

The technology of the production of boli and salt licks with Prussian Blue was given to Ukraine, Russia and Belarus by K.Hove.

In Belarus the boli production and the wide-scaled application of them was organized with the financial support of the Ministry of Agricultural Products and Goskomchernobyl. The Veterinary-Pharmaceutical Council elaborated and approved the scientific-technical documentation on the application of boli, salt licks and concentrates with Prussian Blue.

Boli and concentrates for dairy cows the production of which is paid by the state are used in the private sector the most widely.

In Russia the compounds on the basis of Prussian Blue are given to 11,000 cattle and in Ukraine - for 1,500 cattle.

In Belarus 100,000 boli and 4,000 t of concentrates are produced for 30,000 cattle.

Table 3.

Production of meat with Cs-137 content exceeding TPL (thousand t).

Year	Ukraine	Russia	Belarus
1986	6.41	5.4	21.1
1987	1.28	5.7	6.9
1988	0.168	1.5	1.45
1989	0.064	0.3	0.6
1990	0.017	0.04	0.08
1994	0	0.012	0.0026

One of the most widely applied countermeasures in cattle-farming since the first months of the post-accidental period is the feeding of uncontaminated forage during the final fattening of livestock [13]. The implementation of this countermeasure made it possible to reduce the volumes of contaminated meat in 1988 up to insignificant quantities and during the following years to solve this problem practically (table 3).

In the collective sector to produce milk with the levels meeting the requirements of the permissible levels of Cs-137 and Sr-90 content it is a practice of the diets rate setting according to the radionuclides content.

4. Countermeasures in the chain raw material - ready food produce

The common primary and technological processing of agricultural produce and also the methods of cooking result, as a rule, in the reduction of contamination of ready food products.

Thus, milling of wheat, rye and barley grain to white flour decreases twice the radionuclides content in the final product, the milling of oats gives 3-fold reduction and the processing of cereals for spirits practically excludes the radionuclides content in the final; product.

The peeling of potatoes decreases Cs-137 and Sr-90 content in peeled tubers by 20% and the processing of tubers into starch reduces the radionuclides content in the final product up to 2%.

The processing of milk into butter and into cream decreases Sr-90 and Cs-137 content in the final product up to 1% and 5-7% respectively [14].

The processing of contaminated raw food products can reduce the radioactivity in many kinds of foodstuffs and is especially effective under the conditions of mass production. While estimation of the total result of processing by-products and waste products the use of which for feeding animals must be controlled should be taken into account.

Thus, the whole complex of countermeasures in the technological links of agricultural production provide an opportunity to manage the processes of production and use of agricultural produce for various purposes, ensuring by this the reduction of irradiation doses.

5. Countermeasures efficiency for the individual and collective doses reduction.

As milk makes up to 80% of average annual individual dose from the internal irradiation of the population living in the contaminated areas the most effective countermeasures are those aimed at making of cultivated meadows and pastures.

In table 4 the comparative evaluation of internal irradiation doses from milk after the radical improvement of meadows is presented.

Since the moment of using of herbage from the cultivated pasture during the following 5 years the averted individual doses makes up 0.089 mSv. For all this the milk component contribution during the 5th year is 4 times lower on cultivated pastures as compared with natural ones.

Currently the necessity of countermeasures application is more important on a private farm as the products produced just in this sector condition the internal irradiation dose of a villager.

Table 4.

Average annual doses of internal irradiation conditioned by the milk component
with the contamination level on pastures of 37 kBq/m^2 (mSv/y).

Years	Pasturing of cattle on natural meadow	Pasturing of cattle on a meadow after radical improvement
1989*	0.06	0.06
1990	0.048	0.048
1991**	0.04	0.03
1992	0.035	0.018
1993	0.033	0.013
1994	0.03	0.009
1995	0.028	0.007
Total 1991-1995	0.166	0.077

Note: * -   radical improvement of the meadow has been carried out.

     ** -   the use of one has been started

As a rule, the commodity agricultural produce produced in the contaminated areas is purchased by the state and exported to other regions for consumption if the radionuclides concentration in it does not exceed the permissible levels.

Nevertheless, the total Cs-137 activity in such produce can form the collective dose (table 5).

Table 5.

Efficiency of countermeasures complex for the reduction of the collective dose exported from the typical farm with the agricultural lands contamination level of 370 kBq/m^2

Product	Square, ha	Productivity, kg/ha	TF, kg/m^2	Total Q, MBq	Food Proc. Retention, F	Export, Q, %	Export. dose man-Sv
Grain	740	3000	0.1	82.1	-	50	0.58
Potato	200	20000	0.05	48.1	-	50	0.34
Milk	1200	820	0.4	145.6	-	90	1.84
							2.76
Countermeasures in chain soil-plant							
Grain	740	3000	0.07	58.7		50	0.42
Potato	200	20000	0.03	28.9	-	50	0.20
Milk	1200	2460	0.1	109.2	-	90	1.38
							2.00
Food processing							
Grain	740	3000	0.07	58.7	0.5	50	0.21
Potato (starch)	200	20000	0.03	28.9	0.02	50	0.004
Milk (cream)				109.2	0.05	45	0.035
Milk (butter)				109.2	0.01	45	0.007

The activity of natural products (grain, potato, milk) exported from the farm while countermeasures applying in the chain soil-plant decreases by 28%. The processing of raw materials into other kinds of food products can make the exported collective dose 10 times less. Therefore to reduce the export of activity over the boundaries of the contaminated territory it is expedient to perform the processing in the place of active raw materials production.

Thus, the countermeasures in the chain soil-plant are effective for the reduction of individual doses of the population living in the contaminated areas. For the collective dose reduction the contribution of the processing is the most significant.

References

[1] Fesenko S.V., Alexakhin R.M. *et al.*, "Dynamics of Cs-137 concentration in agricultural products in areas of Russia contaminated as a result of the accident at the Chernobyl NPP". Rad. Prot. Dos., 1995, Vol. 60, No. 2, 155-166.

[2] Prister B.S., Loshchilov N.A. "Agro-industrial production in the area of radioactive contamination on the territory of USSR". Proceedings of the scientific conference, Minsk, 1990, 5-11.

[3] Firsakova S.K. *et al.*, "On the state and aims of the scientific investigations in the sphere of agricultural radiology for agro-industrial production in the contaminated areas in BSSR". Proceedings of the scientific conference, Minsk, 1990, 21-27.

[4] Bondar P.F., Doutov A.I. "Estimate of potassium fertilizers efficiency as a means to reduce the radiocaesium contamination in the yield". Collection of scientific works of UIAR, Kiev, 1991, part 3, 69-83.

[5] "Soil-to-plant transfer and countermeasures", in Final Report "Transfer of radionuclides through the terrestrial environment to agricultural product including agrochemical practices" (ECP-2), 1994.

[6] Firsakova S.K. *et al.*, "The efficiency of agroland-reclamation measures to reduce Cs-137 accumulation by plants in meadows and pastures in the zone of the Chernobyl accident". Reports of AUAAS, 1992, N3, 25-27.

[7] Firsakova S.K. *et al.*, "The efficiency of agroland-reclamation measures used on agricultural lands of Belorussian Polesye to reduce the radionuclides uptake by plants". Proceedings of I.U.R. Soviet Branch Seminar on the Radioecology and Countermeasures. 1992, DOC UIR, 201-207.

[8] Prister B.S. *et al.*, "Efficiency of measures aimed at the reduction of contamination in plant produce in the areas contaminated as a result of the Chernobyl accident". Collection of scientific works of UIAR, Kiev, 1991, part 1, 141-153.

[9] Perepelyatnikov G.P. *et al.*, "Some problems of the feed production technology under the conditions of radioactive contamination". Collection of scientific works, Kiev, 1993, part 3, 115-125.

[10] Ilyin M.I. *et al.*, "The effect of radical improvement of natural meadows in the Ukrainian Polesye on the radiocaesium transfer from soil into the herbage". Agrochemistry, 1991, N1, 101-105.

[11] The use of caesium binders to reduce radiocaesium contamination of milk and meat from the territory of Ukraine, Belarus and Russian federation. Draft tech. doc. version 4.2, 1992.

[12] Hove K. "Chemical methods of reduction of the transfer of radionuclides to farm animals in semi-natural environments". Sci. Total Environ., 137 1-3, 1993, 235-248.

[13] Firsakova S.K. "Effectiveness of countermeasures applied in Belarus to produce milk and meat with acceptable levels of radiocaesium after the Chernobyl accident". J. The Science of the Tot. Environ., 137, 1993, 199-203.

[14] Guidelines for agricultural countermeasures following an accidental release of radionuclides. Tech. reports series No 363, Vienna, 1994.

Management of contaminated forests

A. Jouve1, F.A. Tikhomirov2, A. Grebenkov3, M. Dubourg4, M. Belli5, N. Arkhipov6.

1 Institut de Protection et de Sûreté Nucléaire, IPSN, Centre d'Etudes de Cadarache, 13108,
Saint Paul Lez Durance, France.
2 Moscow State University, Soil Science Faculty, Leninskie gory, 119899, Moscow, Russia.
3 Institute of Power Engineering Problems, Sosny, 220109 Minsk, Belarus.
4 FRAMATOME, Tour Fiat, 92084, Paris La Defense, France, Cedex 16.
5 ANPA, Via Vitaliano Brancati, 48, 00144, Roma, Italia.
6 RIA "Pripyat", Radiology and Land Restoration Department, Libknecht Str. 10, Chernobyl,
Ukraine.

Abstract

This paper examines the main radioecological issues, the consequence of which are the distribution of doses for critical group of populations living in the vicinity of contaminated forest after the Chernobyl accident and the effects on the forestry economy. The main problems that have to be tackled are to avert doses for the population and forest workers, mitigate the economical burden of the lost forestry production and comply with the permissible levels of radionuclides in forest products. Various options are examined with respect to their application, and their cost effectiveness in terms of dose reduction when such attribute appears to be relevant. It is found that the cost effectiveness of the various options is extremely dependant of the case in which it is intended to be applied. Little actions are available for decreasing the doses, but most of them can lead to an economical benefit.

1. Introduction

Although reducing the doses for populations is the aim of implementing remedial actions for sites contaminated during the Chernobyl accident, the possible options are largely dependent of economical issues. As shown by the example of Ukraine, the allocation of financial compensation represent the major part of Chernobyl budgets (figure 1). The general economical situation in the CIS has also an important impact on the existing situation of the consequences of the Chernobyl accident and the management routes which can be implemented. It becomes obvious from the factual behaviour of people living in contaminated areas, that no successful remedial action can be achieved without people involvement. A way of involving people in remedial actions could be found in options likely to provide some additional economical benefit by the introduction of new profitable activities such as for example, the production of biofuel using rape seeds, or the amelioration of crop yields applying high doses of fertilisers. The cost-benefit analysis of the countermeasures already applied, for example in Ukraine showed that the cost of the averted man sievert is very expensive [1]. It may be proposed that the cost of the averted dose could be partially compensated by the additional economical benefit of economy enhancing options. The economical situation in the CIS is also likely to determine the increasing consumption of food from private farms, or the use of food from forest ecosystems such as mushrooms, in spite of bans.

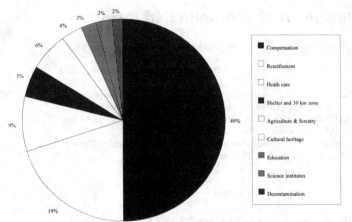

Figure 1: Chernobyl budget in Ukraine in 1993. This budget represents 12% of the national budget. (After [1])

Indeed, except bans, little can be done to mitigate the doses from forest ecosystems. A number of agricultural countermeasures have been applied after the Chernobyl accident, but their application to forest ecosystems may lead to disrupt the delicate equilibrium of forest life. Forest occupy a significant part of terrestrial ecosystems exposed to various kind of pollution. Forest health is therefore becoming an important issue in connection with the emerging concern for global changes on the earth planet. The general problems of forest management, which are pivotal environmental issues, do not disappear after a nuclear accident. On the contrary, the necessity to take care of forests becomes more important with respect to items such as fire preventing, or tree felling for sanitary purposes. Concerning forest fires, it can be enlighten that the areas of forest fires in Russia represent 5,5% of total forest areas, and are 7,5 times larger than harvest areas [2]. This proportion is likely to apply to contaminated forests also. Moreover, the reforestation of contaminated agricultural fields has been recognised as a pertinent option. This also leads to broaden the task of contaminated forest management.

In addition, it becomes sensible in certain countries such as Russia, that the transition to market economy necessitates the reorganisation of the system of forest management. It is thought that the new system could be a combination of market mechanisms and regulations applied to forest use, regeneration and protection which could be financed by forest revenue [2]. Indeed, the evidence that forest possess a substantial economic value that should be not forgotten in evaluating the consequences of radioactive polluted forest was mentioned by earlier authors [3]. Therefore, the role of economy appears to be of primary importance for the management of contaminated forests.

It is appearing that natural and forest ecosystems provide a significant complement to the diet of certain group of populations such as forest workers. Moreover, it was reported that the ratio of the fluxes of radionuclides over the fluxes of energy was higher from natural and forest ecosystems than for agricultural ones [4]. As result, official dose records indicated the critical situation of forested areas, with dominating internal exposures in certain cases. Indeed, the behaviour of radionuclides in forest ecosystems exhibits peculiar features which take a large part in the resulting dose distributions and the subsequent problems to be solved as well as the options of forest management that can be proposed. On the basis of forest radioecology, the purpose of the present study is to review the various problems of forest management, to examine a set of various options with respect to their practical application and some elements of cost benefit analysis.

2. Forest radioecology

It is generally agreed that very little caesium is lost from forest ecosystems, and usually more than 90% of the deposited activity remains distributed in the upper organic horizon, especially in boreal forests.[5-8]. The main cause of ^{137}Cs removal from the ecosystem is the radioactive decay. Strontium-90 is known to exhibit a higher mobility. However, its affinity for organic substrate lead it to remain in the upper organic horizon of forest soils. [9]. This was namely observed during the field campaign carried out by the authors in 1994 in the Belarus part of the 30 km zone of Chernobyl in the framework of the ECP/4 project (Figure 2). This long time residence of radionuclides in the soil has a repercussion on the contamination of mushrooms which exhibit the highest transfer factors (Tf, equation 1) of any kind of forest vegetation [7].

$$\text{Equation 1 : Tf} = \frac{\text{Activity in the plant Bq .kg}^{-1}}{\text{Activity in the soil Bq . m}^{-2}}$$

Mycorrhizae groups are the most effective accumulators. Moreover, the most tasty mushrooms belong to this group which also includes the largest number of edible species. Transfer factors of up to 23 and 396×10^{-3} $m^2.kg^{-1}$ have been recorded on sandy and peaty soils respectively. Using a Tf of 20 $m^2.kg^{-1}$, Tikhomirov has calculated a dose of 1,5 $mSv.year^{-1}$ for a deposition level on the soil of 0,5 $MBq.m^{-2}$ and an annual consumption of mushrooms of 10 $kg.year^{-1}$ which is likely to be the usual diet in the CIS [7]. Balonov mentions that mushrooms may contribute to up to 70% of the dietary intake of ^{137}Cs (cited in [7]).

In comparison, berries are seemed to have a minor contribution to the ingestion dose. Doses of 0,02 to 0,03 $mSv.year^{-1}$ have been calculated for the same areas as those where the doses from mushrooms were of 1,5 $mSv.year^{-1}$[7].

A variable part of ^{137}Cs inventory is distributed in the stands, depending mainly of the soil type and the age of trees. In Sweden, for a mixed Spruce and Scots pine forest of 44 years on podzol, up to 14% of the weapon fallout inventory was found in the trees after a 14 year period [3]. In this case a transfer factor of 0,0014 $m^2.kg^{-1}$ was found for wood This value is consistent with Tf already observed in Belarus or likely to be reached within the next five years for similar ecological conditions [10]. The prediction of radionuclide migration in wood stands is of course of paramount importance for planning the production of contaminated wood. Caesium exhibit a peculiar behaviour in conifer wood in that it trends to concentrate in core wood and contaminates the biomass produced during the time that preceded the contamination of the tree. Weapon fallout ^{137}Cs was found to contaminate the wood formed in 1914 in Japan and the United States of America [11, 12]. Strontium in conifer wood remains located around the rings corresponding to the year of pollution and such a behaviour is observed for both Cs and Sr in deciduous trees.

The contamination of games is also an important question but it is beyond the scope of this paper.

3. Problems to be solved

3.1. Monitoring of contaminated wood

The ^{137}Cs contamination of wood depends mainly of the soil properties. Therefore, it is weakly correlated to the deposition level on the soil. As defined to the equation 1, the transfer factor is not a good predictive parameter. Models involving more relevant parameters have been proposed [10], but so far, the map of contamination is still the main basis used by forest enterprises for choosing the areas where to cut the trees. The authors were told during their visit to the forest enterprise of Khoniky in Belarus that wood exceeding permissible levels is found in relatively clean areas when clean wood is found in areas above 1,5 MBq.m^{-2}. Therefore, due to the uneven distribution of ^{137}Cs deposition, it is thought that monitoring the wood contamination using only sampling procedures and laboratory measurements which are currently used, may lead some contaminated wood to escape from the control and clean wood unnecessarily withdrawn from the market.

3.2. Compliance with permissible levels of radionuclides in wood

The wood which does not comply with permissible levels is currently stored in contaminated areas. In Belarus this wood, amounts 2.10^5 m^3.year^{-1} and the amount of standing wood in mature forest in the zone of 0,5-1,5 MBq.m^{-2} (table 1) is estimated at 2 millions of m^3. The more stringent limitation applies for fuel wood with a permissible level below 740 Bq.kg^{-1} of [137]Cs (figure 4). Wood with ^{137}Cs levels of up to 3700 Bq.kg^{-1} may be used for various purposes such as keeping cases as well as house keeping. However, it seems likely that such a wood may at least end in fire after a time of normal use according to permissible levels, being known that the best way to manage wood waste is the incineration. Although the resulting doses from the uncontrolled use of contaminated wood for fire are thought to be negligible, to make coherent a system of contaminated wood management, it seems that a specific destination may be proposed for such contaminated wood.

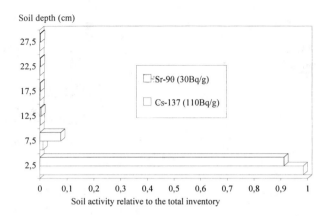

Figure 2 : distribution of radionuclides in a forest soil. Savichy, Belarus, pine forest on podzol, 1994. The absolute numbers in the legend refer to the 5 cm upper layer.

3.3 Doses for workers in forest and industry

The measurements performed during the author's field campaign in Savichy, Belarus, in 1994 gave an average dose rate of about 2,9 µSv.h⁻¹ for a deposition of about 1,1 MBq.m⁻². Assuming that workers would stay 500 h.year⁻¹to this place, this would result in a dose of 1,45 mSv.year⁻¹. This observation is quite consistent with the doses recorded in the Belarus State Radioecological Reserve in 1994 (figure 3). It was noted during the author's visit to the forest enterprise of Khoniky in Belarus in 1995 that trees are mainly felt using chain saws and that tree felling machines are seldom used.

The fluxes of radionuclides in wood pulp industry have been studied in Sweden and some possible dose formation pathways examined [13]. Assuming that similar pulp processing techniques are used in the CIS, the dose received by a CIS worker staying in the conditions described in the Swedish study can be calculated. On the basis of the permissible level for the wood used for pulp i.e. 3800 Bq.kg⁻¹, a dose rate of up to 2,3 µSv.h⁻¹ would be received from a pile of contaminated ashes. It can be noted that this dose rate is similar to this recorded in the forest.

Table 1 : Contaminated forests for restricted activities, i.e. above 0,18 MBq.m⁻²,

Country	Km²	% of forest areas
Belarus	3670	5,45%
Russia	1350	0,02%
Ukraine	1420	1,43%

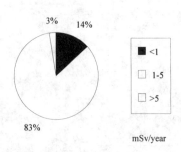

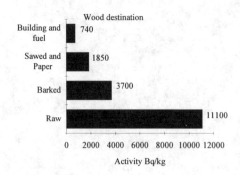

Figure 3 : External doses for forest workers in the Belarus State Radioecological Reserve in 1994 : % of the 400 surveyed workers and doses in mSv.year[-1]

Figure 4 : Permissible levels for wood in Russia

3.4. Economical burden from the lost forest production

The losses of forest production after the Chernobyl accident have been calculated in Belarus [5]. The absolute figures of economical losses are not easy to determine because of the difficult evaluation of currency rates, but it is shown that the losses of non woody forest products are about 16 times as high as this of lost wood products. A significant part of the non woody products are components of the food chain to man, such as mushrooms, or forest pastures. The losses of wood are the consequence of, on one hand, the wood exceeding permissible levels and on the other hand the diminishing forest care because of high doses for workers and subsequent forest production. The losses of mushrooms are quoted to about 1000 tons per year. However this figure may be theoretical since it is observed that contaminated mushrooms are still consumed by the population and lead to an other kind of problem examined in the next paragraph. The same may apply for forest pastures which are known to have a significant contribution to the internal dose through the consumption of contaminated milk. But contrarily to mushrooms, pastures as well as milk can be decontaminated.

3.5. Internal doses after eating contaminated mushrooms and berries

As previously shown in this paper, berries have a minor contribution to the dose and may cause a negligible problem, when mushrooms are among the main pathways of dose distribution. It was reported that for mushroom eating populations, up to 60-80% of the internal dose come from mushrooms [4]. It was proposed [7] to implement a strict interdiction to pick-up contaminated mushrooms, using an accompanying measure which would consist in supplying the population with mushrooms imported from clean areas. But it is seemed that the availability of clean mushrooms may not be guaranteed since mushrooms come from the natural production which cannot be managed on request.

4 Possible options of forest management

4.1. Standing tree monitoring

This option is intended on one hand, to decrease the doses from the use of contaminated wood escaped from control procedures and on the other hand to avoid loosing not contaminated wood which may have been miss classified as contaminated. The evaluation of the dose that can be avoided is difficult to achieve. It is likely that since wood does not inter the food chain, these doses may be very low. However for the need of a coherent policy of contaminated forest management, poor practices of wood control should not exist. A prototype of a counter for standing tree monitoring is currently being developed by the Institute of Nuclear Problems, Belarus State University. But so far, no information is available on the relevance of this option.

4.2. Use of tree felling machines

One tree felling machine can replace 12 loggers cutting the trees by hand using a chain saw machine. For the driver of the tree felling machine, the dose inside the cabin can be evaluated from the measurements performed in the framework of the ECP/4 project. It turned to about 30% of the dose outside the cabin, i.e. the dose received by the loggers. Therefore, using a tree felling machine in contaminated areas instead of 12 chain saw machines may achieve a collective dose reduction factor of about 40. The price of a tree felling machine which can be purchased in Belarus is of 300 kECU. The price of such a machine purchased in France would be of 400 kECU which is a similar figure. Using only the annual investment cost based on the Belarus price with an amortisation over a 5 year period and assuming for 12 loggers an averted individual dose of 1,5 mSv.year^{-1} which is the most frequent dose record in 1994 (figure 3), the cost of the averted man sievert roughly evaluated would be of 3,4 MECU. It may be assumed that using a machine may increase the productivity of felling operations and therefore provide an additional benefit.

4.3. Grow saprophyte mushrooms on clean substrate

The agricultural production of mushroom is currently performed and developed, at least, in western Europe. A professional organisation : the European Mushroom Growers' Group is representing the producers of the European Union. So far, the practical application of mushroom culture is mainly concerning saprophyte mushrooms, i.e. mushrooms growing on compost [14-17].

Saprophytes do not constitute a large part of the mushrooms which are naturally growing in Chernobyl forests. Nevertheless, the controlled culture of symbiotic mushrooms is also being investigated, but would suppose to be implemented in not contaminated forests, when saprophyte mushrooms could be grown on artificial substrate inside contaminated areas, therefore avoiding transport constraints from the production area to contaminated ones. Provided that the populations would adapt their habit to the consumption of new varieties of mushrooms, it may be imagined that off soil mushroom cultures could be organised in forest enterprises in contaminated areas. The mushroom culture necessitates indoors areas which may be available in contaminated areas from the buildings which have been abandoned after the Chernobyl accident. One of the requirements of this culture, which absence sometimes compromises the production under temperate climates, is a sufficient period of cold. This may not be a limiting factor in the CIS.

The cost benefit analysis based on the available economical data shows that for a production cost of mushrooms in France of 3,2 ECU.kg^{-1} and an equivalent of 0,15 averted Man.mSv.y^{-1}.kg^{-1} of mushroom [7], the cost of the averted man sievert would be of 21 kECU. However, it must be stressed that French costs do not satisfy CIS economical comparison. It is likely that this cost could be divided by 5 to 10 to be realistic, the manpower being a significant part of the total cost. Moreover, it should be noted that an additional benefit may be obtained through the economical valorisation of the production of mushrooms, as far as the technique of production would be introduced in the area.

4.4. Valorisation of contaminated wood

This countermeasure is expected to help for mitigating the problem of compliance with permissible levels of ^{137}Cs in wood as well as the problem of the economical burden related to the loss of contaminated wood. So far three options are likely to suit the use of contaminated wood : to burn it in incinerators equipped of a smoke filtration and a contaminated ash removal system, to convert it into paper pulp or into chemicals such as technical alcohol. The incineration may produce energy but the requirements for an economically competitive energy is to care of the availability of fuel as well as the quantity and the form under which the energy can be used. Pulp factories satisfy these requirements in that about 30% of the wood which comes in the plant goes to waste and the need for steam to recycle the chemical reagents is very important.

The decontamination factor (Df) of wood transformed into pulp can be defined by the equation 2

$$\text{Equation 2 : Df} = \frac{\text{Activity of raw wood}}{\text{Activity of the wood derivative}}$$

The Df observed at the industrial scale in Sweden was of about 3 and 58-125 for the acidic process and the alkaline one respectively [13]. These records show that the acidic process is less efficient in decontaminating the wood than the alkaline one. However, is was observed in laboratory experiments performed in the framework of the ECP/4 project that the acidic treatment may be able to achieve a Df of up to 90 [18]. This infers that the pH of the treatment may not be the cause of the different efficiencies of the two processes, but the technology itself.

From the above technical description, it is appearing that pulp factories are pivotal structures for the development of a strategy based on the valorisation of wood capable of providing an economical benefit which may help in the implementation of countermeasures for contaminated forests. But the justification of this countermeasure on the sole basis of a probable benefit in terms of dose reduction, by avoiding the discharge in the environment of radionuclides from wood waste, would necessitate an exhaustive study which results may be affected of large uncertainties, and may not lead to such an evident issue than the need for improving the economy of Chernobyl affected areas.

The production of technical alcohol from contaminated wood may also be one of the possible options to valorise the production of contaminated forests. So far this way has not been thoroughly investigated with respect to contaminated forests. The possibility of producing ethanol from wood to substitute gasoline has been investigated namely in France during the years of petrol crisis. The way involving the enzymatic digestion of cellulose to liberate simple sugars was proven to be more efficient than the chemical one. However, the ethanol was produced at a higher cost than petrol gasoline. But in the peculiar context of energy dependence combined with the need for contaminated forest management, which is at least the case of Belarus, such option could be re-examined.

4.5. Removal of contaminated litter

This option has been studied in the framework of the ECP/4 project on the basis of a field experiment and by other authors as the application of a model of contaminated forest management [19]. The latter study concluded that about ten years after the deposition of radionuclides, the efficiency of removing the litter is questionable. Indeed the direct observation during the field experiment carried out in 1994 in Belarus showed that a significant part of ^{137}Cs is associated with the fine fraction of organic matter still incorporated into the mineral fraction of the soil. It is therefore necessary to remove a part of the mineral soil to achieve a significant decontamination factor. The rotating brush which was tested proved to be of promising issue, but some adaptation are necessary to the existing equipment, namely the independent depth setting of the rotary brush and its rotation speed.

It is unrealistic to think that the litter can be removed on the thousand of km^2 of contaminated forest. However, in some cases such as clear felling operations or the restoration of some high value recreational forest these operations may be undertaken on limited areas. The evaluation of the benefit in terms of dose reduction is difficult to calculate since it is strongly dependant of the practical case of application. One example is given in the case studies performed within the ECP/4 project : the decontamination of the forested area of Karchovka in Russia, close to the city of Novozybkov, Briansk province. This recreational area was very often visited by people before the Chernobyl accident. Using a quite low dose reduction factor of 2 when removing the litter using a rotating brush, the calculated averted dose was of 0,1534 Man.sievert and the subsequent cost of the averted Man sievert was of 3316 ECU. The transport of the 230 m^3 of generated waste was not included in the cost.
Other examples of application may be given, but cannot be presented without an exhaustive sustaining discussion which is beyond the scope of the present paper.

4.6. Reforestation of agricultural land.

This option may have some connection to the option of litter removal in forest which purpose may be to transform forest areas in agricultural land after decontamination, especially for areas where sufficient clean pastures are not available. On the opposite, agricultural land which are not any more suitable for agriculture may be reforested. A technique for planting trees in contaminated areas has been developed by the Belarus Institute of Forestry. The interest of this option is to both increase the productivity of an operation of tree planting, relatively to hand made planting, and to decrease the dose for workers. However, although the reforestation option being supposedly applied in heavily contaminated areas, i.e. above 40 Ci.km^{-2}, the dosimetric cost is likely to be negligible. The benefit of this option cannot be expressed in terms of dose reduction, but only in terms of economical benefit.

4.7. Decontamination of forest pastures

This option has been investigated in the framework of the ECP/4 project for decontaminating permanent pastures. As mentioned in the above chapter on the problems to be solved, it is likely that the losses of forest pastures represent a large part of the total economical burden. On the other hand, leaving private cows to graze on forest pastures is reported to have a significant contribution to the dose [4]. The cost benefit analysis of the use of a turf harvester for decontaminating permanent pastures may be applied to forest pasture, using only the existing small turf harvesters which are capable of operating in complicated conditions of soil topography [20]. The cost of the averted man sievert was of about 80 ECU.

Table 2 : options for the management of contaminated forests

Option	Aim	Cost of the averted man sievert (kECU)	Economical benefit
Standing tree monitoring	Avoid doses, decrease wood losses	nd	mainly
Tree felling machines	Avoid doses for loggers	3400	yes
Mushroom culture	Avoid internal doses	21✦	yes
Valorisation of contaminated wood	Mitigate economical burden	nd	mainly
Litter removal	Avoid external doses	3,32	yes
Reforestation	Use of contaminated land	nd	mainly
Decontamination of pastures	Avoid internal doses	0,08	yes

✦ = Based on west European costs.

5. Conclusion

The figures that have been calculated to evaluate the various options are not intended to do more than to give an idea of the range of cost effectiveness. Indeed, it appears that these cost effectiveness are strongly dependant of the practical case to which the countermeasure is foreseen to be applied. For example, the most expensive of the proposed options (table 2) which is the use of a tree felling machine could probably exhibit a lower cost of the averted Man.sievert if used in very heavily contaminated areas, e.g. at a level of up to 37 MBq.m^{-2} where care felling may be necessary. But one of the major difficulty to carry-out studies on decontamination strategies was to obtain data on case studies, especially with respect to dose distributions. It can be noted that among the various options (table 2), the most cost effective countermeasures is aiming at decreasing the internal dose, but on the other hand, it appeared that data about internal dose was very difficult to collect. Therefore one of the first recommendation would be to organise a systematic monitoring of whole body content of ^{137}Cs of people living in affected areas, and to provide the local health units with a number of whole body counters.

One other important issue is the number of people concerned by the dose intended to be decreased. The contamination of forest lead to the exposure of critical groups of populations such as mushroom pickers and forestry workers, but it is likely that the use of contaminated wood lead to a collective dose for the users of this wood. Moreover, forest industry such as pulp factories may lead to the discharge in the environment of radionuclides which may lead to increase the doses for the concerned populations. Whether these doses have a risk signification is questionable, especially if these doses are added to the Chernobyl background. To answer this question, further radioecological studies are necessary.

Finally it must be stressed that few options are available to decrease the doses, but most of the options are likely to have a positive impact on the economy. In terms of compensation, it seems that all actions leading to revive the local economy and namely forestry activities are likely to be interpreted as remedial actions by the population who is facing in his every day life the decreasing forest economical activities.

6. References

1   Rudy, C., "Accident-generated contamination and remediation technologies in a country with relatively limited ressources," Ministery for environmental protection and nuclear safety, Kiev, Ukraine, 1995.

2   Korovin, G., "Problems of forest management in Russia," *Water Air and Soil Pollution* 82.13-23 (1995): 13.

3   Melin, J., L. Wallberg, and J. Suomela, "Distribution and retention of cesium and strontium in swedish boreal forest ecosystems," *The Science of the Total Environment* 157 (1994): 93.

4   Anonymous, "Fluxes of radionuclides in rural communities in Russia, Ukraine and Belarus : transfer of radionuclides, their comparative importance under different agricultural Ecosystems and appropriate countermeasures," Norvegian Radiation Protection Authority, 1994.

5   Ipatyev, V., ed., Forest and Chernobyl, (Forest Institute of the Belarussian Academy of Sciences, 1994) 1: 252.

6   Bergman, Ronny, *et al.*, "Cesium 137 in boreal forest ecosystem," FOA, 1993.

7   Tikhomirov, F.A., "Optimal management routes for different types of forest contaminated during and after the Chernobyl-4 reactor accident," Laboratory of radioecology, soil Science faculty, Moscow State University, 1995.

8   Strandberg, Morten, "Radiocesium in a Danish pine forest ecosystem," *The Science of the Total Environment* 157 (1994): 125.

9   Tikhomirov, F.A., and A.I Shcheglov,   "The radioecological consequences of the Kyshtym and Chernobyl radiation accidents for forest ecosystems,"   (Luxembourg: Commission of the European Communities, 1990), 867.

10   Dvornik, A., T. Zhuchenko, and V. Ipatyev, "Forestlife is predictive model of radiation contamination of wood," IAEA, 1995.

11   Momoshima, N., and E.A. Bondietti, "The radial distribution of ^{90}Sr and ^{137}Cs in trees," *J. Environ.Radioactivity* 22.2 (1994): 93.

12   Kohno, Masuchika, *et al.*, "Distribution of environmental cesium-137 in tree rings," *J. Environ. Radioactivity* 8 (1988): 15.

13   Ravila, Aaro, and Elis Holm, "Flux and concentration processes of radioactive elements in the forest industry dosimetry, biofueled heating plants, the alkaline and the acidic pulp mill processes," Radiation physics department, Lund University, Sweden, 1992.

14   Guimberteau, J., J.M. Olivier, and M.R. Bordaberry, "Données récentes sur la culture des "pieds bleus" (Leoista sp.)," *P.H.M. Revue Horticole* .298 (1989): 17.

15   Poitou, N., *et al.*, "Mycorhization contrôlée et culture expérimentale au champ de Boletus (=Suillus) granulatus et Lactarius deliciosus," (Braunschweig, Germany: 1987),

16   Delpech, P., and J.M Olivier, "Champignon parfumé ou shii-take, une méthode française pour sa culture," *P.H.M., Revue Horticole* .305 (1990): 25.

17   Anonyme, "Dossier Pleurote," INRA Institut National de la Recherche Agronomique, 1995.

18   Anonymous,   "CEU/CIS agreement on the consequences of the Chernobyl accident; ECP/4, decontamination strategies, report on the work done in 1993.," European Commission, 1994.

19   Schell, W.R., and I. Linkov, "Radiologically-contaminated forests: a modelling approach to safety evaluation and management," (Washington, DC, USA: 1995),

20   Jouve, A.,   "Technical and economical study of the use of a turf harvester for decontaminating permanent pastures in the Chernobyl affected areas," Institut de Protection et de Sûreté nucléaire, IPSN/CDSN, BP6,92265, Fontenay aux Roses, France, 1995.

Water Quality Management of Contaminated Areas and its Effects on Doses from Aquatic Pathways

Oleg VOITSEKHOVITCH
Ukrainian Hydrometeorological Institute, Nauka Avenue, 37 - 252028 Kiev, Ukraine
Umberto SANSONE
National Environmental Protection Agency, Via V. Brancati, 48 - 00144 Roma, Italy
Mark ZHELESNYAK
Inst. of Cybernetics, Ukrainian Acad. of Sciences, Glushkova, 42 - 252187 Kiev, Ukraine
Dimitri BUGAI
Inst. of Geology, Ukrainian Acad. of Sciences, Tchkalova, 55b - 252054 Kiev, Ukraine

Abstract. A critical analysis of remedial actions performed in the Chernobyl close zone are presented in term of effectiveness to dose reduction and money expenditure. The Chernobyl experience proved the need to consolidate the international water protection capacity on the basis of scientific knowledge which should exclude inefficient use of national resource. Strategical and technological interventions on water quality management need to be revised on the base of the experience gained after the 1986 accident.

The lesson learned from the Chernobyl experience has to be used as a key element in the adoption of a strategy of water bodies management. Remedial actions have to be based with an integrated approach considering:

- dose reduction;
- secondary environmental effects of countermeasures;
- synergisms of radionuclides and countermeasure applications with other toxicants;
- social and economical factors of the contaminated areas.

1. Introduction

Surface and ground water resources of Ukraine, Belarus and the Russian Federation represent a vital problem for nearly 150 million population of these regions. After the Chernobyl accident, 8.5 million people living in the Dnieper basin became recipients of radionuclides through direct consumption of water from the Dnieper river-reservoir system and more than 30 million through irrigation and fishery.

Most of the radioactive fallout originated from the Chernobyl accident was deposited within the Dnieper River drainage basin that lies adjacent to the Chernobyl nuclear power plant site. This territory forms an extensive area from which contaminated run-off flows downstream through the Pripyat and Dnieper River systems across the Ukraine to the Black Sea (Figure. 1).

The analysis of water remedial actions carried out to minimize and mitigate water bodies, can provide an unique opportunity for decision makers who are working in other extensively contaminated regions, to optimize their approaches to surface and groundwater protection. Most engineering measures inside the Chernobyl 30-km exclusion zone were focused on prevention of secondary contaminations of the Pripyat river and the Kiev reservoir.

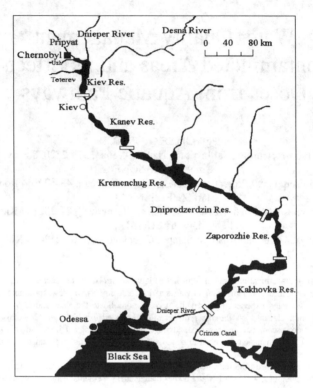

Fig. 1. The Dnieper Cascade

This paper describes nine years of mitigative measures applied on contaminated water bodies surrounding the 30 km exclusion zone. The activities performed demonstrated that effectiveness of mitigative measures depends, not only on proper application of technology, but also on selection of projects offering significant risk reduction potential. In a limited national economy, environmental mitigative projects must maximize risk reduction and cost effectiveness avoiding the risk of ineffective use of national resources.

2. Radioactive Contamination of Water Bodies Located in the Chernobyl Evacuated Zone and of the Dnieper cascade

Most of the radioactive material released during the accident was deposited on the watershed areas of the Dnieper river (as well as the Pripyat catchment), heavy polluting large territories in the vicinity the Chernobyl nuclear power plant [1] [2] [3]. At present, inside the Chernobyl close-in zone, a large amount of radioactive materials is concentrated. Part of this material is constituted by soils of the catchment areas or, by water and bottom sediments from the lakes and the Chernobyl cooling pond, as result of the contamination of 1986. More than $3,7 \times 10^{14}$ Bq of ^{90}Sr and about 7.4×10^{12} Bq of Pu are deposited in the flood plain areas that can be flooded during each spring and in the soils of polder areas in 30 km zone, that are annually inundated during snowmelt and rain-fall periods. Huge amount of radioactive wastes are located in shallow underground waste disposal sites, contacting ground water flows that move in direction towards the Pripyat river and to other directions [4] [5].

The watershed areas of the Pripyat and Dnieper rivers constitute the main potential secondary sources of ^{137}Cs and ^{90}Sr to the Dnieper cascade and to the Black sea [6]. Monitoring data indicate that ^{137}Cs and ^{90}Sr have different behaviour in the Dnieper system. Since 1986 the amount of radiocaesium transported within the system decreased continuously. On the contrary no significant decreasing has been observed for ^{90}Sr and its migration in hydrologic systems is considered to be a major long-term problem [7]. The different behaviour of these radionuclides is attributable to the fact that radiocaesium in soils is strongly sorbed by clay minerals, while strontium is in highly mobile chemical forms.

A second source of radioactive contamination of the Dnieper cascade is due to the groundwater contamination resulting from the wastes buried in shallow sands and trenches in the evacuated zone around Chernobyl. At present the concentration of ^{90}Sr in most of the superficial groundwater tables located in the 30 km zone, exceeds the permissible level (3.7 Bq l^{-1}). More serious situation has been observed in the shallow groundwater tables near the temporary waste disposal sites (at a distance of 5 km from the reactor), where the highly contaminated pine trees of the "Red Forest" were buried in 1987. In this particular case, where groundwater is contacting radioactive wastes, the 1994 monitoring data indicated, mean values of ^{90}Sr concentrations of 100 Bq l^{-1} and in some cases higher than 1000 Bq l^{-1}. Despite these levels of contamination, radionuclide transport to river through the groundwater pathways, contribute only marginally to the off-site radiological risk. In fact, groundwater flow is very slow and its contribution to river contamination is expected to occur not before 10-15 years.

In addition, also the infiltration into the Pripyat river, of contaminated water from the cooling pond of Chernobyl nuclear plant, has as a consequences the increase of river water contamination during low water seasons.

These are the sources of radioactivity that contribute to the contamination of the Dnieper cascade. Since 1986 a radiological monitoring system has been setup by the Ukrainian Hydrometeorological Institute along the Dnieper cascade, to assess the long-term impact of the Chernobyl accident on this system. Samples of water were regularly collected at several transects in the Pripyat river (Chernobyl cross-section), in the Uzh river (before its confluence into the Pripyat river), in the Dnieper river (before its input in the Kiev reservoir) and at the output of the six large artificial reservoirs (Kiev, Kanev, Krementchug, Dneprodzerdzinsk, Zaporozhye, Kakhov) [8].

The highest level of radioactivity in the water of the Pripyat river was observed during the initial period after the accident. The radionuclide contents in water exceeded in some cases the permissible levels. The data show from 1987 to 1994 a general decreasing of the radionuclide content in the river. As the chronological data indicate, ^{137}Cs values (dissolved and particulate phases) are lower than those for ^{90}Sr and usually the deviation of radionuclides from the linear trend is strictly corresponding to water discharge. A close relationship was found between ^{90}Sr and river water discharge (correlation, $r^2=51\%$). Peaks of ^{90}Sr correspond to spring flooding in the areas located within 5-10 km around Chernobyl [8].

Also the data of the radionuclides inflowing into the Kiev reservoir (Fig. 2) suggest a close relationship between ^{90}Sr values and water discharge, while ^{137}Cs values are less dependent on surface hydrology. The increase of ^{90}Sr content in water is evident during the main spring floods, generated by snow melting [8].

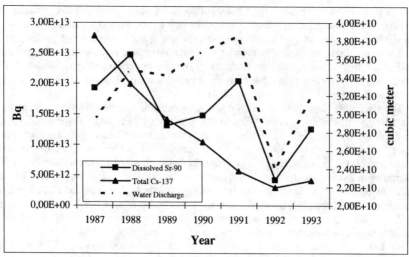

Fig. 2.   Water discharge volume (m³), total ¹³⁷Cs and dissolved form of ⁹⁰Sr (Bq) inflowing into
the Kiev reservoir (1987-1993)

Most of the ¹³⁷Cs originated in the watershed of Pripyat and Dnieper rivers is accumulated in
bottom sediments of the Kiev reservoir. The Kiev reservoir being characterized by low water
velocities and high sedimentation rate has been revealed to operate as trap for ¹³⁷Cs transported
in the particulate phase [9].

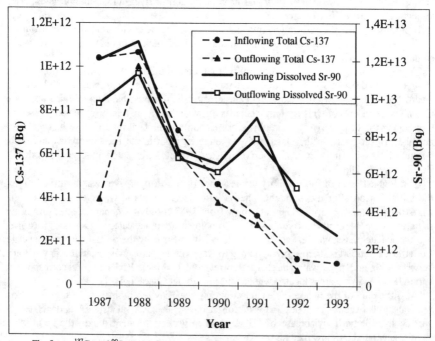

Fig. 3.   ¹³⁷Cs and ⁹⁰Sr (Bq) inflowing and outflowing from the Kakhov reservoir (1987-1993)

Moving downstream to the series of reservoirs along the Dnieper cascade system, the balance of ^{137}Cs and ^{90}Sr show a similar trend than that found in the Kiev reservoir. The data of radionuclides outflow from the last reservoir of the Dnieper cascade (Kakhov reservoir) reported on Figure 3, show that ^{137}Cs values are an order of magnitude lower than those for ^{90}Sr. Two peaks are evident in the ^{90}Sr values, corresponding to 1988 and 1991 spring floods in Pripyat river watersheds. The total ^{137}Cs outflowing from the last reservoir in the period 1987-1993 (2.73 TBq) represents only 3% of the ^{137}Cs that entered the Dnieper cascade (85 TBq) in the same period. In contrast, the dissolved form of ^{90}Sr that reached the Black sea (46.8 TBq) from 1987 to 1993 represents the 43% of the amount of the radionuclide that entered the Dnieper river-reservoirs system (109 Tbq) [8].

These data confirm a long-term input of ^{137}Cs and ^{90}Sr, from the northern heavily contaminated areas, into the Dnieper and Pripyat rivers. From these rivers, most of the ^{137}Cs is distributed along the six artificial reservoirs composing the Dnieper cascade system, while the dissolved form of ^{90}Sr reach more easily the Black sea.

In conclusion, the radioactive contamination of the Dnieper cascade is mainly caused by direct surface water interactions with contaminated floodplain and catchment soils. At present the potential contribution of groundwater sources are expected to account for about 2 to 5% of the total release to the Pripyat river [7]. The present radiological situation of the Dnieper cascade, is the result of the combined effect of natural secondary contamination processes and the effects of the remedial actions performed in the contaminated areas since 1986.

3. Analysis of Remedial Actions Performed along the Dnieper Cascade

Since the Chernobyl accident, engineering and administrative countermeasures have been directed to protect from secondary radioactive contaminations, the main aquatic system of Ukraine (Dnieper river-reservoirs system), in order to reduce radioactivity dose for population living in Dineper basin, downstream the Chernobyl area.

The water protective actions were carried out in the following periods:
- Emergency phase (2-3 months after the accident);
- Early intermediate phase (from summer 1986 to 1988);
- Later intermediate phase (from 1989 to 1991).

3.1. Emergency phase (2-3 months after the accident)

Countermeasures during this period were mainly based on administrative decisions. The main actions are following reported:
- attempts to regulate the flows of contaminated water through the Kiev reservoir dam;
- increasing of the use of ground water sources for municipalities purpose (avoiding where possible the use of contaminated surface water);
- construction of supplementary ground water wells;
- purification of drinking water in municipal water treatment plants.

Most of these actions have been adopted without any cost-benefit analyses and were mainly influenced by social factors as emotional situation and pressure of the population living on affected areas. In this way a lot of national resources were spent, but in term of effectiveness to dose reduction and money expenditure, the results were extremely limited.

Some of the remedial actions were also applied without an adequate interpretation of the monitoring data. In a conflictual situation between scientists and decision makers, the first assessment of the adsorption-desorption kinetic parameters for radionuclide liquid-solid interactions was erroneously determined and consequently, the wash-off coefficients were overestimated. As a result many useless water protective actions were adopted in the first months after the accident. As example, the discharge into the rivers flowing downstrean Chernobyl of unprocessed zeolite material.

On early May 1986, several actions were also done at the Kiev reservoir dam. The surface gates were opened and closed the bottom ones. Unfortunately, these operations were adopted on the base of a lack of hydrologic knowledges, considering that the most clean water were on the surface of the reservoir. But, during the first weeks after the accident, the deeper water of the reservoir were less contaminated and the actions at the Kiev reservoir dam have had as a consequence an increase of the dose to population living downstrean to the reservoir, in the period in which the main exposure took place from drinking water intake.

3.2. Early intermediate phase (from summer 1986 to 1988)

During the 1986 summer, protective dikes of several km were constructed along the Pripyat river with the purpose to trap the run-off of contaminated material from the towns of Chernobyl and Pripyat. These actions were not effective to controll the run-off from the broad landscapes.

In the same period (spring-summer) several canal bed traps were dredged in the bottom bed of the Pripyat river, to increase the river's cross-section and to reduce water velocity. The expected result was to favorize the sedimentation of the contaminated suspended material in the bulk flow according to the Stock's law. Subsequent investigations [10] [11] showed that these traps were ineffective, because the size of these suspended particles were too fine to be settled in a river as Pripyat, characterized by huge water discharges and turbolent flow conditions.

The cooling pond of Chernobyl nuclear power plant was also interested in the remedial actions performed in the early intermediate phase. A special drainage well was built around the damaged reactor, to isolate the cooling pond from the Pripyat river, with the main purpose to trap the infiltrating radioactive water. Up to now, this drainage system is not operating because many doubts arised as a consequence of its operation. Recent researches [6] indicate that pumping water from these wells to the cooling pond could cause problems in the balance of water and dissolved salts in the pond. A lot of national resources were spent for the realization of this project.

During this phase, an underground clay barrier was constructed between the Chernobyl reactor and the Pripyat River. This barrier was built to prevent the migration of contaminated ground water into the river. Additional studies showed that the migration of radionuclides through the ground flows was very slow and consequently the project was stopped.

From 1986 to early 1987, more than 100 zeolite-containing dikes were completed, aiming to adsorb radionuclides from smaller rivers and streams. Subsequent studies indicated that only 5% to 10% of ^{90}Sr and ^{137}Cs transported by the rivers were adsorbed by the zeolite barriers within the dams. In addition, the streams that were dammed during this activity contributed only with a few percent to the total flow of the Pripyat and Dnieper drainage basins. In 1987, the construction of new dams were stopped and it was decided to destroy most of the existing.

In 1987, after the evacuation of the local population and the core of the damaged reactor was extinguished, the burial of contaminated soils, vegetation, debris etc, was used as mitigative measure. This was the case of the highly contaminated pine trees of the "Red Forest" (killed by the high levels of radiation), located at a distance of about 5 km from the reactor. These measures were deemed necessary to protect the "liquidators" and people working inside the exclusion zone, from high radiation doses. To this end all wastes were buried in shallow sands and trenches without any action to prevent the contamination of ground water. As a result, intense local ground-water contamination from these buried trees poses a large, long-term problem for environmental remediation and restoration.

3.3. Later intermediate phase (from 1989 to 1991)

A new phase of hydrologic remediation started after the summer flood of 1988. High water levels covered most of the contaminated flood plain and caused a secondary contamination of ^{90}Sr of the river systems. Modeling of surface hydrology indicated as a realistic, dangerous or "worst-case" scenario, that would cause the highest radionuclide concentration in rivers, a spring flood with a maximum discharge of 2000 m^3 s^{-1} - a 25-percentile probabilistic flood. The model simulation of this "possible worst hydrological scenario", in which the Chernobyl exclusion zone would be significantly flooded, indicated a ^{90}Sr contamination of the water near Kiev and in the last reservoir of the Dnieper cascade (Kakhov reservoir), several times higher than in normal condition. But when in the computed results the washing-out processes were simulated with isolated radioactive sources on the flood plain, the ^{90}Sr concentration decreased in 2-4 times. Several approaches have been proposed and the potential effectiveness of each of them, to reduce radioactive radionuclide concentrations in the rivers, were simulated. Creation of dikes around the contaminated areas on the left (east) bank of the river was chosen as the best protective options. These measures, with additional decontamination of soils of the right bank and with a solution for the infiltration of cooling-pond water could significantly reduce the ^{90}Sr concentration at the downstream boundary. The flooding events of January 1991 and summer 1993 confirmed definitively the simulated results [6] and consequently in 1992 the dikes were built. With this option, during the flooding of summer 1994, more than 3.7 x 10^{12} Bq of ^{90}Sr were prevented to be washed-out from the flood plain of the Pripyat to the Dnieper cascade. This event proved the need to consolidate the water protection strategy on the base of scientific knowledge which should exclude inefficient use of national resources. Engineering works were completed in 1993 in the Chernobyl close-in zone and at present time developement of future technological and strategical capacities on water protection are in progress.

4. Radiation Risk Assessment from Water Usage

The present level of radionuclides content in the Dnieper cascade water does not pose health risks. Analysis of ^{90}Sr migration via groundwater to surface water and down-river population centers shows that, radionuclide transport via the groundwater pathway has potential to contribute only marginally to the off-site radiological risk, which is governed by wash-out of radionuclides from the contaminated river flood plain and catchment areas by surface water during snow melt and rains. Groundwater contributes to rivers contamination for about 2 to 5% of the total release from terrestrial environment [7]. It has been estimated that also after 20 years from the accident, the groundwater will not significantly contribute to surface water contamination.

The technological possibilities to control the existing non point sources of radioactive contamination on a such large catchment scale as the Chernobyl area, are very limited. The optimization of a future strategy of water bodies management can be done only comparing human doses, that could be economized as result of engineering water protection and water remediation actions in the Chernobyl site. To optimize and substantiate a new concept of water protection, a huge research work has been done in term of risk assessment and cost-benefit analyses. The radiological monitoring data and the predicted radionuclide contents in the Dnieper's water up to 2056 year were used for the estimation of the committed collective internal doses from ^{90}Sr and ^{137}Cs to the population of the Dnieper's regions. The structure of committed dose (70 years intake), due to water usage, averaged for all water consumers living along the Dnieper system, is distributed respectively as 35, 40 and 25 % from drinking water, fish and irrigated products. This dose structure is different for the different regions of Ukraine. The studies also showed that annual averaged individual effective internal dose for different regions of Ukraine is very varied. In 1993 for instance, the water pathway contribution to the total radiation dose for the Kiev citizen was about 6% and ranged between 20 and 30% for people living in the southern region of Ukraine [5] [12].

Using the nominal probability coefficient 7.3×10^{-2} 1/Sv (ICRP-60), the number of stochastic cancer effects due to Dnieper water usage during 70 years were estimated. The results indicate about 200 cancer cases on 30 million people (70 years exposure) and about 60 persons during 1986-1992. Recalculation of dose values for all the exposed population, demonstrates that the averaged individual human radiation risk from Dnieper water use cannot be more than 1×10^{-5}. For some critical groups of water users, expected individual risk would be 4-5 and even more times higher. The implementation of the most effective water protective countermeasure can reduce 3-4 times the above radiation risk from water usage.

These results indicate a very low level of radiation risk due to water usage, if compared with others sources of radiation. It is necessary to point out that the stress component for radiation risk of population consuming radioactive contaminated water, is more dominant than the dose component effect. At present, the contribution to the radioactive dose for inhabitants of Kiev (via the water pathway) is likely to be small (a few percent) although it does increase to 10-20% for populations lower down the Dnieper river-reservoir system, where no other water sources are available and much more radioactivity from water enters the diet, as irrigation of food crops is extensive. However, as a recent study has shown, the popular conception within the population of Kiev is that radioactivity in water is equally important as food as far as the dose to man is concerned.

The future strategical and technological interventions on water quality management of the Dnieper cascade need to take into account also the synergisms of radionuclides and countermeasure applications with other toxicants. In fact, the Dnieper reservoirs are situated in industrial and agricultural areas, badly affected by great number of various industrial entreprises with backward technologies including chemical industry, metallurgy, energetics and misuse of pesticides in agriculture. Toxicological investigations showed the presence in water of other toxic substances with carcinogenic and mutagenic properties. In many cases, the origin of these substances is not defined and no information are available about their environmental behaviour. This aspect is a relevant factor to take into account when assessing risk for population on contaminated areas.

5. Conclusions

In spite of low level of exposure for population of Ukraine due to aquatic pathways, the radionuclides transfer by the water flow (mainly from sources situated in the Chernobyl zone) is the most relevant factor of the contamination of the Dnieper aquatic system. In the case of Chernobyl, where the sources of water radioactive contamination are determined, the realisation of a set of priority of water protective actions are recommended. However any countermeasures cannot be implemented without real benefits according with "ALARA" principles. On this basis the Ukrainian Authorities developed from 1993 a scheme of priority chain for water remedial actions. The future devolpment of countermeasures must assure that the ^{90}Sr concentration in the water of Pripyat river, downstream the boundary of the evacuated zone, not exceed the value of 2 Bq l^{-1}. On this basis, the following list of remedial actions has been suggested:

- development of a radioactive monitoring network for ground and surface water, inside and outside the Chernobyl evacuated zone;
- completion of diking on the highest contaminated flood plains, in the left and right side of Chernobyl plant. This action will keep back water and prevent flooding and it is estimated that it will possible to reduce to about 4-5 times the present annual wash-out of radionuclides;
- development of a project for the clean-up of the cooling pond, after the shutdown of the Chernobyl reactors;
- water regulation on the highly contaminated wetlands located in the Chernobyl site. This action has been suggested in agreement with Belarus to keep flooded peat bog areas against fire risk;
- construction of engineering and geochemical barriers around the waste disposal sites located in the Chernobyl zone.

The above mentioned countermeasures will allow to reduce the potential risk from water usage and they will also have a positive effects on the unnecessarily high levels of stress amongst the population living downstream Chernobyl and along the Dnieper cascade. The implementation of these remedial actions will require only several percent of the current financial budget of the Minchernobyl of Ukraine, directed to the liquidation of the consequences of the Chernobyl accident.

References

[1] Vakulovsky S.M., Voitsekhovitch O.V.et.al., Radioactive Contamination of Water Bodies in the Area Affected by Releases from the Chernobyl Nuclear Power Plant Accident. Proc. Environmental Contamination following a Major Nuclear Accident. IAEA, 1990, pp. 231-246.

[2] Vakulovsky S.M., Nikitin A.I and Voitsekhovitch O.V., ^{137}Cs and ^{90}Sr contamination of water bodies in areas affected by releases fron Chernobyl nuclear power plant accident: an overview. Jour. Env. Radioactivity, 23, 1994, pp. 103-122.

[3] Voitsekhovich, O.V., Borzilov V.A.,Konoplyov, A.V., Hydrological Aspects of Radionuclide Migration in Water Bodies Following the Chernobyl Accident. Proceedings of Seminar on Comparative Assessment of The Environmental Impact of Radionuclides Released During Three Major Nuclear Accidents: Kyshtym, Windscale, Chernobyl. Luxembourg, 1-5 October 1990, Vol.2. Commission Of The European Communities, Radiation Protection-53, EUR 13574, 1991, pp.528-548.

[4] Voitsekhovitch O.V., Kanivets V.V., Laptev G.V., Bilyi I.Ya., Hydrological Processes and their Influence on Radionuclide Behaviour and Transport by Surface Water Pathways as Applied to Water Protection after Chernobyl Accident. Proc.UNESCO, Hydrological Impact of NPP. 1992.

[5] Voitsekhovitch 0.V., Nasvit O., Los Y., Present Concept of Water Remedial Activities on the Areas Contaminated by the Chernobyl Accident. Proc. of EC Seminar on Freshwater and Estuarine Radioecology, Lisbon, 1994.

[6] Voitsekhovitch 0.V., Zheleznyak M.I., Onishi Y., Chernobyl Nuclear Accident. Hydrologic Analysis and Emergency Evaluation of Radionuclide Distributions in the Dnieper River, Ukraine, During the 1993 Summer Flood. Rep.Doc. PNL-9980, Contract DOE,USA, Battelle, June 1994, Washington. p. 96

[7] Waters, R., Gibson D., Bugay B., Dzhepo S., Skalsky A., Voitsekhovitch O., A Review of Post-Accident Measures Affecting Transport and Isolation of Radionuclides Released from the Chernobyl Accident. In: Proc. of International Symposium on Environmental Contamination in the Central-Eastern Europe, Budapest-94, Budapest, Hungary, September, 19-24, 1994.

[8] Sansone U., Belli M., Voitsekhovitch O.V. and Kanivets V.V., ^{137}Cs and ^{90}Sr in Water and Suspended Particulate Matter of the Dnieper River-Reservoirs System (Ukraine). Submitted for pubblication, 1995.

[9] Sansone U., Riccardi M., Voitsekhovitch O.V., Kanivets V.V., Osadchiy V., Osadchiy N. and Madruga M.J. Final Report 1991-1995 ECP-3 Project, VII.2.1.4. The Role of Suspended Matter in Radionuclide Transport in Rivers and Reservoirs, 1996.

[10] Voitsekhovitch O.V., Kanivets V.V., Shereshevky A.I., The Effectiveness of Bottom Sediment Traps Created to Cought Contaminated Suspended Materials. Proc. Ukrainian Hydromet Institute, USSR, 1988, Vol. 228, pp. 60-68.

[11] Zheleznyak M.I., Voitsekhovitch O.V. et al., Simulation of Effectiveness of Countermeasures Designed to Decrease Radionuclide Transport Rate in the Pripyat-Dnieper Aquatic Systems. Proc. International Seminar "Intervention Levels and Countermeasures for Nuclear Accidents" Cadarache, France, 1991.

[12] Berkovsky V., Ratia G., Nasvit O., Forming of Internal Doses to Ukrainian Population in Consequences of Usage of the Dnieper Water. 39° Health Physics Society Meeting, June 24-28, 1994, San Francisco, USA, p. 10.

Local Strategies for Decontamination

Philippe HUBERT[1], Valeri RAMZAEV[2], Gennadyl ANTSYPOV[3], Emlen SOBOTOVICH[4], Lubov ANISIMOVA[5]

[1] Institut de Protection et de Sûreté Nucléaire, (IPSN), B.P. 6, 92265 Fontenay-aux-Roses cedex, France.

[2] Branch of Institute of Radiation Hygiene, Karchovka 243000 Novozybkov, Bryansk region, Russia.

[3] Chernobyl State Committee of the Republic of Belarus, 14 Lenin Str., 220030, Minsk, Belarus.

[4] Institute of Geochemistry, Mineralogy and Ore formation, 34 Palladin Av., 252680, Kiev-142, Ukraine.

[5] EMERCOM, Moscow, Russia.

Abstract. The efficiencies of a great number of techniques for decontamination or dose reduction in contaminated areas have been investigated by several teams of E.C. and CIS scientists (ECP4 project). Modelling, laboratory and field experiments, and return from experience allowed to assess radiological efficiencies (e.g. « decontamination factor ») and requirements for the operation of numerous practical solutions. Then, those data were supplemented with data on cost and waste generation in order to elaborate all the information for the optimisation of decontamination strategies. Results will be presented for about 70 techniques. However, a technique cannot be compared to another from a generic point of view. Rather it is designed for a specific target and the best technology depends on the objectives. It has been decided to implement decision analyses on case studies, and the local conditions and objectives have been investigated. Individual doses ranged from 1 to 5 mSv, with contrasted contributions of internal and external doses. The desire to restore a normal activity in a partially depopulated settlement, and concerns about the recent increase in internal doses were typical incentives for action. The decision aiding analysis illustrated that actions can be usually recommended. Results are outlined here.

1. Introduction

In the aftermath of the Chernobyl accident, and the years immediately following the catastrophe, large scale decontamination actions and evacuations of populations have taken place. Meanwhile a series of restrictions were set-up, bearing on economic activities and on daily life. Therefore, six years after the accident, many of the critical areas had been decontaminated, the contamination could be considered as fixed, especially on urban objects and the social situation was felt to be stabilized.

Under those conditions, some questions remained without answer.

What is the efficiency of the « classical » decontamination techniques when contamination has been deposited years ago, or when previous countermeasures were applied?

Which specific techniques can be developed?

Is it sensible to undertake new decontamination actions?

The ECP 4 project was launched in order to look at the opportunities for further dose reduction actions in the contaminated territories of the three republics affected by the accident. The objective was to provide a local decision maker, faced with many alternatives for decontamination, with helpful informations related to the various objectives he may consider. The project involved teams from different disciplines. The techniques themselves, the assessement of their costs, the set up of a methodology for decision aiding and its application to case studies were studied.The main results are presented here.

1 Documenting decontamination techniques

1.1 Scope and organisation of the work

The project had a very broad objective. It surveyed usual techniques, techniques that had been designed within the framework of ECP4 (under assumption that they can be made effective at a full scale level), techniques that had been developed or examined by other Programmes(e.g. ECP2, ECP9, JSP2, JSP5...[1] [2] [3] [4]), and the « do nothing » option.

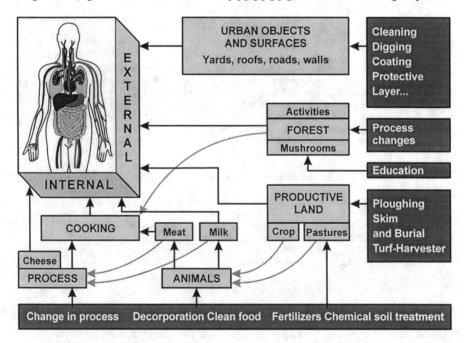

Figure 1 : **Exposure pathways and targets for countermeasures**.

Exhaustivity was searched for. Actions dealing with all sources and pathways were considered, and the tasks were divided in accordance with the type of sources on which it is possible to act (see figure 1). Five sub-projects dealing with techniques were set-up : decontamination of urban environment (walls, roofs, yards, roads...), soil decontamination (arable soils and pastures) with chemical and physical approaches, decontamination of forests, protection and decontamination of machines and the decontamination of domestic and industrial food. They were associated with a sub-project on self-restoration in ecosystems. A specific sub-project was devoted to cost-benefit analysis which eventually made use of the data collected on techniques.

The first years of the project were devoted to laboratory and field experiments, often at a sensible scale. Among other results, it was proved that it is still possible to decontaminate walls and roofs on which Caesium was deposited almost 10 years ago (for example using water under pressure), or fields in which the fertile layer is too thin for ploughing (e.g. turf harvester), and that modifications of cheese making processes can achieve significant decontamination factors (e.g. the Phoenix cheese). During the last year, syntheses were made in order to build up an information that is of practical use in decision analysis.

1.2. Techniques and decontamination yields

Unless otherwise specified, the figures are related to the effectiveness considered several years after an accident and they apply to Caesium. Some figures are close to unity, suggesting a poor efficiency. Most often it traces a countermeasure of current use in the early stages of the accident, which was included here for the sake of completeness.

Urban decontamination and decontamination of machines have been assessed with different techniques such as Ammonium Nitrate, Water Hosing, Vacuuming, Sandblasting, Clay treatment, Polymer coatings, Sorbents, Roofwasher or special digging, in the towns of Pripyat and Vladimirovka (1992-1994) [5][6][7], Novozybkov and surrounding settlements (1993) [6] and Halch (September 1994) [7][8] (see Table 1).

Table 1: Decontamination yields of Techniques applying to urban objects

Technique	Target	D_f	Constraints	Comments
Turning flagstones manually	Flagstones	6	No	-
Set of tools for dismantling	Houses	∞	No	Need to build a new house
Road planing	Road	>100	No	Grinding off surface
Fire hosing	Roads	1.10	No	Water rinsing
Vacuum sweeping	Roads	1.4	No	Dust close to operator
Roof washer	Roof	2	No	Rot. brush, air compressor
Set of tools	Roof (Asbestos)	∞	Two ladders	Manual change of roof cover
ARS-14 with trailer	Roof, Wall	1.2-3.6	No	Dry & collect clay films
Electectric drill, steel wool or sand-paper	Roof (iron), wall (painted)	2-2.3	Possibly scaffolding	Grinding
High pressure turbo nozzle OM-22616	Roof, Wall, Asphalt & Concrete surface	1.3/2.2, 1.7-2.2	No	High pressure water hosing 120 bar
Detached polymer paste	Surfaces (smooth)	4-30	T > +5°C	Transports. Manual work
Sandblasting (dry or wet)	Wall	4 (dry) 5 (wet)	Scaffolding preferable	High-pressure with sand, Whole-body protect/air supply
Ammonium nitrate spraying	Wall	1.3	No	Surface rinsed with clean water
Vacuum cleaner, razors...	Walls (paper)	>100	No	Replacement of wallpaper
Polymer coatings	Walls (not wooden)	4-5	T=20-30°C	Humidity <80%
Manual electric cutting machine	Wooden wall	5	Residual nail remove	Upper layer mechanically removed (dust)

Among the techniques of decontamination of soils (see table 2), mechanical as well as chemical actions were tested. The yields of « classical » actions applying to undisturbed soils were also documented. Among others, the Turf Harvester was tested on meadows and arable lands, in Burakovka and Chistogalovska (1992) [5][6], in Millyachi and Savichi (1993, 1994) [5][6][7][8], Soil fractionation, in Karchovka (1993) [6]; Phytodesactivation in Burakovka (1994)[8].

Table 2:Decontamination yields of Techniques applying to soils in urban/rural environment

Technique	Target	Df	Constraints	Comments
Front loader / Bulldozer	Soil	28 / 10-100	No	Scraping top soil (10-30cm) Removes fertile soil layer
Grader	Top layer of ground	4-10	No	Scraping of soil surface
Shovel	Garden soil	6	Virgin soil	Digging to about 30 cm depth
Turf harvester (small)	Undisturb. grass. soils, priv. & forest pastures, urban grassed lands.	3-20	No or few stones	Removes the 3-5 cm top soil. No further intervention required
Turf harvester (industrial)	Undisturbed grassed soils	20 (grass & milk)	Few stones	Need to build a prototype
Lawn mower (mulcher)	Grassed areas in city	1(9 y after)	No	With other tech. (turf-harv.)
Showel (triple digging)	Garden soil	10-20	Virgin land	Burying top layer 30-40 cm
Mobile equipment for soil fractionation	Soil	4-6	-	Mechanical separation of the soil, Sand, sand clay (20%) soil
Ordinary plough and tractor	Arable soils	9-12(ext.)	Virgin land	Ploughing to 25-45 cm depth
Deep ploughing	Arable soils	2-4(crop)	Virgin land	Plough soil layer (25-35cm)
Skim-and-burial plough	Arable soils	10-20	Virgin land	5 cm topsoil buried at 45 cm
Liming (special trucks for spreading)	Acid arabic land	1.3-3	Soil pH = 4.5-5.5	Requires K addition, Persistent effect during 4-5 y.
Addition of potassium	Arable lands	1.3-3	No	K addition needed
Addition of phosphorus	Arable lands	0.8-1.3	With K,N	-
Organic amendment of soil	Arable soils	1.3	-	Yield and quantity increase
Radical improvement of Pasture (draining, cleaning, disking 3 times...)	Pastures	4-16(peat) 4-9 (podzol)	No	Yield increase
Disking, fertilising, liming and sowing new grass	Pastures	1.4-2.2 (Cs), 1.2-1.4 (Sr)	Repeat disking 4-6 times	Dilution of Cs and Sr in the soil profile, Yield and quantity increase
Liming and fertilising	Forest pastures	1.5	Manual	Poor soils Enriched by Ca, K
Ferrasin bol or Prussian blue	Cow	2-3	3 bolus/3 m.	Where Cs level > 1000 Bq/l
Clean fodder to animals before slaughter	Cow	2-3 on meat	2 m. before slaughter	Organisation of special animal feed before slaughter
Prussian blue salt licks	Cows and bulls	2-3		Salt lick duration= 3 months
Phytodecontamination	Soils (mixed)	1.1-1.3/y	7procedures	
Exchange of food crops with technical (industrial) crops	Arable lands	Exclusion food uptake	Plant /Crop processing	Use of contaminated area for crop production
Ferrasin filters for milk	Milk	ca. 10	Private farm	If milk contamination>400 Bq/l

For the sub-project concerning the rehabilitation of forests techniques (see table 3) techniques such as Brushing Machine, Grinding, Tree Felling Machine, Wood Incineration, Gasification or Thermal and Chemical Treatment, were tested in Kopachi, Burakovka and in the Red Forest(Nov. 1992)[5], in Rovno region, in Kruki and in Savichi (1993, 1994)[6][8].

Table3: Decontamination yields of Techniques applying to forest environment and products.

Technique	Target	Df	Constraints	Comments
Mechanical brush	Forest litter	3.5-4.5		Not wet forest areas or forest < 30 y old. Litter layer removal
Grinding mower	Under-wood forest, shrubs	1.2	∅ wood stem < 8 cm	Not wet forest areas or forest <30 y old. Cleaning of underwood
Wood sawing plant 20-K63-2	Timber	2-4	No wet area	Mec. removal of bark and phloem
Twin-screw extruder	Contaminated wood	50-100	Preparation wood chips	Special wood pulp treatment, extracts Cs, Sr from wood pulp

The study of decontamination of food products was organized in field trips to 3 farms in the Ukraine (1991-1992, 1994)[5][7], in Chistogalovka and in Chernobyl zone (1993)[5] and it also took place in laboratory (see Table 4).

Table 4: Decontamination yields of techniques applying to food processing.

Technique	Target	Df	Constraints	Comments
Replace prod. Tvorog / prod. Phoenix cheese	Domestic and Industrial milk	5	Training required	Phoenix cheese has low transfer factor for Caesium
Exchange milk private plots / milk from collective farm	Industrial milk	6	Transport costs	Use privately produced milk for feeding young animal in coll.farm
Exchange feed(hay) private plots / feed collective farm	Industrial milk	1.3	Transport costs	Use privately produced feed for feeding animals on coll. farm
Use fat fraction, Convert skim to cheese	Industrial milk	complete	Transport costs	Contamination fed to animals does not directly re-enter the food chain
Separate milk into cream & skim milk fractions	Industrial milk	complete	Transport costs	Skim is returned to farm to be used as feed for young animals
Ion-exchange	Industrial milk	10	-	Minimal nutritional implications
Use of sorbents	Industrial milk	>20	-	Minimal nutritional implications
Brine meat (UIAR recommend.)	Industrial meat, Domestic meat	2.5	Training required	Increase Na content,decrease K, B12, Fe, Z, Niacin.
Replace wild / cultured mushrooms	Domestic mushrooms	complete	Training required	Role of education in implementation will be vital
Educate to pick less contaminated mushrooms	Domestic mushrooms	1.8-12.5	Training required	Cs transfer factor can vary greatly between species of mushrooms
Domestic salting of mushrooms	Domestic mushrooms	5	Training	
Parboil potatoes	Domestic potatoes	2	Training	20% of all potatoes in the diet.
Domestic pickle cabbage	Domestic vegetables	2	Training	
Dispose of berries while making compote	Domestic vegetables	2	Training required	Possibly some losses of nutrition associated with disposal of berries
Marinating meat / eating fresh	Domestic/industrial meat	2.5	Training required	Decrease of content of water soluble vitamins and minerals
Wet culinary techniques	Domestic meat	1.6	Training	Losses of Cs are less with dry tech.
Brine carrots/ consume fresh	Domestic vegetables	2	Training	

Self Restoration was followed on Chistogalovska site, Khojniki, Kormyany, and Cherikovskij region (1993)[5][6][8]. Radiological characterisation is an important step in the definition of strategies. Its cost cannot be neglected in the optimisation analysis. Design of measurement tools that are portable and usable in large scale characterisation were worked out and experimented. This was the case of an NaI detector which can at the same time count Caesium activity in the soils and estimate the depth under certain conditions (CORAD device from RECOM). This technique as well as classical methods were documented within the same format as decontamination techniques [9].

1.2. Logistic and economical aspects

Teams working on the techniques have been asked to document them according the needs of optimisation. Components of the costs were depicted in physical terms: direct manpower and overheads, skill requirements for workers, and needs in education for the public, transportation, consumables, loss of value for products, generation of wastes (solid and liquid volumes, activity and toxicity). They were slightly different for techniques bearing on surfaces [9] and for techniques bearing on food stuff [10].

Prices were necessary in order to assess costs. A short synthesis was made in order to know the prices of manpower (from 70 to 100 ECUs per month), consumables (e.g. 0.15 to 0.3 ECUs per liter for gasoline) and products (e.g. 0.25 to 0.4 ECUs per liter for milk). Prices are indeed fluctuant at the local scale but this task was necessary to provide default values and standard values when comparing case studies.

A specific study was then necessary in order to look at the costs associated with waste management, within the framework of Belarus regulations [11]. Wastes are normally in low or very low level (Conventionally Radioactive) categories and the costs were found to depend mostly on handling, transportation and process, which do not vary very much according to the category. The relative costs for disposal of radioactive waste range from about 70 ECU/ m3 for woody waste up to 700 ECU/m3 for liquid waste. Experience in decontamination showed that local options were retained in the past at lower costs (e.g. Kirov with a cost of 17000 Rb 1989 for 3000 m^3 [6]).

2 Principles for the definition of local strategies

2.1. From cost benefit analyses on individual techniques to local strategies

The costs and the savings associated with the application of a technique can be estimated at the scale of an «Intervention Element», that is an element which is homogeneous with respect to radiological characterisation, response to the application of a technique, and contribution to a collective dose (e.g. a particular culture on a given field). The radiological benefit is usually the collective dose, valuated on the basis of the cost of Man Sievert. Other benefits can be considered (e.g. increase in the crop production when fertilizers are used). Very often the cost of the countermeasure is a constant, while the efficiency increases as the contamination increases, for that simple reason that the effect is a multiplicative effect (see table 1 to 4). Thus for a given a priori value of the Man Sievert, one can determine a « Specific Intervention Level » (S.I.L.), that is contamination level above which a countermeasure is justified. Cost benefit ratio allows the ranking of a set of countermeasures that apply to the same target. CIEMAT experimented this approach on the basis of Intervention Elements having the characteristics of some soil types from Kirov [12]. The areas were agricultural areas devoted to pasture and potato production and the S.I.L. ranged from 550 to more than 8000 kBq per sq m (turf harvester, skim and burial, mineral fertilization, liming, organic fertilization, phytodesactivation).

Eventually decision analysis involved a more complex approach. When defining a strategy, a local decion maker needs to determine the best combination of countermeasures that would satisfy a series of criteria. This decision cannot be based solely on the cost-benefit ratio of individual techniques, because some techniques may interact at the level of their efficiencies, or of their costs. The main reason is that all actions must be considered simultaneously when looking at criteria such as individual doses, to which all Intervention Elements contribute simultaneously. Thus optimisation analysis is embodied in a global assessment of the radiological impact of a given situation.

The methodology was tested on case studies. At this level, problems that are encountered are complex, but their size is manageable and the alternatives are limited because the regulatory framework and other national provisions were considered as a given reference. Practical decision on the means for decontamination are also usually taken at this level. Key parameters such as behaviour of populations become meaningfull atthe scale of a « case ». Two type of cases were considered: decontamination of a settlement, reduction of the doses attached to a practice.

Selection of case studies aimed at representing typical situations. Consequently, the choice involved many criteria, such as soil type, importance and nature of previous actions, behaviour of the population, (diet, occupancy factors and professional activities), economic activities, and of course, contamination levels, actual doses and structure of the doses. At last the social situation is of interest, as some settlements are partially evacuated and some others do have the same life-style as before the accident. A second series of criteria was more pragmatic; availability of good quality data and involvement of local authorities.

Eventually 8 cases were developed, 5 of them bearing on settlements, 3 others dealing with the enhancement of the radiological situation in an activity. The settlements were Kirov and Savichi in Belarus, Millyachi and Polleskoe in Ukraine, Zaborie in Russia. Difficulties were encountered in the investigation of the case of Polleskoe (a partially evacuated little town, quite dependant on the forest industry, submitted to heterogeneous fall-out) so that results are not presented here. Some features of the settlements are outlined (see table 5).

Table 5: Some features of the case studies.

Cases	Reference Soil contamination (Ci/sq km)	Present population	% of pre-accident population	Dominant pathway	Previous actions
Zaborie	66	180	20%	External	Extensive
Kirov	30	500	40%	External	Extensive
Savichi	10	160	20%	External	Limited
Millyachi	5	3200	100%	Internal	Numerous

One other case dealt with work in the forest and in the wood industry, an activity of major importance in the Belarus economy. It considered the practical small scale actions that can improve the radiological situation. A recreational activity was investigated, associated with the use of a site of Karchovka, near Novozybkov. At last the approaches for the rehabilitation of a school in Halch were examined. A school is a limited object, but it is an important matter of concern in many areas.

2.3. Implementationof case studies

The work on case study required preliminary steps : identification of the local goals for decontamination (lowering average or group specific individual doses, bringing food or wood contamination level below thresholds, restoring an activity...); collection of data (demography, contamination levels, surfaces according to land use and soil types, productions, diets & occupancy factors); structuration of this information in order to put forward the « Intervention Elements » on which specific actions can be applied.

On those bases, the computation of the effectiveness of countermeasures required models combining both dose prediction and cost computation. In addition, it was necessary to use very flexible tools in order to simulate the efficiencies of countermeasures that could modify almost any of the steps in the chain of dose build-up, and in order to be able to look at very different outputs. For example, individual doses, doses to a critical group, collective doses, « exported » collective dose, contamination of food, volume of food above various levels, time span before reaching desired thresholds have been used in case studies. In contrast with this structural complexity, simplifications were made (e.g. grouping subpopulations or intervention elements whose behaviours are similar, reduction in the number of « in house » localisations ...), and it was decided to use the simplest dose computation models (transfer factors in the food chain, « influence » factors for computing external dose rates in given locations).

Some standardisation took place between the teams (CIEMAT and IPSN), in order to ensure comparability, as was done for costs. For example « recommended figures » worked out by JSP5 for dosimetric predictions have been used as a starting point ([4] and further recommendations), and they were modified only in the case they were contradicted by observations.

Indeed, observed figures in case studies were often available at two levels: foodstuff contaminations and dose rates in various locations on one side, body contents and data from individual dosimeters on the other. They were compared to "predicted doses" and the reasons for the differences were searched for. Whenever a correct explanation was found, basic coefficient such as transfer factors were modified. For example, in Zaborie [15], milk and crop contaminations were lower than predicted by about a factor of two. This was found to be consistent with the fact that previous countermeasures such as ploughing were extensively applied. Nevertheless, alterations were limited because there are many "free parameters" in such models (categorization of areas and populations, diets, occupancy factors, fraction of imported food ...). Therefore, it is always possible to reconstruct observed doses, but non documented modifications, aiming only at improving the fit, may lead to a fallacious accuracy. Preference was given to the robustness of a predictive tool which must be used in simulations. Modelled and observed figures are provided here (see table 6).

Table 6 : Mean individual dose 1993-1995 from Cs 137-134 ($\mu Sv.y^{-1}$):

	EXTERNAL			INTERNAL		
	Values based on predicted model	Dose rate+ Occupancy factor	Individual measurement	Values based on predicted model	Activity in food + diet	Whole Body Counting
Zaborie	4.15[***]	4.264[***]	3.098[***]	1.2[***]	1.05[***]	1.3[***]
Kirov	1.8•	1.570[*]		0.7•	0.96[*]	1.5•
Savichi	0.8•	0.510[*]		0.2•	0.092[*]	0.42♦
Millyachi	0.445[**]	No data	No data	2.83-3.1[***]	5-8	0.23-0.7

*- Predicted by Dr.M. Savkin [14]; **- [15]; ***- Predicted by IPSN in 1995
•- From passport 94 [16][17]; ♦- From passport 93 [17].

The "do nothing" option was the reference, so experimental work took place within ECP4. At first, ECP4 programme was focused on horizontal migrations, in contrast with the numerous studies on vertical migration. In the 1995 programme both aspects were considered [18] [19]. Ideally the self-restoration or, more properly, the reduction in the local exposures, depends on many factors and it is quite difficult to define a generic reduction parameter. Effective half life from 10 (external exposures) to 20 (internal exposures) were suggested by JSP5 [4]. The latest figure was kept for all cases; although this simplification goes along with some uneven treatment depending on pathways (external exposures from natural soils, urban objects and internal exposures).

The time scale for comparison and for computation of "averted collective dose" was taken as 50 years, leading, with the previous assumption to about 25 "effective years". It was shorter when the life time of a source of exposure was shorter, for instance for asbestos roofs which are to be replaced anyway before 50 years. A time scale of 100 years was discarded, because predictions on the behaviour of population, a major parameter for dose computation, was quite uncertain, and because the difference in "effective years" was small.

3. Lessons from case studies

3.1. Goals and criteria for decontamination

When dealing with the settlements, no cases were found in which decontamination was desired in order to meet one unique target. The overall idea is a desire for a return to a normal life style, by suppressions of interdictions in the formally "evacuated" areas (Savichi, Zaborie) or by alleviation of restrictions in villages in the other areas (Kirov, Millyachi).

Individual doses are a matter of concern that was put forward in the four main case studies, but criteria were heterogeneous. Average population dose is a criterion in Zaborie, to be compared with 5 mSv per year. Critical groups of workers have been identified in Zaborie, Kirov; Millyachi, whose doses are still to be compared with 5 mSv (e.g. forestworkers, cowboys,...). Reference to 1 mSv for the average population is also quoted (e.g. Belarus "Passports" on settlements). Most often 1 mSv is not a direct target, rather it is an indication that normal life is possible, and the length of time that is necessary to reach 1 mSv is often estimated. Individual doses are also adressed in Millyachi. It seems that the concern is rather a "potential for exposure" to high internal doses, than a limitation on actual average doses. Indeed, the presence of high transfer factors in peat-bog marshy soils results in observation of high contamination levels in some samples (especially milk). Together with mushroom consumption, internal doses to a family whose diet relies on those particular food chain can reach more than 5 mSv, although body content measurements on the average population failed to show such high figures (see table 6).

Not that the possibility for relatively high individual doses from « extreme » diets is not specific of Millyachi, especially as regards mushrooms and other "gifts of nature". Increases in individual doses in products that were noticed in 1993-1994 in Kirov and Zaborie may result from a lower vigilance of inhabitants.

The need to be less dependant on the restrictions to life style is not a quantified criterion but it is a clearly expressed goal. It has been said that population stayed or came back in evacuated territories. Direct observations also showed that prohibited land can be used either by the people from the settlement or by neighbours.

Improvement of agriculture activity corresponds to another sort of criteria : Diminution of production above standards is a criterion in Kirov an Millyachi. Volume out of range or duration before everything is below standards can be used alternatively. Collective exported dose was also a criterion. Restoration of competitivity for local products was also quoted by local authorities. It can result from the demonstration that activities are permanently low.

More complex goals can be associated with a decision on "rehabilitation". People may have a prejudice on the activity that should be restored, but such a choice is sometimes an expected result from the decision analysis. In this respect, comparison with past situations cannot serve as a reference. Almost 10 years after, a reorganisation took place, abandonned equipments may have been destroyed, the economic context is quite different. It is believed that case studies are a valuable input to such analyses, but discussion with decision makers did not take place yet on this point.

No objectives were assessed in terms of collective dose. Nevertheless, the gain in collective dose was considered as the best criterion to assess the efficiency of a countermeasure. Cost effectiveness was measured by the « cost of averted Man Sievert ». It was admitted that the most cost effective measure should be applied first in the strategy ; so the decision making process is actually an optimisation under the contraints associated with the above mentioned goals and criteria.

The problem faced with the forest industry was close to the one discussed in settlements : need to alleviate restrictions on the activity, issue of individual doses to forest workers (to be brought down to 5 mSv); contamination in wood depending on the use, collective doses associated with the uses. The decontamination of a school in Halch was associated with targets in doses received during attendance. At last, in Karchovka, the goal was to reopen the area for the citizens of Novozybkov city. Contribution to individual doses to members of critical groups (anglers, mushroom collectors and .... leisure) were considered, on the basis of time spent. Collective doses were also considered.

3.2. Elements from the decision analysis

Results from the case studies were contrasted, as could be expected from the differences in radiological situation, soil types and other factors.

In Zaborie, application of countermeasures was limited by the fact that extensive decontamination took place so that options like ploughing were pointless. However, calculated doses being only slightly higher than 5 mSv, many countermeasures would allow to go below this figure ; cleaning yards, education on mushroom consumption, but also combination of agricultural countermeasures. Actions dealing with yards (Front leader or Bulldozer cleaning) or educational programmes yield to a cost of averted Man Sievert below 2000 ECU, and they would be used first in applying a strategy. The most effective action was the cleaning of the yards surrounding houses, provided that a local solution for waste disposal is accepted, as was done in the neighbouring settlement of Yalovka. Should generic costs be applied (see §1) it would be 50 times less effective. Education on mushroom was priced on the basis of one visit per year to each family. Liming, in spite of a poor global impact, has a good cost-efficiency ratio and deserves to be included in the strategy. Costs and impacts on collective dose of individual or combined options are displayed below. Options leading to an individual dose below 5 mSv are underlined (see figure 2). Some options yield very high costs for averted Man Sievert, others are limited in their scale of application, because areas fitted for them are limited (e.g. Turf harvester).

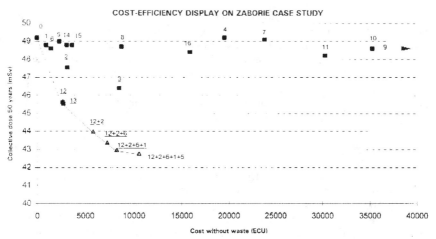

Figure 2: Cost-efficiency display on Zaborie Case study.

Situation in Savichi is such that the effectiveness of countermeasures is usually lower, simply because the contamination levels are lower. Most countermeasures yielded cost of avoided Man Sievert higher than 8000 ECUs. Actions on yards and liming are ahead as in Zaborie, but when benefits are taken into account (i.e. increase in the yields), application of fertilizers such as liming and potassium chloride are indicated. It was said that their efficiency has actually decreased due to previous applications, but they might prove useful again, after they have been stopped. Radical improvement of meadows and pastures can be proposed. In parallel with the cost benefit analysis, the classification of Savichi as a prohibited settlement was reconsidered

The collective farm in Kirov accounts for significant differences with the two previous case studies, as regards goals. Cleaning gardens is efficient (3000 ECUs per Man Sievert), followed by cleaning roofs. Although with a better efficiency due to higher levels of contamination, agricultural countermeasures have rankings similar to those in Savichi.

In Millyachi, the importance of internal dose and the high transfer coefficients increase the efficiency of agricultural countermeasures in comparison with urban countermeasures. Education for the mushroom consumption remains highly efficient. Many agricultural measures have cost efficiency around 1000 ECUs per averted Man Sievert, but cleaning yards still remains efficient. One may also notice that those yields are averaged and that targeting the areas where contamination is the highest and where soils are peat or peat-marshy soils results in even lower costs per averted Man Sievert.

The study of the forest industry appears to be connected with other case studies in which a critical group, was made of forest workers. Tree falling machine, removal of litter targeted on working areas, brushing and grinding equipment were considered for dealing with the dose to forest workers. A figure of 3000 ECUs per avoided Man Sievert was quoted for removal of litter, while a quite high figure applied to tree falling machine (3-4 Million ECUs). In the latest case the question is rather associated with the move to a capital intensive production approach. Other options for wood decontamination were considered, but their goal was not mostly associated with workers doses [20].

The recreational zone of Karchovka is a good example where several decontamination techniques deserve implementation now. For example, the cost of avoided Man Sievert, for litter removal and for litter and soil removal is respectively of 3300 and 3500 ECUs. Deep ploughing has the best efficiency (around 200 ECUs per Man Sievert), although it leads to some reorganization of the beaches that surround the ponds of this area [21].

The case of Halch school is the only one in which the costs of averted dose is high with all options (10^6 to 10^7 ECU per avoided ManSievert). This is due to somewhat low contamination of the walls and roofs [22].

3.3. Generic lessons

In spite of the contrasted situations, generic lessons can be drawn. First, in all cases but one, there are actions with a reasonably low cost for averted dose. Should Man Sievert be valuated to 5000 ECUs or even 2000 ECUs, there are indications for application of countermeasures. Of course there is a decrease in the cost efficiency ratio, and ambitious targets in individual doses (e.g. 1 mSv) would imply the use of techniques of poor economic efficiency. Besides, it has been confirmed that actions on non disturbed land are among the most efficient and that they are limited in the present case studies.

Within a given family of techniques, the hierarchy remains quite unchanged through the various case studies. On the other hand, hierarchies between families of techniques are subject to changes from one case study to another. The techniques bearing on agricultural production are more efficient where internal dose are more important than external doses. This is hardly a surprise but it was worthwhile to quantify this point. The cost effectiveness ratio, vary significantly from one study to an other, even when differences in contamination are accounted for. Differences in diets, or in use of the products can explain those differences , but cost components are also subject to changes. As an example, the cost of averted dose for turf harvester ranged from 100 [23] to 15000 ECUs, and liming ranges from benefit to negative values when the effect on yields is denied.

The case study approach still suffers from four limitations. First, not all techniques were included and domestic and industrial food processes have not been systematically introduced in the simulations. Second, the treatment of wastes was not fully mastered. Generic solutions, local solutions and absence of treatment are three possibilities, actually used and also simulated, but the differences are impressive. In principle, waste treatment should be incorporated in the optimisation as part of the technique. Third, the case studies were performed in areas in which radiological characterization was well advanced. In other areas, the costs assiociated with the acquisition of information must be taken into account. At last, consideration of Strontium would lead to propose new specific actions; however, it would also increase the efficiencies of the present ones.

4. Conclusion

The necessity to undertake new decontamination actions or other dose reduction actions almost one decade after the Chernobyl accident raised serious doubts when the programme was launched. The goals for such actions were unclear, the availability of large scale efficient techniques was questionned and their costs were felt to be prohibitive. Today, a global programme for decontamination of all the contaminated territories is still out of reach. Nevertheless the studies that were undertaken have demonstrated that there are clear indications for action in practical cases. Case studies put forward two aspects : availability of cost effective actions, eventhough limited, in most cases, and confirmation of practical demands and goals at the local level. Moreover they demonstrated that a thorough investigation of a case is necessary before decision, because indications for techniques is highly dependant on a situation.

The feeling that there are opportunities seems to be shared by the local authorities that were involved in the case studies. Some achievements of the programme contributed to this evolution (development of new technologies or adaptation of classical techniques, development of a synthetic information on techniques, practical implementation of case studies). Side aspects of the project cannot be neglected either. A reciprocal training took place between the teams dealing with the technical and decisional aspects of radiation protection and it mitigated the lack of mutual awareness wich was a major difficulty in the development of the project. Implication of authorities did have a similar effect, as shown by the development of decisional aspects in the Radiation Passport in Belarus.

Other incentives for decontamination do not arise from the programme but from the evolution of the situation in the contaminated areas where demands for rehabilitation are getting stronger in "partially evacuated" areas and where it is more and more difficult to maintain the efficiency of restrictions on the life style and activities.

Obviously, some developments are still needed for the improvement of decision. For example, specific questions linked with Strontium were not dealt with extensively, some conditions require adoption of techniques and some differences in the fied experiments must still be explained. The case study approach can be applied to other situations in order to give an exhaustive picture of the possibilities and needs. Among the questions that deserves most interest is the characterisation of contamination. Its global cost must be taken into account in the optimisation, and besides, its efficiency can still be increased.

References

[1] Anonymous, Transfer of radionuclides through the terrestrial environment to agricultural products including the evaluation of agrochemical practices. Final report for the period 1st November 1993 - 31st October 1994 for the Experimental Collaboration Project N°2 (ECP2). COSU-CT93-0040.

[2] Anonymous, Fluxes of radionuclides in rural communities in Russia, Ukraine and Belarus. Transfer of radionuclides to animals, their comparative importance under different agricultural ecosystems and appropriate countermeasures. ECP9 Annual Report. Nov 93-Dec 94.

[3] Anonymous, Joint Study Project N°2, Decision-aiding System for the Management of Post-Accident Situations, 1993-1994 Annual Report, 1995.

[4] Anonymous, Joint Study Project N°5. Pathway analysis and dose distributions. Contract COSU-CT93-53, Report for period Nov 93 to Oct 94. February 1995.

[5] A. Jouve, Decontamination strategies: Project N°4. Agreement on the consequences of the Chernobyl accident. Report for the period 1 November 1992 - 1 Marsh 1994. European Commission. 1994.

[6] Anonymous, CEU/CIS Agreement on the consequences of the Chernobyl accident; ECP/4, decontamination strategies, report on the work done in 1993, part two, CIS contribution. European Commission, 1994.

[7] Anonymous, Agreement on the consequences of the Chernobyl accident; Projet N°4: Strategies of decontamination, Part One, Contribution of EU institutes. Final report on the work done in 1994,. European Commission, 1995.

[8] Anonymous, Agreement on the consequences of the Chernobyl accident; Projet N°4: Strategies of decontamination, Part Two, Contribution of CIS institutes. Final report on the work done in 1994,. European Commission, 1995.

[9] J. Roed, K.G. Andersson, H. Prip, Practical Means for Decontamination 9 Years after a Nuclear Accident. RISØ Report, 1995.

[10] Anonymous, ECP4, Sub-project 5, Decontamination by Food Frocessing, Food Decontamination Methodologies, Radiological Protection Institute of Ireland, August 1995.

[11] A. Grebenkov, K. Rose, Management of Low Level radioactive Waste arising from ECP-4 Decontamination Techniques. Technical Report on the work done in 1995, AEA Technology, England, September 1995.

[12] C. Vazquez, J. Gutiérrez , R.Guardans, B.Robles, M.Savkin, Optimization of intervention strategies for the recovery of radioactively contaminated environments. Application to a case study : Kirov (Belarus). ECP4 Annual Progress Report (Oct. 1992-Oct. 1993) Second version. CIEMAT/IMA/UGIA/M5A 22/02/94, 1994.

[13] L.I. Anisimova, V.P. Ramzaev, V.I. Kovalenko, A.V. Ponomarev, V.A. Trusov, C.V. Krivonosov, A.V. Chesnokov, A.P. Govorun, O.P. Ivanov, B.N. Potapov, S.B. Shcherbak, Zaborie: NonStop Relocation ? A Russian case study, Novozybkov Branch of Institute of Radiation Hygiene, Novozybkov, 1995.

[14] M. Savkin, Savichy case study, General informations, 1995.

[15] Y. Kutlakhmedov, G. Perepelyatnikov, L. Perepelyatnikova, V. Pronevich, I. Los, Millyachi Case Study, A short report, 1995.

[16] S. Firsakova, V.S. Averin, Yu.M. Zhouchenko, Social-Economic and Radiation-Hygienic Passport on Kirov, Belarussian research Institute of Agricultural Radiology, 1995.

[17] S. Firsakova, V.S. Averin, Yu.M. Zhouchenko, Social-Economic and Radiation-Hygienic Passport on Savichi, Belarussian research Institute of Agricultural Radiology, 1995.

[18] V. Davydchuk, Project ECP-4, Geographic Assistance of Decontamination Technologies Trials and Decontamination Strategy Evaluation,Laboratory of Landscape and Ecological Problems of Chernobyl, Kiev, 1995.

[19] G. Arapis, ECP4/ Sub-project 7, Self-restoration of Contaminated Territories, Report 1995, Agricultural University of Athens, 1995.

[20] A. Jouve, F.A Tikhomirov, A. Grebenkov, M. Dubourg, M. Belli, N. Arkhipov, Management of Contaminated forests, Minsk Conference, 1996.

[21] V. Ramzaev, V. Kovalenko, A. Ponomarev, E. Ponomareva, A. Kacevich, V. Trusov, S. Krivonosov, I. Gorbovskaya, J. Roed, K.G. Andersson, Recreation Zone: Karchovka, a russian Case Study, ECP4 Report, 1995.

[22] N. Voronik, A. Bykov, Evaluation of the decontamination efficiency of school in Khalch village, Lazurny meeting, 1995.

[23] Jouve A. Technical and economic study of the use of a turf hrarvester for decontaminating permanent pastures in the Chernobyl affected areas , Rapport IPSN/DPEI/SERE Cadarache 1995.

The Need for Standardisation in the Analysis, Sampling and Measurement of deposited Radionuclides

Yuri S. Tsaturov
Roshydromet, Moscow, Russia

Marc De Cort, Gregoire Dubois
Environment Institute, CEC Joint Research Center, Ispra, Italy

Yuri A. Izrael, Evgeny D. Stukin, Shepa D. Fridman
Institute of Global Climate and Ecology, Moscow, Russia

Leonid Ya. Tabachnyi
Minchernobyl, Kiev, Ukraine

Ivan I. Matveenko, Maria G. Guermenchuk
Republic Centre of Radiation and Environment Monitoring, Minsk, Belarus

Vasilij A. Sitak
Goscomhydromet, Kiev, Ukraine

Abstract: Following the Chernobyl accident in 1986, diverse sampling and measurement methods for radioactivity deposition have been applied by the various European institutes. When compiling these datasets together on the same data platform, in view of preparing the atlas on caesium contamination in Europe, data quality analysis has shown a lack of harmonisation between these various methods. Because of the necessity to dispose of compatible and representative measurements for further analysis, e.g. time series analysis, and the need for better standardisation methods in the event of a future accident with large transboundary release, several suggestions are made of how such harmonisation might be achieved.

Also in view of taking appropriate decisions in case of accidental releases by gaining experience in data standardisation, the variety of the sampling and measurement methods of radioactivity currently used are briefly summarised and the results intercompared.

In order to improve the quality of datasets, GIS, amongst other methods, can be applied as a useful tool to highlight the lack of harmonisation between the various sampling methodologies by indicating the data uncertainty.

1. Introduction

The largest nuclear accident in history took place at the Chernobyl Nuclear Power Plant on the 26th of April 1986. The radioactive release continued for 10 days and was accompanied by a very complicated, unusual meteorological situation. In the initial period of the accident the air masses were transported westward and north-westward. Later onwards, the wind changed direction to the north-east and to the east (through the north), and from 30 April to the south and south-west. As a result, considerable parts of some republics of the former USSR were highly contaminated, namely Belarus, Ukraine and western regions of the Russian Federation; some lower levels of ground contamination were a characteristic feature for the territories of Moldavia, Latvia, Lithuania and Estonia. The central and western European countries were also contaminated by the Chernobyl fallout. Only Portugal, the western regions of Spain and the northern parts of Scandinavia were characterised by low deposition levels. As one could expect, the highest deposition levels were observed in the highlands of northern, central and southern Europe, i.e. in the Alps, the Carpathians, the Balkans, and certain regions of Scandinavia.

The monitoring of environmental objects on the presence of the Chernobyl radionuclides began in May 1986 over almost all European countries. In this study we consider only soil contamination. The data which are further being discussed were collected from European Institutes in view of the preparation of "the Atlas of Caesium-137 Contamination of Europe after the Chernobyl Accident", which was compiled under the Joint Study Project (JSP6) of the CEC/CIS Collaborative Programme on the Consequences of the Chernobyl Accident, implemented into the European Commission's Radiation Protection Research Action.

2. Overview of Sampling and Measurement Methods

From 1986 till 1991 all investigations of terrain contamination in the republics of the former USSR which were contaminated resulting the Chernobyl NPP accident were conducted using the same techniques and under the direct methodical leadership of "Goscomhydromet of the former USSR". The central and western European countries followed their own sampling and measuring strategy in case of nuclear accidental situation. As could be expected, there was no common procedure for soil sampling and analysis of the contamination by the long-lived radionuclides. Besides, there was no any agreed way of approach to measuring of the soil contamination by Cs-134 (caesium-134) and Cs-137 (caesium-137) in situ, i.e., using gamma-spectrometers located directly in the field.

Probably initially - and in some countries also at later stages - the aim of the conducted investigations were not to map the results, since less attention was paid to the regularity of the sampling net.

Nevertheless, mainly three methods were used to investigate the terrain contamination:

- Soil sampling where samples of varying area and depth are taken, eventually mixed, treated (removal of stones and dried), followed by semi-conductor gamma-spectrometer measurements in laboratory;
- Field gamma-spectrometry, where a gamma-spectrometer is mounted at a height of 1 m., and the measurements are carried out in-situ;
- Aerial gamma-spectrometry, with a scintillation or semi-conductor gamma-spectrometer mounted under a helicopter (or plane), flying along fixed routes at an altitude of 50-150 m and at a speed of 60 - 150 km per hour.

At the initial stage of the JSP6 Project (The Atlas of caesium-137 contamination of Europe after the Chernobyl accident), it turned out that no common soil sampling and measurement methodology existed for the various institutes. Table 1 shows the above mentioned differences in the ways of approach for thirteen European countries.

Table 1. Difference in ways of sampling and measurements methods in some European countries

Country		Number of sampling points	measured soil depth (cm)	Method of analysis
Albania		1	5	SAL
Bulgaria		35	deposition on plane	SAL
Croatia		4	5 and 10	SAL
Germany	FRG	250	5, 15-20	SAL, FGS
	GDR	500		
Poland		310	10	SAL
Portugal		4	?	SAL
Romania		47	5	SAL
Slovakia		388	3	SAL
Slovenia		44	12	SAL
Sweden		200,000	15-20	AGS
Switzerland		160	15-20	FGS
Turkey		38	1	SAL
United Kingdom		242	5-15	SAL

SAL: Sample analysis in laboratory
FGS: Field gamma-spectrometry (in-situ)
AGS: Aerial gamma-spectrometry

The main differences can be summarised as following:
1. In the CIS, soil sampling and analysis was carried out from the first days of the Chernobyl NPP accident, and the initial database has been updated up to now both by repeated soil sampling and analysis, and by soil sampling in new regions (previously not covered by the sampling). Two different methods of analysis of the soil contamination by gamma-radiating radionuclides have been used :
   • soil sampling with the following gamma-spectrometry;
   • air-gamma-spectrum survey.
2. In Western Europe and some countries of central Europe, sampling and analysis are not unified by common procedures, and the techniques used are not intercompared by results of intercalibration. Where aerial gamma surveys were conducted at a high level, no reliable data on comparison of these results with groundbased measurements (soil sampling) were available.
3. At the initial stage of the work at "the Atlas of Caesium-137 Contamination of Europe after the Chernobyl Accident" the differences in the approach to the ways and stages of the maps compilation have already been highlighted. A common methodology has been defined and is here briefly presented: the map compilation of caesium-137 deposition are based on manual and automated plotting of isolines of fixed deposition values (e.g. 1, 5, 15, 40, 100 Ci/km2)

on the field of values recorded in the dataset. Therefore nearly every single value has been carefully and repeatedly compared with its neighbouring values taking into account:
- orographical conditions of the territory;
- precipitation maps during the initial period;
- wind field information during the radioactive release.

3. On the Standardisation of Sampling and Measurement Methods

As an example of possible ways of standardisation of soil sampling and analysis, as well as of intercomparison of aerial gamma survey data and corresponding ground-based measurements, the way of approach initially used in the former USSR and then in Russia from 1986 up to now, is considered.

3.1. The requirements to the soil sampling [1, 2, 3]

- Samples have must be taken at a depth not less than 5 cm (during the first two years) and further at a depth not less than 10 cm.
- The place of soil sampling must be located not less than 20 m far from roads, trees, buildings and other obstacles in order to obtain representative results.
- The soil sample has to be taken without being disturbed to avoid the mixing of its layers.
- The gamma-dose rate should also be measured in order to estimate the representation of the soil sample. In this case, if we consider D_0 is the gamma-dose rate at the ground level and D_1 is the gamma-dose taken at the height of 1 m, the ratio D_0/D_1 should be equal to 1,5 - 2,0 to obtain accurate measurements.
- Checking the ratio of Cs-134 and Cs-137 is also an efficient way to determine the quality of the Cs-137 measurements. The ratio Cs-134/Cs-137 was equal to 0,56 immediately after Chernobyl accident and decreased with time.
- The measurements of the radioactivity in the soil samples must be carried out using standard gamma-spectrometry methods defined by international calibration exercises.
- These standardised methods should be defined by international calibration exercises.

3.2 The Requirements to the Aerial Gamma Survey

- The aerial gamma survey must be combined with soil sampling - 5 samples on 100 km of aerial survey. Maps on terrain contamination must be compiled based both on data of aerial gamma survey and of soil sampling;
- The requirements for aerial gamma survey for various Cs-137 deposition levels are described in Table 2;

Table 2. The requirements for the scale of survey [4].

Condition of measurement	Scale of survey	Distance between the routes (km)	One measurement on one route (km)
Preliminary territory survey	1: 1,000,000	10	2.0
Territory survey for Cs-137 levels from 18,5 till 185 kBq/m^2	1: 200,000	2	0.4
Territory survey for Cs-137 levels >1480 kBq/m^2	1: 50,000	0.5	0.1

- Reproducibility of the aerial gamma-spectrometry of Cs-137 by the same way is determined in Table 3;

Table 3. The require to reproducibility of aerial gamma-measurement of Cs-137 at the global level (4 kBq/m^2) [4].

Condition of measurement	Error (%)
on anchored land	< 10
on arable land	< 10
on forest land	< 10
on arable land with 30 cm of ploughing depth	< 15

- Relative mean square errors (rms) of Cs-137 determined by the comparison with sampling must not exceed 30-40%. The values of these errors are shown for the real measurements conducted on the territory of Russia in 1992 (See Table 4).

Table 4. The comparison of the aerial gamma-survey and soil sampling data in various regions of Russia [4].

Region	Deposition interval for Cs-137 (kBq/m^2)	average Cs-137 deposition (kBq/m^2)		nr. of measurements	systematic error (kBq/m^2)	mean square error (σ) (kBq/m^2)	rms errors contents of Cs-137 (%)
		soil sample	aerial gamma survey				
Vorenesh	3.7 - 59.2	20	18.9	103	1.1	7.4	37.0
Belgorod	11.1 - 64.8	28.1	30.0	34	- 1.9	8.9	31.6
Briansk	14.8 - 46.2	31.0	24.8	17	6.2	6.7	21.6
	3.7 - 273.9	45.5	41.1	72	4.4	10.0	22.0
	370 - 1116	540.2	570.0	47	- 29.8	96.2	17.8
	148 - 555	227.2	225.0	60	2.2	37.0	16.3
Rostov	3.7 - 25.5	7.4	6.3	90	1.1	1.8	24.3
Saratov	3.7 - 55.5	7.8	7.0	159	0.8	0.74	10.5
Tambov	3.7 - 85.1	17.5	17.4	106	0.1	7.0	40.0
Ulianovsk	9.2 - 74.0	24.8	32.9	28	- 3.2	10.0	40.3

The main requirements to the investigations of the radioactive contamination of the territories by gamma-radiating radionuclides is the necessary to use these methods of soil sampling and aerial gamma-spectral methods in complex. It is aimed to decrease the enormous mount of hand-work. E.g. in some countries of the CIS where aerial gamma survey was not performed 500,000 soil samples had to be analysed during the last ten years.

4. The Use of GIS as a Tool to highlight the Lack of Harmonisation between the various Sampling Methodologies.

4.1. Introduction

In the course of preparing the Caesium contamination atlas, Geographic Information Systems and the semivariogram were found to be a powerful combination for analysing the quality of datasets which come from different sources. This chapter will briefly discuss their combined use. In order to generate maps of contamination, we usually had to estimate the level of the radioactivity at a non-sampled spatial point, when a set of spatial data is available. One approach was to assume the data to be spatially correlated-normally it would be assumed that the correlation decreases with distance, so that the closer together two points, the more likely it is that they are similar. Formalising the intuition leads to the specification of a stochastic model for the distribution of the pollutants. To narrow down the possible choice of stochastic models, we apply *regionalised variable theory* [5], which assumes that the value of any spatial variable can be expressed as the sum of three terms: a deterministic component with a constant mean value (the drift), and two stochastic error components, a spatially correlated random component and a residual term which is spatially uncorrelated [6]. Such an approach was adopted by Kanevski [7] in his investigation of data pertaining to the Chernobyl accident.

The semivariogram is used to describe the spatial variability of the data, in order to select a theoretical model to be used for the prediction of the unknown points. It is defined as

$$\hat{\gamma}(\mathbf{h}) = \frac{1}{2N(\mathbf{h})} \sum_{i=1}^{N(\mathbf{h})} \left[z(\mathbf{x}_i) - z(\mathbf{x}_i + \mathbf{h}) \right]^2$$

where $z(\mathbf{x})$ is the value of the radioactivity at point $\mathbf{x}$. The distance and direction between $\mathbf{x}$ and $\mathbf{x+h}$, defined by vector $\mathbf{h}$, is termed the lag of the semivariogram. $N(\mathbf{h})$ is the number of pairs of observations separated by the lag.

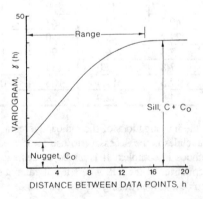

Fig. 1. Example of a spherical semivariogram model [8]

The range is defined as the distance at which the observations become independent.
The sill is the maximum value reached at the range and is equal to the variance of independent observations. Comparing the sills between two semivariograms in nearby regions is a good way

to indicate the different sample sizes and measurement methods and preferential sampling strategies.

The nugget, which should be zero in theory, represents the stochastic component of the variance at a given scale of observation. It includes the variation due to an irregularity of the studied phenomenon at a very small scale and the variation due to errors in measurements. The ratio of the nugget effect to the sill is often referred to as the realtive nugget effect and reflects in a standardised way the variations from the point of view of the sampling and measurement methods between close or overlapping datasets.

Other statistical tools for examining the spatial correlation structure of a dataset include h-scatterplots, correlograms, madograms, crossvalidation techniques and have been described in [9] and [10].

Prior to define the semivariogram model checking for data quality, clusters, trends and discontinuities is essential [11]. This primary analysis has revealed the wide variety between sampling methodologies and measurement methods.

4.2. Data Quality Analysis

In view to improve the effectiveness of this primary analysis, a general quality control procedure was defined: the data were sorted depending on their place of origin to obtain data with identical histories. Those with inaccurate positions or with an extravagant Cs-137/Cs-134 relation were deleted. It was expected that this global filtering would result in homogeneous datasets between which comparisons of relative quality could be made.

4.3. The Variety of Sampling Strategies

Two types of sampling strategies have been encountered during the preparation of the atlas: regular and irregular sampling networks.

Since a regular sampling generally tends not to generate too many problems during primary analysis (if periodicity in the data is not suspected), much more attention has been paid to irregular sampling networks. Random and preferential samplings may lead to spatially clustered data. In general, preferential sampling occurred at strategically sites such as cities and power plants; i.e. areas with higher levels of contamination tended to be oversampled. This is easily shown by displaying cartographic information overlaid with indicator maps, by checking duplicated sample locations in data sets and, by displaying Voronoi polygons [12]. Therefore, in order to obtain representative distributions of data for further analysis, the different weights of these data had to be taken in consideration.

Discontinuities in these irregular sampling scheme could be explained most of the time with the help of additional information provided by cartographic information: the display of the political borders and geographical barriers (seas, lakes, mountains) often explain the lack of data at certain places. Therefore, depending on the required map resolution of the investigated area, interpolating data including these discontinuities had to be done with extreme care.

Another type of discontinuity we have encountered was due to the different sampling and/or measurement methods. Two neighboured datasets can show at their borders a "jump" in the values of the measured variable.

In previous papers, Burgess, Webster, and McBratney [13] have shown how to make better use of data or sampling resources for isarithmic mapping. Reconnaissance samplings should be made in view to obtain the semivariogram and so to define the appropriate sampling strategy. If the semivariogram is not known, the best strategy would be to sample on a regular grid with an interval determined by the number of observations that can be afforded. Irregular sampling strategies appeared in our analysis to have been generally preferred to regular networks, especially in the western countries.

Another important difference between the sampling strategies we have encountered was the density of samples.

4.4. Sampling Density:

As underlined by Burrough [14], it should always be possible for a given aim "to devise a local optimum for sampling that will give the most precise and the most accurate results for a given expenditure". The different sampling densities show that the objectives were slightly different between the countries. Apparently, a common scenario that followed the Chernobyl accident was for every country to perform global monitoring to obtain a first estimate of the radioactive fallout. Based on these measurements new samplings were organised to define zones of higher contamination. Also, because of the very complex spatial structures of the radioactive trails, higher density samplings were often necessary to describe these zones in order to reduce the probability that some regions of higher contamination would be omitted [15].

In consequence, the quality of the estimated radioactivity at unsampled locations is directly related to the amount of data surrounding these locations. Therefore this lack of harmonisation in the sampling density leads to a loss of data quality in zones of low level contamination when estimations are made.

5. Conclusion

When "The European Atlas of Caesium-137 after the Chernobyl accident" project was carried out, analysis of spatial data made clear the need for further collaboration between the countries. The variety of sampling and measurement methods has been shown, due to the lack of harmonisation between the countries. Common requirements to the application of soil sampling by aerial gamma spectrometry and ground-based gamma spectrometry were stated to define future monitoring of the radioactivity. Also, in order to compile maps on caesium-137 deposition, there is a need for deeper analysis of the very local parameters affecting radioactive fallout if complex interpolation methods are required. Finally, after sharing experiences between the members of the CEC and the CIS involved in the JSP6 project, considerable progress has been achieved in developing common methodologies in the fields of radioecology and spatial data analysis. In the future, an optimal monitoring network based on this experience should certainly improve decision making by indicating general contamination patterns based on harmonised measurements, by reducing sampling cost and by conducting further investigations for the contamination assessment at very local scales.

References:

[1] Special Interministry Commission of radioactive control and monitoring of environment (1993) "General direction of investigation, estimation radioactive situation and mapping radioactive contamination on the territory of Belarus republic". Minsk, 1993, 5 p.

[2] Circular of Ministry of extraordinary situation and social protection population after Chernobyl catastrophe, 1993, 5 p.

[3] Report of National Commission of Radioactive Protection of Belarus, Minsk, 1993, 5 p.

[4] Izrael Yu. A, Nazarov I. M., Fridman Sh.D., Kvasnikova E.V. 'The radioactive contamination of the CIS European part in 1992 from the Chernobyl nuclear power plant accident'. Collection of articles, edited by Sh. D. Fridman and A. N. Pegoev, Saint Petersburg, Gidrometeoizdat, 1994, pp 16-51

[5] Matheron, G. (1963) Principles of Geostatistics, Econ. geol. 58, 1246-1266

[6] Burgess, T.M. Webster. r. (1980a) Optimal interpolation and isarithmic mapping of soil properties, 1. The semi-variogram and punctual Kriging, J. Soil Science **32**, 315-331.

[7] Kanevski, M. (1994) Chernobyl fallout: stochastic simulations of the spatial variability and probabilistic mapping, Preprint Nsi-23-94. Moscow: Nuclear Safety Institute, July 1994, 19 p.

[8] Zirschky, J. (1985) Geostatistics for Environmental Monitoring and Survey design, Environment International, Vol. 11, pp. 515-524, 1985

[9] Cressie, N.(1991) Statistics for Spatial Data. John Wiley & Sons, N.Y. 900p.

[10] Isaaks, E.H., Srivastava, R.M. (1989) An Introduction to Applied Geostatistics. N.Y., Oxford University Press 561 p.

[11] Deutsch C., and Journel, A.G. (1992) GSLIB. Geostatistical Software Library and User's guide. Oxford Univ. Press. 340 p.

[12] Kanevski, M., Arutyunyan, R.V., Bolshov, L.A., Linge, I.I., Savel'eva, E., Haas T. (1993) Spatial data analysis of Chernobyl fallout, 1. Preliminary results, Preprint NSI-23-93. Moscow: Nuclear Safety Institute, September 1993, 51p.

[13] Burgess, T.M. Webster, R., McBrackney, A.B. (1981) Optimal interpolation and isarithmic mapping of soil properties, IV. Sampling Strategy, J. Soil Science **32**, 643-659.

[14] Burrough, P.A. (1991) Sampling Designs for Quantifying Map Unit Composition, in Spatial Variabilities of Soils and Landforms, Soil Science Society of America, SSSA Special Publication no. 28, pp 89-125.

[15] The International Chernobyl Project, Technical report, Annex 1: Frequently used sampling procedures for the assessment of Chernobyl fallout in the USSR, p. 191

PSYCHOSOCIAL CONSEQUENCES OF THE CHERNOBYL DISASTER.
A survey of psychological and physical well-being in an exposed
and a non-exposed population sample.

J.M. Havenaar[1], T.J.F. Savelkoul[1], J. van den Bout[2] and P.A. Bootsma[3]

[1] University Hospital Utrecht
[2] Dept. of Clinical and Health Psychology, Utrecht University
[3] National Institute of Public Health and Environmental Protection
The Netherlands.

Abstract. The importance of psychological factors in the aftermath of industrial disasters is being recognized increasingly. Two field studies (total N=3084) were conducted in two regions of the former Soviet Union, to investigate the long-term psychosocial consequences of the Chernobyl nuclear power plant disaster in 1986. A subsample of the respondents (N=449) was studied using a standardized physical and psychiatric examination. The first study took place in the Gomel region (Belarus) in the direct vicinity of the damaged nuclear plant. A control study was conducted in the Tver region (the Russian Federation), about 250 km north-west of Moscow. The results of the study indicate significantly higher levels of psychological distress, poorer subjective health and higher medical consumption in the exposed population. These findings were most prominent in risk groups such as evacuated people and mothers with children. No significant differences in overall levels of psychiatric or physical morbidity were found. Radiation related diseases could not account for the poor health perception in the investigated sample. These results indicate that psychological factors following the Chernobyl disaster had a marked effect upon psychological well being, on perceived health and on subsequent illness behaviour. Fears about future health play a key role in determining this response. The provision of adequate information to the public as well as to the public health services may be important to counteract these fears.

1. Introduction

Although the nature and extent of the physical health damage caused by the Chernobyl disaster continues to be a subject of considerable debate [1-4], there seems to be a consensus that the psychosocial consequences are substantial [5]. So far, however, hardly any studies providing solid evidence to document this have been published in the international scientific literature [6]. In this paper we describe a comparative study about the long term psychosocial consequences of 'Chernobyl'. The study was conducted in 1992-1993 in two regions in the former Soviet Union, one in the direct vicinity of the partly destroyed nuclear plant at Chernobyl and one approximately 1000 km to the north-east well outside of the range of significant fall-out deposits.

The first survey was conducted in the in the Gomel region (Republic of Belarus), one of the most seriously contaminated regions. The survey was set up as the assessment phase of a Belarussian-Dutch humanitarian aid project ('The Gomel project'). It served to provide a solid epidemiological basis for the development of psychosocial and health education services in the region [7]. A second survey was conducted 6 months later in the Tver region (the Russian Federation) about 250 km north-west of Moscow as a control study. The Tver region was selected because of its comparable socio-economic structure and population size. Just as in Gomel, there is a nuclear power plant at about 100 km distance of the regional capital. The Tver region has not significantly been contaminated by fall-out from Chernobyl.

The aim of the study is to test the hypothesis that, in contrast to convictions generally held by the inhabitants of the exposed region, their symptoms and complaints cannot be explained by exposure to ionizing radiation, but rather to psychological distress and subsequent changes in illness behaviour.

2. Methods

2.1. Subjects

Because no reliable records of the inhabitants of the regions were available, it was not possible to draw random samples of the population. As an alternative, we started by taking a sample of employed inhabitants of the exposed region, supplemented with a sample of students and retired or unemployed people. In this way we obtained a broad sample in which all strata of the population were represented (at the time, according to official statistics, more than 99% of the adult population was either employed, retired or student). The study was conducted using a two-phase sampling design [8]. During the first phase the population samples described above were examined using a self-report questionnaire. From these samples stratified subsamples were drawn for participation in the second phase of the study, which consisted of standardized clinical examination. For the recruitment for this second phase, respondents from the population samples were divided into three strata according to their level of psychological distress as measured using the General Health Questionnaire (GHQ; see below). Respondents with high or medium GHQ-scores were given a higher selection rate for phase-2 in order to study respondents with high levels of distress in greater detail. A detailed description of study design and sample characteristics has been published elsewhere [9]. A schematic description of sampling schemes and response rates is given in table 1.

Table 1. Schematic representation of study design and sampling fractions

Phase 1 Screening of population sample using the GHQ-12 and MOS-SF		
	Gomel	**Tver**
Sampled for phase 1	1763	1620
Participated in phase 1	1617 (91.7%)	1427 (88.1%)
Stratified sampling for phase 2		
GHQ-12 score: 0,1	1:10 sampled	1:10 sampled
2-7	1:5 sampled	1:4 sampled
8-12	1:3 sampled	1:2 sampled
Phase 2 Standardized physical and psychiatric examination (ICD9-CM and DSM-R diagnoses)		
Sampled for phase 2	322	284
Participated in phase 2	265 (82.2%)	184 (64.8%)

2.2. Instruments

The phase 1 respondents (N=3084) were examined using a self-report questionnaire to assess psychological well-being, subjective health and health-related behaviours. Psychological well-being was studied using the General Health Questionnaire, 12-item version (GHQ-12), which is a widely used self-report questionnaire for psychological distress [10,11]. Subjective health was assessed using a single item derived from the Medical Outcomes Study questionnaire, Short Form (MOS-SF). This single item, in which the respondent is asked to rate his or her general health on a 5-point scale ranging from excellent to poor, has been shown to be a valid measure for health-related quality of life in a number of studies [12,13]. The phase 1 questionnaire further contained items concerning the number of visits to doctors and the use of prescription drugs during the previous month.

The clinical examination conducted during the second phase of the study (N=449), consisted of a standardized examination by a Dutch physician specialized in internal medicine, who administered a standardized full medical history, evaluation of current complaints, and a physical and basic laboratory examination, including whole blood count, hepatic, renal and thyroid function tests. In addition to establishing clinical diagnoses according to ICD9-CM, the physicians rated the overall health of the respondents on the same 5-point scale ranging from excellent to poor, that was administered to the respondents in phase 1. Using performance status as a guideline, patients with a score of 4 (fair health: good performance status, but with a disorder demanding medical attention) or a score of 5 (poor performance status and clinically ill) were counted as 'clinical case'.

A psychiatric examination was performed by specially trained Byelorussian and Russian psychiatrists, who administered a semi-structured interview for making DSM-III-R diagnoses (the Munich Diagnostic Checklist for DSM-III-R) [14,15]. For all questionnaires a time frame of one month (last four weeks) was used.

2.3. Statistical analysis

In order to reverse the effects of over-sampling of cases with high GHQ scores, prevalence estimates for phase 2 parameters were estimated by weighting the results back to phase 1 proportions, using the observed sampling fractions as weights (i.e. corrected for non-response). Univariate odds ratio's (OR) were calculated to estimate the relative risk associated with living near Chernobyl for the health indices measured in the study. Multivariate logistic regression was performed to calculate adjusted odds ratio's (AOR), correcting for potential confounding by sex, age, civil status and education. The effects of weighting on standard errors were corrected by using the PCCARP computer programme [16]. Bonferroni-Holm correction was performed to correct for the effects of multiple pairwise comparisons [17].

In addition the relative risk for psychological distress was estimated for a number of potential risk factors within the exposed sample. These risk factors were: living in a moderately or severely contaminated zone (resp. > 185 kBq ^{137}Cs/m^2 and > 555 kBq ^{137}Cs/m^2), having been evacuated, having participated in clean-up activities following the disaster ('liquidators'), being a mother with children younger than 18 years old living at home.

3. Results

Table 2 shows the health indices in Gomel and Tver as they were found in both phases of our study. It is apparent that the general health status of both samples is rather poor. There are high rates for all self-reported and clinical health indices collected in our study.

Table 2. Self-reported and clinical health indices in Gomel and Tver

Self-reported health	Prevalence		Univariate OR		Adjusted OR[*]	
	Gomel	Tver	OR	95% CI	AOR	95% CI
Psychological distress (GHQ-12)	64.8	48.1	**1.93**	1.69-2.22	**2.03**	1.75-2.37
Health 'fair' or 'poor' (MOS-SF)	74.5	56.5	**2.25**	1.96-2.58	**2.80**	2.35-3.34
Visited doctor last 4 weeks	47.7	41.1	**1.31**	1.14-1.50	**1.38**	1.18-1.61
Used medication last 4 weeks	69.9	60.4	**1.52**	1.30-1.78	**1.55**	1.33-1.84
Clinical status						
Any ICD-9CM physical diagnosis	63.7	55.1	1.43	0.94-2.17	1.57	0.99-2.49
Any DSM-III-R psychiatric diagnosis	35.8	37.1	0.95	0.64-1.41	1.08	0.70-1.67

[*] Adjusted OR (AOR): odds ratio adjusted for sex, age, marital status and education.

All statistically significant OR's and AOR are typed in bold face.

Around 50% or more of the respondents report unsatisfactory health or psychological well-being. A similar proportion has consulted a doctor and/or used medication during the previous month. More than a third of the sample suffers from a diagnosable medical or psychiatric condition. As may be observed, all self-reported health indices showed substantially higher rates in the exposed population, especially the variable MOS-SF, our measure of subjective health, and the GHQ, our measure of psychological distress (p < .001). For the parameters that indicate medical service utilisation and use of medication differences were less dramatic, but still statistically significant (p < .001).

For the parameters of clinical health, collected in the second phase of the study, differences did not reach statistical significance at a 5% level. There was a trend towards more physical illness in the Gomel sample. This was especially the case with respect to obesity, hypertension, angina pectoris, anaemia and non-malignant thyroid abnormalities. None of these differences reached statistical significance at a 5% level. Importantly, no differences were observed for diseases which could be attributed to the effects of ionizing radiation. Somewhat contrary to expectations, no differences could be demonstrated in the prevalence of psychiatric disorders.

In table 3 the results of the analysis of risk-factors for psychological distress in the Gomel sample are shown. As may be seen, there is no statistical relation between level of contamination of the area of residence and psychological distress. An increased risk for having a high sore on the GHQ was predicted by two risk factors: being a mother with one or more children under 18 years living at home and having been evacuated after the disaster, the latter being only statistically significant after adjusting for sample composition (AOR). No increased risk was observed in the group of liquidators.

Table 3. Risk-factors for psychological distress in the Gomel region

Risk factor	Gomel	Univariate OR		Adjusted OR	
	%	OR	95% CI	AOR	95% CI
Overall	64.8				
living in area > 185 kBq ^{137}Cs/m^2	67.8	1.25	1.00-1.55	1.26	0.99-1.59
living in area > 555 kBq ^{137}Cs/m^2	71.6	1.49	0.88-2.53	1.54	0.89-2.66
evacuated	72.0	1.42	0.85-2.36	**1.45**	1.17-1.80
liquidator	66.9	1.11	0.78-1.57	1.21	0.83-1.75
mother with child <18 yrs.	73.2	**1.80**	1.34-2.42	**1.88**	1.37-2.57

All statistically significant OR's and AOR are typed in bold face.
' Adjusted OR (AOR): odds ratio adjusted for sex, age, marital state and education.

4. Discussion

Perhaps the most striking finding of this study is the fact that more than 50% of a presumably healthy and active sample of adults living in the former Soviet Union perceive it's health to be less than satisfactory. The percentage of respondents with a high GHQ-score (64.8% and 48.1%) has only been matched under extreme conditions, such as in the immediate aftermath of a natural disaster [18]. The alarming prevalence figures for medical conditions in terms of ICD-9CM (more than 50%), are consistent with recent health statistics published by WHO [19]. Over the past 15 years all former Soviet republics have witnessed increasing levels of morbidity and diminishing life expectancy. A prevalence of more than 35% of the population suffering from diagnosable psychiatric illness, meeting internationally accepted criteria, is the highest reported in the literature [20,21].

The results of this study indicate that people in the exposed region report more psychological distress, poorer health related quality of life and higher medical consumption. Our study provides evidence that, even though the people in the Gomel area experience their health as substantially worse than people from in a non-exposed area, these differences may not be attributed to radiation induced diseases, as is their widely held conviction. Rather, it appears that psychosocial factors are responsible for these differences. In this respect, it appears that mothers with young children and to a lesser extent evacuees are the most vulnerable groups.

The psychological distress and the loss of health related quality of life are not merely 'subjective' epiphenomena of a disaster, which in other aspects turned out not as bad as was initially feared. Also, as was clearly demonstrated in this study, people's worries and distress concerning the disaster are not a form of mental disturbance or 'radiophobia'

[22], because psychiatric morbidity was equally high in both regions. Instead, the public distress is a reality with a direct relevance for policy makers. It should be a central focus in national and international relief programmes set up in the area, such as for example the 'Gomel project' [7] and the UNESCO Chernobyl programme [23].

The psychological stress induced by the disaster has profoundly changed health related behaviour patterns in the affected population. In the perception of the population, minor everyday symptoms have become potential heralds of serious radiation induced disease, which needs to be thoroughly examined by a doctor. In this respect the role of the local health professionals is a crucial one. In an earlier study in the Gomel region we have demonstrated that most doctors perceive 'Chernobyl' as the most important threat to health, thereby fuelling their patient's worries and anxieties [24]. Further development of services, which can provide psychosocial support and health education programmes to the public as well as to health professionals is needed.

Acknowledgements

1. This study was conducted in the framework of a Belarussian-Dutch humanitarian aid project to alleviate the consequences of the Chernobyl disaster. The project was sponsored by the government of The Netherlands and executed by the National Institute of Public Health and Environmental Protection in cooperation with the University Hospital, Utrecht, The Netherlands.

2. The authors wish to thank W van den Brink, T Wohlfarth and MWJ Koeter for their valuable advice and practical support in the analysis of these data.

LITERATURE

[1] Dorozynski A: Chernobyl damaged health, says study. BMJ 1994; 309: 1321.

[2] Williams, D. Chernobyl, eight years on. Nature, 1994; 371: 556.

[3] Boice J, Linet M: Chernobyl, childhood cancer, and chromosome 21. BMJ, 1994; 309:139-40. Editorial authors' response. BMJ, 1994; 309: 1300.

[4] Butturini A, Izzi G, Benaglia G, Lloyd D, Pass B, Gale RP: Not all health problems seen close to Chernobyl can be attributed to radiation. BMJ, 1994; 309: 1299.

[5] Coryn P, MacLachlan A. Mental disorders said spreading among Chernobyl-affected people. Nucleonics week, 1995, July 6: 13-15.

[6] Viinamäki H, Kumpusalo E, Myllykangas M, Salomaa S, Kumpusalo L, Komakov S, Ilchenko I, Zhukowsky G, Nissinen A. The Chernobyl accident and mental wellbeing - a population study. Acta Psychiatrica Scandinavica, 1995; 91: 396-401.

[7] Nijenhuis MAJ, van Oostrom IEA, Sharshakova TM, Pauka HT, Havenaar JM, Bootsma PA, Savelkoul TJF. Belarussian-Dutch Health Information Centre Gomel, Belarus. Belarussian-Dutch Humanitarian Aid Project: "Gomel Project". RIVM report no. 801002005. National Institute for Public Health and Environmental Protection, Bilthoven, the Netherlands, 1995.

[8] Duncan-Jones P, Henderson S: The use of a two-phase design in a population survey. Social Psychiatry, 1978; 13: 231-37.

[9] Havenaar JM, Savelkoul TJF, Poelijoe, NW, Kaasjager K, Westermann A van den Brink W, van den Bout J. Report on a health survey in the Tver region (Russian federation). A comparison with the general health status of the population in the Gomel region (Belarus) and the Tver region (Russia). National Institute of Public Health and Environmental Protection, Bilthoven, the Netherlands ,1995

[10] Goldberg D & Williams P. A users guide to the General Health Questionnaire. NFER-Nelson: Berkshire, 1988.

[11] Havenaar JM, Poelijoe NW, Kasyanenko A, van den Bout J, Koeter MWJ. Screening for psychiatric disorders in an area affected by the Chernobyl disaster. The reliability and validity of three psychiatric screening questionnaires in Belarus. Psychological Medicine, 1995 (in press).

[12] Cunny KA, Perri III M: Single-item vs multiple-item measures of the health-related quality of life. Psychological Reports, 1991; 69: 127-30.

[13] Kempen GIJM: The MOS short-form general health survey: single item vs multiple measures of health-related quality of life: some nuances. Psychological Reports, 1992, 70, 608-610

[14] Hiller W, Zaudig M, Mombour W. Development of diagnostic checklists for use in routine clinical care. A guideline designed to assess DSM-III-R diagnoses. Archives of General Psychiatry, 1990; 47: 782-784.

[15] Havenaar, JM, Rumyantzeva GM, Filipenko VV, van den Brink W, Poelijoe NW, van den Bout J, Romasenko L. Experiences with a checklist for DSM-III-R in the Russian Federation and Belarus. Inter-rater reliability and concurrent validity of the Munich Diagnostic Checklist for DSM-III-R in two former Soviet countries. Acta Psychiatrica Scandinavica, 1995 (in press).

[16] Fuller WA. (). Statistical Laboratory. Iowa State University, 1986.

[17] Seaman MA, Levin IR, Serlin RC. New developments in pairwise multiple comparisons. Some powerful and practicable procedures. Psychological Bulletin, 1991; 110: 577-86.

[18] Fairly M, Langeluddecke P & Tennant C. Psychological and physical morbidity in the aftermath of a cyclone. Psychological Medicine, 1986; 16: 617-676.

[19] World Health Organisation. Highlights on health in the Russian Federation, Belarus and Ukraine. Unedited draft. World Health Organization, Denmark, 1992

[20] Wittchen H-U, Ahmoi Essau C, van Zerssen D, Krieg J-C & Zaudig M. Lifetime and six-month prevalence of mental disorders in the Munich follow-up study. Archives of Psychiatry and Clinical Neurosciences, 1992; 41: 247-258.

[21] Kessler RC, McGonagle KA, Zhao S, Nelson CB, Hughes M, Eshleman S, Wittchen H-U & Kendler KS Lifetime and 12-month prevalence of DSM-III-R psychiatric disorders in the United States: results from the National Comorbidity Survey. Archives of General Psychiatry, 1994; 51: 8-19.

[22] Drottz-Sjöberg BM and Persson L: Public reaction to radiation: fear, anxiety, or phobia? Health Physics, 1993; 64, 223-31.

[23] Horich L, Garnets ON, Panok VG. Social-psychological rehabilitation centres conception in regions affected by the Chernobyl NPP accident. Paper presented an the International conference on the mental health consequences of the Chernobyl disaster: current state and future prospects. Association 'Physicians of Chernobyl', Kiev, Ukraine, 1995.

[24] Bout van den J, Havenaar JM, Meijler-Iljina LI. Health problems in areas contaminated by the Chernobyl disaster: radiation, traumatic stress or chronic stress? In: Kleber RJ, Figley C, Gersons B (eds). Beyond trauma: cultural and societal dynamics. Plenum Press, New York, 1995.

THE INFLUENCE OF SOCIAL AND PSYCHOLOGICAL FACTORS IN THE MANAGEMENT OF CONTAMINATED TERRITORIES

G. M. Rumyantseva[1], B.-M. Drottz-Sjöberg[2], P. T. Allen[3], H. V. Arkhangelskaya[4], A. I. Nyagu[5], L. A. Ageeva[6], and V. Prilipko[5]. [1]The Serbsky Center of Social and Forensic Psychiatry, Moscow, Russia. [2]Center for Risk Research, Stockholm, Sweden. [3]Robens Institute, University of Surrey, Great Britain. [4]Institute of Radiation Hygiene, St. Petersburg, Russia. [5]Institute of Radiation Medicine, Kiev, Ukraine. [6]Institute of Sociology, Minsk, Belarus.

ABSTRACT

The paper outlines briefly the evolution of the social and psychological situation in the Soviet Union and the Commonwealth of Independent States (CIS) from the time of the Chernobyl accident in 1986 to 1995. The empirical material presented is based on several survey studies conducted in Russia, Ukraine and Belarus within the Joint Study Project 2 (JSP2) in 1992-95. Investigated topics included e.g. perception of risk, trust in information, everyday worries, level of stress and self-rated health, as well as knowledge of radiation. It is suggested that major political changes combined with a declining standard of living have contributed a background of uncertainty, vulnerability, and distrust in official information and widespread dissatisfaction. Reactions to the radiological situation have included strong expressions of exposure to risk, experiences of helplessness, and dependencies on authorities, but little adaptation with respect to adoption of appropriate behaviors for personal protection. It is recognized that the social and psychological conditions of the populations directly affected by the accident (i.e. people living in contaminated areas, as well as relocated people) differ from others in the long term perspective, although the Chernobyl accident had general and worldwide psychological effects when it occurred. Risk perception increased over time, peaking between 1989-91 in response to political events and intensified mass media coverage. The development in the 1990's included significant differences between affected and non-affected populations with persistent stress reactions in the former group. Health anxiety was a major feature among these reactions and many symptoms and misfortunes were attributed to the accident. Differences in legislation across states after 1991 point to their potentially disparate impact on public reactions, e.g. different intervention levels and degree of voluntariness in the context of relocation were associated with different adaptive behaviors. Financial compensation based on the notion of victimization may have reinforced expressions of helplessness, vulnerability and self-reported low health status. The difficult issue of health consequences related to radiation in combination with extensive medical examinations and communication problems furthermore seem to have enhanced worries about current and long term health effects. The importance of decision makers taking account of social and psychological factors in the management of radiological accidents is emphasized and the central role of correct and continuous information is acknowledged and specified regarding type and focus in the medium and long term time perspectives. Information to populations in areas to which people may be relocated is discussed, as well as the social psychological influences of utilized countermeasures and their relationships to decision making and public reactions. It is suggested that less overall negative psychological impact could be achieved by regular monitoring of public opinions and sentiments, a general availability of information and medical care combined with selective medical examinations focused on vulnerable groups, selected on the basis of predictive studies of the health development, time limited financial compensation and the distribution of compensation or benefits in relation to adaptive protective behavior.

1. Evolution of the social and psychological situation

The Joint Study project 2 was conducted in an historic era; in the very first years of the existence of the Commonwealth of Independent States (CIS). Our task was to investigate the social and psychological effects of the Chernobyl accident - an accident which occurred in 1986 in the former Soviet Union, within a different economic, political and legal framework. The results of our investigations must be related to the changes introduced by the policy of Perestroika and Glasnost, and the intensified social uncertainty which occurred during the mid 1980's and beginning of the 1990's. The Chernobyl accident divides a more than 70 years old social system from a new time. Its significance will therefore remain salient in people's minds for a long time. The accident's special characteristic of spreading an invisible risk over the motherland and the world contributed greatly to its reputation.

The project was not designed to investigate cultural traditions or to compare social systems. It should be noted, however, that the salience and role of social factors are part of the society from which they emerge. Cultural traditions, satisfaction with everyday life and expectations related to current and future events offer a background against which the important social and psychological dimensions arise. The extent and nature of simultaneously occurring societal changes belong to this type of dimensions. Traditions, experiences and expectations are carried and transmitted by human minds and behaviors. Societies differ with respect to acceptability of expression of possible reactions. The type of society is thus another important social factor accounting for public reactions. Such reactions reflect e.g. different expectations of executive authority, responsibility and disaster mitigation. The expectations provide a basis for blame and praise. They influence experiences, and guide behavior in certain directions. Individual and mass psychological reactions adjust to the circumstances and can be expected to emerge in socially traditional forms.

2. Investigations of public reactions

2.1. Studies, designs and methods

The joint studies were initiated in 1991/92 and pilot investigations were conducted by a team of Russian and European researchers in the Bryansk region of Russia. The design of these studies aimed at describing reactions to the Chernobyl accident among people in an affected region, to investigate the feasibility of various methodological approaches, and to examine preliminary hypotheses related to stress reactions and perceived risk. The performance of the task built on previous studies and experiences of Russian researchers [1,2]. In 1993 the work expanded considerably and researchers from Russia, Belarus, Ukraine and Europe collaborated in several major survey studies. The surveys conducted in 1993 to 1995 were mainly designed to investigate psychological reactions, including measures of stress, personal mastery, common sentiments, and perceptions of risk related to various hazards. However, they also included measures of trust in information sources, self-rated knowledge of radiation, perceived degree of radioactive contamination of the home area, and standard of life. Respondents were usually selected by quota sampling methods from affected and non-affected populations. Additional studies included media analyses of number of articles, types of authors and emotional content of Chernobyl related materials, collection of official statistics regarding e.g. health related issues, and a survey among relocated people in the three CIS states[1].

Three general methods were used by the Robens Institute and CIS team. First, a pilot study utilizing tape recorded semi-structured interviewing was designed to elicit the main concerns of people in an affected area. Stress was expected to be problematic [3,4], and measurements of psychological distress were made using the GHQ-28 instrument [5,6]. Second, larger scale surveys within a quasi-experimental framework were conducted (N=1800). They were designed to examine differences between groups with statistical confidence. The third method used the data collected at surveys but relied on regression methods to examine the relative contribution of factors on dependent measures; initially stress and subsequently ingestion dose.

The work conducted by the team of researchers in CIS and the Center for Risk Research included surveys, compilations of official statistics, and media analyses. A pilot questionnaire study was carried out in 1992 in the Bryansk region [7]. A major survey in 1993 (N=3067) included people in three CIS states living in contaminated areas, including a group from the 30 kilometer zone, non-contaminated areas geographically close to those contaminated, resettled people and control groups in distant, non-contaminated regions [8]. The questionnaire was developed on the basis of results and experiences from the pilot study and related work [9-12]. Analyses of newspaper materials were conducted in 1994 [13-15]. The survey of relocated people in 1995 (N=598) was designed to investigate psychological effects of voluntary and involuntary resettlement, and the hypothesized positive influence of the newly introduced UNESCO centers for psychological rehabilitation on perceived health and well-being.

2.2. Reactions to the radiological situation

Previous studies had shown that psychological problems associated with the Chernobyl accident did not decrease with time [2]. Rich qualitative data were obtained from the first interviews and a number of main features emerged in the pilot investigations in Bryansk, 1992 [16,7]. There appeared to be a pervading sense of helplessness and pointlessness in the affected region. Mistrust and anxiety were rife. Worries were focused on health risks due to radioactive contamination, but there was also considerable concern about hardships of everyday life. Concerns about health, material standard and social life correlated between 0.50 and 0.74. The best predictor of personal distress due to the accident [7] was fear of health effects for oneself and the family, followed by self-rated health and worry about social consequences due to the accident[2]. Fear of health effects was also the best predictor of everyday worry, but here attention to media and radiation knowledge also made significant contributions[3].

The Chernobyl accident was followed by massive medical screening of the affected populations. The pilot study showed that although 81 percent of the participants affirmed that they had been subjected to a medical examination due to the radiological situation after the accident, as many as 74 percent also reported that they did not know if they had received a dangerous dose of radiation. The result was interpreted as indicating a lack of efficient communication in relation to the medical screening. Self-rated health showed overall low ratings, and women reported lower values than men. The best predictor[4] of self-rated health was having a disease diagnosed by a doctor, followed by personal distress due to the accident. Trust in various information sources was overall low, but foreign experts gained a higher rating than domestic experts, and health promotion bodies were more trusted than various political bodies. People who paid more attention to media also gave higher ratings of everyday worry. The result inspired further investigations of trust related variables and an interest in what media output people had been exposed to.

2.3. Exposure, distress and self protection

Survey studies in the CIS states showed that distress levels were rated rather highly [4,8,17-19]. And as expected, people in the contaminated and resettled areas, as well as the 30 kilometer zone, gave higher ratings as compared to respondents of the control and the non-restricted areas. People experienced exposure to real risk. Some results are shown in figure 1. Note that resettled people provided the highest risk ratings. The figure also illustrates that people acknowledged some benefits related to the accident. Those may include improved medical services and infrastructure projects.

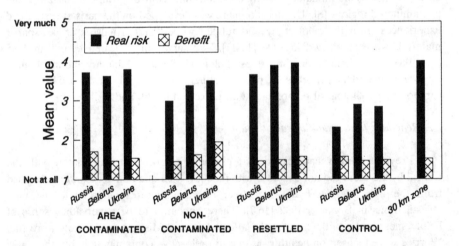

Figure 1. Mean values of rated exposure to real risk, and benefits, due to the Chernobyl accident, by states and types of residential areas.

The response to a question about ability to protect oneself from radioactive contamination was overall discouraging. People indicated an interest, however, in improving their knowledge of radiation and radioactive contamination. There are a number of social cognition models within social psychology which have been applied to health related problems [20-27]. These theories have in common the attempt to predict health related behaviors and have been used particularly to examine protective behaviors. The early results pointed to the importance of the role played by some form of 'fatalism' and the investigation of Locus of Control revealed that people in the affected areas were less likely to believe in the effectiveness of taking responsibility for their own health. These results highlight 'outcome expectancies'. Also a direct measure of fatalism [28] was adopted[5]. In the clean areas personal characteristics and Locus of Control had significant effects; women, those with dependent children and older people had higher stress whilst internal Locus reduced stress. Worries about everyday life increased stress. In the contaminated areas factors specific to radiation also had effects, e.g. self attributed radiation knowledge had an effect in reducing stress. Similarly, the extent to which people believed they could affect the amount of dose they received also reduced stress. However, trust was negatively indicated. It is clearly having some effect on stress and, since it is known to be low, this indicates an exacerbation of the stress problems.

Although knowing the effects of various factors on stress is useful, e.g. in revealing what needs to be addressed when attempting to improve countermeasures, it is also probable that the behavior of people will affect their dose uptake. A working hypothesis was adopted that some characteristics of people, including their general outlook, might lead them to act in different ways and hence to receive relatively more or less dose. A number of whole body Cs137 measures were taken in conjunction with psychological questionnaires [29]. The results showed e.g. that men had a higher ingestion dose, after standardization, than women. More interesting, however, was the finding that Fatalism had an effect, in the direction expected by the general hypothesis[6]. The result indicates that people who tend to believe that things are determined by fate are somewhat more likely to have a higher dose compared to others.

2.4. Reactions among relocated groups

The effects on the stress of people in the relocated areas were somewhat different and highlighted the continuing problems that these people have [30,31]. In addition to the factors which predicted stress in the clean areas the relocated populations were still subject to concerns about contamination, perhaps reflecting the fact that some relocations have simply placed people back into contaminated regions, or maybe a result of experiences due to life in contaminated areas. The effect of having some radiation knowledge was lower for these people. It is also worth noting that the everyday worries which helped predict stress in the normal populations of the control areas did not disappear in the affected regions but were added to by the special concerns of those people. Trust was also a factor amongst the relocated people but appeared to be much more salient for them than for other groups[7]. This suggests that whilst issues of dose control are important for people in the contaminated areas this role is replaced by concerns over trust for relocated people.

A survey of relocated people was conducted in late summer of 1995 in the CIS states. The level of reported distress due to the Chernobyl accident was highest in Ukraine, followed by Belarus and Russia. 63 percent in the Russian sample, 52 percent in the Belarussian and 34 percent in the Ukrainian samples indicated that their resettlement was voluntary[8]. People who had resettled voluntarily *and* found the relocation justified indicated the lowest level of distress. Those who resettled involuntarily *and* who did not find the relocation justified reacted the strongest. Figure 2 illustrates the results for each state. Note the strong overall reaction in the Ukrainian sample, and the strong reaction of distress among Russians who were resettled involuntarily and who did not think the relocation was justified. Young people who had experienced an involuntary resettlement rated higher perceived real risk due to the accident, as well as risk due to the resettlement, as compared to young people who resettled voluntarily. Persons over 55 years who moved voluntarily experienced a higher perceived real risk due to the accident in comparison to those of the same age group who moved involuntarily, but a *lower* real risk due to the resettlement itself.

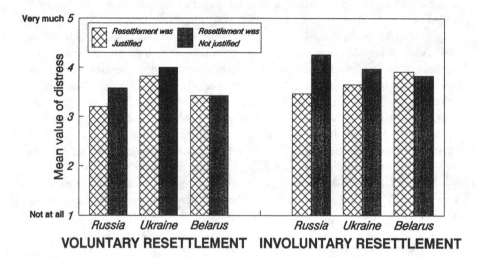

Figure 2. Mean values of distress due to the Chernobyl accident in samples from Russia, Ukraine and Belarus for groups of people who resettled voluntarily vs. involuntarily, and who indicated that the resettlement was justified vs. not justified.

2.5. Newspaper output

The number of articles[9] related to the Chernobyl accident in major republican newspapers and local or regional papers was determined in each state[10]. The months of June and November, from 1986 to 1993, were chosen to facilitate the time-consuming task[11]. A common feature in the material from all states was that a great majority of the articles were written by journalists. Material written by experts, specialists and authorities was much less frequently published, but appeared to some extent in 1986 and then again from 1989-90 and onwards. Another common feature was that the newspaper materials in 1986 and 1987 tended to have a soothing emotional content. Emotionally strained materials appeared more frequently around 1990 and onwards. Data from Ukraine revealed an intensive article production in 1986. It was characterized as mainly containing emotionally soothing materials. The Belarussian data instead showed an information peak between 1989-91, written within a strained or neutral emotional context. The Russian results showed similar information peaks in 1986 and 1990, with a mainly soothing emotional framework in 1986 and a mainly neutral emotional tone in 1990, although strained materials appeared around 1990, especially in the local press. The appearance of strained material around 1990 has been interpreted as reflecting the uncertain political climate at the time of the creation of the CIS, and the election campaigns. The delay of the information peak in Belarus can be related to the relocation of people from contaminated areas.

3. The role of social and psychological factors

Peoples' reactions to a radiological accident may have lasting emotional, social, and economic effects on a community or society. If these reactions become better understood, the knowledge facilitates improved risk management, effective risk mitigation, optimal use of financial means, as well as relief from unnecessary suffering. Such knowledge would allow a focusing on the most effective countermeasures. Our studies have shown that people were unfamiliar with concepts of radioactive contamination and had problems understanding its nature, as well as how to utilize protective measures. These are obviously problems they share with many people worldwide. However, people were willing to learn. If given appropriate information at the right times we believe that radiation knowledge could improve, and protective actions would become better established. Knowledge creates improved mastery of personal life. The studies have pointed to the importance of perceived personal control for psychological well-being and preparedness for adaptive behavior. In the case of a radiological accident, however, personal control is mediated by information [32]. Trust in information sources and credibility of the information therefore precede mastery. Effective information and communication about risk issues require an active involvement of individuals, organizations and authorities who enjoy or can establish credibility. New information is usually more easily digested in exchange with already trusted persons. Our results indicate that medical personnel may provide such a resource. Other trusted groups include teachers, youth leaders or other local leaders. Such core groups could provide invaluable support to decision makers and risk managers if they are properly informed and supported. Correct information is imperative, however, since obvious discrepancies of rules or recommendations may cause uncertainty and concern. An effective and cooperative relationship with the local press and mass media is likewise necessary for quick, correct and instructive information, as well as in providing continuous updates of e.g. recommendations or changes in the radiological situation.

The study of reactions to the Chernobyl accident has demonstrated different reactions and needs amongst the population, and changes over time. The differences seem to be increasing between directly affected population groups and others. In the short and medium time perspective similar needs of information, behavioral recommendations and health care may appear. In the long term, however, our results point to the importance of preparedness for the emergence of new needs generated by the countermeasures themselves, e.g. relocation. The results suggest that voluntariness or personal choice are associated with less psychological distress. In Russia relocation strategies appear to have altered the psychological outlook as well as reduced stress [4]. The policy was implemented in a staged manner [33], being mainly voluntary, spread over time, accompanied by significant financial benefits, and facilitating the maintenance of social networks[12]. Regarding organized health care it may be that preparedness for immediate and large scale medical screening is a necessity after a radiological accident, but the medium term strategy could include other options moving away from mandatory rules, e.g. services for counselling, for measuring personal dose or food products. The long term health risk management could be tailored to needs of affected and vulnerable groups. Similarly it seems that financial support and compensation strategies become very important in the immediate and medium term perspective, but that the beneficial effects may be threatened in the long time perspective if they create dependency rather than enhance self sufficiency. One lesson of relocation which is seldom addressed concerns information and support to the communities which accommodate new members. Although relocated people may be provided with newly constructed housing they will nevertheless have an impact on the local community. Well-being could be increased in the communities if the long term risk management includes a review of common resources and helps accommodate common needs.

4. Conclusions and contributions of the project

The project has highlighted the social and psychological effects of a radiological accident and the changes of reactions and sentiments over time. The empirical studies contribute new information and knowledge due to the massive scale of the investigations and the detailed analysis of specific groups. Control groups have been involved to adjust for overall major social and political changes. The project has also shown the feasibility of investigating citizens' personal experiences for facilitating an accommodation to specific current needs and future planning. The area of research has gained knowledge due to the international cooperation and the personal exchange of experience and information.

5. Notes

1. Studies conducted by CIS researchers in cooperation with Mutadis Consultants are reported elsewhere and are not included here.
2. The model included eight predictors; the adjusted multiple correlation (R^2) was 0.506.
3. Together these three variables explained 28 percent of the variance in everyday worry.
4. The regression model involved six predictors; the adjusted multiple correlation (R^2) was 0.318.
5. In order to examine the relative effects of various factors on stress an approximate version of this approach was adopted; hierarchical stepwise regressions were conducted in blocks representing the main stages, with GHQ-28 dependent. Regressions were carried out separately for the clean control areas, the contaminated areas and relocated people.

6. The efficiency of the equation, as estimated by the R-square criterion, was 0.05 with Sex alone, rising to 0.11 when Fatalism was included. Thus each factor explained about the same proportion of variance.

7. A convenient way to illustrate the trust situation is to compare correlations between the judgements about contamination that people made themselves with those that they attribute to 'the authorities'. In Russia this correlation was quite high (r=0.72), lower in Ukraine (r=0.50) and lowest in Belarus (r=0.41). These correlations highlight some very real credibility gaps, especially in the affected areas of Ukraine and Belarus.

8. Among those volunteers 68 percent thought the resettlement was justified (responses of "yes, definitely" and "yes, probably"), whereas 32 percent did not think so; 65 percent would take the same decision again if faced with a situation similar to that after the Chernobyl accident; 35 percent would instead remain in their home village. In the group of involuntary relocation 56 percent believed that the move was justified, but 44 percent did not think so; 50 percent would move again if confronting a similar situation and 50 percent would remain.

9. "Articles" is used for some different kinds of newspaper material. It includes also e.g. editorials, anonymous letters and official information.

10. The chosen newspapers were: in Russia: Pravda, Izvetzia, Trud and Brianski Rabochi. In Ukraine: Pravda Ukrainy, Kievskaya Pravda, Krasnoye Znamya and Zytomirskaya Pravda. In Belarus: Sovetskaya Belorussia, Zvezda, Gomelskaya Pravda and Minskaya Pravda.

11. Published material was expected to be unevenly distributed across months, with a larger number in May, at the time of the anniversary.

12. The Government permitted relocated people to use both their former and new places of residence for one year. These measures were expected to make relocation attractive and stabilize the social status of the relocated people.

6. References

[1] H.V. Arkhangelskaya, Population Reaction to the Radiation Risk following the Chernobyl Accident. In: J. Baarli (Ed.), *Proceedings of Conference on the Radiological and Radiation Protection Problems in Nordic Regions*. Tromsö 21-22 November. Nordic Society for Radiation Protection, Oslo, 1991.

[2] G.M. Rumyantseva and A.N. Martyushov, Risk of Psychic Dysadaptation in Nuclear Accidents: Chernobyl Accident as a Model. Paper Presented to Third Conference of the Society for Risk Analysis, European Branch, Paris, December, 1991.

[3] International Advisory Committee, *International Chernobyl Project*. ISBN 92-0-129191-4. International Atomic Energy Agency, Vienna, 1991.

[4] P.T. Allen, Stress and Locus of Control in the Chernobyl Region. Paper Presented at the EC Joint Studies Program Workshop, October, Paris, 1993.

[5] D. Goldberg and V.F. Hillier, A Scaled Version of the General Health Questionnaire, *Psychological Medicine 9* (1979) 1398-1405.

[6] D. Goldberg and P. Williams, *A User's Guide to the General Health Questionnaire*. NFER-NELSON, Windsor, 1988.

[7] B.-M. Drottz-Sjöberg *et al.*, Perceived Risks and Risk Attitudes in Southern Russia in the Aftermath of the Chernobyl Accident, *RHIZIKON: Risk Research Report No. 13*, Center for Risk Research, Stockholm, 1993.

[8] B.-M. Drottz-Sjöberg *et al.*, Public Reactions to the Chernobyl Accident. Report from a Data Collection in 1993 in Russia, Belarus and Ukraine. Center for Risk Research, Stockholm, 1994.

[9] B.-M. Drottz-Sjöberg and L. Sjöberg, Risk Perception and Worries after the Chernobyl Accident, *Journal of Environmental Psychology, 10* (1990) 135-149.

[10] B.-M. Drottz-Sjöberg and L. Sjöberg, Adolescents' Attitudes to Nuclear Power and Radioactive Wastes, *Journal of Applied Social Psychology 21* (1991) 2007-2036.

[11] B.-M. Drottz-Sjöberg and L. Persson, Public Reactions to Radiation: Fear, Anxiety, or Phobia? *Health Physics* (1993) 223-231.

[12] D.L. Collins *et al.*, Chronic Stress from the Goiania 137Cs Radiation Accident, *Behavioral Medicine* 18 (1993) 149-157.

[13] H.V. Arkhangelskaya *et al.*, An Analysis of Publications on the Chernobyl Accident in Russia. Manuscript. St. Petersburg Institute of Radiation Hygiene, St. Petersburg, 1994.

[14] L.A. Ageeva and V.S. Tarassov, The Problem of the Chernobyl Accident in the Mass Media. Manuscript. Academy of Sciences of Belarus, Institute of Sociology, Minsk, 1994.

[15] A.I. Nyagu, Reflections of the Chernobyl Accident in the Press. Manuscript. Science Academy Ukraine, Ukrainian Center of Radiation Medicine, Kiev, 1994.

[16] P.T. Allen and J. Marsden, Social and Psychological Factors of the Chernobyl Aftermath: A Qualitative Analysis of Interview Responses in One Affected Area of Russia. Robens Institute Report R193/PSY/002, University of Surrey, Guildford, 1993.

[17] P.T. Allen *et al.*, Two Roles for Psychological Variables in the Aftermath of Chernobyl: Inputs and Outcomes. Paper Presented at Conference on Psychological Aspects of Chernobyl, Kiev, 1995.

[18] A.I. Nyagu *et al.*, *Social-psychological Consequences of Chernobyl Catastrophe in Ukraine*. Final Report in the CEC-CIS Joint Study Project 2. Ukrainian Center of Radiation Medicine, Kiev, 1995.

[19] L.A. Ageeva *et al.*, *The Investigation of Social and Psychological Factors of the Chernobyl Accident in Belarus*. Report on 1993-1995 Work in the JSP2 Project. The Academy of Sciences of the Republic of Belarus, The Institute of Sociology, Minsk, 1995.

[20] I. Ajzen, The Theory of Planned Behaviour, *Organizational Behaviour and Human Decision Processes* 50 (1991) 179-211.

[21] I. Ajzen and T.J. Madden, Prediction of Goal-directed Behaviour: Attitudes, Intentions, and Perceived Behavioral Control, *Journal of Experimental Social Psychology* 22 (1986) 453-474.

[22] E.F. Bazhin, *et al.*, Methods and Methodology of Research: The Method of Researching the Level of Subjective Control, *Psychological Journal* 5 (1984) 153-162.

[23] E.C. Butterfield, Locus of Control, Test Anxiety, Reactions to Frustration and Achievement Attitudes, *Journal of Personality* 32 (1964) 355-370.

[24] H.M. Lefcourt, *Locus of Control: Current Trends in Theory and Research* (2d ed.). Lawrence Erlbaum Associates, London, 1982.

[25] H. Levenson, Attitudes towards Others and Components of Internal-External Locus of Control, *Psychological Reports* 36 (1974) 209-210.

[26] J.B. Rotter, Generalized Expectancies for Internal versus External Control of Reinforcement, *Psychological Monographs 80*, No. 1, 1966.

[27] R. Schwarzer, Self-efficacy in the Adoption and Maintenance of Health Behaviours: Theoretical Approaches and a New Model. In: R. Schwarzer (Ed.), *Self-efficacy: Thought Control Action*. Hemisphere, Washington, 1992.

[28] K. Dake, Myths of Nature: Culture and the Social Construction of Risk, *Journal of Social Issues 48* (1992) 21-37.

[29] P.T. Allen and G.M. Rumyantseva, The Contribution of Social and Psychological Factors to Relative Radiation Ingestion Dose in Two Russian Towns Affected by the Chernobyl NPP Accident. Paper Presented at SRA Europe Conference, May, Stuttgart, 1995.

[30] M. Fried, Grieving for a Lost Home: Psychological Costs of Relocation. In: L. Duhl (Ed.), *The Urban Condition*. Basic Books, New York, 1963, pp. 151-171.

[31] T. Lee, Moving House and Home, In: S. Fisher and C.L. Cooper, *On the Move: The Psychology of Change and Transition*, John Wiley & Sons, Chichester, 1990, pp. 171-189.

[32] B.-M. Drottz-Sjöberg, Medical and Psychological Aspects of Crisis Management during a Nuclear Accident. In: B. Stefenson, P.A. Landahl and T. Ritchey (Eds.), *International Conference on Nuclear Accidents and Crisis Management*, Crisis Management Project Group, Stockholm, 1993, pp.33-48.

[33] G.M. Rumyantseva, The History of Relocation Caused by Chernobyl Accident in Russia. Robens Institute, University of Surrey, Guildford, 1994.

Decision Support Systems for the Post-Emergency Management of Contaminated Territories

Mary Morrey*, Stanislav Dovgiy#, Boris Yatsalo§, Neil Higgins*, Ilya Likhtariov†,
‡Mona Dreicer, ‡Jacques Lochard, •Michael Savkin, •Vladimir Demin,
•Pavel Khramtsov, #Leonid Grekov, °Tatyana Utkina

*NRPB UK, #Topaz-Inform Ukraine, §RIARAE Russia, †USCRM Ukraine,
‡CEPN France, •Institute of Biophysics Russia, •Kurchatov Institute Russia, °INTEH Belarus

Abstract

The work implemented within the framework of the project was directed towards understanding the conceptual basis for the organization of intervention strategies after the accident at the Chernobyl nuclear station. Based on the situation in regions of Belarus, Russia and Ukraine that suffered the consequences of the accident, this project was directed towards the provision of support for decision makers. The work will assist in the choice of proper strategies to protect the population from the effects of environmental contamination, taking into account the available resources. The experience gained, both of the problems of decision aiding in this context and their solution, will be of use in post-emergency planning for possible future accident situations.

At present there are several prototype computer systems which provide the following: access to a wide range of information gathered after the accident at the Chernobyl nuclear station in the CIS; support in complex evaluations of the post-accident situation for a wide range of parameters; analysis and forecast of how the situation may develop using mathematical models and algorithms; support in choosing strategies at each level of decision making taking into account the possibilities of applying a wide range of countermeasures; exploration of multifactor interdependence and the consequences of resource and other limits; the integration of experience in social and psychological factors into the decision making process.

Calculations made by the computer modules are based on actual data from contaminated territories, including structure of soils, age/sex structure of the population, and dietary habits. At present the models for the calculation of doses and radionuclide migration in soil are specific to the regions contaminated after the Chernobyl accident. They are based on a large amount of experimental data ranging from whole body measurements of the population to data about radionuclide transfer from soils to plants, milk, and meat.

A risk module has been developed which includes a database of demographic data on health protection for different territories, and which can calculate risks for any population structure. It can be used for risk estimation for different age groups in any region for which data is provided.

A module on indirect countermeasures has been developed to assist in the selection of countermeasures that will improve conditions for a population that inhabits a contaminated area, in a way that is distinct from a direct reduction in the effective contamination level of the environment and its products. These types of countermeasure can be taken both as separate actions or in combination with direct countermeasures to increase the efficiency of the latter.

The tools and interfaces developed within JSP2 enable the decision maker to estimate consequences and analyze the post-emergency situation according to his own criteria and in a user-friendly manner. The available analyses include: forecast calculations (concerning the estimated contamination of agricultural products, the levels of doses in the local population, and the associated radiation risks); the identification of the critical factors that influence the health of the affected population; a simulation of human intervention taking into account the countermeasures chosen or a time ordered set of countermeasures; an estimation of the influence such factors as dose, risk, cost/benefit and time, etc. have on decisions. The results can be presented in the form of maps, diagrams, and tables.

In this paper, work carried out on the development of computer based decision support systems for the post-emergency management of contaminated territories is discussed, together with possibilities for further development of such systems.

Introduction

Following an accidental release of radionuclides to the environment, there may be a need for countermeasures to be implemented to protect the public. These countermeasures can be classified either as emergency or as longer term countermeasures, depending on their urgency of introduction and their duration. In general, emergency countermeasures are short term actions, such as sheltering, evacuation (for a few days) and the administration of stable iodine. Their primary role is to protect the public from exposures during the release and initial dispersion of the radionuclides. During the emergency phase of an accident, there will, in general, be insufficient time for many detailed measurements to be made or thorough evaluations of the impact of the accident to be undertaken. Therefore, emergency countermeasures are usually introduced according to pre-determined criteria, and based on rapid and approximate calculations. However, because such countermeasures are short term, the scale of any undesirable or unintended consequences is necessarily limited. Longer term countermeasures, by contrast, may continue for many years. Their purpose is to protect the public against continuing, relatively low level exposure from radionuclides in the environment. The risk from short term exposure to these environmental sources is unlikely to be serious: it is the cumulative risk from prolonged exposure which is considered unacceptable. Because such countermeasures may be in force for long periods, the consequences of non-optimal intervention may be very undesirable or costly. It is therefore of particular importance that the design and implementation of such countermeasures should be tailored to the exact post-accident situation taking into account the potential for long term adverse effects. Decisions on such countermeasures are therefore likely to require a large amount of detailed measurement and supporting information. The longer decision time available to work out the details of such countermeasures provides the opportunity to consider the information more fully than during the emergency phase. However, the difficulty is in assessing and synthesising this information, so that the best decision can be made on a reasonable timescale. This synthesis of large amounts of disparate information in the post-emergency phase of an accident is ideally suited to computer management. Within the Joint Study Project 2 (JSP2) of the EC/CIS collaborative programme recent computer developments have been explored in terms of their potential application to post-emergency management, and example models and modules developed. This is the subject of this paper.

The post-emergency phase of an accident is likely to be characterised by the emergence of a wealth of information from a wide variety of sources. For example, an increasing number of radiological measurements should become available, both those requested by the authorities and those made by independent groups. Widely varying opinions and expert advice are likely to be promulgated regarding the most likely future course of environmental contamination, exposures, health effects, and the possibilities and costs of countermeasures for averting these. Pressure groups may emerge to present the interests of particular population groups, and the affected populations will be subject to significant stress. Despite, or perhaps partly because of, the growing volumes of information, the situation will be full of uncertainty. It is likely that anomalies in the measurements and predictions will appear, due to a combination of error, misunderstanding and the sheer complexity of the environmental situation. Moreover, since exposure of the population to the contamination will partly be dependent on behaviour patterns, the prediction of the interaction between social-psychological factors and behaviours which can affect exposure is necessary, if countermeasures are to be evaluated. The emergence of pressure groups and the spread of opinions and advice will tend to require at least some protective decisions to be taken quickly, even though, from a solely radiological risk perspective, this would not be necessary. However, decision makers have a responsibility for considering the wider implications of their decisions: the need for equity between individuals and communities, the need for continuity (frequent changes in countermeasure policy is likely to increase stress and mistrust of the authorities) and the need to limit the commitment of resources to a level commensurate with the scale of the problem. Moreover, there are different levels of decision making required after an accident. At the highest level, a global policy is required, eg generic criteria for initiating the implementation of major classes of countermeasures, such as relocation. But in areas where such policies indicate that countermeasures are needed, the detailed implementation needs to be worked out, including timescales for implementation, practical methods, degree of involvement of those affected in planning the detail and so on. Different levels of decision require different types and amounts of information. One task of a computer based post-emergency support system would be to select the information relevant to each level of decision maker. All these issues are discussed in the paper.

Needs of the Decision Maker

A post-emergency computer decision support system should exist to meet the needs of the decision maker. It is important to recognise that the purpose of such a system is not to replace the decision maker but to facilitate his job by presenting what is known or can be predicted in a relevant and straightforward manner. As mentioned in the introduction, the post-emergency management of an accident is likely to involve a hierarchy of decision making, ranging from generic policy statements at a national level, down to specific local implementations of national policy in a settlement or at community level. Clearly, decision support should be tailored to meet the needs of particular types of decision and decision maker. Thus, a single system, with the inherent advantages of continuity and data sharing, may require several presentational forms if it is to try to meet all needs. However, whether a multifaceted single system or a series of independent systems, it is possible to identify factors and requirements that are common to most levels of a decision hierarchy. These are: the need to store large amounts of heterogeneous data, to keep this database up-to-date, and to facilitate the selective retrieval of these data and their presentation in a helpful form; the need to enable the consequences of possible protective strategies to be explored ('what-if' type questions); the need for possible strategies to be suggested, ranked, rejected; the need for a wide range of non-radiological (eg demographic, geographic, economic, regulatory, and social-psychological) data to be stored; the need for the system to incorporate a range of models/algorithms for interpreting data, and for the system to select the appropriate ones in response to a particular request; and the need for a flexible interface with the user, so that questions do not need to be posed in potentially obscure scientific terms. These common requirements are discussed below.

An intervention strategy may consist of a combination of one or more countermeasures, or no action. In order for such a strategy to be developed, it is necessary not only to evaluate the likely consequences of each individual countermeasure, but also the consequences of any interaction between countermeasures that may take place. The range of factors that need evaluation can be very large, although they will depend on the level and type of decision being made. Clearly, the prime purpose for introducing countermeasures is to reduce adverse health effects in the population. These health effects comprise both those resulting directly from exposure to the radionuclides released in the accident, and those resulting from increased stress caused either by the accident itself, or the subsequent accident response measures. For example, such factors could include: characteristics of the release or environment such as radionuclide transfer factors, initial deposition levels and soil types; geographic factors such as land use; and social psychological factors such as lifestyle and the likely level of compliance with official advice. In addition, all decisions would need to consider temporal and spatial aspects of the proposed strategy. For example, where the proposal is to implement a generic countermeasure such as relocation in areas where the average ground deposition level exceeds a specified criterion, the decision maker requires, amongst other things, information on the extent of areas contaminated above given levels, the practical constraints on timescales for introducing countermeasures (which in turn requires demographic data and knowledge of the spatial distribution of resources), and the possible reception areas for the relocated population. Moreover, where local decision makers need to interpret national policy for the local situation, it is important that they should have access to information on the legal framework within which they are operating. An important aspect of a computer decision support system is therefore the capacity to store large amounts and varieties of data which may be rapidly and flexibly retrieved within a spatial and temporal context.

There is another way in which temporal and spatial considerations influence decision making. Decision makers need advice on the harms and benefits of delaying a decision until additional information has been collected, compared with making the decision based on current knowledge. They also need information on the likely impact of a decision for one location on other locations, both in terms of economic interactions and in terms of social implications. Ideally, the computer system should support such questions as 'what benefit do I gain by postponing the implementation of countermeasures whilst I collect more data?' and 'if I do this in this settlement, what will be the effect on my options for other settlements?'. At one level this requires a sophisticated user interface, but, more importantly, it requires an intelligent and flexible synthesis of different types of information held within the system, and an ability to assess the overall uncertainty associated with this use of the data.

The above examples also illustrate the need for an interface that supports the decision makers' needs, by providing management options and guidance as opposed to a scientifically oriented tool providing descriptive and mechanistic analyses. The questions asked by decision makers need to be re-interpreted in order for them to be answered as a series of specific steps that can be carried out by a computer. The ideal support system will not require this decomposition to be done by the user, in order to ensure that it may be easily used by both local and high level decision makers and their advisors. The system should therefore decompose the more general questions into their constituent parts. This has implications, not only in terms of the interpretation of the primary question, but also in terms of the imposition of constraints on the scope of the analysis, eg in space, time, detail and accuracy, and hence the appropriate selection of models/algorithms and data. For example, a high level decision relating to the whole country requires only approximate measures of the consequences of adopting a given strategy. For this, simple empirical models or those designed to predict average consequences over a period of several years are the most appropriate, together with data aggregated over major geographical units. However, the local implementation of the strategy, eg the decision on how best to cultivate a particular field, given national constraints on maximum permissable radionuclide concentrations in foods, requires site specific models appropriate for that growing season, and data specific to that location.

Models Relevant to Post-Emergency Decision Support

The discussion of the needs of decision makers has indicated a need for a wide range of models and algorithms for processing radiological and other data in the post-emergency phase of an accident. These include models for interpolating between measured data, models for predicting into the future or into areas where there are no, or very limited, measurements, and techniques for combining model predictions/estimates and measurements. In addition, there is a need for models of varying degrees of complexity and precision, depending on the type of decision being taken, and the timescale on which it must be made.

Even in the very long term after an accident, measurement data will never be sufficient to define the full impact of an accident. It will always be necessary to interpolate between and extrapolate from measurements in both space and time. Where measurements are plentiful, the uncertainties introduced by appropriate procedures may be small. Where they are more sparse, it will be necessary to use models to supplement what can be deduced directly from the measurements. These models, which can be empirical or phenomenological, are required to estimate a wide variety of endpoints. Such endpoints include: environmental radionuclide concentrations, doses, risks, costs and effectiveness of countermeasures, costs and effectiveness of medical treatments, the social psychological response to countermeasures, the consequences of indirect countermeasures (ie those not directly addressing the radiation risk from the accident, but used to reduce other risks or to enhance the effectiveness of direct countermeasures). Any model should also include an estimate of the uncertainty associated with the result, both that arising from the data input to the model and that introduced by the model itself. Where a series of models are used, then these uncertainties need to be propagated through the sequence together with the model estimates.

Ideally, a computer decision support system should incorporate, for estimation of a given endpoint, models of differing complexity, scale of application and input data requirements. This is for two reasons. First, the amount and quality of data available to the system will vary depending on the data type, the area of interest and the time after the accident. The models or algorithms best suited to manipulate these data will therefore also vary in accordance with the available data. Second, the nature of the decision being taken will influence the type of model that is most appropriate, in terms of the precision, detail and scope required. The system should therefore hold information about the applicability of the models contained within it and select appropriate models based on the amount and quality of the available data and the type of question being asked.

Computer Techniques and Developments Relevant to Post-Emergency Decision Support

A number of techniques and applications have been developed for computers which are of potential use for a post-emergency decision support system. These are described in this section, and their potential application explored. In the following section, specific applications of these techniques within JSP2 are described.

Databases

Developments in database technology and computing power mean that very large amounts of disparate data can now be stored, accessed and manipulated rapidly. Moreover, this technology has now matured to the point where a degree of standardisation has been achieved, with many database systems using the Structured Query Language (SQL) to retrieve information. Sophisticated databases facilitate the connection of other software applications (for example geographical information systems - see below) and the development of flexible user interfaces, tailored for a specific purpose. An advanced database interface enables prompting for necessary information, screening for erroneous input and the simple construction of complex queries to the database. Information concerning the time of input, links between data and the history of manipulations carried out on the data can also be logged, facilitating quality assurance and retrospective analysis. Finally, triggers can be set, so that the user is alerted if particular data are entered, for example radionuclide concentrations exceeding given levels. Thus, the database can prompt the user to the most relevant information held at any given time.

Geographic Information Systems

At their most basic level geographic information systems (GIS) enable spatial information to be displayed and combined on a map. Even at this level, the GIS offers the decision maker a powerful tool for the communication and comprehension of complex information. The spatial distribution of data can be readily comprehended, and potential interactions between different spatial quantities can be identified. Combined with a relational database, simple map displays on a GIS can be used to present model results and time snapshots, and to assist in the exploration of data interactions (eg comparisons of measurements and model estimates). However, GIS are potentially far more powerful than simple map displays. The more advanced applications of GIS use spatial and temporal analysis techniques to maximise the amount of information that can be extracted from data. Whilst the use of GIS in this more complex manner is still very new in post-emergency management, applications such as automatic averaging of large volumes of data, or the 'intelligent' display of stored information on a scale appropriate to the scale of map being displayed, are examples of the future possibilities of GIS.

Multi-criteria Decision-aiding Techniques

Multi-criteria decision-aiding techniques offer the possibility of comparing protective strategies with competing and widely varying consequences. Implemented in computer systems, they provide a fast and flexible tool for evaluating strategy options according to criteria provided by the user. Implementations exist for ranking options, identifying the most influential attributes of the options, testing the sensitivity of the ranking to changes in the evaluation criteria, screening out options which could never be favoured, and developing hybrid strategies which combine the best features of those originally proposed. By linking such computer applications in a larger decision support system, the software could provide a user interface through which models were run to calculate strategy consequences, and all relevant parameters of the decision could be displayed in map form for ready assimilation.

Computer Networking

The speed and robustness of communication between computers is improving rapidly. The popularity of the Internet is a good example of how effective these links now are. Consequently, networking computers in different organisations and between different countries for post-emergency management is now a realistic goal. At a limited level, this should allow rapid data exchange between organisations with different expertise. However, the possibility of developing distributed databases (ie a single database that has component parts stored on different computers in different organisations) and accessing computer models held on other machines is now emerging. This has the advantage of removing the requirement for a single organisation to maintain an up-to-date compilation of all information and computer models relevant to an accident. Instead, each organisation can make its own data and particular models available to other organisations through the network, thus contributing one component of a larger, distributed decision support system.

Development of Decision Aiding Tools within JSP2

Within JSP2 research has been carried out with a view to: identifying more clearly the requirements of a post-emergency management support system; demonstrating the potential application of computer capabilities and techniques; and developing new models where gaps have been identified. The previous sections have described progress in the first area. In this section the specific computer applications and models developed within JSP2 are discussed.

Use of Measurements in the Post-Emergency Phase

One of the features of the post-emergency phase, compared with the emergency phase, is the existence of a much greater amount of measurement data. This, at the very least, enables predicted consequences to be compared with those observed, and, where measurements are particularly plentiful, for some endpoints to be estimated directly from these measurements. Within JSP2, two models have been developed which derive dose and related quantities empirically from measurement data.

The first[1] has been developed by the Ukrainian Scientific Centre for Radiation Medicine and uses whole body measurements of radiocaesium concentrations to determine the retrospective time profile of ingestion doses and to infer the intakes of radiocaesium in food which have led to these doses. By comparing the inferred actual intakes with the theoretical intakes based on consumption rate statistics and average radiocaesium concentrations in foods, much can be learned about local variations in concentrations and dietary habits and the effectiveness of food countermeasures.

The second[2] has been developed by the Gomel branch of the Institute for Radiation Medicine in Belarus. This model uses the observed relationship between ground contamination level, availability of privately produced milk (as measured by the ratio between the numbers of privately owned cows in a settlement and the number of people in the settlement), and the proximity of the settlement to a forest (ie a measure of the degree of access to forest foods) to average settlement ingestion dose. Using a combination of cluster analysis and linear regression techniques, empirical relationships (dose as a function of average ground contamination) for five types of settlements have been developed. Testing of this model has demonstrated a good correlation between estimated and actual dose (based on whole body measurements); a much better correlation than is obtained by applying a phenomenological model using regional parameters.

Development of a Risk Module and Database

The Kurchatov Institute, Moscow, together with BelCMT, Minsk have developed a risk module, BARD, for application within a post-emergency decision support system[3]. This module is in some ways similar to the computer programmes ASQRAD and SPIDER developed by CEPN (France) and NRPB (UK), but contains a larger intrinsic database and includes non-radiological as well as radiological risks. This allows radiological risks to be compared with other health risks and should help in the allocation of resources devoted to protection. For a wide range of population groups (eg the population of a state, region city or settlement) and specific years, the database holds age distribution densities, which are further sub-divided according to gender and the fraction of population who are either rural or urban. For each type of health risk the database holds age-cause-specific death rates. The health risks considered include: infection and parasite diseases, malignant neoplasms, circulatory system diseases, respiratory system diseases, digestive system diseases, and accidents. For each of the general risk categories there is information on more specific injuries, for example the category malignant neoplasms contains information on leukaemia, respiratory cancer, breast cancer, digestive cancer, etc. For predicting future incidence and death rates, the risk assessment models of ICRP/UNSCEAR, BEIR V, RERF and NRPB are included in the module. Using these, lifetime risk, loss of life expectancy and incidence in specified cohorts and years can be calculated.

Agricultural Countermeasures Module

A regional to farm level module characterising the effects of changes and treatments to soil, crops and husbandry practice enables agricultural countermeasures to be assessed. This module provides information on the change in radionuclide concentration distributions in crops grown in regions of varying contamination level subject to a potentially complex set of user defined countermeasures. Information on the costs of actions are provided and, if required, the user is guided through the selection of countermeasure combinations. The models were developed and tested using data from the Kiev, Zhitomir and Bryansk regions.

Indirect Countermeasures Module

In the past, guidance on intervention after nuclear accidents has focused on countermeasures to reduce exposures to radionuclides released in the accident. However, experience after the Chernobyl accident has indicated that it is pertinent to consider another class of countermeasures - those aimed at improving the overall well-being of the population, without necessarily reducing the direct radiation exposure stemming from the accident[4]. These have been termed indirect countermeasures and have been the subject of investigation within JSP2[5]. Indirect countermeasures can be sub-divided into two categories: risk reducing indirect countermeasures and social action indirect countermeasures. The purpose of risk reducing indirect countermeasures is to reduce other risks the population is exposed to, for example those from radon, medical irradiation or chemical pollutants. Social action indirect countermeasures may be in the form of compensation, eg financial payments or improved amenities, or in the form of supporting actions to other countermeasures, in order to improve their effectiveness (eg information campaigns, local involvement in decisions).

In order to explore the application of indirect countermeasures after an accident, a computer module has been developed by the Institute of Biophysics. The module assesses the costs and benefits of applying countermeasures in three areas, two risk reduction applications, and one social action application. The risk reduction applications consider the costs and benefits of different radon treatment options and the costs and benefits of new equipment and training to reduce the dose received by patients having medical diagnostic examinations. This information can then be assessed by a decision maker who can decided how much weight to attach to these actions when comparing them against alternative direct countermeasures. The social action countermeasure is discussed in the next section.

Social Psychological Aspects

The Chernobyl accident has clearly demonstrated the importance of social and psychological factors after an accident. Research has been undertaken within JSP2 to quantify the influence of these factors, particularly with respect to their interaction with direct countermeasures. Although the complexity of this interaction means that only limited progress has so far been made, it is clear that some social and psychological factors can influence the dose effectiveness for direct countermeasures, whilst others can influence their monetary costs[6]. In future it should be possible to model these interactions explicitly by developing existing social-psychological models[eg 7], and to integrate the modelling within existing computer decision support systems. At present, a simpler approach has been adopted, that of providing the decision maker with additional information on social psychological factors relevant to particular countermeasures. For example, as part of the indirect countermeasures module, an informed assessment of the consequences of implementing a decision to relocate a population group in a variety of alternative ways is presented to the decision maker. The decision maker is then able to consider the dose, cost and health implications of particular relocation choices.

Display Methods

If information is to be assimilated efficiently by the decision maker, then it needs to be presented in a manner which facilitates this. RIARE, USCRM, and Topaz-Inform have all placed emphasis on developing appropriate display methods. For example, information can be presented in context using a GIS. The GIS display can show maps of the areas of interest at scales selected by the decision maker with additional information illustrating the current situation or the predicted consequences of a series of countermeasures. Standard graphic displays are also provided to show the variation in selected properties as a function of time. The decision maker or his representative interacts with the whole system through a series of menus which provide guidance for both an expert and novice user.

Linking between Different Computers

As a result of experimental investigations undertaken as part of JSP2, Topaz-Inform have developed a prototype model of a distributed problem-solving system. This model includes: the user interface (implemented on each computer under X-Windows), separate and different databases and models held on individual computers, and a calculation server to enable the user of one computer access to data and models on other machines. Data exchange between a group of modules takes place independently of other exchanges which are being transacted. The databases work in a multitasking regime and can therefore process several inquiries simultaneously. The system can link different types of computers and operating systems, for example, HP 9000 and PC486 computers and BSD Unix and SMOOTH operating systems.

Analysing the countermeasures at regional and local levels

Decision support should be available at all administrative levels from national to local. By bringing together many of the features and modules discussed above the PRANA (Protection and Rehabilitation of Agrosphere after Nuclear Accident) decision support system has been developed by RIARE[8] to achieve this.

PRANA is a GIS-based system which has been implemented for Novozybkov district in Bryansk region. The database includes detailed information on individual areas (field, settlement, ecology and radiological data including transfer factors) as well as the characteristics of the countermeasures (efficiency, costs etc). The system may be applied on any scale subject to the appropriate information being supplied.

The models used for estimating the contamination of agricultural products and the internal and external doses which result with and without countermeasures, are based on models developed within the framework of ECP9. A particular feature of the implementation of these models within PRANA is the development of procedures to undertake calculations using distributions of parameter values (as opposed to single value parameters).

The main functions of the system are in three areas:

♦ Database analysis: - this allows the selection of fields, settlements, farms etc. according to various criteria, for example, level of contamination, individual dose, collective dose.

♦ Assessment of countermeasures: - definition of a set of countermeasures for a specific area and for a specific duration. The analysis provides a set of radiological and economic indicators such as the cost-effectiveness.

♦ Indirect Countermeasures:- three sub-modules are included, as examples of the application of indirect countermeasures models within post-emergency computer decision support systems. Two enable the modelling of changes in radon exposure and doses from medical diagnostics, whilst the third provides information on some of the social-psychological implications of different implementation strategies for relocation.

By providing a user-friendly interface, and access to the data through Paradox 5.0, PRANA facilitates comparisons both of the consequences of alternative countermeasure strategies implemented at the local or regional levels and of how these consequences change with time.

The Way Forward

The experience of the Chernobyl accident, and the research carried out within JSP2 has improved the understanding of the needs of decision makers and how computer based decision support systems for post-emergency management can assist them. Key lessons are the need for any such system to be centred around a sophisticated relational database, holding a wide variety of data types in their spatial and temporal context. This in turn raises the need for continuously up-dated data: preliminary investigations within JSP2 at Topaz-Inform have indicated that a possible future system would consist of a distributed database, with a number of organisations each holding and being responsible for their own data. This is an area where further research could yield dividends.

Another key lesson is the need for post-emergency models to be developed that utilise collected data efficiently. In the immediate aftermath of an accident, it will usually be necessary to rely on simplified or averaging models. However, once more data become available, it is important that best use is made of these. In addition, there is a need to further develop methods for integrating model estimates, data and expert judgement, so that the best possible picture can be developed of both the accident impact and that of possible countermeasures.

A need has been identified for the inclusion of a number of models for each desired endpoint, so that the system can select and apply the most appropriate model, depending on the needs of the decision maker and the quality and quantity of data available. Finally, the need to develop interfaces that are more accessible to the decision maker has been identified. Both GIS and multi-criteria decision aiding packages have an important role to play in this. However, further research is required towards developing interfaces that enable the decision maker to ask more open-ended and 'natural questions', such as 'what strategies should I consider for this settlement/these settlements/this region/this farm?', 'what additional information could I collect in order to choose between these two strategies?'.

References

1.   Likhtarev, I et al. Assessing internal exposures and the efficacy of countermeasures from whole body measurements. To be presented at the CEC/CIS Minsk Conference, 1996.

2.   Skryabin, A M, Osipenko, A, Morrey, M, Vlasova, N and Podobedov, V. Characterising settlements: a cluster-regression model for improved estimates of ingestion dose in Belarus. Poster presented at an IAEA international symposium on Environmental Impact of Radioactive Releases, Vienna, May 1995.

3.   Demin V F. Methodological recommendations on risk assessment applied to the situation after nuclear weapons tests or accidents. In: Bulletin of the Federal Research Program "Semipalatinsk test site / Altay case study" No 1 (1995).

4.   United state programme on protection of the population of the Russian Federation against the influence of consequences following the Chernobyl catastrophe from 1992-1995 and for period till 2000. Act of Government of Russian Federation Oct 1992 protocol No 29. Adopted 11 July 1993 No 5437-1.

5.   Higgins, N A and Morrey, M. Social intervention and risk reduction - indirect countermeasures. Paper presented at an international workshop on Radiation Risk, Risk Perception and Social Constructions, Oslo, October 1995.

6.   Morrey, M and Allen, P T. The role of social and psychological factors in radiation protection after accidents. Paper presented at an international workshop on Radiation Risk, Risk Perception and Social Constructions, Oslo, October 1995.

7.   Schwarzer, R. Self-efficacy in the adoption and manitenance of health behaviours: theoretical approaches and a new model. In: R. Schwarzer (ed) Self-efficacy: thought control of action. Washington: Hemisphere (1992).

8.   Yatsalo, B et al. PRANA: Decision support for analysis of countermeasures in long term period of liquidation of consequences of nuclear accident (agrosphere). RIARE report (to be published).

Self-restoration of contaminated territories

Gerassimos ARAPIS

Laboratory of Ecology and Environmental Sciences, Agricultural University of Athens, Iera odos 75, 11855 Athens, Greece

Emlen SOBOTOVICH, German BONDARENKO, Igor SADOLKO

Department of Environmental Radiogeochemistry, Institute of Geochemistry, Minerology and Ore Formation, National Academy of Sciences of Ukraine, 34 Palladin av., Kiev 152680, Ukraine

Evgeny PETRAYEV, Galina SOKOLIK

Department of Radiochemistry, Belarus State University, Leninski prosp. 4, Minsk 220080, Belarus

Abstract. This paper illustrates the experience gained in the field of natural restoration of contaminated vast ecosystems. Prior to recommending a large-scale application of any rehabilitation technique, it is important to know the medium- and long- term intensity of self-restoration for most of the affected territories. Three main ways express the process of self-restoration: 1) the natural radioactive decay, 2) the transfer of radionuclides out of natural ecosystems and 3) the ability of some pedological components to fixate the contaminants. The first way is a real decontamination process resulting in the removal from the biosphere of significant quantities of radionuclides. Indeed, during the last years the total activity of short-life-isotopes was decreased by a factor of some thousand and actually, the main contaminants are ^{137}Cs and ^{90}Sr which are decreasing according to their half-life. The two other ways of self-restoration are closely connected with radionuclides migration (vertical or/and horizontal) in soils. The vertical migration velocities of ^{137}Cs and ^{90}Sr in typical soils of contaminated regions in Ukraine and Belarus were evaluated annually during 9 years since the accident. In most of these soils the migration rate of ^{90}Sr seems higher than this of ^{137}Cs and ranges from 0.71 to 1.54 cm/year and 0.10 to 1.16 cm/year respectively. At present time the main part of radionuclides is located in the upper 10 cm layer of soils. The ability of soils components to immobilize the radionuclides was also investigated. From 1989 to 1994 approximately 57% of ^{137}Cs was converted in fixed forms and for the year 2000 it is expected that this percentage will be 80%. Finally, for total contaminated regions, the obtained results on vertical migration velocity of radionuclides as a function of the soil type, are presented under the form of a map in order to help decision makers to determine the feasibility and the methodology for restoration of areas contaminated by ^{137}Cs and ^{90}Sr.

1. Introduction

The scale of the territory contaminated from the Chernobyl accident (approximately 140000 Km^2) makes difficult the implementation of an extensive restoration strategy on account of its economical, technical and social implications.

Before to recommend a large-scale application of any rehabilitation technique, it is important to know the medium- and long- term intensity of self-restoration for the most of the affected territories.

This work presents the experience gained during the last decade in the field of natural restoration of contaminated vast ecosystems.

Three main ways express the process of self-restoration: 1) the natural radioactive decay, 2) the transfer of radionuclides out of natural ecosystems and 3) the ability of some pedological components to fixate the contaminants for long-term. The first way is a real decontamination process resulting in the removal from the biosphere of significant quantities of radionuclides. The two other ways of self-restoration are closely connected with radionuclides migration (vertical or/and horizontal) in soils and their entry to the food-chain. It is obvious that any reduction in time of radionuclide transfer into the surface biota and/or any decrease of the radiation of soils, caused by spontaneous natural processes, brings out a reduction of internal and external irradiation doses.

2. Methodology

The natural behaviour of ^{137}Cs and ^{90}Sr in typical soils of contaminated regions in Ukraine and Belarus was studied for 9 consequent years since the accident. However, the study of natural restoration of contaminated areas must be considered in terms of variable landscape-geochemical complexes and thus the study of radionuclide's behaviour in different media such as soil, vegetation, surface and ground waters, became of great interest for the assessment of the efficacy of decontamination techniques applicable in areas affected by the Chernobyl accident.

Reasoning from the concept that self-restoration of natural landscapes is any removal of radionuclides from the active geochemical cycle of the affected areas due to the active migration and/or durable fixation into the soils, from 1993 to 1995 we considered two main lines of study within the EC/CIS Experimental Collaboration Project n° 4 (ECP-4).

The first one was based on the evaluation of the intensity of radionuclides migration. A study of vertical redistribution of radionuclides in different soil types was conducted in Ukraine and Belarus. An important surface ablation of soils may be occurred in sloping landscapes thus, they can considered as a good topographical model which may give indications of long-term behaviour of radionuclides soon, we studied the efficiency of self-restoration on four sloping sites with different incline, soil types, vegetation character, density of contamination and form of radionuclides.

The second investigation line was based on the dynamics of distribution of ^{90}Sr and ^{137}Cs and their forms of occurrence in different types of soils and the concurrent changes of phytomass contamination.

The main processes of self-restoration resulting to the reduction of external irradiation are the radioactive decay and the vertical or/and lateral migration of radionuclides. The processes leading to the decrease of internal irradiation are the radioactive decay, the vertical or/and lateral migration and the immobilization of radionuclides.

It is necessary to know the main parameters of self-restoration of different types of soils for the recovery of large contaminated areas. This knowledge forms the background for any project and decision associated to the application of countermeasures. Indeed the following information can be delivered:

1. Predict the decontamination of soils as a function of time and thus forecast of natural rehabilitation of contaminated areas. The affected areas in Ukraine, Belarus and Russia being vast, it is essential to know the intensity of their natural self-restoration.
2. Classify the contaminated territories and select the areas and sites for the implementation of countermeasures.
3. Define the secondary effect of the rehabilitation actions and introduce them in the cost-benefit analysis of the countermeasures.
4. Qualify the contaminated areas according to their rate of self-restoration in order to plan their reintegration into the normal uses.

Within the period of our involvement in the ECP-4, studies were conducted related mainly to the two first of the above items. Data referring to the properties of migration (horizontal and vertical) and of immobilization of radionuclides into the contaminated soils of different type are presented below.

3. Results

3.1. Evaluation of the radioecological balances of contaminated territories

The evaluation of the radioecological balance of affected areas in Ukraine and Belarus was made in the base of general quantitative estimations of radionuclides horizontal migration.

The balance of radionuclides in the landscapes reflects the ability of the natural systems to "evacuate" the pollutants. It can be negative or positive, depending of the direction and the intensity of the natural processes of migration of radionuclides.

In addition to the classic geochemical parameters, terrestrial balance of radionuclides depends on the forms of relief, the lithology of soil forming deposits and the vegetation cover. These factors are to be taken into account in order to evaluate the ability of different elements of landscapes to evacuate or to accumulate ^{137}Cs [1,2,3]. Based on the balance of this radionuclide, the landscape elements of 30-km zone (Ukraine) or of Khoiniki region (Belarus) can be classified on three main groups 1) areas with negative balance (where the process of radionuclides evacuation is dominant), 2) areas with neutral balance and 3) areas with positive balance (and accumulation of radioactivity) [4]. We prepared short-term and long-term maps of balance evaluation of ^{137}Cs surface migration in natural landscapes of Chernobyl zone. Short-term balance reflects the evaluation from the present situation up to 20-30 years in future. Long-term balance is a forecast situation 60-80 years from now, when some stable forest succession will arrive, and shows globally small differences compared to the short-term one.

The landscape of extreme-morainic ridge of Chistogalovka, shows both in short-term and in long-term aspects, a clear negative balance of ^{137}Cs. Generally landscapes with negative short-term balance of the radionuclide occupy 36,9% of the estimated territory. Neutral short-term balance is expected for sandy territories of the river terraces and fluvioglacial plains and together with water surfaces and industrial areas cover approximately 38,9% of the territory. Zones of positive short-term balance of ^{137}Cs are identified with closed depressions, watershed catchments, elements of erosional network and rear lowered parts of river terraces and flood plains. Positive balance of the radionuclide is estimated 23,9% of the territory inside the 30-km zone. Similar situation characterizes the ^{137}Cs long-term balance in the divers landscapes. A small differentiation of surfaces is evaluated: 4,1% decrease of the areas showing neutral balance and 4,5% increase of the negative balances or self-restored landscapes.

In a similar way a map of balance evaluation of ^{137}Cs in the natural landscapes of Khoiniki district (Gomel region, Belarus) was prepared. Thus, in correspondence with the work done for the 30-Km zone in Ukraine, we created the cartography of ^{137}Cs balance evaluation for Belarus, in order to cover the most important part of the most contaminated territory of these two republics. Our data show that 47.4 % of the territory of Khoiniki district is characterized by negative balance. 19.3 % of this territory shows neutral balance and 33.3 % is characterized by positive balance of ^{137}Cs.

3.2. Vertical Migration of radionuclides

This part is related to the measurements of the radionuclides vertical migration in the soils of Ukraine and Belarus (30-Km zone and Khoiniki district-Gomel region respectively) and is based experimental data collected from 1986 - year of the accident - up to the present time.

The results of the vertical distribution of ^{137}Cs in soils profiles, from the Chernobyl area, obtained during 8 years after the accident are presented in table 1 as the annual position of the center of ^{137}Cs-reserve.

Table 1. Dynamic of transfer of the center of ^{137}Cs-reserve in soils of Chernobyl area.

Type of soil	Elementary landscape	Biocoenosis	Depth of Cs reserve center, cm*							
			1987	1988	1989	1990	1991	1992	1993	1994
sod-semi-podzolic-sandy	eluvial	pine forest	0.60	0.78	0.90	1.04	1.18	1.30	1.41	1.40
sod-podzolic dusty-sandy	eluvial	pine forest	(0.60)	(0.85)	1.10	1.10	1.25	1.35	1.42	1.40
sod-podzolic sandy, dusty-sandy	eluvial	long-fallow land	0.62	0.82	1.12	1.45	1.75	2.00	2.13	2.20
peat gleic, peat bog	trans-eluvial-accumul.	alder marshy forest	(1.15)	(2.00)	2.91	3.12	3.75	4.35	5.06	5.58
primitive alluvial sandy	trans-super-aquatic	pioneer phytocoenosis	(0.88)	1.25	1.61	(1.95)	2.22	(2.50)	2.69	2.80
alluvial semi-podzolic semi-gleic sandy	super-aquatic	meadow	0.70	1.24	1.70	1.80	2.15	2.50	3.10	3.20
sod-podzolic gleic dusty sandy	super-aquatic	meadow drained	0.80	1.22	1.44	1.72	2.12	2.35	2.56	2.67
sod-podzolic gleic loamy-sandy	super-aquatic	meadow non-drained	0.72	1.30	1.65	2.20	2.65	3.05	3.56	3.83
alluvial gleic loamy	super-aquatic	meadow non-drained	(0.90)	1.30	1.86	2.45	2.87	3.42	3.95	4.20
soddy-gleic	super-aquatic	meadow	0.92	1.25	1.77	2.24	2.74	3.30	4.10	4.40
peat bog	super-aquatic		1.44	(2.00)	2.73	2.95	3.44	4.52	5.65	6.05

Notes: * = error of definition is not more than 10%
( ) = data are obtained on the base of 2-3 profiles

Analysis of the motion of the center of ^{137}Cs-reserve as a function of time, shows that slower speed of the vertical migration of ^{137}Cs (about 0.1-0.2 cm/year) is observed in areas with sod-podzolic sandy, dusty-sandy and sandy-loamy soils of the eluvial landscapes covered by pine and oak-pine forests. Such slow speed is probably due to the presence of litter, where is the main part of radionuclides reserve and thus it plays the role of an organic barrier against the active migration of ^{137}Cs. Areas represented by long-fallow lands and sod-podzolic semi-gleic dusty-sandy, and alluvial dusty-sandy soils, are characterized by higher speed of vertical migration of ^{137}Cs (about 0.2-0.3 cm/year). The velocity of migration is higher (0.3-0.5 cm/year) for areas which represent sod-podzolic gleic loamy-sandy. The same velocity of vertical migration show primitive alluvial sandy soils of river beaches. The highest velocity is observed on areas with soddy gleic and peat-bog eutrophic soils, as well as in alluvial ones. The speed arrives 0.7-1 cm/year. Finally, the velocity of the displacement of the center of ^{137}Cs-reserve for soils of the same type is not constant from year to year. In fact, the difference of depth of the reserve centers between 1989 and 1990 was 0.1 cm while between 1992 and 1993 it was 0.6 cm.

The topsoil of Khoiniki is characteristic of the whole Belarus Polessye which was highly contaminated by the Chernobyl accident. The following dominant soil types were investigated [5,6,7]:

1. Soddy-podzolic soils (sandy, sandy-loam).
2. Soddy-podzolic soils with redundant moisture.
3. Soddy-podzolic-gley soils.
4. Soddy-gley soils.
5. Peaty-marshy soils of a lowland type, as well as alluvial ones.

The swift of the center of the radionuclides total reserve is calculated on the base of rate parameters, estimated according to Konstantinov's and Prokhorov's quasi diffusive models, which were worked out for global radionuclides fall-out [8,9].

Table 2 shows the migration speed of ^{137}Cs and of ^{90}Sr in the five different types of soil.

1) Soddy-podzolic soils (sandy, sandy-loam) and 2) soddy-podzolic soils with redundant moisture:

The speed of radionuclides migration in soddy-podzolic automorphous soil does not differ from that of those with redundant moisture. The average speed of ^{137}Cs in soils of these types seems to be 0.48 cm/year and of ^{90}Sr 0,74 cm/year. In a subgroup of unprocessed soddy-podzolic soils the migration speed of ^{137}Cs is 0.14-0.26 cm/year and of ^{90}Sr1.17-5.70 cm/year.

3) Soddy-podzolic-gley soils:

In soddy-podzolic-gley soils we observed a clear dependence of migration speed of radionuclides from hydrological conditions. So, the average linear speed of migration of ^{137}Cs is found 1.16 cm/year in soils with constant high moisture (the so-called "wet" soils), whereas in "drier" soils of this type the speed of ^{137}Cs migration is considerably lower: 0.43 cm/year. Similar observations are made with the migration speed of ^{90}Sr. In "wet" soddy-podzolic-gley soils the migration is about 1.30 cm/year and in "dry" soils 0.65 cm/year.

4) Soddy gley soils:

In soddy gley soils the speeds of migration for both radionuclides do not seem to differ and were found to be approximately 1 cm/year.

5) Peaty-marshy soils of a lowland type, as well as alluvial ones:

Migration characteristics of radionuclides in peaty-marshy soil of a lowland type were variable. The speed of ^{137}Cs transfer in meliorated peat bogs is approximately 0.4 cm/year and in non-meliorated ones about 0.9 cm/year. ^{90}Sr speed in the above mentioned soils was found 0.52 and 1.54 cm/year respectively.

3.3. Classification of migration velocity

Here in below we show the cartographic classification of the above soils as a function of the observed migration velocities of ^{137}Cs. Two maps are prepared separately for the most contaminated areas of Ukraine and Belarus.

For the 30-Km zone of Ukraine the observed intensity of ^{137}Cs migration as a function of the type of soils, is classified in five groups and presented on the map of figure 1. This map was prepared on the base of definition of groups of soil with similar velocities of dislocation of the center of ^{137}Cs-reserve. Similar cartographic work was made showing the migration speed of the two radionuclides (^{137}Cs and ^{90}Sr) into the most typical soils of the Khoiniki district.

Table 2

Speed of ^{137}Cs and ^{90}Sr vertical migration in soils of different type
(cm/year, average speed and standard deviation)

	TYPE OF SOIL A		TYPE OF SOIL B		TYPE OF SOIL C		TYPE OF SOIL C1		TYPE OF SOIL D		TYPE OF SOIL E		TYPE OF SOIL E1	
	^{137}Cs	^{90}Sr	^{137}Cs	^{90}Sr	^{137}Cs	^{90}Sr	^{137}Cs	^{90}Sr	^{137}Cs	^{90}Sr	^{137}Cs	^{90}Sr	^{137}Cs	^{90}Sr
Average:	0,45	0,71	0,50	0,76	1,16	1,30	0,43	0,65	1,07	1,03	0,92	1,54	0,39	0,52
St. Dev. :	0,24	0,13	0,14	0,21	0,26	0,19	0,01	0,08	0,39	0,11	0,26	0,72	0,10	0,06

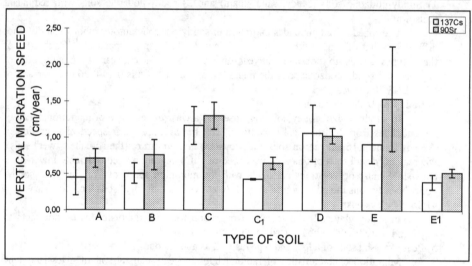

SOIL A = SODDY-PODZOLIC SANDY AND SANDY-LOAM (AUTOMORPHOUS)

SOIL B = SODDY-PODZOLIC SANDY AND SANDY-LOAM WITH MOISTURE

SOIL C = SODDY-PODZOLIC-GLEY ("WET")

SOIL C1 = SODDY-PODZOLIC-GLEY ("DRY")

SOIL D = SODDY-GLEY

SOIL E = PEATY-MARSHY (NON MELIORATED)

SOIL E1 = PEATY-MARSHY (MELIORATED)

Figure 1

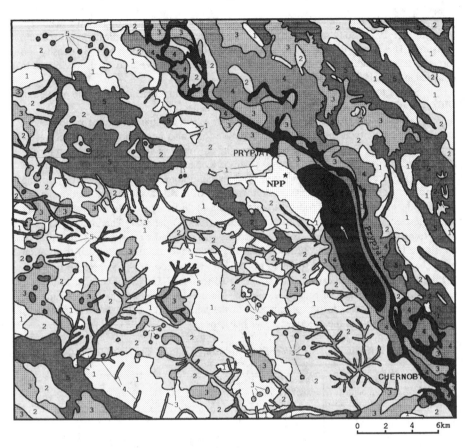

Intensity of Cs-137 vertical migration
in Chernobyl zone (cm/year)

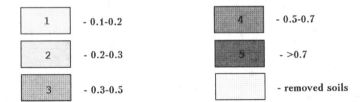

1	- 0.1-0.2	4	- 0.5-0.7
2	- 0.2-0.3	5	- >0.7
3	- 0.3-0.5		- removed soils

3.4. Migration of ^{137}Cs on slopes

The study of natural migration of radionuclides in sloping soils of 30-km zone began in September 1993. Soils from four new sampling areas were systematically analyzed and measured. These areas differ in slope angle, in distance from Chernobyl NPP, in density and forms of fallout, and in some bio-geochemical characteristics. We also used previous data obtained in 1986-1993 from some other slopes..

Self-restoration was expressed by the intensity (I) and efficiency (E) of the process of ^{137}Cs migration which were calculated by the simple equations:

$E=(A_0-A_T)/A_0.100\%$ and $I=dA/dt$, were: $A_0=$ the initial activity of radionuclides and $A_T=$ the activity at the time of sampling. We took also into account the processes of physical decay of ^{137}Cs. During the last 9 years the efficiency of self-decontamination for the studied sites ranged from 7,5-8% (for the site of Kopachy) to about 20% (for the site of Novoselki), or from 1 to 2.5-3% per year. The distribution of ^{137}Cs in the soil of a sloping burnt forest needed a separate examination. Analysis of our data shows that before the fire of 1991, the efficiency (E) of self-restoration during the first 5 years following the accident was estimated 5.2-5.7 % (±1.1 %) or 1.1-1.2 % (±0.25 %) per year. The intensity (I) was found to be 600 Bq/kg per year. Even an increase was detected from the top to the foot of the slope. In two years following the forest fire the above parameters of self-restoration were changed and E increased to 23 % (±4.5 %) - for the top of the slope rose to 36 % (±7%) or 11.5-15 % per year - and I also increased to 3-4.5 KBq/kg per year. In all cases it was observed a tendency of redistribution of ^{137}Cs on the slopes and at present time the maximum activity seems to be concentrated in the middle part of the slope. In table 3 is presented an example of the described above behaviour.

Table 3. Distribution of ^{137}Cs in soil profiles from a sloping site in Savichi (1994)

Sampling point on the slope	Upper layer activity (A_{up}) Ci / km^2	Total activity (A_{tot}) Ci / km^2	A_{tot} / A_{up}
1 (upper)	28.7	32.1	1.12
2	24.0	24.0	1.00
3	35.7	40.5	1.13
4	29.6	35.6	1.20
5	25.7	31.0	1.21
6	16.6	24.1	1.45
7	18.4	26.0	1.41
8	25.3	29.1	1.15
9	19.4	24.4	1.26
10	24.8	31.2	1.26
11 (down)	14.6	20.9	1.43

3.5. Existence form of radionuclides

In order to determine the form of radionuclides in soils, we investigated the site which is situated on the landscape of Shepelichi, at the first over-valley terrace of Prypiat river. Sampling point was situated on a shallow-wavy and relatively dry terrace. Sandy soddy-podzolic soils are lain on aged-alluvial sands. A humus layer of soils reaches 25-31 sm. This site was ploughed before 1986.

A selective leaching of elements was achieved by consecutive treatment of soils using solutions with different composition. For the watersoluble form we used distilled water, for the exchanging ones 1M ammonium acetate solution and for the acidsoluble ones 1M nitric acid solution. The leaching was made at the ratio of solid and liquid phases equal to 1:5, at 200° C temperature and without intensive mixing.

The distribution of radionuclides' forms of existence (occurrence) in soil's profile is presented in table 4.

Table 4. Content (%) of mobile form of radionuclides in vertical profile of soil from the sampling site of Chystogalov.

Layer	Form	^{144}Ce %	^{106}Ru %	^{134}Cs %	^{137}Cs %	^{90}Sr %
0-1	water-soluble	0.40	0.61	0.22	0.15	2.19
	exchangeable	1.39	-	4.5	3.86	6.25
1-2	water-soluble	-	1.09	0.18	0.13	3.23
	exchangeable	1.65	1.21	6.07	4.38	41.3
2-3	water-soluble	-	1.2	0.31	0.11	1.46
	exchangeable	2.8	3.6	5.0	5.56	27.7
3-4	water-soluble	-	1.79	0.31	0.14	1.73
	exchangeable	1.95	1.27	5.6	4.0	30.0
4-5	water-soluble	-	4.24	0.96	0.24	3.02
	exchangeable	3.05	2.42	12.5	9.04	53.5
5-6	water-soluble	-	5.71	-	0.29	1.67
	exchangeable	5.35	4.43	8.92	5.11	45.9
6-7	water-soluble	7.19	10.0	1.11	0.38	7.22
	exchangeable	5.16	13.3	8.22	5.64	58.3
8-9	water-soluble	23	18.9	7.80	4.67	1.19
	exchangeable	-	45.7	71.2	70.8	50.0
10-11	water-soluble	-	-	17.4	8.89	2.13
	exchangeable	75	-	17.8	16.0	87.0
12-14	water-soluble	-	-	-	6.30	28.9
	exchangeable	-	-	-	9.26	45.6

In land biomass ^{137}Cs and ^{90}Sr contents vary from 10 to 200 Bq/Kg. ^{137}Cs and ^{90}Sr mobile forms contents (sum of water-soluble and exchangeable forms) in soils and relation of this nuclides in grass presented in table 5.

Table 5. Mobile forms of ^{137}Cs and ^{90}Sr in soddy-podzolic soil and ratios of these radionuclides in grass.

Year	Mobile forms in soil			^{137}Cs/^{90}Sr
	^{137}Cs, %	^{90}Sr, %	Cs-137/Sr-90	in grass*
1987	8	15	0,35	0,95
1988	7,2	24	0,3	-
1989	8,1	30	0,27	0,39
1991	4,9	37	0,13	0,35
1992	3,5	40	0,09	0,12
1993	-	-	-	0,15
1994	2,2	49	0,036	0,047

* calculated on data from Arhipov A.N., Arhipov N.P., Gorodetzkiy D.V., Meshalkin G.S. "Dyinamika radioekologicheskoy obstanovky naselhosugodjah 30-km zone". Preprint NPO "Pripyat", 1994, Zelony mys (Russian).

4. Conclusions and discussion

Related to the radioecological balances of the contaminated territories, the 30-km zone of Chernobyl NPP can be divided in three parts that are approximately equal, one of them is self-restored. The self-restored part which has negative balance seems to increase in time, extended to landscapes with neutral balance, and it contaminates the second part which is obviously characterized by a positive balance. The percentage of the surfaces with positive balance seems not to differentiate in time. The last part of the territory is characterized by a neutral balance and seems to decrease in time transformed slowly in self-restored territory. In Khoiniki region approximately the half of the territory is self-restored, 33% shows positive balance and only 19% has a neutral character.

Related to the vertical migration and distribution of radionuclides into soil profiles our study showed that at present time the main part (up to 90%) of radionuclides is in the upper 10 cm layer. The rate of vertical migration essentially depends on type of soil and natural landscape. The rate of penetration of the ^{137}Cs-reserve is within 0.1-0.15 cm per year for grass-podsol sandy and sandy loam soils covered by pine forests, and 0.7 cm (or more) per year for grass-gley and peaty-boggy soils.

The obtained data allowed us to classify the soils of some highly contaminated areas in Ukraine and Belarus according to intensity of vertical migration. Our calculations on external dose reduction due to the 10-year natural vertical migration of ^{137}Cs, results in a decrease arising up to 30%.

For the sloping areas, the intensity of self-restoration depends on type of soils, character of the cover vegetation and inclination of slope. For soils of the same type, the factor of acceleration of self-restoration is estimated to be 1.5-3 when the slope incline increases from 8-10° to 30-40°. However, in future self-restoration intensity and efficacy will decrease due to the increasing long-term fixation of radionuclides by different soil components. Nevertheless, the damages of the surface layer caused by erosion or removal of vegetation will become of great importance.

Since the accident more than 57% of ^{137}Cs was converted in fixed forms and for the year 2000 it is expected that this percentage will be 80%. The distribution of the activity of mobile forms of ^{137}Cs and ^{90}Sr in vertical section of soil coincides with that of activity of radionuclides' reserve. 94-97% of the total ^{137}Cs activity is accumulated in the upper 5-cm soil layer. From 1987 till 1994 the steady increasing of ^{90}Sr mobile form content in soils and contrary the decreasing of that of ^{137}Cs were observed. Such effect can be explained by the immobilization of ^{137}Cs in soil. Dynamics of the ratios of ^{137}Cs to ^{90}Sr mobile forms reflects the rate of ^{137}Cs immobilization. The time of transformation of 50% of mobile ^{137}Cs into immobilized form is evaluated to be 3-5 years for soils of 30-Km zone.

Finally, the trends of ratios of ^{137}Cs to ^{90}Sr concentrations in over ground parts of plants are conform to that of mobile forms of that radionuclides in soil. Intensive vertical migration results in a decrease of radionuclide transfer into the phytomass because of their wash-out the root layer.

ACKNOWLEDGEMENTS

The authors wish to thank Dr. Vassili Davydchuk, Ukrainian Institute of Geography, for definition of radioecological balances and for cartographic assistance, and the European Commission for financial support (contract N° COSU-CT94-0080).

REFERENCES

[1] Arapis G., Davydchuk V., Koutlahmedov Y., Sadolko I. (1994): Estimation of self-decontamination of ecosystems to optimize the strategy of decontamination. *Fourth International Scientific and Technical Conference on the Problems of Chernobyl Accident Clean-Up*, Chernobyl-Zeleny Mys, 24-28 October (in press).

[2] Davydchuk, V. (1992): Landscape approach to the estimation of radionuclides migration conditions in the Chernobyl accident zone. *Seminar on the Radioecology and Counter-Measures*. Kiev, 27 April-2 May, Proceedings of I.U.R. Soviet branch, p. 117-123.

[3] Grinevetskiy V., Davydchuk V., Marinich A., Micheli S., Rudenko L., Shevchenco L. (1991): Regional ecological problems of the Ukraine; theoretical and methodological aspects. *Soviet Geography*, v.XXXII, p.533-537.

[4] Davydchuk V., Arapis G. (1995): Evaluation of 137Cs in Chernobyl landscapes: mapping surface migration balance as background for application of rehabilitation technologies. *Journal of Radioecology* (in press).

[5] Arapis G., Petrayev E., Shagalova E., Zhukova O. Sokolik G., Ivanova T. (1995): Effective migration velocity of ^{137}Cs and ^{90}Sr as a function of the type of soils in Belarus. *Journal of Environmental Radioactivity* (submitted paper).

[6] Petrayev E.P., Leinova S.L., Sokolik G.A. (1993). Composition and properties of radioactive particles detected in Southern districts of Belarus, *Geochemical International J.*, 7, 930-939.

[7] Petrayev E.P., Sokolik G.A., Ivanova T.G. (1994). Forms of occurrence and migration of Chernobyl radionuclides in Belarussian soils, *Proceeding of SPECTRUM 94*, Nuclear and Hazardous Waste Management International Topical Meeting, August 14-18, Atlanta, Georgia, USA.

[8] Konstantinov I.E. Skotnikova O.G., Soldayeva L.S. (1974). The migration forecast of Cs-137, *Pochvovedeniye*, 5, 54-58 (in Russian).

[9] Prokhorov V.M. (1981). Migration of radioactive contamination in the soils, *Moscow: Energyizdat*, (in Russian).

Geochemistry of Chernobyl Radionuclides

E. SOBOTOVICH, G. BONDARENKO, E. PETRIAEV
Department of Environmental of Radiogeochemistry of National Academy of
Sciences, Kiev, Ukraine
National University, Minsk, Belarus

The accident at the Chernobyl NPP caused contamination of the most of Ukrainian and Belorussian territory and of Briansk region (Russia). Over 80 radionuclides with half-life more then 5 hours and total activity amounted to almost $1,9 \cdot 10^{18} Bq$ were released into the environment.

Solid-phase are characteristic for the contaminated areas of Ukraine and condensational ones for those of Belorus. About 90% of solid-phase radioactive deposits are parts with radionuclide composition close to that of irradiated nuclear fuel. In the first post-accidental months the main mechanism of vertical dislocation of radionuclides of the Chernobyl fuel fallout was migration of radionuclides in a form of solid particles. In soil radioactive deposits are subjected to influence of soil solutions. The pace of transfer of ^{90}Sr into the mobile form is measured by years another radionuclides release from particles with the same rate but relatively quickly transfer into immobile form. Owing to difference in immobilization rates in soils ^{90}Sr is found at present on the whole in mobile form, 80-95% of activity of another radionuclides are found in immobile form.

Grading of radionuclides caused by forms of nuclides occurs the river system. Owing to that, ^{90}Sr transfers into soluble state and depletion of bottom sediments. Regional evacuation of ^{90}Sr from contaminated drainage system into the river system of Dnieper is 5 times higher than of ^{137}Cs.

General assessment of the fallout was originally made basing on the amount estimation of radionuclides on the contaminated areas. Till present, in the damaged 4th reactor there have been found only about 100 tons of uranium contained in solidified silicate lava which fused through two ferroconcrete overlaps. Uranium content in the lava is within 2,5-10%. Taking into account the possibillity of finding new uranium stores in the adjoining premises, the amount of nuclear fuel released from the reactor is assessed to be over 1/3 of the whole bulk of it. The major part of fuel radionuclides was ejected by steam and hydrogen explosions over the NPP site and the northern part of the cooling-pond. About 3,5% of fuel, i.e. $1,5 \cdot 10^{18} Bq$ of activity is considered to be released outside the site.

Similar quantity of radionuclides was released into the environment during burning of graphite. In this case temperature should exceed 2500°C. In the jet over the 4th reactor the temperature decreased causing condensation of evaporated radionuclides together with atoms and combinations of materials droped into the reactor well. That is the reason that radioactive condensates on the contaminated areas have various matrix-from zirconium and iron to silicate ones.

Radionuclide composition of the fallout is presented by vast spectrum of splinter nuclides and products of other nuclear reactions with half-lives from 24 hours (^{99}Mo) to 30 years (^{137}Cs and ^{90}Sr) as well as by long-living transuranics. The total area contaminated, over 1 Ci/km^2 by ^{137}Cs is more than 100000 km^2, though there is only 11 kg of ^{137}Cs isotope on this territory. Radionuclide composition in the zone enclosed by 3700 Bq/m^2 of ^{239}Pu isoline is close to the fuel one, isoline 110000 Bq/m^2 of ^{90}Sr encloses the area where the ratio of nuclides of ^{90}Sr, ^{137}Cs, ^{239}Pu and ^{144}Ce varies within relatively narrow limits.

Among ecological, medical, biological, biochemical problems arisen due to the accident at the Chernobyl NPP, the most important are those of redistribution of radionuclides in the environment during their dispersion and secondary localization. Decrease of radioactive contamination density in the course of time is determined by radioactive decay and migration.

Radioactive released from nuclear plants, emissions of radioactivity from liquid radioactive waste storages and to some extent global radioactive fallouts are mostly in ion-dispersed form. Radionuclides being in ion form are relatively mobile, easily assimilated by biosphere and included into geochemical and biochemical cycles. Behaviour of such radionuclides is adequately studied in laboratory and natural conditions.

Distinctive feature of the radioactive deposits arisens owing to the accident at the Chernobyl NPP is subdivision of the bulk of radionuclides into two main form:

- solidphase radioactive deposits (dispergated fuel, parts of constructions, graphite, etc) ;
- products of condensation of volatile radionuclides.

Solid-phase are characteristic for the contaminated areas of Ukraine and condensational ones for those of Belorus. About 90% of solid-phase radioactive deposits are fuel parts with radionuclide composition close to that of irradiated nuclear fuel and various content of mobile fission products.Condensational parts, created due to destruction of heat-generating elements and burning of nuclear fuel and containing considerable amount of volatile fission products, caused lower surface contamination as compared to the fuel ones. Both radioactive deposit forms, though to different extent, are present on all contaminated areas. Within the 30-km zone, the main activity is presented by fuel particles although in the North-East and North-Western parts of the zone-mostly by condensational products.

Presence of low-dispersed hot particles on the contaminanted areas is verified by sampling of the particles from soils, analysis of their elementary and radionuclide composition, autoradiography of soil and building surfaces and also special features of distribution of activity in soil profiles in the first months after the accident. So, radioactive micro-spots not typical for homogeneous distribution radionuclides in ion-dispersed state were found in the near-zone by autoradiography. Distribution of high activity was shown on photographs in form of separate grains and their clusters. It was found that the particles are of 2 main sizes - 30 μm and 1-2 μm. Distribution of ^{144}Ce activity in particles depend on degree of dispersion.

Physico-chemical state of radionuclides from accidental release is unique. The very accident lead to creation of a unique test site by virtue of the specific features of radionuclide form and unprecedented extension of contamination.

The problem of hot particles consist of two aspects: medico-biological, i.e. influence of hot particles over a human organism via intake with air and food, and radiational-geochemical treating hot particles as source of mobil radionuclide forms in the environment. Peculiarity of radioactive contamination by solid-phase deposits lies in the fact that the environment, particular, food chains, are contaminated not by activity of the particles, but by radionuclides released from hot particles by leaching and dissolution of the material of particles on their destruction. That is why physico-chemical stability of hot particles in the environment and their ability to relain radionuclides are of greatest importance.

Radionuclide composition of the hot particles sampled in "Red" forest area corresponded to that of irradiated fuel in an accidental phase. Particles enriched in ^{144}Ce (6,3% of total amount of particles), ^{144}Ce+^{106}Ru (5,4%), ^{144}Ce+^{137}Cs (3,6%), ^{106}Ru (1,8%), ^{134}Cs+^{137}Cs (1%) were found. It is worth mentionin₅ ṭnat hot particles comprise combinations of different nature: from inert to radioactive ones. In the hot particles sampled near Janov station, 1,8 km from the Chernobyl NPP, radioactivity was as follows: 65-75% of ^{144}Ce, 22-35% of ^{106}Ru, 3-8% of radiocaesium. In some particles depleted in ^{144}Ce, activity was caused mainly by ^{106}Ru; in other particles ^{144}Ce was not found; in some particles the content of ^{144}Ce was 80-94%.

The investigations indicated that there are no separate particles outside 1 km area around the Chernobyl NPP. The particles are always included into a conglomerate and it presents difficulties to extract them.

When leaching particles of fuel radionuclide composition from conglomerate by distilled water 4 months after the accident, extraction of ^{144}Ce was minimum (0-7,3%), of ^{106}Ru maximum (5,3-33%), ^{134}Cs - 5,45-15%, ^{137}Cs - 1,9-12,6%. In particles enriched in one of the radionuclides (^{144}Ce, ^{106}Ru) we observed maximum water leaching of contained in smoller quantities. From 30 to 55% of ^{106}Ru and ^{134}Cs were leached from particles containing over 90% of ^{144}Ce and 19% of ^{144}Ce and only 6,1% of ^{106}Ru from those containing mostly ^{106}Ru. Water leaching of ^{134}Cs prevailed over that of ^{137}Cs in particles of different types. it can be explained by difference of nuclear genesis of the caesium isotopes.

The ratio of liquid and solid phase volumes upon leaching was rather high (S : L - 1:10^4). It accounts for large proportion of watersoluble forms of radionuclides. The obtained results give notion of radionuclide`transfer into liquid phase on exposure of hot particles to precipitation. To estimate potential bulk of mobile forms of radionuclides in solid and silts`we studied transfer of radionuclides into solution by acting with salt and acid solutions over individual agregates of hot particles. Tolerance of particles to influence of water solutions should be emphasized. At the same time in a number of cases hight leaching degrees were obtained, as a rule, of radiostrontium and radiocaesium, radioruthenium. They were maximum for annealed particles, enriched in ^{144}Ce. Release of ^{144}Ce (up to 33%) and ^{106}Ru (up to 37,4%) was observed in some particles.

Tolerance of hot particles to influence of acid is supported by experiments on treating soils with muriatic, nitric, perchloric acid and mixtures $HNO_3+H_2O_2$, $HClO_4+HF$. Complete dissolution of hot particles was observed only in some cases for samples annealed at 650^{0}C in nitric asid or perchloric and hydrofluoric acids mixture.

Tolerance of hot particles to influence of aggressive media is supported by experiments on treating soils samples with aqua-regia. In this case ^{144}Ce is leached completely. ^{106}Ru is the most stable to aqua-regia influence. Leaching degree of it is about 31%.

The particles are of dark-brown (81%), black (9,4%), grey (3,6%), light-brown (2,4%), brown colours (2,4%) and colourless. The colour is caused by intrusion of luminophores into the crystal lattice of the matrix. Almost all (97%) brown particles contain iron, the black ones contain U^{+4} and the grey ones-lead.

Particles of condensational type are most often ball-shaped. Their shape is often complicated by adhering of smaller particles of different shape. Fuel particles are irreqularly-shaped. Majority of the particles are fragile, especially particles of lamellar habit.

Matrix of the majority of hot particles contains oxides of U^{+4} and U^{+6}, in some cases iron, lead, titanium, silicium, etc.

Microscopic and primary sounding test showed that the matrix of a number of particles is represented by coaly structureless mass, probably of organogenous or predominantly organogenous nature. At the same time the general high background is attributable to presence of radionuclides of ^{144}Ce, ^{106}Ru, ^{137}Cs, ^{134}Cs, ^{90}Sr. In some particles these nuclides are located close to outside zones (may be on the surface) when in others they are distributed over the whole particle.

So, experimental data on the content of mobile forms of radionuclides in agregates of hot particles permitted to conclude that hot particles are able to limit build up of mobil forms of radionuclides in contaminated soils and silts due to stability to influence of natural solutions. By this means they are able to hold radionuclide transfer into surface and ground waters as well as into biotic chains in the first months after an accident.

According to the pilot observations and studies of radionuclide form of being, a scheme of transformation was determined and parameters of transformation of radionuclides of fuel particles into mobile forms and immobiligation of them in soil and disposal sites were established. Half-removal of ^{90}Sr from particles is measured in years.

The data on physico-chemical state of radionuclides suggest that the pace of transfer of ^{137}Cs into the mobile form in soils correlates with the rate of extraction of from ^{90}Sr from particles; ^{137}Cs, isotopes of plutonium and other nuclides, in this case, relatively quickly transfer into immobile (fixed) form. Owing to difference in immobilization rates of radionuclides in soils, ^{90}Sr is, at present, in mobile form, when other nuclides are in immobile form.

Mobilization rates of dispergated fuel radionuclides is one-two orgers lower for air-dry state of media than for soil medium. It allowed us to predict the state of radionuclides in the premises of the 4th reactor. Dust-suppression in the premises caused increasing of dampness with consequent increasing of radionuclide mobilization processes. In this case the fuel as a radionuclide matrix is not dependable even in the first past-accidental ten years.

Slightly reduced mobilization rates of radionuclides is observed in radioacive waste srorages of the Chernobyl NPP near-zone than in soils-due to lower partial pressure of oxygen.

Migration of radionuclides is determined by their forms of being in deposits and soil. Study of spatial-temporal features of migration of radionuclides of Chernobyl fallout includes investigation of lateral and vertical structure of distribution of radionuclides and their forms of being as well as study of changes of these forms in the course of time.

Distribution of radionuclides of fuel deposits in soil profiles resembles logarithmic dependence on the depth of the layer, excepting the surface (0-1) and low horizons. Over 90% of radioactivity, at present, is in 0-10 cm surface layer. Average rate of vertical migration of the bulk of radionuclides is 0,2-1 cm per year. The highest rates of penetration were observed in the first months after the accident due to radionuclide migration in a form of solid particles. In all cases, lower soil horizons contained more mobile radionuclide forms than the upper ones.

Comparison of radionuclide behaviour in global and accidental fallouts testifies that the latter were in the form resistable to influence of the environment. Only insignificant quantity of radionuclides of fuel deposits was in mobil state in the first months after the accident and exhibited radiochemical and geochemical behaviour characteristic for them.

In the first post-accidental months radionuclide migration in soil had the following features:

1. The main mechanism of vertical dislocation of radionuclides of the Chernobyl fallout was migration of radionuclides in a form of solid particles.

2. This process was followed by grading of radioactive particles and predominat penetration of smaller particles. Increasing of specific surface of radioactive particles (which depends quadratically on their size) in lower horizons favours leaching and dissolving of radionuclides.

3. In soil radioactive deposits are subjected to influence of soil solutions, containing dissociating non-organic and organic salts and exhibiting properties of strong electrolytes. They are also influenced by vital activity of the vegetation and soil organisms. Leaching of radionuclides from solid-phase deposits and creation of new physico-chemical form of radionuclides characteristic for the soil complex.

4. Soil regime characteristic for Polessie area makes penetration of watersoluble forms down the soil profile possible. It increases relative content of mobile forms in the lower horizons.

These processes are unidirectional and provide constant but limited removal of radionuclides from mantle of soil.

Distribution of activity of condensational deposits along profiles of soil cuts testifies that penetration of radionuclides was followed by grading on early stages of migration and it was characterized by more intensive penetration of activity in comparison with that observed for fuel deposits. For example, less than 0,1% of radionuclides of fuel deposits penetrated to the depth of 6-8 cm during half a year. In soils contaminated by condensational deposits, there were 0,1-2,3% of ^{90}Sr, 0,05-2% of ^{137}Cs and 1-6% of ^{106}Ru at the same depth. The cited data testifies for larger input of mobile radionuclide forms of these deposits into vertical migration. Experimental data on water leaching from soil supported the following conclusion: 2,5-5% of ^{90}Sr, 0,45-2% of ^{137}Cs and 0,8-2,7% of ^{106}Ru were leached from the upper (0-2 cm) layer of soil contaminated by condesational deposits. 30-70% of watersoluble ^{90}Sr, 5-10% of ^{137}Cs, 20-50% of ^{106}Ru was in layers deeper than 5 cm. ^{106}Ru of condensational deposits penetrated deeper than ^{137}Cs and ^{90}Sr. Water leaching rate of ^{106}Ru from soils of the far-zone was higher than for ^{90}Sr. Contrastingly, ^{106}Ru of global deposits behave like such low-mobile radionuclides as ^{144}Ce and ^{137}Cs.

^{106}Ru behaviour in soils of condensational contamination is similar to that of liquid radioactive wastes in soils, where it is characterized by higher migrational ability than other products of uranium fission. This behaviour of ^{106}Ru is explained by build-up of stable isotopes of ruthenium (masses 99-102 and 104) during operation of the reactor and storage of fuel.Stable isotopes of ruthenium are isotope carriers of ^{106}Ru in the environment encouraging in this way migrationfl processes.

So, essential difference of penetration mechanism of radionuclides of fuel and condensational radioactive deposits in the first post-accidental months lies in the fact that radionuclides of fuel deposits migrated withourt grading of fission products. Whereas in contaminated soils of the far-zone, where aerosol condensational deposits prevailed, penetration of radionuclides was followed by their grading during the first year of exposition. The content of mobile radionuclide forms in them was one-two orders higher than in fuel deposits.

Because of the exposive character of radionuclide release, difference of build-up coefficients was observed for the same plants on similar podsol sandy loam soils, but in various distances from the Chernobyl NPP: for maize it was from 0.1 to 2.5 (factor 25), for lupin - from 0.3 to 4 (factor 13). These data agree with spatial distribution of dispersion of radioactive particles, forms of deposits and content of mobile radionuclide forms in soils.

Transformation of physico-chemical forms of radionuclides and vertical migration influenced sufficiently over the dynamics of radionuclide release into the river system. Decontamination of Ukrainian and Belorussian territories from ^{90}Sr at the expense of surface runoff was about 1.4% /year in 1980, 0.6% / year in 1987, 0.8% /year in 1988, 0.42%/year in 1989, i.e. in the average, about 0.5% annually, except for rthe first post-accidental year. Now, about 80% of ^{90}Sr activity and 20% of ^{137}Cs activity transported by the river Pripyat into the Dnieper Cascade is released from the 30 km zone and 20% of ^{90}Sr activity and 80% of ^{137}Cs activity is flowing from Belarus.

Grading of radionuclides caused by forms of nuclides and their physico-chemical properties occurs in the river system. Owing to that, ^{90}Sr transfers into soluble state and depletion of bottom sediments in this nuclide takes place. Regional evacuation of ^{90}Sr from contaminated drainage system into the river system of Dnieper is 5 times higher than of ^{137}Cs.

Considerable redistribution of radionuclides in the ponds on the river Dnieper is observed. Spotting character of radionuclide contamination of bottom sediments in the ponds is a fundamental feature of radioecological processes in the water ecosystems. High (up to 10^3-10^4) radionuclide build-up coefficients in some objects located in the exclusion zone of the Chernobyl NPP had to be temporaly localized and disposed. The bulk of radionuclides is concentrated on the site of the Chernobul NPP, including "Ukrytie" object and in other premises in the zone. A part of radioactive substances is stored in PVLROS (storage sites) having no hydroisolation. To solve the problem of radioactive waste management in the exclusion zone, uncontrolled release of radionuclides into the environment should be excluded.

Preliminary assessment indicates that radioecological risk caused by PVLROs and PZROs ($1.4 \cdot 10^{16}$ Bq) and the risk caused by surface contamination ($1 \cdot 10^{16}$ Bq) are similar. Reasoning from this data the question on advisability of redisposal of radioactive waste not supported by thorough study of different variations of their storage seems to be debatable. When solving problems of minimization of radioecological impact of the disposal sites, such factors as the quantity of radioactive matter, present and predictable physico-chemical state of radionuclides in the deposit sites, potential evacuation of radionuclides by ground water, etc. should be taken into consideration. At present , we obtained optimistic predictions on radionuclide releases into the river system.

The experience gained in the course of studies of radionuclide behaviour of the Chernobyl deposits in the environment allows to advocate that results of geochemical studies are basing for solving any technical, agro-industrial and medical-biological problems.

The first forecast of contamination of the river system after accident was based on geochemical behavior of radionuclides of the Chernobyl fallout and subsequent events bore out it. Now forecast of natural decontamination of contaminated lands and contermeasured are based on radiochemical knowledge.

II. ENVIRONMENTAL ASPECTS
OF THE ACCIDENT
C. Management of contaminated territories

Posters

Migration of Radionuclides in the Soil-Crop-Food Product System and Assessment of Agricultural Countermeasures

I. BOGDEVITCH, V. AGEYETS

Belarussian Research Institute for Soil Science and Agrochemistry

Abstract Studies on dynamics of redistribution of radionuclides through of profile of the different soils on uncultivated agricultural lands of Belarus during the 1986-1995 period show that vertical migration occurs with low rate. In arable soils the radionuclides are distributed in comparatively uniform way through the whole depth of the 25-30 cm cultivated layer. Investigations on migration of radionuclides with wind erosion on the drained series of wet sandy and peat soils and water erosion on sloping lands show that one should take into consideration the secondary contamination of soils while forecasting a possible accumulation of radionuclides in farm products.

1. Introduction

Radionuclides in the topsoil are potentially available for uptake by plant roots, although plant uptake of many radionuclides is controlled by variety of factors. The rate of uptake differs substantially for different radionuclides and soil types and depends on the physicochemical processes in the soul that govern their availability and on the physiological requirements of the plant.

The amount of information for various radionuclides concerning their behaviour in soil varies greatly. Studies on the dynamics of redistribution of radionuclides through the profile of the different soils, on the transport of radionuclides with wind and water erosion and the effect of soil moisture on the processes of migration of ^{137}Cs and ^{90}Sr in the soil-crop system are of actuality to the population of Belarus.

2. Results and Discussion

2.1 Vertical migration of radionuclides in soils

Studies on the character and rates of migration processes of the long-lived radionuclides ^{137}Cs and ^{90}Sr were conducted on the main typical soil series of the Gomelskaya and Mogilevskaya regions, uncultivated areas of pastures, hayland and lealand. On each site the initial characteristics of morphological properties, granulometric, mineralogic and bulk chemical composition and of agro-chemical properties of all genetic horizons of soil cross sections have been established. Annual measurements of the exposure dose and determinations of the levels of radionuclides in relation to 1 cm and 5 cm soil layers were performed.

Vertical and horizontal migration of ^{137}Cs and ^{90}Sr in arable soils is studied in field experiments. The forms of radionuclides in soils and accumulation of those in the yields were determined. Studies on dynamics of redistribution of radionuclides through the depth of a soil profile on uncultivated agricultural lands during the 1986-1995 period show that vertical migration occurs with a low rate. Practically all the radionuclides are in the upper active part of a root layer of soil humus horizons. The greatest quantity of ^{137}Cs and $^{238\text{-}240}$Pu (80-90%) appears to be in the upper 5 cm layer, i.e. in the sod. And only on the soils being cultivated before the accident (ploughing, grassland establishment) a more appreciable movement of radionuclides to the 5-15 cm depth has been found (Fig. 1).

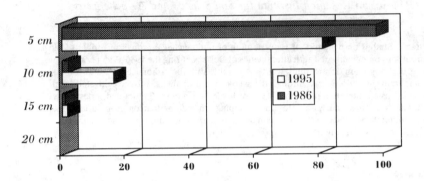

Fig. 1. Dynamics of migration of ^{137}Cs through the profile of sod—podzolic loamy sands (in % of the total level)

The mean rate of migration of ^{137}Cs from the upper 5-cm layer was within a range of 1.6-5.2% per year. ^{90}Sr is more mobile, the level in the 0-5 cm layer decreasing with a rate of 2.8-6.4% per year. In all uncultivated soil the 20-25 cm layer is characterised by background level of radioactivity. Notwithstanding some peculiarities, vertical distributions of ^{137}Cs, ^{90}Sr and $^{238\text{-}240}$Pu are much similar in character; this fact is probably conditioned by a predominantly biomechanic transfer of radionuclides with soil particles.

In arable soils ^{137}Cs and ^{90}Sr are distributed in a comparatively uniform way through the whole depth of the 25-30 cm cultivated layer. Their movement to 30-50 cm depth is significant.

2.2   Horizontal transport radionuclides with water and wind erosion

Transport of radionuclides with processes such as wind erosion has been studied on the drained series of wet sandy and peat soils used for growing field crops. According to the direction of prevailing winds profiles were laid down, the space between them being 100 metres. Over a four-year period since 1991 in spring, summer and autumn from each profile in accordance with the motion of an air flow, composite soil sample have been taken and analysed for ^{137}Cs. The results obtained were compared with the initial levels of radiocaesium in soils according to the data of the 1987 measurements. It has been found that various sites of the fields investigated differ greatly for the levels of radiocaesium in the arable horizon of soils. The levels of ^{137}Cs increased normally by 1.5-2 times in the direction from the central part of the field to the tract of forest and at the edge of the forest the levels of ^{137}Cs were 2-3 times higher than the initial levels of contamination. This fact is due to the transport of contaminated dust particles; it is confirmed by the presence of deflation material in dusters designed by Bagnold and in plotting boards installed according to the motion of an air flow.

Studies on transport of radionuclides with water erosion were conducted on sloping lands. these studies are underlaid by quantitative indices of a solid flow during 1991-1994 in the Gomelskaya region. It has been found that most of the lands subjected to washout are represented by areas having a potential average annual washout that ranges between 2.5 and 10 t/ha. Thus, in Vetkovsky district one can observe a natural increase in the levels of ^{137}Cs in the middle and particularly in the lower part of the slope up to 1.5-2.0 times comparing with the upper part while cultivating grain and tilled crops. On the lands under four-year old perennial grasses, where practically a solid flow was absent, there was no significant differentiation in the density of contamination of soils according to units of a slope. On the sloping lands, with an intensity of a solid flow of 10-20 t/ha per year, a more significant accumulation of ^{137}Cs in the arable horizon of soils of the lower part of the slope has been found.

Thus, one should take into consideration the secondary contamination of soils while forecasting a possible accumulation of radionuclides in farm products. It is obvious that there is a need for taking regular measures focused on soil conservation, because in the zone of radioactive contamination more than half of the arable land represented by sandy and drained peat bog soils and also by sloping soils can be subjected to erosion.

2.3 The degree of soil moistening and processes of migration of radionuclides in the soil-crap system

Due to contradictions of data in the literature a need has appeared for studying the effects of soil moisture on the processes of migration of radionuclides in the soil-crop system. As the degree of soil moistening increased the share of radiocaesium fractions available to plans increased by a factor five and the transfer of radionuclides into hay, by a factor twenty. Taking into account the natural increase in yield on sod-gley sandy soils the total removal of ^{137}Cs has increased by a factor 76 in comparison with overmoistened soils.

The availability of ^{137}Cs to plants is significantly declining in the course of time due to the processes of "ageing" and its fixation in soil. During the period 1987-1992, the share of mobile radiocaesium decreased in different soils from 29-74% too 5-29% of the total or on an average more than 3 times. At the same time a corresponding reduction in the transfer of ^{137}Cs into the yields occurred.

On the contrary, ^{90}Sr is characteristics of predominance of easily available exchangeable and water-soluble forms that as a whole account for 50-87% of the total level. As for the share of the fixed fraction extracted by 6M HCl it is not high and varies within a range of 2 to 19%. Overmoistening of soils also results in the increase in the share of mobile fractions of ^{90}Sr. But this process is accompanied by a significantly lower increase in the transfer of ^{90}Sr into perennial grasses than for ^{137}Cs. Accumulation of strontium in plants to a higher degree depends upon the agrochemical properties of soils than on the moisture regime.

2.4   Assessment of agricultural countermeasures

The protective measures in farming industry are carried out in two stages : the 1st stage -- 1986-1991 and the 2nd stage -- since 1992. The highly contaminated lands, where obtaining farm products having permissible levels of radionuclides have been removed from crop rotations; there liming and fertilisation of acid soils using phosphorous and potash fertilizers have been performed on waterlogged soils, drainage and deep ploughing of the sod were done, grass establishment and regrassing of hayland and pastures was done. All the measures made it possible to reduce the entry of radiocaesium into farm products by 3.5 times.

There exists only random monitoring concerning the levels of ^{90}Sr in farm products. Approximately, the entry of radiostrontium into food products over the postaccident period has decreased twice. The availability of ^{90}Sr to plants remains high with a trend towards increase.

Conclusion

In Belarus one can see apparent comparatively high effectiveness of large-scale protective agricultural measures that have been taken mainly in the public sector. However, the problems of safe residence for population and obtaining quality food products are still far from being solved.

Geographic Assistance of Decontamination Strategy Elaboration

Vassili DAVYDCHUK
Inst. of Geography, National Ac. of Sci., Volodymyrska 44, Kiev 252034 Ukraine
Gerasimos ARAPIS
Agricultural University of Athens, Iera Odos 75, Botanikos 11855 Athens, Greece

Those who elaborates the strategy of decontamination of vast territories is to take into consideration the heterogeneity of such elements of landscape as relief, lithology, humidity and types of soils and, vegetation, both on local and regional level. Geographic assistance includes 1) evaluation of efficacy of decontamination technologies in different natural conditions, 2) identification of areas of their effective application and definition of ecological damage, 3) estimation of balances of the radionuclides in the landscapes to create background of the decontamination strategy.

1. Introduction

Geographical assistance is directed to take in consideration of heterogeneity of environmental conditions which determine the efficacy and the ecological consequences of the technologies of decontamination and elaboration of decontamination strategy.

2. Obtained results

2.1. Field trials and evaluation of decontamination technologies

Geographic assistance of experimental trials and of decontamination technologies evaluation includes identification and characterisation of testing plots, evaluation of efficacy of decontamination technologies in different natural conditions and definition of ecological damage.

The program of field trials and ecological evaluation of the decontamination technologies includes identification and characterisation of testing plots, evaluation of efficacy in different natural conditions and definition of ecological damage. Soils, surface deposits, relief and vegetation cover of field experimental plots were identified, and efficacy of the technologies depending of natural conditions was evaluated. Thus the areals of most effective expected application of the technologies tested were defined.

On the basis of geosystem analysis of the field trial results special map shoving the application areals of turfcutter soil decontamination technology, urban decontamination technologies and rotating brush removal of the forest litter is prepared for the evacuation zone of Chernobyl NPP. The map shows natural units (areals) with certain combinations of relief, lithology, soil and vegetation conditions, which are supposed as areals where effectiveness of decontamination by the technologies mentioned will correspond to results discovered during experimental field trials.

Best natural conditions for application of turfcutter technology are obtained in non-ploughed grasslands with thick turf layer and composed mineral or organic soil horizon of high and intermediate humidity. Among them gleic soddy-podzolic, soddy and alluvial loamy and sandy-loamy soils, covered with meadows or long-fallow grass vegetation and especially drained organic (peat bog) soils of flood plains and lowered river terraces. Good results of this technology are corresponding to semi-gleic and non-gleic soddy and soddy-podzolic sandy-loamy soils. Damage of the ecosystems is minimal because of fast vegetative selfrestoration of ground cover.

At the same time results of the turfcutter technology on poor, dry sandy podzolic and soddy soils with sparse grass cover and thin mat are not very good: turf mat can't be rolled. Simultaneously areals of some special types of forests, swamps and sandy areas, which are shown on the map, can't be decontaminated using a technology evaluated in the frame of ECP-4.

2.2. Evaluation and mapping of local balances of the radionuclides

Any decontamination strategy for large territories has to take into consideration the direction and intensity of natural processes of migration of the radionuclides which influence the formation of their regional and local balances and therefore self-decontamination (or self-contamination) of the natural areals. Processes of redistribution of the radionuclides vary depending of such stable components of landscape as relief, lithology and soils, and variable ones as vegetation cover. Areals with negative, neutral, or positive balances of the radionuclides are defined on the maps.

The balance of radionuclides in the landscapes reflects the ability of the natural systems to "evacuate" the pollutants. It can be negative, positive or neutral. Landscapes which occupy high levels of relief, normally are characterised with negative balance. They are geochemically autonomous, and their balance depends only of the intensity of natural evacuation of the pollutants. The landscapes with positive balance (depressions, valleys etc.) are under geochemical influence of surrounding territories, which belong to the same watershed basin.

Attention should be taken also on tendencies and intensity of natural evolution and self-restoration of the landscapes in Chernobyl zone which started after the evacuation of population and sufficient modification and limitation of human activity. It caused intensive development of plant succession processes which modify initial ploughed lands, meadows and cultivated forests into relatively stable forest ecosystems according to edaphical conditions of territory. Self-restoration of vegetation, accumulation of phytomass, deconsolidation and self-restoration of soils decrease washing-off and increase infiltration.

Maps of local balances of Cs-137 in ecosystems of Chernobyl zone are created by overlaying the map of natural landscapes and maps of vegetation successions. Maps were elaborated by computer using PC View Color and Adoba Photoshop 2,5 softwares.

Long-term balance of Cs-137 was evaluate taking into account the direction and velocity of the plant successions and landscape selfrestoration processes which are obtained in evacuation zone. According to our elaborated succession model, this processes, in collaboration with human activity, shall create dense forest cover at present long fallow grasslands, dry meadows territories and settlements during 50-80 years.

Influence of vegetation on washing-off balance of the radionuclide is expected semi-homogenous for the territory of evacuation zone after restoration of the forest cover. Expected long-term changes of spatial structure of balance of Cs-137 in landscapes of the territory shall be connected with decreasing of role of washing-off processes as result of fixation of the soil surface by forest vegetation. Proportion of areals of Chernobyl zone with different balances in the short- and long-term aspects is presented on the Figure 1.

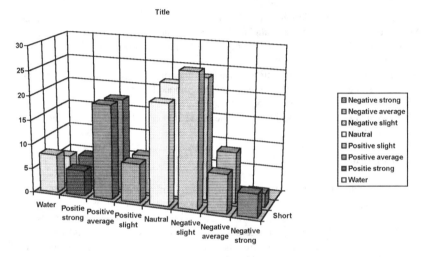

Fig. 1. Landscape areals (% of territory) with negative, neutral and positive short-term (up to 50 years after the accident) and long-term (more than 50 years) balances of Cs-137

The figure shows that evolution of the situation is under influence of natural landscapes restoration. Areals with **strong negative balance** which cover 4,7% of territory for sort-term evaluation decrease to 2,8% for long-term one. **Average negative balance** of Cs-137 is expected for 7,9% of the territory in short-term aspect and for 10,4% in long-term. **Slight negative balance** is expected for 27,1% of territory in short-term evaluation and for 24,7% - in long-term. Generally landscapes with negative short-term balance of the radionuclide occupy about 39,7% of estimated territory, and with long-term - 37,9%. It means that more than one third part of Chernobyl zone proclives to lose Cs-137 by washing-off.

Neutral balance is expected for territories of the river terraces, which are characterised with sandy deposits, flat or hillock surfaces and very few washing off. Together with water surfaces and industrial areas, neutral balance areals cover 20,6% of the territory for short-term evaluation and 23,1% for long-term one. The increasing of this category and decreasing of **water surface** from 7,9% to 5,9% in future is connected mainly with future drainage of cooling pond of NPP and annexation of its territory to ones with neutral balance.

Areals of **positive balance** of Cs-137 are identified in Chernobyl accident zone with closed depressions, watershed catchments, elements of erosional network and rear lowered parts of river terraces and flood plains. **Positive short-term balance** of the radionuclide is expected for 32,3% of territory of 30-km zone, and long-term one - for 33,1%. Landscapes with **strong positive balance** cover about 5,2% in short-term evaluation, and about 6,4% - in long-term one. Expected increasing of this category and corresponding decreasing of category of **average positive balance** from 19,2 to 18,9% from short-term to long-term evaluation can be explained by intensive restoration of the forest vegetation during future 30-50 years.

Thus, territory of 30-km zone of Chernobyl NPP is divided in three approximately equal parts. One of them is self-decontaminating. Radionuclides are directed from here to second third by natural processes of washing-of. The last third does not participate this processes. Countermeasures can be planned for the last one without taking into consideration processes of self-decontamination. The estimations presented here are in good correspondence with field radiogeochemical experimental data.

The proposed above evaluation is qualitative one. Using experimental data its quantitative interpretation can be made. The physical decay of the radionuclides is also to be taken into account.

3. Discussion

Maps of short-term and long-term balances of radionuclides 1) show localisation of places, 2) explain the factors of their concentration and therefore stability of the radioecological situation in 30-km zone of Chernobyl NPP, 3) reflect the increasing of this processes in time. This conclusion combined to prepared maps can be used as background for evaluation of intensity and direction of processes of natural redistribution of the radionuclides. Also it can be used for elaboration of general conception of decontamination strategy and for preparation of decisions related to application of the countermeasures.

Depending of the development of natural processes, long-term rehabilitation strategy can be recommended for application at landscapes with positive long-term balance of the radionuclides. These landscapes can be proposed also as depositories for radioactive soil material generated from the application of countermeasures.

Moreover, evaluation of long-term efficacy of the decontamination technologies has to take into consideration both regional (local) balances of radionuclides, and intensity of natural processes of self-decontamination, especially for the zones of negative balance.

On the stage of elaboration of spatial aspects of decontamination strategies the maps of balance of the radionuclides in the natural landscapes can be useful to determine places of recommended soil (litter) removal and their relocation. One of possible strategies can be based on artificial relocation of the radionuclides to the zones of natural accumulation.

In addition to well known economical and dosimetrical criteria of cost-benefit analysis, some ecological criteria are to be considered. First one is to evaluate the cost of ecological consequences of decontamination, if damage of the ecosystems takes place. Second one is to extend the analysis of cost-benefit proportions for the period of total restoration (artificial or natural) of initial landscapes and ecosystem conditions.

CEC-CHECIR ECP-4
OPTIMIZATION OF INTERVENTION STRATEGIES FOR THE
RECOVERY OF RADIOACTIVE CONTAMINATED ENVIRONMENTS.

C. Vázquez [a], J. Gutiérrez [a], C. Trueba [a] and M. Savkin [b]

[a] CIEMAT, Av. Complutense 22, 28040 Madrid, Spain
[b] Institute of Biophysics, Zhivopisnaya St. 46, 123182 Moscow, Russian Federation

Abstract. The goal of this work is to evaluate different options of intervention for the recovery of contaminated environments. It will consider not only the efficiency of the countermeasures in terms of dose reduction, but also in terms of costs, wastes and other possible secondary consequences, in order to obtain the best possible strategy for each particular circumstance. This paper summarizes the methodology of optimization of intervention, which has been carried out in the framework of CEC-CHECIR ECP-4 Project.

1. Introduction

Following a nuclear accident with environmental consequences, intervention, leading to the recovery of the contaminated environment to as close to normality as possible with the lowest social cost, could be necessary. The reduction of the damage from the existing contamination should be justified and optimized. This means that the best strategy for applying recovery actions must be selected from a set of potential alternatives, analyzing the positive and negative effects related with their applicability.

2. Methodology

Two main branches of activity on which the strategy analysis is based are identified. The first one deals with the scenario of intervention which requires actions leading to its characterization, its classification and the evaluation of its radiological impact. The second one is related with the different decontamination procedures, and their relationship with the scenario. Both of them will converge in an evaluation process under different criteria. Figure 1 shows the relationship between the two branches and the general sequence of operations involved.

The characterization of the scenario consists in a complete physical, radio ecological and socioeconomic description. This will allow its classification in different intervention elements (IE), defined as class elements of any scenario where similar activity concentrations lead to similar radiological risks and similar response to the same recovery actions. Once classified, it is possible to calculate the normalized (per unit of deposited activity) radiological impact in terms of dose rates and integrated doses derived from each IE.

The analysis of the decontamination procedures consists of an assessment of their performance and applicability for the different identified IEs. The *performance* represent its radiological and economic behaviour on each IE (in relation to each specific radionuclide). The cost will include the operation costs and the costs concerning the waste management and

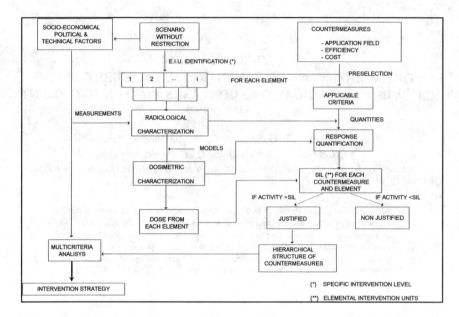

Figure 1. Schematic diagram of the methodology for strategy analysis.

disposal as a consequence of the intervention. The *applicability* incorporates other factors such as the availability of resources, effort involved in the implementation, and all other possible constrains related with the scenario. Both performance and applicability will determine their *practicability* for the real case under analysis.

Several categories of Intervention Scenarios can be immediately envisaged, but this work deals only with two of them:

1. Urban scenarios: dealing with all the environments where people normally live. The radiological risk will be evaluated only through the external irradiation from the different locations of deposition. Typical IEs in these scenarios are gardens, yards, streets, roofs, walls.

2. Agricultural scenarios: where the most significant risk contribution will come through the food chain, directly from crops, or through cattle, mainly through milk, but also external doses to the farmers are evaluated. Typical IEs are soils used for pastures, haylands, and arable lands.

A data base of countermeasures developed in the frame of ECP4 Project and others, usually applied to recover Chernobyl scenarios, has been prepared. For each countermeasure, radionuclide and IE, the following factors of performance have been included: frequency of application; decontamination factor (in terms of transfer factor to certain crops and/or in terms of external dose); man power; depreciation of equipment; consumables; overheads; secondary effects on the IE, such as changes in productivity or quality; restriction time after intervention and amount and activity of wastes.

Factors determining the applicability are: scale of application; number of operators; equipment and consumables and constrains, if any.

Using adequate radio ecological and dosimetric models, the following items are evaluated: radiological risk from each IE in terms of collective and individual doses and length of restriction, if legal levels are applicable. The data base of countermeasures make it possible to calculate: the averted and residual collective dose from each IE after applying each countermeasure; the cost of application using all cost factors, included wastes and secondary consequences and the volume, specific activity and management cost of the generated wastes.

The developed procedure, using the evaluated dosimetric and cost factors, can make possible a cost effectiveness analysis or a complete cost benefit analysis, if a preestablished value for the collective dose is introduced; in this case a specific intervention level (SIL) is calculated for each countermeasure on each IE, below wich the procedure would not be justified. According to the type of analysis and taking into account external restrictions such as available budget, machinery, man power..., it is possible to decide the final strategy of intervention.

3. Case study

As a demonstration of the usefulness of the proposed methodology, a case-study concerning a local strategy of intervention is exemplified on Savichy, a large rural settlement in the Southern-Eastern part of Belarus. After Chernobyl accident all the population was evacuated but in 1987 part of it came back without permission. The objective is to analyze the radiological situation of the population, at present and in future, supposing that the settlement would recover the former population ("shadow population") and to provide criteria to decide about the possibility of applying some decontamination to improve the situation. For both, urban and agricultural scenarios, the impact evaluation and the applicable countermeasure's behaviour have been analyzed using the two branches defined in the methodology. This paper only shows the result of the urban analysis.

Figure 2 shows the external dose distribution on different locations from all IE, calculated using models where the inputs are the dose rate on undisturbed land (33 μR h⁻¹), the relative distribution of activity on the different urban elements and the permanence factors for the population. The total yearly individual dose rate was 1,05mSv. The contribution of the different urban IEs to the dose in each collective farm is shown in Figure 3 for wooden wall houses.

The results obtained for the "shadow population" are very similar.

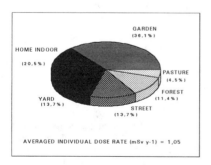

Figure 2. External dose distribution on different locations from all IEs.

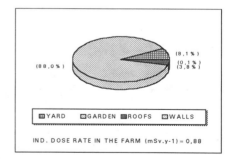

Figure 3. Wooden houses: contribution of the different urban IEs to the total dose.

The response of the different countermeasures applicable to farms with brick and wood walls under three different criteria is shown on Table 1 and 2. For the cost-benefit analysis the preestablished value for the collective dose used was 15.000 ECU Sv^{-1}.

Both the cost-benefit and the cost efectiveness analysis give as best decontamination procedure the spade on yard. The countermeasure which avoids a higher value of the collective dose is the spade on garden.

Table 1. Decontamination of farms with brick homes.

INTEGRATION TIME (y)				50			
DOSE RATE INDOORS (nGy/h)				40,3556			
DOSE RATE OUTDOORS (nGy/h)				281			
OPTIONS OF DECONTAMINATION	Decontam. factor	Individual dose from the IE doing nothing Sv/y	Resi.Ind.dose from the IE after decont Sv/y	Collective averted dose per home man.Sv/home	SIL micro R/h	Cost Effectiv. ECU/Sv	Proc. cost/ waste cost ECU/home
SPADE/YARD	6	1,2E-05	2,0E-06	2,0E-03	16	7209	14
SPADE/GARDEN	6	1,3E-04	2,2E-05	2,2E-02	19	8631	188
SPECIAL DIGGING/YARD	13	1,2E-05	9,2E-07	2,2E-03	30	13973	31
SPECIAL DIGGING/GARDEN	13	1,3E-04	1,0E-05	2,4E-02	36	16728	404
ARS-14/ROOF	1,9	4,7E-06	2,5E-06	4,5E-04	143	65742	29
SET OF TOOLS/ROOF	total	4,7E-06	0,0E+00	9,5E-04	445	204303	193
HAMMER NAIL-TAKER/ROOF	total	4,7E-06	0,0E+00	9,5E-04	563	258367	244
SANDBLASTING DRY/WALLS	4	1,2E-07	3,0E-08	1,8E-05	2207	1012505	18
ROOF WASHER	2	4,7E-06	2,4E-06	4,7E-04	2836	1301073	615
TURBO NOZZLE HP/ROOF	2,2	4,7E-06	2,1E-06	5,2E-04	3910	1793602	925
POLYMER COATINGS/WALL	4,5	1,2E-07	2,7E-08	1,9E-05	4132	1895399	35
ARS-14/WALL	2,4	1,2E-07	5,0E-08	1,4E-05	6349	2912547	41
AMMONIUM NITRATE SPRAY/wall	1,3	1,2E-07	9,2E-08	5,5E-06	139790	64123943	356
SANDBLASTING WET/walls	5	1,2E-07	2,4E-08	1,9E-05	329280	151046062	2906
TURBO NOZZLE HP/walls	1,3	1,2E-07	9,2E-08	5,5E-06	411348	188691552	1047

Table 2. Decontamination of farms with wood homes.

INTEGRATION TIME (y)				25			
DOSE RATE INDOORS (nGy/h)				75			
DOSE RATE OUTDOORS (nGy/h)				309			
OPTIONS OF DECONTAMINATION	Decontam. factor	Individual dose from the IE doing nothing Sv/y	Resi.Ind.dose from the IE after decont Sv/y	Collective averted dose per home man.Sv/home	SIL micro R/h	Cost Effectiv. ECU/Sv	Proc. cost/ waste cost ECU/home
SPADE /YARD	6	5,9E-05	9,8E-06	1,7E-03	18	8437	14
SPADE /GARDEN	6	6,4E-04	1,1E-04	1,9E-02	22	10122	188
SPEC DIGGING/GARD.	13	6,4E-04	4,9E-05	2,1E-02	43	19618	404
ARS-14 /ROOFS	1,9	2,7E-05	1,4E-05	4,5E-04	143	65386	29
SPECIAL DIGGING/YARD	13	5,9E-05	4,5E-06	1,9E-03	36	106022	31
SET OF TOOLS/ROOF	total	2,7E-05	0,0E+00	9,5E-04	443	203196	193
HAMMER NAIL-TAKER/ROOF	total	2,7E-05	0,0E+00	9,5E-04	560	256968	244
ROOF WASHER/ROOF	2	2,7E-05	1,4E-05	4,8E-04	2821	1294024	615
TURBO NOZZLE HP/ROOF	2,2	2,7E-05	1,2E-05	5,2E-04	3889	1783885	925
MANUAL ELECTRIC CUTT/WALLS	5	8,0E-07	1,6E-07	2,2E-05	12667	5810531	130

References

[1]   C. Vázquez, J. Gutiérrez, R. Guardans, B. Robles and M. Savkin, ECP4 1993 Annual Progress Report, "Optimization of Intervention Strategies for the Recovery of Radioactively Contaminated Environments. Application to a Case-Study: Kirov (Belarus)." CIEMAT/IMA/UGIA/M5A12/01-93. December 1993.

[2]   C. Vázquez, and C. Trueba, ECP4 1994 Annual Progress Report, "Optimization of Intervention Strategies for the Recovery of Radioactively Contaminated Environments. Application to a Case-Study: Kirov (Belarus)." 2 nd Version CIEMAT/IMA/UGIA/M5A22/02-94. December 1994.

[3]   C. Vázquez, C. Trueba and N. Rodríguez, CHECIR ECP4 Subproject 6: Modelling and Cost-Benefit Analysis. January-August 1995 Progress Report. CIEMAT/IMA/52F21/02 95. August 1995.

OPTIMAL SYSTEMS OF MEANS AND METHODS AND UNIVERSAL ALGORITHM OF DECONTAMINATION OF RADIONUCLIDE'S CONTAMINATED SOILS.

Y.KUTLAKHMEDOV, N.ZEZINA, A.MICHEEV
Institute of cellular biology and Genetic engineering National Academy of Ukraine

A.JOUVE
Institute de Protection et de surete Nucleare /DPEI/ CEA, Cadarache, France

G.PEREPELYATNIKOV
Ukrainian Radiation Training Centre, Kiev

Abstract. This paper represents our data of comparative analysis of efficacy of different countermeasures in decontamination of soils in Ukraine in total and in case study Milyachi. On this base it was created of optimal algorithm of strategy of decontamination of soils which is based on method of usage turf harvester for unploughed soils and method of phytodesactivation for ploughed soils of Ukraine after Chernobyl accident.

1. Ten years ago after Chernobyl accident the problems of decontamination and rehabilitation of contaminated territories are removing on first place. We believed that 30-km zone must be return in normal usage. Other contaminated territories of Ukraine needs in rehabilitation. It is clear that all contaminated lands of Ukraine must be returned for normal usage.

2. We provided of comparative analysis of efficacy methods and means decontamination of soil which were contaminated of radionuclide's after accident on ChNPP. It was calculated Cd (coefficient of decontamination) of different countermeasures on criteria of decreasing individual and/or collective doses for population (table 1).From data of table 1 it is seen that efficacy of realised and planned countermeasures (CM) distinguishes. Most part of CM are able to decrease of individual doses but only small part can decrease collective doses.

Table 1

Effectively of different methods of radionuclide contaminated soil's desactivation

Desactivation methods	Coefficient of desactivation by individual dose (Cd -1)	Coefficient of desactivation by collective dose for population(Cd-2)	Time for realisation (years)
Territory fixation	1,2	1,2	1

Turf removal	20	20	1
Removal of soil with plough, buldozer, screper	6-8	2	1
Deep ploughing	2-3	1	1
Change of agriculture	2-3	1	1
Including of fertilisers	2-3	1	2-3
Phytodesactivation	3-5	3-5	4-5

3. On territory of Ukraine were realised any effective countermeasures. In table 2 summary results of estimation efficacy of any CM,s represented with calculation of territories and years.

Table 2.

Summary data about realised CM in agricultural collective farms on Ukraine (1986-1994)

Countermeasures	Square th.ha	Coast th.Ecu	Economy of collective dose th. man-Sv	Benefit th.Ecu	Benefit-coast th.Ecu
The liming of soils	578	7029	34,67	22189	15160
The fertilising of soils	803	10130	48,19	30792	20702
Improvement of pastures	536	22032	42,86	22471	13346
Total	1917	30191	125,72	75392	49208

It is good seen how high efficacy of agricultural CM which were realised on Ukraine. From total sum of waited collective dose for Ukraine 19-20 mln man-rem for 9 years nearly 12,6 mln man-rem was economised owing to realised CM,s(63%).It is seen that waited benefit-coast consists 49,2 thousands ECU.(Price of one man-em for Ukraine is 6,4 ECU).

4. We have based the high potential efficacy technology of decontamination of turfed soils with help turf harvester, which can to cut of upper layer of turf (1-5 sm.).It is shown in our experiments in Rivno district on peat bog soils, that removing of upper layer of turf (5 sm.) promises in 20 times to decrease contamination of crops and milk on this pastures(table 3).

Table 3.

Cs-137 contents in cow's milk of experimental and control group of animals in experiments on turf removal by turf harvester (Bq/l)

Selection data	Cow number	Cs-137 content in milk(experiment	Cs-137 content in milk (control)	Cd
21.07.93	227	59,2	723	9,9
	246	80 (aver- 72,7)		
	920	79		
22.07.93	246	85		9,2
	227	81 (aver-78,3)		
	920	69		
23.07.93	246	74		10,1

	227	67 (aver-71,7		
	920	74		
24.07.93	920	45		15,3
	246	48 (aver-47,3)		
	227	49		
25.07.93	227	38,2		19,9
	920	34(aver-36,4)		
	246	32		

We estimated possibilities of usage turf harvester technology on territory of Ukraine for pastures which did not plough after accident. It is shown that on nearly 30 thousands Ha of pastures for private cattle can be used this technology of turf harvester. This can bring nearly 180 thousands man-rem economy of collective dose and relation of benefit-coast nearly 668 thousand dollars of USA. In our investigations on polygon "Buryakovka" we were based the high efficacy Phytodesactivation of soils with help of optimal system crop rotation of plants, with usage :the special processing of seeds; the including of fertilisers; the including in soil any micro-organisms and etc.It is shown that for 4-5 years coefficient of decontamination (Cd) for phytodesactivation can reach 4 units. 5. For case study Milyachi (Rivno district, Dubrivitsa region) was anylized concrete system of realised CM,s(table 4).On collective farms in result of usage CM collective dose was decreased on 50-70% but small quantity of CM,s for private farms showed the decreasing of collective dose for population only 20-30%. It means that for population of this case study individual dose can compose significant value 3-8 mSv/year. The using of turf harvester on pastures in this case study could bring economy of collective dose in 1800 man-rem. The usage of CM in this case study brought real economy collective dose in 376 man-rem only.

Table 4.

The general characteristics of realised CM in case study Milyachi in 1988-1993

Countermeasures	Square (ha)	Quantity of heads	Including in tons	Cd (on milk)
On collective farms				
Deep ploughing	990			1,5-2
The raised dose of fertilisers	720		360	2-2,5
The liming of soils	420		1260	1,5-2,5
Pasture improvement	250		75	2,5-3
The including of manure and sapropell	440		13200	1,7-1,9
On private farms				
The using of boluses		80	240 pieces	2,2-2,8
The feeding of humolit		150	45	1,5-1,9
The feeding of ferrasin		50	7 kg	2-3
The turf-cuttering	0,5	3		18-20

6. All possible means and methods decontamination can be divide on the two groups: the first - for ploughed soils(after accident);the second - for unploughed soils and territories. For unploughed turfed soils is most effective the usage of turf harvester, and for ploughed of soils is more effective the usage phytodesactivation. It was created universal algorithm for election of means and methods decontamination of soil for different conditions. The optimal choice of means and methods of decontamination depends from next circumstances: a) Ploughed or unploughed the soils; b)The characters of distribution of radionuclides on profile of soils; c) The aims of decontamination (agricultural usage of territories, usage of territories for recreation. The decreasing of individual and/or collective doses for population); d) The relation of benefit - cost for concrete countermeasures; e) The limit of time which can be use for decontamination of territories. It was shown significant difference in systems of countermeasures which were oriented on decreasing of individual and/or collective doses for population on local territories and system of countermeasures which was oriented on decreasing of exported collective dose for population.

Aspects of risk analysis application to estimation of nuclear accidents and tests consequences and intervention management

Demin V.F.,[*]   Hedemann-Jensen P.,[†]   Rolevich I.V.[‡]

Schneider T.,[§]   Sobolev B.G.[¶]

Abstract For assessment of accident consequences and a post-accident management a risk analysis methodology and data bank (BARD) with allowance for radiation and non-radiation risk causes should be developed and used. Aspects of
these needs and developments are considered. Some illustrative results of health
risk estimation made with BARD for the Bryansk region territory with relatively
high radioactive contamination from the Chernobyl accident are presented.

1   Introduction

In accordance with regulatory documents adopted the decisions on post-accident off-site
protection and restoration measures have been made with allowance for only radiological consequences. In this decision making and for an assessment of radiation accident
consequences the concept of the effective dose D_E has been often used especially in the
practice in the FUSSR and CIS after the Chernobyl accident.

As it follows from the experience in the assessment and analysis of the consequences
of nuclear accidents or nuclear weapon tests as well as in the implementation of the
protection and restoration measures, there are some reasons, on the one hand, to go
beyond the scope of the radiation protection and to consider the non-radiation causes of
risk as well. On the other hand, remaining in the framework of the radiation protection,
it is not enough to base oneself on the concept of the effective dose, even though only the
stochastic effects of exposure are under consideration. The value of risk determined by
D_E is

- integrated over the whole duration of the radiobiological stochastic effect after an
  exposure (tens of years for carcinogenesis and all generations to come for genetic
  consequences of the exposure);

- averaged over the human age at exposure and over the population of different
  countries.

[*]RRC "Kurchatov Institute", Moscow, Russia
[†]RISØ, Denmark
[‡]Emergency Management Ministry, Minsk, Belarus
[§]CEPN, France
[¶]USCRM, Kiev, Ukraine

As a result, the D_E-based assessment of the radiation consequences does not involve the time factor. No data on the radiological risk can be obtained for the different intervals of time after a nuclear accident or test. The value of D_E cannot make allowance for the local and age features of population cohorts (or personnel) for which the aftereffects are estimated.

The necessity of estimating non-radiation risks goes from the following:

- Some countermeasures being implemented can have negative side consequences of a non-radiological nature for a population; for example, the relocation, as follows from the experience available, may adversely affect the human health because of changing the social and other living conditions;
- Some possible trouble with the health of the population caused by local or national-wide social living conditions requires, in the context of the most efficient investitions in health protection, to assess by a unified way—through the risk analysis—the state of health as a whole and the background radiation and non-radiation risk causes;
- Taking into account the acute, at all times, need for the socio-psychological substantiation of the countermeasures (interaction with the local population, authorities and mass media), a substantiated scientific-methodical basis must be available to perform the comparative assessments and analysis of various risks;
- As it follows from the present-day methodology of estimating the radiological risk, the background values of carcinogenic risk must be known for the application of this methodology (models of relative risk).

All these points show that for assessment of consequences of nuclear accidents or tests and decision making on their mitigation health risk assessment from various radiation and non-radiation risk causes should be developed and used.

2 Methodology and data bank

In the frame of the CIS state research programmes (Chernobyl and Altai case studies) and the international (EU–CIS) project JSP2 the research subproject "developing the methodology (MAR) and data bank (BARD) on risk analysis" started in 1994. The first version of MAR is published in the international (EU–CIS) project JSP2 report for 1995 [1]. Main functions of BARD are:

- assessment of the radiological and non-radiological consequences of nuclear tests and accidents,
- assessment of the health of a population in terms of risk indices.
- analysis of effectiveness of radiation and social protection measures.

BARD includes
1. Service and calculation codes realizing the methodology mentioned;
2. Health-demographic data (HDD: the age-cause-specific death rates and the age distribution density) which are neccessary for radiological and non-radiation risks assessment.

HDD have been prepared for population of many regions of Russia for different years, for some regions of CIS and some countries around the world.

Input data for calculations with BARD are: 1) values of absorbed or equivalent doses (short-term exposure) and dose rates (chronic exposure) of different human body organs due to radiation exposure from a source considered; these doses and dose rates should be given in their dependence on age, time at exposure, countermeasures adopted etc.; 2) respective HDD from the internal BARD data base; 3) primary radiation or non-radiation risks models.

BARD to a certain extent analogous to the computer codes ASQRAD and SPIDER being developed by CEPN (France) and NRPB (UK). BARD differs from them by the large intrinsic HDD data base, the possibility of calculating non-radiological risks, areas of application etc.

BARD is constantly supported and developed in the local and distributed versions. The last one can be accessible through Internet (http://144.206.130.230/) at RRC "Kurchatov Institute" (Moscow, Russia).

A specific version of BARD is developed as a module of the decision support system for a post-accident activity (DSS JSP2).

3 Conclusion

Some results of health risk estimation made with BARD for the Bryansk region territory with relatively high radioactive contamination from the Chernobyl accident ($> 30Ci(Cs-137)/km^2$) are presented in Figure. They are illustrative to the ideas expressed above on the usefulness of detailed risk assessment for decision making.

Pecularities of excess health risk due to the radiogenic thyroid cancer, leukemia and other cancers should be taken into account in preparing and performing the health protection and rehabilitation programme for the current and following years on the territories suffered from the accident. Using risk assessment results can change notions about consequences of the nuclear accidents or tests and effectiveness of countermeasures. It is important in this assessment to take into consideration both radiation and non-radiation risks, radiation as well as social and medical health protection.

References

[1] Demin, V.F., Hedemann-Jensen, P., Schneider, T. Methodological recommendations on risk assessment in application to situations after nuclear accidents or weapon tests, the 1995 report of the international (EU - CIS) project JSP2.

[2] Demin, V.F., Hedemann-Jensen, P., Khramtsov, P.B., Okeanov, A.A., et al., Risk assessment data bank for JSP2 DSS, the 1995 report of the international (EU - CIS) project JSP2.

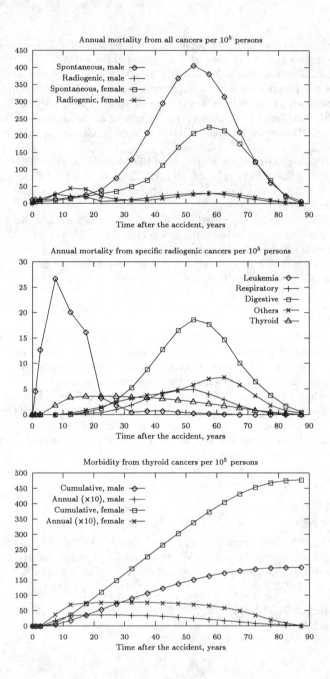

Figure 1: Mortality and morbidity (thyroid cancer) from spontaneous and radiogenic (due to the Chernobyl accident) cancers (per 100 000 persons, rural, age 0 - 18 years at the accident; male for specific cancers), as a function of time t after the accident (calculation with BARD; exposure doses without countermeasures).

Radionuclide Transfer To Meadow Plants

Natalya Sanzharova[1], Maria Belli[2], Andrei Arkhipov[3], Tamara Ivanova[4],
Serguei Fesenko[1], Georgei Perepelyatnikov[5], Olga Tsvetnova[6]

[1]RIARAE, Obninsk, Russia; [2]ANPA-DISP, Rome, Italy; [3]RIA "Pripyat", Chernobyl,
Ukraine; [4]BSU, Minsk, Belarus; [5]UIAR, Kiev, Ukraine; [6]MSU, Moscow, Russia

Abstract. Experimental data on ^{90}Sr and ^{137}Cs transfer to plants of natural and semi-natural meadows selected in the main CIS region contaminated due to the ChNPP accident are discussed. The highest TF's in grass stand are obtained for peatlands, and minimal ones - for dry meadows. ^{137}Cs content in plants decreased after the accident, on average, by a factor of 2-4. The dynamics of ^{137}Cs uptake by plants depends on meadow and soil properties. The first half life of ^{137}Cs transfer to plants change from 2,0 to 2,2 years and the second (slower) period half life change from 4,0 to 12 years for different meadow types. ^{90}Sr TF's are higher than those obtained for ^{137}Cs. The correlation between soil parameters and TF's are shown. ^{137}Cs TF's in grass stand depend on meadow type and decrease in the following order: peatlands> flood plain and wet (lowland) meadows> dry meadows.

1. Introduction

Meadow ecosystems are the main source of radionuclide transfer into animal products. After a nuclear accident one of the critical pathway of radionuclide transfer in animal products is consumption of grasses derived from contaminated pastures and natural meadows used for hay production. Natural and semi-natural ecosystems are characterised by a high variability of geobotanical composition and soil characteristics. As a result of the ChNPP accident meadows of different types - dry, flooded, lowland and peaty - were contaminated. The radionuclide uptake in meadow stand differs up to 2 order of magnitudes [1].

2. Dynamics of transfer factors of ^{137}Cs in plants

The estimation of the dynamics of internal radiation doses during the post-accidental period requires consideration on the decrease of radionuclide content in food products with time. This decrease is attributable to the sorption processes of radionuclide in soil components that decrease the biological availability for uptake by plant [2]. For comparative analysis of radionuclide bioavailability in food chains it is very reasonable to use the half-life periods of decrease of radionuclide levels in components of natural ecosystem, including meadows. For 9 regions of Russia, contaminated after Chernobyl accident, a data base of coefficients of ^{137}Cs accumulation in grass stand of meadows of different types was developed. Analysis of the data show that ^{137}Cs content decreased from 1987 till 1994 on average by a factor of 2-4. The character of this decrease differs considerably for various meadow types (Fig.1). Two periods characterised by different velocity of ^{137}Cs decreasing in grass stand can be recognised:

T_{ec}^1 represents ^{137}Cs half-life in the first period from 1987 to 1989 when the processes of radionuclide sorption in soil are more intensively;

T_{ec}^2 represents ^{137}Cs half-life in the second period since 1989 when the process of decreasing ^{137}Cs content in grass stand decelerated.

The parameters given above were estimated by trend values using the method of non-linear regression. The first half-life period for dry meadows on soils with light mechanical composition was 2,2 years and the second one - 4,0 years. For dry meadow on soils with heavy mechanical and for wet (lowland) meadows on peaty soils the first period is shorter and amounted 2,0 years. The second periods for these meadows are higher up 9 and 12 years, respectively. The differences in the half-lives are connected with influence of soil properties and various mechanisms of radionuclide fixation.

These data show that 5 years after the accident the amount of radiocaesium transfer annually from soil to plant reach a value that after this period show little variation with time.

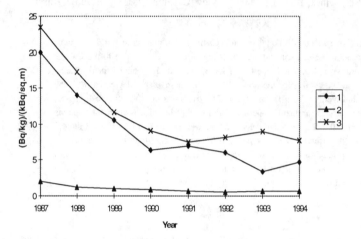

Fig. 2 Dynamic of ^{137}Cs TFs for different meadow types. 1 - dry meadows with sandy and loamy sand soils; 2 - dry meadows with heavy loam and clay soils; 3 - wet (lowland) meadows with peaty and peaty-gleyed soils

3. Accumulation ^{90}Sr and ^{137}Cs in grass stand of different type meadows

Experimental data on ^{90}Sr and ^{137}Cs transfer to plants of natural and semi-natural meadows selected in the main CIS regions, including the 30-km zone, contaminated due to the ChNPP accident are discussed. Accumulation of ^{90}Sr and ^{137}Cs in grass stand depends on meadow type and decrease in the following order: peatlands> flood plain meadows and lowland (wet) meadows>dry meadows (Table 1). ^{90}Sr transfer factors are higher than those obtained for ^{137}Cs from 2,1 to 5,2 times. ^{137}Cs TFs for meadows of different types vary in a wide range - from 0,2 to 58,2. ^{90}Sr TFs vary from 5,4 to 45,1. Maximum TFs are characteristic for grass stand of meadows on peats and minimum ones for dry meadows which cover is presented by mineral soils of heavy mechanical composition. The relation between soil parameters and transfer factors are shown. There is the interaction between mechanical composition of mineral soils and ^{137}Cs transfer factor (Fig. 2). ^{90}Sr and ^{137}Cs TFs decrease in the following order: sandy and sandy-loam soils > light loam and middle loam soils > heavy loam soils.

4. Correlation between soils characteristics and ^{137}Cs TF's in grass stand different type meadows

The effect of soil parameters on the soil-to-plant transfer factors (m^2 kg^{-1}) of ^{137}Cs was assessed using a non-parametric method (relative comparison method). The parameters considered are: soil pH; soil physical clay content (particle size<0.001mm); soil organic Carbon content; Calcium content; cationic exchange capacity; K content in soil. The main results are: pH, in the range considered (from 3.9 to 7.5), does not influence uptake from vegetation; soils with low content of clay (peaty and sandy soils) have a relative transfer factor 5 times higher than those with higher content of physical clay (more than 20%); the analysis discriminate mineral soils from organic soils giving a relative transfer factor two times higher for soil with high organic Carbon content than for mineral soils.

Table 1. Transfer factors of ^{90}Sr and ^{137}Cs in grass stand of different meadow types, 1993-1994

Soil type	Mechanical composition	TF, Bq/kg plant kBq/m² soil					
		^{137}Cs			^{90}Sr		
		average	min	max	average	min	max
DRY MEADOW							
Mineral	Sandy, loamy sand	4.9	0.4	10.1	25.6	5.4	45.1
	Light loam, middle loam	3.0	0.2	14.4			
	Heavy loam	0.5	0.3	0.7			
FLOOD PLAIN MEADOW							
Mineral	Sandy, loamy sand	5.1	0.9	10.1	11.7	5.8	22.9
	Light loam, middle loam	4.4	0.4	11.9			
Organic	Peaty	17.3	-	-			
LOWLAND (WET) MEADOW							
Mineral	Sandy, loamy sand	7.2	2.6	11.1			
	Light loam, middle loam	6.4	1.7	11.0			
Organic	Peaty, peaty gleyed	8.7	3.4	23.6	18.7	9.8	42.0
PEATLAND							
Organic	Peaty	32.7	14.3	58.2			

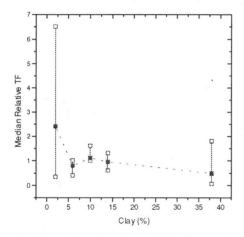

Fig. 2 Median relative TF's values (with 10-90% interval) as a function of soil clay content

5. Accumulation ^{137}Cs in different plant species

Content of ^{137}Cs in feed which was received from natural meadows in depending mainly on plant species composition of grass stand. Species composition of grass stand depends on meadow type. Radiocaesium concentrations, measured in different species, show high variability between individual plant species. It was shown that for different species (in the same meadows) the transfer factors range from 6,5 to 115. Differences in ^{137}Cs accumulation by the same species of plants on different meadow types are from 1,7 (Trifolium pratensis) to 57,8 (Poa pratensis).

Table 2. ^{137}Cs transfer factor for different plant species on different meadow type, $(Bq/kg)/(kBq/m^2)$, 1993-1994

PLANT SPECIES	MEADOW TYPE					
	Dry		Lowland (wet)		Flood plain	Peatland
	Soil type					
	Mineral		Mineral	Organic	Mineral	Organic
	Sandy, loamy sand, light loam, middle loam	Heavy loam				
Agropyron repens (L.) Beauv.	1.9		10.0	42.4		
Agrostis gigantea Roth.					3.0	6.6
Achillea millifolium L.	17.7	0.75			1.3	29.6
Carex acuta, caespitosa L.	6.8		34.4	22.0		46.9
Dactylis glomerata L.	1.1		2.8	34.8	8.4	
Deshampsia caespitosa (L.) Beauv.	3.3	2.6				
Erigeron danadensis Ster.	1.8		1.4			
Festuca pratensis Huds.	1.2	0.07	1.1	6.1		
Hypericum perforatum L.	1.7		0.3			
Poa pratensis L.	5.2	0.09	3.5	4.9		
Potentilla anserina, argentea, L.	14.9	0.49	6.1	8.3	4.9	
Phleum pratense L.	1.6			7.8		44.6
Stellaria graminea L.	8.5			7.1		
Taraxacum officinale Wigg.	6.4	0.42				
Trifolium pratensis L.	14.8		8.7			

6. Conclusion

For meadow ecosystems a wide variety of soil conditions, species composition of stand is characteristic what determines differences in contamination of fodder for animal breeding on meadows of various types. The results obtained allow the conclusion that behaviour ^{90}Sr and ^{137}Cs depend on meadow type and soil properties. The analysis of received data shows, that the type of meadow influence to speed of migration. The maximum significance TFs of ^{137}Cs are determined for grass stand of meadow on transient peat, in 5-8 times higher accumulation on flood plain and wet meadows. The minimum accumulation ^{137}Cs is marked on dry meadow where soil cover is submitted heavy soils. ^{137}Cs content decreased from 1987 till 1994 years on average by a factor of 2-4 as a result of sorption its by soil. The first half life period was 2-2,2 years and the second one 4-12 years for the different type meadows.

REFERENCES

[1]. Alexakhin, R.M. and Korneyev, N.A. (Eds) Agricultural Radiology (Moscow: Ecologia Publisher) (in Russian) (1992)

[2]. Sanzharova, N.I., Fesenko S.V., Alexakhin, R.M., Anisimov, V.S., Kuznetsov V.K., Chernyayeva L.G. Changes in the form of ^{137}Cs and its availability for plants as dependent on properties of fallout after the Chernobyl nuclear power plant accident. The Science of Total Environment 154 (1994) 9-22

Study of Decontamination and Waste Management Technologies for Contaminated Rural and Forest Environment

Alexandre GREBENKOV[1], Vassily DAVYDCHUK[2], Slava FIRSAKOVA[3],
André JOUVE[4], Yuri KUTLAKHMEDOV[5], Kelvin ROSE[6], Tatiana ZHOUCHENKO[7],
Gennadz ANTZYPAW[8]

[1]-*Institute of Power Engineering Problems, Minsk, Belarus*
[2]-*Institute of Geography, Kiev, Ukraine*
[3]-*Belarus Institute of Agricultural Radiology, Gomel, Belarus*
[4]-*Institut de Protection et de Sûreté Nucléaire, Cadarache, France*
[5]-*Institute of Cell Biology & Genetic Engineering, Kiev, Ukraine*
[6]-*AEA Technology, Abingdon, England*
[7]-*Institute of Forestry, Gomel, Belarus*
[8]-*Chernobyl State Committee, Minsk, Belarus*

Abstract. Pilot and demonstrative scale *in situ* trials of several decontamination technologies proposed in the framework of ECP-4 project were carried out in real conditions of Chernobyl Zone. Their results proved that industrial scale decontamination of various types of land is feasible. The management of radioactive waste arising from decontamination techniques can be provided by ecologically sound and efficient technologies.

1. Introduction

Except for some countermeasures such as the evacuation of populations, a universal method to solve a great deal of Chernobyl related problems does not exist to our knowledge. Only a set of various methods adapted to the different areas of the environment can be proposed. The application of these technologies will not only be profitable for restoring the Chernobyl Zone, but also for the preparation to possible accident situations which probability must not be excluded.

The main research activity in the framework of the Experimental Collaboration Project N°4 entitled "Decontamination Strategies" was focused on pilot and demonstrative scale *in situ* trials of several soil decontamination technologies in order to obtain real characteristics of their efficiency and cost in terms of economy, waste management and occupational exposure. Results of this study are presented in the paper

2. Target Surfaces and Methods of Decontamination

For decontamination of pastures and meadows, as well as of forest, whose lands have not been ploughed since an accident, a simple method of removal of the top layers of soil can be applied with sufficient effectiveness, since the major part of deposited radionuclides is still localised in the 4-7 cm of surface horizons [1, 2, 3]. This sort of decontamination is supposed to provide sufficient dose reduction. In order to prove that, the series of experiments have been carried out at real conditions including several test plots of the Chernobyl zone in the Ukraine and Belarus (Fig. 1), involving soil of different type and deposited radioactivity of different level.

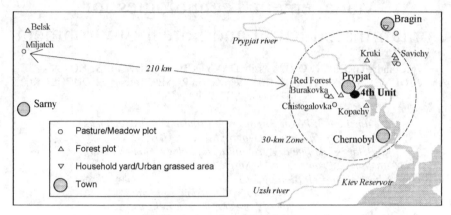

Fig. 1. Location of the plots

Soil type and other characteristics of the experimental plots, where pilot and demonstrative scale trials were performed, are given in Table 1. Typical distribution of radionuclides along the soil profiles in these plots is shown in Fig. 2 and Fig. 3.

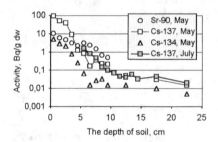

Fig. 2. Radio-nuclides distribution in soil profile. Savichy. Point N° 2

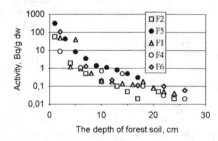

Fig. 3. Distribution of Cs-137 in the forest soil of different sampling plots

Among other essential countermeasures the surface soil layer removal remains as the only method that can provide a high effectiveness for a long term. For meadows which have not been ploughed since accident this method can be applied using turf harvester and attached technologies that was initially developed within the European RESSAC Programme and showed a good result with acceptable amount of waste [4]. In case of contaminated forest a similar method [3, 5] of removal of decomposed litter and top layer of humus accumulating the significant portion of activity can be applied. Shrubs and underwood have to be ground before.

The principle of the turf harvester is to cut out a mat layer of 1 to 6.5 cm thick and remove it using the root network of plants. The machine is able to follow the soil irregularities, therefore the quantity of waste produced is minimised. Turf-cutter is equipped with a blade moving back and fro between two depth settings and behind driving wheels, that results in cutting the strips of turf mat of about 40 cm wide.

Table 1. Description of plots

Location	Soil type	Dominant vegetation	Activity of soil, kBq/m²	Landscape
1992 Burakovka (Meadow) Chystogalovka (Former arable)	Sod-podsol gleic sandy-loamy	*Setarina viridis* *Agropyrum repens*	3700(^{137}Cs)	
1993 Miljatch (Meadow)	Drained peat-bog	*Betula pubescens Eheh. et*	174 (^{137}Cs)	Drained flat terrace
1993 Savichy (Meadow)	Sod-podsol gleic sandy-loamy	*Nardus stricta L., Lusula multiflora (Retz) Lej.,*	1036(^{137}Cs) 355 (^{90}Sr) 1.3 (^{239}Pu)	Poorly drained flat terrace
1994 Savichy (Haushold yard)	Sod-podsol gleic sandy-loamy		925 (^{137}Cs)	
1994 Bragin (Wet meadow)	Peat bog		225 (^{137}Cs)	Non drained terrace
1994 Savichy (*Pinetum cladinosum*)	Sod-podsol gleic sandy-loamy	*Pinus silvestris L.* 40% *Pleurozium schreberi Mitt.*	1370 (^{137}Cs)	Automorfic flat and hillock terrace
1994 Savichy (*Pinetum - dicranosum*)	Sod-podsol sandy-	*Pinus silvestris L.* 90% *Dicranum scoparium Hedw.*	2440 (^{137}Cs)	Hydromorphic flat and hillock terrace

The mechanical brush collecting litter consists of the rotor with frequent firm elastic cores located on its cylindrical surface. The rotor is driven by hydraulic engine with a reductor placed inside the rotor cavity. This mechanical brush is assembled on the frame together with a storage bin having a volume of about 0.4 m^3 where the litter is collected and then can be easily unloaded into a trailer. The soil depth is controlled by means of a couple of wheels. The grinding mover to shred underwood consists of head equipped by rotor with free hanging incisors and cutting blades. It rotates by means of hydro-mover connected to tractor's hydro-driving system. Width of the head is about 1.1 m.

3. Results of Experiments *in situ* and Discussion

Two first small-scale tests carried out in summer 1992 (Table 1) on abandoned lands showed rather promising results regarding the efficacy of turf harvester with decontamination factor ($D_F = A/A_0$) ranging from 13.7 to 27.5 in terms of ^{137}Cs specific activity before (A) and after (A_0) decontamination. Our next attempts, each at the scale of 2500 m² were made in 1993 on two different soil types (Table 1). The peat bog soil was a fertile soil with a peat layer of more than 50 cm. The soddy podzolic soil was a poor soil with an organic layer of only 12 cm. After applying the turf harvester technology amount of ^{137}Cs and ^{90}Sr in the soil of Savichy reduced with an average D_F of 38.4 $\pm$ 17.3 and 6.33 $\pm$ 2.04 respectively. The exposure rate in the middle of the test plot after decontamination reduced by a factor of 3.4. In case of peat bog of Miljatch the efficacy is considerably less and D_F does not exceed 14. The next trials of turf harvester technology in 1994 (Table 1) confirmed again that D_F looks to reach higher values for the soddy podzolic soil than for the peat bog soil. This observation is in agreement with the higher migration rates in organic soils which are due to the presence of fulvic acids in peat bog soils. In is shown [1, 2] that removed top soil is a little proportion of the arable layer in both cases and does not compromise the re-growth of natural vegetation and agricultural crops.

In summer 1995 the demonstrative trials of grinding mover and then litter removal machine were carried out in the real conditions of forest plots chosen (Table 1). The litter layer of about 5-6 cm thick, underwood and shrubs of up to 8 cm diameter were removed from the area of about 0.1 hectare by means of brushing machine and grinding mover. The D_F of 5.0-26.6 was achieved in ideal conditions while removing the whole layer of 5-6 cm thick. In real conditions an effect of separation of litter matter into the fractions by brushing machine leads to decreasing the effectiveness of this technology up to D_F of about 3.5-4.0. It is proved that more than 45% of mass containing fine organic fractions with only 15% of radioactivity could be returned back to the soil owing to design of brush.

4. Management of Resulting Waste

The distinctive features of the above wastes are connected with (i) variation of specific activity of the wastes (mainly from 0.4 to 400 kBq/kg), and (ii) large amounts of very low level radioactive waste of different compositions that are difficult to segregate. Most of related waste can be disposed nearby the place of decontamination. In the case described in [2] a simple on-surface disposal site was arranged that occupies about 5.1% of field area decontaminated. During the very wet 1993 summer season, leaching of 0.05-0.1% of the radio-caesium from the pile to the collectors was observed. A natural grass cover will encapsulate the turf mat with radionuclides. The overall cost of management and disposal of non-compacted waste (e.g. soil) and woody waste was evaluated as 160 ECU/m^3 and 70 ECU/m^3 respectively.

5. Conclusion

The results of pilot and demonstrative scale trials of the proposed decontamination technologies performed during three years in the ECP-4 framework proved the real possibility of design and efficient application of industrial scale technologies for remediation of contaminated site.

References

[1] Jouve, H. Maubert, R. Millan-Gomez, Y. Kutlakhmedov, Rehabilitation of Soils and Surfaces After a Nuclear Accident: Some Techniques Tried in the Chernobyl zone, Proc. of the 1993 International Conference on Nuclear Waste Management and Environmental Remediation, Prague, P.-E. Ahlstroem, C.C Chapman, R. Kohout, J. Marek editors, 2 (1993), pp. 391-393.
[2] Grebenkov, A. Jouve, S. Firsakova et al, Turf Harvester Technologies for Soil Remediation After Chernobyl Accident: Field Trial in Belarus. Proc. of Nuclear and Hazardous Waste Management International Topical Meeting "SPECTRUM '94", Atlanta, GA, USA, Aug. 14-18, 3 (1994), pp. 2102-2106.
[3] Grebenkov, A. Jouve, I. Rolevich et al, Possible Technologies for Belarus Forest Site Remediation After Chernobyl Accident, ibid, pp. 1640-1644.
[4] A.Jouve, E.Schulte, P.Bon, A.L. Cardot, Mechanical and Physical Removing of Soil and Plants as Agricultural Mitigation Techniques, The Science of the Total Environment, 137 (1993), pp 65-79
[5] Jouve, A. Greben'kov, J. Tormos, R. Zanon, Forest Decontamination and Generating Economic Value from Wood: a Case Study on the Belarus Forest, Symposium on the Remediation of Radioactive Contaminated Sites in Europe, 11-15 October, Antverpen, Belgium (1993), p. 67.

Natural Sorbents for Decontamination of Objects of Urban Territories

N. MOVCHAN, YU. FEDORENKO, B. ZLOBENKO, A. SPIGOUN
Department of Environmental of Radiogeochemistry of IGMOF,
National Academy of Sciences of Ukraine

Abstract. This paper gives an information about the use of film coverings, based on natural sorbents, in decontamination of buildings, contaminated after the Chernobyl accident. This method has incontrovertible advantages in the beginning period after the accident and can be used for cleaning considerable areas of urban territories.

The works in decontamination of large territories, that were polluted by the Chernobyl accident, show the complexity of the problem, the relationship between the methods of decontamination and the character of contamination, the type of building material and others.

After the Chernobyl accident a considerable quantity of dispersed nuclear fuel and products of its decay was gushed out in the environment. All that leads to radioactive contamination of neighboring territories, including buildings and structures, and areas, that are far from the emergency block (hundreds and thousands kilometers). Two types of radioactive contamination predominate - wet (rain) contamination from a condensation phase and contamination by the dispersed nuclear fuel. The role of these contamination types is measured depending on a distance from the emergency block, a direction of wind, a wind power and others. The deposition of fuel particles prevails near the block; at great distances a condensation phase predominates. Combined contamination is observed also.

From the beginning of the accident the measures conducted in the clearing of buildings showed an insufficient development of methods of decontamination of construction material of surfaces. During the work within the ECP4 project three groups of methods of decontamination of buildings were approved: 1) a mechanical removal, using vacuum, sandblasting, and the removal of layers of the contaminated material [1], 2) physical-chemical methods of decontamination, using partial dissolving and washing away the radioactive contamination by different water and organic solutions [2]. The third group includes the combination of physical-chemical and mechanical methods [3].

The researches show that in the first period after the accident (the first months and during the first 2-3 years) it is more effective to use a method, employing natural sorbents for creating a film covering on surfaces and removing the adhered and partially adsorped radioactive contamination [4, 5]. In Table I there is formation about the decontamination of surfaces of rural house in Vladimirovka, Polesskoe district, Kiev region, three years after the accident. The village was seriously contaminated by Cs-137 in 1986 (now it is more than 40 Ki/sq.m) and the population of the village was evacuated.

Putting the coverings was performed with the aid of the autopouring station ARS-14, which is installed on the lorry ZIL-152 with a trailer.

A complex like this allows to process on average 90 m²/h of decontaminated roof and 70 m²/h of walls. The labor costs is 0,006-0,007 people-days/m² of decontaminated surface; the tentative value of works is 0,7 ECU/m²

Table 1.

The decontaminated object	Contamination β-particles/sm²min before treatment	Contamination β-particles/sm²min after treatment	Coefficient of decontamination (Fd)
The asbocement roof			
eastern slope	390 ± 70	83 ± 12	4.7 ± 1.5
western slope	495 ± 78	92 ± 15	5.4 ± 1.7
The walls from silicate bricks			
eastern side	353 ± 25	98 ± 8	3.67 ± 0.5
south side	108 ± 6	52 ± 4	2.1 ± 0.3
western side	85 ± 5	48 ± 3	1.8 ± 0.25
northern side	148 ± 10	55 ± 4	2.7 ± 0.15

The collection of waste of decontaminating clay covering is performed by the individual vacuum cleaner (type KU5S) with accumulating capacity.

The decontamination wastes are the removed decontaminating coverings. An average amount of waste represents nearly 0.25 kg/m² of decontaminating surface with an activity from $1*10^3$ to $2.5*10^5$ Bq/m³. A dose reduction of about two near the object from 0,01 mSv/h to 0,0045 mSv/his observed. In Chernobyl similar works in the cleaning a rural house were performed. The house had a combined type of contamination. The information about the decontamination is shown in Table 2.

Table 2.

The decontaminated object	Contamination β-particles/sm²min before treatment	Contamination β-particles/sm²min after treatment	Coefficient of decontamination (Fd)
The asbocement roof			
eastern slope	417 ± 87	214 ± 43	1.95 ± 0.3
south slope	395 ± 62	193 ± 40	2.0 ± 0.3
western slope	380 ± 78	182 ± 36	2.1 ± 0.5
northern slope	814 ± 78	271 ± 42	3.0 ± 0.5
The walls from silicate bricks			
eastern side	192 ± 8	48 ± 4	4.0 ± 0.45
south side	116 ± 6	42 ± 3	2.8 ± 0.26
western side	206 ± 9	76 ± 9	2.7 ± 0.3
northern side	121 ± 6	42 ± 3	2.9 ± 0.36

The collection and the nature of wastes are the same as above. After the decontamination it was observed a dose redution of 1,7 was observed, from 0,0044 mSv/h to 0,0026 mSv/h. In Table 3 the average coefficients of decontamination (Fd) are given .

Table 3.

	1987	1988	1989	1990	1991	1992	1993	1994
Average value of FD	15	4.5	2.9	2.5	2.0	1.7	1.5	1.4

The depth of penetration of contamination by Cs-137 was different for building materials. Ninety five % of contamination concentrated in the layers:
- for the absocement slate   0.1 - 0.3 mm
- for the tiles   0.2 - 0.3 mm
- for the red brick   0.3 - 0.5 mm
- for the white silicate brick   0.5 - 0.8 mm
- for the rusty iron   0.1 - 0.15 mm

Therefore we propose to clean surfaces by the cleaning-decontaminating paste "Cleadecon" with abrasive, which was tested and gave the results presented in Table 5.

Table 5.

The material type	Contamination before treatment	Contamination after treatment	Coefficient
The tile	240 ± 18	89 ± 6	2.7 ± 2
The slate	280 ± 20	85 ± 4	3.3 ± 0.25
The iron	185 ± 12	67 ± 3	2.7 ± 0.16
The red clay brick	135 ± 7	58 ± 4	2.32 ± 0.12
The white silicate brick	148 ± 10	55 ± 4	2.7 ± 0.15

Conclusion

1. A relationship between the character of radioactive contamination and f methods of decontamination is observed.

2. In the first period after the accident a method based on the use of the film-covering of natural sorbents is very effective and worth-while.

3. If the decontamination penetrated into the material, it is necessary to use methods removing the contaminated layers of building materials from surfaces.

4. Using the decontaminating compositions based on natural sorbents is very worth-while, because of:
- the low costs of ingredients ;
- high effectivity;
- low labor costs (0,096-0,007 people-days/sq.m);
- rather high productivity (70-90 sq.m/h);
- costs for decontamination 1 sq.m are near 0,07 ECU.

5. If the radioactive contamination penetrated into surface layer of building material, it's effective to use the decontaminating pastes like the "Cleadecon" with abrasive.

References

[1]   A.D. Simon. Decontamination (in Russian). M., Atomizdat, 1975.
[2]   C.Z. Wood. A review of the application of chemical decontamination technology in the United States. Progress in Nuclear Energy, 1990, v.23 N1 pp. 35-80.
[3]   A.D. Simon. Picalov V.K. Decontamination (in Russian), M, Izdat, 1994.
[4]   I.F. Vovk et al. "clay Radiological Barrier and Countermeasure Research Related to the Accident at the Chernobyl NPP", Proceeding of U.I.R. Soviet Branch Seminar on the Radioecology and Countermeasures, Kiev, pp. 67-82, 1991.
[5]   Using clay to clean up in the Ukraine Nuclear engineering international. London, Heywood-Temple Industrial Publication Ltd. v.39 N481, aug. 1994, p. 36.

The Removal of Radionuclides from Contaminated Milk and Meat

Long, S.C., Sequeira, S. and Pollard, D.
Radiological Protection Institute of Ireland (RPII),
3 Clonskeagh Square, Clonskeagh Road, Dublin 14, Ireland

Fedin, F.A. and Krylova, N.V.
Ukrainian Institute of Agricultural Radiology (UIAR),
7 Machinostroiteley Street, Chabany, 255205, Kiev, Ukraine

Abstract

The transfer of radiocaesium and radiostrontium from milk to a number of commonly consumed European and Ukrainian cheeses was quantified. The F_f[1] for radiocaesium in these cheeses was typically between 0.5 and 0.7, while that for radiostrontium ranged from 0.7 to 8.3. Those steps during the cheese making process that resulted in removal of radiocaesium were then identified. The production of cheeses by modified technologies succeeded in reducing the radiostrontium transfer by a small amount. Acidification of the milk (to a pH < 5.1) prior to coagulation resulted in a significantly smaller transfer of radiostrontium to the cheese. The coagulation process varied by up to 20% for duplicate experiments.

At least 50% of radiocaesium may be removed from contaminated meat by soaking in NaCl solution (brine). Greater removal (up to 90%) may be achieved by varying treatment parameters. The losses of vitamins and minerals associated with this treatment were quantified.

1. Introduction and background

The Chernobyl accident has shown that the ingestion of contaminated food and drink is a significant exposure pathway, both in the immediate aftermath of a nuclear accident and in the long term. Even today, in some of the regions most contaminated by the Chernobyl accident, a large proportion of the dose received is due to the ingestion of contaminated food and drink [1]. The consumption of milk, milk products, meat and meat products accounts for a large proportion of this ingestion dose. However, considerable decontamination of these and other food types may be achieved through the use of specific food decontamination techniques or through the adoption of particular food processing strategies.

In terms of ingestion dose, radiocaesium and, to a lesser extent, radiostrontium are the nuclides of greatest significance in the long term. The behaviour of these nuclides during culinary preparation and food processing is dictated by their chemical properties. Caesium is a highly soluble ion and thus tends to concentrate in the aqueous fractions during food processing. Strontium is less soluble and tends to follow the behaviour of calcium. Thus, during milk processing radiocaesium tends to be transferred to those products with a high water content, while radiostrontium is transferred to high calcium products. The soluble nature of radiocaesium means that it can be removed from contaminated meat with liquids used during culinary preparation or processing. Radiostrontium contaminated meat is rarely a problem as this nuclide tends to accumulate in the bones rather than in the muscle.

Although the chemical properties of caesium and strontium are quite different, the distribution of these nuclides between milk products is similar. An exception to this is their proportionate transfer from milk to cheese. In general, approximately 50% of the raw milk content of radiostrontium is transferred to the cheese, while about 10% of radiocaesium is transferred. This, combined with the low processing efficiency (P_e)[2] for cheese (approximately 0.1), can result in a high F_f value for radiostrontium in cheese. It is, however, possible to reduce the transfer of both radiocaesium and radiostrontium to cheese through the application of modified cheese making procedures. A number of experiments were carried out to determine:

1. The transfer of radiostrontium (*in vitro* contamination) and radiocaesium (*in vivo* contamination) from milk to a range of Ukrainian cheeses. Those steps during the cheese making process that resulted in removal of radiocaesium were also identified.

2. The transfer of radiostrontium and radiocaesium from milk (*in vivo* contamination) to a range of Ukrainian and European cheeses. The effects of modifications designed to reduce the radiostrontium transfer were examined.

3. The variability of the coagulation process and the effects of coagulation pH on the transfer of radiostrontium from milk to cheese.

One of the most effective ways of removing radiocaesium from contaminated meat is by soaking the meat in brine. More than 90% decontamination can be achieved in this way, depending on the applied treatment. A series of experiments were carried out in order to assess the effects of varying a number of treatment parameters. Losses of soluble vitamins and minerals during treatment were also determined.

2. The transfer of radionuclides from milk to cheese

1. Materials and methods

Experiment #1: Milk was contaminated *in vivo* with Cs-137 and *in vitro* with Sr-90. The milk was then kept at 4° C for 48 hours to allow equilibration of radiostrontium with the binding proteins [2]. Standard techniques were used to produce a number of commonly consumed Ukrainian cheeses: dutch, susaninsky, brynza, suluguni, dneprovsky and adygeisky.

Experiment #2: Milk contaminated *in vivo* with Cs-137 and Sr-90 was obtained from cattle grazing some of the highly contaminated pastures close to the Chernobyl reactor. Four European cheeses: edam, cheddar, camembert and quarg, and three Ukrainian cheeses: adygeisky, suluguni and brynza, were produced using standard and modified procedures. Modifications were designed to reduce the radiostrontium transfer by producing a more acidic cheese. This was achieved by adding more starter culture and less rennet during production.

Experiment #3: Milk was contaminated *in vitro* with Cs-137 and Sr-85 and then stored at 4° C for 3 days. The pH of coagulation was varied from 0.1 to 6.7 by using different amounts of starter culture. The variability of the coagulation process was assessed by setting up duplicate samples for each set of conditions and by carrying out each experiment twice.

In each experiment the Cs-137 and Sr-85 activities in the curd, whey and final cheese was measured by high resolution gamma spectrometry, while the Sr-90 content was determined with standard radiochemical techniques.

2. Results and discussion

The F_f values for radiocaesium and radiostrontium for the European and Ukrainian cheeses produced during experiments #1 and #2 are summarised in Table 1. The specific activity of cheeses was found to be closely related to their moisture content. The low F_f for radiostrontium in quarg can be attributed to its acidic nature. During standard cheese making, those steps found to result in most removal of radiocaesium were; soaking the cheese in saline solution, the complete replacement of whey with water and the removal of whey at each stage of the process. The production of cheeses incorporating these steps, or the addition of these steps to other cheese making procedures will significantly reduce the final radiocaesium content of the cheese. More details of the results of these experiments are reported in Fedin *et al.* [3].

Small reductions in the transfer of radiostrontium to cheese were achieved through the application of modified procedures, for example, the F_f value for standard camembert was 4.2, while that for the modified cheese was 3.7. Nutritional analysis of the cheeses produced during experiment #2 has shown that the composition of cheeses (i.e. the protein, fat, moisture, calcium and potassium contents) produced by modified technologies did not differ significantly from those produced by the standard methods.

Table 1. The transfer of Cs-137 and Sr-90 from milk to European and Ukrainian cheeses

Cheese type	F_f - Cs-137	F_f - Sr-90
Edam - fresh	0.49	6.91
Cheddar - fresh	0.52	5.82
Camembert - fresh	0.48	4.15
Quarg - fresh	0.59	0.68
Dutch - fresh	0.57	8.30
Dutch - salted in brine and matured for 60 days	0.49	8.30
Susaninsky - fresh	0.16	8.00
Susaninsky - salted in brine and matured for 15 days	0.11	8.00
Brynza - fresh	0.76	4.70
Brynza - salted in brine and matured in brine for 20 days	0.11	4.70
Suluguni - fresh	0.70	6.80
Suluguni - after cheddaring, salting in solution and 2 days maturing	0.19	4.00
Dneprovsky	0.74	3.60
Adygeisky	0.66	4.50

The sensitivity of the cheese making process to a number of variables means that the F_f values may vary considerably between duplicate experiments. Due to time limitations, it was not possible to duplicate the experiments carried out with *in vivo* contaminated milk. Thus it is difficult to draw any firm conclusions regarding the reduction in radiostrontium transfer from the results of these experiments. The experiments did however illustrate the criticality of the coagulation pH on the transfer of radiostrontium to cheese and the variability of radionuclide transfer during cheese making. This dependence on coagulation pH and variability in coagulation was further investigated in the third set of experiments.

It was found that approximately 75% of the radiostrontium remained in the curd at a pH of 6.5, compared to 20% at a pH of 4.5. Thus transfer of radiostrontium is strongly dependant on the pH of the milk at coagulation. This is due to the fact that acidification of milk to a pH of 5.1 (the isoelectric point of the milk protein) results in the liberation of the strontium from the protein into the aqueous phase. The proportion of radiocaesium present in the curd remained between 10% and 17% for the pH range studied. Thus the behaviour of radiocaesium during coagulation is independent of pH.

For radiocaesium the variability both within-batch and between-batch ranged from 1% to 20%. For radiostrontium the within-batch variation ranged from 0% to 20% and the between-batch variation from 0% to 10%.

3. Conclusions

Under standard production conditions F_f values for the transfer of radiocaesium from milk to cheese are approximately 0.5 to 0.7, while those for radiostrontium are approximately 5.0 to 8.0. A considerable reduction in the radiocaesium transfer is possible through the introduction of specific steps during production. Reduction of the radiostrontium transfer is more difficult, as it is necessary to acidify the milk to a pH of less than 5.1 prior to coagulation. The pH of the milk at coagulation affects the radiostrontium transfer critically, its effect on radiocaesium transfer is negligible. The variability of the transfer of radionuclides from milk to cheese is considerable, thus replicate experiments are necessary for conclusive results. The nutritional composition of modified cheeses do not differ significantly from that of standard cheeses. It should be noted that the production of cheese by modified technology results in a cheese with different characteristics, i.e. a new type of cheese is produced.

3. The removal of radiocaesium from meat

1. Materials and methods

The effects of varying treatment parameters on the removal of radiocaesium from contaminated meat were studied. The parameters investigated were:
- meat size (minced meat, 1.5-2 cm cubes and 250 g pieces).
- the duration of treatment (1 hour, 2 hours, 1 day, 2 days, and 3 x 2 days).
- concentration of brine solution (2.5%, 5%, 10%).
- brine-to-meat ratio (1:1, 3:1 and 5:1).

In addition, losses of water soluble nutrients (B_{12}, niacin, iron and zinc) during treatment were investigated. The radiocaesium content of meat, before and after treatment, and of the brine solution was determined by high resolution gamma spectrometry. All experiments were duplicated.

2. Results and discussion

The smaller the meat pieces, the greater the surface area exposed to the brine solution and consequently, the greater the decontamination achieved. However, the usefulness of minced and cubed meat following treatment is less than that of 250g pieces. In general, the longer the treatment time, the greater the decontamination achieved. For example, a 1 hour treatment [3] resulted in 47% decontamination, a 2 hour treatment, with replacement of brine after 1 hour, resulted in 61% decontamination, while a 1 day treatment resulted in 82% decontamination. The NaCl concentration was not found to have a significant effect on the removal of radiocaesium. For minced meat, the brine-to-meat ratio had a significant effect on radiocaesium removal; a 5:1 ratio [4] resulted in 72% decontamination, while a 1:1 ratio resulted in 33% decontamination. In the case of cubed meat the brine-to-meat ratio did not affect decontamination.

In summary, at least 50% decontamination of radiocaesium from meat can be easily achieved, using the treatment conditions detailed in footnote 1. Greater decontamination may be achieved by changing various parameters. For example, increasing treatment time to 24 hours will increase removal to 80%, treatment of minced meat will increase decontamination to 70% and a 2 hour treatment with replacement of solution will increase decontamination to 60%. Some of the longer treatment times recommended by other studies (up to 6 days) [4] did not achieve significantly greater decontamination.

Meat is a major dietary source of the water soluble vitamins B_{12} and niacin and of the trace metals iron and zinc. These treatments resulted in losses of these nutrients to the salt solution. The longer the treatment times, the greater these losses. In addition, longer treatments resulted in higher Na+ and lower K+ concentrations in the treated meat, which is undesirable from a dietary point of view. The results of nutritional measurements are summarised in Table 2.

3. Conclusions

At least 50% of radiocaesium may be easily removed from meat by soaking in brine. Losses of vitamins and minerals may be minimised through careful choice of treatment conditions. If necessary, greater than 90% decontamination may be achieved by altering appropriate treatment parameters. Losses of soluble vitamins and minerals are considerable, however, these losses are relatively inexpensive to replace.

Table 2. The effects of salting on the Na, K, Fe, Zn, B_{12} and niacin content of meat

Treatment	Na	K	Fe	Zn	B_{12}	Niacin
As in footnote 3	+ 148%	- 62%	-34%	-11%	-26%	-39%
As in footnote 5	ɪ 408%	-98%	-60%	-74%	-66%	-45%

4. Acknowledgements

The authors gratefully acknowledge the assistance of the State Laboratory, Dublin, where the nutritional analysis of meat was carried out.

5. Footnotes

1. F_f = Food processing retention factor $= \dfrac{\text{Activity (fresh wt.) in processed food (Bq/kg)}}{\text{Activity (fresh wt.) in raw materials (Bq/kg)}}$

2. P_e = Processing efficiency $= \dfrac{\text{Weight of processed food (kg)}}{\text{Weight of raw material (kg)}}$

3. 1.5-2 cm cubes soaked in a 2% NaCl solution, with a brine-to-meat ratio of 3:1, for 1 hour at room temperature, followed by washing in tap water for 10-15 minutes

4. Minced meat soaked in a 5% NaCl solution, for 2 days at 2 ˚C, followed by washing in tap water for 10-15 minutes

5. 250 g piece of meat soaked in a 5% NaCl solution, with a brine-to-meat ratio of 4:1, for 3 x 2 days, with replacement of solution after 2 days and 4 days, at 6˚C, followed by washing in tap water for 10-15 minutes

6. References

[1] Lazarev, N. Personal communication.

[2] Edmonson, L.F., Keefer, D.H., Douglas, F.W., Harris, J.Y. and Dodson, E. Comparison of the removal of radiostrontium from *in vivo* and *in vitro* labelled milk by ion exchange resins. *J. Dairy Sc.* 46:1362-1366, 1963.

[3] Fedin, F.A., Dolgin, N.L., Krivtsov, I.L. and Krylova N.V. Effect of technological factors on radionuclide migration from milk into natural cheese. In: Proc. All-Union Conference on Agricultural Radiology, Moscow, Russia, 1199-1200 (in Russian), 1989.

[4] Petäjä, E., Rantavaara, A., Paakkola, O. and Puolanne, E. Reduction of radioactive caesium in meat and fish by soaking. *J. Environ. Radioactivity* 16, 273-285, 1992.

The Ways of Reduction of People's Psychological Distress During the Post-Accidents on the Atomic Plants

Ludmila AGEEVA

Institute of Sociology (IS), Surganova str. 1-2, 220072 Minsk, Belarus

Abstract. The study of people's status in Belarus after the Chernobyl accident shows that independently from the place of residence they experienced a strong psychological distress which was caused by fears before the accident consequences. This paper gives the characteristics of fears influencing on the growth of the given phenomenon and proposals how to eliminate the same situations in the future.

1. Method.

The data for this paper were taken from the sociological surveys. They included the questionnaire study of people, the expert's evaluations, statistical data collection and etc. The questionnaire study concerned 218 persons from the restricted area with the radioactive contamination 15-40 Cu/km², 250 persons from the non restricted area, 255 resettled persons from the 30 km area and 262 residents from the clean (controlled) area. Respectively there were 42% of men and 58% of women. The subjects of study were representatives of all main occupational groups: pensioners, housewives and pupils in the age from 16 to 75 years.

2. Data.

The sociological study showed that 55.2% of respondents considered the Chernobyl accident the reason of the rather much and very much psychological distress, 33.3% indicated rather little and very little psychological distress and only 11% gave a negative answer (0.8% did not answer). 48.4% of the respondents considered that they and their families experienced the real radiation risk due to accident in rather high and very high extent, 45.0% reported very little and rather little extent of the risk, 4.9% indicated the absence of any kind of risk (1.8% did not answered). The distress was connected first of all with people's worry about personal and children health. For example, 64.5% of the respondents worried rather and very much about personal and their children health, 64.5% were rather little and very little worried, only 3.7% were not at all worried (0.7% did not answer). People's fear for personal and their children health was caused at that time and later by their ignorance of the radiation affects on human body and environment and by the firm belief in its harmful consequences for all living creatures. The long term concealment of the trustworthy information about the real scale of the accident, its hushing up by the mass media had led to the belief that "there are something to be withheld". For example, 53.6% of respondents in September 1993 supposed that the experts and medical doctors knew the information which had been withheld from the population, 27.9% of respondents were not quite sure, and only 17.6% supposed that such an information did not exist (0.9% did not answer). Taking into account such a situation people acquire the attitude towards the perception of any information from the non official sources as trustworthy, because they need the information very much. Thus 31.4% of respondents reported that they talk about the Chernobyl accident almost every day, 13.4% at least once a week, 9.4% - at least once a month, 42.9% - seldom. Only 2.3% did not talk at all (0.4% did not answer).

Being sure in the destructive radiation affect on person people try to explain all their diseases recently appeared and health disorders to the Chernobyl accident consequences and demand with insistence from the authorities to recognize them the victims and to give them a privilege established for this group of citizens. Since the connection of the disease with the radiation affect is not proved, they are inclined to distrust the competence of people, who investigate the radiation affects on the health. For example, while the questionnaire study 15.1% did not trust at all the competence of specialists, 62.9% - had very little and neither much nor little trust, and 20.1% trusted almost completely and completely. Therefore, many people are coming to the capital for the medical examination and hope to find the competent specialists, but they are usually disappointed by these visits that strengthens the tension.

The other strong distress factor is the people's confidence that they consume dirty foodstuffs and are not able to change something in their life because the economic difficulties which endure people in Belarus, the low standard of life do not allow them to consume guaranteed clean foodstuffs (delivered from the clean areas and from abroad) and force them to consume the local ones, but people do not know how to make it less dangerous.

Thus, the distress due to the Chernobyl accident could be marked less, if people had been ready to the eventual consequences and management structures knew exactly how to act in such situations and first of all how and about what to inform people in order to avoid rumours and panic.

These knowledge should be the next:

- openness of the information about the work of the atomic plants, their construction, reliability;
- acquaintance with devices for measurement of the environment, foodstuffs radioactive contamination and their accessibility to people;
- knowledge of the radiation affect on the human health and preventive measures for reduction of its destructive affects;
- exhaustive information of people about the eventual consequences of their residence in the contaminated areas and the modes of risk reduction;

- elaboration of such measures of people's social protection, which guarantee their future.

4.Conclusions

The study revealed the high level of people's worry, their complete distrust towards the power structures, mass media, experts in the field of radiation and public health. People reported that the local authorities did not take care of them. 77.1% of the respondents were worried about their future. 87.4% of the respondents were helpless to change something in this situation. All these circumstances create the background for people's chronicle distress and require the carrying out of the rehabilitations measures throughout the republic.

DYNAMICS OF SOCIAL-PSYCHOLOGICAL CONSEQUENCES 10 YEARS AFTER CHERNOBYL

*G.M.Rumyantseva, ** H.V.Archangelskaya,
**I.A.Zykova, *T.M.Levina

*State Scientific Center of Social and Forensic Psychiatry,
3 Poteshnaia St., 107258, Moskaw, Russia
Phone a FAX:009-7-095- 200-7308

**Institute of Radiation Hygiene, 8, Mira St.
St.Petersburg, 197101 Russia
Phone a FAX: 009-7-812-232 -7025

Abstract

The study has been carried out according to the long-term JSP2 in comparison with the results of data acquired by the authors in previous years in other programs in 1988-95 for more then 5 thousand people. In working out the strategy of post-catastrophe situation it is necessary to have a joint effort of the population and authority. The studies have showed that cooperation has not been achieved in this case. Hence, the effect of protective measures have been seriously decreased. Countermeasures taken after the catastrophe have had not only a positive, but in some cases a negative impact. The results of many previous studies as well as JSP2 program have shown serious social and psychologicalconsequences of Chernobyl Accident. There is a constant year-to-year comprehension among population anxiety concerning their health. The main result of the study is that social and psychological consequences of the Chernobyl Accident include nonradiological risks as seriously as the radiation risk.

General Background.

The Chernobyl Accident in 1986 has become a cause of stress for the general public possessing all features of informational stress. The situation is characteristic for almost any global ecological stress. The cause of stress was information about radioactive contamination. Approximately 5 mln people had not seen with their own eyes the fire and the accident. they personally, by personal direct (without devices) perceptions could not evaluate the radiation risk, but they had to change their habits because of countermeasures.

Under these above-mentioned conditions stress and worry in the affected people is formed and prevails within a different time framework. They have their own distinctive features, channels and forms of incidence, which differ from the described classic cases when stress appears immediately after an accident, catastrophe or other extreme event for any direct participant of these events. A process of worry dissemination acquires an epidemic process character.

The aim of the study.

The results of many previous studies as well as JSP2 program have shown serious social and psychological consequences of Chernobyl Accident. For working out a long term strategy it is necessary to study the effects in actual dynamics of the factors that influence the social and psychological consequences.

Method.

The study has been carried out according to the long-term JSP2 in comparison with the results of data acquired by the authors in previous years in other programs. The basis for this study were questionnaires of the population. The study was carried out on radionuclide contaminated territories with different levels of contamination, on adjacent territories, on territories of relocation and on distant clean territories. Study period:1988-1995; number of polled participants: more then 5 thousand persons. Computer processing: SYSTAT program.

Results.

The level of psychological stress and worry is independent of the level of radiation contamination of the territories (Table 1).

Table 1
RADIATION RISK EVALUATION BY THE POPULATION IN 1988

Question	% of respondents with positive answer on contaminated territories with the level of contamination (Cu/km2)			
	1-5	5-15	15-40	>40
The radiation situation is dangerous for health	68.4	74.2	78.5	71.4
Mentioned cases of berth restriction	64.1	66.2	64.0	64.2
Thinking of moving to a different region	33.8	28.9	47.5	33.8
Definitely decided to move	4.6	2.0	3.7	0.0

Table 2
ATTITUDE TO RADIATION

Question	% of respondents with positive answer on contaminated territories in a year		
	1988	1992	1993
The radiation situation is dangerous for health	76.5	97.8	89.0
Aware of radiation influence on the health	87.5	69.2	89.2
The radiation has influence on the health	65.9	45.2	60.0

In contrast to other catastrophe the Chernobyl induced stress has no tendency of decrease with time (Table 2). In 1993 89% of the inhabitants of radionuclide contaminated territories thought that the situation of contamination is dangerous for their health against 78% of the respondents in 1988.

Subject of the psychological stress during the years after Chernobyl is a worry to radiation influenced to the health his and his child. The main content of the sickness stress has been the serious concern about personal health and the health of the children. More then 80% of the respondents are afraid to be affected by a radiation related diseases or think they have already been affected by such a disease. Radiation risk is evaluated as being at rather high level equal to risk of the vital social and economic factors (Table 3).

Table 3
ANXIETY OF THE POPULATION CONCERNING EXTERNAL FACTORS
INFLUENCE UPON HEALTH IN 1993

| Question | % of anxiety respondents on the territories | | | | | |
| | contaminated | | | control | | |
	no	little	very	no	little	very
Have you been exposed to a dangerous radiation dose?	4.0	30.4	65.6	30.0	59.4	10.7
Are you worried about radiation effect?	2.0	38.1	59.8	19.4	46.6	34.0
Are you worried of the economical changes?	6.9	30.7	62.5	30.8	33.5	33.8
Are you worried of the social changes?	2.6	37.6	52.8	35.9	40.1	24.1

The population has a tendency to link all its health problems with the accident and its consequences. Up to 70% of the respondents on territories with different levels of radionuclide contamination are sure their health has been influenced by the catastrophe. All the other every day factors that usually caused health problems are either ignored or pushed into the background (Table 4).

Table 4
POPULATION RISK EVALUATION IN 1993

| Factor | % of anxiety respondents on the territories | | | | | |
| | contaminated | | | control | | |
	no	little	very	no	little	very
Traffic	18.7	55.2	36.2	4.3	62.8	32.9
Alcohol	44.3	35.7	20.0	44.1	39.7	16.2
Smoking	45.3	33.0	21.8	43.0	43.5	14.5
Drugs	58.0	12.2	29.9	66.3	12.8	20.9
Chemical pollution induced disease	7.7	32.6	59.6	9.1	32.7	58.2
Radionuclide pollution induced disease	2.1	21.7	76.2	13.4	34.9	51.7

Moreover, the population is definitely expecting radiation impact. Up to 90% of the population on the radionuclide contaminated territories knows of cases of thyroid diseases, and more 70% of the respondents have heard about sudden deaths of young people. Even in remote control areas a number of these respondents is great - 44% to 52% of the total number (Table 5).

Table 5
EVALUATION OF THE RADIATION IMPACT ON THE HEALTH
OF OTHER PEOPLE IN 1993

Question	% of respondents with positive answer on territories			
	contaminated	adjacent	relocation	control
I am acquainted with someone who has fallen seriously ill	47.8	38.4	55.6	18.2
In my immediate surroundings there are people with thyroid disease	92.0	71.1	89.1	51.6
I have heard of young people's sudden deaths with no apparent reason	71.2	73.6	72.0	44.1

Countermeasures taken after the catastrophe have had not only a positive, but in some cases a negative impact. Studiesperformed years after the tragedy showed that such as an overall medical check, disactivation, radiation measurements enhance the stress in about 40% of population. Stress level in villages were cattle has been expropriated turned out to be much higher then in those were this countermeasure was not performed - even 7 years after the catastrophe. Apparently, cattle expropriation was a notable sign of the accident, and was not regarded by the population as a means of defence.

The main source of knowledge was information. During resent years the trust in official information decreased dramatically. In 1988 70% of the respondents trusted official information. In 1993 their number fell to 5%. At the same time 50% of the population still trust specialists. Special trust is given to rumors; 30% to 60% of the population consider them the most reliable source of information.

The study of social-psychological consequences of one of the most important countermeasures relocations has shown, that character and specific features of it's organization may determine level of psychological response of the population.

While 1988 52% of the population wished to move from radionuclide contaminated territories, only 31% expressed such an intention in 1993.

Discussion.

There is a constant year-to-year comprehension among population anxiety concerning their health. During the polls in protective measures priority is given to medical treatment support. This is also true for the relocated in 1995.

Dynamic observation over the psychological consequences with the relocated have shown that stress level has grown in the period from 1992 till 1995. This coincides with decrease of social standards and lack of financial support in the medical sphere, construction and with the developing of the infrastructure. All these have been mentioned in the answers to open-ended questions by the relocated in 1995.

Results of the poll in the dynamics for the period of 1992-95 have shown a marked tendency of dependence growth of the population on the local authority. More then 80% of the population think state authority should be responsible for the liquidation of the catastrophe results; 50% demands it of the local authority and 50% of the doctors. Only 20% of the population think that they should try to deal with the present situation.

In developing post-catastrophe management guidelines it is necessary to take into account the forms of overcoming thestress that the population itself uses. It would have been possible to imagine that people from radionuclide contaminated territories would be worried to a great extend by the possibility of dangerous radiation influence of the health (as it has been shown earlier), that they would take measures to prevent getting another dose of radiation. In 1993 locus-control (adapted for slavic population) poll showed that 69% of the exposed told that their routine life-style has changed. Nevertheless, protective measures are being used only 22% from the exposed. In other words, everyday life and subjective image do not consider. Apparently this shows a double standard in approach of the population to the problem.

In working out the strategy of post-catastrophe situation it is necessary to have a joint effort of the population and authority. The studies have showed that cooperation has not been achieved in this case. Hence, the effect of protective measures have been seriously decreased. Stress among population increases with time; it is enhanced by the fact that a 45% of the population did not trust the authority (the data does not change with time) and only 14% of population believe at present (1995) that local authority takes definite steps to change the situation to the better.

If we compare the social-psychological consequences of the Chernobyl catastrophe in the USSR and Tri-Mail-Ailend accident in the USA we can point out the following positive features of the Chernobyl Accident:

1. The population on the contaminated territories in Russia has not been stricken by panic immediately after the catastrophe.

2. For the long time after the accident a trust to the specialists has been maintained; more then 50% of the respondents prefer to receive information from the specialists up till now.

3. People still believe in atomic power as a source of energy. Only 30% of the population suggest to close down all the existing atomic power stations, 95% of the population consider to build atomic power stations with the better protection.

Conclusion.

The study showed, that radiation risk perception and as a result of this, population anxiety appeared to be the result of the following factors:
- situation (changing of the radiation situation and introducing of the protection measures);
- information (trusting the sources, availability, urgency);
- social and economic (changing the social structure of the social and difference of economic situation);
- political (authorities' stability, functioning of the legal system)
- geographical (close location to contaminated area).

The basis of the social and psychological rehabilitation should be found in strengthening personal attitudes and consideration of personal responsibility. Social and psychological consequences include nonradiological risks. Their influence should be taken into account as seriously as the influence of radiation, for psychological distress takes its toll both in diseases and in deaths.

Information System "Chernobyl" of EMERCOM of Russia

L. Bolshov, I. Linge, R. Arutyunyan, A. Ilushkin, M. Kanevsky,
V. Kiselev, E. Melikhova, I. Ossipiants, and O. Pavlovsky
IBRAE - Nuclear Safety Institute of Russian Academy of Science
Bolshaya Tulskaya 52, 113191, Moscow, Russia

Since 1991, USSR (Russian, upon 1991) State Programs on Elimination of Consequences of the Chernobyl Accident included the section "System and information support of works". That section made provisions for funding two trends of the work:

Development of Information System "Chernobyl" of EMERCOM of Russia;

System and analytical support of the State Program [1].

The paper presents results of the first trend of the work. Information System "Chernobyl" of EMERCOM of Russia included the following: Central bank of generalized data, Bank of models, Information systems for federal and local authorities. The analysis of many phenomena demanded retrospective data collection. In that way, banks of primary data were created and experience of analysis of directly accident information was acquired.

The main element of the system-analytical support is the administrative informational system (AIS) of the Department for elimination of consequences of radiological and other disasters of EMERCOM of the Russian Federation. AIS is intended for providing specialized program-technical complexes and systematized data related to the Chernobyl- accident effects and measures on their elimination for heads and specialists of the Central staff and territorial and regional administrative bodies, all other interested ministries, departments and organizations.

The AIS software comprises a number of specially developed topical electronic erence books as well as a stock of algorithms and programs intended for modeling the spreading of radionuclides in the environment, calculations of exposure doses, and estimations of the additional risk due to the population exposure on contaminated territories, and others.

Compatibility of information comprised in the above-mentioned data banks is provided (within network) through the use of unified classifiers and data formats.

An urgent need is interdepartmental information integration and adjustment of constant interdepartmental data exchange. In this connection works on interbranch information exchange with numerous ministries and departments were carried out.. Over 40 organizations were drawn in data collection works. The management of works on elimination of the Chernobyl consequences is carried out by regional departments, centers, divisions, and administrations of corresponding departments. In all regions affected by the radioactive contamination special bodies of regional administrations were established; they differed not only in their names, but also in place and subordination level within the administration structure as well as in the scope of questions being solved. In connection with a diversity of organizational forms that can be used for elimination of management of accident consequences at localities, individual elements of the system for information support of regional administrative bodies were developed.

In several territorial-management subdivisions of the affected regions elements of the administrative information system for regional administrative bodies were put into service.

For successful functioning of AIS and implementation of interregional and interbranch data and message exchanges the organization of software/hardware means for computer reception, processing and transfer of data via communication channels is of decisive importance.

The main component of the software/hardware complex for AIS telecommunications, namely, network communication unit was developed and introduced into practice. Such a unit provides reception and transmission of data and messages from remote subscriber's stations of AIS. Developed hardware/software provides an access to data bases of scientific organizations working on the Chernobyl problem.

The Central Bank of Generalized Data (fig. 1) comprises the several subdivisions: factual, documentary and personal data, and electronic map bank. For example the brief characteristics of two data banks are listed below.

In the data bank on radiological and sanitary situation the information has been accumulated for about 10,000 settlements belonging to 17 regions of Russia such as Bryansk, Kaluga, Tula and others. The number of data types (fields) is about 250 for each settlement. These data have been collected from more than 300 information sources. The main data groups include populated area name, administrative code and geographical coordinates; social and economic development data; data on types of protection measures used; data on soil and milk contamination by radionuclides (cesium, strontium, iodine and plutonium); data on measured population internal and external exposure doses; estimates of expected -population exposure doses; data on doses for making decision. Data collected cover the period from 1986 till 1995. All the data are accompanied by bibliographic erence data fields.

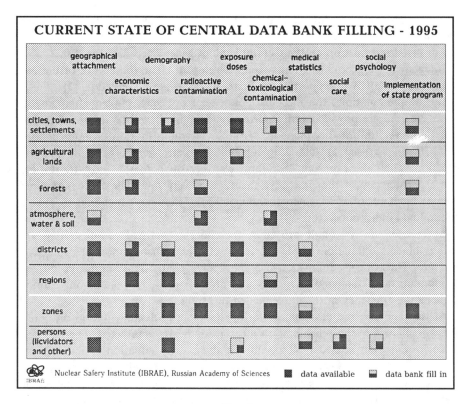

Fig. 1

The data bank on demography and medical statistics embraces plenty of data that are grouped as follows: population size and structure (age, sex, town, country), natural changes in population including migration, births and deaths, mortality on different causes, cancer incidence in children and adults, medical service. Data are available for the period 1982-1994 for almost all regions of the Russian Federation (for about 80 regions). Naturally the most detailed statistics was collected for 15 regions affected by the Chernobyl accident.

For retrospective data analysis, development and validation of models, analysis of statistical distributions of different data types local data banks on primary measurement were created. Among them there are whole body counter measurements, agricultural products and food-stuffs contamination, dose rate measurements, individual external dose control, iodine thyroid activity measurements, radionuclide concentrations in samples of soil, surface water and some others. These primary databases contain information not only for Russian contaminated regions but also for Belarus and Ukraine. Various kinds of data covered the period from 1986 till present time.

The model bank consists of numerous calculating and simulating computer codes, including programs for calculation of atmosphere transfer, migration of radionuclides, estimates for internal and external doses, assessments of radiation risk, and others.

For detail information on the Central bank of generalized data see .[2].

Integrated and geoinformation systems developed in the framework of the information and analytic support of the State programs are presented in [3].

By the present time, the most part of special problems implemented in the State Program are provided with information support. It will be improved in future. However, information support of the local administration is more important. It can be used for choosing rehabilitation arrangements in the present situation, when the exposure doses are relatively small in the most territories, and the efficiency of the protective measures is very low. The State programs provide hundreds of arrangements, and erection of thousands buildings in social, cultural and industrial sector.

Absence of the systematic procedure for choosing optimal administrative solutions hinders estimation of the efficiency of given rehabilitation or protective measure. Thus, local administrative bodies, solving large-scale problems on rehabilitation of the suffered territories and protection of the population, cannot evaluate the efficiency of the implemented measures.

At the moment, the development of methodological documentation on comparative analysis of the rehabilitation measures is nearly completed. The urgency of such evaluation is evident, especially at the present situation, when contradictions aggravate between the available resources (financial and material) and wide range of needs. The essence of the method of comparative estimation of measures is an hierarchical presentation (decomposition) of the process of making administrative decisions on rehabilitation of the territories (for example, as follows: goal - action - arrangements - subjects - specific goals - development scenario) and expert estimation of the extent to which low-level elements exert effect on each high-level element. The developed algorithms allow one to analyze consistently any complicated system of arrangements, and determine the most effective measures for the final goal, and set the optimal combination of measures.

References

[1]L.Bolshov, I.Linge, R.Arutyunyan et al. Experience of information-analytical support of the State program of elimination of the Chernobyl accident consequences and its further application (this issue).
[2]L. Bolshov. I. Linge. R Arutyunyan et al. Chernobyl Experience of emergency data management. In proc. of NEA Workshop Emergency Data Management, 12 - 14 September, 1995, Zurich, Switzerland (to be published).
[3]R. Arutyunyan, L. Bolshov, V. Demianov, A. Glushko et al., Environmental Decision Support System on Base of Geoinformational Technologies for the Analysis of Nuclear Accident Consequences (this issue).

Environmental Decision Support System on Base of Geoinformational Technologies for the Analysis of Nuclear Accident Consequences.

R.V.Arutyunyan, L.A.Bolshov, V.V. Demianov, A.V.Glushko, S.A.Kabalevsky,
M.F.Kanevsky, V.P.Kiselev, N.A.Koptelova, S.F.Krylov, I.I.Linge,
E.D. Martynenko, O.I.Pechenova, E.A.Savelieva, and A.N.Serov
IBRAE - Nuclear Safety Institute, B.Tulskaya 52, 113191 Moscow, Russia
Fax: (007) 095 2302065, E-mail: mkanev@ibrae.msk.su
T. C. Haas
School of Business and Administration, University of Wisconsin-Milwaukee, USA.
M.Maigan
Lausanne University,Switzerland

Abstract. The report deals with description of the concept and prototype of environmental decision support system (EDSS) for the analysis of late off-site consequences of severe nuclear accidents and analysis, processing and presentation of spatially distributed radioecological data. General description of the available software, use of modern achievements of geostatistics and stochastic simulations for the analysis of spatial data are presented and discussed.

1. Environmental Decision Support Systems.

Modern information technologies: data bases and data base management systems (DBMS), decision support systems (DSS) and intelligent DSS; expert systems; geographical information systems (GIS), integrated systems and information support systems enable us to create new generation of computer models (CM) and computer environments that may be widely used in the analysis of off-site consequences (ecological modeling, risk analysis, emergency preparedness planning, decision making etc.).

Classical Decision Support Systems (DSS) include Data bases, Bank of models and User's interfaces [1]. At present time DSS are essential part of any good ecological project. Intelligent DSS includes also expert systems or knowledge bases. With the help of DSS it is possible to manage both data and models in order to prepare support for decision making process. Environmental DSS (EDSS) deal with spatially/geographically distributed data. It means that EDSS have to be connected Geographical Information Systems (GIS).

The main parts of Environmental Decision Support Systems are as follows[2]:

- Data bases and data base management systems. Data bases contain information concerning environment, demography, contamination, economic information and others needed for the description of the region of interest and decision-making process.
- Specialized geographical informational system, based on electronic maps bank. Cartographic database includes electronic maps about most of contaminated zones due to Chernobyl and South Ural. Attributive data for EDSS are from NSI Central Data Bank. Information has been accumulated for about 10,000 populated areas belonging to 15 regions of Russia. Original software (RVS) was created with use 'user-friendly interface". By this system users can create efficient queries and to make some thematic maps (Fig.1).
- Bank of Models and ModelBase management system. Bank of Models includes computer models on different aspects of radionuclides behavior in environment and their impact on people. All coefficients and parameters of models are organized as a model-dependent data bases. At the moment, Model bank consists of a number of models on various subjects: short and long range atmospheric dispersion models, compartment and advection-diffusion models on radionuclides migration in soils, compartment and dynamic models on radionuclides migration in food chains, dosimetric models, dose-effect models, risk models, etc. An important phase of model development is models' verification ands validation. These were done for majority of models by using different kinds of data, also data on Chernobyl accident consequences.
- Information support systems. They include different kind of information concerning computer science, radioecological modeling, decision-making process, intervention levels and countermeasures, etc.
- User interfaces or dialogue modules. This is an important part of any DSS and helps user to prepare, process, present, and understand results.

The off-site analysis includes different subjects: atmosphere, hydrosphere, soil, food chains, dose calculations and dose-effect analysis, etc. First step is to prepare scenario compatible with available information. For example: source term-atmospheric dispersion-migration of radionuclides in food chains-dose calculations — countermeasures. Database management systems controls flows of information between models. The main results of the EDSS are decision-oriented maps, legends, analysis (reports and recommendations) (Fig.2).

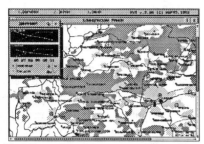 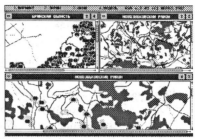

Fig.1.Examples of data presentation with use of RVS software.

Fig.2. Multywindow system for radionuclides migration analysis on maps.

2. Analysis of spatially distributed radioecological data.

Environmental data, as a rule, are discrete, spatially distributed and time dependent. Before using data have to be analyzed and prepared, correspondingly. The results of analysis are strongly dependent on both input data (how do they represent reality) and data processing methods used.

One of the first and important question is: what is a spatial and dimensional resolution of the network and which phenomena can be detected by the given network? These problems are related to the qualitative and quantitative description of monitoring network and its clustering (spatial inhomogeneity). Clustering of network (preferential measurements of some regions) leads to wrong calculations of mean value, variances, histograms that is essential in many applications. We have used several indicators in order to describe network's clustering and its spatial resolution: The most frequently used are Voronoi polygons (area-of-influence polygons) and Morishita index. Voronoi polygons along with Delauney triangulation are used also for network structure visualization. If a monitoring network is self-similar, the network can be characterized by its fractal dimension $D_f[T]$, where D_f is dimensional resolution [3].

There are several approaches for spatial data analysis and processing: statistics, interpolations, geostatistics, stochastic simulations. All methods can be considered as deterministic (ordinary interpolation techniques) or statistical (geostatistics and some kinds of modeling). They differ in degree of preliminary data preparation, underlying mathematics and in used interpretation models and obtained results. Statistical methods need more deep preliminary analysis of spatial/temporal data correlation (including anisotropy) and they estimate errors of prediction as well. Geostatistical models can be and are used for the monitoring network optimization and redesign. This problem is also essential for remediation and restoration of contaminated sites.

At the moment, more and more linear and non-linear geostatistical models (various kinds of kriging: simple, ordinary, disjunctive, indicator, moving window regression residual) are used in the analysis of ecological data. Methods of structural analysis allow to discover and model spatial correlation and take into account existing trend, which is a common phenomenon in environmental data. It brings a significant influence especially on a large scale. Moving window approach provides non-linear trend model and builds separate correlation models for every set of local data (within window). Using artificial intelligence methods is another way to estimate spatial trend. Residuals obtained from neural network estimations of global trend represent local spatial structures modeled by variography. Geostatistical prediction methods allows to analyze prediction uncertainty by building "thick" estimation contours with a certain probability to cover the real (unknown) isoline. Such maps are more useful for decision making then pure estimation ("thin") isolines [4]. It is possible to explore spatial structures of different data populations and using non-linear methods to model cumulative distribution function. This leads to risk analysis, maps showing probability of exceeding certain levels of contamination assigned by a decision maker. Stochastic simulations provide different possible variation of pollutant's distribution. Different methods give different both quantitatively and qualitatively decision-oriented maps (Figs. 3, 4).

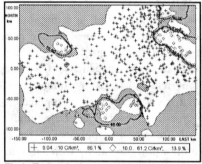

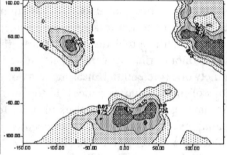

Fig.3. Estimation uncertainty contours for Cs137 contamination at a level of 10 Ci/km^2 Gomel region.

Fig.4 Probability map of exceeding the level of 10 Ci/km^2 contamination by Cs 137, Gomel region. Contours at 0.05, 0.25, 0.5, 0,75.

References

1. Tom Beer "Bushfire - control decision support system. Environment International, v. 17, p. 101-110 (1991).
2. M.Kanevsky, V.Kiselev, P.Fache, J.Touche. Decision-oriented mapping in emergency and post-accident situations. Radioprotection, Special Issue, February 1993, p. 441-445.
3. S. Lovejoy, D. Schertzer, P. Ladoy. Fractal chracterization of inhomogeneous geophysical measuring networks. Nature, 319, 1986. pp. 43-44.
4. M. Kanevsky, R. Arutyunyan, L. Bolshov, V. Demyanov, T. Haas, L. Savel'eva. Environmental Data Analysis. Case Study: Chernobyl Fallout. 10th International Congress on Mathematical and Computer Modelling and Scientific Computing. Boston, July 5-9, 1995.

"Hot" Particles - The Source of Radionuclides on the Territory of Belarus

E. PETRYAEV[1], M. BELLI[2], G. SOKOLIK[1], S. LEYNOVA[1], V. DUKSINA[1]

[1] Radiochemistry Department, Belarussian State University, Minsk, Belarus

[2] ANPA (Environmental Protection Agency), Rome, Italy

Abstract. Investigations of radioactive fallouts on the territory of Belarus indicate that a substantial part of radioactive isotopes is found on all the studied territory in the form of "hot" particles, which according to the mechanism of formation can be conditionally classified into two types: fuel and condensing. The analysis of change of "hot" particles distribution, activity, stability with time show, that "hot" particles are potential source of radionuclides, which penetrate into soil and redistribute in the biological-geochemical chains..

1. Introduction

Investigations of radioactive fallouts on the territory of Belarus indicate that a substantial part of radioactive isotopes is found on all the studied territory in the form of "hot" particles. According to the mechanism of formation the detected "hot" particles can be conditionally classified into two types: fuel, representing finally divided fuel and condensing, being the consequence of condensation of products in steam-gas phase over the reactor.

Using the calculated-experimental method suggested in scientific paper [1], has shown, that for sites within 10-20 km from ChNPP principal part of fallout have fuelling nature, for sites remote within 40 km from ChNPP 65-85 % of particles have fuelling nature and for sites within 200, 250 km from ChNPP principal part of fallout have condensed origin.

Conditions of formation of "hot" particles, determine their composition and physico-chemical properties and, as a consequence, dynamics of changing of radioactive contamination of territories and the level of radiological influence on biological objects.

2. Results and Discussion

2.1. Distribution of "hot" particles in soil

Experimental investigation of "hot" particles properties was carried out on the sites remote from the place of the accident from 10 to 250 km. Such methods as radiography, granulometry, α-, γ-spectrometry, β-radiometry, radiochemical analysis were made use of.

As it is seen from the data given in Table 1, a decrease of the content of "hot" particles in upper layers of soil and decrease of the share of beta-activity connected with them are going on. The character of distribution of "hot" particles, along the territory of the Republic of Belarus has practically not been changing during all the years of investigations: the quantity of particles in the soil and their activity decrease depending on remoteness from ChNPP.

Table 1.   The content of "hot" particles in upper soil layers and the share of beta-activity, connected with them (1987-1993).

Years	Quantity of "hot" particles/m^2	Share of β-activity of "hot" particles in %	Quantity of "hot" particles/m^2	Share of β-activity of "hot" particles in %
	40 km from ChNPP		200, 250 km from ChNPP	
1987	3.9×10^5 - 1.2×10^6	28 - 48	1.6×10^4 - 8.1×10^5	9.8 - 43
1988	2.7×10^4 - 9.0×10^5	27 - 47	1.8×10^3 - 7.0×10^4	9.0 - 42
1989	1.9×10^4 - 8.9×10^5	26 - 45	1.1×10^3 - 4.0×10^4	8.5 - 40
1990	4.9×10^3 - 9.0×10^5	21 - 40	1.2×10^3 - 7.0×10^4	8.0 - 21
1991	3.7×10^4 - 8.9×10^5	9.0 - 35	-	-
1992	4.2×10^3 - 5.0×10^4	8.0 - 33	1.9×10^3 - 1.6×10^4	8.0 - 20
1993	1.8×10^2 - 2.8×10^4	5.3 - 28	1.8×10^2 - 4.8×10^3	5.0 - 18

The decrease of beta-activity associated with "hot" particles during the period from 1987 to 1994 in all soil studied could be attributable to the destruction of "hot" particles and to their chemical solubility in the environment. Comparison of the results obtained in the investigation have shown, that the beta-activity, connected with "hot" particles, decreases 1.2 - 1.5 times in one year. It has been noticed that the velocity of lowering of this value depends on the properties of soils and on the own properties of the "hot" particles.

Insignificant migration of "hot" particles against time because of their mechanical redistribution along the depth of soil slits is observed in all types of soils only in the limits of 2-3 one-centimeter layers. Nevertheless, in all investigated stations a decrease of the share of β-activity of "hot" particles contained in all depth of soil profiles was registered during the time. This evidence of the destruction and partial chemical solution of "hot" particles, going on under the influence of various natural conditions.

2.2.   Physico-chemical properties of "hot" particles

2.2.1   Dispersity of "hot" particles
The analysis of distribution of "hot" particles in soil fractions with different sizes has shown, that relative content of "hot particles in smallest fraction increases in a course of time. The share of "hot" particles in fractions sized <100 μm in soils from sites, remote from different place of the accident at present constitutes 55-90 % of their total content in the whole sample in soils. The share of "hot" particles in fractions sized less than 5 μm in separate cases reaches 20 % of their total content in the sample.

2.2.2 Stability of "hot" particles

The results of investigation of mechanical and chemical quality of "hot" particles separated from soil have shown that they depend on the location of the investigated site: stability of particles in near sites is 1.2-2 times higher than the stability of particles in remote sites. This is probably connected with different composition of matrixes of the deposited "hot" particles, which is in turn conditioned by the nature of "hot" particles origin nature.

2.2.3 Composition of "hot" particles

Radionuclide composition of "hot" particles spotted at different distances from ChNPP considerably differs: particles Ce, Ru and Cs enriched were revealed [2]. It must be noted that the quantity of Cs enriched particles increased when moving away from the place of the accident, and content of plutonium in separate "hot" particles decreases. Its quantity in particles revealed in soils of sites remote from the place of the accident 200, 250 km did not exceed 0.1 Bq/part.

Investigations of the "hot" particles composition by X-Ray fluorescent method and by neutron-activation analysis had shown that "hot" particles contained Ca, Pb, Cu, Al, Cr, Br, V, Na, Mn, Rh, Mo, Te. In separate particles Au, Ag, Zr are registered. The content of Ca in some particles exceeds 50 weight percent (w.%), of iron - 17 w.%, of lead - 12 w.%.

The presence of Si, Al, Mg, Cl, Na in particles can point to the fact that fine-dispersive natural particles acted as aerosols at the moment of formation of particles and also to the fact that there existed an interaction of particles with natural substances.

2.2.4 Mobility of radionuclides from "hot" particles

In order to assess the mobility of radionuclides from "hot" particles sequential selective leaching of Cs, Sr, Pu and Am by NH_4COOCH_3 solutions, 1M HCl and 6M HCl was conducted. As a result solutions containing this radionuclides in exchangeable, mobile and fixed forms were separated. A determination of the quantity of radionuclides was conducted. Values showing the content of radionuclides in exchangeable and mobile forms are interpreted as the release of radionuclides from "hot" particles.

In Table 2 the content of different forms of Cs, Sr, Pu and Am in "hot" particles sampled in 1993 in soils of different remoteness from ChNPP is shown. It is seen from the data the mobility of analyzed radionuclides from "hot" particles increases, when moving away from the place of the accident. Increases of radionuclides mobility from "hot" particles was registered in the course of time: so in 1987 this value constituted for caesium 12-25 % and for strontium 8.0-31 %, in 1993 this value constituted 22-33 % and 30-38 % accordingly.

Table 2.   Occurrence forms of Cs, Sr, Pu and Am in "hot" particles found at a different distance from ChNPP (1993).

Occurrence forms	Distance from ChNPP											
	10 km				40 km				250 km			
	Cs %	Sr %	Pu %	Am %	Cs %	Sr %	Pu %	Am %	Cs %	Sr %	Pu %	Am %
exchangeable	10	10	2.0	15	11	12	9.0	15	15	16	10	15
mobile	12	20	6.0	20	14	19	13	30	18	22	20	37
fixed	20	18	22	3.0	25	17	13	5.0	26	22	27	4.0
heavily soluble	58	52	70	62	50	52	65	50	41	40	43	44

Conclusion

In a course of time destruction of "hot" particles, decrease of their activity and increase of their dispersity are observed. Decrease of particles stability and increase of the share of radionuclides in exchangeable and mobile states let us foresee their further decay; This makes "hot" particles a potential source of radionuclides, which penetrate into soil and than redistribute according biogeochemical chains. The presence of both fuel and condensed particles in soils of the Republic has formed zones with different sources of radionuclides.

References

[1]   A.A. Ter-Saakov, M.V. Glebov, S.K. Gordeev, Physico-Chemical Characteristics of Fuel and Condensed Particles and Their Inhalation Intake Into Arespirator Organs of Man. In: Working Material of Conference on the Radiobiological Impact of Hot-Beta Particles from Chernobyl Fallout: Risk Assessment, Vienna, 1992, p. 1-37.

[2]   E.P. Petraev, S.L. Leinova, G.A. Sokolik et al., Composition and Properties of Radioactive Particles detected in the Southern Districts of Belarus. Geochemical International Journal 7 (1993) 930-939

Hot Particle Factor in Radiation Dose Formation after the Chernobyl Accident

O. BONDARENKO[1], V. DEMCHUK[2], V. TEPIKIN[3] and V. NAGORSKY[3]

[1] *Radiation Protection Institute, 53 Melnikov, 253050 Kiev, Ukraine*
[2] *Radioecology Institute, 14 Tolstoj, 252033 Kiev, Ukraine*
[3] *Research Industrial Association Pripyat, Chernobyl, Ukraine*

Abstract. The necessity to apply original data about the size and the activity distributions of hot particles has been arising at many post-Chernobyl research. Such researches include first of all (i) studying of migration processes at soil-water complex, (ii) retrospective inhalation dose reconstruction for the population, and (iii) validation different scenarios of the Chernobyl accident deployment. Results of this research show that the fuel matrix in soil can be considered as constant with accuracy 20-30 % for transuranic nuclides and major of long-living fission products. Temporal stability of hot particles at the natural environment gives a unique possibility to use the hot particle size distribution data and the soil contamination data for retrospective restoring of doses even 10 years after the Chernobyl accident. In present research the value of the integral of hot particle activity deposited into the lung was calculated using a standard inhalation model which takes into account the hot particle size distribution. This value normalized on the fallout density is equal to 0.6 $Bq/(Bq\,m^{-2})$ for areas nearby the Chernobyl NPP.

1. Introduction

The peculiarity of the Chernobyl accident was that most part of radioactivity released from the emergency reactor was in the hot particle form [1]. By hot particles are meant all classes of radioactive particles (i) of condensation type (caesium and ruthenium), and (ii) of fuel type having the radionuclide contents similar to that of burnt out fuel.

Hot particles represent a potential hazard to humans, for principally two reasons. First, after the deposition in (or on) the human body, hot particles produce huge local doses. This can make a micro injury inside different organs (or on the skin) as well as contribute to late (carcinogenic) effects. Second, hot particles are "repositories" of a large amount of transuranic radioactivity which is encapsulated inside them. The solubility of hot particles and the degree of mechanical destruction determine the biologically accessible fraction of transuranic nuclides. Thus it forms the additional source of transuranic elements at intake through the food chain or at dissolution of inhaled hot particles.

2. Methods

Identification of hot particles and measurements of their size and activity distribution are carried out by the well known and worked out methods of emulsion autoradiography (X-ray film PM-B, PM-1) for the beta component [2] and solid-state nuclear track detection (SSNTD) (films of LR-115 and CR-39 types) for the alpha component [3].

The lower level of detection (LLD) for emulsion autoradiography is about 0.1 Bq/particle after an exposure time of 96 hours. In this way the minimal registered size of a hot particle is about 0.05 μm. The SSNTD method is able to identify a hot particle and provides simultaneous measurement of its size and activity, and also gives information about its radionuclide contents without destroying the particle. The LLD of the latter method is 0.1 mBq/cm^2 after one week exposure time.

Contrary to the above methods a standard radiochemical analysis destroys particles and is only able to give the total amount of radionuclides in a sample. Additional radiochemistry of soil is rather laborious and can be hardly done as an automatic procedure.

3. Formation of radioactive contamination

Primarily the radioactive pollution of the environment took place by air transport of radionuclides and their deposition onto the surface. In the bulk radionuclides were released to the air in a form of dispersed fuel particles and their mixture with inactive carriers. The fuel component of the release comprised dispersed fuel with a matrix of UO_2, UO_3, U_3O_8 and other. Inactive components contained a complex mixture made from graphite, building materials of the reactor, materials used for roofing in the devastated areas of the reactor.

Gases released from the active zone of the reactor were mostly isotopes of Kr and Xe. Some elements sublimated from the active zone and then deposited onto the surface as mononuclide particles. Among those there were mostly isotopes of I and partially of Cs and Ru. formation of such mononuclide particles occurred at condensation centres which may have been particles of dust, smoke etc. Caesium particles prevail in the region of less than 1-2 μm whereas fuel particles comprise mostly the region above. The ratio of the fuel components to the condensation components is in the range 5-8 for the near zone [4].

Fuel particles mainly were forming the contamination of the near zone whereas gaseous and vapor components and partially fine fraction of fuel particles were contributing to the contamination of the far zone. After the analysis of the structure, forms and mapping density of radioactive pollution of the near zone, it can be assumed that the hot particles were a main radiation factor of polluting the atmosphere and of forming inhalation doses of the population after the Chernobyl accident.

The pollution of the zone in vicinity of the NPP was formed basically by hot particles which radioactive contents coincides with the structure of burnt fuel. In addition ruthenium (1-3 %) and caesium (10-20 %) types of particles were met.

By the eighth year after the accident the portion of the caesium activity aggregated in hot particles is 40-80 %. It should be taken into account that at the moment of the initial explosion the released particles were already depleted by caesium. Changes of the hot particle dimensions structure in time is insignificant. Seven years or research from 1989 to 1994 indicate these changes to be about 10-20 % (fig. 1). Processes of hot particle mechanical destruction in soil and selective leaching of radionuclides from them, have been insignificant relatively to the rate of vertical migration. Thus the fuel matrix in soil can be considered as approximately constant (with accuracy 20-30 %) for transuranic nuclides and the majority of long-living fission products.

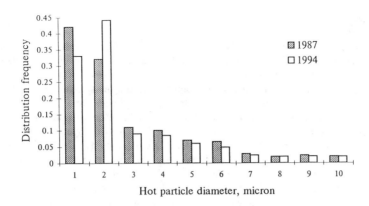

Fig. 1.   Typical distribution of hot particles in soil at vicinity of the Chernobyl NPP

The vertical activity distribution in the soil eight years after the accident is basically (90-95 %) concentrated in the 5-10 cm ... layer. Only trace amounts of caesium and strontium are detected below 10 cm. The activity distribution depends locally on soil type and water regime. The number of hot particles varies from 70-80 % in the top 0-2 cm layer down to 1-3 % in the 6-8 cm layer. Below 10 cm hot particles are hardly observed.

4. Retrospective assessment of hot particle intake into the lung

The spatial distribution of airborne hot particles during the earliest fallout with regard to HP dimensions and activities is very important and in the same time is the least investigated side of the entire HP problem. Information about this distribution based on direct measurements is almost completely absent. Such a situation forces to look for indirect methods to make HP lung intake estimation [5].

In the simplest ways, the intake model looks as follows. Suppose a human living in the damaged territory all the time of radioactive ground track information (with accuracy about "human behavior coefficient" approximately equal to 0.5). This human lung may be considered as a filter system for depositing radioactive dust.

The amount (or the integral activity) of radioaerosols of a given size l deposited in the respiratory tract can be derived from the contamination level at a given location and the size distribution of hot particles . The approach considered [5] allows to fold this complex dependence to the following view $l = AS_{eff}$, where A - soil contamination level (Bq m^{-2}) and S_{eff} - effective retention capacity (m^2).

Data of soil fallout may be approximated by a log normal distribution. For the particular area in the vicinity of the 4th block an average hot particle size $X = 2.56\mu m$ was found (fig. 1). It means that for normal breathing rate of 30 m^3 day^{-1}, the value of effective retention capacity S_{eff} obtained after integration is equal to $S_{eff} = 0.6$ m^2. Thus the integral I of HP activity deposited into the lung normalized on the fallout density A numerically equals I/A = 0.6 Bq/(Bq m^{-2}).

5. Discussion

Results of this research show that the fuel matrix in soil can be considered as constant with an accuracy of 20-30 % for transuranic nuclides and the majority of long-living fission products. Temporal stability of hot particles at the natural environment gives a unique possibility to use the hot particle size distribution data and the soil contamination data for retrospective dosimetry even 10 years after the Chernobyl accident.

The whole period of inhalation doses formation could be divided into two stages: 1) a short-term period of primary air pollution directly from the destroyed active zone (April-May 1986), and 2) a period of secondary atmospheric pollution as a result of re-suspension.

The first period is distinguished by a high concentration of primary radioactive particles and availability of plenty of short-living radionuclides in air.

An important feature of that stage was that hot particles massively irradiated (by beta emitters) the lung of many people. They could subsequently not found inside their bodies because of the relatively fast removal from top compartments of lung (during several days). However, calculations show that these hot particles could make a main contribution to the inhalation dose for a broad group of the population. In this research the value of the integral of hot particle activity deposited into the lung was calculated using a standard inhalation model which takes into account the hot particle size distribution. This value normalized on the fallout density is equal to $0.6\ Bq/(Bq\,m^{-2})$ for areas nearby the Chernobyl NPP.

The second stage is stretched in time by now. It has been characterized by a considerably lower intensity of radioaerosol sources (some orders of magnitude). Main dose forming radionuclides have been caesium, strontium and actinides. The hot particle size distribution has been conditioned by soil grain properties. The source of secondary airborne hot particle has been the soil surface, plants, buildings, etc.

References

[1] Hot Particles from the Chernobyl Fallout. Proc. of Intern. Workshop held in Theuern 28-29 October, 1987, Ostbayern, Theuern, 1987, p. 147.

[2] V.V. Demchuk, N.V. Viktorova, V.V. Morosov and E.B. Ganja. *On Study of Disperse and Radionuclide Characteristics of Chernobyl Fallout Particles by Means of Macroradiography.* In: Proc. II Intern. Workshop on Solid-State Nuclear Tracks Detectors, March 24-27, 1992, E-7-93-61, JINR, Dubna, 1992.

[3] O.A. Bondarenko, D.L. Henshaw, P.L. Salmon and A.N. Ross. *The Method of Simultaneous Size and Activity Measurement of Alpha Emitting Hot Particles Using Multiple Track Analysis of Solid State Nuclear Track Detectors.* Radiat. Meas., Vol. 25, Nos 1-4, pp. 373-376, 1995.

[4] V.V. Demchuk, N.V. Viktorova, V.V. Morosov and E.B. Ganja. *Migration and Transformation of Fuel Particles in Soil-Water System in Near Zone NPP.* In: Proc. Intern. Symposium on Radioecology, 12-16 October 1992, Znojmo, European Branch of IUR, 1992.

[5] I.A. Likhtariov, V.S. Repin, O.A. Bondarenko and S.Ju Nechaev. *Radiological Effects After Inhalation of Highly Radioactive Fuel Particles Produced by the Chernobyl Accident.* Radiat. Protect. Dosimetry, 1995, Vol. 59, No. 4, pp. 247-254.

III. HEALTH EFFECTS
FOLLOWING THE ACCIDENT
A. Treatment of accident victims

EXPERIENCE IN TREATMENT OF THE RADIATION SYNDROME IN ACCIDENT VICTIMS EXPOSED WITH NON-UNIFORM DISTRIBUTION OF THE DOSE WITHIN A BODY.

Angelina Guskova and Anjelika Barabanova,
SRC - Institute of biophysics,Moscow,Russia

Summary. Experience in diagnosis and treatment of radiation accident victims undergone to radiation exposure with non-uniform distribution of the dose within a body is presented and the most significant features of medical management of such patients are discussed. The term "compound radiation injury" is propoused to use for this form of radiation disease. Treatment of compound radiation injury demands a participation of very qualified specialists. The first medical aid and management should include careful body surface monitoring. Beside daily haematological observation and cytogenetic study with corresponding treatment, careful observation and registration of skin reaction are necessary. Some features of treatment are the following: more early administration of antiinfectious means, including isolation in sterile room, timely surgicall intervention, prophylacsis and treatment of endogenic intoxication improving of microcirculation, long time follow up study with pathogenic therapy.

Accidental exposure to people in most cases characterized with non-uniform distribution of absorbed dose within a body. The difference between the local and total dose (or dose to bone marrow) can be 20 - 30 times and more. Such type of exposure brings to development several syndromes, among which skin injury and especially necrosis of the skin are of greate importance. Clinical course of these cases is much more severe than in classical form of acute radiation disease.We use the term "compound radiation injury" for that. The frequency of such cases among all cases of acute radiation disease is close to 50 %.

Non-uniformity of distribution of dose can be conditioned by several factors,among which short distance or close contact a source to the body is the most important. Another significant factor is an energy of radiation, its penetrating ability. Gamma - beta or gamma-neutron exposure, typical for reactor accident, leads to specifical type of dose distribution with significant overexposure to skin and underlying tissues.

For successful management of compaund radiation ingury (CpRI) good knowledge on condition of exposure with all details is quite important. This can help in estimation of actual dose distribution, evaluation of local and whole body dose. All means must be used : careful study of the accident circumstances, simulation of the accident with using different type of dosimeters, calcullation, biological dosimetry (chromosome aberration analysis of blood cells and bone marrow from several parts of the body), electron spin resonance dosimetry of biological tissues and clothing.The last one sometimes can be especially useful.

Clinical course of CpRI characterized with more short latent period, since severe skin radiation injury develope earlier than cytopenia. The period of main manifistations is more severe because of often septical complications and very severe pain syndrome. More long period of high fever, significant decrease of mass of body, more early and deep anaemia, disproteinaemia, high tachycardia and neurological disorders all together form the syndrome of endogenic intoxication(SEI), too typical for CpRI. The SEI can be a cause of death, as it was in most of chernobyl lethal cases.

As it was shown in special study, the degree of the SEI depends on the size of skin lesions, especially on the mass of skin in zone of necrosis, and less depends on the severity of bone marrow syndrome.Repeated plasmaferesis was one of the most effective therapy procedure for detoxication.

Late period of CpRI is also much more severe: amputative defects and cataracts are the main consequences of the CpRI.These patients need more long careful surveillance and treatment. Some of them are invalides.

Planning of treatment in acute phase of the CpRI and in period of late effects should be based on early diagnostic and prognostic evaluations. For diagnosis of severity of bone marrow syndrome the rulles ellaborated previously [1] for bone marrow exposure appeared to be the most useful and help to predict the time and degree of cytopenia. The significance of evaluation of the size of skin injury has already been mentioned.For this the registration of early erythema considered to be important. Prognostical evaluation of the severity of skin syndrome is more difficult, since affected by many factors: type of radiation, its energy, size and place of skin lesion and so on...

However, in any case, careful day-by-day observation of skin reaction with photo and/or video registration is very useful. In our practice we use also the registration of skin reaction as a function of time by means of arbitrary skin reaction scores elaborated for this purpose.

Day-by-day development of symptomes of injury, as well as, the process of healing can be registered and presented in form of curve (the scale of scores and two examples of its use is shown on the fig 1).These examples demonstrate two different situations with various values of the rate of skin reaction (RSR), designated as index "J" .The analysis of a number of such curves brought us to the conclusion that the RSR could serve as a prognostical criterion. This is because the analysis has shown that there was some critical value of index "J" beyond which surgical intervention became a necessity. Moreover, it was found from retrospective analysis that some dependance pertained between this value and the time of operation; the higher the RSR, the earllier the operation became essential;(fig 2) this means that estimation of the RSR can help to predict the need and time of surgical intervention[2]. Timely operation is essential to prevent the severe SEI and to evoid secondary changes in exposed and adjacent tissues.

However, observation and registration of skin reaction is not enough to make a choice of the type of operation (amputation, its level, necroectomia, type of grafting and so on).It was shown by T.Protasova (unpabllished data) that pathomorphological findings in skin exposed to radiation can serve as a some sort of biological dosimetry and hence as a prognostic test. Three zones of character changes were considered. The first - was a zone of total necrosis, the second - the zone of injury with possible reparation, the third - the zone of relatively safe tissues. Pathomorphological findings in these zones were discribed for various periods of time elapsed since exposure. The main recomendation was to use skin biopsy for diagnosis of severity of skin lesion, and to perform operation, if necessary, in the periphery of the third zone. The best time for operation was indicated between the third week and third month.

In conclusion we would like to summarise, that:

The first medical aid in case of CpRI should include careful radiomonitoring of skin surface; careful dosimetric study with using all of possible means and methods; careful observation and registration of skin reaction; more early administraation of antiinfectious means; prophylacsis and treatment of endogenic intoxication; early prophylacsis of skin necrosis, improving of microcirculation; timely and radical surgical treatment; long time follow up study of patients.

References
1. Baranov, A. E. Dose estimates and the prediction of the dynamics of the peripheral-blood neutrophyl count according to haematological indicators in human gamma irradiation. Med. Radiologia 26,11-16 (1981).
2. Barabanova,A. V., A. E. Baranov, A. K. Guskova et al.Acute radiation effects in man. USSR State Commettee on the Utillization of Atomic Energy.Moscow-TSNII Atominform,(1986).

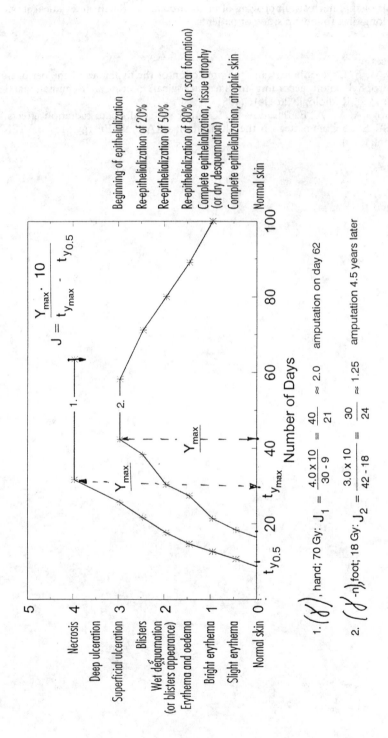

Fig. 1: Description of the scale of scores and two examples of its use

- 556 -

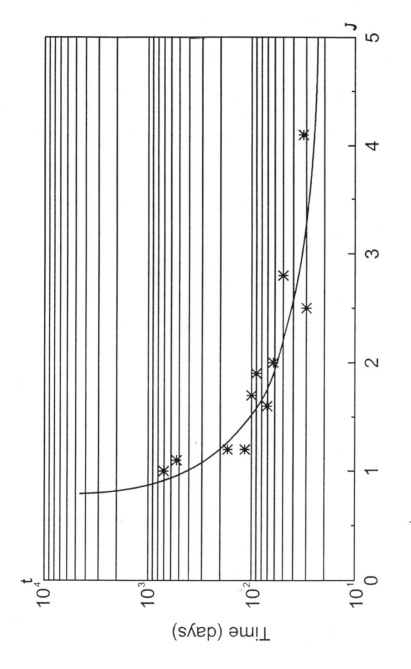

Fig. 2: The dependence of the time (t) at which amputation becomes essential on the rate of skin reaction (J)

CRITICAL ASSESSMENT OF EFFICIENCY OF IMMEDIATE INTERVENTION

Anatoly E. ROMANENKO

Research Center for Radiation Medicine,Academy of Medical Sciences of Ukraine

Abstract. Critical analysis is given in this report on measures conducted after the Chernobyl NPP accident fo provision of health care of inhabitants of suffered areas. Among shortages of the first stage of the accident it is pointed out: lack of modern instruments, dispersion of establishmental forces and means and dyscoordination of actions of different establishments having taken part in the accident elimination, excessive confidential;ity, unsifficient level of knowledge in the field of radiation emdicine, low efficiency of antiradiation measures, in particular, iodsine prophylkaxis. Thyroid irradiation doses were not taken into account in full measure for decision making about evacuation.

Almost ten years have passed since the day of tragic events at the Chernobyl NPP, but painful feeling about this event is still alive in memory of our nation. Many people are anxious about one question - what are consequences of this accident for our country and first of all for health of present and future generations.

Answering this question we should first of all point out unprecedental character and complexity of the situation appeared as a result of the accident at the nuclear reactor number four of the Chernobyl NPP and expansion of radioactive substances at relatively large areas. That's why, speaking about elimination of consequences of the Chernobyl accident, as a rule, we use such terms as "unprecedental", "first in the world practice" etc. Large-scale state measures were required, adoption of principally new decisions, implementation of the whole medical experience to provide health of inhabitants of affected areas. We should recognise that not everybody was ready to adopt such decisions, and, moreover, for their practical implementation.

One of the lessons of the Chernobyl accident became evident character of development of necessary general standards which consider radiation levels when different kinds of food products should be prohibited for consumption - intervention levels. Wide and non-justified variability of standards existed adopted in different countries. it impeded orientation in rapidly changing radiation situation.

Scales of the Chernobyl NPP accident, despite of experience of previous accidents in other countries, appeared to be unexpected. That's why in the health care system urgent and very complex problems and difficult tasks appeared on health care of population of territories involved into the accidental situation. They required both scientific consideration and rapid responce and operative decisions.

For planned and purposeful activities of authorities and health care system establishments during elimination of any catastrophe situation, taking into account huge experience of military medicine and using its terminology, first of all scientifically based imaginations are needed about possible magnitude and structure of military casualties, and also about basic peculiarities of conditions under which it is necessary to act giving medical aid to victims.

Forces and means of health care services, amount, stage character and continuity of medical aid and organisational principles of activities should comply to these basic initial data.

As for the Chernobyl situation such initial parameters was data about development and assessment of radiation situation, amount of medical and sanitary measures taking into account forthcoming population evacuation, areas and terms of population accomodation, sources of possible damage to population health and cosequently assessment and prognosis of levels of population irradiation. Only on such basis it is possible to provide efficient planning and organisation of adequate medical and sanitary measures, development of substantiated proposals on provision of population radiation safety.

But the experience of implementation of measures on population radiation protection has clearly shown that under condition of radiation contamination of territories calculated by thousands of square kilometers fighting against such factor as external gamma-irradiation, especially after relative stabilisation of radiation situation, represents considerable difficulties.

In contrast to comparatively "controlled" factor of internal irradiation methods of fighting against the factor of external gamma-irradiation are determined, first of all, by decontamination of radioctively contaminated areas and subjects of human residential environment. It is understandable that in contrast to a character of decontamination activities on the site of the Chernobyl NPP and areas directly adjacent to it decontamination activities in settlements, in forests, on fields and farms can be mainly accomplished only by methods of mechanical removal of radionuclides and, certainly, without implementation of special decontamination solutions (prescriptions) containing either active chemical components. It decreases considerably efficiency of decontamination.

The following conditions impeded in-time revealing and assessment of radiation situation:

First, peculiarities of the accident and unstable metheorological situation determined very complex dynamics and character of development of radiation situation which mainly stabilised only in two weeks after beginning of the accident. It impeded detection of the accident not only in acute period but also at subsequent time.

Second, tremendous deficiency of modern instruments and automatised systems of radiation monitoring in this country as the consequence of "stagnation" period and loss of interest to radiation problems after halting of nuclear weapons tests in three media. So, in particular, initially there was no modern equipment for automatic gamma-shooting of terrain from a board of an aircraft or a vehicle neither in the headquarters of the civil defence of Ukraine, nor in the Hydrometheorology Department. There was no not only modern, but also obsolete dosimeters, type DK-02, which fitted for establishment of group dosimetry by range of doses measured, because sets of dosimeters designated for combat actions (type DP-22, DP-23) appeared to be unfit for such situation. In number of ministries and establishments there was no also any instruments for technological radiation control during processing of food raw materials and production of food products. As a result there was also lack of trained personnel of specialists (dosimetrists, radiometrists, radiochemists and spectrometrists).

Third, dispersion of forces and means available separated by excessive confidentiality, establishmental interests and interestablishmental barriers, different methodic basis of instrumental measurements conducted, uncompetence of headquarters of civil defence as coordinators of such activities.

Separation and lack of coordinated actions between organisations and establishments were expressed mostly during implementation of radiometric control of food products, water, environmental subjects not only by general activity, but also by radionuclide content. Paradoxal situation appeared: there was lot of ministries and establishments responsible for production, procurement, processing, storage, transportation and sale of food raw materials and food products and also for water supply. Whereas their equipment and trained personnel available was so unsifficient that initially bulk of control was responsibility of health care authorities.

Miscoordination in other matters was also felt: in some civil defence headquarters Standards of radiation Protection (NRB-76) were ignored and during assessment of radioactive contamination of people, vehicles etc. standards of wartime were used. Because such standards in average two orders higher of acceptable standards mentioned above, in some places dosimetric points were cancelled. Consequently secondary radioactive contamination was brought into cities and settlements by vehicles. Decision making about order of vehicles decontamination was delayed, mainly because of uncoordinated actions of economic managers, existing establishmental barriers.

First attempt to unite efforts and put in good order implementation of control of radioactive contamination and irradiation of people was made under initiative of the Health Ministry of Ukraine. The Health Ministry has prepared a draft adopted by the order of the Council of Ministers from 13, June, 1986 No 332 pc about distribution of functions between Ministries and establishments on radiometric and dosimetric control.

Only by the order of the Council of Ministers of the USSR from 31, October, 1986 No 2204 pc functions of head organisation on control, assessment and prognosis of radiation situation in this country, contaminated territories around the Chernobyl NPP first of all, were given to the State USSR Committee on Hydrometeorology.

The health Ministry of the Ukrainian Soviet Socialist republic was not informed about the decision of the Governmental Commission of the

USSR to implement evacuation of first turn of the city of Pripyat' and settlements of the 10 km zone on the territory of the Ukrainian Soviet Socialist Republic.

Decision about relocation of inhabitants was made in process of development and worsening of radiation situation and under situation when not only prediction of expected dose, but even assessment of already received (realized) dose dose was sufficiently complex task. Difficulties in development of such decision are conditioned mainly by serious character of scales of the accident and exceptional complexity of configuration and structure of sources and fields of irradiation. Naturally, for decision making about evacuation it is necessary to have information at least about dynamics of external irradiation doses or thyroid irradiation doses. Comparing these doses with Criteria having been existed for that period it is necessary to provide timely population protection against irradiation. It should be taken into account that in concordance with these Criteria two levels of radiation influence have been established - A and B.

A system of urgent measures was established in such manner that if levels of radiation influence reach level A, but are within the range of A-B, a decision on population protection, including evacuation, is made taking into account concrete situation. However, if prognosis of radiation situation is such that reaching of level B can be real, a decision about evacuation is made immediately. Factual decisions were made with some reserve, if to rely upon reliability of that information which was available at the Governmental Commission and specialists at that acute period of time. In particular, the city of Pripyat' was evacuated when there was threat of reaching only of level A - 250 mSv, and upon completion of evacuation from the whole 30-km zone only little part of population exceeded level A, but did not reach level B, equal to 750 mSv.

Thus, it is possible to declare that for overwhelming majority of inhabitants of the 30-km zone external gamma-irradiation dose did not exceed 250 mSv.

However, by the same Criteria implementation of urgent countermeasures was foreseen for thyroid irradiation dose. Here if threat existed for excession of level of 2500 mSv (level B for thyroid), decision about evacuation could be made indepently. By all means, it was more difficult to use second criterium. Control of thyroid radioactivity or milk contamination during acute period of the accident could not be implemented in such operative and efficient way as external irradiation doses control. That's why mass radioiodine measurements in thyroid glands of inhabitants, including evacuated population, were implemented later.

During recent studies we managed to conduct more detailed and correct reconstruction of external irradiation doses and thyroid irradiation doses. Based on these works there is an opportunity to compare rates of factual received doses with levels pointed out in Criteria for evacuation and to evaluate critically efficiency of decisions made (Fig. 1).

Rates of factual doses are drawn on the Figure by rectangle, and medianes of dose distributions - by black stripe.

On Figure it is seen that maximum values of factual external irradiation doses do not reach level B, whereas by thyroid doses upper limit of 2500 mSv is closed. It means that criterium of evacuation on external irradiation doses for inhabitants of the 30-km zone around the Chernobyl NPP appeared to be less conservative and not adequate to thyroid irradiation doses.

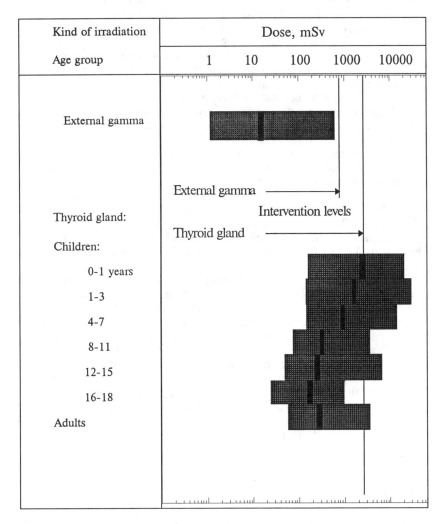

Fig. 1. Comparison of external irradiation doses and thyroid irradiation doses for inhabitants of the 30-km zone around the Chernobyl NPP with levels adopted as criteria for evacuation.

As subsequent events of step-by-step population evacuation firstly from the 30-km zone and then from the Western trace and in the outskirts of the 30-km zone during detection and adjustment of radiation situation, deployment of forces and means of health care system in Ukraine and Belarus was implemented not on basis of previously developed plans and prevention assessment of situation, but following the events. That's why decisions were made in rush, without appropriate calculations and substantiations. It was the main source of errors, failures and delays. As an example we may point out evacuation of inhabitants of some settlements of the Chernobyl raion (district) into other settlements of the same raion with high radionuclides contamination densities.

Traditional information of population about consequences of implementation of nuclear energy for military purposes has caused its negative influence on formation of exaggerated, erroneous imaginations about real danger of radiation situation appeared at population. Whereas population has the most superficial imagination about that fact that not only appearance of life, but also all the evolution of biosphere was accompanied by ionising irradiation.

Elementary knowledge about constant presence of radiation in biosphere, its biological influence, methods of its measurement was not given to each individual, and it is a serious error.

Pointing out huge work made by medical workers for provision of population safety under condition of elimination of consequences of the Chernobyl NPP accident we should not also keep silence about negative phenomeni. Relative well-being in exploitation of atomic energy objects has played its appropriate role in self-calmness appeared at some managers of health care system and medical establishments. System of actions of authorities and establishments of health care system has not been worked out quite clearly under condition of large-scale accident. Even under condition of clinical establishments physicians had no appropriate education in the field of radiation pathology, did not know in appropriate degree methods of diagnostics and treatment.

Not all the establishments of sanitary and epidemiological service were completely ready for activities under condition of large-scale radiation accident. Standardisation documents which regulated order of implementation of control for food products, were mainly oriented on determination of background contents of radioactive substances in them. Radiological service did not have sufficient reserve of dosimetric and radiometric instruments.

On first stage there was no clear coordination in actions of links of all the levels of sanitary and epidemiological service. In most degree medical and sanitarian situation in acute period was complicated by the following conditions:

1. Defects of systems of warning and information. As a result neither population nor management authorities, both local and republican ones, during a few days did not have any information about this event and scale of the accident.

2. Lack of data about irradiation doses of both personnel and participants of the accident elimination and population. External radiation control instruments on the territory of the NPP were broken, and individual dosimeters available, not fitting to high irradiation doses, showed readings out of scale. Neither individual nor group dosimetric

control was foreseen for population, that's why population was not provided with dosimeters.

First preliminary and very oriented assessments and prognoses of irradiation doses were made only in June, 1986.

3. Accident of such scale was not prognosed and expected, that's why all the medical measures in case of the Chernobyl NPP accident were authorised to medical and sanitarian department of the NPP, and engagement of forces of local health care system was not foreseen. In this connection a system and organisational principles of population medical aid was not developed timely, not fitting to scale of evacuation measures and number of population involved into the accidental situation, including both first medical and qualified medical aid. A system of medical service of civil defence and possibilities of its units and also possibilities of medical and sanitarian department of the Chernobyl NPP appeared to be non-adequate to the real situation. That's why main burden of medical provision laid on authorities and establishments of health care system of UKraine and Belarus.

Taking into account this fact urgent necessity of covering quantitative and qualitative deficiency of necessary forces and means of health care for enforcement and specialisation of local health care system had to be satisfied in urgent order, without sufficient substantiation.

Conditions mentioned above sufficiently impeded goal-directed activity of health care system and decreased efficiency of large-scale and laborious measures.

More than 135 000 of population was evacuated into other areas. As a result loading on existing treatment and prophylaxis establishments has exceeded considerably.

4. Very low level of knowledge in the field of radiation emdicine at overwhelming majority of practical physicians, and also at managing staff and specialists of health care authorities. They were limited by superficial study and separate knowledge about injuries by nuclear weapons and standards of war time obtained at lessons on civil defence. It was a source of unsubstantiated anxiety, tendencies to radiation hyperdiagnostics. In combination with low level of knowledge on medical psychology it established favourable environment for wide expansion of radiophobic reactions and iatrogeniae. It was also promoted by deficiency of and defects of social information about the situation.

5. Lack in period of mass population evacuation of standards of acceptable contamination of surface of human body and clothes, and also uncompetence of authorities and establishments of Ministries of Well-Being of republics on organisation of sanitary processing of population, despite that fact that it was foreseen by plans of civil defence. That's why health care authorities, together with mass medical measures delt with functions not inherent to them - sanitary processing of population. The task on provision of population by exchangeable fund of clothes also was solved in rush and gave bulk of anxiety to managers of medical establishments.

6. Unpreparedness to in-time and mass implementation of one of the most efficient measures of radiation protection - iodine prophylaxis,

especially for protection of critical subpopulations of population - children, pregnant and feeding mothers. Basic causes having decreased dramatically efficiency of this measure is lack of sufficient quantities of iodine preparations on spot, late issue of "Temporary instruction onextremum iodine prophylaxis of injuries by radioactive iodine", which was issued on 7, May, 1986.

7. Weak material basis and deficiency of specialists for conduct of complex mass instrumental investigations on determination of incorporated radionuclides, especially such urgent ones as thyroid radiometry. As a result on territory of Belarus time for it was lost.

8. Lack of primary medical documentation adequate to the situation, which disrupted succession in registration of radioactive irradiation and contamination, in giving medical aid. It required appropriate local self-activities and not always considered decisions. Then it impeded analysis and generalisation of investigation results.

Retrospectively assessing medical aspects of the Chernobyl catastrophe it is possible to make the following basic general conclusions:

- despite measures undertaken after Chernobyl on increase of safety of NPP of technical and technological character, so called "primary protection level", NPP are becoming inherent element of general system of safety of nuclear energy, where measures are considered to be priority ones directed on prevention or maximum possible dewcrease of unfavourable influence of the whole complex of accidental situation on people, including organisational and medical measures;

- all the previous accidental plans, not taking into account possibility of the accident of such scale and involvement of such huge amount of population into accidental situation, were oriented on comparatively limited (so called hypothetic) accident and thus did not foresee the whole complex of population protection measures;

- the Chernobyl accident by its scale and mass of population involved in it and also number of participants of its elimination has no precedent and it is difficult to imagine a catastrophe of larger sclae, that's why it is expedient to take expeience of its elimination as a basis for a system of suggested organisational and medical measures of population antiradiation protection;

- its principal points are possibilities of adequate reconstruction of existing health care system with operative, subsequent inclusion of rapid reaction forces of oblast (regional), republican and the All-Union levels for necessary enforcement and specialisation of forces and means of local health care system on territories involved into accidental situation with their timely preparation in potentially dangerous zones around functioning NPP and NPP under construction and designed.

Multiple character of possible versions of accidrnts by scale, radiological peculiarities and dynamics of development in time, and also considerable difference and peculiarities of territorial location of forces and means of health care system in different areas of the country and zones of potential danger do not allo to give recommendations on development of specific accidental plans, but give opportunity to develop organisational principls, general scheme of population protection measures.

Based on everything mentioned above it is possible to suggest the following recommendations.

Timely information of health care authorities about radiaion situation and prognosis of its development have decisive meaning foe goal-directed conduct of medical and sanitarian measures. That's why it is expidient during first day to have preliminary data about borders of contaminated territory, revealing of number and composition of population (number of children, pregnant women, feeding mothers) residing on this territtory etc.

Efficiency of antiradiation measures in decisive degree depends on timely preparation and realisation coordinated under general management of local authorities (councils) from one hand, and appropriate training of population to perception and preparedness to their implementation from the other hand.

The basic ones are:

- sheltering of population, children first of all, pregnant women and feeding mothers in antiradiation shelters or residential houses and industrial premises, children's and medical establishments with their maximum hermetisation, and also limitation of stay of adult population out of hermetised premises, including industrial ones, without necessity and without means of protection of respiratory organs. Doors closed tightly, windows, coverings of ventilation holes, chimneys should be subject of attention of communal services under timely preparation of settlements on territories included into plans of of antiradiation measures. It should be noted that timely sheltering of population in hermetised premises or antiradiation shelters is the most efficient protective measure and can completely prevent inhalation (internal) intake and sufficiently decrease external irradiation dose;

- iodine prophylaxis, for children's population first of all. Consumption of stable iodine preparations in six hours after intake of radioiodine into huma organism can decrease thyroid dose at 50-60%, and in 24 hours conduct of iodine prophylaxis is practically useless. Operativeness of such work is determined in many aspects by timely preparation of reserves of iodine preparations packed for rapid distribution to population and clear order of this distribution taking into account necessary forces distributed by concrete objects which should be corrected periodically.

Evaluating retrospectively medical consequences of Chernobyl it is necessary to recognise that pronounced radiogenic effects are connected just with lack of timely warning, delays and uncomplete involvement by iodine prophylaxis of all people who need it, unpreparedness of population to actions under accidental situation - it everything mainly has determined high thyroid irradiation doses at considerable part of population, children first of all, and also higher external irradiation doses because of stay of bulk of population out of shelters.

Evacuation of population is the most urgent measure and should be conducted only in case if it is reliably known that radiation levels will condition excess of quote of external irradiation dose limit established for first year after the accident. Assessment of other criteria for evacuation within the limits of first day will not be possible.

Here it should be taken into account that extremum evacuation without its sufficient preparation and clear organisation in maximum short terms can lead to considerably higher dose loadings in comparison with timely sheltering of population in hermetised premises with subsequent evacuation.

At least we should point out such important question as establishment of protective means of accidentald esignation, in particular, unified complexes of means of individual protection designated for implementation during conduct of hard accidental works, organisation of serial production of prospective means of individual protection, including filtrating autonomous individual protective means with inflation, universal afcial parts of means of individual protection not limiting vision, disposable special clothes from non woven materials etc.

Methods for Assessing the Extent of Acute Radiation Injury

T. M. Fliedner

Department of Clinical Physiology, Occupational and Social Medicine of the University of Ulm, 89081 Ulm/Do., Germany

1. INTRODUCTION

Previous radiation accidents have shown that the medical management of exposed persons cannot be performed without the use of "biological indicators" of effect and of repair. For the clinical management of a patient with the acute radiation syndrome, it is essential to obtain information on the subjective symptomatology as well as on laboratory parameters, especially during the first 3 to 6 days after exposure. The medical doctor responsible for the clinical care of patients has to rely on the use of what has been described as "sequential diagnosis" (1). This approach consists essentially of the determination of a limited number of parameters as a function of time. From the analysis of the pattern of the determined and evaluated signs and symptoms in the first hours and days, one is able to characterize patients according to type and severity of symptomatology. This has been clearly demonstrated in the Moscow - Ulm Radiation Accident Database (MURAD) developed in a collaborative project between the Institute of Biophysics in Moscow and the Department of Clinical Physiology and Occupational Medicine of the University of Ulm (2). On the basis of the radiation accident clinical response pattern observed early after irradiation, one is able to develop a first approach for therapeutic strategies.

It is the purpose of this contribution to outline the diagnostic and prognostic significance of blood cell changes and to discuss the following problem areas:

1. Significance and elements of a sequential diagnosis
2. Significance of blood lymphocytes for radiation accident diagnosis
3. Significance of blood granulocyte changes for the prognosis of the acute radiation syndrome

4. Analysis of granulocyte changes by means of feedback regulated system models
5. Utilization of indicators of response and repair for planning therapeutic options

2. SIGNIFICANCE AND ELEMENTS OF THE "SEQUENTIAL DIAGNOSIS"

For the medical doctor, who is called to attend a radiation accident patient, it is important and decisive to come as soon as possible to a clinical evaluation of the situation of the patient on the basis of symptoms and laboratory findings. The following questions require an immediate answer: 1. Has a radiation exposure occurred in a patient or are the symptoms characteristic for alternative causes (for instance skin burns, chemical poisenings, physical injuries)? 2. If there is a high probability for the radiation aetiology of the clinical symptomatology, what is the extent of the radiation injury: Does one have to assume the consequences of an injury to the central nervous system, to the gastrointestinal system or to the hemopoietic system?

The answer to these questions can be derived only to a limited extent through the evaluation of the radiation exposure history. More important are the diagnostically relevant laboratory findings, which show and indicate the extent of strain to the human organism. The reason for this basic consideration is very simple: (table 1) In the case of accidental whole body irradiation exposure it is almost impossible to obtain a sufficiently precise evaluation about a "radiation dose". This is not necessarily due to the missing or insufficient dosimetry. Rather this impossibility to determine quickly the exact dose to the organism is due to the fact that the radiation exposure of the human being is usually not uniform and hence not all parts of the body receive the same radiation dose.

The evaluation of radiation accidents from all over the world (2) indicate that a homogeneous total body exposure usually does not occur in the real world. Rather radiation accidents result as a rule in an inhomogeneous radiation exposure. The pathophysiology of the radiation damage indicates that any inhomogeneity of the exposure of the human being works in favour of the regeneratory potential of the organism. In the clinical setting an estimated radiation dose does not contribute significantly to the consideration of the therapeutic approaches to be taken.

A "sequential diagnosis" supports the attempt to obtain information about the extent of the injury to the critical organ systems within the first hours and days after an acute radiation exposure in order to derive from this assessment conclusions for the therapeutic management. It is not the purpose of a sequential diagnosis to calculate an "exposure dose". The determination of an exposure dose is usually only possible with the help of radiation accident reconstructions. Such dose measurements and estimates may take months or even years or are debatable for many decades as seen in Hiroshima and Nagasaki.

The relationship between the assessment of stress and strain in the case of radiation accident medical management is symbolized in table 1.

The organ system that is most important for executing a meaningful sequential diagnosis in the case of an acute radiation exposure, is the blood cell formation. A second organ system that is of great importance for assessing the consequences of radiation exposure is the skin. In case of skin injury (for instance thermal or radioactive exposure of Chernobyl patients) it is essential to not only assess the effects on the blood cell forming tissue, but also the extent of damage to the skin. In these cases the diagnosis and treatment of skin injuries must take preference and requires classical approaches in order to assure the immediate survival of the patient.

As far as the assessment of injury to the blood cell forming tissues is concerned, one has to find as soon as possible an answer to the following question: Has the stem cell pool of blood cell formation been injured "irreversibly" or "reversibly"? If the damage to the stem cell pool would be diagnosed as being "irreversible", then in principle a stem cell transplantation, be it autologous or allogeneic, needs to be considered. If the damage is most likely "reversible", then one would have to answer the question in what way therapeutic measures are capable to bridge the expected transitory hemopoietic failure and/or to use possibilities to shorten such a period of hemopoietic failure. In parallel to the evaluation of the blood cell forming tissues, one would naturally have to consider other important organ systems, such as the gastrointestinal system, the metabolic organs, but also the lung, the cardiovascular system and the central nervous system. Based on the results of the hematological "pacemaker"-evaluation, it is quite possible and the rule that other special diagnostic and therapeutic measures are indicated.

3. THE SIGNIFICANCE OF BLOOD LYMPHOCYTES IN A "SEQUENTIAL DIAGNOSIS"

If one is analyzing blood lymphocytes as a function of time after an acute radiation exposure, one can after exposure give an answer to the question "Has a radiation exposure occurred and what clinical symptomatology might be expected?" within a few hours. In figure 1 a-d the lymphocyte curves of 52 persons that were exposed in radiation accidents can be seen. One fact is common to all curves: Within less than 10 days a new level of lymphocyte concentration is reached that does not change drastically in the next days or weeks. In other words, the regeneration is slow. If one correlates the lymphocyte patterns with the severity of the clinical courses observed, one can show that the clinical courses of the patients can be correlated with a particular lymphocyte pattern.

In the case of a "very severe" clinical course one can show that there is a decrease of blood lymphocytes within 2 to 4 days to concentrations of less than $200/mm^3$ blood (group d). In the case of "severe" forms of the acute radiation syndrome, the lymphocytes fall within 4 days to values between 200 and $800/mm^3$ blood (group c). In the case of "moderate" effects (group b) the lymphocytes decrease to concentrations between 800 and $1.100/mm^3$ blood. If there is a minimal injury, then the lymphocyte changes are very unspecific, but as a rule do not decrease below $1000/mm^3$ blood (group a). Therefore, it is possible to derive from this assessment that lymphocyte changes in the blood beyond 10 days after the radiation accident are of low diagnostic value and cannot necessarily predict whether a spontaneous recovery can be expected or not. The regeneration of the lymphocyte system can be shown in all radiation accident cases to take many weeks or even months.

However, a primary diagnostic answer is possible within hours after accidental radiation exposure. One can say with high probability whether a clinically significant radiation exposure has occurred or whether one is dealing most likely with a moderately severe course of the acute radiation syndrome. In figure 2 one can see the lymphocyte course in patients that were exposed in major radiation accidents. In the case of the radiation exposure of Marshall Island victims (1954) the slow lymphocyte decrease in the first 3 days indicates a light or moderate degree of radiation effects on the victims. In case of the accident in Los Alamos (1946) one can see that the lymphocyte concentration decreases within 1 day to less than 10 % of normal. In this case, the patient died within a few days from severe hemopoietic failure. The Oak Ridge patients and the Chernobyl patients shown in the graph had a moderate decrease within the first 3 days to about 30 % of normal. In these cases a spontaneous hemopoietic recovery occurred.

In conclusion it is suggested that the lymphocyte changes in the blood as a function of time after accidental whole body radiation exposure allow one to predict with high probability whether there is a causal relationship between the clinical symptomatology and the radiation exposure and whether a very severe or a more moderate course of the radiation syndrome can be expected.

The reason for the fact that one cannot draw clinically significant prognostic conclusions from the initial lymphocyte decline, is seen in the fact that an initial decrease of the lymphocyte count may also be observed after an inhomogeneous or large volume partial body irradiation. Under these circumstance a spontaneous recovery of the hemopoietic system is quite possible, regardless of the fact that a severe damage of several organ systems has to be assumed. From the viewpoint of pathophysiology the lymphocyte pattern is not only determined by the radiation sensitivity of these cells, but also by the sensitivity of the lymphocyte recirculation from blood via the lymph to the lymph nodes and back to the blood.

4. THE SIGFICANCE OF BLOOD GRANULOCYTES IN THE "SEQUENTIAL DIAGNOSIS"

The changes of granulocyte concentration in the peripheral blood as a function of time after a single acute radiation exposure have found to be an important, if not decisive indicator for the performance of a sequential diagnosis. The granulocyte changes can be considered as the key indicator for determining the prognosis of a patient and for the therapeutic planning. It should be mentioned at this point that the assessment of changes of platelets, of erythrocytes or reticulocytes result in a very similar evaluation in comparision to granulocyte changes. However, granulocyte changes can be determined more quickly and precisely even with automated equipment and are therefore the preferred indicators.

For the sequential diagnosis the granulocyte concentrations in the first 6 days after radiation exposure are of particular importance. If one finds an initial granulocytosis with values well above 10.000 mm^3 in the first 2 days and especially, if one finds beween the 4th and the 6th day after exposure, a progressive decrease of cell numbers to values below 200 mm^3 blood one can take this pattern as an indicator for the most severe form of the acute radiation syndrome resulting in an essentially "irreversible damage" to the stem cell system of hemopoiesis, which is distributed in the skeletal system of the organsim (see fig. 3).

If, however, one finds in the first 6 days after radiation exposure an initial moderate granulocytosis (values most likely between 5.000 and 15.000 mm^3 blood) and if one finds beyond the 4th day a moderate decrease of granulocytes to values between 200 and 1.000/mm^3 blood (which can be clearly depicted, if one is performing a graphic demonstration of granulocyte values), one can predict significant damage to the hemopoietic system. However, such a pattern is compatible with a reversible damage to the hemopoietic stem cell system. In this case, one would aim for a therapy of "bridging the hemopoietic failure" or "stimulating" hemopoiesis for a more enhanced regeneration. The analysis of radiation accident case reports, which are collected in the Moscow - Ulm Radiation Accident Database (MURAD), indicates that there are very clearcut granulocyte patterns to be associated with the notion of a "reversible damage" of the stem cell pool (2) (see fig. 4).

In these cases of a severe form of the acute radiation syndrome with however reversible blood formation damage (fig. 4) one can recognize a granulocyte pattern which is characterized by an initial granulocytosis, a moderate, but not critical granulocyte decrease towards the 10th day, a transitory increase or a "inbetween plateau" until the 20th to the 25th day and by a final regeneration beyond the 30th day after exposure. In the patients with an even lighter form of the acute radiation syndrome the pattern of decrease, the abortive regeneration, the inbetween plateau and the regeneration beyond the 30th day is very characteristic. This course is compatible with a significant, but nevertheless benign course of the acute radiation syndrome. The patients with a light form of the acute radiation syndrome show such a pattern only in a mild form. A statistically significant difference between the groups can most likely be obtained, if one evaluates the blood counts about 30 days after exposure.

A chromosomal analysis in such patients is of importance not necessarily to calculate a "dose", but to document conclusively that irradiation exposure has occurred and not alternative causes of injury. Therefore, under any circumstances, a chromosomal analysis should be attempted even for forensic reasons.

In conclusion one can state that lymphocyte changes after total body radiation exposure can be considered as an important "retrospective indicator" (Does one have to assume a significant radiation exposure or not?"), but that granulocyte changes (and in parallel the changes of platelets and reticulocytes) need to be considered as a "prospective indicator" of effect and repair. This is most important for the planning execution and continuous evaluation of the therapy. The pathophysiological justification for this conclusion is given in other publications.

5. QUANTIFICATION OF GRANULOCYTE COURSES WITH THE HELP OF FEEDBACK REGULATED SYSTEM MODELS

The question was whether it is possible to simulate the granulocyte changes after total body exposure with the help of a biomathematical systems model. Such a quantification and characterization was considered to be decisive in order to determine the number of "virtual" stem cells in the bone marrow necessary to produce a sufficient number of granulocytes to keep their level in the blood constant. Such a quantification of the stem cell number would be essential to predict whether a autochthonous regeneration of the stem cell pool and thus, of the entire hemopoietic system can be expected or not. In other words, if it is possible to assess the damage to the hemopoietic system is reversible or irreversible.

The basic model of granulocytopoiesis was developed in several steps due to an intensive collaboration between the Department of Clinical Physiology and the Department of Measurement Regulation and Microtechniques of the University of Ulm (3). The model (fig. 5) that is presently used consists of 7 cell and 2 regulatory compartments and is based on the assumption, supported by experimental evidence, that in a homeostatic equilibrium between cell production and cell removal, which exist in hemopoiesis, each granulocyte that is moving out of the blood by ageing or immigration is replaced by a granulocyte from the bone marrow (compartment F). For each granulocyte delivered to the blood stream one has to assume a netto increase of one cell by cell division (compartments S, CBM, P). One knows the life span of the granulocytic precursor cells and the frequency of cell division and one has evidence for a humoral regulation of the entire system. The present system considers 2 regulatory compartments REG I und REG II. All in all it is the question whether the stem cell compartment is capable to replace the cell loss in the blood by cell production in the proliferative compartments. In order to model such a system, 37 differential equations were necessary. The details of this biomathematical compartment model were published elsewhere (3). The pattern of granulocyte changes in the different categories of the course of the acute radiation syndrome can be simulated, if one assumes that there is not only a destruction of stem cells in the system by radiation exposure, but that some of the stem cells are injured, but apparently capable of repair. It has been shown previously that radiation response data of the hemopoietic system can only be explained, if one assumes that the radiation exposure results in stem cells that are completely repaired and hence are able of an unlimited replication and differentiation and other stem cells in which the replicative capability is limited and hence the clon may die out. These assumptions were published and discussed elsewhere (4). With the help of such a biomathematical simulation model it is now possible to assess the damage of radiation to the stem cell pool on the basis of granulocyte changes. Each granulocyte pattern can be correlated to a virtual stem cell number. Important is the number of uncommitted virtual stem cells (S) that are capable to initiate a final granulocyte recovery beyond 25 to 30 days.

In fig. 3 the radiation accident granulocyte response patterns are simulated in which the granulocyte values decrease within 5 to 6 days to minimal values (below 300 per mm^3 blood). This pattern, as indicated before, is correlated with an irreversibility of the damage to the stem cell pool.

If one takes a look at the course of the Sor-Van accident and finds that the granulocyte regeneration was found beyond the 14th day, one can underline the thesis: in this particular case there was a bone marrow cell transfusion on day 5, which led within 10 days to the signs of a "take". In the accident case Moscow 1991 one did not want to use a bone marrow transplantation. This patient died 3 months after the accidental exposure inspite of intensive cytokine therapy and without a sign of a permanent regeneration.

In fig. 4 four curves are demonstrated which are typical for severe, but nevertheless reversible courses of the radiation accident syndrome.

In table 2 the numbers of virtual stem cells that were calculated on the basis of the model are demonstrated. One can see that the values for intact stem cells should be above 6/10.000, in order to be able to predict a reversible course of the acute radiation syndrome. Values below 6/10.000 are compatible with an irreversible damage of the stem cell system. This evaluation can be made within a very few days (maximally 5 to 6 days after radiation exposure) and forms a rational basis for further diagnostic and therapeutic measures. It goes without saying that, in addition, one has to consider the status of other organ systems, such as CNS, heart, circulation, renal system, skin and so on. It is of course necessary to perform more research in order to develop this model and models for other cell systems even further in order to be able to improve clinical decision making in radiation accidents.

It should be mentioned that an examination of the bone marrow by particle smears and by histological section can greatly assist in the evaluation of the radiation exposed patient. The bone marrow smear will show within 12 to 24 hours the effects of the radiation exposure on marrow cells. This requires, however, bone marrow particle smears and this should be done by an experienced hematologist. He would distinguish between cell pyknosis and cell edema from cells with mitotically connected abnomalities (.....). One would also expect that within this time one would find an excessive increase of apoptotic cells.

Most recently, further studies in our group have shown that the granulocyte regeneration simulation can also be used in stem cell transplanted patients. Under these circumstances one can find that granulocyte regeneration patterns are directly proportional to the number of CD34$^+$ cells transfused. Thus, one has now a way to correlate the biomathematically calculated "virtual stem cells" with the number of remaining intact stem cells as determined by biological means (CD34$^+$ cells essential to be in a transfusate to induce hemopoietic regeneration).

6. CONSEQUENCES FOR THE PLANNING OF THERAPY

It is not the purpose of this presentation to discuss extensively the present therapeutic concepts for the treatment of the acute radiation syndrome. However, it is important to point out the following facts: If the diagnosis is that of an "irreversible damage" to the stem cell system based on the analysis and the calculation of the granulocyte course, it is highly recommended to treat such patients with stem cell transplantation. This should be done as soon as possible. Of large significance would be an autologous stem cell transplantation. However, this is of course only possible, if one would take blood stem cells by means of continuous cell centrifugation before radiation exposure. This leads to the question whether it might be advisable to convince "persons at particular risk" to set up a "stem cell bank" of their own stem cells. Such a approach might be used in persons who may have an increased risk of accidental exposure due to clean-up or rescue operations.

However, as a rule, one has to find appropriate stem cell donors essentially similar to the approach that is being used in the oncological services for the therapy of leukaemic patients or of patients with other systemic neoplasias. Related donors are preferred to non-related donors. In general, the treatment of such an "irreversible damage" to hemopoiesis is identical to the treatment used in cases of severe aplastic anemia.

If the initial diagnosis with the help of the analysis of the granulocyte pattern as an indicator shows that hemopoiesis was damaged, but that a spontaneous regeneration might occur ("reversible damage"), one has to discuss in what way one can bridge the transitory pancytopenia and shorten it.

Such a bridging therapy aims at the prophylaxis and therapy of granulocytopenia as a course of infection. This results in the recommendation of the treatment of such a person in the isolationbed-system under germfree conditions (gnotobiotic therapy) and to the treatment of these patients after careful evaluation of the sensitivity of the microorganisms to the therapy with antibiotics, antimycotics and virostatic substances. Most recently one is trying to shorten the phase of transitory pancytopenia. This can be done by the application of hemopoietic stimulation factors, for instance IL-3. It can be shown that these factors are capable of shortening the duration of pancytopenia in cases of pancytopenic conditions. This therapeutic form is only effective, if the stem cell pool contains a sufficient number of intact hemopoietic stem cells that have survived the radiation exposure. Only, if there is such a sufficient number, then this pool can react to such recombinant factors. One has, however, also to consider the problem of stem cell competition.

If one considers the platelet course, then the therapeutic approaches have to result in a platelet concentration above 15.000 - 20.000/mm^3 blood. This can be achieved by platelet transfusion, with histocompatible platelets. Such a thrombocyte transfusion therapy has to be continued until the spontaneous regeneration of hemopoiesis has occurred. At the present time many studies are on their way to try to shorten the period of thrombocytopenia by the application of recombinant stimulatory factors. One can show that the factor IL-3 alone or in combination with IL-6 may play a very good role.

7. CONCLUSIONS AND SUMMARY

The questions posed at the beginning of this contribution can be answered - in summary - as follows:

1. Using the approach of "sequential diagnosis", it is feasible to assess the damage to the organism, its organs and functional systems after accidental whole body radiation exposure within a very few days (4-6 days at the most) to such an extent that a prognosis is possible as a basis of planning the therapeutic options. A "sequential diagnosis" is based on the determination of a few biological indicators of effect and repair as a function of time (for instance every 6 hours for 6 days) and their evaluation on the basis of the pathophysiology of the acute radiation syndrome.

2. The canges observed in the concentration of lymphocytes in the peripheral blood can be used only as a "retrospective" indicator of effect: If there is a decrease of lymphocytes below 50 % of normal within 24 hours after exposure, one has to assume the development of a severe course of the acute radiation syndrome. However, one cannot predict whether a spontaneous hemopoietic recovery can be expected or not.

3. In contrast, the pattern of granulocyte concentration changes in the blood stream can be considered to be the most important "prospective" indicator of effect and repair. Within 4 - 6 days after an exposure, the pattern of

granulocyte changes and the extent of decrease between days 4 and 6 allows one to predict whether the blood cell forming bone marrow has been damaged to an essentially "irreversible" extent or whether the clinical course is most likely a "reversible" one. The therapeutic measures to be taken are quite different based on such an assessment.

4. Using a biomathematical simulation model of granulocytopoiesis it has been possible to determine the number of "virtual" stem cells that have remained after radiation exposure in the hematopoietic tissue and that can be associated with a particular granulocyte response pattern. Due to the collection of clinical case histories of radiation accident patients in the Moscow-Ulm Radiation Accident Database (MURAD) it was possible to determine the number of "virtual" stem cells that are essential to allow an autochthonous hemopoietic recovery. This number can be used as an important prognostic indicator.

5. If the number of remaining intact "virtual" stem cells is below ... % of normal, the hematopoietic damage is essentially irreversible. Under these circumstances a stem cell transplantation has to be performed using the same approach as in the treatment of severe aplastic anemia. If the number of remaining intact "virtual" stem cells is well above 95 % of normal, one can assure that appropriate cytokines can act and shorten significantly the autochthonous hemopoietic recovery. Even under these circumstances, a gnotobiotic therapy in a germfree setting and systemic antibiotics and platelet transfusions may be essential.

REFERENCES

1) Fliedner, T.M.: Strategien zur strahlenschutzmedizinischen, ambulanten Versorgung von "Betroffenen" bei kerntechnischen Unfällen. In: Medizinische Erstmaßnahmen bei kerntechnischen Unfällen. Georg Thieme Verlag Stuttgart 1981.

2) Baranov, A.E., Densow, D., Fliedner, T.M. and H. Kindler: Clinical pre computer proforma for the international computer database for radiation exposure case histories. Berlin, New York (...): Springer, 1994. ISBN 3-540-57596-0 (Berlin), ISBN 0-387-57596-9 (New York).

3) Hofer, E.P., Tibken, B. und Fliedner, T.M.: Modern Control Theory as a Tool to Describe the Biomathematical Model of Granulocytopoiesis. In: Möller, D.P.F. und Richter, O. (Hrsg.): Analyse dynamischer Systeme in Medizin, Biologie und Ökologie, 4; Ebenburger Gespräch, Bad Münster, April 1990, Informatik-Fachberichte, Springer Verlag, Berlin 275 (1991) 33-39.

4) Fliedner, T.M., Weiss, M. Hofer, E.P., Tibken, B. und Y. Fan: Blutzellveränderungen nach Strahleneinwirkung als Indikatoren für die ärztliche Versorgung von Strahlenunfallpatienten. In: F. Holeczke, Chr. Reiners, O. Messerschmidt: Strahlenexposition bei neuen diagnostischen Verfahren. Biologische Dosimetrie - 6 Jahre nach Tschernobyl. Strahlenschutz in Forschung und Praxis, Band 34, 137 - 154, Gustav Fischer Verlag Stuttgart, 1993.

Table 1

Stress	Strain
(as a consequence of Ionizing radiation)	
Exposure	Biological consequences (CNS, organ systems)
(external, internal)	influencing variables
Type	Genetic factors
Quality	Age
Duration	Previous health impairments
Rate	Sex (?)
Etc. (to be determined by	Etc. (to be determined by physical examination
physical and chemical	and tests to evaluate extent of reparable and
indicators)	irreparable damage)

———————→ !

←——— //——— ?

Table 2

	Remaining Intact Stem Cells % (Cell Number)	Remaining Injured Stem Cells % (Cell Number)	Destroyed Stem Cells % (Cell Number)
Patient ID: 4	0,06 $(7,5 \cdot 10^5)$	5,6 $(7,0 \cdot 10^{10})$	94,34
Patient ID: 5	0,26 $(3,25 \cdot 10^6)$	8,0 $(1,0 \cdot 10^8)$	91,74
Patient ID: 1	0,0004 $(5,0 \cdot 10^3)$	5,28 $(6,6 \cdot 10^7)$	94,72
Patient ID: 3	0,0006 $(7,5 \cdot 10^3)$	9,12 $(1,14 \cdot 10^8)$	90,88
Brescia Case	0,0 (0)	0,0 (0)	100,0
Norway Case	0,0 (0)	0,0 (0)	100,0
Sor-Van Case	0,0 (0)	0,0 (0)	100,0
Moscow Case	0,0 (0)	0,001 $(1,25 \cdot 10^4)$	99,999

Figure 1

LYMPHOCYTES AFTER ACUTE IRRADIATION
Mean and Standard Deviation (n = 52)

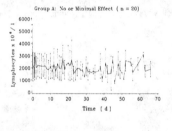

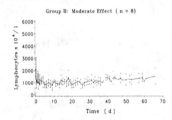

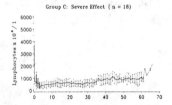

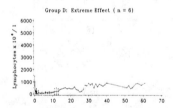

Figure 2

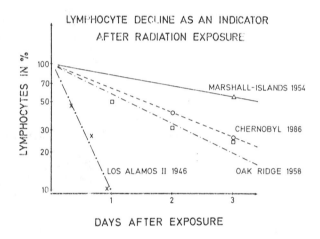

LYMPHOCYTE DECLINE AS AN INDICATOR
AFTER RADIATION EXPOSURE

Figure 3

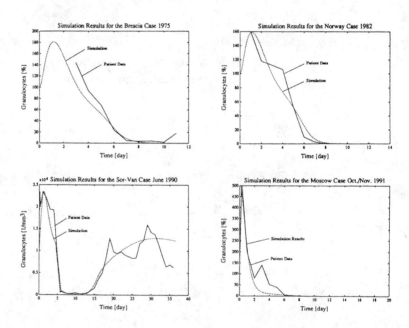

Figure 4

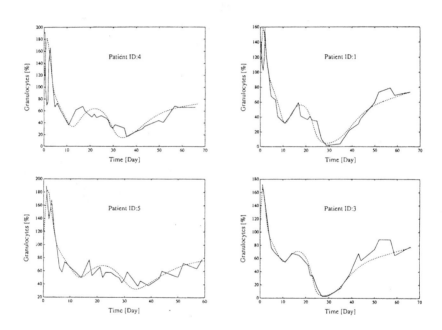

Figure 5

MODEL OF GRANULOCYTOPOIESIS

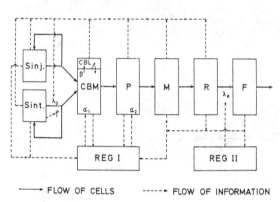

FLOW OF CELLS ----► FLOW OF INFORMATION

New Developments in the Treatment of the Acute Radiation Syndrome.

Gerard WAGEMAKER

Institute of Hematology, Erasmus Universiteit Rotterdam, The Netherlands

Abstract. Radiation accidents associated with the use of nuclear power, radiation devices and industrial applications have resulted in a number of individuals with significant partial or total body exposures, in a limited number with lethal outcome. Such exposures generally result in life-threatening suppression of bone marrow and immune functions and may be accompanied by high doses to the skin and underlying tissues. In recent years, new methods of diagnosis and treatment of such patients have entered development, while experience with relatively large-scale accidents, such as the Chernobyl cases, have clearly demonstrated the limitations of previous approaches. In particular bone marrow transplantation has met with limited success, mostly due to the inhomogeneous nature of accidental exposures as well as the frequent occurrence of other injuries other than bone marrow damage. Present day technology allows for direct estimates of the number of residual bone marrow stem cells to enable a rapid decision on the type of treatment indicated, while some twenty-five recombinant hemopoietic growth factors are under investigation to accelerate the recovery of white blood cells and platelets, as well as immune functions. If applied appropriately, such growth factor treatment will successfully replace bone marrow transplantation in the majority of accident cases. Advances in hemopoietic stem cell biology, both in terms of source of stem cells as well as their isolation, will on the other hand make stem cell infusion a much less risky operation than bone marrow transplantation has been in the past.

1. Introduction

Radiation accidents, which have occurred during the last half century associated with the use of nuclear power, radiation devices and industrial applications of ionizing radiation, have resulted in significant partial or total body exposures with lethal outcome in a number of individuals. Exposures resulting from these and similar accidents are usually associated with non-homogenous or partial body irradiation, resulting in suppression of bone marrow and immune functions and with high, sometimes extreme, doses to parts of the skin. Bone marrow and immune functions are the most radiation sensitive vital functions of a mammalian organism and diagnostic procedures must be rapidly deployed, followed by a detailed plan of treatment in conjunction with the treatment of other injury.

A number of radiation accident victims have been treated with bone marrow transplantation, originally thought to be a life-saving therapy for severe radiation bone marrow damage. Although this may be a valid and successful approach in modern cancer treatment, it has met with only limited success for radiation accident victims. This failure is due to the inhomogeneous nature of accidental exposures, resulting mostly in insufficient immunosuppression for graft acceptance, as well as to the frequent other injuries, either or not radiation inflicted, characteristic for the accident situation. The emergence and large-scale availability of recombinant cytokines, in particular the hemopoietic growth factors, which stimulate the recovery of bone marrow hemopoietic of stem cells and the production of mature blood cells, have conceptually largely replaced bone marrow transplantation in the treatment regimens for radiation accident victims, although only very limited experience exists. In particular, the optimal combination of hemopoietic growth factors required has not been worked out in detail and has neither been tested for efficacy nor for risks in large outbread experimental animal models. However, it is certain that at relatively high exposures to radiation the number of residual hemopoietic cells will be insufficient for an adequate growth factor response, indicating that following high doses of irradiation only transplantation of hemopoietic cells would be life-saving.

2. Radiation sensitivity of hemopoietic stem cells

It has been recognized that immature hemopoietic stem cells are heterogenous with respect to radiation sensitivity, the most immature stem cells, which are responsible for long-term hemopoietic and immune reconstitution, being less sensitive to radiation than previously thought on the basis of LD_{50} data and the classical spleen colony test. (1,2). Using the LD_{50} for total body irradiation X-rays, autologous or syngeneic bone marrow rescue numbers and estimates of total bone marrow cell numbers in different species, D_0 values between 0.5 and 0.75 Gy were derived (3). Although this approach may still be valid to characterize the radiation sensitivity of hemopoietic cells required for rescue from radiation induced bone marrow failure in an untreated patient, more recent evidence has accumulated to demonstrate that the heterogeneity of immature hemopoietic cells is also reflected by a heterogeneity in radiation sensitivity with practical consequences for the treatment of radiation accident victims.

In rhesus monkeys (4,5), we noted some ten years ago survival from "lethal" total body irradiation without or with very small numbers of autologous bone marrow grafts, provided adequate supportive care was given, consisting of gastrointestinal decontamination, fluid and electrolyte substitution, treatment of infectious complications and (irradiated) thrombocyte and whole blood transfusions (6). In monkeys which had not received autologous bone marrow grafts, hemopoietic reconstitution occurred within 6 weeks to 2-3 months. Although some of these monkeys died eventually from infectious complications and interstitial pneumonitis, all reached normal values of granulocytes, lymphocytes, monocytes and red cells. Characteristically, in some of these monkeys, similar to those which rejected allogeneic grafts and recovered from endogenous hemopoietic stem cells, longstanding thrombopenia was observed. Thus, the effect of supportive care on 100-day mortality of rhesus monkeys is virtually indistinguishable from that of autologous bone marrow grafts.

This observation prompted questioning of the values reached for the radiation sensitivity of hemopoeietic stem cells calculated on the basis of the LD_{50} for the bone marrow syndrome and autologous rescue numbers of bone marrow cells, as the D0 values derived on this basis would require that a rhesus monkey has at least a log more bone marrow cells than could possibly be contained in a 3 kg monkey, Indeed, when, following graded doses of TBI, the endogenous regeneration time of reticulocytes, found to be the most reliable cell type for quantitative for evaluation of hemopoietic reconstitution (6), was compared with that following transplantation of graded numbers of bone marrow cells in lethally irradiated monkeys, we could establish a dose-effect relationship in the dose range of 5 - 10 Gy and calculate a D0 value, which was estimated to be approximately 1.3 Gy for 6 MV X-rays and 1.1 Gy for 300 kV X-rays (7). However, the extrapolated N_0 was significantly lower than the most reliable estimates for the total number of rhesus monkey bone marrow cells, on which basis we had to assume that in the lower dose range of 1 to 5 Gy, hemopoietic reconstitution is mainly derived from a more radiosensitive cell type. In fact, such a cell type is perfectly in agreement with the LD_{50} for the untreated bone marrow syndrome and the D_0 value of 0.5 - 0.75 Gy reported earlier (3). We concluded that, even following doses of TBI which are lethal due to gastrointestinal damage, sufficient protracted hemopoietic stem cells survived for ultimate hemopoietic reconstitution, be it at the expense of a prolonged period of profound pancytopenia in addition to longstanding immunodeficiency.

Several murine studies, which enabled cell separation and syngeneic transplantation experiments with congenic markers, have established the cellular basis for these findings, in that level and kinetics of peripheral blood reconstitution following total body irradiation appeared to be dependent on functionally distinct stem cell subsets (8-11) with differential sensitivity to ionizing radiation.. Cells with marrow repopulating ability which generate many secondary spleen colony-forming cells, appeared to be considerably less sensitive to X-rays than the more mature colony-forming cells (12). The D_0 value derived for the MRA cells ranged between 1.1 and 1.2 Gy, strikingly similar to the independently derived D_0 value for cells from which hemopoietic reconstitution originated in rhesus monkeys exposed to 5-9 Gy X-rays. This was similarly shown for long-term repopulating cells in comparison to spleen colony-forming cells (13) and confirmed by transplantation in mice in which donor and recipient cells could be distinguished by a congenic marker (14). In the latter studies, the ratio a/ß of the linear quadric (LQ) radiation response model (15) demonstrated that the radiation sensitivity of the more immature, long-term repopulating cells are characterized by a low a/ß. This indicates that these immature stem cells have a considerable capacity to repair radiation damage, which was confirmed by fractionated irradiation (14), in contrast to the spleen colony-forming cells, which are characterized by a high a/ß.

Table 1. Radiation sensitivity of hemopoietic stem cell subsets

stem cell type	D_0 (Gy)	α/β (Gy)
short-term, rapidly regenerating	~ 0.7	~ 10^2
long-term, slowly regenerating	1.1 - 1.3	~ 5

Thus, these results demonstrate that the heterogeneity of immature stem cells is also reflected in a heterogeneity of radiation sensitivity among the subsets, the most immature hemopoietic stem cells, from which long-term hemopoietic reconstitution originates following cytotoxic reduction of bone marrow cells, being considerably less sensitive to radiation and characterized by a much greater repair capacity than assumed until recently (Table 1). The reappraisal of the radiation sensitivity of immature hemopoietic stem cells may have several consequences for the future of whole body irradiation accident victims. It is likely that even following doses of exposure approaching lethally from gastrointestinal damage, sufficient numbers of hemopoietic stem cells will remain available for endogenous hemopoietic recovery. Treatment by allogeneic bone marrow grafts may therefore aim more at temporarily abridging pancytopenia than a replacement of damaged endogenous bone marrow. This may also very well mean, that in the high dose range of exposures just compatible with survival, hemopoietic growth factor treatment should aim at acceleration of stem cell regeneration rather than at an accelerated production of peripheral blood cells, which may in this situation have adverse rather than beneficial effects.

3. Transplantation of hemopoietic cells: identification of stem cells

Bone marrow transplantation finds wide-spread application as a routine procedure in hematology/oncology centers as treatment for a variety of acquired and hereditary diseases, although the majority of transplant recipients are leukemia patients. Basic to the further progress and saftely of bone marrow transplantation is the analysis and identification of the cells exclusively responsible for hemopoietic reconstitution, i.e., the hemopoietic stem cells. Immature hemopoietic stem cells are known to be heterogenous in repopulating capacity, physical properties, hemopoietic growth factor receptor expression and responsiveness, and sensitivity to ionizing radiation (9-13, 16-22). It has become well established that among immature stem cells, a subset is capable of rapid, but transient repopulation following high dose total body irradiation and another subset appears to be responsible for sustained hemopoietic reconstitution. For the future development of the diagnosis of stem cell damage following whole body exposure to ionizing radiation, it is of importance to establish methods that directly measure the extent of cell reduction and damage at the level of the most immature stem cell subsets, in particular those that have the life-saving rapid reconstitution capacity. For this purpose, methods need to be developed that distinguish immature stem cell subsets and identify those which have rapid and those with sustained repopulating capacity. In addition, effective and safe hemopoietic growth factor therapy is probably not only dependent on growth factors that stimulate the reconstitution of granulocytes and thrombocytes, such as G-CSF and thrombopoietin, but also on those growth factors which influence reconstitution of hemopoietic stem cell numbers, in particular the type of stem cells required for sustained hemopoietic reconstitution.

Our preclinical studies in subhuman primates are directed at identification of immature stem cells by surface antigens and growth factor receptors in an attempt to identify those cells that are best suitable tor rapid reconstitution, as well as the design of optimal hemopoietic growth factor combination therapy to be used in case of accidental over-exposure to ionizing radiation. Although much of such analyses can be done by in vitro methodology, the ultimate test for pluripotent stem cell capacities, as has been well established by mouse studies, can only be provided by transplantation experiments, preferably in so called competition assays, in which transplanted cells compete with residual endogenous hemopoietic stem cells following TBI (19). If such an approach is combined with selection on the basis of surface hemopoietic growth factor receptors, it would in principle be possible to establish which growth factors are relevant to accelerate reconstitution of hemopoietic stem cells.

The heterogeneity of hemopoietic stem cells has been best studied in mice, although the results have not been unambiguous between laboratories (23-28). We demonstrated recently, that stable multilineage hemopoietic chimerism, using sublethally irradiated a-thalassemic recipient mice for easy longitudinal monitoring of donor-type red cells, originated from a stem cell subpopulation devoid of the majority of CFU-S-12 (19). It was simultaneously shown that the highly purified day-12 CFU-s have only short term (i.e. weeks to a few months) repopulating capacity. Following accidental high dose whole body exposure to ionizing radiation, one might argue that the latter, more mature stem cell population, might be more important for rescue from bone marrow failure than the former, which, either by their low numbers or by their slow activation, gave rise to sustained regeneration only after months. Obviously, damage to both subsets should be assessed in case of accidental whole body exposures. If similar subsets can be isolated from human donor bone marrow, one may contemplate to transplant allogeneic short term repopulating stem cells to abridge a period of prolonged pancytopenia in cases of accidental high dose exposure.

In rhesus monkeys, we established earlier that autologous, CD34 positive cells reconstitute hemopoiesis as effectively as transplanted unfractionated bone marrow cells and demonstrated quantitatively that the repopulation capacity of bone marrow grafts resides in CD34 positive cells. CD34 positive cells are heterogeneous and represent one to a few per cent of bone marrow cells (29), whereas in mice the frequency of the long term repopulating stem cell subset has been estimated at 1 to 2 per 10^5 bone marrow cells (19). Although transplantation of CD34 positive cells in the present clinical practice may be considered as a major advance, isolated CD34 positive cells are most likely still a factor 10^3 remote from pure stem cells. Further subfractionation of CD34 positive cells is therefore necessary to establish markers (i.e. the surface antigen and growth factor receptor phenotypes) for immature stem cell subsets relevant for radiation protection.

A marker associated with activation and, thereby, with maturity of stem cells, is the major histocompatibility antigen RhLA-DR. Three subsets of CD34 positive cell can be distinguished on the basis of DR staining, of which the CD34 bright, DR dull cells contain, apart from committed progenitor cells, immature multipotent cells as established by single cell cultures (22) as well as transplantation experiments. Transplantation of limited numbers of the latter fraction in lethally irradiated rhesus monkeys resulted in rapid hemopoietic reconstitution, which quantitatively related very well to hemopoietic reconstitution following transplantation of unfractionated bone marrow (unpublished observation). This finding might indicate that most, if not all of the repopulating capacity of a bone marrow graft is contained within the tiny CD34 bright, DR dull fraction, which contains about 0.05% of all bone marrow cells. An experiment in which CD34 bright, DR bright cells were transplanted, did not result in rapid hemopoietic reconstitution; instead, the monkey died aplastic. This preliminary experiment strongly corroborated the notion, that the DR dull fraction among the CD34 positive cells contain both the rapid as well as the long term repopulating stem cells, although the latter can formally only be assessed in an autologous setting following gene marking. Further subfractionation focussed therefore on the CD34 bright, DR dull cells.

Presently, we have assessed the growth factor receptor distribution of GM-CSF (21), IL-3 (22, 30), Il-6 (20) and c-Kit. This was done by directly binding the biologically active, biotinylated ligand to the functional, (hetero-)dimeric receptor, and subsequent binding of fluorescently-labelled streptavidin. This enabled flow cytometric receptor detection with the advantage of a simultaneous identification of subsets on the basis of more conventional markers, and avoided detection of low-affinity or non-functional single receptor chains as occurs with antibodies directed at individual receptor sub-units. In addition, the cells can be sorted and analyzed for their capacities in vitro as well as following transplantation in irradiated animals. The CD34 bright, DR dull fraction, which by all criteria contains the most immature stem cells, hardly expressed the GM-CSF receptor, whereas the IL-3 and IL-6 receptor are moderately expressed, most likely on the majority of the cells, and c-Kit expression appears to be bright. The latter observation is in agreement with the finding in mice, that selection of bone marrow cells for c-Kit expression selects for repopulating stem cells as assessed by transplantation in irradiated recipients (31). In our hands, an at least 100-fold enrichment was obtained for mouse stem cells, measured in competitive transplantation experiments, by selecting for c-Kit alone (unpublished observations). Therefore, c-Kit would appear to be a growth factor receptor marker for stem cells. IL-3 and IL-6 receptors would appear to be strong candidates for expression on immature stem cell subsets, but formal proof by sorting cells followed by transplantation will be difficult to achieve due to the low expression levels. Along these lines we expect to identify a combination of surface antigens and growth receptors that may serve as biological markers to detect immature stem cell subsets, which then may enable a direct and quantitative analysis of residual stem cell numbers following whole body exposure to ionizing radiation, provided these cells can be obtained from a representative and reproducible site.

4. Transplantation of hemopoietic cells: alternative sources.

Alternative sources to bone marrow cells for transplantation have become available in various stages of clinical development. There is a long-standing experience to establish the "blood stem cell" as an alternative source to the use of bone marrow derived stem cells, which has been tested in extensive canine studies and has been transferred to clinical practice some 10 years ago (31a). In the meantime the use of mobilized peripheral blood stem cells in leukemic and cancer patients has expanded tremendously, most recently supplemented with the introduction of the use of blood stem cells for allogeneic transplantation. The use of blood derived stem cells as part of the management of bone marrow failure seen after accidental whole body radiation exposure and comparison of its effectiveness and efficiency with other stem cell sources such as bone marrow and cord blood cells should receive priority. In addition, several studies have shown that one single umbilical cord blood contains sufficient hematopoietic stem cells to fully reconstitute myeloablated patients treated for malignant and nonmalignant disorders (32,33). Following the analysis of the first 50 children treated by cord blood transplants, research has expanded throughout the world on the properties of the immature hematopoietic stem cells and the immune system in the new born, and on the standardization of methods of cord blood banking for various therapeutic and epidemiological purposes. Both developments should lead to the establishment of a "stem cell bank" for clinical availability, using stem cells of the blood especially under allogeneic conditions and umbilical cord as an essential element for the immuno-hemopoietic reconstitution of the accidentally high dose irradiated individual. In this respect, its is of great priority to compare the properties of the hemopoietic stem cells of these various sources, their engraftment potential and the risk of graft-versus-host disease in direct experimental studies.

Table 2. Sources of Hemopoietic Stem Cells

stem cell source	advantages	disadvantage(s)
bone marrow	high concentration	donor comfort and risks
	well characterised	relatively slow reconstitution
mobilised in peripheral blood	large numbers	exposure of donors to growth factors
	donor comfort	
	relatively rapid reconstitution	
	experience rapidly accumulating	
umbilical cord blood	readily available	relatively small numbers
	banks feasible	relatively immature stem cells
		experience limited

5. Use of circulating CD34 positive cells to assess stem cell regeneration.

Bone marrow is dispersed over numerous sites in the body and its distribution within a single site is not homogenous. Bone marrow aspirates are furthermore subject to variable admixture with peripheral blood and, for these reasons, are not a very reliable source to measure the residual number of stem cells and radiation damage of stem cell subsets. It is well known that following cytotoxic insult to bone marrow, hemopoietic stem cells start to circulate during the process of hemopoietic reconstitution. Taking the CD34 antigen as a marker for immature hemopoietic cells, we investigated the numbers of these cells which appear in the circulation and the relationship to numbers of bone marrow CD34 positive cells following total body irradiation. Already on the first day after 5 Gy TBI (which eliminates approximately two logs of bone marrow stem cells), measurable frequencies of CD34 positive cells enter the circulation to reach a relatively high percentage within 5 days. The CD34 positive cells in the first days after TBI reach numbers of $1\text{-}2 \times 10^4$ per ml, which means that, in principle, there are sufficient numbers of cells in one ml blood for enumeration and further analysis. Weekly small bone marrow punctures enabled us to relate the number of circulating CD34 positive cells to those in the bone marrow. A clear, but complex relationship was observed during the first three weeks after irradiation, which equals the pancytopenic period. Although it may be concluded, that circulating CD34 positive cells reflect bone marrow regeneration, more data are needed to investigate the predictive value of CD34 positive cells that appear early after TBI (i.e. within the first three days) for endogenous hemopoietic recovery and to estimate stem cell reduction. CD34 analysis is sufficiently simple to recommend measurement of circulating CD34 positive cells in each case of accidental whole body exposure to high doses of ionizing radiation. We noted that circulating CD34 positive cells can be readily detected in human bone marrow transplant recipients.

The presented data demonstrate that immature hemopoietic cell analysis combined with transplantation assays may eventually yield information on characteristics of stem cells, such as surface antigens and growth factor receptors, which allow for rapid enumeration of residual numbers of cells. By the same token, sufficient information on the growth factor responsiveness of the immature stem cell subsets required for rescue from bone marrow insufficiency following whole body irradiation may be obtained to design rational growth factor regimens to stimulate reconstitution of stem cells. In addition, we demonstrated a quantitative relationship between numbers of circulating CD34 positive cells and bone marrow regeneration of such cells, which may eventually lead to an early, predictive test for residual numbers of stem cells. The ready accessibility of those residual immature cells may also allow for clonal expansion in vitro and cytogenetic analysis to assess stochastic radiation effects of the hemopoietic system directly at the stem cell level.

6. Growth factor therapy

In the area of growth factor therapy, several new developments have occurred recently. The long search for thrombopoietin has met with success in 1994 The identification of thrombopoietin is perhaps the most important event in the growth factor arena of the last few years. Thrombopoietin appeared to be the ligand for c-mpl and its gene has been cloned as such(34-39). The gene product has been produced by recombinant DNA technology under various names, i.e., c-Mpl ligand, MGDF and thrombopoietin (TPO). Thrombopoietin is the first unique and potent stimulator of megakaryocytopoiesis presently known: it specifically stimulates megakaryocytopoiesis in vivo and in vitro and improves platelet production (34-41), while c-Mpl deficient mice are thrombopenic (42). It is conceivable that thrombopoietin will play an important role in the future in the treatment of thrombocytopenia, for instance after high dose radio-/chemotherapy and following stem cell transplantation (39). Preliminary results have demonstrated that TPO is more effective in stimulating platelet production following radiation-induced myelosuppression than any other growth factor thus far studied (Table 3). It was highly effective in rhesus monkeys at the LD50 dose of 5 Gy X-Rays, and stimulated platelet produuction recovery far above the level where transfusions become necesary. Similar to all other growth factors, however, TPO was not effective at doses greater than 7 Gy TBI due to limited numbers of progenitor cells available for stimulation. As such, TPO can be expected to be a highly effective agent in the mitigation of thrombopenia following radiation accidents, although in the higher dose range still hemopoietic cell infusions should be considered as life-saving. It remains to be investigated as to whether thrombopoietin treatment will alleviate the long-standing thrombopenia that frequently occurs following hemopoietic reconstitution from limited numbers of stem cells.

Table 3. Effectiveness of hemopoietic growth factors

stem cell reconstitution in vivo (only partly explored):

IL-6 >> IL-3 > placebo

platelets:

TPO > IL-6 > IL-3 > GM-CSF = GCSF

neutrophils:

G-CSF = GM-CSF >> IL-3 > IL-6 = placebo

immune system (unexplored)

T cells IL-2
B cells IL-3 (early), IL-6 (late)
various candidate growth factor untested

The number of identified growth factors, besides the established growth factors G-CSF andGM-CSF to stimulate reconstitution of neutrophilic grnaulocytes, and erythropoietin, has been extended to a total of some twenty-five known to influence stem cell proliferation and/or the production of mature blood cells, and the search for suitable combinations of growth factors, guided by in vitro studies at the level of immature stem cells, to be used for radiation accident victims should receive high priority. To date, its is known that growth factors such as IL-3 and IL-6 will stimulate stem cell regeneration, but in vivo acceleration of stem cell reconstitution is largely an unexplored area. The posssibility to monitor immature stem cell regeneration by assessing peripheral blood CD34 poitive cells might change this in the near future. Also growth factor stimulated immune reconstitution in humans and primates is largely unexplored; scanty data suggest that IL-3 and IL-6 may promote B cell recovery, and it has been known for some time that IL-2 will stimuate T lymphocytes. Whether such growth factors may improve the fate of radiation accident patients, however, remains to be investigated.

Acknowledgements

This work was made possible by grants of the Netherlands Organization for Scientific Research NWO, the Dutch Cancer Foundation and contracts of the Commission of the European Communities.

References

[1] JE. Till, EA. McCulloch, A direct measurement of the radiation sensitivity of normal bone marrow cells. *Radiat Res* 14 (1961) 213-222.
[2] JH. Hendry, A. Howard, The response of haemopoietic colony-forming units to single and split doses of gamma-rays or D-T neutrons. *Int J of Radiat Biol* 19 (1971) 51-64.
[3] HM. Vriesendorp, DW. Van Bekkum, Role of total body irradiation in conditioning for bone marrow transplantation. *Immunol. of Bone Marrow Transplantation* (1980) 345-364.
[4] G. Wagemaker, HM. Vriesendorp, DW. Van Bekkum, H Balner, Successful bone marrow transplantation across major histocompatibility barriers in rhesus monkeys. *Transplant. Proc. XIII*   (1981) 875-880.
[5] G. Wagemaker, PJ. Heidt , S. Merchav, DW Van Bekkum, Abrogation of histocompatibility barriers to bone marrow transplantation in rhesus monkeys. In: SJ. Baum, GD. Ledney, S Thierfelder (Eds.) Experimental Hematology Today, S. Karger, Basel , 1982, pp 111 - 118.
[6] WR. Gerritsen, G. Wagemaker, M. Jonker, JJ. Wielenga, JHK. Kenter, G. Hale, H. Waldmann, DW. Van Bekkum, The repopulating capacity of autologous bone marrow grafts following pretreatment with monoclonal antibodies against T lymphocytes in rhesus monkeys. *Transplantation* 45 (1988) 301 - 307.
[7] JJ. Wielenga, FCJM. Van Gils, G. Wagemaker, The radiosensitivity of primate haemopoietic stem cells based on in vivo measurements. *Int J Radiat Biol* 55 (1989) 1041.
[8] G. Wagemaker, KJ. Neelis, AW. Wognum, Surface Markers and Growth Factor Receptors of Immature Hemopoietic Stem Cell Subsets. *Stem Cells* 13 (1995) 165-171.
[9] MC. Magli, NN. Iscove, N. Odartchenko, Transient nature of early haemopoietic spleen colonies. *Nature* (1982) 295: 527.
[10] I. Bertoncello, GS. Hodgson, TR. Bradley, Multiparameter analysis of transplantable hemopoietic stem cells. I. The separation and enrichment of stem cells homing to marrow and spleen on the basis of rhodamine-123 fluorescence. *Exp. Hematol.* 13 (1985) 999-1006.
[11] RE. Ploemacher, NHC. Brons, Separation of CFU-S from primitive cells responsible for reconstitution of the bone marrow hemopoietic stem cell compartment following irradiation: evidence for a pre-CFU-S cell. *Exp Hematol*; 17 (1989) 263-266.
[12] EIM. Meijne, AJM. Van der Winden-van Groenewegen, RE.Ploemacher, O. Vos, JAG. Davids, R. Huiskamp, The effects of X-radiation on hematopoietic stem cell compartments in the mouse. *Exp Hematol* 19 (1991) 617-623.

[13] RE. Ploemacher, R. Van Os, CAJ. Van Beureden, JD. Down, Murine hemopoietic stem cells with long term engraftment and marrow repopulating ability are less radiosensitive to gamma radiation than are pleen colony forming cells. *Int J Radiat Biol*1 61 (1992) 489-499.

[14] JD. Down, A. Boudewijn, R. Van Os, HD. Thames, RE. Ploemacher, Variations in radiation sensitivity and repair among different hemopoietic stem cell subsets following fractionated irradiation. *Blood* 86 (1995) 122-127.

[15] HD. Thames, JH. Hendry, Fractionation in radiotherapy. Taylor and Francis, London, 1987.

[16] NS. Wolf, GV. Priestly, Kinetics of early and late spleen colony development. *Exp. Hematol* 14 (1986) 676.

[17] Y. Imai, I. Nakao, *In vivo* radiosensitivity and recovery pattern of the hematopoietic precursor cells and stem cells in mouse bone marrow. *Exp Hematol* 15 (1987) 890-895.

[18] MC. Baird, JH. Hendry, NG Testa, Radiosensitivity increases with differentiation status of murine hemopoietic progenitor cells selected using enriched marrow subpopulations and recombinant growth factors. *Radiat Res* 123 (1990) 292-298.

[19] JCM.Van der Loo, C Van den Bos, MRM. Baert, G. Wagemaker, RE. Ploemacher, Stable multilineage hemopoietic chimerism in a-thalassemic mice induced by a bone marrow subpopulation that excludes the majority of day-12 spleen colony-forming units. *Blood* 83 (1994) 1769-1777.

[20] AW.Wognum, FCJM. van Gils, G. Wagemaker, Flow cytometric detection of receptors for interleukin-6 on bone marrow and peripheral blood cells of humans and rhesus monkeys. *Blood 81* (1993) 2036.

[21] AW.Wognum, Y. Westerman, TP Visser, G. Wagemaker. Distribution of receptors for granulocyte/macrophage colony stimulating factor on immature CD34-positive bone marrow cells, differentiating monomyeloid progenitors and mature blood cell subsets. *Blood* 84 (1994) 764.

[22] AW.Wognum, TP. Visser, MO. De Jong, T. Egeland, G. Wagemaker, Differential expression of receptors for interleukin-3 on subsets of CD34 expressing hemopoietic cells of rhesus monkeys. *Blood* (1995) in press.

[23] GJ. Spangrude, S. Heimfeld, IL. Weissman, Purification and characterization of mouse hematopoietic stem cells. *Science* 241 (1988) 58.

[24] BI. Lord, TM. Dexter, Purification of haemopoietic stem cells - the end of the road? *Immunol Today* 9 (1988) 376.

[25] IL. Weissman, S. Heimfeld, G. Spangrude, Haemopoietic stem cell purification. *Immunol Today* 10 (1989) 184.

[26] N. Iscove, Searching for stem cells. *Nature* 347 (1990) 126.

[27] I. Weissman, G. Spangrude, S. Heimfeld, L. Smith, N. Uchida, Stem cells. *Nature* 353 (1991) 26.

[28] N. Iscove, Reply to Weissman et al. *Nature* 353 (1991) 26.

[29] MR. Loken , VO. Shah, KL. Dattilio, CI. Civin, Flow cytometric analysis of human bone marrow. II. Normal B lymphocyte development. *Blood* 70 (1987) 1316.

[30] FCJM. Van Gils, MEJM. Van Teeffelen, KJ. Neelis, J. Hendrikx, H. Burger, RW. Van Leen, E. Knol, G. Wagemaker, AW. Wognum, *Blood* (1995) in press.

[31] D. Orlic, R. Fischer, S-I. Nishikawa, AW. Nienhuis, DM. Bodine, Purification and characterization of heterogeneous pluripotent hematopoietic stem cell populations expressing high levels of c-*kit* receptor. *Blood* 82 (1993) 762-770

[31a] M. Körblin, B. Dorken, AD. Ho, A. Pezzutto, W. Hunstein, TM. Fliedner, Autologous transplantation of blood-derived hemopoietic stem cells after myeloablative therapy in a patient with Burkitt's lymphoma. *Blood*. 67 (1986) 529-532.

[32] JE. Wagner, NA Kernan, M. Steinbuch, HE. Broxmeyer, E. Gluckman,E Allogeneic sibling umbilical-cord-blood transplantation in children with malignant and non-malignant disease. *The Lancet* 346 (1995) 214-219.

[33] JE. Wagner, HE. Broxmeyer, RL. Byrd, B. Zehnbauer, B. Schmekcpeper, N. Shah, C.Griffin, PD. Emanuel, KS. Zuckerman, S. Cooper, C. Carow, W. Bias, GW. Santos. Transplantation of umbilical cord blood after myeloablative therapy: analysis of engraftment. *Blood* 79 (1992) 1874-1881.

[34]. FJ. Sauvage, PE. Hass, SD. Spencer, BE. Malloy, AL. Gurney, SA. Spencer, WC. Darbonne, WJ. Henzel, SC. Wong, SC. Kuang, KJ. Oles, B. Hultgren, Jr LA. Solberg, DV. Goeddel, KL. Eaton, Stimulation of megakaryocytopoiesis and thrombopoiesis by the c-mpl ligand. *Nature* 369 (1994) 533.

[35] S. Lok, K. Kaushansky, RD. Holly, JL. Kuijer, CE. Lofton-Day CE et al., Cloning and expression of murine thrombopoietin cDNA and stimulation of platelet production in vivo. *Nature* 369 (1994) 565.

[36] K. Kaushansky, S. Lok, RD. Holly, VC. Broudy, N. Lin et al., Promotion of *Nature*, 369, 568, 1994.

[37[F. Wendling, E. Maraskovsky, N. Debill, C. Florindo et al., C-mpl ligand is a humoral regulator of megakaryocytopoiesis. *Nature* 369 (1994) 571.

[38]. TD. Bartley, J. Bogenberger, P. Hunt, Y-S. Li, HS. Lu, et al., Identification and cloning of a megakaryocyte growth and development factor that is a ligand for the cytokine receptor mpl. *Cell* 77 (1994) 1117.

[39] D. Metcalf, Thrombopoietin - at last. *Nature* 369 (1994) 519.

[40] DC. Foster, CA. Sprecher, FJ. Grant, JM. Kramer, JL. Kuijper, RD. Holly, TE. Whitmore, MD. Heipel, LA Bell, AF. Ching, et al., Human thrombopoietin: gene structure, cDNA sequence, expression, and chromosomal localization. *Proceedings of the Nat. Ac of Sciences of the USA* 91 (1994) 13023-13027.

[41] FC. Zeigler, F. De Sauvage, HR. Widmer, GA. Keller, C. Donahue, RD. Schreiber, B. Malloy, P. Hass, D. Eaton, W. Matthews, W. In vitro megakaryocytopoietic and thrombopoietic activity of c-mpl ligand (TPO) on purified murine hematopoietic stem cells. *Blood* 84 (1994) 4045-4052.

[42] AL. Gurney, K. Carver-Moore, FJ. De Sauvage, MW. Moore, Thrombocytopenia in c-mpl-deficient mice. *Science* 265 (1994) 1445-1447.

Diagnosis and prognosis of radiation injury to skin.

A.Barabanova, N.M.Nadejina
SRSC Institute of Biophysics, Moscow

Abstract. One of the most distressing complications of the clinical course of acute radiation syndrome in persons affected in a result of the Chernobyl accident was the development of severe skin injury. This form of injury was due to beta radiation from distant beta-gamma sources and from skin contamination. Of the 108 persons who developed ARS and were treated in the clinical department of the Institute of Biophysics, 56 experienced radiation induced skin injuries. Injuries to skin significantly complicated the clinical course and prognosis of the patients and were the only cause of death in some cases. Treatment of these injuries was difficult. Doses of radiation to skin were not known at first.This together with impossibility to use special methods of diagnosis (such as thermography, rheovasography and other) created significant difficulties for prognostic evaluations. Only careful day-by-day observation and registration of skin reaction were the base for diagnosis and prognosis.

Clinical manifestations served as the prognostic tests were:

- primary erythema showed the size of the most affected areas;
- duration of latent period;
- expression clinical manifestations in acute period of injury;
- degree and rate of recovering of pathological process (outcomes of acute period of local radiation injuries).

Histological study of tissues in case of surgical intervention was very important. Late outcomes could be prognosed on the base of clinical course in early period and confirmed by using such special methods, as ultrasound investigations and duplex scanning and other.

Treatment and Follow-up of Patients Suffering from the Cutaneous Radiation Syndrome

Ralf U. Peter[1,2], Petra Gottlöber[1], Marc Heckmann[1], Otto Braun-Falco[1], Gerd Plewig[1]
1-Department of Dermatology Ludwig-Maximilians-University Frauenlobstraße 9-11, 80337 Munich
2-Institute of Radiobiology, FAF Medical Academy, Neuherbergstrasse 11, 80937 Munich
Germany

Abstract

The hazards of acute radiation exposure are commonly addressed with respect to total body gamma or neutron irradiation, resulting primarily in bone marrow failure as the main clinically relevant aspect of the acute radiation disease. Under conditions of inhomogeneous exposure, as they are characteristic for many accident scenarios, other organ systems, such as the skin may become more important in determining clinical prognosis. This became especially obvious in the two worst radiation accidents since 1945, the Chernobyl accident in April 1986 and the Goiania accident in September 1987.

The characteristic chronic sequelae of accidental cutaneous radiation exposure and therapeutic results have been described based on own clinical experience with treating patients with acute and late cutaneous effects after therapeutic irradiation, and a distinct group of patients having survived the Chernobyl nuclear power plant accident of April 26, 1986.

Apart from clinical examination, histological analysis and high-frequency (20MHz) ultrasound as well as a variety of functional tests have been used to determine the extent of radiation fibrosis and to exclude malignant transformation of keratoses and ulcers.

Treatment included, apart from dermatosurgical procedures and plastic surgery for disabling contractures or ulcers, argon laser treatment of telangiectasias, topical tretinoin 0,005% (Epi-Aberel[R], Cilag, Frankfurt), etretinate and acitretin (Tigason[R], Neotigason[R], Hoffmann LaRoche, Grenzach) for radiation keratoses, partly combined with a novel, nonatrophogenic steroid, Mometasonefuroate (Elocon[R], Schering-Plough, New Jersey) to antagonize inflammatory reactions, and low-dose interferon-gamma (Polyferon[R], Rentschler, Laupheim) for extensive radiation fibrosis. Basic dermatotherapy was performed with an ointment containing linoleic acid (Linola[R], Wolff, Bielefeld). With this combination treatment, transepidermal water loss could be sustained, progression of keratoses and inflammation were stopped. The most remarkable result was the reduction of radiaton fibrosis: sonographically determined skin thickness partly returned to normal levels after treatment of 18 months with Interferon gamma 50μg s.c. 3x/week. Based upon therapeutic experience with patients undergoing radiation therapy, symptomatic relief of pruritus can be achieved by administration of nonsedating antihistamines, such as Loratadine. In the chronic stage of the CRS after radiation therapy, where interferon gamma cannot be used, reduction of radiation-induced cutaneous fibrosis can be reached by a combination of Vitamin E and pentoxyfiline even two decades after radiation exposure. It is concluded that under accidental partial body exposure with high doses of beta and gamma irradiation, the predominant involvement of the skin, described as the cutaneous radiation syndrome, may become the characteristic trait of this increasingly probable accident pattern. Though treatment is complex and requires dermatologic and radiobiological expertise, it results in marked clinical improvement of the affected patients. These results demonstrate that considerable progress has been achieved during the last years regarding diagnosis and treatment of cutaneous radiation sequelae. Surgery has to be considered the treatment of choice for radiation carcinoma, but is no longer the only available therapeutic option. However, systematic controlled clinical trials are still missing for these options.

Introduction

As radiation accidents basically are very rare incidents, standardization of treatment is difficult to achieve. Additionally, the effects of ionizing radiation on skin have long been neglected in defining guidelines for treatment and follow-up. Therefore, the status of treatment of cutaneous radiation sequelae apart from surgical methods is so far not satisfactory (1,2). An established treatment scheme which would go beyond recommendations, based on anecdotal observations, does not exist.

Additionally, differring procedures in documentation of accidents further worsen the intercomparability of therapeutic efforts in accident situations which generally are seperated by large gaps in space and time. This situation may improve after implementation of appropriate data base systems (3) where all radiation accidents can be reported to.

Generally, the clinical course of cutaneous radiation reactions follows a distinct clinical pattern, for which the term „cutaneous radiation syndrome" (CRS) has been coined (4). Within minutes to hours after exposure an erythematous reaction develops, which may be associated with a burning itch. This prodromal stage is transient in nature, and is followed after upto 36 hrs by a clinically inapparent latency stage. The manifestaional stage is characterized by occurrence of an intensively erythematous skin which may show some scaling. In more severe conditions subepidermal blisters and even ulcerations may develop. Though resembling skin lesions produced by thermal injury, the time course and underlying processes involved in the development of the CRS are so different from thermal burns (5), that the terms „radiation burns" or „β-burns" are considered inappropriate for this clinical condition and misleading and should therefore be abandoned.

In the chronic stage of the CRS three clinical symptoms are dominating the further course:

- radiation keratoses may develop in all exposed areas. They have to be considered precancerous lesions and should be monitored thouroughly.

- radiation fibrosis is caused by an increase of collagenous tissue in dermal and subcutaneous fibroblasts which may lead to pseudoatrophy of fatty tissue. Fibrosis may lead to an occlusion of blood vessels and can thus cause secondary ulceration.

- telangiectasias are a characteristic sign of the chronic stage of the CRS in humans (they may be absent in many laboratory animals, among them pigs). Apart from cosmetic disfiguring, they may cause a permanent itching sensation and a feeling of warmness, which is reported to be disturbing by the affected patients.

Treatment of the CRS and Discussion

Treatment has to focus on symptomatic relief and the avoidance of additional risk to the patients. The applied therapies, dosages, numbers of patients to which the treatment had been applied and the therapeutic outcome are summarized in table 1.

With regard to the manifestational stage of CRS our experience is limited to radiotherapy patients. In these conditions, an erythematous and erosive condition occasionally occurs, which is associated often with a burning itch. Treatment with loratadine, a nonsedating and mast-cell stabilizing antihistamine, induced a marked relief of these symptoms and a shortening of the erythematous phase as compared to untreated patients (Peter et al., submitted). Topical steroids generally can be used with good success, as reported earlier by other authors (6).

Topical dressings of Tetrachlorodekaoxide (TCDO) induce considerable granulation and reepithelization in erosive skin conditions (Braun-Falco and Landthaler, unpublished observations). These data confirm the radioprotective properties of TCDO reported by other authors in irradiated mice (7), rats (8) and regenerative capacities in complicated wounds (9).

Additional treatment modalities which have been reported to be of value in the manifestational stage are cleansing of the oral cavity and administration of pilocarpine for prevention of mucositis (10, 11) as well as heparinization and antibiotic prophylaxis for bacterial and viral infections (1).

In some older studies beneficial effects of hydroxyethylrutosides (12) and bovine blood extracts (13) on radiation-induced acute skin reactions have been reported. It should be noted, however, that these reports have been based animal experiments or on open trials with comparatively small patient groups, respectively.

Our therapeutic experience with the chronic stage of the CRS comes, apart from the survivors of the Chernobyl accident, from patients suffering from chronic cutaneous sequelae following therapeutic irradiation. These experiences shall be discussed consecutively. All Chernobyl patients who have been treated by us responded well to a basic therapy with a specific ointment containing linoleic acid (Linola Fett[R], Wolff, Bielefeld), which led to a marked decrease of the initially severely increased transepidermal water loss, as determined by evaporimetry. In the meantime, this beneficial effect of linoleic acids has been demonstrated in animal experiments as well (14). Teleangiectasias, though primarily a cosmetical problem, in some localizations, such as the ankles of the knees, caused discomfort due to sensations of a burning itch and heat, which disappeared after treatment of telangiectasias by Argon laser.

Tretinoin cream 0,005% (Epi Aberel[R], Cilag, Frankfurt), applied once daily, led to clearance of focal and patchy radiation keratoses, as it has been reported for solar keratoses (15). In more extensive lesions, oral application of the retinoid Acitretin (0,1-0,2 mg/kg/d) was used, analogous to the reported treatment of radiation-induced keratoacanthomas (16). However, on radiation-exposed skin tretinoin cream appeared to be more irritant than it is known from patients with actinic keratoses; therefore, intermittent antiinflammatory treatment with topical steroids was necessary. To avoid additional damage of the atrophic skin caused by steroid atrophy, a novel steroid preparation, which has been proven to be nonatrophogenic in a variety of clinical trials (Mometasonefuroate, Elocon[R], Schering-Plough, New Jersey), was applied intermittently.

The most striking results were reached with subcutaneous administration of interferon (IFN) gamma (Polyferon[R], Rentschler, Laupheim), in eight patients with severe and extensive radiation fibrosis: in six patients receiving IFN gamma for 18 months in a dosage of 50µg s.c. 3x/week according to a protocol used in scleroderma patients (17) fibrosis could be reduced almost to the level of uninvolved contralateral skin, as determined by cutaneous 20 MHz sonography (cp. also the article of P. Gottlöber et al, this issue). Side effects which were noted included elevation of body temperature upto 38,5°C after the first two injections, and a reduced frequency of herpesvirus infections, which recurred in three patients after discontinuation of interferon treatment. In two patients who rejected interferon injections after the first injection, and who were followed up together with the other six patients, an increase of fibrosis occured during the observation period.

The efficacy of IFN gamma in radiation fibrosis may be explained in part by its antagonizing effect towards the cytokine TGF-beta, which is of eminent importance for the induction of radiation fibrosis (18)

Another therapeutic option for radiation fibrosis is the combined administration of Pentoxyfilline (PTX) 3x400mg/d and Vitamin E 1x400mg/d. By this regimen, applied for a mininimum of six months, radiation fibrosis persisting and being progressive for more than 20 years could be reduced (19).

Controversial data exist in the literature for superoxide dismutase with regard to its efficacy in reducing acute radiation-induced tissue reactions. In muscular and subcutaneous fibrosis however, there is good evidence of a beneficial effect (20)

Surgical procedures followed in general reported guidelines (21) and included excision of ulcers and contractures; wound closure was performed with split and full thickness skin grafts, on certain instances vascularized flaps were used. All grafts healed without complications, even in localizations where the surrounding tissue was affected by late radiation effects. This is a specifically interesting aspect of the CRS in the Chernobyl survivors, as due to the primary cause of cutaneous lesions, namely short-range nuclides causing cutaneous contamination, only the upper parts of the dermis and subcutis were affected, whereas the larger vessels penetrating the muscle fascia were not or only partially harmed. Therefore the surgical experience derived from patients with skin fibrosis after deeply penetrating radiation therapy, that skin grafts do not heal if not the complete surrounding affected tissue is removed, proved to be inadequate for the survivors of the Chernobyl accident.

In conclusion it can be stated that the last decade not only brought a substantial gain in understanding about the mechanisms underlying cutaneous radiation reactions, but this knowlegde could also be transformed into several novel therapeutic approaches for the benefit of those patients who suffered so badly from the most severe accident in the history of the civilian use of nuclear energy so far.

References

1) A. Barabanova, A.K. GuskovaThe diagnosis and treatment of skin injuries and other non-bone- marrow syndromes in Chernobyl victims. In: Ricks R C Fry S A (eds.) The medical basis for radiation accident preparedness II clinical experience and follow-up since 1979. Elsevier, New York 1990, pp. 183-189

2) Angelina K. Guskova. Basic principles of the treatment of local radiation injuries. *Brit J Radiol Suppl* **19** (1986) 122-124

3) H. Kindler, D. Densow, T:M. Fliedner. RADES-medical assistance system for the management of irradiated persons. In: Proceedings of the 2nd DEXA 91. Springer Wien 1991

4) Ralf U. Peter. The Cutaneous Radiation Syndrome. Proceedings, 2nd Consensus Conference on the Treatment of Radiation Injuries, Bethesda, April 14-17, 1993 (in press)

5) M.J. Hoekstra, R.W. Kreis, M.H.E. Hermans, *et al.* The management of burn patients. *Med Corps Int* **3** (1987) 49-52

6) J. Chung, C.W. Song, T. Tamaguchi, J. Tabachnick Effect of anti-inflammatory compounds on -irradiation induced radiodermatitis. *Dermatologica* **144** (1972) 97-107

7) S. Ivankovic, S.R. Kempff. Regenerative effects of tetrachlorodecaoxid (TCDO) after total body irradiation with y-rays in BD IX rats. *Radiat Res* **115** (1988) 115-123

8) T. Sassy, N. Breiter, K.R.Trott. Die Wirkung von Tetrachlorodecaoxid (TCDO) auf den chronischen Strahlenschaden am Rattenkolon. *Strahlenther Onkol* **167** (1991) 191-196

9) R.U. Peter, A. Bleyl Beschleunigung des Heilungsverlaufs bei Operationen an der Fußsohle nach Anwendung von Tetrachlordekaoxid. *Wehrmed Monatsschr* **1** (1988) 38-39

10) C. Hasenau, B.P.E. Clasen, D. Roettger. Anwendung einer standardisierten Mundpflege zur Prophylaxe und Therapie einer Mukositis bei Patienten während der Radiochemotherapie von Kopf-Hals-Malignomen. *Laryng Biol Otol* **67** (1988) 576-579

11) I.H. Valdez, A. Wolff, J.C. Atkinson, *et al.* Use of pilocarpine during head and neck radiation therapy to reduce xerostomia and salivary dysfunction. *Cancer* **71** (1993) 1848-1851

12) H. Fritz-Niggli, E. Fröhlich. Reduktion der strahleninduzierten Frühschäden der Haut (Mauspfote) durch O-(ß-Hydroxyäthyl)-rutoside. *ROFO* **133** (1994) 316-321

13) W.H. Pohl Über die Behandlung von Strahlenschäden der Haut mit Actihaemyl. *Strahlenther* **117** (1962) 3-10

14) J.W. Hopewell, M.E.C. Robbins, G.J.M.J. Vandenaardweg, *et al.* The modulation of radiation-induced damage to pig skin by essential fatty acids. *Br J Cancer* **68** (1993) 1-7

15) G.D. Weinstein, TP Nigra, P.E. Pochi, *et al.* Topical tretinoin for treatment of photodamaged skin. *Arch Dermatol* **127** (1991) 659-665

16) J.C. Shaw, F.J. Storrs, E. Everts. Multiple keratoacanthomas after megavoltage radiation therapy. *J Am Acad Dermatol* **23** (1990) 1009-1011

17) Hein R, Behr J, Hundgen M, Hunzelmann N, Meurer M, Braun Falco O, Urbanski A, Krieg T (1992) Treatment of systemic sclerosis with gamma-interferon. Br J Dermatol 126: 496-501

18) M. Martin, J.-L. Lefaix, P. Pinton, *et al.* Temporal modulation of TGFb1 and b-actin gene expression in pig skin and muscular fibrosis after ionizing radiation. *Radiat Res* **134** (1993) 63-70

19) P. Gottlöber, G. Krähn, H.C. Korting, R.U. Peter. Behandlung der kutanen Strahlenfibrose mit Pentoxifyllin und Vitamin E - ein Erfahrungsbericht. *Strahlenther Onkol* (im Druck)

20) J.-L. Lefaix, S. Delanian, J.-J. Leplat, *et al.* La fibrose cutaneo-musculaire radio-induite (III): efficacite therapeutique majeure de la Super Oxyde Dismutase Cu/Zn liposomiale. *Bulletin du Cancer* **80** (1993) 799-807

21) Eric F. Hirsch. The role of surgery in the management of acute local radiation accidents. In: Ricks R C Fry S A (eds.) The medical basis for radiation accident preparedness II Clinical experience and follow-up since 1979. Elsevier, New York (1990) pp. 191-194

Table 1 Symptom-oriented treatment of the Cutaneous Radiation Syndrome

Symptom	Treatment	Application	Dosage	Result	Side Effects.
Pruritus	Antihistamines	p.o.	10mg q.d.	relief of itch	-----
Erythema	Steroids	topical b.i.d.		alleviation	none if used less than 3 wks
Blisters	Steroids TCDO	wet dressing t.i.d.		alleviation	-----
Dryness	Linoleic acid cream	topical 1 x /day		inhibition of water loss	----
Keratoses	Tretinoin	topical	1 x /day	clearance	irritation
„	Acitretin	oral	0,1-0,3 mg/kg	moderate	dryness of lips
Inflammation	Mometasone	topical	3-4 x /week	alleviation	- - -
Fibrosis	IFN gamma	s.c.	50 µg 3x/w	reduction	Fever
	PTX+Vit E	p.o.	400mg t.i.d. + 300mg q.d.	reduction	-----

Long Term Follow-up of Irradiated Persons: Rehabilitation Process

Vladimir BEBESHKO, Alexander KOVALENKO, David BELYI

Scientific Center of Radiation Medicine, Melnikov St. 53, Kiev, Ukraine

Abstract. In patients after acute radiation syndrome as result of Chernobyl accident a gradually forming of late radiation pathology was observed in following years. It is connected with destructive changes in the tissues with low proliferative activity. Among some of these patients a deviation of biochemical data and different clinical variants of displasia of haemopoiesis have been found. Taking in account the specialities of development and evolution of nonstochastic effects the system of rehabilitation and prophylactics has been developed and improved. This system was directed on the reduce of the late radiation pathology development and it's clinical manifestation. This system is characterised by the complex of different medical and rehabilitation measures.

During post-accidental period somatic status of the patients who suffered from acute radiation syndrome (ARS) was characterised by increase of a frequency rate of the digestive, circulatory diseases and it's chronic flow, decrease of physical and mental capacity, transforming neurovegetative and psychoneurological disorders into organic neurovascular pathology (hypertensive disease, discirculatory encephalopathy, psychoorganic syndrome). Steady changes of brain's functional status with non-linear "dose-effect" correlation were revealed. Disablement reaches of approximately 90%.

In 40% of patients after ARS changes in the haemopoiesis (transitory and/or stable leukopenia, leukocytosis) and in 1/3 of the patients immune disorders were observed.

Changes in enzymes activity of antioxidant defence, activation peroxide oxidation of lipids processes, weakening of antiradical defence, affects of membranes and lizosomal enzymes (hydrolyses) activity increase were revealed in erythrocytes of patients after ARS. It is shown cells' destruction, severity of pathology and bad prognosis.

There were observed changes in hormonal supply of adaptation and reproduction processes (steady hypercorticism, hypohonadism, increase of polyamines in serum, atherogenic changes of lipid metabolism).

Six cases of hypothyreosis and seven cases of radiation cataract took place.

It is known that in pathogenesis of different organs and systems defeat the main roles are belong to activation of processes of peroxide oxidation of lipids (POL) and non-adequate antioxidant system (AOS).

So rehabilitation must be are based on following principles:

- to weaken lipid peroxide oxidation;
- to increase the activity of organism antioxidant system;
- correction of vitamin and microelements balance;
- to reduce the immunological insufficiency;

- to stabilise the vegetative status;
- to normalise the lipid metabolism (dislipidemia correction);
- to treat and prevent chronic disease of different organs and systems;
- to increase the physical and mental capacity;
- to propagandise the health way of life and refuse from bad habits and professional harmfulness;
- to have rational and balance nutrition;
- to treat in sanatoriums;
- to make succession in hospital, out-patient and sanatorium treatment.

The rehabilitation system was elaborating and improving during whole post-accidental period.

On the hospital stage patients received different combinations of drugs that contained radioprotective, antioxidant, antitoxin, membrane defence, haemostimulative, immunomodulative, vasoactive, metabolic, hepatotropic and sedative components. The rehabilitation, prophylactic and physiotherapy measures have the main role.

The rehabilitation system must be directed to prove oxidation homeostasis, normalise regulation and metabolism, prevent cells' membrane defeat (see Figure 1).

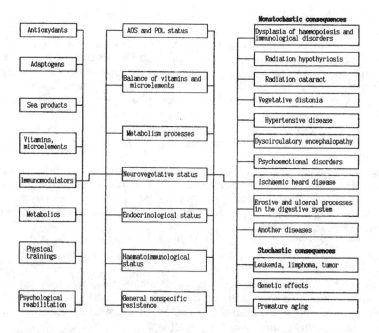

Figure 1. The main rehabilitation measures for the patients after ARS

On the outpatient department stage the main role belongs to the drugs treatment, physiotherapeutic treatment, massage and physical training.

In sanatoriums natural and climatic factors of treatment were used.

It is understandable that 10 year's term is not sufficient to do conclusions about effectiveness of the rehabilitation programme. Our experience shows positive effect of the rehabilitation in most of the patients after ARS. But 2 cases of myelodysplastic syndrome ended to death on 7th and 9th year after Chernobyl accident prove that the problem of radiobiology stochastic effects' prophylactic especially haematological oncopathology is a very significant problem till now.

An introduction of elaborated rehabilitation system directed to reduce late radiation pathology in patients after ARS has some difficulties. It is connected with deformation of mental, psychological and social status of these patients. Some of these patients continue to smoke, overeat and refuse physical training. So rehabilitation includes permanent explanatory and sanitary education.

The Effects of Single Dose TBI on Hepatic and Renal Function in Non-human Primates and Patients

J.J. BROERSE, B.BAKKER, J. DAVELAAR, J.W.H. LEER, M.M.B. NIEMER-TUCKER and
E.M. NOORDIJK

Department of Clinical Oncology, University Hospital, PO Box 9600, 2300 RC Leiden,
The Netherlands

Abstract. Total body irradiation (TBI) and bone marrow transplantation
(BMT) are common procedures in the treatment of severe combined
immune deficiency syndromes, leukemia, non-Hodgkin lymphoma and
other hematological disorders. Improved results following TBI and BMT
have increased the number of patients in long term follow up. Late
detrimental effects of TBI have been investigated in non-human primates
and patients with emphasis on vital organs like liver and kidney. The
response of monkeys to radiation is not significantly different from that in
man. Long term effects of TBI could be studied by keeping 84 monkeys of
different ages under continuous observation for a period up to 25 years.
Effects on hepatic and renal function were demonstrated using serological
and histological parameters. The values of the liver function parameters
such as alkaline phosphatase and gamma glutamyl transferase in the
irradiated group are significantly increased after TBI. Also the parameters
of kidney dysfunction, e.g., Ht and urea show a significant change in the
irradiated old aged cohort with respect to the controls.
Between 1967 and 1993, 336 bone marrow transplantations were performed at the University Hospital Leiden. The present study was restricted
to those patients who survived at least 18 months after transplantation.
This retrospective analysis consequently amounts to 120 patients. The
monkey data indicated subclinical organ damage for postirradiation intervals exceeding 15 years. However, up to the present time, the human data
do not support these findings since the follow up time is still restricted to
a median survival of 4,5 years. Detrimental effects in liver and kidney function at a later stage can not be excluded yet, and careful examinations of
the patients remain indicated.

1. Introduction

High dose total body irradiation in combination with intensive chemotherapy
followed by bone marrow transplantation has shown to be of benefit for the
treatment of hematological malignancies, some non-malignant disorders of the
hemopoietic system and selected solid tumors. Presently, many thousands of
patients per year are being treated with high dose total body irradiation (TBI)
followed by autologous or allogeneic bone marrow transplantation. The number of
patients in long term follow up after bone marrow transplantation has increased
Delayed effects of this treatment modality, including high dose total body
irradiation, become therefore of more importance. The radiation doses involved are
at the threshold for acute radiation damage to lung, kidney and liver.
Radiation experiments in primates are of relevance since the response to radiation
of monkeys does not seem to be significantly different from that in man. This has
been demonstrated for a number of acute effects including those on the hemopoietic

and intestinal system and also for some late effects such as tumor induction [3, 4, 5, 6]. Furthermore an outbred species as the Rhesus monkey (Macaca mulatta) is more representative as an animal model to assess the effect of total body irradiation as a single toxic factor than inbred rodents. The dose distribution of TBI and the number of circulating T lymphocytes in Rhesus monkeys are comparable to those in man. Moreover the life expectancy of the Rhesus monkey in captivity can exceed 35 years and is therefore long enough for simulating late radiation effects in man [7, 8, 9].

For fractionated irradiation the approximate limits of radiation tolerance of different human tissues including the kidneys and the liver are clinically defined [10, 11, 12]. The tolerance dose after fractionated exposure, which is the dose associated with 1 to 5% complications occurring within 5 years of treatment is assumed to be 20 Gy for the kidneys and 30 Gy for the liver. There is no universal agreement on the exact tolerance dose of many tissues after single dose TBI including the kidneys and liver. Rubin et al. [13] mention values between 11 and 19 Gy and 15 and 20 Gy, respectively as tolerance doses for kidney and liver after single dose irradiations. Van Rongen [14] observed radiation induced alterations in different renal functional and histological parameters in the rat after single irradiations with doses in excess of 6 Gy. He observed that deviations from control values start earlier and increase more rapidly with increasing radiation dose.

On the basis of the results for fractionated irradiation in man, an iso-effect dose for kidney damage could be derived under the assumption of an α/β ratio of 2 Gy [15, 16]. The resulting extrapolated total dose, ETD, of 40 Gy would imply a tolerance value of 8 Gy for single dose irradiation.

We investigated the long term effect of TBI as a common toxic agent in Rhesus monkeys on the incidence and time course of hepatic and renal function. In addition we studied the risk of radiation induced late organ damage in a retrospective analysis using the medical records and laboratory data of the adult bone marrow transplant patients of the Leiden University Hospital.

2. Materials and Methods

The risks of total body irradiation with large doses of X-rays could be investigated by keeping long-term surviving monkeys from an experiment on acute effects [17] under continuous observation for a period up to 25 years. This group of irradiated animals is a unique population to study late deterministic effects due to irradiation in relation with age. The control group consisted of untreated monkeys with comparable age distribution. This group was maintained under identical housing conditions and they were fed commercial food pellets (Hopefarms) and a diet of fresh fruit and vegetables. Information on the age distribution of the group of Rhesus monkeys, the irradiation conditions and the morphological assessment can be found elsewhere [18].

During all examinations the monkeys were anesthetized by an intramuscular injection of 10 mg/kg of ketamine hydrochloride. All animals underwent a general physical examination. Clinical signs of organ failure like jaundice, hepatomegaly and edema were registered. Blood samples were withdrawn from the femoral vein. The long term hepatic and renal toxicity of the TBI treatment were investigated through the parameters summarized in Table 1. In some animals the assessment of the parameter was performed twice with a time interval of several months, therefore the total number of blood samples investigated is larger than the total number of animals. Hematocrit was tested in an additional investigation and therefore the number of animals equals the number of laboratory results.

Table 1. Parameters used to monitor hepatic and renal function in Rhesus monkeys.

Hepatic function	γ-glutamyl transpeptidase
	alkaline phosphatase
	aspartate aminotranspherase (ASAT)
	alanine aminotranspherase (ALAT)
	total bilirubin
	LDH
Renal function	serum creatinine
	blood urea nitrogen (BUN)
	hematocrit
	albumin

Between January 1967 and March 1993, 336 bone marrow transplantations were performed in the Department of Hematology at the Leiden University Hospital. The indications for BMT including total body irradiation, were investigated in these patients. The evaluation was restricted to patients who survived at least a period of 18 months after transplantation.

Consequently, the total number of evaluable cases of this retrospective study amounts to 120. The median follow up time was 43.1 months (range 12.1 - 218., mean 53.3). The series included 68 males and 52 females. Fifty patients were treated with an autologous BMT, 69 patients were treated with an allogeneic BMT and 1 patients with an isologous BMT.

Eighty-three (69%) out of the evaluated long term surviving patients were conditioned with chemotherapy (cyclophosphamide 2 x 60 mg/kg body weight) followed by total body irradiation. Patients with treatment other than TBI (31%) received cyclophosphamide alone or in combination with cytosin-arabinoside, etoposide, BCNU (Carmustine) or busulfan. Total body irradiation was performed in a single fraction with a dose of 8 to 9 Gy in the majority of the cases (97.6%).

The long term hepatic toxicity of the TBI treatment was evaluated by the assessment of serological parameters of the liver function: alkaline phosphatase (reference interval 15-60 U/L), total bilirubin (reference interval 1-17 µmol/L), ALAT (reference interval 1-15 U/L) and ASAT (reference interval 1-15 U/L). Parameters for long term nephrotoxicity of the TBI treatment included blood urea nitrogen (BUN) (reference interval 2.5 - 7.5 mmol/L) and serum creatinine levels (reference interval 50 - 110 µmol/L).

3. Results in Rhesus monkeys

With respect to the hepatic function the parameter γ-glutamyl transpeptidase demonstrated a significant difference between the irradiated and non-irradiated animals in the old cohort (p<0.0001). The difference in alkaline phosphatase levels between the TBI group and the age-matched control group was significant for the young cohort (p=0.0021) and the old cohort (p=0.0009). The levels of alkaline phosphatase levels (AP) in the middle aged cohort did not show any significant difference. The increase in AP levels in the old cohort corresponds with the increase of γ-glutamyl transpeptidase in the same cohort. In the non-irradiated control group the levels of γ-glutamyl transpeptidase and AP did not show any increase with age.

The liver parameters alanine aminotranspherase (ALAT), aspartate aminotranspherase (ASAT) and bilirubin did not show any significant difference between the irradiated and non-irradiated groups. An interesting observation was an increase in LDH with age, in both the irradiated and non-irradiated groups with p-values of respectively 0.0013 and 0.0020. Also a significant difference was found between the TBI group and the control group with regard to the average levels of LDH in the young, the middle aged and old cohort with p values of respectively 0.024, 0.029 and 0.0015. The isoenzymes LDH_{1-5} did not show any statistical changes in the irradiated animals compared to the control animals.

With respect to renal function the results of the present study demonstrated a decrease in renal clearance capacity. Since serum creatinine levels are related to muscle mass the data are expressed in terms of creatinine/body weight. Changes in bodyweight due to loss of lean body mass should be accompanied by a subsequent decrease in creatinine levels in absence of renal failure.

A significant increase in serum creatinine/body weight index was found in the irradiated group indicating significant renal impairment for post-irradiation intervals exceeding 10 years. The control animals did not show such an increase of creatinine/body weight, even in the very old animals (age > 30), the creatinine/body weight index levels seemed to be stable.

Renal impairment is affirmed by a significant decrease of the hematocrit (Ht) levels with post-irradiation time in the irradiated group (p=0.0083). In contrast the control group showed a significant increase of Ht with age (p=0.0061). There was a significant difference of Ht levels between the irradiated and non-irradiated animals of the old cohort (p=0.0035). There was no significant difference between the Ht values of the irradiated and non-irradiated animals of the young and middle aged cohort.

The blood urea nitrogen (BUN) levels in the old cohort showed a significant difference between the TBI and control animals (p=0.0005). The BUN levels demonstrate an increase with post-irradiation time in the irradiated groups only (p=0.0041). Similar to the liver parameters, a late radiation effect is demonstrated on the renal function which becomes only detectable after a latency period of more than 10 years after TBI.

Albumin levels did not show a significant difference between the irradiated and non-irradiated monkeys and in both groups a similar decrease was found with age (p <0.0001). This finding can not be attributed to radiation because in both groups the same tendency was found.

Table 2. Summary of alterations in hepatic and renal functional parameters in irradiated animals versus age matched controls.

parameter	postirradiation interval					
	< 5 years		10-15 years		> 15 years	
hepatic function						
bilirubin	=		=		=	
γ-glutamyl transpeptidase	=		↑	NS* p=0.09	↑	p<0.0001
alkaline phosphatase	↑↑	p=0.002	↑	NS* p=0.13	↑	p=0.0009
ASAT and ALAT	=		=		=	
LDH	↑	p=0.02	↑	p=0.03	↑	p=0.0015
renal function						
creatinine/body weight	=		↑	p<0.0001	↑↑	p<0.0001
blood urea nitrogen (BUN)	=		=		↑	p=0.0005
hematocrit	=		=		↓↓	p=0.0035
albumin	=		=		=	

*NS: not significant;
=: no change

From the summary of the results in Table 2 it becomes evident that some deterministic effects become only apparent at long time intervals after irradiation. The alterations in liver and kidney function which we found were not related to any clinical symptoms of organ damage.

4. Patient results

For clarity the numerical values are collected for two time intervals, respectively before and after 18 months, to demonstrate the disparity between early and late changes of the parameters of the various organ functions.

After application of the exclusion criteria such as GVHD, the total number of patients eligible for evaluation for their liver function was 102. At 18 months posttransplantation the total bilirubin levels remained stable both in the TBI and non- irradiated group. During the early phase after transplantation in the TBI group however, the bilirubin levels demonstrate a slight decrease with time (p=0.01). Only 4.8 % of the total bilirubin values of the TBI group exceeds the normal range of 17 mmol/L within the first 18 months after TBI. Elevations of the total bilirubin level above 50 mmol/L were never seen in the TBI group and only once in the non-TBI group. A well known early liver toxicity called veno-occlusive disease was never diagnosed in the group of long term (> 1 year) surviving patients.

During the early phase the ALAT levels show a significant increase in the irradiated group (p<0.0001); the mean ALAT level in the irradiated group was 26.5 ± 1.4 U/L and 16.9 U/L in the non-irradiated group. The ASAT levels show a similar behaviour. After 18 months the ASAT and ALAT levels remain stable in the TBI and non-TBI group, but slightly elevated for the mean ALAT level in the TBI group as compared to the non-irradiated group (p=0.01), at respectively 16.6 ± 1.0 U/L and 12.4 ± 0.9 U/L. The alkaline phosphatase levels gradually declined with time and show no statistical significant difference between the irradiated and non irradiated group, both in the early phase as well as the period beyond 18 months post transplantation.

After application of the exclusion criteria such as the use of nephrotoxic drugs or previous known renal failure, the total number of patients eligible for evaluation of the kidney function was 112. The BUN levels (reference value 2.5 - 7.5 mmol/L) were found to be elevated once or more in course of time in 18/81 = 22.2 % of the patients in the TBI group and in 11/31 = 35,5 % of the non-irradiated group during the total post-transplantation period. However, clinically significant renal failure was never reported in either of the groups. Neither was there a significant change of BUN levels after 18 months post-transplantation between the TBI and non-irradiated group and the BUN levels never exceeded 15 mmol/L in both groups. Beyond 18 months post transplantation also no significant increase of creatinine was found in either of the groups.

During the early and late phase post-transplantation no statistically significant difference of albumin levels could be demonstrated between the TBI and non-TBI group.

5. Discussion

To our knowledge delayed hepatic damage following TBI has not been reported for post-irradiation intervals longer than 10 years, neither in humans nor monkeys. We observed a significant increase in AP and γ-glutamyl transpeptidase levels in the old cohort of Rhesus monkeys. Since γ-glutamyl transpeptidase is a parameter exclusive for liver tissue, it can be concluded that after a post-irradiation interval of more than 10 years an increase of this liver parameter might be due to irradiation. The increase of the LDH level however, in the irradiated group should be attributed to a radiation effect on other organs since liver cell decay could not be confirmed with the additionally performed investigations of the iso-enzymes of LDH. The observation of a significant increase of AP levels in the young irradiated group is likely to be due to alterations in the bone-metabolism, as the other liver parameters in this age group did not show any alterations.

The alterations of the levels of alkaline phosphatase and γ-glutamyl transferase could not be explained by pathology of the bile ducts neither by changes in the liver tissue resembling nodular regenerative hyperplasia, as described by Snover et al. [19]. However, the incidence of periportal fibrosis and mild pericentral and capsular fibrosis was slightly increased in the irradiated animals.

Our study demonstrates a significant alteration in the parameters of renal function in the irradiated group of old and middle aged monkeys. Since no other toxic agents were involved in the treatment of the monkeys it is reasonable to conclude that the radiation (TBI) was the nephrotoxin responsible for the relative elevation in serum creatinine levels. Also a decrease of hemoglobin and hematocrit levels could be demonstrated in the irradiated group of monkeys after a post-irradiation interval of more than 5 years.

In the patient study we tried to identify the role of TBI in the occurrence of late organ damage. The total body irradiation is carried out to achieve immunosuppression in order to allow full engraftment and to increase tumour cell kill. In our present analysis nearly all irradiated patients were treated with single dose TBI (97.6%) and no major late liver or renal toxicities were found in this group of long term (>1 year) BMT survivors. These results are encouraging and suggest that unfractionated total body irradiation up to a dose of 9 Gy and a high instantaneous dose rate of about 0.23 Gy/min in this population is well tolerated. The present data did not show any significant difference between the group of patients with a conditioning regimen including single dose TBI and the other conditioning regimens without irradiation with regard to the incidence of liver and kidney dysfunction. In our experience veno-occlusive disease (VOD) was never diagnosed among the group of BMT patients while this complication is common according to the literature.

In summary, no late adverse effects of TBI, as a part of the conditioning treatment preceding bone marrow transplantation, on hepatic or renal function have been found in man. Studies with non-human primates with post-irradiation intervals of more than 15 years have provided evidence for subclinical liver and kidney damage. In patients detrimental effects can not be excluded after longer survival periods and careful examination of the patients remains indicated.

References

[1] Sanders, J.E., Late effects in children receiving total body irradiation for bone marrow transplantation, *Radiotherapy and Oncology*, 1990, Suppl. 1, **18**, 82-87.

[2] Zapatero, A., Martin de Vidales, C., del Cerro, E., Vazquez, M.L., Dominguez, P. and Perez Torrubia, A., Total body irradiation before allogeneic bone marrow transplantation for leukemia, Radiotherapy and Oncology, 1994, **32**, Suppl. 1, 167.

[3] Ainsworth, E.J., Leong, G.F. and Alpen, E.L., Early radiation mortality and recovery in larger animals and primates. Response of different species to total body irradiaton. Edited by: J.J. Broerse and T.J. Macvittee, Martins Nijhoff, Boston, Dordrecht, 1984, pp. 87-111.

[4] Broerse, J.J, van Bekkum, D.W., Zoetelief, J. and Zurcher ,C., Relative biological effectiveness for neutron carcinogenesis in monkeys and rats. *Radiation Research*, 1991, **128**, 128-135.

[5] Carsten, A.L., Acute lethality, the haemopoietic syndrome in different species, Response of different species to total body irradiaton. Edited by: J.J. Broerse and T.J. Macvittee, Martins Nijhoff, Boston, Dordrecht, 1984, pp. 59-86.

[6] Vriesendorp, H.M. and Van Bekkum, D.W., Susceptibility to total body irradiation, Response of different species to total body irradiaton. Edited by: J.J. Broerse and T.J. Macvittee, Martins Nijhoff, Boston, Dordrecht, 1984, pp. 43-57.

[7] Bowden, D.M., Aging in non-human primates, New York, Van Nostrand Reinhold, 1979.

[8] Chambers, K.C. and Phoenix, C.H., Sexual behavior in response to testosterone in old long-term-castrated rhesus males. *Neurobiology of Aging*, 1983, **4**: 223,227.

[9] Kaufman, P.L. and Bito, L.Z., 1982, The occurence of senile cataracts, ocular hypertension and glaucoma in rhesus monkeys. *Experimental Eye Research*, 1982, **34**, 287-291.

[10] ICRP- 41, Nonstochastic effects of Ionizing Radiation. *International Commission on Radiological Protection*, 1984, Vol 14, no. 3 (Pergamon Press, Oxford UK).

[11] UNSCEAR, Sources, Effects and Risks of Ionizing Radiation. United Nations Scientific Committee on the Effects of Atomic Radiation, 1988, (United Nations, New York).

[12] Fajardo, L.F. and Berthrong, M., Radiation injury in surgical pathology. *The American Journal of Surgical Pathology*,1978, **2**, 159-199.

[13] Rubin, Ph., Constine, L.S. and Nelson, D.F., Late effects of Cancer treatment: Radiation and Drug Toxicity, In: *Practice of Radiation Oncology*,1992, Perez, C.A. and Brady, L.W., Edited by: JB Lippincott Company, Philadelphia. pp. 124-161.

[14] Van Rongen, The influence of fractionation and repair kinetics on radiation tolerance. Studies on rat lung and kidney, Thesis, University of Amsterdam., 1989.

[15] Barendsen, G.W., Dose fractionation, dose rate and iso effect relationships for normal tissue responses. *International Journal of Radiation Oncology, Biology, Physics*, 1982, **8**, 1981-1999.

[16] Thames, H.D. and Hendry, J.H., 1987, Fractionation in Radiotherapy. Taylor and Francis, London, New York, Philadelphia, 1987.

[17] Broerse, J.J., van Bekkum, D.W., Hollander, C.F. and Davids, J.A.G., Mortality of monkeys after exposure to fission neutrons and the effect of autologous bone marrow transplantation. *International Journal of Radiation Biology*, 1978, **34**, 253-264.

[18] Niemer-Tucker, M.M.B., Sluijsmans, M.M.J.H., Bakker, B., Davelaar, J., Zucher, C. and Broerse, J.J., Long-term consequences of high dose total body irradiation on hepatic and renal function in primates, *Int. J. Radiol. Biol.* 1995, **68**, 83-96

[19] Snover, D.C., Weisdorf, S., Bloomer, J., McGlave, Ph. and Weisdorf, D., Modular Regenerative Hyperplasia of the liver following bone marrow transplantation. *Hepatology*, 1989, **9**, 443-448.

III. HEALTH EFFECTS
FOLLOWING THE ACCIDENT
A. Treatment of accident victims

Posters

Developing Diagnostic Guidelines for the Acute Radiation Syndrome

D. Densow[a], H. Kindler [b, a], and T. M. Fliedner[a]

[a]Institute of Occupational Medicine, University of Ulm, D-89070 Ulm, Germany, WHO
Radiation Emergency Medical Preparedness and Assistance Network (REMPAN) Centre
[b]Institute of Applied Knowledge Processing, University of Ulm, D-89010 Ulm, Germany

Abstract. Diagnostic guidelines seem to be promising for improving medical care. One aspect of a diagnostic guideline for the acute radiation syndrome has been tested against an extensive case history database. Subsequently, the guideline has been optimised for a small set of case histories. The improved performance has then been proven by a test against the rest of the case history database.

1. Introduction

Clinical guidelines created by medical bodies are systematically developed statements to render decision support to physicians for specific clinical circumstances. Generally, guidelines aim at enhancing the quality of care by reduction of commonplace errors, by diffusion of optimal care delivery, through increased efficiency of diagnosis, and by allowing for scientific evaluation through means of feedback and statistical tests. Even though, the above advantages seem very much obvious, clinical guidelines have but scarcely been tested. In this presentation one guideline will be tested against reality, i. e., case histories in a database.

2. Research objectives

Nowadays, clinical guidelines are developed on the basis of experts' knowledge. The necessary formalisation process may corrupt the expert's knowledge, leading to a lack of generality and sub-optimal performance. Clinical guidelines, however, should perform well for the majority of possible case types in an application domain. A means to improve clinical guidelines could be the case-oriented refinement based on tests against real cases ascertaining the correctness of the guideline. This would add more objectivity to the experts' opinion.

A diagnostic guideline for the acute radiation syndrome (ARS) has been tested against case histories to get an impression of its correctness. As a first step, the test was limited to the granulocyte concentration changes which serve as the most decisive indicator for the grading of the ARS. Since the results have shown a mediocre performance of the guideline, it has been carefully considered how the performance could be improved. The guideline has been optimised for a small set of test cases and evaluated against a large number of real cases.

3. Methods

In the last years a Clinical Pre Computer Proforma (Baranov 1994) has been developed for the standardised structured documentation of ARS case histories. This questionnaire has been accepted by the 8 WHO Collaborating Centres on Radiation Emergency Medical Preparedness and Assistance (WHO 1993). Hundreds of case histories have since been collected world-wide and inserted into the International Computer Database for Radiation Exposure Case Histories. These case histories served as test set.

In 1979 the International Commission on Radiological Protection (ICRP) devised a guideline for the medical management of persons accidentally exposed to ionising radiation (ICRP 1980). Concerning the haematological disorders only qualitative descriptions were included. In 1985 the German Commission on Radiological Protection (BMI 1986) decided to create more precise recommendations by defining leukocyte concentration intervals between day 4 and 7 to predict the clinical outcome of the disorders related to the haemopoietic system (see figure 1).

	degree	BMI-ICRP-classification
transient mild granulocytopenia	1	I, II
mild depression of haemopoiesis	2	III
spontaneous recovery of haemopoiesis	3	IV
no spontaneous recovery due to lack of stem cells	4	V
no spontaneous recovery due to insufficient stroma	5	VI

Figure 1.   Degrees of the clinical outcome of the haemopoietic disorders of the ARS.

The leukocyte concentration intervals between day 4 and day 7 are given in figure 2 as time-concentration-rectangles. The degrees of the clinical outcome are marked in or immediately below the rectangles. The leukocyte concentrations of one real case are depicted as circles. In the guideline no information has been given how to interpret the measurements, provided more than one degree can be taken into account. The best estimates for real cases are obtained deciding for the highest degree. Thus, the clinical outcome of the case in figure 2 is classified as of degree 2.

In order to evaluate the ICRP based guideline it was tested against 159 real case histories of the International Computer Database for Radiation Exposure Case Histories.

The correct estimates for each degree of clinical outcome are indicated by the red characters. The guideline underestimates the degree of severity of the haemopoietic syndrome systematically as can be seen by the high percentages on the left hand side of the highlighted numbers.

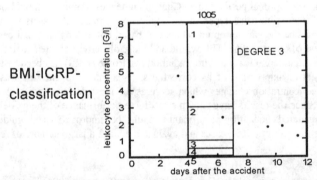

Figure 2.   Classification schema based on the ICRP recommendations applied on one case history.

Less measurement noise can be obtained by taking into account the granulocytes as the main fraction of leukocytes only. Their correlation to the clinical outcome is even better.

After radiation leukocytes and granulocytes do not decrease on straight-lines, but, on curves. Therefore, curves yield better classification. Instead of polynomials exponential functions as output functions of a bi-compartmental models have been used as discriminating curves.

The bi-compartmental model is shown in figure 4. The model is a simplification of the well-known structure of granulocytopoiesis (Fliedner 1988).

<table>
<tr><td></td><td>0</td><td>1</td><td>2</td><td>3</td><td>4</td><td>5</td></tr>
</table>

	0	1	2	3	4	5
0	0	0	0	0	0	0
1	6	21	0	0	0	0
2	2	67	8	0	0	0
3	0	75	34	0	0	0
4	0	3	24	9	7	0
5	0	0	0	0	0	0

- clinical course vs. ICRP-BMI-guideline
- κ-coefficient (Cohen 1968) = - 0.02
- correspondance is inadequate

Figure 3.   Results of the ICRP based classification schema tested against 159 case histories.

It is composed solely of a maturing and a blood granulocyte (functional) pool. The predecessors of the maturing cells were neglected, due to the radio sensitivity of dividing cells.

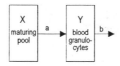

Figure 4.   Simplified physiological bi-compartmental model of granulocytopoiesis.

They will only if at all come into play after a longer period of time. A full maturing and functional pool were assumed as initial conditions. Assuming the maturing and functional pool to be completely filled is justified since their cellular content is not very sensitive to ionising radiation. These initial conditions and the loss of granulocytes b, which is not significantly changed by radiation, have already been measured in reality.

classification based on the bi-compartmental model

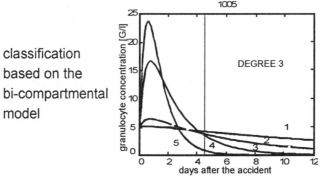

Figure 5.   Classification schema based on the physiological model applied on a real case history .

Ionising radiation changes the transit time a, because substances (cytokines) will be liberated according to the intensity of the radiation to mobilise the cells of the maturing pool. The stress causing the change of transit time a is strongly correlated to the prognosis of the future granulocyte level over time.

Taking the parameters $X(0)$, $Y(0)$, and b from experimental results four output functions optimally differentiating between the 5 degrees of clinical outcome for 39 cases have been estimated by adapting the parameter a. (see figure 5). The discrimination curves correctly partition off the different degrees from day 4.5 onward due to the overlap during the first days. If a granulocyte measurement lies in one of the spaces then its degree indicated by the numbers

1 to 5 is assumed. E. g., all the granulocyte measurements (circles) of a real case with a clinical outcome of degree 3 lie between two discrimination functions in the space marked with (degree) 3.

Even after optimisation of the parameter a for the four discrimination functions the partitioning between the degrees was not a hundred percent successful.

4. Results

The above presented diagnostic guideline to predict the severity of the haemopoietic lesions after total body irradiation has been tested against 120 real cases. The classification results are shown in figure 6.

Nevertheless, classification errors have occurred. The confusion of degree 3 with degree 4 during the diagnostic process will be hazardous for the patient. The guideline is quite good in differentiating between degree 3 and degree 4. Only 2% of degree 3 were misclassified as degree 4 and only 6% vice versa.

	1	2	3	4	5
• clinical course vs. exponential classifier 1	21	0	1	0	0
• only cases 2	5	26	6	0	0
• κ-coefficient = 0.74 3	0	7	35	2	0
• correspondance is 4	0	0	1	15	0
good 5	0	0	0	0	1

Figure 6.   Results of the exponential classificator tested against 120 case histories.

Already existing guidelines and the experts' formalised knowledge were too weak to achieve a proper problem-solving behaviour as has been shown in figure 3. The partitioning curves can be applied as a relatively simple grading scheme, working like a nomogramm, on paper. The guideline cannot only be justified by the authority of experts, but, by physiology and statistics.

5. Conclusion

Only one aspect of the ARS could be discussed in this short paper. Even for the haematopoietic syndrome other parameters like lymphocytes, reticulocytes have to be taken into account. The same should be performed for other affected organs systems to create a general guideline for the ARS. The results must and will be revised by an expert panel later-on.

One necessary prerequisite for this approach is a standardised collection of a large number of comprehensive case histories. Nowadays only few international standardised case history databases exist. Introducing the electronic medical record may facilitate the collection of large numbers of case histories.

6. References

[1] Baranov A. E., Densow D., Fliedner T. M., Kindler H.: "Clinical Pre Computer Proforma for the International Computer Database for Radiation Exposure Case Histories", Springer, Heidelberg, 1994
[2] BMI (ed.): "Medizinische Maßnahmen bei Kraftwerksunfällen", Veröffentlichungen der Strahlenschutzkommission 4, Gustav Fischer, Stuttgart, 1986
[3] Fliedner T. M., Steinbach K. H.: "Repopulating Potential of Hemopoietic Precursor Cells", Springer Verlag, New York, 1988, Blood Cells 14, 1988, 393-410
[4] ICRP (ed.): "Annals of the ICRP - The Principles and General Procedures for Handling Emergency and Accidental Exposures of Workers", ICRP Publication 28, Pergamon Press, Oxford, 1980
[5] World Health Organisation (ed.): "Fourth Co-ordination Meeting of WHO Collaborating Centres in Radiation Emergency Medical Preparedness and Assistance", report PEP/93.2, confidential, 1993

A Multi-Centre Clinical Follow-Up Database as a Systematic Approach to the Evaluation of Mid- and Long-Term Health Consequences in Chernobyl Acute Radiation Syndrome Patients

Fischer B.[1,2], Belyi D.A.[3], Weiss M.[1], Nadejina N.M.[4], Galstian I.A.[4], Kovalenko A.N.[3], Bebeshko V.G.[3], Fliedner T.M.[1]

[1]*Department of Clinical Physiology and Occupat. Medicine, University of Ulm, Germany*
[2]*Research Institute of Applied Knowledge Processing FAW, Ulm, Germany*
[3]*Scientific Center of Radiation Medicine, Academy of Medical Science, Kiev, Ukraine,*
[4]*Institute of Biophysics, Clinical Department, Moscow, Russian Federation*

Abstract: This paper describes scope, design and first results of a multi-centre follow-up database that has been established for the evaluation of mid- and long-term health consequences of acute radiation syndrome (ARS) survivors. After the Chernobyl accident on 26 April 1986, 237 cases with suspected acute radiation syndrome have been reported. For 134 of these cases the diagnosis of ARS was confirmed in a consensus conference three years after the accident. Nearly all survivors underwent regular follow-up examinations in two specialised centres in Kiev and in Moscow. In collaboration with these centres we established a multi-centre clinical follow-up database that records the results of the follow-up examinations in a standardised schema. This database is an integral part of a five step approach to patient evaluation and aims at a comprehensive base for scientific analysis of the mid- and long-term consequences of accidental ionising radiation. It will allow for a dynamic view on the development of the health status of individuals and groups of patients as well as the identification of critical organ systems that need early support, and an improvement of acute and follow-up treatment protocols for radiation accident victims.

1. Introduction

The multi-centre clinical follow-up database described in this paper is an integral part of a methodological framework for the evaluation of mid- and long-term health consequences in patients surviving the acute radiation syndrome (ARS). After the Chernobyl nuclear power plant disaster on 26 April 1986, a total of 237 persons were suspected of having an ARS due to irradiation. Following acute care in hospitals mainly situated in Moscow and Kiev the 199 surviving patients (28 died during the acute disease) were under regular follow-up investigation in Hospital No. 6 of the Institute of Biophysics in Moscow, and the Clinical

Department of the Scientific Center of Radiation Medicine of Ukraine. Additionally, 14 of the patients with the most severe skin injuries had repetitive follow-up examinations in the Dermatological Department of the Ludwig-Maximilians-University, Munich, Germany.

2. Methodology: Five Step Approach to Patient Evaluation

The multi-centre clinical follow-up database is an integral part of a general methodological approach to patient evaluation. This five step approach (Fig. 1) consists of the following five steps: (1) Design of a follow-up questionnaire reflecting the reality of repeated follow-up examinations, time flow and the implication of different centres in following-up a single patient (Weiss et al. 1995). (2) Design of the conceptual, logical and physical database schema (SQL for Microsoft Access 2.0 and/or Oracle 7); design of a form-based question-naire-like user-interface for putting-in and browsing patient data (Microsoft Access 2.0); assuring data quality by data validation through integrity con-straints that are reflecting the seman-tics of the clinical data. (3) Po-

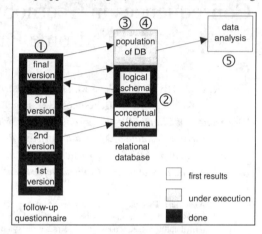

Fig. 1: Overview on Methodology

pulation of the database; integrating the data from different participating centres. (4) Further data integration for a common view on acute-care (Baranov et al. 1994) and follow-up clinical data; user interface design. (5) Analysis of the data in order to improve the knowledge on mid- and long-term consequences of ARS and give recommendations for their prevention in acute care and subsequent follow-up as well as recommendations for standardised follow-up examinations for patients after ARS. Data analysis techniques include the visualisation of individual health dynamics, evaluation of health trouble rates in different groups of patients and in relation to the initial exposure situation.

3. Population under Investigation

After the Chernobyl accident on 26 April 1986, 237 persons were suspected of suffering from an acute radiation syndrome (ARS). A later revision of all cases resulted in a confir-mation of 134 cases while 103 could not be confirmed. 28 victims died from acute compli-cations, 10 died during the follow-up period (October 1995). 199 persons are still alive and most of them are under regular follow-up. The patients undergoing follow-up examinations performed by the three centres in Kiev, Moscow and Munich form the study population of the follow-up study and the computer database. Tab. 1 shows the number of follow-ups performed so far.

Tab. 1: Numbers of patients and follow-up examinations in the three examining centres

Centre	approx. nb. of performed follow-up examinations	nb. of patients followed-up in 1994
Kiev	> 1200	168
Moscow	> 350	18
Munich	~ 60	15

3. Results

After filling-in the data from the original patient records into the questionnaire data can be entered into the computer database by the use of an Microsoft Access based set of computer forms. Putting the data directly from the patients records into the database is also possible but needs further steps for data validation. Since the beginning of 1995, the clinical follow-up database is in the phase of population after having been tested by several case examples in the year before.

Follow-Up Data Available	patients	exams.
Kiev	168	>1200
Moscow	18	>350
Munich	14	~60
total	199	>1600

⇒

Follow-Up Questionnaires	patients	exams.
Kiev	2	20
Moscow	14	135
Munich	10	10
total		165

⇒

Included in Follow-Up Database	patients	exams.
Kiev	2	20
Moscow	9	69
Munich	0	0
total	11	89

Fig. 5: Number of patients and number of examinations (available, reported, in database; October 1995)

4. Discussion

The clinical data collected in the three examination centres represent a base of enormous scientific value. Our effort has the objective to make these data available for scientific analysis. (1) An increased availability to a larger scientific community can be achieved by using English language and a common standardised terminology for the documentation of the clinical data. (2) If the same standardised and structured schema is used for the documentation of all follow-up examinations later comparison of case histories will be much easier. (3) The availability for scientific analysis can be increased by storing the clinical data in a computer. Compliance to database standards (SQL, ISO/IEC 9075) enhances the interoperability. Additionally, the relational database model allows for flexible data access paths and to answer a large variety of research questions. (4) The data from all participating centres are stored by the help of the same database schema. This allows for an integration into a single clinical database and provides research and practice with the complete history of follow-up examinations during the last 10 years even if a patient has been followed-up by different centres.

Together with the possibility of integrating the rapidly growing follow-up database with another database comprehensively documenting the acute course of acute radiation syndrome widens the range of questions that can be answered by the help of this well-structured multi-centre clinical follow-up database.

During the database development process and the following database population we were faced with the following problems. (1) Clinical data are person-related and require measures protecting the privacy. The gold standard is only to store anonymised data. Since data from different follow-up examinations of the same patient have to be connected to each other the examining centres agreed on a common table of patient codes that is used to fill in the follow-up questionnaires and allows to link all data of an individual patient. (2) The data reported in the questionnaires is not perfect. A sound quality control procedure was required to get reliable data. The first measure in this direction was feed back from the data entry personnel (medical students). A second were the build-in integrity constraints of the database.

5. Conclusion and Outlook

The multi-centre clinical follow-up database lays a sound foundation for patient evaluation after acute radiation syndrome (ARS). Once populated with a larger number of cases, this valuable research tool will be at the disposal of an international group of scientists to be used as a source for new knowledge on mid- and long-term consequences of this rare disease including the generation of hypothesis by explorative data analysis and testing formerly stated hypotheses.

The continuation of regular follow-up examinations and the input of their results into this multi-centre database seems necessary to us. Continuous analysis should accompany this process. An extension to include cases from others than the Chernobyl accident - as already performed for the acute period database - should be discussed.

The comprehensive database schema and its relational design have shown to be difficult to handle by novices. It seems to be useful to supplement the existing forms-based graphical user interface for data entry and browsing with an easy-to-use graphical user for flexible data visualisation. Possible functions are: (1) to present the health dynamics of an individual case, (2) to find similar cases from the database, and (3) to present multiple cases for comparison.

Acknowledgements
This research was carried out within the framework of the "EC/CIS Agreement for International Collaboration on the Consequences of the Chernobyl Accident". The work was partially funded by the European Atomic Energy Community as project number COSU-CT94-0089 (JSP3).

References
Baranov A.E., Densow D., Fliedner T.M., Kindler H.: *Clinical Pre Computer Proforma for the International Computer Database for Radiation Exposure Case Histories*. Berlin, New York [...]: Springer, 1994. ISBN 3-540-57596-0 (Berlin), ISBN 0-387-57596-0 (New York).

ISO/IEC 9075:1992(E)   Information technology - Database languages - SQL, 1992.

Weiss M., Nadejina N.M., Galstian I.A., Belyi D.A., Fischer B., Bebeshko V.G., Fliedner T.M.: *Epicrisis on the Primary Health Effects in Radiation Exposed Persons*. University of Ulm, Institute for Occupational and Social Medicine, 1993 - 1995.

Weiss M., Nadejina N.M., Galstian I.A., Belyi D.A., Fischer B., Bebeshko V.G., Fliedner T.M.: *Questionnaire for the Clinical, Laboratory and Functional Follow-Up of Radiation Exposed Persons*. University of Ulm, Institute for Occupational and Social Medicine, 1993 - 1995.

Evaluation of Mid- and Long-Term Consequences, Clinical and Social Performance in Chernobyl Acute Radiation Syndrome Patients in A Multi-Centre Clinical Follow-Up Study

Weiss M.[1], Bebeshko V.G.[2], Nadejina N.M.[3], Galstian I.A.[3], Belyi D.A.[2], Kovalenko A.N.[2], Fischer B.[1], Fliedner T.M.[1]

[1]Department of Clinical Physiology and Occupat. Medicine, University of Ulm, Germany
[2]Scientific Center of Radiation Medicine of Ukraine, Kiev, Ukraine
[3]State Research Center, Institute of Biophysics, Moscow, Russia

Abstract: Since the Chernobyl accident in 1986 nearly all survivors (n=199) of 237 patients with suspected acute radiation syndrome (ARS) underwent regular follow-up investigations in the scientific centres in Kiev and in Moscow. In a close collaboration with these centres we investigate the health status of this population in a five step approach. An integral part of this approach to patient evaluation and analysis of the mid- and long-term consequences of the Chernobyl accident is a „Questionnaire for clinical, laboratory and functional follow-up of radiation-exposed persons", developed with these centres. Beyond this project we report as an interim some results of analyses performed by the scientific centers in Kiev and in Moscow about disorders of the cardiovascular system and the digestive tract, formation of cataract, generalised and local skin injuries and/or disorders as well as for a subpopulation (n=89) the Karnofsky performance score and working ability.

1. Introduction

As a result of human error partial fusion of the reactor core and loss of its confinement by explosion occured at the Chernobyl nuclear power plant (Ukraine) on April 26 in 1986. In a multi-centre clinical follow-up study we investigate the health status of 199 survivors with suspected acute radiation syndrom (ARS). In a five step approach (Fischer et al. 1995) we focus on following aims: (1) Assessment of the health status in dynamics. (2) Description of occurance and development of neoplastic and non-neoplastic late effects. (3) Assessment of critical organ systems. (4) Dose-effect relationship. (5) Clinical and social performance.

Furthermore, we present as an interim some results of analysis of mid- and long-term health consequences in patients that survived the acute radiation syndrome after the Chernobyl accident and are under regular follow-up examinations in Moscow and in Kiev.

2. Patients and Methods

The investigations are based on patients described in Tab. 1 showing the total number of survivors of the Chernobyl accident that were suspected of suffering from ARS. According to the degree of severity patients are grouped to ARS I - IV and a group of patients for whom ARS could not be confirmed.

Tab. 1: Patients involved in the JSP3 Study (Survivors)

Degree of Acute Radiation Syndrome (ARS)	total nb. of "Chernobyl survivors"
ARS I	41
ARS II	47
ARS III	13
ARS IV	1
ARS not confirmed	97
total	199

38 patients died during the 9 years after the accicent (28 during the acute period of ARS and 10 in the follow-up period).

We developed a "Questionnaire for clinical, laboratory and functional follow-up of radiation exposed persons" (Weiss et al. 1995) and a database as major instruments for our patient evaluation. The data will be recorded retrospectively from the original patient charts into the Questionnaires.

3. Results

3.1 JSP3 Project

Fig. 1 shows exemplarily an individual clinical course in a day-to-day analysis by the visualisation of the course of neutrophils, lymphocytes and thrombocytes of a male patient exposed to 4,9 Gy (based on chromosomal aberrations in peripheral blood lymphocytes).

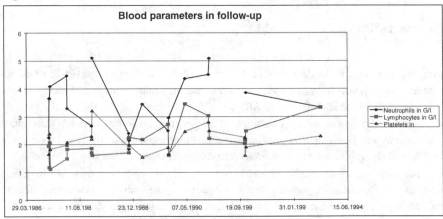

Fig. 1: Blood parameters in follow-up

In the same way health dynamics for an individual patient and for a group of patients can be presented. This data can also be used for decision support by providing case histories for comparision when a quick decision is necessary e.g. whether the patient has sufficient residual haematopoietic capacity to survive with conventional supportive care or requires intensive, sophisticated treatment in specialised centers, which may include bone marrow transplantation and/or treatment with hematopoietic growth factors.

3.2 An excerpt of data analysed in Moscow and in Kiev.

Eyes
In a subgroup of 89 patients cataract formation of different etiology has been found in 14 patients, 7 patients suffered from a radiation cataract.

Cardiovascular disorders
There is an increase in diseases of the cardiovascular system like hypertension and/or ischaemic heart disease for every group and in both centers. Fig.1 shows data of patients beeing treated in Kiev.

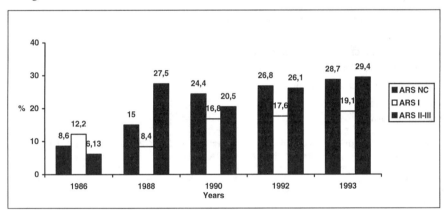

Fig. 2: Dynamics of cardiovascular system disorders

Digestive disorders
In both centers a steady increase of digestive disorders like gastritis, gastroduodenitis, cholecystitis, erosions, ulcers have been noticed for all groups. The highest increase was found in the group with the lowest radioactive burden.

Skin disorders
After exposure to beta and gamma irradiation local (atrophy, hyper-/hypopigmentation, teleangiectesia, peeling) and generalised skin injuries (severe dermatofibrosis, atrophy, hyperkeratosis, chronic ulcers, hypo-/hyperpigmantation) have been found. Malignant transformations did not occur so far.

Clinical and social performance
The clinical performance was rated according to the Karnofsky score. Information was also requested for social activity as work attendance. These data are based on last known data

(1993) of a subpopulation of 89 patients. Patients from the non-confirmed group and patients with ARS III have a very low Karnofsky performance score of about 60%. Patients with ARS II and ARS III recovered well with a Karnofsky performance score of nearly 80%. Additionally, working ability has been in sequence 35%, 70%, 60%, 35% (part-time) and 0%, 37%, 13%, 0% (full-time).

Conclusions

With continued observations it becomes apparent that radiation exposure through the Chernobyl accident caused significant acute, delayed and chronic toxicity. In this paper we present our approach for scientific analysis of the health status of the described population. Further investigations are needed to elucidate the mechanisms and aetiology of the described health effects and to improve consequently the outcome of radiation exposed persons.

As consequence conclusions in following fields can be drawn:

- Understanding of pathophysiological mechanisms
- Determination of biological indicators that are of predictive value for development and occurance of neoplastic and non-neoplastic late effects
- Improvement of existing and development of new approaches to assess the extent of radiation injury to every organ systems, using molecular, cellular and cell system oriented methods established in different Institutes in Western and Eastern Europe
- The extensive data will be collected and made available in a case-history oriented database for irradiated individuals

Acknowledgements

This research was carried out within the framework of the "EC/CIS Agreement for International Collaboration on the Concequences of the Chernobyl Accident. The work was partially funded by the European Atomic Energy Community as project number COSU-CT94-0089 (JSP3). The success of this project depends on the continuation of the close collaboration with the scientific centers in Moscow and in Kiev.

References

Weiss M., Nadejina N.M., Galstian I.A., Belyi D.A., Fischer B., Bebeshko V.G., Fliedner T.M.: *Epicrisis on the Primary Health Effects in Radiation Exposed Persons.* University of Ulm, Institute for Occupational and Social Medicine, 1993 - 1995.

Weiss M., Nadejina N.M., Galstian I.A., Belyi D.A., Fischer B., Bebeshko V.G., Fliedner T.M.: *Questionnaire for the Clinical, Laboratory and Functional Follow-Up of Radiation Exposed Persons.* University of Ulm, Institute for Occupational and Social Medicine, 1993 - 1995.

Fischer B., Belyi D.A., Weiss M., Nadejina N.M., Galstian I.A., Kovalenko A.N.,Bebeshko V.G., Fliedner T.M.: A Multi-Centre Clinical Follow-up Database as a systematic Approach to the Evaluation of Mid- and Long-Term Health Concequences in the Chernobyl Acute Radiation Syndrome Patients.

Weiss M., Bebeshko V.G., Belyi D.A., Fischer B., Fliedner T.M. et al. *Late effects, clinical and social performance of long term survivors after the Chernobyl accident.* In: Hagen U., Jung H., Streffer C (ed): Radiation Research 1895-1995, Würzburg, 1995, p. 331.

Hemoblastoses of the Chernobyl Accident Clean-Up Workers and the Population in Belarus

G.V. TOLOCHKO[1], E.P. IVANOV[1], A.E. OKEANOV[2], V.G. LAVRENYUK,
L.P. SHUVAEVA
[1]*Research Institute of Hematology and Blood Transfusion, 160 Dolginovski Trakt,
223059 Minsk, Belarus*
[2]*Belarussian Centre for Medical Technologies, Information Computer Systems,
Health Care Administration and Management, 7a Brovki str., 220600 Minsk, Belarus*

Abstract. The frequency and structure of hemoblastoses among Chernobyl nuclear power plant disaster liquidators and the population of Belarus are studied. The nosology structure of morbidity are documented; the dependence of the latter on the location of a work place in the hazard zone and of the permanent place of inhabitancy are noticed. No excessive increase of liquidators morbidity levels as compared with the one of the whole population is discovered. However, a certain increase in the last years is determined.

1. Introduction

"Clean-up workers" or "liquidators" of the consequences of the Chernobyl nuclear power plant disaster are a group of a higher risk for oncohematological morbidity. Despite the majority of liquidators are of the age of active ability to labor, a communication already appeared that among the reasons of their primary labor disability the 2nd rank is of oncohematological diseases (after the diseases of cardiovascular system) [1]. Japanese follow up observations of the A-bomb survivors documented that in Hiroshima the excess of leukemia cases appeared after a 5-year period following the A-bombing and in Nagasaki after a 10-year period. It was also demonstrated that the duration of latent period of leukemia is shorter for the cases exposed to ionizing radiation before 30 years of age [2;3].

2. Materials and methods

The data on the Chernobyl nuclear power plant liquidators and on their oncohematological morbidity were obtained by means of screening the two registers; the Republican Register of the population suffered as a result of Chernobyl nuclear power plant Accident and of the Republican Register of blood diseases. The first one contains information of about 63.5 thousand liquidators. That is approximately not more that 50 % of the entire number of persons who worked in the 30 km area from the epicenter of the catastrophe and who, due to various reasons, were registered as regular patients of specialized dispensaries. The comparison of these data with the data of the second - hematological - register made it possible to obtain the most full and verified information about oncohematological cases among liquidators and about the analogous morbidity of the Belarus population (the latter being used as a control group of the present study) as well as of the population of its separate administrative regions.

3. Results

As a result of the screening, 67 liquidators were found to have leukemia or other hemoblastoses. About 79 % of these persons worked in the hazard zone in 1986, 13 % in 1987, the rest in 1988-89. 6 were diagnosed - Hodgkin's disease, true polycythemia etc. - prior to the disaster and these cases were excluded from the further analysis. 61 liquidators were diagnosed at first from May 1986 till 1994 (Table 1).

Table 1.   Characteristics of hemoblastoses morbidity among liquidators - Number of cases according to nosologies

	LEUKEMIA					LYMPHOMA	
Year	Acute	Chronic lympho-leukemia	Chronic myelo-leukemia	Erythremia	Multiple myeloma	Hodgkin's disease	Non-Hodgkin's
1986	0	0	0	2	0	0	0
1987	1	0	0	0	0	1	0
1988	2	1	0	1	0	2	0
1989	1	1	0	1	0	1	2
1990	2	2	0	1	0	1	0
1991	4	2	3	0	0	0	0
1992	1	1	3	2	0	0	1
1993	1	4	2	3	1	0	1
1994	1	2	4	0	1	0	2
Total	13	13	12	10	2	5	6

They consist in 8 women and 53 men, 11 rural and 50 urban habitants. The age structure of this group is plotted in Table 2.

When the morbidity of hemoblastoses of the population of the regions of the Belarus is compared with the quantity of the liquidators cases diagnosed in these territories, one is able to suppose that the inhabitants of the Mogilev, Gomel and Grodno Oblasts were sent to work in the comparatively less contaminated zone than the ones ordered to the duty trips from the City of Minsk, Minsk and Vitebsk Oblasts.

Table 2.   Age distribution of cases among Chernobyl nuclear power plant disaster liquidators in 1986 - 1994

Number of liquidators	Age (years)						
	19<25	25<30	30<40	40<50	50<55	55<60	=>60
At the moment of work in the zone	7	4	14	17	10	7	2
At the moment of morbidity	3	5	10	16	7	10	10

The revealed in the present study excess of the hemoblastoses levels among the liquidators of the Vitebsk and Minsk Oblasts (Table 3), as compared with the one of the population is associated exactly with the character of their work in the zone and with the radionuclide contamination of the zone of their duties fulfillment, but not with the particularities of the dissemination of hemoblastoses among the Republic and the location of their permanent inhabitancy. This supposition found its confirmation during the following analysis that revealed that the most radiation-dependent nosological forms of hemoblastoses - acute leukemia, chronic myeloleukemia - more often are diagnosed among the liquidators who fulfilled the work in the territories, closest to the epicenter of the disaster: in Chernobyl, Bragin and Khoiniki Raions. It should also be noted [Table 1.] that the hemoblastoses quantity among liquidators is being step-by-step increases with the time lag extension from the moment of the accident.

4. Conclusions

The level of morbidity of various forms of hemoblastoses (leukemias and lymphomas) among liquidators has not exceeded during 9 years after the disaster (1986-1994) the analogous level of the one of the whole population of the Republic of Belarus.

There is a trend of increase of the number of diagnosed cases in the last years.

Table 3.   Morbidity levels magnitudes (per 100,000) among the population of Belarus
            (aged 20-59 years) and the liquidators

Location of inhabitancy	Hemoblastoses	
	among the population	among the liquidators
Minsk City	16.1	16.1
Brest Oblast	13.8	11.3
Vitebsk Oblast	14.9	19.5
Gomel Oblast	13.3	7.9
Grodno Oblast	14.5	4.5
Minsk Oblast	15.9	22.6
Mogilev Oblast	14.8	2.6
In the Republic of the Belarus in toto	14.8	10.7

5. References

[1]   Zubritskiy M.K. et al. Some medical and social aspect of disability caused by the Chernobyl
        nuclear power plant disaster among the inhabitants in Mogilev Oblast; in: Abstracts of the
        IVth Republican Scientific Conference, Mogilev, 1994, Part 1, pp. 104-105 (in Russian).
[2]   Ichimaru M. et al. Cancer in atomic bombs survivors, NY, 1986 - pp. 113-129.
[3]   Oguma N. et al. Func. Biol. Med., 1985, Vol. 4, N 1.-1-7..

The Hematopoietic System of the Acute Radiation Syndrome Reconvalescents in Post-Accidental Period

V. KLIMENKO, I. DYAGIL, L. YUKHIMUK, N. BILKO, V. BEBESHKO,
S. KLIMENKO, O. OBERENKO
Scientific Center of Radiation Medicine, Melnikov str. 53, Kiev, Ukraine

Abstract. The state of hemopoietic system has been studied since 1986 up to now in 145 patients who had acute radiation sickness after the Chernobyl accident. We studied clinical, morphofunctional, histological, ultrastructural, biophysical, cultural, cytochemical indexes of the hematopoietic elements. The connection between hemopoietic microenvironment and hemophoiesis state was put up. The realization of the hematological disorders as myelodysplastic syndrome testified the most important problem in future.

1. Introduction

The Chernobyl Accident led to special situation when some people taking part in elimination of its consequences of Chernobyl accident were irradiated in high doses. We are interested in the hemopoietic system reaction among 145 patients with acute radiation sickness (ARS) of different degrees. Some scientific publications show a high sensitivity of hemopoietic tissue to ionizing radiation [1,2,3].

2. Materials and methods

One hundred and five patients after ARS were studied. The morphofunctional, clinical, biophysical, cultural, ultrastructural methods were used. The clinical features, the quantitative and qualitative changes, the ultrastructural organization of blood and bone marrow cells were evaluated. Besides, the colony-forming activity of the progenitor cell in the tissue culture in vitro and the influence of hemopoietic microenvironment on the stromal elements and its role in the forming of hematopoietic pathology were studied.

3. Main results

The dynamic investigation of hemopoietic system state in the patients after ARS I-II degrees showed that in acute period, changes in bone marrow were typical. The myelokaryocytes number varies from 8 to 138 x 10^9/l. The quantity of megakaryocytes in 50 % patients in the first 3-5 days after irradiation stayed normal. In some cases its number decreased to full disappearance. Simultaneously, the karyorrhexis and nuclears' fragmentation of hematopoietic cells were intensified. Also, the number of granulocytes and erythrocytes were decreased and the number of lymphoid elements was increased. In the same time the number of reticular and plasma cells in bone marrow was enlarged. 3-5 days after irradiation the active macrophages with ability to phagocytosis appeared. We found the hypoplastic state of bone marrow in this group of patients by using morphological, histological and ultrastructural methods. The cultural investigations showed changes of ability of hemopoietic elements to colony-forming with predominance of eosinophilic-neutrophilic colony. In the same time there was reduction of stromal elements ability to colony-forming in monoline culture to its disappearance in some cases.

In the first period after irradiation in most of the patients leukopenia, thrombocytopenia, erythrocytosis, leukocytosis and lymphocytosis were found. If in the acute period the qualitative changes (hyper- and hyposegmentation, nuclear fragmentation, karyorrhexis, karyolysis, pyknosis, basophilic cytoplasm, nuclear and cytoplasm vacuolization, bilobal nuclear) were observed in most of the patients, some 30 days later these changes decreased. It was revealed in 30-40 % patients.

Ultrastructural investigations have shown that most of the endothelyocytes are characterized by the modifications of the cells' sizes and forms, the increase of electron density of cytoplasm and other changes.

9 years after the Chernobyl accident in the reconvalescents ARS changes in the hemopoietic system were found. The clinic-laboratory examination of the patients showed more often stable moderate anemia, leukocytosis, lymphocytosis and leukopenia than in the first period after irradiation.. Hypersegmentation, fragmentation of nuclears, the toxical granulation, hair- and basophilia of the lymphocytes cytoplasm, the vacuolization of cytoplasm and nuclears were observed in the blood and bone marrow cells.

We found the decrease of bone marrow cellularity to 10,0 - 13,0 x 10^9/l in some cases of examined groups after myelogram analysis. For the others the number of myelocaryocytes was normal and the hypercellularity was found only in 7 patients. The number of megacaryocytes fluctuated, as a rule, on the level of physiological standard, but the increase of the number of megacaryocytes were with lesser size of nuclear and shape modification. The number of cells with active production of platelets was decreased.

The correlation between the hematopoietic lineages was normal in the most cases in spite of the number of myelocaryocytes which were decreased. In some cases the narrowing of erythroid lineage from 5,5 to 10,5 % took place.

The membrane permeability of blood erythrocytes increased in the investigating group according to the data of osmotic and mechanical resistance.of erythrocytes. The thorning and spherical forms of erythrocytes were found in the peripheral blood of 40 % observed patients.

9 years later the ultrastructural changes of the hemopoietic elements remained the same. The histological investigations showed the predominance of the fat marrow over hemopoietic one with essential fibrosis.

We observed the increase of activity of antibacterial enzymes and decrease of the number PAS-positive lymphocytes in the bone marrow and peripheral blood.

Assessment of functional activity of hemopoietic progenitor cells (CFU-GMdc) was determined in bone marrow by quantity colonies - clones on 1e10 cultivating cells and their morphological analysis. The cloning efficiency of bone marrow in 1986 was low 12,6 + 0,6 for ARS degree I and 9,4 + 2,6 for ARS degree II. Those colonies consisted of blasts, metamyelocytes and maturing neutrophils, but rarely monocytes and macrophages. Eosinophilic and eosinophilic-granulocytic colonies predominated. Presence of eosinophilic's progenitor cells correlated with rising eosinophilic leucocytes in peripheral blood and bone marrow.

Cloning of progenitor cells 5-10 years after accident has shown that independent of the degree of ARS, in 19 patients with ARS (degree I) and 4 patients with ARS (degree II) quantitative changes in colony-forming activity in vivo were 38,5 - 3,6 and 30,0 - 5,4, correspondingly. The average number of colonies in the control was 36,5 + 3,8. Tendency to lowering this index in 5 patients was accompanied by morphological changes manifested by predominance of eosinophilic and eosinophil-neutrophilic colonies.

So, after some years in the majority of cases no significant differences in colony-formation in comparison with control were seen. However, predominance eosinophilic and eosinophilic-neutrophilic colonies in culture of 80 % examined patients was noticed though no eosinophilia in peripheral blood was observed.

The cultural research of stromal progenitor cells showed that the stromal progenitor cell activity varied from normal indexes to total proliferative potential depression. For three cases (ARS of degree I) grew none cluster and colony in the culture after 14 days of the bone tissue cultivation in vitro. The fibroblasts were absent. That testified to the presence of total depression of bone marrow stem cell stromal activity.

The research of paramagnetic centers carried out by using electron paramagnetic resonance method let us find the CU^{2+}- ceruloplasmin level decrease for the reconvalescents of ARS. The methemoglobin paramagnetic signal efficiency was normal and the free radical efficiency was greater than of the control group. The results of bone marrow paramagnetic centers studying (8 patients) appeared very interesting. The research carried out showed that the bone marrow paramagnetic centers efficiency was different from that of blood for stable modifications of peripheral blood (leukocytosis, leukopenia, anemia).

During the observation period we found the development of the hematological disorders after high doses of irradiation: 3 cases of myelodysplastic syndrome.

4. Conclusions

The results of our study show that the hemopoietic system of the reconvalescents ARS remain damaged. The presence of the qualitative and quantitative changes in the hematopoietic elements is acknowledged. The connection between hemopoietic microenvironment and hemopoiesis state was put up. The realization of the hematological disorders as myelodysplastic syndrome testified the importance of this problem in future.

References

[1]   V.G. Baryakhtar (1995), The Chernobyl Accident, Kiev, 560 pages.
[2]   S.P. Yarmonenko (1988), Radiobiology of men and animals, Moscow, 424 pages.
[3]   V.G. Bebeshko, V.I. Klimenko, L.N. Yukhimuk, I.S. Dyagil (1994), The hematopoietic system and bone marrow microenvironment state of the persons who were heavily irradiated as a result of the Chernobyl Accident. Proceedings of the International Read Table "Chernobyl: Never Again", Venice, Italy, 148 pages.

Therapy of combined radiation injuries with hemopoietic growth factors

Boudagov R., Oulianova L.

Medical Radiological Research Centre, Obninsk, Russia

Abstract.

Radiation accidents of the 5-7th levels according to IAEA scale lead to life-threatening acute radiation syndrome and many patients will probably suffer from additional thermal burns. These combined injuries (CI) will be among the most difficult to achieve survival. Present therapeutic means need to augment with new approaches to stimulate host defence mechanisms, blood system recovery and to enhance survival. The evaluation of therapeutic properties of human recombinant G-CSF, IL-1, IL-2 and other so called "biological response modifiers" on survival and blood recovery after CI was the purpose of this work. Experiments carried out with mice CBA x C57BL6 receiving 7 Gy total body irradiation followed by a full-thickness thermal burn of 10% of body surface. It established that G-CSF does not exhibit a positive modifying action on the damage level and on hematopoietic recovery. I.p two-four/fold infusion of IL-2 during the initial 2 days has provided a significant statistically survival increase from 40% (untreated mice with CI) to 86%. Single s.c IL-1 injection resulted in abrupt deterioration of the outcome when dealing with CI; three/fold administration of IL-1 in 2,4 and 6 days after CI did not increase survival. Extra-cellular yeast polysaccharides resulted only a 15 to 30% increase in survival if given 1 h after CI. The best results obtained when mixture of heat-killed L.acidophilus injected s.c immediately after CI - survival has increased from 27% (untreated mice) to 80%. Revealed beneficial effects of IL-2 and biological response modifiers did not accompany by a corresponding correction of depressed hematological parameters.

1.Introduction.

It is well known that in the event of radiation accidents many patients will probably suffer from thermal burns in addition to acute radiation syndrome (ARS). The outcome of these combined injuries (CI) is worse than for ARS alone. Even benign by themselves thermal burns become hazardous when a combination of trauma occurs in conjunction with total body irradiation at minimal-lethal or midlethal dosages. Exposed patients with additional thermal burn die during the first two or three weeks after irradiation mainly due to sepsis. In spite of long standing study of CI pathogenesis, management of infectious complications after CI remain as very hard medical problem [1-4]. Early supportive therapies of patients who have been exposed to high doses of radiation and trauma include antimicrobial therapy, platelet transfusion, fluids, electrolytes, immunoglobulin therapy, and surgical debridement or cleansing of wounds [5]. Recently several authors have reported that so called "immunoregulatory biological response modifiers" (glucan, trehalose dicorynomycolate, heat-killed Lactobacillus casei, recombinant cytokines et al.) can enhance hematopoietic cell recovery and increase survival when given after gamma or fission neutron irradiation alone [6-10]. Published data proved that agents of this group may increase macrophage's activity and secretion of cytokine such as hemopoietic growth factors. Efficacy of this new therapeutic approach for CI treatment did not study as a matter of fact. Madonna et al.[11] showed that injection of immunomodulator significantly augments resistance to infection and increases survival of irradiated mice. However, this treatment did not increase survival of mice with sepsis following irradiation and wound trauma. Recent work with mice in our laboratory showed that bacterial polysaccharide pyrogenal, thymus preparations, heterologic immunoglobulines don't modify in fact the low values of 30-day survival under CI. Single injection of prodigiozan, zymozan and some other yeast polysaccharides in 1 h after CI resulted at unimportant increase of survival [12-13]. The evaluation of therapeutic properties of human recombinant IL-1, IL-2, G-CSF and other biological response modifiers on survival and blood system state after CI was the purpose of this work.

2. The effectiveness of the hemopoiesis regulatory cytokines in CI models.

Recombinant IL-1-β was kindly supplied by Dr Ketlinski (Institute of High-Purity Biopreparation, St.-Petersburg, Russia). IL-2 (Biotech, St.-Petersburg, Russia) was a generous gift from Dr Jurkevich. G-CSF was supplied by Institute of Radiation Medicine (Chine).

CBA x C57BL6 male mice used in all studies. Animals held in quarantine for 2 weeks then irradiated from ^{60}Co source at a dose-rate 0.45 Gy/min. The midline absorbed dose was 7.0 Gy. Non-lethal per se full-sickness thermal burn 10% of body surface inflicted immediately after irradiation by means of powerful light of halogen lamps. This model of CI characterised by sharp decrease of 30-day survival in compare with only irradiated mice; untreated animals died mainly from sixth to twelfth days after CI.

The most attention we spared to therapeutic use of IL-1 in murine model of CI. It was taken into account that IL-1 act as an essential molecular master switch for secretion of GM-CSF, G-CSF, M-CSF, IL-3, IL-6 and other hemopoietic growth factors [14,15]. Several authors [16,17] have reported that IL-1 given 1-4 hour's postradiation (50-100-200 micrograms/kg, subcutaneously or intraperitoneally) increased survival of mice exposed to radiation alone. Our studies showed that rIL-1-β given once in 4 hr after CI (100 micrograms/kg, s.c.) caused a higher and earlier rate of mortality in 2-3 day after CI. In particular, 28 from 40 "treated" mice died instead of 100% survival untreated animals. Analogous results obtained when IL-1 dose reduced to 150 ng/mouse (40% of "treated" mice died at early phase of CI). Single s.c injection 150 ng IL-1 accelerated lethal outcomes of CI even followed 24 hr after CI. Repeated i.p. injection of smaller dose IL-1 (100 pg/mouse) in 2-4 and 6 days after CI did not influence on disease development and outcomes. It should be stressed that single administration of "high" dose IL-1 (100 micrograms/kg) and repeated small dose injection of this cytokine to only irradiated mice did not modify the early phase of ARS. Moreover slightly therapeutic effect took place and survival rate in 30-day period increased up to 15 or 30%. Aggravating effect of IL-1 on outcome of CI when injected in 4 or 24 hr are apparently connect with "burn component" of CI. Really, IL-1 administration in 4 h after burn alone resulted in 55% dearth rate during the first 2-4 days after non-lethal and non-shockogenic trauma per se. Possibility exist that at early phase of thermal burn or CI exogenous IL-1 interacts synergistically with endogenously produced TNF-α and IL-1-β. As result severe hypotension or other toxic effects of these cytokines may occur and potentially beneficial action of IL-1 on the radiosensitive systems (immunity and hemopoiesis) proves masked.

The role of interleukin-2 has been shown to be of great significance in the modulation of immune response and host resistance to sepsis in thermally injured mice [18]. On the other hand IL-2 may increase survival of irradiated mice and dogs [19]. These data lead us to investigate therapeutic properties of recombinant IL-2 for combined injury's treatment. Animals received intraperitoneal injections of rIL-2 in dose 5000 U/mouse. Three experimental schemes used: a) 4 injections in 15 min, 4, 24 and 48 hours after CI; b) 2 injections in 15 min and 4 hour after CI; c) 2 injections in 24 and 48 hours after CI. Results showed that IL-2 provides statistically significant survival increase from 40% (untreated control group) accordingly to 86%, 82% and 89%. Comparative study effects of "unstriking" cytokine IL-1 and "beneficial" cytokine IL-2 on hematology made too (Tabl. 1). Both cytokines did not correct severe cell devastation of bone marrow, leucopenia and anemia during the critical phase of CI. Only IL-2 administration increased E-CFU number per spleen and slightly raised blood platelets count.

Table 1

The influence of IL-1 and IL-2 treatment on mice blood system state after combined injury

Parameters	CI + saline	CI + IL-1	CI + saline	CI + IL-2
Number of bone marrow cells	2.25 ± 0.25	2.29 ± 0.36	2.00 ± 0.20	2.70 ± 0.46
Endogenous CFU/spleen	0.95 ± 0.18	1.95 ± 0.23	2.89 ± 0.74	6.40 ± 1.00
White blood cells	289 ± 44	329 ± 55	200 ± 28	183 ± 24
Granulocytes	50 ± 11	79 ± 29	45 ± 16	85 ± 22
Platelets	41 ± 4.4	51 ± 3.0	42 ± 2.2	79 ± 7.2
Erythrocytes	3.67 ± 0.12	3.95 ± 0.08	3.64 ± 0.10	3.67 ± 0.15

IL-1 injected in 2-4 and 6 days after CI. IL-2 injected in 15 min and 4 hour after CI. Blood system state registrated in 8 day after CI. Mean ± standard error of values presented. Bone marrow cells - x 10^6 per femur, erythrocyte - x 10^6 μl, platelets - x 10^3 μl.

Recombinant human G-CSF was investigated for the ability to accelerate bone marrow regeneration and to decrease the severity of leukopenia after irradiation only or CI (we thank O.Semina and T.Semenets for the participation in this experiments carry out). Mice were exposed to sublethal dose 4 Gy. G-CSF (2.5 micrograms/day) or saline administered on days 0-4 post-irradiation. Bone marrow cellularity, exogenous CFUs and white blood cell's number evaluated in 8 days after irradiation or CI. Results demonstrated that therapeutic G-CSF increased number of CFUs per femur from 640 ± 91 (untreated only irradiated mice) to 1030 ± 165. When animals exposed to CI this parameter changed insignificantly from 517 ± 71 (untreated mice with CI) to 541 ± 54. G-CSF did not strongly modify lowered bone marrow cellularity and leukocytes score in 8 day after irradiation or CI.

3. Therapeutic properties of new biological response modifiers in CI models.

Extra-cellular yeast polysaccharides of Bullera alba (B-678) and Sporobolomyces albo-rubescens (Sp-50) prepared by Prof. Elinov et al. (St.-Petersburg, Russia) and heat-killed Lactobacillus acidophilus prepared by Dr. Pospelova et al. (Moscow, Russia) have been investigated for CI treatment. Single i.p. injection of B-678 and Sp-50 (20 mg/kg) in 1 h after CI increased 30-day survival from 3% (untreated mice) accordingly to 23 and 20%. Heat-inactivated L.acidophilus (10^8 microbes per 1 ml growth media) injected s.c. following CI in volume of 0.1 ml/mouse. In this experimental group survival increased from 27% to 80%. None of the studied preparation rendered any beneficial action on the scores of bone marrow nucleated cells, white blood cells and CFUs as compared to control untreated groups (Tabl.2).

Therefore our studies demonstrated that short course of rIL-2 therapy or single subcutaneous injection of heat-killed L.acidophilus may increase survival of irradiated mice inflicted with thermal burn. Curative efficacy does not accompany by corresponding correction of depressed blood system state. Further research needs to establish the mechanisms of revealed beneficial effect on survival under combined injuries.

Table 2

Effects of biological response modifiers on some hematological parameters after CI

Experimental group	Bone marrow cells, x 10^6	Endogenous CFU/spleen	Leukocytes
1. CI + saline	1.10 ± 0.09	0.8 ± 0.33	292 ± 50
CI + B-678	1.27 ± 0.13	2.5 ± 0.94	345 ± 60
CI + Sp-50	1.44 ± 0.22	1.70 ± 0.39	291 ± 50
2. CI + saline	1.80 ± 0.21	2.24 ± 0.86	200 ± 41
CI + L.acidophilus	1.35 ± 0.20	1.14 ± 0.53	240 ± 49

Hematological parameters registrated in 8 day after CI. Mean ± standard error of values presented.

Acknowledgement

This investigation was made in framework of "EC/CIS Agreement for International Collaboration on the Consequences of the Chernobyl Accident" and was supported in part by the Commision of the European Communities. The authors thank T.Kliachina, S.Isaeva and T.Eriomina for their assistance in this work..

References

[1] Conklin, J.J. et al. Current concepts in the management of radiation injuries and associated trauma. *Surgery, Gynaecology and Obstetrics*, **156** (6): 809-829 (1983).

[2] Gruber, D. et al. (Eds). *The pathophysiology of combined injury and trauma. Management of infectious complications in mass casualty situation..* Orlando, Florida, Academic Press,Inc, 1987.

[3] Tsyb, A. & Britun, A. (Eds). *The pathogenesis and treatment of combined radiation-thermic injuries.* Moscow, Medicine, 1989 (in Russian).

[4] Tsyb, A. & Farshatov, M. (Eds). *Combined radiation injuries: the pathogenesis, clinic and treatment.* Moscow, Medicine, 1993 (In Russian).

[5] Brook, I. Use of antibiotics in the management of postirradiation wound infection and sepsis. *Radiation Research*, 115 (1): 1-25 (1988).

[6] Gallicchio, V.S. Accelerated recovery of hematopoiesis following sub-lethal whole body irradiation with recombinant murine interleukin-1 (IL-1). *Journal of Leukocyte Biology*, **43** : 211-215 (1988).

[7] Madonna, G.S. et al. Trehalose dimycolate enhances resistance to infection in neutropenic animals. *Infection and Immunity,* **57** (8): 2495-2501 (1989).

[8] Ledney, G.D. et al. Therapy of infections in mice irradiated in mixed neutron/photon fields and inflicted with wound trauma: a review of current work. *Radiation Research*, **128** : S18-S-28 (1991).

[9] Nomoto, K. et al. Radioprotection of mice by a single subcutaneous injection of heat-killed Lactobacillus casei after irradiation. *Radiation Research*, **125** (3): 293-297 (1991).

[10] Patchen, M.L. et al. Therapeutic administration of recombinant human granulocyte colony-stimulating factor accelerates hemopoietic regeneration and enhances survival in a murine model of radiation-induced myelosuppression. *International journal of cell clonning*, **8** (2): 107-122 (1990).

[11] Madonna, G.S. et al. Treatment of mice with sepsis following irradiation and trauma with antibiotics and synthetic trehalose dicorinomycolate (S-TDCM). *The Journal of Trauma*, **31** (3):316-325 (1991).

[12] Boudagov, R. et al. Rationale for the methods of prevention and treatment of toxic infectious complications of combined radiation and thermal injuries. *Medical Radiology*, **37** (9-10):41-44 (1992). In Russian.

[13] Khlopovskaya, E. et al. Application of drugs of protozoan origin as means of prophylaxis and early therapy in combined radiation-thermal injuries. *Science Academy's news. Biological series*, (2):227-234 (1993). In Russian.

[14] Bagby, G.C. Interleukin-1 and hematopoiesis. *Blood Reviews*, 3:152-161 (1989).

[15] Ketlinski, S.A. & Kalinina, N.M. Cytokines of mononuclear phagocytes in the regulation of inflammation and immunity reactions. *Immunology*, (3): 30-44 (1995). In Russian.

[16] Rogacheva, S.A. et al. Study of the antiradiation effect of interleukin-1-β in experiments. *Radiobiology. Radioecology*, **34** (3): 419-423 (1994). In Russian.

[17] Legeza, V.I. et al. Experimental study of IL-1-β efficacy in radiation injury. *Hematology and Transfusiology*, (3): 10-13 (1995). In Russian.

[18] Gough, D.B. et al. Recombinant interleukin-2 (rIL-2) improves immune response and host resistance to septic challenge in thermally injuried mice. *Surgery*, **104** (2):292-299 (1988).

[19] Rogacheva, S.A. et al. Curative action of human recombinant IL-2 at the acute radiation disease in experiment. *Radiobiology. Radioecology*, **35** (2): 237-243 (1995). In Russian.

Assessment of Cutaneous Radiation Fibrosis by 20 MHz-Sonography

P. GOTTLÖBER[1], N. NADESHINA[2], O. BRAUN-FALCO[1], G. PLEWIG[1],
M. KERSCHER[1], R.U. PETER[1]

[1]*Department of Dermatology, Ludwig-Maximilians-University, Munich, Germany*
[2]*Institute of Biophysics, Clinical Department, Moscow, Russia*

Abstract. Radiation fibrosis is the cardinal symptom of the chronicle stage of the cutaneous radiation syndrome. The degree of cutaneous fibrosis can clinically be estimated by palpation. High-frequency 20 MHz-sonography is an established, noninvasive procedure, which renders an exact determination of skin thickness and additionally densitometry is possible. We investigated 15 survivors of the Chernobyl accident in 1986, who developed symptoms of the chronical stage of the cutaneous radiation syndrome. We determined skin thickness and echogenicity of skin areas clinically suggestive of radiation fibrosis before, during and after treatment. 20 MHz-sonography showed a distinct enlargement of the echorich corium and a reduction of the subcutaneous fatty tissue in comparison with the unaffected, contralateral skin, here demonstrating typical features of radiation fibrosis, namely dermal fibrosis and reactive pseudoatrophy and fatty tissue. The histology presented an increase and swelling of the collagen fibers and atypical fibroblastic cells. The patients received treatment with low-dose interferon y (Polyferon [R], 3 x 50 µg s.C., three times per week) up to 30 months. A marked reduction of skin thickness and echogenicity reaching nearly normal values could be observed. We conclude that 20 MHz-sonography is an easy to apply, noninvasive, well established procedure to quantify cutaneous radiation fibrosis and to assess therapeutic outcome.

1. Introduction

Cutaneous radiation fibrosis, a symptom of the chronicle stage of the cutaneous radiation syndrome [1,2], is caused by an increase of collagenous connective tissue.. It is of clinical importance as severe fibrosis can lead to functional impairment, contractures and secondary ulcerations [3]. The degree of cutaneous fibrosis can clinically only be estimated by palpation. Histologic examination is of limited applicability, as fibrotic skin has a decreased vascularization and does not heal easily.

Other procedures for determination of cutaneous radiation fibrosis like nuclear magnetic resonance tomography and positron emission tomography are time-consuming and expensive. Computer tomography is associated with an additional radiation load for the patients.

Therefore a noninvasive method which allows a quantitative determination of the extent of cutaneous radiation fibrosis after exposure to ionizing radiation is needed.

The 20 MHz-B-Scan-ultrasound is a well established procedure for three-dimensional imaging of morphology and topography of cutaneous structures, which proved to be highly reproducible in the hands of experienced examinators [4]. It has been used for skin thickness measurements in patients with localized and progressive systemic scleroderma [5] and for evaluation of drug effects on the skin [6]. Furthermore, sonometric measurements of tumor thickness in patients with basal cell carcinoma [7] and malignant melanoma [8] have been performed.

Here we report on the experience in the follow-up of survivors of the Chernobyl accident with severe radiation-induced skin damage by this diagnostic method.

2. Patients and methods

15 male survivors of the Chernobyl nuclear power plant accident in 1986, suffering from the chronical stage of the cutaneous radiation syndrome, were examined in our department in 1991 and during the following four years. Telangiectases, radiation keratoses, radiation ulcers, atrophy, hypo- and hyperpigmentation, hemangiomas and hematolymhangiomas were noted (2). Additionally, a marked induration on the exposed skin areas in 8 of the 15 survivors could be determined. According to the clinical investigation fibrotic skin areas were selected and examined by 20 MHz-B-scan-ultrasound before during and at the end of treatment with low-dose interferon y (Polyferon ^{R}, 3 x 50 µg per week). Furthermore, biopsies were taken of selected cutaneous areas at the beginning and the end of therapy.

Histologic examination revealed the typical histological signs of cutaneous radiation damage like orthohyperkeratosis, flattening of epidermis, i.e. loss of papillary folds and an increase of collagenous fibers with atypical fibroblastic cells. The collagenous fibers were swollen.

Ultrasound measurements including A-scans, sum A-scans and B-scans were performed by the DUB 20 S, a digital ultrasound imaging system (tpm, Lüneburg, Germany) The 20 MHz-scanner with an axial resolution of about 80 µm and a lateral resolution of 200 µm is suitable to investigate epidermis, corium and subcutaneous fat tissue up to a depth of about 10 mm. All fibrotic areas of the examined patients were sonographically compared with contralateral, nonaffected sites. Skin thickness was determined as distance extending from the entry echo to the output echo representing the border between the deepest coral region and the subcutaneous fat tissue.

3. Results

The sonographic image of the involved skin showed under the entry echo a very echorich, corium considerable increased thickness. Furthermore, some isles of echorich spots were seen in the subcutaneous fat tissue, presenting enhanced collagenous septs. Additionally the echolucent space between the echo of the corium and echo of the muscle fascia was markedly reduced, presenting pseudoatrophy of fatty tissue caused by proliferative coral fibrosis.

Ultrasound B-mode images of radiation fibrosis revealed increased thickness of the corium from 27 % to 98 % in comparison with the uninvolved skin before treatment . The skin density was increased to 40 % of the normal value.

In two patients who didn't tolerate the therapy and topped the treatment after the first injection, a progression of cutaneous radiation fibrosis from 67 % to 200 % could be determined sonographically during 30 months. In these patients treated with interferon y there was a reduction of skin thickness during therapy and an approximation to the normal skin thickness. The values decreased from 26 % to 64 % compared with initial thickness before treatment after 30 months of therapy. Furthermore, there was a normalization of the skin density. The details of the therapeutic experience are published elsewhere (Peter et al submitted).

Conclusion

Radiation fibrosis develops after exposure to ionizing radiation in various tissues and is of eminent clinical significance, as the extent and progression of radiation fibrosis determines the clinical outcome [9].

Information about changes of the corium, subcutis and muscle fascia represent important data with regard to quality and quantity of the fibrotic process which has not been available until now. The 20 MHz-B-scan-ultrasound is an available, noninvasive and well reproducible procedure for determining the thickness and density of the fibrotic skin. There is a good agreement of the sonometrically and histometrically obtained skin thickness values. The 20 MHz-B-scan-ultrasound should be considered in the routine follow-up procedures for determination of the clinical course of cutaneous radiation damage. Furthermore, it should be used to assess the clinical course and to evaluate the therapeutic results.

References

[1]   R.U. Peter, Klinische Aspekte des kutanen Strahlensyndroms nach Strahlenunfällen-Erfahrungen nach Goiania und Tschernobyl, Akt Dermatol 19 (1993), pp. 364-367.

[2]   R.U. Peter, O. Braun-Falco, A. Birioukov, N. Hacker, M. Kerscher, U. Petersheim, T. Ruzicka, B. Konz, G. Plewig, Chronic cutaneous damage after accidental exposure to ionizing radiation: The Chernobyl experience, J Am Acad Dermatol 30 (1994), pp. 719-723.

[3]   O. Braun-Falco, G. Plewig, H.H. Wolff, R.K. Winkelmann, Dermatology. Berlin, Springer, 1991.

[4]   M.J. Stiller, J. Driller, J.L. Shupack, C.G. Gropper, M.C. Rorke, F.L. Lizzi, Three-dimensional imaging for diagnostic ultrasound in dermatology, J Am Acad Dermatol 29 (1993), pp.171-175.

[5]   K. Hoffmann, U. Gerbaulet, S. el Gammal, P. Atlmeyer, 20 MHz-B-mode ultrasound in monitoring the course of localized scleroderma (morphea), Acta Derm Venereol 164 (1991), pp. 3-16.

[6]   M. Kerscher, H.C. Korting, Topical glucocorticoids of the non-fluorinated double-ester type: Lack of atrophogenicity in normal skin as assessed by high-frequenced ultrasound, Acta Derm Venereol 72 (1992), pp. 214-216.

[7]   K. Hoffmann, K. Winkler, S. el Gammal, P. Altmeyer, A wound healing model with sonographic monitoring, Clinical and Experimental Dermatology 18 (1993), pp. 217-225.

[8]   K. Hoffmann, J. Jung, S. el Gammal, P. Altmeyer, Malignant melanoma in 20 MHz-B scan sonography, Dermatology 185 (1992), pp. 49-55.

[9]   E. Scherer, C. Streffer, R. Trott: Radiopathology of organs and tissues. Berlin, Springer, 1990.

Evaluation of Lioxasol for the Treatment of Accidental Local Radiation Injuries: an Experimental and Clinical Study.

Natalia M. Nadejina[1], John W. Hopewell[2], Igor A. Gusev[1], Mohi Rezvani[2], Gerard M. Morris[2], Tatiana G. Protasova[1], Tatiana E. Chelmodaeva[3], Nina I. Fetisova[3] and Lev B. Shagalov[3].

[1]Scientific Research Center - Institute of Biophysics, Moscow, Russia
[2]Research Institute of Oxford University, The Churchill Hospital, Oxford, Great Britain
[3]"Pharmazashita" Scientific Productive Association, Moscow, Russia.

Abstract: The Chernobyl accident caused the development of Acute Radiation Syndrome (ARS) in 134 individuals, these were either treated at Hospital 6 (Moscow) or in hospitals in Kiev. Local radiation injuries (LRI) were found in 54 patients from the 108 ARS patients treated in Moscow over the acute period; 2 additional patients from this group had combined radiation and thermal skin injuries (the total number of LRI patients was 56).

The effectiveness of Lioxasol, an ethyl alcohol based product containing 2-alliloxoethanol, was investigated in these patients. The treatment group was composed of 8 survivors of ARS with a second degree LRI caused by relatively uniform gamma-beta exposure. The control group was composed of 8 patients suffering from ARS also of second degree (7 patients) or first degree (1 patient) reactions caused by external, relatively uniform, gamma-beta exposure between 1956 and 1970.

The time of of re-epithelisation in the treated group was 25.4 ± 3.1 days after irradiation. This was slightly shorter than the 28.3 ± 4.9 days in the control group. However, this difference was not statistically significant (p>0.05).

The effectiveness of Lioxasol was further studied on pig skin. Multiple sites in the same animal were irradiated with 22.5 mm diameter ^{90}Sr/^{90}Y plaques. The time of onset of moist desquamation and the subsequent healing times were used as end points. Following a single dose of 35 Gy, a dose known to produce moist desquamation in all irradiated sites, Lioxasol was applied topically twice a day. Lioxasol treatment (twice daily), which started the day after irradiation, delayed the time of onset of moist desquamation significantly from 5.1 ± 0.2 weeks to 5.5 ± 0.2 weeks. However, the most marked effect was on the number of sites that healed within 3 weeks of the first appearance of moist desquamation. This was 80 ± 10.3% for sites treated with lioxasol whereas in untreated sites only 26.7 ± 11.4% of the irradiated fields were healed by this time (p<0.001). The possibility that this might be explained by enhanced proliferation of cells in the basal layer of the epidermis was tested by examination of the effects of prolonged Lioxasol treatment on unirradiated pig skin. Twice daily applications of lioxasol for 2 weeks resulted in a 28.6 ± 1.7% increase in the ^{3}H-thymidine labelling index of basal cells. This level of elevation in the labelling index was maintained until 10 weeks; when Lioxasol treatment was stopped. There was no quantifiable evidence of hyperplasia in the epidermis but the number of labelled cells, fibroblasts and endothelial cells, was markedly increased in the papillary dermis. Twice daily application of lioxasol for 2-10 weeks produced a 43-58% increase in the number of labelled cells per mm^2. These observations provide a possible mechanism for the actions of Lioxasol in ameliorating localized radiation injury to the skin.

1. Introduction

Non uniform exposure to individuals has been a feature of the numerous radiation accidents that have been associated with the use of nuclear power plants, radiation devices in medicine and industrial applications of ionizing radiation, since the end of the second world war. Partial body exposures have been associated with extremely high doses to the skin and mucous membranes and relatively lower non-uniform doses to the bone marrow. Extremely high doses to the skin, from beta emitters, have caused severe skin effects which were the primary cause of death in some cases.

Initial reports [1] on the Chernobyl nuclear power plant accident in 1986, have accounted for 115 patients, with clinical signs of general acute radiation syndrome (ARS), who were admitted to the Hospital of the Institute of Biophysics, Moscow. Among these patients 56 had a varying degree of radiation-induced skin lesions. Lymphocyte analysis has subsequently shown that two of these cases had not received a total body dose sufficient to result in a general ARS. Skin lesions in these two individuals were caused by the combined effects of radiation and thermal burns.

Initially, the victims of ther Chernobyl accident who developed skin were classified into four groups [1] according to the following criteria: The contribution of the ß- and γ-radiation components to the total dose, the prevalence of distant or contact exposure to the skin and the nuclide composition of the radioactive sources to which an individual was exposed. In the light of additional information, obtained during the last few years, the number of cases allocated to each group has been revised and are slightly different to the initial values. These new values are given in Table 1. In some cases, skin lesions were so severe that they were not compatible with patient recovery even in the absence of any ARS. Deaths in 26 patients in the first 3 months of the exposure were associated with skin lesions involving >50% of the total body surface area (Table 2). This left 28 survivors who had acute local injuries and developed the long term consequences of these injuries. Of these 25 remain alive 10 years after the accident.

Table 1. Dependence of clinical signs of radiation-induced damage to the skin on the depth-dose distribution.

Group	Number of Patients current (original)	γ doses (Gy)	Skin doses (Gy) at: 70 μm	150 μm	Time of onset of severe damage (days) 1st wave	2nd wave	Maximal clinical signs (time of onset in days)
I	17 (15)	2-5.8	8-10	3-4	-	-	Erythema (15)
			60-90	20-30	25-28	~60	Small erosions (35 & 65)
			100	>30	20	45	Ulcerations (30 and 50)
II	6 (6)	4.0-12.7	100	10	10	-	Small erosions (20)
			250	~30	5-7	35	Necrosis (20-30)
			360†				
III	6 (6)	9.0-1 4	>200	>50	5-10	-	Necrosis (15)
IV	25 (29)	2.0-11.5	150	50	8-10	~30	Ulceration & necrosis (25)
			300†	100	5-7	~30	Total necrosis of skin (20-25)
Total	54 (56)						

†The most exposed cases

Group I: exposed to distant high energy ß-γ sources; Group II: exposed to γ-radiation from smoke constituents of the plume and γ skin contamination from particles of fall out; Group III: high dose ß-γ radiation from smoke and steam (firemen); Group IV: varied group, some with total or local wetting of clothes to generate a thick radiation source on the skin surface (ß-γ irradiation).

Table 2. Number of deaths over the acute period in patients showing a varying severity of radiation-induced skin lesions and the number still alive after 10 years.

Group	No. of patients	No. of acute deaths	Alive at 10 years
I	17	1	14
II	6	4	2
III	6	6	0
IV	25	15	9
Total	54	26	25

Clinical management of localised radiation injuries to the skin involves the development of more effective therapeutic measures for both the acute and for the late periods as well as the elaboration of methodologies for diagnosis and surveillance of the dynamics of the pathological processes at different times after an accident. Main clinical manifestations of different grades of radiation induced skin lesions in radiation accident victims are given in Table 3.

Table 3. The main clinical manifestations and levels of absorbed doses (short-term gamma irradiation) for different grades of local radiation injuries to the skin.

Local radiation injury	Grade of lesion (approximate dose, Gy)			
	I (8-12)	II (12-20)	III (20-25)	IV >25
Initial erythema and time of appearance	Continues for few hours, may be absent	Seen from a few hours to 2-3 days	Seen in all; duration from 3-6 days	Seen in all; its severity does not decrease prior to manifestion of main reaction
Main Reaction - latent period - clinical signs	15-20 days erythema, 'dry' desquamation	10-15 days erythema, oedema, blisters, moist desquamation	7-14 days erythema, pain, blisters, erosions	Oedema, pain necrosis, local haemorrhages

Cases involving exposure doses of more than 20 Gy (Grade III-IV lesions) are likely to develop late radiation ulcers even where there is initial healing of acute lesions. This is illustrated by an analysis of representative cases with either Grade I, II or III LRI (Figure 1). Patients with Grade I lesions show full re-epithelialisation with skin scarring or atrophy. The majority of patients with Grade III lesions do, on the other hand, develop late ulceration. This confirms the necessity for early (20-30 days) surgical intervention after local irradiation with doses higher than 20 Gy. After such doses it is important to perform surgical intervention in the period of ulceration and necrosis to reduce the risk of

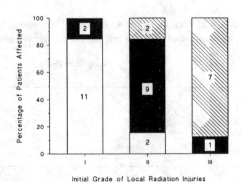

Figure 1: Incidence of late consequences of exposure in patients with an initial diagnosis of either Grade I, II or III LRI. ▨▨ Late radiation ulcers, ▮▮ Tissue scarring or atrophy ☐ full re-epithelisation.

the development of anatomical abnormalities and defects of damaged tissues. The following types of operation are performed:

1. The simultaneous dissection of all the damaged tissues, with healing of the postoperative wound. This operation prevents late pathological processes, including the development of late trophic ulcers. Usually such therapy is possible when the LRI are not localised in regions of large blood vessels, nerves and tendons.

2. Transplantation of a skin flap onto an ulcerated defect without any preliminary dissection of the ulcer. This operation is recommended when LRI is localised on the face, fingers or if major arteries, nerves or tendons are located at the base of the ulcer. For the successful survival of a skin flap, the surface of the ulcer should be prepared for transplantation (treatment with vasoactive drugs and local application of agents to clean the surface of the ulcer). The best time for such an operation is after full ulcer cleansing and the appearance of bright-red granulation tissue. Sutures are removed on days 12-14 after surgery and the flap stem is dissected on day 20. According to available experience a free divided skin flap does not always prevent the development of late radiation ulcers.

3. Amputation of damaged part of extremities is carried out in cases of dry gangrene, large necroses of soft tissues with bones opening. The level of amputation depends on the localisation and area of destruction and on the possibility of creating the most favourable conditions for the formation of a viable amputation stump, considering future prosthetics.

4. Necrotomy with skin transplants.

    a) free skin graft
    b) multilayered pedical skin flap
    c) skin-muscular flaps on vascular pedicule (microsurgical method)

Since 1990 the Department of Plastic and Reconstructive Microsurgery of the Russian Academy of Medical Sciences have used microsurgical techniques in association with operations on 7 patients with LRI who had ulcerative and necrotic skin defects at late times after exposure. In 4 patients, LRI resulted from accidental gamma-irradiation. The absorbed radiation doses varied from 20 to 60 Gy. As the patients had multiple skin lesions, operations were performed in stages. The results of the operations were evaluated by defects healing and recovery of hand function. Substantial increases in hand functional activity were seen as well as the social rehabilitation of the patients.

LRI of the mild and moderate degree (Grades I and II) can be treated using conservative approaches. Such conservative therapies are subdivided into general and local methods. It is the view of the Moscow group that the most effective means of local therapy in cases of LRI, of Grade I-II, are preparations giving the opportunity to use 'blister cover' as a natural barrier to secure against wound infection and to allow the re-epithelization of the eroded surfaces situated under the skin blister. Therefore, we have investigated the effectiveness of a Lioxasol preparation in victims of LRI of the mild and moderate degree of severity.

2. Treatment of LRI with topical application of Lioxasol:

A Lioxasol preparation, in the form of an alcohol aerosol, has been developed at Hospital 6 in Moascow. This compound has been widely used, topically, in the treatment of acute local radiation injuries during the last decade. This was carried out as part of the routine treatment in that hospital and unfortunately, there is no independent study to demonstrate the effectiveness of Lioxasol. This paper reports the results of the clinical course of LRI in patients who received Lioxasol as part of their treatment compared with those patients who did not receive Lioxasol. The analysis of the clinical course of LRI includes the time of occurrence, time to re-epithelisation of radiation ulcers and the duration of the clinical manifestation of lesions.

There were some difficulties in finding appropriate control patients for some historical reasons. Developments in the nuclear industry and improvements in radiation protection for radiation workers has produced a change in the type of radiation accidents during recent years. In the early periods (1955-1977) local exposures were dominated by accidents with unsealed sources (compounds of radium and nuclear fuel fission products), whereas in the later periods (1980-1992) cases of accidental local exposure were caused by sealed industrial radiography sources (Ir-192, Cs-137). These different types of casualties caused some difficulties in compiling comparable study groups with similar exposure characteristics. The selection of patients was carried out using the following criteria:

- radiation type (taking into account the radiation quality and energy).
- exposure dose or degree of severity of the lesion: if the severity of the lesion was not compatible with the calculated dose clinical assessment was accepted.
- localisation: area effected by local radiation injuries (hands, upper extremities).

1. Effectiveness of Lioxasol in the treatment of LRI caused by local gamma exposure

The multiple origin of the criteria given above caused relatively small numbers of patients to be selected for the control and Lioxasol treated groups. Lioxasol group consisted of 11 patients who suffered from local radiation injury of Grade I-III. The

Table 4. Characteristics of patients with LRI caused by local gamma exposure, who received Lioxasol as part of their treatment.

Patient No.	Patient age at accident (years)	Year of accident	Source	Radiation	Exposure dose (Gy) calculated	Exposure dose (Gy) clinical	Reaction grade	Localisation of LRI	Latency before clinical manifestation (days)	Time to re-epithelisation (days)	Duration of clinical manifestation (days)
1	52	1988	Sr-Y-90	gamma-beta	-	20	I-II	1st, 2nd finger right hand	12	34	22
2	31	1982	Cs-137	gamma	25	25	II-III	1st - 3rd finger right hand	8	59	51
3	18	1991	Cs-137	gamma	16	16	I-II	1st - 3rd finger right hand	16	30	14
3	45	1982	Co-60	gamma	30	30	II-III	1st, 2nd finger left hand	11	46	35
5	24	1989	Ir-192	gamma	50	25	II	1st - 3rd finger right hand	10	60	50
6	43	1982	Co-60	gamma	25	25	I-II	1st, 2nd finger right hand	10	30	20
7	50	1990	Ir-192	gamma	7	15	I-II	right hand	18	43	25
8	60	1992	Ir-192	gamma	10	15	II	2nd, 3rd finger left hand	13	32	19
9	57	1982	Co-60	gamma	20	20	I-II	1st, 2nd finger right hand	10	62	52
10	45	1984	Ir-192	gamma	280	20	I-II	1st - 3rd finger right hand	10	32	22
11	49	1984	Ir-192	gamma	65	18	I-II	1st, 2nd finger right hand	10	54	44
								Mean (± SE)	11.6±0.9	43.8±3.7	32.2±4.2

control group also included 11 patients who suffered from local radiation injury of Grade I-III caused by gamma radiation between 1951 and 1974.

Lioxasol group: local application of Lioxasol from day 6 to day 26 (100% of cases), Ozocerite (45% of cases), Hippophae oil (9% of cases), Rivanolum (9% of cases), anti-burn ointment (9% of cases), DNA with Dioxidinum (9% of cases), electric puncture (9% of cases), systemic applications of: Solcoseril i.m. (45% of cases), Trental i.v. (45% of cases), Hyperbaric oxigenation (36% of cases), Complamin i.v. (9% of cases), vitamins B-1, B-6, and C i.m. (18% of cases), H Hemodez i.v. (18% of cases), Contrical i.v. (9% of cases), Troxevasinum p.o. (9% of cases), Curantilum p.o. (9% of cases), Cinnarisinum p.o. (9% of cases). A description of these patients is given in Table 4.

Control group: local applications of Cintomicinum emulsion (36% of cases), Sintoson (27% of cases), Rivanolum (18% of cases), peach oil (18%), Lanolin (18% of cases), Locacorten (9% of cases), glycerine (9% of cases), Hippophae oil (9% of cases), Furacilinum (9% of cases), paraffin baths (9% of cases); systemic application of: vitamin B-1 i.m. (36% of cases), polivitamins p.o. (27% of cases), vitamin C i.m. (18% of cases), Nikoshpan p.o. (18% of cases), Galidore p.o. (18% of cases), Glivenol p.o. (9% of cases), hyaloid i.m. (9% of cases), Neocompemsan i.v. (9% of cases), vitamin A p.o. (9% of cases). A description of these patients is given in Table 5.

The latency period for the clinical manifestation of LRI was 11.6 ± 0.9 days in the Lioxasol group. This was found to be longer than 7.5 ± 1.4 days in the control group. This difference was statistically significant ($p<0.05$). The time taken for the re-epithelisation of skin lesions was 43.8 ± 3.7 days and 54.9 ± 8.3 days in Lioxasol and control groups, respectively. Although the time taken for the re-epithelialisation of skin lesions after exposure was shorter in Lioxasol group the difference did not prove to be statistically significant ($p = 0.2$). This was also true for duration of the clinical manifestation of the reaction ($p = 0.1$).

2. Effectiveness of Lioxasol in LRI caused by local beta exposure:

Here the Lioxasol treatment group consisted of patients with acute radiation syndrome (ARS) from the Chernobyl accident whose lesions were caused by beta irradiation. A major difficulty associated with patient selection was the assessment of physical dose to the skin of these patients. This was due to the following reasons:

- beta-radiation intensity was unknown and/or rapidly changing depending on the time and distance from the radiation source. These included those changes caused by decontamination procedures.
- the presence of short-lived beta emitters, which were not taken into account.
- uncertainty associated with the exposure time.

Therefore, patient selection was based mainly on the clinical assessment and the severity of skin lesion.

Lioxasol group comprised of 8 patients with Grade II ARS caused by relatively uniform gamma exposure. All patients were observed to have LRI of the skin of Grade I-II caused by beta-gamma exposure. The characteristics of this group of patients are given in Table 6.

Table 5. Characteristics of patients with LRI caused by local gamma exposure who did not receive Lioxasol as part of their treatment.

Patient No.	Patient age at accident (years)	Year of accident	Source	Radiation	Exposure dose (Gy)		Reaction grade	Localisation of LRI	Latency before clinical manifestation (days)	Time to re-epithelisation (days)	Duration of clinical manifestation (days)
					calculated	clinical					
1	25	1951	Uranium fission products	gamma	10	25	II-III	hands	5	100	95
2	24	1958	Br-82	gamma	10	15	II	right hand	13	41	28
3	31	1974	Rn-106	gamma	20	20	II	1st finger right hand	1	20	19
3	26	1969	Uranium fission products	gamma	15	15	II	hands	10	54	44
5	30	1973	Co-60	gamma	30	30	I-II	1st - 3rd finger left hand; 1st, 2nd finger right hand	5	31	26
6	28	1967	Co-60	gamma	20	20	II-III	hands	5	90	85
7	28	1957	Ra-224	gamma	20	20	II	1st, 2nd finger left hand	1	38	37
8	23	1974	Uranium fission products	gamma-beta	20	20	I-II	hands	13	33	20
9	34	1968	Sc-46	gamma	15	15	II-III	left hand	14	87	73
10	22	1971	Uranium fission products	gamma	30	30	II	hands	6	30	24
11	34	1968	Brem-stralen	gamma	20	20	II	1st finger left hand	10	80	70
								Mean (± SE)	7.5±1.4	54.9±8.3	47.4±8.1

Table 6. Characteristics of patients with ARS caused by relatively uniform gamma exposure and with LRI caused by local gamma-beta exposure (Chernobyl, 1986) receiving Lioxasol.

Patient No.	Patient age at accident (years)	Radiation	Exposure dose (Gy)		ARS grade	Acute LRI grade and area of skin damage (%)	Localisation of LRI	Time to re-epithelisation of radiation ulcers or start of pilation (days)
			calculated	clinical				
1	37	gamma-beta	10	25	II	1 - 15%	face, hands, lower legs	34
2	36	gamma-beta	10	15	II	1 - 20% 2 - 1%	face, breast, lower legs, buttocks	19
3	64	gamma-beta	20	20	II	1 - 5%	face, neck	17
3	27	gamma-beta	15	15	II	1 - 20% 2 - 2%	face, neck, breast, feet, lower legs, thighs	39
5	55	gamma-beta	30	30	II	1 - 12%	face, neck, breast, lower legs, buttocks	25
6	39	gamma-beta	20	20	II	1 - 12%	face, neck, breast, lower legs	12
7	35	gamma-beta	20	20	II	1 - 10%	neck, face, hands, forearms, lower legs	23
8	30	gamma-beta	20	20	II	1 - 6% 2 - 6%	breast, hands, forearms, feet	34
							Mean (± SE)	25.4±3.1

Table 7. Characteristics of patients with ARS caused by relatively uniform gamma exposure, and with LRI caused by local gamma-beta exposure, not receiving Lioxasol as a component of their treatment.

Patient No.	Patient age at accident (years)	Year of accident	Radiation	ARS degree of severity	Acute LRI, degree of severity and area of radiation skin damage	Localisation of LRI	Time to re-epithelisation of radiation ulcers or start of pilation (days)
1	37	1956	gamma-beta	II	1 - 25% 2 - 3 %	face, inner hand, forearms, inner elbows	40
2	36	1970	gamma-beta	II	1 - 30% 2 - 5%	face, hands, forearms, thighs	30
3	64	1956	gamma-beta	II	1 - 6%	face, inner hands	18
3	27	1956	gamma-beta	II	1 - 14%	hands, lower thirds of forearms, face	20
5	55	1956	gamma-beta	II	1 - 15%	face, neck, breast, hands	20
6	39	1956	gamma-beta	II	1 - 8%	face, hands	19
7	35	1970	gamma-beta	II	1 - 35% 2 - 15%	neck, face, right hand, back, abdomen, breast, legs	60
8	30	1956	gamma-beta	II	1 - 9%	face, hands, elbows	19
						Mean ± SE	28.3 ±4.9

The control group was also comprised of 8 patients suffering from Grade II ARS (7 patients) and Grade I (1 patient) caused by a relatively uniform external gamma-beta exposures between 1956 and 1970. All patients entered in control group also had gamma-beta skin lesions of Grade I and II (Table 7). Average area of Grade I and II injured surface of skin was similar in both groups.

The appearent difference between times to re-epithelisation of 25.3 ± 3.1 days in Lioxasol group and 28.3 ± 5.3 days in control group was not significantly different (p>0.05). However, it must be borne in mind that both control and Lioxasol patients received other treatments too. The true effect of Lioxasol could be better determined in an experimental set up where comparable groups of animals could be treated with Lioxasol only.

3. Experimental investigation of the effectiveness of topical application of Lioxasol on radiation damage:

1. Animal model

Acute radiation-induced damage to pig skin can best be quantified by assessing the dose-related incidence of moist desquamation, an easily recognised clinical change in the appearance of the skin. After beta-irradiation from standard 22.5 mm diameter $^{90}Sr/^{90}Y$ plaques (dose-rate ~3 Gy/min), the dose associated with a 50% incidence of the effect (ED_{50} ±SE) following an acute single exposure is 27.3 ± 0.5 Gy [2, 3]. The latency period for the first appearance of this effect (4.75 ± 0.16 weeks) at least for doses of 22 - 45 Gy, is independent of the level of the radiation dose. The severity of moist desquamation and hence the time taken for a desquamated area to heal is dose-related. This model was used to study the effectiveness of the topical application of Lioxasol.

2. Topical application of Lioxasol:

For these studies 22.5 mm diameter sites were irradiated with a single dose of 35 Gy or 65 Gy of β-rays from $^{90}Sr/^{90}Y$ plaques.

After irradiation skin sites were sprayed twice daily with the aerosol form of Lioxasol and sites allowed to dry. Treatment with Lioxasol was stopped when skin sites developed moist desquamation. The effects of Lioxasol treatment, as compared with untreated skin sites, on the opposite flank of two pigs, are shown in Table 8.

Table 8. Effects of Lioxasol treatment as the latency and healing of moist desquamation in pig skin after irradiation with 35 Gy of $^{90}Sr/^{90}Y$ ß-rays.

	Lioxasol sites	Control sites
Latency (wks)	5.53 ± 0.18	5.07 ± 0.17
Healing time (wks)	>2.0 ± 0.46	>4.37 ± 0.64
Sites unhealed at 11 wks	1/15	5/15
Sites healed after 3 wks	12/15	4/15

Treatment with Lioxasol slightly, but significantly, delayed the time of onset of moist desquamation by 3-4 days (p<0.05). The delay in the clinical manifestation of skin damage was also approximately 4 days in accident cases after gamma-irradiation. The marked reduction in the healing time for moist desquamation was somewhat difficult to interpret since some sites remained unhealed 11 weeks after irradiation when the study was completed. However, assessment of the number of sites, healed after 3 weeks, which was 80 ± 10.3% in the Lioxasol treated group and only 26.7 ± 11.4% in the control group was highly significant. (p<0.01).

Following the higher dose of 65 Gy to pig skin Lioxasol had no significant effect in the latency for the appearance of moist desquamation after irradiation and on the very delayed healing of this reaction.

4. Effects of Lioxasol on the cell proliferation kinetics of pig skin:

In addition to quantifying the effectiveness of the above treatment protocol, it was also envisaged that studies in animals might be used to determine possible mechanisms of action of the therapeutic procedure. In this respect, experiments were designed to determine the effects of Lioxasol on the proliferation kinetics of basal cells in the epidermis of pig skin. The response of this cell population is the key to responses of the skin. Counts of the number of labelled cells, endothelial and fibroblasts, were also made in the papillary dermis.

Lioxasol was applied topically, twice daily, to unirradiated skin. The experimental findings related to labelling index of the epidermal basal cell layer and the number of labelled cells/mm^2 in the papillary dermis are given in Table 9.

Table 9. Basal layer labelling index (±SE) of pig epidermis and in the papillary dermis at various times after the topical administration of Lioxasol.

Time after compound administration (weeks)	Basal layer labelling index (%)	Labelled cells/mm^2 Papillary dermis
0	8.4 ± 0.4	14.3 ± 0.6
2	10.8 ± 0.4	22.6 ± 1.5
4	11.2 ± 0.3	21.7 ± 1.8
6	10.9 ± 0.6	20.4 ± 1.2
10	11.3 ± 0.7	20.5 ± 1.2

A clearly defined enhancement of cell proliferative activity in the basal layer was demonstrated in pig skin after the administration of Lioxasol. This peaked at a level that was ~29% higher than controls, at 2 weeks after the start of administration. Thereafter, the labelling index remained relatively constant. There was no quantifiable evidence of hyperplasia in the epidermis after the topical application of Lioxasol.

Cell proliferative activity was also accentuated in the papillary dermis. Changes in number of labelled papillary dermal cells paralleled those for the epidermal labelling index (Table 9). Peak values for labelled papillary dermal cells were in the range ~20-22 per mm^2.

The time taken for radiation-induced moist desquamation to heal (4 to 5 weeks in control pigs) was reduced in pigs receiving Lioxasol. This enhanced healing response was probably related, at least in part, to the acceleration in epidermal cell proliferation induced by the compound.

References

[1] A. Barabanova & D.P. Osanov: Int. J. Radiat. Biol. 57, 775-782, 1990
[2] G.J.M.J. van den Aardweg, J.W. Hopewell and R.H. Simmonds: Radiotherapy & Oncology 11:73-82, 1988
[3] J.W. Hopewell and G.J.M.J. van den Aardweg Int. J. Radiat. Oncol. Biol. Phys. 14:1047-1050, 1988

Cardiovascular System and Physical Working Capacity in Patients Who Had Acute Radiation Syndrome as the Result of Chernobyl Accident

David BELYI, Oleg GERGEL, Alexander KOVALENKO
Scientific Center of Radiation Medicine, Melnikov St. 53, Kiev, Ukraine

Abstract. The functional state of cardiovascular system has been studied since 1986 in 168 patients who had acute radiation syndrome as the result of Chernobyl accident. There was revealed a progressive increase of cardiovascular system pathology. The number of patients with pathological signs at ECG increased from 4.8 % in 1987 to 11.3 % in 1994 and with myocardial hypertrophy from 1.2 % to 22.6 %. The number of patients with coronary heart disease increased on 17.2% and with essential hypertension on 15.5 %. The physical working capacity reduced to 50-60 % of a due level for healthy persons. Two patients suffered from acute myocardial infarction during this period of observation. Thirteenth patients died from 1987 to 1995. Among them 4 patient died in a result of acute cardiac failure. The development of cardiovascular pathology has no any correlation with a dose of exposure. Three factors of cardiovascular pathology growth are supposed.

1. Introduction

In a framework of treatment of Chernobyl patients we are interested in the cardiovascular system reaction on exposure that courses the acute radiation syndrome (ARS). Some scientific publications are favoured to this interest: we know about acceleration of arteriosclerosis [1, 2, 3] and cardiovascular pathology growth in irradiated body [3, 4, 5], including a clinical observation after Chernobyl accident [6, 7, 8].

2. Materials and Methods

Since 1986 the dynamic observation was doing for 168 patients from total number of 237 patients who have endured ARS in result of Chernobyl accident. The 168 investigated patients divided on 86 persons with ARS NC (index Not Confirmed is marked persons who were irradiated in doses 0.7-1.0 Gy with the signs of primary reaction on radiation exposure but without bone marrow syndrome), 35 with ARS 1, 40 with ARS 2 and 7 with ARS 3. The last two cohorts were combined in one group with ARS 2-3 for more reliable statistics and as very closed in dose levels and in probability of stochastic and non stochastic consequences development.

The past medical documentation had shown that most patients were healthy persons before the accident. Only 4 patients with ARS NC, 1 with ARS 1 and 2 with ARS 2-3 suffered from coronary heart disease (CHD) before accident. The average age of the patients was 36.3±10.1 years in group of ARS NC, 34.2±8.1 years in ARS 1 group and 35.6±11.9 years in ARS 2-3 group.

As a control were investigated 20 healthy persons who had never undergone to ionising radiation (age 31.9±12.1 years).

As the methods of investigation we used ECG (Mingograph 720, Simens Elema, Sweden), cycle ergometry (KE-12, Medicor, Hungary), spiroergometry (Oxicon 4, Mingdhardt, Holland), echocardiography (SSD 630, Aloka, Japan).

3. Main Results

The clinical investigation 1 year later the Chernobyl accident revealed in all groups of patients the great number of individuals with subjective and objective signs of cavardio-vascular neurosis (ICD-9 code is 306.2) or vegeto-vascular distonia (as in Ukrainian classification). The main symptoms were cardialgia, headache, transitory jumps and drops of systolic blood pressure, tachyrhythmia, arrhythmia, tachypnoea that combined with neurological and sexual disorders. At that time we did not observe any acute pathological findings or coronary insufficiency with ECG.

In 1988-1989 was revealed the increase of vegeto-vascular distonia (VVD): on 11.1% in patients with ASR NC (initial number 72.1%), 10.0% - with ARS 1 (initial - 85.7%) and 11.9% - with ARS 2-3 (initial - 78.7%) (P<0.05). In next years the number of VVD decreased. One of the features of VVD tendency was the appearance on the stage of 1991-1993 years the hypothalamic crises: in 8.1% patients after ARS NC, 25.7% - after ARS 1 and 12.8% - after ARS 2-3. In half cases from this number we observed the VVD transformation in hypertension disease (HD).

The neuro-reflecting and neuro-endocrine disorders that take place with VVD, are that pathogenic mechanism of CHD and HD development. It is shown early in the scientific literature [9, 10]. These facts are confirmed by our data of progressive growth of HD and CHD in patients after ARS (P<0.01) (see Figures 1 and 2).

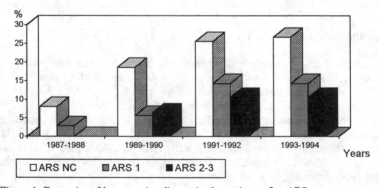

Figure 1. Dynamics of hypertensive disease in the patients after ARS

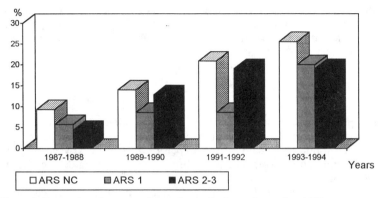

Figure 2. Dynamics of coronary heart disease in the patients after ARS

The development of cardiovascular system pathology combined with myocardial hypertrophy. We also revealed the increase of complete or partial atrioventricular (His) bundle's blockade or deceleration of electric impulse flow in heart muscle.

The analysis of cardiovascular pathology as a reason of death has shown it's high rate. Four persons from 13 (30.8 %) died in result of acute cardiac failure.

The cycle ergometry test has been done since 1989 year. During this study we found a decrease of physical capacity level in all patients after ARS in comparison with control level that was 184.6±25.2 Watt (P<0.01) (see Figure 3).

In all indexes that characterise aerobic and anaerobic metabolism of the body (anaerobic threshold, oxygen debt, including its alactic and lactic fraction and the constants of debt's recovery speed) did not find any reliable differences from control or published data early by others' authors. All these indexes varied within the valid levels.

An analysis of stress-test stopping reasons had shown that in all groups prevailed objective reasons: not increased systolic blood pressure (BP) or heart rate (HR) during increase load or very high increase of BP and HR, or segment ST depression. There were not any significant changes of these indexes during 1989-1993. We revealed only growth of ST segments depression from 4.2% in 1989 to 7.2% in 1993.

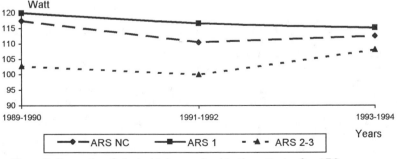

Figure 2. Dynamics of physical tolerance level in the patients after ARS

In all groups we found a combination of objective and subjective (cardialgia, headache, tachypnoea) reasons of test stopping. Only 15.1% of patients were limited in their physical capacity by physiological reasons (adequate muscular fatigue and reach of submaximal HR).

4. Conclusions

1. The progressive growth of cardiovascular pathology is observed in patients after different degrees of ARS severity from 1987 to 1994.

2. The development of cardiovascular pathology has no any correlation with the dose of irradiation. This fact is confirmed by less pathology in the patients after ARS 2-3.

3. It is supposed three factors of pathogenic mechanism of cardiovascular pathology development: a) changes of neuro-humoral regulation processes after ARS that expresses in vegeto-vascular distonia development that in next order transforms to coronary heart disease or hypertensive disease; b) radiation affects at vessels' endothelium; c) high frequency of dislipoproteinaemia that takes place in patients after ARS and has atherogenic character [11].

4. The adequate aerobic and anaerobic metabolism of body in spite of cardiovascular pathology shows a good compensation of circulation and microcirculation that have to have an arteriosclerosis prevention.

5. The natural tendency to growth of cardiovascular pathology during ageing and especially after irradiation requires permanent medical prophylaxis of cardiovascular pathology development.

References

[1] Dzuravlev V.F. (1982) Toxicology of radioactive substances, Moscow, 128 p.

[2]. Yarmolenko S.P. (1988) Radiobiology of men and animals, Moscow, 424 p.

[3] Vorobiev E.I., Stepanov R.P (1985) Ionizing radiation and vessels, Moscow, 296 p.

[4] Gillette E.L., McCheseney S.L., Hoopes P.J. (1985) Isoeffect curves for radiation-induced cardiomyopathy in the dog, J.Radiat.Oncol., 11, 12, p.2091-2097.

[5] Kodama K., Okumura Y., Kasagi F. et al. (1995) Incidence of Myocardial Infarction in Atomic-bomb Survivors, Tenth International Congress of Radiation Research, Wursburg, Germany, August 27-September 1, 1995, Radiation Research 1895-1995, Volume 1: Congress Abstracts, p.168.

[6] Shishmarev Y.N., Alekseev G.I., Nikiforov A.M. et al. (1992) Clinical Aspects of Chernobyl Accident Consequences, Radiobiology, 32, 3, p.323-332.

[7] Kordysh E., Goldsmith J.R., Quastel M.R. et al. (1994) Cardiovascular Findings and Exposure to Chernobyl Radiation in Immigrants to Israel from Former USSR, Int. Congress Environ. Med., Abstracts of Papers, p.207.

[8] Shidlovsky P.R. (1992) Dynamics of the General Morbidity of Byelorussia Population Before and After the Chernobyl Atomic Station Disaster (1985-1989), Vrachebnoe Delo, 2, p.20-22.

[9] Vein A.M., Solovieva A.D., Kolosova O.A. (1981) Vegetovascular distonia, Moscow, 318 p.

[10] Makolkin V.I., Abakumov S.A. (1985) Neuro-circulatory disthonia in internal practice, Moscow, 192 p.

[11] Chayalo P.P., Chobotko G.M. (1995) Four Years' Study of Blood Lipid and Lipoprotein Amount Changes in Liquidators of Chornobyl Accident Aftereffects, Ukrainian Radiologycal J., 1, p.40-43.

III. HEALTH EFFECTS
FOLLOWING THE ACCIDENT
B. Thyroid cancer in children living near Chernobyl

Human Thyroid Cancer Induction by Ionizing Radiation: Summary of Studies Based on External Irradiation and Radioactive Iodines.

Roy E. Shore, PhD, DrPH
New York University Medical School
Department of Environmental Medicine
550 First Avenue, New York, NY 10016

Abstract

To provide a context for the Chernobyl thyroid cancer experience, a summary of the findings from other studies is given. The data on external radiation and thyroid cancer come primarily from studies of children irradiated for a variety of benign medical conditions and the Japanese atomic bomb cohort. Unfortunately, only small amounts of data are currently available on thyroid cancer following radioactive iodine exposure in childhood. In order to predict the risk of thyroid cancer in the Chernobyl experience, a number of radiation-related factors need to be considered: the magnitude of radiation risk from available studies; shape of the dose-response curve; variations in risk by gender, time since irradiation, and age at irradiation; the effects of dose fractionation or dose protraction. Other considerations pertaining to the frequency of thyroid cancer and its outcome are thyroid-tumor surveillance effects and background iodine intake. The data to date suggest that ^{131}I produces less thyroid cancer than a comparable dose of external radiation, but the Chernobyl experience will provide extensive new information on this issue. Principles are discussed as to how to maximize the scientific validity and informativeness of Chernobyl thyroid studies.

This paper will consider what is known about the risk of thyroid cancer from ^{131}I exposure and what gaps still exist in our knowledge. The approach will be first to examine what is known about thyroid cancer risk from acute, external irradiation in order to provide a context regarding ^{131}I exposure. Then studies of ^{131}I exposure and thyroid cancer risk will be reviewed and compared to external radiation risk. Finally, several scientific criteria will be considered which will help assure that future studies are valid and provide the maximum scientific information.

A number of studies have documented that acute, external irradiation of the thyroid gland in childhood produces an appreciable thyroid cancer risk [1]. The thyroid gland and bone marrow (leukemia) are perhaps the most radiosensitive cancer sites. A summary of the results of major thyroid irradiation studies is shown in Table 1. Each study found a statistically significant dose-related excess of thyroid cancer. There is a fair degree of variation in the risk coefficients; some of it may be attributable to cell-killing effects, different ages at irradiation, different population susceptibilities, etc., while some may be simply sampling variation.

A recent pooled analysis [2] found an average excess relative risk of 7.7 per Gy (95% confidence interval= 4.9-12), while the excess absolute risk was 4.4 per 10^4 person-year Gy (95% CI= 1.9-10). A linear dose-response function was found to fit the data well; there was no significant quadratic curvature. When dose fractionation was examined in these studies, it was found that the risk was about 30% lower for fractionated doses than for unfractionated doses. However, the degree of fractionation was small and the dose per fraction fairly large, so these data may underestimate the degree to which dose fractionation or protraction reduce thyroid cancer risk.

It is well known that females have a higher incidence of thyroid cancer than males. Most studies have found that the excess relative risks per Gy are roughly similar for males and females, and a formal test of a gender x radiation interaction in five major studies did not show a statistically significant effect, although the risk estimate was nearly twice as high for females than males [2].

Table 1. Thyroid Cancer Excess Relative Risk (ERR) per Gy and Excess Absolute Risk (EAR) per Gy for Cohort Studies with Exposure to Acute, External Ionizing Radiation before the Age of 20 Years

Study (Reference)	# Irradiated Subjects	Mean Dose (cGy)	Observed / Expected Cancers	ERR / Gy (95% CI) [a]	EAR / 10^4 Person-year Gy [a]
A-bomb (<15 y at exposure) [2, 4]	~13,000	23	40/19.2= 2.1	4.7 (1.7,11)[c]	2.7
Enlarged thymus [12]	2,475	136	37/2.77= 13.4	9.1 (3.6,29)[c]	2.6
Tinea capitis [15]	10,834	9	44/11.2= 3.9	32.5 (14,57)[b,c]	7.6
Enlarged tonsils [16]	2,634	59	309/125= 2.5 [d]	2.5 (0.6,26)[c]	3.0
Skin hemangioma [17]	14,351	26	17/7.46= 2.3	4.9 (1.3,10)	0.9
Skin hemangioma [18]	11,807	12	15/7.98= 1.9	7.5 (0.4,18)	1.6
Lymphoid hyperplasia [19]	1,195	24	13/~2.2= 5.9	~20 (9.5,37)	15.1
Childhood cancer [3]	9,170	1,250	23/0.4= 53	1.1 (0.4,29)[c]	0.4

[a] Both the ERR and EAR estimates are based on dose-response analyses.

[b] When an indicator variable for irradiated vs. control group was included, the ERR dropped to 6.6/Gy [2].

[c] Risk estimates based on dose-response analyses by Ron et al [2].

[d] This study had no unirradiated control group; the expected value was estimated from the dose-response intercept, not from general population rates.

~ Value estimated for this tabulation from the data available.

Studies in the literature have reported almost no thyroid cancers prior to 5 years after irradiation [1]. The finding of apparent increases before 5 years in Chernobyl-irradiated populations may be due to the extensive thyroid screening programs in Ukraine and Belarus. One of the effects of screening is to cause earlier detection of the disease than would occur otherwise. The duration of thyroid radiation effects is not well characterized. The pooled analysis of five studies suggested that the ERR per Gy was greatest about 15 years after exposure, but it was still elevated 40 or more years after irradiation [2].

Some of the classic ^{131}I studies were based on patients who received ^{131}I treatment for hyperthyroidism, where the thyroid doses were typically 60-100 Gy. One may question whether these data are meaningful for risk assessment. To address this, it would be valuable to know what the external thyroid irradiation data tell us about the dose-response at high doses. The data seem to indicate an approximately linear dose-response up to about 10 Gy, but the one available study with high doses [3] suggests that above roughly 10 Gy the carcinogenic effect begins to level off, so that the risk per unit dose is lower at very high doses. This implies that ^{131}I studies of hyperthyroid patients may be misleading for inferring risk at lower doses.

Unlike studies of external radiation where there is much thyroid cancer data from childhood irradiation and little from adult irradiation, for ^{131}I most of the available data are based on adult exposures. Hence, changes in thyroid radiation sensitivity by age are important to examine. Two years ago, for the first time, a good comparision of thyroid cancer induction across the age range became available from the Japanese atomic bomb study [4]. There is a striking decrease in radiation sensitivity for adult irradiation as compared to childhood irradiation; the ERR coefficients per Gy for those irradiated at ages 0-9, 10-19, 20-39 and ≥40 were 9.5, 3.0, 0.3 and -0.2 respectively. It is clear that in order to characterize thyroid cancer risk from ^{131}I we must have information on childhood exposures, because studies of adult exposure will probably provide negative information that is not necessarily indicative of childhood exposure effects.

Table 2 summarizes the data for juvenile (i.e., under age 20) exposure to [131]I. It is evident that the data are very sparse, i.e., the juvenile populations are small and the number of thyroid cancers few. Most of the ^{131}I data suggest that the risk is less than for external irradiation; however, interpretation is clouded by the small numbers of cancers and by the fact that in some studies a fraction of the subjects were being evaluated for thyroid conditions and in others the doses were very high, in the cell-sterilization range.

Hence, it is clear that we need more scientific information on thyroid cancer in children and young people exposed to [131]I. The Chernobyl accident provides the premier opportunity to obtain such information because of the fact that a large population received relatively high doses. For example, among those with thyroid measurements made just after the accident, preliminary dose calculations in Ukraine and Belarus (prepared by Drs. Likhtarev and Minenko respectively) show about 1,800 juveniles with estimated thyroid doses over 5 Gy and another 5,000 with doses between 2 and 5 Gy. These numbers exceed the total children with ≥2 Gy in all the other completed or ongoing studies of ^{131}I put together.

Table 2. Thyroid Cancer Excess Relative Risk (ERR) per Gy and Excess Absolute Risk (EAR) per Gy for Studies with [131]I Exposure before the Age of 20 Years

Study (Reference)	# Irradiated Subjects	Mean Dose (cGy)	Observed / Expected Cancers	ERR / Gy (95% CI)	EAR / 10⁴ Person-year Gy
Swedish diagnostic [131]I [20, 21]	2,408	150	3/1.78= 1.7	0.25 (<0, 2.7)	0.15
FDA diagnostic [131]I [22]	3,503	~80	4/3.7= 1.1 [a]	0.10 (<0, 2.0)[a]	0.05 [a]
Utah ^{131}I fallout [23]	2,473	17	8/5.4= 1.5	7.9 (<0, 41)	3.3
Marshall Islanders [24]	127	~1,240 [b]	6/1.2= 5.0	0.32 (0.1-0.8)	1.1
Juvenile hyperthyroidism, combined series [25-33]	602	~8,800	2/~0.1= 20	0.3 (0-1.0)	0.1

~ Estimated for this tabulation from the data available.

[a] Expected value based on SEER registry data [34]. If the analysis is instead based on the internal control group, among whom one thyroid cancer occurred, then the expected value is 1.4, and the nonsignificant ERR and EAR values are 3.1 and 0.6 respectively. An approximate dose-response analysis showed an ERR of 0.85/Gy (95% CI= 0.04, 16) and an EAR of 0.2 per 10⁴ person-year Gy (95% CI = 0, 1.9)

[b] Over 80% of this dose was from short-lived radioiodines and external irradiation rather than [131]I.

However, given that Chernobyl populations in principle provide the best opportunity to quantify the effects of [131]I , we need to address the question of how we can design and execute studies that will be of the highest quality and will provide scientifically valid risk estimates. Let us review several of the criteria against which to measure thyroid irradiation studies. Comments will be directed towards three broad types of studies: *aggregate studies* (often called "ecological" studies) in which thyroid cancer incidence rates for various geographic regions and/or various time periods are compared; *case-control studies* in which thyroid cancer cases are compared to a selection of persons (typically matched to the cases on age and sex) without thyroid cancer with respect to estimated thyroid doses and perhaps other characteristics; and *cohort studies* in which a large number of persons with a range of defined doses are followed up for some period of time to determine the frequency with which they develop thyroid cancer and to relate this to the ascribed doses. The following paragraphs list some pitfalls regarding methods of study and suggest how to maximize the scientific information we can glean from future Chernobyl studies.

Aggregate studies are useful as preliminary scoping studies, i.e., one may use them to get an idea of whether, at least crudely, there appears to be an association between thyroid cancer incidence and approximate gradations of ^{131}I exposure. However, in such studies one does not know the actual doses received by the cancer cases or other individuals. A serious limitation of aggregate studies is that it is difficult, and usually impossible, to control for the effects of possible confounding factors, most notably the intensity of thyroid screening which could seriously bias any quantitative estimates of risk one might attempt to derive from such studies. A number of weaknesses and limitations of aggregate studies have been identified [e.g., [5-9]]; the upshot of the literature is that the potential undetectable and unavoidable biases of aggregate studies are so great, they cannot be relied upon for quantitative estimates of risk.

A *cohort or case-control* study needs reasonably accurate estimates of doses to the individuals in the study. Obviously, doses that are systematically biased on the low or high side will provide over- or underestimates of risk respectively. However, random error in dose estimates will tend to produce an inaccurate estimate of risk as well. It is preferable to derive individual doses with realistic estimates of the amount of uncertainty attached to those doses, so that the risk estimates can potentially be corrected for dose uncertainty using methods that are beginning to become available to epidemiologists.

The issue that the thyroid dose estimates in the Chernobyl experience may have very limited accuracy and precision needs to be taken seriously, and thought should be given to how to validate the dose estimates and how to realistically determine the degree of uncertainty in doses estimated by various methods/instruments.

It is important that *cohort or case-control* studies of ^{131}I have a substantial number of study subjects with relatively high thyroid doses so that there is a wide range of doses. Having a wide range of doses is essential if the study is to have adequate statistical power; having high doses also permits one to derive more precise estimates of risk. As discussed above, Chernobyl studies have the potential to excel in this respect.

In a *cohort study* it is important to carefully define the study subjects in a standard way that is completely independent of disease status, so that a biasing correlation between cancer incidence and thyroid dose will not inadvertently be built into the study. The safest way to do this is to ensure that the investigator who defines the cohort does not have knowledge of the potential study subjects' thyroid disease status (i.e., the investigator is "blinded") and that sufficient effort is made so that the participation rate is high. The principle of blinding also applies to those who work at trying to enlist subjects into the study.

For a *cohort study*, a standardized protocol for thyroid examinations is essential, so that examination procedures and frequency will be identical for individuals at all dose levels. In addition, it would be desirable to keep the examiners blinded to the thyroid doses of the individuals being examined so that unconscious biases will not creep into the thyroid evaluations. However, since blindedness may be impossible to maintain, in that dose levels are largely determined by geographic area, an alternative is to use a strictly objective criterion as to which tumors to include, e.g., only tumors ≥1 cm on ultrasound examination.

For a *case-control study*, it is imperative to try to control for the degree of thyroid surveillance. Studies have shown that as much as seven times more thyroid cancers will be found with thyroid screening than without it [10]. There naturally has been more intense thyroid screening in high-exposure areas as compared with lower exposure areas, so this correlation of dose with screening intensity can seriously bias the results. Any case-control study therefore needs to be designed so that the cases and controls have had similar degrees of thyroid surveillance which will minimize this bias.

Another aspect of increasing the scientific value of a *case-control or cohort study* is to elicit information on other possible risk factors for thyroid cancer in a standardized fashion, so that these other variables can be controlled for if they are confounders or examined as possible risk modifiers. For instance, studies have suggested that familial thyroid cancer, Jewish ethnicity and childbearing may be risk factors for thyroid cancer [1, 11, 12]. Animal studies have indicated that goitrogens increase thyroid tumor induction by [131]I [13, 14], so thyroid volumes and possible goitrogenic factors such as low iodine intake may be worth investigating as co-factors in thyroid cancer induction.

In conclusion, while the epidemic of childhood thyroid cancer around Chernobyl is first and foremost a public health crisis, it also affords an extraordinary opportunity to obtain much-needed scientific information on thyroid cancer risk from [131]I. However, we need to maximize the the scientific quality of such studies by ensuring that the study design and methodology meet the highest standards of epidemiological and clinical research.

References

1. Shore R. Issues and epidemiological evidence regarding radiation-induced thyroid cancer. Radiat Res 1992;131:98-111.
2. Ron E, Lubin J, Shore R, et al. Thyroid cancer after exposure to external radiation: a pooled analysis of seven studies. Radiat Res 1995;141:259-277.
3. Tucker MA, Jones P, Boice J, et al. Therapeutic radiation at a young age is linked to secondary thyroid cancer. Cancer Res 1991;51:2885-2888.
4. Thompson D, Mabuchi K, Ron E, et al. Cancer incidence in atomic bomb survivors. Part II: Solid tumors, 1958-1987. Radiat Res 1994;137:S17-S67.
5. Piantadosi S, Byar D, Green S. The ecological fallacy. Am J Epidemiol 1988;127:893-904.
6. Greenland S, Morgenstern H. Ecological bias, confounding, and effect modification. Int J Epidemiol 1989;18:269-274.
7. Greenland S. Divergent biases in ecologic and individual-level studies. Stat Med 1992;11:1209-1223.
8. Brenner H, Greenland S, Savitz D. The effects of nondifferential confounder misclassificatin in ecologic studies. Epidemiol 1992;3:456-459.
9. Greenland S, Robins J. Ecologic studies--biases, misconceptions, and counterexamples. Am J Epidemiol 1994;139:747-760.
10. Ron E, Lubin J, Schneider A. Thyroid cancer incidence (letter). Nature 1992;360:113.
11. Perkel VS, Gail M, Lubin J, et al. Radiation-induced thyroid neoplasms: Evidence for familial susceptibility factors. J Clin Endocrinol Metab 1988;66:1316-1322.
12. Shore RE, Hildreth N, Dvoretsky P, Andresen E, Moseson M, Pasternack B. Thyroid cancer among persons given x-ray treatment in infancy for an enlarged thymus gland. Am J Epidemiol 1993;137:1068-1080.
13. Doniach I. The effect of radioactive iodine alone and in combination with methylthiouracil upon tumor production in the rat's thyroid gland. Br J Cancer 1953;7:181-202.
14. Axelrad AA, Leblond C. Induction of thyroid tumors in rats by a low iodine diet. Cancer 1955;8:339-367.
15. Ron E, Modan B, Preston D, Alfandary E, Stovall M, Boice J. Thyroid neoplasia following low-dose radiation in childhood. Radiat Res 1989;120:516-531.

16. Schneider AB, Ron E, Lubin J, Stovall M, Gierlowski T. Dose-response relationships for radiation-induced thyroid cancer and thyroid nodules: Evidence for the prolonged effects of radiation on the thyroid. J Clin Endocrinol Metab 1993;77:362-369.

17. Lundell M, Hakulinen T, Lindell B, Holm L-E. Thyroid cancer after radiotherapy for skin hemangioma in infancy. Radiat Res 1994;140:334-339.

18. Lindberg S, Karlsson P, Arvidsson B, Holmberg E, Lundberg L, Wallgren A. Cancer incidence after radiotherapy for skin haemangioma during infancy. Acta Oncologica 1995;34:735-740.

19. Pottern LM, Kaplan M, Larsen P, et al. Thyroid nodularity after childhood irradiation for lymphoid hyperplasia: A comparison of questionnaire and clinical findings. J Epidemiol 1990;43:449-460.

20. Holm LE, Wiklund K, Lundell G, et al. Thyroid cancer after diagnostic doses of iodine-131: A retrospective cohort study. J Nat Cancer Inst 1988;80:1132-1138.

21. Hall P, Mattsson A, Boice J. Thyroid cancer following diagnostic iodine-131 administration. Radiat Res 1995;In Press.

22. Hamilton P, Chiacchierini R, Kaczmarek R. A follow-up of persons who had iodine-131 and other diagnostic procedures during childhood and adolescence. Rockville, MD: CDRH-Food & Drug Admin., 1989:

23. Kerber RA, Till J, Simon S, et al. A cohort study of thyroid disease in relation to fallout from nuclear weapons testing. J Am Med Assoc 1993;270:2076-2082.

24. Robbins J, Adams W. Radiation effects in the Marshall Islands. In: Nagataki S, ed. Radiation and the Thyroid. Amsterdam: Excerpta Medica, 1989:pp. 11-24.

25. Safa AM, Schumacher O, Rodriguez-Antunez A. Long-term follow-up results in children and adolescents treated with radioactive iodine (I-131) for hyperthyroidism. New Engl J Med 1975;292:167-171.

26. Dobyns BM, Sheline G, Workman J, Tompkins E, McCohaney W, Becker D. Malignant and benign neoplasms of the thyroid in patients treated for hyperthyroidism: a report of the cooperative thyrotoxicosis therapy. J Clin Endocrinol Metab 1974;38:976-998.

27. Kogut MD, Kaplan S, Collipp P, Tiamsic T, Boyle D. Treatment of hyperthyroidism in children. New Engl J Med 1965;272:217-221.

28. Crile G, Schumacher O. Radioactive iodine treatment of Graves' disease. Am J Dis Child 1965;110:501-504.

29. Hayek A, Chapman E, Crawford J. Long-term results of thyrotoxicosis in children and adolescents with radioactive iodine. New Engl J Med 1970;283:949-953.

30. Goldsmith RE. Radioisotope therapy for Graves' disease. Mayo Clin Proc 1972;47:953-961.

31. Sheline G, Lindsay S, McCormack K, Galante M. Thyroid nodules occurring late after treatment of thyrotoxicosis with radioiodine. J Clin Endocrinol Metab 1962;22:8-18.

32. Starr P, Jaffe H, Oettinger L. Late results of I-131 treatment of hyperthyroidism in 73 children and adolescents: 1967 followup. J Nuc Med 1969;10:586-590.

33. Holm L-E. Malignant disease following iodine-131 therapy in Sweden. In: Boice J, Fraumeni J, eds. Radiation carcinogenesis: epidemiology and biological significance. New York: Raven Press, 1984:263-271.

34. SEER. Cancer Incidence and Mortality in the United States, 1973-81. Bethesda, MD: U.S. DHHS, NIH Publ. No. 85-1837, 1985.

Thyroid cancer in children in Belarus

E.P. DEMIDCHIK
Thyroid Cancer Center, Skoriny av. 64, 220600, Minsk, Belarus
I.M. DROBYSHEVSKAYA
Health Ministry of Belarus, Miasnikova str. 39, 220097, Minsk, Belarus
E.D. CHERSTVOY
Minsk Medical Institute, Dzerzhinsky av. 83, 220116, Minsk, Belarus
L.N. ASTAKHOVA
Institute for Radiation Medicine, Aksakovshina, 223032, Minsk, Belarus
A.E. OKEANOV
Medical Technology Center, Brovky 7a, 220600, Minsk, Belarus
T.V. VORONTSOVA
Institute for Radiation Medicine, Aksakovshina, 223032, Minsk, Belarus
M. GERMENCHUK
Center for Radiation Control, Skoriny av. 110a, 220023, Minsk, Belarus

Abstract

Paediatric thyroid cancer was diagnosed in 390 patients in Belarus after the Chernobyl accident. The morbidity rates increased by 55.7 times as compared with the 10 year pre-accident period. Thyroid cancer in children is highly aggressive disease accompanied by surrounding tissues and metastatic involvement of lymph nodes.

A sufficient increase of thyroid cancer incidence has become one of the serious consequences of the Chernobyl accident. For a nine year period after the catastrophe this disease was diagnosed in 210 children more as compared with the same pre-accident interim (Table 1).

Table 1

Thyroid cancer incidence in Belarus

Pre-accident period			Post accident period		
Years	Adults	Children	Years	Adults	Children
1977	121	2	1986	162	2
1978	97	2	1987	202	4
1979	101	0	1988	207	5
1980	127	0	1989	226	7
1981	132	1	1990	289	29
1982	131	1	1991	340	59
1983	136	0	1992	416	66
1984	139	0	1993	512	79
1985	148	1	1994	553	82
Total	1131	7	Total	2907	333

This perceptible increase in the incidence rate started 5 years after the accident in 1990 and by now the thyroid cancer has been diagnosed in 390 children (Table 2). 76.1% of them were from Gomel and Brest regions where the highest soil contamination with I^{131} has been observed. Boys and girls ratio was 1÷1.5 (Table 3).

Table 2

The incidence of thyroid cancer after the Chernobyl accident

Region	Years			
	1986-89	1990-94	7 months of 1995	Total
Brest	2	74	9	85
Vitebsk	0	7	0	7
Gomel	7	172	33	212
Grodno	5	14	3	22
Minsk	3	15	1	19
Mogilev	0	16	3	19
Minsk-city	1	17	8	26
Belarus	18	315	57	390
%	4.6	80.8	14.6	100

Table 3

Sex ratio for children

Region	Boys	Girls	Total
Brest	28	57	85
Vitebsk	3	4	7
Gomel	83	129	212
Grodno	12	10	22
Minsk	11	8	19
Mogilev	10	9	19
Minsk-city	10	16	26
Belarus	157	233	390
%	40.3	59.7	100

Radioactive iodine uptake by thyroid has become a certain cause of thyroid cancer in children. Childhood thyroid has a high capacity to accumulate radioactive isotopes and it is less tolerable to ionizing irradiation as compared with adults.

Thyroid cancer has mainly appeared in the children born before (380 cases) or at the time of the accident (6). "Spontaneous" carcinoma was diagnosed only in 4 children after I^{131} disintegration after the catastrophe.

The children irradiated at the age of 0-4 were at the highest risk of cancer promotion (Table 4). 17.9% patients of this group at the time of the accident were younger that one year of age.

The risk of radiation induced thyroid cancer developing in the children affected by the accident remains for long years and, possibly, for the whole life. Thyroid cancer was dignosed in 39 (10%) patients who were evacuated and migrated from the contaminated zones.

Table 4

The age of children at accident

Region	Number of cases	Age group		
		0-4	5-9	10-14
Brest	85	59	24	2
Vitebsk	7	4	3	0
Gomel	210	137	69	4
Grodno	21	9	11	1
Minsk	19	13	5	1
Mogilev	18	12	5	1
Minsk-city	26	17	8	1
Belarus	386	251	125	10
%	100	65.0	32.4	2.6

The most frequent histological type was papillary carcinoma (94.9%). Follicular tumors were verified in 4.3% and only two patients had anaplastic carcinomas. Only in one patient medullary tumor was diagnosed.

At the time of diagnosis carcinomas manifested themselves as nodules, while diffuse lesions were less common.

Of 380 children operated on in the Thyroid Cancer Center solitary carcinomas were in 64.5% cases and multifocal in 35.5%.

Tumors of more than 4 cm in size were rare. Tumor size didn't impact significantly the frequency of multifocal lesions. In the majority of patients the primary tumor had a high potential for invasion. Even small carcinomas (up to 8 mm) involved the thyroid capsule and the surrounding tissues. 48.4% of children had extracapsular spread of the tumor which corresponds to pT4. Capsular involvement formed mainly 3 months after a nodule was diagnosed.

Thyroid cancer in children frequently results in metastatic involvement of the regional lymph nodes. Approximately in one third of the patients (31.6%) metastatic disease of both neck sides was diagnosed before surgery. In many cases the rate of metastatic growth was significantly higher than the rate of primary tumor growth in the thyroid. The frequency of the regional nodes involvement depended directly on the tumor spread to the thyroid capsule and the surrounding tissues. The number of multifocal lesions was significantly higher in patients with extracapsular tumors. Thus, in our studies thyroid cancer was a rapidly developing disease.

When treating thyroid cancer in children it is not always clear whether thyroidectomy should be performed in every case. After a surgery of this kind a patient needs a high-doze thyroxine replacement therapy during all his life. It is doubtful that a long-term thyroxine therapy can completely replace the removed thyroid. It is this fact that in certain cases makes many surgeons perform smaller surgeries, such as subtotal resection and hemithyroidectomy. After these surgeries a small-doze thyroxine replacement therapy is quite effective.

It is not clear either whether wide neck dissection is necessary in cases when lymph node metastases in the neck and mediastinum are not routinely identified.

Our experience is based on the results of the follow-up of 380 children in the Center for Thyroid Cancer after the Chernobyl accident. In every case the diagnosis was histologically proved. The results of surgeries were reviewed and analyzed for 292 patients who were operated on 2 to 8 years ago.

In the first 5 years after the Chernobyl accident the surgical strategy was complete tumor removal and, where possible, preserving a part of the thyroid tissue to avoid heavy hypothyroidism difficult to treat with thyroxine (there was no L-thyroxine in Belarus at that time). Not only total thyroidectomy but also subtotal resection and hemithyroidectomy were performed. These surgical procedures resulted in paratracheal lymph nodes dissection. If metastases were diagnosed in other neck regions, wide lymphadenectomy on one or both sides of the neck was performed.

In the following years hemithyroidectomy and subtotal resection of the thyroid was performed only in case of T1aNOMO, when a solitary carcinoma was 3 to 7 mm in size and there were no metastases in lymph nodes.

Of 292 children followed up during the period of 2 to 8 years after surgery in 56 (19,1 %) recurrences were observed. In 8 cases cancer was found in the parts of the thyroid saved at surgery. In 48 cases there were metastases in regional lymph nodes (Table 5). The interval between the first surgery and the time when carcinoma was observed in the thyroid remnants was 16,8 months. Lymph node metastases appeared 14 months after surgery. All 56 children were operated on again.

Table 5

Patterns of failure

Operation	Cases	Relapse	Regional metastases
Total thyroidectomy	106	0	12
Subtotal thyroidectomy	57	3	17
Hemithyroidectomy	129	5	19
Total (abs. and %)	292 (100%)	8 (2.7%)	48 (16.4%)

The most common postoperative complication was parathyroid insufficiency (Table 6). There were no lethal outcomes.

Table 6

Surgical morbidity

Complications	For primary surgery	For secondary surgery
N.recurrence damage:		
-unilateral	7	0
-bilateral	1	0
Parathyroid insufficiency	30	2
Ductus thoracicus damage	0	1
Total	38	4

Scintigraphy was usually performed after total thyroidectomy. When lung metastases appeared (55 cases) the patients had radioiodine therapy in the clinic of Professor Ch.Reiners (Essen, Germany).

Clinical and epidemiological studies showed that of the children affected by ionizing radiation during the accident those younger than 4 years of age were at the greatest risk in terms of developing thyroid cancer.

Thyroid cancer in children is of highly aggressive nature. Lymphogenic and hematogenic metastases often develop in cases of occult thyroid tumors and are hard to identify even through modern diagnostics. That is why complete removal of the thyroid and lymphadenectomy on both sides of the neck and anterior mediastinum can be considered as a curative surgical procedure. After such a surgery there is no source left for a relapse in the thyroid and regional lymph nodes. A smaller surgery (total thyroidectomy with no neck lymphadenectomy on both sides) is only relatively radical as it leaves a certain degree of risk of tumor developing. Nevertheless, this kind of surgery is very often acceptable. In a number of cases for pT1aN0M0 hemithyroidectomy can be performed. In such cases residual carcinoma is identified in 2,7% of patients.

Total thyroidectomy does not always save the parathyroid gland. Grafting of the removed parathyroid into a muscle does not eliminate parathyroid insufficiency. Short duration of the follow-up period of the patients who developed pulmonary and bone metastases and underwent radioiodine therapy does not allow to draw conclusions on the results of this strategy.

Besides it does not seem to be possible to evaluate the consequences of a long-term treatment of cancer in children with L-thyroxine which is a synthetic substance.

Thyroid cancer in children requires an early diagnosis as well as an adequate treatment procedure to get better long-term results without possible abnormalities.

Thyroid Cancer in Children and Adolescents in Ukraine after the Chernobyl Accident (1986-1995)

N. TRONKO, T. BOGDANOVA, I. KOMISSARENKO, E. BOLSHOVA,
V. OLEYNIK, V. TERESHCHENKO, Y. EPSHTEIN, V. CHEBOTAREV
*Institute of Endocrinology and Metabolism, Academy of Medical Sciences of Ukraine,
Vyshgorodskaya Str. 69, Kiev 252114, Ukraine*

Abstract

The increase in the incidence of thyroid cancers in children and adolescents in Ukraine following the Chernobyl accident made it necessary to compile a clinical morphological register of respective cancers. In 1986-1994 there were 339 cases registered in children and adolescents, of them 211 children (who were operated at the age under 15 years) and 128 adolescents (who were operated at the age of 15-18 years). Before the Chernobyl accident (1981-1985) in Ukraine 59 cases of thyroid cancer in children and adolescents were reported: 25 cases in children and 34 cases in adolescents. This increase has been observed since 1990. In 1981-1985 the incidence rate (number of thyroid cancers per 100 000 children population) ranged 0.04 - 0.06. In 1990 this estimate was 0.23 and in 1992-1994 0.36 - 0.43, thus a 7-10 fold increase exceeding the pre-Chernobyl level. In the 5 most contaminated northern regions of Ukraine (Kiev, Chernigov, Zhitomir, Cherkassy, Rovno oblasts) and the city of Kiev the incidence rate was much higher. For example, in 1984 it was 3.8 in Chernigov oblast, 1.6 in Zhitomir oblast. The total "contribution" of the above-mentioned regions to the incidence of thyroid cancer in children after the Chernobyl accident makes more than 60%. It has been noted that in 1990-1994 there was an increase in the number of children operated at the age under 10, it means that these children were under 6 years at the time of the accident and were most sensitive to radioiodine exposure. As for the sex ratio, there has been a shift to males : in 1981-1985 F:M = 1.8:1, in 1990-1994 F:M = 1.4:1. Morphologically, 93.4% of 196 carcinomas resected from children and adolescents at the Institute of Endocrinology from 1986 to August 1st, 1995 were papillary carcinomas. They manifested high invasive and infiltrative growth, signs of intraglandular spread. Regional lymph node metastases were found in 59% of cases, distal lung metastases observed at various periods after surgery were noted in 23.7% of cases.

The increase in the incidence of thyroid cancers in children and adolescents in Ukraine following the Chernobyl accident made it necessary to compile a clinical morphological register of respective cancers at the major centre of endocrinology and endocrine surgery in Ukraine, Institute of Endocrinology and Metabolism in Kiev. This register includes all cases of thyroid cancer in patients who were under 18 years of age at the time of the Chernobyl accident. In the period of 1986-1994 there was a total of 531 such cases. There were 177 such patients operated at the surgical department of our Institute. The data on other patients are included into the register based on statistical information from 27 regions of Ukraine. In addition, we have developed an individual record form of clinical and morphologic data which includes the following : the duration of the disease, doses of radiation exposure, peculiarities of clinical course, surgical intervention, metastases, description of light microscopic and electronmicroscopic pathology. The compilation of a number of findings in each case for an extended length of time will allow for improved future analysis of the accumulated data [1].

This report will review the cases of thyroid cancer in Ukraine in children and adolescents aged 0-18 years at the moment of surgery. During the 5 years preceding the accident a total of 59 cases of thyroid cancer were registered in this age group, while during the period of 1986-1994, 339 additional cases were discovered. Between 1981-1985 the average number of thyroid cancer cases per year was 12. Following the Chernobyl accident between 1986-1990 there was an average of 22 cases per year. During the 1991-1994 period this number jumped to 57 cases per year.

In the group 0-14 years of age at surgery the number of new cases was the highest. In the 1981-1985 period there was a total of 25 cases of thyroid cancer among children in Ukraine whereas in 1986-1994 they numbered 211 cases. The most significant increase was observed beginning from 1990 : in 1990 : 26 cases, in 1991 : 22 cases, in 1992 : 47 cases, in 1993 : 43 cases and in 1994 : 39 cases. By August 1st, 1995, at the Institute of Endocrinology alone 29 children under 15 were operated, i.e. no decrease in the number of thyroid cancer cases can be anticipated.

Analysis of thyroid cancers in adolescents, who were operated at the age of 15-18 years showed that in the period of 1981-1985 34 cases of thyroid cancers were found, while in 1986-1994, 128 cases were found. Although it was more pronounced in the younger groups, a considerable increase in malignant thyroid tumors was also observed in adolescents : in 1991 : 20 cases, 1992 : 18 cases, 1993 : 18 cases and in 1994: 21 cases.

It should also be mentioned that between 1981-1985 the number of childhood thyroid cancers per 100.000 children population ranged 0.04-0.06 yearly. In 1990 this index increased up to 0.23, 1991 : 0.19, 1992 : 0.43, 1993 : 0.39 and 1994 : 0.36, thus 6.5-10 times exceeding the "pre-Chernobyl" level. As for several of the most contaminated regions of Ukraine this index was much higher. In 1993, it was 2.3 in the Kiev oblast, 1.5 in the Cherigov oblast, 1.3 in the Cherkasy oblast, 1.0 in the Zhitomir oblast and 0.7 in the Rovno oblast. There has also been marked increase of the incidence rate of 1.0 in the city of Kiev. In 1994 this index jumped to 1.6 in Zhitomir oblast and 3.8 in Chernigov oblast.

In absolute numbers in the period 1986-1994 there were 35 thyroid cancers registered in the Kiev oblast, 23 in the Chernigov oblast, 8 in the Rovno oblast, 13 in the Zhitomir oblast and 14 in the Cherkasy oblast. In the city of Kiev there were 27 cancers registered during this time. According to the preliminary data for 1995 the greater incidence was observed in Kiev oblast (8 cases), the city of Kiev (7) and Zhitomir oblast (4). It should be emphasized that in 1981-1985 period no cases of thyroid cancer were reported in the above-mentioned areas with the exception of the Cherkasy oblast. In general, the "contribution" of these highly contaminated regions to the incidence of thyroid cancer in children in Ukraine in the period of 1990-1994 makes more than 60%.

As for the age distribution of patients 0-18 years, operated during 1990-1994 it is obvious that in 1992-1994 there was an increase in patients under 10 years of age. This translates to these patients being under 6 years of age at the time of the Chernobyl accident. This age group was the most vulnerable to radioactive iodine exposure.

The female to male ratio has changed only slightly : 1.8:1 before the accident to 1.4:1 during the 1990-1994 period with a recent shift to males.

Therefore, a pronounced increase in thyroid cancer cases in children and adolescents in Ukraine have been obvious following the Chernobyl accident.

It is most probable that excessive cases of thyroid cancer in children in Ukraine are of radiation genesis since more than 60% of all cases were registered in the 5 most contaminated oblasts out of 25 oblasts of Ukraine.

Radiation thyroid doses in children and adolescents operated in 1994 for thyroid carcinoma were estimated at the department of Dosimetry and Radiation Hygiene of the Scientific Centre of Radiation Medicine, Academy of Medical Sciences of Ukraine, under the guidance of Prof. I.A. Likhtarev.

As in the previous years in the majority of cases (79%) the absorbed thyroid dose did not exceed 30cGy, in 10.5% it was 30-100 cGy and in 10.5% - 100 cGy and more. The number of excessive cases of thyroid carcinoma as compared with the calculated spontaneous level per 100.000 children and adolescents was much higher in areas where thyroid radiation dose was 50-100 cGy and 100-200 cGy [2].

Analysis of the health state of patients aged 0-18 years with malignant thyroid tumors who were treated at the Department of Surgery of the Institute of Endocrinology and Metabolism in 1986-1994 showed that when admitted to the hospital, children had no complaints. Some patients complained of tumour-like formation in the neck, compression, hoarseness, problems with swallowing. In some children thyroid cancer was diagnosed when they were examined for submaxillary and cervical lymphadenitis.

Clinical thyroid cancer in children examined was characterised by cervical tumour-like formation of different size. Palpation showed solitary nodule, multiple nodules, diffuse tumour with vague contours. The size of palpable and ultrasonically identified nodules was 0.5-3.6 cm in diameter. In the majority of cases thyroid tumors were of extreme density, even stone-like, in this they differ from ordinary nodular goitre. The surface of the tumors, depending on the spread, was elastic, rugged or uneven.

One of differential diagnostic criteria of thyroid cancer is limited tumour nodule mobility. The degree to which this sign is manifested depends on the tumour size and its spread over the surrounding tissues. However, a soft nodule does not exclude thyroid cancer.

The treatment of thyroid cancers in children and adolescents in Ukraine is total thyroidectomy. Diagnostic scans with radioactive iodine-131 are performed 6 weeks after total thyroidectomy to reveal whether residual thyroid tissue or metastases are present. When areas accumulating radioagents are found they are ablated with therapeutic dose of radioiodine. The postoperative management of thyroid cancer also involves full-scale hormonal rehabilitation of patients.

A morphological study was carried out on 196 cases of thyroid cancer removed in 1986-1995 (by August 1st) at the Department of Surgery, Institute of Endocrinology, from children and adolescents aged 0-18 years. The male to female ratio was 1:1.3. In these cases the diagnosis of thyroid cancer was made by 4 pathologists in Kiev, according to the WHO classification [3]. 130 of these cases were additionally reviewed by the American expert in thyroid pathology Prof. V. LiVolsi in 1994 and 125 such cases were reviewed by the EU expert Prof. D. Williams (UK) and by Prof. B. Egloff (Switzerland). In 98.5% of cases the diagnoses were confirmed.

The results were the following: 183 cases of papillary carcinoma (93.4%)
6 cases of follicular carcinoma (3.1%)
4 cases of medullary carcinoma (2.0%)
2 cases of anaplastic carcinoma (1.0%)
1 case of malignant lymphoma (0.5%)

In patients who were operated in other Ukrainian clinics the diagnosis of thyroid carcinoma was verified by the local pathologist. Pathologic data for these cases will be presented later after we and other experts confirm the analysis of these specimens. Thus, it is clear that following the Chernobyl accident papillary forms of carcinoma are most common in children and adolescents in Ukraine.

Histological analysis of papillary carcinoma revealed the following :
. Prevalence of non-encapsulated tumors : 91.0%
. Infiltration of adjacent soft tissue : 50.5%
. Signs of intraglandular tumour spread when tumour sites are detected in relatively unchanged tissue : 55.3%

Typical papillary carcinoma was observed in 12.2% of cases. Papillary structures in the tumors studied were characterized by typical clear nuclei, usually oval or rounded. Nuclear chromatin was located mainly in the periphery and, as a rule, was absent in the central part, this gave the nucleus the "ground-glass" appearance. In addition, the papillary carcinoma cells were characterized by grooves, intranuclear inclusions and evidence of nuclear overlapping.

Electromicroscopically in papillary areas of carcinomas studied well differentiated thyrocytes were observed. They differed in the degree of functional activity. Cells with well developed cytoplasmic organellas prevailed showing granular endoplasmic reticulum, mitochondria, and the Golgi complex. The nuclei of the tumour cells were indented with low heterochromatin content and cytoplasmic "inclusions" into nucleoplasm, which is typical for this carcinoma pattern. These "inclusions" are not true intranuclear inclusions, but just reflect deep and complex nucleolemmal invaginations. Changes in the nucleus are very important diagnostic criteria of papillary carcinoma for cytologists and pathologists when studying the specimens with light microscope [4-5].

Follicular variant of papillary carcinoma was observed in 38.3% of cases. Nuclei in follicular areas are also clear and poor with chromatin. Typical papillary structures were not numerous or absent.

Solid or mainly solid variant was observed in 27.7%. Tumors with the alveolar-solid pattern prevailed, and tumour sites were divided by connective tissue.

Electromicroscopy revealed solid areas in the parenchyma consisting of poorly differentiated cells, when solid variants of papillary carcinoma were studied. Special attention was paid to details such as : smaller size, changed forms, presence of processes, and a decrease of organoid content [1-6]. Often only separate small canals of granulated endoplasmic reticulum, flocks or ribosomes, mitochondria and large vacuoles were observed, findings that are typical of the early stages of embryogenesis of endocrine glands. It should be pointed out that clusters of poorly differentiated cells were detected in the so-called uninvolved part of the thyroid gland thus indicating multifocal tumour growth.

Diffuse sclerotic variant of papillary carcinoma was observed in 3.7% and characterized by diffuse tumour growth, signs of sclerotic manifestations, lymphoid infiltration, prominent invasion of the tumour cells in the lymphatic vessels and a large number of psammoma bodies.

Mixed variant with papillary, follicular and solid areas was found in 18.1% of all cases.

It should be emphasized that irrespective of the variant of papillary carcinoma, solid clear-cell clusters in the areas of infiltrating tumour growth were found in 55.3% of cases. In addition psammoma bodies were present in 80.3% of cases, sclerotic stromal changes in 79.9%, and signs of background thyroiditis (9.6%) or lymphoid infiltration in 32.4% 67.6% of cases showed tumour invasion into lymphatic and 23.9% into blood vessels.

Tumour vessels in children were characterized by oedematous endothelium protruding into the lumina and by reduced micropinocytotic activity on a background of thickened basement membranes. Such changes in the components of the vascular wall were described for radiation exposure [4].

From our point of view, ultrastructural criteria of the initial stages of metastatic spreading of tumour cells is most interesting. These tumour cells lose their intercellular bonds, become rounded and may be found both in the capsule of the gland and occasionally in the lumina of tumour vessels.

In lymph nodes, metastases were observed in 59.0% of cases. Metastases were characterised by typical papillary structures or follicular and solid areas. In the solid areas cell clusters had preformed nuclei similar in structure to the focus of origin.

In general, the described morphologic signs indicate comparatively rapid growth and highly invasive properties of these tumors, especially when patients' ages are taken into consideration. This aggressiveness is revealed by frequent distant metastases. At our institute 23.7% of cases in this age group showed metastases to the lungs at the time of surgery and at various periods after surgery.

A true comparison with pre-Chernobyl papillary carcinomas is difficult because in the period of 1981-1985 there were only 8 these tumors removed in the age group 0-18 years. Of these 75% showed a typical variant of papillary carcinoma, 12.5% a follicular variant and 12.5% a solid variant. Although both groups are not strictly comparable, there seems to be in the post-Chernobyl period, an increase in the solid and follicular variants of papillary carcinoma, the presence of diffuse sclerotic variant and evidence of highly invasive tumour properties (multifocal growth, capsular and vascular invasion, lymph node metastases and infiltration of the adjacent tissues).

It should also be mentioned that a great number of solid and follicular variants of papillary carcinoma on a background of aggressive tumour behaviour was registered in thyroid carcinomas in the children of Belarus [7,8].

The increase in the incidence of solid and follicular variants was noted in comparison to those described papillary thyroid carcinomas in children from Philadelphia [9], and United Kingdom [10].

The comparison of the thyroid cancers in Ukrainian children with the British series [11] shows that in the period of 1990-1994 175 children aged 0-14 years were operated in Ukraine (on average, 35 cases per year). During the 30-year period (1963-1992) 154 children were operated in the United Kingdom (on average, 5 cases per year), that is it was 7 times more common in Ukraine.

There were major differences in the age structure of the two series. In the United Kingdom there was a smooth increase with age after about 6 years, with a rapid increase after 10. This contrasts with the Ukrainian series, were the peak of incidence occurs at the age of 8, with a subsequent decline.

In British series the female/male ratio was 2.7:1, while in Ukraine it was 1.3:1, that is in this age group a shift to males was obvious in the Ukrainian series.

In the British series papillary carcinoma in children made only 69%, whereas in Ukrainian series they numbered 95% with a sex ratio of 3.8:1 (F:M) in the UK compared with 1.4:1 in the Ukrainian series. At that in the United Kingdom typical variant of papillary carcinoma was 2.4 times more frequent and on the contrary, the most unfavourable solid variant of papillary carcinoma was 2.3 times less common.

Thus, these observations have confirmed the occurrence of a very large number of cases of thyroid carcinoma in children in Ukraine since 1990, and have shown that the

Ukrainian tumors differ from thyroid cancer in children in the United Kingdom by being more common in younger children and in boys. Analysis of the subtypes of thyroid cancer shows that the increase is specifically in a solid/follicular type of thyroid carcinoma..

Molecular-genetic and immunomorphologic study of malignant tumors is one of the most promising topics of these times. Changes in numerous oncogenes, growth factors, tumour suppressor genes are observed in malignant tumors of different organs [12-14]. For poorly and well differentiated carcinomas various genetic pathways of carcinogenesis may exist [15,16]. Some of carcinomas may develop due to the accumulation of a number of genetic changes which, as a rule, closely correlate with tumour progression and metastasis.

There are many publications about the oncogene and growth factor expression in thyroid cancer [12-22]. These problems have not been investigated in the sphere of childhood thyroid cancers so far.

That is why with the Experimental Collaborative Project-8 (EC/CIS Collaboration project) in cooperation with Cambridge University together with Prof. D. Williams, Dr. G. Thomas and Dr. R. Harach, thyroglobulin, calcitonin contents were studied with the help of immunohistochemistry and in situ hybridisation. Besides, met, ret and p53 oncogene expression, in tumour cells, lymph node metastases and extratumoral microscopically unchanged thyroid tissue were studied using immunoperoxidase technique.

Immunochemistry showed uneven thyroglobulin distribution in tumour cells. In the solid areas and in the lymph node metastases thyroglobulin reaction was focal. The same was noted in thyroglobulin biosynthesis in solid areas, lessening of their differentiation [6], and it causes anxiety as far as the postoperative prognosis is concerned.

Calcitonin-containing cells were identified in tumour tissue only in 5% of cases while C-cell hyperplasia in the thyroid extratumoral tissue was revealed in 41% of cases thus proving a definite role of C-cells in oncogenesis of the thyroid gland. We had already paid attention to this phenomenon following our electromicroscopic investigations [1,6].

When studying oncogene expression the most pronounced reaction was shown for the oncogene ret, it was detected in 87% of cases and located both in papillary and follicular and solid structures. The most intensive reaction was observed in cells infiltrating the capsule or stroma of the tumour, these were mainly solid cell clusters. In positively stained cells reaction product was mostly seen near the basement membrane. Only in some cases there was positive staining in the luminal sites of tumour cells.

The oncogene met was detected in 74% of cases, mainly with diffuse cytoplasmic staining. Only in some sections membrane staining was obvious. The reaction was much less intensive than for the oncogene ret.

Data on identification of p53 gene also seem to be of interest. Its mutation in carcinoma and its visualisation indicate lessening of tumour differentiation. As far as we know this gene was detected in thyroid tumours only in anaplastic thyroid carcinoma [15,16]. In our study p53 gene was found in 56% of cases but in the majority of cases (65%) the number of immunopositive nuclei was very small, on average only 5-10%. But at the same time in 24% of cases the number of immunopositive nuclei was about 20% or more.

Thus, the data obtained once more confirm highly aggressive properties of the studied papillary carcinomas in children and indicate the necessity to continue the work we have begun.

Further research in this field is necessary and could elucidate the mechanisms of carcinogenesis of the thyroid gland in children after the Chernobyl accident. A better understanding of the role of oncogenes in carcinogenesis of the thyroid gland may be helpful in determining specific tumour markers and be used for the earlier detection of these malignancies.

So we conclude that : After the Chernobyl accident in 1986, there has been an increase of thyroid cancer cases in children and adolescents in Ukraine. The incidence rate of thyroid cancer in these groups has increased most considerably since 1990.

These additional cases of thyroid cancer in children and adolescents, are most probably of radiation genesis, since more than 60% of all cases have been registered in the 5 oblasts most contaminated by the Chernobyl accident out of a total of 25 oblasts in Ukraine.

Thyroid cancers in children and adolescents in Ukraine have been mainly of the papillary carcinoma pattern. The aggressive biological behaviour is supported by frequent signs of multifocal growth, high invasive properties, intensive reaction to various oncogenes and growth factors. Such aggressive behaviour is also confirmed by regional and distal metastases.

References

[1]   N. Tronko, Epstein Ye., Oleinik V., Bogdanova T.I., Likhtarev I., Gulko G., Kairo I., Sobolev B., Thyroid gland in children after Chernobyl accident (yesterday and today). In:Nagasaki Symposium on Chernobyl : Update and Future, Sh. Nagataki editor (ed), Excerpa Medica Intern. Congress Series 1074:Elsevier (1994) 31-46

[2]   I.A. Likhtarev, B.G. Sobolev, I.A. Kairo, N.D. Tronko, Bogdanova T.I., V.A. Oleinic, E.V. Epshtein, V. Beral, Thyroid cancer in the Ukraine, Nature (1995) 375:365

[3]   Hedinger Chr., Williams E.D., Sobin L.H., Histological typing of thyroid tumours, WHO, 2nd Edn. Berlin:Springer, 1988

[4]   LiVolsi V.A., Surgical pathology of the thyroid, Philadelphia:WB Saunders, 1990

[5]   Rosai, J., Carcangiu M.L., Dellelis R., Atlas of Tumor Pathology, Tumors of the Thyroid Gland, Washington:Armed Forces Institutes of Pathology, 1992

[6]   Bogdanova T.I., Pathomorphologic Characteristics of Malignant Thyroid Tumors in Children. In: Treatment of Thyroid Cancer in Childhood, Ed. J. Robbins, National Institutes of Health, Bethesda, Maryland, (1994) 51-59

[7]   Furmanchuk A.W., Averkin J.I., Egloff B., Ruchti C. et al., Pathomorphological findings in thyroid cancer of children from Repubic Belarus. Histopathology (1992) 21:401-408

[8]   Nikiforov Y., Gnepp D.R., Pediatric thyroid cancer after the Chernobyl disaster. Pathomorphological study of 84 cases (1991-1992) from the Republic of Belarus, Cancer (1994) 74:748-766

[9]   Peters S.B., Chatten J., LiVolsi V.A., Pediatric papillary thyroid carcinoma. In: Abstracts of the Annual Meeting of the United States and Canadian Academy of Pathology, San Francisco (1994) 55A

[10]   Williams E.D., Thyroid Cancer in United Kingdom children and in children exposed to fall-out from Chernobyl. In:Nagasaki Symposium on Chernobyl : Update and Future, Sh. Nagataki (ed.), Excerpta Medica Intern. Congress Series 1074:Elsevier, (1994) 89-94

[11]   Bogdanova T., Bragarnik M., Tronko N.D., Harach H.R., Thomas G.A., Williams E.D., Childhood Thyroid Cancer after Chernobyl, In:Pathological Society of Great Britain and Ireland, Amsterdam (1995) 37

[12]   Wynford-Thomas D., Molecular-genetics of thyroid cancer. Trends in endocrinology and metabolism (1993) 4:224-232

[13]   Fabien N., Fusco A., Santoro M., Barbier Y., Dubois P.M., Paulin C., Description of a human papillary thyroid-carcinoma cell-line morphologic study and expression of tumoral markers, Cancer (1994) 73:2206-2212

[14]   Basolo F., Fugazzolo L., Fontanini G., Elisei R., Pepe S., Bevilacqua G., Pinchera A., Pacini F., Markers of cell-proliferation as prognostic factors in differentiated thyroid cancer, International J. of Oncology (1993) 3:1077-1081

[15]   Nakamura T., Yana I., Kobayashi T., Shin E., Karakawa K., Fujita S., Miya A., Mori T., Nishisho T., Takai S., p-53 gene mutation associated with anaplastic transformation of human thyroid carcinomas, Jpn J. Cancer Res. (1992) 83:1293-1298

[16]   Fagin J.A., Matsuo K., Karmakar A., Chen D.L., Tang S.H., Koeffler H.P., High prevalence of mutations of the p53 gene in poorly differentiated human thyroid carcinomas, J. Clin. Invest (1993) 91:179-184

[17]   Hall W., Hall E., Oncogenes and growth factors in thyroid carcinogenesis, Endocrinol. Metab. Clin North Am. (1990) 19:479-493

[18]   Enomoto T., SUgawa H., Inoue D., Miyamoto M. *et al*, Establishment of a human undifferentiated thyroid cancer cell line producing several growth factors and cytokines, Cancer (1990) 65:1971-1979

[19]   Imyaninov E.N., Chernitsa O.I., Nikiforova I.F., Socolov S.I., Valdina E.A., Plutalov O.V., Berlin Y.A., Knyazev P.G., Low frequency of alterations of ERBB-1 (HER-1), ERBB-2 (HER-2/neu), C-MYC, HRAS-1 oncogenes, RB-1, p53 antioncogenes and deletions of some loci of chromosome-17 in human thyroid carcinomas, Eksperementalnaya onkologiya (1993) 15:37-41

[20]   Shimizu T., Usuda N., Yamanda T., Sugenoya A., Iida F., Proliferative activity of human thyroid tumors evaluated by proliferating cell nuclear antigene/cyclin immunohistochemical studies, Cancer (1993) 71:2807-2812

[21]   Basolo F., Pinchera A., Fugazzola L., Fontanini G., Elisei R., Romei C., Pacini F., Expression of P21 RAS protein as a prognostic factor in papillary thyroid-cancer, European Journal of Cancer (1994) 30A:171-174

[22]   Bertheau P;, De La Rosa A., Steeq P.S., Merino M.J., NM23 protein in neoplastic and nonneoplastic thyroid tissues, Am? J. Pathol? (1994) 145:26-32

Thyroid Cancer in Children and Adolescents of Bryansk and Kaluga Regions

A.F. TSYB, E.M. PARSHKOV, V.V. SHAKHTARIN, V.F. STEPANENKO,
V.F. SKVORTSOV, I.V. CHEBOTAREVA

MRRC RAMS, Obninsk, Russia

Abstract

We analyzed 62 cases of thyroid cancer in children and adolescents of Bryansk and Kaluga regions, the most contaminated as a result of the Chernobyl accident. The data on specified radiation situation as well as probable radiation doses to the thyroid are given. It is noted that the development of thyroid cancer depends on the age of children at the time of accident (0-3, 7-9, 12-15 years). They are the most critical periods for the formation and functioning of the thyroid, in particular, in girls. It is suggested that thyroid cancer develops in children and teenagers residing in areas with higher Cs-137 contamination level at younger age than in those residing in less contaminated regions. It is shown that the minimal latent period in the development of thyroid cancer makes up to 5 years. The results of ESR method on tooth enamel specimen indicate that over postaccident period the sufficient share of children has collected such individual radiation dose which are able to affect on their health state and development of thyroid pathology.

For a long period of time Russia unlike Belarus and Ukraine was considered to be "favourable" by the development of thyroid cancer in children and adolescents after the Chernobyl accident. Such a fact appeared to be a dissonance in the common concept on the possible radiation induction of thyroid tissues to malignancy when received relatively low doses of iodine radionuclide. In addition, it was incomprehensible why Briansk region directly adjacent to Gomel region in Belarus, which had practically the same soil contamination, showed no increase in thyroid cancers in children. To settle this problem in 1991 a systematic medical examination of population residing in south-west districts of Bryansk region started under the concept of 3-stage prophylactic examination.

At the same time specification of radiation situation in Bryansk and Kaluga regions as well as probable radiation doses to the thyroid was performed.

The total area in Bryansk region contaminated by Cs-137 over 37 kBq/sq.m makes up about 12.000 sq.km. 1.392 settlements with 470.000 population are located in this area. Among this population children and adolescents total 120.000. More than 93.000 individuals reside in the areas with Cs-137 contamination over 555 kBq/sq.m.

Mean doses of thyroid irradiation by iodine radionuclides ranges from 120 mGy for adults to 500 mGy in children aged under 7. In settlements with contamination level over 555 kBq/sq.m, the mean doses to the thyroid in children under 7 achieve 2200 mGy and individual doses several thousand cGy.

In Kaluga region the total area contaminated by Cs-137 over 37 kBq/sq.m is equal to 5.000 Sq.km. 424 settlements with 94.000 population including about 24.000 children and adolescents are located on this area. Kaluga region has no settlements with contamination density over 555 kBq/sq.m. The mean doses of thyroid irradiation by iodine radionuclides ranges from 50 mGy for adults to 250 mGy in children aged under 7. In settlements with contamination density over 185 kBq/sq.m the mean doses in children under 7 are equal to 500 mGy and individual doses in some children achieve 10.000 mGy.

The estimation of collective doses of thyroid irradiation by iodine radionuclides in children and teenagers residing in the regions with soil Cs-137 contamination density over 37 kBq/sq.m shows that during the whole life period one may expect about 240 cases of radiogenic thyroid cancer in Bryansk region and 30 in Kaluga region. In addition to this rate 30% increase in thyroid cancers may be expected due to external and internal Cs irradiation.

To date, thyroid cancer has been verified morphologically in 62 patients who were children and adolescents at the time of the accident (Fig. 1).

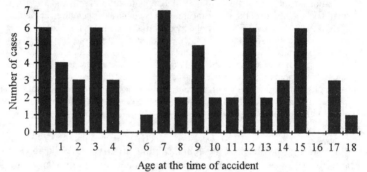

Fig. 1 Number of thyroid cancers in children and adolescents of Bryansk and Kaluga regions depending on the age at the time of accident

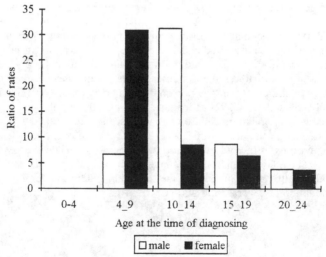

Fig.2 Ratio of Standardazed rates of thyroid cancer morbidity in children and adolescents of Bryansk Region to the same rates in Russia

This figure shows that thyroid cancer developed in 19 children who were irradiated in-utero and at the age of 4, in 14 children at the prepubertal age and 17 at the pubertal age.

In fact, thyroid cancer development depends on children's age at the time of accident (0-3, 7-9, 12-15). They are the most critical ages for the development and functioning of the thyroid, in particular, girls. Of 62 registered thyroid cancers 42 cases occurred in females and 20 in males.

The standardized rate of thyroid cancer incidence in residents of Bryansk region, who were at the age of 24 at the end of 1994, is much higher than the same rate in Russia on the whole (Fig. 2). The data presented indicate that the highest risk of thyroid cancer occurs in girls aged 5-9 and boys aged 10-14. By 15-19 and especially 20-24 the rate of thyroid cancer incidence is practically similar in groups of both sexes.

To our mind, it is interesting that thyroid cancer develops in children and teenagers residing in the regions with soil Cs-137 contamination density over 185 kBq/sq.m at younger age (Fig. 3, 4). Figure 3 shows that the average age rate at the time of accident and diagnosis makes up 4.8 and 12.3 years, respectively, whereas this rate is equal to 10.6 and 18.1 years (Fig. 4) for individuals residing in less contaminated territories (up to 185 kbq/sq.m).

The medical and dosimetric investigations of the verified cases of cancer showed that the most probable reconstructed thyroid doses of I-131 irradiation ranges from 20 to 2400 mGy, these values being in the direct dependence on the residence (Table 1).

Estimating the obtained results one may suggest that a basic etiological factor in the development of thyroid cancer appears to be radioactive iodine. It is quite possible that low doses (20-200 mGy) to thyroid tissues only provoke the disease available and, on the contrary, higher doses result in the development of malignant tumors de novo.

Currently, it is impossible to answer the question on the contribution of other radiation sources (external and internal Cs-137) to the development of malignant tumors although our results and the results of other researchers indicate their participation in the development of thyroid cancer. It is difficult to estimate the significance of other sources due to their long-term exposure (years) at low dose rate.

Children residing in the regions where soil contamination density is equal to or over 555 kBq/sq.m have the whole body average collective dose of external and internal irradiation much higher for the period of 1986-94 (90-140 mGy) than those residing in the territories with lower soil contamination (20-90 mGy).

The results obtained using ESR method for tooth enamel specimen show that for the period of time after the accident a sufficient part of children have such individual collective doses of radiation which are able to affect on their health state including the development thyroid pathology and thyroid cancer. In 41% of cases this collective dose makes up 120-150 mGy and in 12% of cases even more than 250 mGy.

One may suggest that the aggressive course of thyroid cancer especially in children of younger age group is associated with the suppression of immunity which occurs in the examined groups. Moreover, the humoral link is more suppressed in children residing in the areas with soil Cs-137 contamination density over 555 kBq/sq.m. the most number of thyroid cancer has been found just in these regions.

The clinical manifestation of thyroid cancer is known to have a latent period which may range from 5 to 20 years [1, 2]. Our results indicate that in fact the minimal latent period makes up 5 years having no dependence on the age and, probably, dose of thyroid radiation (Table 2). The table shows that the pathological process takes practically the same time in all age groups. On the basis of these results one may conclude that if this tendency is observed in adults, the same increase in thyroid radiogenic cancers should be expected in them.

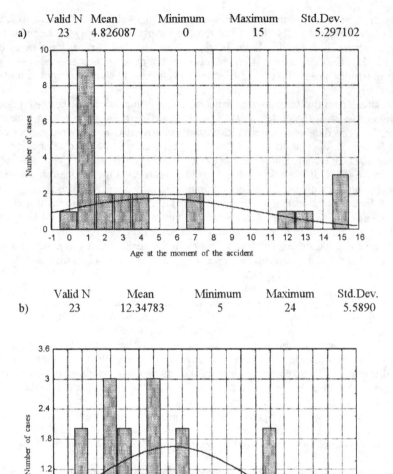

	Valid N	Mean	Minimum	Maximum	Std.Dev.
a)	23	4.826087	0	15	5.297102

	Valid N	Mean	Minimum	Maximum	Std.Dev.
b)	23	12.34783	5	24	5.5890

Fig.3   Age-specific distribution of thyroid cancer cases in children and adolescents residing in the territory of Bryansk regon with Cs-137 surface soil contamination over 185 kBq/m^2 (at the moment of the accident - a; at the moment of diagnosis establishment - b)

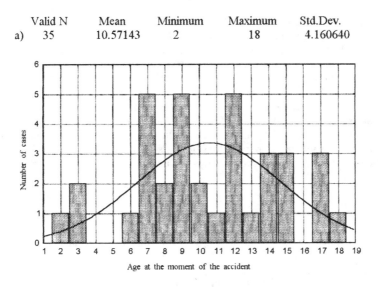

	Valid N	Mean	Minimum	Maximum	Std.Dev.
a)	35	10.57143	2	18	4.160640

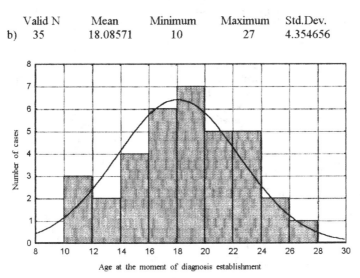

	Valid N	Mean	Minimum	Maximum	Std.Dev.
b)	35	18.08571	10	27	4.354656

Fig.4 Age-specific distribution of thyroid cancer cases in children and adolescents residing in the territory of Bryansk regon with Cs-137 surface soil contamination before 185 kBq/m^2 (at the moment of the accident - a; at the moment of diagnosis establishment - b)

The availability of the prolonged latent period of clinical manifestation of thyroid cancer shifts "children's cancer" to teenager and adult groups. Without taking into account this factor, the problem of child's thyroid cancer would be artificially avoided during next 5 years. In this connection it is necessary to focus a great attention on the study of thyroid pathology in individuals born in 1968-86. This cohort of persons should be studied as those who received external and internal thyroid irradiation as child and adolescent. Control for them should be individuals selected by the same age and sex from the "clear" areas.

The analysis of thyroid cancers in children and adolescents taking into account sex and age at the time of accident and diagnosis, time and clinical manifestation and frequency of the disease depending on soil contamination density by Cs-137 and radioactive iodine enables to come to the conclusion that most revealed thyroid cancers could be considered as radiation-induced ones.

Table 1

Results of medical and dosimetric investigation of thyroid cancer in children and adolescents of Bryansk Region

Range of Cs-137 contamination, Ci/km^2	Plase of residence at the moment of the accident	Number of cases	The most probable dose to the thyroid, mGy
0.1 - 1.0	Bryansk, Dyatkovo, Karachev, Seltso, Unecha, Trubchevsk, Vygonichckiy, Kamarichsky and Suzemsky districts	14	15.0 ± 2.9 (10 - 50)
1.0 - 5.0	Klintsy, Navlya, Klimovsky, Starodubsky and Trubchevsky districts	10	229.0 ± 74.4 (15 - 700)
5.0 - 15.0 and over 15.0	Novozybkov, Novozybkovsky, Klintsovsky, Klimovsky and Krasnogorsky districts	11	1290.9 ± 210.4 (400 - 2700)

Table 2

Time of klinikal manifestation of thyroid cancer in children and adolescents after the Chernobyl accident (Russia, 1995)

Age at the moment of the accident	Time of the thyroid cancer detection (years)										Number of cases
	1986	1987	1888	1889	1990	1991	1992	1993	1994	1995	
1					1		3	3	2	1	10
2									2	1	3
3								1	2	3	6
4								1		2	3
5											
6									1		1
7							1	1	5		7
8								1	1		2
9		1			1			1	2		5
10								1	1		2
11					1				1		2
12						2		2		2	6
13						1			1		2
14					1		1		1		3
15						2	1		2	1	6
16											
17						1	1	1			3
18										1	1
Total		1			4	4	8	13	21	11	62

Literature

1. Protection of the thyroid gland in the event of releases of radioiodine: Report No. 55/NCRP. Washington, 1977, p. 60.

2. Robbins J., Adams W. H.   Radiation effects in the Marshall Islands. In: " Radiation and thyroid. [Sh. Nagataki, ed.]". Excerpta Medica, Amsterdam, Tokyo, 1989, p. 11.

Interaction of Pathology and Molecular Characterization of Thyroid Cancers

E.D. WILLIAMS, E. CHERSTVOY, B. EGLOFF, H. HÖFLER,
G. VECCHIO, T. BOGDANOVA, M. BRAGARNIK, N.D. TRONKO

Abstract. This paper presents the results of joint studies of thyroid cancer in children under 15 years of age between departments in Cambridge, Brussels, Naples and Munich in the European Union, and departments in Minsk, Kiev and Obninsk in the newly independent states of Eastern Europe.

The pathology of 264 cases of childhood thyroid cancer out of 430 that have occurred since 1990 in the 3 countries in which high levels of fallout from the Chernobyl accident occurred has been restudied by NIS and EU pathologists. The overall level of agreement reached was about 97%. The diagnosis was supported by immunocytochemistry and ISH for the differentiation markers, thyroglobulin and calcitonin, and the tumours were classified according to the WHO, with papillary carcinomas being further subclassified. 99% of the 134 Belarussian cases were papillary carcinomas, as were 94% of the 114 Ukrainian tumours. All 9 of the Russian cases available for study were papillary in type. 76 of 154 cases of childhood thyroid cancer reviewed over a 30 year period in England and Wales and were also studied, 68% of these were papillary carcinoma. Histological study showed that a subtype of papillary carcinoma, rarely found in adults, with a solid/follicular architecture occurred in children. It was found in 72% of the Belarussian papillary carcinomas, 76% of the Ukrainian cases, but only 40% of the England and Wales cases.

Molecular biological studies showed that the proportion of cases of papillary carcinoma expressing the ret gene was not significantly different in the exposed and the unexposed tumours, studies of the type of translocation leading to ret gene expression are not yet conclusive. Ras gene mutations were found as expected in follicular carcinoma, but were absent from any papillary carcinoma, whether from exposed or unexposed cases. TSH receptor mutations, normally found in follicular tumours were not found in any papillary carcinomas, nor were any p53 mutations identified. All these results conform to the proposal that papillary carcinomas only are associated with ret translocation and that ras and TSH receptor mutations occur in follicular tumours, and p53 mutations in undifferentiated carcinomas. The results support the diagnosis of nearly all of these cases as papillary carcinoma.

The major differences between the tumours from the exposed areas and those from England and Wales, lie in their frequency, the proportion of papillary carcinomas, and within these the proportion of the solid / follicular subtype. The high level of agreement between the different centres in establishing the diagnosis allows all the data to be studied. The number of tumours shows a very great increase, particularly noticeable in Southern Belarus and Northern Ukraine. The birth dates show that virtually all children with thyroid carcinoma in the areas exposed to fallout were born before the accident, very few occurred in children born after the accident. Analysis of the age at operation showed that it differed greatly in the exposed children as compared to children in England and Wales, and further analyses suggest that there is a very great increase in sensitivity relating to the age at exposure with the youngest children showing the greatest sensitivity. Overall the findings suggest that the great increase reported in childhood thyroid carcinomas in the areas exposed to fallout from Chernobyl is currently reported, that it is not due to over-ascertainment, that the incidence is almost certainly due to exposure to fallout from Chernobyl with the most likely cause being isotopes of iodine, and that this exposure to radiation has led to the development of an unusual subtype of papillary carcinoma.

1. Introduction

It is now generally accepted that cancer is derived by an interactive process of somatic mutation and clonal selection, and that a number of steps are involved, perhaps 6 or more in many cancers. The interaction of the consequences of these mutations with each other, and with the pattern of expression of the genome of the cell of origin of the cancer determine the structure and behaviour of the tumour; both would be expected to correlate with the somatic mutations present. In studying such correlations it must be remembered that, particularly in slowly growing tumours such as thyroid carcinoma, clones with different mutations may coexist, and a mutation conferring malignancy in a small clone is unlikely to produce instant metastases. Correlation of the structure of a tumour and of its behaviour is based upon a very large body of accumulated knowledge, and it is now important to correlate structure with molecular biological changes, to build upon and add to the ability to predict behaviour and determine appropriate treatment. Different mutagens are known to produce differing spectra of mutations, for example liver cancer induced by aflatoxin is characterized by a particular mutation in the p53 oncogene (1) and radiation is known to be more likely to induce translocations and deletions than point mutations, in contrast to some chemical mutagens which produce point mutations (2, 3).

The accident in the nuclear power station at Chernobyl in April 1986 led to the exposure of large numbers of the population in southern Belarus, northern Ukraine, and adjacent western areas of the Russian Federation to high levels of radioactivity in fallout. The radioactivity released from the reactor contained large amounts of isotopes of iodine, mostly ^{131}I, but the population was also exposed to significant amounts of short lived isotopes, including ^{133}I and ^{132}I. The distribution of the dose from these isotopes is not known with precision, and is likely to have diminished more rapidly with distance from the reactor than did the dose from ^{137}Cs. An increased incidence of childhood thyroid cancer in Belarus and the Ukraine was first noticed in 1990, four years after the accident, it was drawn to the attention of the West in 1992 (4, 5) and has continued up to the present. The purpose of the present paper is to describe the pathomorphology of the tumours in the three countries exposed to high levels of fallout from Chernobyl, to correlate these findings with a study of the molecular biology of selected oncogenes in thyroid cancer from the exposed population, and to compare and contrast both the pathological and the molecular biological findings with studies in a series of childhood thyroid cancers from an unexposed population in England and Wales.

2.   Pathomorphology - general

430 cases have been diagnosed as childhood thyroid cancer in the three republics of Belarus, Ukraine and the Russian Federation since 1990; 266 of these have been studied jointly with pathologists in the UK, and a total of 76 cases of childhood thyroid cancer diagnosed in the UK have been studied jointly with pathologists from the CIS countries.   The overall level of agreement in the diagnosis of malignancy has been high, 97% in the CIS countries and 96% in the UK.   The tumours from all 4 countries could be divided into the major groups of papillary, follicular and medullary carcinomas, and the papillary carcinomas subdivided into classic, solid follicular, diffuse sclerosing and other types.   The follicular and medullary carcinomas showed the expected features, all medullary carcinomas were confirmed by calcitonin immunocytochemistry and all follicular carcinomas showed capsular or vascular invasion with one exception in which invasion was not seen in the section available, but metastasis occurred. The classic papillary carcinomas showed the typical branching papillary architecture commonly found in adult papillary carcinomas, and the typical ovoid grooved nuclei, together with intranuclear cytoplasmic inclusions.   The solid follicular papillary carcinomas were characterized by a solid and/or small follicular pattern, occasionally with a minor papillary component.   They commonly contained psammoma bodies, but in general showed irregular round nuclei, lacking prominent nuclear grooving.   Diffuse sclerosing papillary carcinomas showed a diffuse pattern of invasion through thyroid tissue, usually the whole of one lobe, with a marked fibrous and lymphoid response.   The distinction between these subtypes was not absolute, and some cases showed features of more than one type. These were classified by the dominant pattern in the primary tumour.

2.1   Belarus

298 cases of thyroid cancer have been diagnosed in children under the age of 15 in the Pathology Institute in Minsk between 1990 and 1994 inclusive. A total of 134 of these (45%) have been seen also in Cambridge, and the diagnosis agreed in consultation with pathologists from Minsk. Agreement was reached in 98% of cases. The cases included 67 from the first 100 cases seen during the period, and 65 from the last 100, the selection being made on the basis of those cases where blocks were available for study. There was no change in the level of agreement in the diagnosis between the two periods, with disagreement between the original and the revised diagnosis present in only 1 case from each period. Virtually all of the Belarussian cases (99%) were papillary carcinomas, with 1 medullary carcinoma seen in the cases available for study. Because of the very high level of agreement on diagnosis in the sample examined from the earlier and from the later cases, the sex and age data from the whole series has been

examined. There was an overall peak age of 9 years, with a mean age of 10.4 years, and a sex ratio of 1.7:1 F:M. When the cases occurring in the early part of the outbreak were compared with the last group seen there was a change in both peak and mean age from 8 years with a mean of 9.1, to 9 years with a mean of 10.9 years (Fig. 1). The diagnoses in the whole group of 298 were 293 papillary carcinomas (98%), 4 follicular carcinomas (1.3%) and 1 medullary carcinoma (0.3%). Of the 134 cases studied histologically in Cambridge, 2 were excluded because on the material available the diagnosis could not be confirmed, 4 were papillary microcarcinomas, and 2 were malignant but unclassifiable. 124 papillary carcinomas were subclassified, 72% were of the solid follicular type, 14% were classic, 8% diffuse sclerosing and 6% others. The sex ratio, F:M was 1.2:1 for the solid follicular type, 1.25:1 for the classic type and 8:1 for the diffuse sclerosing tumours.

Because of the high level of agreement with the diagnosis from the Pathology Institute in Minsk, the figures for all 298 cases were used to demonstrate the changes in age at operation with time (Fig. 2). It can be seen that there is a sharp cut off in the number of younger patients, but that this increases with time, and corresponds closely to the age of a child born 6 months after the Chernobyl accident. The data has also been analysed by cohorts based on the age of the child at the moment of the accident (Fig. 3). It can be seen that overall the number of cases within each cohort continues to rise with time.

2.2 Ukraine

122 cases of thyroid cancer have been diagnosed in children under the age of 15 at the Institute of Endocrinology in Kiev between 1990 and 1994 inclusive. 114 of these cases have also been studied in Cambridge, and the diagnosis agreed in correlation with pathologists from Kiev. The diagnosis of malignancy was agreed in 97.4% of the cases, with the small number of disagreements probably being due to lack of appropriate material for restudy. The great majority (94%) of the cases were papillary carcinomas, there were 2 medullary carcinomas and 5 follicular carcinomas. The papillary carcinomas could be subdivided into 80 of the solid follicular type (76%) 7 classic papillary carcinomas (7%), 9 diffuse sclerosing carcinomas (9%) and 7 others (7%) 1 occult papillary carcinoma is excluded. The sex ratio (M:F) for the different types was 1.2:1 for the solid follicular type, 2.3:1 for the classic type, and 3:1 for the diffuse sclerosing variant. The peak age for the Ukraine cases was 8 years, and the overall sex ratio F:M was 1.2:1. The age at operation rose with time, (Fig. 4) and the change in the lower limit corresponded to the age of a child born 6 months after the Chernobyl accident.

2.3 Russian Federation

Study of the thyroid carcinomas that have occurred in children in the exposed areas of the Russian Federation is more difficult than studies in Ukraine or Belarus because the exposed areas form only a relatively small part of the very large Federation, and because no single central hospital has treated all or the great majority of the cases. Pathological material has been made available from 10 cases of childhood thyroid cancers by the RAMS Institute in Obninsk, from material originally treated and diagnosed in different hospitals serving the contaminated areas. The diagnosis of malignancy was confirmed in 9 of the 10 cases, the tenth contained no tumour in the material available in Cambridge. All 9 cases were papillary carcinomas, 5 were of the solid follicular type, 2 were classic papillary carcinomas and 2 oxyphil carcinomas. The overall sex ratio was 0.7:1, F:M., the mean age was 11 years. One of the classic papillary carcinomas was 3 mm in diameter in the section available. None of the children were born after the Chernobyl accident, and all came from contaminated regions in Bryansk, Kaluga or Tula Oblasts.

2.4 England and Wales

The occurrence of thyroid carcinomas in children in England and Wales was studied over a 30 year period, 1963 to 1992 (6). Information was obtained from the United Kingdom Childhood Cancer Registry, which had recorded 154 cases of thyroid cancer in children under the age of 15 during the study period. Tissue blocks were requested for all cases, and material was made available for 81 of the cases. The review diagnosis confirmed the original diagnosis of malignancy in 76 of the cases (94%); the cases where there was a disagreement included several earlier cases when the diagnosis of papillary carcinoma was made on the basis of architecture rather than cytology. One case of malignant lymphoma of doubtful primary origin in the thyroid was excluded. The remaining 75 cases included 1 teratoma, 53 papillary carcinomas, 8 follicular carcinomas, 12 medullary carcinomas and 1 other. The 53 papillary carcinomas were divided into 21 solid follicular tumours, 27 classic papillary carcinomas, and 5 of the diffuse sclerosing variant. The sex ratio (F:M) for these tumours was 6:1 for the solid follicular carcinomas, 3.5:1 for the classic papillary carcinomas, and 4:1 for the diffuse sclerosing carcinomas.

The overall sex incidence for the whole series was 2.3:1 F:M, and the age incidence rose rapidly with age, and was still rising at the age of 14 (Fig. 5).

3. Immunocytochemical and Molecular Biology studies

Immunocytochemical investigations have been carried out on the great majority of these tumours from all 4 countries. They have included studies of the differentiation markers thyroglobulin and calcitonin which were used to confirm the cell type in all the carcinomas studied, and were in most cases used together with *in situ* hybridisation to demonstrate the specific mRNA. No specific changes were seen between the tumours from the different countries with these techniques. Immunohistochemical techniques have also been used to demonstrate the presence of ret and met peptides in the tumours. The great majority of papillary carcinomas were positive for ret, using two antibodies. In most the distribution of the peptide was unusual, presenting as a dot-like deposit of reaction product between the nucleus and the basement membrane. The distribution of the met oncogene product (the hepatocyte growth factor receptor) was mostly on the cell membrane, it was found in approximately 80% of the papillary carcinomas, but was absent from the follicular carcinomas.

Molecular biological studies were carried out in a collaboration between the CIS countries and groups from Cambridge, Brussels, Naples and Munich.

The three ras genes have been studied by PCR and direct sequencing in formalin fixed material from 14 childhood papillary carcinomas from the CIS. No mutations have been observed in the commonly mutated codons (12,13 and 61). Similar studies carried out on a control group of childhood thyroid papillary carcinomas from England and Wales also showed no mutation at these sites. However, mutations of these genes have been identified in 3 of 10 adult follicular carcinomas using the same approach. This suggests that ras gene mutation is probably not normally involved in the genesis of thyroid papillary carcinomas, whether induced by radiation or not.

Expression of the ret gene has been studied using RT-nPCR for a 90 base pair sequence within the tyrosine kinase domain and direct sequencing. Ret expression was identified in 6 of 18 (33%) Chernobyl associated childhood papillary carcinomas so far studied. This was a significantly lower frequency than that found in a study which used the same technique on adult papillary carcinomas from England and Wales, and was also lower than the frequency observed in 20 childhood papillary carcinomas from England and Wales. However, due to the smaller number of carcinomas from children so far studied, it is not yet possible to say whether the irradiated series shows a significant reduction in the frequency of ret expression. A small study carried out on frozen material from the Ukraine showed ret expression in a similar proportion (3/11) of childhood thyroid papillary carcinomas, suggesting that the low frequency found in the paraffin embedded material was not due to a decreased sensitivity of the system. There does not appear to be a correlation between the expression of the ret oncogene as observed by RT-nPCR analysis and morphological subtype of papillary carcinoma. Positivity for actin amplification was used as a control for quality of the RNA extracted from the sections.

Twenty five cases of childhood thyroid carcinomas from the CIS have been used in a study of the type of ret gene translocation; only 11 yielded sufficient RNA for further analysis. Two controls for RNA quality were used: amplification of actin mRNA by RT-PCR and Northern blot with an 18S RNA probe. Using primers which allow detection of the three individual translocations of the ret oncogene so far identified, three papillary carcinomas with ret translocation have so far been identified by PCR and Southern blotting. All 11 cases which provided sufficient RNA have been analysed for the PTC1 translocation; only one case was positive. Two of the 7 cases so far analysed have been found to be positive for the PTC3 translocation. Analysis of the presence of PTC1 in a small series of adenomas from the Ukraine has also been carried out and shown to be absent. Interestingly one of the carcinomas found to be positive for ret expression by RT-PCR was not found to possess one of the 3 known translocations. Further studies on the remaining papillary carcinomas for PTC2 and PTC3 expression and to identify other translocations involving the ret oncogene are underway.

As part of a study of TSH receptor mutations in thyroid carcinogenesis, DNA extraction has been performed on material from 41 cases of childhood thyroid carcinoma from the CIS and regions of interest in exon 10 of the TSH receptor have been amplified using PCR. The regions studied include the third intracellular loop and the third transmembrane segment of the TSH receptor gene. Single stranded conformational polymorphism (SSCP) in all cases followed by sequencing (in 15 papillary carcinomas) has been used to identify TSH receptor mutations in exon 10. So far no mutations have been observed in 41 cases of papillary thyroid carcinoma or in 18 follicular adenomas and 3 follicular carcinomas from children and adolescents from the radiation exposed population. Cases known to be positive for mutations in exon 10 have been used as control for the technique used. The absence of mutations in the TSH receptor correlates with the morphological observations that the childhood thyroid tumours identified in the Ukraine after Chernobyl are of the papillary subtype.

P53 is a gene which is widely involved in human neoplasia, and although it is not usually involved in the early stages of thyroid carcinogenesis, a study has been performed to look for p53 mutations in possible radiation induced tumours. Exons 5 and 7 and 8 have been successfully amplified from 23 papillary carcinomas using nested PCR. SSCP analysis under four different running conditions has so far been applied to all 23 samples, but no aberrant cases have been found. p53 mutation involvement in thyroid carcinoma is usually a late phenomenon, at the interface between differentiated and undifferentiated carcinoma. None of the tumours so far studied has been un-differentiated and none show widespread positivity for p53 on immunocytochemistry. However, exposure to radiation has been reported to increase the frequency of mutation in the p53 gene in other tumours. From the results presented here p53 mutation does not appear to play a major role in papillary thyroid carcinogenesis post Chernobyl.

4.   Discussion

These studies have involved cases of childhood thyroid cancer from Belarus, Ukraine, the Russian Federation, and England and Wales. The countries obviously differ in that parts of the first three were exposed to high levels of fallout from Chernobyl, while in England and Wales only very small amounts of fallout took place, and the cases can effectively be regarded as unexposed. England and Wales differ from the CIS countries in other ways, particularly in that they show no significant iodide deficiency, while mild iodine deficiency is prevalent in parts of the CIS. The rate of occurrence of childhood thyroid carcinomas in England and Wales is about 0.5/million children/year(6), a rate that is towards the lower end of the spectrum of figures for childhood thyroid carcinoma. Most countries would expect to have a rate of about 1/million/year, and a few rise as high as 3/million/year. The rates calculated for Belarus and the Ukraine are very much higher than any of the reported international figures, and this study has confirmed the diagnostic accuracy on which the figures are based, and excluded papillary microcarcinoma as a significant contribution. There are major differences between the findings in the CIS countries and those in England and Wales (Tables 1 & 2).

Table 1:

Childhood thyroid carcinoma in Belarus, Ukraine and England and Wales

	Belarus	Ukraine	England and Wales
Total no. of carcinomas registered (years studied)	298 ('90-'94)	122 ('90-'94)	154 ('63-'92)
Numbers available for review	134 (45%)	114 (93%)	81 (53%)
% of diagnosis confirmed	98%	97.4%	96%
% of papillary carcinomas in reviewed cases	99%	96%	68%

In this discussion the figures from Belarus and the Ukraine will be used, the numbers available for study from the Russian Federation are currently too few for adequate comparisons. The most obvious difference between the CIS tumours and those from England and Wales lies in the proportion of papillary carcinomas and the different subtypes of papillary carcinoma. Well over 90% of the CIS tumours were papillary carcinomas, as compared to 68% of the England and Wales tumours. Within the papillary carcinoma group over 70% of the CIS tumours were of the solid follicular type compared to only 40% in England and Wales. The difference was even more marked with the classic type of papillary carcinoma, which formed over half the England and Wales series, but only 12% of the CIS cases. There was no significant change in the proportion of the different type of papillary carcinoma between the earlier and the later cases, or between the younger and the older CIS cases. These findings suggest that the aetiological agent causing the increase in thyroid cancer in children in the CIS countries is specifically leading to an increase in papillary carcinoma of the solid follicular type.

Table 2:

Subtypes of papillary carcinoma of the thyroid in Belarus, Ukraine and England and Wales

	Belarus	Ukraine	England and Wales
Solid/follicular papillary carcinoma	72%	76%	40%
Classic papillary carcinoma	14%	7%	51%
Diffuse scelerosing variant	8%	9%	9%

Mutations in the ras oncogenes were absent from both the exposed and the non-exposed childhood papillary carcinomas, and from the non-exposed adult papillary carcinomas. Recent work suggests that in the thyroid ras mutations are much more restricted to follicular carcinomas than was previously thought (7) and this study confirms this. It shows that the exposed group of tumours do not differ from non-exposed in their lack of ras gene mutations. The study of the

TSH receptor gene gives a similar negative result; mutation in this gene is generally found in follicular rather than papillary tumours, and the papillary tumours from the exposed population show the lack of TSH receptor mutation expected in non-exposed areas. The position with the ret gene is more complex. There are at least 3 translocations of the ret oncogene in papillary carcinoma in adults; the commonest is PTC1. The papillary carcinomas in the exposed group could differ from unexposed tumours in their overall frequency of ret translocation, or in the type of ret translocation involved. These studies show that there is no increase in the overall frequency of ret activation in these tumours, but that compared to adults there may be a reduction in the cases due to PTC1. However the frequency of the various ret translocations in childhood thyroid carcinoma is not yet known, so that no conclusion can yet be drawn from these or from other results suggesting an increase in PTC3 in these cases. The p53 results again conform to the result expected in non radiation exposed papillary carcinomas.

The age and sex structure is interesting. In England and Wales there were more males in the younger but not the older age group. In the CIS countries the female to male sex ratio was closer to equality than it was in the England and Wales, and the difference was not solely due to the younger age of the cases. There was a marked difference in the age structure of the children with thyroid cancer in the CIS compared to England and Wales. The Chernobyl related cases showed a peak age of about 8-9, while in England and Wales the incidence continued to rise with increasing age. The peak age in Belarus showed a change with time, suggesting that there might be a cohort affect, with one group of children showing a higher sensitivity than others. Analysis by age at exposure to the accident shows that the cohorts showed a continuing increase in incidence with age. When this is compared with the expected incidence derived from the England and Wales figures, it can be seen that there is a great increase in observed as compared to expected tumours in the youngest cohort at exposure, and a steady drop in sensitivity with increasing age (Fig. 6).

These results lead to a number of conclusions. The pathological diagnosis and number of cases reported from Belarus and the Ukraine have been broadly confirmed, and the number of microcarcinomas shown to be very small. The relation of the tumours to exposure to fallout from the Chernobyl accident is strongly supported by the great reduction in the numbers of tumours operated on from children born more than a few months after the accident; the rapidity of the drop in the number of cases suggests that the causative agent did not persist at high levels in the environment. This, together with the apparent restriction of the increase in malignancy to thyroid carcinomas strongly suggests that exposure to isotopes of iodine from fallout is the major, if not the only cause of the increase.

The molecular biology results do not distinguish the papillary carcinomas in the exposed children from those in an unexposed population. It is possible that the ret oncogene translocations differ in type, or perhaps in the exact break points of the translocation, although these are difficult to ascertain because they take place in an intron. Although there is not a relative increase in the proportion of papillary carcinomas with a ret translocation in the CIS cases, as increase is almost exclusively in papillary carcinoma, there is a very large absolute increase in cases with translocation of the ret oncogene. The genes involved in the ret negative papillary carcinomas are not known, except for trk, which is translocated in a small proportion of adult papillary carcinomas. Radiation is known to give rise to more translocations than point mutations.

The age structure of the children with thyroid carcinoma in the CIS and a comparison with cases in England and Wales suggest that the children aged under 1 at the time of the Chernobyl accident have shown the greatest susceptibility to the carcinogenic effect on the thyroid of exposure to fallout, and that older children have a gradually reducing sensitivity. In summary these results overall lead to the hypothesis that the great increase in thyroid carcinomas recorded in CIS countries is due to exposure to radiation from isotopes of iodine causing translocation of the ret oncogene and translocation or other mutational events in other genes, leading to the development of papillary carcinoma of the thyroid. The chance of developing cancer is heavily influenced by age at the time of exposure, leading to a cohort which carries a much greater risk than the other exposed population.

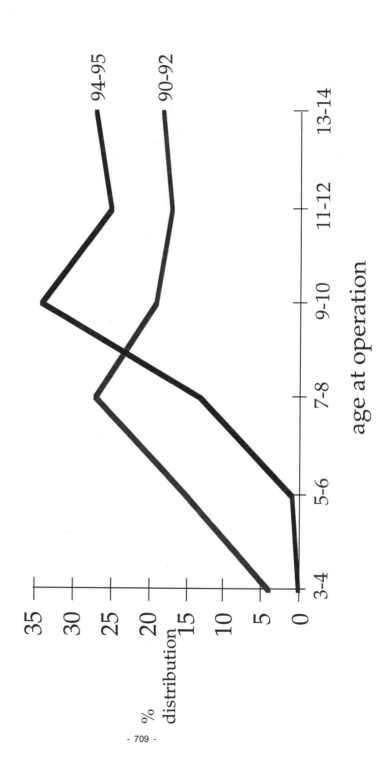

Figure 1: Change in age distribution of cases of childhood papillary carcinoma in Belarus with time

Figure 2

Childhood thyroid cancer in Belarus Jan 1990 - Dec 1994

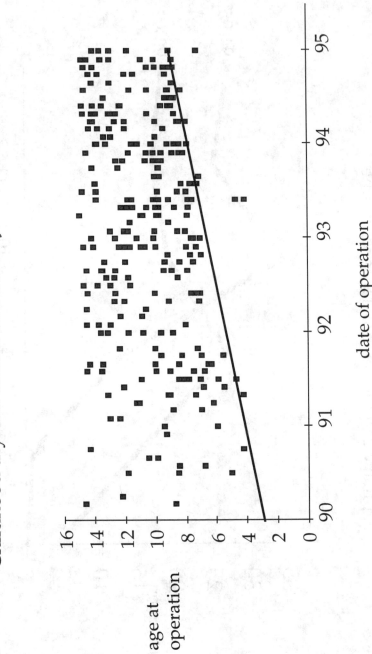

The line corresponds to the age of a child born on November 26th 1986 - that is 3/12 intrauterine age at the time of the Chernobyl accident

Figure 3

Relationship between age at exposure and age at operation for childhood
thyroid carcinomas occurring in Belarus between 1990-1994

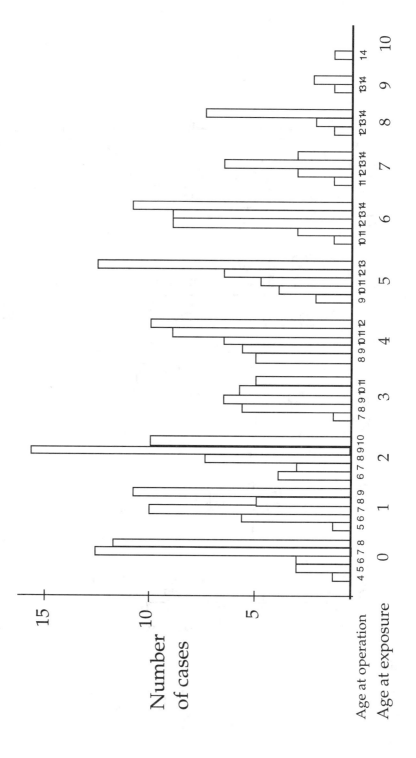

Figure 4

Childhood thyroid cancer in Ukraine Jan 1990 - Dec 1994

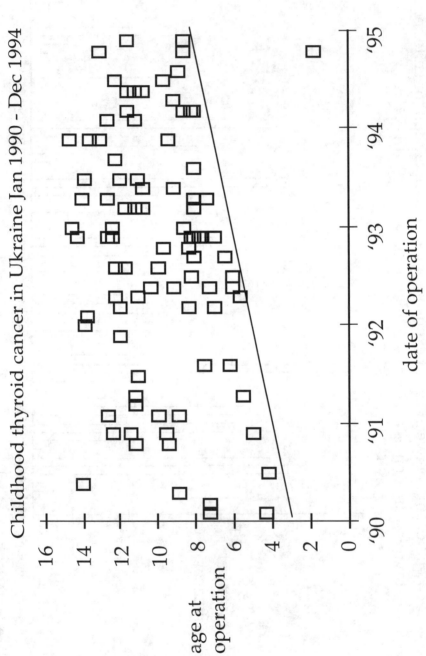

The line corresponds to the age of a child born on November 26th 1986 - that is 3/12 intrauterine age at the time of the Chernobyl accident

Figure 5

Age Distribution of thyroid tumours in children aged 15 and
under at the time of operation in England and Wales 1963-1992

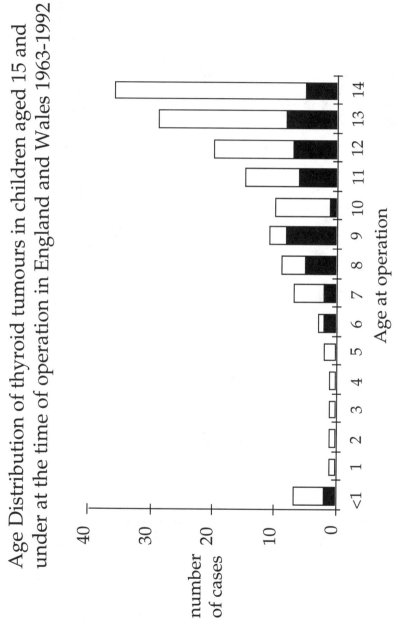

Figure 6

Risk of development of thyroid carcinoma relative to age at exposure.
Acomparison of rates in Belarus and rates in England and Wales

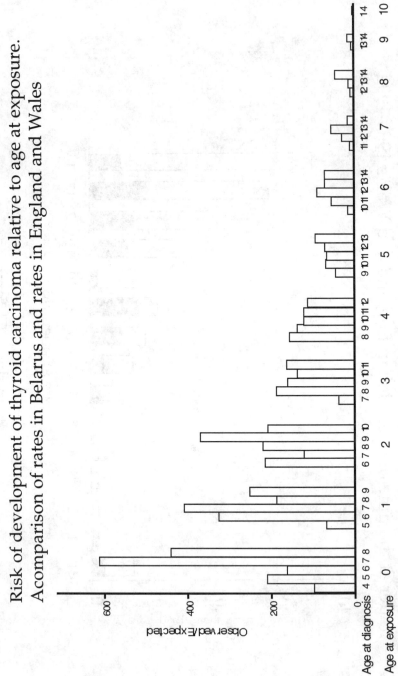

The WHO Activities on Thyroid Cancer

Keith BAVERSTOCK[1] and Elisabeth CARDIS[2]

[1]WHO ECEH, Rome Division, Via G.V. Bona, 67, I-00156 Rome, Italy
[2]IARC, 150 Cours Albert Thomas, F-69372 Lyon, France

Abstract The WHO has been involved in activities related to thyroid disease in populations exposed to Chernobyl fallout since 1991. The International Programme on the Health Effects of the Chernobyl Accident, based in Geneva, undertook a pilot project on screening for thyroid disease and the WHO European Centre for Environment and Health responded to claims from Belarus of an increase of childhood thyroid cancer. Since then the WHO has been developing the public health response in partnership with scientists and physicians in Belarus and a number of centres of excellence outside the CIS specialising in the disciplines relevant to the problem. In 1993 the International Thyroid Project was initiated in partnership with the International Agency for Cancer Research. The activities developed with scientists and physicians in Belarus to respond to the increase are described. The increase in thyroid cancer and its implications for future accidents have been addressed. Revised advice on stable iodine prophylaxis has been formulated.

1. Historical Preview

The first reports of an increase in childhood thyroid cancer were made at a scientific meeting in October 1991 in Munich by Professor L. Astakhova and Dr. V.M. Drozd of the Institute of Radiation Medicine in Minsk. As a direct result of these claims the WHO European Centre for Environment and Health convened, jointly with the CEC, an expert panel, to meet in Munich in January 1992, to discuss the claims directly with Professor Astakhova and Dr. Drozd. This meeting recommended that an expert mission should visit Minsk to review, at first hand, the pathological evidence and discuss with the physicians and surgeons the case notes and laboratory and diagnostic evidence.

Following initiatives by the CEC to strengthen capabilities for the preparation and examination of histological specimens in Minsk, the WHO organised a small group[1] which visited Minsk in July 1992 and saw evidence of substantial increase in the normally very rare childhood thyroid cancer. Indeed in a single day, at two hospitals in Minsk, eleven patients, recently operated for thyroid cancer, were seen. In the United Kingdom, with a population of some 50 million, there were 154 cases in the thirty years since 1960, i.e. some 5 cases per year (E.D. Williams: personal communication). Belarus, in contrast, has a population of only 10 million.

As a result of this first visit to Minsk, two letters were published in the journal *Nature* [1,2] in order to attract the attention of the international scientific community to what appeared to be an extraordinary situation. The letters gave the numbers of cases

---

[1] Consisting of Professor Aldo Pinchera, Pisa, Italy, Professor Sir Dillwyn Williams, Cambridge, United Kingdom and Dr. Keith Baverstock, WHO ECEH, Rome, Italy

diagnosed and the stage of the tumour at diagnosis and raised the hypothesis that this increase was due to exposure to radioactivity released from the stricken reactor. Given the exposure to the radioactive isotopes of iodine that was known to have taken place in the three months after the accident in the regions surrounding Chernobyl, an increase in thyroid cancer was not unexpected. What was surprising was the large number of cases that had appeared so early, five years, after the accident. The letters in *Nature* provoked considerable scepticism regarding the origin of the increase, the principle criticism being that screening programmes might have increased ascertainment through the use of ultrasound screening of affected populations, a possibility discussed in one of the original letters [2]. In addition, the finding was thought to contradict the results of a study of thyroid morbidity in contaminated territories compared to uncontaminated regions [3]. In fact, in that study, only 323 exposed children were examined and it thus had too little statistical power to detect an increase of the order reported in Belarus (3-10 per 100,000 children). Further reasons for scepticism were the widespread belief that the isotopes of iodine did not cause thyroid cancer and that a much longer latency, at least ten years, was expected. In fact little or no experience of the effects of the isotopes of iodine on the child's thyroid was then available although it was well known that it was sensitive to external x-rays. However, the available studies were typically of only a few thousand children, in contrast to about 1 million children exposed to significant level (ranging from several mGy to several Gy) of the isotopes of iodine in Belarus.

The principal, but not the only reason, to think that the increase was not due entirely to enhanced case ascertainment was the observation that the tumours were unusually aggressive, metastasising to the perithyroid tissue and the lung [1]. It was reasoned that these tumours, even if they had been diagnosed by screening of the exposed population, would quickly have led the patient to consult a doctor. It was also far from clear that screening by ultrasound was in sufficiently widespread use to have detected the number of cancers reported given the still low incidence rates that were being observed. The continuing increase in the number of cases up to the end of 1994 [4] confirms that ascertainment bias is not a major factor.

In 1993 the CEC [5] published a consensus report of experts[2] which confirmed the increase as most probably due to exposure to the isotopes of iodine in the fallout cloud, further focusing attention on the role of these isotopes in an environment contaminated by fission products.

There are several radioactive isotopes of iodine in reactor generated fallout as well as ^{132}Te (which decays to ^{131}I), and two major routes of entry to the body, namely inhalation and ingestion. These constitute a very complex exposure regime. While the measurements of activity in the thyroids of exposed persons made in the few months after the accident, before the iodine had decayed, provide rough and ready estimates of dose from ^{131}I to a small minority of individuals, heavy reliance has to be placed on dose reconstruction for the majority not so measured. Given the long duration of the release, about 10 days, and the changing composition of the fallout cloud as the more volatile components, such as iodine, were driven off by the heat from the fire, reconstruction is not straightforward : there is no simple ratio between deposited iodine and the much longer lived and more systematically measured caesium isotopes.

However, a number of factors, including the geographical and temporal

---

[2] Consisting of Professor Sir Dillwyn Williams, Cambridge, United Kingdom, Professor Aldo Pinchera, Pisa, Italy, Professor Bruno Egloff, Kantonsspital Winterthur, Switzerland, Professor C. Reiners, Essen University, Germany, Dr. P. Fragu, INSERM, France, Dr. Keith Baverstock, WHO, ECEH, Rome, Italy

distributions of cases that is emerging, suggest that a simple association with levels of ^{131}I and a minor contribution from enhanced case ascertainment may not fully explain the increase and thus other factors, which may be influential, have to be considered. Although a large increase in the numbers of cases was observed in 1990 in the heavily contaminated and screened Gomel region, no similar increase has been observed in the equally heavily contaminated and screened regions of Mogilev region, or in the Bryansk and Kaluga regions of the Russian Federation, although more recently an increase is reported from Bryansk region [6]. In contrast, the Brest region of Belarus has shown an increase second only to that observed in Gomel region although both contamination levels and screening intensity were not so great. Factors which it was thought might contribute to this pattern were exposure to the short lived isotopes of iodine and [132]Te, and iodine deficiency.

The circumstantial evidence for implication of radiation exposure, principally ^{131}I from the Chernobyl accident, the increase in the number of cases among children in Belarus was, however, compelling: the distribution of the cases under the path of the initial cloud of fallout, the small number of cases diagnosed in those not born at the time of the accident (the incidence of these is consistent with the pre 1990 rates) and the cumulative incidence of the cases in the exposed cohort of children living in the Gomel region, up to the end of 1993, which appears to be consistent with an average dose to the thyroid equivalent to 0.7 Gy of x-rays [7].

As a result of these observations, and in response to a request from the Minister of Health of Belarus, the WHO proposed the establishment of the International Thyroid Project. The overall objective, specific aims and approach, taken from the project document, are given in the Table 1.

This project, managed from the WHO ECEH in collaboration with the IARC under the "umbrella" of the WHO International Programme on the Health Effects of the Chernobyl Accident (IPHECA) builds on the expertise and achievements of its predecessor, the IPHECA Pilot Project on Thyroid, which developed screening capabilities for thyroid disease by providing diagnostic equipment and capabilities, medial supplies, advanced training for specialists, computers for data collection and facilitating collaboration between the states most affected by the Chernobyl accident - Belarus, the Russian Federation and Ukraine - and the international community. The project was instrumental in confirming the incidence of thyroid disease through a program of ultrasound screening and determination of thyroid hormone and antibodies concentration. As the result of this work incidence rates for malignant and benign thyroid nodules, auto immune diseases and hypothyroidism have been determined in children living in territories contaminated by caesium-137 above 555 kBq/m² (the so-called "Strict Control Zones") in the three states. It was completed in 1994 and the results have been published [8].

Given the uncertain situation regarding the dosimetry of exposure to the isotopes of iodine after the accident, it was acknowledged that some of the aims and objectives of the project would have to be achieved largely without recourse to reliable doses.

To assist in achieving the aims and objective of the ITP, a network of collaborators has been set-up, including WHO Collaborating Centres (in Minsk, Berne, Cambridge, Munich, Nagasaki, Pisa, Poznan, Stockholm, and Villejuif) and research centres (IARC and ECEH).

Table 1 - The ITP : objective, specific aims and approach

Overall objective :
• to provide a comprehensive response to the increased thyroid disease in Belarus in terms of research and training directed to the public health issues raised by the outbreak of childhood thyroid cancer and the increasing evidence that older age groups are being affected;

Specific aims :

- to understand the origin of the outbreak of thyroid cancer;
- to identify the groups most at risk
- to determine the likely extent of the outbreak, both in time and geographically and the factors that may influence risk and
- to optimise treatment through a programme of co-ordinated collaborative research studies centred on the WHO collaborating Centre in Minsk.

The project is composed of two parts: part A, relating to children aged 0 to 14 at diagnosis and part B, relating to adolescents and adults. The reason for this division is to reflect work being undertaken by other programmes, in particular the bilateral collaboration between Belarus and the National Institutes of Health (NIH) in the USA on long-term epidemiology of children exposed to known doses of radiation and the collaboration between CIS states and the European Commission on pathology and treatment of childhood thyroid cancer.

Several activities under parts A and B of the project are presently being elaborated between the staff at the Minsk Collaborating Centre and those in the international centres. It is recognised that close coordination between activities and co-operative interaction between the centres is essential to the ongoing development and success of the project. An integral part of the project is the exchange visits between Minsk and the international centres for the purposes of collaborative research and training in advanced techniques. In the first instance, the project is envisaged to last 5 years and to study morbidity over a 10 year period, i.e. 5 years retrospectively and 5 years prospectively.

Detailed objectives of parts A and B of the project are given in:
"Green Book" International Programme on the Health Effects of the
Chernobyl Accident, Thyroid cancer in Belarus after Chernobyl: International
Thyroid Project, WHO European Centre for Environment and Health, Rome
Division

The Minsk Centre, in collaboration with the others, has developed a series of activities, some of which are solely humanitarian and others which have a significant research component. Most activities under the ITP have public health implications in the short or medium term.

2. Past Activities of the ITP

2.1 Survey of 50 cases

In 1993-94 a survey of 50 childhood thyroid cancer cases was carried out in Belarus [9] in preparation for a possible case control study. The objectives of the survey were to better characterise the reported increase in childhood thyroid cancer and to test mechanisms for identifying and interviewing cases. Detailed information about the behaviour of children around the time of the accident and about the circumstances of diagnosis of thyroid cancer was obtained by questionnaire from the parents, particularly the mothers.

Although the results of this survey must be interpreted with caution because of the lack of the information on suitable controls, a number of behaviourial factors that would enhance exposure to the short-lived isotopes were observed. More than half (34/50) of the cases drank milk from a local cow, thus potentially minimising the time between milking and consumption of the milk. In addition, most (64%) spent more than eight hours outdoors every day and about half the cases slept with open windows during the immediate aftermath of the accident, thus maximising their exposure to inhaled radioactivity.

Twenty-four cases had reconstructed dose estimates. Levels of doses were relatively low : only half of these were 0.25 Gy and only one above 2 Gy.

About half the cases (23/50) were identified through medical examination other than for thyroid problems and about a quarter (12/50) through formal screening programmes. The diagnosis was made on the basis of clinical examination in 35 out of 43 children for which this information was available, and on ultrasound screening in eight.

Only eight cases reported that they had received stable iodine shortly after the accident.

The survey resulted in the identification of important hypotheses concerning factors which may modify the association between radiation dose and thyroid cancer risk. There are a possible genetic predisposition, iodine status and exposure to the short-lived isotopes of iodine and ^{132}Te.

The hypothesis of a possible genetic predisposition to radiation induced thyroid cancer arose from an incidental observation that out of the 50 cases included in the study, at least two had a sibling who also had an operated cancer. Anecdotal evidence from other centres in Europe appears to indicate than expected incidence of non medullary thyroid cancers among first degree relatives.

The second hypothesis concerns the role of stable iodine in radiation induced thyroid carcinoma. It is thought that iodine deficiency, which appears to have been - and to remain - common in some of the areas most affected by the Chernobyl accident, may affect the risk of radiation induced thyroid cancer both by increasing the uptake of radioactive iodine and by increasing the production of thyroid stimulating hormone (TSII), which increases cell proliferation in the thyroid gland. On the other hand, systematic prophylaxis with large doses of stable iodine, which the survey revealed has occurred for several years after the accident in a number of schools, may influence the frequency of thyroiditis.

2.2 Mission to review treatment of thyroid cancer in Minsk

In May 1994, at the request of the Ministry of Health of Belarus, the WHO convened an expert panel[3] to review the treatment being offered in Minsk. The panel was greatly impressed by the way in which the medical authorities had responded to an unprecedented situation, especially in the light of the lack of facilities for nuclear medicine, and the "scattered" nature of the institutes that had combined their expertise and resources to address the problem. Their recommendations included the proposal for a network of computers to make case notes available to physicians and for a detailed treatment protocol. This protocol has been used in developing a new treatment protocol which is being discussed between the three affected countries.

3. Current activities of the ITO

3.1 Rationalisation of Thyroid Hormone Testing

The increase in childhood thyroid cancer has increased the demand for laboratory testing of thyroid hormone levels, both directly, as a result the increased disease and the need for diagnostic tests and follow-up of patients, and directly by sensitising the population to the possibility of radiation induced thyroid disease. In the past, reliance has been placed on imported test kits from Western Europe and the USA, which are expensive and require foreign currency.

At a meeting attended by scientists and clinicians from the three most affected countries in October 1994 it was decided, on the basis of the findings of the IPHECA Pilot Project on Thyroid, that the use of thyroid hormone tests in diagnostic screening was not cost effective and should be discontinued. This enabled attention to be focused on the use of such tests on diagnosis and treatment. In association with the Collaborating Centre in Pisa, a protocol for optimal use of thyroid hormone tests was prepared. It was recommended that TSH and FT3 measurements to be used for monitoring effectiveness of suppressive therapy with l-thyroxine in patients already treated by total thyroidectomy and that serum Tg and anti-thyroglobulin antibodies be used for monitoring tumour recurrence. There was no evidence that any of these tests was useful for the diagnostic evaluation of thyroid nodules.

To further the cost-effectiveness of thyroid hormones testing, encouragement was given to the development of local production of test kits, starting with kits for thyroglobulin (Tg) and antibodies to antithyroglobulin (Tg_{ab}). These locally produced kits are now being tested for reliability and quality against commercially available products. It has proved possible to produce such kits at about 20% of the price of imported kits.

3.2 Survey of Iodine Deficiency

Dietary iodine deficiency is a long-standing problem in regions surrounding Chernobyl, leading to a high prevalence of thyroid goitre. This condition can be rectified by the supplementation of the diet by sodium iodine, normally in salt used for table and cooking or, as in the former Soviet Union, through adding stable iodine to bread. By the end of

---

[3] Consisting of Professor J.F. Leclère, Nancy, France, Professor Furio Pacini, Pisa, Italy, Professor Hans Rocher, Dusseldorf, Germany, Professor Martin Schlumberger, Paris, France and Professor Norman Thompson, Ann Arbour, USA

the 1970's Belarus had largely eliminated iodine deficiency disorders. There was, however, a re-emergence of iodine deficiency in the early eighties. It seems likely that by 1986, the time of the accident, little or no supplementation was taking place, although there are presently no reliable data upon which to assess the degree of iodine deficiency then. Following the accident, and the recognition that exposure to the radioactive isotopes of iodine was widespread, sporadic attempts to reinstate supplementation were made, especially among children in schools, by the distribution of sodium iodine tablets. Iodised salt is currently available in shops but is far from universally used by the population.

A survey of iodine deficiency in Belarus is being planned. It will involve measurements of iodine in urine and thyroid volume in more than 11,000 children in 30 schools distributed across the territory of Belarus. Measurements will be made according to a strict protocol. Five age group will be investigated, between 6 and 18 years. The aim is to obtain information, in the form of a country-wide map, on the current status of iodine deficiency, and on the status at the time of the accident. The map of current iodine deficiency will be used as a basis for directing iodine supplementation to the areas where it is most needed; the information on iodine deficiency in 1986 will be valuable for studies of the aetiology of the observed thyroid cancers.

Iodine deficiency may play to important roles in the development of thyroid cancer: it influences the dose, from the isotopes of iodine, taken up by the thyroid; it may also modify the expression of "latent", i.e. initiated, but not clinically diagnosable, cancer. These two factors are not fully understood. While the iodine deficiency thyroid tends to have an increased uptake of iodine in the circumstances of a single exposure, the complex kinetics of uptake and excretion of iodine, and the larger organ size, do not allow a ready estimate of the total effect on dose to the thyroid. The iodine deficient thyroid expresses a higher level of thyroid stimulating hormone (TSII) and thus latent cancer may be accelerated in its expression by continued iodine deficiency and concomitantly suppressed by correcting the deficiency. Both features, i.e. in the extent of dose modification and accelerated expression of cancer, need careful examination.

At the time of writing, a pilot phase involving measurements at two schools has been completed and the results are being evaluated. A preliminary evaluation of the results from 430 children and teenagers in Minsk showed that about 60% have palpable goitre, grade 1B and 2 (or grade 2 according to the new classification). Only about 30% of children excreted more than 100 μg of iodine per litre. The others excreted much less and in some cases less than 20 μg/l.

The Collaborating Centres in Minsk and in Poznan, Poland, are the main participants in this activity.

3.3   Case Control Study

The objective of the study is to identify, and if possible quantify, genetic and environmental factors which modify the risk of radiation induced thyroid cancer. The primary factors of interest are genetic predisposition to radiation induced thyroid carcinoma (perhaps associated with some DNA repair deficiency), iodine status (deficiency and overdose) and intake of very short lived isotopes of iodine. Secondary factors which will be considered include age at exposure and screening intensity.

The cases are all incident cases, and a random sample of prevalent cases, of thyroid carcinoma occurring in the study population (defined as all persons who were children or adolescents at the time of the Chernobyl accident) during the study period

(1990-1997) and operated in Belarus. All cases will be independently verified by two pathologists, one from Belarus and one from Western Europe. The primary source for case ascertainment will be the Belarus Cancer Registry and the records for the Republican Surgical Centre.

For each case, four controls will be selected as follows. Two controls will be matched closely on age an sex and on village/city at the time of the accident. The other two controls will also be matched closely on age and sex and on administrative region at time of accident. Matching on village or city at the time of the accident is proposed in order to study the modifying effect of short lived isotopes of iodine while controlling for ^{131}I dose and iodine deficiency. Matching on region will ensure variability of ^{131}I exposure and of iodine deficiency, while controlling for other geographical or life factors which could confound the associations under study. The proposed mechanisms and source for control selection will vary depending on the age of the subject and the mode of diagnosis of the case.

Information on variables of interest will be obtained using the following approaches, depending on the variable, including:

- questionnaire administered by a trained interviewer; which will include questions about:
  - family history of cancer and thyroid disorders;
  - the behaviour of the subject at the time of the accident, in the following days and the first two months (this information will be useful for deriving gradients of probable exposure to ^{131}I and short lived isotopes);
  - stable iodine prophylaxis and thyroid hormone administration.
- medial and school records to confirm the information on stable iodine prophylaxis and thyroid hormone administration;
- results of current and past geographical surveys (particularly for iodine deficiency) and surveys of countermeasures (iodine prophylaxis, evacuation, sources of clean food) being carried out in particular within the framework of the ITP;
- analysis of biological samples.

Information on radiation dose from ^{131}I will be obtained in two ways. First, a simple gradient will be construed on the basis of the information on behaviour in the two months following the accident obtained by questionnaire. In a second phase, individual dose reconstruction will be carried out using a protocol similar to that developed for Belarus/US collaborative studies.

Information on a possible genetic predisposition will be first obtained by the questionnaire. If a subject reports a history of cancer, thyroid disorders, congenital malformation or mental retardation among first or second degree relatives, an investigation will be carried out and the pedigree will be drawn; microsatellite typing of selected regions of the genome will be performed. In a second phase, once genes have been identified in a parallel project, blood samples obtained from all study subjects will be analysed to screen for mutations of these genes and thus evaluate the risk of radiation induced thyroid cancer associated with the genetic predisposition.

This activity is being set-up in as a collaboration between the Minsk Collaborating Centre, IARC, the Collaborating Centres in Pisa and Cambridge and WHO/ECEH Rome Division.

3.4 Ecological Epidemiology of Thyroid Cancer in Belarus

In the study of diseases related to environmental pollution, such as occurred in the aftermath of the Chernobyl nuclear accident, personal characteristics of individuals are often not sufficient to explain regional differences of disease incidence, because individuals within areas may be similarly exposed. The Ecological Epidemiology Project was set up to document and better understand the ongoing epidemic of thyroid cancer in belarus in terms of its geographic spread, its relationship with the distribution of radioactive isotopes, and of time trends and ages at highest risk.

This is achieved by combining detailed and quality checked information on patients with that on the demographic, ecological and socio-economic characteristics of the areas from which they come. A computer-based relational data bank has been installed for this purpose at the Lesnoj Oncological Institute in collaboration with the Berne Collaborating Centre, and data analyses are ongoing. Results so far have allowed to respond convincingly to doubts about whether the increase is primarily the consequence of methodological artefacts.

3.5 Registry of Surgically Treated Thyroid Diseases

Although a national cancer registry exists in Belarus, it would be useful from the points of view of public health planning and evaluation of possible preventive strategies (such as iodisation of table salt), as well as for research purposes, that a number of non-cancerous thyroid pathologies be systematically registered. In preparation for a possible registry of thyroid diseases, a pilot study of operated thyroid pathologies has been carried out. This study revealed that, among adults, a large proportion of thyroid pathologies, including carcinomas, have been misclassified, particularly before the accident. It demonstrated that a registry of operated pathologies, with careful verification of diagnoses, would therefore be useful to ensure complete ascertainment of cancer cases and that it is feasible to set one up, both prospectively and retrospectively. As a result, such a registry is now being set-up in phases. In a first step, it is being established both prospectively and retrospectively (to 1982) in Minsk region and city. A data collection facility has been created at the Lesnoj Oncological Institute in collaboration with the Berne Collaborating Centre, in which information on all surgeries in children, adolescents and young adults will be reported.

3.6 Thyroid Pathology Pilot Project

Hospitals outside Minsk do not all benefit from the most up-to-date and expert pathology facilities; diagnoses, therefore, do not always conform to international nomenclature and may not be uniform. This has not posed a problem for children for whom all surgery is likely to be carried out in Minsk. So far, however, there has been no international validation of diagnoses of thyroid cancer in adolescents or adults operated in the regional hospitals.

The objectives of the thyroid pathology project is to investigate the level of diagnostic integrity in regions outside Minsk and to assess the needs for training and equipment to cope with the increased thyroid disease resulting either from the accident or from the accident or from an increased awareness of thyroid problems. Visits to regional hospitals will be made by pathologists from Minsk and the Cambridge Collaborating Centre; reviews of past cases will be conducted and the quality of clinical records assessed. Particular emphasis will be placed on studying the morphology of adenomas and collecting material for molecular biological study of adenomas and carcinomas in adolescents.

3.7   Descriptive Epidemiology

Data from the Belarus cancer registry are being used to study trends and features that may assist in developing the public health strategy to, for example, optimise diagnostic efficiency. An early analysis of the cumulative incidence of the childhood thyroid cancer in Gomel region up to 1993 [7] suggested, assuming the increase was induced by radiation, that the increase was consistent with that which would be expected if the children in the region had received an average dose equivalent to 0.7 Gy of external X-rays. More recent analysis [in press], using data up to 1994 and including childhood, adolescent and young adult cases in four regions of Southern Belarus, shows clearly the start of the increase to have been 1989/90 in all four regions. With the help of more recent data on the age dependence of sensitivity to radiation induced thyroid cancer, average doses equivalent to about 1.2 Gy in Gomel region, just over 0.6 Gy in Brest region and of the order of 0.2 Gy in Minsk region and city, of x-rays are inferred. These levels of doses are more-or-less consistent with thyroid doses resulting from exposures to the isotopes of iodine after the accident provided it is assumed that iodine isotopes are as efficient at inducing cancer in children as are externally delivered x-rays.

These on-going studies are carried out as collaboration between the Minsk Collaborating Centre and the WHO/ECEH (Rome Division).

3.8   Computer Networked Case Records for Minsk

Concerned that the centres in Minsk involved in the treatment of children were so widely 'scattered' across Minsk, so making it difficult for the physicians to communicate and review case notes, the expert group on treatment (see paragraph 2.2) recommended the implementation of a computer network between participating physicians so that case notes could be readily exchanged. Plans to implement such a network are in progress in collaboration between the Minsk Collaborating Centre and WHO/ECEH (Rome Division).

4.   Nuclear Emergency Preparedness

In addition to the International Thyroid Project, the WHO Regional Office for Europe has a programme concerned with the public health aspects of the response to nuclear emergencies. Activities come under two headings, namely response in the event of an accident and the development and maintenance of a harmonised region-wide contingency plan. Prophylaxis by stable iodine is a part of the latter activity.

The situation in Belarus has highlighted the substantial potential health risk, especially to children, posed by the accidental release of radioactive iodine to the environment.

The hazard to health posed by radioactive iodine has been well recognised for some time; it was the major preoccupation after the Windscale accident in the UK in 1957 [10]. It has also been long recognised that the administration of stable iodine, before, or at least as soon after exposure commences as possible, is an effective way of blocking the further uptake of iodine including radioactive iodine. Stimulated by the

- 724 -

Chernobyl accident the WHO issued guidelines for the use of stable iodine prophylaxis after reactor accidents [11]. In brief these recommended that sodium iodine or iodate tablets be available, either pre-distributed, or ready for rapid distribution, in what was described as the near field for nuclear facilities where it was possible that very doses might be received.

Following the Chernobyl accident the authorities in Poland distributed stable iodine in solution to a large part of the child population of Eastern Poland. Subsequently the results, in terms of undesirable side effects, have been assessed ([12]. In the light of the situation in Belarus and the experience in Poland the WHO's Regional Office for Europe reconsidered and revised their advice. The new advice is given in the recent published Manual on Public Health Action in Radiation Emergencies [13]. It is recommended that stocks of suitable preparation of stable iodine are available in such a way as to be distributed at very short notice to pregnant women, neonates and children more-or-less anywhere where thyroid doses could exceed a level at which the risk of taking the countermeasure exceeded the risk of inducing thyroid cancer. This dose turns out to be of the order of 1 mSv to the thyroid very much lower than the currently accepted levels for intervention which are in the range of 20 to 50 mSv. Such a low dose makes it impractical to implement such a procedure and a more practically applicable level might be 10 mSv [14].

5.   Conclusion

Following the reports of a dramatic increase in the number of cases of childhood thyroid cancer in Belarus after the Chernobyl accident, the WHO has implemented, in partnership with Belarussian scientists and physicians, a programme of activities to address the public health consequences of exposure to radioactive fallout and to learn from the experience. The project is of medium to long term and will require sustained support over the next few years to achieve its goals. The WHO guidelines of stable iodine prophylaxis are in the process of revision based on the experience gained in Belarus.

References

[1]   S. Kazakov et al, Thyroid cancer after Chernobyl, *Nature*, 359 (1992) 21.
[2]   K.F. Baverstock et al, Thyroid cancer after Chernobyl, *Nature*, 359 (1992) 21-22.
[3]   A. Mettler et al, Thyroid nodules in the population living around Chernobyl, *JAMA*, 268 (1992) 616-9.
[4]   Y. Averkin, Th. Abelin, J. Bleuer, Thyroid cancer in Belarus:ascertainment bias? *Lancet*, 34 (1995) 1223-4.
[5]   Commission of the European Communities, Thyroid cancer in children living near Chernobyl: expert panel report on the consequences of the Chernobyl accident, ISBN:92 826 5515 6, Office for Official Publications of the European Communities, Luxembourg 1993 (Report EUR 15248 EN)
[6]   V.A. Stsjazhko, A.F. Tsyb, N.D. Tronko, G. Souchkevitch, K.F. Baverstock, Childhood thyroid cancer since the accident at Chernobyl, BMJ Vol. 310, 25 March 1995, p. 108
[7]   K.F. Baverstock, Thyroid cancer in children in Belarus after Chernobyl, *World Health Statistics Quarterly*, 46 (1993) 204-8.
[8]   Report of the International Project for the Health Effects of the Chernobyl Accident, WHO (World Health Organization), Geneva, 1995.
[9]   L.N. Astakhova, E. Cardis, L.V. Shafarenko, L.N. Gorobets, S.A. Nalivko, K.F. Baverstock, A.E. Okeanov, C. Lavé, Additional documentation of thyroid cancer cases (Belarus), IARC, Internal report 95/001, International Agency for Research on Cancer, Lyon and Environmental Health in Europe Report No. 3, World Health Organization Regional Office for Europe, Copenhagen, 1995.

[10] K.F. Baverstock, J. Vennart, Emergency Reference Levels for Reactor Accidents Health Physics, **30** (1976) 339.

[11] World Health Organisation Guidelines for iodine prophylaxis, FADI, Publishers, Copenhagen, 1989.

[12] J. Nauman, Potassium Iodine Prophylaxis in Poland: Review of far field experience in: Rubery E., Smales E. (ed.), Iodine Prophylaxis following nuclear accidents Proceedings of a joint WHO/CEC Workshop, July 1988, Pergamon Press 1990 p. 135-140.

[13] Manual on Public Health Action in Radiation Emergencies, World Health Organisation European Centre for Environment and Health, Rome Division EUR/ICP/EHAZ9413/PB 01, 1995

[14] W. Paile, "Optimal Intervention level for stable iodine prophylaxis", WHO International Conference on the Health Consequences of the Chernobyl and Other Radiological Accidents 20-23 November 1995, Geneva SWI, World Health Statistical Quarterly, in press

Thyroid cancer in Belarus: The epidemiological situation

*Abelin Theodor, **Averkin Juri I, Okeanov Aleksei E, *Bleuer Jürg P

*WHO Collaborating Centre for Epidemiology of Radiation and Thyroid Disease, Dept. of Social and Preventive Medicine, University of Berne, Switzerland; **State Research Institute of Oncology and Medical Radiology, Lesnoj/Minsk, Belarus; Belarus Centre for Medical Technology, Minsk, Belarus.

Abstract. Starting in 1990, an increasing number of children were diagnosed as suffering from thyroid cancer in regions close to the Chernobyl nuclear accident site, and this increase is continuing. But still today, doubts about the significance of this increase are being voiced.

Using data from the Belarus epidemiological cancer registration system up to 1994, the geographic distribution, time and cohort trends, age distribution and other characteristics of this epidemic are reviewed. Results show that the geographic distribution is similar to that of iodine-131 following the accident; that when looking at cohorts of children born in the same years incidence has steadily increased since 1990; and that deviations from this pattern might be explained by active case finding.

The most likely interpretation of these results is that of a causal association with radiation exposure related to the Chernobyl accident, but possible modifying factors should be examined closely. The most likely future course of the epidemic is an increasing number of cases among those exposed in childhood, and public health measures should take this into account.

1. Introduction

The Chernobyl nuclear accident occurred on the 26th of April, 1986. In November, 1991, Prisyazhiuk et al [1] were the first to report on a possible increase in incidence of childhood thyroid cancer following this accident. Their communication was based on three well examined cases having occurred in 1990 in three districts of the Ukraine close to the accident site and a doubling of cases from nine to seventeen cases in the country as a whole. A year later, in a letter in 'Nature', Kazakov, Demidchik and Astakhova [2] reported that in Belarus an increase from five and six cases in 1988 and 1989 to 29 in 1990, 55 in 1991 and 30 for the first half year of 1992 had occurred. According to a letter published in the same issue of 'Nature' [3], the the clinical and histopathological diagnosis was confirmed by an international group of experts.

These reports were met by great skepticism on the part of a number of prominent scientists [4-6]. In particular, questions were raised about whether the observed cases were indeed thyroid cancer, or whether they represented clinically irrelevant occult carcinoma discovered through screening activities in areas with radioactive contamination.

The purpose of this presentation is to review what is presently known about the epidemic of thyroid cancer following the Chernobyl accident, and to use this evidence to answer the question whether the facts support the belief that a real epidemic of thyroid cancer is occurring, which is caused by the Chernobyl nuclear accident, or whether the observations support the suspicion that the reported increase represents an artefact - in which case the treatment of hundreds of children would have been unjustified.

Some of the reasons for doubt that the increase of childhood thyroid cancer reported in 1991 and 1992 was real [4,5,6], are:

Artefact due to active case finding: Active case finding by methods such as ultrasound screening could have led to the detection of what is known as occult carcinoma. These are very small focuses of cancer, which remain without malignancy throughout a person's life. Such occult cancer has been found in autopsy studies of persons having died for reasons having no connection with the thyroid gland. Their diameter is usually around one millimeter, and they never go beyond 15 millimeters.

Previous underreporting: The increase reported from Belarus could also have been an artefact, if childhood thyroid cancer had been grossly underreported in the years preceding the reported increase.

Unexpected geographic distribution: Based on assumptions formulated in an expert report of the International Atomic Energy Agency (IAEA) [7], an increase of thyroid cancer was expected in all areas highly contaminated by Cesium 137, but the distribution of reported childhood thyroid cancer differed from this expectation.

Short latency period: The fact that in some of the affected regions, childhood thyroid cancer incidence was considerably increased in 1990 already, i.e. after a latency period of only four years, was considered as biologically improbable.

Decrease of childhood thyroid cancer after initial increase: After a few years of increase, a decrease of childhood thyroid cancer was observed in the affected regions, which was interpreted as evidence for an effect of detection of occult carcinoma.

These reasons for doubt will be discussed on the basis of epidemiologic considerations, and conclusions will be drawn about the nature and cause of this epidemic.

In addition, a number of questions of practical relevance for the *planning of health services and public health action* have been asked, and an indications will be given on how these can be answered. They concern:

Screening strategy: What is the optimal strategy for programmes of thyroid cancer screening in high risk regions?

Iodine supplementation strategy: Are there regions where iodine deficiency has contributed to an increased risk of thyroid cancer, and where iodine supplementation is needed to reduce the risk in case of future radioactive contamination?

Planning of therapeutic services: If the number of thyroid cancer cases is expected to further increase, what will be the need for specific services, and where would be the most cost effective location for such services?

Preventive distribution of stable iodine: up to what distances from nuclear reactors should preventive distribution of stable iodine be planned?

Although the three Republics of Belarus, the Ukraine and Russia are affected, the focus will be on Belarus, where the radioactive contamination covered the largest area, and the increase of incidence rates was most marked. The analyses presented here are based on all cases of childhood thyroid cancer reported to the Belarus Cancer Registry during the years 1986 to 1994. In addition, given the long time since the date of the accident, information on cases among adolescents and young adults were added to the data bank.

For computation of incidence rates, population data were used by age and sex groups based on the last Belarus census.

2. Is there a real increase or an artefact?

In order to objectively evaluate the possible risks of nuclear accidents, the health consequences of disasters have to be established beyond doubt. Critical questions have to be examined on the basis of well documented evidence. Since skeptical comments continue to be published, this is particularly important in the context of the observed increase of thyroid cancer incidence following the Chernobyl nuclear accident.

Artefact due to active case finding: If the cases diagnosed in Belarus are occult carcinomas discovered through active case finding, then the characteristics of these tumours should be those generally known for occult carcinoma. Tumour sizes should be very small, and signs of malignancy such as invasion beyond the organ capsule should be rare or not be found. In reality, assessment by experienced pathologists shows invasion beyond the capsule in 61 percent of cases [8], and measurement of tumour diameters revealed that 12.8 percent of surgically removed tumours were smaller than 1 cm, against 96 percent in a combined series of eight studies including 1759 cases. Finally, the frequency of distant metastases was that known from clinical series of childhood thyroid cancer patients [9]. Therefore, the cases diagnosed in Belarus are not occult carcinomas, and if active case finding played a role, it was by earlier detection of clinically relevant thyroid cancer only.

Previous underreporting: Examination of childhood thyroid cancer incidence rates from cancer registries across the world shows that these are in the order of magnitude of those observed in Belarus before 1990. Therefore previous underreporting cannot explain the reported increase following the Chernobyl accident [9].

Unexpected geographic distribution: One of the arguments used against a causal relationship between radioactive contamination following the Chernobyl accident and childhood thyroid cancer was that there were hardly any cases of thyroid cancer in the region of Mogilev, which, however, was known for very high radioactive contamination. What was not considered was that in that region radioactivity is from cesium-137 rather than from radio-iodines, which due to the strong affinity of iodine to the thyroid gland can be expected to be involved in thyroid cancer. On the other hand, maps of radio-iodine contamination show a high correspondence to the geographic distribution of childhood thyroid cancer cases.

Short latency period: Latency periods between exposure to radioactivity and the development of thyroid cancer of five years had been observed before, but what was striking was the large number of cases after such a short period. This does not speak against a causal relationship, but may indicate a particularly high risk, in which case a large number of cases is yet to be expected after longer latency periods.

Decrease of childhood thyroid cancer after initial increase: As suggested by earlier publications, after a few years of increase, a flattening or even decrease of the curve of childhood thyroid cancer was observed for Belarus and its the affected regions. An important reason is that among children aged 0 to 14 years, there is an ever decreasing number of children who had been alive at the time of the Chernobyl accident. For example, among those aged 0 to 14 years in 1996, only those aged 10-14 had been born when the accident occurred. Therefore for epidemiological purposes, it is no longer sufficiently informative to use statistics on incidence rates by the age at diagnosis, but rather cohorts of persons defined by their age at the time of the accident should be followed up. This was done in *figure 1*, which shows that within all cohorts exposed between the ages of zero and twenty, a steady increase of cases has continued at least from 1990 through 1994. The ups and downs between 1991 and 1994 in the cohort aged 0 to 5 at the time of the accident could perhaps be explained by varying intensities of case finding activities [10], but further research on this hypothesis is needed.

Figure 1. Thyroid cancer incidence 1986 to 1994 for cohorts defined by age in 1986.

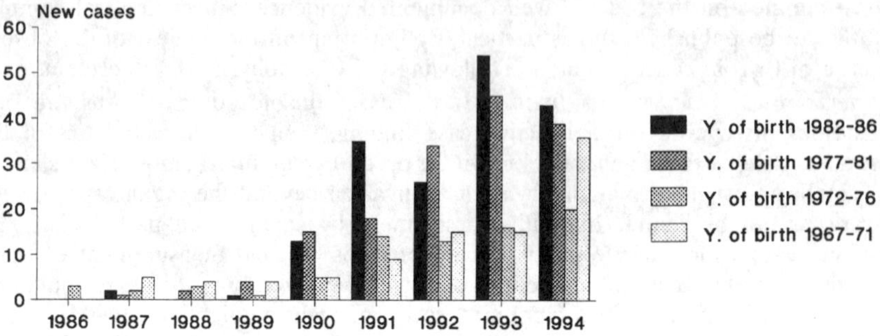

3. Discussion and conclusions

In summary, the examined evidence is well compatible with a real increase of thyroid cancer related to the Chernobyl nuclear accident. Projections suggest that numbers among those exposed in childhood will still increase, and based on experience with longer latency periods, this increase may still continue for years. Preventive activities, in particular screening programs should therefore be concentrated on those who were children when the accident occurred, and who lived in areas contaminated by radioactive iodine. To facilitate decisions of public health relevance, research is needed to identify these areas more precisely, and to better understand the role of iodine deficiency as a modifying factor.

We thank the Swiss Federal Office of Public Health for financial support.

References

[1] A. Prisyazhiuk *et al*, Cancer in the Ukraine, post-Chernobyl. *Lancet* 338 (1991) 1334-5.
[2] V.S. Kazakov *et al*, Thyroid Cancer after Chernobyl. *Nature* 359 (1992) 21.
[3] K. Baverstock *et al*, Thyroid Cancer after Chernobyl. *Nature* 359 (1992) 21-2.
[4] V. Beral, G. Reeves, Childhood thyroid cancer in Belarus. *Nature* 359 (1992) 680-1.
[5] I. Shigematsu, J.W. Thiessen, Childhood thyroid cancer in Belarus. *Nature* 359 (1992) 681.
[6] Ron E *et al*, Thyroid cancer incidence. *Nature* 360 (1992) 113.
[7] The International Chernobyl Project, Technical Report. Assessment of radiological consequences and evaluation of protective measures. Report by an international advisory committee. International Atomic Energy Agency, Vienna, 1991.
[8] A.W. Furmantchuk *et al*, Pathomorphological findings in thyroid cancers of children from the Republic of Belarus: a study of 86 cases occurring between 1986 ('post-Chernobyl') and 1991. *Histopathology* 21 (1992) 401-8.
[9] T. Abelin *et al*, Thyroid cancer in Belarus post-Chernobyl: Improved detection or increased incidence? *Soz. Praev.med.* 39 (1995) 189-197.
[10] J.I. Averkin *et al*, Thyroid cancer in children in Belarus: ascertainment bias? *Lancet* (in print).

Epidemiologic Studies
of Thyroid Cancer in the CIS

Gilbert W. BEEBE, Ph.D.
National Cancer Institute, Bethesda, Maryland 20892, USA

Abstract. Despite the great international interest in Chernobyl and the need for quantitative risk information on the carcinogenic effectiveness of the radioiodines, there has been relatively little epidemiologic research on thyroid cancer following the Chernobyl accident. The reasons for this are many, diverse, and difficult to eliminate, although some progress is being made. Among them are the natural priority of public health concerns, a weak infrastructure for conducting studies in chronic disease epidemiology, and the difficulty of assigning thyroid dose estimates to individuals for study. In spite of the difficulties a number of significant studies have been begun or are planned, and several valuable reports have appeared. From the descriptive studies it is now known that the latent period for thyroid cancer in children exposed to radioiodines is not 5 to 10, but probably three years, that the magnitude of the increase in thyroid cancer among children is beyond anything previously experienced or expected, and that there is a strong correlation between thyroid cancer and environmental radiocesium contamination levels in the Gomel region of Belarus, and between thyroid cancer and average regional levels of I-131 dose to the thyroid in Ukraine. However, even today, there is very little hard scientific information on the relation of thyroid cancer in children and their exposure to the radioiodines in the fallout from the Chernobyl accident. This is information that only well-designed scientific epidemiologic studies, based on firm dose estimates, could be expected to provide. With that purpose in mind, the US has planned with Belarus and Ukraine long-term cohort studies of many thousands of subjects with thyroid activity measurements.

1. Introduction

For the Post-Accident Review meeting at the International Atomic Energy Agency (IAEA) in August, 1986, the documentation provided by the USSR State Committee on the Utilization of Atomic Energy [1] revealed considerable concern among Soviet authorities about the effect of the accident on thyroid disease, especially cancer. In the medical annex, where this concern is most clearly evident, plans for the medical surveillance of the exposed population are briefly described. Included there (p 68 in the English translation) are two significant statements:

"A special study will be made of the functioning of the thyroid gland and over extended periods the frequency of development of adenoma and malignant neoplasms," and

"In dose ranges presenting even minimal risk of dysfunction of the thyroid gland (the pediatric) examination will be complemented by special dynamic observation of the endocrinology of the thyroid gland, using hormones--thyroxine, triiodothyroxine, thyroid stimulating hormone and others."

In May, 1987, there were two meetings on epidemiologic studies of the post-Chernobyl experience, one at the World Health Organization (WHO) Regional Office in Copenhagen, the other at the IAEA in Vienna [2], both chaired by Professor Albrecht Kellerer. At both meetings it was assumed that epidemiologic studies of thyroid cancer would be undertaken, and specific methodologic suggestions were made. In neither report, however, is there an expectation that thyroid cancer could become the problem that confronts the CIS today, especially Belarus.

In October, 1987, the Nuclear Energy Agency of the Organization for Economic Cooperation and Development held a workshop in Paris entitled "Epidemiology and Radiation Protection" at which the principles and practice of radiation epidemiology were extensively reviewed. The only reference to Chernobyl was contained in a paper by Kaldor and Parkin detailing plans to study leukemia through cancer registries throughout Europe, including Belarus and parts of the Russian Federation; the first paper on this study was published in 1993 [3], and a second is in process of publication [4].

In the spring of 1990 the WHO began planning its International Program on the Health Effects of the Chernobyl Accident (IPHECA), based on a request from the Minister of Health of the USSR. Thyroid disease was one of five radiation effects initially targeted for pilot studies, the others being leukemia and related hematologic disease, brain damage from pre-natal exposure, epidemiologic registers, and oral health in Belarus. The protocol for the thyroid survey visualized a case-control study built on a population screening effort in heavily contaminated (Cs-137 deposition greater than 550 kBq m^{-2}) areas with controls selected from the same areas and from less contaminated areas in anticipation of obtaining adequate dose information [5].

In the fall of 1990 the IAEA carried out an epidemiologic field survey in 7 villages in strictly controlled zones and 6 villages in lightly contaminated areas in Belarus, the Russian Federation, and Ukraine. Although by late 1990 the increase in thyroid cancer among children had begun in both Belarus and Ukraine, the careful thyroid screening by IAEA teams revealed no difference between heavily and lightly contaminated villages with respect to the prevalence of thyroid nodules [6]. The samples of children were too small for the IAEA survey to demonstrate a difference that had just begun to emerge.

Also in 1990, under an earlier US-USSR agreement to cooperate in the field of civilian nuclear reactor safety, thyroid cancer and leukemia were selected as subjects for collaborative epidemiologic investigation. In 1995 thyroid studies are just beginning in both Belarus and Ukraine.

In short, the catastrophe set in motion an intensive effort to screen children for nodules and cancer, national and international groups have attempted to organize epidemiologic studies with scientific as well as public health objectives. The effects of the radioiodines released during the accident on thyroid disease, especially cancer, has caused considerable concern and interest. Although the radioiodines are known to have caused thyroid cancer among the Marshallese, the cases were few, doses were difficult to establish, and the respective roles of I-131 and short-lived radioisotopes could not be clarified [7]. Exposure to both x and gamma rays has shown the thyroid to be highly sensitive to their carcinogenic potential; indeed, the thyroid is regarded as one of the most sensitive organs, especially in children [8-10].

2. The Need for Scientific Information on the Carcinogenic Effect of Radioiodine

Radiation protection strategies depend in part on scientific risk estimates derived from medical exposures and from the atomic bombing of Hiroshima and Nagasaki. Thus far information obtained from the medical use of I-131 has been inconclusive, partly because it rarely pertains to children or juveniles and little or no risk has been demonstrated among adult patients treated with I-131 [8]. Although the Marshallese experience does demonstrate the carcinogenic potential of radioiodine, it is estimated that the thyroid dose was principally due to short-lived radioiodines, especially I-132 and I-133. Animal experiments have confirmed that I-131 can induce thyroid cancer but without producing a sure basis for adjusting the human risk estimates for x and gamma rays. In sharp contrast to this paucity of information on the risk of radioiodine exposure to children is the fact that the extent of their vulnerability in case of nuclear accidents such as Chernobyl remains unknown and may be great if Chernobyl is any guide. Nuclear power reactors have large inventories of radioiodine and their accident rate, although low, is not zero. Also, if, as many believe, I-131 is much less effective than x or gamma radiation, we need to know why this is so, whether it depends on the protraction of dose within the gland or on some other mechanism. From the available data Shore has estimated that I-131 may be only 20-25 percent as effective as external radiation in inducing thyroid cancer among juveniles [9].

3. The Opportunity for Epidemiologic Study

However tragic the Chernobyl accident, it demonstrated the need for additional information on radiation protection and provided an unparalleled opportunity to create the information on which radiation protection must depend. Comparative figures on accidental releases of I-131, recently summarized by Becker et al. [10] are: Three Mile Island, 15-20 Ci (about 0.6 TBq); Windscale, 20,000 Ci (0.7 PBq); Hanford, from 1944-1947, 690,000 Ci (26 PBq); and Chernobyl, 40-50 million Ci (1-2 EBq). The Chernobyl accident exposed hundreds of thousands of people living within a relatively small area to doses that ranged well above a Gray. Soviet physicists were well aware of the potential hazards of radioiodine fallout and made literally hundreds of thousands of direct thyroid radioassay measurements, especially in Belarus and

Ukraine. These measurements, used in conjunction with personal histories of residence and nutritional sources, offer promise in dose reconstruction. Endocrinologists were not slow to initiate systematic screening programs. Within a comprehensive epidemiologic framework that includes information on individual dose, these programs could be expected to provide scientific information hitherto unavailable. Although only one area (Belarus) had a cancer registry in operation at the time of the accident, the highly organized and standardized medical care system in place at the time, and subsequently, constitute important assets in the conduct of large-scale epidemiologic studies.

4. The Effort To Establish Epidemiologic Studies of Thyroid Cancer

According to the preliminary WHO catalogue issued in the fall of 1994 [11], in each of the three now separate countries with significant fallout an indigenous cohort study of some size, 6,000 in the RF, 50,000 in Ukraine, and 20-30,000 in Belarus, was begun in 1986. In early 1990, following extensive collaboration between the USSR and WHO on matters relating to Chernobyl, the Ministry of Health of the USSR asked WHO to develop an international program, "to mitigate the health consequences of the Chernobyl accident." Although the thyroid component of the program was mainly a disease-detection effort in children living in highly contaminated areas and under 16 at the time of the accident, there was hope that the prevalence of thyroid disease could be related to thyroid dose from the Chernobyl accident. The case-control approach would be used, and dosimetric investigations were prescribed. In addition to the case-control approach a number of descriptive and ecologic methods were to be used [5].

Meanwhile, systematic case-detection continued in response to public health concerns, and additional studies were planned or begun in all three countries. In 1990 the Japanese Sasakawa Health and Medical Foundation began its extensive program of humanitarian aid to affected areas of all three countries, concentrating on case-finding by the most modern means. By the time the WHO issued its first draft Catalogue on Studies on the Human Health Effects of the Chernobyl Accident, depending on the criteria one uses for an "epidemiologic" study, there were 4 or more such studies in the RF, 6 or more in Ukraine, and 8 or more in Belarus. The criterion used here is that a study seeks, by comparison, to relate a health outcome to some measure of radiation exposure [12]. Most of the studies are of the "cohort" type, only a few, "case-control." It remains to be seen what results these studies will produce.

5. US-Sponsored Epidemiologic Studies

Probably the most ambitious plans for epidemiologic studies of thyroid disease following the Chernobyl accident are those being sponsored by the US National Cancer Institute with the support of the US Department of Energy and Nuclear Regulatory Commission in collaboration with the Ministries of Health of Belarus and Ukraine. They are planned with explicit scientific objectives, with concrete plans for dosimetry, with sampling plans based on existing files of subjects with thyroid measurements in 1986 and on power calculations related to the scientific objectives, with precise descriptions of clinical and laboratory methods, and with an emphasis on quality assurance. Procedures for data collection, storage, and retrieval, and for storage of specimens, are specified. Equipment and supply needs are detailed together with the pattern of bi-national collaboration, a plan for the management of the project, and guidelines for the preparation and release of reports on the work. Scientific objectives include the estimation of risk coefficients, as functions of dose, for nodules, for cancer, and for hypothyroidism, a search for ancillary risk factors, and the detection of hyperparathyroidism and lymphocytic thyroiditis. On the basis of estimated dose distributions, calculations were made to determine the power of each study not only to detect a dose-response relationship but also to discriminate among various effectiveness ratios for radioiodine in comparison with risk estimates derived for external radiation.

6. The Difficulties

The difficulties are many, diverse, and hard to eliminate, but some progress is being made.

6.1. Infrastructure.

A serious difficulty stems from the academic and research experience of investigators and allied personnel in the countries of the former USSR. Although many aspects of public health, including infectious disease epidemiology, appear to have been essential parts of the medical school curriculum in the USSR, little or no attention seems to have been paid to chronic disease epidemiology. And opportunities for training overseas were minimal to absent prior to the break-up of the USSR. In contrast, chronic disease epidemiology has flourished in Europe and the US for the past 50 years, is widely taught in the academic environment, and increasingly utilized in the investigation of public health and scientific problems. In the US alone there are professional societies and journals concerned largely with the methods of epidemiology, as well as more specialized societies and journals concerned with areas of application. The broad use of epidemiology as a research tool means that not only universities but also government agencies and private companies are engaged in the collection and analysis of information bearing on the distribution of disease in the population and its links to environmental and genetic causes. There has been very little of this in the CIS, and the absence of suitably trained and experienced personnel has slowed efforts to mount effective epidemiologic studies. It is to the credit of WHO, the European Community, private Japanese organizations, and others in the international community that they have organized training programs under which personnel from the CIS have become familiar with various aspects of chronic disease epidemiology. But if a beginning has been made in this way, the problem is far from solved.

6.2. Supplies and Equipment

As the WHO found in planning the thyroid phase of the IPHECA project it was necessary to provide not only large items of equipment such as radioimmunoassay analysers and ultrasound scanners, but also kits for the various assays, pipettes, needles, etc., etc. These needs reflected not merely the scale of the projected studies, but also the paucity of up-to-date equipment and the lack of funds for expendable supplies. Similarly, in planning its collaborative studies with Belarus and Ukraine the US National Cancer Institute developed pages of needed items of supply and equipment.

6.3. Communication

Communication networks have been slow and inefficient, partly because necessary equipment such as facsimile machines and computers have been in short supply, and telephone systems inadequate for both domestic and international communication. Communication has also been difficult because of habitual attitudes toward the ownership of information and toward collaborative work among different administrative entities. Epidemiologic work on the scale necessary to realize the potential of the opportunity presented by the Chernobyl accident requires teams from different disciplines, collaboration among independent institutes, and ready sharing of information. Interpersonal communication has been hampered by language differences and an insufficient supply of interpreters skilled in medical and scientific concepts and terminology, but also by differences in training and experience. Library resources have needed supplementation in all three countries, and inter-library loans have had to be encouraged. As part of this effort the US National Library of Medicine has established e-mail connections with national medical libraries in the Russian Federation, Belarus and Ukraine.

Cultural differences in program development have been especially difficult for US, Belarussian, and Ukrainian investigators and have greatly impeded the progress of their efforts to obtain approval for, fund, and initiate, projects of interest. In part this is because US investigators are required to formulate very explicit research plans and to have them favorably peer-reviewed as a condition of funding, and because of stringent official US regulations to protect the privacy and safety of human subjects of study. These requirements have stressed negotiators on both sides, Belarussian and Ukrainian investigators feeling restrained by them, and US negotiators probing for facts in an unfamiliar environment.

6.4. Dosimetry

All the scientific questions to be addressed on the basis of the post-Chernobyl experience with thyroid disease require adequate dosimetry if they are to be answered. The development of a basis for adequate individual dose estimates seems likely to be a slow and difficult process, despite the large number of direct thyroid measurements made in 1986 shortly after the accident. These measurements were not made in all cases under highly standardized conditions and with the same, uniformly calibrated, equipment. It may be helpful to remember that the present DS86 dosimetry for the A-bomb survivors was developed in 1986 and replaced the T65 dose estimates generated in 1965, and that in 1995 there is dissatisfaction with the neutron components of the DS86, 50 years after the bombings. The two dose reconstruction tasks are quite different, of course, but environmental dose reconstruction is never an easy task. Environmental measurements of radioiodine were few and the plume carrying radioisotopes was subject to the vagaries of the winds. Environmental measurements of radiocesium are much more numerous, but the relation between radiocesium and radioiodine is variable. Moreover, the Marshallese experience warns that all radioiodines may not be equal, and that estimates of I-131 dose alone may not suffice. Also, the ultimate thyroid dose to an individual derives from both inhalation and ingestion, depending on location, calendar time, shielding, and food contamination. This means that the contribution of the various radioiodines to the total dose of an individual is also a variable, and one on which very little information is likely to be forthcoming. If studies are confined to subjects with good direct measurements, as seems necessary, there remains the problem of estimating the temporal variation in the ingestion of contaminated food. Obtaining good dietary information retrospectively must be much more difficult than obtaining the shielding histories for the A-bomb survivors. On the positive side is the fact that all these difficulties are well known to dosimetrists and that some highly competent people are working to ameliorate or solve them.

6.5. The Worsening Economic Situation

In 1988, when the US-USSR agreement was made, it was possible to plan on an equitable sharing of the economic burden of specific research projects. In-country expenses for indigenous personnel, space, communication, etc. was to be covered by the host country. In 1995 this is no longer true: to put a project in the field local assistance must be provided to the host country and the in-country expenses of visiting collaborators must be borne by the collaborator. Local scientists with language skills are tempted to emigrate, and sending personnel abroad for training enhances the temptation. Official salaries are often so depressed by inflation that physicians and scientists have been forced to work at outside jobs in order to maintain their living standard. All these factors work against the development of secure, ongoing relationships necessary for the maintenance of a research organization. Finally, at least one international funding agency has expressed concern over the share of the national budget dedicated to the amelioration of the effects of the Chernobyl accident. Such pressure would seem inevitably to depress research budgets for Chernobyl research.

6.6. Locating Subjects for Study

Screening endeavors need not be limited to fixed cohorts. In fact, the public health interest driving them favors regional case-finding where all subjects of interest may be examined. It is possible, on this basis, to build case-control studies of considerable rigor, but difficulties are experienced at the level of dose. In this approach one finds the 1986 files of individuals with direct thyroid measurements to be of little help. In the recent case-control study of Astakhova et al. [13] only 12 among its 107 cases were in the thyroid measurement file for Belarus. With the cohort approach, which is the preferred design for tight inferences and for mapping the long-term consequences of an exposure, locating subjects for study may pose difficulties. In Belarus, for example, Voronetsky et al. conducted a pilot study [14] of a representative sample of 600 children drawn from the thyroid measurement file for Belarus. They found that the original measurement file lacked sufficient identifying information to make tracking its members an easy task. It would be much more efficient to select subjects by mapping large external files of dispensaries, etc. into the measurement file. This would be at some sacrifice of numbers and also affect the integrity of the sample as a basis for making inferences about the period from 1986 to the date of selection.

7. Progress to Date of Epidemiologic Studies of Thyroid Cancer in the CIS

Perhaps very little of the progress to date can be attributed to epidemiologic studies. Certainly the most important contribution has been that made by the pathologists [15-17] in validating the overwhelming majority of diagnosed cases in Belarussian children. The best epidemiologic report, although not based on a specific epidemiologic study, is that of Abelin et al. [17] who combined the evidence from pathology and incidence. They showed, as had Kazakov et al. earlier [18], that incidence rates by oblast in Belarus correlated well with estimated fallout deposition, and that they were well in excess of expectation based on the thyroid cancer rates in the better national cancer registries in Europe and the US. By the end of 1992 there could be little doubt that the Chernobyl accident had caused much of the rise in thyroid cancer incidence in Belarus. The first dose-specific demonstration of the relationship of thyroid cancer to thyroid dose appeared in 1995 in a letter to Nature by Likhtarev et al. [19] in which they showed a correlation between incidence and estimated average rates of thyroid cancer in children for 7 zones of Ukraine. In their small case-control study of thyroid cancer in children under age 15 in 1986, Astakhova et al. have shown the dose-specific relationship to be true at the level of the individual thyroid dose [13].

Taken together the various lines of investigation have established that: 1) the cases of thyroid cancer in children meet international clinical and pathology standards of diagnosis, 2) the minimal latent period for radiogenic thyroid cancer in children is on the order of 3-4 years, not 5-10 as previously believed, 3) the excess incidence is unprecedented, 4) the excess incidence is related to the Chernobyl accident, and 5) the dose-response relationship is so strong that individual high-dose cases can be called radiogenic with a very high probability, a fact that lays the groundwork for molecular comparisons between radiogenic and presumably non-radiogenic cancers.

References

[1] USSR State Committee on the Utilization of Atomic Energy: The Accident at the Chernobyl Nuclear Power Plant and its Consequences. Working document for the Post-Accident Review Meeting, Vienna, August, 1986.

[2] World Health Organization: Nuclear Accidents and Epidemiology, Reports of Two Meetings. Environmental Health Series No 25, Copenhagen, 1987, 108 p.

[3] D. Parkin, E. Cardis, H. Masuyer et al. Childhood leukemia following the Chernobyl accident: The European childhood leukemia-lymphoma incidence study (ECLIS). 1993, Eur J Cancer 29A:87-95.

[4] D. Parkin, R. Black, E. Masuyer et al. Childhood leukemia in Europe after Chernobyl: Five-year follow-up. To be published.

[5] World Health Organization: International Program on the Health Effects of the Chernobyl Accident: Protocol for the Pilot Project "Thyroid." PEP/93/9, Geneva, 1994.

[6] F. Mettler, M. Williamson, H. Royal et al. Thyroid nodules in the population living around Chernobyl. 1992, Jour Amer Med Assn 268:616-619.

[7] E. Lessard, N. Miltenberger, R.Conard et al.: Thyroid absorbed dose for people at Rongelap, Utirik and Sifo in March 1, 1959. BNL 51882, Upton, N.Y.: Brookhaven National Laboratory, 1985.

[8] United Nations Scientific Committee on the Effects of Atomic Radiation: UNSCEAR 1994 Report to the General Assembly, with Scientific Annexes, New York, United Nations, 1994.

[9] R. Shore: Issues and epidemiological evidence regarding radiation-induced thyroid cancer. Radiation Research 131:98-111, 1992.

[10] D. Becker, J. Robbins, G. Beebe et al.: Childhood thyroid cancer following the Chernobyl accident: A status report, Endocrinology and Metabolism Clinics of North America (In press).

[11] WHO European Center for Environment and Health: Catalogue on Studies on the Human Health Effects of the Chernobyl Accident (Draft), Rome, Italy, 1994.

[12] G. Beebe: Epidemiologic studies based on the Chernobyl accident. Proceedings of the 31st Annual Meeting of the National Council on Radiation Protection and Measurements, April, 1995 (In press).

[13] L. Astakhova, L. Anspaugh, G. Beebe et al.: Thyroid cancer in children following the Chernobyl accident: A case-control study in Belarus. (To be published)

[14] B. Voronetsky: (Personal communication)

[15] A. Furmanchuk, J. Averkin, B. Egloff et al.: Pathomorphological findings in thyroid cancers of children from the Republic of Belarus: A study of 86 cases occurring between 1986 ("post-Chernobyl") and 1991. Histopathology 21:401-408, 1992.

[16] K. Baverstock, B. Egloff, A. Pinchera et al.: Thyroid cancer after Chernobyl (letter). Nature 359:21-22, 1992.

[17] T. Abelin, J. Averkin, M. Egger et al.: Thyroid cancer in Belarus post-Chernobyl: Improved detection or increased incidence? Soz Praventivmed 39:189-197, 1994.

[18] V. Kazakov, E. Demidchik, L. Astakhova: Thyroid cancer after Chernobyl (letter). Nature 35: 21, 1992.

[19] I. Likhtarev, B. Sobolev, I. Kairo *et al.*: Thyroid cancer in Ukraine (letter). Nature 375:365, 1995.

RADIATION RISK ASSESSMENT OF THE THYROID CANCER IN UKRAINIAN CHILDREN EXPOSED DUE TO CHERNOBYL[1]

Boris SOBOLEV, Iliya LIKHTAREV, Irina KAIRO,
The Research Center of Radiation Medicine, AMS of Ukraine,
53, Melnikov St., 252050 Kiev, Ukraine
Nicolai TRONKO, Valeriy OLEYNIK, Tatiana BOGDANOVA
The Research Institute of Endocrinology and Metabolism, AMS of Ukraine,
69, Vyshgorodskaya St., 254114 Kiev, Ukraine

Abstract. The children's thyroid exposure to radioiodine is one of the most serious consequences of the Chernobyl accident. The collective dose to children aged 0–18 in the entire Ukraine was estimated to be 400,000 person-Gy. The dose estimates were calculated on the basis of measurements of thyroid content of ^{131}I for about 108,000 people in Ukraine aged 0–18 years in May–June 1986. Up to the end of 1994, 542 thyroid cancers throughout the Ukraine have been reported in children and young adults who were aged 0–18 at the time of the accident. Rates of thyroid cancer have climbed, from about 0.7 per million children aged 0–14 in 1986 to more 7 per million in 1994. Rates increased most in raions closest to Prypiat'. Between 1990 and 1994, 9 of the 14,580 people who had been children at the time of the accident in Prypiat' developed thyroid cancer. This corresponds to an annual incidence of 123 cases per million persons. The estimated average thyroid dose in Ukrainian children varies by several orders of magnitude. There is a more than 30-fold gradient in thyroid cancer incidence rates corresponding to the gradient in thyroid doses from ^{131}I. A preliminary investigation shows an excess in the annual incidence rate of thyroid cancer, throughout the northern territory of Ukraine, corresponding to the average doses to thyroid from ^{131}I. Coefficients of regression of excess cancers versus thyroid dose have been calculated.

1. INTRODUCTION

This paper presents some preliminary results in risk assessment of thyroid cancer among people exposed in childhood due to Chernobyl accident. The children's thyroid exposure to radioiodine is one of the most serious consequences of the accident. The collective dose to irradiated subjects of ages 0–18 when exposed in the entire Ukraine was estimated to be 400,000 person-Gy [1]. The dose estimates are calculated on the basis of direct measurements of thyroid content

---

[1] This study was supported the Ukrainian Radiation Protection Institute. The authors thank Dr. Andre Bouville for his help in text preparing. We also thank Miss Irina Plakhotnik and Miss Anna Semenchenko for their help in text and illustrations preparing.

of ^{131}I for about 150,000 people in Ukraine and are the results of a large scale dose reconstruction.

Under the Order of the Ukrainian Health Care Ministry, the Institute of Endocrinology and Metabolism and the Research Center for Radiation Medicine jointly have been collecting data on cases of the thyroid cancer in Ukraine since 1992. The objective is to estimate the influence of children's thyroid exposure to radioiodine due to Chernobyl accident on thyroid cancer incidence.

Cancers have been observed among Ukrainians who were born between 1968 and 1986 years, that is who were children at the time of the accident. We have collected evidence about 542 surgically treated patients which developed thyroid carcinoma during the 1986–1994 time period.

In the analysis of these data, we have considered 3 crucial questions.
1. Has the thyroid cancer incidence increased in the entire Ukraine?
2. What are birth cohort effects?
3. What is the relationship between the increase in the thyroid cancer incidence and the radiation exposure due to Chernobyl accident?

2. HAS THE THYROID CANCER INCIDENCE INCREASED IN UKRAINE?

We selected 4 two-year and 1 one-year intervals of observation (1986–1987, 1988–1989, 1990–1991, 1992–1993, and 1994) in order to have an appropriate number of cases for the analysis. Besides, we have assumed that the consequences due to the Chernobyl accident had not manifested themselves during the first two time intervals (1986–1987, 1988–1989)[7]. Table 1 gives the distribution of cases over the considered time intervals among the population born between 1968 and 1986 in the entire Ukraine.

The distribution over time shows a considerable increase in the number of cases since 1986. Changes in the annual incidence rate further demonstrate this increase. Rates of thyroid cancer in children aged 0–14 in 1986 have climbed from about 0.7 per million to more 7 per million in 1994. Rates increased most in raions closest to the Chernobyl nuclear power plant. Between 1990 and 1994, 9 of the 14,580 people who at the time of the accident were children in Prypiat'[2] – a town 3.5 kilometers from Chernobyl – developed thyroid cancer. This corresponds to an annual incidence of 123 cases per million persons.

Table 2 shows the increase of an annual incidence of thyroid cancer in people aged 0–18 at the time of the accident. However, it cannot be readily determined whether this increase is related to exposure due to the accident.

Table 1. Distribution of thyroid cancer cases observed between 1986 and 1994 among people born between 1968 and 1986.

	86–87	88–89	90–91	92–93	94
males	12	16	32	66	27
females	21	41	87	155	85
total	33	57	119	221	112

---

[2] In this paper we have used latin transliteration from Ukrainian language in names of settlements and regions, except name *Chernobyl*.

Table 2. Annual incidence of thyroid cancer during the 1986–1994 time period among people born between 1968 and 1986, per 10^6 persons.

	86–87	88–89	90–91	92–93	94
males	0.86	1.14	2.27	4.64	3.80
females	1.57	3.01	6.36	11.24	12.34
total	1.21	2.06	4.28	7.89	8.00

It can be argued that the observed increase of thyroid cancer is due to an age-dependent increase, as exposed population is aging over time and the age-specific baseline incidence rate increases from 0.5 per million for children to 10 per million for young adults [3].

3. EFFECTS OF BIRTH COHORT ON THYROID CANCER INCIDENCE

When the data were subdivided into 9 birth cohorts (<3, 3–4, 5–6, 7–8, 9–10, 11–12, 13–14, 15–16, and 17–18 years old at the time of the accident), we studied case series of thyroid cancer for every cohort in order to examine the assumption that the observed increase in the entire population is only due to variation with age of the baseline incidence rate.

Table 3 shows the observed annual incidence of the thyroid cancer in the entire Ukraine during the period of 1986–1994 according to two-year age groups and two-year birth dates. The table contains 9 columns corresponding to 9 birth cohorts and 19 rows corresponding to 13 age groups which were observed between 1986 and 1994.

Table 3. Annual age-specific incidence of thyroid cancer from 1986 to 1994 among people born between 1968 and 1986, per 10^6 persons.

Age group	Year of birth								
	1968–1969	1970–1971	1972–1973	1974–1975	1976–1977	1978–1979	1980–1981	1982–1983	1984–1986
0–2									–[a]
3–4								–	0.22
5–6							0.34	–	1.93
7–8						0.35	0.69	2.61	7.47
9–10					1.39	2.08	3.79	3.58	6.45
11–12				1.03	1.73	4.15	9.18	7.31	
13–14			2.38	1.72	3.10	6.50	8.03		
15–16		2.82	4.38	5.78	6.49	8.10			
17–18	3.45	3.87	4.99	7.46	6.81				
19–20	5.39	5.65	5.71	3.43					
21–22	8.41	9.80	10.45						
23–24	16.09	7.68							
25–26	15.07								

[a] No cases.

For example, people born in 1978 and 1979 had attained 7 years by the moment of the accident. Annual incidence rate of thyroid cancer for this birth cohort was 0.35 per million in 1986–1987. It was 2.08 per million in 1988–1989 when attained age was 9; 4.15 per million in 1990–1991 when attained age was 11; 6.50 per million in 1992–1993 when attained age was 13; 8.10 per million in 1994 years when attained age was 15.

Every column has 5 non-empty cells corresponding to 5 intervals of observation. Moving along a column, we can see how the disease rate in a given birth cohort changes as a function of time after the accident. Moving along a row, we can compare the incidence rate in different birth cohorts who have attained the same age. One can see the incidence rate for same age is higher in younger birth cohort.

We assumed that the two first time intervals (1986–1987, 1988–1989) for every cohort correspond to a time period when exposure consequences have not been expressed yet ("latent period")[7]. Therefore the disease rate in these time intervals provide an estimate of the baseline rate for the corresponding ages. Since the "latent period" for every cohort falls on different ages, disease rates in two first time intervals collected jointly for all cohorts could be an estimate of age-specific baseline rates. Table 4 gives the estimation of base-line disease rates obtained by fitting a power function of age to data from Table 3. We have chosen the power function as a simple guess.

Fig. 1 shows the estimate of baseline disease rate for different ages. Thus, we have an opportunity to investigate how the disease rate for cohort in question deviates from spontaneous disease rate in different ages. Deviations from the spontaneous rate in the birth cohorts are summarized graphically in Fig. 2. A considerable increase in the thyroid cancer incidence rate is observed in the birth cohorts under 8 years at the time of the accident, by a factor of 2–11.

This means that the cohort effects are strong. Since the age distributions of different birth cohorts are extremely different also, the observed disease rates for the pooled population will not correspond to the expected baseline rates of any particular cohort. Consequently, the observed increase of incidence since 1986 within the population born between 1968 and 1986 cannot be explained by age-specific changes only.

Table 4. Annual age-specific incidence of thyroid cancer observed in 1986–1987 and in 1988–1989 and estimate of spontaneous rates for different ages, per 10^6 persons.

Age, year	Observed in 1986–1987	Observed in 1988–1989	Estimate
3	$-a$	0.22	0.096
5	0.34	–	0.288
7	0.35	0.69	0.597
9	1.39	2.08	1.027
11	1.03	1.73	1.585
13	2.38	1.72	2.275
15	2.82	4.38	3.100
17	3.45	3.87	4.063
19	5.39	–	5.168

a No cases.

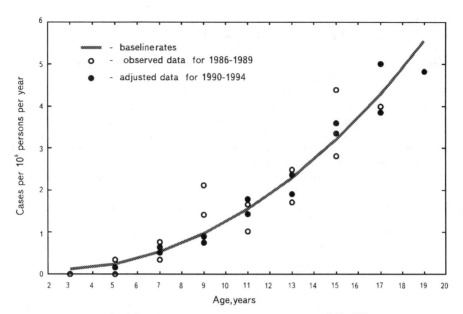

Fig.1. Baseline rates of thyroid cancer during the 1986-1989 time
period in Ukrainian population born between 1968 and 1986

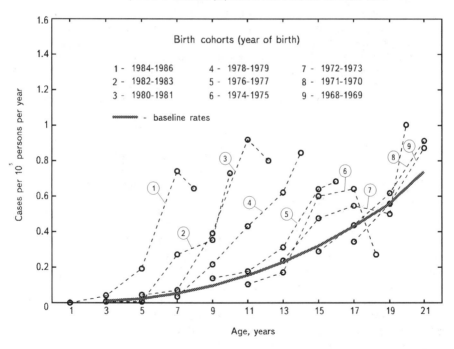

Fig.2. Thyroid cancer incidence for the 1986-1994 time period
in birth cohorts born between 1968 and 1986 in Ukraine

4. EFFECTS OF EXPOSURE TO RADIOIODINE ON INCREASED THYROID CANCER INCIDENCE

A preliminary investigation shows that the average doses to thyroid from ^{131}I throughout the territory of Ukraine correspond to the excess annual incidence of thyroid cancer. There is a more than 30-fold gradient in thyroid cancer incidence rates (about 4 cases per million per year in southern Ukraine and 123 cancers per million per year among Prypiat's children). This gradient corresponds to the gradient in thyroid doses from ^{131}I [1]. The results of the large scale thyroid dose reconstruction started in 1993 allowed to mark territories of Ukraine according to average levels of thyroid exposures from ^{131}I.

We determined the average dose from ^{131}I to the population born between 1968 and 1986 in the administrative raions of the northern Ukraine. Territory in question contains raions from the five most affected oblasts of Ukraine – Chernihivs'ka, Kyivs'ka, Zhytomyrs'ka, Vinnyts'ka, Cherkas'ka, cities Chernihiv, Kyiv, Zhytomyr, Vinnytsia, Cherkasy and a town of Prypiat'. We considered this territory because of reliable thyroid dose estimates for its population. The dose estimates were calculated on the basis of direct measurements of thyroid content of ^{131}I for about 150,000 people in the Ukraine in May–June 1986, including about 108,000 aged 0–18 years at that time [4]. In addition, about 7000 settlements from these oblasts have gotten certificates of thyroid exposure as a result of the large scale thyroid dose recovery [5]. Certificate contains the estimates of average thyroid dose from ^{131}I for seven age groups.

Besides, about 42% of all cancers since 1986 and more than 45% of all cancers during the 1990–1994 time period fall on these 5 oblasts out of a total of 25 oblasts in Ukraine. The estimated average thyroid dose in children varies by several orders of magnitude within these five oblasts. Therefore we combined raions with close average doses in five dose clusters. Fig. 3 shows the distribution of raions' average thyroid dose to people aged 0–18 in 1986.

Then we examined age-specific and age-adjusted incidence of thyroid cancer in pooled population born between 1968 and 1986 for every dose cluster. The analysis was restricted to cases occurring 5 years and more after exposure. Table 5 shows the age-adjusted disease rates in pooled population for 1990–1994 years. The age standardization procedure used demographic data of the Census of Ukraine, 1989 [2]. We used the preliminary estimate of the age-specific baseline disease rates from Table 4 for calculating the age-adjusted annual spontaneous rate. Its value is 3.56 per million for pooled population born between 1968 and 1986 in calendar time 1990–1994. We studied the changes of the excess thyroid cancer incidence over dose clusters. Table 5 shows the excess absolute risk (EAR) and the excess relative risk (ERR) in 1990–1994 calculated in pooled population born between 1968 and 1986 for every dose cluster. EAR is obtained by formula EAR = observed incidence – spontaneous rate. ERR is obtained by formula ERR = observed incidence : spontaneous rate – 1.

We calculated the coefficients of linear regression between the excess cancer and the thyroid dose from ^{131}I: EAR = k_{AR}×dose – for the excess absolute risk; ERR = k_{ER}×dose – for the excess absolute risk, where k_{AR}, k_{ER} – regression coefficients. This allows us to obtain the first reliable estimate of the relationship between thyroid doses received as a consequence of the accident and increased thyroid cancer incidence.

SCALE 1:2900000

Boundaries	Inhabited localities	Average dose, Gy

Boundaries

——— - of Oblast

——— - of Raion

Inhabited localities

⊙ - less than 30 000

○ - 30 000-100 000

◉ - 100 000-300 000

⊗ - 300 000-1 000 000

- over 1 000 000

Average dose, Gy

1.618

0.713

0.342

0.106

0.033

Fig.3. Thyroid doses from ^{131}I to population born between 1968 and 1986

Table 5. Average thyroid dose (cGy), age-adjusted disease rate (per 10^6 PY),
excess absolute risk (per 10^6 PY), and excess relative risk in dose clusters

Interval	Dose	Disease rate	EAR	ERR
1–4.9	3.3	5.19	1.63	0.46
5–19.9	8.8	6.20	2.64	0.74
20–59.9	32.1	8.94	5.38	1.51
60–119.9	71.1	17.15	13.60	3.82
120–200	165.1	23.81	20.25	5.69

We will interpret obtained regression coefficients k_{AR} and k_{ER} with some care as radiation risk per unit of dose [8]. So, the excess absolute risk per 10^4 PY per Gy (EAR/10^4 PY Gy) was 0.1350 (95% CI = 0.0971, 0.1728). The excess relative risk per Gy (ERR/Gy) was 3.79 (95% CI = 2.73, 4.86). Estimates and confidence intervals were calculated in procedures from [6].

5. CONCLUSION

In this paper we presented the materials which allow to formulate the following inferences.

1. The observed increase of thyroid cancer in Ukrainian population exposed in childhood due to the Chernobyl accident exceeds the increase of baseline disease rate, even when the variation of baseline rate as a function of age is taken into account.

2. The excess of thyroid cancer in population aged 0–18 at the time of the accident from the 5 most exposed oblasts of Ukraine could be related to the thyroid exposure to radioiodine from the Chernobyl accident.

References

[1] I.A.Likhtarev, B.G.Sobolev, I.A.Kairo, N.D.Tronko, T.I.Bogdanova, V.A.Oleinic, E.V.Epshtein, V.Beral. Thyroid cancer in the Ukraine. *Nature*. (1995) 375: 365.
[2] Склад населення Української РСР за статтю та віком на 12 січня 1989 року (за данними Всесоюзного перепису населення 1989 року). – Міністерство статистики УРСР, Київ, 1991, 310 с.
[3] W.Jacobi, H.G.Paretzke, D.Chmelevsky, M.Gerken, K.Henrichs, F.Schindel. Risiken Somatischer Spaetschaeden durch Ionisierende Strahlung. GSF-Bericht 15/89. – Munchen 1989.
[4] I.A.Likhtarev, N.K.Shandala, G.M.Gulko, I.A.Kairo, N.I.Chepurny. Ukrainian thyroid doses after the Chernobyl accident. *Health Physics* (1993) 64, 6: 594–599.
[5] I.A.Likhtarev, G.M.Gulko, B.G.Sobolev, I.A.Kairo, N.I.Chepurnoy, G Proehl, K. Henrichs. Thyroid dose assessment for the Chernigov region (Ukraine): estimation based on ^{131}I thyroid measurements and extrapolation of the results to districts without monitoring. *Radiat. Environ. Biophys.* (1994) 33:149–166.
[6] N.Draper, H.Smith. Applied Regression Analysis. 2nd edition. Wiley, 1981
[7] National Council on Radiation Protection and Measurements (1985). Induction of thyroid cancer by ionizing radiation. NCRP Report No 80, Bethesda, Md.
[8] E.Ron, J.H.Lubin, R.E.Shore, K.Mabuchi, B.Modan, L.M.Pottern, A.B.Schneider, M.A.Tucker, J.D.Boice, Jr. Thyroid Cancer after Exposure to External Radiation: A Pooled Analysis of Seven Studies. *Radiat. Res.* (1995) 141: 259–277.

SCREENING FOR THYROID CANCER IN CHILDREN

Shigenobu Nagataki and Kiyoto Ashizawa
The First Department of Internal Medicine,
Nagasaki University School of Medicine.

Abstract

In the screening of the thyroid diseases in the radiation exposed cohort, it is essential to make correct diagnosis and to measure radiation dose in every subjects in the cohort and to analyze the dose response relationship by the most appropriate statistical method. Thus, thyroid cancer, thyroid adenoma and autoimmune hypothyroidism were confirmed to be radiation-induced thyroid diseases among atomic bomb survivors.

A group of investigators from Nagasaki university have been working in the thyroid part of Chernobyl Sasakawa Health and Medical Cooperation Project, and more than 80,000 children were screened in 5 diagnostic centers (Mogilev, Gomel, Kiev, Korosten and Klincy).

In order to make correct diagnosis, thyroid echo-tomography, measurements of serum levels of free thyroxine, TSH, titers of anti-thyroid antibodies were performed in every children in the cohort and aspiration biopsy was performed when necessary. Whole body Cs-137 radioactivity was also determined in every subjects.

Children with thyroid cancer confirmed by histology (biopsy or operation) were 2 in Mogilev, 19 in Gomel, 6 in Kiev, 5 in Korosten and 4 in Klincy (until 1994). Since children screened in each center were less than 20,000, prevalence of thyroid cancer was remarkably high (lowest 100 and highest 1000/million children) when compared to the other parts of the world (0.2 to 5/million/year). However, there was no dose response relationship between the prevalence of cancer or nodule and whole body Cs-137 radioactivity.

Although a significant correlation between thyroid cancer and reconstructed thyroid I-131 dose was presented, there are no previous reports to prove that I-131 produces thyroid cancer in human. Investigation on external radiation and short lived isotopes along with I-131 may be important to elucidate the cause of thyroid cancer.

1. Introduction

The explosion of nuclear weapons, nuclear tests and accidents at nuclear power plants can affect the thyroid gland, and radiation-induced thyroid diseases are due to exposure to radiation at the time of explosion or to radioactive fallout (external radiation) and exposure to radioactive iodine that accumulates within the thyroid (internal radiation). Patients with radiation-induced thyroid disease can survive for a long time with treatment. Therefore, a survey of subjects with radiation-induced thyroid diseases can be conducted using the same protocol to examine all subjects in a certain cohort at the same time. In normal controls, the prevalence of thyroid diseases is higher than that of other radiation-induced diseases, which is an advantage for epidemiological studies. For these reasons, the thyroid is an ideal organ for the investigation of the health effects of the atomic bomb explosions, nuclear tests and accident at nuclear power plants [1].

In the investigation of radiation-induced thyroid disease, the followings are essential to obtain a significant results:

1) Determination of the exact thyroid radiation dose;
2) Correct diagnosis of thyroid diseases;
3) Analysis of the results by the most appropriate statistical method.

Thus we confirmed the radiation-induced thyroid diseases among atomic bomb survivors in Nagasaki [2]. We examined 2856 subjects who were cohort members of the Nagasaki Adult Health Study from October 1984 to April 1987. Thyroid radiation doses in each subjects was estimated with Dosimetry System 1986 (DS 86) [3]. Significant correlation between the thyroid radiation dose and the prevalence of thyroid diseases was found in thyroid solid nodules and thyroid cancer in female and antibody positive spontaneous hypothyroidism. The prevalence of thyroid nodule increases monotonously with thyroid radiation dose and the prevalence is significantly higher as the age at the time of bombing is decreased. However, the prevalence of antibody positive spontaneous hypothyroidism (autoimmune hypothyroidism) displayed a convex dose response relationship with maximum level of 0.7 Sv. Forty five years after the explosions of atomic bomb autoimmune, hypothyroidism was confirmed to be the radiation-induced disease.

Based on these experiences of atomic bomb survivors in Nagasaki, children around Chernobyl area were examined.

2. Subjects and Methods

The subjects are children born between April 26, 1976 and April 26, 1986 (age at the time of accident: 0-10 year old) and examined in the period from May 15, 1991 to December 31,

1994. The number of subjects is as follows: Mogilev (17,927), Gomel (14,054), in Belarus, Kiev (18,848), Korosten (18,792), in Ukraine and Klincy (17,467) in Russia respectively. The course of health examination includes the followings; (1) collection of individual history and biographical information; (2) anthropometric data; (3) determination of serum hormone levels (free thyroxine, TSH), (4) titers of anti-thyroid antibodies, (5) ultrasonography of the thyroid, (6) echo-guided fine needle aspiration biopsy (FNAB) & cytological diagnosis when necessary, and (7) whole body 137-Cs radioactivity.

Estimation of thyroid volume was performed with an arch-automatic scanning ultrasonographic instrument (Aloka-SSD-520) [4] . Images of 11cross sections of the thyroid were recorded on optic disc, then the total thyroid volume was calculated. Diagnosis of thyroid diseases was established on the basis of the following criteria of thyroid images: (1) position, (2) structure, (3) echogenity, (4) presence of nodules and cysts, and (5) volume. The criteria for goiter is a thyroid volume exceeding the volume calculated by the following formula:

$$\text{LIMIT} = 1.7 \times 10^{0.013 \times age + 0.0028 \times height} \times (\text{body weight})^{0.15}$$

Three hundred ten children showing echographic thyroid abnormalities were selected for FNAB, and a sample was successfully obtained for cytological diagnosis from 272 cases [5] . Ultrasonographical abnormalities over 5mm in diameter such as nodular lesions, cystic lesions and abnormal echogenity were chosen for FNAB.

3. Results

The results of investigations is summarized in Table 1. Children with thyroid cancer confirmed by histology were 2 in Mogilev, 19 in Gomel, 6 in Kiev, 5 in Korosten and 4 in Klincy (until 1994). In Gomel region prevalence of thyroid cancer (19 out of 14054) was especially high. All 19 cases were histologically papillary carcinoma and the characteristics of thyroid cancer were highly invasive which were similar to childhood thyroid cancers in other parts of the world [6] .

Since children screened in each center were less than 20,000, prevalence of thyroid cancer in not only Gomel region but also in other region was remarkably high (lowest 100 and highest 1000/million children) when compared to that of U.S.A., Europe and Japan (0.2 to 5/million/year) [7-9] .

Combined with the accurate thyroid examination data, we also compared thyroid disease prevalence with whole body Cs-137 radioactivity. The results of screening, however, did not show any significant correlation between the prevalence of thyroid cancer and whole body Cs-137 radioactivity which is closely related to Cs-137 in the soil [10] .

Table 1. Incidence of thyroid diseases around Chernobyl (%)

Thyroid disorders	BELARUS		RUSSIA	UKRAINE	
	Mogilev	Gomel	Klincy	Kiev	Korosten
Goiter (range)	6-31	4-56	31-53	38-75	12-49
Nodule(s)	0.12	1.81	0.59	0.16	0.28
Cancer	0.01	0.14	0.02	0.03	0.03
Hyperthyroidism	0.43	0.11	0.05	0.09	0.07
Hypothyroidism	0.07	0.28	0.09	0.04	0.13
ATG	1.1	0.9	1.2	1.2	3.1
AMC	1.9	2.4	1.8	2.1	3.5

The incidence of goiter, nodule(s) was analyzed by ultrasound images.

Hyperthyroidism: (free thyroxine $>$ 25.0 pmol/L, TSH $<$ 0.24 μ IU/mL)

Hypothyroidism: (free thyroxine $<$ 10.0pmol/L, TSH $>$ 2.90 μ IU/mL)

Cancer: All cases were histologically confirmed by operated tissue.

ATG: anti-thyroglobulin antibody, AMC: anti-microsome antibody

4. Discussion

Since atomic bombs in Japan, more than thousands of nuclear weapon tests and nuclear plants accidents took place in the world and human beings are still at the risk of exposure to radiation. However, the health effects of many of these tests and accidents are not clarified yet. Chernobyl accident was one of the most serious nuclear plants accidents in the world. In the Chernobyl accident various investigations are supported by many international organizations and many groups in the world. Among them our project supported by the Chernobyl Sasakawa Health and Medical Cooperation is the biggest and the most international, contributive medical aid .

It is confirmed that there are many children with thyroid cancer in Belarus and Ukraine and diagnosis was confirmed in virtually all cases. A part of Russia was contaminated as highly as in some areas in Belarus and Ukraine, but childhood cancer was not increased in the contaminated areas except some report which found 14 cases of thyroid cancer in 1994. In this project we demonstrated very high prevalence of thyroid cancer in 3 Republics including Russia (Klincy: 0.02%) using the same protocol screening. However, no dose response relationship between the prevalence of thyroid cancers and whole body Cs-137 radioactivity at the time of examination [10] . Since many children were evacuated from highly contaminated areas, it is necessary to investigate the relation with Cs-137 in the soil at the time of shortly after the accident.

Even many international experts believe that it is very likely the increase of thyroid cancer is due to the radioactive fallout by the Chernobyl accident [11-12], it is not known what kind of radioactivity is the cause of thyroid cancer. It is essential to show the dose response from causal radioactive materials. Of course the possible radioactive materials are, first of all, various radioactive iodine which accumulate within the thyroid. However, it would be very difficult to pick up one specific radioactive iodine.

A map of I-131 contamination measured in 1986 was presented recently and it was reported that a significant dose response was found between the prevalence of thyroid cancer and I-131 in the soil or the reconstructed thyroid I-131 dose [13]. Although thyroid I-131 dose measured directly within several months after the accident in majority of operated case in Ukraine was less than 30cGy, it was reported that the prevalence was higher in the population with thyroid dose greater than 50-100 and 100-200 cGy. These are very important findings and have to be confirmed by international experts. It should be noted that there are no previous publications which showed that I-131 in any dose produced thyroid cancer in human. Therapeutic dose of I-131 clearly induce hypothyroidism within several weeks after the radiation. However, no reports that I-131 induced thyroid cancer at least in humans were published. And several reports showed that no significant thyroid diseases were induced by the diagnostic dose of I-131. In Chernobyl accident, we have to notice the special situations which were different from previous publications on I-131 in many respects; 1) Prevalence of thyroid cancer was very high in especially children. 2) Around Chernobyl are iodine deficient areas. 3)Various iodine prophylaxis was given. Therefore, I-131 could be the cause of thyroid cancer and investigation on the clear dose response relationship must be encouraged.

Investigation on radiation from other types of radioactive materials have not been reported. Radiation by short-lived isotopes of iodine and tellurium, which may contribute a large percentage of the absorbed thyroid dose by inhalation may be more carcinogenic than I-131 and could be the cause of thyroid cancer. And external radiation by any type of isotopes also can produce thyroid cancer. Radiation to atomic bomb survivors was mainly external radiation at the time of explosion of atomic bomb. And thyroid external radiation dose by atomic bomb was estimated for each atomic survivor by DS 86 system [4]. A large dose of external radiation induces hypothyroidism. However, it should be emphasized that thyroid cancer was induced in children by low dose of medical external radiation (0.06 to 1.4 Gy) for enlarged thymus, tinea capitis, skin hemangioma [14-18]. Therefore, investigation on external radiation and short lived isotopes along with I-131 may be important to elucidate the cause of thyroid cancer.

References

[1]   S. Nagataki, Radiation and the Thyroid: A Model of Investigation. In : S. Nagataki (Ed.), Nagasaki
      Symposium on Chernobyl Update and Future ISBN 0-444-81953- 3. Excerpta Medica, Amsterdam,
      1994, pp 265-270.

[2]   S. Nagataki et al., Thyroid Diseases Among Atomic Bomb Survivors in Nagasaki. *J.A.M.A.*. 272
      (1994) 364-370.

[3]   W.C. Roesch, US-Japan Joint Reassessment of Atomic Bomb Radiation Dosimetry in Hiroshima
      and Nagasaki Ⅰ. Radiation Effects Research Foundation, Hiroshima, 1987.

[4]   N. Yokoyama et al.., Determination of the Volume of the Thyroid Gland by a High Resolutional
      Ultrasonic Scanner. *J. Nucl. Med.* 27 (1986) 1475-1479.

[5]   M. Ito et al ., Childhood Thyroid Diseases Around Chernobyl Evaluated by Ultrasound Examination
      and Fine Needle Aspiration Cytology. *Thyroid* (1995) (in press).

[6]   M. Ito et al ., Histopathological Characteristics of Childhood Thyroid Cancer in Gomel, Belarus.
      International Journal of Cancer. (1995) (in press)

[7]   H. Campbell et al ., The Incidence of Thyroid Cancer in England and Wales. *Br. Med. J.* 2 (1963)
      1370-1373.

[8]   Danish Cancer Registry. Incidence of Cancer in Denmark 1973-1977. Copenhagen. Danish Cancer
      registry, Kraeftens Bekaempelse, Danish Cancer Society, 1982.

[9]   J.L.Young, JL., Percy, CL., Asire, AJ (eds). Surveillance Epidemiology and End Results: Incidence
      and Mortality Data, 1973-1977. National Cancer Institute, Monograph 57, 1981.

[10] S. Yamashita et al ., Chernobyl Sasakawa Health and Medical Cooperation-1994. In : S. Nagataki (Ed.),
      Nagasaki Symposium on Chernobyl Update and Future ISBN 0-444-81953- 3. Excerpta Medica,
      Amsterdam, 1994, pp 63-72.

[11] E.D. Williams, Chernobyl, Eight Years On. *Nature*, 371 (1994) 556.

[12] T. Abrlin, Belarus Increase was probably Caused by Chernobyl. *B.M.J*, 309 (1994) 1298.

[13] I.A. Likhtarev et al ., Thyroid Cancer in the Ukraine. *Nature*, 375 (1995) 365.

[14] R.E. Shore et al ., Thyroid Cancer Among Persons Given X-ray Treatment in Infancy for an Enlarged
      Thymus Gland. *American J. Epidemiology*, 137 (1993) 1068-1080.

[15] E. Ron et al ., Thyroid Neoplasia Following Low Dose Radiation in Childhood. *Radiation Research*, 120
      (1989) 516-531.

[16] R. Shore et al ., Carcinogenic Effects of Radiation on the Human Thyroid Gland. In: Radiation
      Carcinogenesis, New York, Elsevier, 1986.

[17] C. Furst, et al ., Cancer Incidence after Radioteraphy for Skin Hemangioma: A Retrospective Cohort
      Study in Sweden. J. of the National Cancer Institute, 80 (17) 1988, pp 1387-1392.

[18] M. Lundell, et al ., Thyroid Cancer after Radiotheraphy for Skin Hemangioma in Infancy. *Radiat.
      Research*, 140 (1994) 334-339.

Diagnosis, Surgical Treatment and Follow-up of Thyroid Cancers

F. Pacini, T. Vorontsova*, E. P. Demidchik**, F. Delange°, C. Reiners#, M. Schlumberger°°,
and A. Pinchera.

Istituto di Endocrinologia, University of Pisa, Italy; *Institute of Radiation Medicine and
**Oncology Center of Thyroid, Minsk, Belarus; °Departement de Pediatrie et des
Radioisotopes, ULB, Bruxelles, Belgium; #University Clinic for Nuclear Medicine, Wurzburg,
Germany; °°Institute Gustave Roussy, Villejuif, France.

Abstract. This paper reports the activities and the results of the research carried out by the Centers participating to the JSP4 project, within the framework of the EU program on the consequences of the Chernobyl disaster. The project was aimed to develop and to control the application of basic principles for the diagnosis, treatment and follow-up of thyroid carcinoma, with special attention to the peculiar requirement of children and adolescents. To this purpose, training in Western European Centers was offered to a number of scientists from Belarus, Ukraine and Russia. Several official meetings were organised to share views and to discuss the progress of the project. A basic protocol for the diagnosis, treatment and follow-up of thyroid carcinoma has been developed and approved by all participating Centers. Hopefully, it will be applied to the new cases and to those already under monitoring. A large part of the protocol is dedicated to the post-surgical treatment with thyroid hormones for the suppression of TSH and with calcitriol for the management of surgical hypoparathyroidism. A detailed protocol to asses iodine deficiency and, eventually, to introduce a program of iodine supplementation has been proposed.

The collection of control cases of childhood thyroid carcinoma in non-radiation exposed European countries has been initiated in Italy, France and Germany. This data will be used as control for the post-Chernobyl childhood thyroid carcinomas. Here is reported a preliminary comparison of the clinical and epidemiological features of almost all (n=368) radiation-exposed Belarus children who developed thyroid carcinoma (age at diagnosis <16 years), with respect to 90 children of the same age group, who, in the past 20 years, have received treatment for thyroid carcinoma in two centers in Italy (Pisa and Roma). Finally, by molecular biology, genetic mutations of the RET proto-oncogene have been found in several samples of thyroid carcinomas provided by the Belarus partners.

All together, this project has provided the CIS partners with the most useful information and knowledge in order to allow an effective treatment of the diseased population and, possibly, the prevention in those not yet affected.

1. INTRODUCTION

Thyroid cancer is a rare condition in childhood. When it does occur it is most usually of the papillary type, and is in general susceptible to effective treatment and cure, provided that early diagnostic and therapeutic measures are undertaken (1-3). General guidelines for the management of thyroid carcinoma in children are shared by most international centers, but protocols may differ in relation to local conditions. The unique nature of the Chernobyl accident (4-7), the pathology and the number of thyroid cancers registered to date (8,9) have outlined the importance of investigating the relationship between thyroid cancer and the causing event. This need has obtained a prompt answer from the European Community who financed the present and other projects (10).

The aim of this project was to develop and to control the application of basic principles for the following objectives: a) developing and implementing a protocol for the diagnosis, treatment and follow-up of childhood differentiated thyroid cancer, including the correct use of l-thyroxine suppressive therapy; b) developing new diagnostic imaging techniques for the post-surgical monitoring of thyroid carcinoma, including the possible use of recombinant human TSH; c) studying the autoimmune phenomena in radiation-induced childhood thyroid cancer; d) evaluating the biological and clinical behaviour of radiation-induced thyroid carcinoma as compared with non-radiation-induced thyroid carcinoma; e) assessing iodine deficiency and the need for iodine supplementation and thyroid suppressive therapy in the prevention of radiation-induced thyroid cancer.

Several exchange of visits and meetings have taken place between the Western and the CIS participating institutions. Thanks to these initiatives, several objectives and results have been reached in the years covered by this project, as summarised in the present report.

2. RESEARCH ACTIVITY AND TRAINING OF BELARUS SCIENTISTS IN EUROPEAN CENTERS.

Several scientists from Belarus had training at the Institute of Endocrinology and at the Department of Surgery of the University of Pisa. They performed an extensive number of endocrine and immunological assays in a large series of children with or without thyroid cancer or with other thyroid disorders. These children were from Belarus areas exposed or unexposed to radiation. The results obtained so far, suggest that there is a significant increase in autoimmune phenomena both in children with thyroid cancer and in unaffected children. The Belarus partners provided several samples of thyroid carcinomas which have been studied by molecular biology for the presence of genetic mutation of the RET proto-oncogene.

Other investigations on oncogenes (ras and gsp) in thyroid tumors occurring either spontaneously, or after neck external irradiation or as a consequence of the Chernobyl accident have been performed in the center of Villejuif.

Two senior physicians from Ukraine spent two weeks at Würzburg University Clinic for Nuclear Medicine to become more familiar with recent developments in the treatment and follow-up of childhood thyroid cancer. On return these colleagues were able to improve the local protocols for I-131 therapy with the equipment delivered by a CEC ECHO-2 project.

Training of medical staff from the three CIS countries in the Brussels Center is already scheduled for January 1996. The object of this visit is to became acquainted with the basic technique for the epidemiological and laboratory assessment of iodine deficiency.

3. DEFINITION OF A COMMON PROTOCOL FOR DIAGNOSIS, TREATMENT AND FOLLOW-UP OF CHILDHOOD THYROID CARCINOMA.

31. Protocols.

A protocol for diagnosis, treatment and follow-up of thyroid cancer in children has been approved by all partners and will be applied to the new cases and to those already under monitoring. A large part of the protocol is dedicated to two major problems encountered up to now by the teams involved in the treatment of the affected children in the CIS countries. These include the post-surgical treatment with thyroid hormones for the suppression of TSH and with calcitriol for the management of surgical hypoparathyroidism.

A detailed protocol for iodine supplementation and thyroid suppressive therapy has been proposed. The objectives of this protocol are two fold and are as follows:

1. Evaluating the past and present amount of iodine in the nutrition of the areas under investigation.

2. Implementing thyroid suppressive therapy and evaluating its effects. The type of therapy will be decided on the basis of the results obtained by objective 1.

32. Treatment.

In the center of Würzburg (previously in the one in Essen), 120 courses of I-131 therapy were administered to 30 children from Belarus with advanced thyroid cancer. This joint project was carried out together with the Centre for Thyroid Tumours in Minsk. The children came to Essen in groups of 4 where they stayed for a week. The group was accompanied by a physician who was trained during her/his stay in the radioiodine treatment of thyroid cancer, in the follow-up with sonography, scintigraphy and laboratory tests as well as in hormonal replacement therapy.

In addition, several new assays for thyroglobulin and TSH were tested in the laboratory of Würzburg University Clinic for Nuclear Medicine. It could now be demonstrated that the implementation of such recent techniques improves the management of patients with thyroid cancer.

A separate project focused on pulmonary fibrosis as a possible side effect of high dose I-131 therapy on children with disseminated pulmonary metastases of thyroid cancer. High resolution CT and lung function tests were used for follow-up. In 2 out of 45 children with lung metastases signs of pulmonary fibrosis developed after more than 20 GBq of I-131.

4. CLINICAL AND EXPERIMENTAL STUDIES OF THYROID CARCINOMA IN IRRADIATED BELARUS TUMORS AND IN UNIRRADIATED WESTERN EUROPEAN CONTROL TUMORS.

4.1. Data base.

The collection of control cases of childhood thyroid carcinoma in non-radiation exposed European countries has been initiated in Italy, France and Germany. In France a total of 150 cases have been registered. In Italy, a total of 120 cases ageing 16 years or less have been analyzed. These cases were compared from the epidemiological and clinical point of view with all the cases that occurred in Belarus from the year 1986 to the first 6 months of 1995.

4.2. Clinical features of radiation-induced childhood thyroid cancer from Belarus, as compared with non-irradiated childhood Italian cancer.

We studied the clinical and epidemiological features of almost all (n=368) radiation-exposed Belarus children who developed thyroid carcinoma (age at diagnosis <16 years) , compared to 90 children of the same age group, who, in the past 20 years have received treatment for thyroid carcinoma in two centers in Italy (Pisa and Roma) .

Patients

We had access to the records (from 1986 to June 1995) of most Belorussian children (n=368) who were younger than 16 when the diagnosis of thyroid carcinoma was made. The number of cases per year is shown in Table 1.

Table I. Number of Childhood thyroid cancer registered yearly from 1986 to 1995.

YEAR	NUMBER OF CASES
1986	1
1987	2
1988	4
1989	6
1990	28
1991	58
1992	65
1993	80
1994	80
1995	44 (First 6 months)

In a subgroup of 63 children (35 females, and 28 males) serum samples were available for biochemical determinations. All the patients had a histologically proven differentiated thyroid carcinoma and at the time of the Chernobyl reactor accident, were living in regions of Belarus heavily contaminated by radioactive fall outs: 201 children were living in Gomel, 81 in Brest, 41 in Minsk, 20 in Mogilev, 20 in Grodno, and 7 in Vitebsk (Table II). The diagnosis of thyroid carcinoma was done either by screening programs or by referral from the family doctor.

Table II. Distribution of cases by region.

REGION	NUMBER OF CASES
Gomel	201
Brest	81
Minsk	41
Mogilev	20
Grodno	20
Vitebsk	7

Initial treatment of all patients was carried out in Minsk and consisted in total (or near-total) thyroidectomy in 134 patients, sub-total thyroidectomy in 57, and lobectomy in 172.

The epidemiological and clinical pattern of Belarus tumors was compared with that of 88 Italian children (females, males) of the same age group treated for differentiated thyroid cancer in the past twenty years at the Institute of Endocrinology of the University of Pisa (n=33), and at the Nuclear Medicine Department, University of Roma (n=55). Initial treatment was of near-total thyroidectomy in 88 children, and sub-total thyroidectomy in 2.

Results

Belarus versus Italian tumors.

The female to male ratio was lower, but not significantly different, in Belarus children (F:M=1.6:1) compared to Italian children (F:M=2.7:1). The mean($\pm$SD) age at the diagnosis was significantly younger in Belarus (10.0$\pm$2.5 years) than in Italy (11.5$\pm$3.2 years; p<0.0001 by unpaired Student's t test).

Most Belarus tumors (n=351, 95.3%) were papillary thyroid carcinomas, with or without follicular component, 15 were pure follicular carcinomas (4.1%), 1 was medullary thyroid carcinoma (0.3%), and 1 was Hurthle cell carcinoma (0.3%). In the Italian series, 72 tumors (81.8%) were papillary and 16 (18.2%) were follicular. According to TNM classification, in the Belarus cases, tumors extending outside the thyroid gland were 178 (48.3%); lymph node involvement was present in 241 (65.4%) cases, and distant metastases were found in 33 (8.9%).

Belarus children: age at radiation exposure and age at diagnosis.

As shown in Fig. 1, the mean age of the children at the time of the Chernobyl accident and at the time of diagnosis was 3.57±2.6 years (range 0-13.0) and 10.0±2.5 years (range 3.1-14.1 years), respectively

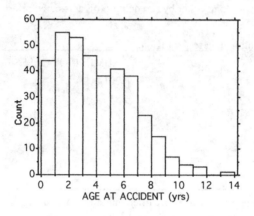

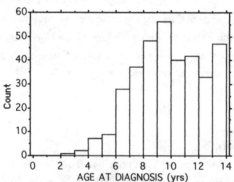

Fig. 1. Age distribution at the moment of the accident (upper panel) and at diagnosis (lower panel).

The mean time elapsed between exposure and diagnosis was of 6.47±1.6 (range 1-9) years with a strong positive correlation (r=0.79; p<0.0001; Fig. 2) between the two parameters, indicating that the latency period between exposure and diagnosis of thyroid cancer was similar in all children.

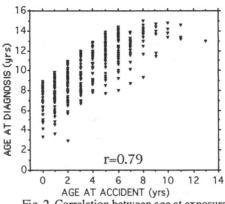

Fig. 2. Correlation between age at exposure
and age at diagnosis.

Lymphocytic infiltration

Lymphocytic infiltration was assessed in 52 Belarus tumors and was present in a total of 27 (50.9%) cases. In 5 cases it was defined as diffuse infiltration typical of lymphocytic thyroiditis, and as focal in the other cases.

As shown in Fig. 3, circulating anti-TPO autoantibodies were more frequently found in Belarus patients (44.4%) than in Italian patients (23.3%; p<0.01), while anti-Tg autoantibodies were equally found in both populations (11.7% and 9.6%, respectively).

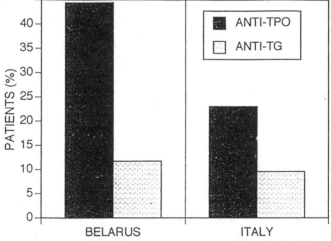

Fig.3   Incidence of anti-TPO and anti-TG autoantibodies in children with thyroid carcinoma from

Belarus (radiation exposed) compared to children from Italy (unexposed). The incidence of anti-

TPO, but not of anti-TG, antibodies is significantly higher (p<0.001) in Belarus children.

4.3. ONCOGENES in Belarus papillary thyroid carcinomas. (11,12)

Through different molecular techniques, six papillary thyroid carcinomas of children living in Belarus at the time of the Chernobyl accident were studied (These experiments were carried out in the laboratory of Dr. M. Pierotti, Istituto Nazionale Tumori, Milano, Italy). This was in order to identify tumor-specific gene rearrangements of the proto-oncogenes *RET* and *TRK*, previously found activated in papillary thyroid carcinoma. As shown in Table III, using the Southern blot analysis, in four cases we detected specific rearranged bands indicating an oncogenic activation of *RET* .

TABLE III. RET rearrangements in 6 Belarus children with papillary thyroid carcinoma after Chernobyl.

Patient/Sex	Age at accident	Age at diagnosis	Rearrangement
1/F	14 mo	6 yr	Ret/PTC3
2/F	7 yr	13 yr	None
3/F	8 yr	14 yr	None
4/M	5 yr	11	Ret/PTC2
5/F	8 yr	14	Ret.PTC3
6/F	7 yr	13	Ret/PTC3

The DNA of 3 tumors was also able to induce transformation of the NIH-3T3 cells after the DNA-mediated transfection assay, and the respective NIH-3T3 transfectants were found to express the oncogenic fusion transcripts. These results, although preliminary, support the possibility that the *RET* oncogenic activation could represent a major genetic lesion associated with radiation-induced thyroid carcinoma. The question of whether, and to what extent, young age *per se* rather than radiation, or the combination of both, is responsible for a *RET* activation remains to be established.

5. CONCLUSIONS

The JSP4 Project was mainly concerned with the clinical aspects of the post-Chernobyl childhood thyroid carcinoma. A protocol for diagnosis, treatment and follow-up has been devised. It is expected that the adoption of such protocol in its final version by all treatment Centers will greatly facilitate an optimisation of the management of thyroid cancer. Because of the limited experience acquired in this disease in most hospital Centers, it was important that the specific expertise of major European referral institutions could be made available to Belarus and other CIS countries in order to cope with the multiple and difficult aspects of the post-Chernobyl outbreak of childhood thyroid cancer. The program for training of CIS medical specialists and technicians in EU Centers has been successfully carried out and is expected to contribute to the effectiveness of the diagnostic and therapeutic procedures and to the standardisation of laboratory procedures. Clinical and molecular aspects of Belarus childhood thyroid carcinoma have been investigated and compared with unirradiated control populations.

The question of whether and to what extent the greater aggressiveness and the higher frequency of the RET oncogene activation observed in the post-Chernobyl thyroid cancer cases is related to radiation remains to be established. Furthermore, the impact of iodine deficiency on the development of the post-Chernobyl thyroid carcinoma is under evaluation. Iodine prophylaxis corrects iodine deficiency and thus may be considered an effective preventive measure for decreasing the risk of radiation-induced thyroid cancer.

REFERENCES

1. Ceccarelli C, Pacini F, Lippi F et al. Thyroid cancer in children and adolescents. Surgery 104: 1143-48, 1988.
2. Schlumberger M, de Vathaire F, Travaglia JP et al. Differentiated thyroid carcinoma in childhood. J. Clin. Endocrinol. Metab 65: 1088-94, 1987.
3. Treatment of thyroid cancer in childhood. Edit by Robbins J. Proceedings of a workshop held at the N.I.H, Bethesda, Maryland, September 10-11, 1992.
4. Baverstock K, Egloff B, Pinchera A et al. Thyroid cancer after Chernobyl. Nature 359:21-22, 1992.
5. Kazakov VS, Demidchlk EP, Astakhova LN. Thyroid cancer after Chernobyl. Nature 359:21, 1992.
6. Nagasaki Symposium on Chernobyl Update and Future. Edited by S. Nagataki; Excerpta Medica, Elsevier Science BV, 1994.
7. Thyroid cancer in children living near Chernobyl. Edited by D. Williams, A. Pinchera, A. Karaoglou, KH. Chadwick; CEC , Report EUR 15248 EN, 1993.
8. Furmanchuk AW, Averkin JI, Egloff B, Ruchti C, Abelin T, Schappi W, Korotkevich EA. Pathomorphological findings in thyroid cancers of children from the Republic of Belarus: a study of 86 cases occurring between 1986 ("post-Chernobyl") and 1991. Histopathology 21:401, 1992.
9. E.D. Williams, F. Pacini, A. Pinchera. Thyroid cancer following Chernobyl. J Endocrinol Invest. 18:144, 1995.
10. A. Pinchera, F. Pacini, ED Williams. Chernobyl in the future: European Community project. In "Nagasaki Symposium on Chernobyl: Uptade and Future", Nagataki S. (Ed), Elsevier Science B.V., 133-144, 1994.
11. Ito T, Seyama T, Iwamoto KS, et al. Activated RET oncogene in thyroid cancers of children from areas contaminated by Chernobyl accident. Lancet 344:259, 1994.
12. Fugazzola L, Pilotti S, Pinchera A, Vorontsova TV, Mondellini P, Bongarzone I, Greco A, Astakhova L, Butti MG, Demidchich EP, Pacini F, and Pierotti M. Oncogenic rearrangement of the RET proto-oncogene in papillary thyroid carcinomas from children exposed to the Chernobyl nuclear accident. Cancer Reserach, 1995 (in press).

Radioiodine Treatment in Children with Thyroid Cancer from Belarus

Chr. Reiners, J. Biko, L. Geworski, M. Olthoff, E. P. Demidchik,
C. Streffer, H. Paretzke, G. Voigt, Y. Kenigsberg, W. Bauer,
G. Heinemann, H. Pfob

Clinic and Policlinic for Nuclear Medicine University of Würzburg,
Centre for Thyroid Tumors Minsk,
Institute for Medical Radiation Biology University of Essen,
Institute for Radiation Protection GSF Munich-Neuherberg,
Institute for Radiation Medicine Minsk
Coordination Office Minsk

Abstract. Between 1st of April 1993 and 15th of November 1995, 95
children from Belarus with most advanced stages of thyroid cancer have
been treated totally 305 times with radioiodine in Germany. In spite of a
high frequency of advanced tumor stages pT4 (82 %), lymph node
metastases (95 %) and distant metastases (55 %) in those selected children,
the preliminary results of radioiodine treatment are promising. In 55 % of
the children complete remission and in 44 % partial remission of thyroid
cancer could be achieved. In no case progressive disease under treatment
has been observed.

1. Introduction

The frequency of thyroid cancer in children from Belarus is increasing since
1990. The relative incidence of thyroid cancer per 100.000 children below age of
15, which amounted to 0.1 - 0.3 between 1986 and 1989 increased to 1.2 - 3.5 between
1990 and 1994. In the region of Gomel, which has been most heavily contaminated
after the Chernobyl accident by radioactive fallout containing I-131 and short-
lived radioisotopes of Iodine, the relative incidence increased from 0.3 - 1.0
between 1986 and 1989 to 3.3 - 11.7 between 1990 and 1994. On the whole, 333 cases
of childhood thyroid cancer have been detected and operated by the Centre for
Thyroid Tumors in Minsk between 1986 and 1994. 54 % of those children lived in
the Gomel region. At the time of surgical intervention 78 cases had to be classified
in Minsk as stage pT1 (tumors of less than 1 cm of diameter), 98 as stage pT2
(between 1 and 4 cm of diameter), 4 as stage pT3 (more than 4 cm of diameter
without invasion of surrounding tissue) and 145 as the most advanced stage pT4
(tumors invading soft tissue surrounding the thyroid gland). In 72 % of tumors
stage pT1-2 and 84 % of tumors stage pT4 cancer growth were classified
histologically as multicentric. In 53 % of the cases stage pT1-2 and 81 % of patients
stage pT4 lymph node metastases have been observed during surgery. The relative
frequency of distant metastases to lungs diagnosed in Minsk postoperatively
amounted to 8 % in tumor stages pT1-2 and 23 % in tumor stage pT4.

Since 1991, many international organisations and institutions from abroad offered
their help to screen for thyroid cancer in children from Belarus and the Ukraine.
Those screening programmes now allow the early detection of thyroid cancers in
children living in regions affected by the reactor accident of Chernobyl. In
contrast to those screening projects, no comparable actions had been undertaken
before 1993 to offer international help for treatment of children with thyroid
cancer diagnosed after Chernobyl accident.

2. The Project „Scientists Help Chernobyl Children"

In April 1993, the German project „Scientists help Chernobyl Children" has been started by the Joint Committee for Radiation Research. This committee consists of 7 German Scientific Societies namely the Society for Physics, for Biophysics, for Medical Physics, for Nuclear Medicine, for Radiation Protection, the Röntgen-Society and the Association of Radiation Protection Physicians. The project is supported by the German Minister of the Environment, Radiation Protection and Reactor Safety as well as by the Minister of Health of Belarus. Generous funds for this project amounting to more than 3 Million German Marks have been raised from German Electricity Suppliers.

The project has been established in the framework of a bilateral cooperation between the Centre for Thyroid Tumors in Minsk and the University Clinic for Nuclear Medicine in Essen (later Würzburg), the Institute for Medical Radiation Biology of the University of Essen and the Institute for Radiation Protection in Munich-Neuherberg and the Institute for Radiation Medicine in Minsk. Surgical resection of thyroid tumors and removal of lymph nodes have been performed by the surgical team of the Centre for Thyroid Tumors in Minsk. Radioiodine treatment and staging with nuclear medicine procedures was the task for the Clinics for Nuclear Medicine in Essen and Würzburg respectively. The Institute of Medical Radiation Biology in Essen studied mutations of tumor suppressor gene p53 and the relative frequency of micronuclei for biological dosimetry. The Institute for Radiation Protection in Munich-Neuherberg and the Institute for Radiation Medicine in Minsk reconstructed thyroid doses retrospectively to correlate those with the frequency of thyroid cancer in children of Belarus. The results of these research projects will be reported elsewhere. The coordination office in Minsk organized - in close cooperation with the Centre for Thyroid Tumors - selection and transport of children for treatment in Germany.

In addition to those activities, training of 15 physicians and physicists of Belarus in Germany in the field of diagnosis and treatment of thyroid cancer with nuclear medicine procedures has been arranged in the framework of the project „Scientists Help Chernobyl Children". Theses activities have been carried out in close connection with training projects sponsored by the European Community.

3. Patients

Between the 1st of April 1993 and the 15th of November 1995 95 children from Belarus with most advanced stages of thyroid cancer have been selected for treatment in Germany. The children travelled by plane in groups of 4 and were accompanied by a physician from Minsk. They stayed for 1 week in the Clinic and Policlinic for Nuclear Medicine of the University of Essen (since the 1st of January 1995 of the University of Würzburg). Totally 309 courses of I-131 therapy have been carried out until November 15th, 1995.

58 % of the children originated from the Gomel region. Their age at the time of surgery ranged from 7 - 18 years with a mean age of 9.4 +/- 2.8 years. 59 % of the children were female, 41 % male. 98 % of the cancers were typed histologically as papillary thyroid and 2 % as follicular carcinomas. 82 % of the cases selected for treatment in Germany because of the aggressiveness of tumors had to be classified as stage pT4. In 95 % of the cases lymph node metastases and in 55 % of the children distant metastases had been detected. With the exception of one case, distant metastases were localized in the lungs. Nearly all of the cases with lung metastases presented as disseminated miliary spread, only one case showed localized multinodular lesions. In 25 of the 95 children radioiodine treatment had been performed with different activities in Minsk previously (mainly low activities up to 1 GBq), 18 of the children had been irradiated percutaneously with

mainly low radiation doses (up to 20 Gy). In 5 children chemotherapy with different drugs had been performed in Minsk.

4. Protocol

The diagnostic protocol included ultrasonography and scintigraphy of the neck, thorax X-ray, computer tests of pulmonary function, determinations of thyroglobulin, TSH, free T4 and free T3 in serum as well as measurements of Calcium, Phosphate and differential blood cell counts. Additionally, wholebody counter measurements of incorporated radionuclides and biological dosimetry using the micronucleus test have been performed on a routine base.

For treatment 50 MBq of I-131 per kg of bodyweight have been applied for the elimination of thyroid remnants. For ablation of metastases, 100 MBq of I-131 per kg of bodyweight were given. Simultaneously, antiemetica and emulsions for the protection of gastric mucosa were given to reduce gastrointestinal side effects. 2 days after treatment, replacement therapy with levothyroxine which had been withdrawn 4 weeks before treatment was restarted. The mean dose amounted to 2.5 µg of levothyroxine per kg of bodyweight. For staging, wholebody scans were performed 4 days after the application of radioiodine.

5. Results of Treatment

In 80 of 95 children more than one course of radioiodine treatment has been performed in Germany. In those cases, the results of treatment could be checked by follow-up with I-131 scintigraphy, ultrasonography of the neck, X-ray of the thorax and determinations of thyroglobulin in serum.

Table 1: Results of treatment in 80 children with thyroid cancer from Belarus treated more than once in Germany

N=80			CR	PR	NC	PD
pT_{1-3}	N_0	M_0	2	--	--	--
		M_1	1	1	--	--
	N_1	M_0	7	1	--	--
		M_1	2	1	--	--
pT_4	N_0	M_0	1	--	--	--
		M_1	--	--	--	--
	N_1	M_0	21	3	--	--
		M_1	10	27	3	--

CR=complete remission, PR=partial response, NC=no change, PD=progressive disease

In 44 out of 80 children (=55 %) complete remission of thyroid cancer could be achieved up to now. In 44 % we were able to recognize partial remission defined as decrease of tumor volume, tumor marker serum level or intensity of radioiodine uptake for at least 50 %. In additional 3 cases we saw a response to radioiodine treatment, however, the changes were not sufficient to be classified as partial remission so that we had to range them as „no change". Fortunately, in no case progressive disease has been observed. It is important to mention that the results given here are not the final results of treatment since in all of the cases without complete remission further courses of radioiodine are planned.

For explanation of the positive effects of radioiodine treatment on tumor activity, figure 1 shows the individual courses of thyroglobulin in serum. Thyroglobulin is a specific tumormarker for thyroid cancer, which has been measured before each therapeutic activity of I-131.

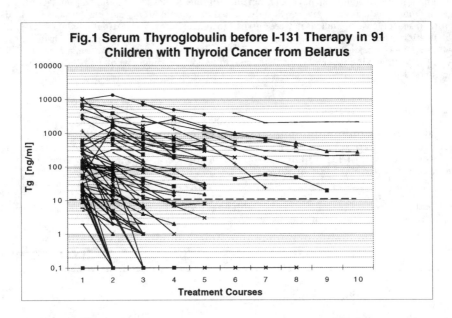

Fig.1 Serum Thyroglobulin before I-131 Therapy in 91 Children with Thyroid Cancer from Belarus

6. Discussion

According to data published by the Minister of Health of Belarus in 1993, children living in the Gomel region have received mean thyroid doses of 3.1 Gy after the Chernobyl accident (Kazakov et al. 1993). There is no doubt that exposure to radiation may induce thyroid cancer in children. According to a recent publication of Ron et al., „The thyroid gland in children has one of the highest risk coefficients of any organ and is the only with convincing evidence for risk at about 0.1 Gy". For childhood exposures, the authors specify the pooled excess relative risk per Gy to 7.7 (95 % confidence interval 2.1-28.7). The excess relative risk seems to be higher for females than for males. Linearity best described the dose response in children exposed to radiation before age of 15. Risk decreased significantly with increasing age at exposure, with little risk apparent after age of 20 (Ron et al. 1994).

The biggest series of thyroid cancer in young children below age of 16 has been studied at the Mayo-Clinic in Rochester, Minnesota (Moir and Telander 1994). According to the statistics of the Mayo-Clinic based on 58 patients (40 girls and 18 boys) only papillary cancer has been seen below age of 16. This corresponds well to our patient material where papillary cancer has been observed in 98 % of 95 children from Belarus. Surprisingly, the sex ratio in children from the Mayo-Clinic with a female predominance of 2.2 in considerably higher than the sex ratio in children from Belarus showing a female predominance of only 1.4. This observation may be explained by the fact that mean age is low in children from Belarus and that a female predominance is usually seen only after age of 10 (Nikiforov and Gnepp 1994).

The authors from the Mayo-Clinic report that the average tumor size with 3.0 cm for children was greater than with 2.1 cm for adults (Moir and Telander 1994). In their series, 18 % of the children had tumors with diameters greater than 4 cm. Unfortunately, the publication of Moir and Telander gives no data about local tumor invasion. Taking all the data from the Centre for Thyroid Tumors in Minsk collected in 333 children, tumors bigger than 4 cm and/or showing local invasion of tissue surrounding the thyroid gland have been observed in 54 % of the cases. This behaviour, which has been confirmed by independent pathomorphological reviewers (Nikoforov and Gnepp 1994), seems to be a sign of special aggressiveness in tumors diagnosed in children after the Chernobyl accident.

In contrast to this, the frequency of lymph node metastases in children from Belarus with 53 % in stage pT1-2 patients and 81 % in stage pT4 patients is not higher than in children from the Mayo-Clinic (reported as almost 90 %). A high incidence of lymph node metastases is typical for childhood papillary thyroid cancer (Moir and Telander 1994).

In 7 % of children from the Mayo-Clinic distant metastases have been reported. The frequency of distant metastases in children from Belarus with 8 % in stages pT1-2 and 23 % in stage pT4 seems to be higher. Taking this all together, thyroid cancers in children from Belarus perhaps are more aggressive because of frequent local invasion and higher frequency of distant metastases.

With respect to treatment, the authors of the Mayo-Clinic recommend near-total thyroidectomy when the tumor is widespread or multifocal as well as removal of lymph nodes in a node-picking procedure. In patients with (local or distant) metastases, radioiodine treatment should follow surgery. Afterwards sufficient thyroid hormone replacement to inhibit TSH-secretion has to be given. Only in cases of low-risk tumors of less than 2 cm in size that are not multifocal and have no evidence of cervikal lymph nodes and distant metastases a less aggressive approach seems to be justified (Moir and Telander 1994).

Prognosis of children with papillary carcinoma treated and followed at the Mayo-Clinic was excellent. Children generally had a better prognosis than adults. Of 95 children with papillary cancer who were followed at the Mayo-Clinic for 30 years, 7 % died as compared with 3 % expected in the general population. Even in the subgroup of children with pulmonary metastases, only 1 child (3 %) died of thyroid cancer (Moir and Telander 1994).

In conclusion, the aggressiveness of thyroid cancer in the subgroup of children from Belarus which has been selected for treatment in Germany justifies the aggressive treatment strategy which has been chosen. The preliminary results of treatment are promising. According to the literature, prognosis for those children is expected to be good.

References:

[1] Kazakov V S, Asthakova L N, Demidchik E P, Matyukhin V A, Kenigsberg Y M, Gavrilin Y I, Minenko V F, Drozd V M, Tochitskaya S I, Polynskaya O N, Khmara I M, Zelenko S M, Davydova E V, Dardynskaya I V, Markova S V, Dubovtsov A M: Characteristics of a Developing Thyroid Pathology in Children Exposed to the Effects of Radionuclides in Connection with the Accident in the Nuclear Power Plant Chernobyl. Workshop „Radiation Epidemiology after the Chernobyl Accident", Gesellschaft für Strahlen und Umweltschutz Neuherberg, 23.-25.10.1991. *BFS-Schriften 9/93, Grosche B, Burkart W (Hrsg.) München (1993) 129-146.*

[2] Moir C R, Telander R L: Papillary Carcinoma of the Thyroid in Children. *Seminars in Pediatric Surgery 3 (1994) 182-187.*

[3] Nikiforov Y, Gnepp D R: Pediatric Thyroid Cancer after the Chernoby Disaster, Pathomorphological Study of 84 cases (1991-1992) from the Republic of Belarus. *Cancer 74 (1994) 748-766.*

[4] Ron E, Lubin J H, Shore R E, Mabuchi K, Modan, B, Pottern L M, Schneider A B, Tucker M A, Boice Jr J D: Thyroid Cancer after Exposure to External Radiation: A Pooled Analysis of Seven Studies. *Radiation Research 141 (1995) 259-277.*

PERSPECTIVES OF DEVELOPMENT OF THYROID CANCERS IN BELARUS

J. Kenigsberg[a], E. Buglova[a], H.G. Paretzke[b], W. Heidenreich[b]

[a] Research Institute of Radiation Medicine, Masherov ave. 23, 220600 Minsk, Belarus
[b] GSF-Institut für Strahlenschutz, Ingolstädter Landstr. 1, 85758 Neuherberg, Germany

Abstract. This paper gives an overview on the total number of thyroid cancers observed in Belarus after the Chernobyl accident among children, discusses possible sources of the observed increase over expected cases and compares these observations with predictive calculations using different risk coefficients published in the literature. To this purpose exposure estimates of the thyroid are made for children living in three selected areas. Different radioecological, dosimetric and other reasons make it very difficult to obtain reliable dose estimates for these victims, and the use of published risk coefficients for the assessment of future developments of the thyroid cancer incidence rates results in predictions which do not agree too well with the observations.

1. Introduction

A significant increase of the incidence of thyroid cancer in the Belorussian population was registered after the Chernobyl accident, mainly among those exposed in childhood [1,2]. This has stimulated great scientific interest, because it is well known that ionizing radiation, as it was received by Belorussian population due to the Chernobyl accident, can induce thyroid cancer [3,4,5]. The thyroid gland of children is especially sensitive to the biological action of radiation exposure, and the thyroid doses among children were higher than among adults of Belarus. These may be two of the main reasons of especially high expression of thyroid cancer incidence among Belorussian children. There are several other causes that could have influenced the development of thyroid tumors in Belarus after the Chernobyl accident. To quantify the contribution of radiation exposure to the increased thyroid cancer incidence rates it was therefore necessary to study

- the relationships between radiation exposure of the thyroid gland in childhood and the probability of the subsequent development of a thyroid cancer,

- the role of screening - intensity, - technique, etc. carried out in the territory of Belarus, in the revealing of the thyroid cancer cases,

- the role of stable iodine deficiency in Belarus, which is an an endemic goiter area, in the development of radiation-induced cancer,

- the influence of stable iodine application during the acute phase of the accidental exposure on the development of thyroid cancer.

The purposes of this paper are a discussion of the possible causes of this increase in thyroid cancer incidence and a comparison between epidemiological observations of excess cancer cases in three selected areas in Belarus with calculations of predicted cases using published risk coefficients.

2. Thyroid Cancer Cases Observed in Belarus

From the time of the accident until the end of 1994 422 thyroid cancer cases were registered among children exposed under fourteen years of age in different regions of Belarus. 51 % of these cases were revealed among persons who were living in Gomel region during the accident, and more than a quarter of these 217 cases were registered among persons who were exposed in the city of Gomel. On the whole, 25% of all 422 cases registered during 1986-1994 were found among children who were exposed in the cities of Minsk, Gomel, Mogilev, Brest, Vitebsk and Grodno.

Only 22 % of the thyroid cancers occured in persons who were older than 14 years at the time of diagnosis. 62 % of all cases were observed among females. The ratio between cancer cases among females and to those among males is 1.64.

3. Thyroid Exposures in Selected Reference Areas

The analysis of the contribution of the radiation exposure to the increase of the thyroid cancer incidence requires data about thyroid doses for exposed populations. In this study the doses to the thyroid are estimated from the results of direct measurements carried out in May-July 1986 for persons living in the contaminated regions of Gomel, Mogilev and Brest and in the city of Minsk.This preliminary analysis shows that 2.3% of children exposed between 6 months and 2 years and 0.03% of adults might have received doses higher than 10Gy. A comparison of average thyroid doses calculated for adults and for persons aged under 18 years shows that doses received by children and juveniles are likely to have been 3-10 times higher than those for adults [6].
 Unfortunately, the poor quality of the input data for the dose reconstruction is responsible for a high level of uncertainty in the dose assessment.The derived values of calculated doses are strongly depending on the method of evaluation and on the assumed parameters. For the purpose of the present investigation we selected three so-called reference areas for which the

Table 1   Characteristics of the selected reference areas

Area	Distance from the Chernobyl station, km	Density of iodine 131 in soil, kBq per m^2	Collective dose, person-Gy
Minsk-city	320	185 - 370	37140
Gomel-city	120	1850 - 5550	57900
Hoiniki-district	<60	1850 - >37000	12210

values of the average thyroid doses were derived by Gavrilin et al. [6,7] on the basis of direct measurements (Table 1). These reference areas differ in several characteristics; they represent settlements at different distances from the nuclear power station and with different levels of doses received. In addition, they reflect two categories of exposed populations, namely urban and rural populations.

4. Observed Thyroid Cancer cases in the Reference Areas

After the accident in these reference areas a significant number of thyroid cancer cases were registered for people exposed in childhood (table 2).

Table 2   Thyroid cancer incidence rate for children exposed under 14 years
of age in the reference areas (cases/10^5 children)

Region	1986	1987	1988	1989	1990	1991	1992	1993	1994
Minsk-city	0	0	0.27	0	0.81	0.54	1.35	2.15	1.62
Gomel-city	0	1.73	0.86	0.86	2.59	14.68	8.61	7.77	15.5
Hoiniki-district	0	0	0	18.0	0	72.17	54.1	36.0	18.0

It is known that the thyroid cancer incidence rate among children and young adults was relatively small. According to WHO data, the incidence rate for children under 15 years before the Chernobyl accident was $0.1/10^5$ per year during 1983-1987, for young adults it was higher, namely $0.41/10^5$ per year.

These data show that a significant increase of the thyroid cancer incidence rate was observed in different regions and in all Belarus earlier in time after exposure than expected on the basis of some previous studies [3,4,5].

5. Possible Factors of Influence on the Reported Cancer Incidence Rates

As mentioned at the beginning of this report, the role of screening in the observed increase of the thyroid cancer incidence rate after the accident was evaluated. On the basis of investigations carried out in Belarus the following conclusions were obtained. Children and juveniles in the Ushachi district of the Vitebak region (which was almost not contaminated) and in the Hoiniki district of the Gomel region were examined in 1993 by groups of physicians with the same qualification of personal, the same equipment and the same method of investigation. In the Hoiniki district they found 4 thyroid cancer cases and none in the Ushachi district [9]. The role of screening in the increase of the revealing may therefore be estimated as small, because if there were no developed cancers, they were not revealed.

Another serious question is the potential influence of the traditional iodine deficiency in Belarus, particularly, in the Southern regions. Because of this, the uptake can be increased but the size of an affected gland may also be different, namely bigger than the average assumed in the dose assessment. As a result,the iodine deficiency may actually lead to small a decrease of calculated dose values. Besides that, the effectiveness of radioiodine in its carcinogenic action on the modified gland may be different from the expected level. The induction of thyroid cancers by radiation exposure may be changed by prior application of stable iodine. After the Chernobyl accident the application of stable iodine was not carried out in necessary terms. The evidence of this fact is confirmed by the differences in levels of thyroid doses particularly in regions near the nuclear power station. The question about the influence of stable iodine application after the first week after the accident on the thyroid cancer appearance is disputable now. 662 evacuated children were investigated, who were examined in Minsk clinics because of their high dose rate above the thyroid gland. 552 of these children received stable iodine with an average dosage of 2.89 g (range: 0.02 - 27 g). During 9 years after exposure no thyroid cancer cases occurred among these children. This study can be a first part of an evaluation of the influence of stable iodine application on the development of thyroid cancer. However, no final answer can be given already now.

6. Calculated Thyroid Cancer Incidence Rates

To investigate the ability of well-known published coefficients of radiation-induced thyroid cancer risk to calculate the observed incidence of thyroid cancer in Belarus, the observed and calculated thyroid cancer cases were compared for these selected reference areas.

According to the NCRP-model [8] we assumed that the latency period for thyroid cancer is 5 years and that the distribution of the cases during years at risk is practically the same. The first value of an absolute risk coefficient adopted here is 2.5 cases per 10^4 PYGy [8]; this value was derived from a combination of several studies. According to another risk estimation, which was carried out on the basis of seven previously published studies of thyroid cancer after acute external irradiation, the absolute risk coefficient might be 4.4 per 10^4 PYGy (95% CI = 1.9, 10.1) in this case [9]. These both values were used for comparison with the observed data; the results are shown in Table 3.

Table 3   Comparison between observed and predicted thyroid cancer cases after the Chernobyl accident in selected areas of Belarus

Area	Collective dose (person-Gy)	Observed cases for 1990-1994	Predicted cases for 1991-1994	
			Absolute risk = $2.5 \times 10^4 (PYGy)^{-1}$	Absolute risk = $4.4 \times 10^4 (PYGy)^{-1}$
Minsk-city	37140	21	37	65
Gomel-city	57900	54	58	102
Hoiniki-district	12210	20	12	22

The result of this comparison between observed and predicted number of cases using these two coefficients shows that for the Hoiniki district with high levels of average thyroid doses the number of observed cases is higher than predicted with the NCRP-risk coefficient and compares better to the higher value of ref. 9. For the cities of Minsk and Gomel the number of observed cases are less than predicted. The lack of regularity in the results of this comparison makes it difficult to draw general conclusions. However, it appears that the smaller risk coefficient of 2.5 x 10^4 $(PYGy)^{-1}$ leads to results which agree better with the observed numbers than the higher risk coefficient.

7. Conclusions

In general, it must be admitted that different radioecological, dosimetric and other reasons cause significant difficulties of correct prognosis for the thyroid cancer incidence rates in Belarus after the accident. Using published risk coefficients for these predictions leads to considerable uncertainties (at least about a factor of 2) in this prognosis.

References

[1]   V.S. Kasakov., E.P. Demidchik, L.N Astahova, Thyroid cancer after Chernobyl. Nature 359 (1992) 21-22.

[2]   Th. Abelin, Ju. Averkin, M. Egger, et al, Thyroid cancer in Belarus post-Chernobyl: Improved detection or increased incidence? Soz Praventivmed. 39 (1994) 189-197.

[3]   R.E. Shore, Issues and epidemiological evidence regarding radiation-induced thyroid cancer. Radiat. Res. 131 (1992) 98-111.

[4]   L.H. Hempelmann, W.J. Hall, M. Phillips, et al, Neoplasms in persons treated with x-rays in infancy: fourth survey in 20 years. J. Natl. Cancer Inst. 55 (1975). 519-530.

[5]   H.R. Maxon, E.L. Saenger, S.R. Thomas, et al, Clinically improtant radiation-associated thyroid disease. J.Am.Med.Assoc. 244 (1980) 1802-1805.

[6]   Ju. Gavrilin, V. Khrusch, V. Ivanov, et al, Particularities and results of determination of the internal radiation dose to the thyroid for the population of contaminated regions of Belarus. Vestn. Akad. Med. Nauk SSSR, 2 (1992) 35-43.

[7]   Ju. Gavrilin, V. Khruck, S. Shinkarev, Internal irradiation of the thyroid in the residents of regions contaminated with radionuclides in Belarus. Med. Radiol., 6 (1993) 15-20.

[8]   Induction of Thyroid Cancer by Ionising Radiation. Recommendation of the National Council on Radiation Protection and Measurement. NCRP Report - 80, Bethesda, 1985.

[9]   E. Ron, J.H. Lubin, R.E. Shore, et al. Thyroid cancer after exposure to external radiation: a pooled analysis of seven studies. Radiat. Res. 141 (1995) 259-277.

[10]   E.V. Davidova, L.N Astachova, S.V Gnipel, et al. Results of endocrinological screening of children and juveniles for Hoiniki and Ushachi district of Belarus. Proceedings of 4-th Republic conference. Part I. Mogilev, 1994, pp. 65-66.

III. HEALTH EFFECTS
FOLLOWING THE ACCIDENT
B. Thyroid cancer in children living near Chernobyl

Posters

THE PATHOLOGY OF CHILDHOOD THYROID CARCINOMA IN BELARUS.

E Cherstvoy[1], V Pozcharskaya[1], HR Harach[2], GA Thomas[2], ED Williams[2].
[1]Department of Pathology, Minsk State Medical Institute, Minsk, Belarus and
[2]Department of Histopathology, University of Cambridge, UK.

Abstract

We have studied data on the sex and age distribution of 293 cases of thyroid carcinoma in children operated in Belarus between January 1990 and December 1994. We have also reviewed the histology of 134 cases and performed immunocytochemistry for calcitonin, thyroglobulin, ret, met and p53 and in situ hybridisation for thyroglobulin and calcitonin on a sample of these cases. We have compared the data derived from this series with those obtained from a similar series of 122 cases operated in Kiev, Ukraine over the same time period and those from 154 cases operated in England and Wales over a 30 year period. There was agreement on the diagnosis of malignancy in 132 of the 134 Belarussian cases (98%). In 2 of the cases there was no evidence of malignancy in the material seen in Cambridge, but not all the original pathological material was available for review. In 7 cases there was evidence of malignancy, but inadequate material to determine the subtype of malignancy. The papillary carcinomas were classified as of the classic type when they showed a papillary architecture and the nuclear features typical of adult papillary carcinoma, or of the solid follicular type as described in the series studied in England and Wales (1). Four were papillary microcarcinomas. The age and sex distribution of all cases from Belarus showed a markedly different pattern from that observed in England and Wales,. In Belarus the peak was at age 9, while the England and Wales series showed a smooth rise in incidence with increasing age. Virtually all the cases from Belarus were papillary carcinoma (99%) compared with only 68% in England and Wales. In addition, there was a higher proportion of papillary carcinomas of the solid/follicular type (72% in Belarus, 35% in England and Wales). The frequency of this subtype did not change significantly with age in Belarus, whereas there was a relative decrease from 62% in the 0-9 year age group to 23% in the 10-14 year age group in England and Wales. The data from Belarus concerning sex, age and pathology of the tumours does not differ significantly from that obtained from a similar study carried out in Ukraine over a similar time period. The frequency of the solid/follicular subtype of papillary carcinoma may therefore be related to the aetiological agent involved. There is also a clear change in the age threshold for development of thyroid carcinoma over time, consistent with a causative agent at the time of the Chernobyl accident, and suggesting that the causative agent does not persist in the environment. These findings provide confirmation of the great and continuing increase in the incidence of thyroid carcinoma in children in Belarus, provide strong evidence that exposure to fallout from Chernobyl was the cause, and are compatible with radioisotopes of iodine as the main thyroid carcinogenic agent in fallout.

1:   Introduction

The increase in childhood thyroid cancer reported in Belarus following the Chernobyl accident (2,3) is unprecedented in size and the brevity of the latent period.

The increase started in 1990, 4 years after the accident, and in the 5 years from 1990 to 1994 inclusive 293 cases of thyroid cancer have been diagnosed in children under the age of 15 at the Pathology Institute in Minsk. The aim of this study was to determine whether the tumours that have occurred in Belarus post Chernobyl show any differences in age or sex incidence or in histological type when compared to tumours that have occurred in an area unaffected by fallout from Chernobyl. Diagnosis, treatment and follow-up of thyroid carcinoma arising in the 2,347,000 children in Belarus is centralised in Minsk, and all the pathology specimens are referred to the Pathology Institute for diagnosis. Data on age at operation, date of birth and diagnosis on all children operated in Minsk was made available, and a histological review was undertaken of 69 of the first 100 cases operated in 1990-1991, and 65 from the last 100 cases in 1993-1994. Data on a control non irradiated population (10,448,000 children) was obtained from the United Kingdom Childrens Cancer Registry which collects data on all UK childhood cancers. Sex and age data was available on the 154 childhood thyroid cancers reported in England and Wales in the period 1963-1992. Tissue for histological review was available on 81 of the 154 cases (53%). A more detailed report of the England and Wales series has already been published (1). Where sufficient material was available, serial sections from material from both series were taken for analysis of calcitonin, thyroglobulin, met, ret and p53 peptides by immunocytochemistry and calcitonin using commercially available antibodies and thyroglobulin mRNA by in situ hybridisation using digoxigenin labelled oligoprobes.

2: Results

On the material available, the diagnosis of malignancy was agreed in 75 (94%) of the England and Wales cases, and in 68 (99%) of the cases operated in Belarus between 1990-1991 and 64 (98%) of the 1993-1994 Belarussian cases. The lack of confirmation of the diagnosis does not mean that the original diagnosis was wrong, as in some cases not all the material on which the original diagnosis was made was available for review. Because of the very high level of agreement in both the early and late cases from Belarus, the sex and age distribution analysis has been based on all cases recorded in the Pathology Institute.

2.1: Age and sex distribution

The sex and age distribution was very different in the two series, with the England and Wales data showing a smooth increase with age from about 5 years up to 14, the limit of the study, while the Belarus cases showed a bell shaped curve. When all 293 Belarus cases were analysed there was a peak age at 9 with an overall mean of 10.41. The first 100 cases, showed a peak at 8 years with a mean of 9.1, while the last 100 cases showed a peak age of 9, and a mean age of 10.9 years (Figure 1).

The sex ratio of the two series was also different with an overall ratio of 1.7:1 (F:M) for Belarus and 2.3:1 (F:M) for England and Wales.

2.2: Histological type

A histological review was undertaken on 132 thyroid carcinomas from Belarus and 81 cases from England and Wales. In 7 of the Belarus cases there was insufficient material in the sections studied to confirm the histological type of tumour present, and in 6 cases from England and Wales malignancy was not confirmed on review. There were significant differences between the two series with respect to tumour type (see Table1). Papillary carcinomas comprised almost all of the Belarussian cases (99%), but only 68% of the England and Wales series. Only one of the Belarussian tumours reviewed was found to be a medullary carcinoma; only four follicular carcinomas were reported in the total 293 tumours; sections from these tumours were not included in those available for review.

Figure 1:

Comparison of the number of cases by age at operation of all thyroid carcinomas operated in Belarus in 1990-91 and 1993-1994

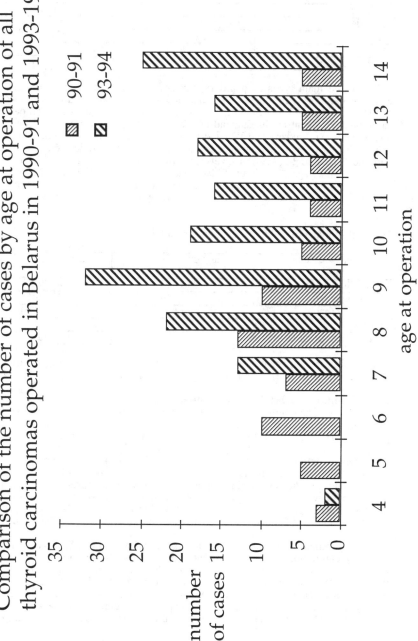

Table 1:
Review diagnosis of 75 carcinomas of childhood thyroid carcinoma from England and Wales and 125 carcinomas from Belarus

Histological diagnosis	England & Wales 1963-92	Belarus 1990-1994
Medullary Ca	12 (16%)	1 (1%)
Follicular Ca	8 (11%)	0
Papillary Ca	51 (68%)	124 (99%)
Others	4 (5%)	0
Total number of cases	75	125

Three main types of papillary carcinoma were present in both series - a classical papillary carcinoma, a solid/follicular variant and a diffuse sclerosing papilary carcinoma (Table 2). In Belarus these formed 14, 72 and 8% of all papillary carcinomas, by contrast in England and Wales the figures were 41, 35 and 10% respectively. In England and Wales there was a relatively greater frequency of the solid/follicular type in children aged 9 years and under at operation (62%) compared with those aged 10-14 (23%). However, the frequency of this subtype (72%) did not drop in the older age group in the Belarussian series. Only 4 of the Belarus papillary carcinomas studied showed the features of "occult" or microcarcinomas; the majority frequently showed direct invasion of adjacent muscle and connective tissue.

Table 2:
Different subtypes of papillary carcinoma in children.

Papillary Ca subtype	England & Wales 1963-92	Belarus 1990-1994
Solid/follicular	18 (35%)	89 (72%)
Classic type	21 (41%)	18 (14%)
Diffuse sclerosing	5 (10%)	10 (8%)
Others	7 (14%)	7 (6%)
Total number of cases	51	124

These results are consistent with another study comparing the sex and age distribution and pathological features of childhood thyroid cancer in the Ukraine post Chernobyl.

2.3: Immunocytochemical and in situ hybridisation studies
Calcitonin and thyroglobulin mRNA was localised on serial sections using digoxigenin labelled oligoprobes and a sensitive in situ hybridisation (ISH) protocol (4). Calcitonin, thyroglobulin, ret and met peptides were also localised using indirect peroxidase techniques and commercially available rabbit polyclonal antibodies. Control dilution profiles for each antibody or probe were carried out on appropriate human material and appropriate concentrations selected for this study.
Control sections were either hybridised with an inappropriate probe (for ISH) or incubated with an inappropriate antibody used at the same concentration for immunocytochemistry (ICC). Other controls included omission of probe or primary antibody. These techniques were carried out on all available tumours from both series.
All follicular and papillary carcinomas from England and Wales and all but 6 papillary carcinomas from Belarus were positive for thyroglobulin mRNA and peptide. In the papillary tumours studied there was considerable intercellular variation on both ISH and ICC; papillary tumours also showed decreased and a more heterogeneous expression of thyroglobulin mRNA relative to normal background thyroid (Figure 2). All follicular and papillary carcinomas were negative for calcitonin mRNA and peptide. Normal C cells where present in the background

thyroid were positive for both calcitonin peptide and mRNA. Both series showed greater than 80% positivity for both ret (Figure 3) and met protein in papillary carcinomas. The ret staining showed a characteristic dot like pattern at the base of the cell; in solid areas this altered to circumferential in some tumours. The met staining was generally diffuse, some membrane staining was also observed mainly at the periphery of the tumours. 11 (19.3%) of the Belarussian cases showed weak positivity in occassional nuclei for p53. No significant differences were observed between the results presented in this paper and those observed in a similar study of 104 Ukraine childhood papillary carcinomas.

3: Conclusions

The large increase in childhood papillary carcinoma in Belarus post Chernobyl reported in previous studies (5,6) has been confirmed and shown to be continuing in this study. All diagnoses have been confirmed by ISH and ICC for thyroglobulin and calcitonin. No major differences between the irradiated population from Belarus and the non radiation exposed population of children from England and Wales were observed in the proportion of papillary carcinomas positive for ret, met or p53. The Belarus cases were younger than the England and Wales cases, but show an increasing peak and mean age with time since the Chernobyl accident. The increase in childhood thyroid carcinoma in Belarus is confined to papillary carcinoma, and when compared with the England and Wales series there is a relatively greater frequency of the solid/follicular subtype, which is particularly marked in the older age group in Belarus. This suggests that development of this morphological subtype may be related to the causative aetiological agent.

The increase in the peak age of children with thyroid carcinoma with time suggests that the increase is related to the Chernobyl accident, and that the aetiological agent does not persist in the environment. These findings provide strong evidence that exposure to radioisotopes are the cause of the considerable increase in thyroid carcinoma in children from Belarus post Chernobyl.

4: References

1: Harach HR, Williams ED (1995) Br J Cancer 72: 777-783
2: Kazakov VS, Demidchik EP, Astakhova LN (1992) Nature 359: 21
3: Baverstock K, Egloff B, Pinchera A, Ruchti C, Williams D (1992) Nature 359: 21-22
4: Thomas GA, Davies HG, Williams ED (1993) J Clin Pathol 46: 171-174
5: Furmanchuk A W et al., (1992) Histopathology 21: 401-408
6: Nikiforov Y, Gnepp DR (1994) Cancer 74: 748-766

Legends to photomicrographs

Figure 2:

2A: Section from a solid/follicular papillary carcinoma in a female of 8. Intercellular variability of thyroglobulin mRNA expression is seen on in situ hybridisation.

2B: Semi serial section to 2A showing similar intercellular variability of thyroglobulin protein expression on immunocytochemistry.

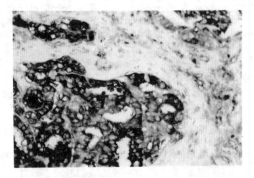

Figure 3:

Section from a solid/follicular papillary carcinoma from a male of 13. Dot like positivety for ret protein is observed in tumour cells on immunocytochemistry.

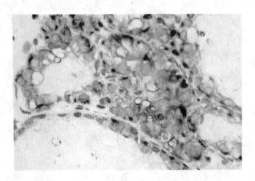

The Pathology of Thyroid Cancer in Ukraine Post Chernobyl

T Bogdanova[1], M Bragarnik[1], ND Tronko[1], HR Harach[2], GA Thomas[2], ED Williams[2],
[1]Institute of Endocrinology, Kiev, Ukraine and [2]Department of Histopathology,
University of Cambridge, Cambridge, UK CB2 2QQ

Abstract

We have analysed data on the sex and age distribution of 122 cases which have been operated at the Institute of Endocrinology and Metabolism in Kiev, Ukraine during the period January 1990 to December 1994 and compared these to information on 154 cases recorded by the UK Childhood Cancer Registry in England and Wales over the period 1963-1992. The histology has also been reviewed in 114 cases from Ukraine and in 81 cases in England and Wales. In addition immunocytochemistry for calcitonin,thyroglobulin, ret, met, IGF1 receptor and p53 and in situ hybridisation for thyroglobulin, calcitonin, and IGF1 mRNAs has been performed on a sample of cases from each of the two series. Our results show that there are clear differences between the sex and age distributions of the two series. In England and Wales there is a smooth rise with increasing age, but in Ukraine there was a peak incidence at eight years of age. The sex distribution was closer to equivalence in Ukraine then in England and Wales. The majority of thyroid carcinomas were papillary in type in both series, but Ukraine showed a higher frequency (96% compared with 68%). In addition, there was a particularly high incidence of the solid/follicular subtype of papillary carcinoma in children from Ukraine. There is a clear change in the age threshold for development of thyroid carcinoma over ime, consistent with a causative agent at the time of the Chernobyl accident, and suggesting that the causative agent does not persist in the environment. These findings provide strong evidence for exposure to radioisotopes of iodine as the cause of the considerable increase in the incidence of childhood thyroid cancer in the Ukraine.

1: Introduction

The accident at Chernobyl Nuclear Power Station took place in April 1986, and during 1990 a clear increase was noticed in the numbers of thyroid cancers in children from areas exposed to fallout in northern Ukraine (1,2) and southern Belarus (3,4). We have set out to compare some of the features of the tumours seen in the Ukraine with those of childhood thyroid carcinomas in the United Kingdom (5). The Ukraine cases were all seen at the Institute of Endocrinology in Kiev between 1990 and 1994 inclusive. The Institute is the major centre for Endocrinology in the northern half of Ukraine, it drains a population of approximately 30 million people, with approximately 6.5-7 million children.

Approximately 80% of the cases from the areas heavily contaminated by fallout from Chernobyl, a population with approximately 2 million children, have received treatment for thyroid problems in the Institute. The UK cases were derived from the UK Childhood Cancer Registry, information was obtained on all cases registered in England and Wales (population approximately 50 million, with 10 million children) during a 30 year period, pathological material was available on slightly over half of the cases. The comparison was carried out to determine whether the tumours that have occurred in the Ukraine in areas exposed to fallout from Chernobyl show any differences in age and sex incidence or in histological type from tumours that have occurred in areas remote from Chernobyl.

2: Results
2.1: Age and sex incidence

122 cases had been diagnosed as thyroid carcinomas in children under the age of 15 in the Institute of Endocrinology in Kiev between 1990 and 1994, while 154 cases had been

diagnosed as thyroid cancers in children under the age of 15 in England and Wales between 1963 and 1992 inclusive. The sex and age distribution was very different in the two series, with the England and Wales data showing a smooth increase with age from about 5 years up to 14, the limit of the study, while the Ukraine cases showed a peak incidence at the age of 8, with a subsequent decline in numbers (Fig. 1). The sex incidence also differed, being 1.3:1 F:M, for the Ukraine, and 2.3:1 for England and Wales. This was partly related to the differing age structure, as in children under 10 the difference was much less (1:1 Ukraine, 1.3:1 England and Wales). It was not entirely due to age difference, in the 13-14 year olds it was still present (1.5:1 Ukraine, 4.1:1 England and Wales).

2.2: Histological type

Material was available for study from 114 of the 122 thyroid carcinomas from Ukraine, and for 81 of the 154 thyroid carcinomas from England and Wales. The diagnosis of malignancy was agreed in 95% of all Ukraine cases and 94% of the England and Wales cases. Most of the small number of disagreements were due to lack of appropriate material or change in diagnostic criteria. The Ukraine cases were virtually all (96%) papillary carcinomas, while only 68% of the England and Wales cases were papillary carcinomas, most of the remainder were medullary or follicular carcinomas (Table 1).

Table 1:
Review diagnosis of 75 carcinomas of childhood thyroid carcinoma from England and Wales and 108 carcinomas from Ukraine

Histological diagnosis	England & Wales 1963-92	Ukraine 1990-1994
Medullary Ca	12 (16%)	2 (2%)
Follicular Ca	8 (11%)	2 (2%)
Papillary Ca	51 (68%)	104 (96%)
Others	4 (5%)	0
Total number of cases	75	108

Three main types of papillary carcinoma were present in both series - a classical papillary carcinoma, a solid/follicular variant, and diffuse sclerosing papillary carcinoma (Table 2). In England and Wales these formed 41, 35 and 10% of all papillary carcinomas, in contrast in the Ukraine the figures were 7, 76 and 9% respectively. The great excess of the solid/follicular type was in part age related; - they formed 62% of the papillary carcinomas in children from England and Wales under the age of 10, compared to 23% of older children, but there was no significant difference in frequency of the solid/follicular type with age in the Ukraine (80% in under 10's compared with 74% in those aged 10-14).

Table 2:
Different subtypes of papillary carcinoma in children.

Papillary Ca subtype	England & Wales 1963-92	Ukraine 1990-1994
Solid/follicular	18 (35%)	80 (76%)
Classic type	21 (41%)	7 (7%)
Diffuse sclerosing	5 (10%)	9 (9%)
Others	7 (14%)	8 (8%)
Total number of cases	51	104

The papillary carcinomas frequently showed direct invasion of adjacent muscle and connective tissue. The single papillary microcarcinoma is included in the category "others".

2.3: Immunocytochemical and in situ hybridisation studies

Calcitonin, thyroglobulin and IGF1 mRNA was localised on serial sections using digoxigenin labelled oligoprobes and a sensitive in situ hybridisation (ISH) protocol (6). Calcitonin, thyroglobulin, ret and met peptides were also localised using indirect peroxidase

Figure 1:

Comparison of the age and sex distribution of all differentiated thyroid cancer in children from England and Wales 1963-1992 with children from Ukraine 1990-1994

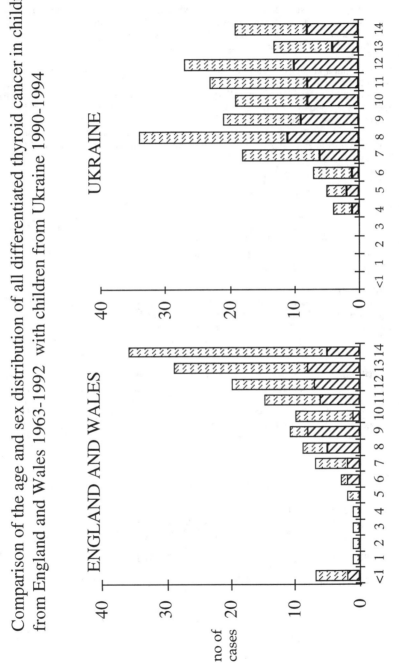

techniques and commercially available rabbit polyclonal antibodies. IGF1 receptor was localised using a mouse monoclonal antibody. Control dilution profiles for each antibody or probe were carried out on appropriate human material and appropriate concentrations selected for this study.

Control sections were either hybridised with an inappropriate probe (for ISH) or incubated with an inappropriate antibody used at the same concentration for immunocytochemistry (ICC). Other controls included omission of probe or primary antibody. The these techniques were carried out on all available tumours from both series.

All follicular and papillary carcinomas from both series were positive for thyroglobulin mRNA and peptide. In the papillary tumours studied there was considerable intercellular variation with both ISH and ICC; papillary tumours also showed decreased and a more heterogeneous expression of thyroglobulin mRNA relative to normal background thyroid where present in the section examined. All follicular and papillary carcinomas were negative for calcitonin mRNA and peptide, although some (8.7%) of the Ukrainian papillary carcinomas showed included normal C cells which were positive for calcitonin on both ISH and ICC (Fig 2). Forty-three percent showed positive C cells in the background normal thyroid. All medullary carcinomas were positive for calcitonin mRNA and peptide, but negative for thyroglobulin peptide and mRNA (Fig 3).

Over 80% of papillary carcinomas from both series were positive for ret expression by ICC with a similar figure for met expression. The ret staining showed a characteristic dot-like pattern at the base of the cell; in solid areas this pattern altered to circumferential membrane staining in some tumours (Fig 4). Met staining was diffuse; occasional membrane staining was observed. The majority of papillary cancers (90%) showed weak positivity for both IGF1 receptor protein and IGF1 mRNA (80%) in the majority of tumour cells, similar to that observed in adult papillary carcinomas (7).

3: Conclusion

The major increase in childhood thyroid cancer reported from the Ukraine in children exposed to fallout from Chernobyl has been documented, the tumours classified, and compared to childhood thyroid cancer cases from England and Wales.

All diagnoses were confirmed by ICC and ISH. No major differences between the two series were observed in the ICC or ISH studies, in particular the proportion of ret positive cases were approximately the same in both series.

The Ukraine cases are younger and more frequently male than the England and Wales cases. The Ukraine cases are nearly all papillary carcinoma; medullary or follicular carcinomas show little or no increase compared to the frequency found in the UK. In contrast papillary carcinoma is greatly increased in frequency in the Ukraine compared to the UK, particularly in the areas exposed to the highest fallout. The increase is most marked in the solid/follicular variant of papillary carcinoma, which may be specifically related to exposure to fallout.

4: References

1: Tronko N et al., (1994) In: Nagasaki Symposium on Chernobyl: Update and Future. Nagataki S (ed) Excerpta Medica Intern Congress Series 1074. Elsevier pp31-46
2: Likhtarev IA et al., (1995) Nature 375: 365
3: Kazakov VS, Demidchik EP, Astakhova LN (1992) Nature 359: 21
4: Baverstock K, Egloff B, Pinchera A, Ruchti C, Williams D (1992) Nature 359: 21-22
5: Harach HR, Williams ED (1995) Br J Cancer 72: 777-783
6: Thomas GA, Davies HG, Williams ED (1993) J Clin Pathol 46: 171-174
7: Takahashi MH, Thomas GA, Williams ED (1995) Br J Cancer 72: 813-817

Legends to photomicrographs

Figure 2:
Section from a diffuse sclerosing variant of papillary carcinoma from a male of 9. Included normal C cells, shown staining positive for calcitonin peptide on immunocytochemistry are found within the tumour.

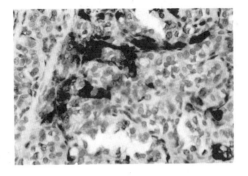

Figure 3:

3A: H and E stained section of a medullary carcinoma from a male of 9.

3B: Semi serial section showing strong positivity for calcitonin on in situ hybridisation

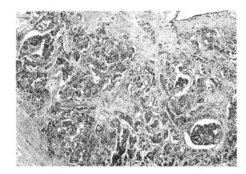 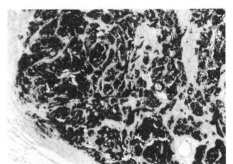

Figure 4:
Section from a solid/follicular papillary carcinoma in a male of 11 showing strong positivity for ret peptide on immunocytochemistry.

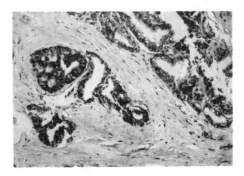

The Pathology of Childhood Thyroid Tumours in the Russian Federation after Chernobyl

A.Yu.ABROSIMOV

MRRC RAMC, Obninsk, Kaluga Oblast, Russian Federation

E.F.LUSHNIKOV
MRRC RAMS, Obninsk , Kaluga Oblast, Russian Federalion

AF.TSYB
MRRC R,AMS, Obninsk, Kaluga Oblast, Russian Federation

HR.HARACH
Department of Histopathology, University of Cambridge, UK

G.A.THOMAS
Department of Histopathology, University of Cambridge, UK

E. D. WILLIAMS
Department of Histopathology , University of Cambridge, UK

Abstract The histological verification of thyroid carcinoma that have occurred in children in the contaminated areas of the Russian Federation after Chernobyl has been performed by pathologists from Obninsk and Cambridge. Formalin fixed material and paraffin blocks of 10 cases of childhood thyroid cancer were received from different hospitals in Russia during 1993-1995. 4 of the cases were female, and 6 male. In one of these cases the material available in Cambridge unfortunately showed no tumour. Of the other 9 cases, all were papillary carcinomas. 5 showed the solid follicular pattern, predominant in younger children in the UK and forming the great majority of the recent childhood cases in both Belarus and the Ukraine. 2 were predominantly oxyphil carcinomas which were classified with papillary carcinomas on both architectural and cytological grounds, and 2 showed the features of the classic type of papillary carcinoma, predominant among the older children in the UK. All children came from areas contaminated by fallout from the Chernobyl accident, with 6 from Bryansk 1 from Kaluga and 3 from Tula. All cases were confirmed by immunohistochemistry and in situ hybridisation for thyroid differentiation markers. The oncogenes ret, met and p53 were also studied by immunohistochemistry.

1.   Introduction

Thyroid carcinoma is relatively rare in childhood [1]. However it is currently assuming greater importance, because of a greatly increased frequency in children exposed to fallout in the areas around Chernobyl [2-5]. We performed histological verification of 10 cases of thyroid carcinoma in children from contaminated areas of Russia and compared the histologic features with those from children in the UK.

2. Materials and methods

Paraffin blocks and 10% formalin fixed material of primary thyroid tumours, surrounding thyroid tissue and metastases to regional lymph nodes were requested from different hospitals in Russia. Histological sections of material available were examined by using routine haematoxylin-eosin staining, immunohistochemistry and in situ hybridisation to confirm the histological diagnosis. Primary antibodies for calcitonin and thyroglobulin were applied, (Dako anti-human calcitonin 1:2,000 dilution and Dako anti human thyroglobulin 1:10,000 dilution), followed by the indirect peroxidase technique. In situ hybridisation methods included the application of digoxigenin labelled probes for calcitonin (0.lng/ml) and thyroglobulin (0.2ng/ml). ret, met and p53 oncogene expression was also investigated by immunohistochemistry. Anti-bodies to ret and met (1:100 dilution) were followed by indirect peroxidase technique and p53 (1:200 dilution), followed by the avidin-biotin technique. All antibodies were from Santa Cruz Biotechnology.

Morphological verification of diagnosis was carried out in accordance with the criteria of the WHO classification of thyroid tumours [6].

3. Results

Among 10 patients 6 were male and 4 female. The age of the patients was 8, 9, 10 (2 cases), 11, 12 (2 cases), 13 and 14 years (2 cases). Papillary carcinomas were verified in 9 cases. In one case the material available unfortunately showed the surrounding thyroid tissue only. Five tumours showed the solid follicular pattern of papillary carcinoma. Two were predominantly oxyphil carcinomas. They were classified with papillary carcinomas on both architectural and cytological grounds. Two tumours showed the features of the classic type of papillary carcinoma; in one of these cases the histological feature of papillary microcarcinoma was found. The diameter of this tumour was less than 1 cm. Lymph node metastases were found in 7 cases. Histological conclusions were confirmed by immunohistochemistry and in situ hybridisation in all cases of papillary carcinomas. All 9 tumours showed follicular cell differentiation, with irregular accumulation of thyroglobulin in different parts of the tumour and irregular distribution of cells which show thyroglobulin synthesis in different tumour areas. Focal collections of C-cells were found in the surrounding thyroid tissue of 4 cases. Oncogene investigation showed expression of met oncogene by single and groups of carcinoma cells in 5 primary and 3 metastatic thyroid tumours. Only tumours with cells showing membrane staining for met were considered to be positive. Four tumours were positive by immunohistochemistry for ret protein. These tumours showed cells with either cytoplasmic or membrane positivity. No tumour showed positivity on immunohistochemistry for p53 protein.

4. Discussion

We have performed histological verification of cases of thyroid cancer that have occurred in children in the contaminated areas of Russian Federation after Chernobyl and confirmed the diagnosis of papillary thyroid carcinoma in 9 cases. In one case material available were presented only by surrounding thyroid tissue. The solid follicular pattern of thyroid carcinoma is predominant in younger children in the UK and in the great majority of the recent childhood cases in both Belarus and the Ukraine. This variant formed 55% of our cases. In two cases tumours were predominantly oxyphil cell in type. They were classified as papillary carcinomas on both architectural and cytological grounds which included intranuclear cytoplasmic invaginations, nuclear grooves and ground glass nuclei. Two cases showed the features of the classic type of papillary carcinoma, predominant among the older children in the UK. The age of the Russian children in these cases was 12. In one of these cases papillary microcarcinoma of thyroid was diagnosed. Its maximal size was less than 1 cm. The tumour was revealed by screening ultrasound examination.

Analysis of met oncogene expression showed positivity in more than half of all primary tumours and in one third of metastatic tumours. It is known [7], that the c-met proto-oncogene encodes a transmembrane tyrosine kinase, identified as the receptor for a Hepatocyte Growth Factor (HGF). HGF is a potent mitogen for epithelial cells and promotes cell motility and invasion. It is suggested [7], that the overexpression of c-met oncogene may play a role in the pathogenesis and progression of thyroid tumours derived from the follicular epithelium. Four tumours showed positivity for the ret oncogene. Translocation of this protooncogene has only been demonstrated in papillary thyroid carcinomas [8]. Further studies are needed to determine the frequency and type of ret translocation in these tumours.

References

[I]   H.RHarach and E.D.Williams, Childhood thyroid cancer in England and Wales, *British Journal of Cancer (1995,) 72, 777-783.*

[2]   Yu.Nikiforov and D R.Gnepp Paediatric Thyroid Cancer After the Chernobyl Disaster Pathomorphologic Study of 84 Cases (1991-1992) From the Republic of Belarus, Cancer (1994) 74, 748-766

[3]   A Furmanchuk et al., Pathomorphological findings in thyroid cancers of childrᴄ.. ᴦom the Republic of Belarus a study of 86 cases occurring between 1986 (post-Chernobyl} and 1991, *Histopathology* (1992) *21,* 401-408.

[4]   Yu.N Zumadzhi et al., Morphofunctional characteristics of thyroid malignant tumours in children from different areas of Ukraine affected by the Chernobyl accident after effects, *Archiv Patologii,* (1993) 55, 55-60.

[5]   Williams ED, Chernobyl, eight years on. *Nature,* (1994), *371* , 556.

[6]   C Hedinger et al., Histological Typing of Thyroid Tumours. In: International Histological Classification of Tumours, 2nd Edn, WHO Springer: Berlin, 1988

[7]   MF Di Renzo et al, Overexpression of the c-met oncogene in human thyroid carcinomas derived from the follicular epithelium, *Journal of Endocrinology Investigation,* (1994), 17, 75

[8]   M.Santoro et al, ret oncogene activation in human thyroid neoplasms is restricted to the papillary cancer subtype, *Journal of Clinical Investigation,* (1992) 89, 1517-1522.

The Molecular Biological Characteristics of Childhood Thyroid Carcinoma

Belarus: E Cherstvoy, A Nerovnya, Ukraine: L Voskoboinic, T Bogdanova, ND Tronko,

Brussels: M Tonnachera, JE Dumont, F Lamy

Munich: G Keller, J Boehm, H Hoefler

Naples: GC Vecchio, G Viglietto, G Chiappetta,

Cambridge: GH Williams, GA Thomas, ED Williams (Coordinator)

Collaborating Institutes: Institute of Endocrinology and Metabolism, Kiev, Ukraine, Department of Pathology, Minsk State Medical Institute, Minsk, Belarus, IRIHBN, Campus Hopital Erasme, Belgium, Brussels, Insitute of Pathology, Technical University, Munich, Germany, Department of Biology and Pathology, University of Naples, Naples National Tumour Institute, Italy, and Department of Histopathology, University of Cambridge, UK

Abstract:

We have used molecular biology to study mutation and expression of key oncogenes in childhood thyroid carcinomas from Belarus and Ukraine. All cases were histologically verified by two or more pathologists including at least one from the CIS and one from the EU. We chose to study six genes which have been shown to be involved in thyroid carcinogenesis in adults: ret, Ha, Ki and N ras genes, p53 and the TSH receptor. Expression of the ret oncogene, which has been shown to be activated by translocation in a proportion of papillary carcinomas has been studied by two independent methods. The first, used by the Cambridge group uses RT-PCR to identify the expression of the tyrosine kinase domain of the gene; as the gene is normally silent in follicular cells, this approach allows demonstration of activation of ret, but does not identify the particular translocation involved. The second approach, used by the Naples group, also uses RT-PCR, but amplifies across the breakpoint of each of the three translocations already identified to provide information on the proportion of tumours which express the individual translocations of this gene. Mutations in the TSH receptor, a key modulator of thyroid follicular growth have been sought by the Brussels group using SSCP and direct sequencing. The Munich group have analysed the samples for presence of mutation in p53, which is believed to play a role in genetic instability which is a features of carcinomas derived from may different tissues. Mutations in the common sites of the ras oncogenes have been studied by the Cambridge group.

Analysis of 26 papillary carcinomas so far studied has shown that mutations in the TSH receptor and in p53 do not play a significant role in the genesis of the tumours studied. The proportion of tumours showing ret expression does not differ significantly from that found in a control non exposed population from the UK. However, the pathological study shows that nearly all the increased number of thyroid carcinomas found in children exposed to fallout from Chernobyl are of the papillary type, and although the proportion showing ret activation does not increase, the numbers of tumours in which ret activation plays a role in carcinogenesis has very greatly increased. We therefore consider that radiation leading to ret translocation is a major feature in the increased number of childhood thyroid carcinomas in the population exposed to fallout from Chernobyl.

1: Introduction

The thyroid is an organ which is extremely sensitive to the carcinogenic effect of radiation. It has been known for many years that radiation causes an increase in thyroid cancer in animals (1) and that exposure to external radiation in childhood, usually as a result of clinical treatment increases the risk of developing thyroid cancer in later life (2). More recently, there has been a dramatic increase in juvenile thyroid cancer reported in the areas of Belarus and Ukraine exposed to fallout from the Chernobyl nuclear disaster (3-6). This may be related to the fact that the thyroid is the only organ in the body that actively takes up and then binds iodine. Large amounts of radioiodine were released during the Chernobyl accident. However, radioiodine is widely used in man for the treatment of Graves Disease, and it has not been shown to be carcinogenic to man in the doses used in adults.

There are three types of differentiated thyroid cancer, medullary carcinoma, derived from the minority C cell component of the thyroid, follicular and papillary carcinomas which both derive from the thyroid follicular cell. Thyroid follicular cells require iodine in order to synthesise the iodide containing hormones T3 and T4, which are essential for metabolic regulation in vertebrates. Although both follicular and papillary carcinomas increase in frequency after radiation exposure, it is papillary carcinoma which shows a greater relative increase (7). There is little evidence to suggest that medullary carcinoma is associated with exposure to radiation.

Molecular biological studies have recently suggested that papillary and follicular carcinomas in humans show different oncogene involvement; ras genes being more frequently mutated in follicular carcinoma (8) and translocations of the ret and trk oncogenes being more frequent in papillary carcinomas (9). So far three different translocations of the ret oncogene have been identified, two intrachrosomal and one interchromosomal. Interestingly the ret oncogene has been shown to be frequently mutated, rather than translocated, in medullary carcinoma (10).

2: Material studied

Sections of paraffin embedded material were obtained from 26 cases of thyroid cancer from children under 15 at the time of operation from Ukraine and 50 cases from Belarus. All cases were histologically verified and all except one (a medullary carcinoma) were papillary carcinoma. Sections were circulated to each of the collaborating centres for analysis.

3: Results
3.1: Ras gene mutation

The three ras genes have been studied in 14 papillary carcinomas by PCR and direct sequencing. No mutations have been observed in the commonly mutated codons (12,13 and 61). Similar studies carried out on a control group of childhood thyroid papillary carcinomas from England and Wales also showed no mutation at these sites. However, mutations of these genes have been identified in 3 of 10 adult follicular carcinomas using the same approach. This suggests that ras gene mutation is probably not normally involved in the genesis of thyroid papillary carcinomas, whether induced by radiation or not.

3.2: Ret gene expression

Expression of the ret gene has been studied using RT-nPCR for a 90 base pair sequence within the tyrosine kinase domain and direct sequencing. Ret expression was identified in 6 of 18 (33%) Chernobyl associated childhood papillary carcinomas so far studied. This was a significantly lower frequency than that found in a study which used the same technique on adult papillary carcinomas (11), and was also lower than the frequency observed in 20 childhood papillary carcinomas from

England and Wales. However, due to the smaller number of carcinomas from children so far studied, we are not yet able to say whether the irradiated series shows a significant reduction in the frequency of ret expression. A small study carried out on frozen material from the Ukraine showed ret expression in a similar proportion (3/11) of childhood thyroid papillary carcinomas, suggesting that the low frequency found in the paraffin embedded material was not due to a decreased sensitivity of the system. There does not appear to be a correlation between the expression of the ret oncogene as observed by RT-nPCR analysis and morphological subtype of papillary carcinoma. Positivity for actin amplification was used as a control for quality of the RNA extracted from the sections.

3.3: Ret gene translocation

Twenty five cases of childhood thyroid carcinomas have so far been studied; only 11 yielded sufficient RNA for further analysis. Two controls for RNA quality were used: amplification of actin mRNA by RT-PCR and Northern blot with an 18S RNA probe. Using primers which allow detection of the three individual translocations of the ret oncogene so far identified, three papillary carcinomas with ret translocation have so far been identified by PCR and Southern blotting. All 11 cases which provided sufficient RNA have been analysed for the PTC1 translocation; only one case was positive. Two of the 7 cases so far analysed have been found to be positive for the PTC3 translocation. Analysis of the presence of PTC1 in a small series of adenomas from the Ukraine has also been carried out and shown to be absent. Interestingly one of the carcinomas found to be positive for ret expression by RT-PCR was not found to possess one of the 3 known translocations. Further studies on the remaining papillary carcinomas for PTC2 and PTC3 expression and to identify other translocations involving the ret oncogene are underway.

3.4: Mutations in the TSH receptor

DNA extraction has been performed on material from 41 cases and regions of interest in exon 10 of the TSH receptor have been amplified using PCR. The rgions studied include the third intracellular loop abd the third transmembrane segment of the TSH receptor gene. Single stranded conformational polymorphism (SSCP) in all cases followed by sequencing (in 15 papillary carcinomas) has been used to identify TSH receptor mutations in exon 10. So far no mutations have been observed in 41 cases of papillary thyroid carcinoma or in 18 follicular adenomas and 3 follicular carcinomas from children and adolescents from the radiation exposed population. Cases known to be positive for mutations in exon 10 have been used as control for the technique used. The absence of mutations in the TSH receptor correlates with the morphological observations that the childhood thyroid tumours identified in the Ukraine after Chernobyl are of the papillary subtype.

3.5: p53 mutation

Exons 5 and 7 and 8 have been successfully amplified from 23 papillary carcinomas using nested PCR. SSCP analysis under four different running conditions has so far been applied to all 23 samples, but no aberrant cases have been found. p53 mutation involvement in thyroid carcinoma is usually a late phenomenon, at the interface between differentiated and undifferentiated carcinoma. All of the tumours so far received have been well differentiated and do not show widespread positivity for p53 on immunocytochemistry. However, exposure to radiation has been reported to increase the frequency of mutation in the p53 gene in other tissues. From the results presented here that p53 mutation does not appear to play a major role in papillary thyroid carcinogenesis post Chernobyl.

4: Conclusions

The results obtained from this study so far suggest that contrary to a previous report (10) there is no increase in the frequency of ret translocation in thyroid carcinomas post Chernobyl. However, we cannot yet exclude the possibility that there may be a difference in the frequency of the three known translocations of ret when compared to the data from an adult population. Whether any such difference is due to radiation or is due to a differing frequency in children requires study of the proportion of PTC 1, 2 and 3 positive tumours in a non radiation exposed population of children. Previous reports had also suggested that there was an increased frequency of mutation in Ki ras in human follicular thyroid carcinomas following radiation (13). However, we have been unable to demonstrate any Ki ras mutations in papillary carcinomas whether from children exposed to radiation or not. Other work has failed to find any ras mutation in adult papillary carcinomas (8). The papillary carcinomas we have studied showed evidence of aggressivity, but we were also unable to demonstrate mutations in N ras. N ras mutations have recently been shown to be more common in aggressive thyroid tumours (14). The lack of mutations in the TSH receptor is related to the fact that the tumours examined so far have all been papillary carcinomas, and the lack of p53 mutations is not surprising as all tumours analysed so far have been well differentiated. p53 immunopositivity has been noted only in occasional nuclei in sections from the same tumours suggesting that alterations in p53 gene expression have not played a major role in the development of these tumours.

The molecular biological results so far obtained provide additional evidence to support the view that ret translocations are one of the key events in papillary carcinogenesis, while ras mutations are one of the key events in follicular carcinogenesis. The post Chernobyl tumours do not appear to differ significantly in their pattern of mutations from a non radiation exposed comparison group of children, although the frequency of different types of ret translocation have not yet been adequately studied. Overall there has been a very great increase in the frequency of childhood thyroid cancer in the children exposed to fallout from Chernobyl, this increase has been very specifically in the papillary subgroup of thyroid tumours, and the oncogene changes have been those found in papillary tumours from unirradiated patients, not those found in follicular carcinoma whether associated with radiation exposure or not. They suggest therefore that exposure to radiation has led to a great increased in tumours showing a ret translocation, and also a great increase in the ret negative papillary carcinomas, where the molecular biological changes have yet to be characterised.

5:   References
1: Doniach I (1974) Br J Cancer 30: 487-495
2: Hemplemann LH (1968) Science 160: 159-163
3: Baverstock K,et al., (1992) Nature 359: 21-22
4: Furmanchuk AW et al.,(1992) Histopathology 21: 401-408
5: Tronko N et al., (1994) In: Nagasaki Symposium on Chernobyl: Update and Future. Nagataki S (ed) Excerpta Medica Intern Congress Series 1074. Elsevier pp31-46
6: Likhtarev IA et al., (1995) Nature 375: 365
7: Shore RE (1992) Radiation Res 131: 98-111
8: Manenti G et al., (1994) Eur J cancer 30A 987-993
9: Santoro M et al., (1993) Br J Cancer 68: 460-464
10: Eng C et al., (1994) Human Mol Genet 3: 237-241
11: Williams GH et al., (1995) submitted
13: Ito T et al., (1994) Lancet 344: 259
14: Wright PA et al., (1991) Oncogene 6: 471-473
15: Hara H et al., (1994) Surgery 116: 1010-1016

Analysis of the Results on the Use of Frozen Sections and Smears–Imprints in Express Diagnosis of Tumoral Processes in Children and Adolescents in Ukraine After the Chernobyl Accident

V.G.KOZYRITSKY, T.I.BOGDANOVA, L.G.VOSKOBOINIK,
M.N.BRAGARNIK, I.V.KOMISSARENKO, S.I.RYBAKOV,
A.Ye.KOVALENKO, N.P.DEMCHENKO, N.D.TRONKO
Institute of Endocrinology & Metabolism, Academy of Medical Sciences of Ukraine
Vyshgorodskaya Str. 69, Kiev 254114, Ukraine

Abstract. It has been carried out a comparative analysis of the conformity of the results of instant diagnosis made on frozen sections (152 biopsies), with the final conclusion on fixed preparations of thyroid tumors in children and adolescents of Ukraine. The bening nodular and multinodular hyperplastic processes on frozen slices are verified in 100% of observations (48 cases). In case of follicular adenoma the coincidence in diagnosis made 90.1% (55 cases), in case of papillary carcinoma it made 91.1% (45 cases). In case of follicular (3 cases) and medullary (1 case) thyroid carcinomas, presence of carcinoma was suspected but the accurate diagnosis was made on fixed preparations. Thus, instant diagnosis on frozen sections of thyroid tumors allows, in more than 90%, to make the accurate diagnosis of a tumoral process in the course of operation and to choose adequate tactics of surgical treatment.

The increase in thyroid cancer cases among children and adolescents in Ukraine after the Chernobyl accident makes it necessary to carry out morphological studies of thyroid nodules at three stages: preoperatively (cytological study of the puncture material with fine–needle aspiration biopsy), intraoperatively (express diagnosis using frozen sections and smears–imprints) and postoperatively (histological study of paraffin–embedded sections). Among the methods mentioned express diagnosis using frozen sections plays a special role since it helps surgeons to choose a proper surgical treatment and define the extent of surgery at the time of operation. Modern equipment provided to us by the European Commission within the International Scientific Project ECP–8 as well as methods of accelerated specimen staining allow a conclusion on the form of pathological process in the thyroid gland within 15 minutes after the biopsy material has been taken to the laboratory of morphology.

Analysis of the material that required intraoperative verification of the diagnosis in 1994–1995 (152 biopsies by 1.07.95) showed that benign nodular and multinodular hyperplastic processes in children and adolescents (48 cases) are verified on frozen sections in 100% of cases. In multinodular thyroid lesions sections from all nodules should be studied to exclude the presence of carcinoma in any of them. The histologic structure of benign nodes may be different — microfollicular, normo-follicular or mixed. A part of nodes may undergo a cystic degeneration or massive fibrotization of central part. Some nodes in multinodular lesion of thyroid gland have a capsule of irregular thickness and resemble adenomas as to their structure. In such cases, in our opinion the histologic conclusion must reflect this feature of their structure and be called "multinodular adenomatous goiter" of thyroid gland. In the case of folicular adenoma (55) intraoperative and postoperative diagnosis coincided in 90.1% of cases. In 9.9% of instances the conclusion was indefinite because solitary encapsulated tumors often require an additional study of subcapsular areas and capsule in paraffin—embedded sections to exclude minimally invasive follicular carcinoma or follicular variant of papillary carcinoma. In the latter case it is advisable to perform an additional analysis of cytological characteristics of tumor cell nuclei using smears—imprints. The main nuclear peculiarities in neoplastic cells are: their large size, enlightenment and dust—like chromatin. Nuclear polymorphism is moderate. Intranuclear cytoplasm inclusions were found in more than half bioptates under study. The major peculiarities of tumor cell nuclei in papillary carcinoma are low optical density (clear nuclei), dust—like chromatin, grooving and intranuclear cytoplasmic inclusions [1].

A pathologist who studies follicular adenoma (and other thyroid lesions) on frozen (and paraffin—embeded) sections should know whether fine—needle aspiration biopsy was performed because this assay can imitate capsular invasion and cause false diagnosis of "minimally invasive follicular carcinoma" [2]. The main difference between "puncture" capsular invasion and the true one is presence of hemorrhages in the parenchyma of the gland and inside seperate follicles at the site of puncture, existence of inflammatory reaction at the site of cell entry as well as normal cell structure.

Follicular carcinoma causes particular difficulty when the diagnosis is made on frozen sections. As a rule it is a solitary nodule in dense capsule of varions thickness showing follicular pattern. Usually in children and adolescents minimally invasive follicular carcinoma is diagnosed and brings about additional difficulties with its identification. The probability to find capsular or vascular invasion on 3—4 frozen sections during express—biopsy is very low, that is why often the diagnosis was rather ambiguous being "follicular neoplasm" [2]. Such a diagnosis, to a certain extent, made surgeons to perform a more radical operation than that performed for follicular adenoma. In children follicular carcinomas are extremely uncommon (3 cases over 1.5 years), that is why certain difficulties in diagnosis of this pathology in express diagnosis cannot diminish the value of this method in differential diagnosis of neoplastic processes in the thyroid gland.

In papillary carcinoma (45) the diagnoses coincided in 91.1% of cases, in 2.2% of instances (one case of encapsulated follicular variant of papillary carcinoma) the result was false—negative and in 6.7% of cases carcinoma was suspected but the precise diagnosis was made on paraffin—embedded specimens. Unfortunately, one cannot rely upon nuclear cytologic feature in frozen sections since the majority of nuclei often demonstrate swelling, become clear and contain false (artefact) intranuclear inclusion. In this respect imprint smears are very helpful preserving nuclear cytologic features of papillary carcinoma. On frozen sections papillary carcinoma is diagnosed by the presence of true papillary structures with prominent fibrovascular core, tumor invasion to the adjacent thyroid tissue as well as tumor cell invasion to the lymphatic and blood vessels and the capsule. Sometimes transcapsular invasion forming mushroom—like structures or tumor cell spreading over the adjacent soft tissues is observed. Finding of psammoma bodies on the frozen sections of the thyroid gland is an important diagnostic sign. Psammoma bodies can be located among tumor cells, within papillary structures, in the stroma of the tumor and in the lumina of lymphatic vessels. Psammoma bodies shoud be distinguished from calcificates which may be found in other thyroid lesions as well, benign ones included. Usually psammoma bodies demonstrate regular rounded shape and laminated structure which is well seen at the extreme low position of the microscope condenser. The amount of psammoma bodies in different papillary carcinomas vary considerably: commonly, there are single instances in follicular variant of papillary carcinoma and they are numerous in diffuse sclerosing variant. It shoud be mentioned that psammoma bodies may be absent in some papillary carcinomas (at least on the tumor sections available) but they may be found in regional lymph node metastases.

Anaplastic carcinoma (1) is easily diagnosed on frozen sections by the growth of small roundish or angular cells with a dense round nucleus, by marked invasion into blood vessels and capsule, by presence of mitosis, invasion of capsule, presence of metastases into adjacent soft tissues. The anaplastic carcinoma must be differentiated from medullary carcinoma which is noted more often in children and adolescents. The latter is usually represented by spindle cells gathered in isolated small islets surrounded with thick layers of connective tissue [3]. Mitosis in medullary carcinoma are rarely determined, invasion is noted into blood vessels and capsule.

Further accumulation of the material and its analysis will make it possible to determine more exactly diagnostic value of the signs mentioned and to unify diagnostic approaches so as to reduce the percentage of errors in preoperative diagnosis.

Thus it is concluded that the study of frozen sections and smears is an important additional method of differential dignosis of tumoral processes in the thyroid gland. It allows morphological diagnosis in more than 90% of cases at surgery as well as selection of the adequate tactics of surgical treatment.

References

[1] S.R.Kini, Thyroid. Guides to Clinical Aspiration Biopsy. Igaku—Shoin, N.Y.—Tokyo, 1989, 362 p.
[2] V.A.LiVolsi, Surgical Pathology of the Thyroid. WB Saunders, Philadelphia, 1990.
[3] J.Rosai, M.L.Carcangiu, R.Delellis, Atlas of Tumor Pathology. Tumors of the Thyroid Gland. Armed Forces Institute of Pathology, Washington, 1992.

Comparative Light— and Electromicroscopic Characteristics of Thyroid Carcinoma in Children and Adolescents in Ukraine Following the Chernobyl Accident

T.I.BOGDANOVA, V.G.KOZYRITSKY, N.D.TRONKO,
G.V.PETROVA, I.L.AVETESYAN
Institute of Endocrinology & Metabolism, Academy of Medical Sciences of Ukraine
Vyshgorodskaya Str. 69, Kiev 254114, Ukraine

Abstract. 190 thyroid carcinomas in children aged up to 15 (154 cases) and adolescents aged 15 to 18 (36 cases) operated at the Institute's Clinic from 1986 to the 30th of June 1995, have been studied using light and electron microscopy. It has been found in 93.2% papillary, in 3.2% — medullary, in 1% — anaplastic carcinomas. A typical papillary carcinoma was revealed in 11.5%, follicular variant — in 39.0%, solid variant — in 28.1%, diffuse and sclerosing variant — in 3.8%. In cases of solid variant low—differentiated cells prevailed in the tumor, what manifested itself the most obviously by electron micro-scopic analysis. The thyroid carcinomas studied in children and adolescents of Ukraine are characterized by high invasive properties, that is confirmed by a high percentage (66.5%) of regional metastases.

One hundred and ninety thyroid carcinomas removed from children under 15 years and adolescents aged 15—18 years at the clinic of the Institute of Endocrinology and Metabolism of the Academy of Medical Sciences of Ukraine following the Chernobyl accident (1986—30.06.1995) were studied morphologically. Of them, 154 carcinomas were resected from children, 36 cases from adolescents. The diagnoses were verified according to the WHO classification [1]. One hundred and thirty carcinomas were additionally verified by the international experts. In 98.5% of instances the diagnoses were confirmed. In 177 cases (93.2%) the experts have found papillary thyroid carcinoma, in six cases (3.2%) follicular, in four cases (2.1%) medullary, in two cases (1%) anaplastic thyroid carcinoma and in one case (0.5%) malignant lymphoma was revealed. It is obvious that excluding a small number of cases thyroid carcinomas in Ukrainian children and adolescents after the Chernobyl accident display mainly papillary forms.

As histopathological analysis of papillary carcinomas has shown non—encapsulated tumors prevail (91.2%). The capsular invasion was registered in 50.0% of cases and infiltration of the surrounding soft tissues in 37.4% of cases. In 54.9% signs of intraglandular spreading were observed when tumor sites or psammoma bodies are found in relatively unchanged tissue. Classical papillary carcinoma in children and adolescents was found in 11.5% of cases. It consists of numerous characteristic papillae. Papillary structures in the tumors studied were characterized by typical clear nuclei, usually oval or rounded. Nuclear chromatin was located mainly in the periphery and as a rule, was absent in the central part, this gave the nucleus the "ground–glass" appearance. In addition, the papillary carcinoma cells were characterized by grooves, intranuclear inclusions and evidence of nuclear overlapping. Electromicroscopically in papillary areas of carcinomas studied well differentiated thyrocytes were observed. They differed in the degree of functional activity. Cells with well developed cytoplasmic organellas prevailed showing granular endoplasmic reticulum, mitochondria, and the Golgi complex. In the apical part of such cells a large amount of secretory granules is accumulated, numerous microfilli protrude into the follicle lumina. Desmosomal contacts are most common among intercellular junctures. There are also dense, low–active thyrocytes with unevenly broadened (even lacunae are formed) small canals of granular endoplasmic reticulum and pycnomorphous perishing cells in the state of desquamation. Nuclei of tumor cells have markedly uneven, cut nucleolemma outline. The nuclei of the tumor cells were indented with low heterochromatin content and cytoplasmic "inclusions" into nucleoplasm, which is typical for this carcinoma pattern. These "inclusions" are not true intranuclear inclusions, but just reflect deep and complex nucleolemmal invaginations. Cytoplasmic inclusions are often represented by so–called M–bodies [2]. Their presence proves distinct malignant character. Changes in the nucleus are very important diagnostic criteria of papillary carcinoma for cytologists and pathologists when studying the specimens with light microscope [3,4].

In the majority of tumors studied (39.0%) follicular pattern was predominant. Nuclei in follicular areas are also clear and poor with chromatin. Typical papillary structures were not numerous or absent. The main distinctive features of such carcinoma are cytologic peculiarities in the nucleus of tumor cell [4].

In solid variant (28.1%) the alveolar–solid pattern prevailed, and tumor sites were divided by connective tissue. Electromicroscopy revealed solid areas in the parenchyma consisting of poorly differentiated cells, when solid variants of papillary carcinoma were studied. Special attention was paid to details such as: smaller size, changed forms, presence of processes, and a decrease of organoid content [2,5]. Often only separate small canals of granulated endoplasmic reticulum, flocks of ribosomes, mitochondria and large vacuoles were observed, findings that typical of the early stages of embryogenesis of endocrine glands. It should be pointed out that clusters of poorly differentiated cells were detected in the so–called uninvolved part of the thyroid gland thus indicating multifocal tumor growth. The cells may adjoin each other or may be separated by wide lacuna–like intercellular spaces which is a typical sign of small–cell carcinoma simplex. The remaining desmosomal contacts (though not numerous) allow to differ these cells from transformed lymphocytes on ultrastructural level, i.e. to make differential diagnosis of poorly differentiated thyroid tumor and malignant lymphoma [5,6]. In the described solid sites preformed Askanazy–Hurthle cells are revealed. The latter may be di– and multinuclear with distinct nucleolemma invaginations. Mitochondria in the cytoplasm are numerous, as usually, but polymorphism in their form and size, matrix osmophilia, cryst content is observed.

Diffuse sclerosing variant was observed in 3.8% and in 17.6% tumors showed combined papillary—follicular—solid pattern and characterized by diffuse tumor growth, signs of sclerotic manifestations, lymphoid infiltration, prominent invasion of the tumor cells in the lymphatic vessels (66.5% of cases) and a large number of psammoma bodies. In 23.6% of cases invasion in blood vessels was demonstrated. In carcinomas from children under 15 these figures were higher: 79% for invasion in lymphatic vessels and 38.8% for invasion in blood vessels. Ultrastructure of the tumor vessels is characterized by edematous endothelium protruding into the lumina and by reduced micropinocytotic activity on a background of thickened basement membranes. Such changes in the components of the vascular wall were described for radiation exposure [7]. From our point of view, ultrastructural criteria of the initial stages of metastatic spreading of tumor cells is most interesting. These tumor cells lose their intercellular bonds, become rounded and may be found both in the capsule of the gland and occasionally in the lumina of tumor vessels. Lymph node metastases were registered morphologically in 59% of cases (in children in 63.3%). As a rule tumor metastases contain characteristic papillae, follicular and solid areas. In the solid areas cell clusters had preformed nuclei similar in structure to the focus of origin.

Thus light— and electromicroscopically studied papillary carcinomas from children and adolescents in Ukraine are characterized by highly invasive properties and signs of aggressive behavior, this being confirmed by a high percentage of regional metastases. The role of radiation factor in the increase of thyroid malignancy in children and its contribution to pathomorphism of the studied tumors needs deep analysis and further investigations.

References

[1] Chr.Hedinger, E.D.Williams, L.H.Sobin, Histological Typing of Thyroid Tumours. WHO, 2nd Edn. Berlin, Springer, 1988.
[2] D.W.Henderson et al., Ultrastructural appearances of tumours Diagnosis and classification of human neoplasia dy electron microscopy. Edinburgh, London, Melbourne and N.Y., 1986, 150—156
[3] V.A.LiVolsi, Surgical Pathology of the Thyroid. WB Saunders, Philadelphia, 1990.
[4] S.R.Kini, Thyroid. Guides to Clinical Aspiration Biopsy. Igaku—Shoin, N.Y.—Tokyo, 1989, 362 p.
[5] J.V.Johannessen, V.E.Gould, W.Jao, The fine structure of human thyroid cancer. Human Pathology 9 (1978) 385—400.
[6] J.V.Johannessen, M.Sobrino—Simoes, Well differentiated thyroid tumors. Problem in diagnosis and understanding. Pathology Annual 18 (1983), pt.1, 255—285.
[7] J.Rosai, M.L.Carcangiu, R.Delellis, Atlas of Tumor Pathology. Tumors of the Thyroid Gland. Armed Forces Institute of Pathology, Washington, 1992.

Thyroid Cancer Following Diagnostic Iodine-131 Administration

Per HALL,* John D. BOICE, Jr.,‡ and Lars-Erik HOLM*

*Department of General Oncology, Radiumhemmet , and ‡Epidemiology and Biostatistics Program, Division of Cancer Etiology, National Cancer Institute, Bethesda, MD 20892, USA

Abstract. To provide quantitative data on the risk of thyroid cancer following ^{131}I exposure, 34,104 patients administered ^{131}I for diagnostic purposes were followed for up to 40 years. Mean thyroid dose was estimated as 1.1 Gy, and 67 thyroid cancers occurred in contrast to 49.7 expected [standardized incidence ratio (SIR) = 1.35; 95% confidence interval (CI) 1.05-1.71]. Excess cancers were apparent only among patients referred because of a suspected thyroid tumor and no increased risk was seen among those referred for other reasons. Further, risk was not related to radiation dose to the thyroid gland, time since exposure, or age at exposure. The slight excess of thyroid cancer, then appeared due to the underlying thyroid condition and not radiation exposure. Among those under age 20 years when ^{131}I was administered, a small excess risk (3 cancers *vs* 1.8 expected) was about 2-10 times lower than that predicted from A-bomb data. These data suggest that protraction of dose may result in a lower risk than acute x-ray exposure of the same total dose.

1. Introduction

Thyroid cancer has been convincingly linked to ionizing radiation only after childhood exposure [1-5]. Despite some reports that adult exposure might increase the risk, the evidence is weak [6-8]. In the most recent follow-up of A-bomb survivors, thyroid cancer risk was significantly increased only among individuals under age 20 years at exposure [9]. Other than radiation dose, age at exposure appears to be the most important determinant of future risk, and differences in reported risk estimates might merely reflect differences in age distribution.

2. Material and Methods

The patients were less than 75 years of age when examined with ^{131}I during the period 1950-69 and characteristics are given in Table I. A total of 2,408 individuals were exposed before 20 years of age and 316 before the age of 10 years.

Table I. Characteristics of patients exposed to ^{131}I in relation to reason for referral. The first 5 years after exposure were excluded.

| | Reason for referral | | |
	Suspicion of thyroid tumor	Other reasons	All
No. of patients	10,785	23,319	34,104
Males/Females, %	14/86	22/78	20/80
Mean age at exposure (range), years	44 (1-75)	42 (1-75)	43 (1-75)
Patients <20 years of age at exposure, %	6	8	7
Mean follow-up period (range), years	23 (5-38)	24 (5-39)	24 (5-39)
Mean 24-hour thyroid uptake (range), %	40 (0-96)	40 (0-96)	40 (0-96)
Mean administered activity (range), MBq	2.4 (0.04-37)	1.6 (0.04-37)	1.9 (0.04-37)
Mean dose to the thyroid (range), Gy	1.3 (0.0-25.7)	0.9 (0.0-40.5)	1.1 (0.0-40.5)

The individual absorbed thyroid dose was estimated from the amount of ^{131}I administered, the physical half-life of ^{131}I, the ^{131}I uptake in the thyroid gland, and the thyroid gland size.

The follow-up period started at the time of first ^{131}I administration or if examined prior to 1958, at January 1, 1958. The end of follow-up was the date of thyroid cancer diagnosis, death, emigration, or December 31, 1990.

The cohort was matched with the Swedish Cancer Register for the period 1958-90 to identify thyroid carcinomas. The expected number of thyroid cancers were calculated using incidence data from this register and indirect standardization with adjustment for sex, attained age at exposure, and calendar period. Thyroid cancers occurring during the first 5 years of follow-up were excluded because any thyroid cancer occurring shortly after examination would be likely related to referral or increased medical surveillance and not ^{131}I exposure.

3. Results

Between 1958 and 1990, 67 thyroid cancers were identified more than 5 years after ^{131}I administration. Forty-two of the 67 patients who developed thyroid cancer were referred because of clinical indications that a thyroid tumor might be present. The mean time after ^{131}I administration to the diagnosis of thyroid cancer was 15 years. There were 36 papillary, 18 follicular, 11 anaplastic or giant cell thyroid cancers found, and one sarcoma of the thyroid gland. In one case the exact histopathology was not given.

The significant overall risk for thyroid cancer more than 5 years after exposure was 1.35 (Table II). A significantly higher risk was seen for the 10,785 patients referred under the suspicion of a thyroid tumor (SIR=2.86) compared to those referred for other reasons (SIR=0.75; Table II).

Table II. Observed number of cases (Obs.) and thyroid cancer risk (SIR) in relation to estimated thyroid dose. The first 5 years after exposure were excluded.

Dose, Gy	Obs.	SIR	95% CI
All			
≤0.25	11	1.03	0.51-1.83
0.26-0.50	16	1.84	1.05-2.98
0.51-1.00	9	0.46	0.38-1.57
>1.00	31	1.60	1.09-2.27
All	67	1.35	1.05-1.71
Referred for suspicion of thyroid cancer			
≤0.25	6	3.57	1.31-7.77
0.26-0.50	12	4.30	2.22-7.51
0.51-1.00	4	1.39	0.38-3.56
>1.00	20	2.72	1.66-4.20
All	42	2.86	2.06-3.86
Referred for other reasons			
≤0.25	5	0.55	0.18-1.29
0.26-0.50	4	0.68	0.18-1.73
0.51-1.00	5	0.47	0.20-1.46
>1.00	11	1.04	0.52-1.86
All	25	0.75	0.48-1.10

No dose-response relationship was noticed regardless of the reason for referral The highest risks were seen during the period 5 to 9 years after ^{131}I administration. Only 3 thyroid cancers occurred among the 2,408 patients exposed before the age of 20 years (SIR=1.69; 95% CI 0.35-4.93). Men had a significantly higher relative risk than women among those with a suspicion of a thyroid tumor. No cases of thyroid cancer were found among men referred for other reasons.

4. Discussion

The thyroid gland of children appears to be one of the organs most susceptible to radiation carcinogenesis with relative risk estimates at 1 Gy ranging from 4 to 12 [10]. Since 93% of the

34,104 patients were over age 20 years when ^{131}I was administered, the apparent absence of an overall effect might be attributable in part to the lower sensitivity of the adult thyroid gland.

Iodine-131 delivers nearly all its radioactivity within the first 6 weeks of exposure and it is conceivable that this time was sufficient to allow repair of DNA damage to occur. Lifetime exposure to elevated levels of natural background radiation in China has not been associated with an increase in thyroid tumors [11]. A recent parallel analysis of most major studies of thyroid irradiation concluded that spreading dose over time may lower the risk of subsequent thyroid cancer [12].

The influence of the underlying thyroid condition, real or suspected, could influence the observations in several ways. Persons under surveillance for a suspected thyroid condition might be more likely to have a thyroid cancer detected because of increased medical surveillance. We attempted to address these possibilities by excluding all thyroid cancers that were reported within 5 years of the initial ^{131}I administration. The decreasing risk of thyroid cancer with time after ^{131}I examination is consistent with the possibility that the underlying condition and medical screening contributed to some of the early thyroid cancers. On the other hand, exclusion of individuals who had been examined because of a suspected thyroid tumor may have depleted the exposure group of persons who would likely have developed thyroid cancer. To address this possibility, we conducted dose-response and time-response evaluations. No trend of increasing risk with increasing dose or time since exposure was suggested.

A recent report from cancer registry data in Belarus purports high rates of thyroid cancer to be associated with radioactive fallout, mainly radioactive iodine, including ^{131}I from the Chernobyl accident [13]. This opinion was shared by an expert panel formed by the Commission of the European Communities, although they emphasized that the influence of screening should be carefully considered in assessing the results [14]. The time between exposure and appearance of the thyroid cancer is remarkably short and the dramatic increase in thyroid cancers most likely is, at least in part, related to the intense screening, increased awareness and changed referral routines [15]. Ongoing studies with estimated thyroid doses to individuals should help clarify the causal factors associated with the increase.

References

[1] C.J. Fürst et al., Cancer incidence after radiotherapy for skin hemangioma: a retrospective cohort study in Sweden, J Natl Cancer Inst 80 (1988) 1387-1392.

[2] E. Ron et al., Thyroid neoplasia following low-dose radiation in childhood, Radiat Res 120 (1989) 516-531.

[3] M.A. Tucker et al., Therapeutic radiation at a young age is linked to secondary thyroid cancer, Cancer Res 51 (1991) 2885-2888.

[4]   A.B. Schneider *et al.*, Dose-response relationships for radiation-induced thyroid cancer and thyroid nodules: evidence for the prolonged effects of radiation on the thyroid, *J Clin Endocrinol Metab* 77 (1993) 362-369.

[5]   R.E. Shore *et al.*, Thyroid cancer among persons given x-ray treatment in infancy for an enlarged thymus gland, *Am J Epidemiol* 137 (1993) 1068-1080.

[6]   J.D. Boice, Jr *et al.*, Radiation dose and second cancer risk in patients treated for cancer of the cervix, *Radiat Res* 116 (1988) 3-55.

[7]   J.-X. Wang *et al.*, Cancer incidence among medical diagnostic X-ray workers in China, 1950 to 1985, *Int J Cancer* 45 (1990) 889-895.

[8]   S.L. Hancock, R.S. Cox, and I.R. McDougall, Thyroid diseases after treatment of Hodgkin's disease, *N Engl J Med* 325 (1991) 599-605.

[9]   D.E. Thompson *et al.*, Cancer incidence in atomic bomb survivors. Part II: Solid tumors, 1958-87, *Radiat Res* 137 (1994) S17-S67.

[10]   United Nations Scientific Committee on the Effects of Atomic Radiation (UNSCEAR). Sources and effects of ionizing radiation. 1994 Report to the General Assembly, with scientific annexes. United Nations, New York, 1994.

[11]   Z. Wang *et al.*, Thyroid nodularity and chromosome aberrations among women in areas of high background radiation in China, *J Natl Cancer Inst* 82 (1990) 478-485.

[12]   E. Ron *et al.*, Thyroid cancer after exposure to external radiation: a pooled analysis of seven studies, *Radiat Res* 141 (1995) 259-277.

[13]   V.S. Kazakov, E.P. Demidchik, and L.N. Astakhova, Thyroid cancer after Chernobyl, *Nature* 359 (1992) 21.

[14]   D. Williams *et al.* Thyroid cancer in children living near Chernobyl. Expert panel report on the consequences of the Chernobyl accident. Report EUR 15248 EN. Luxembourg: Commission of the European Communities, 1993.

[15]   E. Ron, J. Lubin, and A. Schneider, Thyroid cancer incidence, *Nature* 360 (1992) 113.

Iodine Deficiency Disorders (IDD) in Regions of Russia Affected by Chernobyl

G.GERASIMOV, G.ALEXANDROVA, M.ARBUZOVA, S.BUTROVA,
M.KENZHIBAEVA, G.KOTOVA, B.MISHCHENKO, A.NAZAROV,
N.PLATONOVA, N.SVIRIDENKO, T.CHERNOVA, E.TROSHINA and
I.DEDOV
Russian Endocrinology Research Centre, D.Ulyanova, 11, Moscow 117036, Russia

Abstract. The present article provides an update on IDD in the Western regions of Russia (Bryansk, Kaluga, Tula and Orel) which were contaminated by radioactive fallout after the Chernobyl accident in 1986. These surveyed areas meet the criteria of ICCIDD/UNICEF/WHO for mild and moderate IDD. Higher iodine excretion and smaller goiter prevalence (mild level of IDD) were more typical for urban sites, while lower iodine levels and higher goiter endemicity (moderate level of IDD) were found in rural areas. IDD control programmes should be developed and implemented in Chernobyl areas and iodine excretion should be monitored continuously to minimize future thyroid abnormalities.

1. Introduction

Elimination of iodine deficiency disorders (IDD) has been defined as a primary goal for international action. After the dissolution of the USSR in 1991, IDD became a common problem of nearly all Newly Independent States. It has been known for years that four provinces (oblast) of Western Russia (Bryansk, Tula, Kaluga, and Orel) were iodine-deficient areas and had a high prevalence of endemic goiter [1]. Now these provinces are one of the main targets for IDD control programmes in Russia because they experienced the highest levels of radioactive fallout shortly after the Chernobyl accident in 1986. The density of radioactive cesium-137 contamination of the soil in these areas varies from 5 to more than 40 Ci/km2 [2].

The present article provides an update on IDD in the Western regions of Russia, which were contaminated by radioactive fallout after the Chernobyl accident in 1986.

2. Materials and methods

Since 1991 several surveys have been performed in Western regions of Russia (Bryansk, Kaluga, Tula and Orel oblast) by research teams from Russian Endocrinology Research Centre with the assistance of the International Council for Control of Iodine Deficiency Disorders (ICCIDD), WHO and UNICEF. Special attention was given to Bryansk oblast since it had relatively high levels of radioiodine fallout and an increased incidence of childhood thyroid cancer [2]. More than 10,000 persons have been surveyed in these areas. Nonexposed regions of Tambov oblast served as a control. No iodine supplementation took place in these regions before or during the study.

Casual urine samples were obtained from 1,450 children and young adults. The iodine analyses were performed after wet digestion of the urine, using a modification of the ceric-arsenite analytical method of Wawschinek et al. [3]. Results were expressed as micrograms of iodine per 100 ml of urine (mcg/dl).

Goiter incidence in the population under study was evaluated by ultrasonography, using a portable instrument (Philips SDR 1200) with 5.0 a MHz transducer. Thyroid volumes calculated from the study population were compared with those from populations with sufficient iodine intake [4].

3. Results and discussion

The data are summarized in table 1, showing minimal and maximal values of diffuse goiter in children and adolescents, nodular goiter in adults, median urinary iodine levels and the percent of samples below 10 mcg/dl in each area.

These surveyed areas meet the criteria of ICCIDD/UNICEF/WHO [5] for mild and moderate IDD. The level of iodine deficiency in nonexposed Tambov oblast appeared to be even higher than in the other surveyed areas. Higher iodine excretion and smaller goiter prevalence (mild level of IDD) were more typical for urban sites, while lower iodine levels and higher goiter endemicity (moderate level of IDD) were found in rural areas. The difference can probably be explained by fact that the rural population is more self sufficient for food, especially milk, vegetables, meat and poultry.

These results confirm earlier observation of IDD in regions of the former USSR, contaminated by radionuclides at the time of Chernobyl [6]. Recently, a group from Byelarus reported urinary iodine excretion in children from Gomel oblast, which received considerable radioiodine fallout in 1986. Most of the median

TABLE 1. GOITRE PREVALENCE IN BRYANSK, KALUGA, TULA AND OREL PROVINCES OF RUSSIA.

	BRYANSK	KALUGA	TULA	OREL	TAMBOV (control area)	Reference data from nonendemic areas
Diffuse goiter (in children and adolescents)	12 - 30 %	18 - 30 %	18 - 35 %	20 - 45 %	35-50%	< 5%
Nodular goiter (in adults)	3 - 5%	3 - 6%	4 - 6 %	up to 8%	6-10%	< 3%
Median urinary iodine level (mcg/dl)	6.9 - 8.4	5.4 - 8.9	5.2 - 7.5	3.7 - 5.5	2.8-5.6	> 10.0
Percent of samples below 10 mcg/dl	59 - 82%	57 - 80%	55 - 78%	73 - 92%	75-95%	< 50%

values for urinary iodine in the surveyed sites indicate low iodine intake. The data also suggest the presence of relatively isolated "pockets" of severe iodine deficiency [7]. It is not known to what extent, if any, iodine deficiency in Russia, Byelarus and Ukraine may have contributed to the increase in thyroidal irradiation and thyroid cancer after the Chernobyl accident [8].

IDD control programmes should be developed and implemented in Chernobyl areas and iodine excretion should be monitored continuously to minimize future thyroid abnormalities. Iodized salt is still not available in these areas. Implementation of alternative methods (iodization of bread, iodized oil) can be an immediate action. In 1992, more than 300,000 capsules of iodized oil (Lipiodol) donated by UNICEF, were distributed among children in Bryansk and Tula oblast of Russia. A new project with iodized oil supplementation in the areas of Russia near Chernobyl is being launched in 1995 with the assistance of the International Red Cross.

References

[1] A. Nikolaev, Endemic goiter. Meditsina, Moscow, 1956. (in Russian)

[2] A. Tsyb, E. Parshkov, V. Ivanov, V. Stepanenko, E. Matveenko, Yu. Skoropad, Disease indices of thyroid and their dose dependence in children and adolescents affected as a result of the Chernobyl accident. In: S.Nagataki (Ed.), Nagasaki Symposium on Chernobyl: Update and Future - ISBN: 0 444 81953 3. Elsevier, Amsterdam, 1994, pp. 9-19.

[3] O. Wawschinek, O. Eber, W. Petek, P. Wakonig, A. Gurakar, Bestimmung der Harnjodausscheidung mittels einer modifizierten Cer-Arsenitmethode, *Ber. OGKC* **8** (1985) 13-15.

[4] R. Gutekunst and H. Martin-Teichert, Requirements for goiter surveys and the determination of thyroid size. In: F. Delange, J. Dunn and D.Glinoer (Eds.), Iodine Deficiency in Europe: A Continuing Concern - ISBN: 0-306-44410-0. Plenum Press, New York and London, 1993, pp. 109-115.

[5] J Dunn and F. Van Der Haar, A practical guide to the correction of iodine deficiency - ISBN: 90 70785 12 9, WHO 525, Washington, DC, 1990.

[6] G. Gerasimov, Update on IDD in the Former USSR, *IDD Newsletter* **9** (1993) 43-48.

[7] T. Mityukova, L. Astakhova, L. Asenchyk, M. Orlov and L. VanMiddlesworth, Urinary iodine excretion in Byelarus children. *European J. of Endocrinology* **133** (1995) 216-217.

[8] V. Kazakov, E. Demidchik, L. Astakhova, Thyroid cancer after Chernobyl [letter], *Nature* **359** (1992) 21.

Acknowledgments

The authors would like to acknowledge encouragement and financial and material assistance of the Ministry EMERCOM, Russia, the International Council for Control of Iodine Deficiency Disorders (ICCIDD), the World Health Organization (WHO), the United Nations Children's Fund (UNICEF), Henning Berlin GmbH (Germany) and Boots Pharmaceuticals, Inc. (USA). We are grateful to Dr. John Dunn for revision of the English of this manuscript and Vassily Gerasimov for excellent typing.

Differentiation Between Malignant and Benign Thyroid Tumors by X-ray Fluorescent Analysis - Comparison of Cases from Russia and Albany, New York.

I.TOMASHEVSKY, G.GERASIMOV, K.TROSHINA, G.ALEXANDROVA,
S.SERPUKHOVITIN, M.BRONSTEIN and J.FIGGE (*)
Russian Endocrinology Research Centre, D.Ulyanova, 11, Moscow, 117036, Russia
(*) Albany Medical Center, 47 New Scotland Ave., Albany, NY 12208 and State
University of New York, Albany, NY, USA

Abstract. Intrathyroid iodine level in different types of thyroid neoplasms in Russia (126 cases) and USA (37 cases) were investigated by X-ray fluorescent analysis in vitro. A decrease in intrathyroid iodine concentration is associated with the stepwise loss of differentiation in thyroid tumors. In colloid goiter tissue from Russia, the intrathyroid iodine level is increased. The intrathyroid iodine level was markedly increased in microfollicular adenomas and colloid goiters from American patients which may reflect a higher iodine supply in the USA. X-ray fluorescent analysis together with careful clinical appraisal can be used for management of thyroid patients with suspicious nodules which should be treated by surgery.

1. Introduction

The evaluation of thyroid nodules represents a common and important clinical problem. The distinction among thyroid hyperplastic nodules, adenomas and follicular carcinomas is often difficult [1]. Thyroid cancer is the most common of the endocrine malignancies and represents 0.6-1.6% of all malignant tumors in humans [2]. Typically, the tumor develops slowly within the thyroid gland. Local metastases and invasion appear several years after the formation of primary foci. Thyroid carcinomas can present as thyroid nodules and may be difficult to differentiate from nodular goiters, adenomas, authoimmune thyroiditis and other benign thyroid lesions. The differential diagnosis in such cases is based mainly on cytological examination of thyroid tissue after fine needle aspiration. In certain cases, particularly in the case of follicular lesions, cytologic evaluation often cannot differentiate benign from malignant neoplasms [3].

In patients with thyroid neoplasms, normal tissue is replaced by functionally inactive and iodine-deficient tissue. Iodine content in the thyroid tissue can be measured *in vivo* by X-ray fluorescent analysis (RFA). This method is non-invasive, rapid and inexpensive; it does not require injection of any radioactive compound and the level of external irradiation is negligible [4]. It has been found that iodine content in thyroid lesions as measured by the above method is dependent upon the type of neoplasm. For example, in thyroid cancer, tissue iodine content is low (50-70 mcg/g tissue) due to the decreased synthesis of thyroglobulin, a major iodine-containing species in the thyroid gland [5]. Intrathyroidal iodine content also depends on iodine consumption: in the Moscow area, which is iodine- deficient, the iodine content in thyroid tissue, measured *in vivo*, was 380 +/- 50 mcg/g, while in the iodine-sufficient American population iodine content was 1000 +/- 80 mcg/g [6,7]. The intrathyroid iodine level can be also measured by RFA *in vitro* in paraffin-embedded thyroid tissues obtained after surgery. In this case, iodine levels can be compared with the surgical histology and correlated with the type of thyroid lesion.

The aim of the present study was to evaluate whether the iodine concentration in thyroid tissue can be used as a factor to predict the probability of malignancy in thyroid nodules. We also compared iodine levels in thyroid nodule tissue originating from three geographic regions: Moscow, the Arsenyevo region of the Tula province (which was contaminated by radioactive fallout after the Chernobyl accident in 1986) and Albany, New York, USA.

2. Materials and methods

One hundred twenty six patients with thyroid nodules underwent thyroid surgery in the Surgery Division of the Russian Endocrinology Research Center in Moscow. Twenty four patients came from the Arsenyevo region of the Tula province, contaminated by radioisotopes at the time of the Chernobyl accident in 1986. The density of cesium-137 pollution in this area is 5 - 15 Ci per km2. The diagnosis of thyroid nodule in these patients was made by palpation, ultrasonography and fine needle biopsy in the course of the field survey conducted directly in the region. Since no difference between iodine concentration in the thyroid tumours from Moscow and Arsenyevo has been found, we combined these data into one group. Thirty-seven patients with thyroid nodules underwent surgery at Albany Medical Center, New York, USA. Twenty cases of normal thyroid tissue (e.g., tissue surrounding nodules) from Russia served as additonal controls. Thyroid tissue specimens derived from surgical resection were fixed with 10% formalin solution and imbedded in paraffin blocks. Morphological laboratories in both Moscow and Albany used the same formalin and paraffin to prepare blocks.

RFA has been developed for measurement of the iodine concentration in thyroid tissue *in vivo* and *in vitro*. Equipment was produced in Russia by the Research Scientific Institute of Technical Physics and Automation. This technique involves the use of a collimated source of radiation (241-Am) and a high-resolution detector to obtain an intensity map of the stable iodine distribution in the thyroid [4,6].

3. Results and discussion

Intrathyroid iodine levels in different types of thyroid neoplasms in Russia and USA are listed in the Table 1. The lowest iodine levels were found in malignant thyroid neoplasms (papillary, mixed papillary-follicular and follicular) from Russia. In benign thyroid lesions from Russia a statistically significant decrease in iodine levels was found in fetal and embrional adenomas (p<0.02), while in microfollicular and B-cell adenomas no difference in intrathyroid iodine levels was detected compared to normal tissue. These results are consistent with the idea that a decrease in iodine concentration is associated with the stepwise loss of differentiation in thyroid tumors. In colloid goiter tissue from Russia iodine level was even higher than those in normal tissue (p<0.001).

Table 1. Intrathyroid iodine levels in different types of thyroid neoplasms in Russia and USA

(Number of group) Pathology	Region (number of cases)	Iodine concentration (mcg/g tissue)	Statistical difference between groups
(1) Papillary carcinoma	Russia (13)	206+/-5.0	1,9 p<0.01
(2) Papillary-follicular carcinoma	Russia (15)	186+/-8,5	2,9 p<0,001
(3) Follicular carcinoma	Russia (12)	181+/-8,7	3,9 p<0,01
(4)Microfollicular adenoma	Russia (23)	376+/-24	4,9 NS
(5) Fetal adenoma	Russia (16)	306+/-21	5,9 NS
(6) B-cell adenoma	Russia (14)	360+/-46	6,9 p<0,02
(7) Embrional adenoma	Russia (13)	252+/-17	7,9 p<0,02
(8) Nodular colloid proliferative goiter	Russia (20)	848+/-8.9	8,9 p<0.001
(9) Normal tissue	Russia (20)	389+/-6,0	----
(10) Papillary carcinoma	USA (22)	102+/-5,0	1,10 p<0,01
(11) Microfollicular adenoma	USA (10)	1112+/-47	11,4 p<0,001
(12) Nodular colloid proliferative goiter	USA (5)	1584+/-103	12,8 p<0,001

Intrathyroid iodine level was markedly increased in microfollicular adenomas and colloid goiters from American patients compared to patients from Russia (p<0.001). The higher iodine concentration in cases from the USA (benign tumours) might result from differences in environmental factors between these countries (e.g., iodine deficiency in Russia [8]. Strangely enough, iodine levels in papillary carcinomas from the USA were even lower than in Russian specimens. The reason for that difference remains obscure.

RFA appears to be a useful noninvasive diagnostic test in differentiating between malignant and benign thyroid nodules. Intrathyroid iodine levels decrease in thyroid adenomas and carcinomas and increase in benign nodular goiter. Carcinomas have the lowest levels, differentiating them from adenomas and other benign lesions. At the same time, this marker is not specific: iodine levels in thyroid lesions can be influenced by iodine supply. For that reason, reference values should be developed for specific countries and/or regions.

For populations in iodine-deficient areas, a localized area of low iodine content in the thyroid can be specific for malignancies or poorly differentiated adenomas, which both need surgery. Patients with lesions characterized by a low thyroid iodine level should be carefully worked up to exclude thyroid malignancy. At the same time, high or normal iodine levels are signs of benign conditions which may still require follow up and therapeutic intervention.

4. Conclusions

1. A decrease in intrathyroid iodine concentration is associated with the stepwise loss of differentiation in thyroid tumours. Localized areas of decreased iodine content within the thyroid gland may represent poorly differentiated adenomas or carcinomas. In colloid goiter tissue from Russia, the intrathyroid iodine level is increased.

2. The intrathyroid iodine level was markedly increased in microfollicular adenomas and colloid goiters from American patients which may reflect a higher iodine supply in the USA.

3. X-ray fluorescent analysis together with careful clinical appraisal can be used for management of thyroid patients with suspicious nodules which should be treated by surgery.

References

[1] J. Ruschoff, Diagnostic and Prognostic Significance of Histochemical and Cytochemical methods in Thyroid Tumours, *Exp. Clin. Endocrinol.* **101** (1993) 11-16.
[2] A. Schneider, Carcinoma of Follicular Epithelium. In: L.Braverman and R.Utiger (Eds) The Thyroid, ISBN 0-397-51205-8, J.B.Lippincott Company, Philadelphia, 1991, pp.1121-1165.
[3] E.C.Ridgway, Clinical Evaluation of Solitary Thyroid Nodules. In: L.Braverman and R.Utiger (Eds) The Thyroid, ISBN 0-397-51205-8, J.B.Lippincott Company, Philadelphia, 1991, pp.1121-1165.
[4] I.Tomashevsky, X-Ray Fluorescent Analysis of the Stable Iodine Content in Diagnosis of Thyroid Diseases, *Med.Radiologia,* (1991) N 6, 17-20. (In Russian)
[5] T.Tadros, M. Medisey, F.Tavy, P.Turner, The Iodine Concentration in Benign and Malignant Thyroid Nodules. Measurement by X-Ray Fluorescence, *Br. J. Radiol.* **54** (1981) 656-629.
[6] I.Tomashevsky, G.Gerasimov, Intrathyroid Iodine Content and Thyroid Function in Nontoxic Diffuse Goiter, *Probl. Endokrinol.* **41** (1995), in press. (In Russian)
[7] J.Patton, J.Sandler, C. Partain, Prediction of beningnancy of the solitary "cold" thyroid nodule by fluorescent scanning. *J.Nucl. Med.* **26** (1985) 461-470.
[8] G. Gerasimov, Update on IDD in the Former USSR, *IDD Newsletter* **9** (1993) 43-48.

Acknowledgements

Authors would like to acknowledge the Administration of the Tula province for support of field surveys in Arsenyevo region.

Immunologic Status of Children with Thyroid Cancer Living near Chernobyl
(flow cytometric and electron microscopic study)

K.P.ZAK, M.A.GRUZOV, E.V.BOLSHOVA, V.V.AFANASYEVA,
V.S.SHLYAKHOVENKO, O.A.VISHNEVSKAYA, N.D.TRONKO

*V.P.Komissarenko Institute of Endocrinology & Metabolism, Acad.Med.Sci.,
254114, Vishgorodska,69, Kiev, UKRAINE*

Abstract. It has been carried out a light, electron microscopic and flow cytometric study of blood leucocyte of children with malignant tumors(papillary carcinoma) of thyroid gland who were living at the moment of the accident near Chernobyl.The results obtained point out the presence of some disturbances of immune status of theese children.

1. Materials and Methods.

It has been carried out light microcsopic, electron microscopic and flow cytometric study of blood leucocytes of 30 children aged 8 to 15 with malignant tumors of thyroid gland (papillary carcinoma), who were living at the moment of the accident near Chernobyl. The control group consisted of children of the same age residing on territories not subject to radioactive contamination.

Blood for examination was sampled from ulnar vein using a heparinized syringe. Differential blood count was performed in blood smears stained by Pappenheim. When performing count of 200 leucocytes large granular lymphocytes (LGL) were considered as separate group.

Flow cytometric analysis of surface phenotype of lymphocytes was performed with a cytofluorimeter FACStar plus ("Becton Dickinson",USA) using separated by Ficoll-Hypaque cells and monoclonal antibodies to CD3, CD4, CD8 and CD16 surface antigenes ("Sorbent",Russia and "Becton Dickinson",USA).

For electron microscopic study the buffy-coat isolated from blood was fixed with a 2.5% glutaraldehyde ("Merck", Germany) and postfixed with 1% osmium tetroxide and after dehydratation was included into Araldite ("Fluka",Germany). Ultrathin slices obtained on ultratome LKB-8800 ("LKB",Sweden) were examined under electron microscope JEM-100C ("JEOL", Japan).

2. Results.

The investigations showed that the children with thyroid cancer had a statistically significant increase in the absolute and relative contents of eosinophils, and a part of patients had a decrease in blood content of LGLs - the morphologic homologue of natural killer (NK) cells. Cytologic study of leucocytes in blood smears in patients as compared to control group, a significant increase in the number of hypersegmented form of neutrophils with needle-shaped prominences. A certain part of lymphocytes had an atypical cconfiguration of nuclei. Sometimes, lymphocytes with two-blade nuclei were noted. In certain patients a marked vacuolation of cytoplasm of monocytes was observed.

Flow cytometric analysis of surface antigens of lymphocytes revealed in the patients examined a some decrease of in total number of T-lymphocytes (CD3$^+$) at the expense of fall in the amount of both T-inductors/helpers (CD4$^+$) and T-suppressors/killers (CD8$^+$). In some patients with the low number of LGLs a decrease in CD16$^+$cells was noted.

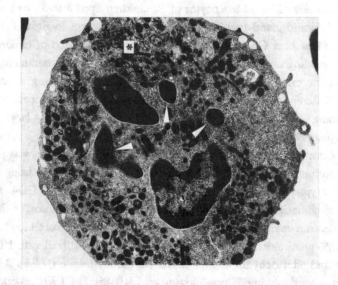

Fig.1.Ultrastructure of blood neutrophil in an child with thyroid cancer. Hypersegmentation of nuclei(arrow) and swelling and destruction of granule contents (*). X 10000.

Electron microscopic studies of blood cells in children with thyroid tumors revealed marked changes in the ultrastructure of neutrophil nuclei: hypersegmentation, whimsical prominences, loops, pockets, numerous small segments (Fig. 1). Swelling and distrucction of granule contents are also noted. There was an increase in the amount of the so-called "aberrant" forms of lymphocytes, i.e. cells with a disturbed structure of chromatin, similar to that in apoptosis, nuclear prominences, additional micronoclei, what seems to refleecct a genome disturbance. In many LGLs marked distruction of azurophilic granules has been noted, what suggests a disturbance of their secretory and cytolytic function.

The results obtained point out the presence of significant disturbances of immune status in Chernobyl's children with malignant tumors.

Malignant Neoplasms on the Territories of Russia Damaged Owing to the Chernobyl Accident

L.V. REMENNIK, V.V. STARINSKY, V.D. MOKINA, V.I. CHISSOV,
L.A. SCHEPLYAGINA, G.V. PETROVA, M.M. RUBTSOVA
Moscow A. Herzen Cancer Research Institute,
2nd Botkinsky proezd, 3, 125284 MOSCOW, Russia

Abstract. The work presents the results of descriptive analysis of development of oncoepidemiological situation in six of the most polluted regions owing to the Chernobyl accident in 1981-1994. The growth of malignancies incidence is marked in all territories as well as in the Russian Federation as a whole. The most adverse tendencies have been revealed in the Bryansk, Orel, Ryazan regions. It is marked that the formation of a structure, levels and trends of the malignancies incidence has been occurring under influence of a complex of factors usual up to the accident. The analysis of the data from the specialized cancer-register evidences that the incidence of thyroid malignancies is actively growing in the population of the Bryansk region. The probability of connection of growth of the thyroid cancer incidence in children of the Bryansk region with the Chernobyl accident is reasonably high, but should be confirmed through the application of methods of analytical epidemiology.

1. Introduction

The increase in risk of development of malignancies is one of the most probable effects of ionizing radiation on populations living on radioactively polluted territories. In the Russian Federation the most polluted by radionuclides owing to the Chernobyl accident are the territories of the Bryansk, Kaluga, Orel, Tula, Ryazan, Kursk regions. The study of dynamics of the oncological incidence in the population of these territories in view of the background tendencies that were determined up to the accident appears to be urgent.

2. Materials and methods

Methods of the descriptive epidemiological analysis to study the development of oncoepidemiological situation over the period 1981-1994 in the Bryansk (BR), Kaluga (KR), Orel (OR), Tula (TR), Ryazan (RR), Kursk (KuR) regions and Russia as a whole were used. Base of research were materials of official oncological statistics and data personal cancer registers.

The following registers have been generated: the population cancer register of BR, the specialized cancer registers of the children incidence in the BR and TR, the nosological registers (for thyroid cancer) in the BR and TR.

3. Results and discussion

3.1.   Structure and dynamics of malignancy incidence

There is the steady growth of malignant neoplasm incidence (MNI) in the population of these regions as well as in whole Russia. However, since 1987, the MNI in the 6 above-mentioned regions has become higher than in Russia. In 1994 the rates of the general oncological incidence were: in BR -333.2, KR -294.3, OR -313.3, TR -339.3, RR -357.1, KuR -300.0, whereas the average rate in Russia was 278.0 per 100,000 of population. Along with it the rate incidence increments in 1994, as compared to 1981, were: in BR -38.6 %, KR -29.6 %, OR -40.1 %, TR -21.4 %, whereas an increment of the average rate in Russia was 19.1 %. The analysis of the sex- and age- standardized rates evidences high incidence rates in males of the BR and the RR. In 1994 the incidence rates in the male population of these regions were 304.9 and 305.2, respectively (in Russia -272.4). High rates of the incidence, which tend to grow, are in the junior group (0-29) of male population of the BR. Over the scrutinised period the incidence rates in female population of the BR and RR were higher than in whole Russia and in 1994 they were in BR -178.2, KR -145.5, OR -171.5, TR -163.4, KuR -148.3, RR -180.2, whole Russia -169.2.

After the accident the structure of the MNI did not significantly change. In order of prevalence rates everywhere over the territories, excluding the BR, the first place goes to lung, trachea, and bronchus cancers, the second place - to stomach cancer. In the BR, first goes to stomach cancer, whereas cancer of the respiratory organs goes behind. The contribution of the stomach, hemopoietic and lymphatic tissue, thyroid, pharynx malignant neoplasms to the total incidence of cancer in the BR is more than in Russia. As for the OR the contribution of thyroid cancer to the total incidence of cancer in this region is significantly more than Russia. Dynamics of the incidence of lung, trachea, bronchus cancers in the population of the investigated territories has a positive trend, the highest rates being for the KR, TR, RR. However, the damages territories were characterized by the higher incidence rate of stomach cancer than in the whole of Russia as before the accident as after it along with decreasing indices. In 1994 the BR occupied the second and third places in the incidence rate of stomach cancer in males and in women, respectively. The incidence rate of hemopoietic and lymphatic tissue malignant neoplasms in the BR, TR and RR is higher than in whole Russia. In 1994 the highest incidence rates were registered in the RR: for males -18.0, for females -13.5.

In 1994 the BR was marked highest in the Russian Federation for a level of malignancy incidence in children. The standardized rates for boys and girls made up 21.8 and 24.2, respectively. Appropriate parameters for the Russian Federation were 10.9 and 9.0.

3.2.   Thyroid malignancies

Accounting for a probable increase in the thyroid cancer (TC) incidence on the territories damaged owing to the Chernobyl accident, dynamics of this process has been estimated.

The growth of thyroid malignant neoplasms incident rate appears to be everywhere over the damaged territories and its increment is significantly higher than in whole Russia (Table 1.) and the highest ones in Russia were registered in 1994 in the BR, OR, RR and they were 7.8, 7.7, 7.7 per 100,000 of population, respectively.

Table 1. *Dynamics of thyroid malignancy incidence in the populations of the Bryansk, Kaluga, Orel regions and of Russia for the period 1981-1994 - Standardized rates per 100 thousand of population of appropriate sex.*

Territory	Sex	Years						
		1981	1982	1983	1984	1985	1986	1987
Bryansk region	m	1.2	1.7	0.4	0.5	0.5	1.4	1.3
	f	3.2	2.3	2.2	2.0	1.6	2.9	6.6
Kaluga region	m	0.6	0.2	0.3	0.2	0.0	0.2	0.6
	f	0.9	1.8	1.4	1.0	1.5	0.9	1.4

Territory	Sex	Years						
		1988	1989	1990	1991	1992	1993	1994
Bryansk region	m	1.4	1.0	0.8	2.3	1.8	2.1	2.6
	f	5.5	7.2	6.7	5.6	8.9	10.0	11.4
Orel region	m	*	1.1	1.4	2.5	1.9	1.5	2.1
	f	*	4.1	4.2	5.5	7.2	11.0	10.7
Kaluga region	m	*	0.6	0.6	0.3	1.0	0.6	1.3
	f	*	2.0	1.5	1.8	1.5	3.0	2.4
Russia	m	*	1.0	1.0	1.0	0.9	1.0	1.1
	f	*	3.1	3.0	3.0	3.4	3.6	4.0

* = no data

In 1994 the age-standardized rates of the TC incidence in males and in females of the BR and OR were 1.7 and 11.4, 2.1 and 10.7 per 100,000, respectively, whereas the average rates in Russia were 1.1 and 4.0, respectively. The average annual rates of increment of the TC incidence for the last five years in the BR (15.8 %) and OR (24.9 %) were much higher than that in whole Russia (6,4 %). Up to now the contribution of TC to the total incidence of cancer in both regions is more than twice as much as that for Russia as the whole. The registered increase in the TC incidence for these regions is more pronounced in young and middle-aged persons. The average annual rates of increment of the 0-29 age group of the male and female populations in the BR over the period 1987-1994 were 35.5 % and 22.2 %, respectively, whereas those for the region as the whole were 13.4 and 10.1, respectively.

The main feature of the TC age group is the high contribution of children to the total number of patients with this disease. So in 1994 7.0 % and 2.9 % of all the patients with TC in the BR and OR, respectively, were children under 14, whereas in Russia as the whole they were 1.3 %. The contribution of TC to the total incidence of cancer in children of the BR (12.4 %) in 1994 was significantly higher than in OR (8.7 %) as well as in Russia (1.9 %) as the whole. Every fifth solid tumour from those registered in children of the BR in 1994 was TC. The contribution of the TC to the whole number of solid tumours in children of the OR was 11.9 % compared with 4.0 % in Russia.

Along with it in 1994 the TC incidence rates in children were: in BR -2.5, in OR -1.1, in Russia -0.2 per 100,000 of children population. Before the Chernobyl accident in the period 1981-1986 the TC incidence rate in children of the BR (0.1 per 100,000 of children population) was not higher than that Russia average rate. In 1987-1994 there were 14 children with TC in the BR. Of them 9 live in the western most polluted areas, 3 in the areas with a medium rate of pollution, 2 in pure areas, therewith 11 cases were morphologically diagnosed papillary cancer.

In analysis of the thyroid cancer incidence in adults of the BR living in areas with various rates of radioactive pollution there was no decrease in the ratio of the follicular forms of cancer to the papillary ones as compared with 1981. For this region as the whole it was 0.7. The incidence rates in adults of the BR do not correlate with the rates of radioactive pollution. In 1994 the incidence rates in residents of most polluted, pooly polluted, and pure areas were as follows: 11.7±4.4, 6.6±3.4, 7.1±1.7 per 100,000 of population.

3. Conclusion

Formation of the current rates of malignant neoplasms incidence, its structure and the trends of its dynamics can be traced to the complex of factors that had been established before the accident. Unfavourable trends in development of the oncoepidemiological situation appear to be for the BR and the RR. To monitor the incidence one must use population cancer registers. The significant growth of the thyroid cancer incidence in children in the BR in 1994, probably is connected to the accident but requires study through the application of analytical methods. Further monitoring should be carried out of the TC incidence in the population on the damaged territories using screening programs based on wide application of the ultrasound thyroid examination, accounting for the regional structure of thyroid pathology as well as non-radioactive cancerogenic factors.

What the Educated Public Wants to Know From Chernobyl

Ralph Lapp*, Alexander Shlyakhter[#], Richard Wilson[#0]
*Alexandria, Virginia, U.S.A.
[#] Department of Physics, Harvard University, MA, U.S.A.
[0] Advisory Committee, ISIR, Minsk, Belarus

Conferences of this sort tend to concentrate on what the experts in the field, especially those toiling in the vineyard, know and have recently discovered, often with jargon or esoteric detail. In this paper we will take a different approach and present some thoughts that we have encountered in our many interactions with the public.

Most of what the public is told about the effects of radiation is based upon the excellent work of Radiation Effects Research Foundation (RERF). This has a couple of logical deficiencies. Firstly, the effect is of a large dose received in a short period of time, whereas public interest is mostly about the effects of small doses spread out over a long period. Secondly, the way the dose at Hiroshima and Nagasaki are calculated retrospectively is still a matter of discussion - and the uncertainty is still perhaps as much as a factor of three.

The total exposure of the people exposed at Chernobyl to more than 30 Rems - measured in person-Rems - is considerably greater than the comparable exposure to the RERF cohorts. This leads to the possibility of addressing these issues.

It is now 10 years since the accident, and it should now be possible to make a preliminary assessment without impeding the more careful work that will take 50 or more years to complete. The public needs preliminary indications now.

In particular, we would like to see the following items addressed:

1.   On Thyroid Cancers

a)   How can one reconcile the incidence of thyroid cancer among children exposed near Chernobyl [1,2] with other data from

(i)   Fallout from weapons testing [3]

(ii)   Exposure during radiotherapy for skin hemangioma in infancy [4] where no significant excess was found.

(iii)   Other radiotherapy for benign diseases [5,6]

(iv)   Exposure to the Hiroshima/Nagasaki survivors

b)   Is the thyroid cancer incidence larger or smaller among those who took potassium iodide (KI) pills? Is there a difference between those who took KI pills before or after the exposure? Since it is likely that there will not be reliable information on the point for all the affected populations, can we reliably define a subset for which it is known? [We note that there is prior evidence that KI pills taken before exposure may protect against exposure but KI pills taken after exposure may "lock in" the iodine and accentuate the dose.] There is also some indication that KI itself may be carcinogenic.

c)   Are there any data before 1986 on iodine deficiency that can be used to try to relate the thyroid incidence to thyroid deficiency? See, for example, Mityukova *et al.*, [7] who found mild thyroid deficiency <u>after</u> the Chernobyl accident.

d)   It is important to continue to study thyroid cancer incidence among those who were children at the time of the Chernobyl accident. One possibility is that the incidence of childhood thyroid cancer will diminish with time - and be close to zero eight years or more after exposure. This would be expected if thyroid cancer incidence were similar to leukemia incidence.

e)   How large is the population in which thyroid dose was measured in the first month after exposure? Is it large enough for a sensible study and a large enough fraction of those exposed to be sure that no biases exist? Does the work of Balonov and others from St. Petersburg in the region around Briansk enable us to get a fairly reliable measure of dose? [Noting here that if the dose is known to within a factor of 3, then the precision is comparable to that of the RERF data.]

It is widely believed that it is impossible to obtain good dose data because of the fact that in only a small proportion of exposed individuals was the dose measured. But it is important to realize that even if a group of children with tumors as small as 20 were well categorized, and this was a good statistical (random) sample of the population, one could obtain an accuracy greater than the factor of three that still plagues the RERF data.

f)   What is the death rate among the unfortunate victims of thyroid cancer? Are the cancers similar to those naturally occurring in the USA? This would be a vital conclusion for assessment of risks of a nuclear power accident since most safety studies have assumed that 90% of thyroid cancers are non-fatal and can be cured. The results of the SEER program [8] are attached in Table 1 which suggests that few naturally occurring childhood cancers are fatal. Preliminary indications [9] are that the cancers in Belarus and Ukraine are more <u>aggressive</u>. Can this important detail be confirmed?

g)   The above may have an important implication for the optimal <u>treatment</u> of these cancers and should therefore be discussed soon. There seems to be two contrasting viewpoints. One that radiation treatment is needed [10]. The other suggesting that aggressive treatments such as thyroid replacement should no longer be regarded as harmless therapy.

Table 1

Ages	Deaths (D)	Cases (Incidence I)	Ratio I/D
0-9	0	11	--
10-19	1	171	171
20-29	1	748	748
30-39	5	799	160
40-49	28	758	27
50-59	55	689	12.5
60-69	112	506	4.5
70-79	142	339	2.4
80-89	97	153	1.6
ALL	440	4,174	9.5

2. On Leukemias

h) In 1993 Ivanov *et al.*, [11] did not find an increase of leukemia attributable to Chernobyl. In 1995 Ivanov (private communication) reports a 30% overall increase in cancer in Belarus since 1986 compared to before 1986. If the increase is due to the Chernobyl accident, one expects it to be primarily in high dose areas. Is there any consistency? Are the data consistent with the numbers expected from the RERF data?

i) Can one categorize the leukemias by type (AML, CML, CLL and ALL)? If this cannot be done before 1986, can it be done for leukemias since 1986?

j) AML cases that occur subsequent to radiotherapy have been identified as having a specific DNA marker. Have any such markers been seen among those exposed at Chernobyl? Can one preserve tissue cultures to analyze for DNA markers at some time in the future?

k) In view of the fact that Chronic Lymphocytic Leukemia (CLL) has not been identified as being caused by radiation, it is of crucial interest to know whether CLL in Belarus has been observed to increase as much as other leukemias. We note that CLL is a slowly progressing disease and is believed to have a very long latency [12].

l) Even if the identifying of leukemia types before 1986 was not reliable, what is the identification since 1986? Does the CML/AML ratio vary with dose? Is this variation the same as seen in RERF?

m) Have other cancers been seen in excess? Would any have been expected by now on the basis of previous data?

3. Other Comparisons

n) How do the whole-body doses compare to the whole body doses from the exposures to the workers at the MAYAK plant at Kyshtym, to the villagers along the Techa River exposed to the releases from Lake Karachai, or to the population in the Altai exposed to nuclear testing fallout?

References

[1]   V.S. Kazakov, E.P. Demidchik, L.N. Astakhova, Thyroid Cancer After Chernobyl, *Nature* **359** (1992) 21.

[2]   V.A. Stsjazhko, A.F. Tsyb, N.D. Tronko, G. Souchkevitch, K.F. Baverstock, Childhood Thyroid Cancer Since Accident at Chernobyl, *BMJ* **310** (1995) 801.

[3]   Kerber et al; *JAMA* 270 (1993) 2076-2082.

[4]   M. Lundell, T. Hakulinen and L-E Holm, *Radiat. Res.* **140** (1994) 334-339.

[5]   M.P. Mehta, P.G. Goetowski, T.J. Kinsella, *Int. J. Radiation Oncology Biol. Phys.*, Radiation Induced Thyroid Neoplasms 1920 to 1987: A Vanishing Problem? **16** (1989) 1471-1475.

[6]   C. Silverman, Thyroid Tumors Associated with Radiation Exposure, *Public Health Reports* **99** (1984) 369-373.

[7]   T.A. Mityukova, L.N. Astakhova, L.D. Asenchyk, M.M. Orlov, L. VanMiddlesworth, Urinary Iodine Excretion in Belarus Children, *European J. Endocrinol. 133 (1995)* 216-217.

[8]   J.L. Young, *et al.*, Cancer Incidence and Mortality in the United States 1973-77, Surveillance Epidemiology and End Results, *NIH Publication*, 82-2435 (1982).

[9]   L. Baschieri, A. Antonelli, M. Ferdeghine, B. Alberti, M. Puccini, G. Boni, P. Miccoli, Thyroid Cancer in Children After Chernobyl Nuclear Disaster, *Thyroid*, **5** (supplement. 1) (1995) S-30.

[10]   T. Delbot, L. Leenhardt, A.Moutet, D. Leguillouzic, B. Gameiro, M.L. Simonet, A. Aurengo, Management of Childhood Thyroid Cancer Following the Chernobyl Accident, *Thyroid*, **5** (supplement. 1) (1995) S-30.

[11]   E. Ivanov *et al.*, *Nature* **365** (1993) 702.

[12]   E. Montserrat, *et al.*, Lymphocytic Doubling Time in Chronic Lymphocytic Leukemia, *Clinics in Haematol.* **6** (1977) 185-202.

III. HEALTH EFFECTS
FOLLOWING THE ACCIDENT
C. Epidemiology of exposed populations

What is Feasible and Desirable in the Epidemiologic Follow-up of Chernobyl

Elisabeth CARDIS[1] and Alexey E. OKEANOV[2]

[1]IARC, 150 cours Albert Thomas, 69372 Lyon Cedex 08, France and
[2]Belarus Centre for Medical Technologies, Information Computer Systems, Health Care Administration and Management. 7-a Brovki St, 220600, Minsk, Belarus

Abstract. This paper summarises the results of pilot studies carried out to evaluate the feasibility of long-term epidemiological studies of cancer risk among populations exposed to radiation from the Chernobyl accident.

These studies demonstrated that it is feasible to carry out a study of radiation-induced risk of specific cancers among "Chernobyl liquidators" included in the State Chernobyl registries using a case-control approach in Belarus and Russia. Careful large-scale studies of liquidators will provide important information concerning the effects of exposure protraction and perhaps of radiation type in the relatively low dose (0-500 mSv) range. Protocols for case-control studies of leukaemia and thyroid cancer risk among liquidators are being prepared in Belarus and Russia.

Non-specific studies of cancer risk among the general population exposed in the contaminated regions are unlikely to be informative for radiation risk estimation because of the generally lower doses received by the majority of these populations, the difficulties in estimating these doses and following populations. An exception is the study of thyroid cancer risk in children, the incidence of which has been observed to increase dramatically in the first years following the accident. A careful study appears to be feasible and may provide a unique opportunity to increase our understanding of factors which modify the risk of radiation induced cancer and thus have important consequences for the radiation protection of patients and the general population.

1. Introduction

The accident which occurred in the Chernobyl reactor in Ukraine underlined the fact that radiation accidents - and nuclear safety in general - are a global concern, and cannot be restricted to a single region or country. From the point of view of radiation protection, there were some uncertainties concerning the possible health effects of radiation of the types received by populations as a result of the Chernobyl accident. Indeed, although ionising radiation is one of the best studied environmental carcinogens, a number of questions remain in radiation research and radiation protection today. Some of these questions could, in principle, be answered by epidemiological studies of the consequences of the Chernobyl accident. They include:

- the effects of protracted exposures, which could be examined in studies of liquidators;
- the effects of relatively low doses, such as those resulting from environmental exposures in large areas contaminated by the Chernobyl accident;

- the effects of different radionuclides and different radiation types;
- the effects of factors which may modify radiation induced risks (including age at exposure, sex, possible genetic predispositions, other host and environmental factors).

To provide answers to these questions, however, studies of the consequences of the Chernobyl accident must fulfil several important criteria: they must cover very large numbers of exposed subjects; the follow-up must be complete and non-selective and precise and accurate individual dose estimates (or markers of exposure) must be available. In particular, the feasibility and the quality of epidemiological studies largely depend on the existence and the quality of basic population-based registers, and on the feasibility of linking information on a single individual from different data sources.

When preliminary contacts were made between scientists from the three most affected countries and from Western Europe concerning the possibility of carrying out epidemiological studies of the Chernobyl accident jointly, the feasibility of such studies was unclear to the international research community. By then, following a 1987 directive of the Ministry of Public Health of the USSR, much effort had been invested in the CIS countries into registration and active follow-up – with obligatory annual examinations – of the most affected populations in the State Chernobyl Registries [1]. Little work was published, however, which demonstrated the feasibility of carrying out long-term, large-scale passive follow-up of populations. Thus, the first order of work was the assessment of the feasibility and informativeness of studies of populations exposed as a result of the accident. To this end, a number of activities were started within the framework both of Experimental Collaboration Project 7 (ECP-7) between the CIS states and the European Commission, and of the WHO International Thyroid Project (ITP).

1.1. Liquidators

Efforts within ECP-7 concentrated on the evaluation of the feasibility of studies of "liquidators" or clean-up workers (i.e. the participants in the "liquidation" of the consequences of the Chernobyl accident) since it was thought that this population was likely to be the most informative for the study of radiation effects and the easiest to follow.

In the former USSR, several hundred thousand inhabitants, the majority residing in the Ukraine, the Russian Federation and Belarus, were drafted - or in some cases volunteered - to undertake clean-up activities within the polluted areas. On average these workers spent between one and six months in the region of Chernobyl, and were officially allowed to receive up to about 0.25 Gy external radiation before being discharged from the area. It is expected that during the early phases after the accident, workers may have received a higher dose, in particular those workers having particular expertise, who could not easily be replaced by others. Table 1 presents the information concerning the distribution of doses among liquidators which had been published by the time the ECP-7 programme began [2].

By a decree of the Ministry of Public Health of the USSR in 1987, liquidators must be registered in the Chernobyl Registry of their country of residence at the time of the accident and must undergo a yearly medical examination. They are also entitled to a number of social, health and financial benefits and can be identified by a special certificate and/or a stamp in their military passport.

Table 1 Distribution of levels of radiation dose among populations exposed as a result of the Chernobyl accident (dose estimates from Ilyin et al. [2])

Exposed group	Approximate size	Exposure type	Exposure level	
Evacuees from 30 km zone	135 000	Whole body γ-rays	range	30-500+ mSv
			average	120 mSv
		Thyroid , ^{131}I, children	average	0.3 Gy
			range	0.1->2.5 Gy
Inhabitants of SCZ's[1]	270 000	CEDE[2] from γ-rays	average:	60 mSv
			4%	> 100 mSv
			0.3%	> 200 mSv
		Thyroid , ^{131}I, children	range	0.1->10 Gy
Liquidators	600 000	Whole body γ-rays	45%	< 100 mSv
			47%	100-250 mSv
			8 %	250-500 mSv
			0.02%	> 500 mSv
European part of USSR	75 000 000	Total CEDE	average:	6-7 mSv

Because of their number, their levels of radiation exposures and the fact that they are entitled to benefits - and hence may be easier to follow than the general population -, the population of liquidators from the start was seen as one of the most likely to be informative for the epidemiological study of the effects of relatively low doses of radiation received at different exposure rates.

Concerns existed, however, about the feasibility of obtaining precise and accurate individual dose estimates: official dose estimates were not available for all liquidators. Dosimetric practices varied, moreover, depending on the type of liquidators (civilians, military, radiation specialists, etc.) and estimates were sometimes based on group (i.e. an individual dosimeter was assigned to one member of a group of liquidators) rather than individual monitoring. Moreover, errors may have occurred in the entering of these doses, which had been measured in various different units, on the liquidators' certificates of work, as well as in the Chernobyl Registries. There were therefore questions about the accuracy and comparability of available dose estimates.

1.2. The general population

The information on environmental radiation exposure levels of the general population available at the time (Table 1) was mainly based on tens of thousands of direct measurements of external and internal dose made mainly for persons residing on the most contaminated territories [2] and on models of environmental transport of radionuclides and of behaviour of individuals at the time of the accident. Given the levels of exposure, the absence of systematic individual exposure estimates in populations with different contamination levels, and the amount of population movement thought to have taken place since the accident, it was judged from the start that non-specific mortality and/or cancer morbidity follow-up of the general population living in contaminated territories would be more difficult, and potentially less informative, than studies of liquidators. ECP-7 activities

---

[1] SCZ: strict control zones, defined as areas with Cs-137 contamination above 40 Ci/km^2

[2] CEDE: committed effective dose equivalent

were therefore concentrated at first on the assessment of the feasibility of epidemiologic studies of liquidators and not of the general population.

Specific studies in the general population may be important and feasible, however. Reports [3,4] of a large increase in thyroid cancer incidence in children, particularly in the Gomel area of Belarus, were initially met with scepticism, because of the early occurrence of the increase (as early as four years after the accident) and the very high incidence of the disease reported (annual age specific incidence in children under the age of 15 in 1991-94: 9.6 per 100 000 in Gomel region and 2.9 per 100 000 in Belarus as a whole [5]). In addition, little increase was observed in other regions thought to be nearly as highly contaminated as the Gomel area (Mogilev, in Belarus, Bryansk in Russia), while relatively large numbers of cases were seen among children residing in regions thought to be less contaminated (Brest and Grodno). Since then, increased incidences have, however, also been reported in the Ukraine [6] and, more recently, in the part of Russia most contaminated by the accident [5,7]. The observation of such a large and early increase in incidence raised the possibility that host and environmental factors may be playing a role in the risk of radiation induced cancer. Because of the rarity of this disease, this situation provided a unique opportunity to identify such factors and quantify their effect. Activities within the ITP therefore focused on additional documentation of this increase and on the evaluation of the feasibility of a study of thyroid cancer risk in children [8].

2. Materials and methods

2.1. Liquidators

In October 1993, three pilot studies aimed at assessing the feasibility of a long-term epidemiological study of cancer risk among "liquidators" of the Chernobyl accident (sometimes referred to as "emergency accident workers" or "clean-up workers") were started in Belarus and the Russian Federation [9] within the framework of ECP-7:

- Pilot study I   Test of follow-up mechanisms
- Pilot study II   Case-control study of leukaemia among liquidators
- Pilot study III   Calibration/validation of dosimetry

These three studies were developed in parallel to test different epidemiological approaches and provide the information necessary to identify which studies are feasible, how they should be designed and what are the requirements, in terms of expertise, equipment, existence of population registries and funds, for them to be informative for the understanding and quantification of radiation induced cancer risk.

The first two studies were carried out in both Belarus and the Russian Federation, the third in Russia alone as the sample size required was small. No effort was made to set-up similar studies in the Ukraine since plans for such activities were already underway in collaboration with scientists from Canada and the USA. The study population for these studies was defined as all 1986-87 liquidators having worked in the 30 km zone around the Chernobyl reactor and included in the State Chernobyl Registries of Belarus and Russia.

2.1.1. Test of follow-up mechanisms

The specific objectives of the first study, the test of follow-up mechanisms, were to assess the feasibility of long-term passive mortality and incidence follow-up of liquidators in

Belarus and in Russia and to examine available mechanisms for tracing individual subjects and collecting relevant data pertaining to these individuals.

The study covered 500 liquidators in each country, selected at random from a computerised list of the study population. In Russia, due to difficulties in carrying out a State wide follow-up, the study was restricted to three regions - Rostov area, Moscow city and region and St-Petersburg city and region - and the random sample was stratified on region. The following step-by-step approach was used to ascertain vital status at the study cut-off date (1 October 1993 in Belarus and 31 December 1992 in Russia):

- a subject was assumed alive at the end of the study if a recent (within the last 3 months in Belarus and 6 months in Russia) medical visit had been recorded in the Chernobyl Registry;

- where a recent medical visit was not registered, the address bureau of the area of last known place of residence (based on last address in Chernobyl Registry) was to be contacted in order to confirm the address; if the person had moved, a visit was made to the address bureau of that locality to ascertain the subject's new address; if the person had moved outside Belarus, the date and country of emigration were obtained;

- if the subject's current vital status could not be ascertained through the relevant address bureaus, the records of the local Chernobyl Registries and hospitals and polyclinics were consulted. If no information could be obtained from these sources, the records of the population registry (ZAGS - buro zapicii akta grazhdanskovo sostoyania) at the last known place of residence were to be consulted to determine whether the person had died; if so, a copy of the death certificate was to be obtained;

- for subjects who could not be traced at all, the date of last known vital status (and the source of this information) was to be noted.

2.1.2. Pilot case-control study of leukaemia

The objective of the second study, the case-control study of leukaemia among liquidators, was to assess the feasibility of carrying out case-control studies of cancer risk among Chernobyl liquidators by examining available mechanisms for identifying cases of leukaemia among liquidators and for selecting appropriate controls and by testing a questionnaire (adapted from the Estonian-US-Finnish collaborative study of liquidators) to collect information about work as a Chernobyl liquidator as well as about possible non radiological risk factors for cancer.

In Belarus, the study was restricted to all cases diagnosed in the period 1991-92 in the study population, while in Russia, it included all cases diagnosed to the end of 1993 in two regional centres of the State Chernobyl Registry, that of the North Caucasus and that of the Central-Chernozem region.

The mechanism for case ascertainment differed between Belarus and Russia. In Belarus, the primary source for case ascertainment was the Registry of Haematological Diseases; the Cancer Registry and Chernobyl Registry were used as complementary sources to ensure completeness of ascertainment. In Russia, the sole source was the State Chernobyl Registry as no cancer registry exists at present. To be included in the study, all cases had to be confirmed by a haematologist in Minsk or Obninsk.

Controls were drawn at random from the study population on the basis of the computerised listings of the State Chernobyl Registries. The questionnaire was administered in person, usually at the home of the study participants, by a trained interviewer.

2.1.3. Validation/calibration of biological dosimetry

The third study, calibration/validation of dosimetry, was designed to assess the feasibility of using biological markers of radiation damage, namely stable chromosome aberrations scored through fluorescent in-situ hybridisation (FISH), to estimate individual radiation doses, or at least to provide a gradient of individual radiation doses, which could be used in epidemiological studies of cancer risk among liquidators.

For this purpose, a random sample of liquidators residing in the Obninsk area of Russia was selected, stratified on the total official dose registered in the Chernobyl Registry as follows: 31 with total official doses below 100 mGy (4 had 0 dose), 18 with official dose between 100 and 200 mGy and 13 with official dose above 200 mGy.

The study subjects were asked to present themselves to a clinic in Obninsk where a blood sample was drawn and a simple questionnaire was administered. Questions concerned their work in the 30 km zone and exposures (such as smoking and viral infections) which may influence the level of stable aberrations. Blood samples were sent to the NRPB in the UK for cell culture and preparations were distributed to partner laboratories in the CIS and Western Europe for FISH and conventional dicentric analyses (in the CIS only).

2.2. General population studies - thyroid cancer in children

In 1993-94 a survey of 50 childhood thyroid cancer cases was carried out in Belarus [10] within the framework of the ITP, in preparation of a possible case-control study. The objectives of the survey were to better characterise the reported increase in childhood thyroid cancer and to test mechanisms for identifying and interviewing cases, in preparation of a possible case-control study.

The study subjects were the first 50 cases of childhood thyroid carcinoma who attended the clinic of the Institute of Radiation Medicine between December 1993 and April 1994. A trained nurse interviewed the parents - preferably the mother - of the cases to obtain detailed information about the circumstances of diagnosis of thyroid cancer (screening vs. self presentation) and the behaviour of children around the time of the accident.

3. Results[3]

3.1. Test of follow-up mechanisms

Follow-up was virtually complete (99.2%) in both countries (Table 2). In Belarus, one hundred and thirty eight (27.6%) subjects had had a recent medical examination registered in the Chernobyl Registry; the vital status of 253 subjects (i.e. 69.9% of those who had not had a recent medical visit registered in the Chernobyl Registry) was ascertained passively, through the address bureaus, and the remainder through the local registries (100 subjects) and hospitals and polyclinics (Table 3).

---

[3] Detailed results of the studies presented here are available elsewhere [9,10].

Table 2: Trace rates and distribution of vital status

Follow-up	, Belarus	Russia			
		Total	Rostov	Moscow	St Petersb.
Number of subjects	500	500	300	100	100
Vital Status:					
Lost to follow-up	4 (0.8%)	4 (0.8%)	3	0	1
Alive	489 (97.8%)	457 (91.4%)	267	94	96
Dead	5 (1.0%)	17 (3.4%)	13	2	2
Emigrated	2 (0.4%)	22 (4.4%)	17	4	1

Table 3: Sources of information used to determine vital status

Source	Belarus	Russia			
		Total	Rostov	Moscow	St Petersb.
Chernobyl Registry	138 (27.6%)	299 (59.8%)	211	57	31
Passport department or address bureau	253 (50.6%)	53 (10.6%)	0	9	44
ZAGS	0	0	0	0	0
Other and combination	109 (21.8%)	152 (30.4%)	86	43	23
Unknown	0	3 (0.6%)	3	0	0

In Russia, 59.6% of the subjects had had a medical examination registered in the Chernobyl Registry in the previous 6 months (Table 3). For the remainder, however, the main source of information was active tracing through the local registries, hospitals and other medical facilities; 4% were traced by personal contact. Only 53 were traced passively, through passport departments or address bureaus. It is noted that, in a full-scale study, the resources associated with direct contact with all but a very small proportion of liquidators could be considerable. It would therefore be highly important to concentrate on the most probable avenues of tracing (e.g., the Chernobyl Registry, passport department and address bureau, local registries, etc.) before turning to this option.

Among subjects who were traced successfully in Belarus, five (1%) were found to have died - and cause of death was obtained from the Chernobyl Registry and the ZAGS - and two (0.4%) emigrated (place unknown) (Table 2). In Russia, the corresponding figures were 17 (3.4%) and 22 (4.4%); the place of emigration was known for all twelve of these; all but two had moved to other parts of Russia. In Belarus, the cohort was also linked to the Cancer Registry and one case of cancer was identified. Information on cancer diagnosis was obtained in Russia from the Chernobyl Registry: two cases were found.

Since the samples of liquidators in this pilot study were chosen at random, descriptive analyses of the data collected provide some comparison between characteristics of Belarus and Russian liquidators. The former tended to be younger (median age at the time of the accident 29 compared to 35) and to have received substantially lower doses (when known) than the latter (median dose 48 compared to 130 mSv), even though the majority of Belarus liquidators had worked in the first six months following the accident. This difference in age and doses may be due to many Russian liquidators having been specialists sent to Chernobyl for specific missions. The majority (74%) of Belarus liquidators in the sample had missing dose estimates, moreover, possibly reflecting the incomplete personal monitoring performed in the first few weeks after the accident.

3.2. Pilot leukaemia case-control studies

The means of identifying leukaemia cases differed between Belarus and Russia. As discussed above, in Belarus, the primary source of case ascertainment was the Republican Registry of Haematological Diseases; secondary sources were also used to ensure completeness of ascertainment. As indicated by Storm *et al.* in another paper in this conference [11], the Registry of Haematological Diseases appears to provide adequate coverage of recent haematological malignancies diagnosed in Belarus.

In Russia, details of cases recorded on the State Chernobyl Registry were sent to the corresponding regional centres and departmental registries of the Chernobyl Registry in order to confirm or disprove the leukaemia diagnosis. Out of 117 cases on the Chernobyl Registry, only 60 were confirmed; 48 were not confirmed and nine were neither confirmed nor disproved. This low level of confirmation may be explained by the fact that diagnostic information in the Chernobyl Registry is primarily obtained from local clinics; thus, if the diagnosis of a patient referred to a regional or national level medical centre for confirmation of diagnosis and treatment is changed, the information is not necessarily sent to the Chernobyl Registry. Within a case-control study, this is of concern mainly from the viewpoint of the resources required for the checking of potential cases that are ultimately not confirmed. This also raises questions about the completeness of cancer data recorded in the Chernobyl Registry since some cancers may be missed if the initial diagnosis is not cancer. It is therefore important to perform cross-checking against data held in regional oncological dispensaries where, as indicated in [11], the completeness of cancer data is likely to be high. The low level of confirmation of cancers registered in the Chernobyl Registry must however be kept in mind, when interpreting results of other types of studies of disease patterns based solely on that Registry.

Some of the other problems encountered in this study were common to both Belarus and Russia. For example, it was generally considered that the questionnaire used was too long and that some questions were not informative (e.g., most liquidators did not know whether they had consumed locally produced food since they ate in canteens). These points are valuable in deciding how to rearrange the questionnaire for a full-scale study.

There were also logistical problems associated with the tracing of cases and controls: the current address given in the Chernobyl registries was often out of date and additional efforts were required to locate the new address. On occasion, time was expended visiting the old address in order to determine the new one. It would be preferable in future to check other sources such as the address bureau before resorting to more labour and time-expensive options such as visits to last known addresses. Time and transport problems arising from travel to the liquidators' homes were also cited by the interviewers as a problem.

Owing to the small number of cases included in the study (6 in Belarus and 16 in Russia), it was not possible to make meaningful comparisons between cases and controls in this pilot study. However, various other types of comparison were possible. For example, some data were common to the questionnaire and to the database of the Chernobyl Registry. In Belarus it is difficult to examine the consistency of information from these two sources, since the Chernobyl Registry was used as the primary source in most instances. However, such comparisons were possible using the data from Russia. There was generally good agreement between the Russian Chernobyl Registry and the questionnaire data on variables such as the reason for the liquidator's presence in or near Chernobyl, the dates of entry to and departure from this region, and the external radiation dose. However, it would be worth determining whether there is an alternative data source that could be used in deciding whether to utilise the Chernobyl Registry or the questionnaire as the main source of information on the official dose for a full-scale study.

Comparisons were also made of characteristics of subjects in Belarus and Russia. The proportion of liquidators in the regular army and the proportion of subjects who entered the Chernobyl region before the end of June 1986 were higher in Belarus than in Russia. As indicated in Study I above, this latter finding explains in part the higher proportion of missing doses in the military passport for Belarus liquidators. More Russian subjects worked in the special zone and for a longer period than their Belarus counterparts. This may again be explained in part by the Russian liquidators having worked at a later time than those from Belarus, as well as which the Russians tended to be radiation specialists.

The most common activities performed by the subjects were earth removal, work in the forest and transport. The majority of subjects in both countries reported having used breathing masks or respirators often or always, but few wore gloves or were given iodine tablets. Based on the data from Russia, no association was found between the official radiation dose estimates and aspects of work at Chernobyl for which information was available such as date of entry or reason for being in the zone.

3.3. Calibration/validation of dosimetry

There was no correlation between official radiation dose, as registered in the Chernobyl Registry, and the percentage of translocation of the genome as estimated by FISH or conventional dicentric analyses in any of the participating centres (Fig. 1). There was also no correlation between the results of biological analyses in any of the participating centres or between these results and characteristics of work as a liquidator. The number of cells scored for each individual is, however, relatively small (range: 74 to 1506; few individuals had more than 1000 cells scored) and it is thought that, in the range of doses received by the liquidators under study, several thousand of cells per individual would need to be scored for results to be meaningful.

FISH analyses are very labour intensive and time-consuming, however, and even in the framework of this small-scale pilot study - and despite the fact that analyses were redone to increase the number of cells analysed - it was not possible to score sufficient numbers of cells per individual. It therefore appears that, in the range of exposures received by the majority of liquidators, studies of chromosomal aberrations - whether conventional or through FISH - are unlikely to provide accurate and precise individual dose estimates for use in large scale epidemiological studies, at least without prohibitive costs.

The association between official doses and characteristics of work as a Chernobyl liquidator was also considered in this study. Official dose estimates were found to correlate with duration of work as a liquidator and with the fact of having worked in the zone of 500 m around the reactor, on the roof of the reactor or on the construction of the sarcophagus. This information, together with that obtained in pilot studies I and II above, and with information of the type of work done in the zone and the likely level of exposure in different jobs, locations and time periods, may prove useful for the construction of a gradient of exposure which could be used in epidemiologic studies.

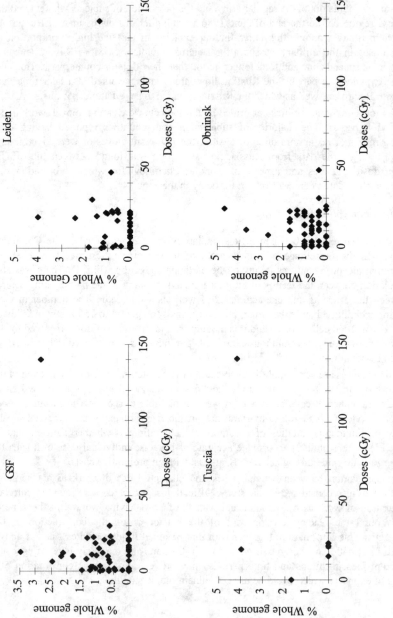

Fig 1. Distribution of percentage of translocation of whole genome by dose and laboratory.

3.4. Additional documentation of childhood thyroid cancer

Table 4 shows the distribution of circumstances in which the diagnosis of thyroid cancer was made among the 50 cases. The majority of participants were diagnosed as having had thyroid problems through one form of screening or another - this includes: systematic screening programmes, routine medical examinations, medical examination for other reasons or medical consultation unrelated to any symptom or illness, because the mother was worried (Table 4). Out of 50 children, only 10 were diagnosed when the mother consulted a physician because they were unwell. The diagnosis was made on the basis of clinical examination in 35 out of 43 children for which this information is available, and on ultrasound screening in eight.

The survey showed that the majority of the cases had behaviours around the time of the accident which would have maximised their exposure to iodine isotopes: 66% were in the habit of drinking milk from local cows; the majority spent all day (eight hours or more) out of doors everyday in the period 26 to 30 April 1986 and slept with open windows in the days following the accident. Only eight of the children were reported to have received stable iodine in the weeks following the accident. This information must, however, be interpreted with caution as no information is available on the behaviour of controls.

Available ^{131}I dose estimates on 24 of the 50 study participants indicate that doses among these children were not as high as might be expected given the fact that several hundred children in Belarus received doses of 10 Gy or more to the thyroid (Table 1). The available dose estimates in this study were relatively low, ranging from 0 to 2 Sv, with an average of 0.55 Sv. No information on dose from short lived isotopes was available.

There was a strong correlation between age at the time of the accident (or year of birth) and age at diagnosis, with younger children at time of exposure being diagnosed at earlier ages (Figure 2), in particular for girls. Within the range of ages at exposure in the survey (0 to 9 years old), the length of time between exposure and diagnosis does not appear to be correlated with age at time of the accident; the average time between exposure and disease is of the order of 6 years. The distribution of ages at exposure and diagnosis among boys was fairly uniform. In girls, however, peaks were observed around the ages of 1 and 2 years at the time of exposure and 8 and 12 years at the time of diagnosis.

Both the interviewee and the interviewer were given the opportunity to provide additional information which they judged relevant to the tumour occurrence or the situation at the time of the accident. Two out of the 50 children were reported to have a sibling who was also operated for thyroid cancer. Even in some areas of Gomel oblast, where the iodine contamination is likely to have been the greatest and the prevalence of thyroid cancer in children has been estimated to be as high as 1 per 1 000 up to the end of 1992, the probability of observing two or more sibling pairs with thyroid cancer is extremely low. Additional information suggests that at least one other sibling pair of thyroid cancer cases exists in Pinsk (Brest oblast). The possibility that there is some genetic predisposition to radiation induced thyroid cancer in children should therefore be assessed in an analytical epidemiologic study.

Three children were also reported to have received "antistrumin", a preparation containing 1 mg of potassium iodine, on a regular basis (once per week) at school for periods ranging from 6 months to 5 years. It is thought that iodine deficiency, which appears to have been - and to remain - common in some of the areas most affected by the

Table 4. Circumstances of diagnosis of thyroid problem

Circumstances of diagnosis	Ultrasound diagnosis			Total
	unknown	no	yes	
Formal screening programme	2 (28.6%)	7 (20.0%)	3 (37.5%)	12
Incidental to other examination	2 (28.6%)	18 (51.4%)	3 (37.5%)	23
Consulted since unwell	1 (14.3%)	8 (22.9%)	1 (12.5%)	10
Consulted even though well	2 (28.6 %)	2 (5.7%)	1 (12.5%)	5
Total	7	35	8	50

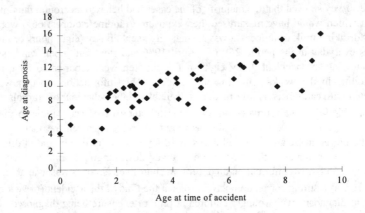

Fig 2. Distribution of age at diagnosis by age on 24/04/86

Chernobyl accident, may affect the risk of radiation induced thyroid cancer both by increasing the uptake of radioactive iodine and by increasing the production of thyroid stimulating hormone (TSH), which increases cell proliferation in the thyroid gland [12]. On the other hand, systematic prophylaxis with large doses of stable iodine, which the survey revealed has occurred for several years after the accident in a number of schools, may influence the frequency of thyroiditis. The role of iodine status in the observed increases in thyroid cancer must therefore be assessed in a careful epidemiological study.

4. Discussion

4.1. What is feasible

4.1.1. Liquidator studies

The results of the pilot studies of liquidators indicate that it is feasible to carry out case-control studies of specific cancers in liquidators using the Chernobyl Registry as the source population. Approaches for ascertaining cases, selecting controls and interviewing study subjects have been examined and a detailed questionnaire tested.

With sufficient financial and human resources, it is also possible to carry out a long-term mortality follow-up of cohorts of liquidators in Belarus and in Russia. The circumstances for follow-up differ, however, between the two countries. Since there are plans to centralise the address bureau information at the Republican level in Belarus, follow-up for vital status ascertainment based on that source together with the Chernobyl registry appears to be feasible. In Russia, the difficulties of performing a centralised follow-up meant that the ascertainment of vital status in the pilot study was performed mainly actively, via either the main Chernobyl Registry or the regional registries and their contacts with local hospitals and polyclinics. This approach, within a full scale study, would require considerable resources and its success would depend both on the continued support of the Chernobyl Registry over time and on the continued participation of liquidators in the annual medical examinations.

At present, therefore, no cohort study is planned in Russia and two case-control studies are being set up in Belarus and Russia as a result of the ECP-7 pilot activities. They are studies of leukaemia and of thyroid cancer risk among 1986-87 liquidators having worked in the 30 km zone and included in the State Chernobyl Registries. The detailed protocols for these studies are being prepared. The approach for ascertaining cases and controls will be similar to that used in the pilot studies. The questionnaire is being modified in the light of the responses to the pilot study.

Dosimetry still poses an important problem. The results of the pilot studies indicate that some characteristics of the liquidators work, such as duration of work in the 30 km zone, may not be correlated with radiation dose, while others, such as the fact of having worked on the construction of the sarcophagus, are associated with doses recorded in the Chernobyl Registry. Work is now underway to gather more detailed information on the types of work performed by liquidators and the likely level of exposure in different jobs, locations and time periods from persons familiar with the work in the 30 km zone (liquidators themselves and persons responsible for co-ordinating the work). The aim of this work is to formulate a short and simple questionnaire, which liquidators can still answer 10 years following the accident, and which will allow the construction of a gradient of probable individual exposures.

4.1.2. Case-control studies of the general population

The pilot survey described here has shown that it is feasible to obtain detailed information about behaviour of children around the time of the accident and about the circumstances of diagnosis of thyroid cancer from the parents, particularly the mothers, of cases seven to eight years after the Chernobyl accident. Given the very large increase in thyroid cancer incidence in Belarus, Ukraine and the contaminated territories of Russia, one or more important risk factor for this disease must be present and the pilot survey has provided important hypotheses to identify them.

Information derived from the pilot studies indicate that case ascertainment is feasible using the records of the Belarus Cancer Registry and those of the Centre for Thyroid Pathology (where virtually all the thyroid cancers in young persons are operated in Belarus). Selection of controls from the general population (rather than from the Chernobyl Registry as was done in the liquidator studies) has not been tested in the pilot projects described here. It has, however, been carried out in a Belarus-US case-control study of radiation and childhood thyroid cancer also reported in this conference [13]. Although it posed some logistic difficulties, because of the lack of a centralised population list from which to draw, it was shown to be feasible.

A case-control study of factors which are modifying the risk of 131-I induced thyroid cancer in children is therefore feasible and a detailed protocol for this study is now being developed in Belarus in close collaboration with pathologists, dosimetrists, geneticists and molecular biologists.

4.2. What is desirable

As mentioned above, there are a number of outstanding questions in radiation research and radiation protection which could in principle be answered by epidemiological studies of the consequences of the Chernobyl accident.

On the basis of the work described here, we believe that case-control studies of cancer risk among liquidators are desirable in order to study the effects of exposure protraction in the relatively low dose range and the effects of different types of radiation. Given the distribution of known doses among the liquidators, the power of these studies in individual countries is relatively low, however, and it is important that studies be set up in the three mainly affected countries (Belarus, Russia and Ukraine)[4] and that the protocols be similar so that the results can ultimately be compared and combined. At this time, case-control studies of leukaemia and of thyroid cancer are particularly important: leukaemia because in high dose studies it is among the most radiogenic cancers and its latency is relatively short (increases have been observed as early as two to five years after exposure) compared to solid tumours. Studies of this disease could therefore be informative in the short to medium term (3-5 years). Thyroid cancer is also of interest as sporadic reports of increases in the incidence of this disease among liquidators have appeared. In the future, it may also be appropriate to set-up case-control studies of other cancer types depending, in part, on the results of the studies of leukaemia and thyroid cancer. Once the approach has been set-up for one cancer type, however, the difficulty and cost of additional studies should be minimal.

The observation of a very early and large increase in the incidence of thyroid cancer in children and young adults in Belarus, and later in the Ukraine and Russia, raises the important possibility that host and environmental factors may be playing a role in the risk of radiation induced cancer. This is confirmed by the preliminary results of a case-control study of thyroid cancer in relation to dose from ^{131}I carried out in Belarus [13]: although a dose-related increase in risk has been observed in this study, a large number of cases received low thyroid doses (estimated to be below 0.3 Gy). Our survey has resulted in the identification of important hypotheses concerning factors which may modify the association between radiation dose and thyroid cancer. These are a possible genetic predisposition and iodine status. In addition, it is important to assess the role of short lived isotopes in the observed increases and whether exposure to these isotopes implies a greater risk per unit of radiation dose than external irradiation or contamination with ^{131}I.

Because of the rarity of this disease, the observed increase in thyroid cancer incidence in children and adolescents is a unique opportunity to identify such factors and quantify their effect. It is noted that such factors may be important not only for thyroid cancer but also, possibly, for other radiation induced cancers. This work, in addition to contributing to the understanding of the mechanisms of carcinogenesis, may therefore have important

---

[4] Note: studies of cancer risk among liquidators are already underway or planned in the Baltic countries.

public health implications, in Europe and elsewhere, for the protection of patients treated with radiotherapy, of radiation workers as well as of the general population in the event of further nuclear accidents.

5. Conclusion

The results of pilot studies indicate that it is feasible to carry-out long-term epidemiological studies of cancer risk among Chernobyl liquidators included in the State Chernobyl registries using a case-control approach in Belarus and Russia. The cohort study approach is also feasible, but would require substantial financial and human resources, except in Belarus where circumstances exist for relatively inexpensive systematic passive follow-up of populations. Given the distribution of known doses among the liquidators, the power of such studies in individual countries is low, however, and it is essential that studies carried out in different affected countries be similar so that the results can ultimately be compared and combined. Efforts must be made to estimate radiation levels from detailed questionnaires of activities in the Chernobyl area, as the official dose estimates are incomplete and biological estimation is too imprecise in the range of doses received by the majority of liquidators. If these conditions are met, studies of liquidators will provide important information concerning the effects of exposure protraction and perhaps of radiation type in the relatively low dose (0-500 mSv) range.

Non-specific studies of cancer risk among the general population exposed in the contaminated regions are unlikely to be informative for radiation risk estimation because of the generally lower doses received by the majority of these populations, the difficulties in estimating these doses and following these populations. An exception is the study of thyroid cancer risk in children, the incidence of which has been observed to increase dramatically in the first years following the accident. A careful case-control study appears to be feasible and may provide a unique opportunity to increase our understanding of factors which modify the risk of radiation induced cancer and thus have important consequences for the radiation protection of patients and the general population.

Acknowledgements

The authors are indebted to the Nuclear Fission Safety Programme of the European Commission, to the WHO European Centre for Environmental and Health in Rome and to the Ministries of Health and Chernobyl affairs of Belarus and Russia for financial support of these studies and of the Chernobyl Registries. They also wish to acknowledge the help of the other ECP-7 partners, from the CIS and Western Europe, in the preparation and conduct of these studies.

References

[1]   A.F. Tsyb and et al. Registry material, Radiat. Risk 1 (1992) 67 - 131.
[2]   L.A. Ilyin, M.I. Balonov, L.A. Buldakov, V.N. Bur'yak, K.I. Gordeev, S.I. Dement'ev et al. Radiocontamination patterns and possible health consequences of the accident at the Chernobyl nuclear power station, J. Radiol. Prot. 10 (1990) 3 - 29.
[3]   V.S. Kazakov, E.P. Demidchik, and L.N. Astakhova, Thyroid cancer after Chernobyl, Nature 359 (1992) 21.

[4]   E.P. Demidchik, V.S. Kazakov, L.N. Astakhova, A.E. Okeanov, and Y.E. Demidchik, Thyroid cancer in children after the Chernobyl accident: clinical and epidemiological evaluation of 251 cases in the Republic of Belarus. In: S. Nagataki (Ed.), Elsevier Science, B.V. Amsterdam, 1994, pp. 21 - 30.

[5]   V.A. Stsjazhko, A.F. Tsyb, N.D. Tronko, G. Souchkevitch, and K.F. Baverstock, Letter to the editor, British Medical Journal 310 (1995) 801.

[6]   N.D. Tronko, Y. Epstein, V. Oleinik et al. Thyroid gland in children after the Chernobyl accident (yesterday and today). In: S. Nagataki (Ed.), Elsevier Science, B.V. Amsterdam, 1994, pp. 31 - 46.

[7]   A.F. Tsyb, E.M. Parshkov, V.K. Ivanov, V.F. Stepanenko, E.G. Matveenko, and Y.D. Skoropad, Disease indices of thyroid and their dose dependence in children and adolescents affected as a result of the Chernobyl accident. In: S. Nagataki (Ed.), Elsevier Science, B.V. Amsterdam, 1994, pp. 9 - 19.

[8]   K.F. Baverstock and E. Cardis, The WHO activities on thyroid cancer. First International Conference of the European Commission, Belarus, the Russian Federation and the Ukraine on the radiological consequences of the Chernobyl accident (Minsk, Belarus, 18-22 March 1996). IOS Press, Amsterdam, 1996.

[9]   A.E. Okeanov, V.C. Ivanov, E. Cardis, E. Rastoptchin, A. Sobolev, C. Lavé, and A. Mylvaganam, Study of cancer risk among liquidators, Report of EU Experimental collaboration project 7: Epidemiologic Investigations including Dose Assessment and Dose Reconstruction. IARC Internal Report 95/002 International Agency for Research on Cancer, Lyon, 1995.

[10] L.N. Astakhova, E. Cardis, L.V. Shafarenko, L.N. Gorobets, S.A. Nalivko, K.F. Baverstock, and A.E. Okeanov, Additional documentation of thyroid cancer cases (Belarus): Report of a survey, International Thyroid Project. IARC Internal Report 95/001 International Agency for Research on Cancer, Lyon, 1995.

[11] H.H. Storm, A.E. Prisyazhniuk, A.E. Okeanov, V.K. Ivanov, and L. Gulak, Development of infrastructure for epidemiological studies in the three CIS republics. First International Conference of the European Commission, Belarus, the Russian Federation and the Ukraine on the radiological consequences of the Chernobyl accident (Minsk, Belarus, 18-22 March 1996). IOS Press, Amsterdam, 1996.

[12] Thomas GA and Williams ED. Evidence for and possible mechanisms of non-genotoxic carcinogenesis in the rodent thyroid. Mutat Res 1991;248(2):357-70.

[13] G.W. Beebe, Epidemiological studies of thyroid cancer in the CIS. First International Conference of the European Commission, Belarus, the Russian Federation and the Ukraine on the radiological consequences of the Chernobyl accident (Minsk, Belarus, 18-22 March 1996), IOS Press, Amsterdam, 1996.

Health Status and follow-up of the
liquidators in Belarus

Alexey E. OKEANOV[1], Elisabeth CARDIS[2], S.I. ANTIPOVA[1],

Semion M. POLYAKOV[1], Alexander V. SOBOLEV[1], Natalya V. BAZULKO[1]

[1]*Belarus Centre for Medical Technologies, Information Computer Systems, Health Care
Administration and Management. 7-a Brovki St, 220600, Minsk, Belarus and* [2]*International
Agency for Research on Cancer, 150 Cours Albert Thomas, 69372 Lyon Cedex 08, France*

Abstract. This paper presents information on the organisation of the follow-up of
Chernobyl liquidators in Belarus. The characteristics of the liquidators cohort and
results of preliminary analyses of their health status, including cancer incidence and
general morbidity, are presented.

1. Introduction

The data presented in this paper were obtained from two health registries functioning in
Belarus: the State Registry of Individuals Exposed to Radiation as a Result of the
Chernobyl Accident (which is referred to as the "Chernobyl Registry") and the Belarus
Cancer Registry.

The Chernobyl Registry has functioned in Belarus since 1987. It was established by a
directive of the Ministry of Public Health of the USSR as a comprehensive registration and
follow-up system for the persons most affected by the Chernobyl accident. The directive
identified four groups of subjects - the groups of "primary registration" - for whom
registration and follow-up was mandatory. Group 1 consists of participants in the
"liquidation" of the consequences of the Chernobyl accident, the so-called "liquidators" or
clean-up workers. Information contained in the Chernobyl Registry includes demographic
("passport") variables, including the group of primary registration, information on location
and behaviour (food and milk consumption, time spent in contaminated zones) around the
time of the accident and work in the Chernobyl area, dosimetric information (when available)
and medical information, updated periodically to include the results of the obligatory annual
medical examination and all diagnoses and treatments made at the raion (district) and oblast
(region) levels (from local dispensaries and clinics), as well as at the state level (from
republican clinics and institutes).

The central Cancer Registry began functioning in Belarus in 1973. Data are available on
magnetic tapes since 1978. Until 1985, however, the information stored in the computer
files did not include the names and addresses of the cancer patients. In 1985, a computer

Table 1 Distribution of Belarus liquidators by place of work, sex and duration of stay in contaminated area

Place of work	Duration (days)	Men	Women
30 km zone	less than 30	15 541	2 117
	more than 30	12 424	1119
	Total	27 965	3 236
Area with Cs^{137} contamination level greater than 555 kBq/m^2	less than 30	3 953	1 182
	more than 30	8 110	1 228
	Total	12 063	2 410

system of dispensary control for cancer patients was set up in the oncological dispensaries of Belarus. This system facilitated long-term prospective data collection on, and follow-up of cancer patients. Since 1991, this system has functioned on personal computers in all the oncological dispensaries in the country. The Belarus Cancer Registry registers all cases of malignant neoplasms, including diseases of lymphatic and hematopoietic tissues, and carcinoma in situ.

2. Characteristics of the Belarus Liquidator Cohort

At the beginning of 1995, more than 63 thousand liquidators were registered in the Belarus Chernobyl Registry. Place of work in the Chernobyl region was known for 45 674 of them (the results presented in this paper are restricted to this sub-cohort). Among these, 31 201 liquidators worked in the evacuation area around the Chernobyl power station (30 km zone), 14 473 worked in the areas of primary or further relocation (with 137-Cs contamination levels higher than 15 Ci/km^2, i.e. 555 kBq/m^2). Fifteen percent were women (10 284). Table 1 shows the distribution of liquidators by place of work, sex and duration of stay in the contaminated area. The distribution of liquidators by age in 1994 is shown in Table 2.

3. Follow-Up

3.1 Test of Passive Follow-up in the Framework of ECP-7

In 1993-1994, a liquidator follow-up study was carried out in the framework of Experimental Collaboration Project 7 (ECP-7) - "Epidemiological Investigations Including Dose Assessment and Dose Reconstruction"- between CIS States and the European Union [1]. The objectives of this study were to assess the feasibility of long-term passive mortality and incidence follow-up of liquidators, and to examine available mechanisms for tracing individual subjects and collecting relevant data pertaining to these subjects.

The study covered 500 liquidators, selected at random from the population of 1986-87 liquidators having worked in the 30 km zone and included in the Chernobyl Registry.

Table 2 Distribution of Belarus liquidators by age in 1994

Age range in 1994	Men	Women
<30	2 010	549
30-40	17 964	1 539
40-50	11 816	1601
50-60	5 737	1 413
60+	1 903	468
Total	39 430	5 570

A step by step approach was used to ascertain vital status on 1 October 1993:

- a subject was assumed alive on 1 October 1993 if a recent (within the last 3 months) medical visit had been recorded in the Chernobyl Registry;

- where a recent medical visit was not registered, the address bureau of the area of last known place of residence (based on last address in Chernobyl Registry) was contacted in order to confirm the address; if the person had moved, a visit was made to the address bureau of that locality to ascertain the subject's new address; if the person had moved outside Belarus, the date and country of emigration were obtained;

- if the subject's current vital status could not be ascertained through the relevant address bureaus, the records of the local Chernobyl Registries and hospitals and polyclinics were consulted. If no information could be obtained from these sources, the records of the population registry (ZAGS - buro zapicii akta grazhdanskovo sostoyania) at the last known place of residence were to be consulted to determine whether the person had died; if so, a copy of the death certificate was to be obtained;

- for subjects who could not be traced at all, the date of last known vital status (and the source of this information) was to be noted.

Follow-up was virtually complete (99.2%). One hundred and thirty eight (27.6%) subjects had had a recent medical examination registered in the Chernobyl Registry; vital status of 253 subjects (i.e. 69.9% of those who had not had a recent medical visit registered in the Chernobyl Registry) was ascertained through the address bureaus and the remainder through the local registries (100 subjects) and hospitals and polyclinics.

Among subjects who were traced successfully, five (1%) were found to have died - and cause of death was obtained from the Chernobyl Registry and the ZAGS- and two (0.4%) emigrated (place unknown). The cohort was also linked to the Cancer Registry and one case of cancer was identified. The diagnosis was Hodgkin's disease (ICD 9 code 201) and morphological confirmation was obtained.

An analysis of data contained in the Chernobyl Registry for this cohort was carried out. The mean age at the time of the Chernobyl accident was 31.3 years (53% of the liquidators were below the age of 30; 31.5% between 30 and 39 and 15.5% were 40 or above). Information on social group was available for all study subjects: most were employees (68.8%) or industrial workers (27.4%). The overwhelming majority were sent to the 30 km zone around the Chernobyl reactor on mission, following the accident (97.6%); five subjects were actually residing in the area at the time of the accident and three were employed there.

Most liquidators worked in the 30 km zone in 1986 (82.6%); 46% first entered the 30 km zone between April 26 and 30 June of that year; 41.8% of liquidators were in the 30 km zone for less than one month. Thirty-one liquidators were sent to the 30 km zone a second time.

Dose estimates were missing in the Chernobyl Registry for 73.8% of subjects. The mean registered dose of external radiation among those for whom estimates were available was 56.6 mSv (median 48 mSv); registered doses ranged between 1.4 and 185 mSv.

The results of this pilot study confirmed the feasibility of carrying out a passive long-term follow-up of liquidators in Belarus. As there are plans to centralise the passport department information at the Republican level, follow-up for vital status ascertainment based on passport department information appears to be feasible. For subjects identified as dead, cause of death information would be obtained from local ZAGS and verified using data from hospitals and, in the case of cancer, from the Cancer Registry.

3.2 Active Follow-up of Liquidators in Outpatient Clinics

As a consequence of the 1987 directive of the Ministry of Public Health of the USSR, all liquidators must undergo an obligatory annual medical examination in which he or she is seen by a general practitioner, an endocrinologist, an ophthalmologist, a neuro-pathologist, an othorhinolaringologist and a gynaecologist (for women). The liquidator is also directed, as appropriate, for additional examinations to oncologists and other specialists.

All data on diseases diagnosed during the annual medical examination, as well as at any other time during the year, is entered in the personal outpatient card at the regional outpatient clinic from which the liquidator depends and is also sent to the Chernobyl Registry for inclusion in the registry data base. As mentioned above, the Chernobyl Registry includes information on place of residence, social group and on internal and external radiation exposure, in addition to data on medical conditions. This information is also updated, as appropriate, at the time of the annual examination.

In Belarus, several official directives regulate the activities of groups which are in charge of controlling the special dispensary examinations and the local registries. According to them, if a person does not present him or herself for the obligatory annual examination, the cause must be ascertained: personal refusal, emigration outside of the area covered by the health institution activities or outside of Belarus, or death. For this purpose, a nurse from the outpatient clinic visits the subject at home; if he or she has moved, a query is sent to the passport department where all population is registered to confirm the move and information is entered into the database of the Chernobyl Registry. In case of death, a nurse visits the district ZAGS bureau and ascertains the date and cause of death, which are also entered into the registry data base.

4. Preliminary Results on the Health Status of the Liquidators

4.1 General Morbidity

In Belarus, regional outpatient clinics systematically collect information on disease diagnoses on all the residents of the region they cover (not only on those included in the Chernobyl Registry). This information is summarised locally and is sent on special statistical reporting forms at yearly intervals to the Ministry of Health. These forms contain information about the number of cases of acute and chronic diseases diagnosed in a given year in the population in all areas of Belarus. This information is not broken down by age or sex. The number of acute and chronic diseases thus reported to the Ministry of Health has been increasing in recent years in the population of Belarus as a whole. This passive system of collection of morbidity data on the population contrasts with the active follow-up carried out, as described above, for the liquidators.

In this section, we present preliminary analyses of the morbidity of liquidators both over time and in comparison to the general population of Belarus. It should be noted that comparisons to the general population may not be appropriate and should be interpreted with caution, as no age-adjustment could be made and as the method for ascertainment of disease differs.

Preliminary analyses of the morbidity of liquidators, as registered in the Belarus Chernobyl Registry indicate an increase in the crude (i.e. not age-adjusted) incidence of diseases over time between 1990 and 1994. This finding may at least partly be explained by the ageing of the population (as indicated above, nearly half the liquidators are now over the age of 40 and close to 20% is above the age of 50) and needs to be investigated in more detail. The increase in morbidity is seen for many disease classes, in particular diseases of the endocrine and digestive system and of metabolism and immunity (2 508 diagnoses per 100 000 in 1990 compared to 4 787 per 100 000 in 1994), mental disorders and diseases of the blood and circulatory system. Morbidity for these disease classes tends to be higher among the liquidators than among the adult population of Belarus (Table 3). By the beginning of 1995, the cumulative prevalence of chronic diseases registered in the Chernobyl Registry since the accident was greater than 50%.

The greatest apparent differences in morbidity between liquidators and the general population are registered for endocrine diseases, and disorders of the digestive system, metabolism and immunity (Table 3). The ratio of morbidity from thyroid diseases in liquidators to that of the entire adult population is particularly high. These results must be interpreted with caution as these ratios are not adjusted for age or sex and the intensity of screening of liquidators resulting from the annual medical examinations may have artificially increased the number of disorders diagnosed.

Table 3 Crude incidence of disease diagnosed among liquidators and the general Belarus population in 1993 and 1994

Class of diseases (ICD 9 codes)	Year	Rates per 10^5	
		Liquidators	Belarus
Diseases of the endocrine system, disorders of the digestive system, metabolism and immunity (240-279)	1993	2 560.6	630.9
	1994	2 862.0	667.6
Diseases of the thyroid gland (240-246)	1993	329.8	329.8
	1994	391.4	391.4
Diabetes mellitus (250)	1993	316.7	101.1
	1994	313.1	94.5
Mental disorders	1993	1 466.2	1 014.0
	1994	2 438.7	1 098.6
Diseases of the circulatory system	1993	4 960.2	1 626.2
	1994	5 974.6	1 646.1
Diseases of the digestive system	1993	5 318.6	1 937.6
	1994	6 411.2	1 889.1
Cataracts	1993	281.4	136.2
	1994	420.0	146.1

An apparent difference in morbidity from nervous and perception organs diseases including cataract is also observed (morbidity ratio 2.6), as well as an increase in digestive system disorders and benign tumours (morbidity ratio 1.7). Again, the depth of medical examination of liquidators may account for at least part of this difference.

Among liquidators, an apparent increase over time in the frequency of non-specific diseases of the vegetative nervous system (ICD-9 code 337.9) – 997.9 per 100,000 in 1994 compared to 599.8 in 1991 – and of chronic respiratory diseases is also noted. Morbidity from a number of diseases associated with older age has also been registered in younger individuals (diabetes mellitus, ischemic heart disease, hypertension, atherosclerosis etc.).

Differences in the morbidity of 1986-1987 liquidators and 1988-1989 liquidators have been observed recently. In particular, morbidity from thyroid diseases, diabetes mellitus, diseases of blood and hematopoietic tissues, circulation system, respiratory organs, digestive, urinary and genital systems, skin and subcutis appear to be higher in the first group. The rate of disease and trauma related disability and mortality appears to be approximately 1.5 times higher among those who worked in 1986-1987 than among those who worked later. The possibility that these differences may at least be partly explained by a different distribution of age and of participation in the annual medical examination between these two groups needs to be studied further.

The preliminary observations reported here may reflect a real increase in morbidity following the Chernobyl accident. On the other hand, they may at least be partly explained by a bias introduced by the active follow-up of liquidators and by failure to take into account the effects of age and sex in the analyses. They must therefore be interpreted with caution.

4.2 Cancer Incidence in Liquidators in 1993-1994

The Belarus Cancer Registry which, as described above, was already functioning before the Chernobyl accident, is an important tool for the study of cancer incidence among liquidators. Cancer diagnoses among liquidators can be systematically identified by linkage of the Chernobyl Registry and the Cancer Registry using full name (last, first, patronymic), year of birth and current (last known) address [2]. Age and sex-specific cancer incidence rates among liquidators can then be calculated and compared with the incidence in the general population of same sex and age.

Dosimetric data reported to date in scientific papers show that, in general, the liquidators received the highest doses among the populations exposed to radiation from the Chernobyl accident (with the exception of smaller groups of populations residing in the most contaminated areas, particularly Pripyat, and evacuated in the days following the accident). It has been estimated that 15% of the Belarus liquidators have received whole body radiation doses below 50 mSv, 30% between 50 and 100 mSv, 48% in the range 100-250 mSv and 7% in the range 250-500 mSv [3]. In principle, therefore, studies of liquidators should be more informative for estimating radiation risks than studies of comparable populations having been exposed environmentally [4]. For studies to be informative radiation risk estimation, however, it is important that individual radiation dose estimates be available; at present, dose estimates are available for less than half of all liquidators in the cohort.

Table 4 Distribution of observed and expected numbers of cancers in 1993-94 among liquidators, by sex and cancer type.

ICD-9	Site	Men				Women			
		O	E	SIR	95% CI	O	E	SIR	95% CI
151	Stomach	19	12.1	157	94-244	1	2.5	39	98-218
153	Colon	11	6.8	161	80-288	1	1.2	81	2-453
162	Lung	33	51.0	65	45-91	0	0.7	0	0-502
173	Skin	5	10.5	48	15-111	3	2.3	128	27-375
174	Female breast	0	_[1]	-		4	7.4	54	15-139
188	Urinary bladder	15	6.9	219	123-361	0	0.2	0	0-2165
189	Kidney	8	8.2	98	42-192	2	0.9	226	27-817
193	Thyroid	4	1.7	241	66-617	5	1.33	376	122-878
204-208	Leukaemia	12	6.5	184	95-321	1	0.8	124	3-692
140-208	All sites	155	202	77	65-90	27	30.0	90	59-131

Table 5 Distribution of observed and expected numbers of cancers in 1993-94 among male liquidators by duration of work in the 30 km zone.

ICD-9	Site	Duration of work in the 30 km zone							
		Less than 30 days				1-6 months			
		O	E	SIR	95% CI	O	E	SIR	95% CI
151	Stomach	7	12.1	58	23-119	5	7.3	69	22-160
153	Colon	7	2.9	241	97-497	2	1.7	117	14-423
162	Lung	14	22.1	63	35-106	9	12.0	75	34-143
173	Skin	2	4.4	45	5-163	0	2.7	0	0-139
188	Urinary bladder	5	3.0	167	54-390	4	1.6	245	67-628
189	Kidney	3	3.4	88	18-257	4	2.1	189	52-485
193	Thyroid gland	1	0.7	151	4-844	3	0.5	625	129-1826
204-208	Leukaemia	3	2.7	111	23-325	6	1.8	342	126-746
140-208	All sites	61	85.2	72	55-92	41	50.4	81	58-110

The following analyses are based on age standardised cancer incidence rates in the period 1993-94. The standardisation was carried out according to the age and sex distribution of the liquidators population. Expected numbers are based on the incidence in the general population of Belarus of same age and sex. It is noted that, for cancer types with small number of cases, the rates are very unstable and are thus difficult to interpret.

Table 4 shows the distribution of observed (O) and expected (E) numbers of cases and the standardised incidence ratio (SIR) and its 95% confidence interval (CI) for 9 cancer types by sex. The incidence of cancer in general appears to be lower in liquidators, both in men – SIR: 77 – and in women – SIR: 90 –; only for the former is this difference statistically significant. Among men, a significant increase in the incidence of urinary bladder cancers is seen among liquidators (SIR: 219, 95% CI 123-361) compared to the general population. Non significant increases are also seen for cancers of the colon and thyroid and for leukaemia; the number of cases on which these comparisons are based, however, are small, particularly for thyroid cancer (Table 4). Among women, an increased incidence of cancers of the skin, kidney and thyroid gland was observed on the basis of even smaller numbers of cases.

---

[1] not applicable

Table 5 presents the distribution of observed and expected numbers of cases and SIR's for the same cancer types among male liquidators by duration of work in the 30 km zone: less than a month (30 days) and 1-6 months. The SIR for all cancers together is slightly higher among those who worked in the 30 km zone over 30 days (SIR: 81, 95% CI 58-110) than among those who worked for shorter times (SIR 72, 95% CI 55-92). In particular, a higher incidence of tumours of the bladder, kidney, thyroid and of leukaemia is observed among those who worked in the 30 km zone more than 30 days. None of these SIR's is significantly elevated compared to the general population, however, apart from thyroid cancer among those who worked more than 30 days. The number of cases in these categories is, moreover, quite small and these comparisons must be interpreted with caution.

5. Discussion

It is difficult at present to evaluate with certainty the health of the liquidators. Results of analyses of the general morbidity of the liquidators appear to indicate an increase in the prevalence of a number of acute and chronic diseases, both over time and compared to the general population. These results, however, are preliminary and must be interpreted with caution. Analyses were not adjusted for age or sex. Moreover, the observed increases may at least partially be explained by the different approaches for ascertaining diseases in the general population and the liquidators. Among the liquidators themselves, although participation in the annual medical examination is obligatory, only about 80% of liquidators actually participate in it in any given year. The possibility that participation rates vary with the age and health status of the liquidators - and may, in addition, vary over time - cannot be rejected.

The results of cancer incidence analyses are less subject to these biases, except for some cancer types such as thyroid cancer and chronic leukaemia where the depth of screening may greatly influence the observed incidence. Overall, in 1993-4, the liquidators appeared to have a lower incidence of cancer than the general population of Belarus. Among male liquidators, the incidence of cancer of the urinary bladder was significantly increased compared to the general population and non-significant increases in cancers of the colon, thyroid and leukaemia were observed. These results are, however, based on relatively small number of cases and on active follow-up of the liquidators. Further studies of this population is needed to confirm or reject this observation.

The population of liquidators in Belarus, because of the level and type of radiation exposure they received, is an important population to study the health consequences of radiation exposure, in particular the effects of relatively low doses (compared to atomic bomb survivors, for example) and the effects of exposure protraction.

We have shown that a passive mortality follow-up of liquidators is possible in Belarus using existing population registration structures. Furthermore, the existence of a long-established cancer registry, and the ongoing work aimed at improving the registry and the possibility of linking [5] it to other sources of data – in particular the Chernobyl Registry – ensures the feasibility of systematic cancer morbidity follow-up.

The population of 1986-87 liquidators having worked in the 30 km zone, although large, is still relatively small for studying directly the effects of low doses of radiation. It is therefore important that studies of liquidators be carried out in parallel in other countries, using similar and compatible protocols, in order to maximise their informativeness for radiation risk estimation.

Many of the liquidators were young at the time of the accident; today, 10 years later, the mean age among the Belarus cohort is approximately 40 and 15% of the liquidators are over the age of 50. As the population ages, and enters the age range where cancer incidence increases rapidly, the informativeness of the follow-up for cancer risk is increasing.

6. Conclusions

The population of Belarus liquidators is important for the epidemiologic study of the health consequences of the Chernobyl accident. The distribution of doses and exposure patterns in this population and the demonstrated feasibility of carrying out passive follow-up for mortality and cancer morbidity make this a particular important population for the study of radiation effects. Although the total number of subjects in the cohort is large (over 45 000), this is still a relatively small number of subjects for the study of the effects of low doses. It is therefore important that this population be followed-up over time, and that similar studies be carried out in parallel in other countries with substantial numbers of liquidators, particularly Russia and the Ukraine, but also the Baltic countries, in order to maximise the information which can be obtained about radiation risks.

References

[1]   A.E. Okeanov, V.C. Ivanov, E. Cardis, E. Rastoptchin, A. Sobolev, C. Lavé, and A. Mylvaganam, Study of cancer risk among liquidators, Report of EU Experimental collaboration project 7: Epidemiologic Investigations including Dose Assessment and Dose Reconstruction. IARC Internal Report 95/002 International Agency for Research on Cancer, Lyon, 1995.

[2]   A.E. Okeanov, S.M. Polyakov, H.H. Storm, A. Sobolev, and R. Winkelman, Development of cancer registration system in Belarus. First International Conference of the European Commission, Belarus, the Russian Federation and the Ukraine on the radiological consequences of the Chernobyl accident (Minsk, Belarus, 18-22 March 1996), IOS Press, Amsterdam, 1996.

[3]   Chernobyl Accident: health aspects (Collection of scientific papers). Health Ministry of the Republic of Belarus, Minsk, 1994.

[4]   E. Cardis and A.E. Okeanov, What's Feasible and Desirable in the Epidemiologic Follow-Up of Chernobyl. First International Conference of the European Commission, Belarus, the Russian Federation and the Ukraine on the radiological consequences of the Chernobyl accident (Minsk, Belarus, 18-22 March 1996), IOS Press, Amsterdam, 1996.

[5]   H.H. Storm, A.E. Prisyazhniuk, A.E. Okeanov, V.K. Ivanov, and L. Gulak, Development of infrastructure for epidemiological studies in the three CIS republics. First International Conference of the European Commission, Belarus, the Russian Federation and the Ukraine on the radiological consequences of the Chernobyl accident (Minsk, Belarus, 18-22 March 1996). IOS Press, Amsterdam, 1996.

Health status and follow-up of liquidators in Russia

Victor IVANOV

Medical Radiological Research Center of RAMS, Obninsk, Russia

Abstract. In 1986 immediately after the Chernobyl accident the USSR Ministry of Health adopted the large-scale program on the establishing of All-Union Distributed Registry of persons affected by radiation at the Medical Radiological Research Center of RAMS, Obninsk. To 1992, to the time of dissolving of the USSR, the database of the Registry comprised medical and dosimetric information on 659 thousand of affected persons including 284 thousand of liquidators. At present time data on 435 thousand of affected persons, citizens of Russia, including 152 thousand of liquidators are kept in Russian National Medical and Dosimetric Registry. Officially registered average doses of external irradiation of liquidators are as follows: liquidators of 1986 - 15,9 cGy; 1987 - 9,0 cGy; 1988 - 3,3 cGy; 1989 - 3,2 cGy. The prognosis of excess mortality from malignant tumors among liquidators in 20 years after the irradiation has been made with the account of age distribution of liquidators (average age of liquidators at the moment of their work within the 30-km zone was 33 years) and dosimetric data. Attributive risk of mortality from all malignant tumors could be 2,8%, from leukemia - 23,6%. The prognosis is in good agreement with the actual data on mortality from malignant tumors among liquidators gathered for the period from 1986 to 1995. We have stated that liquidators of 1986 and 1987 comprise the group of the especially high risk by the incidence and disability rates. These rates for the liquidators several times exceed average rates for the whole cohort. The number of essential studies among them cancer risks in liquidators have been performed within the framework of EEC Experimental Collaboration Project 7 "Epidemiological investigations including dose assessment and dose reconstruction". The findings can be used as basis further long-term epidemiological studies on the assessment of radiation risks of leukemia among liquidators.

1. Introduction

In 1986 immediately after the Chernobyl accident the Ministry of Public Health of the USSR adopted a large-scale program to establish in the country the All-Union Distributed Registry of persons exposed to radiation. Towards 1992 (by the time of collapse of the USSR) the data base of the Registry comprised medical and dosimetric information for 659292 people including that for 284919 emergency workers (EWs) (liquidators). Into establishment of the Registry all republics of the former Soviet Union as well as a wide range of scientific and practical institutions were involved [1, 2].

At present (in accordance with the Decree of the Government of Russia № 948 of 22.09.93) the National Radiation and Epidemiological Registry (NRER) operates in the country. The general customer of works on the Registry is Ministry on Emergency Situations. The leading organization is the Medical Radiological Research Center of RAMS responsible for sampling primary medical and dosimetric data through 24 regional centers.

The NRER involves 3 main bases: the Registration List of persons exposed to radiation which is established on special dosimetric criteria from the Decree of the Government of Russia № 948); the Chernobyl Registry since 1992 the Russian National Medical and Dosimetric Registry (RNMDR); the Registry on Interdepartmental Expert Councils.

In this paper we will enlarge on radiation-epidemiological analysis of the Chernobyl Registry of Russia (RNMDR).

2. Current status of the Russian National Medical and Dosimetric Registry

Fig. 1 presents the information on the dynamics of the RNMDR registrant number growth in 1986-1995. As the Fig. 1 shows during all these years of its existence the data base of the Federal level of the RNMDR kept accumulating medical dosimetric information and as of 1.09.95 comprises data on 435276 people from throughout the Russian Federation. All the RNMDR registrants are divided into five primary registration groups (PRG):

PRG 1 - emergency workers - 152325 (35,0 %);
PRG 2 - evacuated and resettlers - 12889 (3,0 %);
PRG 3 - residents (persons living or lived in monitoring territories) - 251246 (57,7 %);
PRG 4 - children born of emergency workers of 1986-1987 - 18816 (4,3 %).

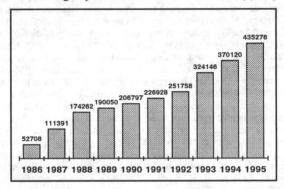

Fig. 1. Dynamics of RNMDR registrant number growth in 1986-1995.

Fig. 2 demonstrates the distribution of persons registered in the RNMDR on their representation in regional centers.

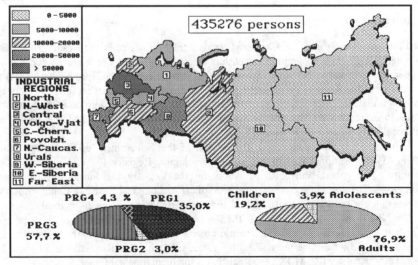

Fig. 2. Number of persons recorded in the RNMDR.

3. Cancer morbidity and mortality among Chernobyl emergency workers: radiation risk assessment

3.1. Cohort and radiation doses

The cohort of Chernobyl emergency workers (EWs) discussed in this paper includes 143032 people and represents the subgroup of EWs registered in the RNMDR. The EW cohort studied in this paper conforms to the data contained in the RNMDR database as of the beginning of 1994. This subgroup contains EWs with no cases of cancer registered before the Chernobyl accident and for which the following information is available in the RNMDR: date of birth, date of arrival to the 30-km zone around the Chernobyl NPP, period spent in the zone, data on medical examinations and their results.

Among 143032 EWs included into the cohort under study, 113936 (about 80%) people have documentary verified external radiation doses. The term "dose" here implies a documentary verified amount of external radiation received by each EW. Distribution of EWs by external radiation dose is shown in Fig. 3.

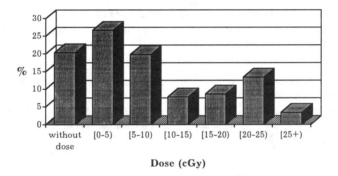

Fig. 3. Distribution of EWs by external radiation dose.

Men constitute the majority of EWs (about 98%), 85% of which are in the age group of 20-40 years. The average age of EWs at the moment of arrival to the Chernobyl zone is 34,3 years. The majority (82%) of EWs has taken part in recovery works in 1986-1987. The average period of work in the exposure zone amounts to 2,7 months and the average dose of EWs from Russia amounts to 10,7 cGy.

3.2. Medical information collection system

The acquisition of data in RNMDR is performed in the following way. Special Registry record papers [1] containing the information about radiation doses and health status of the person under examination are filled in on the basis of unified medical protocols during regular medical checkups in central district hospitals. The papers filled in are verified and passed on to the regional level. On this level the data are recorded on magnetic media (floppy disks), checked and passed on to the Regional Center. RNMDR encompasses 24 Regional Centers collecting information throughout Russia [1].

RNMDR database is annually updated with information from Regional Centers and from specialized Registries of the Defence Ministry of Russia, Ministry of Home Affairs, Ministry of Security, Ministry of Transport and Ministry of Atomic Power Engineering and Industry of Russia.

3.3. Methods of the analysis

The term "case of disease" is used here to denote the registration of a diagnosis of a particular class of diseases by a health care institution. The diagnoses for EWs implied here are the first ones after his/her arrival to the 30-km zone. For each person the time spent at risk to develop a disease of a particular class is calculated as the difference between the date of registration of the primary diagnosis for this class of diseases and the date of arrival to the 30-km zone. Therefore, the term "incidence rate" used below is determined as the ratio of the total sum of cases to the sum of time at risk, measured in person-years.

The mortality rate is defined in a similar way, i.e. as the ratio of the sum of deaths to that of person-years.

To find the dependence of incidence (mortality) rate individual data on EWs are divided into 5 strata: according to age: (18-20), (20-30), (30-40), (40-50), (51+), 4 strata according to the date of arrival to the 30-km zone: 1986, 1987, 1988, 1989+, 4 strata according to the period spent in the zone: 1, 2, 3, 3+ months, 6 groups according to dose: (0-5), (5-10), (10-15), (15-20), (20-25), (25+) cGy.

Let i be the stratum in a data array aggregated by age, dose, arrival date and period of work in the zone. Let Y_i be the number of cases of diseases (deaths), P_i - person-years, M_i - the incidence (mortality) rate in the stratum. In these terms the incidence (mortality) rate for a given class of diseases can be defined as:

$$M_i = Y_i / P_i \qquad (1)$$

It is reasonable to assume that Y_i values are independent Poisson random values with mathematical expectation $E(Y_i) = P_i M_i$. To determine the dependence of M_i on any parameter x (age, year of arrival etc.) it is necessary to present M_i in the form of parametric function f and its parameters are to be determined via maximization of likelihood function:

$$L = \sum Y_i \ln(P_i M_i) - P_i M_i, \qquad (2)$$

where $M_i = f(x_i)$.
Simple functions are used in the paper:

$$f(x_i) = M = const \qquad (3)$$
$$f(x_i) = M_0 + M_1 x_i \qquad (4)$$
$$f(x_i) = M_0 (1 + M_1 x_i) \qquad (5)$$

Function (3) is used to determine crude mortality and incidence rates, whereas functions (4) and (5) can be used to determine linear trends for mortality and incidence rates according to risk factor x.

3.4. Analysis of cancer morbidity and mortality among EWs

Among 143032 EWs studied in this paper, 1026 cases of cancer diseases were diagnosed and 341 deaths from malignant neoplasms were registered. The most widespread among them are malignant neoplasms of digestive system (ICD-9: 150-159) - 251 cases of diseases and 125 deaths; and malignant neoplasms of the respiratory system (ICD-9: 160-165) - 186 cases of diseases and 112 deaths.

Shown below are dependences of incidence (IR) and mortality rates (MR) on various factors (external radiation dose, age, date of arrival to the zone and duration of being there) for all cancer diseases (ICD-9: 140-208.9). Estimates of relative risk for cancer morbidity and mortality of EWs will be also shown.

As is seen from Fig. 4, IR increases with age (p<0,05), external radiation dose (p<0,05) and decreases (p<0,05) with the date of arrival to the 30-km zone for all cancer diseases

(ICD-9: 140-208.9). As for the dependence of the incidence rate on the period spent in the zone no conclusion about its statistical significance can be made (p>0,05).

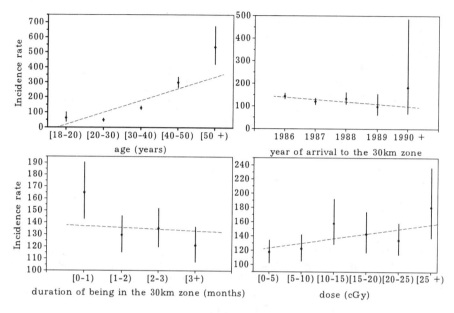

Fig. 4. Incidence rates for EWs (per 100000 person-years) for cancer diseases (ICD-9: 140-208.9) depending on different factors: age, year of arrival at the zone, period spent in the zone and external radiation dose for EWs. Dashed line shows the linear trend corresponding to the regression function (4).

Both the cancer mortality and incidence rates among EWs (see Fig. 5) increase with age, external radiation dose and decrease with the date of arrival to the 30-km zone. However, in contrast to the incidence rate the mortality rate shows statistically significant dependence (p<0,05) only on age and external radiation dose of EWs. As for the dependence of MR on the period spent in the zone and the date of arrival at it, we cannot make conclusions about its statistical significance (p>0,05).

Therefore, the radiation dose fits into the group of the basic "external" factors of risk for EWs, i.e., factors independent of "natural" increase in incidence and mortality rates, as is the case with age.

Consideration of dynamic dependences of incidence and mortality rates allows to make a conclusion about significant (p<0,05) increase in incidence and mortality rates with time. The incidence and mortality rates for cancer diseases among EWs (ICD-9: 140-208.9) as of calendar year are shown in Fig. 6. As mentioned above, radiation exposure level, in particular, the external radiation dose of EW fits into the group of basic factors of risk for EWs. Relative risk coefficients (RR)- the ratio of incidence (mortality) rates in different dose intervals to the corresponding value in the dose group of [0-5) cGy were estimated for quantitative assessment of radiation factor influence on the state of health of EWs. Fig. 7 demonstrates the RR coefficients for cancer morbidity and mortality of EWs. When modelling the dependence of the relative risk on the dose by regression function the following excess relative risk values per 1 cGy with 95% confidence limits are obtained: 0,017 (0,003; 0,031) - for cancer morbidity of EWs and 0,025 (-0,004; 0,05) - for cancer mortality.

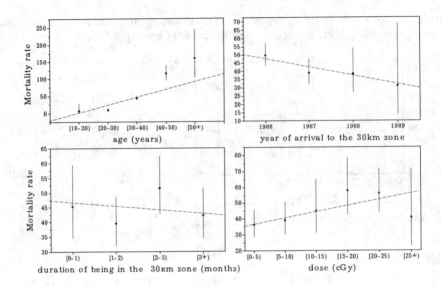

Fig. 5. Mortality rate for EWs (per 100000 person-years) for cancer diseases (ICD-9: 140-208.9) depending on different factors: age, year of arrival at the zone, period spent in the zone and external radiation dose for EWs. Dashed line shows the linear trend corresponding to the regression function (4).

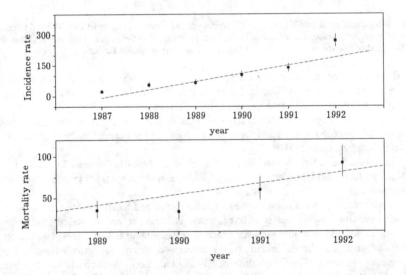

Fig. 6. Incidence (upper part) and mortality (lower part) rates for EWs (per 100000 person-years) for cancer diseases (ICD-9: 140-208.9) depending on calendar years. Dashed line shows the linear trend corresponding to the regression function (4).

For qualitative assessment of dynamic tendencies in the behaviour of the relative risk the crude relative risk values were evaluated, i.e. - the ratio of cancer incidence (mortality) rates in the dose interval over 5 cGy to the corresponding value in the background dose group of [0,5) cGy. Analysis of the variables obtained (Fig. 8) does not allow to make a statistically significant conclusion (p>0,05) that time dependences of relative risk exist either for cancer morbidity or cancer mortality of EWs.

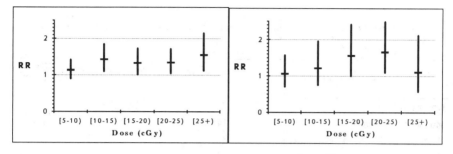

Fig. 7. Relative risk (RR) for morbidity (on the left) and mortality (on the right) among EWs from cancer diseases in different dose ranges.

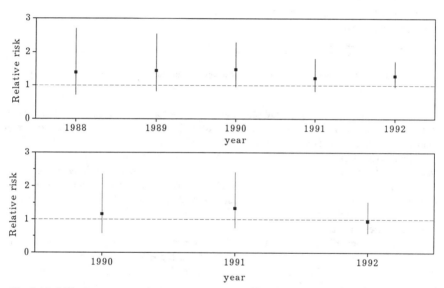

Fig. 8. Morbidity (upper part) and mortality (lower part) crude relative risk for cancer diseases of EWs (ICD-9: 140-208.9) depending on calendar year.

4. Morbidity and disability of emergency workers

Table 1 demonstrates the comparison of morbidity rates per 100 thousand people on general classes of diseases both for population of Russia as a whole and for EWs. It is clear from the Table 1 that morbidity rates of EWs in a series of cases do repeatedly exceed the analogous ones for population of Russia. Undeniably, level, completeness and quality of prophylactic medical examination of EWs differ much from the All-Russian practice. Really, peculiarities and quality of prophylactic medical examination of EWs are that for their examination the most currently available methods of diagnosis of diseases are applied, in so doing, the works listed are carried out by trained and competent specialists. In such a manner, according to the data of the MRRC of RAMS the establishment of primary registered diseases by specialists of this institution is several times higher than by local physicians. In this situation it is very difficult to choose an adequate review control group of comparison.

Table 1
Comparison of morbidity rates per 100000 persons on general classes of diseases for population of Russia as a whole and emergency workers on 1993

Classes of diseases	Population of Russia	Emergency workers	Relationship among the indices
Neoplasms	788	747	0,9
Malignant neoplasms*	140	233	1,6
Diseases of the endocrine system	327	6036	18,4
Diseases of the blood and blood-forming organs	94	339	3,6
Mental disorders	599	5743	9,6
Diseases of the circulatory system	1472	6306	4,3
Diseases of the digestive system	2635	9739	3,7
All classes of diseases	50785	75606	1,5

* - For malignant neoplasms the standardized index on age distribution of emergency workers as of 1993 is given.

It is known that in forming of pathologic state and morbidity among EWs the factors of social and psychological character connected with the Chernobyl accident are of great importance. All this in combination with radiation effect can be defined as "Chernobyl syndrome". Attempts to outline the significance of role of radiation factor in this complicate syndrome complex are very important. Therefore, we have estimated morbidity and disability rates on the following groups: 0-5 cGy, 5-20 cGy and over 20 cGy based on dosimetric data for EWs included in the RNMDR. In this case, the contingents of EWs exposed in the range of 0-5 cGy were used as an interval control group. As it is seen from Table 2, the morbidity rates on a series of classes of diseases in the dose groups of 5-20 cGy and over 20 cGy are statistically significant higher than those in the dose group of 0-5 cGy. The dose group of over 20 cGy was demonstrated to consist by 99,1% of the EWs of 1986-1987. In the dose group of 5-20 cGy the EWs of 1986-1987 are presented by 91,2%. In the 1st dose group (0-5 cGy) the EWs of 1986-1987 comprise less than a half of the cohort (48,9%). As a consequence within the frame of standard multifactor analysis we discussed two competitive factors from the viewpoint of influence on morbidity rates: dose one (with 3 gradations of 0-5 cGy, 5-20 cGy and over 20 cGy) and the second - on the date of arrival in the zone of radiation influence (with 3 gradations as well: EWs of 1986, 1987 and 1988-1990). Using the analysis of morbidity rates on 3 classes of diseases (of the endocrine system, blood and blood forming organs as well as mental disorders) as an example, it was established that the factor of the date of arrival in the zone of radiation influence (1986, 1987, 1988-1990) is by far the determining one relative to the dose factor with respect to its influence on morbidity. By this is meant that the state of health of EWs first and foremost of 1986 and 1987 is of particular alarm.

Table 3 presents the disability rates among EWs dependent on received doses of external radiation. In the 2-nd and 3-rd dose groups these rates exceed significantly corresponding coefficients for the 1st dose group (0-5 cGy). It should be noted that the disability rates for the EWs as a whole exceeds 2,8-3,2 times the analogous control one on Russia.

Table 2
Comparison of morbidity rates per 100000 persons on general classes of diseases
among emergency workers of different dose groups in 1993

Classes of diseases	[0,5[cGy	[5,20[cGy	over 20 cGy
Neoplasms	690	648	747
Malignant neoplasms	217	232	225
Diseases of the endocrine system	5270	6120 *	6075 *
Diseases of the blood and blood-forming organs	213	354 *	450 *
Mental disorders	5178	5490	5472
Diseases of the circulatory system	5287	6090 *	6648 **
Diseases of the digestive system	9106	9743	9515
All classes of diseases	69831	75346 *	75785 *

* - indices reliably (p<0,001) differ from corresponding ones in the dose group of 0-5 cGy; ** - indices reliably (p<0,01) differ from corresponding ones in the dose group of 5-20 cGy. Degrees of significance were calculated with Fisher fi-criteria and the criteria of the ratio of two Poison parameters; the degree of significance of the less powerful criteria was used in each index of the Table.

Table 3
Dynamics of disability rates (per 1000 persons) in dose groups in 1990-1993 (the data of the RNMDR)

Year of observation	[0,5[cGy	[5,20[cGy	over 20 cGy
1990	6,0	10,3	17,3
1991	12,5	21,4	31,1
1992	28,6	50,1	57,6
1993	43,5	74,0	87,4

Thus, on the cohort of EWs registered in the RNMDR two main conclusions may be done:
- factual evidence for the period just ended and prognostication of total mortality rate as well as that from malignant neoplasms made on the basis of radiation risk coefficients by ICRP are in a good agreement with observed rates which do not exceed corresponding control values on the Russian Federation;
- on morbidity and disability rates the EWs of 1986 and 1987 comprise the group of higher risk.

5. EEC Experimental collaborating project 7: epidemiological investigations including dose assessment and dose reconstruction

In the end of 1993 the following 3 pilot studies of CEC-CIS project titled "Study cancer among liquidators" were started in Russia:
Study 1. Test of follow-up mechanism.
Study 2. Cases-control study of leukemia among liquidators.
Study 3. Validation/calibration of biological dosimetry.

As a result of the implementation of the first study the detailed medical and dosimetric information on 500 participants of recovery works at the Chernobyl NPP randomly selected from the data base of the Russian medical and Dosimetric registry was collected. For the period of the implementation of the Project (6 months) 496 specially developed questionnaires containing information on health status and the character of performed work within the 30-km zone of the Chernobyl NPP by those included in the cohort under study were filed in. So the loss in the "follow-up" was 0,8%. It allows us to make conclusion that the project may be continued in larger scale if financial and human resources are available.

Within the framework of the second study of the investigation of leukemia cases among liquidators of Russia the following work has been done:

1) identification of leukemia among liquidators living in Russian Federation (58 cases of leukemia among liquidators have been detected, 48 of the cases have been detected among those worked in the Chernobyl zone in 1986-1987);

2) filling in questionnaire forms and collecting blood samples for the biological dosimetry of leukemia cases (17 cases) among male liquidators worked in the zone of Chernobyl in 1986-1987 delivered from two Regional Centers of the Registry - North Caucuses and Central Chernoziom Region;

3) analysis of the obtained medical and dosimetric information.

It should be noted that though the dependence of morbidity with leukemia among liquidators on the dose of exposure has been established nevertheless the continuation of the investigation is of special concern.

References

[1] V.K.Ivanov et al., Planning of long-term radiation and epidemiological research on the basis of the Russian National Medical Dosimetric Registry. Nagasaki symposium on Chernobyl update and future. Elsevier, Amsterdam, 1994, pp. 203-216.
[2] V.K.Ivanov et al., Information systems and modelling: data and organizational aspects of the Chernobyl State registry. Proceeding of the First Conference of International Simulation Societies. Zurich, 1994, pp. 579-583.

CHERNOBYL NPP ACCIDENT CONSEQUENCES CLEANING UP PARTICIPANTS IN UKRAINE HEALTH STATUS EPIDEMIOLOGIC STUDY MAIN RESULTS

V.Buzunov, N.Omelyanetz, N.Strapko, B.Ledoschuck, L.Krasnikova, G.Kartushin

Radiation Medicine Scientific Centre of Ukrainian Medical Sciences Academy, Kiev, Ukraine

Summary

The Epidemiologic Studies System for Chernobyl NPP Accident consequences cleaning up participants (CNPP ACCP) health status was worked out and than improveing in Ukraine after the CNPP Accident. The State Register of Ukraine both with several other Registers are the organizational, methodological and informational basis here. The ACCP health status worsening was registered in dynamics through the post-accidental period i.e. the nervous system, digestive system, blood circulation system, respiratory system, bone-muscular system, endocrine and genitourinary systems chronic non-tumoral pathology both with mental disorders amount increase. In cohort study the differences of morbidity formation were fixed among emergency workers with different radiation exposure doses. The dependence of leukemia morbidity on presence in 30-km zone duration was noticed, it's access manifested 5 years after the participance in ACC. The ACCP invalidisation increase with main reason of general somatic diseases, and annual mortality growth are registered. But that doesn't exceed the mortality rate among population of working age in Ukraine.

1. The Accident consequences cleaning up participants (ACCP) contingent general characterization.

Nearly 300,000 residents of Ukraine took part in Chernobyl NPP ACC. Among all the affected population just the ACCP contingent received the most high radiation exposure doses, that may lead to major prevalence of radiation negative effects here. Some other antropogenic harmful factors besides the radiational one effected the mentioned contingent during the period of activity. The psycho-emotional strain, related to the fact of Accident was of significant meaning.

174,812 ACCP are at present under the survey of State Register of Ukraine (SRU) created after the Accident. More than 77% of them participated in emergency works during 1986 - 1987. (Fig.1.)

The ACC participants distribution according to the year of participation beginning

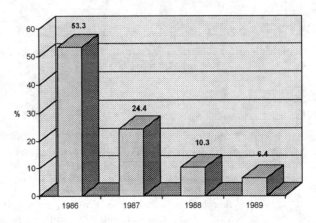

Fig. 1

The ACCP contingent is mainly represented by males (96%) of young and medium age. 80% of them are persons 20-39 years of age and 99.1% - 20-59 years respectively. At the same time there are some differences in age structure distribution between ACCPs of 1986-1987 and 1988-1990. (Table 1).

Table 1.

The ACCPs distribution among 1986-1987 and 1988-1990 according to age at the moment of ACC participance (in %)

Study Group	Age Groups (years)					
	up to 20	20-29	30-39	40-49	50-59	60 and over
ACCP of 1986-1987	1.58	40.0	40.2	13.7	3.9	0.49
ACCP of 1988-1990	0.49	4.4	77.7	16.5	0.81	0.08

Depending on duration of presence in 30-km zone the ACCPs of 1986-1987 were distributed as following:

worked more than for 3 months in 30-km zone - 15.2%,

up to 3 months - 20.1%,

up to 2 months - 26.8%,

less than 1 month - 37.9%.

That for ACCPs of 1988-1990:

53.6% worked in 30-km zone for more than 3 months,

30.1% - up to 3 months,

11.4% - up to 2 months,

and only 4.9% - less than one month.

The ACCPs external irradiation SRU data analysis revealed that persons working in 1986-1987 received the higher dosed compared to that among those in further time person. (Table 2).

Table 2.

ACCPs distribution according to extarnal irradiation doses depending on year of participance in ACC (data from SRU)

Study Group	External irradiation dose mGy					
	no data	up to 50	50-99	100-249	250-500	over 500
ACCP of 1986-1987	48.1	5.9	15.4	25.1	5.4	0.20
ACCP of 1988-1990	17.0	67.2	14.2	1.3	0.23	0.00

The ACCPs of 1988-1990 external irradiation doses not exceeded 50 mGy in 82% of cases. At the same time the dosed reconstruction and verification remain actual for emergency workers 'cos of data absence in base of Register.

2. Accident consequences cleaning up participants morbidity.

Basing on existing prognoses of radiation-induced delayed medical effects, the accident survivors' health status was not to change significantly during the post-accidental period, as far as the main somato-stochastic effects latent period consists 5-10 years. But the survey data prove the strong negative tendencies presence in ACCPs health status changes. The quota of healthy persons among them decreased from 78% to 26.1% as the result of general somatic pathology manifestation, that are at present the main reasons of disability and mortality growth among ACC participants.

The ACCP health status epidemiologic monitoring results showed the annual general and initial morbidity growth, that increased more than twice up to 1992 'cos of nearly all classes chronic pathology. Especially significant was the nervous system and sensoric organs pathology increase, both with that of blood circulation system, digestive system. All the mentioned is of prior meaning in diseases structure. The pathology of bone-muscular system, mental disorders, blood system pathology growth was also meaningful. The autonomous nervous system - vascular disorders possessed the heavy place in nervous system pathology structure, the prevalence grew from 29.5 up to 90.7 here for 1000 examined persons. The endocrine pathology and genitourinal system diseases growth was registered also.

The differences in chronic diseases morbidity growth among previously healthy contingent of ACCPs depending on the year of participance in works of ACC were detected during special cohort epidemiologic studies (Table 3).

Table 3.

Age-standartized frequency of newly registered diseases in ACCP of 1986-1987 and 1988-1989 five years after the ACC participance (per 1,000 persons).

Diseases Class IDC - 9	ACC participants of 1986 - 1987	ACC participants of 1988 - 1990
All diseases	356.5	204.8
II. Neoplasmas	2.40	1.96
III. Endocrine system	41.3	29.8
IV. Blood and hemopoetic organs	2.69	0.71
V. Mental disorders	30.2	6.2
VI. Nervous system and sensoric organs	110.8	51.9
VII. Blood circulation system	52.9	34.9
VIII. Respiratory system	19.2	24.5
IX.Digastive system	56.3	29.3
X.Genitourinary system	5.7	3.1
XII.Skin and subskinal tissue	3.4	1.0
XIII.Bone-muscular system	27.2	18.5

The received data prove the non-neoplasmal pathology chronic forms growth in CNPP ACC participants is of definite dependence on external irradiation dose. The ACCPs morbidity level in persons with external radiation total exposure dose exceeding 250 mGy is significantly higher than in those with received doses up to 250 mGy (Table 4).

Table 4.

Age-standartized frequency of obtained during 1986-1991 chronic non-neoplasmal morbidity in ACC participants of 1986-1987 period, depending on external irradiation dose (per 1,000 persons).

Diseases Class IDC - 9	Dose less than 250 mGy	Dose over 250 mGy
All diseases	580.6	685.1
III. Endocrine system	74.3	72.5
IV. Blood and hemopoetic organs	4.64	6.57
V. Mental disorders	50.8	72.7
VI. Nervous system and sensoric organs	168.6	190.5
VII. Blood circulation system	91.8	93.2
VIII. Respiratory system	30.2	38.8
IX.Digastive system	93.1	128.5
X.Genitourinary system	8.8	9.8
XII.Skin and subskinal tissue	6.3	10.0
XIII.Bone-muscular system	42.3	52.0

The analysis results of held in 1993-1994 complex transversal study with profound clinic-epidemiological monitoring standartized program revealed the high level of pathology prevalence in ACC participants (Table 5).

The profound complex medical examinations results are concordant to health status self-estimations data, received during the ACCPs questioning. The questioned marked the health status rough worsening compared to that of pre-accidental period. 93% of ACCPs estimated the own health before the Accident as the "good" one; at present only 4% consider that as "good" one, 42% - as "satisfactory" and nearly 50% - as "poor" one. The ACC participants fixed the bad cold catching frequency growth both with present before Accident chronic diseases run exacerbation and new chronic pathology arisement. From the point of view of questioned persons the factors related to Accident injured them mostly of all. Approximately 20% pointed out to the unsatisfactory social-living conditions, residence ecological conditions, medical service lacks role.

Table 5.

Level (per 1,000 persons) and structure (in %) of pathology prevalence in ACC participants according to profound clinic-epidemiological study results of 1993-1994.

Diseases Class (according to IDC-9)	IDC - 9 code	Level per 1000	%
Infections and parasitic diseases	001-139	147.3	1.69
Neoplasmas	140-239	236.4	2.71
Endocrine system diseases and nutrition disorders	240-279	304.5	3.49
Blood and hemopoetic organs diseases	280-289	37.4	0.43
Nervous system and sensoric organs diseases	320-389	2817.3	32.33
Blood circulation system diseases	390-459	1121.8	12.87
Respiratory system diseases	460-519	463.6	5.32
Digestive system diseases	520-579	2687.3	30.83
Genitourinary system diseases	580-629	377.3	4.33
Skin and subskinal tissue diseases	680-709	67.3	0.77
Bone-muscular system and joining tissue diseases	710-739	311.8	3.58
Congenital malformations (development viciums)	740-759	80.9	0.93
Symptoms, signes	780-799	49.1	0.56
Traumas and poisonings	800-999	11.8	0.14
All diseases		8713.6	100.0

Held according to this program the social-hygienic studies enabled to receive data for harmful to health risk factors prevalence before Accident and at the moment of examination. On the background of rather objective estimation, among the significant number of ACC participants the such health risk factors presence was determined: hardness and strainness of labour, harmful environment at the place of residence, smoking, alcohol consumption, unproper dietary mode, social-psychologic stress.

In connection to the ACCPs health status significant worsening the requirement of health-improveing measures increased. During the questioning 46% of ACCPs pointed out to necessity of hospital treatment, 72% - of sanatorium - health resourt treatment, 38% - of dietal nutrition, 40% - of more prolonged rest, 27% - of curative physical training and sports, 10% require psychotherapy. The presented data convince, that the life quality of Chernobyl Society members posesses the significant and maybe the leading role in health risk factors estimation and Accident medical consequences optimization system.

For ACCP morbidity with hemoblastoses epidemiologic studies run the Special Register was formed, that contains information concerning detected cases of disease. In 1987-1993 the 141 case were registered among ACCPs, and 86 cases from those - with leukemias. The studies revealed that major number cases of leukemia were newly registered among emergency workers, who participated in ACC during April - June 1986 i.e. just after the Accident. The ACC participants of 1986 year period morbidity intensive parameters summarizes for 1987-1992 consisted 13.35 cases per 100,000 persons of studied cohort (Table 6). The morbidity access was confidentionally fixed among this group in 1991, i.e. 5 years after exposure.

Table 6.

Leukemia morbidity in ACC participants of 1986-1987 years period (per 100,000 males).

Years of survey	Morbidity level	
	ACCP of 1986	ACCP of 1987
1987	13.33 + 4.71	0
1988	6.42 + 3.21	6.32 + 4.47
1989	14.06 + 4.69	4.41 + 3.12
1990	14.50 + 4.59	5.32 + 3.07
1991	18.13 + 4.84	7.74 + 3.46
1992	12.59 + 3.98	12.02 + 4.25
Total	13.35+1.80	7.04+1.57

For ACCP of 1987 and further years period the leukemic morbidity tolally for 1987-1992 consixyed 7.04 cases per 100,000. The confidential morbidity peak was registered in 1992, i.e. 5 years since presence in 30-km zone.

Further years according to prognoses are to be the time of irradiation delayed consequences outcome. According to calculations of RMSC specialists the cancer fatal forms increase during life under risk in ACC participants of 1986 year period will consist 12%, and that of 1987 - 4.2% respectively.

3. The Accident consequences cleaning up participants disablement and mortality.

The ACCP disablement dynamics, level and reasons study results revealed the significant annual parameters increase already during the first post-accidental years and rapid growth rate here in further period. The disablement parameter grew from 2.71 per 1,000 persons in 1988 to 130.0 in 1994, with the same parameters for working population of Ukraine - from 3.7 to 4.9 per 1,000 persons. The disablement high level and rapid growth is registered both among males and females (Fig.2). The disablement level in ACCP of 1986-1987 years period is meaningly higher than that in ACCP who worked in 30-km zone later.

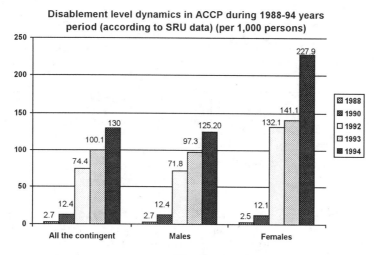

Fig. 2

The disablement reasons structure changed in post-accidental period dynamics. The major quota among disablement reasons in 1988 was possessed by traumas and poisonings, including both the pathology connected to radiation effection (24.8%). The main reason of disablement in ACCP at present are the general somatic diseases - that of nervous system and sensoric organs, blood circulation system, digestive system. The part of these three leading reasons of ACCP disablement consisted 84.4%

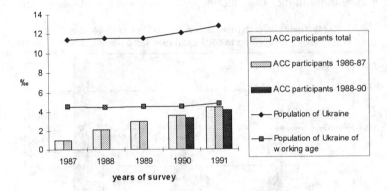

Fig. 3

The ACCP mortality analysis revealed it's level increase in dynamics, with growth rate dependence on terms of emergency works participation manifestation in 1991 (Fig.3). Both with that, mortality among the ACC participants is not exceeding those parameters among Ukrainian population of working age. The age-parameters of ACCP mortality were lower than those among similar age groups of Ukrainian population. The medium age of the deceased ACC participants in 1987-1988 consisted 43.8-43.9 years old. In mortality structure reasons the leading places were possessed by traumas and poisonings, blood circulation system diseases and neoplasmas. The average personal external irradiation dose of the deceased ACC participants of 1986-1987 consisted 190 mGy, of 1988-1990 - 60 mGy respectively.

4. Conclusion

All the presented above prove the necessity of further carrying out of profound epidemiological studies aimed at prognosing and risk determination of prior pathology arisement in ACC participants, that lead to disablement and life duration shortening; and also of radiational and non-radiational factors degree of effection on their formation here. These studies will enable not only to make the existing knowledge in the field of radiation medicine and epidemiology more deep and wide, but also to work out and applicate the rehabilitation-prophylactic measures, form the correct medico-social policy in relation to these most affected contingents. The basis for such the studies carrying out may be the number of functioning special scientific subregisters in the system of State Register of Ukraine.

Development of infrastructure for epidemiological studies in Belarus, the Russian Federation and Ukraine.

Storm H. H. [1], Winkelmann R. A. [1], Okeanov A. E. [2], Prisyazhniuk A. E. [3], Ivanov V. K. [4], Gulak L. [5], Yakimovitch G. [2]

1   Department of Cancer Epidemiology, Danish Cancer Society, Copenhagen, Denmark
2   Belarussian Centre for Medical Technologies, Information Computer Systems, Health Administration and Management, Minsk, Belarus
3   Laboratory of Epidemiology Radiation Stochastic Aftereffects, Ukrainian Scientific Centre of Radiation Medicine, Kiev, Ukraine
4   National Radiation and Epidemiological Registry, Medical Radiological Research Centre, Obninsk, Russian Federation
5   Laboratory of Medical Computer Systems, Ukrainian Research Institute of Oncology and Radiology, Kiev, Ukraine

Abstract

The Chernobyl accident in 1986 raised worldwide concern about the health effects of the radiation fallout. International collaborations were established between scientists to investigate the long-term consequences of the accident. However, lack of knowledge about the mechanisms of data collection and the quality of basic epidemiological tools, such as mortality and cancer incidence, has been recognised as a major limitation for the conduct of epidemiological investigations according to international standards in the Newly Independent States (NIS). In the framework of a collaboration which aims to develop and implement epidemiological infrastructure in Belarus, the Russian Federation and the Ukraine, a survey on cancer registration techniques was conducted. A system of compulsory reporting of all new cases of cancer was introduced in 1953 throughout the former Soviet Union for health planning purposes. This cancer registration system, however, was developed entirely independent from similar activities in other parts of the World. In each of the countries surveyed, a nationwide network of regional dispensary-based cancer registries exists. Cancer registration in the NIS relies on passive reporting from hospital and laboratory sources. Death certificates are searched actively. Whereas in Belarus and the Ukraine computerised cancer registration has been developed in recent years, cancer registration in most areas of the Russian Federation is still a manual operation. Although computerisation was identified as the major objective in all three countries, further efforts are required to assess the completeness and the quality of the information collected. The introduction of internationally recognised classifications would considerably improve the comparability with registries in other parts of the World. In addition to preparing annual statistics for health planning purposes, cancer registries should consider reporting cancer incidence for research purposes following international standards. Relating cancer registration more closely to research would allow better and continuous monitoring of the epidemiological situation, which in turn could serve as a basis for public health purposes in the priority setting and the improvement of the quality of health care in the NIS.

Acknowledgements

We thank Prof. M. Rahu, Institute of Experimental and Clinical Medicine, Tallinn, Estonia for his valuable comments and suggestions.

Introduction

The Chernobyl accident on April 26, 1986 caused great international concern about the short- and long-term health effects of the radiation fallout. In order to adequately study health effects in both the general population and in particularly exposed population groups - such as the clean-up workers (liquidators) -, basic tools for epidemiology of diseases must be viable for decades to come. These tools include reliable registration of mortality, morbidity and of the population. Epidemiological investigations rely on correct and complete identification of target populations as well as on individual health status indicators such as mortality or cancer incidence to asses the effect of the exposure in these populations. Since individual exposure may vary considerably, and, since data in epidemiological studies may be combined from a variety of data sources, the basic information must be available at the individual level. A common and unique identifier of individuals must therefore be present. Such tools also exist in the Newly Independent States (NIS) based on principles developed in the former Soviet Union. As part of a vital registration system deaths are systematically recorded and cancer registration is compulsory in all Newly Independent States. However, these demographic and epidemiological registers were developed entirely independent from the experience gained in other parts of the World and in many instances deal with aggregated data using locally developed coding and classification systems. To assure the quality of epidemiological investigations in the Newly Independent States, it is therefore of primary importance to understand the mechanism by which health outcome measures are collected and to asses the quality of the information collected. In the framework of our activities in the development of the epidemiological infrastructure in Belarus, the Russian Federation and the Ukraine, under the European Union supported and initiated project (ECP-7), our efforts were concentrated on the establishment or the improvement of cancer registration techniques, with provision of support in terms of equipment and know-how.

The concept of cancer registration[1] and its techniques developed empirically since the beginning of this century. During the early 1940s population based cancer registration developed in North America[2] and in Western Europe[3]. About one decade later cancer registration was also introduced in countries of Central and Eastern Europe[4] (Jensen et al., 1991). The rapidly increasing number of population-based cancer registries throughout the World raised the problem of the comparability of the strategies elaborated in the registration process and, among the initiatives towards the standardisation of the methods and definitions, the "WHO[5] Handbook for standardized cancer registries" was the first one in 1976 to recognise the need for standardisation of both, hospital-based and population-based cancer registries, with regards to their nomenclature and classifications used as well as to the evaluation and presentation of their results. A more detailed manual of cancer registration methodology was published by the International Agency for Research on Cancer (IARC) in 1978, followed by an updated version, including new developments, in 1991 (MacLennan et al., 1978; Jensen et al., 1991). Whereas this methodology was largely adhered to in Western Europe and North America, cancer registration techniques developed almost independently in Eastern Europe.

---

[1] The term "cancer registration" refers to population-based cancer registration throughout this document, unless stated otherwise.

[2] New York State, USA (1940), Connecticut, USA (1941), Saskatchewan, Canada (1944).

[3] Denmark (1942), South West Region, England and Wales (1945).

[4] Slovenia, Yugoslavia (1950), different areas in Hungary (1952), German Democratic Republic (1953).

[5] World Health Organization.

Compulsory cancer registration was introduced in the Union of Soviet Socialist Republics (USSR) in 1953. Death certificates and autopsy reports as information sources were included in 1961 (Rahu, 1992). Registration of leukaemias only became compulsory in 1965. Since 1966 international cancer incidence statistics are published in regular intervals in the series entitled "Cancer Incidence in Five Continents" (Doll *et al.*, 1966, 1970; Waterhouse *et al.*, 1976, 1982; Muir *et al.*, 1987; Parkin *et al.*, 1992). In the first five volumes of this series no data were included for any of the republics of the Soviet Union. Only the most recent Volume VI, published in 1992, provides cancer incidence statistics for selected areas of the former Soviet Union: Belarus, St. Petersburg, Kyrgyzstan, Latvia and Estonia. It was noted, however, that the quality of these cancer incidence could not be evaluated in detail as it is usually the case for other cancer registries. In the absence of any detailed knowledge on cancer registration techniques in the Newly Independent States, a survey was conducted in selected regional cancer registries in Belarus, the Russian Federation and the Ukraine. The study aimed to provide detailed insight into the local cancer registration techniques, to identify potential differences with cancer registries in other parts of the World and to suggest areas for further improvements towards the ultimate goal of developing a network of computerised cancer registries in the Newly Independent States according to internationally accepted and recommended standards.

Methods

In the absence of any knowledge on the details of the cancer registration process in the NIS, the survey methodology was adopted as the most appropriate approach to collect information according to a sufficient level of detail. Guidelines for the basic items to be covered during the survey were developed *a priori* following the available internationally recognised methodology for cancer registration (Jensen *et al.*, 1991).

In each of the Newly Independent States a nationwide network of regional population-based cancer registries forms the basis for cancer registration. Each individual state is administratively subdivided into *oblasts* (regions) covering a population of approximately 1.5 million inhabitants. In each of these oblasts one or more oncological dispensaries are responsible for diagnosis, treatment and registration of cancer occurring in patients resident in a defined catchment area. In each of these regional oncological dispensaries one department is responsible for cancer registration, which we will refer to as the *regional cancer registry* hereafter.

The survey was restricted to regional cancer registries in Belarus, the Russian Federation and the Ukraine. Taking into account the geographical variety of conditions in which regional cancer registries operate, the study aimed to include at least one registry functioning in an urban environment, one in a predominantly rural area and one registry operating in a contaminated region for each of the countries covered by the survey. In addition, institutes responsible for cancer registration at the national level in each of the three countries were included as further information sources. The survey was carried out during a period of three months in spring 1995. Due to logistic difficulties and time constraints, the initial inclusion criteria could not entirely be fulfilled in the Russian Federation. The list of institutions which were visited during the survey is presented in Table 1.

The concept and details of cancer registration were discussed extensively with the heads of the National Cancer Registries as well as with their staff directly responsible for the data co-ordination and the elaboration of the final summary statistics. Contacts concerning the visits of foreign scientists at the regional oncological dispensaries and other health care institutions were initiated by the respective National Cancer Registry in Belarus and Ukraine. In Russia, the need for crossregional computerised cancer registration was first recognised by the Medical Radiological Research Centre, where developments towards a cancer registry for all contaminated areas were initiated. At the local cancer registry, the contact persons always included the principle oncologist as well as the head of the cancer registration department of the local oncological dispensary. Discussion were primarily held in Russian as the working language and we had the opportunity to work with professionals at different levels, and to gain a comprehensive picture of cancer registration practices.

Table 1: Institutions visited in the framework of the survey on cancer registration techniques in the Newly Independent States.

Country	Region	Name of the institution (Location)	Type of institution	Survey category
BELARUS	Minsk city	Belarussian Centre for Medical Technologies, Information Computer Systems, Health Administration and Management (Minsk)	Central cancer registry	Central cancer registry
	Grodno oblast	Grodno oblast oncological dispensary (Grodno)	Regional cancer registry	Rural, contaminated
	Minsk city	Minsk city oncological dispensary (Minsk)	Regional cancer registry	Urban - non-contaminated
	Minsk oblast	Minsk oblast oncological dispensary (Lesnoe)	Regional cancer registry	Rural non-contaminated
RUSSIAN FEDERATION	Moscow city	Cancer Research Centre, Institute of Carcinogenesis (Moscow)	Research Institute	---
	Kaluga oblast	Medical Radiological Research Centre (Obninsk)	Future central cancer registry for the contaminated areas	Future central cancer registry for the contaminated areas
	Kaluga oblast	Research clinics No. 1 and 2 (Obninsk)	Research Institutes	---
	Kaluga oblast	Kaluga oblast oncological dispensary (Kaluga)	Regional cancer registry	Rural, contaminated
	Moscow city	Moscow children dispensary (Moscow)	Specialised health care institution	---
	Moscow city	Moscow city oncological dispensary (Moscow)	Regional cancer registry	Urban, non-contaminated
UKRAINE	Kiev city	Ukrainian Research Institute of Oncology and Radiology (Kiev)	Central cancer registry	Central cancer registry
	Kiev city	Kiev city oncological dispensary (Kiev)	Regional cancer registry	Urban, non-contaminated
	Zhitomir	Zhitomir oblast oncological dispensary (Zhitomir)	Regional cancer registry	Rural, contaminated
	Zakarpat	Zakarpat oblast oncological dispensary (Uzhgorod)	Regional cancer registry	Rural, non-contaminated

Results

Principles and particularities of the health care system

In the NIS free medical care is provided for the entire population. In addition to diagnosis and treatment, health care in the NIS also includes systematic early detection of diseases and follow-up of the patient, which is generally known as *dispensarisation*. One of the particularities of the health care system in the NIS compared to that of other countries today is the complete separation of adult and child health care in different institutions. Initial medical education is divided into either adult- or child care as from the third year of the medical studies onwards and students graduate as adult general practitioner or paediatrician.

Health care is provided at the official place of residence. Primary health care is provided in medical district units, serving 3-4,000 inhabitants each. A district unit consists of a group of physicians (e.g. general practitioner, paediatrician, gynaecologist, obstetrician, etc.) and nurses responsible for primary health care of the resident population. In the past, private activity of physicians and family doctors were practically unknown in the NIS. The basic health care unit is the rayon hospital, always comprising an out-patient department (or polyclinic) as well as a hospital (or inpatient section). Highly specialised health care institution are primarily located in oblast centres, such as health care institutions providing specialised care for particular diseases or population groups, which are locally termed *dispensaries* (e.g. oncological dispensaries, tuberculosis dispensaries, etc.). Furthermore, clinics attached to medical educational establishments and research institutes are mainly located in the capitals. The various levels of health care providers in urban areas are summarised in Figure 1.

Figure 1: Organisational structure of medical care in urban areas.

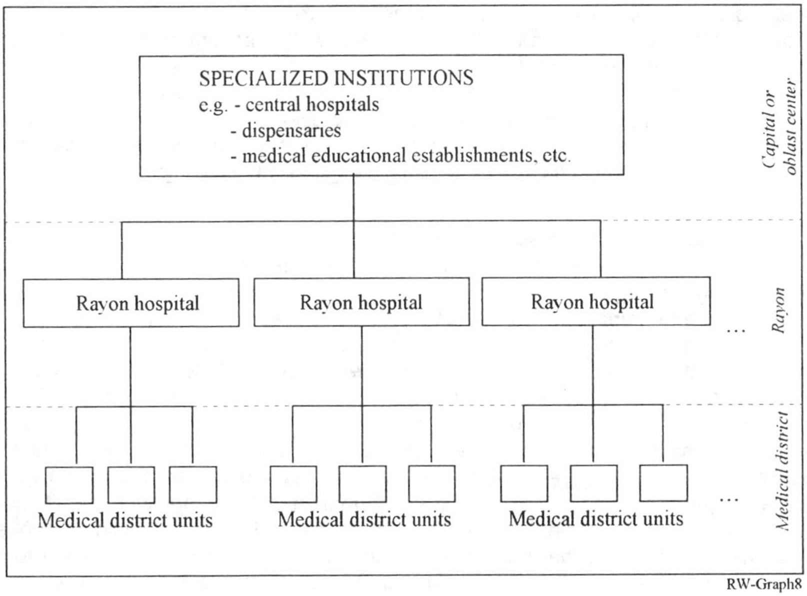

Adapted from: Participants in a study tour organized by the WHO (1960) Health services in the USSR. Public Health Papers No. 3, WHO, Geneva.

However, in rural areas an additional level of health care units exists. In the latter, medical districts are equipped with small hospitals and primary health care is provided by feldsher units or feldsher-midwife posts in rural communities or farming areas. Feldshers are medical assistants, who, following a paramedical education of three years, provide primary health care under the direct supervision of a district doctor.

Cancer care relies on two principal functional units: the oncological consulting room at the rayon level and the oncological dispensary at the oblast level. The rayon oncologist is responsible for (early) diagnosis of cancer and follow-up of cancer patients. The oncological dispensary is responsible for diagnosis and treatment of cancer cases. The large majority of the cancer cases are treated at the regional oncological dispensary. Cancer registration is performed at both, the rayon and the oblast level.

Basic documents of the cancer registration process

Six different documents are either directly or indirectly involved in the cancer registration process (Table 2). Each patient ever admitted to a medical establishment, irrespectively whether to the out-patient department or to the in-patient section, will have an outpatient card in the central register of the admission office. The out-patient card contains information on the subjects identification (surname, name, patronymic, date of birth, sex, occupation and place of work), preventive examinations, diagnosis, treatment, short medical history as well as follow-up information. Each diagnostic procedure performed results in the establishment of a special form (loose leaf) which is systematically inserted in the out-patient card. The out-patient card is the lifetime patient record which is systematically updated at subsequent discharges.

All patient admitted for in-patient treatment will have a record of medical history. This includes patient identification information as well as diagnosis, treatment, treatment complications, results of diagnostic examinations and prescriptions. For in-patients, both, the medical history and the outpatient card will follow the patient through the various departments. The medical history is admission-specific and is stored in archives after discharge of the patient. The medical history has a special *epicrisis* section which provides a summary of diagnosis and treatment. The latter represents the document by means of which cancer cases diagnosed and treated in the oncological dispensary are notified to the cancer registration department (oncological dispensary internal notification).

Two additional forms are specifically designed for reporting of cancer cases from peripheral institutions to the regional oncological dispensary: the notification and the extract. The "notification of a patient with newly diagnosed cancer or other neoplasm", include personal and demographic information (surname, name, patronymic, sex, year of birth, occupation and place of work) as well as date of diagnosis, diagnosis, method of diagnostic verification and circumstances of diseases detection. This must be "completed by all medical doctors of the general and special health care network for every patient with newly diagnosed malignant neoplasm, including patients detected during surgery, preventive examinations, during autopsy, etc.". The extract is used for notification of cancers first diagnosed and treated in a medical establishment other than the regional oncological dispensary. The extract records information on the patients identification, complete diagnosis and treatment as well as a short medical history including diagnostic investigations and course of the disease. Whereas the notification has to be send to the relevant regional oncological dispensary within three days of the completion of the form, the extract is send at the time of discharge of the patients. Both forms are send by regular mail.

Table 2: Documents directly or indirectly involved in the cancer registration process in the Newly Independent States.

Document type	Type of information recorded	Directly involved in the cancer registration process	Used at/by
Out-patient card	Lifetime out-patient record	NO (source document)	All out-patient facilities
Medical history	Medical history for a single admission	YES (source document)	All in-patient facilities
of which: Epicrisis	Summary of diagnosis and treatment	YES	
Notification	Notification following diagnosis	YES	All medical doctors of the general or special health care network
Extract	Notification following diagnosis and treatment	YES	All cancer diagnosing and treating medical institutions
Control card of dispensary follow up for patients with malignant neoplasms	Lifetime cancer patient follow up record	YES	Oncological dispensaries
Death certificate	Death record, including date and cause of death	YES	By all medical doctors for submission to the vital registration department

At the regional oncological dispensary, the cancer registration department systematically receives all out-patient cards, medical histories after discharge of the patient and the notifications and extracts from other medical establishments. The cancer registry maintains a card register on cancer cases in the resident population. For each patient a "control card of dispensary follow-up of patients with malignant neoplasms" (control card) exists in this register. For each of the documents reaching the cancer registry the corresponding card will be searched and updated or for new patients a new control card will be established. In regular monthly intervals, the cancer registration staff actively searches the information on cancer deaths in the corresponding vital registration department. For cancer deaths not previously known to the cancer registry, trace-back is initiated and notifications and extracts for previous contact with the health care system are requested retrospectively. A control card will be established for each cancer case first identified through death certificate.

Approximately 30 different data items are on record for each cancer case during the manual process of registration. The introduction of computerised cancer registration doubled the number of data items collected. In addition to the identification and the medical information, a number of basic characteristics (e.g. dispensary, rayon, oblast, nationality, occupation) are included. Information is available in chronological order. Diagnosis is recorded according to the International Classification of Diseases 9th Revision (ICD-9). Additional information concerning diagnosis includes method of diagnostic confirmation, a local stage code and TNM classification, histological type (local 1 or 2-digit classification) and clinical group. Furthermore basic items on treatment are gathered. Follow-up status is evaluated at the end of each year. Heads of cancer registration departments are predominantly medical doctors. The coding is performed by the assistants under the supervision of the medical doctor.

Manual cancer registration procedures

Health care is provided at the official place of residence. Whereas in the event of specialised diagnosis or treatment there may be exceptions to this principle, cancer cases are <u>always</u> registered at the official place of residence. The complete cancer registration process consist of the joint action of the oncological dispensary internal reporting process and the peripheral reporting from other health care institutions to the regional oncological dispensary (Figure 2).

Figure 2: Reporting of cancer cases to the regional oncological dispensary

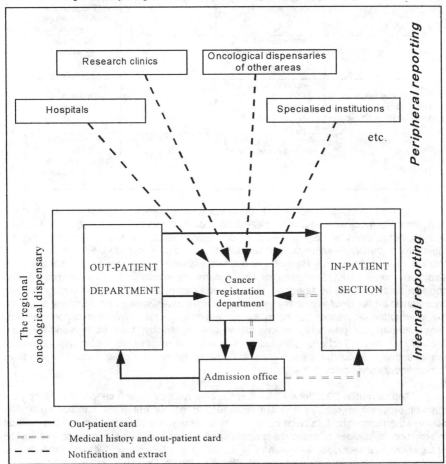

RW-Graph14

At discharge from the oncological dispensary, all documents are transferred to the cancer registration department. After registration, the out-patient card is kept at the admission office, whereas the medical history is stored in archives. The process of reporting of cancer cases from peripheral health care institutions consists of establishing a notification or extract which is mailed to the oncological dispensary of the place of residence. Within a given catchment area, the rayon hospitals and the regional oncological dispensary mutually report cancer cases to each other.

At monthly intervals both the regional cancer registry staff and the rayon oncologist independently gather information on vital status and cancer deaths from the vital registration department and the respective card registers are updated accordingly. In the event a cancer case is first known from death certificate, the cancer registry traces back its history and retrospectively requests a notification or an extract wherever previous contact with a health care institution can be identified. Furthermore, the rayon oncologist visits the regional oncological dispensary quarterly to manually verify the completeness of the information in the two cancer registers. The latter enables the rayon oncologist to maintain and update his (her) rayon cancer registry for the follow-up of the cancer patients and, at the same time, represents an independent assessment of the completeness of the regional cancer registry.

Routine reporting of cancer data

Two different types of statistical reports are prepared annually: the "Report on cancer incidence by site sex and age of the patient" (the cancer incidence report) and the "Report on patients with malignant neoplasms" (the patient report). The information source are the control cards for all patients first diagnosed during the previous calendar year. The cancer incidence report provides number of patients by site, sex and five year age groups, and, since 1989, includes 44 individual- or groups of sites. The patient report mainly provides prevalence figures according to different criteria (e.g. stage, treatment, etc.) as well as mortality among cancer patients.

The layout of the statistical reports defines the amount of detail according to which cancer data are available at rayon-, oblast- and country level. Wherever computerised technology is not available to the local cancer registry, these forms are prepared manually. The latter particularly applies to the large majority of the areas of the Russian Federation.

Statistical reports are prepared annually as requested by the Ministries of Health. The reporting activity starts at the rayon level sometime in January. Cases diagnosed during December of the previous year for which the notification or extract is received at the cancer registration department after the date of start of the reporting process will be included in the January contingent of the subsequent year. Cases diagnosed before December of the previous year and registered after January of the subsequent year will not be included in any annual statistical report on cancer incidence. The rayon reports are submitted to the regional oncological dispensary, which is responsible for elaborating a summary report. The latter is submitted to the central institution responsible for the elaboration of the country report, usually the Institute of Oncology, and the final report is requested by the Ministry of Health during March of the same year. Statistical reports are considered final and they are not revised thereafter. In certain well established cancer registries, however, research is based on cancer incidence statistics elaborated specifically for research purposes, including cases omitted in the official reporting process.

The limitations of the manual cancer registration process were first recognised in Belarus, where efforts towards computerised cancer registration started in 1973 at the Institute of Oncology and Medical Radiology and computerised cancer registry data exist since 1978. However, computerisation was first introduced to assist in the statistical reporting. During the early periods only coded information was computerised. This resulted in cancer registration files without identification information such as surname, name and patronymic of the patient between 1978 and 1985. Development towards decentralised Personal Computer (PC)-based cancer registration recording all items of the control card started in 1985. During a period of almost five years the twelve oncological dispensaries in Belarus were equipped with the PC technology and the cancer registration software. Since 1990 a network of computerised oncological dispensaries provides the basis for nation-wide cancer registration, coordinated by the Belarussian Centre for Medical Technologies, Information Computer Systems, Health Administration and Management.

Similarly, initiatives towards a computerised national cancer registry started in 1990 in the Ukraine at the Ukrainian Research Institute for Oncology and Radiology. An original concept aimed at simultaneously developing both, computerised hospitals- as well as population-based cancer registration. The hospital cancer registration system uses the out-patient card and the medical history as information sources, whereas the population-based cancer registration system is based on the notification, the extract and the epicrisis. The population-based cancer registry could be optimised by the exchange of already computerised information between the two types of registries. To date, approximately one quarter of the 46 regional- or city oncological dispensaries in the Ukraine are equipped with the population-based cancer registration technology.

Unlike Belarus and the Ukraine, no initiative for the development of a computerised national cancer registry was taken to date in the Russian Federation. Most of the 119 oncological dispensaries across the 71 oblasts and territories of the Russian Federation still rely on the manual cancer registration process. Following individual localised initiatives however, computerised regional cancer registries operate in certain areas. During the Soviet period, the cancer registration software developed in the Belarus was exported to a number of regional oncological dispensaries in the Russian Federation (i.e. Kaluga oblast, etc.). Other oncological dispensaries developed their own computerised cancer registration systems (i.e. Moscow city, St. Petersburg city, etc.). Following the Chernobyl Accident the need for computerised cancer registration was recognised and the Medical Radiological Research Centre (MRRC) in Obninsk is currently developing the relevant technology to introduce computerised cancer registration in the four contaminated oblasts (Oriol, Tula, Kaluga and Briansk). A revised control card was specifically introduced for this purpose. Both the official and the revised control card currently are used in parallel in the Kaluga oblast. However, an order by the Ministry of Health of the Russian Federation may render the use of the revised control card compulsory, at least for the use in the contaminated areas.

A modernised infrastructure

Cancer registration techniques as currently in operation in the NIS developed entirely independent from the experiences gained in the rest of the World. In view of the design and implementation of future international epidemiological studies to investigate, among others, the health effects of the Chernobyl accident, it is important to highlight the areas which could result in measurable differences in the comparison of cancer incidence statistics from the NIS to those of other countries.

Classification and coding of neoplasms

Since the 1980s, all Newly Independent States code cancer diagnosis according to the 9th Revision of ICD. Coding was first performed at the three digit level and in the early 1990s the four-digit level coding was introduced in most of the countries. Histology is coded in all Newly Independent States, but the classifications used vary in their level of detail, ranging from one to three digits. For the purpose of comparability of the cancer data, however, internationally accepted and recommended classifications should be used consistently. Belarus, the Russian Federation and the Ukraine have agreed to implement this International Classification of Diseases for Oncology (ICD-O) for all computerised cancer registration systems in the future. As a first step, the ICD-O second edition (1990) is currently translated into Russian in the framework of the current collaboration. Subsequently the cancer registration software will be modified to allow for inclusion of the relevant codes and dictionaries. The success of the implementation and the quality of the future data, however, will primarily depend on the understanding of the cancer registration staff of the need for this transition as well as the provision of the staff with adequate training in the new coding procedures.

Further diagnosis-related information systematically collected by cancer registries in the NIS include stage, TNM code and clinical group. It should however be mentioned, that, although many cancer registries throughout the world record stage and TNM code, these data items are generally subject to substantial variability between hospitals (Jensen *et al.*, 1991). Similar and even stronger limitations apply to the NIS local definition of clinical group.

Completeness and quality of cancer registration data

Monitoring of the completeness and quality of the data collected should be part of the regular activities of the cancer registry, rather than measured on an ad-hoc basis. Various methods have been proposed to measure the completeness of registration, mostly based on death certificates or samples of hospital records (Jensen *et al.*, 1991). In the NIS, death certificates are regularly screened and trace-back is initiated for cancer deaths not previously registered. However, detailed results of these activities have never been reported. Another important aspect is the completeness and accuracy of the individual information items. Incompleteness of data items may introduce severe bias in statistical analyses, if the non-reporting is not a random process. Furthermore, even if complete, a data item may not necessarily be accurate. Errors in detail may occur during abstraction, transcription or coding of the information. While some of these errors are detectable by range and consistency checks, others may not. The more data items collected by a cancer registry the higher is the probability of incompleteness and inaccurateness of the individual items. Some of the cancer registries in the NIS gather a quite substantial level of detail. Following international cancer registration practices, a common set of procedures for assessing the completeness and accuracy of the cancer registration data should be developed for the systematic use in all NIS cancer registries. As a part of the ECP-7 program, standard software for data entry and data checking developed by the International Association of Cancer Registries and the International Agency for Research on Cancer was translated into Russian and distributed within the NIS.

Particular attention has to be devoted to the retrospective recording of cancer incidence data. Computerisation of cancer patient data usually starts from a given point in time, where the manual process is converted into a computerised process and all patient data will be directly entered into the database thereafter. In addition, most cancer registries put special efforts into the computerisation of retrospective data. In this context it should be remembered that patients who die are eliminated from the card file in the cancer registration department (and are likely to be stored in archives). The same applies to patients with basal cell carcinomas who survive more than five years and who are considered to be cured thereafter. Special efforts should be devoted to assess the completeness of the retrospective recording of cancer patient data.

Another area of particular concern in cancer registries of the NIS is the quality and thus the comparability of the coding of diagnoses. In each oblast cancer registry some three to five persons are involved in the coding of cancer diagnoses. A special study should be designed to evaluate the comparability of the coding between areas of the NIS as well as by comparing the latter to coding practices in well established cancer registries in Europe.

Computerised record linkage

Record linkage technology is used in a variety of circumstances in cancer registries. It may be used to identify data from different sources (e.g. hospital, death certificates, pathology laboratories etc.) relating to the same person. Similarly, record linkage is commonly used for merging the information from two or more different registries (e.g. incidence registry with mortality registry, etc.). In the particular case of the cancer registries in the NIS, the principle of linkage technology is also used at the data collection level for the identification of patients already registered (e.g. finding the name of a patient) as well as for duplicate checking (e.g. match of name, year of birth, etc.). The identification information of all patients in the NIS cancer registries is a combination of surname, name and patronymic and all record linkage developed so far is based on exact matches. There is a particular problem in case of text strings, since actual matches may not be identified due to spelling errors or abbreviations in one of the elements entering the comparison. Many cancer registries throughout the World have already developed probabilistic linkage procedures. It might be advisable to review the available linkage procedures to propose appropriate algorithms for direct implementation in or their adjustment to the specific situation of the NIS cancer registries or to suggest to the authorities a population monitoring system based on a unique number as in the Nordic Countries and other places. The NIS countries involved in the ECP-7 discussed and adopted common formats for variables to be recorded on cancer patients. In order to facilitate record linkages in the future to in particular the established Chernobyl Registers a common identification number between these registers "the Chernobyl number" will be included.

Reporting and analyses

Statistical reporting of cancer data in the NIS is regulated by deadlines of the respective Ministries of Health. In general, reporting starts mid-January at the lowest territorial level (the rayon) for cases diagnosed during the preceding year and they are submitted to the next territorial level, responsible for the elaboration of summary reports. The final national statistical reports are expected at the relevant authorities by the end of March of a given year. The initial lag-period of 1-2 weeks before summarising the data at the lowest level was considered sufficient by the health authorities. However, a certain proportion of the patient admitted for treatment in December of the previous year may not have been discharged by early January of the following year. Furthermore, the information on cancer deaths is gathered once a month. For those cancer deaths identified for the month of December, the trace-back procedure is unlikely to be completed before the statistical reporting starts. It is common practice in most Western cancer registries to allow for a lag-period of one to two years before publishing the cancer incidence data, in particular when death certificates are used. Thus the official cancer incidence reports from the NIS are not directly comparable to those from other countries. Therefore, researchers often base their studies on specially elaborated cancer incidence statistics. The effect and the magnitude of the difference related to the reporting procedure in the official statistics should be evaluated. Depending on the evidence, modifications of the reporting procedure may then be suggested to the relevant authorities. Whereas the current reports from the registries lack detail and accuracy, the ECP-7 program included exchange of scientists, software and collaboration on data analysis and interpretation, in order to improve the situation. More descriptive epidemiological papers already occurred or will occur in international scientific journals as a result of this effort.

Mortality data

Mortality statistics, represent another major outcome measure commonly used in epidemiological studies. Throughout the world and also in the NIS, registration of death is the responsibility of the vital statistics department. A death certificate is established for each death, on which information concerning the identification of the deceased as well as concerning the cause of death is recorded. Only after registration of the death an "authorisation for burial" is issued. Death certificates are coded according to a slightly modified version of the ICD-9 B-list ("Special tabulation list"). Summary mortality statistics are produced annually, the correctness however is unknown and as elsewhere in the world they are believed to be of low quality. Independently, cancer registries sometimes prepare "corrected" mortality data, taking into account the information on diagnosis. In establishing the quality of the necessary tools for the conduct of epidemiological studies into the health effects of the Chernobyl accident, the mortality data should be surveyed as we did for cancer registration and the feasibility of developing a computerised mortality database, including information on all deaths (personal identification, socio-demographic characteristics as well as the underlying and associated causes of death) should be investigated.

Population data

Evidently, to produce rates, population data specified for age and sex are needed on a regional basis. These data, normally are made available from the central statistical office of the country. Although such data in the past was not readily released in the Soviet Union, it is known that such data is timely and available at a sufficient level of detail for epidemiological purposes. Whether the information is based on censuses, or on other mechanism is however not clear from the data we collected. The population is often given as of 1. January in official populations.

Conclusion

Compulsory cancer registration in the Union of Soviet Socialist Republics (USSR) was introduced in 1953 as part of the generalised health planning activities. The principle of cancer registration has remained unchanged and still applies today to the Newly Independent States (NIS). Annual cancer statistics related to incidence and prevalence pattern are requested by the Ministries of Health to allow the preparation of future health plans. Research based on these statistics however has remained extremely limited and mechanisms of collection and the quality of cancer incidence data as one of the most important outcome measures in epidemiological studies are virtually unknown.

The survey on cancer registration confirmed the existence of a nationwide network of regional dispensary-based cancer registries responsible for elaborating cancer incidence statistics for health planning purposes in each of the countries surveyed. The underlying cancer registration process relies on passive reporting using hospital records (including laboratory test results) as the primary source. In addition, death certificates are actively searched by the cancer registration staff in the records of the relevant authorities.

However, major differences with cancer registries in other parts of the World were observed in particular in coding and classifications, in quality control and in many areas cancer registration is still a manual operation. Special efforts should be devoted to assess the completeness and the quality of the information collected and the introduction of internationally recognised classifications would improve the comparability considerably. A review of the current record linkage methodology could form the basis for the development of probabilistic concepts suitable for cancer registration systems in the NIS, where identification is primarily based on names. Furthermore, NIS registries should begin to report cancer incidence rates for research purposes following international standards. Relating cancer registration more closely to research would allow better and continuous monitoring of the epidemiological situation, which in turn could serve as a basis for public health purposes in the priority setting and the improvement of the quality of health care in the NIS.

The ECP-7 program was instrumentive in bringing together the researchers who will be responsible for following the long term effects of the Chernobyl accident. A common platform and collaboration in cancer registration in the NIS and with the rest of Europe was established. A training program, including courses and training in western Europe was initiated and the necessary tools in registration and epidemiology developed for a Russian speaking audience. All the human building blocks and networks are thus established. Also some equipment (computers) and software was provided. If the development and maintenance of an infrastructure for epidemiological studies and public health should continue to be successful, so far supported by the European Union with the ECP-7 project, full support (including financial) and collaboration between Ministries of Health, - Chernobyl, and the Interior within and between the NIS must be secured as data for follow-up of the consequences of the accident is developed under the auspices of these ministries. Evidently the future effort is huge. We cannot afford competition for funds, data and power between different groups, but must seek collaboration at all levels of competence.

References

1.   Coleman M.P., Demaret E. (1988) Cancer registration in the European Community. *Int J Cancer*, 42(3):339-45.

2.   Doll R., Payne P., Waterhouse J., *eds.* (1966) Cancer incidence in five continents, Vol. I. International Union Against Cancer, Springer Verlag, Berlin.

3.   Doll R., Muir C., Waterhouse J., *eds.* (1970) Cancer incidence in five continents, Vol. II. International Union Against Cancer, Springer Verlag, Berlin.

4.   Jensen O.M., Parkin D.M., MacLennan R., Muir C.S., Skeet R.G., *eds.* (1991) Cancer registration: principles and methods. IARC Scientific Publication No. 95, IARC, Lyon.

5.   MacLennan R., Muir C., Stenitz R., Winkler A., *eds.* (1978) Cancer registration and its techniques. IARC Scientific Publication No. 21, IARC, Lyon.

6.   Ministry of Health of the USSR (1974) The training and utilization of feldshers in the USSR. Public Health Papers No. 56. WHO, Geneva.

7.   Muir C., Waterhouse J., Mack T., Powell J., Whelan S., *eds.* (1987) Cancer incidence in five continents, Vol. V. IARC Scientific Publication No. 88, IARC, Lyon.

8.   Napalkov N.P., Tserkovny G.F., Merabishvili V.M., Parkin D.M., Smans M., Muir C.S. (1983) Cancer incidence in the USSR. IARC Scientific Publication No. 48, Second Revised Edition. IARC, Lyon.

9.   Parkin D.M., Muir C.S., Whelan S.L., Gao Y.-T., Ferlay J., Powell J., *eds.* (1992) Cancer incidence in five continents, Vol. VI. IARC Scientific Publication No. 120, IARC, Lyon.

10.   Parkin D.M., Chen V.W., Ferlay J., Galceran J., Storm H.H., Whelan S.L. (1994) Comparability and quality control in cancer registration. IARC Technical Report No. 19, IARC, Lyon.

11.   Participants in a study tour organized by the World Health Organization (1960) Health services in the USSR. Public Health Papers No. 3. WHO, Geneva.

12.   Participants in a study tour organized by the World Health Organization (1970) Postgraduate education for medical personnel in the USSR. Public Health Papers No. 39. WHO, Geneva.

13.   Percy C., Van Holten V., Muir C. *eds.* (1990) ICD-O - International classification of diseases for oncology. Second Edition. WHO, Geneva.

14.   Rahu M. (1992) Cancer Epidemiology in the former USSR. *Epidemiology*, 3(5):464-470.

15.   Waterhouse J., Muir C., Correa P., Powell J., *eds.* (1976) Cancer incidence in five continents, Vol. III. IARC Scientific Publication No. 15, IARC, Lyon.

16.   Waterhouse J., Muir C., Shanmugaratnam K., Powell J. (1982) Cancer incidence in five continents, Vol. IV. IARC Scientific Publication No. 42, IARC, Lyon.

17.   World Health Organization (1976) ICD-O - International classification of diseases for oncology. First Edition. WHO, Geneva.

18.   World Health Organization (1976) WHO Handbook for standardized cancer registries. WHO Offset Publication No. 25. WHO, Geneva.

19.   World Health Organization (1977) International classification of diseases. Revision 1975. Volume 1. WHO, Geneva.

20.   World Health Organization (1977) International classification of diseases. Revision 1975. Volume 2. WHO, Geneva.

Problems & Peculiarities of Medical Service of the liquidators of the Bryansk Region, who were engaged in the Clean-Up after the Chernobyl Accident in 1986-1987

M.M. DUKOV, V.M. SAMOILENCKO*
V.V. DOROKHOV, A.D. PROSHIN, G.A. ROMANOVA, N.B. RIVKIND**
*Bryansk Department on Chernobyl Problems, MES of Russia
Karl Marx Sq. 2, 241000 Bryansk, Russia
** Bryansk Regional Medical Diagnostic Center
Bezhitskaya str. 2, 241007 Bryansk, Russia

Abstract

Investigation of the radiological effects of the Chernobyl accident's consequences among the liquidators engaged in the clean-up in 1986-1987, represents, from our point of view, special interest. It is associated with the fact of multiply received high radiation doses as well as by the following radiological consequences, related to particular radiocontamination over the Bryansk region. This paper contains data analyses of medical, dosimetric, demographic and other health status characteristics of this group of liquidators during recent years.

Peculiarities of radiation effects among the Bryansk region liquidators, engaged in the clean-up after the Chernobyl accident in 1986-1987.

Radioactive cloud, formed as a consequence of the Chernobyl accident on 28 April 1986, passed over the whole Bryansk region territory, that lead to contamination of the soil by such short-lived radionuclides, as I-131, I-132, as well as by the main dose-forming radionuclides : Cs-137, Sr-90, Pt-240 and determined the rate of both internal and external exposure. According to the data of direct measurements, made in the region area during May-June 1986, average doses of the thyroid gland exposure from I-131 ranged between 10 and 220 rad.

Population of 22 districts of 27 was affected by radioactive contamination with Cs-137. The ground contamination with Cs-137 ranged between 1 and 71 Ci/km². So the persons, who were engaged in the accident's aftercare activities within the 30 km area from the Chernobyl Nuclear Power Station, received considerable doses. The value of such dose varied in accordance with place of residence and character of job.

During aftercare activities they have received doses, ranging from 5 to 25 Rad. Having come back to the areas, contaminated with radionuclides, these persons still have being irradiated.

Cytogenetic study had been performed among 987 liquidators. We detected evidence of multiaberrant cells in 10% of them; the rate of dicentrics and rings, as markers in peripheral blood lymphocytes, is still high : 2.2 +/- 0.4 per 1000 cells and differs from control figures (0.9 +/- 0.5 per 1000 cells). This data shows the significant values of doses received by the liquidators of the Bryansk region, but the question of it's value is still open.

Analyses of morbidity during 4 years reflects a trend of increase : 1991 : 78.202 cases per 100.000 of inhabitants (average value over the Russian Federation : 71.709), 1994 : 120.548 cases per 100.00 (over the Russian Federation in 1993 : 75.585). So, the rate of morbidity during the last 4 years came to 65% (Table 1.).

Table 1

Dynamics of morbidity among the liquidators in 1990-1994
(per 100.000 people)

Indicators / Years	1991	1992	1993	1994
Number of subjects examined	1491	1527	1572	1640
All types of diseases	78202	74263	88295	120548
Malignant tumors	536	131	254	793
Cardiovascular diseases	1006	1048	1336	2073
Endocrine diseases	5768	6158	6488	6585
Digestive diseases	3622	2816	4835	8841
Genito-urinary diseases	536	720	891	1341
Neuro-psychic disorders	6439	6811	8015	10488

In the structure of morbidity in liquidators which had been examined at the regional dispensaries, the leading position in recent years have had neurotic and psychic disorders with a trend of growth (Table 2). Nearly 90% of subjects have neurotic reactions of different stage of dissociated changes, and 20% of cases are psychosocial disadaptation.

The second position takes diseases of gastro-intestinal tract. Particular feature is a high frequency of "mute" ulcers, slow process of cicatrization. Hyperplasia of the thyroid gland of 2-3 stages and nodule forms are the most frequent. A trend of increase of nodule forms is noticed: from 22 (1.6%) in 1991 to 144 (7%) in 1994. Diseases of endocrine, cardio-vascular and genito-urinary systems have a trend to grow too, but their rates are still within the values, registered in the areas with contamination levels higher than 15 Ci/km^2.

Table 2

Morbidity rate among adults of the Bryansk region and the liquidators
(per 100 000 people)

Diseases	Population		Liquidators	
	Total in the Bryansk region	in the con-taminated areas ($>5Ci/km^2$)	in the Bryansk region	total in Russia
Cardiovascular diseases	11438	14287	7555	6306
Endocrine diseases	3159	5100	6585	6036
Digestive diseases	7179	9215	8841	9739
Genito-urinary diseases	6075	6351	1341	no data
Mental disorders	4329	4929	10488	5743
All types of diseases	97586	107705	120548	75606

Correlative analyses of morbidity rate among the liquidators of Russia having determined dose and those of the Bryansk region shows that the latter can be equated with liquidators having dose higher than 20 cGy.

An increased number of persons with detected pathology have been followed by an increased number of subjects requiring various health care services (Table 3.).

Table 3

Dynamics of treatment and rehabilitation services among the liquidators

	1991	1992	1993	1994
Out-patient treatment	103	154	178	220
In-patient treatment	41	53	69	76
Sanatory treatment	19	20	28	62

In spite of 2-3 times increase in the scope of various forms of medical service (in-patient, out-patient, sanatory) it should be noted about it's insufficient effects that is caused by the still high trend of increase of such health status characteristics, as morbidity and disability.

The number of persons with temporary disability requiring out-patient medical care was 103 (6.9%) in 1991 and 200 (13.4%) in 1994 with increase level of 1.9 times.

In comparison with 1990, when the number of persons required in-patient care was 27, in 1994 the number was already 76 and the growth of hospitalisation came to 2.8 times. The number of subjects requiring sanatory treatment was 18 in 1990 and 62 in 1994 with increase level of 3.4 times.

However, there are still a growing number of persons getting certificates of disability. The number of liquidators who received certificates of disability for the first time was 46 in 1990 and 79 in 1994 with growth level of 1.7 times (Table 4.).

Mortality rate among the liquidators has permanently exceeded average regional values 2.4-4.5 times. The structure of mortality causes has not changed during these years, including 1994. Injuries and poisonings are still on the first stage : from 50 to 70% of cases, neoplasms - from 15 to 75% of cases.

Table 4

Health status characteristics in liquidators of the Bryansk region in 1990-1994
(per 1000 population)

Characteristics	1991	1992	1993	1994
Morbidity	78.2	74.3	88.3	120.5
Disability	27.8	38.6	39.4	48.1
Mortality	3.9	4.0	5.9	7.8

Analyses of disability characteristics among the liquidators of Russia having determined dose values, and those of the Bryansk region shows that the liquidators of the Bryansk region can be equated with emergency workers having dose from 5 to 20 cGy.

Morbidity rate among the liquidators of the Bryansk region and those of the central areas of Russia where the Bryansk region is in is practically the same.

Conclusion

The liquidators of the Bryansk region represent a unique group among the other liquidators of Russia in relation to multiple exposure doses : during passing the radioactive cloud over the Bryansk region; during aftercare activities at the Chernobyl Nuclear Power Station; living in the territories contaminated with radionuclides.

Exposure doses among this group of inhabitants are probably within 5-20 cGy, outgoing from the data of correlative analyses by criteria of general morbidity and disability. The data of cytogenetic studies shows, that 10% of these persons have significantly higher individual doses. But absence of such characteristics, as individual accumulated doses, enables to estimate dose-effect and to evaluate it's input to morbidity level of this group of inhabitants of the Bryansk region. It is necessary to conduct detailed examinations of the liquidators, engaged in the aftercare activities at the Chernobyl Nuclear Power Station in 1986-1987, and of their children, having markers of received high exposure doses, that could allow to obtain additional data of health status peculiarities among them.

WHO - IPHECA: Epidemiological aspects

G. Souchkevitch, Radiation Scientist, World Health Organization, Geneva

1. Introduction

In May 1991 the World Health Assembly endorsed the establishment of the International Programme on the Health Effects of the Chernobyl Accident (IPHECA) under the auspices of WHO.

Five pilot projects have been carried out within IPHECA in the study territories of Belarus, Russian Federation and Ukraine in a period from 1991 to 1994. This pilot projects dealt with the detection and treatment of leukemia and related diseases (Haematology Project), thyroid disorders (Thyroid project), brain damage during exposure *in-utero* (Brain Damage *in-Utero* project) and with the development of the Chernobyl registries (Epidemiological Registry Project). A fifth pilot project on oral health was performed only in Belarus.

Epidemiological investigations have been an important component of all IPHECA pilot projects. Within "Epidemiological Registry" Project such investigations have been the principal activity. But with respect to other IPHECA projects it was carried out in addition to main objectives relating to medical monitoring, early diagnosis and treatment of specific diseases included in project protocols.

To support the epidemiological investigations within IPHECA, WHO supplied 41 computers in Belarus, Russian Federation and Ukraine and provided training for specialists from these countries in internationally recognized centres. The training programmes and host countries were as follows: (1) Standardization of epidemiological investigations (United Kingdom), (2) Radiation epidemiology (Russia), (3) Development of software (United Kingdom), (4) Principles of Epidemiological investigations (The Czech Republic), (5) Cohort investigations (Japan).

2. Activities within "Epidemiological Registry" Project

The "Epidemiological Registry" pilot project aimed at the development and introduction of data basis, for obtaining of health indices and identifying the causes of morbidity and mortality among the population of territories with the level of ^{137}Cs contamination more than 555 kBq/m^2 (Table 1). It should be noted that some rayons had not been included initially in the project protocol but added later when the new information regarding soil contamination became available.

The Epidemiological Registry project served as a focal point for epidemiological investigations carried out within other IPHECA pilot projects. Therefore main attention was paid to:

- the development of standardized protocols and formats for computerized data collection;

- the creation of Software for all pilot projects;
- the determination of optimal methods for epidemiological studies.

Table 1.   **Territories in Belarus, Russian Federation and Ukraine within IPHECA Epidemiological Registry project**

Country	Oblast[1]	Contaminated rayons[2] (more than 555 kBq/m^2)	Clean territories (reference data)
Belarus	Mogilev Gomel	Slavgorodski Hoynikski Checherski Kormyanski Gomelski	Ushachski rayon of Vitebsk oblast
Russian Federation	Bryansk Kaluga	Novozybkovski Zhizdrinski	Borovski rayon of Kaluga oblast
Ukraine	Kiev Zhitomir	Ivankovski Polesski Ovruchski	Yagotinski rayon of Kiev oblast

Three specialized institutions and national project coordinators supervised the fulfilment of the IPHECA "Epidemiological Registry project" in situ:

Belarus   -   Belarus Centre of Medical Technology, Informatics   and Economics of Health care, (Drs. E. Okeanov and G. Chernikov);

Russian Federation   -   Medical Radiological Research Centre (Dr. V. Ivanov);

Ukraine   -   Institute of Epidemiological and Prevention of Radiation Injuries (Dr. V. Buzunov).

2.1   *Development of standardized documents*

In order to standardize epidemiological investigations within all IPHECA pilot projects, specialists from the three States in cooperation with international experts have developed basic documents as follows:

---

[1]   Oblast-   a larger administrative and territorial unit, includes several rayons

[2]   Rayon-   an administrative and territorial unit in the former USSR

(1) Epidemiological Registry Project Protocol - for the determination of the common technology for collection, storage, transmission and analysis of demographic, medical, dosimetric and epidemiological data obtained in radiocontaminated territories in Belarus, Russian Federation and Ukraine. The protocol has clearly outlined main procedures in territorial units (rayon, oblast, national level).

(2) Recommendations how to integrate soft and hardware with National systems.

(3) Special IPHECA software package (SISP) - for harmonization of IPHECA with National programmes.

(4) The General Record Chart (OKSO) which includes the following main sections:

- General information;
- Dosimetry data on exposure of the whole body and thyroid;
- Data on exposure to hazardous factors;
- Data on health states, disability and death;
- Parents chronic diseases (for children examined);
- Data on diseases of the examined persons before and after radiation exposure.

(5) Unified general examination chart (for all IPHECA projects and National Programmes in Belarus).

(6) General endocrinological examination form.

(7) General haematological examination form.

(8) Form for epidemiological investigations in haematology.

2.2 Special IPHECA Software Package (SISP)

Collection of data within IPHECA Epidemiological Registry was started with the completion of the OKSO chart, which was used as the outpatient's medical card and/or the child's development history. It was then entered in the passport section of the chart, each patient was assigned a unique registration number. The passport section in the unified database was common for all pilot projects (all pilot projects also required completion of the OKSO). This facilitated collection all data collected on a patient in each pilot project, preventing duplication of information in the data bank, removing ambiguities and ensuring complete analysis an individual's health status. In order to computerize and standardize the collection of primary information and the operation of the database with common technology and joint epidemiological analysis within IPHECA, specialists from Belarus, Russian Federation and Ukraine developed Special IPHECA Software Package (SISP). The main application and advantages of IPHECA software are:

(1) The construction of the database and the verification and analysis of data is uniform in all the IPHECA Pilot projects;

(2) The use of common computer technology for the input editing, verification and correction of data in all IPHECA Pilot projects at all levels;

(3) The use of standardize reference information (unified codes for settlements, ICD-9 codes for diseases, etc);

(4) The use of common statistical packages;

(5) Facilitation of information exchange between observation levels and between National programmes and IPHECA and compatibility of research results;

(6) Guarantees of reliability and scope for modification of software;

(7) Effective use of computer technology and minimization of repeated data entry.

The following user - characteristics were taken into consideration when creating the software:

(1) Requirement to gather large amount of primary data simultaneously in all IPHECA Pilot projects;

(2) Need to be user - friendly for medical workers and operators;

(3) Necessity for information checking at all stages of processing.

The implementation of SISP has been supported by WHO through providing the three states with computers for the national, oblast and rayon levels (Table 2).

2.3 Indicators of the health status used in the Project

Morbidity indices of the population studied within IPHECA have been evaluated according to 15 classes of diseases indicated in ICD-9. In addition, distribution of children and adults into different health groups was used for analysis of their health status. The following health groups were classified:

Group I -   Healthy persons without deviations according to all health indicators and who were never sick during the observation period;

Group II -   Persons with functional disorders, with a risk of contracting a chronic pathology and with increased morbidity;

Group III, IV and V - Sick persons with chronic pathology in the states of respectively compensation, subcompensation and decompensation.

Table 2   **Characteristics of the computers used at the various observation levels**

Observation Level	Type of Computer	Technical Specifications	Assumed Population
National (in each country)	PS/2 8595- VO1	16 MB RAM 3 hard disks (400 MB), Nape (2.3 GB)	70-130,000
	PS/2 70 - 161	4 MB RAM Hard Disk (160 MB)	
Oblast (in each oblast)	PS/2 8595-AKF	8 MB RAM Hard Disk (400 MB)	30-70,000
	PS/2 70 - 161	4 MB RAM Hard Disk (160 MB)	
Rayon (in each rayon)	PS/2 70 - 161	4 MB RAM Hard Disk (160 MB)	20-50,000

PS/2 L40 SX laptop computers with 2 MD RAM and a 60 MD hard disk are used when working away from the office.

2.4   *Main results within "Epidemiological Registry" Project*

During the pilot phase of IPHECA, about 80,000 records related to OKSO were collected in IPHECA Registries of the three States. Having used these records morbidity indices have been studied in selected cohorts of population of radiocontaminated territories included in the Epidemiological Registry Project Protocol. In Belarus, the selected study cohort consisted 9846 people. This number includes 6798 residents of radiocontaminated territories in Gomel oblast and 3048 in Mogilev oblast. The level of ^{137}Cs soil contamination in these territories is more than 555 kBq/m^2. The control cohort (reference data) compiled 1068 residents of clean territories in Vitebsk oblast. The major part of completed OKSO has been related to children and Fig. 1 shows the distribution by health groups of this category of residents of contaminated and noncontaminated rayons. As it can be seen the health status of children in a clean rayon is significantly better than that in the contaminated rayons. Childhood morbidity data analysis from contaminated rayons is compared with the control rayon in Table 3.

In Russian Federation, by the end of 1993, over 13,600 OKSO charts had been completed in the Novozybkovski rayon of Bryansk oblast and Zhizdrinski, Ulyanovski and Khvastovichski rayons of Kaluga oblast. All these records were on individuals who were children or adolescents at the time of the accident and for whom information on the iodine dose to the thyroid was available.

Table 3   **Childhood morbidity (per 1000 children) in strictly controlled zones in comparison with clean rayon of Vitebsk oblast**

NN	Classes of diseases	1000 children		
		Gomel oblast	Mogilev oblast	Vitebsk oblast
1.	Infectious and parasitic disease	141.1	65.59	149.02
2.	Malignant Neoplasms	2.49	4.05	2.81
3.	Endocrine diseases	133.81	191.9	23.43
4.	Mental disorders	8.29	17.41	2.81
5.	Diseases of the nervous system and sense organs	191.01	117.41	55.3
6.	Diseases of the circulatory system	32.66	34.41	30.93
7.	Diseases of the respiratory system	486.98	364.37	779.76
8.	Diseases of the digestive system	390.98	446.56	60.92
9.	Diseases of the genitourinary system	11.11	12.15	5.62
10.	Congenital abnormalities	23.38	6.48	1.87
11.	Symptoms, signs and ill-defined conditions	92.19	46.15	6.56

In addition, about 40,000 OKSO charts had been completed in Borovski rayon of Kaluga oblast (clean rayon) and on almost 20 000 children living in the contaminated areas of the Bryansk oblast. These results provided a picture of the health status in the controlled areas and will serve as the basis for future analytical studies..

Of the whole number of registered individuals with completed OKSO and known thyroid doses, 5694 residents of Zhizdrinski, Ulyanovski and Khvastovichski rayons of Kaluga oblast who were exposed to radioiodine in the period April-June 1986, were included in the sub-cohort for in depth analysis of morbidity indices and assessment of radiation risks. The mean thyroid dose in this subcohort was 510 mGy. The data showed that a statistically significant dose relationship was not observed for the majority of diseases. However, the excess relative risk for nononcological thyroid diseases was significantly greater than 0. The Relative Risk for

Fig 1.   **Distribution of children (%) by health groups**

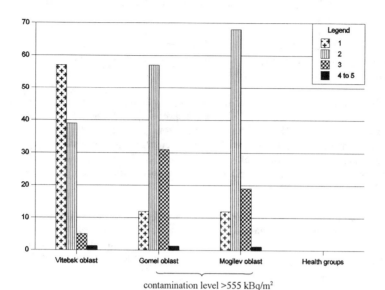

contamination level >555 kBq/m²

this class of diseases within the cohort was calculated and it was 1.1102 (95% CL:1.0422-1.1918).

In Ukraine, the primary demographic and epidemiological characteristics of contaminated territories under the IPHECA Epidemiological Registry Project have been studied on a basis of records included in the State Registry of Ukraine (SRU). Observation cohort numbered 37,409 people residing in Polessko-Ivankovski rayon of Kiev oblast and Ovruchski rayon of Zhitomir oblast.

The morbidity structure of the children examined under the Epidemiological registry protocol showed the greatest weight of respiratory diseases (45.8%). The second place was taken by digestive organs diseases (10.6%), then circulatory system (3.1%), musculoskeletal system (2.3%), blood and haematopoietic organs (2%). The contribution of the above pathology to the morbidity structure of the child group under examination exceeded 60% which tallies with SRU data and age regularities in morbidity formation.

The analysis of OKSO information has been performed in two selected cohorts. The first one included 1.563 adults residing in Polesski and Ivankovski rayons of Kiev oblast and the second one 1814 children living in Ovruchski rayon of Zhitomir oblast. Primary evaluation of the health status of adults revealed a low proportion of fit and practically fit persons. Among the males those in the first health group were only 13% of the total number, and those in the second 5.8%, i.e. 18.9% in both groups. The indicators for females were found to be lower:

the first and second health groups included 8.4 and 5.8% respectively, i.e. 14.2% in both. The morbidity level in this cohort in 1992 was lower as compared to SRU statistical data for Polessko-Ivankovski rayon.

The highest contribution to morbidity in children of the selected cohort was from diseases of the endocrine system (not for 0-3 age group), respiratory tract (in particular diseases of the ear, throat and nose); nervous system and sensory organs.

It should be outlined that results of the OKSO records analysis obtained in the three States are preliminary. Due to restricted funds and a short period of time it was not possible to collect and analyse more OKSO data in clean rayons and carry out intercomparison investigation of morbidity indices in contaminated and noncontaminated territories. Moreover, the initial selection of Borovski rayon of Kaluga oblast in Russian Federation as a reference territory for comparison analysis with contaminated rayons was not fully appropriate due to local endemic peculiarities.

Without valid reference data on morbidity in each of the three States it is too early to make any conclusions about health hazard of residing in radiocontaminated territories. But, according to the prognosis performed by specialists from Russian Federation within the Epidemiological Registry Project, the population residing in areas contaminated to 555-740 kBq/m^2 will live under conditions of "unacceptable" risk for 10 years.

2.5 Accident Recovery Workers (Liquidators)

Despite the fact that the IPHECA pilot phase did not provide for a study of health effects among the Chernobyl accident recovery workers, the morbidity and mortality study among that group of people was in practice within the framework of the Epidemiological Registry Project. The more comprehensive analysis has been done in Russian Federation. The Russian State Medical and Dosimetry Registry compiled records on 324,146 liquidators by the end of 1993. These records were used for analysis of morbidity and mortality indices, causes of death, assessment of dose-dependent morbidity risk and prognosis of long-term stochastic effects of radiation for liquidators. The results of the analysis point to an increasing trend in the incidence of malignant tumours from 97.6 per 100,000 in 1989 to 180.7 per 100,000 in 1992. Morbidity for endocrine system, blood and blood-forming organ diseases, psychiatric disorders, circulatory and digestive organ diseases exceeded those in control groups. The mortality in liquidators for 1990 and 1991 did not exceed that of the controls. The assessment of dose-dependent morbidity risk was done in 99,344 liquidators who had documented proof of having received a radiation dose and had undergone an additional examination to verify a diagnosis.

It was shown that for malignant tumours, blood and blood-forming organ diseases, endocrine system diseases, psychiatric disorders, nervous system and sensory organ diseases a statistically significant (P <0.01) linear increase in the relative risk values was observed dependent on dose. There is also a significant (P <0.05) positive trend for digestive organ diseases and infectious and parasitic diseases. The results obtained in Russian Federation suggest there will be a continuing increase in morbidity and mortality among the liquidators.

3. Epidemiological approaches in other IPHECA pilot projects

In addition to Epidemiological Registry Project, epidemiological investigations, as it has been already mentioned, have been carried out within other IPHECA pilot projects.

3.1 Thyroid project

Results obtained within Thyroid Project provided data on incidence rate of thyroid cancer in the three States. Before the Chernobyl accident the annual incidence of childhood thyroid cancer in Belarus was about one per million, in Russian Federation and Ukraine about 0.5 per million in each. After the accident the incidence rate has increased significantly (Table 4). In the most contaminated Gomel oblast, the annual disease incidence reached more than 100 per million children in 1994.

3.2 Haematology Project

In the course of implementing of the Haematology Project, it was established that in the areas of Belarus, Russian Federation and Ukraine under the project protocol general morbidity for leukaemia and related diseases did not significantly differ in levels although there was a slow upward trend in both contaminated and noncontaminated areas. Nevertheless, the morbidity level did not exceed that which is typical for many other European countries. Before the accident, in a period from 1979 to 1985, the annual mean incidence of leukemia and related diseases in controlled territories of Belarus, Russian Federation and Ukraine was 58 per 100,000. The annual incidence of these diseases in a period from 1986 to 1993 was 64 per 100,000. Comparison of leukemia morbidity in the areas with different levels of radioactive contamination did not produce any significant differences. Childhood leukemia incidence did not change its level after 1986. There were no deviations from age distribution standards or any correlations of disease types with dose. Thus, the results obtained so far show no changes in morbidity which could be linked to the effects of radiation.

3.3 "Brain Damage in-Utero" Project

Epidemiological investigations within this project aimed at detecting cases of mental retardation and other brain disfunctions in a cohort of children born between the 26 April 1986 and 26 February 1987 and who were exposed to radiation in-utero due to the Chernobyl accident. The total number of the children examined in the three States was 4210. The results obtained are different to interpret and require verification. Well planned epidemiological investigations should be continued.

3.4 "Oral Health" Project

The incidence of some stomatological diseases in contaminated territories has been studied in the framework of IPHECA "Oral Health" Project in Belarus. Investigations under the project were conducted in the areas with a mean radionuclide contamination of 185 kBq/m^2 and 555 kBq/^{2} and also in clean zones. Under observation were more than 2,686 residents.

Table 4.   **Incidence (rounded) per million children**

	1986	1987	1988	1989	1990	1991	1992	1993	1994
Belarus	0.9	1.7	2.2	3.0	13	26	28	34	36
Russian Federation	0.0	2.0	0.0	0.0	4.0	0.0	8.0	12	22
Ukraine	0.7	0. 6	0.7	0.9	2.2	1.8	3.9	3.5	3.1*

NOTE*:   = incomplete number
Belarus hasd 2.3 million children, Ukraine has 12 million and the Bryansk and Kaluga oblasts of the Russian Federation have 500,000.

A practically identical incidence of diseases of the oral mucosa, periodontal and dental tissues was detected among the residents examined from the contaminated and clean rayons of Belarus.

4.   Development of epidemiological investigations within IPHECA after pilot phase

In 1994, the IPHECA Management Committee endorsed the "Accident Recovery Workers" Project and recommended to continue "Thyroid" Project. Epidemiological aspects are included in both these projects. An estimated 100,000 liquidators and their families will have their health data recorded in central registries of Belarus, Russian Federation and Ukraine and form the follow-up cohort. The following epidemiological indicators will be studied : mortality (by cause of death); incidence of various malignant and nonmalignant tumours, incidence of other various diseases, frequency of disability or not returning to work. In addition, it is intended to conduct special studies on cohorts consisting of the higher exposed liquidators in order to evaluate health risks associated with radiation exposure.

The study of thyroid cancer incidence will be continued within the "Thyroid" Project. Having used the case-control method, it is also intended to study more carefully the role of radiation factor in the increase rate of thyroid cancer and in the development of each case of leukemia and related diseases in contaminated territories. For this reason thea "Dosimetry" project has been established within IPHECA. One of the objectives of this project is to reconstruct individual doses of radiation for each case of thyroid cancer in children and leukemia and related diseases.

The international experts recommended also to search for excess occurrence of cancers other than leukemia and thyroid cancer, to identify the structure and time trends of prevalence, morbidity and mortality in the residents of radiocontaminated territories. But significant additional funding would be needed for completion of all planned activities.

Epidemiology of Cancer in Population Living in Contaminated Territories of Ukraine, Belarus, Russia after the Chernobyl Accident

Anatoly Prisyazhniuk[1], Zoya Fedorenko[2], Alexey Okeanov[3], Victor Ivanov[4],
Valery Starinsky[5], Vladimir Gristchenko[1], Lyubov Remennik[5]

[1]Research Centre for Radiation Medicine, 53, Melnikov str., Kiev, 254050, Ukraine;
[2]Ukrainian Institute of Oncology and Radiology, 33/43, Lomonosov str., Kiev 252022,
Ukraine; [3]Belarussian Centre for Medical Technologies, Computer Systems, and
Management of Health, 7a, P.Brovka str., Minsk, 220600, Belarus; [4]Medical Radiological
Research Centre AMS of Russia, 4, Korolev str., Obninsk 249020, Russia; [5]Moscow
Oncological Institute of P.A.Gerzen, 2 Botkinsky bystr., 3, Moscow, 125284, Russia

Abstract. Statistical data of oncologic service of Ukraine, Belarus, and Russia
on the number of new patients with cancer and leukemia in 1980-1994 in 12
regions adjacent to the Chernobyl NPP are generalized. Spatio-temporal for
incidence of malignant diseases in population are developed. The analysis of
possible connections between the effective dose and incidence of cancer in
population living in the area contaminated by radionuclids is performed.
Spatio-temporal models for the incidence of cancer including leukemias and
lymphomas are found to be the same in the pre- and postaccidental periods.
Nine years after the Chernobyl accident there are no scientific evidence for the
excess of incidence of malignant tumors, except thyroid cancer, attributed to
radiation factor, even in the most contaminated areas. Appearance of
previously unregistered thyroid cancer cases in children living in the territory,
where considerable amount of radioactive iodine was deposited, can indicate
stochastic radiation effects in thyroid.

1. Introduction

The identification of possible stochastic effects from relatively small doses of ionizing
radiation requires both accurate determinations of radiation doses and a thorough
study of the spontaneous incidence of malignant neoplasms. The latter is especially
important because increased levels of tumor incidences attributed to radiation from the
Chernobyl plant have appeared recently in both the social and scientific literature.

Results of our previous studies [1, 2, 3] have been carried out in 1980-1993 in
adjacent to the Chernobyl territories (large regions and small areas) and did not show
significant increase of cancer incidence due to the radiation exposure in the first years
after the accident, except thyroid cancer. So there are continuous needs for long term
study of possible radiation effects.

2. Materials and methods

To estimate radiation doses received by inhabitants, data were used from the paper by Ilyin et al. Based on data on expected collective effective dose equivalent during 70 years and population size, we determined the expected average dose per inhabitant in the adjacent areas of the Ukraine, Belarus, and Russia. According to the dosimetric model, the expected dose received over 70 years in linearly related to the dose received in the first several years after the accident.

The average per capita lifetime dose for large regions is relatively small. It is variated from 0.23 cSv in Orel region to 4.0 cSv in Gomel region. In contrast to these numbers, the value of average per capita lifetime dose for population of the most contaminated with radionuclids areas differ by an order of magnitude: from 17.62-24.24 cSv in Bryansk region to 20.19-34.62 cSv in Kiev region.

Ilyin et al. [4] also presented collective doses from ^{131}I in the thyroid of inhabitants of the contaminated regions in 1986. The data shows that average thyroid doses were significant in some cases, and that the average thyroid dose to children below 7 years of age was 3 to 5 times higher than that to adults. Together with the known susceptibility of the thyroid to induction of radiogenic tumors, especially in young people, this justifies extensive epidemiological investigations of thyroid cancer in the affected regions.

We used data from oncology studies performed in the Ukraine, Belarus, and Russia on the number of cancer cases from 1980 through 1994 to determine the incidence of cancers. The available data included information on ages of the affected individuals, which allowed us to account for differences in age distributions of the populations in different regions and time periods. We assumed as the reference age distribution the population of the former Soviet Union in 1979 (the year of the all-union census). We analyzed data on tumor incidence in six areas in the Ukraine (Vinnitsa, Zhitomir, Kiev, Cherkasy, and Chernigov regions and the city of Kiev), two areas in Belarus (Gomel and Mogilev regions), and four areas in Russia (Bryansk, Kaluga, Orel, and Tula regions) totally numbering 19.13 mln people at the moment of the accident. Collective doses for populations in most of these areas (excluding Vinnitsa and Cherkasy) were reported by Ilyin [4].

The investigation also included the study of cancer incidence rate in small areas subjected to maximal radionuclide contamination. Such investigation was performed in the four areas of Ukraine closely adjacent to the Chernobyl: Naroditchy, Ovrutch (Zhytomir region), and Ivankov, Polesskoye (Kiev Region). In addition, data were reconstructed on all cancer patients in former Chernobyl district for the preaccidental period 1981-1985. All medical records were collected, including the regular notifications on new cancer cases as well as death certificates from all the institutions and hospitals where patients had been diagnosed and treated. Every record was cross-checked before the final register was compiled. A total of 8550 new cases of cancer have been registered since 1980. At the time of the accident, the five districts including former Chernobyl district accounted for a total population of 274,000, of which 59,200 children of age up to 15. In 1994 the four districts without the now unpopulated Chernobyl district had a population of 146,000, including 26,000 children of age up to 15.

Age-specific and age-adjusted incidence rates were calculated for each year between 1980 and 1994, and were compared to relevant data for the entire Ukraine, as well as for Kiev and Zhytomir regions which include the study districts.

Regression-trend-models of dynamics of cancer incidence rate in male and female population for both large regions and small areas were performed for investigated 15-

years period. The value of dispersion of regression coefficient and signifficance of the slope of regression line, and 95% confidence limits were determined [5]. Correlation analysis of factorial (expected per capita exposure dose) and effective (incidence rate) attributes, and of incidence before (1980-1995) and after (1986-1994) the accident has been carried out.

3. Analysis of materials

After the Chernobyl accident there were observed increased incidences of cancer in each region and compared to the preaccidental period. For males the greatest increases were observed in the Gomel (32.8%), Chernigov (24.2%), Mogilev (22.5%), Orel (22.4%), and Bryansk (22.0%) regions. For females the greatest increases were observed in the Gomel (22.4%), Orel (20.6%), Bryansk (16.3%), and Mogilev (14.2%) regions.

Comparison of the increased incidence with the radiation doses received by the populations indicates no correlation. Analysis of tumor incidence during the complete 15-yr period in each area indicates that convergence of the level of tumor incidence took place. This could be attributable to the fact that during the observation period the quality of diagnostics and accounting of this pathology became consistent among all regions.

The investigated areas increased cancer incidence were observed in all regions following the Chernobyl accident. However, in the regions with initially high relative incidences the increases were lower than in the regions where the level of incidence was relatively low at the beginning of the period studied.

The analysis of incidence rates during the two periods (1980-1985 and 1986-1994) indicate that there is good correlation between rates in individual regions for the two periods. For male inhabitants the value of the correlation coefficient between incidence rates during the investigated time periods was 0.83 ($P<0.01$); for female inhabitants the value was 0.95 ($P<0.01$). Thus, there is evidence that the factors influencing cancer incidence rates in 1980-1985 were still present in 1986-1994.

The study of dynamic models of incidence of cancer in the most contaminated with radionuclids areas, where expected average per capita dose of radiation will exceed the value for population of large regions by 20-30 times and presents great interest.

Between 1980 and 1994 the age-adjusted incidence rates in the study districts were lower than those in the entire Ukraine, Kiev and Zhytomir regions. The time-trends, however, were the same (fig. 1, 2).

To test for different trends before and after the accident, linear regressions are performed separately for the periods 1980-1985, 1986-1994, and 1980-1994 (fig. 3). It is found that the trends are similar with slopes that do not differ with statistical significance before and after accident. The generally rising trend represents increases in incidence rates of cancer of the oral cavity, colon, rectum, gall bladder, pancreas, larynx, lung, skin, breast (in women), corpus uteri, ovary, prostate, kidney and urinary bladder.

Analysis separately for different age groups - the annual incidence rates for all cancers except leukemia, and thyroid cancer (fig. 4) has shown that the predominant change in the time-trend is an increased cancer incidence in the oldest age group (65+) that begins one year after the accident; it amounts to nearly one third and has remained at about this level since then. In spite of deviations in some years, no significant increase of the frequency of cancers is seen in the younger age groups.

Because radiation-induced cancer occur in all age groups, especially in children, it can be assumed that the indicated increase of incidence in the eldest age group is not attributable to radiation. Thus, a study of the frequency of malignant neoplasms in the most contaminated areas before 1995 provides no bases for the conclusion that the observed malignant neoplasms are attributable to radiation doses.

A separate study of the distribution of specific malignant neoplasms that have short latent periods (approximately 5 years) is important because radiation-induced tumors of these types could reasonably be expected to be observed at this time. Such types include tumors of lymphatic and hemopoetic tissues and thyroid cancer.

During the analyzed period in large regions there were registered increase of leukemia and lymphoma incidence (except of Kiev city). The most significant increase in males was observed in the Gomel (46.6%), Zhytomir (45.7%), Kaluga (44.5%) regions and in females in Bryansk (56.1%), Zhytomir (54.4%), Gomel (42.3%) regions. Overall, the pattern observed for all cancers was observed for this subject: the increased rates in the regions with initially high incidence levels were considerably lower than the increases in the areas where the incidence levels in the initial period was low. This results are in convergence of the observed incidence rates.

As in case of all forms of cancers there were no found correlation between increased incidence of hemoblastosis and radiation doses received by population of different regions.

Detailed study of trends of incidence of leukemias and lymphomas in the most contaminated areas and their comparison with the data in large regions (Fig.4 and 5) has shown that trend models have no significant difference. Even when the male and female population are analyzed together (Fig.6) the trends remains insignificant.

The analysis of the dynamics of special forms of leukemia and lymphoma show significant increase in incidence rates in 1986-1994 compared to 1980-85 only for lymphatic leukemia (P<0.05). Small number of cases of myeloid leukemia do not allow to draw conclusion about real time trends. Another forms of hemoblastosis (lympho- and reticulosarkoma, Hodgkin and non-Hodgkin lymphomas, multiple myeloma, unspecified cellular types of leukemia) do not show significant increase of incidence (P>0.05).

As for different age groups it was noted that the rates of leukemia (ICD-9, 204-208) in the eldest age group (65+) increased in 1987 sharply and has remained two or three times higher than the pre-accident rates. It was due to chronical lymphatic leukemia which is not belong to radiation dependant form of malignancy. In the intermediate age groups the observed rates before and after accident were broadly similar. A seeming increase for children is not, at this point, statistically significant.

Thus, through analysis of epidemiological data on leukemia and lymphoma in populations residing in territories subjected to the highest contamination levels provided no evidence of radiogenic tumors of lymphatic and hemopoetic tissues.

Malignant neoplasms of the thyroid were included in official statistics only since 1989, i.e. in connection with the Chernobyl accident. Therefore, the data on this pathology for inhabitants of the Ukraine were available mainly from our previous investigations in 1962-1964,1969-1972 and 1977-1979 and from official medical reports for 1989-1991.

Fig. 7 illustrates that, as a whole, moderate increases in thyroid cancer incidence rates were registered in the Ukraine during the period from 1962 to 1979. The increase in women was greater than in men (30% vs.less than 20%).

Fig.8 shows incidence rates and linear regression analysis for 1980-1994 in the CA where a through investigation of incidence rates has been performed.

The time trends for the over-all thyroid cancer incidence rates indicate a marked increase soon after the reactor accident.

The most notable aspect of the data on thyroid cancer is the occurrence of 5 thyroid cancers in children and 3 - in teenagers which contrasts sharply with the fact, that no such cases have been observed in children and only one in teenage in years before the accident. Observed number of thyroid cancer in children exceeded expected one in 13.9 times (confidence interval: 4.5 - 32.2). In 1994 no one case of thyroid cancer was found in children in age group 0-4 in Ukraine at whole. It is very evidencable because this age group was born in 1990-94 and did not affected by radioactive iodine.

4. Discussion and conclusions

The analysis of possible radiation effects in the population should be based on observable population markers. An observed excess of cancer incidence in the youngest age groups, and an excess of special forms of malignant neoplasms such as myeloid leukemia, multiple myeloma, and thyroid cancer could be considered such population markers. If excess of these cancers were observed in the most vulnerable age group, this would provide evidence of radiogenic effects.

Among the three categories of malignancies that have been considered, only the thyroid cancer show a statistically significant increase that suggests a relation to the radiation exposures after the accident of Chernobyl.

In view of small number of cancer in different age groups especially in children, and long latent period for radiation-induced cancer, there will be continued need to monitor the cancer incidence in the contaminated areas.

References

[1] Prisyazhniuk A. et al. Cancer in the Ukraine, post-Chernobyl. *The Lancet* 1991; **338:** Nov 23, 1334-1335.
[2] Prisyazhniuk A.Ye. Spatio-temporal models for incidence of malignant neoplasms in the area subjected to radioactive contamination after the Chernobyl accident. In:*The Chernobyl Papers, Vol.1:* S.E.Merwin, Michail I.Balonov, eds. *Doses to the Soviet population and early health effects studies,* 1993, 399-423.
[3] Prisyazhniuk A. et al The time trends of cancer incidence in the most contaminated regions of the Ukraine before and after the Chernobyl accident. *Radiat Environ Biophys* 1995; **34:** 3-6.
[4] Ilyin L.A. et al. Radiocontamination patterns and possible health consequences of the accident at the Chernobyl nuclear power station. *J Radiol Prot* 1990; **10** (1): 3-20.
[5] Armitage P. Statistical methods in medical research. Third printing 1974, John Wiley and Sons, New York.

Age-adjusted all cancer incidence rates in Ukraine (1), Kiev (2), Zhitomir (3) regions and strict control areas (4), males, 1980-1994

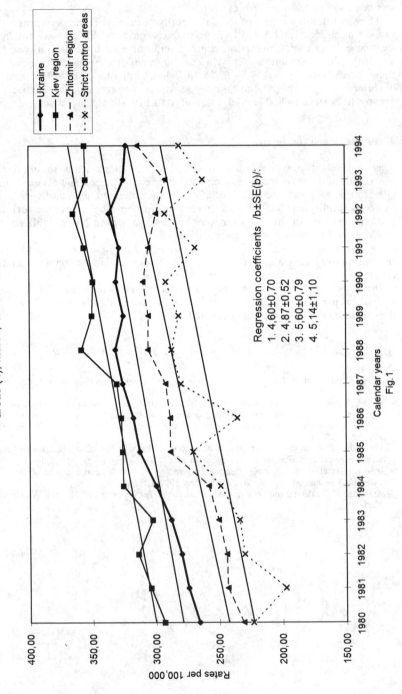

Calendar years
Fig.1

Age-adjusted all cancer incidence rates in Ukraine (1), Kiev (2), Zhitomir (3) regions and strict control areas (4), females, 1980-1994

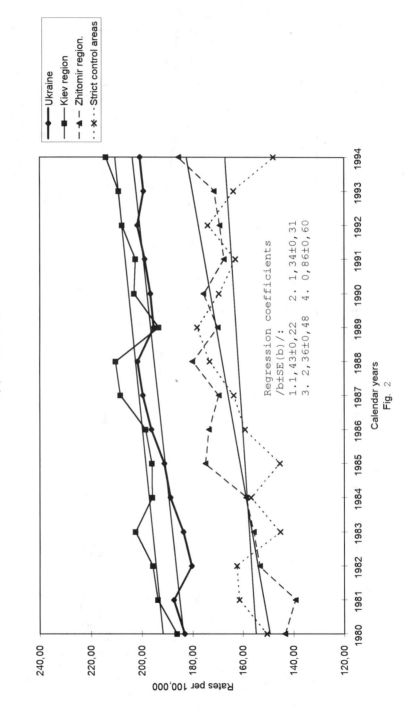

Fig. 2

Regression analysis of age-adjusted cancer incidence rates in strict control areas (males and females)
for periods: 1980-85, 1986-94, 1980-94

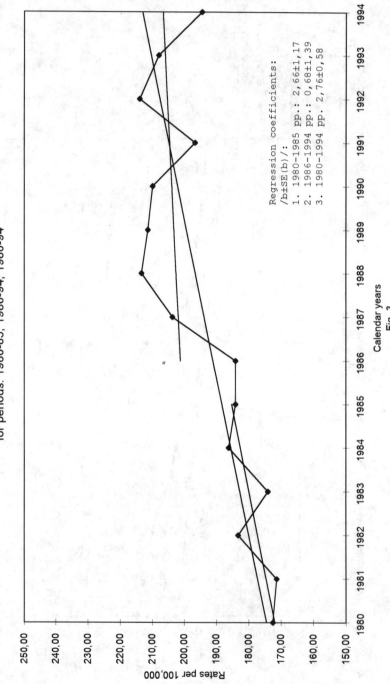

Rates per 100,000

Calendar years

Fig. 3

Regression coefficients:
/b±SE(b)/:
1. 1980-1985 pp.: 2,66±1,17
2. 1986-1994 pp.: 0,68±1,39
3. 1980-1994 pp. 2,76±0,58

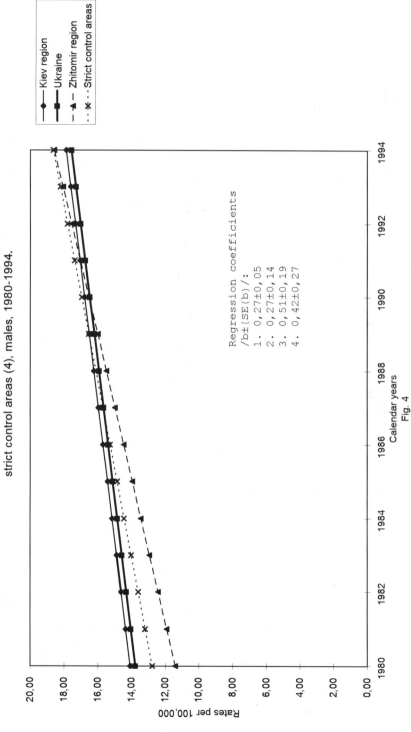

Age-adjusted leukemia and lymphoma incidence rates in Ukraine (1), Kiev (2), Zhitomir (3) regions and strict control areas (4), males, 1980-1994.

Fig. 4

Age-adjusted leukemia and lymphoma incidence rates in Ukraine (1), Kiev (2), Zhitomir (3) regions and strict control areas (4), females, 1980-1994..

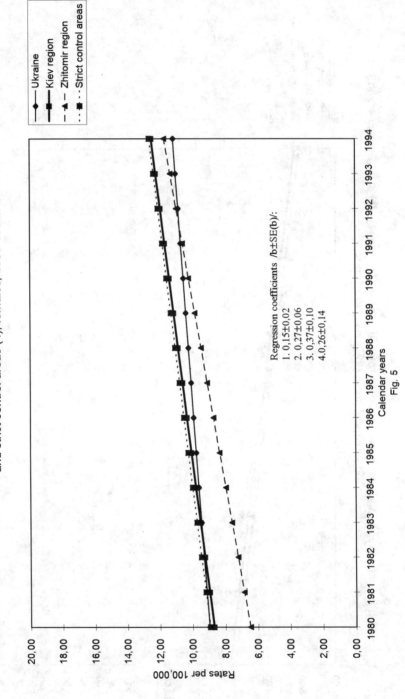

Rates per 100,000

Calendar years

Regression coefficients /b±SE(b)/:
1. 0,15±0,02
2. 0,27±0,06
3. 0,37±0,10
4. 0,26±0,14

Fig. 5

— ◆ — Ukraine
— ■ — Kiev region
— ▲ — Zhitomir region
- - ■ - - Strict control areas

Regression analysis of age-adjusted leukemia and lymphoma incidence rates in strict control areas (males and females) for periods: 1980-85, 1986-94, 1980-94

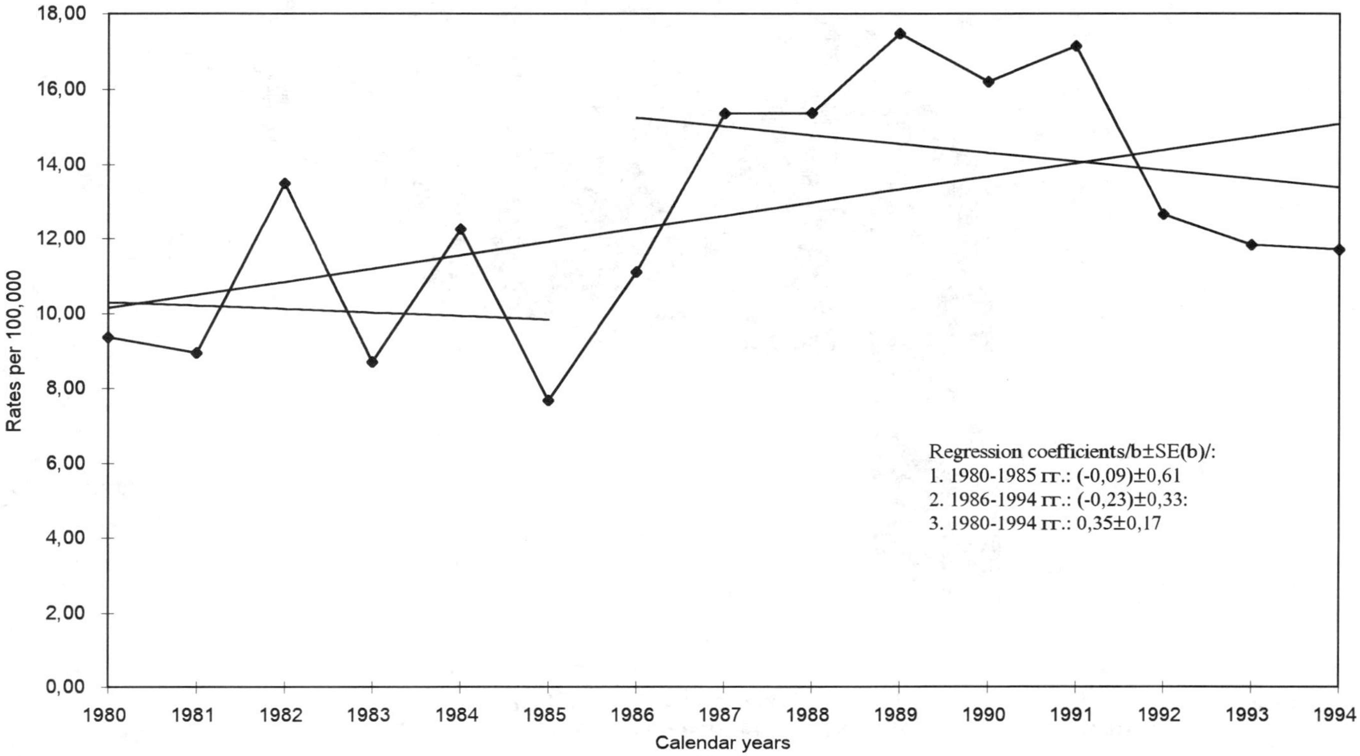
18,00
16,00
14,00
12,00
10,00
8,00
6,00
4,00
2,00
0,00
Rates per 100,000
1980 1981 1982 1983 1984 1985 1986 1987 1988 1989 1990 1991 1992 1993 1994
Calendar years
Regression coefficients/b±SE(b)/:
1. 1980-1985 гг.: (-0,09)±0,61
2. 1986-1994 гг.: (-0,23)±0,33:
3. 1980-1994 гг.: 0,35±0,17

Fig. 6

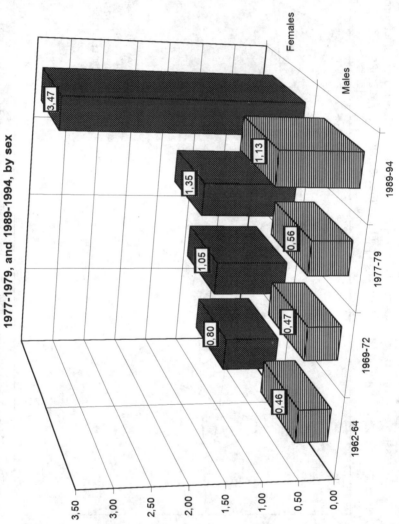

Mean annual age-adjusted thyroid cancer incidence rates in the Ukraine: 1962-1964, 1969-1972, 1977-1979, and 1989-1994, by sex

Fig. 7

Regression analysis of age-adjusted thyroid cancer incidence rates in strict control areas (males and females) for periods: 1980-85, 1986-94, 1980-94

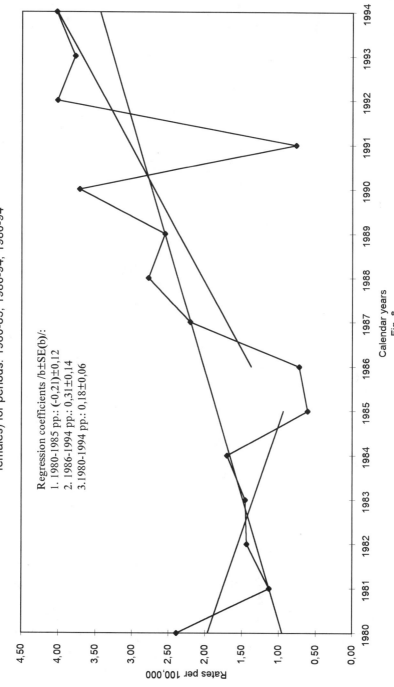

Regression coefficients /b±SE(b)/:
1. 1980-1985 pp.: (-0,21)±0,12
2. 1986-1994 pp.: 0,31±0,14
3. 1980-1994 pp.: 0,18±0,06

Calendar years
Fig. 8

Rates per 100,000

III. HEALTH EFFECTS
FOLLOWING THE ACCIDENT
C. Epidemiology of exposed populations

Posters

Development of the Cancer Registration System in Belarus

A.E. OKEANOV[1], S.M. POLYAKOV[1], A.V. SOBOLEV[1],
R.A. WINKELMANN[2], H.H. STORM[2]

[1] *Belarussian Centre for Medical Technologies, Information Computer Systems,
Health Care Administration and Management,
7a Brovski street, 220600 Minsk, Belarus*
[2] *Danish Cancer Registry, Danish Cancer Society,
Strandboulevarden 49, Box 839, DK-2100 Copenhagen, Denmark*

Abstract. Cancer registration was established in Belarus in 1953, however was not complete until the 1970's. In 1973 a computerized central cancer registry was established (files available only from 1978) based on coded and anonymous information received from each of the 12 oncological dispensaries in the country. In 1985 a computer system of dispensary control for cancer patients was set up in the oncological dispensaries in Belarus, whereby identification of individual cancer patients in the cancer registry was made possible. The Belarussian cancer registry records all cases of cancer including those of the lympho-hematopoietic system, and carcinoma in situ. The registry is person-based with information on all tumours and their treatment in a given individual. Coding and classification is carried out in accordance with ICD-9. For histology a local classification is used. Currently the registration system is under modernization in order to achieve full correspondence with internationally accepted standards and for the purpose of easy linkage to the Belarussian Chernobyl Registry.

1. The history of cancer registration in Belarus

Cancer registration started in 1953 in Belarus based on information in the oncological hospitals. Due to the small number of oncological hospitals population coverage and completeness was not reached until the beginning of the 1970's.

In 1973 a central computerized cancer registry was established. It was based on the special cards of dispensary control, which are filled-in and coded at each of the 12 oncological dispensaries recording available information on a given cancer patient irrespective if treated at the oncological dispensary. Annually the forms were delivered to the central registry in Minsk, where the cards were computerized, controlled and after completed checking procedures loaded into the computer memory, however without any person identifying information. The annual cancer incidence, prevalence and death in period files were created and stored on magnetic media from 1978 onwards.

The lack of person identifying information made any verification and/or correction of the entered data virtually impossible. It was thus a major advantage when a computer system of dispensary control for cancer patients were established in the oncological dispensaries in 1985.

This system allow identification of individual patients, whereby both follow-up, collection of continuous data and corrections to previous collected data is facilitated. Also linkages to other external files, such as the Chernobyl Registry is possible. Since 1991 the entire system has functioned on personal computers in all Belarussian oncological dispensaries.

2. The Belarussian Cancer Registry 1995

Reportable diseases
The cancer registry collects data on all malignant neoplasms, including malignancies of the lympho-hematopoietic tissues and carcinoma in situ. Carcinoma in situ is not included in the tabulations for cancer incidence. If tumours of the brain and CNS are reported without a clear indication whether malignant such cases are included. Only invasive bladder tumours are recorded.

Sources of information
Reports to the cancer registry are received from the registration department at each of the oncological dispensaries. This department collects and processes (coding and computerization) the data on all cancer cases of the region, including those appearing on death certificates. The data is primarily collected from 3 main sources: the oncologist at the rayon or city policlinic level; from the oncological dispensary (more than 85 % of the cases); from death certificates checked monthly at the regional population registry. If an unreported cancer case is found on a death certificate, a query is mailed to the treating hospitals requesting confirmation and more detailed information.

The medical care system of Belarus is established with an oncologist in each rayon (administrative subdivision) or city polyclinic. These are responsible for the cancer patients of their area, and are supposed to admit all patients to the oncological dispensary for specialized diagnosis and treatment. However, irrespective of this, any medical institution diagnosing a cancer case must fill a special form and submit this to the registration department at the oncological dispensary. This also apply to pathologists who unexpected find a cancer case at autopsy.

Identification, coding and classification
The collected information is arranged by person, with separate records of each tumour diagnosed. Each individual person is to day identified by names, birthday a region code, a dispensary code, date of registration and a registration card number. For each tumour is recorded the ICD-9 code (4 digit), stage, the treatment and the method of diagnose as well as the histology according to a local classification. A case of multiple cancer is accepted if arising in a different organ or different parts of organs as defined by the 4th digit of the ICD-9 classification. Multiple skin cancers are always counted separately if appearing in different anatomic regions.

Results
The registry is able to analyse data by administrative areas, and to study trends in incidence. This can be done by ICD-9 categories which are primarily topographic, and only to a limited extent by relevant histological groups. The data is produced annually, for submission to the Ministry of Health in accordance with the national regulations. Usually data for a year must be available in March April after the turn of the year, a period that is far too short to achieve completeness and high quality of data reported late in the year or found at death certificates. Nevertheless a dataset from Belarus appears in the international

monograph on Cancer Incidence in Five Continents although the data quality in the past have not been checked sufficiently, and the presented incidence for for instance leukaemia is lower than expected. Irrespective of these weaknesses in the data, a clear incidence in thyroid cancer in the young and adolescent was observed, and no clear pattern changes in leukaemia incidence.

Future development of the registration system
The current cancer registration system has a number of problems. Firstly the classification of tumours (in particular with respect to morphology) do not adhere to the recommended international standard. Secondly, the data quality procedures are not well developed. Thirdly, the data capture from the specialized unit on hematology and collaboration with this Institute and the institute of oncology needs to be strengthened and fourth, the system does not facilitate linkages and data-exchange with external files, such as the Chernobyl registry.

In order to solve these problems the following is planned or in process:

1. Introduction of the second edition of the ICD-O in the registry, as an addition to the current system to maintain comparability with previous recorded data and possibilities for evaluating trends in incidence etc.

2. A modernization of the registration software with implementation of quality control measures in accordance with international standards.

3. Include identifying information such as date of birth, place of residence and address as recorded by the Chernobyl registry.

4. Introduce a field for registration of migration.

5. Introduce a flagging field for persons also registered in the Chernobyl registry.

Health Status and Follow-up of the Chernobyl Nuclear Power Plant Accident Liquidators in Latvia

E. ČURBAKOVA, B. DZĒRVE, M. EGLITE, I. FRICKAUSA, T. ZVAGULE
Center of Occupational and Radiological Medicine of P. Stradiņs State clinical Hospital
13 Pilsoņu Street, LV-1002 Riga, Latvia

1.   Introduction

The accident at the Nuclear Power Station in Chernobyl created a new problem for health professionals in Latvia due to the fact that 6475 inhabitants (mainly healthy and men of reproductive age) of Latvia took part in clean-up works in Chernobyl within the period 1986-1991. Chernobyl clean-up workers were exposed to external γ-radiation and they also incorporated radionuclides. The doses documented for the clean-up workers are variable; they are estimated to be between 0.01 -0.5 Gy although the specialists working on the precision of received doses think that they could be even 2 or 3 times higher. Many data show that these doses could be classified as small doses but they may cause somatic and genetic biological effects (2, 3, 4, 5).

Therefore all the above-mentioned factors require very serious attitude in the evaluation of the Chernobyl clean-up workers health status as well their further treatment and rehabilitation. There is a difference between Latvian Chernobyl accident liquidators and liquidators in other countries contaminated with radionuclides because Latvia Chernobyl clean-up workers have only had a low ionizing exposure during an exact period of time. The aim of our work is to evaluate the health status of liquidators investigating them on a long-term basis:
- to create the correct system of health status evaluation of Chernobyl clean-up workers;
- to improve the register of Chernobyl clean-up workers and of their children;
- to analyze the data about the incidence of different diseases and mortality gained from follow-ups;
- to evaluate health status and clinical picture within the period of time;
- to work out and use adequate methods of treatment.

2.   Materials and methods

A system of Chernobyl clean-up workers registration and observation is organized in our republic. There is a screening system in every district of Latvia which means that local hospitals make a follow-up of every clean-up worker at least once a year including a general and complete check-up.

The registration and more detailed examinations of their health condition are carried out in the P. Stradiņs State Clinical Hospital at the Center of Occupational and Radiological Medicine where examinations of the thyroid gland, the immune status, endoscopic examinations when indicated (fibrogastroscopy, firbrocolonoscopy, bronchoscopy), functional tests (doplerography, veloergometry, respiratory function) are done. Scientific investigations in this field are carried out in the Institute of Labour Medicine of Medical Academy of Latvia. Since June of 1994 the register have been officially recognized as *Latvian State register for patients with occupational disease and for patients that have received ionizing radiation in Chernobyl and for the children of Chernobyl clean-up workers.* The register also contains data about occupational disease because about 80 % of Chernobyl clean-up workers were people working in difficult working conditions (welders, tractor drivers, and etc.). At the 01.09.1995 we have registered 4663 Chernobyl clean-up workers, 116 persons evacuated from Chernobyl, 790 children from the families of clean-up workers (born after the accident) and also 3000 occupational disease patients. It is important to note that many of patients registered have both occupational diseases and effects of ionizing radiation. We have collected the information about liquidators place and condition of work, duration of their stay and work in Chernobyl, received doses of radiation, protective measures used, diseases before, during and after the accident, occupation before and after the accident, habits and data from objective clinical investigations from 1986 till today. Latvian citizens have been working in Chernobyl from 1986 till 1991. More than half of them are documented as ones having received radiation exposure (Table I.)

Table 1. - Doses of radiation received by Latvian Chernobyl accident clean-up workers

Doses of radiation (Gy)	Amount of clean-up workers
0.5 and >	3
0.4 - 0.49	2
0.3 - 0.39	4
0.2 - 0.29	515
0.1 - 0.19	909
0.01 - 0.1	1218
Not measured	2012
	Total: 4663

1320 workers worked close to the reactor and turbine hall, 1130 worked at installations next to reactor site, 2213 worked outdoors. These workers were mainly between the age of 18 and 48 at that time. Those employed between 1986 and 1987 worked for 1 to 3 months in average, while those employed later worked for 4 to 6 months in average. The workers employed in 1988 to 1989 were mainly occupied with building the town near the Chernobyl reactor and did not get involved in the works on the site of the reactor itself. Their doses are, therefore, lower on average than the doses of the earlier clean-up workers.

We are also analyzing the incidence of different diseases of Chernobyl clean-up workers in comparison with control cohort. Control cohort consists of 3.000 workers from 29 to 50 years old with similar working and life conditions and habits. The analysis covered the period 1991 to 1994. In order to evaluate the health status, clinical, physiological, biochemical and immunological methods were used and reasons of death cases among the Chernobyl clean-up workers were also analysed. All the data collected have been processed with statistical methods.

3. Results

Morbidity analysis of 4.500 Chernobyl clean-up workers in comparison with control cohort showed that Chernobyl clean-up workers suffer more often than the persons in control group from infections, diseases of the nervous, cardiovascular and respiratory system and also from bone and muscle diseases.

Elevated number of infectious diseases could be connected with the changes in immune system. We have determined the immune state of 833 clean-up workers during the period 1986-1989. The most characteristic changes in the individuals involved in the clean-up were the increased level of IgM in 14.5 - 26.4% of cases, the increased level of IgA in 14.7 - 26.0%, the decreased number of T-lymphocytes in 58.4% and also large number of granular lymphocytes in 75%. The ability to produce interferons (IFNs) was lower in 22.1 - 50.4% of the individuals (1). The deficiency in IFNs as well as in IgM and IgG might determine the frequent infection complications. Between 1991-1994 the diseases that mainly occured were diseases of bones, muscles and connective tissues, diseases of digestive and respiratory organs as well as diseases of nervous system and organs of sense.

During last three years the number of diseases of digestive organs, nervous system, organs of sense and physical disorders have increased significantly (table 2). We also have analyzed the mortality of Chernobyl clean-up workers. The number of deaths among the clean-up workers are higher than among the rest of the population.

The most important cause of their death have been different diseases (particularly of the cardiovascular system), but the number of suicides and accidents are also significantly higher than among the rest of the population (table 3).

Table 2. - Structure of diseases of Chernobyl clean-up workers.

Years	1991	1992	1993	1994
Registered diseases	5174 - 100 %	6003 - 100 %	5553 - 100 %	4342 - 100 %
Registered at first in % (here and after - first number - total number of diseases, second - %)	3794 - 73.3 %	3277 - 54.6 %	1902 - 34.3 %	1214 - 28.0 %
1. Infections and parasitic diseases	2.6 - 0.5 %	319 - 5.3 %	29 - 0.5 %	13 - 0.3 %
2. Neoplasms:	10 - 0.2 %	19 - 0.3 %	20 - 0.4 %	16 - 0.4 %
malignant neoplasms	9 - 0.17 %	9 - 0.14 %	15 - 0.3 %	8 - 0.2 %
3. Endocrine, Nutritional and Metabolic diseases and Immunity disorders	397 - 7.7 %	182 - 3 %	225 - 4 %	275 - 6.3 %
4. Diseases of the blood and blood forming organs	6 - 0.1 %	17 - 0.3 %	13 - 0.2 %	6 - 0.13 %
5. Mental disorders	355 - 6.8 %	795 - 13.2 %	885 - 16 %	599 - 13.8 %
6. Diseases of the nervous system and Sense organs	737 - 14.2 %	918 - 15.3 %	763 13.7 %	695 - 16 %
7. Diseases of the circulatory system	488 - 9.4 %	494 - 8.2 %	563 - 10.1 %	424 - 9.7 %
8. Diseases of the Respiratory system	933 - 18 %	880 - 14.7 %	675 - 12.2 %	452 - 10.4 %
9. Diseases of the digestive system	577 - 11.2 %	1030 - 17.2 %	1030 - 20.5 %	810 - 18.6 %
10. Diseases of the Genito-Urinary system	65 - 1.3 %	117 - 1.9 %	123 - 2.2 %	99 - 2.3 %
11. Diseases of the Skin and subcutaneous tissue	37 - 0.7 %	61 - 1 %	64 - 1.2 %	38 - 0.8 %
12. Diseases of the Musculoskeletal system and connective tissue	1085 - 21 %	1001 - 16.7 %	959 - 17.3 %	782 - 18 %
13. Accidents, Injuries and Poisonings	458 - 8.9 %	160 - 2.7 %	96 - 1.8 %	133 - 3.0 %

Table 3. - Number and causes of death among Chernobyl clean-up workers

Year	Number of deaths
until 1990	52
1990	16
1991	18
1992	16
1993	36
1994	17
Total	155

Causes of deaths	Number of deaths
1. Diseases	51
2. Accidents	33
3. Suicides	13
4. Intoxication with alcohol	8
Total	105

Analysis of diseases which caused death

Causes	Number
1. Diseases of cardiovascular system	34
2. Oncological diseases	8
3. Radiation damage	3
4. Other	6
Total	51

Evaluating the morbidity and changes of clinical picture of clean-up workers in the period of time we have collected data set about the pathologies induced by small doses of ionizing radiation.

Chernobyl clean-up workers most commonly have complaints of headaches (70%), dizziness and weakening of memory, prostration, lowering of working capacities (76%). Characteristic attacks with loss of consciousness and impotence are observed. About 10% of clean-up workers have broken up marriages or they suffer from infertility. There are complaints of changeable arterial blood pressure, excessive nervousness, local cramps and pains in the bones (70-72%). In this stage we mostly observe chronic inflammatory processes in respiratory and in digestive organs (the number of diseases like erosive gastroduodenitis, stomach ulcers, chronic hepatitis have increased) as well as diseases like atherosclerosis, ischemic heart disease, hypertension, osteoporosis, cataract and also endocrine pathologies: diabetes mellitus and euthyroid goiter.

Neurologic investigations proved objectively various degrees of organic neurologic symptoms. Neuropsychological investigations show depressive moods, increased excitability, asthenia, coordination disturbances, polyneuropathy and encephalopathy. Electroencephalograms showed disorganization of cortical bioelectrical activity with signs of diffuse irritation.

After the complex hematological examination following abnormalities have been found: during all the years after the accident we are detecting increased Hb content (appr. for 25%) and elevated erythrocyte count, during first years after the accident we also observed slight leukopenia but during the last few years the amount of Ly have increased and now we observe slight leukocytosis (about 10 000 - 14 000 Ly with proportional Ly formula).

We have been currently trying to establish new methods for treatment and rehabilitation of clean-up workers using our knowledge about the etiology and pathogenesis. We widely use the therapy of disintoxication and methods of correction of metabolic processes and normalization of the circulatory disturbances.

The problem of Chernobyl clean-up workers is not only a medical problem but also a social and financial problem. They are mainly connected with the economical crisis because of the unemployment and poverty. Most of the clean-up workers were qualified workers at the time of the accident but now many of them are disabled and without any chance for normal life due to the fact that we still don't have special services and companies providing these young, professional people with work. Due to the financial problems in our country the question of rehabilitation is also still under discussion.

4. Conclusions

During these years the registry of Chernobyl clean-up workers and their children have been created and proper analysis of health status, morbidity and mortality have been done. The following facts are of high importance: Chernobyl clean-up workers more often than the control group suffer from diseases of the nervous, the endocrine and the metabolic and immune system. They also have higher rate of incidence for diseases of digestive and respiratory system and for diseases of bones, muscles and connective tissue higher rates of accidents and suicides.

Now, ten years after the accident there are Chernobyl clean-up workers who are chronically ill and their health status is expected to be worse in the next few years.

Regular follow-up and medical examination of Chernobyl clean-up workers and their children should be carried out every year.

Regular rehabilitation of Chernobyl clean-up workers should be provided by the government.

References
1. Bruvere R., Heisele O., Zvagule T., Churbakova E., The immune state of Latvia's inhabitants involved in the clean-up of radioactivity in Chernobyl, Acta Medica Baltica, vol. 1., 1994., pp. 30-37.;
2. Владимиров В.Т., Биологические эффекты при внешнем воздействии малых доз ионизирующих излучений, Военно-медицинский журнал, 1989., N° 4., стр. 44-46.;
3. Моекаев Ю.И., Отдаленные последствия воздействи ионизирующих излучений, Москва, 1991., 462 стр.;
4. Shigematsu J., Ito Ch. et al., A - Bomb radiation effects diggest, Bunkodo Co., LTD., 7-2-7 Hongo, Bunkyo-ku, Tokyo 113, Japan, 1993., 37 p.;
5. Summer D., Wheldon T., Watson W., Radiation risks, The Tarragon Press, Glasgow, 1988., 236 pp.

III. HEALTH EFFECTS
FOLLLOWING THE ACCIDENT
D. Methods and techniques for dose reconstruction

The reconstruction of thyroid dose following Chernobyl

V. Stepanenko[1], Yu. Gavrilin[2], V. Khrousch[2], S. Shinkarev[2],
I. Zvonova[3], V. Minenko[4], V. Drozdovich[4], A. Ulanovsky[4],
K. Heinemann[5], E. Pomplun[5], R. Hille[5], I. Bailiff[6],
A. Kondrashov[1], E. Yaskova[1], D. Petin[1], V. Skvortsov[1], E. Parshkov[1].,
I. Makarenkova[2], V. Volkov[2], S. Korneev[4], A. Bratilova[3], J. Kaidanovsky[3].

[1] - Medical Radiological Research Centre (MRRC), Obninsk, Russia
[2] - Institute of Biophysics (IBPh), Moscow, Russia
[3] - Institute of Radiation Hygiene (IRH), St. Petersburg, Russia
[4] - Institute of Radiation Medicine (IRM), Minsk, Belarussia
[5] - Research Centre Juelich GmbH, Juelich Germany
[6] - Durham University, Durham, England

The report presents the overview of several approaches in working out the methods of thyroid internal dose reconstruction following Chernobyl.

One of these approaches was developed (IBPh, Moscow; MRRC, Obninsk; IRM, Minsk) using the correlations between the mean dose calculation based on I-131 thyroid content measurements and Cs-137 contamination of territories. The available data on I-131 soil contamination were taken into account. The lack of data on I-131 soil contamination was supposed to be compensated by I-129 measurements in soil samples from contaminated territories. The semiempiric model was developed for dose reconstruction. The comparison of the results obtained by semiempiric model and empirical values are presented. The estimated values of average dose according semiempiric model were used for individual dose reconstruction.

The IRH (St.-Petersburg) has developed the following method for individual dose reconstruction: correlation between the total I-131 radioiodine incorporation in thyroid and whole body Cs-137 content during first months after accident. The individual dose reconstruction is also mentioned to be performed using the data on individual milk consumption during first weeks after accident. For evaluation of average doses it is suggested to use the linear correlation: thyroid dose values based on radioiodine thyroid measurements vs Cs-137 contamination, air kerma rate, mean I-131 concetration in the milk. .

The method for retrospective reconstruction of thyroid dose caused by short-living iodine nuclides released after the Chernobyl accident has been developed by Research Centre, Juelich, Germany. It is based on the constant ratio that these nuclides have with the long-living I-129. The contamination of soil samples by this nuclide can be used to assess thyroid doses. First results of I-129 contamination values and derived thyroid doses are to be presented.

1. Introduction

At present time about four hundred thyroid cancer cases among children and adolescents (of this age at the moment of irradiation) have been stated in the regions of Belarussia and Russia. These areas suffered the contamination caused by the reactor accident of Chernobyl. The observed increase in the number of thyroid cancer incidents have to be explained. Therefore it is necessary to determine the radiation exposure in the first few weeks after the accident.

The special interest to the thyroid organ is caused by the fact that its cancer probably derives from the radiation exposure by the short living iodine nuclides (I-131 to I-135; longest half life 8.04 d) and high radioiodine releases which was significantly higher than Cs-137 release.

A lot of radiation measurements (thyroid and whole body burden, environmental) have been done beginning after the reactor accident until now including about 160 000 thyroid burden determinations in Belarussia and Russia. These data may be used for dose calculations and for development of the models of the retrospective dose estimation. Before the data can be used for any calculations, they have to be carefully validated in some way.

There are also many settlements for which no radioiodine measurements are available. Fore these sites retrospective dose assessments have also to be realised. Such retrospective calculations may be based on the determined ratio of I-131 to Cs-137 contamination of the soil. Unfortunately there are only a small number of settlements with known I-131 contamination data. One of the possible ways to estimate I-131 contamination is to use the data on I-129 contamination. So the special program of I-129 measurements in the soil must be developed. The net I-129 contamination has to be taken into account because there are also I-129 contaminations due to the atmospheric atomic bomb tests, the value of which is not very well known; but both are very low. Further considerations on the error of such assessments have to be included into the calculations.

This means retrospective dose assessment is a wide field using empirical, half empirical and more or less pure theoretical calculations. In the frame of ECP 10 such investigations are done. The results are described in this report.

2. Thyroid dose reconstruction modelling on the base of the results of radioiodine thyroid content measurements vs soil radioactive contamination data.

2.1. According to the data of the Institute of Experimental Meteorology (SPA "Typhoon", Obninsk, Russia) released I-131 activity to the atmosphere from the Chernobyl accident is estimated about $6,3 \cdot 10^{17}$ Bq and it was much higher than released Cs-137 activity.

As distinguished from uniform caesium distribution in body about 30% of I-131 intake accumulates in thyroid. Thus it is obvious to state that after the Chernobyl accident thyroid exposure to population was one of the main factor or radiation safety due to the short living radioisotopes of iodine (mainly I-131).

By September 1995 about 400 thyroid cancer cases among children up to 14 years old were revealed in Republic of Belarussia (1) and 62 cancer cases among children and adolescents up to 18 years old at the time of irradiation - in Russia (2) (Bryansk , Kaluga and Tula oblasts, mainly in Bryansk). The observed increase in the number of childhood thyroid cancer incidents in these countries and Ukraine as well needs yet to be severely explained. But it doesn't contradict to the hypothesis of radiation causation of cancer, which is being confirmed indirectly by ascertained correlation between the weighed number of cases and average thyroid dose in the groups of the settlements (2).

Thus the dosimetry aspects in epidemiological studies have special importance.

2.2. First of all these aspects of the dosimetry investigations are aimed to clarify the individual thyroid dose estimates, calculated on the basis of direct thyroid measurements of people, and also to assess the average thyroid doses to the residents in any settlement independently whether the residents had been measured or not.

At the same time it should be taken into account that total number of the people with thyroid dose calculations based on the direct measurements of appropriate quality in Belarus and Russia is approximately 160 000. This number includes:

- 130 000 - in Belarus (3);

- 30 000 - in Russia (28 000 in Kaluga oblast (4) and 2 000 in Bryansk oblast (5)).

It is possible to estimate mean thyroid doses on the basis of these data for some settlements where representative number of direct measurements were made in May and at the beginning of June 1986 .

However much more contaminated settlements were without thyroid radioiodine content measurements. For inhabitants of these settlements the thyroid dose reconstruction can be done on the basis of ascertained tendencies of thyroid dose formation. The tendencies were revealed due to comparison of known average thyroid doses for adult population (D_{jx}) in the settlements with I-131 ground deposition densities (q_{jx}) in the vicinity of the respective settlements (j), located in area (x) (6,7). The calculations for other age groups can be done by known age dependencies.

Matching D_{jx} to q_{jx}(I-131) and the following estimation of the unknown doses on the basis of available I-131 ground deposition density in accordance with the ascertained tendencies is the best way. Unfortunately the data on I-131 ground deposition density practically are lack for Russia and are not enough for Belarus.

The most simple, reliable and accessible way is to use the available data relating to Cs-131 ground deposition density. But direct Cs-137 data usage can result in additional uncertainties because the ratio of I-131 to Cs-137 activity in the Chernobyl fallout's were able to vary in wide range. The main reasons for it are the distinction in physical-chemical properties of iodine and caesium as well as the distinction in time of fall-out for different parts of Belarussia and Russia.

I-129 (half life $16 \cdot 10^6$ years) has just the same chemical and physical (excluding the different half life) properties as I-131. Thus it is possible to accomplish the results with less uncertainty when one calculates q_{jx}(I-131) using q_{jx}(I-129) rather then q_{jx}(Cs-137). However the correct determination of average I-129 ground deposition density q_{jx}(I-129) even in the vicinity of one settlement is rather complicated technical task. Resolving this task by traditional techniques such as neutron - activation analysis and accelerator-mass spectrometry requires a lot of time and expenses.

Thus the solution of the task to determine the values of unknown thyroid doses on the basis of q_{jx}(I-131), or q_{jx}(Cs-137), or q_{jx}(I-129) has both the advantages and disadvantages.

2.3. More than 160 000 thyroid radioiodine content direct measurements for 350 settlements in some rayons in Belarus as well as for 62 settlements in three rayons in Kaluga and Bryansk oblasts in Russia were used for ascertaining the tendencies in thyroid dose exposure to population after the Chernobyl accident. Besides the mentioned data the values of ground deposition density of Cs-137 and I-131, which had been measured by specialists from Goscomgydromet of Russia and Belarus, Institute of Nuclear Physics (Minsk) (8), Institute of Biophysics (Moscow) were used in the analysis.

On the basis of this information the dependencies of D_{jx}/q_{jx} versus q_{jx} were drawn with grouping of the settlements (6).

The analysis of this sort of dependencies for all the available groups of settlements in Belarus and Russia showed that as a whole they can be approximated satisfactorily with the formula:

$$D'_{jx} = C_0 \times R_x \times q_x(Cs\text{-}137) + B_0 \times R_{jx} \times q_{jx}(Cs\text{-}137) \qquad /1/$$

where: D'_{jx} - calculated average thyroid dose for adults in area (x) and in the settlement (j), Gy; $q_x(Cs\text{-}137)$ and $q_{jx}(Cs\text{-}137)$ - average Cs-137 ground deposition density, respectively in area (x) and in the settlement (j), $Bq \times m^{-2}$; $R_x = (q_I/q_{Cs})_x$ and $(q_I/q_{Cs})_{jx}$ - average ratio of I-131 to Cs-137 ground deposition density, respectively in area (x) and in the settlement (j), rel.unit;

$C_0 = 3.6 \times 10^{-8}$ $Gy \times m^2 \times Bq^{-1}$;

$B_0 = 1.3 \times 10^{-8}$ $Gy \times m^2 \times Bq^{-1}$.

Equation /1/ is just for the groups of the settlements, located in selected areas. Each of these selected areas is characterised with approximately the uniform radionuclide concentration in radioactive cloud during the main deposition.

The following assumptions have been done:

1). Peculiarities of radioiodine intake by residents in the settlement (j) were absent. (Such kind of peculiarities as life style; relationship between levels of milk consumed from private cows and from shops; taking pills of potassium iodide; pasture period of cows and so on.)

2). The same level of standing of crop biomass.

3). The absence of essential contribution to the total activity of the fallout on the part of the radionuclides containing in the fuel parts of fallout.

Any deviation from typical variant is taken into account by inserting the additional coefficients.

Distinctions in the levels of standing of crop biomass, interception abilities of vegetation, feed intake rates and fractions of cow's intake from pasture grass are taken into consideration by inserting to formula /1/ the special coefficient too.

The most correct assessment of the values of D'_{jx} on the basis of q_x and q_{jx} can be reached in the case of adequate selection of areas (x), when each of these selected areas is characterised with the approximately uniform radionuclide concentration in radioactive cloud during the main deposition. As first step it was suggested that area (x) coincide with administrative rayon.

The parameters R_x and R_{jx} in the eq. /1/ are very variable among the others due to the distinction in physical-chemical properties of iodine and caesium and distinction in time of fall-out between different contaminated spots. That is why the data relating to I-131 and Cs-137 content in environment especially in milk, grass, and soil are of great importance. Unfortunately such data are available only for some settlements in Belarussia. There are not so much data for the settlements in Russia.

The missing data can be restored by analysing I-129 and Cs-137 content in soil samples and following assessment the values of ratio of $q_{jx}(I\text{-}131)/q_{jx}(Cs\text{-}137)$ on the basis of estimated values of ratio of $q_{jx}(I\text{-}129)/q_{jx}(Cs\text{-}137)$. It must be noted that for calculations we need most correct estimation of the ratio of I-129 to I-131 in Chernobyl reactor at the moment of accident.

At present time the traditional techniques for analysing I-129 content are the neutron-activation analysis and accelerator-mass spectrometry. However due to their high cost it is desirable to develop more cheap method. With this purpose the techniques of beta-x-ray coincidence may be used together with iodine extraction from the sample. This method provides the detection limit for I-129 not less than 10^{-3}

Bq/sample. The sensitivity of the method can be increased by using the greater amount of soil.

2.4. Individualised doses D'_{ijx} of internal thyroid irradiation for resident (i) living in settlement (j) located in territory (x) are reconstructed according to the following relationship:

$$D'_{ijx} = D'_{jx} \times k_i \qquad\qquad /2/$$

where D'_{jx} - see /1/ , k_i - is an age coefficient.

The value of the k_i coefficient depends on the age of individual (i) and his daily consumption of unskimmed milk during the early period after accident.

For persons who were not consuming milk are used k_i values corresponding to the zero milk consumption (inhalation intake only).

For persons in whom the information on milk consumption is unknown are used the k_i values that correspond to the values of mean daily milk consumption taken for different age groups. The values of k_i coefficient are presented in (9).

It was assumed that the concentration of radioiodine in consumed milk was similar for different persons in the settlement. It was taken into account age differences in mean daily milk consumption, breath speed, mass of thyroid gland and the speed of radioiodine turnover (7).

With the purpose to retrospectively evaluate individualised doses in persons a special questionnaire has been devised. The questionnaire helps to take into account the following factors related to the period from 26 April to 15 June 1986 influencing the value of the evaluated dose, age of an examined person, duration and time of stay in the zone of radioactive contamination, regimen of life in that period, protective measures, daily consumption of unskimmed milk, the beginning of the pasturing season in the locality. The data for infants and those exposed to radiation when in uterus are found by the mother's questionnaire (the data on breast-feeding are included). In this case thyroid dose in a child is estimated using the data of his mother by the relationships presented in (10).

2.5. The Table 1 presents the thyroid doses for adults calculated using semiempirical model /1/. Age dependence for estimation of children doses was taken from (9). The parameters q_x, q_{jx}, R_x, R_{jx}, were estimated basing on data of (11). The values of R_x, R_{jx} correspond to 1 May, 1986. The calculations relate to Nikolaevka settlement of Krasnogorsky rayon and to Novozybkov settlement of Novozybkov rayon (Bryansk oblast). It was supposed that areas (x) in /1/ coincide with administrative territories of Krasnogorsky and Novozybkovsky rayons. No countermeasures and no departure of inhabitants took place.

As it follows from Table 1 the dose values for age 3-6 years old are close to the corresponding estimations from (5) for the levels of the settlements contamination by Cs-137 in the range 570-2800 kBq/m^2.

Table 1. Results of the retrospective thyroid dose calculation according semiempirical model /1/ for settlements in Bryansk oblast.

Rayons	Settlements	Parameters of the model /1/				Thyroid doses by /1/, mGy			Thyroid doses by (5), mGy
		q_x, kBq/m^2	q_{jx}, kBq/m^2	R_x	R_{jx}	age <1 yr	age 3-6 yrs	adult	age 3-6 yrs
Krasnogorsky	Nikolaevka	460	2800	7	7	2800	1800	370	2150
Novozybkovsky	Novozybkov	640	570	7	7	1600	1000	210	750

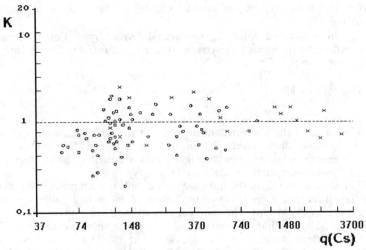

Fig 1. K = D$_e$/D$_c$ - ratio of the empirical dose values, D$_e$, to the results of calculation according model /1/, D$_c$; q(Cs) - Cs-137 contamination of the settlements, kBq/m². Circles - data for Bragyn rayon, crosses - data for Krasnopolsky rayon in Gomel and Mogilev oblasts of Belarussia.

It is interesting to compare the results of calculation according model /1/ with empirical values in the wide range of contamination levels. The comparison of the dose calculation performed using model /1/ with empirical estimations based on the radioiodine thyroid content measurements is presented on the Fig. 1.

It is a evident from Fig. 1 the proposed semiempirical model /1/ obviously brings the results of calculations close to the empirical values.

2.6. The developed semiempirical model /1/ was used for evaluation of mean dose values in inhabitants of every Russian settlements having Cs-137 contamination more then 3,7 kBq/m2.

Overall collective thyroid dose in 3,674,000 residents of contaminated territories of 8 oblasts of Russia (Bryansk, Tula, Kaluga, Orel, Riasan, Kursk, Leningrad) is estimated as 234,000 person·Gy (only due to internal irradiation by radionuclides of iodine) (12).This collective dose may cause about 700 cases of radiation induced thyroid cancer for all period of life after irradiation. The highest values of collective dose are in Bryansk and Tula regions, where expected numbers of radiation-induced thyroid cancer cases are about 280 and 270, respectively. The majority of cancer cases belongs to those who were children and adolescents in April-May 1986 (2).

The linear correlation is established between thyroid doses and rates of thyroid cancer cases diagnosed by the end of 1994 in children and adolescents of Bryansk oblast (12).

The retrospective evaluation of individual thyroid doses was performed in 30 of 49 cases of thyroid cancer found in those residents of contaminated territories of Bryansk oblast who were children and adolescents in April-May 1986 (2).

In 14 of 30 cases (47%) the most probable values of individual doses ranged from 200 to 2700 mGy. In the rest cases (53%) the most probable values of individual doses were 50 mGy or less. The attention attracts the fact that the typical distribution of individual doses is different for overall population of children residing in contaminated territories but not having the diagnosis of thyroid cancer: the majority of children have relatively low doses, less than 200 mGy, and the minority - the doses 200 mGy and more (4,5).

3. Dose reconstruction modelling vs parameters of radioactive contamination, milk consumption and Cs-137 WB content.

3.1. The problem of thyroid dose reconstruction for inhabitants of contaminated regions who were not measured on I-131 content in thyroid in May-June 1986 have been solved in two stages. At first, average doses for inhabitants of different age groups in a studied settlement, then individual doses based on average values were determined (5).

It was used linear proportional model for regression analysis of connection between the mean thyroid dose in the age group 3-6 years and different parameters of radioactive contamination of the environment in settlement (surface Cs-137 activity on soil, kerma rate in the air on May 10-12 1986, mean I-131 concentration in milk on May 10-12). The results of this linear regression analysis were used for the reconstruction of the mean dose in settlements. Dose values for regression analysis were calculated from the direct measurements of radioiodine content in inhabitant's thyroids in May-June 1986. Besides, the correlation between the average thyroid dose and the average Cs-134,-137 whole-body content in adult measured in July-August 1986 was studied.

It was shown (5) that the connection between parameters was reliable in all cases: correlation coefficients were within the range from 0.86 to 0.95 (number of measured pairs for each correlation dependence were from 6 to 13). The difference in the linear regression coefficients for Cs-137 soil contamination and the kerma rate in air in Bryansk oblast is less than 50 % comparatively with Tula oblast.

It was found from analysis of connection between the thyroid dose and the Cs-137,134 content in the whole body that it should be better to consider the Bryansk and Tula regions separately, presumably because of the difference in transfer coefficients from soil into plants. The regression equation obtained for the Bryansk region can be applied to the areas with poor "podzol", peat and sandy soils, and the equation for the Tula region - for the areas with "chernozem" soils.

The regression equations provide estimates of the maximum thyroid dose in the absence of protective actions. For each settlement the package of the countermeasures performed there was analysed. Finally the dose was evaluated by comparison of values obtained with different regression equations taking into account the reliability of initial parameters and introducing the correction for the performed countermeasures. The accuracy of final mean values was assumed not worse than 50%.

3.2. As far as the dose in thyroid of inhabitants was formed basically because of consumption of contaminated food of local produce, it is natural to suppose that the individual dose is proportional to the amount of the consumed milk - the basic source of the I-131 intake in the human body.

To check this suggestion the results of the thyroid control in May 1986 and the data of the poll on the nutrition regime were analysed for about 600 inhabitants of the Bryansk region. The dependence of individual dose in thyroid on the daily milk consumption appeared statistically reliable with the correlation coefficient within the limits of 0.6-0.8 for all age groups.

One can change over from the mean age dose in the settlement D (mGy) to evaluation of the individual dose D_i on the basis of these data, if the information on individual V_i (l/day) and mean V (l/day) consumption of milk of local produce in the specified age group, on the duration of its consumption by the given person t_i (days) and on the average in the village t (days) is present, with the help of the formula:

$$D_i = D \times [2 + 8 \times V_i \times t_i/(V \times t)] \qquad /3/$$

Such method for evaluation of the individual dose can be applied to groups of persons with similar behaviour and nutrition modes after the accident. If in the settlement people consumed milk from different sources (from private ownership, from a collective milk farm, from a state shop) the correction for the mean ratio of the I-131 concentrations in milk from different sources must be introduced in formula /3/.

The other opportunity of evaluation of individual thyroid dose is based on the correlation between the total I-131 intake in a body (or the dose in thyroid) and the content of caesium radionuclides in it measured later, for instance, in August-September 1986. It was shown that statistically reliable linear connection (r=0.7-0.9) between the total iodine incorporation in thyroid and the caesium content in the body exists for inhabitants of the controlled area of the Bryansk region - children over 3 years and adults, which in the main ceased the consumption of food products of local produce in May 1986 and kept these restrictions during the summer. In this case the individual dose of the I-131 irradiation in thyroid can be evaluated with the formula:

$$D_i = K(T) \cdot A_i \ , \ mGy \qquad /4/$$

where K(T), mGy/kBq, is the age dependent coefficient, its values are presented in Table 2; T - age, years; A_i, kBq, is the Cs-137,137 content in the whole body of the investigated person in July-September 1986.

It was evaluated that possible deviation of the reconstructed by this way individual doses from the real ones by the factor of 2-3 to the both sides.

Table 2. Coefficients for determination of thyroid dose according eq. /4/.

T, years	3 - 6	7 - 11	12 - 17	>18
K(T), mGy/kBq	8	4	2	1

A retrospective evaluation of the individual thyroid doses for about 60 thousands inhabitants of the contaminated areas of the Bryansk oblast was made on the base of the developed techniques. For 2 300 children from the most contaminated settlements of the Tula oblast the same estimations were made.

4. Reconstruction of thyroid dose following Chernobyl accident on the base of I-129 measurements in the soil.

4.1. To reconstruct retrospectively the radiation dose of the thyroid received by the short-living iodine nuclides, the former concentration of these nuclides in air and food must be determined. This is tried by measuring the soil contamination of the long-living nuclide I-129 which is in known correlation with the short-living iodine nuclides produced in the nuclear reactor. Parallel to the experimental investigation of the soil activity a computer- aided model has been established to assess the infant thyroid dose. This model is guided along the idea that the I-129 analysis resulting in the determination of the integral soil contamination must be completed by the determination of the uncertainty of the intake function. First results are given.

4.2. Soil samples were taken from undisturbed ground around villages or towns in the western Bryansk region (Russia),which is situated about 200 km north east of Chernobyl. These sites had been selected because in 1986 I-131 contamination were measured in this area. Since the general distribution of I-129 with depth is unknown the soil samples had been divided into 2 cm thick layers. For details of taking soil samples see (13). To determine the I-129 soil contamination caused by

Chernobyl only, one has to subtract the background soil contamination due to atomic weapon test explosions in the atmosphere. Therefore soil samples were taken also from low-contaminated areas similar in the kind of soil and climate. In the background areas the Cs-137 soil contamination was less by three orders of magnitude. The different layers showed an increase of mass with depth; this may be explained by higher densities resulting from the pressure increasing with depth. From each layer of each village 100 g dry soil were used for analyses purpose. This means that for each layer the aliquots are not constant and correction factors had to be introduced with regard to the mean mass of all layers. The radioactivity, mainly of the nuclide Cs-137, was determined by gammaspectrometry. According to the radioactivity content of the samples the measuring time changed from 5 to 900 minutes. The concentrations of I-129 and I-127 have been measured by neutron activation analysis.

To assess the uncertainty of the intake function three theoretical cases have been modelled for different contamination patterns resulting from different time-dependent contamination and precipitation rates. All these patterns are based on the same total amount of deposited activity per square meter. For the total grass activity a value of $3 \cdot 10^5$ Bq/m^2 has been derived from measurement performed in the contaminated regions of Bryansk in 1986 (14). From the three differential contamination patterns corresponding intake functions have been evaluated. This function describes the time-dependent intake of short-living iodine nuclides via milk consumption. To determine the intake function the time distribution of contamination and intake is necessary, at least for short-living nuclides as I-131. The contamination of milk is directly proportional to the one of grass. The total grass contamination at a specific day was obtained by summing up all the activity deposited until this day minus the amount which has been already vanished. The assessment of the daily thyroid dose, D, of the infants has been performed according to

$$D = F \times (I_{cow} / Y) \times T \times I_{infant} \times g, \text{ Gy/day or Sv/day} \qquad /5/$$

where F represents the contamination of the grass in Bq/m^2 , Y - the yield of grass from the meadow (0.85 kg/m^2), I_{cow} - the grass consumption of the cow (65 kg/d), T - the transfer factor (grass/milk: 3×10^{-3} d/kg), I_{infant} the milk consumption (infant) 0.55 kg/d, g - the thyroid dose factor ($g_{thyr} = 3,5 \times 10^{-6}$ Gy/Bq) or effective dose factor ($g_{eff} = 1,1 \times 10^{-7}$ Sv/Bq). For the first assessments the German values of the parameters are used. Inserting Russian values may lead to some differences in the result.

4.3. For each village or town the normalised depth profiles for Cs-137 are determined including the "background" villages. The mean slope of the background depth profiles is smaller than the other ones. The so-called relaxation length is 7.5 cm in the background areas compared to about 4 cm in the regions contaminated by the reactor accident of Chernobyl. This may be explained by the longer contamination time of the atomic bomb tests. The Cs-137 contamination of the background areas is less by two or three orders of magnitude in regard to the Chernobyl effected regions.

The Cs-137 contamination for each village is calculated from the gammaspectrometrical data, the correction factor and the known area of sample taking. The results are in good agreement either with parts of the Russian average values or with parts of the Russian measurements in the year 1986 (14). The latter one is in general the result of only one measurement at one point near the villages or towns.

Until now, the determination of the I-129 concentration in soil had not been finished (October 1995). There is no clear dependence of the iodine content with depth. However, to draw a general conclusion a much broader data base is necessary. But for two villages the preliminary dose reconstruction will be given.

4.4. In general it can be assumed that contamination of soil or grass may happen only during the passing time of the plume. If the time of contamination is small compared with the effective half life of the regarded nuclide on vegetation, the influence of the time distribution of contamination is negligible. One other parameter, influencing the contamination of grass and in consequence the intake function, is the mechanism of contamination: washout or fallout. The very inhomogenious contamination in Russia and Belarus shows that washout has been the decisive process. Three different contamination patterns have been considered here. In the first case it is assumed that all the activity is deposited at the first day after the accident. Since there is no further contamination during the following days the activity concentration is reduced with time in dependence of the effective half-life of the I-131 on grass ($t_{1/2}$ = 5.9 d, (15)). The grass contamination will decrease continuously and reach half of the initial contamination at the end of the week assuming an effective half life of one week. For the calculation it is supposed here that the deposition takes place at the beginning of each day. The second function represents the case that the activity is contaminating the grass over ten days with the same amount of activity each day. This means a continuous but - due to the decay - sub-linear increase of the activity until day '10', the end of the supposed deposition. In addition to these quite simple assumptions a third function has been used here, namely a hyperbolic- like trend of the activity deposition based on literature data for the time dependence of the nuclide release (16). In this case the first six days show a slight increase of the activity followed by a very steep one until day '10' before again the decay characteristic controls the curve.

The grass contamination resulting from these patterns and the related thyroid dose have exactly the same shape since all parameters in the above equation except the contamination F are constant in time.

A summation of the daily thyroid doses results in an accumulated dose for infantile thyroids. Among the three considered cases the situation is worst for the total activity deposition at the first day. In particular between day '5' and day '10' the thyroid dose for the first case is about 0.3 Gy above the values for the two other contamination patterns (deposition over ten days) which result in nearly identical values for the integrated thyroid dose. In the period of more than 25 or 30 days there is no great difference between the three cases, and an accumulated dose of about 1 Gy has been evaluated by these model calculations which are based on the assumption that no counter- measures have been initiated. A dispersion calculation (17) shows that in these areas there was one main washout contamination during only one day.

The results of these three cases considered demonstrate very clearly that e.g. the prohibition of milk consumption from the contaminated areas would reduce the dose very effectively. In the cases of a contamination over ten days the supply with non-contaminated milk at day '5' would reduce the thyroid dose by about a factor of 5. Even in the worse case of total activity deposition at the first day this countermeasure would let expect a dose of 0.5 Gy instead of 1 Gy.

4.5. This report shall be finished by preliminary results of the retrospective dose reconstruction for the two settlements Novozybkov and Nikolaevka. The I-129 concentration in soil for the town of Novozybkov has been determined by neutron analyses for the 2 cm layers from 0 to 10 cm depth. For the two layers 6 - 8 and 8 - 10 cm the I-129 concentration has the constant value of about 1 mBq/kg. The soil measurements of the nearly uncontaminated village of Kostenichi for the 0 -2 cm layer gives about the same value, which therefore is regarded as background value.

Subtracting this background the I-129 contamination is assessed to

Novozybkov	0,481 Bq I-129/m^2
Nikolaevka	0,446 Bq I-129/m^2

According to the ratio of $5,5 \times 10^{-8}$ for I-129 to I-131 (17) these values can be converted into a I-131 soil contamination of:

Novozybkov	8,7 (0,603 / 4,76)	MBq I-131/m^2
Nikolaevka	8,0 (3,18 / 25,1)	MBq I-131/m^2

(For a longer working time of the reactor the I-131 soil contamination will decrease). The values in the brackets are Russian measuring values. The first one is a measurement of one single soil sample for each settlement (14) at the 22 May 1986. The dispersion calculation (18) shows that the main contamination by washout happened at the 28 April 1986. The correction for a 24 day decay of I-131 results in the second values of the bracket. This results show that our assessments gives at least the same order of magnitude for the I-131 contamination as the Russian measurements of the soil contamination at the end of May 1986.

According to the very sparse grass vegetation in the contaminated area only 1/10 of the German retention factor (0.33) has been used here for the model calculations which result in the following values for the integrated thyroid doses for infants caused by ingestion of I-131 contaminated food (without counter measurements) :

Novozybkov	1,0 Gy
Nikolaevka	0,9 Gy

5. Conclusion.

5.1. The results of I-131 content measurements in thyroid gland as well as Cs-137 and I-131 soil contamination data, results of I-129 measurements in soil samples were used to develop the different models of reconstruction of thyroid doses for those settlements, where direct estimation were not performed in 1986.

5.2. It must be taken into account that ratio of I-131 to Cs-137 activity in the Chernobyl fallout's was able to vary in wide range. It may be the result of the distinction in physical-chemical properties of iodine and caesium as well as the distinction in the time of fall-outs for different contaminated territories. Very small quantity of available data concerning I-131 contamination can be restored due to analysing I-129 content in soil samples. Then the values of ratio of I-131 to Cs-137 should be assessed on the basis of estimated ratio of I-129 to Cs-137. For correct calculations it is extremely desirable to make a proper estimation of the I-131 to I-129 activity ratio in the Chernobyl reactor at the time of accident.

References.

1. A.E. Oceanov, E.P. Demidchik, M.A. Ankudovich et al. Thyroid cancer in Belarus before and after Chernobyl accident. WHO/EOS/94.26., Geneva, 1994. (Russian).
2. E.M. Parshkov, A.F. Tsyb, V.F. Stepanenko. Analysis of the thyroid pathology and childhood/adolescents thyroid cancer cases as a result of Chernobyl accident. In: "Radiological, medical and social/economic consequences of the Chernobyl accident". Moscow. 1955. p.57 (Russian).
3. Yu. I. Gavrilin, K.I. Gordeev, L.A. Il'in et al. Results of thyroid dose assessment for contaminated territories of Belarussia. "The Bulletin of AMS of USSR", 1991, N8, p. 35 (Russian).
4. Stepanenko V.F., Tsyb A.F., Matveenko E.G. et al. Thyroid exposure doses among the population of contaminated territories: methodology and results of measurements. In: Deutch- Russische Konferenz fur Messprogramm in Russland. Moskau, 1992, s. 53.
5. Zvonova I.A., Balonov M.I. Radioiodine dosimetry and prediction of consequences of thyroid exposure of the Russian population following the Chernobyl accident. In: "The Chernobyl papers", Washington, 1993, p.71.
6. A.F. Tsyb, V. F. Stepanenko, Yu. I. Gavrilin et al. The problem of retrospective dose estimation for inhabitants as a result of Chernobyl accident: peculiarities of dose-forming, structure and levels of

irradiation according results of direct measurements. Part 1: Doses of internal irradiation of the thyroid. WHO/EOS/94.14, Geneva, 1994. (Russian).

7. Yu. I. Gavrilin, V.T. Khrousch, S.M. Shinkarev et al. Passportisation of the settlements in rayons of Gomel oblast in hard contaminated zone. Report of IBPh, M., 1991. (Russian).

8. Yu.V. Dubina, Yu.K. Schekin, L.N. Guskina. Systematisation and verification of the spectrometrical analysis data of soil, grass, milk and milk products samples with the results of 131-Iodine measurements. Minsk, 1990. (Russian).

9. V.T. Khrousch, Yu. I. Gavrilin, S.M. Shinkarev, V. F. Stepanenko. The estimation of the thyroid doses on the base of I-129 measurements in the environment. Goskomsanepidnadzor of Russia, Moscow, 1995 (in press, Russian).

10. Evaluation of absorbed doses from iodine radionuclides in thyroid glands of persons exposed to radiation as the result of the Chernobyl accident (Methodical instructions). Ministry of Public Health of the USSR, Moscow, 1987. (Russian).

11. Data on radioactive contamination the territory of Russian Federation by Cs-137, Sr-90, Pu-239-240, I-131. Radiation and Risk. Bulletin of the All-Russia Medical and Dosimetric State Registry. Issue 3, Moscow-Obninsk. 1995 (Russian).

12. V.F. Stepanenko, A.F. Tsyb, Yu.I. Gavrilin et al. Thyroid dose irradiation in population of Russia as a result of Chernobyl accident (retrospective analysis). Radiation and Risk. Bulletin of the All-Russia Medical and Dosimetric State Registry. Issue 7, Moscow-Obninsk. 1995 (Russian, in press).

13. Forschungszentrum Juelich GmbH, Reconstruction of the thyroid doses obtained by man during the Chernobyl accident, COSU - CT93 - 0051, December 1994

14. I.A. Zvonova, personal communication, 1994

15. K. Heinemann and K.J. Vogt, Measurements of the deposition of iodine onto vegetation and of the biological half-life of iodine on vegetation, Health Physics 39, 463 - 474, 1980

16. Gesellschaft fuer Reaktorsicherheit, Neuere Erkenntnisse zum Unfall im Kernkraftwerk Tschernobyl, GRS-S-49, 1986

17. Kolobashkin V. M., Rubtsov P. M., Ruzhansky P. A., Sidorenko V.D., Radiation characteristics of irradiated nuclear fuel. Handbook, Energoatomizdat, 1983, 620 pp.

18. H. Haas, H. J. Jacobs, S. Elbel, Dispersionsrechnungen fuer Radioisotope in der Folge des Reaktorunfalls von Tschernobyl, Bericht des Foerdervereins des Instituts fuer Geowissenschaftliche Umweltforschung und Analytik Koeln e.V. an der Universitat zu Koeln, Januar 1995.

Retrospective individual dosimetry using EPR of tooth enamel

V. Skvortzov [1], A. Ivannikov [1], A. Wieser [2], A. Bougai [3], A. Brick [4],
V. Chumak [5], V. Stepanenko [1], V. Radchuk [5], V. Repin [5], V. Kirilov [6]

1) *Medical Radiological Research Center of RAMS, Obninsk, Russia*
2) *GSF-Forschungszentrum fur Umwelt und Gesundheit, Munchen, Germany*
3) *Institute of Semiconductor Physics of UAS, Kiev, Ukraine*
4) *Institute of Geochemistry, Mineralogy and Ore Formation of UAS, Kiev, Ukraine*
5) *Scientific Center of Radiation Medicine of UAMS, Kiev, Ukraine,*
6) *Research Institute of Radiation Medicine of MHB, Minsk, Belarus.*

Abstract. The results of joint investigations (in the framework of ECP-10 program) aimed on the improvement of the sensitivity and accuracy of the procedure of dose measurement using tooth enamel EPR spectroscopy are presented. It is shown, what the sensitivity of method may be increased using special physical-chemical procedure of the enamel samples treatment, which leads to the reducing of EPR signal of organic components in enamel. Tooth diseases may have an effect on radiation sensitivity of enamel. On the basis of statistical analysis of the results of more then 2000 tooth enamel samples measurements it was shown, what tooth enamel EPR spectroscopy gives opportunity to registrate contribution into total dose, which is caused by natural environmental radiation and by radioactive contamination. EPR response of enamel to ultraviolet exposure is investigated and possible influences to EPR dosimetry is discussed. The correction factors for EPR dosimetry in real radiation fields are estimated.

1. Introduction

The large area in central zone of Russia has been contaminated by radioactive fallout as a result of the Chernobyl accident. The large-scale individual dosimetric control of population was not performed because of objective circumstances. Only averaged estimations of collective doses for separate settlements were carried out as a rule. The essential variance of real individual doses of the averaged dose even for population of separated settlements should be expected, because of the high heterogeneity of radionuclide contamination and the difference in behavior of individuals. That is why the individual doses assessment using analytical methods, which are based on information about level of radioactive contamination, encounters the principal difficulties. But nevertheless, the knowledge of individual accumulated doses is necessary for formation of population groups of high risks for medical observation and rehabilitation.

Tooth enamel exhibits the features of individual dosimeter and it may be used for the assessment of individual accumulated doses in accidental and uncontrolled situations. The stable paramagnetic centers in tooth enamel are formed after its exposing to ionizing radiation. These centers may be detected using the electron paramagnetic resonance spectroscopy (EPR-spectroscopy), and the individual accumulated dose may be estimated using the intensity of radiation induced (RI) EPR signal . The method of tooth enamel EPR-spectroscopy becomes one of the most useful techniques for the dose evaluation in uncontrolled and accident situations. Nevertheless, there are the lot of problems concerning the samples preparation and spectra interpretation in order to achieve the sufficient sensitivity and accuracy for the reliable dose evaluation. The investigations in framework of ECP-10 program were carried out as a joint work in the following directions:

- Optimizing the procedure of sample preparation and physical-chemical treatment of samples in order to achieve the maximal sensitivity.

- Estimation of the influence of the line shape used for simulation of native background signal on the results of dose evaluation.

- Influence of tooth diseases to dose response.

- Estimation of the method sensitivity using the statistical analysis of the results of wide-scale measurements of doses by tooth enamel EPR-spectroscopy.

- EPR response to ultraviolet (UV) exposure and possible influences to EPR dosimetry

- Correction factors for EPR-dosimetry in real radiation fields.

2. Optimizing the procedure of sample preparation and physical-chemical treatment of enamel

2.1. Sample preparation

The lowest measurable absorbed dose in tooth enamel was found to be not limited primarily by the sensitivity of the EPR spectrometer but by the existence of background EPR signals in the sample. In competition to the EPR signal of the carbonate radicals also the background EPR signals were found to increase by radiation. In dentine the background was found more intensive than in tooth enamel. The background radicals have insufficient dosimetric properties and prevent precise dosimetry below 1 Gy for tooth enamel. The lowering of the detection threshold for dosimetry requires an effective separation of the attached dentine from the tooth enamel. Furthermore, the concentration of the background radicals in tooth enamel needs to be reduced by suitable sample treatment.

2.2. Chemical and physical procedures for effective separation of dentine from tooth enamel

Precise EPR dosimetry in the dose range below 1 Gy requires very pure samples of tooth enamel for the measurements. The incomplete removal of dentine from the tooth enamel deteriorate drastically detection threshold and precision of EPR dosimetry. The grinding of dentine from tooth enamel by a dentist drill was found to be very time consuming and ineffective. The interface between dentine and tooth enamel can not visually be recognized easily. Therefore the separation of dentine is not necessarily complete. Further purification is required, e.g., by density separation in heavy liquid after mortaring the sample. A procedure was developed for effective removal of dentine from enamel. The procedure makes use of the softening of dentine by treatment with sodium hydroxide solution in an ultra sonic bath. After 3 hours treatment the interface between dentine and tooth enamel can be easily recognized visually by the difference in color and hardness of the two tooth components. The dentine becomes bright white in contrast to the yellowish color of tooth enamel. The ultra sonic treatment with sodium hydroxide solution destroys the structure of dentine and allows for its easy removal. This method of tooth preparation provides very pure samples of tooth enamel.

2.3. Influence of pre-treatment on the background EPR signal of tooth enamel

The intrinsic EPR signal in tooth enamel affects the reliability of dosimetry. In the case of bone and tooth dentine the intrinsic signal was found to result mainly from organic components. The procedures developed for removing the organic fraction of bone was found to be less effective in the case of tooth enamel. The removal of the intrinsic EPR signals in tooth enamel requires different chemical treatment than bone or tooth dentine.

The treatment of bone in a Soxhlet apparatus with diethylenetriamine at 200 C was found to result in a 5-fold reduction of the organic EPR signal component [1]. Similar results by the Soxhlet treatment were found also with tooth dentine. However, in the case of tooth enamel only little reduction of the intrinsic EPR signal was achieved after this kind of treatment. The reasons for the difference in behavior of the intrinsic EPR signal in bone and tooth enamel by chemical treatment might be due to the very low organic content of tooth enamel. The most effective removal of the intrinsic EPR signal in tooth enamel was found by successive treatment with sodium hydroxide solution and acetic acid at room temperature in an ultra sonic bath. Chips of tooth enamel were first treated with concentrated sodium hydroxide solution for 40 hours. After grinding the chips grains in the range 0.1 - 0.5 mm were etched with acetic acid (25%) for 10 minutes. The total treatment was found to result in a 3-fold reduction of the intrinsic EPR signal.The efforts in optimizing the procedure for the chemical treatment of tooth enamel resulted not in a complete elimination of the background. However, further reduction of the background is required to improve the precision at absorbed doses below 0.3 Gy. In addition mathematical methods for spectrum evaluation need to be developed in order to eliminate the disturbing background.

3. Comparison of background signal from teeth of different locations . Difference of real and modelled background signals and its consequences to dose evaluation

The physical sense of the EPR tooth enamel dosimetry is due to the generation of free radicals in the crystal lattice of enamel under the irradiation. The concentration of radicals is proportional to the absorbed irradiation dose and can be measured with the EPR . One of the problems of the correct measurement of a concentration of radiation induced free radicals is the presence of non radiation radicals which are generally present in enamel and the EPR spectrum of which is superimposed on the radiation induced spectrum. The non radiation EPR line in tooth enamel is so-called background line. To measure the concentration of the radiation induced radicals one must subtract the background spectrum from the experimentally observed superposition of lines.

So there are questions about the correct line shape of background spectrum, its possible variations from one sample to the another and from one laboratory to another and possible influence the background line shape on the value of obtained in such a way dose.

We used different background signals obtained from some participants of intercalibration in the framework of ECP-10 program which have agreed to carry out the intercomparison of their background lines. Background lines were compared with Gaussian and Lorentzian line shapes.

To compare all background lines we display them at the same scale of magnetic field and the vertical size was adjusted   so that the first extreme of line shape derivative coincides for all lines. The central parts of all lines are very close in the region where is actual to EPR dosimetry. Wings of different lines outside the restricted central part have individual peculiarities but they don't influence significantly the radiation disturbed central part of the spectrum.

Another important feature of the background line is its asymmetry in compare with the Gaussian and Lorentzian line shapes. It means that the use of symmetric background lines should give an overestimated results. The use of Gaussian, Lorentzian and real background line for the same set of the dose dependent spectrum sequence gives three different values of dose : 0.70, 0.48 and 0.20 Gy respectively.

On the ground of this analysis we offer as a first approximation the background line which is the average of all 5 real line shapes. It is represented in digital form for 48 points of its main region.

4. Influence of tooth diseases to dose response

Tooth enamel is a typical representative of the minerals of biological origin. It is very textured system, because microcrystalls of hydroxyapatite in tooth enamel have not chaotic distribution in space [2]. At disease of teeth the degree of texturing and properties of microcrystalls can change. So it is necessary to study influences of diseases of teeth at the results of retrospective dosimetry.

We have established that properties of tooth enamel can be characterized through anisotropy effects by studying EPR at plates of tooth enamel [3]. It were shown that anisotropy effects in healthy and sick teeth are quite different. We have introduced the anisotropy coefficient. It gives quantitative characteristic of state tooth enamel and it is a ratio of different components in EPR spectra in plate of tooth enamel [3].

The state and properties of tooth enamel can be evaluated not only by plates of tooth enamel, but and by powder samples. In this case we had introduced the coefficient of line shape that is ratio of different components of EPR spectra in powder samples [3]. The determination of this coefficient for different teeth had shown that EPR signal in tooth enamel is a superposition from different kind of radiation induced centers. The temperature stability of these centers is different as well. It is incorrectly to say about one half life time of these centers. In tooth enamel there is a distribution of radiation induced centers with different half life time [3].

The chemical composition, detentions, orientation and other properties of hydroxyapatite crystallites are changed at disease of teeth. For example the quantity of carbonate groups can increase from (1-3)% to (5-10)%. Because radiation induced centers are associated with carbonate groups, so it is appeared the problem about change radiation sensitivity of tooth enamel at disease of teeth. We have established that at high level at doses (D > 10 Gy) the sensitivity of carious teeth can be more then healthy teeth approximated at (20 - 50)%. But a low level irradiation (D > 5 Gy) this difference is reduced [3]. We have described theoretical model in the frame of band theory of solid state, which had permitted to explain this experimental results.

From the described above investigations we have assumed that for increase the reliability of retrospective EPR dosimetry it is necessary as a rule previously investigations of state and properties of tooth enamel. At the same time it is important to underline that at present EPR dosimetry by tooth enamel is most effective method for dose reconstruction of population.

6. Development of EPR methods based on statistical analysis of experimental data

We have applied tooth enamel EPR-spectroscopy method for individual accumulated dose assessment during the wide inspection of population of radioactive contaminated after Chernobyl accident territories of Bryansk region. Over 2000 tooth enamel samples, collected during the ordinary dentist practice, were analyzed. All doses were estimated with the use of the standardized methodical approach.

The methodical approach for sample preparation and dose measurements was described in details in our recent publication [4] and briefly it may be described as follows. Tooth enamel samples were prepared by removal of dentine with the dental drill. The enamel was crushed into pieces about 1 mm in size. EPR spectra were recorded at a room temperature using the spectrometer ESP-300E with the standard rectangular resonator in X-band. Microwave power was set 10 mW. Time duration of spectrum measurement with accumulations was equal to 45 min.

Subtraction of background native signal was performed using computer, built-in spectrometer, and its basic software for spectra processing. For simulation of background signal we used the spectrum of mixed enamel of children milky teeth. Enamel samples for this spectrum were selected with the criteria of minimal radiation induced signal. Shape of simulated signal was fitted to background native signal by varying its amplitude, field shift and width using software under operator control. Subtraction gave as a result the RI signal. Its intensity was measured as amplitude of low-field component.

Irradiation doses were determined from the intensity of RI signal in enamel with the use of the universal calibration coefficient. The value of this coefficient was measured as the averaged slope of the linear dependence of RI signal intensity on additional irradiation dose for the mixed enamel sample. The collimated beam of gamma radiation from source Co-60 was used for additional irradiation.

Individual accumulated doses were measured with the use of enamel samples, collected for population of radioactive contaminated south-west territories of Bryansk region and of control radiation-free territories of Kaluga region. Measured dose values with information about patients were entered into computer data bank for subsequent statistical processing. The overall quantity of determined and analyzed values of doses was more than 2000.

The frequency distributions of individual dose values for confined territories and large settlements were described by logarithmic normal function. The mean dose values for control territories for adult population were in the range of 100-120 mGy (with dispersion 70-80 mGy). For the radioactive-contaminated territories the mean dose values were up to 150-200 mGy (dispersion 80-120 mGy), depending on the level of radioactive contamination. These results include all analyzed teeth. The histograms of dose distribution were published in previous work [5].

The observed data scattering of dose values is caused by several reasons. Undoubtedly, essential contribution into data scattering gives experimental error of method used. This contribution is difficult to estimate because of unknown absolute accuracy and systematic errors of method. Possible sources of errors will be discussed in this paper.

The part of scattering is caused by different age of patients and, consequently, by different accumulated dose caused by natural background radiation. We have found, that the measured doses are tend to grow with the age of patients both as for control and radioactive-contaminated territories. The slope of appropriate linear regressions is equals to 1.4+/-0.5 mGy/year. Intercept is dependent on the level of radioactive contamination. For example, for radioactive contaminated territory of Gordeevsky rayon (radioactive contamination up to 40 Ci/sq.km), the value of slope is equal to

113+/-23 mGy, for radiation-free control territory   35+/-30 mGy. The difference between these two intercept values is due to 9-year lasting (since the Chernobyl accident) additional irradiation resulting from radioactive contamination. The determined slope value for dose-age dependence is corresponding to annual accumulated dose resulted from natural environmental radiation with dose rate about 0.18+/-0.06 uSv/hour.

Some part of observed data scattering may be attributed to heterogeneity of local radioactive contamination. Difference between dispersion values for control and radioactive-contaminated territories may be attributed namely to this heterogeneity. Nevertheless, there is clear tendency of averaged dose to grow with the level of radioactive contamination of settlements. The dependence of averaged for some settlements doses on Cs-137 contamination of appropriate territories is described by linear regression with the slope equals to 2.4+/-0.5 mGy/Ci/sq.km and the intercept equal to 113+/-20 mGy. The slope value of linear regression for dose-contamination dependence is in a good agreement with those, obtained by model calculation .

Thus, the statistical analysis of results obtained allowed us to make a conclusion, that tooth enamel EPR spectroscopy method made possible to registrate contribution into total dose, which was caused by natural background radiation and by radioactive contamination.

Some possible sources of systematic errors of method used are described by the following contributions. The main error source is the uncertainty in values of some intrinsic signals, which may be existing in enamel EPR spectrum at the same location as RI signal. These signals may be originated by some undetermined reasons, for example, by some paramagnetic impurities, by medical X-ray diagnostic and so on. It should be appointed, according to our latest results, what IR centers in enamel may be formed under the action of UV exposure and, as it is not excluded, by the action of UV component of solar illumination.

7. EPR response to UV exposure and possible influences to EPR dosimetry

It was found for some patients with several collected extracted teeth, what teeth of front part of jaw (incisors and canines, location 1-3) exhibits, as a rule, more intensive RI signal, than teeth of rear part of jaw (molars and premolars, location 4-8).   We have examined the dependencies of enamel RI signal intensity on age of patients for different located teeth. It was founded, what RI signal intensity values for teeth with location 1-3 are high scattered, whereas data scattering observed for teeth with location 4-8 is much less. For teeth with location 4-8, the clear tendency of RI signal intensity to grow with the age of patient is observed. The observed RI signal intensity mean value for front teeth equals to 250 mGy with dispersion 102 mGy, whereas for back teeth mean value is much less and equals to 120 mGy with dispersion 42 mGy.

In order to explain the observed differences in the RI signal intensity for front and back teeth we supposed, that RI signal in front teeth may be formed under UV component of solar irradiation. To approve this proposition we have measured EPR spectra of tooth enamel after exposing it to different light sources. The EPR spectrum of illuminated during 20 hours by low pressure bactericide mercury lamp tooth enamel is stabile and is similar to spectra of RI centers in gamma-irradiated enamel with dose about 50 Gy. Exposing enamel to solar illumination during 6 days gives rise to similar RI signal having intensity equivalent to 200 mGy. It is clearly seen that sun light illumination leads to forming of signal similar with RI signal. The observed high dose values measured for front teeth and high doses scattering may be explained namely by influence of UV component of solar illumination. Thus, this phenomena should be taken into account on dose measurements using tooth enamel EPR spectra.

Because of possible action of solar illumination on forming RI signal in enamel of front teeth, we have excluded teeth with location 1-3 from statistical analysis. That has lead to reducing of dispersion and to some decreasing of mean values. The dose-age dependencies of corrected doses were tend to be linear with slope value laying in diapason from 1.0 to 2.5 mGy/year, depending on concrete settlement.

8. Correction factors for EPR dosimetry in real radiation fields

Some systematic error may be originated from the significant dependence of enamel sensitivity on gamma-photons energy. As shown [6], tooth enamel exhibits maximum sensitivity in photon energy region below 200 keV. Thus, the presence of low energy component in real acting spectrum of natural environmental radiation and/or scattered component of radiation from Cs-137 contamination may give increased readings for tooth enamel dosimeter compared with direct radiation. This systematic error should be taken into account by the entering a correction factor. We have made experimental and theoretical estimations for this correction factor for some cases of scattered radiation. For example, the presence of high scattered low-energy component of Cs-137 gamma radiation behind the 0.4 m of water shielding results in increasing of the readings of tooth enamel dosimeter by factor 2.4+/-0.5 in comparison with the readings of the LiF thermolumeniscent dosimeter (experimental result). The accounting for energy distribution of natural environmental radiation gives the theoretical estimation of correction factor for tooth enamel dosimeter about value of 1.5.

References

[1] A. Wieser *et al.*, EPR dosimetry of bone gains accuracy by isolation of calcified tissue, *Appl. Radiat. Isot.* **45** (1994) 525-526.
[2] A. Brick and O.Gaver, Principles of building of structure and radiation characteristics of biological origin. Abstracts of the 16th General Meeting of Intentional Mineralogist Association, Pizza, Italy, 1993
[3] A. Brick et al., Metamorphic modifications and EPR dosimetry in tooth enamel. Abstracts of 4th International Symposium on ESR Dosimetry and Applications, Munich, GSF, 1995
[4] V.G. Skvortzov et al., Assessment of individual accumulated irradiation doses using EPR spectroscopy of tooth enamel. *J. Molec. Struct.*, **347** (1995) 321-329
[5] V.G. Skvortzov et al., Results of investigation by EPR-dosimetry of population of radioactive contaminated territories of Bryansk oblast. Reprint of WHO/EOS/94.12, Geneva, 1994
[6] J. Aldrich and B. Pass, Determing radiation exposure from nuclear accidents and atomic tests using dental enamel. *Health Physics,* **54** (1988) 469-471.

International intercomparison of dose measurements using EPR spectrometry of tooth enamel

A. WIESER[1], V. CHUMAK[2], I. BAILIFF[3], N. BARAN[4], A. BOUGAI[4], A. BRIK[5],
S. DUBOVSKY[6], V. FININ[7], E. HASKELL[8], R. HAYES[8], A. IVANNIKOV[9], G. KENNER[8],
V. KIRILLOV[6], S. KOLESNIK[4], G. LIIDJA[10], E. LIPPMAA[10], V. MAKSIMENKO[4],
M. MATYASH[5], A. MEIJER[11], V. MINENKO[6], L. PASALSKAYA[2], J. PAST[10],
J. PAVLENKO[2], J. PUSKAR[10], V. RADCHUK[12], O. SCHERBINA[5], S. SHOLOM[2],
V. SKVORTSOV[9], V. StEPANENKO[9], Ü. VAHER[13]

[1] GSF-Forschungszentrum für Umwelt und Gesundheit, Institut für Strahlenschutz,
Postfach 1129, D-85758 Oberschleißheim, Germany
[2] Scintific Centre of Radiation Medicine, Academy of Medical Sciences of Ukraine,
Melnikova str. 53, 252050 Kiev, Ukraine
[3] University of Durham, Old Shire Hall, Durham DH1 3HP, UK
[4] Institute of Semiconductor Physics, National Academy of Sciences of Ukraine,
Pr. Nauky 45, 252650 Kiev, Ukraine
[5] Institute of Geochemistry, Mineralogy and Oreformation of Nat. Acad. of Sci. of Ukraine,
Palladina prosp. 34, 252180 Kiev, Ukraine
[6] Research Institute of Radiation Medicine, Ministry of Health of Byelarus,
Masherova ave. 23, 220600 Minsk, Republic of Byelerus
[7] Byelarus State University, Scariny pr. 4, 220080 Minsk, Republic of Byelerus
[8] University of Utah, 825 North 300 West #107, Salt Lake City, UT 84103, USA
[9] Medical Radiological Research Centre, Russian Academy of Medical Sciences,
Koroleva str. 4, 249020 Obninsk, Russia
[10] Institute of Chemical Physics and Biophysics, Estonian Academy of Sciences,
Ravala puiestee 10, EE-0100 Tallinn, Estonia
[11] Sedish Radiation Protection Institute, 17116 Stockholm, Sweden
[12] Ministry of Chernobyl, M. Raskovoy str. 12, 253167 Kiev, Ukraine
[13] Institute of Experimental Biology, Estonian Academy of Sciences, Harku, EE-3051, Estonia

Abstract. Electron paramagnetic resonance (EPR) dosimetry with teeth is the only
solid state dosimetry method that allows for direct measurement of the individual
dose. It is considered to be a very promising tool for retrospective individual
dosimetry after accidental radioactive releases. It will help to make a reliable
assessment of the radiation risk. A number of laboratories are engaged in
retrospective EPR dosimetry with teeth. There is consequently a need to develop a
programme of intercalibration and intercomparison to check whether the results
produced by different laboratories are either consistent or accurate. The
Commission of the European Communities has initiated the project ECP10 entitled,
Retrospective Dosimetry and Dose reconstruction. Within the joint EU/CIS project
the '1st International Intercomparison of EPR Dosimetry with Teeth' was started in
1994. Nine research laboratories were involved from Germany, Russia, Byelarus,
Ukraine, Estonia and USA.

1. Introduction

Materials of the human body are of special interest for retrospective individual dosimetry after radiation exposure. At present, electron paramagnetic resonance (EPR) dosimetry using teeth is the only solid state dosimetry method which allows for retrospective dose assessment of the individium. Tooth enamel is privileged for dosimetry due to its high content of hydroxyapatite [1]. By irradiation CO_2^--radicals are created out of the CO_3 impurities which are located inside the hydroxyapatite crystals [2]. The resulting EPR absorption of the radicals is proportional to the absorbed dose in tooth enamel. It has the advantage that the EPR signal in the enamel for the accumulated dose is very stable and no corrections need to be performed due to the time between the radiation exposure and the analysis of the enamel. At 25°C a life time of 10^7 years was determined for the CO_2^--radicals [3]. Therefore, EPR dosimetry of tooth enamel is suitable for individual dosimetry over long periods of exposure and for long time after the exposure. It has currently the disadvantage that a tooth must be extracted. First applications of EPR spectrometry of tooth enamel were done for dose evaluation for atomic bomb survivors [4], nuclear workers [5] and residents close to Chernobyl [6].

Many institutes are engaged in producing results of retrospective EPR dose evaluation with teeth. There is a growing interest by other institutes in using this results. No comparisons have been performed to check whether the results produced by different laboratories are either consistent or accurate. Consequently, there is a great need for developing programmes of intercalibration and intercomparison. The Commission of the European Communities has initiated the project ECP10 entitled 'Retrospective Dosimetry and Dose reconstruction'. Within the joint EU/CIS project the '1st International Intercomparison of EPR Dosimetry with Teeth' was started in 1994. Nine research laboratories were involved from Germany, Russia, Byelarus, Ukraine, Estonia and USA. The results of this first international intercomparison are presented in this paper.

EPR dosimetry with tooth enamel includes four main processes: sample preparation, EPR spectra recording, spectra evaluation and finally dose reconstruction. In the 1st intercomparison programme different methods of dose evaluation were applied. The methods are characterised by differences in handling of the main processes. The process of sample preparation was identical for all applied methods. The aim of the intercomparison was the confrontation of results obtained by different dose evaluation techniques. It is considered by the participants as the base for critical analysis and future refinement of techniques. The final goal of this and following intercomparison programmes will be the development of standardised techniques for EPR dosimetry with teeth by the incorporation of the best features selected out of all currently applied methods.

2. Materials and Methods

The participating laboratories were supplied with powdered and homogenised samples of tooth enamel in order to ensure uniform properties of all samples. The samples were all prepared by one laboratory. Then all powder samples were forwarded to the IAEA laboratories. One fifth of the samples was kept unirradiated and the others were irradiated with four different doses in the range between 100 - 1000 mGy. A set of unirradiated and irradiated samples was distributed to each laboratory. The irradiation doses and the code number of the unirradiated samples were unknown to the laboratories. Only information about the approximate range of the applied doses was provided. Each laboratory used its own evaluation method for dose reconstruction. The applied methods are identified by the numbers A - H. Its characteristics are summarised in the tables 1 - 3.

Table 1. Essential EPR parameters used in the different dose evaluation methods

	A	B	C	D	E	F	G	H
Spectrometer	ERS230	Bruker ESP300E	Bruker ESP300E	Varian E-12	n.i.	Bruker ECS106	Bruker ECS106	ERS231
Microwave power, mW	14	10	8	10	5	5	10	10
Scan range, mT	5	8	5	5	12	10	10	5
Modulation amplitude, mT	0.4	0.3	0.5	0.2	0.25	0.25	0.23	n.i.
Conversion time, ms	0.1	164	20.5	n.i.	n.i.	164	164	n.i.
Time constant, ms	n.i.	164	164	300	n.i.	164	328	n.i.
Number of scans	16000	16	10	120	n.i.	16	15	n.i.
Recording time, min	28	45	4.2	120	60	45	45	n.i.

n.i.: no information provided

Table 2. Features of dose reconstruction applied in the different dose evaluation methods

Method	A	B	C	D	E	F	G	H
Determination of calib. factor	calibration factor additive dose to 2 samples	calibration factor additive dose to 1 samples	calibration factor additive dose to 8 samples	calibration factor additive dose to 4 samples	additive dose	additive dose	additive dose	n.i.
Number of add. irradiations	1	n.i.	7	5	2	2	5	n.i.
Radiation source	^{60}Co and ^{137}Cs	^{60}Co	^{60}Co	^{137}Cs	^{60}Co	^{60}Co	^{137}Cs	n.i.
Add. dose, mGy	1000 and 460	n.i.	37, 74, 148, 296, 592, 1184, 2368	200, 700, 1700, 2700, 4700	250, 500	250, 500	steps of 174 and 348	n.i.

n.i.: no information provided

2.1 Samples

The teeth for the intercomparison were taken in the student clinic of Kiev. The samples were represented mainly by wisdom teeth, i.e., the natural background dose for these teeth can be considered to be negligible. In this set of samples the dentine was removed mechanically using a dental steel drill. No chemical treatment was applied during the complete preparation process. The obtained samples of tooth enamel were crushed in an agate mortar to grains of the size 0.1 - 0.25 mm. At this stage, the individual samples were measured to assure the absence of radiation induced signals. Only samples without any signs of radiation signals were included for the intercomparison.

In total 40 teeth were available. The samples of tooth enamel from all teeth were joined and homogenised. The yield of powdered tooth enamel was more than 10 g. The material was divided into 108 samples of 100 ± 1 mg weight, each.

2.2 Irradiation of samples

The samples were irradiated at the IAEA Laboratories in Seibersdorf, Austria, with dose level of 100, 250, 500 and 1000 mGy. The doses up to the 500 mGy level were applied with a ^{137}Cs source with a dose rate of 800 µGy/min. The 1000 mGy level was applied by a ^{60}Co source with a dose rate of 200 mGy/min. The doses were given in terms of air kerma. All samples were irradiated under a 3 mm Plexiglas cover. They were fixed in the beam on the surface of a polystyrene block. No correction for backscatter from this block was applied, because it was considered to be negligible. The air kerma dose rate at the reference point used for irradiation of the samples was determined by a standard ionisation chamber calibrated in terms of air kerma by BIPM and PTB. The irradiations were done with a source detector distance of 100 cm and 85 cm and a field size of 12.5 x 12.5 cm^2 and 10 x 10 cm^2 with the ^{137}Cs and ^{60}Co source, respectively. The standard combined uncertainty of the air kerma is estimated to be smaller than 3% and 1% for ^{137}Cs and ^{60}Co, respectively.

2.3 Evaluation methods

In all evaluation methods X-band EPR spectrometers were used and the spectra were recorded at room temperature. The essential EPR parameters used by the different methods are given in table 1. The following values of the parameters were covered. The applied microwave power was varied from 5 - 14 mW. The spectra were recorded using a magnetic field scan range from 5 - 12 mT with a variation of the modulation amplitude from 0.2 - 0.5 mT. The recording time was in minimum 4.2 minutes and in maximum 120 minutes.

The main features of dose reconstruction applied in the different methods are summarised in table 2. About half of the methods have reconstructed the doses by individual calibration of the samples. Additional radiation doses were applied to the samples in varying steps and dose increments. The originally absorbed dose was obtained by least-squares-fitting of the signal to dose relation to a linear function for each sample. The other half of the methods used a mean signal to dose calibration factor for dose evaluation. The calibration factor was evaluated by the mean of the fitted signal to dose relation of a few selected samples.

The features of the EPR signal evaluation procedures which are applied in the different methods are given in table 3. Reference samples were measured in half of the methods in order to adjust the EPR signal positions obtained from different measurements. In all of the methods a separation of background and radiation induced signal was performed. The reference background signals were measured or modelled by pure Gaussian or combined Gaussian and Lorentzian functions. Real reference background spectra were used in 4 methods. In two of them the signals were adjusted in width, position and intensity to obtain best congruence with

Table 3. Features of EPR signal evaluation applied in the different dose evaluation methods

	A	B	C	D	E	F	G	H
Reference for signal position	CaO	not used	not used	not used	MnO	MnO	MnO	n.i.
Subtracted background	Modelled	Spectrum of child tooth. Variation of position, width and intensity for best visual agreement with actual spectrum.	Spectrum of the sample from the intercomparison with lowest dose.	Spectrum of pig tooth.	Modelled	Modelled	Spectrum of mixed samples of 18-25 years old students. Variation of intensity for best visual agreement with actual spectrum.	n.i.
Evaluation of radiation signal	Modelling with 4 Gaussians and linear baseline functions. Relative signal positions and width are fixed. 5 fit parameters.	Manual measurement of low-field component of radiation signal.	Manual measurement of intensity difference at positions of high and low field extrema of radiation signal.	Fit by variation of position, width and intensity of standard background and radiation signal. 6 fit parameters.	Convolution of spectrum by Gaussian and Lorentzsian functions. n.i. fit parameters	Convolution of spectrum by Gaussian and Lorentzsian functions. n.i. fit parameters	Manual measurement of amplitude of $g_\perp$ component of radiation signal.	n.i.

n.i.: no information provided

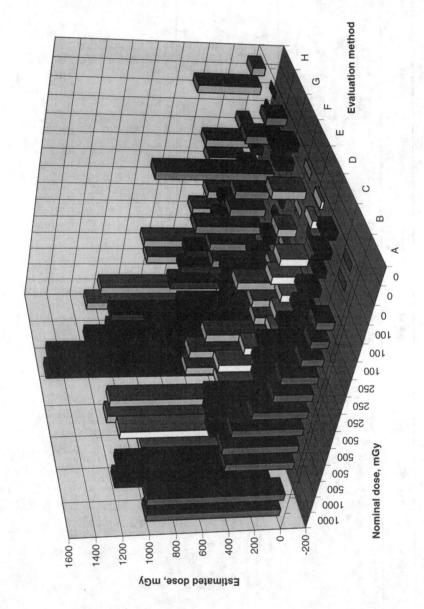

Figure 1. Results of the 1st international EPR intercomparison

the spectrum of the actual sample. In the other two, no or only adjustment of intensity was done. The background spectrum was manually subtracted in three of the methods while in all others subtraction was done by least-squares-fitting procedures. The amplitude of the radiation induced signal was measured manually in three of the methods. It was obtained by fitting to a modelled or reference radiation signal in three and one of the methods, respectively. The reference radiation signal was obtained experimentally by measurement of a 100 Gy irradiated sample. The models used pure Gaussian or combined Gaussian and Lorentzian functions.

3. Results

Each laboratory has reconstructed the absorbed doses on its provided set of samples by applying one evaluation method identified by the numbers A - H. The results of dose reconstruction of this 1st international intercomparison programme are shown in figure 1. For each sample and evaluation method the reconstructed dose is traced versus the nominal dose.

In an integral survey over all methods the reconstructed doses were found to reflect the nominal doses delivered to the samples. However, the precision and accuracy of the results produced by the different methods were found to defer significantly. The dose reconstruction of 1000 mGy and 500 mGy irradiated samples was within ±25% agreement for 6 and 5 methods, respectively. An agreement within ±100 mGy was found for 5 methods at a dose level of 250 mGy. Four of the methods have reconstructed the dose within ±100 mGy for unirradiated and 100 mGy irradiated samples. Non of the applied methods has reconstructed the dose of 100 mGy irradiated samples within ±25%.

4. Conclusions

The presented results of the first international intercomparison programme confirm that EPR spectrometry of tooth enamel is a promising method for retrospective individual dosimetry. It is demonstrated that EPR dosimetry with tooth enamel has the potential to measure absorbed doses as low as 100 mGy. All of the evaluation methods applied in this programme were found to need further refinements for providing precise dosimetry at the 100 mGy dose level.

EPR dosimetry with tooth enamel includes several single procedures with mutual interference. Refinements and optimisation of each procedure require the fixation of all others. The methods applied in this first intercomparison programme were found to be very complex. The obtained results have not the potential to extract in detail the most suitable sub procedures for providing most precise dosimetry. Further intercomparison programmes of the sub procedures are necessary.

Acknowledgements

The authors wish to express their gratitude to Kishor Mehta and Ladislav Czap, IAEA, Vienna, Austria for their kind cooperation in providing the irradiation of the tooth enamel samples. Their support was essential for the successful implementation of the intercomparison programme.

The intercomparison programme was part of the EU/CIS project: Retrospective Dosimetry and Dose Reconstruction. The project was supporeted by the Comission of the European Communities under the contract of COSU-CT93-0051.

References

[1] F.C.M. Driessens, The mineral in bone, dentine and tooth enamel. *Bulletin de la Société Chimique de Belgique* **89** (1980) 663-689.

[2] P.D.W. Moens, F.J. Callens, R.M.H. Verbeeck and D.E. Naessens, An EPR spectrum decomposition study of precipitated carbonated apatites (NCAP) dried at 25°C: Adsorption of molecules from atmosphere on the apatite powders. *Appl. Radiat. Isot.* **44** (1993) 279-285.

[3] H.P. Schwarcz, ESR studies of tooth enamel. *Nucl. Tracks* **10** (1985) 865-867.

[4] M. Ikeya, J. Miyajima and S. Okajima, ESR dosimetry for atomic bomb survivors using shell buttons and tooth enamel. *Jpn. J. Appl. Phys.* **23** (1984) 697-699.

[5] A.A. Romanyukha, D. Regulla, E. Vasilenko and A. Wieser, South Ural nuclear workers: Comparison of individual doses from retrospective EPR dosimetry and operational personal monitoring. *Appl. Radiat. Isot.* **45** (1994) 1195-1199.

[6] H. Ishii, M. Ikeya and M. Okano, ESR dosimetry of teeth of residents close to Chernobyl reactor accident. *J. Nucl. Sci. Technol.* **27** (1990) 1153-1155.

Retrospective Dosimetry by Chromosomal Analysis

D.C. Lloyd[1], A.A. Edwards[1] A.V. Sevan'kaev[2], M. Bauchinger[3], H. Braselmann[3],
V. Georgiadou-Schumacher[3], K. Salassidis[3], F. Darroudi[4], A.T. Natarajan[4],
M. van der Berg[4], R. Fedortseva[5], Z. Fomina[6], N.A. Maznik[7], S. Melnov[8],
F. Palitti[9], G. Pantelias[10], M. Pilinskaya[11], I.E. Vorobtsova[12]

1. National Radiological Protection Board, Chilton, UK
2. Medical Radiological Research Centre, Obninsk, Russia
3. GSF, Institut für Strahlenbiologie, Neuherberg, Germany
4. University of Leiden, Netherlands
5. Centre of Ecological Medicine, St. Petersburg, Russia
6. Institute of Inherited and In-born Diseases, Minsk, Belarus
7. Institute of Medical Radiology, Kharkov, Ukraine
8. Institute of Radiation Medicine, Minsk, Belarus
9. University of Tuscia, Italy
10. National Centre for Scientific Research 'Democritos' Athens, Greece
11. Centre of Radiation Medicine, Kiev, Ukraine
12. Research Institute of Roentgenology and Radiology, St. Petersburg, Russia

Abstract. The joint EU/CIS project ECP-6, was set up to examine whether cytogenetic dosimetry is possible for persons irradiated years previously at Chernobyl. The paper describes the possibility of achieving this by the examination of blood lymphocytes for unstable and stable chromosome aberrations; dicentrics and translocations. Emphasis was placed on the relatively new fluorescence *in situ* hybridization (FISH) method for rapid screening for stable translocations. In a collaborative experiment *in vitro* dose response calibration curves for dicentrics and FISH were produced with gamma radiation over the range 0-1.0 Gy. A pilot study of about 60 liquidators with registered doses ranging from 0-300 mSv was undertaken to determine whether the chromosomal methods may verify the recorded doses. It was concluded that the dicentric is no longer valid as a measured endpoint. Translocations may be used to verify early dosimetry carried out on highly irradiated persons. For the vast majority of lesser exposed subjects FISH is impractical as an individual dosemeter; it may have some value for comparing groups of subjects.

1.   Introduction

Since the 1960s biological dosimetry by the study of chromosomal aberrations in blood lymphocytes has been used to assess persons over-exposed to ionising radiation. The method has been developed into a routine procedure in a number of centres worldwide where it performs a useful role in radiological protection programmes. Traditionally the method has used the dicentric aberration as the indicator of exposure and the dose to an individual is determined by referring the aberration yield to an appropriate calibration curve produced *in vitro*. An account of this method is given by IAEA [1]. The dicentric method works well when a) exposure is acute, ie. delivered within about 30 minutes; b) the irradiation is uniform over the whole body; and c) the blood sample is taken soon after exposure. Departures from this ideal situation increase the uncertainties on dose estimates, such that eventually the method may no longer be reliable as a dosemeter. Experimental studies have shown how one may take into account the non-uniformity of exposure [2] and the protraction of exposure [3] but there is greater uncertainty still associated with delayed sampling.

2.   The problem

The aberrations are induced in the peripheral lymphocytes from which they are sampled and used as an indicator of dose to all cells in the body. The principal problem is that there is a natural turnover of mature lymphocytes with new cells being formed by mitotic division of stem cells, notably within the bone marrow. The dicentric, together with its accompanying acentric fragment, are sometimes termed unstable aberrations, which means that they represent such gross changes in chromosomal morphology that a dividing cell which contains them is unlikely successfully to complete division to produce viable daughter cells. Therefore, in the normal replacement of blood lymphocytes, the newly formed cells comprise a population from which unstable damage has been selectively eliminated with time. The result is that blood samples taken late after exposure will exhibit a time dependent reduction in the dicentric yield and hence that their frequency will underestimate the magnitude of the dose originally received. For any individual we have no means of determining the rate at which he or she has replaced lymphocytes. It is acknowledged that there are likely to be inherent individual variations in this process. Moreover, there may be perturbations in the underlying rate of replacement caused by the radiation itself, particularly if the exposure is sufficiently high as to cause deterministic effects.

3.   The solutions

Two approaches have been employed. These are 1) to continue to use dicentrics and nevertheless make some allowance for their disappearance, acknowledging that this introduces uncertainties which are usually unquantifiable; and 2) to analyse lymphocytes for stable chromosomal aberrations, which are technically much more difficult to assay, and to assume that their level in the cells is constant with time. This means that as fast as cells containing them are eliminated from the circulation, they are replaced by cells bearing similar aberrations which have successfully negotiated stem cell division and maturation with no selection.

3.1   Dicentrics

The PHA responsive T-cells, which are the types used for dosimetry, fortunately seem to have a sufficiently long life-span such that delays of the order of a few weeks generally do not compromise early biological dosimetry based on the dicentric. This assumption may fail in the case of very high exposures when one may detect rapid changes of blood cells counts. Here, more prompt sampling is desirable and this is usually performed because it is apparent that a high dose has been received. This was certainly the case at the Chernobyl accident but, by contrast, at Goiania [4] there was some delay before the authorities became aware that a radiological incident had occurred. At Chernobyl those subjects such as fire fighters and reactor crew who exhibited acute radiation syndrome were quickly evacuated to hospital where their medical surveillance included blood sampling for cytogenetics. The results of this early dosimetry proved to be the most reliable method of quantifying their doses, giving generally credible values when set against other clinical parameters. Biological dosimetry for this important group of persons was a valuable aid in the management of these patients and in the more scientific evaluation of their exposures. Unfortunately, there has not been a case by case account published of these patients including the individual chromosomal data but the cytogenetic work performed has been summarised and reviewed [5,6].

Since 1986 a number of CIS laboratories have carried out dicentric analyses on less heavily irradiated subjects such as liquidators and inhabitants of contaminated territories [eg. 7,8]. These studies have tended to commence at times, as discussed above, where some reduction in the initial dicentric yield would be expected. More protracted exposures received by these persons can be accommodated by considering that their induced aberration yields conform to the linear $y = c + \alpha D$ model of dose response [1]. Their exposures from both external penetrating gamma radiation and internally incorporated radionuclides have generally resulted in more or less whole body exposures. For the internal emitters this is because much of the dose was due to isotopes of caesium which are widely distributed around the body.

Sevan'kaev et al [7] made a generalised assumption of a single exponential decline in dicentrics with a three-year half life. When applied to several cohorts of liquidators, classified by year of working at Chernobyl, they found that the resultant mean dose for each year group matched fairly well with the mean doses that had been ascribed for all liquidators in each of these years. It was noted however that liquidators who had not been issued with a personal film badge and hence must be assumed to have less reliable recorded doses tended to have higher dicentric yields than their badged colleagues. This was particularly evident for liquidators who worked in the first three months after the accident where personal physical dosimetry data are less complete or reliable. These persons probably comprise the most highly exposed of all liquidators and the time-adjusted chromosomal data suggested an average dose for them of about 300 mGy. Sevan'kaev et al acknowledged however that the three-year half life, which had been suggested previously [1], is a value for which there is no strong scientific foundation. For example, it has been suggested that seven years may be a more realistic value [9]. Other workers have questioned this single exponential decay suggesting more complex models. Some examples are two exponential decay terms [10,11], a time hyperbolic function [12] or even a shoulder with the onset of decline not commencing until about a year after exposure [13].

Another means of calculation by which the observed yield of unstable aberrations can be used to estimate dose long after exposure is the Qdr method [14]. This considers the yield of dicentrics and rings only in those cells which contain any type of unstable aberration and assumes that these cells were present at the time of irradiation. It is therefore a dose dependent parameter which is independent of time delay. The method can only be applied to persons for whom an appreciable number of cells with unstable damage are still observed. Thus, in practice it seems to give reasonable estimates for persons who experienced high doses. Salassidis et al [15] have used this method successfully on a group of Chernobyl victims who were selected for treatment of radiation-induced skin reactions and obtained dose estimates in the approximate range 1-6 Gy. These were in good agreement with values estimated by parallel analysis of stable aberrations using the FISH method (see later).

It is certain that now ten years after the accident the simple and traditional method of analysis for dicentric aberrations is invalid as a means of estimating dose, not only for persons exposed briefly in 1986, but also for long term continuous exposure spanning the intervening period. In future, any biological dosimetry using chromosomal damage as the endpoint must rely on analysis for stable aberrations. However, there exists a large amount of dicentric data that were accumulated by many CIS laboratories in the few years post-1986 and these data are potentially very useful to the Chernobyl follow-up. When the ECP-6 group of laboratories was formed it was realised that several of the collaborating CIS partners did not possess

Table 1

In vitro dicentric dose response curves for cobalt-60, 0-1.0 Gy

Lab No.[*]	$C \pm SE$ $\times 10^{-4}$	$\alpha \pm SE$ $\times 10^{-2} Gy^{-1}$	$\beta \pm SE$ $\times 10^{-2} Gy^{-2}$	chi^2	DF	P
1	5.57 ± 7.56	1.33 ± 1.30	7.45 ± 2.04	6.47	3	0.09
2	5.76 ± 6.98	1.32 ± 1.24	9.22 ± 1.98	5.32	3	0.15
5	6.91 ± 7.26	-0.45 ± 0.88	9.58 ± 1.43	2.62	3	0.45
6	5.85 ± 5.78	4.56 ± 1.22	5.45 ± 1.79	2.46	3	0.48
7	8.61 ± 6.43	3.08 ± 1.11	6.27 ± 1.60	1.62	3	0.66
8	4.91 ± 5.00	5.77 ± 1.31	2.07 ± 1.90	1.85	3	0.60
11	9.56 ± 13.75	1.17 ± 2.03	6.60 ± 3.17	12.72	3	0.0005
12	5.06 ± 5.69	3.92 ± 1.26	8.91 ± 1.98	3.39	3	0.34

*As numbered in the authors list

appropriate *in vitro* dose response curves to evaluate their data and for those that did, there had been no systematic attempt at inter-laboratory comparisons. It is well established that variability between laboratories exists particularly in microscope scoring and for that reason the recommendation has been made that laboratories undertaking biological dosimetry should develop their own *in vitro* calibration data [1]. Accordingly, within ECP-6, one laboratory undertook to irradiate blood *in vitro* from two donors with cobalt-60 gamma radiation to a number of doses in the range 0-1.0 Gy, to culture the lymphocytes by a standard method and to supply replicate coded slides to the other participating labs in the CIS for scoring. A full account detailing all the results and statistical analyses of intercomparisons is given in the final report of the ECP-6 group [16]. The data were collated and statistically evaluated by one laboratory. Table 1 shows the yield coefficients and the parameters of fit obtained for the dicentric results for each laboratory fitted to the linear quadratic model of dose response by the iteratively reweighted least squares method using well verified software. It is evident from Table 1 that there is a considerable variability between the laboratories and this is a quality control issue that the respective groups will have to address for the future evaluation of their accumulated dicentric data.

3.2   Stable translocations

These may be detected by banding methods although to use this assay as a biological dosemeter is prohibitively time consuming because of the need to examine cells by full karyotyping for random non-constitutional changes. As with the dicentric analysis, several hundred cells would need to be analysed per individual to detect doses less than 1.0 Gy with confidence limits sufficiently small to be useful. Even with the assistance of semi-automated karyotyping systems this is still a daunting task.

The fluorescence *in situ* hybridization (FISH) method provides the possibility for more rapid screening of metaphases for translocations. The technique and its possible application as a dosemeter has been reviewed comprehensively elsewhere [17] There are considerations of cost and technically demanding protocols to produce the fluorescent images, but more importantly the assay, being only recently developed, still needs validation as a biological dosemeter. Its limits of applicability are not yet properly defined. Some limitation at low doses is imposed by the higher background of stable translocations, particularly in older adults [18], compared with the more transient dicentric. Additionally, the procedure, as currently used, specifically highlights a few pairs of chromosomes and this limits the analysis such that generally about only 30% of all translocations present will be detected. If one needs to intercompare or pool data where different chromosome probe mixtures have been used, then genome equivalence calculations have to be made. These are based on the formula developed by Lucas et al [19] which makes the simplifying assumption that the relative involvement of the different chromosomes in translocation processes is a function of their DNA contents. Whilst useful as a first approximation, there is growing evidence that the situation is more complex [20].

Another important requirement for the use of FISH as a retrospective dosemeter is that the induced translocations do persist over many years. How is stability of aberrations to be judged? Clonal expansion may occur but in principle this may be detected and allowed for [21]. More importantly however, translocation yields should not decline with time [10]. As yet there are no data where prompt and then later FISH analyses have been carried out and very few studies of prompt dicentric yields compared with later FISH. Lucas et al [22] have published data for one case involving an accidental intake of tritiated water (~0.4 Gy) showing a good correlation between dicentrics and later FISH. Salassidis et al [15] have shown, using the skin burn subjects, that at higher doses FISH generally gave dose estimates comparable with those derived by the Qdr method. Natarajan et al [23] have presented data for Goiania victims exposed to <1.0 Gy showing that retrospective FISH translocation yields give comparable results to dicentric analyses carried out within a month of the accident. However, for persons exposed to >1.0 Gy the translocation yields declined with time and so that a time correction would need to be applied.

Partial body irradiation is an important consideration for the unchanging persistence of translocations. It is stated by Pyatkin et al [5] that the highly exposed subjects at Chernobyl generally experienced homogeneous penetrating radiation exposure. Localised skin lesions, as suffered by the patients studied by Salassidis et al [15], might point to partial body irradiation but if these lesions were due primarily to beta radiation, this would not be experienced by the marrow stem cells. An over-dispersed distribution amongst the residual dicentrics was noted but this might be a consequence of lymphocyte kinetics rather than an initial inhomogeneity of penetrating radiation. Following uniform exposure of stem cells the random distribution of lesions would mean that the formation of dicentrics and translocations is independent. Thus, the subsequent selection against cells carrying dicentrics would not affect the yield of translocations in cells passing into the circulation. If however there is non-uniform exposure, as was generally experienced at Goiania where local gamma radiation burns were induced, then a non-uniformity of marrow exposure would cause a correlation between the distribution of cells bearing stable and unstable damage. Accordingly, the subsequent elimination of dicentrics would also result in a reduction in translocations. This is clearly an area where more data need to be developed.

In the ECP-6 programme material from the *in vitro* blood irradiations with cobalt-60 described above, was also used to investigate the dose response for FISH translocations over the range 0-1.0 Gy. The reader is again referred to the ECP-6 final report [16] for the basic data and statistical evaluations. So far, data from only three CIS laboratories are available and as different chromosome probe mixtures were used a combined dose response was derived with genome equivalence. This was based on the scoring of reciprocal and terminal translocations. Table 2 shows the pooled dose response from the three laboratories and for comparison, data from Lucas et al [24] and Bauchinger et al [25] are shown. The former group used cobalt-60 but only considered reciprocal translocations, whilst the latter used caesium-137 and reported complete and incomplete symmetrical translocations. In order to

Table 2

In vitro FISH dose response curves for γ radiation, 0-1.0 Gy
adjusted for full genome equivalence

Laboratory	C ± SE x 10^{-3}	α ± SE x 10^{-2}Gy^{-1}	β ± SE x 10^{-2}Gy^{-2}	chi^2	DF	P
2 + 7 + 12[*]	5.56 ± 2.37	4.05 ± 2.34	6.87 ± 2.99	5.65	3	0.13
Lucas et al 1995	4.68 ± 0.77	2.62 ± 1.45	6.13 ± 2.74	0.17	2	0.92
Bauchinger et al 1993	5.05 ± 1.21	3.70 ± 2.04	2.53 ± 3.18	0.03	2	0.98

[*]As numbered in the authors list

make the comparison, only the published data up to 1.0 Gy and adjusted for genome equivalence, were fitted to the linear quadratic model. In Table 2 it may be seen that the fitted control values are very similar for all three curves. The α coefficient of yields are not statistically different because they carry rather large standard errors but the value from Lucas et al being the lowest, might reflect that only reciprocal translocations were scored. With doses only up to 1.0 Gy the β coefficient also carries large uncertainties. In the context of the Chernobyl accident where dosimetry for persons probably exposed to <1.0 Gy is required, it is clear that considerably more *in vitro* data are required to establish firmly the dose response.

The ECP-6 group undertook a pilot study of about 60 liquidators, who had worked in 1986 or 1987, to assess the possibility of applying retrospective dosimetry on blood samples taken in 1994. These subjects were chosen as a stratified random sample representative of the majority of liquidators in that their recorded doses were mostly in the range 0-300 mSv. We are indebted to colleagues in ECP-7 for identifying suitable subjects from the Obninsk

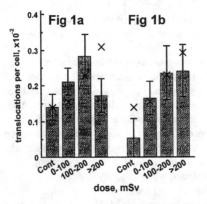
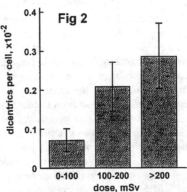

Fig 1.
    Observed FISH translocation yields in liquidators sampled in 1994, having worked in 1986 or 7. Data are pooled into dose cohorts based on their officially registered doses. The crosses are expected yields for the group weighted average doses referred to the published dose response curve [25]. Fig. 1a chromosome probe mixture 1, 4 and 12; Fig. 1b mixture 2, 3 and 8. Data from GSF and Leiden respectively.

Fig. 2.
    Residual dicentric yields in liquidators sampled 7-8 years after exposure. Data are pooled into dose cohorts based on their officially registered doses. Data from Obninsk.

register and applying questionnaires to them. The reader is again referred to the ECP-6 final report [16] for tabulated data on each individual and the evaluation of the results. Figures 1a and b show the results of FISH analyses in two laboratories on lymphocytes from the subjects divided into three dose cohorts based on their officially recorded doses of <100, 100-200 and >200 mSv. Figures 1a and b also show values for local controls produced by those laboratories using the same chromosome probe mixtures. In Fig. 1a there appears to be a slight trend whereby the translocation yield for the dose group 100-200 mSv is higher than that for the lower group or controls. This trend however does not extend to the highest group. Figure 1b shows a progressive increase in yields. However, as shown by the standard errors in Figures 1a and b, the differences between groups fall short of significance and this was confirmed by more rigorous statistical tests on the data from each laboratory and combined.

From the standpoint that the registered doses to liquidators are correct the FISH data have not been able to distinguish between the cohorts. This is despite 50,000 metaphases having been scored. However, it should be borne in mind that there is considerable uncertainty associated with the ascribed individual doses. Thus in the absence of another independent source of evidence it could be equally concluded that the subdivision into the 3 cohorts is inaccurate. Nevertheless, assuming the ascribed doses are reliable an expected yield of translocations indicated by crosses in Figs. 1a and b has been calculated from the local control plus linear yield coefficient for *in vitro* γ radiation published by laboratory No. 3 [25]. In Fig. 1b the points have been corrected for equivalence of the different probe mixtures. Overall there is a good agreement between the observed and expected yields except for the highest cohort in Fig. 1a where, as a group, the observed translocations indicate that the ascribed doses may have been too high. Generally the FISH work has produced results compatible with statements from IAEA that exposure levels for liquidators did not exceed about 250 mSv.

In contrast to the FISH results, Figure 2 shows data on residual dicentrics scored conventionally on lymphocytes from the same blood samples and here around 19,000 cells were examined. An upward trend in aberration yield was found and the yield from the lowest dose group, which is similar to often quoted control levels of ~1 in 1,000 cells, fell significantly below the other two.

4. Conclusions

Now, ten years after the accident, it is not feasible to use the dicentric aberration as a retrospective dosemeter to apply to large numbers of persons thought to have been irradiated to low doses. The FISH method needs to be carefully evaluated with respect to its practical applicability for retrospective dosimetry. An evaluation requires reliable physical dose measurements for comparison and these are not available for the Chernobyl population. A sufficient number of analysed cells is also needed and the present study has not achieved this for individual estimates of dose or for group comparisons of persons officially logged to have been in the range 0-300 mSv. FISH may more readily be applied retrospectively to verify early biological dosimetry performed on highly irradiated persons.

References

[1]   IAEA. Biological Dosimetry: Chromosomal Aberration Analysis for Dose Assessment. Tech. Report No. 260, International Atomic Energy Agency, Vienna, 1986.

[2]   D. Lloyd et al. Biological Dosimetry Applied to *in vitro* Simulated Partial Body Irradiation, Report No. EUR 12558, CEC, Luxembourg, 1991.

[3] M. Bauchinger et al. Calculation of the Dose Rate Dependence of the Dicentric Yield after Co-60 γ-irradiation of Human Lymphocytes. International Journal of Radiation Biology 35 (1979) 229-233.

[4] IAEA. The Radiological Accident in Goiania, International Atomic Energy Agency, Vienna, 1986.

[5] E. Pyatkin et al. Absorbed Dose Estimation According to the Cytogenetic Investigations of Lymphocyte Cultures of Persons who Suffered in the Accident at the Chernobyl Atomic Power Station. Radiation Medicine 4 (1989) 52-57.

[6] A. Sevan'kaev and A. Zhloba. Early and Late Radiation Cytogenetic Effects in Man. Acta. Oncologica 12 (1991) 201-204.

[7] A. Sevan'kaev et al. A Survey of Chromosomal Aberrations in Lymphocytes of Chernobyl Liquidators. Radiation Protection Dosimetry 58 (1995) 85-91.

[8] A. Sevan'kaev et al. Chromosomal Aberrations in Lymphocytes of Residents of Areas Contaminated by Radioactive Discharges from the Chernobyl Accident. Radiation Protection Dosimetry 58 (1995) 247-254.

[9] H. Braselmann et al. Chromosome Aberrations in Nuclear Power Plant Workers: The Influence of Dose Accumulation and Lymphocyte Life-time. Mutation Research 306 (1994) 197-202.

[10] A. Ramalho et al. Lifespan of Human Lymphocytes Estimated during a six year Cytogenetic Follow-up of Individuals Accidentally Exposed in the 1987 Radiological Accident in Brazil. Mutation Research 331 (1995) 47-54.

[11] K. Bogen. Reassessment of Human Peripheral T-Lymphocyte Lifespan Deduced from Cytogenetic and Cytotoxic Effects of Radiation. International Journal of Radiation Biology 64 (1993) 195-204.

[12] M. Bauchinger et al. Time-Effect Relationship of Chromosome Aberrations in Peripheral Lymphocytes after Radiation Therapy for Seminoma. Mutation Research 211 (1989) 265-272.

[13] W. Scheid et al. Chromosome Aberrations Induced in Human Lymphocytes by an X-Radiation Accident: Results of a 4-year Post-irradiation Analysis. International Journal of Radiation Biology 54 (1988) 395-402.

[14] M. Sasaki and H. Miyata. Biological Dosimetry in Atomic Bomb Survivors. Nature (London) 220 (1968) 1189-1193.

[15] K. Salassidis et al. Dicentric and Translocation Analysis for Retrospective Dose Estimation in Humans Exposed to Ionising Radiation during the Chernobyl Nuclear Power Plant Accident. Mutation Research 311 (1994) 39-48.

[16] Biological Dosimetry for Irradiated Persons. EUR Report No. 16532. CEC; Brussels, 1996.

[17] J. Gray et al. Fluorescence in situ Hybridization in Cancer and Radiation Biology. Radiation Research 137 (1994) 275-287.

[18] J. Tucker et al. On the Frequency of Chromosome Exchanges in a Control Population Measured by Chromosome Painting. Mutation Research 313 (1994) 193-202.

[19] J. Lucas et al. Rapid Translocation Frequency Analysis in Humans Decades after Exposure to Ionising Radiation. International Journal of Radiation Biology 62 (1992) 53-63.

[20] S. Knehr et al. Analysis for DNA-proportional Distribution of Radiation-Induced Chromosome Aberrations in Various Triple Combinations of Human Chromosomes Using Fluorescence in situ Hybridization. International Journal of Radiation Biology 65 (1994) 683-690.

[21] K. Salassidis et al. Chromosome Painting in Highly Irradiated Chernobyl Victims: A Follow-up Study to Evaluate the Stability of Symmetrical Translocations and the Influence of Clonal Aberrations for Retrospective Dose Estimation. International Journal of Radiation Biology 68 (1995) 257-262.

[22] J. Lucas et al. The Persistence of Chromosome Translocations in a Radiation Worker Accidentally Exposed to Tritium. Cytogenetics and Cell Genetics 60 (1992) 255-256.

[23] A. Natarajan et al. Biological Dosimetric Studies in the Goiania Radiation Accident. In Proceedings of the International Workshop on the Scientific Basis for Decision Making after Radioactive Contamination of an Urban Environment. IAEA Vienna, in press.

[24] J. Lucas et al. Dose Response Curve for Chromosome Translocations Measured in Human Lymphocytes Exposed to ^{60}Co Gamma Rays. Health Physics 68 (1995) 761-765.

[25] M. Bauchinger et al. Radiation-Induced Chromosome Aberrations Analysed by two-colour Fluorescence *in situ* Hybridization with Composite Whole Chromosome-Specific DNA Probes and a Pancentromeric DNA Probe. International Journal of Radiation Biology 64 (1993) 179-184.

Reconstruction of the External Dose of Evacuees from the Contaminated Areas Based on Simulation Modelling

Reinhard MECKBACH

GSF-Forschungszentrum für Umwelt und Gesundheit, Institut für Strahlenschutz, Postfach 1129, D-85758 Oberschleissheim, Germany.

Vadim V. CHUMAK

Ukrainian Scientific Center of Radiation Medicine, Department of Dosimetry and Radiation Hygiene, 252050 Kiev-50, Ukraine.

Abstract. Model calculations are being performed for the reconstruction of individual external gamma doses of population evacuated during the Chernobyl accident from the city of Pripjat and other settlements of the 30-km zone. The models are based on sets of dose rate measurements performed during the accident, on individual behaviour histories of more than 30,000 evacuees obtained by questionnaire survey and on location factors determined for characteristic housing buildings. Location factors were calculated by Monte Carlo simulations of photon transport for a typical housing block and village houses. Stochastic models for individual external dose reconstruction are described. Using Monte Carlo methods, frequency distributions representing the uncertainty of doses are calculated from an assessment of the uncertainty of the data. The determination of dose rate distributions in Pripjat is discussed. Exemplary results for individual external doses are presented.

1. Introduction

The Chernobyl accident led to heavy radioactive contaminations in the nearby city of Pripjat and in other settlements of the 30 km zone, with the potential to deliver significant radiation exposures to the affected population prior to its evacuation. A detailed reconstruction of individual radiation doses of the evacuees is needed for epidemiological investigations that could lead to an increased knowledge of the relationship of exposure and the somatic effects of radiation.

It can be expected that the external gamma component contributed substantially to the radiation exposure of the evacuated population. In the time between accident and evacuation, gamma dose rate measurements were conducted at different points of the 30 km zone, particularly in the city of Pripjat. This data, together with individual behaviour data obtained by a wide scale public questionnaire survey performed three years after the accident, allows for a reconstruction of external doses of the population. Using deterministic models, a first assessment of individual external doses to more than 30,000 evacuees has been performed [1]. The mean external effective dose of Pripjat residents, evacuated about 36 hours after the accident, is estimated to be 11.5 mSv. The largest part of the population of the 30-km zone was evacuated after eight to ten days, and a mean effective dose of 18.2 mSv was determined, with 644 persons having received external doses higher than 100 mSv. The results revealed large variations between external doses of different individuals, caused by the heterogeneity of the dose rate pattern and the variability of individual behaviour, showing the importance of performing individual external dose reconstruction.

In the present contribution, developments and results of a second stage of individual external dose reconstruction of the evacuated population are reported. Stochastic models are developed, which allow for the determination of uncertainty ranges and distributions of the calculated individual external exposures from an assessment of the uncertainty of the data, using Monte Carlo methods [2] for calculating the propagation of the uncertainties. Location factors for a specific typical housing block of Pripjat and for village houses were computed by Monte Carlo simulations of photon transport for different source energies, various configurations of radioactive clouds, specific deposition areas of the radionuclides and indoor contaminations, for numerous positions inside and outside of the buildings. The results allow to determine location factors for specific radiation scenarios and to quantify the influence of the uncertainty of source configuration, spectral energy distribution of the emitted radiation and position of the individual in the building on the uncertainty of the location factors.

From the data set of dose rate measurements in Pripjat, dose rates relative to reference area were determined from information on the environment of each measurement point and uncertainties due to measurement errors and environment were assessed. Using the Monte Carlo sampling program PRISM [3] and incorporating kriging methods [4] for interpolations in space, uncertainty distributions for dose rates relative to reference area averaged over different sectors in Pripjat were obtained. Finally, exemplary results for individual external doses were calculated.

2. Overview of data

Direct dose rate measurements in Pripjat started two hours after the accident in the nuclear power plant at 1:26 a.m. on April 26, 1986 and lasted until after the completion of the evacuation two days later. The location of the 31 measurement sites in Pripjat and dose rates measured 30 hours after the beginning of the accident are shown in Figure 1. On the average, the measurements were made in intervals of about 3.5 hours, using Geiger Mueller counters at 1 m above ground. The dose rate distribution over the city of Pripjat was highly heterogeneous, with the highest levels in the hospital area in the south-eastern part of the city. Residential areas in the north-western part were affected to a much smaller extent. A detailed description of the radiation situation in Pripjat after the accident is given in [1].

In the 30 km zone, dose rate measurements began on April 26 at 12 settlements close to the nuclear power plant, and were continued on April 28 at 23 measuring sites and the next day at 55 sites. After May 1, nearly daily measurements were made at a total of 84 observation points in settlements. The spatial distribution of the measured dose rates was highly heterogeneous, the dose rates at some of the more strongly affected villages was more than a factor of 100 larger than at other areas of the territory. Also the time dependence of the measured dose rates showed an irregular behaviour, reflecting the various releases of radionuclides and changes in wind directions.

Information on the individual pre-evacuation behaviour of a large fraction of the population was obtained by a wide scale questionnaire public survey launched in 1988 and performed during routine medical examinations. After the data was checked for consistency, records of 16,193 individuals evacuated from Pripjat and of 19,605 from the settlements of the 30 km zone were entered in a computer data base.

For the survey of evacuees from Pripjat, the city was divided, as indicated in Figure 1, into eight sectors according to the open air doses accumulated in the 36 hour period until evacuation. In a questionnaire, the evacuees were asked to reconstruct their location (indoor or outdoor, and in which sector) hour by hour from the time of the accident until their evacuation. Furthermore, the storey of their dwelling was recorded. In a similar questionnaire,

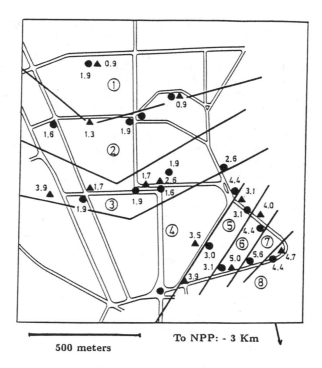

Figure 1. Plan of the city of Pripjat showing the points where dose rate measurements were performed by two independent teams (triangles respectively circles). Also indicated are dose rates in mGy/h measured 30 hours after the accident and the subdivision of the city in sectors for the questionnaire survey.

the evacuees from settlements of the 30 km zone were asked to state in which kind of house they lived (brick house, wooden house or multi-storey building) and to remember, day by day, in which settlement they were staying and the fraction of time they had spent outdoors.

3. Calculation of location factors

Location factor is defined as the ratio of dose rate at a particular indoor or outdoor location to a reference dose rate. For radiation from radioactive clouds, the dose rate 1 m above an infinite smooth air-ground interface irradiated by the same cloud configuration is taken as reference dose rate. For radiation from deposited radionuclides, it is given by the dose rate over an infinite smooth lawn under the same deposition conditions as the other urban surfaces [5]. Location factors for different building types have been reported in the literature (see [5] and references therein). However, in order to account for construction characteristics of Ukrainian buildings and the radiation situation during the accident and to allow for an estimation of uncertainty and variability, location factors were calculated by Monte Carlo simulations of photon transport using the code SAM-CE [6] for a specific five storey building and for two types of rural houses typical for the 30-km zone.

The information available on the radionuclide composition of the radiation sources and the source configurations on the first days after the accident is limited and uncertain. Therefore, separate Monte Carlo calculations were made for a range of source energies and for various definite source configurations. Then, combining the results of these calculations, location factors for different scenarios of radioactive contaminations can be determined. Location factors were calculated for source energies of 80 keV, 140 keV, 365 keV, 662 keV and 1.6 MeV. Simulations were made for radiation from an homogeneous semi-infinite radioactive cloud and from finite clouds with different orientations relative to the buildings. Separate calculations were made for different urban areas on which the radionuclides were deposited, such as roofs, walls, windows, ground, trees and indoor surfaces, allowing to determine location factors for different relative contaminations of these areas.

The simulated five storey building is very typical for Pripjat, several nearly identical buildings of this type are located in the residential area in the southern sector of Pripjat, an area relatively more strongly exposed to radiation after the accident. A complete description of the building was available and an edge section was simulated in full detail, including the urban environment with neighbouring buildings, streets, lawns and trees. In order to determine the variability of the location factors at different positions, a total of 200 indoor detection volumes and several detection volumes at outdoor positions were defined.

In Table 1, examples of location factors calculated for the building in Pripjat are presented for a source energy of 662 keV and different source configurations. Finite clouds were simulated at a distance of 200 m parallel to the long side of the building. For radionuclides deposited outdoors, a dry deposition was assumed with source strengths relative to the source strength on lawns of 10% for streets, walls, windows and balconies, 30% for roofs, 80% for the ground, taken to extend to a distance of 500 m, and a factor of 5 for trees.

Table 1 Location factors for different storeys of a housing building in Pripjat, for radiation from radionuclides deposited outdoors, from an homogeneous semi-infinite cloud and for two finite cloud configurations. Also indicated are the 5 and 95 percentiles of the distribution of location factors resulting from the different positions in the respective storey. The source energy is 662 keV.

| Location | Source | | | |
	Deposited radionuclides	Semi-infinite cloud	Finite cloud sideways	Finite cloud above
Ground floor	0.046	0.057	0.050	0.025
	0.016 - 0.10	0.021 - 0.11	0.004 - 0.16	0.005 - 0.07
Second floor	0.035	0.061	0.058	0.023
	0.012 - 0.08	0.023 - 0.12	0.005 - 0.16	0.005 - 0.06
Fourth floor	0.017	0.066	0.063	0.040
	0.005 - 0.04	0.025 - 0.13	0.007 - 0.17	0.016 - 0.08

It can be seen that the location factors depend, in general, on the source configuration. At the fourth floor the location factor for a semi-infinite cloud is more than three times larger than for radionuclides deposited outdoors. They also depend on the floor, particularly for radiation from radionuclides deposited outdoors, with the uppermost floor shielded more than two times better than the ground floor. The location factors within one floor can vary by up to an order of magnitude, and even more for finite cloud configurations. Location factors for indoor contaminations are around 0.022 for a relative source strength of the indoor surfaces of 2% and become as large as 0.11 if one assumes a relative indoor contamination of 10%. Similar results were also obtained for the other source energies considered. One finds that for the lowest source energy of 80 keV the location factors are generally about 35% lower and for the highest energy of 1.6 MeV about 35% higher than for 662 keV. For indoor contamination the dependence on energy is less than 10%.

The results provide an data base which allows to determine location factors and their uncertainty ranges appropriate for the radiation situation in Pripjat during the accident. An analysis shows that the lack of detailed knowledge on the spectral energy distribution of the emitted radiation does not contribute very strongly to the location factor uncertainty. The dependence of location factors on the storey of buildings is opposite for radiation from clouds and from radionuclides deposited outdoors, and in a mixed radiation situation the dependence on the storey becomes less important. Important sources of uncertainty are the lack of knowledge on indoor contaminations, the variability of building characteristics and the dependence on the position in a particular storey of the building, particularly for radiation from finite cloud configurations.

Effective location factors are established by accounting also for the movements of individuals at indoor respectively outdoor locations. From a first analysis, for Pripjat an effective location factor of 0.05 with an uncertainty range with a 5 percentile of 0.01 and a 95 percentile of 0.2 was obtained for indoor locations and a factor of 0.8 with an uncertainty range between 0.5 and 1 for outdoor locations.

4. Stochastic models

The stochastic models developed for the calculation of the individual external exposures of the evacuated population are based on the individual behaviour obtained by the questionnaire survey. They are designed with a modular structure, separating the calculation of parameters which do not depend on the individual behaviour from the final computation of the individual external dose. The main parameters used are the dose rates for the corresponding positions, expressed with respect to a reference area and the location factors estimated for the respective indoor and outdoor locations.

The uncertainties of the model parameters are represented by frequency distributions of the parameter values. These uncertainty distributions are computed from the assessed uncertainty of the data and from the results of the calculations of location factors. Monte Carlo methods, coded in the program PRISM [3], are used to sample values from the parameter uncertainty distributions. With these values, the external dose for each individual is repeatedly (for example, 500 times) calculated, obtaining a frequency distribution representing its uncertainty.

The individual external effective dose distributions E for the population evacuated from Pripjat are calculated according to

$$E = C \cdot \sum_{h=1}^{h_{evac}} \overset{.}{D}(h,s) \cdot \Delta t \cdot G(h,s) \cdot L(h)$$

with $\Delta t = 1$ hour, where the summation extends over each hour h, starting from the time of the accident to the time h_{evac} the individual was evacuated, and

C   is the conversion factor from absorbed dose in air to effective dose, appropriate to the individual according to its age;

$\overset{.}{D}(h,s)$ is the uncertainty distribution of dose rate with respect to the reference area, for the hour h and the sector s of Pripjat (see Fig. 1) in which the individual was staying at this hour;

G(h,s) is a distribution representing the uncertainty due to the variation of the dose rate over the sector s the individual was at the hour h;

L(h)   is the current distribution for the location factor. Depending where the individual had been at the hour h, it is taken to be equal to the location factor distribution for the particular storey of his home housing building, for a specific other building (like school, hospital), for unspecified other buildings or for outdoor locations.

For the population evacuated from the 30 km zone, the individual external effective dose distributions are calculated according to

$$E = C \cdot \sum_{d=1}^{d_{evac}} \overline{D}(d,v) \cdot \{G(d,v) \cdot L_{out}(d) \cdot t_{out}(d) + K(d,v) \cdot L_{in}(d) \cdot [24 - t_{out}(d)]\}$$

where the summation extends over each day d, starting from the day of the accident to the day d_{evac} the individual was evacuated, and

C   is the conversion factor from absorbed dose in air to effective dose, appropriate to the individual according to its age;

$\overline{D}(d,v)$ is the uncertainty distribution of dose rate with respect to the reference area, averaged over one day, for the day d and for the particular village v of the 30 km zone in which the individual was staying at this day;

G(d,v) is a distribution representing the uncertainty due to the variation of the dose rate over the area the individual moved around outdoors, for the day d and the village v;

$L_{out}(d)$ is the distribution for the outdoor location factor for each day, according to the radiation source configuration assumed for this day;

$t_{out}(d)$ is the uncertainty distribution for the time spent outdoors by the individual on day d;

K(d,v) is a distribution representing the uncertainty due to the variation of the dose rate at the position of the particular house in the settlement in which the individual was staying with respect to the point where the dose rate was measured;

$L_{in}(d)$ is the distribution for the indoor location factor for each day, for the specific home house type of the individual respectively for unspecified other buildings, for the radiation source configuration assumed for this day;

The uncertainty distributions for the parameters of the models are computed in separate modules and then used to compute the uncertainty distributions for the individual external doses, based on the individual histories. In this way, improvements in the modules, leading to a reduction of the uncertainties, can be made without altering the overall structure of the models.

5. Modelling of dose rate distributions in Pripjat

For the calculation of individual external doses of evacuees from Pripjat, dose rates relative to reference area averaged over each sector of the city and frequency distributions representing their uncertainty need to be determined for each hour after the accident from the dose rate measurement data and an assessment of measurement errors and the uncertainty due to the environment of the measurement point. The steps involved can be detailed as follows:

1. Assessment of the uncertainty of the dose rate measurements due to instrumental, calibration and recording errors. Separate uncertainty distributions are assessed for statistical and systematic errors.
2. Assessment of the location factors for the different measuring points and estimation of their uncertainty distributions. Dose rates can be modified significantly by the urban environment due to shielding by neighbouring buildings and different relative contaminations of the urban surfaces. In Pripjat, the measurement points were known precisely and their respective environments surveyed, allowing for an assessment of the corresponding location factors and an estimation of their uncertainty distributions.
3. By Monte Carlo sampling from these distributions with the code PRISM, uncertainty distributions of dose rates relative to reference area are obtained for each measurement point and time.

4. Interpolation in time of the dose rate distributions by linear interpolation. The uncertainty due to the lack of information on the temporal dose rate behaviour between measurement times is accounted for by sampling for these times from appropriate distributions representing this uncertainty.
5. Kriging interpolation [4] combined with Monte Carlo sampling are used for interpolations in space in order to obtain dose rates at positions where no measurements had been made and uncertainty distributions for dose rate averaged over the respective sectors of Pripjat.

In simple terms, kriging interpolation [4] is a means of estimation of interpolated values in which the interpolation weights are determined from information on the correlation of the measured values in dependence of the distance between the measurement points. In the framework of a probabilistic model, the interpolation weights are determined by the condition that the expectation value of the estimation error is zero and its variance minimal. In this sense, the method is unbiased and optimal. It automatically incorporates declustering for inhomogeneous distributions of measurement points and allows for an estimation of the interpolation error. Results of kriging interpolations for dose rates relative to reference area in Pripjat are shown in Figure 2.

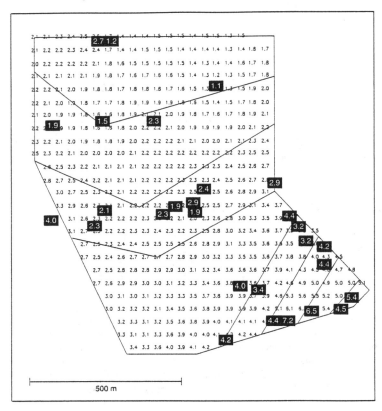

Figure 2. Dose rates relative to reference area in mGy/h 30 hours after the accident at the measurement points in Pripjat (numbers in black fields) and interpolated values on a grid obtained by kriging interpolations. Also indicated are the different sectors in Pripjat (compare Figure 1).

Uncertainty distributions for dose rates relative to reference area averaged over the respective sector were calculated for each of the first 50 hours after the accident. Furthermore, for each hour the variability of dose rate relative to reference area over each sector was assessed from the interpolated values. As an example, the dose rate distribution for sector 3 of Pripjat 30 hours after the accident is shown in Figure 3. For this example, the uncertainty distribution for the averaged dose rate was combined with the variability distribution over the sector. The results obtained provide the set of dose rate parameter distributions needed for the individual external dose reconstruction model for Pripjat discussed in the previous section.

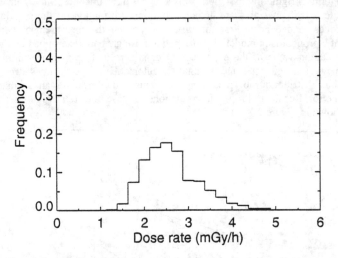

Figure 3. Frequency distribution of dose rate relative to reference area in sector 3 of Pripjat (compare Figure 1), 30 hours after the accident. The distribution reflects the uncertainty of the average dose rate of the sector and the variability of dose rate within the sector.

6. Examples results for individual external doses

Based on the distributions of dose rates relative to reference area determined for Pripjat and the assessed location factors, individual effective external doses of 12,653 evacuees from this city are being calculated. In a first step, deterministic calculations were performed with this data, obtaining a mean external effective dose for this population of 11 mSv, which is only slightly lower than the value of 11.5 mSv reported in [1]. The resulting frequency distribution of individual external effective doses of the population considered is shown in Figure 4.

The calculation of uncertainty distributions for individual external doses of population evacuated from Pripjat is being performed. On the other hand, preliminary results of uncertainty distributions of individual external doses have been obtained for individuals evacuated from the village of Paryshev, one of the settlements of the 30-km zone, situated 18 km to the south-east of the power plant and having a population of 678 inhabitants, of which 335 were surveyed. The village was evacuated eight days after the accident.

As an illustration, the frequency distribution of external effective dose calculated for a female worker from this village, who lived in a wooden house and stayed mostly outdoors during the day is shown in Figure 5. The uncertainty distribution is close to lognormal, with a mean value of 7.34 mSv, a 95% percentile of 17.2 mSv, a geometric mean of 6.04 mSv and a geometric standard deviation of 1.85. In this case, uncertainties in the assessment of dose rates were the main source of uncertainty. Such frequency distributions of individual external doses were computed for the whole surveyed population of the village.

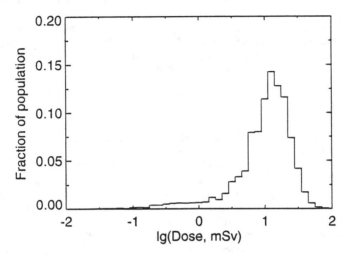

Figure 4. Relative frequency distribution of individual external effective doses for 12,653 evacuees from Pripjat

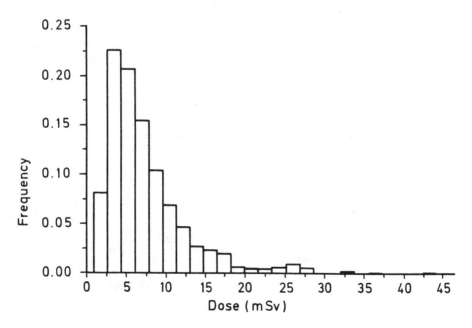

Figure 5. Frequency distribution of external effective dose calculated for an individual from the village of Paryshev in the 30 km zone.

The result is only of an exemplary nature, as it is based on preliminary assessments of the uncertainty distributions for the corresponding dose rates. For the settlements of the 30-km zone, interpolations in time over two to three days are necessary for some measuring points, particularly for the time span of the first days after the accident. Geostatistical conditional simulation methods [4] allow to simulate the time dependence of the dose rate between two measurement times, using information on the covariance of the dose rate between different times. These methods are being combined with the Monte Carlo sampling calculations with the PRISM code in order to obtain more realistic uncertainty distributions for time averaged dose rate distributions for the settlements of the 30-km zone.

8. Conclusions

In this study, the development of stochastic models for the calculation of individual external doses of population evacuated from the contaminated territories after the Chernobyl accident has been discussed. It was shown that Monte Carlo sampling methods can be used for determining frequency distributions representing the uncertainty of dose quantities. The models were designed with a modular structure; distributions for dose rates and location factors, which do not depend on the individual histories, are calculated separately and then used as parameters for the calculation of individual external doses.

Location factors were calculated by Monte Carlo simulations, allowing to estimate effective location factors and their uncertainty ranges to be used for the population of Pripjat. It was shown how dose rates relative to reference area averaged over sectors in Pripjat are determined from the measurement data and how distributions reflecting their respective uncertainty and variability are determined by Monte Carlo sampling incorporating kriging interpolation procedures.

Using these results, individual external doses were calculated for evacuees from Pripjat. A mean external dose of 11 mSv was obtained, which agrees closely with the result that had been reported from a first stage of dose reconstruction [1]. The corresponding uncertainty distributions are being calculated.

References

[1] I.A. Likhtarev, V.V. Chumack and V.S. Repin, Retrospective Reconstruction of Individual and Collective External Gamma Doses of Population Evacuated after the Chernobyl Accident, *Health Phys.* **66** (1994) 643-652.
[2] K.A. Rose, E.P. Smith, R.H. Gardner, A.L. Brenkert and S.M. Bartell, Parameter Sensitivities, Monte Carlo Filtering, and Model Forecasting under Uncertainty. *Journal of Forecasting* **10** (1991) 117-133.
[3] R.H. Gardner, D. Rojder and U. Bergström, PRISM: A Systematic Method for Determining the Effect of Parameter Uncertainties on Model Predictions. Tech. NW-83/555. Studsvik Energietek AB, Nykopping, Sweden, 1983.
[4] A.G. Journel and Ch.J. Huijbregts, Mining Geostatistics. Academic Press, London, 1978.
[5] R.Meckbach and P. Jacob, Gamma Exposures due to Radionuclides Deposited in Urban Environments. Part II: Location Factors for Different Deposition Patterns, *Radiat. Prot. Dosim.* **25** (1988) 181-190.
[6] H. Lichtenstein, M. Cohen, H. Steinberg, E. Troubetzkoy and M. Beer, The SAM-CE Monte Carlo System for Radiation Transport and Criticallity Calculations in Complex Configurations (Revision 7.0.). A Computer Code Manual (Mathematical Application Group Inc., 3 Westchester Plaza, Elmsfort, NY10523, USA), 1979.

The use of luminescence techniques with ceramic materials for retrospective dosimetry

I.K. Bailiff

Luminescence Dosimetry Laboratory, Environmental Research Centre, University of Durham, South Road, Durham DH1 3LE, UK.

Abstract. Luminescence techniques are being used with ceramic materials to provide evaluations of integrated external gamma dose for dose reconstruction in populated areas contaminated by Chernobyl fallout. A range of suitable ceramics can be found associated with buildings: on the exterior surfaces (tiles), within walls (bricks) and within the interiors (porcelain fittings and tiles). Dose evaluations obtained using such samples provide information concerning the time-averaged incident gamma radiation field, average shielding factors and, with the aid of computational modelling techniques, dose estimates at external reference positions.

1. Introduction

Previous dose reconstruction studies in Japan [1] and the USA [2] have demonstrated the value of using thermoluminescence (TL) techniques with ceramic materials for the retrospective measurement of gamma dose many years after the event. Similar promise has been demonstrated in a preliminary study of the use of ceramic materials from the Chernobyl region [3,4].

Since the Chernobyl accident the need to apply a range of techniques for dose reconstruction has become increasingly evident because of the complexity and scale of the fallout and the sporadic nature of monitoring procedures, particularly in populated rural regions. One of the elements within the project ECP10, *Retrospective Dosimetry and Dose Reconstruction*, has been to further develop the application of luminescence techniques. The experimental studies have been combined with the use of computational modelling to aid the deconvolution of dose evaluations to ceramics and to test predicted variations of accrued dose with location. This paper discusses current progress in development of the experimental approach including, the assessment of the types of sample available in contaminated regions of Belarus, Russia and Ukraine, the levels of transient fall-out dose which can be measured and the relationship between dose in ceramic to dose at an external reference point.

2. Scope for sampling

A range of contaminated regions has been examined in programme of field work, including: Pripyat and settlements within the Exclusion Zone, Southern Belarus and the Kaluga and Briansk regions of Russia. While Pripyat has served as a study site for methodological development because of its proximity to the power plant and the extensive scientific work which has been undertaken in the vicinity of the town, the examination of more distant populated (and evacuated) settlements has provided important information concerning the nature of the fall-out, the type of buildings and potential ceramic samples.

2.1 The availability of ceramic samples

For retrospective dosimetry measurements, ceramic samples from fixed locations are sought from the exterior surfaces of buildings, from interior shielded locations and ideally throughout the walls (Fig. 1).

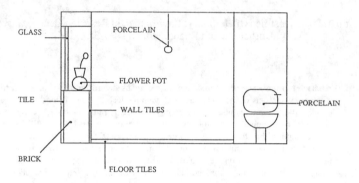

Figure 1
Location of potential ceramic samples

The construction of dwellings and buildings differ widely; in urban areas they may be of concrete with additional ceramic fittings/decorations. Most urban buildings of the 1970's, particularly multiple storey apartment blocks of the type found in Pripyat, however, are constructed of concrete. Although concrete is not, so far, suitable for luminescence dosimetry, glazed terracotta or earthenware decorative tiles fixed to the exterior walls are suitable and have been used in this study. In Zone 1 of Pripyat (the most heavily contaminated region; see [3]) there is a small number of substantial brick buildings with outer tile cladding in Pripyat which provide the opportunity to extract samples at a number of depths, although it is to be noted that mortar is unavoidably present in a region of substantial absorption for external gamma rays (Fig. 2).

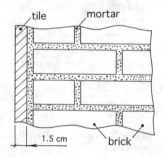

Figure 2. Location of mortar in brick wall cross-section

The evacuated and populated settlements surveyed for potential samples present a more diverse range of dwellings in age (up to ~50 years) and style. Older traditional houses are constructed of timber frame and board with plastered interiors and have a substantial brick oven with chimney stack. One or two courses of external bricks are often used as part of the foundations between the timber building and an underlying a concrete plinth; however where they are close to soil, adsorption of contamination is likely to have occurred. For more recent houses concrete bricks have been frequently used. As in Pripyat, the use of porcelain fittings and components is commonplace.

Within the interiors of buildings, high fired terracotta floor, wall and skirting tiles and porcelain sanitary and electrical fittings provide a ubiquitous source of samples, present in most rooms and available in abundance within bathrooms. Porcelain which is also available in exterior locations (usually light fittings and as electrical insulators) has an important advantage over exposed brick because the glazed surface of porcelain inhibits the ingress of fall-out products.

The measurement of accrued dose for interior, shielded and exterior samples provides information on the time-averaged shielding factors. The more difficult experimental tasks lie with samples where the transient dose is comparable with the accrued natural background dose, such as interior samples and in areas of low contamination. Given the wide availability of porcelain in rural areas in external positions and within the interior, work so far has concentrated on developing the use of porcelain for dose evaluations.

Where buildings are of brick construction, the opportunity to obtain samples of ceramic throughout the wall thickness allows the determination of dose as a function of depth (referred to here as a dose-depth characteristic). Additionally it provides dose evaluations for exterior and interior positions of the wall, indicating the effective shielding factor. Such dose profiles can provide information on the nature of the (time-averaged) incident gamma radiation field which is related to the form of the profile. In a series of experiments with laboratory photon Cs and Co sources the feasibility of obtaining dose-depth curve experimentally using ceramic material has been demonstrated (Section 3.1.1).

The investigation within ECP10 has focused on aspects which are concerned with matching capability of the method to dosimetric requirement, including:

i) The dose response characteristics of ceramic samples, in particular the minimum dose which can be resolved. There is little previous experience in the use of ceramics of the type found in the contaminated areas. Thus basic characterisation needs to be performed (e.g. linearity of response with dose, changes of sensitivity with secondary laboratory heating etc.).

ii) The location of ceramics and their suitability for providing appropriate information for use of computational modelling has been used to identify the position and orientation which could be used to differentiate different source configurations.

iii) The level of accumulated natural background dose within samples at the time of measurement.

iv) The accuracy with which the transient dose can be determined.

Table 1 Sample types

Fired clay or composites:	Brick; terracotta tiles; glazed wall and floor tiles; cladding tile; skirting tile; flower pots.
Porcelain:	Electrical insulators including, switches, lamp holders, domestic and industrial fuses, power insulators.

3. Experimental methods

The dose evaluated by luminescence techniques, D_L, corresponds to the sum of dose arising from: a) naturally occurring radionuclides within ceramic materials and the surrounding medium and b) transient fall-out contamination. In the same way that TL dosimetry phosphors can be used to register the administration of absorbed dose and, at some later time, to yield a quantitative measure of that dose, minerals present in ceramics such as quartz and feldspar can perform the same task, albeit at generally higher dose levels. The transient gamma dose arising from exposure due to an accident, D_X, determined by luminescence techniques, is given by

$$D_X = D_L - A(\dot{D}_\alpha + \dot{D}_\beta + \dot{D}_\gamma + \dot{D}_c) \qquad (1)$$

where, D_L = accrued dose determined by luminescence measurements (TL or OSL); A = sample age in years; $\dot{D}_\alpha$, $\dot{D}_\beta$, $\dot{D}_\gamma$, $\dot{D}_c$ = effective annual alpha, beta, gamma and cosmic dose, respectively, due to natural sources of radioactivity.

Several luminescence (using either thermoluminescence, TL, or optically stimulated luminescence, OSL) techniques are available to evaluate D_L. Apart from the means of stimulating luminescence, they are differentiated by the mineral composition of the sample extracted for measurement, the grain size, the procedures employed to determine the accrued dose and, most pertinent to dosimetry, their sensitivity. These aspects and the various components of the natural background dose-rate are determined using direct and indirect analytical techniques. The details of the experimental work are given in a companion paper [5]. Three of the techniques used are referred to as the *fine-grain*, *quartz inclusion* and *pre-dose* techniques which were originally developed for dating archaeological ceramics [6]. The pre-dose technique [7] generally offers the highest sensitivity and has also been applied to porcelain from interior locations. Porcelain is usually prepared as cut slices rather than crushed and sized material in the case of bricks and tiles.

During the measurement of TL, the release of the trapped charge carriers and consequent emission of luminescence is achieved using thermal stimulation; for OSL measurements the release of trapped charge is achieved by stimulation using light within particular wavelength regions. Under constant illumination of a sample and using suitable optical filters in the detection system to reject the stimulating light, an OSL decay curve is obtained. The wavelengths used for stimulation differ according to the mineral(s) within the sample; in the case of quartz green light (e.g. 514 nm, Ar ion laser) is used for stimulation and for feldspars it has been found that infra-red wavelengths are also suitable, enabling simpler light sources to be used. Under infra-red stimulation the emission is referred to as infra-red stimulated luminescence (IRSL). Procedures employing optical stimulation are a recent introduction to the field and as such are under development [8-11].

For samples receiving little or no transient fall-out dose, D_L is substantially due to natural radiation dose. The annual dose in equation 1 arises from the decay of the ^{238}U, ^{232}Th series and ^{40}K within the sample and the surrounding medium. For typical brick structures the accrued dose due to natural sources of radiation is expected to be in the region of 2.5 - 3.5 mGy/a with contributions of approximately 60%, 25% and 5% for beta, gamma and cosmic components respectively (using inclusions, where the internal radioactivity and external-grain alpha dose contribution is assumed to be negligible). Thus the accrued dose due to natural sources of radiation is expected to be roughly 30 mGy per decade - this forms the baseline above which further transient dose contributions are made. In areas of low fallout or in heavily shielded locations the uncertainty associated with D_X will be strongly influenced by the uncertainty associated with the determination of the age and annual dose-rate. Consequently for heavily shielded interior locations, both high luminescence sensitivity and accurate annual dose evaluations are sought. The precision (and accuracy) of D_X achieved will depend on an assessment of the particular material and location tested. A full appraisal of the levels of minimum resolvable transit dose in areas of low fall-out/high shielding awaits more extensive testing.

3.1 Sampling and testing

One of the potential rôles for luminescence is the provision of benchmark dose evaluations which can be used in conjunction with computational modelling techniques for dose reconstruction. In populated areas the nature and pattern of the fall-out is generally complex due to the variability of the weather, building structures, clean-up operations, resuspension etc. Multiple sampling is consequently desirable to test for variability and also multiple testing by means of inter-laboratory comparisons to evaluate consistency. As a first step towards this objective sites have been sought where samples are readily available and where extensive monitoring has been performed since the accident to allow comparisons of integrated dose obtained by luminescence and by computational modelling.

Field samples from Pripyat and Southern Belarus have been examined to: i) test samples from a common location in Pripyat by different laboratories using both TL and OSL techniques; ii) measure dose as a function of depth for a roof-top location and at ground level for a house in a rural district.

Evaluations of dose were obtained at the site of three apartment blocks constructed of concrete in Pripyat which have been the subject of modelling calculations [12] . Samples of external glazed tile (at ca 1m above ground level) and porcelain from interior locations were analysed [5] from the apartment blocks 8, 18 and 28 in Lenin Prospect. The mean values of the accrued dose obtained by four laboratories using TL and OSL techniques for the exterior samples (as part of the first stage of the intercomparison completed during 1995) ranged from 1.5±0.2 Gy (SD 4) to 2.0±0.3 Gy (SD 4) for four locations. Porcelain samples from a selection of locations within rooms on each floor of one of the apartment blocks were also tested, from which a mean transient dose of 100±10 mGy (SD 9) was evaluated for all locations. The ratio of external to interior transient dose (0.06±0.01) is consistent with that expected on the basis of shielding calculations [13]. A detailed calculation taking into account each sampled location is now needed to test whether the experimental results are consistent with expectation on the basis of modelling. Nonetheless, the broad agreement in the overall shielding factors is very encouraging, given that the level of integrated dose for the interior is within the lower dose range of the method.

3.1.1 Dose depth profiles

The measurement of accrued dose as a function of depth is a potentially powerful tool in employing luminescence in the field. While, as discussed below, dose determinations produced for use in dose reconstruction are likely to be based on the outer 1 cm or less of an external ceramic, the measurement of accrued dose at greater depths in the absorber medium provides verification of the presence of an external photon field in the past and is directly related to the time-averaged energy spectrum of the external radiation (assuming that any bremsstrahlung generated by external beta radiation forms a minor component of the accrued dose).

The dose depth characteristic requires either cores or whole bricks to be extracted in the field. During recent fieldwork, the use of a portable percussive drill with chisel bit has been found to be convenient to remove mortar and extract the whole brick. This usually restricts sampling to one brick depth. Several experimental approaches have been used to determine the accrued dose as a function of depth in samples: i) cutting thin slices; ii) drilling with a fine masonry drill; iii) OSL scanning of the cut surface of the ceramic [5].

As part of the evaluation of the use of different experimental procedures to determine the dose depth characteristic in ceramic [5, 14, 15], measurements have been performed with tile stacks (Fig. 3a) irradiated with either Cs or Co photon beams in the laboratory. Good agreement has been obtained (Fig. 3b) between the experimental and calculated dose-depth characteristics [16].

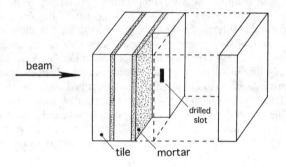

Figure 3a

Modern terracotta tiles (each of 25 mm thickness) were assembled in stacks of varying depth and irradiated (dose at surface ~2 Gy) with ^{137}Cs and ^{60}Co photon beam facilities at GSF Neuherberg. IRSL was used for measurements, giving stronger emission than TL.

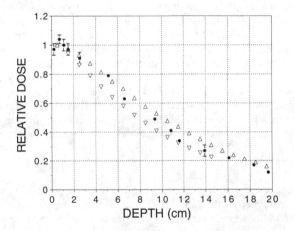

Figure 3b

Open triangles: Monte Carlo calculation of relative kerma in terracotta tile due to parallel irradiation with a Cs photon beam. The average kerma is calculated for a volume element (1cm thick x 2 cm x 2 cm) at various depths located: 1) along the central axis of the tile stack (upper symbols) and 2) along an axis parallel to the central axis located such that the volume element is located at the (same) corner of each tile (lower symbols). Filled circles: Measured absorbed dose vs depth in stack measured using IRSL. Calculation by R. Meckbach, GSF.

A dose depth characteristic obtained for a sample of brick from house in the village of Masani (north of reactor on the Belarus border) is shown in Fig. 4 - the reduction in dose with depth is consistent with an external isotropic photon field dominated by the presence of Cs. Although there are limitations to what may be interpreted concerning the specific composition of the time-averaged field [15], it provides confirmation of the energy range of the incident field. Such confirmation is important in the subsequent use of the luminescence data, as discussed further below.

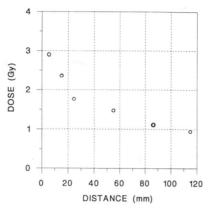

Figure 4
Dose-depth profile in brick: Masani, Belarus.

2.2.1 Roof of Polyclinic, Pripyat

The variation of dose with height was tested using samples from the upper regions of the 4-storey building in Pripyat. In the absence of full scaffolding or special working platforms, the extraction of bricks from the walls of upper levels clear of the roof was not possible. Dose evaluations for the bricks obtained from the roof (using red TL emission as discussed in [14]) indicate a substantial contribution from contamination on the roof (Fig. 5), while the dose in the outer brick is significantly lower. These results illustrate the care needed in selecting samples - for such elevations, samples from an exterior face more distant from the roofline are required to avoid the contribution of the (intermittent) roof contamination .

5. Deployment of luminescence results

Dose evaluations performed with ceramics (or other natural materials) taken from fixed structures provide determinations of the time integrated dose for the sampled location. Thus no temporal resolution is possible unless the sample had been removed from the contaminated area at a known time. A procedure of deconvolution is required to relate the dose in ceramic to dose in air at a reference point. and this can be achieved with the aid of computational modelling techniques. The luminescence determinations are consequently not entirely independent of modelling, but an analysis of uncertainty and sensitivity to variation is being developed in the computational analysis as part of ongoing work .

Calculations by Meckbach [17] have been made to relate dose in ceramic at various locations on buildings to dose at the reference position for: a) a semi-infinite cloud source and b) a semi-infinite ground source based on a source energy of 662 keV (Fig. 6). Sampling at Pripyat has provided the opportunity to investigate the predicted variations of dose with elevation (multiple storey buildings), orientation and source type. The dose to the first 2 cm of external ceramic of a wall at 1 m from the ground is calculated to be about half that at the same height at a reference position. This provides a convenient 'ready reckoner' (i.e. doubling the dose evaluation) in preparing luminescence measurements for use in dose reconstruction. At greater depths in the ceramic, a greater dependency on the energy of the incident radiation [17] and the source configuration is predicted. Thus the dose-depth characteristics perform an important role in providing confirmation of the assumptions made concerning the incident radiation when relating the dose in ceramic to the dose at the reference position.

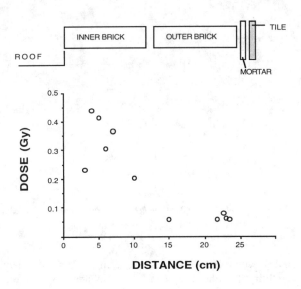

Figure 5
Accrued dose, evaluated by TL, as a function of depth in bricks located on the roof of the polyclinic, Pripyat. The upper schematic diagram indicates the location of the bricks and tiles relative to the (flat) roof.

To seek evidence to test this, however, we have seen that measurements on rooftops need to be considered with some care, since the flat bitumen roofs of multiple storey apartment blocks in Prypiat have a complex history of retention and loss of contaminants, as discussed above.

The comparison of dose estimates based on modelling calculations and experimental dose determinations using luminescence is one of the objectives of this project. In Pripyat there are extensive monitoring data which relate to the period shortly after the accident until evacuation and following re-occupation during 1987. So far there appear to be important gaps in data during the first year following the accident which will introduce additional uncertainties in calculating a value for the integrated dose since the accident at a location/area sampled for luminescence measurements. Such estimates based on integration and including an analysis of uncertainty are not trivial and will form part of future work.

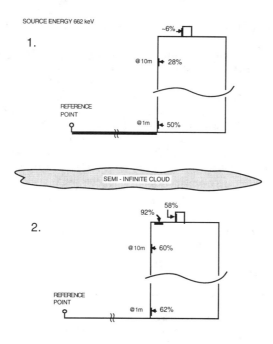

SOURCE ENERGY 662 keV

1.

~6%

@10m 28%

REFERENCE
POINT

@1m 50%

SEMI - INFINITE CLOUD

2.

92% 58%

@10m 60%

REFERENCE
POINT

@1m 62%

Figure 6

The calculated ratio of dose in ceramic (0-2 cm depth) at the locations indicated to dose in air at the reference point for two source configurations (1) distributed on the ground and (2) a homogeneous radioactive cloud. The source energy is 662 keV.

6. Capability and future directions

Ceramic materials, including porcelain, terracotta and bricks, have been sampled from populated areas contaminated by Chernobyl fall-out and, in general, have been found suitable for retrospective dose evaluation. The existing techniques are capable of measuring total integrated dose from ~10 mGy (at best) to well beyond 10 Gy. Substantial improvement of sensitivity has not been a major issue in this project, although it is of importance in testing heavily shielded samples. The main body of developmental work lies in the appropriate deployment of the method and the optimization of the particular techniques.

The use of porcelain inside buildings provides a widely available dosemeter material for measurements in shielded locations and its glazed surface also provides the advantage of low fall-out retention when used in exterior locations. The use of brick from walls to obtain a dose-depth characteristic provides important information concerning the nature of the time-averaged incident field. New luminescence techniques, based on optical stimulation, are promising and may offer advantages in routine dose evaluations and in the evaluation of dose-depth profiles using cut brick sections [15].

The development of computational methods has provided an important adjunct to the experimental determinations enabling i) luminescence dose evaluations in ceramic to be related to reference points; ii) dose-depth profiles to be interpreted in terms of incident radiation; iii) the variation of dose with location and orientation of external samples with source configuration.

This project has enabled some of the key factors in deploying luminescence for retrospective dosimetry to be put into place and the potential for the method is clearly established in this work. An important aspect for future use of the method is in bringing it from a research base into routine use since a large number of determinations will be required for each settlement selected for study. There is much work to be done in settlements, in examining levels of resolvable transient dose, the degree of spatial variation and in seeking correlations with dose estimates based on other data such as soil contamination and monitoring, if available. The establishment of experimental capability in 'novice' laboratories within the CIS has commenced under ECP10 and their full development will be an essential part of completing this work.

References

[1] Maruyama, T., Kumamoto, K., Ichikawa, Y., Nagatomo, T., Hoshi, M., Haskell, E. and Kaipa, P. (1987) Thermoluminescence Measurements of gamma rays. In US-Japan Joint Reassessment of Atomic Bomb Radiation Dosimetry in Hiroshima and Nagasaki. Final Report (Ed. W.C. Roesch) Vol. 1, The Radiation Effects Research Foundation, 143-184.

[2] Haskell, E. H., Bailiff, I. K., Kenner, G. H., Kaipa, P. L. and Wrenn, M. E. (1994) Thermoluminescence measurements of gamma-ray doses attributable to fallout from the Nevada Test Site using building bricks as natural dosemeters. *Health Physics* **66**, 380-391.

[3] Hütt G., Brodski L., Bailiff I.K., Göksu Y., Haskell, E., Jungner H. and Stoneham D.(1993) TL accident dosimetry using environmental material collected from regions downwind of Chernobyl: a preliminary evaluation. *Radiat. Prot. Dosim.* **47**, 307-311.

[4] Vischnevekii I. N., Drozd I. P., Koval, G. N., Fominych V. I., Baran N. P., Bartchuk V. I., Bugal, A. A., Maksimenko M. and Baryachtar V. G. (1993) The use of quartz inclusion thermoluminescence for the retrospective dosimetry of the Chernobyl area. *Radiat. Prot. Dosim.* **47**, 305-306.

[5] Stoneham, D., Bailiff, I.K., Bøtter-Jensen, L., Göksu, H.Y., Jungner, H. and Petrov, S. Retrospective Dosimetry: the development of an experimental methodology using luminescence techniques. These proceedings.

[6] Aitken M. J. (1985) *Thermoluminescence Dating.* Academic Press, London.

[7] Bailiff I. K. (1994) The pre-dose technique. *Radiation Measurements* **23**, 471-479.

[8] Godfrey-Smith, D. I. and Haskell, E. H. (1993) Application of optically stimulated luminescence to the dosimetry of recent radiation events involving low total absorbed doses. *Health Physics* **65**, 396-404.

[9] Poolton N. R. J., Bøtter-Jensen, L. and Jungner H. (1994) An optically stimulated luminescence study of porcelain related to radiation dosimetry. *Radiation Measurements* **24**, 543-550.

[10] Göksu, H.Y., Bailiff, I.K., Bøtter-Jensen, L., Hütt, G, and Stoneham, D. (1995) Inter-laboratory beta source calibration using TL and OSL with natural quartz. *Radiation Measurement s* **24**, 479-484.

[11] Göksu, H.Y., Wieser, A., Stoneham, D., Bailiff, I.K. and Figel, M. EPR, OSL and TL spectral studies of porcelain. Int. Symposium on EPR Dosimetry and Applications, 15-19 May 1995, Munich. Accepted for publication.

[12] Meckbach R. and Jacob P. (1988) Gamma exposures due to radionuclides deposited in urban environments. Part II: Location factors for different deposition patterns. *Radiat. Prot. Dosim.* **25**, 181-190.

[13] R. Meckbach, private communication.

[14] Bailiff, I.K. (1995) The use of ceramics for retrospective dosimetry in the Chernobyl Exclusion Zone. *Radiation Measurements* **24**, 507-512.

[15] Bøtter-Jensen L., Jungner H. and Poolton N. R. J. (1994) A continuous OSL scanning method for analysis of radiation depth dose profiles in bricks. *Radiation Measurements* **24**, 525-530.

[16] Meckbach, R. Bailiff, I.K. Göksu, H.Y., Jacob, P. and Stoneham, D. Calculation and measurement of dose-depth distribution in bricks. 11th Int. Symposium on Solid-state Dosimetry, Budapest 8-11 July, 1995. Submitted for publication.

[17] Final Report, Experimental Collaboration Project ECP10, Retrospective Dosimetry and Dose Reconstruction. CEC/CIS Chernobyl Collaborative Programme.

WHAT IS DESIRABLE AND FEASIBLE IN DOSE RECONSTRUCTION FOR APPLICATION IN EPIDEMIOLOGICAL STUDIES?

André Bouville
National Cancer Institute, Bethesda, Maryland 20892, USA

Lynn Anspaugh
Lawrence Livermore National Laboratory, 7000 East Ave., Livermore, CA 94550, USA

Gilbert W. Beebe
National Cancer Institute, Bethesda, Maryland 20892, USA

Abstract. Formal epidemiologic studies are intended to increase scientific knowledge about the quantitative risk that is associated with radiation exposure. Dosimetric data are needed for such studies.
What dosimetric data are desirable? Doses are needed for a large number of people with a large gradation of radiation exposures in order to ensure a sufficient power for the epidemiological study. The characteristics of the desirable doses are, in some respects, different from those calculated for radiation protection purposes. The desirable data are: (1) absorbed doses to the individual organs or tissues of interest, instead of effective doses; (2) absorbed doses delivered over limited time periods, instead of committed doses; (3) doses specific to the individuals that are subjects in the epidemiological studies, instead of average doses over population groups; and (4) very accurate and precise doses.
What dosimetric data are feasible? Most of the characteristics of the desirable dosimetric data are usually achievable. However, uncertainties can be fairly large and estimated with a large degree of subjectivity. Also, for practical reasons, it may not be feasible to estimate individual doses for all subjects.

1. Introduction.

Epidemiological studies of populations are of two general forms, monitoring or formal, and serve several possible purposes. Monitoring studies inform members of potentially affected population groups of the nature and magnitude of the risks that might have been imposed on them. Formal epidemiological studies can increase scientific knowledge about the quantitative risk that attends exposure. Risks of human health due to radiation exposure are most appropriately estimated by means of formal epidemiological studies.

Dosimetric data are essential for any epidemiological study, but the detail and accuracy needed depend on the purposes to be served. If the need is for a monitoring study, then general information about doses will suffice. However, a formal study that is expected to contribute to scientific information about quantitative radiation risk requires careful individual dose estimation.

This paper is devoted to the discussion of dosimetric data needed for formal epidemiological studies of populations exposed as a result of nuclear power operations. The recommendations made by the National Research Council [1] have largely been followed. The examples used in this paper are relevant to the Chernobyl accident, which caused a large number of people to be exposed at relatively high doses and provided an opportunity for formal epidemiological studies to be initiated. The studies that are singled out are those of thyroid cancer among children who resided in Belarus and in Ukraine at the time of the accident, and those of leukemia among workers involved in the mitigation of the accident and in clean-up operations.

2. What dosimetric data are desirable?

The characteristics of the desirable dosimetric data are presented and discussed.

2.1. Magnitude of the dose and dose distribution

First of all, formal epidemiological studies require a large number of people to be studied, with a large gradation of doses among those people. The potential informativeness of a formal study often is measured in terms of statistical power, which can roughly be defined as the probability of rejecting the hypothesis of no effect (null hypothesis), when in fact it is false and the alternative hypothesis is correct. An example would be that one would conclude that there is no increase rate of disease in the exposed population when, in fact, the radiation exposure does have an effect of the expected magnitude. For the studies considered here, there are two aims that are of primary interest. The first pertains to the probability of detecting a dose-related effect if an effect is present, given the expected size of that effect as derived from the risk estimates that form the basis of radiation protection standards [2-6]. Second, even when a study does not detect an effect, it can yield valuable information if it establishes an upper bound on risk that will increase confidence in the standard radiation risk estimates.

The sample sizes required to achieve adequate statistical power increase greatly as the magnitude of the dose decreases. Under the assumption of linearity between dose and effect, the sample size required to be likely to detect the effect varies roughly as the inverse of the square of the average dose. For example, if it is assumed that each member of a population with an age and sex distribution typical of the United States as a whole was exposed to a specific whole-body dose and followed over a lifetime, the size of the exposed population needed to have an 80% chance of seeing an excess of total cancer mortality against general population rates is estimated to be 20,000 for a whole-body dose of 100 mGy, 220,000 for a dose of 30 mGy, and 2,000,000 for a dose of 10 mGy [1].

Extrapolating these results to higher doses would result in a population size of 2,200 for 30 mGy and of 200 for 1 Gy. However, several caveats should be made regarding these sample size calculations. Normally, one would have a range of doses in the population rather than a uniform dose. Performing a dose-response analysis would create some gain in statistical power (and a corresponding reduction in the required sample size) relative to the simple comparison of the total exposed group to the general population used above; it can be shown that having a modest fraction of more highly exposed persons in the population considered increases the statistical power appreciably and that the greater the spread in dose the greater the increase in statistical power that can be achieved if a dose-response analysis is used [7,8]. Also, the number of persons required is smaller for types of cancer such as leukemia, thyroid and breast cancer if the population considered consists only of children, who are more susceptible to the effects of ionizing radiation than adults for those types of cancer and who have lower baseline rates of disease [7]. On the other hand, several factors would tend to diminish statistical power. First, the uncertainty in estimating individual doses tends to diminish statistical power and increase the required sample size [9,10]. Second, the calculations above assume a full lifetime follow-up, whereas most studies have a much shorter average follow-up. Third, it is generally agreed that beta or gamma irradiation delivered at low doses and low dose rates is less effective, when normalized to unit dose, than the high, acute doses that were used to derive the population sizes given above. Considering all these factors together, the sample sizes given above probably err on the side of underestimating the required sample sizes.

2.2 Type of dose

In contrast to radiation protection purposes, for which the committed effective dose averaged over a population group is often the preferred dose quantity because it provides a measure of the total radiation impact resulting from a given exposure to the population group considered, the type of dose that is desirable for application in formal epidemiological studies is specific of the disease considered, specific of each individual considered in the study, and includes as much information as possible on the physical and temporal characteristics of the dose.

Taking as an example an epidemiologic study of thyroid disease among children exposed as a consequence of the Chernobyl accident, it is only the thyroid dose that is of interest. That thyroid dose should be estimated for all children included in the study. In addition, it is important to separate the contributions to the thyroid dose of external irradiation, of internal irradiation due to short-lived radioiodines (mainly I-132 and I-133) and of internal irradiation due to I-131 because those three components of the thyroid dose may not have the same effectiveness to cause thyroid cancer. Also, even though most of the thyroid dose is expected to have been delivered within a few weeks after the accident, it is helpful to provide information on the protracted dose, due to the long-lived radiocesiums, which have been and will continue to be delivered at low rate, both from external and from internal irradiation.

If leukemia is the disease that is considered, then the bone-marrow dose should be estimated. For members of the public exposed as a result of the Chernobyl accident, it is desirable to estimate separately the contributions to the bone-marrow dose arising from external irradiation and from internal irradiation as those two components are of similar magnitude but with different temporal distributions. The contribution from external irradiation resulted from short-lived radionuclides soon after the accident, whereas it is currently due essentially to the radiocesiums that were deposited on ground surfaces; the contribution from internal irradiation is mainly due to the consumption of foods contaminated with radiocesiums and radiostrontiums. So far, the internal doses from the radiocesiums have been greater, on average, than those from the radiostrontiums, but that trend is expected to be reversed in the future. For workers, however, only the dose from external irradiation would need to be carefully estimated for most individuals, as occupational bone-marrow doses arise for the most part from external irradiation; however, a rough assessment of the bone-marrow dose from internal irradiation would be needed to make sure that it can be neglected.

It is usually sufficient to estimate for all individuals the integrated dose from the beginning of the radiation exposure considered until the time when the disease is diagnosed. However, information on the time dependence of the protracted doses is also desirable.

2.3 Uncertainties attached to the dose estimates

Uncertainties have to be estimated for all individual doses. It is desirable to reduce the uncertainties to the lowest possible level, to separate the random uncertainties and the biases, and to calculate these uncertainties in an objective manner that does not leave any room for subjective judgments. Also, the extent to which these uncertainties are correlated across subjects must be understood. Although complex, statistical methods to account for uncertainties in dose estimates in the analysis of epidemiologic studies have been developed [11-13].

3. What dosimetric data are feasible?

What is feasible in dose reconstruction studies is often much less than what is desirable. At best, individual doses can be estimated with subjective, usually large, uncertainties. Sometimes, it is necessary to resort to group doses because resources are not sufficient to estimate individual doses for all subjects or because the database available does not allow for the identification of individual-specific parameter values.

3.1 Individual doses

In the estimation of individual doses, every effort should be made to use data that are specific of the persons that are considered.

In the case of internal irradiation, the best data are those that are related to the measurement of the contents of radioactive materials in the body. For example, following the Chernobyl accident, several hundreds of thousands of measurements of exposure rates against the neck of individuals were made for the purpose of estimating the I-131 content of the thyroid [14-17]; also, a very large number of whole-body burden measurements have been made in order to evaluate the internal doses due to intake of radiocesiums [18]. Even though the measurements of exposure rates against the neck (usually called "thyroid measurements") were not made in all cases under highly standardized conditions and with the same, uniformly calibrated equipment [12], they provide the best basis from which the thyroid doses can be estimated.

It is important to note that those measurements of radiation originating from the body provide only information on the dose rates received at the time of measurement. Unless the individuals considered were measured many times during the time period when the dose was delivered, the variation of the dose rate as a function of time must be estimated using other sources of information. For example, the additional information needed to reconstruct the thyroid dose resulting from the intake, by inhalation or by ingestion, of I-131 is: (1) the history of I-131 intake by the measured individual both before and after the thyroid measurement, and (2) the metabolic data to permit the conversion from I-131 intake to thyroid dose. Because of the knowledge of the thyroid dose rate at the time of the thyroid measurement, only relative intakes of I-131 from inhalation and from ingestion are needed.

The intake rates of I-131 from inhalation depend on the I-131 concentrations in the air that was breathed and on the breathing rates of the individuals considered. There is very little information on the air concentrations of I-131, which varied both in time and in space. The assumption that has been made so far is that of a single intake during the first day after the accident. Large uncertainties are associated with the estimation of the uncertainties attached to the inhalation doses. However, for the populations of Belarus, Ukraine, and Russia who were affected by the Chernobyl accident, the I-131 intake by inhalation was, in general, much smaller than that from ingestion. Notable exceptions are the early evacuees who inhaled contaminated air before and during evacuation, did not consume contaminated foodstuffs before or during evacuation, and were provided with uncontaminated foodstuffs after evacuation.

Intake of I-131 from ingestion arose from the consumption of contaminated milk and, to a lesser extent, of other contaminated foodstuffs, such as leafy vegetables. Unfortunately, the database on I-131 concentrations on foodstuffs is very limited and has to be complemented by means of environmental transfer models that take into account, among other factors, the influence of the official milk ban in some of the contaminated areas. Information on (1) the consumption rates of the contaminated foodstuffs, (2) the origin of those foodstuffs, and (3) the possible intake of iodine pills or solutions in order to block the thyroid uptake can be obtained by means of personal interviews. It is acknowledged that the data obtained in these interviews are highly uncertain and possibly biased.

Personal metabolic data are also largely unknown in the absence of individual measurements of retention of iodine in the thyroid. However, the use of literature data does not lead to large uncertainties for individuals because of the low range of variability of the effective half-time of retention of I-131 in the thyroid.

In addition to the doses resulting from the intake of I-131, doses arising from the presence of short-lived radioiodines (mainly I-132 and I-133) in the thyroid, from internal irradiation of the thyroid due to the presence of radiocesiums in the body, and from external irradiation originating from the deposition of radioactive materials on the ground and on the clothes, skin, and hair need to be estimated. It is not feasible to estimate the magnitude of these dose contributions for all individuals, as most of the relevant information is not available. These dose components can only be obtained by means of models developed using information from a very small database. Fortunately, the contribution of those dose components to the total thyroid dose is relatively small for most individuals.

In summary, it is clear that, even for individuals with thyroid measurements, the estimation of the thyroid dose over the time period of interest is fraught with large uncertainties, which are very difficult, if not impossible, to quantify in an objective manner.

In the case of external irradiation, individual doses can be recorded by means of personal dosimeters when the radiation exposure occurs, or be derived from biological dosimetry, if the doses are high enough, after the exposure. For example, in the case of the clean-up workers (also called "liquidators") involved in the mitigation of the Chernobyl accident, a measure of the dose from external irradiation is provided from the processing of the badges that they were required to wear. Unfortunately, not all liquidators wore badges during the first few weeks after the accident and the badges worn by highly exposed people during the night of the accident were overexposed. Concern has also been expressed about the validity of some of the recorded doses. There are at least two ways to verify the validity of the recorded doses. One is to perform biological dosimetry, using for example the fluorescent in situ hybridization (FISH) technique to measure the frequency of certain stable chromosome aberrations in blood samples [19] or the Electron Paramagnetic Resonance (EPR) technique to measure the radiation dose accumulated in tooth enamel. These techniques, however, are still in the experimental stage and are not helpful for doses below 0.1-0.2 Gy. The other method that can be used to verify the validity of the recorded doses is to perform detailed time and motion studies, in which the liquidator reconstructs as best as possible where and when he was during the time period he was exposed and the knowledge of the radiation field in various reactor locations allows an estimate of the radiation dose received by the liquidator to be made. Both methods present large uncertainties and may have to be used in conjunction in order to improve the accuracy of the dose estimates.

3.2 Group doses

Although the estimation of individual doses is highly recommended in formal epidemiological studies, it is sometimes necessary to resort to group doses because resources are not sufficient to estimate individual doses for all subjects or because the database available does not allow for the identification of individual-specific parameter values. Group doses are estimates of average doses for population groups presenting similar characteristics. For example, for thyroid studies related to the Chernobyl accident, the same dose could be assigned to children of the same village and the same age group presenting similar dietary and lifestyle habits. Also, liquidators without badges could be assigned the same dose if they were members of a team that carried out a given operation. Uncertainties attached to group doses are, as a rule, larger than those associated with individual doses when they are assigned to individuals in that group.

4. Discussion

Although much could be learned from the Chernobyl accident, it must be emphasized that deriving new scientific knowledge will be difficult. As is the case in most studies of exposed populations around nuclear facilities, the radiation doses received by the individuals exposed cannot be quantified precisely. The dosimetric data that are currently feasible do not present all of the desirable characteristics requested by the epidemiologists. In particular, a substantial effort will have to be made by the dosimetrists in order to provide dose estimates with reasonably low uncertainties.

References

[1]   National Research Council. Radiation Dose Reconstruction for Eoidemiologic Uses. National Academy Press; Washington, D.C.; 1995.

[2]   International Commission on Radiological Protection. 1990 Recommendations of the International Commission on Radiological Protection. Ann. Intl. Comm. Radiol. Protect. 21:1-201; 1991.

[3]   National Research Council, Committee on the Biological Effects of Ionizing Radiations. Health effects of exposure to low levels of ionizing radiation (BEIR V). Washington, D.C.; National Academy Press; 1990.

[4]   United Nations Scientific Committee on the Effects of Atomic Radiation. Sources, effects, and risks of ionizing radiation. UNSCEAR 1988 Report to the General Assembly, with scientific annexes. New York, NY; United Nations; 1988.

[5]   United Nations Scientific Committee on the Effects of Atomic Radiation. Sources and effects of ionizing radiation. UNSCEAR 1993 Report to the General Assembly, with scientific annexes. New York, NY; United Nations; 1993.

[6]   United Nations Scientific Committee on the Effects of Atomic Radiation. Sources and effects of ionizing radiation. UNSCEAR 1994 Report to the General Assembly, with scientific annexes. New York, NY; United Nations; 1994.

[7]   Shore, R.E. Epidemiological issues related to dose reconstruction. Paper presented at the 1995 annual NCRP meeting. Arlington, VA; 1995.

[8]   Shore, R.E., Iyer, V., Altshuler, B., Pasternack, B. Use of human data in quantitative risk assessment of carcinogens. Impact on epidemiologic practice and the regulatory process. Regul. Toxicol. Pharmacol. 15: 180-221; 1992.

[9]   De Klerk, N.H., English, D., Armstrong, B. A review of the effects of random measurement error on relative risk estimates in epidemiological studies. Int. J. Epidemiol. 18: 705-712; 1989.

[10]   Lubin, J.H., Samet, J., Weinberg, C. Design issues in epidemiologic studies of indoor exposure to Rn and risk of lung cancer. Health Phys. 59: 809-817; 1990.

[11]   Clayton D.G. Models for the analysis of cohort and case-control studies with inaccurately measured exposures. Pages 301-331 in: Dwyer, J.H., Lippert, P., Feinleib, M., Hoffmeiser, H., editors. Statistical models for longitudinal studies of health. New York, NY; Oxford University Press; 1992.

[12]   Gilbert, E.S. Accounting for bias and uncertainty resulting from dose measurement errors and other factors. Br. Inst. Radiol. Rpt. 22: 155-159; 1991.

[13]   Pierce, D.A., Stram, D., Vaeth, M. Allowing for random errors in radiation dose estimates for the atomic bomb survivor data. Radiat. Res. 123: 275-284; 1990.

[14]   Gavrilin, Yu.I., Gordeev, K.I., Ivanov, V.K., Ilyin, L.A., Kondrusev, A.I., Margulis, U.Ya., Stepanenko,V.F., Khrouch, V.T., Shinkarev, S.M. The process and results of the reconstruction of internal thyroid doses for the population of contaminated areas of the Republic of Belarus. Vestn. Acad. Med. Nauk SSSR 2: 35-43; 1992.

[15]   Likhtarev I.A., Shandala, N.K., Gulko, G.M., Kairo, I.A., Chepurny, N.I. Ukrainian thyroid doses after the Chernobyl accident. Health Phys. 64: 594-599; 1993.

[16]   Likhtarev, I.A., Sobolev, B.G., Kairo, I.A., Tronko, N.D., Bogdanova, T.I., Oleinic, V.A., Epshtein, E.V., Beral, V. Thyroid cancer in the Ukraine. Nature 375: 365; 1995.

[17]   Zvonova, I.A., Balonov, M.I. Radioiodine dosimetry and predictions of consequences of thyroid exposure of the Russian population following the Chernobyl accident. Pages 71-125 in: Merwin, S.E. and Balonov, M.I., editors. The Chernobyl papers. Vol. I: Doses to the Soviet population and early health effects studies. Richland, WA; Research Enterprises; 1993.

[18]   Ilyin, L.A., Pavlovsky, O.A. Radiological consequences of the Chernobyl accident in the Soviet Union and measurements taken to mitigate their impact. Pages 149-166, Vol. 3 in: IAEA International Conference on Nuclear Power Performance and Safety. IAEA-CN-48/33; 1987.

[19]   Straume, T., Lucas, J.N. A comparison of the yields of translocations and dicentrics measured using fluorescence in situ hybridization. Int. J. Radiat. Biol. 64: 185-187; 1993.

United States-Assisted Studies on Dose Reconstruction in the Former Soviet Union*

Lynn ANSPAUGH

Lawrence Livermore National Laboratory, 7000 East Ave., Livermore, CA 94550, USA

André BOUVILLE

National Cancer Institute, 6130 Executive Blvd., Rockville, MD 20854, USA

Abstract. Following the Chernobyl accident, the US and the USSR entered into an agreement to work on the safety of civilian nuclear reactors; one aspect of that work was to study the environmental transport and health effects of radionuclides released by the accident. After the break-up of the USSR separate agreements were established between the US and Ukraine, Belarus, and Russia to continue work on dose reconstruction and epidemiologic studies of health effects from exposure to external radiation and the incorporation of radionuclides. Studies in Belarus and Ukraine related to the Chernobyl accident now emphasize epidemiologic studies of childhood-thyroid cancer and leukemia, and eye-lens-cataract formation in liquidators. Supporting studies on dose reconstruction emphasize a variety of ecological, physical, and biological techniques.

Studies being conducted in Russia currently emphasize health effects in the workers and the population around the Mayak Industrial Association. As this production complex is an analogue of the US Hanford Works, advantage is being taken of the US experience in conducting a similar, recently completed dose-reconstruction study.

In all cases the primary work on dose reconstruction is being performed by scientists from the former Soviet Union. US assistance is in the form of expert consultation and participation, exchange visits, provision of supplies and equipment, and other forms of local assistance.

1. Introduction

Following the Chernobyl accident the United States and the former Soviet Union (FSU) entered into an agreement in 1988 to work cooperatively on the safety of civilian nuclear reactors. As part of this agreement a Joint Coordinating Committee on Civilian Nuclear Reactor Safety (JCCCNRS) was established, and the formation of twelve working groups soon followed. Most of these working groups were concerned directly with safety aspects of the reactors still operating in the former Soviet Union. One working group, number 7, was concerned with environmental transport of the radionuclides released and with long-term health effects.

The general goals of Working Group 7 were to develop validated models for rapidly projecting doses and health effects in the event of a future reactor accident, to provide the

---

* This work was funded by the Office of International Health Programs; Assistant Secretary for Environment, Safety and Health; US Department of Energy; the National Cancer Institute, and the Nuclear Regulatory Commission. Portions of this work were performed at the Lawrence Livermore National Laboratory under contract W-7405-Eng-48 with the Department of Energy.

basis for the physical dosimetry needed to reconstruct doses for the FSU citizens exposed to the higher doses following the Chernobyl accident, and to study the health effects of the accident. In order to implement these goals many subworking groups were established that included those to study

a. Research on atmospheric dispersion modeling,
b. Wind-driven resuspension of toxic aerosols,
c. External exposure and dose from deposited radionuclides,
d. Transport of radionuclides through terrestrial food chains and the resulting dose to man,
e. Long-term dose from the contamination of aquatic food chains,
f. Modeling the behavior of radionuclides in a soil-aquatic system including rivers and reservoirs,
g. Intercalibration of methods for measuring radioactive contaminants in the environment,
h. Acute radiation syndrome in man,
i. Population registries,
j. Biological dosimetry,
k. Epidemiologic study of thyroid disease, and
l. Epidemiologic study of leukemia.

These tasks were generally agreed upon in 1989-1990, and the work was begun to implement most of the agreed upon research. The National Cancer Institute was asked to lead the epidemiologic studies indicated above, and the US Nuclear Regulatory Commission provided additional funding and support for the studies.

These research plans were changed markedly by three separate events. The first was the implementation of the International Atomic Energy Agency (IAEA) sponsored work at the request of the USSR government by the International Chernobyl Committee, which undertook what can be described as a massive quality-assurance audit of the response by the USSR government to the Chernobyl accident. The work of this group obviated the need for some studies, such as the intercalibration studies indicated above.

The second event was the collapse of the government of the USSR in 1991. Following this collapse (and the generally poor economic conditions in the FSU) the prior agreements could not be adhered to and other separate agreement eventually followed between the US and the individual governments or agencies of Russia, Ukraine, and Belarus.

The third event was the transfer of the Working Group 7 studies from the basic research arm of the Department of Energy (DOE) to the Office of the Assistant Secretary for Environment, Health and Safety (EHS) within DOE. The latter's responsibilities include a focus on the definition of radiogenic health effects in man.

As a result of these changes the US-assisted studies underwent a substantial change. The general research studies were phased out, and a focus was maintained on epidemiologic studies of radiogenic health effects in man. In addition to the studies on leukemia and thyroid disease noted above, studies on eye-lens cataracts have been added more recently. As it is impossible to study meaningfully radiogenic health effects without knowing the doses to which the study enrollees have been exposed, a major component of these continuing studies is dose reconstruction.

In addition to the studies noted above and in keeping with the focus of EHS on the elucidation of radiogenic health effects in man, a separate agreement was negotiated between the US and Russia to study radiation-health effects in general. This work is being implemented through the Joint Coordinating Committee on Radiation Effects Research (JCCRER). Current focus of the JCCRER research is on epidemiologic studies on the workers and the general population in the vicinity of the Mayak Industrial Association in the Urals.

2. Studies of Thyroid Disease in Belarus

The first epidemiologic studies defined under the work of the JCCNRS were studies of childhood-thyroid disease in Belarus. Two separate studies were defined--a case-control study and a cohort study.

2.1. Case-Control Study

The case-control study involved the study of 107 cases and 214 controls of two separate groups. The first control group was drawn from the general population and matched to the cases according to age, sex, and urban/rural type of residence. The second control group was matched as above, but also matched to the cases according to experience with diagnostic procedures that brought the cases to diagnosis.

A few of the enrollees in the study had had their thyroids measured directly during the few weeks immediately following the accident, but most had not. Most of the enrollees had been interviewed in an effort to determine residence and lifestyle information, but most of the questionnaires did not contain lifestyle information that could be considered to be adequate.

Therefore, in order not to bias the results, a "lowest common denominator" approach of dose reconstruction was used. Thus, information that was not generally available for all enrollees in the study was ignored.

The general information used in the method of dose reconstruction included:

a. Country-wide information on the deposition density of ^{137}Cs,

b. Limited information on the deposition density of ^{131}I, and

c. A databank of thyroid measurements of approximately 200,000 people that was used to infer the relationship between adult-mean-thyroid dose and the deposition densities of ^{137}Cs and ^{131}I.

Particular information on residency immediately after the accident was used, but only default values were used for food consumption. Appropriate methods were also used to adjust mean-adult-thyroid doses to doses for children of the appropriate age. Preliminary results of this case-control study were presented at the World Health Organization Conference in November 1995 [1,2], and further details of the general dosimetric methods are provided in Gavrilin et al. [3,4].

2.2. Cohort Study

The cohort study is planned to follow actively 15,000 children. The cohort has not yet been defined, but efforts are being made to include in the cohort only those who had their thyroids measured in 1986. Thus, the children will be drawn from the most heavily contaminated areas in the oblasts near to the Chernobyl Nuclear Power Plant (NPP). In order to maximize the power of the study to detect any increase in thyroid disease, special efforts are being made to enroll as many persons as possible in a high-dose group. Depending upon the results of current investigations, it may be possible to add some people from the Brest Oblast with thyroid measurements to this high dose group. The Brest Oblast is distant from the Chernobyl NPP but is thought to have been rather heavily contaminated by radioiodines. Otherwise, it will be necessary to add persons to this group who are believed to have received high doses, but whose thyroid activities were not measured.

The basis of the dose reconstruction will be the results of thyroid measurements, information on the extent of the radioactive contamination in the subjects' locations, and information on lifestyle as gained from questionnaires. If subjects must be enrolled who did not have their thyroids measured, they will be selected from locations where "passports" of radioiodine doses have been derived on the basis of residents who did have their thyroids measured [3,4].

Some of the important steps that must be completed for this dose-reconstruction study are accurate calibrations for the instruments used for the direct measurements of thyroids and calculation of the radiation doses due to sources in addition to ^{131}I.

The calibration of the instruments is necessary, because survey instruments not designed for this purpose were used for the thyroid-activity measurements. Collimators were not used, and the influence of clothing contamination and the procedures used for "background" subtraction must be investigated and defined.

Other sources of radiation exposure to be calculated include internal irradiation from the intake of short-lived radioiodines and of ^{132}Te; internal irradiation from the intake of ^{137}Cs; and external irradiation from radionuclides deposited on the ground. As these sources of exposure may be more effective per unit dose in inducing thyroid cancer, these inclusions may be particularly important.

3. Epidemiologic Studies in Ukraine

In Ukraine epidemiologic studies of thyroid disease are also approved, and in addition there is a study underway on the induction of eye-lens cataracts in liquidators. An extensive study of leukemia in liquidators has been planned.

3.1. Thyroid Disease

A cohort study has been defined that will consist of those individuals
a. Who were less than 18 years of age at the time of the accident,
b. Who had their thyroid activity measured in 1986 in eight raions and the City of Pripyat, and
c. Who can be located at the beginning of the study.

It is estimated that the number of children satisfying these conditions is 57,000. Individual-thyroid doses will be required for all members of the cohort. The general procedures to be followed have been described in Likhtarev et al. [5]. Primary inputs consist of the deposition density of ^{137}Cs with adjustment according to the polar coordinates from the Chernobyl Nuclear Power Plant and individual data from questionnaires. The sources of exposure in addition to ^{131}I (as named above) will be considered in this study, also.

3.2. Leukemia in Liquidators

A case-cohort study is being planned to study the radiogenic induction of leukemia in liquidators. The underlying cohort for this study consists of all liquidators first employed between 1986 and 1990 and who lived in the City of Kiev and the six oblasts of Dnepropetrovsk, Donetsk, Lugansk, Kiev, and Kharkov at the time of first employment. This target cohort is estimated to contain about 90,000 individuals. A subcohort will be selected from this cohort, and this subcohort will serve as control a control group. The size of the subcohort will be in the range of 1000 to 2000 individuals.

All cases in the target cohort will be identified; the number of cases that will occur is estimated to be approximately 100. Individual doses will be required for the members of the case and the subcohort classes.

Dose reconstruction for these groups will be more difficult than for the children enrolled in the thyroid studies, as the exposures for the liquidators occurred under very individual-specific conditions that are not known with great precision.

It is planned that dose reconstruction will be performed with the following tools:

a. All administrative and other records that provide historical estimates of radiation exposure,

b. Any reconstructed estimates of dose based upon examinations of groups of individuals with a variety of physical dosimetric information,

c. Construction of a matrix (over time and space) of all known measurements of exposure rate or dose,

d. Specification of time and motion steps that led to the exposure of individual liquidators that, when combined with item c above, can be used to estimate doses,

e. Measurements of chromosomal translocations by the fluorescent in situ hybridization (FISH) method (this method will not work for treated cases of leukemia), and

f. Measurements of electron paramagnetic resonance of teeth.

Substantial research, development, and validation of the techniques enumerated above will be required in order to provide the required doses.

3.3. Eye-Lens Cataracts in Liquidators

A cohort of 10,000 liquidators will be defined and actively followed in order to examine the occurrence of eye-lens cataracts. Individuals doses will be required for all 10,000 members of this cohort.

The dose reconstruction for this group will be based upon all of the techniques mentioned above in order to provide a base of underlying information. However, in this case it is significant that the tissue of interest is that of the eye lens. It is possible that beta doses to the eye lens were the dominant sources of exposure that must be quantified. This is a very difficult problem, as there are currently no validated techniques of estimating the possibly critical pathway of deposition of beta emitters on the eye lens itself or of generally specifying the magnitude of beta exposure.

Thus, a substantial effort of research, development, and validation is required in order to produce eventually the required information on individual dose to the eye-lens tissue.

4. Epidemiologic Studies in Russia Related to the Mayak Industrial Association

Feasibility studies of epidemiologic studies of radiogenic disease in the workers at the Mayak Industrial Association and in the general population in the vicinity were begun in March 1994. These one-year feasibility studies will result in reports and recommendations for the conduct of continuing work.

4.1. Workers at the Mayak Industrial Association

The Mayak Industrial Association was opened in 1948 to produce plutonium and other materials related to nuclear weapons [6]. The cohort of workers includes about 20,000 people who started work between 1948 and 1985. High doses of external gamma radiation were received by the workers during the first few years, and at the radiochemical reprocessing plant workers were also exposed to plutonium. Approximately 5000 workers (1700 females) worked at the complex when the exposures were large (more than 1 Sv). External gamma exposure was always monitoring by film-badge or thermoluminescent dosimeters, and plutonium exposure has been monitored by urinary assay and tissue analysis of cadavers.

Recent studies of the worker cohort have shown excess risks of leukemia and solid tumors [7]; a separate study demonstrating excess lung-cancer mortality has also been published [8]. These studies, however, currently lack proper controls and adequate follow up.

The creation of a more refined dosimetry-data base has been proposed This does not require "reconstruction" as much as it requires the assembly of all existing external gamma-dose information into a more refined data base, and the reassessment of all bioassay data for plutonium into a data base consistent with the most recent data on plutonium-excretion rates, etc. This study is considered by the authors very likely to proceed beyond the feasibility phase, due to the generally good dosimetry records and the interest in this very unique population of highly exposed individuals.

Another subcohort of this population will be studied in order to examine the deterministic effects of occupational exposure in this highly exposed cohort. Dosimetry records will be compiled according to the process indicated above.

4.2. Techa River Cohort

Between 1949 and 1956 about 10^{17} Bq of liquid wastes were released into the Techa River from the Mayak facility.[9]. This resulted in substantial exposure of a large number of persons living downstream. Dose-reconstruction activities have been proceeding for some time [9], as have epidemiological studies [10]. A cohort of approximately 29,000 persons has been followed and studies reveal an excess incidence of leukemia and solid tumors. The latter incidence data, however, are complicated by the fact that two distinct ethnic groups are being followed, and these two groups have quite different background rates of cancer incidence.

Dose-reconstruction studies performed so far have depended heavily upon measurements of ^{90}Sr in teeth and upon thermoluminescence measurements of external gamma exposure in environmental materials. Only group doses to village residents have been calculated. For future work it is proposed to consider more general and complete methods of dose reconstruction and to provide individual estimates of doses for the approximately 29,000 members of this cohort. Again, the authors estimate that it is likely that this project will extend beyond the feasibility stage due to the large number of people known to be exposed to moderately high doses at relatively low dose rates.

Dose reconstruction for this cohort will depend upon additional data envisaged to be gained from measurements of chromosomal translocations and from electron paramagnetic resonance measurements of teeth samples.

Additional effort will also be devoted to defining more accurately the source terms for the material released to the Techa River and also released to the atmosphere. It is known that large amounts of ^{131}I (approximately 2×10^{16} Bq), and perhaps other radionuclides, were released to the atmosphere. Once these source terms are defined attempts to model the movement of the radioactive materials in the environment will be undertaken.

4.3. East Urals Radioactive Trace (EURT) Cohort

In 1957 an explosion in a radioactive waste-storage facility (the so-called Kyshtym accident) resulted in the release of approximately 7×10^{16} Bq of radioactive materials into the atmosphere and formed the EURT. About 30,000 people are included in an EURT registry; these people were exposed to both external and internal irradiation. Some work on dose reconstruction and epidemiological follow-up has been done for this group.

Another task of the feasibility studies currently being conducted is to evaluate the possibility and desirability of performing individual-dose reconstructions for this cohort and engaging in a long-term epidemiologic study.

4.4. The Mayak Children Cohort

Another task in the feasibility studies is to identify approximately 40,000 persons born between 1948 and 1973 who lived for at least one year in the vicinity of the Mayak Industrial Association. One of the primary routes of exposure to this cohort was the release of materials to the atmosphere from the facility. An evaluation will be made concerning the possible individual and collective doses and whether a long-term dose-reconstruction and epidemiologic study would be useful.

5. Other Cohorts of Interest

Three other cohorts have been proposed for feasibility studies to be conducted under the auspices of the US-Russia Joint Coordinating Committee on Radiation Effects Research. These proposals will be considered by the JCCRER in April 1996.

One cohort consists of children exposed to high radioiodine doses from the Chernobyl accident and who live in the Bryansk, Kaluga, and Tula Oblasts in Russia The proposal is to conduct a dose-reconstruction effort and to define a cohort to be the subject of an epidemiologic study.

Another subject of general interest is the exposures that have occurred in the Altai Oblast, Russia, from the testing of nuclear weapons at the Semipalatinsk Polygon in the Republic of Kazakhstan. Fairly substantial exposures are known to have occurred, especially from the first test that was conducted in 1949. The proposal is for the US and Russia to work jointly on dose-reconstruction efforts. A primary goal would be to develop a common methodology that could be tested and validated against data from both countries' experiences. If this dose-reconstruction effort is approved and is successful, it might be followed by joint epidemiologic studies of the population of interest in Russia.

6. Pattern of US Cooperation

In nearly all of the situations addressed above the primary work on dose reconstruction is being performed by scientists in the former Soviet Union. US assistance has been in the form of expert consultation and participation; exchange visits; and the provision of medical, computational, and dosimetric equipment. In addition financial assistance to the cooperating institutions has been provided on a limited scale.

A primary goal of the US-assisted efforts is to publish the obtained results in peer-reviewed western journals. This goal is being facilitated by the joint authorship of such articles with the lead authors being those from the FSU responsible for the primary intellectual input.

References

[1] Y.I. Gavrilin, V.T. Khrouch, V.F. Minenko, V.V. Drozdovitch, A.V. Ulanovsky, S.M. Shinkarev, A.C. Bouville, L.R. Anspaugh and T. Straume, Reconstruction of Thyroid Doses for the Population of Belarus following the Chernobyl Accident. Submitted to the Proceedings of the International Conference on the Health Consequences of the Chernobyl and other Radiological Accidents, 20-23 November 1995, Geneva.

[2] L.N. Astakhova, L.R. Anspaugh, G.W. Beebe, A. Bouville, V. Garber, Y.I. Gavrilin, V.T. Khrouch, Y.N. Kuzmenkov, V.F. Minenko, K.V. Moshchik, A.S. Nalivko, J. Robbins, E.V. Shemyakhina, S.I. Tochitskaya, and M.A. Waclawiw, Thyroid Cancer in Children following the Chernobyl Accident: A Case-Control Study in Belarus. Submitted to the Proceedings of the International Conference on the Health Consequences of the Chernobyl and other Radiological Accidents, 20-23 November 1995, Geneva.

[3] Y.I. Gavrilin, V.T. Khrouch, S.M. Shinkarev, A. Bouville, L. Anspaugh, and T. Straume, Chernobyl Accident: Reconstruction of Thyroid Dose for Inhabitants of the Republic of Belarus. Submitted to *Health Physics*.

[4] Y. Gavrilin, V. Khrouch, S. Shinkarev, V. Drozdovitch, V. Minenko, E. Shemyakina, A. Bouville, and L. Anspaugh, Estimation of Thyroid Doses Received by the Population of Belarus as a Result of the Chernobyl Accident. This Volume.

[5] I.A. Likhtarev, G.M. Gulko, B.G. Sobolev, I.A. Kairo, N.I. Chepurnoy, G. Pröhl, and K. Henrichs, Thyroid Dose Assessment for the Chernigov Region (Ukraine): Estimation Based on [131]I Thyroid Measurements and Extrapolation of the Results to Districts without Monitoring. *Radiat. Environ. Biophys.* **33** (1994) 149-166.

[6] A.V. Akleyev and E.R. Lyubchansky, Environmental and Medical Effects of Nuclear Weapon Production in the Southern Urals. *Sci. Total Environ.* **142** (1994) 1-8.

[7] N.A. Koshurnikova, L.A. Buldakov, G.D. Bysogolov, M.G. Bolotnikova, N.S. Komieva, and V.S. Pesternikova, Mortality from Malignancies of the Hematopoietic and Lymphatic Tissues among Personnel of the First Nuclear Plant in the USSR. *Sci. Total Environ.* **142** (1994) 19-23.

[8] V.F. Hohryakov and S.A. Romanov, Lung Cancer in Radiochemical Industry Workers. *Sci. Total Environ.* **142** (1994) 25-28.

[9] M.O. Degteva, V.P. Kozheurov, and M.I. Vorobiova, General Approach to Dose Reconstruction in the Population Exposed as a Result of the Release of Radioactive Wastes into the Techa River. *Sci. Total Environ.* **142** (1994) 49-61.

[10] M.M. Kossenko and M.O. Degteva, Cancer Mortality and Radiation Risk Evaluation for the Techa River Population. *Sci. Total Environ.* **142** (1994) 73-89.

Estimation of Thyroid Doses Received by the Population of Belarus as a Result of the Chernobyl Accident

Yuri GAVRILIN, Valeri KHROUCH, and Sergei SHINKAREV
State Research Center of Russia - Institute of Biophysics, 46 Zhivopisnaya St., 123182 Moscow, Russia

Vladimir DROZDOVITCH, Viktor MINENKO, and Elena SHEMYAKINA
Institute of Radiation Medicine, 23 Masherova Ave., 220600 Minsk, Belarus

André BOUVILLE
National Cancer Institute, 6130 Executive Blvd., Rockville, MD 20854, USA

Lynn ANSPAUGH
Lawrence Livermore National Laboratory, 7000 East Ave., Livermore, CA 94550, USA

Abstract. Within weeks of the Chernobyl accident, about 300,000 measurements of human thyroidal ^{131}I content were conducted in the more contaminated territories of the Republic of Belarus. Results of these and other measurements form the basis of thyroid-dose reconstruction for residents of Belarus. Preliminary estimates of thyroid doses have been divided into three classes:

Class 1 ("measured" doses). Individual doses are estimated directly from the measured thyroidal ^{131}I content of the person considered, plus information on life style and dietary habits. Such estimates are available for about 130,000 individuals from the contaminated areas of Gomel and Mogilev Oblasts and Minsk city.

Class 2 ("passport" doses). For every settlement with a sufficient number of residents with "measured" doses, individual thyroid-dose distributions are determined for several age groups and levels of milk consumption. This action has been called the "passportization" of the settlement. A population of about 2.7 million people resides in the "passportized" settlements.

Class 3 ("inferred" doses). For any settlement where the number of residents with "measured" doses is small or equal to zero, individual thyroid doses are derived from the relationship obtained between the mean adult-thyroid dose and the deposition density of ^{131}I or ^{137}Cs in settlements with "passport" doses presenting characteristics similar to those of the settlement considered. This method can be applied to the remainder of the population (about 7.3 million people).

An approximate estimate of the collective thyroid dose for the residents of Belarus is presented. Illustrative results of individual thyroid dose and associated uncertainty are discussed for rural settlements and urban areas.

The Chernobyl accident concerned most of the population of the world. This tragic accident has led scientists from several countries to undertake epidemiological studies in order to clarify the relationship between thyroid dose and thyroid disease, especially for children. In the dosimetric part of such investigations, the main efforts are aimed at reducing the uncertainties associated with the estimation of individual and collective thyroid doses. It is well known that internal radiation exposure of the thyroid from radioiodine (mainly ^{131}I) was the main contributor to the thyroid doses received by the residents of Belarus within the very first weeks after the accident. A large number of measurements of human thyroidal ^{131}I content were conducted in Belarus from the beginning of May to the middle of June, 1986. The results of these measurements, after they were sorted and verified for approximately 200,000 persons, form the basis of the calculation of individual thyroid doses for Belarussian residents. At present about 130,000 individual thyroid doses have been calculated on the basis of the measured thyroidal ^{131}I contents. These doses are called data of Class-1 reliability ("measured" doses). On the basis of the Class-1 data, average thyroid doses have been calculated for the population of more than 800 rural settlements and "passports" have been established for these settlements. The "passportization" of the settlement is a process that allows the assessment of individual thyroid doses (called data of Class-2 reliability or "passport" doses) for the residents of those settlements who were not measured. Such "passports" also have been established for four cities: Minsk, Gomel, Mogilev, and Mozyr. The thyroid doses received by the other residents of Belarus, who are not provided with "measured" or "passport" doses, can be estimated on the basis of the relationship between the ^{131}I or ^{137}Cs ground-deposition density in the area considered and the average thyroid dose obtained from thyroid measurements among residents of that area. Those thyroid doses are called "inferred" doses, or data of Class-3 reliability.

In this paper, the essential features of Class-1, Class-2, and Class-3 data for Belarussian residents are discussed. The currently available results are the outcome of the first stage of the thyroid dose reconstruction in Belarus. They form the basis of the dosimetric support for the epidemiological studies of thyroid disease among Belarussian children that are conducted jointly with American scientists (a retrospective "case-control" study and a prospective cohort study). Some of the key aspects of thyroid dose reconstruction in Belarus also have been presented, mainly in Russian, in scientific publications and in scientific meetings [1-5]. In addition, an extensive discussion, in Russian, of the work done from 1988 to 1993 can be found in the Appendices to a report of the Institute of Biophysics (Moscow) [6].

1. "Measured" individual thyroid doses (data of Class-1 reliability).

The main steps that were followed in the calculation of "measured" doses are presented in Table 1. The total number of Belarussian residents who were measured in May-June of 1986 is not known with accuracy. This is due in particular to the fact that in some measuring stations the residents with values of dose rate near the thyroid lower than some predetermined level (100, 200, 300, or 1000 $\mu R\ h^{-1}$) were not registered. Also, the reliability of the Class-1 data is not uniform because the uncertainty of the measured thyroidal ^{131}I content depends on the measuring conditions and on the measurement device (DP-5, or SRP-68-01, or DRG3). Therefore, the "measured" doses were divided into 3 subgroups of reliability, according to the level of uncertainty associated with the determination of the ^{131}I content in thyroid, characterized with geometric standard deviations of 1.3, 1.7, and 2.2, respectively. The thyroid dose calculation was performed according to a Guidance report [7], assuming ^{131}I intake with inhalation and with ingestion of fresh milk following a single deposition of fallout on pasture grass. The latter way of ^{131}I intake is considered to be predominant for most individuals. The variation with time of the radioiodine deposition, as recorded in stations of the Gidrometeorology Committee, is shown in Fig.1 [8]. It can be seen in Fig.1 that radioiodine fallout in the cities of Gomel, Mogilev, and Grodno occurred mainly over a period of one to two days. This justifies the assumption of a single radioactive fallout in the first stage of calculation of "measured"

Table 1. Description of the main steps followed in the calculation of the data of Class-1 reliability ("measured" individual thyroid doses).

Dates	Type of activity	Result
End of April and beginning of May 1986	Organization of large scale measurements of human thyroidal ^{131}I content	35 measuring groups; measuring stations in the centers of oblasts
May and June 1986	Monitoring the ^{131}I content in the thyroids to reveal the Belarussian residents that were highly exposed	About 300,000 people were examined
1987 - 1988	Collection of original records of measurements of thyroidal ^{131}I content	About 200,000 records
1988	Interviewing inhabitants on lifestyle and diet (levels of milk consumption)	About 150,000 records
1988 - 1989	Verification of results of in vivo thyroid measurements; development of special ways to correct some original data	Three groups of reliability of available measurements; ways of correcting some original data
1988 - 1991	Creating the database of the direct thyroid measurements; development of computer codes for thyroid dose estimation	Data bank containing "measured" individual thyroid doses for 130,000 residents, who had been measured before June 6, 1986

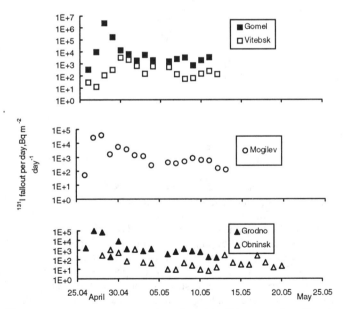

Fig.1. Daily ^{131}I fallout recorded in some observation points of the Gidrometeorology Committee according to the data of SPA "Taifun".

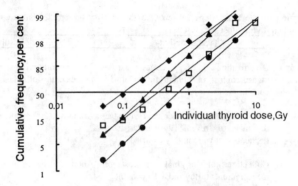

Fig.2 Lognormal probability plot of "measured" individual thyroid doses for adult residents in rural settlements: in Hoiniki raion - Korchevoe (♦), and in Bragin raion - Kolybany (Y); Savichi (▲); Mikhalevka (●)

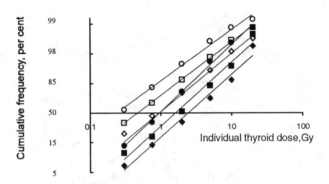

Fig.3 Lognormal probability plot of "measured" individual thyroid doses for the residents in Hoiniki raion: evacuated people - (♦)children up to 7y; (■) children up to 18y; (●) adults. nonevacuated people - (◊)children up to 7y; (Y) children up to 18y; (O) adults.

doses for the residents of the Gomel and Mogilev oblasts. However, it also can be seen in Fig.1 that the radioiodine fallout in Vitebsk (Belarus) and Obninsk (Russia) lasted longer than one or two days; in the future, this protracted period of deposition should be taken into account in the thyroid dose estimation for the residents of those areas.

Differences in lifestyle and dietary habits resulted in different individual thyroid doses even within one settlement and one age group. The obtained individual thyroid dose distributions in general can be described with a lognormal function. Examples of such distributions, which are represented as straights line in a lognormal probability plot, are presented in Fig.2. The lognormal function was shown to describe satisfactorily the distributions of individual thyroid doses for the people from several settlements as well as for the people from the whole administrative raion. Fig.3 shows the lognormal probability

Table 2. Arithmetic mean values of "measured" individual thyroid doses (Class-1 data), calculated for rural residents in some contaminated territories of Gomel and Mogilev oblasts.

| Contaminated territories (raions) | Arithmetic mean of Class-1 data and percentage of the measured residents | | | | | |
| | Children up to 7y | | Children up to 18y | | Adults | |
	D, Gy	%	D, Gy	%	D, Gy	%
Bragin*	2.1	57	1.5	56	0.80	71
Bragin**	1.8	55	1.1	63	0.48	91
Hoiniki*	4.7	59	3.1	56	1.6	50
Hoiniki**	1.8	62	1.1	66	0.53	63
Narovlya*	1.6	13	1.0	15	0.45	34
Narovlya**	1.3	37	1.0	38	0.36	57
Vetka	1.6	9.3	1.2	8.5	0.34	2.6
Rechitsa	1.6	17	1.1	15	0.44	12
Loev	0.87	47	0.72	46	0.32	19
Klimovichi	0.37	6.2	0.25	5.3	0.088	4.8
Kostukovichi	0.45	20	0.35	16	0.21	15
Krasnopolye	0.62	13	0.42	19	0.19	17
Slavgorod	0.22	12	0.15	17	0.10	7.4
Chericov	0.54	28	0.34	25	0.13	20

* - the territory of the raion, which had been left by all the residents before May 5, 1986.
** - the non-evacuated territory of the raion.

plot of "measured" individual thyroid doses for the residents in Hoiniki raion, subdivided into groups.

It can be seen in Fig. 3 that children within one group (nonevacuated people or evacuated people) generally received higher thyroid doses than adults, and also that a fraction of small doses is higher for nonevacuated people than for evacuated ones. Arithmetic mean values of "measured" individual thyroid doses (Class-1 data), calculated for rural residents in some contaminated territories of Gomel and Mogilev oblasts are presented in Table 2 . It is important to note that the highest "measured" doses do not exceed 60 Gy.

2. "Passport" individual thyroid doses (data of Class-2 reliability).

"Passport" doses have been calculated for the rural settlements in Gomel and Mogilev oblasts with sufficient number of available Class-1 data (not less than 10 measured residents). On the basis of the available "measured" doses, average thyroid doses have been estimated for each of 19 age groups (one adult category and 18 age groups for children aged up to 18 y at the time of the accident, with incremental steps of one year) for the residents in the selected settlements. Also, the thyroid doses to nursing infants fed by mother's milk and the thyroid doses to the fetus can be estimated by multiplying the mother's dose by age-dependent coefficients. The "passport" of the settlement presents the data in the form of a Table consisting of 19 rows and 11 columns. Each row corresponds to one age group while each of the 11 columns corresponds to one level of fresh cow's milk consumed daily by the resident, from 0 L d^{-1} (inhalation intake only) to 4.0 L d^{-1}. Also has been included into the Table the special case when the information on the level of milk-consumption rate is absent. The contribution to the thyroid dose due to ^{131}I intake with leafy vegetables is considered to be small for most individuals and has not been included into the Table in this first stage of dose reconstruction.

At present, over 800 rural settlements in the most contaminated areas in Gomel and Mogilev Oblasts have been provided with "passports". In addition, for about 100 settlements, important dosimetric data such as the arithmetic mean of the individual thyroid doses have been determined. Therefore, about 930 settlements (out of a total number of about 23,500 settlements in Belarus) have been provided with collective characteristics (with different degrees of reliability) of thyroid exposure. Taking into consideration the "passportized" large cities (Minsk, Gomel, Mogilev, and Mozyr) about 2,700,000 residents of Belarus (approximately 27% of the Belarussian population in 1986) are provided with Class-1 or Class-2 data of reliability.

3. "Inferred" individual thyroid doses (data of Class-3 reliability).

The large number of Class-1 data and the availability for many areas in Belarus of measured depositions of radionuclides on the ground [9] provided the opportunity to determine empirical relationships between the average thyroid dose received by people in the rural settlements and the ground-deposition density of radionuclides (radiocesium or radioiodine) in this settlement and in the area around the settlement. Using these relationships, thyroid doses can be inferred for residents of settlements with very few, or no, thyroid measurements. These "inferred" doses (or data of Class-3 reliability) can be estimated for the approximately 7,300,000 Belarussian residents without either "measured" or "passport " doses.

Fig.4 shows, for the settlements in three raions (Hoiniki in Gomel oblast, and Kostukovichi and Krasnopolye , in Mogilev oblast), the variation of the quotient of the arithmetic mean "measured" thyroid dose (D_{mjx}) for the adult population in the settlement (j) to the ^{137}Cs ground-deposition density (q_{Csjx}) as a function of the ^{137}Cs ground-deposition density (q_{Csjx}). Such dependencies can be described satisfactorily with the following expression [10]:

$$D_{jx} = 3.5\times10^{-8} \times q_x(^{131}I) + 1.4\times10^{-8} \times q_{jx}(^{131}I)$$
$$= 3.5\times10^{-8} \times R_x \times q_x(^{137}Cs) + 1.4\times10^{-8} \times R_{jx} \times q_{jx}(^{137}Cs) \qquad (1)$$

where D_{jx} - arithmetic mean thyroid dose for adult population in settlement (j) in area (x) in
    the absence of any countermeasures in the settlement and for typical lifestyle and
       dietary habits, in Gy;
    $q_x(^{131}I)$, $q_x(^{137}Cs)$ - average ground-deposition density of ^{131}I (^{137}Cs) in area (x), in Bq m^{-2};
    $q_{jx}(^{131}I)$, $q_{jx}(^{137}Cs)$ - average ground-deposition density of ^{131}I (^{137}Cs) in the settlement (j)
    in area (x), in Bq m^{-2};
    R_x, R_{jx} - average ratio of the ^{131}I to ^{137}Cs ground-deposition densities in area (x) and in
    settlement (j) in area (x), respectively.

The numerical coefficients 3.5×10^{-8} and 1.4×10^{-8} Gy m^2 Bq^{-1} have been estimated by the least square method using the data related to the nonevacuated residents of the Hoiniki raion in Gomel oblast, which had been selected as the "reference" territory. To illustrate the adequacy of using equation (1) for estimating the "inferred" thyroid doses, the ratios of "measured" average thyroid dose (D_{mjx}) for adults from the settlements (j) in Bragin raion (nonevacuated residents) in Gomel oblast and in Kostukovichi and Krasnopolye raions in Mogilev oblast, calculated on the basis of "measured" individual doses, to "inferred" average thyroid dose (D_{jx}) for adults in the same settlements (j), calculated according to eq.(1), have been plotted in Fig. 5 against the ^{137}Cs deposition densities in those settlements

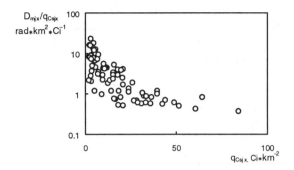

Fig.4. Ratio of arithmetic mean "measured" thyroid dose (D_{mjx}) for the adult population in the settlement (j) to ^{137}Cs ground-deposition density (q_{Csjx}) versus (q_{Csjx}) for the settlements in Hoiniki, Kostukovichi, and Krasnopolye in Belarus.

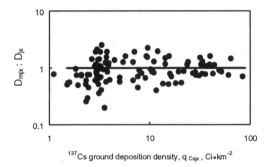

Fig.5. Comparison of "measured" thyroid doses (D_{mjx}) to calculated thyroid doses (D_{jx}) according to the "semiempirical" model.

It can be clearly seen in Fig.5 that the semiempirical model gives reasonable estimates of "inferred" average thyroid dose for all the considered range of ^{137}Cs ground-deposition density.

The procedure of "inferred" thyroid dose assessment on the basis of equation (1) requires information on the ground-deposition density of radiocesium and of radioiodine in the considered settlements and in the area (x) as a whole. That requirement is essential for reconstruction of Class-3 data. Ongoing research may clarify whether equation (1) is generally applicable to all contaminated areas of Belarus and of the other Republics of the former Soviet Union. It is likely that correction coefficients taking into account the role of fuel particles, and parameters such as the standing crop biomass, the fraction of cow's intake from pasture grass, and the date of the beginning of cow's pasture will need to be factored in.

Table 3. Estimates of geometric standard deviation for the parameters used in individual thyroid dose assessment of Class-1, Class-2, and Class-3 data.

Parameter	Geometric standard deviation
Thyroidal ^{131}I content (for Class-1 data)	≤1.3 (Group 1) from 1.3 to 1.7 (Group 2) from 1.7 to 2.2 (Group 3)
Thyroid mass [11]	1.6
Temporal variation of ^{131}I intake	1.3
Daily milk consumption (for each age group)	1.6 ± 0.1
Time-integrated concentration of ^{131}I in milk	2.1 ± 0.3
Uptake of ^{131}I in the thyroid [11]	1.4 - 1.5 (2.0 for newborn)
Distribution of individual thyroid doses for the residents of a given age-group in the settlement	2.7 ± 0.5
Average "passport" dose for the residents of a given age group in the settlement	≤1.5
Average "inferred" dose for the residents of a given age group in the settlement	from 1.5 to 1.8
Class-1 data ("measured" doses)	≤1.7 (Group 1) from 1.7 to 2.0 (Group 2) from 2.0 to 2.5 (Group 3)
Class-2 data ("passport" doses)	from 2.2 to 2.9
Class-3 data ("inferred" doses)	from 2.6 to 3.3

4. Uncertainty of Class-1, Class-2, and Class-3 data.

The current estimates of the main sources of uncertainties as well as of the total uncertainties for Class-1, Class-2, and Class-3 data are presented in Table 3. Future investigations will be devoted to a more thorough evaluation of each parameter uncertainty as well as of the total uncertainty attached to the individual thyroid doses. In particular, a distinction should be made between random and systematic uncertainties, for example for the results of measurement of thyroidal ^{131}I content. The systematic uncertainty of the results of measurement of thyroidal ^{131}I content for the residents of a settlement is transferred in its entirety to the "passport" doses for that settlement, while the random uncertainty decreases as the number of "measured" doses increases. For a number of settlements, the nonrepresentativity of the available Class-1 data, for instance because of the registration of only the highly exposed residents as was the case in Rechitsa raion in Gomel oblast, is a source of systematic uncertainty in the "passport" doses. That source of systematic uncertainty can be removed if information for all other residents is available. As a matter of fact, at the present time, the differences in the uncertainties of Class-2 and Class-3 data result only from the uncertainties in the average thyroidal ^{131}I contents for the adult population. All other sources of uncertainties are taken to be the same.

Table 4. Estimates of the collective thyroid doses for the people of Belarus.

Contaminated territories	Collective thyroid doses for the residents, 10^3 man.Gy		
	Children up to 7 y	Adults	Total
Minsk city	12.2	20.5	39.8
Total in Minsk Oblast (including Minsk city)	*15.0*	*28.2*	*51.7*
Gomel city	22.4	28.3	64.7
Three raions of Gomel Oblast: Bragin, Hoiniki and Narovlya	16.7	37.4	64.2
The other raions of Gomel Oblast	45.4	128.3	197.6
Total in Gomel Oblast	*84.5*	*194.0*	*326.5*
Mogilev city	5.5	11.2	20.9
Five raions of Mogilev Oblast: Chericov, Klimovichi, Kostyukovichi, Krasnopolye, and Slavgorod	8.5	14.8	29.3
The other raions of Mogilev Oblast	9.0	20.2	36.3
Total in Mogilev Oblast	*23.0*	*46.2*	*86.5*
Total in Grodno Oblast	*3.0*	*7.1*	*11.8*
Total in Brest Oblast	*7.3*	*17.6*	*29.0*
Total in Vitebsk Oblast	*1.2*	*2.6*	*4.4*
Total in Republic	*130*	*300*	*510*

5. Collective thyroid doses for the inhabitants of Belarus.

Estimates of collective thyroid doses for the residents of Belarus are presented in Table 4. A large proportion of people were measured in the most heavily contaminated raions of Gomel oblast. The estimates of the collective doses for the populations of those raions have been made on the basis of the available "measured" individual thyroid doses. The same method was used for the urban residents in Minsk, Gomel, and Mogilev cities. A combination of "passport" doses and "inferred" doses, calculated according to equation (1), was used for the raions of Gomel and Mogilev oblasts for which "passport" dose information was incomplete. The collective estimates for the weakly contaminated raions of oblasts other than Gomel and Mogilev were assessed using only equation (1).

References

[1] Gavrilin,Yu.I.; Gordeev,K.I.; Ilyin,L.A.; Khrushch,V.T.; Margulis,U.Ya.; Shinkarev,S.M.; Ivanov,V.K.; Stepanenko,V.F. Irradiation levels on thyroid of inhabitants of Byelorussia after the Chernobyl accident. - Proc. Symp. on the effects on the thyroid of exposed population following the Chernobyl accident. Chernigov, 3-6 December 1990. ICP/CEH 101/13,781r,13 November 1990

[2] Gavrilin,Yu.I.; Gordeev,K.I.; Ivanov,V.K.; Ilyin,L.A.; Kondrusev,A.I.; Margulis,U.Ya.; Stepanenko,V.F.; Khrouch,V.T.; Shinkarev,S.M. The process and results of the reconstruction of internal thyroid doses for the population of contaminated areas of the Republic of Belarus. News Acad. Med.Sci:2:35-43;1992 *(in Russian).*

[3] Yu.I.Gavrilin, V.T.Khrouch, S.M.Shinkarev, V.F.Stepanenko, V.F.Minenko Reconstruction of internal thyroid doses for inhabitants of raions of the Republic of Belarus contaminated because of the Chernobyl accident (state of investigations and tasks for the future).-Abstracts of 3rd Republic Conference "Scientific - practical aspects of health preservation for people exposed to influence of radiation in consequence of accident on Chernobyl NES",Gomel, April 15-17,1992; part 1 pp75-77 *(in Russian).*

[4] Gavrilin,Yu.I.; Khrouch,V.T.; Shinkarev,S.M. Internal thyroid exposure of the residents in several contaminated areas of Belarus. J.Med.Radiol. 6:15-20;1993 *(in Russian).*

[5] Yu.I.Gavrilin, V.T.Khrouch, S.M.Shinkarev, L.N.Astakhova. Radioiodine Irradiation on thyroid glands of people of Byelorussia after the Chernobyl accident. In: Book of abstracts. 4th annual scientific and technical conference of the Nuclear Society: Nuclear energy and human safety. NE-93, June 28 - July 2, Nizhny Novgorod, Russia; v.1, pp.222-224 *(in Russian).*

[6] V.T.Khrouch, Yu.I.Gavrilin, S.M.Shinkarev, U.Ya.Margulis, I.V.Samokhin V.I.Soldatenkov, O.E.Ivanova. Generalization of results of individual thyroid dose reconstruction. Determination of connections between parameters of contamination of people residences and levels of irradiation on thyroid glands. Final report of Institute of Biophysics, Moscow Contract N 7-17/93 with the Ministry of Public Health, Minsk, Belarus. Moscow, 1994, 929pp, including 819 pp of Appendices *(in Russian).*

[7] Arefieva,Z.S.; Badyin,V.I.; Gavrilin,Yu.I.; Gordeev,K.I.; Ilyin,L.A.; Kruchkov,V.P.; Margulis,U.Ya.; Osanov,D.P.; Khrouch,V.T. Guidance on thyroid dose assessment for a man due to his radioiodine intake. Edited by Ilyin,L.A. Moscow: Atomic Energy Publishing House; 1988 *(in Russian).*

[8] S.M.Vakulovsky, K.P.Makhonyko, E.G.Kozlova. Assessment of I-131 ground deposition density and distribution of its external exposure over the territory of the USSR, which was contaminated due to the Chernobyl accident. Obninsk: Scientific Production Association "Taifun"; 1990 (*in Russian*).

[9] State Committee of the USSR on Hydrometeorology. Data on radioactive contamination of settlements of Belorussian SSR by cesium-137 and strontium-90. Moscow: Hydrometeorology Publishing House; 1989 (*in Russian*)

[10] Yu.I.Gavrilin, V.T.Khrouch, S.M.Shinkarev, V.F.Stepanenko. The reconstruction of internal thyroid doses to population of the most contaminated territories of the former USSR after the Chernobyl accident. - Proc.International workshop on environmental dose reconstruction, Atlanta, Georgia, USA, November 15-18, 1994 *(to be published).*

[11] D.E.Dunning,Jr. and G.Schwarz. Variability of human thyroid characteristics and estimates of dose from ingested [131]I. -Health Physics, v.40 (May), 1981, pp.661-675.

RESULTS OF LARGE SCALE THYROID DOSE RECONSTRUCTION IN UKRAINE

Iliya LIKHTAREV, Boris SOBOLEV, Irina KAIRO,
The Research Center of Radiation Medicine, AMS of Ukraine,
53, Melnikov str., 252050 Kiev, Ukraine
Leonid TABACHNY,
The Ukrainian Ministry on Chernobyl Affairs,
8, Lvovska sq. 254655 Kiev, Ukraine
Peter JACOB, Gerhard PRÖHL, Guennady GOULKO
GSF-Forschungszentrum für Umwelt und Gesundheit, Institut für Strahlenschutz,
Neuherberg, D-85764 Oberschleißheim, Germany

Abstract. In 1993, the Ukrainian Ministry on Chernobyl Affairs initiated a large scale reconstruction of thyroid exposures to radioiodine after the Chernobyl accident. The objective was to provide the state policy on social compensations with a scientific background. About 7,000 settlements from five contaminated oblasts have gotten certificates of thyroid exposure since then. Certificates contain estimates of the average thyroid dose from ^{131}I for seven age groups. The primary dose estimates used about 150,000 direct measurements of the ^{131}I activity in the thyroid glands of inhabitants from Chernihivs'ka, Kyivs'ka, Zhytomyrs'ka, and also Vinnyts'ka oblasts. Parameters of the assumed intake function were related to environmental and questionnaire data. The dose reconstruction for the remaining territory was based on empirical relations between intake function parameters and the ^{137}Cs deposition. The relationship was specified by the distance and the direction to the Chernobyl Nuclear Power Plant. The relations were first derived for territories with direct measurements and then they were spread on other areas using daily iodine releases and atmospheric transportation routes. The results of the dose reconstruction allowed to mark zones on the territory of Ukraine according to the average levels of thyroid exposures. These zones underlay a policy of post-accidental health care and social compensations. Another important application of the thyroid dose reconstruction is the radiation risk assessment of thyroid cancer among people exposed during childhood due to the Chernobyl accident.

1. INTRODUCTION

Large areas of Ukraine were contaminated by the radioactive fallout of the Chernobyl reactor accident in 1986. Due to its chemical and physical properties, among the radioisotopes released, ^{131}I contributed significantly to the exposure of the population. Thyroid exposure from ^{131}I, especially to children's thyroid glands, is one of the most important consequences of the Chernobyl accident.

More than 150,000 direct measurements of thyroid ^{131}I activities were carried out by special dosimetric teams under the guidance of emergency group specialists of the Ukrainian Health Care Ministry during May and June of 1986 in Ukraine. Despite a large number of the thyroid activity measurements, 0.2% of inhabitants were measured in oblasts with direct measurements only and wide areas suffered strongly were not covered with the thyroid exposure monitoring at all. Therefore a reconstruction of the thyroid doses due to radioiodine incorporation is of special importance. In 1993, the Ukrainian Ministry on Chernobyl Affairs had initiated large scale reconstruction of thyroid exposure to radioiodine. The objective was to provide the state policy on social compensations aftermath accident with scientific background.

The paper presents some results of the thyroid dose reconstruction in five oblasts from northern Ukraine: Chernihivs'ka[1], Kyivs'ka, Zhytomyrs'ka, Vinnyts'ka, Cherkas'ka. Up to the end of November 1995, 6,773 settlements from these oblasts have gotten certificates of thyroid exposure. Certificate contains the estimates of average thyroid dose from ^{131}I for seven age groups. Table 1 shows the classification and age intervals in the age groups of the thyroid dose certificate.

Table 1. The age groups in thyroid dose sertificate for settlment

Group	1	2	3	4	5	6	7
Year of birth	1986	1983–1985	1979–1982	1975–1978	1971–1974	1968–1970	< 1968
Age at the time of the accident	<1	1–3	4–7	8–11	12–15	16–18	>18

The work on the dose recovery was carried out in series: first, for oblasts with the majority of direct measurements – Chernihivs'ka, Kyivs'ka and Zhytomyrs'ka. The exception was the settlements inside the 30-km zone and city Kiyv that woud be the goal of future investigations.

Then, the dose reconstruction was fulfiled for oblasts more distant from Chernobyl Nuclear Power Plant (NPP) - Vinnyts'ka and Cherkas'ka oblasts. They had only a few direct measurements. The investigation on the thyroid doses recovery is being in process and up to the end of 1995 two additional oblasts – Rivnens'ka and Volyns'ka – will get certificates of thyroid exposure.

2. DATA USED FOR THYROID DOSE RECONSTRUCTION

2.1. Direct measurements of thyroid activities

The primary dose estimates used about 150,000 direct measurements of the ^{131}I activity in the thyroid glands of inhabitants from Chernihivs'ka, Kyivs'ka, Zhytomyrs'ka, and also Vinnyts'ka oblasts. [1]. The settlements of Ukraine in 20 different raions, and also city Kyiv and town Prypiat', with measurements are indicated in Fig 1. The thyroid activity measurements were performed using different types of devices listed in Table 2.

---

[1] In the paper we use latin transliteration from Ukrainian language in names of settlements and regions, except name *Chernobyl*.

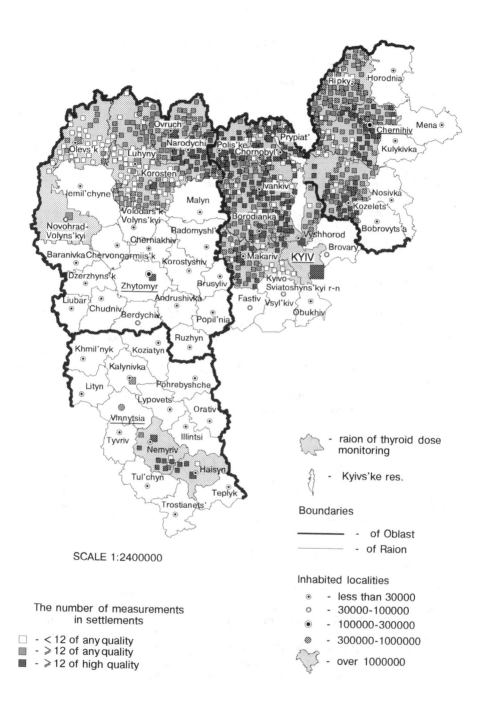

- raion of thyroid dose monitoring

- Kyivs'ke res.

Boundaries

- of Oblast

- of Raion

Inhabited localities

⊙ - less than 30000
○ - 30000-100000
◉ - 100000-300000
⊛ - 300000-1000000

- over 1000000

SCALE 1:2400000

The number of measurements
in settlements

☐ - < 12 of any quality
▨ - ⩾ 12 of any quality
▧ - ⩾ 12 of high quality

Fig.1. Spatial distribution of the thyroid activity measurements
in Ukraine during a period of May-June 1986

Table 2 Devices used for ^{131}I thyroid activities monitoring in Ukraine

Device	Method of measurement	Detector type	Detector dimensions (mm)	Analyzer
SRP-68-01	exposure rate	NaI(T1)	30×25	– [a]
DSU 2-1/DSU-68	spectrometer	NaI(Tl)	63×63	Single-channel
GTRM 01c	spectrometer	NaI(Tl)	40×40	Single-channel
NK 150	spectrometer	NaI(Tl)	25×25	Single-channel
NK 350	spectrometer	NaI(Tl)	25×25	Single-channel
UR 1-1/UR 1-3	spectrometer	NaI(Tl)	63×63a	Four-channel

[a] This device is equipped with two detectors of this size

About 56,000 people from the whole data set were monitored by means of spectrometric techniques used a referent source for calibration. About 35,000 measurements of them were combined in series allowing a statistical analysis of the reliability of these measurements [2]. For the non-spectrometric techniques 4886 measurements were combined in a series with a referent source. Only the serial non-spectrometric measurements with repeated calibration give some indication as to the quality of these non-spectrometric measurements. The analysis of the selected serial reference-source measurements shows that the majority of these measurements (82%) is high or acceptable quality. Therefore the results of non-serial spectrometric technique were defined as high quality individual measurements. The results of the non-serial non-spectrometric measurements were defined as the results of low quality because of the lack of possibility to estimate the reliability of these measurements. .

Table 3 shows the quantity of individual measurements with high and low quality in raions of thyroid monitoring.

2.2. Settlement locations and ^{137}Cs deposition

For the dose reconstruction we use the data on ^{137}Cs soil contamination in every settlement collected by Ukrainian Ministry on Chernobyl Affairs [3]. A special data base has been created to collect the data on ^{137}Cs deposition to soil and geographic coordinates of every settlement for territory under certification. Fig. 2 presents the map of the northern part of Ukraine with izolines of ^{137}Cs deposition.

2.3. Data on behavior in May 1986

The results of large-scale quiz on the individual behavior during May 1986 were used for thyroid dose reconstruction. Fig. 3 presents this questionnaire. The questions especially concerned the following areas: residence (staying in- and outdoors, time and duration of leaving the contaminated territory); the intake of stable iodine; the consumption of fresh milk (source and daily consumption); the consumption of leafy vegetables (origin and daily consumption). The Questionnaire data base contains about 23,000 individual records.

Table 3. Number of the measurements of different quality in raions and cities
with direct thyroid activity measurements in Ukraine

Oblast	Raion	Qua-lity	Number of measurements in age group (according to thyroid dose certificate)						
			1	2	3	4	5	6	7
Vinnyts'ka	Haisyns'kyi	high	16	115	122	110	96	- [a]	16
		low	-	-	-	68	73	-	-
	Nemyrivs'kyi	high	21	227	178	170	120	-	18
		low	-	35	52	40	26	-	-
	Kalynivka town	high	-	-	-	-	-	-	-
		low	-	47	262	18	0	-	38
Zhytomyrs'ka	Zhytomyrs'kyi	high	-	16	104	72	87	-	48
		low	-	-	-	-	-	-	-
	Korostens'kyi	high	-	-	98	111	105	-	-
		low	51	532	747	3024	3119	401	656
	Luhyns'kyi	high	-	-	-	-	-	-	-
		low	-	-	-	-	-	-	155
	Novohrad-Volyns'kyi	high	-	-	-	-	-	-	-
		low	-	-	15	109	109	-	-
	Narodychs'kyi	high	15	111	171	36	57	76	3426
		low	129	1273	1557	1815	1889	953	8000
	Ovruchs'kyi	high	-	17	51	41	60	-	34
		low	199	1900	2534	3747	4019	602	7916
	Olevs'kyi	high	-	-	-	-	-	-	-
		low	-	98	155	164	177	21	252
Kyivs'ka	Borodians'kyi	high	-	-	18	146	105	-	-
		low	-	49	172	1931	2119	564	20
	Vyshhorods'kyi	high	-	-	27	893	1010	366	-
		low	-	177	301	1609	1954	499	79
	Kyiv city	high	-	24	129	1229	1304	12	29
		low	-	83	127	904	711	-	158
	Ivankivs'kyi	high	-	152	214	224	218	49	-
		low	28	196	282	1040	1303	351	233
	Kyivo-Sviatoshyns'kyi	high	-	59	187	1028	1044	-	-
		low	-	12	51	157	185	21	13
	Makarivs'kyi	high	-	-	13	231	195	88	-
		low	-	25	103	1505	1737	363	32
	Polis'kyi	high	-	164	325	438	563	153	-
		low	62	410	446	606	701	337	356
	Prypiat' town	high	-	139	266	613	822	169	17
		low	35	378	627	584	464	142	386
	Chornobyl's'kyi	high	13	166	203	714	926	194	-
		low	29	391	503	443	470	96	500
Chernihivs'ka	Kozelets'kyi	high	37	374	380	307	421	381	1080
		low	138	1555	2090	1878	2175	356	2641
	Ripkyns'kyi	high	-	-	-	15	22	-	-
		low	148	1293	1756	2264	2156	536	3480
	Chernihivs'kyi	high	86	1074	1506	1190	930	145	302
		low	223	2395	3487	4041	3953	657	2355

[a] – less 12 measurements

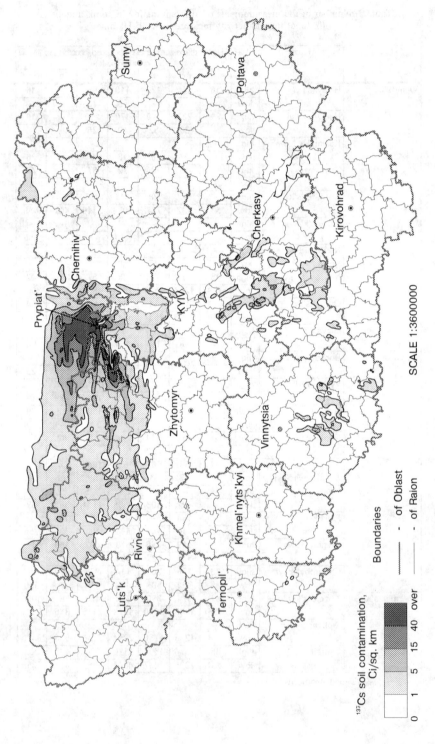

137Cs soil contamination, Ci/sq. km

Boundaries

— - of Oblast
— - of Raion

0   1   5   15   40   over

SCALE 1:3600000

Fig.2. 137Cs soil contamination of the northern part of Ukraine (12 oblasts)

QUESTIONNAIRE

1. Personalities

Surname   Name   Patronymic

Sex [ ]   Date of birth (day, month, year) [ ]

Address at the time of the accident [ ]

Address at present, phone [ ]

2. Peculiar measurement

Were you measured in May-June 1986 in respect to thyroid dose monitoring?- "yes", "no" (underline); if "yes", where? [ ]

3.Countermeasures

3.1 Countermeasures (underline)	wet cleaning, bathing, cloth change,	outdoor time restriction, windows closing

if other (write): _____   Since what time?(date) _____

3.2 Stable iodine administration.

Cross in the table (right) days when you consume stable iodine (even approximately)

TABLE - CALENDAR (APRIL-MAY 1986)

mo	tu	we	th	fr	st	su
					26	27
28	29	30	1	2	3	4
5	6	7	8	9	10	11
12	13	14	15	16	17	18
19	20	21	22	23	24	25
26	27	28	29	30	31	

4. Residence since April 26 to June 1986 (in details)

Residence			Type of a building			Date (day, month, year)	
oblast'	region	settlement	one-floor stone/wooden	many-storied brick	floor	arrival	depature

Fig. 3 Form of questionnaire used for the thyroid dose reconstraction

5.Behaviour

Days	How many hours did you spend (in average)		
	outdoor	indoor with opened windows	indoor with closed windows
26 - 27 April (days off)			
28 - 30 April (working days)			
1 - 4 May (holiday)			
5 - 8 May (working days)			
9 - 11 May (holiday)			
12 - 16 May (working days)			
17 - 18 May (days off)			
19 - 23 May (working days)			
end of May			

6. Consumption of milk (milk products) and of leafy vegetables at the time of the accident.

In table (left side) choose and underline the source of consumed milk (code). In case of absence of appropriate name in the list put it in the row #7.

In table (right side) choose and underline the code of the consumed vegetables. In case of absence of appropriate name in the list put it in the row #7.

Code	Origine
1	not consumed
2	state milk-stores
3	milk from private farm
4	Goat's milk
5	mother's milk
6	baby mixtures
7	(add needed)

Code	Origine
1	not consumed
2	parsley dill
3	lettuce
4	spinach
5	sorrel
6	nettle
7	(add needed)

Fill the following table (use choosen code). *Put the code* into appropriate column " Daily milk consumption " and " Daily consumption of leafy vegetables "

Days	Settlement where you stayed	Daily milk consumption					Daily consumption of leafy vegetables			
		1 l	0.5 l	1 glass	0.5 glass	Other (enter)	50 g	100 g	1-2 leafs	Other (enter)
26 - 27 April (days off)										
28 - 30 April (working days)										
1 - 4 May (holiday)										
5 - 8 May (working days)										
9 - 11 May (holiday)										
12 - 16 May (working days)										
17 - 18 May (days off)										
19 - 23 May (working days)										
end of May										

7. How do you remember your behaviour in May-June 1986 good (underline):

1.Good	2.Satisfactory	3.Badly	4.Hardly

Fig. 3, continued

3. RESULTS OF DOSE RECONSTRUCTION

Two approaches to the dose reconstruction procedure were used. The first one was applied for three oblasts closest to Prypiat' which contained the majority of measurements carried out in May and June of 1986. They were Chernihivs'ka, Zhytomyrs'ka and Kyivs'ka oblasts. Technique has been reported explicitly in [4]. Fig. 4 shows the scheme of this method. The essence of this approach is an using the relationships between the thyroid doses and the ^{137}Cs deposition as well as the location relative to the Chernobyl NPP. Those empirical relationships was estimated for the territory with direct thyroid activity measurements and then they were applied to the territory without measurements.

Questionnaire data were used to check whether the behavior of the population in raions without direct measurements is comparable with that of the rest of the population.

The results of such dose reconstruction were the age-specific average thyroid exposure doses for every settlement in three oblasts (except the settlements of 30-km zone and the city of Kyiv). The obtained data reveal the spatial dose distribution in different age groups of the population. Fig. 5 shows the spatial dose distribution for children born between 1983 and 1985 who dwelt in Chernihivs'ka, Zhytomyrs'ka and Kyivs'ka oblasts.

Estimation of thyroid doses from individual measurement

Age-dependent model of thyroid doses

$$D(age) = K \cdot a^t, \text{ where } t = exp(-b \cdot age)$$

$D(age)$ - mean thyroid dose for given age
K - parameter of the radioiodine intake
a - scaling parameter
b - age-specific parameter

Estimation of model parameters (K,a,b)

---

*Correlation of thyroid dose with ^{137}Cs deposition
and coordinates of settlements*

$$K = K(\sigma, \rho, \varphi)$$

σ - ^{137}Cs soil contamination level
ρ, φ - spheric coordinates of settlement relative to Prypiat'

Calculation of age-specific thyroid doses for settlements

Fig.4. Scheme of thyroid dose reconstruction for population of
Chernihivs'ka, Kyivs'ka, and Zhytomyrs'ka oblasts

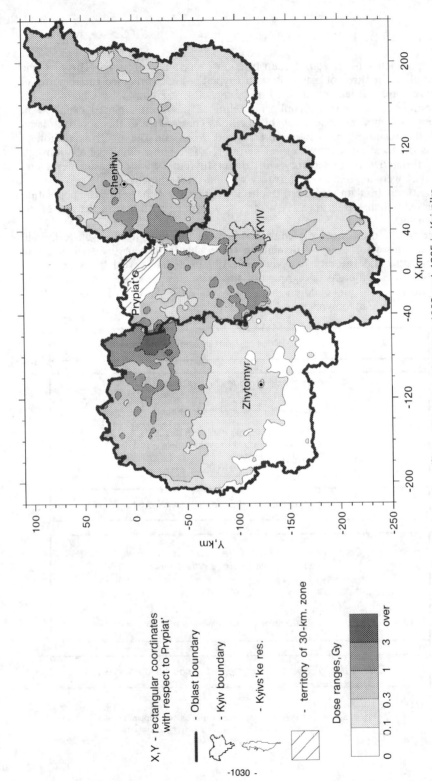

X,Y - rectangular coordinates
with respect to Prypiat'

— - Oblast boundary

— - Kyiv boundary

— - Kyivs'ke res.

[///] - territory of 30-km. zone

Dose ranges, Gy

| 0 | 0.1 | 0.3 | 1 | 3 | over |

Fig.5. Thyroid doses from ^{131}I to children born between 1983 and 1985 in Kyivs'ka,
Chernihivs'ka and Zhytomyrs'ka oblasts

The second approach was applied for oblasts beyond the area of the thyroid activities monitoring - Vinnyts'ka and Cherkas'ka oblasts. Fig. 6 shows the main points of this approach.

Dose reconstruction for this territory was based on empirical relationship between the parameters of the ^{131}I thyroid content model, which indicate the intensity of the accidental environmental contamination, and ^{137}Cs local deposition. Relationship was specified by distance and azimuth to the Chernobyl NPP, by atmospheric transportation routes of radioactive releases, by the day of the iodine stage of the accident.

According to this approach the territory in question was subdivided by the sectors and segments with ^{131}I intake function, suggested to be the uniform one. The subdivision was based on ^{137}Cs soil contamination, the source term of the accident [5], the quantity of the direct thyroid activity measurements, distance and azimuth to Chernobyl NPP.

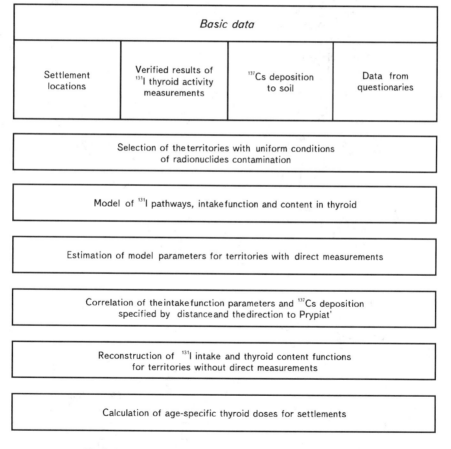

Fig.6. Scheme of the thyroid dose reconstruction for population of Vinnyts'ka and Cherkas'ka oblasts

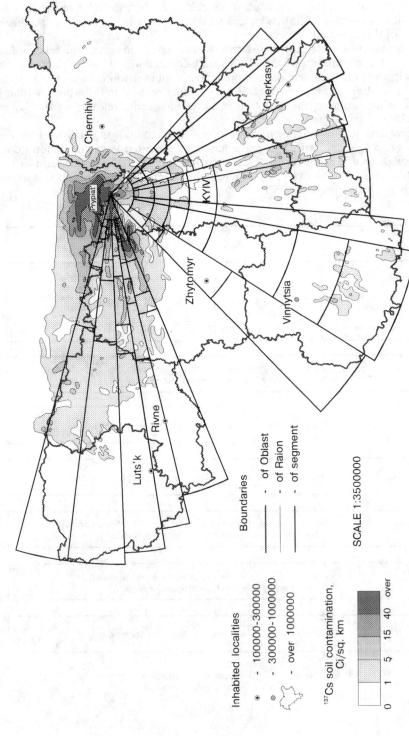

Inhabited localities

⊙ - 100000-300000

⊗ - 300000-1000000

- over 1000000

Boundaries

—— - of Oblast

—— - of Raion

—— - of segment

SCALE 1:3500000

^{137}Cs soil contamination, Ci/sq. km

| 0 | 1 | 5 | 15 | 40 | over |

Fig.7. Sectors and segments on territory of northern part of Ukraine with the uniform ^{131}I thyroid intake function used for thyroid dose reconstruction

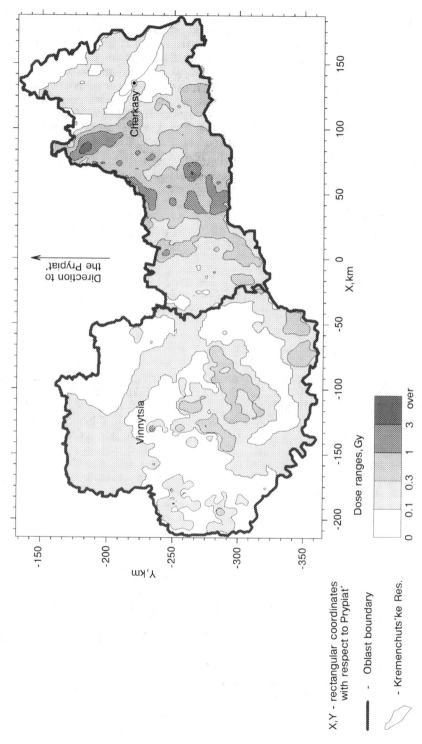

X,Y - rectangular coordinates
with respect to Prypiat'

—▬▬▬▬▬— - Oblast boundary

⬡ - Kremenchuts'ke Res.

Fig.8. Thyroid doses from ^{131}I to children born between 1979 and 1982 in Vinnyts'ka and Cherkas'ka oblasts

Fig.7 shows the location of the sectors and the segments on the territory of the dose recovery. The relationship was found primarily for territory with direct measurements and then it was spread on other areas. The direct measurements performed in three raions of Vinnyts'ka oblast were used for estimation of the model parameters. Questionnaire data allowed to estimate milk and leafy vegetables age- and region-specific consumption rates used as parameters for the thyroid ^{131}I intake function model. Fig. 8 shows the spatial dose distribution for children born between 1979 and 1982 in Vinnyts'ka and Cherkas'ka oblasts. The results of the dose reconstruction reveal also the frequency thyroid dose distribution in different age groups of the population. Table 4 shows the dose distribution in seven age groups for the population of 5 oblasts listed above.

Table 4. Thyroid dose distribution in children of 5 oblasts*(%)

Age at the time of the accident	Dose range, cGy					
	0 – 4.9	5.0 – 9.9	10.0 - 29.9	30.0 – 99.9	100.0 -199.9	≥200
< 1	2.8	10.4	55.5	23.3	7.4	0.5
1-3	3.2	10.3	58.6	24.9	2.1	0.9
4-7	10.8	27.2	47.7	13.4	0.9	0.0
8-11	19.3	39.6	34.0	6.1	0.9	0.0
12-15	35.3	35.2	24.4	4.3	0.7	0.0
16-18	45.7	26.8	23.5	4.1	0.0	0.0
> 18	54.6	21.0	21.6	2.8	0.0	0.0

* – except 30-km zone and Kiyv city

4. CONCLUSIONS

By means of the thyroid dose reconstruction in Ukraine 6673 settlements with about 8,150,000 of inhabitants (2,223 thousand of children aged 0-18 in 1986) received the certificates with the average age-specific thyroid doses. Results of the thyroid dose recovery have revealed dose distributions in different age groups of the population. This allowed to estimate the values of age specific collective doses in different oblasts and raions. Results of reconstruction have also revealed spatial dose distribution. This allowed to seperate dose zones on the territory of Ukraine in according to average levels of thyroid exposure. Dose zones underlay a policy of the post-accident health care and social compensations. Other important applying of the reconstruction is radiation risk assessment of thyroid cancer in people exposed in childhood due to the Chernobyl accident.

Reference

[1] I.A.Likhtarev, N.K. Shandala, G.M.Gulko, I.A.Kairo, N.I.Chepurny. Ukrainian thyroid doses after the Chernobyl accident. *Health Physics* (1993) 64, N 6: 594-599.
[2] I.A.Likhtarev, G.M.Gulko, B.G.Sobolev, I.A.Kairo, G Pröhl, P. Roth, K. Henrichs. Evaluation of the 131I Thyroid-Monitoring Measurements Performed in Ukraine During May and June of 1986. *Health Physics* (1995) 69, N 1: 6 - 15.
[3] Ukrainian Ministry of Chernobyl (1992) Dosimetric characterization of the Ukrainian settlements affected by the radioactive contamination of the Chernobyl accident (in Russian) Kiev, Ukrainian Ministry of Chernobyl Affairs
[4] I.A.Likhtarev, G.M.Gulko, B.G.Sobolev, I.A.Kairo, N.I.Chepurnoy, G Pröhl, K. Henrichs. Thyroid dose assessment for the Chernigov region (Ukraine): estimation based on ^{131}I thyroid measurements and extrapolation of the results to districts without monitoring. *Radiat. Environ. Biophys.* (1994) 33:149-166
[5] Chernobyl: Radiation Contamination of Environment. Edited by Y.A.Izrael, Leningrad (1990).

III. HEALTH EFFECTS
FOLLLOWING THE ACCIDENT
D. Methods and techniques for dose reconstruction

Posters

RETROSPECTIVE DOSIMETRY: THE DEVELOPMENT OF AN EXPERIMENTAL METHODOLOGY USING LUMINESCENCE TECHNIQUES

D. Stoneham[1], I.K. Bailiff[2], L Bøtter-Jensen[3], Y. Goeksu[4], H. Jungner[5], and S. Petrov [2].

[1] Research Laboratory for Achaeology, Oxford University, 6 Keble Rd, Oxford OX1 3QJ, UK.
[2] University of Durham, Woodside Building, South Rd, Durham DH1 3LE, UK
[3] Risø National Laboratory, Postbox 49, DK 4000, Roskilde, Denmark
[4] GSF Forschungszentrum, Ingolstadter Landstr. 1, D8042, Nueherberg, Germany
[5] University of Helsinki, POB 11, SF00014, Finland

Abstract

In a related paper at this conference, the role of luminescence methods in dose assessment in areas contaminated by fall-out from Chernobyl is discussed. This paper presents details of luminescence techniques developed to evaluate doses to ceramics in locations affected by fallout from the Chernobyl accident. The accrued dose is a combination of fallout dose and background dose from naturally occurring radioactivity. For reliable evaluation of the transient dose arising from the accident, the luminescence properties of fired bricks, tiles and porcelain from the contaminated areas has been studied, in particularly sensitivity to dose. Both thermoluminescence (TL) and the newer optically stimulated luminescence (OSL) have been successfully applied. Dose-depth studies have been made on both laboratory irradiated and field samples and results compared with modelling. Measurements on interior and exterior samples have been made to assess shielding; doses to internal locations approaching the minimum resolvable levels of dose (10mGy).

1.   Introduction

Luminescence methods were first used to evaluate fall-out from the Nevada test bomb site[1] as well as in the DS86 assessment[2]. The materials suitable for retrospective measurement are ceramics associated with urban and rural settlements such as bricks, tiles and porcelain fittings. In the earlier studies the main technique used was the pre-dose technique[3]. In the present study, methods such as fine-grain[4], inclusion[5] and pre-dose techniques are used, as well as newer techniques based on optical stimulation using green light[6] and infra-red stimulation[7].

2.   Dose Evaluation

The transient gamma dose, D_X, arising from exposure due to an accident, is given by

$$D_X = D_L - A(D_a + D_b + D_g + D_c)$$

where D_L is the accrued dose determined by luminescence, A is the sample age, and Da, Db, Dg, Dc are the effective internal annual alpha (U, Th), beta (U,Th,^{40}K), gamma (environment before accident) and cosmic dose rates respectively. To determine the accident dose, two groups of measurements have to be made; D_L and A(Da + Db + Dg + Dc). D_L is measured using various luminescence techniques. For the internal dose, the value of A, the age must be known as accurately as possible and the other components are measured using the various techniques described below.

Thermoluminescence (TL) methods

Samples from bricks, tiles, plantpots and other terracotta objects have been measured using several techniques [8,9]:
(i) The fine-grain technique. Samples of powder in the size range 2-8μm are extracted by either drilling or crushing and deposited onto aluminium discs for measurement.

(ii) The quartz inclusion technique. Quartz grains of aproximately 100μm diameter are extracted from the crushed sample by etching in concentrated hydrofluoric acid followed by density separation.
(iii) The pre-dose technique. This technique exploits sensitivity changes observed in the 110°C peak of quartz. It is performed on 100μm grains of quartz prepared as in (ii).
(iv) Porcelain is prepared by coring with a 1cm diamond core drill and slicing into 200-300μm thick slices using a slow speed diamond wheel cutter. The pre-dose technique is the main method used for accident dosimetry [10]. OSL using green light stimulation is also viable[11].

Luminescence Measurements

TL measurements: Measurements were made on a Riso automated reader[12], using a photomultiplier fitted with appropriate filters, a vacuum pump and a source of high purity nitrogen. Laboratory irradiations were performed with 40mCi Sr-90 beta sources, calibrated using sensitized quartz and sensitized porcelain slices so that the calibrations related directly to the materials being measured [13]. OSL measurements: Measurements were made on the same Riso automated reader as above with appropriate OSL attachments.

Background dose-rate measurements

Four techniques are used routinely to determine the background dose-rates arising from naturally occurring radionuclides in the ceramic body: thick source alpha counting for uranium and thorium, flame photometry for potassium content and all three isotopes can alternatively be measured using either neutron activation or gamma spectometry. The mean value for porcelain samples was a contribution of about 3-4mGy per annum. This means that in the 10 years after the accident, 30-40 mGy of the dose consists of the internal component. If the fitment is older, then the contribution is correspondingly more. It is therefore very important for samples from low dose situations such as internal light fittings, where the accumulated dose from all sources is less than 200mGy, that only fitments which were manufactured as close to 1985 as possible are selected.

3. Applications

Properties of materials

The properties of the different materials and methods have been summarised in Table 1. It will be observed that there is a difference between the minimum field dose which can be accurately determined, and the minimum sensitized dose (Min. Sens. Dose in Table 1). All samples containing quartz (which includes porcelain) can be sensitized by dosing and heating several times and this renders them considerably more sensitive than the natural field samples so that the minimum measurable dose can be considerably smaller than that of the field sample. This has important application for their use as background in situ monitors when it is necessary to monitor the ambient dose-rate after removal of the sample.

Table 1. Methods, Sample Types and Minimum Achievable Dose Evaluation

Method	Sample Type	Min. Dose	Field Min Dose	Sens.	Notes	Precision
FG	Pottery	10mGy			Both red and blue emission	<5%
Q inc	Pottery	10-50mGy	100μGy			5-10%
PD	Porcelain	< 10mGy	50μGy		Remove outer 500μm	5-10%
IRSL	Pottery	10-50mGy				<5%
GLSL	Porcelain	<0.5Gy	1mGy		Remove outer 3mm	5-10%

Dose-depth studies

Laboratory studies were made on a modern brick annealed at 500°C and then exposed to a ^{137}Cs photon radiation field from one side only. The dose-depth profile was studied using green light stimulated luminescence on slices (Fig.1), a core scanner[14] along the length of a cut surface (Fig.2) and pre-dose measurements on fine-grains obtained by drilling at intervals along the length of the core (Fig.3) . Field studies were made on samples from two types of location: (i) Rural settlements with wooden houses on a brick foundation where, apart from the foundation, the only other ceramic samples are normally lamp holders and fuses. Fig.4 shows the profile obtained from a foundation brick from a contaminated region in Belarus. The profile was obtained using the pre-dose technique and is consistent with that obtained from the ^{137}Cs laboratory irradiation.

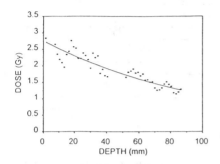

Fig.1 Laboratory irradiated core:
GLSL on slices cut from core

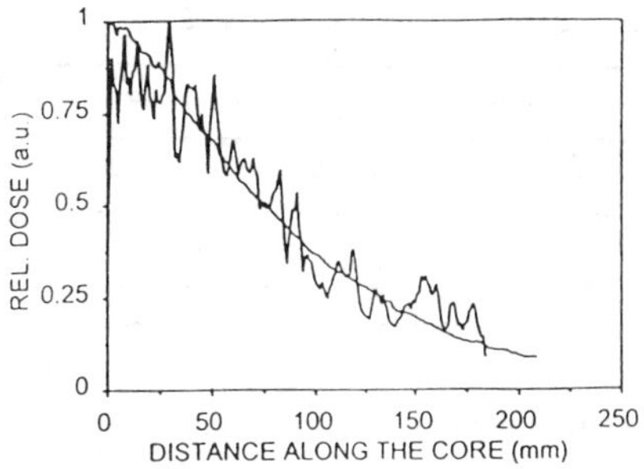

Fig.2 Laboratory irradiated core:
Continuous scanning along length

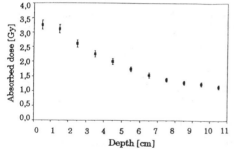

Fig. 3. Laboratory irradiated core:
Fine-grain Pre-dose measurements

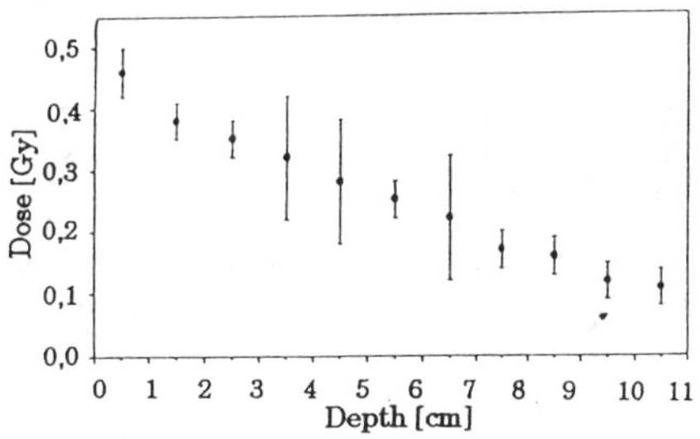

Fig. 4 Dose-depth profile through a brick
from a contaminated region.

(ii)   Urban dwellings may be made from concrete or other synthetic materials but are frequently covered with tiles outside in addition to porcelain fitments. This type of dwelling offers good shielding which can be studied using luminescence technques. Examples of internal samples are porcelain lamp fittings, fuses, sanitary ware and tap fittings as well as stoneware floor and wall tiles. A study was carried out in Pripyat at 18 Prospect Lenina to investigate the shielding given by solid walls. External doses were evaluated for wall tiles (Table 2) and internal doses for porcelain (Table 3). Results from Table 2 show that there is good agreement between TL and OSL evaluations. Mean values of fallout dose from Table 3 are the same for both the 1st floor and the 5th floor, 0.102±0.01Gy, indicating that there is no detectable difference in dose with height.

Table 2. 18 Prospect Lenina: Dose Estimates from External Locations

TILE NO.	LOCATION	TL DOSE (GY)	OSL DOSE (GY)
3	Rear	2.0±0.3	2.1±0.1
8	Front	1.6±0.2	1.5±0.05
44	Front	1.7±0.1	1.4±0.1
45	Front	1.5±0.2	1.4±0.3

Table 3. 18 Prospect Lenina: Dose estimates from Internal Locations

FLOOR	TOTAL DOSE (GY)	INTERNAL DOSE (GY)	FALLOUT DOSE (GY)
1	0.15	0.05	0.10
5	0.17	0.06	0.11
5	0.16	0.06	0.10
1	0.16	0.05	0.11
5	0.13	0.04	0.09
1	0.17	0.05	0.12
1	0.14	0.05	0.09
1	0.14	0.05	0.09
5	0.16	0.05	0.11

7. Conclusions

Luminescence techniques have been successfully applied to a variety of field locations for dose evaluations, dose-depth profiles and shielding studies. Doses of less than 10mGy can be measured - the limiting factor to evaluation of the minimum fallout dose being the background dose due to the internal radionuclides in the ceramic rather than the technqiues. Luminescence has proved a powerful tool of for dose-depth determinations and shielding studies.

8. References

[1] Haskell, E.H., Kaipa P.L. and Wrenn, M.E. (1988). Pre-dose TL characteristics of quartz inclusions removed from bricks exposed to fallout readiation from atmospheric testing at the Nevada test site. Nuc. Tracks Radiat. Meas.14, 113-120.

[2] Maruyama T., Kumamoto Y., Ichikawa Y., Nagatomo T., Hoshi M., Haskell E. and Kaipa P (1987) Thermoluminescence of Gamma Rays in The Dosimetry of Atomic Bombs at Hiroshima and Nagasaki, 2, Ch. 4

[3] Bailiff I.K. and Haskell E.H. (1984) Use of the Predose Technique for Environmental Dsoimetry. Rad. Prot. Dosim. 6, 245-248

[4] Zimmerman D.W. (1971) Thermoluminescence using fine grains from pottery. Archaeometry 13,, 29-52

[5] Fleming S.J. (1970) Thermoluminescence Dating: Refinement of the quartz inclusion method. Archaeometry 12, 133-145

[6] Botter-Jensen L. and Duller G.A.T. (1992) A new systsem for measuring optically stimulated luminescence from quartz samples. Nucl. tracks Radiat. Meas. 20, 549-553

[7] Botter-Jensen L, Ditlevsen C. and Mejdahl V. (1991) Combined OSL (infra-red) and TL studies of feldspars. Nucl. Tracks Radiat. Meas. 18, 257-263

[8] Stoneham D., Bailiff I.K., Goeksu Y., Haskell E.H., Hutt G., Jungner, H. and Tagatomo T (1993) TL Accident dosimetry measurements on samples from the town of Pripyat. Nucl. Tracks Radiat. Meas. 21, 195-200

[9] Hutt G., Brodski L., Bailiff I.K., Goeksu Y., Jungner H. and Stoneham D. (1993) Accident Dosimetry using environmental materials collected from regions downwind of Chernobyl: A preliminary evaluation. Rad. Prot. Dos. 47, 307-311

[10] Stoneham D., (1985) The Use of porcelain as a low dose background dosimester. Nucl. Tracks Radiat Meas. 10, 509-512

[11] Poolton N.R.J., Botter-Jensen, L. and Jungner H., (1995) An optically stimulated luminescence study of porcelain related to radiation dosimetry. Nucl. Tracks Radiat. Meas. 24, No. 4, 543-549

[12] Botter-Jensen L. (1988) The automated Riso TL dating reader system. Nucl. tracks Radiat. meas., 14, 177-180

[13] Goksu H.Y, Bailiff I.K., Botter-Jensen L, Brodski L, Hutt G, Stoneham D. (1995) Interlaboratory source calibration using natural quartz for retrospective dosimetry. Nucl. Tracks Rad Meas. 24, No.4, 479-483

[14] Botter-Jensen L., Jungner H. and Poolton N.R.J. (1995) A continouos OSL scanning method for analysis of radiation depth dose profiles in bricks. Radiat. Meas. 24, No.4, 525-529

^{137}Cs Radiation Burden on Children from a highly contaminated area of Belarus

B. Kortmann[1], R. Fischer[2], V. F. Shaverda[3], H. Wendhausen[1], P. Nielsen[2] and
O. Wassermann[1]

[1]Inst. of Toxicology, Christian-Albrechts-University Kiel, Germany.
[2]UKE, University Hamburg; Germany.
[3]Rayon Hospital, Volozhin; Belarus.

Abstract. The radiation burden from ^{137}Cs sources on 22 children from a small belorussian village was studied from 1992 to 1994. Foodstuff, whole body burden and urinary excretion of ^{137}Cs were measured, intake rates, biological half-lives and doses were calculated. The median value for the ^{137}Cs whole body incorporation level was found to be 124 Bq, the biological half life was calculated as 68 d for girls and 50 d for boys. The internal dose caused by ^{137}Cs was found to be negligible in comparison to that from external sources. No deviation from normal values could be shown in simultaneously studied clinico-chemical parameters.

1. Introduction

In late April 1986 the belorussian village Pershaj (Rayon Volozhin, 70 km north-west of Minsk, 700 inhabitants) was afflicted by the discharges from Chernobyl. A greater proportion of its surface was contaminated with 37 kBq/m^2 ^{137}Cs, to a lesser extent - including some pasture grounds - a contamination of up to 370 kBq/m^2 ^{137}Cs was found [1]. Radiocaesium is known to be the only longterm γ-ray emitting component, that can be detected in man. A study of 22 school children was begun in 1992 to examine their radiation burden caused by incorporated and external ^{137}Cs sources and the influence on the children's health.

2. Material and Methods

Samples of a variety of foodstuffs (1992) and milk (1994) were collected from local farmers and brought to Germany. In order to save time and to emulate belorussian cooking procedures, no reduction of fluid was undertaken. The samples were spectroscopied in 1 L Marinelli beakers in a germanium detector (HPGe p-type: $\Delta E/E$=1.9keV, ε_{rel}.(1331 keV) = 28.3 %; Inst. of Toxicology, Kiel), the detection period being 24 h each time. In 1993 and 1994 the amount of incorporated ^{137}Cs in a total of 22 children (15 females, 7 males) aged 10 to 14 years was measured during a recreational visit to Germany (Oldenburg/Holstein). 19 (1993) and 9 (1994) days after arrival the detection took place at University Hospital (UKE) Hamburg, Div. Med. Biochemistry, in a

4πLSC/HPGe whole body counter (liquid scintill. counter 4πLSC: $\varepsilon_{absol.}(^{137}Cs)$ = 40 %, 2 HPGe p-type: $\Delta E/E$ = 2.1 keV, $\varepsilon_{rel.}$(1331 keV) = 94 %). During the measurement procedure (25 min. each), the children were in contact with a russian-speaking person and were watched by video. Presuming steady-state conditions for intake and excretion in Pershaj, the ^{137}Cs total body activity was corrected for day of arrival, and constant intake rates were calculated from the daily excretion rate constants [%/d] found (*see table 2a*). Simultaneously, the individual biological half-lives were determined by 24-hour urinary excretion measurements of ^{137}Cs (Inst. of Toxicol. Kiel, detection period: 24 hours). The 1-year committed effective dose caused by incorporated ^{137}Cs was also calculated using a factor of $1.56 \cdot 10^{-4}$ $\mu Sv \cdot Bq^{-1} \cdot d^{-1}$ (ICRP 30). After their return to Belarus, the children wore highly sensitive TLD dosimeters (TLD-100, STI/Harshaw, LiF, ($\Delta D/D$ (0.13mGy) = 3.4 %) for 30 days (December 1994) in order to monitor the external radiation dose. To assess the children´s physical situation, hematological parameters were controlled (whole blood count; T3, T4, TSH), the thyroid gland was examined by ultrasound, and they underwent a physical examination.

3. Results

As expected, milk showed the highest contamination of all measured foods (*see table 1*) with a median value of 3.8 Bq/l (range 0.5 - 9.2 Bq/l). The ^{137}Cs total body activity

Table 1: ^{137}Cs contents of different food from Pershaj

Foodstuff	n	Median [Bq/l]	range [Bq/l]
Apples	9	0.67	0 - 1.2
Red Beet	10	0.13	0 - 3.4
Potatoes	14	0.75	0 - 4.0
Milk	9	3.8	0.9 - 3.8

varied from 43 Bq to 2670 Bq with a median value of 124 Bq in girls and 107 Bq in boys (*see table 2*). In the four children with the highest ^{137}Cs body burden (278 - 2670 Bq) ^{134}Cs was found. The biological half-lives were calculated as 68 days for girls and 50 days for boys. From this, the median ^{137}Cs daily intake rates were found to be 1.5 Bq/d for boys and 2.0 Bq/d for girls (range 0.6 - 17.2 Bq/d) under steady-state conditions in Pershaj.

Table 2: Median values of ^{137}Cs incorporation in children from Pershaj

sex	n	age [y]	^{137}Cs total body activity [Bq]	^{40}K total body activity [Bq]	24h-^{137}Cs excret. rate [%/d]	$t_{1/2}$-biol. [d]	^{137}Cs intake rate [Bq/d]
f	15	11.6	124	1840	1.1	68	2.0
m	7	11.4	107	1869	1.2	50	1.5

The median effective equivalent dose caused by incorporated ^{137}Cs was 10 μSv/y with a range from 6 μSv/y to 21 μSv/y. The radiation dose from external sources was found to be 1.4 ± 1.1 mGy/y. No deviation from normal values has been found in the investigated physical and clinico-chemical parameters.

4. Discussion

The calculated low intake rates correspond with daily consumption of milk as the main intake pathway for [137]Cs. Due to its inproportionately high value for the village population and (in comparison with milk) its relatively small portion in children´s food, meat was not included in the study. The high contamination level of the four children mentioned above (max. 78,5 Bq/kg ^{137}Cs in one girl), also indicated by detected [134]Cs, could be explained by consumption of self-harvested mushrooms (*see table 3*) or by the possibility of regular beef intake (mother woking at a meat proceeding facility). The calculated excretion rate constants

Table 3: Maximum contamination level found (girl, regular mushroom consumption)

age [y]	[137]Cs total body activity [Bq]	[40]K total body activity [Bq]	[134]Cs total body activity [Bq]	24h-^{137}Cs excret. rate [%/d]	[137]Cs intern. rad. dose [mSv/y]	[137]Cs intake rate [Bq/d]
11.7	2670	1620	115	0.96	0.21	28.0

i. e. biological half-lifes for boys and girls show the same tendencies as previously described by other authors [2].

In 1993/94 the incorporated activity of ^{137}Cs added an effective equivalent dose of less than 5 % of the natural radiation burden from [40]K (0.1 - 0.2 mSv/y). The radiation burden from external sources is comparable to control TLDs stored in Kiel during December 1994. In extrapolation of data from [1], a dose up to 4.3 mGy/y could be expected as Pershaj is situated near a belt of relatively high external dose rates with hot spots (0.5 - 1.0 µGy/h). One explanation for the reduced external radiation dose could be the limited time spent outdoors in winter. Nevertheless, a direct comparison of doses shows, that, after the decay of short-living isotopes from Chernobyl as ^{131}I, the dose caused by incorporated radiation sources, especially ^{137}Cs, seems to be negligible during the time studied (*see fig. 1*).

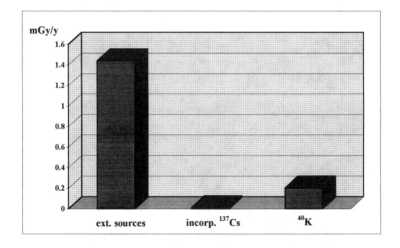

Fig. 1: Comparison of doses from external sources, incorporated ^{137}Cs and [40]K

The methods used to screen the children´s health condition do not represent the most radiosensitive (compared with dicentric chromosome counts or bone-marrow samples), but they are inexpensive, widely established and acceptable to the patient.

This study may give a more detailed insight in the actual local environmental radiation burden in a village of Belarus suffering from the aftermath of Chernobyl, thus, helping to improve also the psychological situation there.

References

[1]: USSR Hydrometeorology Department: „Chart about the radiation burden on the territory of Volozhin, Oblast Minsk, BSSR, in July 1990". Minsk 1990

[2]: Karches, G. J.; Wheeler, J. K.; Helgeson, G. L. and Kahn, B.: Cesium-137 body burdens and biological half-life in children at Tampa, Florida and Lake Bluff, Illinois. Health Physics, 16, 301 - 313, 1969

Monitoring of post-Chernobyl contamination in the Czech Republic

Ivan Bučina[1], Dana Drábová[2], Emil Kunz[1], Irena Malátová[1]
1) National Radiation Protection Institute, 2) State Office for Nuclear Safety
Prague, Czech Republic

Abstract. The results of monitoring of radiation situation in the period 1986-1994 are presented together with the estimation of doses to the Czech population due to Chernobyl accident.

1. Introduction

Large scale monitoring program [1] was implemented in the then Czechoslovakia immediately after the first passage of contaminated air masses from Chernobyl over the territory of the country on 30 April 1986 [2]. Likewise in other countries the first institutions that detected the increase of radioactivity in the air were nuclear power plants during their routine monitoring, then laboratories of hygienic service, hydrometeorologic service and some research institutes started the measurements. On the basis of the Chernobyl experience the Czechoslovak government decided in July 1986 to set up the Czechoslovak Radiation Monitoring Network (RMN) to provide competent authorities with timely information on any changes in radiation situation and that enable them to take appropriate protective measures and to inform the public. Up to the present the data on the activity of radionuclides and on dose rates in the environment are collected and evaluated by the Centre of Czech RMN working in National Radiation Protection Institute in Prague (former Centre of Radiation Hygiene of National Institute of Public Health), monitoring being performed according to monitoring plan set up by the Centre.

2. Methods

In 1986 several thousands of samples were measured to obtain representative values of radionuclide content in all parts of environment and to find out possible extreme values. In following years about one thousand samples (aerosols, fallout, milk and other foodstuff, etc.) were measured annually. All samples were measured by gamma spectrometry using mainly well shielded HPGe detectors. Radiochemical separation followed by alpha and beta spectrometry was used for determination of transuranium radionuclides and ^{90}Sr content in some samples. Additional measurement of surface contamination were performed by means of in-situ and airborne gamma spectrometry. The extensive databases exist about the contamination of environment, food chain and people. The aim of the monitoring in the first days after the accident was in particular the protection of population. Realistic estimation of the average dose and new experimental data on the transfer of different radionuclides in the environment are also important results of long-term monitoring. The average values of radionuclide content in environmental samples and values of doses up to the end of 1992 are valid for the former Czechoslovakia as a whole, since 1993 only the results from the Czech republic were evaluated.

3. Results and discussion

3.1 Activity concentrations in the air

Over the territory of Czechoslovakia three passages of markedly contaminated air masses were observed i.e. on April 30, from May 3 to May 4 and on May 7, 1986 (Figure 1). Activity concentration of radionuclides during those passages differs often even by an order of magnitude depending on locality. The first and the third passage was recorded in all 10 stations which monitored radioactivity in the atmosphere over the Czechoslovak territory. The second passage was not recorded in stations east of 19°E. Up to 20 radionuclides was identified in the aerosol filters in the first days after the accident [1]. The time course of activity concentration of ^{137}Cs and ^{134}Cs up to the end of 1994 is in Figure 2 [3]. Information on particle size distribution was obtained using sampling with cascade impactor [4], from which origin of particles (evaporation, dispersion of fuel) was deduced .

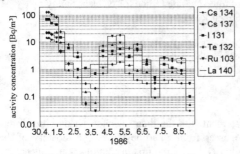
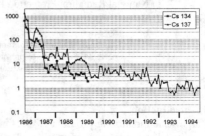

Figure 1 Time course of activity concentration of six most important radionuclides in the air in Prague in the first days after the accident

Figure 2 Time course of monthly averages of activity concentration of ^{137}Cs and ^{134}Cs in the air in Prague

3.2 Concentration of radionuclides in the soil [2, 4, 6]

For the evaluation of fall-out and soil contamination a nationwide survey was organized between June 16 and 18, 1986. Altogether 1300 bare soil samples were collected on sites not shielded by buildings, shrubs and trees, with no grass surface, preferably on agricultural land not tilled since April 26, 1986, with the slope less than 3° , principally not on sandy soil. Results of measurements are listed in Table 1 and for ^{137}Cs displayed in Figure 3. The arithmetic mean of ^{137}Cs deposition in the Czech Republic calculated from the lognormal distribution is 6.5 kBq/m^2. Maximum value of ^{137}Cs deposition found was 95 kBq/m^2 . Some soil samples have been remeasured in 1987 and 1988 to find out radionuclides of minor presence. Mean deposition of ^{110m}Ag (70 Bq/m^2) and ^{125}Sb (140 Bq/m^2) were estimated on the basis of correlation with ^{137}Cs and that of ^{106}Ru (1350 Bq/m^2) on the basis of correlation with ^{103}Ru. Arithmetic means in Table 1 could be underestimated by about 20% as the dry deposition by impaction and interception on the surface of the vegetation was not included.

Table 1. Summary of deposition of individual radionuclides on the Czechoslovak territory as estimated from lognormal distribution of surface activities (reference date June 16, 1986)

Radionuclide	Most Probable Value [Bq/m^2]	Median [Bq/m^2]	Arithmetic Mean [Bq/m^2]	Geometric Standard Deviation
^{137}Cs	600	2190	4200	3.1
^{134}Cs	200	930	1980	3.5
^{103}Ru	490	1560	2800	2.9
^{140}La	0.014 to 1.4	5.18 to 49	101.4 to 294	11.4 to 6.6
^{95}Zr	0.021 to 1.1	7.88 to 20	48.6 to 87	11.4 to 5.5
^{95}Nb	2.92 to 14.5	35.4 to 97	123.3 to 252	4.9 to 3.8
^{141}Ce	3.71 to 7.9	20.63 to 44	48.5 to 103	3.7
^{131}I	9.20 to 14.3	82.74 to 108	248.1 to 295	4.4 to 4.1

3.3 Food contamination [2, 3, 7, 8]

Data about radionuclide content in milk came either from five nationwide surveys which included all large dairies in Czechoslovakia or from regular milk sampling performed from the

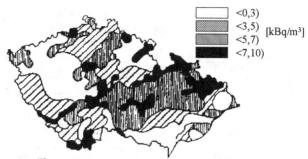

Figure 3 Distribution of ^{137}Cs deposition in soil over the territory of the Czech Republic

beginning of May, 1986 till middle of June, 1986 daily in most of 130 dairies, then weekly, later monthly on a smaller scale covering especially the biggest dairies. The most important radionuclides were ^{131}I, ^{134}Cs and ^{137}Cs. The samples of meat and other important foodstuff have been measured since the accident, too. Time course of specific activity of ^{137}Cs in milk, beef and pork is in Figure 4.

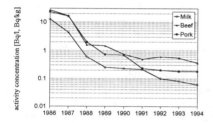

Figure 4 Annual average activity concentrations of ^{137}Cs in milk, beef and pork

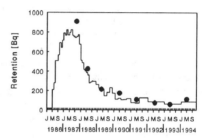

Figure 5 Time course of retention of ^{137}Cs in the Czech population (● nation wide survey by means of mesurement of ^{137}Cs in urine)

3.4 Internal contamination of people [5, 6, 9]

After the Chernobyl accident long term study of the internal contamination has been started. The group of about 30 persons living in Prague has been repeatedly measured by whole body counter. The measurements started on May 1, 1986 and were repeated monthly, later on in longer intervals. From 1993 measurements are performed once a year. Within a short period after the accident internal contamination by ^{131}I, ^{132}Te+^{132}I, ^{103}Ru, ^{137}Cs and ^{134}Cs was measurable. Since the second half of June, 1986 only ^{137}Cs and ^{134}Cs were measurable, since1990 only ^{137}Cs has been measurable. The time course of the ^{137}Cs retention in the refence group in years 1986-1994 is in Fig. 5. In addition to it, as an indirect method, the measurement of cesium daily excretion rate by urine was used, too. Nationwide surveys, that enable to gain information on internal contamination from the whole territory have been performed once a year since 1987. The results are included in Figure 5.

3.5 Doses [3, 6, 9]

The external dose was calculated both for contribution from the radioactive cloud and from the ground deposition. Occupancy factors were taken into account for the calculation of shielding correction. The calculation of effective dose equivalent commitments was as much as possible based on the data from whole body counting. For evaluation of contribution of short lived radionuclides and for assessment of inhalation dose also supplemental information about radionuclide content in air and most important types of foodstuff was used. The assessment of the effective dose equivalent commitments and effective doses during the period 1986-1994 is given for both the external and the internal exposure in Table 2.

Table 2 Summary of doses [μSv] to the Czech population due to Chernobyl accident

	1986	1987	1988	1989	1990	1991	1992	1993	1994
Internal	213	41	9.5	7.5	4.1	3.4	2.9	2.2	4
^{103}Ru	2	-	-	-	-	-	-	-	-
^{131}I	159	-	-	-	-	-	-	-	-
^{132}Te+^{132}I	16	-	-	-	-	-	-	-	-
^{134}Cs	14	14	2.5	1.5	0.6	0.5	0.2	0.1	-
^{137}Cs	22	27	7	6	3.5	2.9	2.6	2.1	4
External	49	31	23	17	14	11.8	10.1	7.9	5.3
Total	262	72	32.5	24.5	18.1	15.2	13	10.1	9.3

4. Conclusions

Immediately after the first reports about the Chernobyl accident the Governmental Commission for Radiation Accidents was activated and started to coordinate all the following activities. All the data concerning radiation situation were summed up and interpreted in terms of the actual and the potential radiation hazards by the Centre of RMN. The main shortcoming of the response of the authorities was the delay of the first official information of the public and the limited will to release fully open information afterwards. On the other hand no substantial technical countermeasures were needed. Even the most conservative estimates of doses to population were substantially bellow the internationally recognized intervention levels. The only exception was the administration of stable iodine to shepherds in mountain region of Slovakia since the activity of ^{131}I in sheep milk they consumed was in some cases several tens of kBq/l. There were, however, a few countermeasures not directly affecting the everyday life of people. Feeding cattle with stored fodder was recommended instead of fresh grass. Milk was daily sampled and measured in all dairies with the intention to discard from direct consumption deliveries exceeding 1000 Bq/l of ^{131}I. The use of fresh milk for production of baby formulae was temporarily stopped and then milk from selected farms from the areas less exposed to fallout was used.

References:
[1] Bučina, I., Kunz, E., Metke, E. Morávek, J. : Basic Principles Covering Environmental Monitoring in the Event of a Radiation Accident at a Nuclear Power Plant in Czechoslovakia, in Emergency Planning and Preparedness for Nuclear Facilities (in Russian), IAEA-SM-280/69, Vienna, 1985
[2] Report on Radiation Situation in ČSSR after Chernobyl accident. IHE-CHZ, Prague (1986). Report to UNSCEAR.
[3] Report on Radiation Situation on the territory of the Czech Republic in the year 1994. National Radiation Protection Institute, Prague. (In Czech).
[4] Rulík,P., Bučina,I., Malátová,I.: Aerosol Particle Size Distribution in Dependence on the Type of Radionuclide after the Chernobyl Accident and the NPP Effluents. Proc. of XVth Regional Congress of IRPA, The Radioecology of Natural and Artificial Radionuclides, Visby, 11-14 Sept 1989, pp.102 -1078 .
[5] Malátová,I., Drábová,D., Češpírová, I. : Internal Contamination of Czechoslovak Population by Cesium Radioisotopes. ibid.pp 433-437.
[6] Validation of models using Chernobyl fallout data from the Central Bohemia region of the Czech Republic. Scenario CB. First Report of the VAMP Multiple Pathways Assessment Working Group. IAEA-TECDOC-795. April 1995.
[7] Drábová,D. et al.: Monitoring of Fallout Radionuclides in Milk in Czechoslovakia after the Chernobyl Accident. Environ. Contamination Following a Major Nuclear Accident, Vol.2 .IAEA Vienna 1990, pp.93 - 96.
[8] Kliment,V., Bučina,I.: Contamination of Food in Czechoslovakia by Caesium Radioisotopes from the Chernobyl Accident. J.Environ.Radioactivity 12, 1990, pp.167 -178.
[9] Malátová, et al.: Effective Dose Equivalents from Internal Contamination of Czechoslovak Population after the Chernobyl Accident. Rad. Prot. Dosimetry, Vol 28, 4, (1989) pp.293-301.

ESR/tooth enamel dosimetry application to Chernobyl case: individual retrospective dosimetry of the liquidators and wild animals

A.Bugai,V.G.Baryakchtar, N.Baran,V.Bartchuk, S.Kolesnik, V.Maksimenko, M.P.Zakcharash, A.B.Bereznoy, A.N.Ostapenko, V.Gaitchenko, V.Radchuk
Institute of Semiconductor Physics, prospect Nauki, 45, 252650 Kiev, Ukraine.

Abstract. ESR / tooth enamel dosimetry technique was used for individual retrospective dosimetry of the servicemen who had worked in 1986-1987 at the liquidation of consequences of the Chernobyl accident. For 18 investigated cases, the values varied from 0,10 (sensitivity limit) to 1,75 Gy. The same technique was used for individual dosimetry of wild animals boarses, red deers, elks) hunted at contaminated 30-km area around the Chernobyl Power Plant. Measured values varied from 0,20 to 5,0 Gy/year and were compared with calculated for external and internal irradiation.

1. Introduction

Many people were irradiated in consequence of the Chernobyl accident. Real individual doses for them may be calculated using known data of external dosimetry but these values are average and have to be verified by independent methods.

The contaminated area around the Chernobyl Power Plant is occupied now by the variety of wild animals and the question is what irradiation doses they usually obtain.

Electron spin resonance of tooth enamel is the only method that allows the direct measurements of individual dose in the case of considerable accidental irradiation of people or animals [1].

2. Experimental technique

Presently there is no generally accepted and unified in all details technique of the ESR/tooth enamel dosimetry; in each case the details are of importance to estimate the correctness of the procedure. In this research Varian ESR E12 spectrometer was used. It was modified with the up-to-date digital technique for multiple field sweep and PC IBM. Because of very weak radiation-induced ESR signals the maximal sensitivity of spectrometer has to be achieved.

To increase the signal/noise ratio the microwave amplifier was used. The important is the field modulation amplitude: it mustn`t exeed 2.5 mT to prevent the overmodulation effects and the loss of the radiation-induced line. Other parameters of the ESR experiment are as follows: microwave frequency 9,3 Ghz; microwave power 10 mW; magnetic field 0,33 T;amplitude of magnetic field sweep 10 mT; ESR spectra were observed at room teperature.The signal was divided into 1024 channels and was accumulated during 36 of magnetic field sweeps, the duration of one sweep was 60 sec. The ESR tooth enamel spectrum was recorded together with the Mn:MgO marker.

The samples for ESR dosimetry were prepared from tooth enamel. The dentine was removed mechanically using the dentists steel and diamond drills. Then pieces of enamel were crushed in agate mortar in order to obtain the grains of size about 0,1...0,3 mm and were treated in the ultrasonic alkaline solution bath for ten hours. Impurities of non-enamel components were removed by hand under the microscop.

Teeth have indvidual sensitivity to radiation and because of this the calibration procedure must be performed. After the careful record of ESR spectrum each sample was in addition iradiated with the calibrated source of gamma radiation. We have used the 137-Cs secondary standard irradiator of the Institute of Nuclear Researches of the National Academy of Sciences of the Ukraine. The additional irradiation was carried out in few steps to obtain the dose dependence of ESR for each sample of enamel. Usually 5 steps of irradiation were used.

Radiation-induced ESR signal in tooth enamel usually is superimposed over the so-called background spectrum of non-irradiated enamel. To find the radiation induced part of the spectrum one must subtract the background spectrum from experimentally observed spectrum of the irradiated tooth. To carry out the procedure the special computer codes LABCALC and SEPARAT were used. For correctness of the subtraction procedure one must use the correct line shapes of radiation induced and background spectra.

The correct line shape of radiation induced free radicals in enamel can be obtained in comparatively simple way and there is no considereble discrepancies about its form.

The problem of the background signal is very important for ESR/tooth enamel dosimetry because of individual peculiarities of this spectrum and impossibility to find this spectrum in the case when the accidental dose must be measured. Because of this the special research of the backgroun line shape was performed.

The limit of sensitivity of the ESR/tooth dosmetry constitutes now about 0,1 Gy. The experimental error is about 30% for doses >0,2 Gy.

3. Background line shape

The background ESR spectrum is inherent to non-irradiated teeth. But one must take into account the natural source of irradiation such as cosmic rays and radiation from natural radionuclei which constitute about 1-3 mGy/year. Because
of this the teeth of young people must be used to obtain the real background line shape. But as a rule sick teeth are extracted and the chemical composition and physical structure of an enamel can be changed and the line shape can be accordingly disturbed. We have proposed to use the pigs teeth. Their age is minimal and thus the natural background irradiation is negligible and unhealthy changes are absent.

We have studied the ESR line shapes of 10-12 years old children and pigs and compared them with standard Gaussian and Lorentzian line shapes. Teeth of children and pigs are very similar. The main central part of all background lines coinsides in details and considerably differs from those of Gaussian and Lorentzian due to substantial asimmetry. For our calculations we have used the averaged line shape of children and pigs.

4. Retrospective ESR/tooth enamel dosimetry of the liquidators

A lot of people were engaged in works at the Chernobyl area after the accident in 1986. A part of them are servicemen and they are under the regular medical control at the Military Medical Board of Security Service of the Ukraine. Extracted bad teeth of some of these people were used for retrospective ESR dosimetry. For some of these people the "official" irradiation doses are known from their medical cards. As a rule the records were made on the grounds of external dosimetry at the places of their work at the Chernobyl area in 1986-1987.

Results of our measurements are shown in Table 1. A part of them is published in [2,3].

Table 1. Indiviual doses of servicemen-liquidators measured with ESR/tooth enamel technique

SampleNo	"Official" dose, Sv	ESR/tooth dose, Gy	Sample No	"Official" dose, Sv	ESR/tooth dose, Gy
3119	-	< 0,10	6662	0,035	0,20
3213	0,10	0,25	6826	0,03	0,19
3331	-	< 0,10	11331	-	< 0,10
4170	-	< 0,10	11956	-	0,15
4824	0,019	0,80	R797	-	0,30
4994	0,013	0,60	G95K	-	< 0,10
5706	0.009	0,30	NN-1	-	1,75
6221	0,036	< 0,10	NN-4	-	0,20
6600	-	< 0,10	NN-5	-	0,30

5. ESR/tooth enamel dosimetry of wild animals

Now the 30-km area around the Chernobyl power plant became the reserve territory and as a result wild animals population is large. The surface density of contamination with radionuclei corresponds to the surface activity from several units to several hundreds Ki per square km. During their life (usually several years) the animals are irradiated and cumulative dose can be large enough. Individual variations of the dose can be found only with ESR/tooth enamel techniques.

Results of our measurements are shown in Tables 2 and 3. Partially they were published in [3,4].

Table 2. Individual doses of wild boars at Chernobyl area

No	Date of hunt	Place of hunt	Age, years	Activ.of muscl., Bq, 137 Cs	Dose, Gy
190	26.02.91	Ilyinysy	2	59000	0,60
191	21.02.91	Zimowistche	?	4700	0,40
192	21.02.91	Maschevo	2	83000	1,10
9201	05.06.92	Opachichi	3	2600	0,35
9202	06.05.92	Kriva Gora	4	40500	0,95
9203	08.06.92	Opachichi	5	270	0,95
9313	19.05.93	Jampol	2	23000	1,50
9316	21.05.93	Korogod	1	4400	0,60
9410	21.05.94	Zimowistche	?	8000	5,60
9411	21.05.94	Usov	4	201000	16,40
9412	22.05.94	Opachichi	?	1400	2.70
9415	14.10.94	Ivanovka	1	?	<0,10
9416	30.10.94	Maschevo	2-3	?	6,10
95**	27.02.95	Krasno	-	?	7.70
95**	-. 95	Buda	-	?	0,85
95**	-. 95	Kriva Gora	-	?	0,50
95**	16.02.95	Opachichi	4-5	?	0,20

Table 3. Individual doses of other wild animals

Animal	Place and date of hunt	Age, years	Dose, Gy
Elk	Opachichi	3-4	0,60
Elk	?	?	0,80
Elk	Koschevka	?	0,70
Red deer	Ilovnitsa	?	1,15
Red deer	Tscherevatch,16.10.94	2-3	0,25
Roe	Jampol, 17.02.95	3	0,25
Roe	Kopatchy, 16.02.95	?	0,65

References

[1] M.Ikeya, New Applications of Electron Spin Resonance - Dating, Dosimetry and Microscopy. World Scientific, Singapore, 1993.

[2] A.A.Bugai et al., The EPR Investigation of Tooth Enamel for Measurements of Absorbed Gamma Doses of People Irradiated in Chernobyl Accident. Reports of the Ukrainian Academy of Sciences (1993) 170-172.

[3] A.A.Bugai et al., Experimental Retrospective Dosimetry and Chernobyl Problems. Reports of the Ukrainian Academy of Sciences (1994) 67-72.

[4] A.A.Bugai et al., External Irradiation Doses of Wild Animals in 30-km Area of Chernobyl Power Plant. Reports of the Ukrainian Academy of Sciences (1994) 149-152.

Reconstruction of the Accumulated Dose in Oncohematological Patients in Belarus

O.V. ALEINIKOVA, A.V. ALEKSEICHIK & V.S. FININ

National Children's Oncohematological Centre, Belarus University
pr. Scorini 64, 220013 MINSK, Belarus

Abstract. The method of the accumulated dose (AD) measured by electronic paramagnetic resonance spectre (EPR) of tooth enamel is the most effective way to reconstruct the accumulated dose for patients.

1. Introduction

For the first time reliable data on the concrete level of the radiation dose have been received. The given doses can be an etiological factor of acute leukaemia in children on the territory of Belarus after the Chernobyl accident. The aim of the study is to investigate the accumulated radiation dose as well as to establish the correlation of the dose value and the incidence of haemoblostosis in children on the territories with different density of radioactive contamination.

2. Materials and methods

At the National Children's Oncohematological Centre 20 EPR-spectroscopy tests of tooth enamel obtained on medical grounds and during autopsy from hospitalised children were carried out. Eight samples of tooth enamel were obtained from the patients with haemoblostosis within the period of time close to the moment of making the diagnosis. Four samples from the patient at the final stage of their treatment which included γ-therapy. Eight tooth enamel samples obtained from the children with various hematological diseases other then haemoblostosis were used as a control group.

3. Results

In the investigated group of three patients with haemoblastosis the accumulated doses from 15 to 30 cGy were found in children from the Brest and Mogilev oblasts the territories of which were contaminated with radioactivity. The AD in five of the children was from 0 to 10 cGy. These children were from Minsk and Vitebsk oblasts which are less contaminated. In the group of the children who underwent γ-therapy in the region of the head for neuroleukaemia prophylaxis the AD was from 30 to 250 cGy. These data prove the sensitivity of the method. The AD in the control group was about zero.

4. Conclusions

The preliminary results on the accumulated radiation dose reconstruction in the children with haemoblastosis make it possible to suggest a possible influence of small radiation doses on the origin of acute leukaemia in children. The present investigations should be carried on, to get statistically, reliable data.

Metabolism in Tooth Enamel and Reliability of Retrospective EPR Dosimetry connected with Chernobyl Accident

A.BRIK, V.RADCHUK, O.SCHERBINA, M.MATYASH, O.GAVER
Institute of Geochemistry, Mineralogy and Oreformation of National Acad. of Sci. of Ukraine, 34 Palladina prosp., 252180, Kiev, Ukraine

Abstract. It is shown that the results of retrospective EPR dosimetry by tooth enamel are essentially determined by the fact that tooth enamel is the mineral of biological origin. The structure of tooth enamel, properties of radiation defects and the role of metabolism in tooth enamel are discussed. It is shown that at deep metamorphic modifications tooth enamel don't save information about its radiation history. The reliability and accuracy of retrospective EPR dosimetry are discussed. Because after Chernobyl accident have passed 10 years the application of tooth enamel for reconstruction of doses which are connected with Chernobyl accident need care and additional investigations.

1. Introduction

It is well known that tooth enamel is one of very potential objects for retrospective EPR dosimetry [1-3]. There are a lot of articles on physical properties of radiation defects in this object [2-6]. Tooth enamel is the typical representative of minerals of biological origin or minerals which have grown in alive organism. The peculiarities of structure and properties of tooth enamel as mineral of biological origin had been discussed by various authors [2,6-9].

Retrospective EPR dosimetry based on tooth enamel can be used for decision medical, social and other problems connected with Chernobyl accident [3]. Using tooth enamel for dose reconstruction of people from Chernobyl accident zone were described in papers [10,11].

But there are a lot of discussing problems in field of retrospective EPR dosimetry based on tooth enamel. They are connected with reliability of reconstructed doses, half life time and stability of radiation defects, changing of radiation sensitivity at diseases of teeth and problems about influence of metabolism and metamorphic modification at the results of retrospective EPR dosimetry.

2. Samples and experimental methods

The experiments were carried out for different kind of human teeth, which were obtained by us in different medical Institutions of Kiev. We have used both powder samples and thin plates from enamel. We coupled to these plates the coordinate system with vertical (v), horizontal (h) and perpendicular (p) axes. These axes were corresponded to the real position of tooth in space. A samples were prepared using stomatological instruments and by gravitation method. This method is grounded at the fact that density of enamel is equal to (2.9-3.0) g/cm^3 and dentin less than 2.8 g/cm^3.

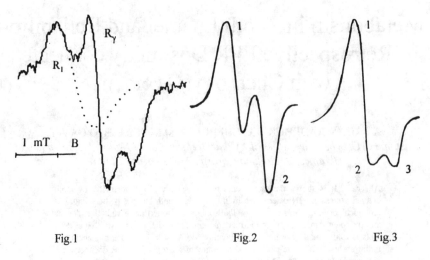

| Fig.1 | Fig.2 | Fig.3 |

The EPR spectra were recorded at T=300K by spectrometer PS-100.X (produced in Minsk, Belarus) and by other spectrometers of X-band. The irradiation of samples were performed using isotopes Co-60 and X-ray tubs. The heating of samples were made in muffle furnace in air atmosphere.

3. Experimental results

As a rule in non irradiation samples of tooth enamel only so called background signal R_1 from organic radicals can be registered. At irradiated samples signal from radiation centers can be registered as well. Fig.1 represents typical EPR spectrum from tooth enamel. This spectrum is related to the sample irradiated to dose about 1 Gy. The parameters of background and radiation signals were described in a number of papers [1-3]. The R_1 signal have asymmetric shape and line width about 0.9 mT. Its g factor is equal approximately to 2 .0045. The radiation signal can be described by axial g tensor with principal values $g_1=g_2=2.002$ and $g_3=1.997$.

We will term radiation centers R_γ centers. Radiation signal R_γ is superposition of signals from CO_2^-, CO_3^{3-} and other radicals. It is important to underline that R_γ centers are mainly connected with carbonate groups. It can be proved by registration of week signals which are connected with magnetic C-13 isotope. This doublet signal (with hyperfine splitting roughly 18 mT) have intensity approximately 1% from intensity of the main signal.

Tooth enamel, as well as other minerals of biological origin, is very textured system. So EPR signals in tooth enamel have anisotropy at rotation of the sample in magnetic field. Fig. 2 represents EPR signal of the plate of tooth enamel (irradiated to dose about 12 Gy) at orientation of magnetic field B parallel to axis p. In article [12] the effects of anisotropy of EPR signals in tooth enamel and effect of reorientation crystallites of hydroxyapatite at the heating of sample were described in detail.

We have established that this reorientation effect and effect of anisotropy of EPR signals in tooth enamel depend on metamorphic modifications and have correlation with disease of teeth. The power of texturing of tooth enamel can be described by texturing coefficient $K_t=I_2/I_1$. Where I_1 and I_2 are intensity of EPR signals of R_γ centers, for which g factor value corresponds to g_3 and g_2 at the orientation of magnetic field B along p axis, see Fig.2. These date were found for samples of sound enamel K_t=(0.9-1.0), but for enamel from carious teeth K_t=(0.5-0.6). It is interesting to underline that for diseased teeth the value of K_t decrease not only for affected part, but for all parts of diseased teeth. In samples from carious teeth the reorientation effect at heating of samples is reduced because the power of texturing for diseased teeth is decreased as well.

Fig.3 is shown EPR signal from R_γ centers in powder sample irradiated by dose about 10 Gy. The shape of this signal can be described by coefficients of shape $k_{ij} = I_i / I_j$, here i,j = 1,2,3 and I_i, I_j are intensity of EPR signal in points 1,2,3 , see Fig.3. For typical signal represented at Fig.3 k_{12}=2.34, k_{13}=2.17 and k_{32}=1.08.

We have established that coefficients of shape for radiation signal in tooth enamel depend on type of teeth dose irradiation, microwave power, time and temperature of heating. For carious teeth the coefficient k_{12} decreased and k_{13} increased and can be equal k_{12}=(0.6-0.8), k_{13}=(2.5-3.0).

We have studied dependence intensity of EPR signals from R_γ centers versus the time of heating of the sample at different value of temperature. It was established that there is distribution of the centers with different half-life time. For example at T=430 K approximately 20% of centers have half-life about 100 min. and at T=450K approximately 40% centers about 70 min.

4. Discussion

It is well known that tooth enamel contains (95-97)% of mineral substances, (2-3)% of organic substances and approximately the same amount of water. Mineral substances are represented mainly by hydroxyapatite, by carbonate hydroxyapatite and by other minerals with similar structure. Mineral substance consists of individual microcrystals - crystallites with size from dozens to hundreds nanometers [8,9]. Crystallites are separated by thin water organic interlayers and they are grown epitaxialy on these interlayers. Since the size of crystallites is small , the surface energy is essential in total energy of crystallites. Thus organic matrix controls, via surface energy of crystallites its properties as well as the properties of biomoneral as a whole [7,8].

We have assumed [7,8] that the properties of radiation centers R_γ in different type of crystallites and crystallites with different size are different. The above mentioned experimental result and other experimental results described in papers [8,12] can be explained from this point of view.

As far as properties of crystallites and correspondently radiation defects are controlled by water-organic matrix, changing properties of this matrix cause to changes in system of radiation defects in tooth enamel. Changing the properties of organic matrix can take place at heating of the sample, at metabolism and at diseases of teeth. The degree of regulating of crystallites for diseases teeth is reduced. It lead to decrease of the texture coefficient k_t accordingly to the above described data. It is obvious that at disease and metamorphic modification of teeth amorphisation of crystallites take place. Because the component 3 in EPR spectrum of R_γ of centers (see Fig.3) is more exiting by spin-orbital interaction, it lead to reducing of this very component of spectrum in carious teeth and increase correspondently the k_{13} coefficient.

It is known that in case of teeth disease essential changes of chemical composition, dimensions, orientation and other properties of crystallites in tooth enamel take place. In carious teeth the quantity of carbonates groups can increase from 5% to 15%. Since the radiation defects in tooth enamel used for dose reconstruction, are associated with carbonate groups and the properties of crystallites are changed at disease of teeth, tooth enamel can not save information about its radiation history at deep metamorphic modification.

At metabolism and modifications of tooth enamel its radiation sensitivity and the shape line of R_γ and R_1 centers can change as well. The latter, in particular, create difficulties at computer processing of experimental spectra.

As far as the teeth used for EPR dosimetry, as a rule, are extracted for medical reasons and if a lot of time passed after irradiation (as in case of Chernobyl accident) it is necessary to conduct additional investigations of dose reconstruction by tooth enamel.

References

[1] M. Ikeya, J.Migajima, S.Okajima, ESR dosimetry for atomic bomb survivors using shell buttons and tooth enamel, *Jap. J.Appl.Ppys.* 24 (1984) L 697-L699.

[2] M.Ikeya, New Applications of Electron Spin Resonance. Dating, Dosimetry and Microscopy. World Scientific, Singapure, 1993.

[3] A.Brik, V.Radchuk, Instrumental retrospective dosimetry, *Priroda*, 2 (1994) 3-17.

[4] F.Callens, R.Vereeck, P.Matthys, L.Martens, E.Boeshman, The contribution of CO^-_2 and CO^{3-}_3 to the ESR Spectrum of tooth enamel. *Calcified Tissue International,* 41 (1987) 124-129.

[5] P.Cevc, M.Schara, Electron Paramagnetic Resonance Study of Irradiated Tooth Enamel *Radiation Research.* 51 (1972) 581-589.

[6] G.Cevc, P.Cevc, M.Shara, U.Snaleric, The carious resistance of human teeth is determined by the spatial arrangement of hydroxyapatite microcrystals in the enamel, *Nature.* 286 (1980) 425-426.

[7] A.Brik, O.Gaver, Principles of building of structure and radiation characteristics of biological origin, *Abstract5s of the 16 th General Meeting of International Mineralogist Association*, Piza, Italy, 1993.

[8] A.Brik, V.Radchuk, O.Scherbina, Metamorphic modifications and EPR dosimetry in tooth enamel, *Abstracts of 4th International Symposium on ESR Dosimetry and Applications*, Munich, GSF, 1995.

[9] E.Borovsky, Terapevtichesnaya Stomatologiya. Medizina, Moscow, 1988 (in Russian).

[10] H.Ishii, M.Ikeya, M.Okano, ESR dosimetry of teeth of residents close to Chernobyl reactor accident, *J. Nucl. Sci. Technol.* 27 (1990) 1153-1155.

[11] N.Baran, V.Barchuk, V.Baryachtar, A.Bugai, ESR investigation of tooth enamel for dose reconstruction connected with Chernobyl accident. *Dokladi Akademii nauk Ukraini* 7 (1993) 170-170, Kiev (in Russian).

[12] A.Brik, N.Saduev, A.Larikov, N.Bagmut, About changing of orientation the crystallites of hydroxyapatite in tooth enamel at heating. *Mineralogicheski journal* 15 (1993) 85-86 , Kiev, (in Russian).

Some Results of the Retrospective Dose Reconstruction for Selected Groups of Exposed Population in Ukraine

Vadim V. CHUMAK, Sergey V. SHOLOM, and Ilia A. LIKHTAREV

Scientific Center of Radiation Medicine AMS of Ukraine,
252050, Melnikova 53, Kiev-50, Ukraine

Abstract. A review of the problem of retrospective dosimetry of some groups of the population of Ukraine is presented. The dose reconstruction efforts are focused now on the cohorts exposed to the highest doses: liquidators and evacuated population. Both analytical and instrumental methods are used for assessment of individual doses of those exposed due to Chernobyl accident. Deterministic and stochastic dosimetric models based on the behavior/migration histories and dose rate data were used for calculation of individual dose of more than 30,000 evacuees. In case of liquidators special fuzzy logic approach was used for assessment of doses with respect to the dose rate data and route lists. However, in some cases, analytical procedures fail to reconstruct dose. ESR dosimetry is used as a tool for instrumental determination of doses to people exposed top doses in excess of 100 mGy. The network for acquisition of teeth is being established in Ukraine in order to provide sufficient coverage of the liquidator population with material for ESR dosimetry. So far, more than 400 teeth from exposed individuals were collected and analyzed by means of ESR. One of the highly exposed groups of these tooth donors are the workers of the Reactor 4 entombment - "Object "Ukrytije". Some results of dose reconstruction for those being exposed in years 1988-1994 are presented in the paper.

1. Introduction

A large fraction of the population, exposed due to Chernobyl, is residing now in Ukraine. Unfortunately, by different reasons, the doses to those are unknown and, therefore, a problem of dose reconstruction for this cohort appears to be of great importance in Ukraine. An acute need for individual dosimetry is constituted by medical, sociological and scientific demands. The efforts in the post-Chernobyl retrospective dosimetry are focused on these two most exposed cohorts - evacuees and liquidators - because of the highest relevance of this population to the biomedical follow up. The final point of the dosimetry of this population is to provide those with the reliable individual dose assessments. Both analytical and instrumental methods are employed in Ukraine for the achievement of this goal. State-of-art approaches and methods are developed in order to perform serial reconstruction of doses and provision of dosimetric registry with necessary data.

2. Evacuated population

About 90,000 citizens of Ukraine, who were residents of the near zone of the Chernobyl nuclear power plant were evacuated during the first after the accident due to the heavy contamination of the environment. Since none of residents wore dosimeters, doses of this cohort were unknown.

Analytical dose reconstruction of this cohort is based on compilation of the results of the direct dose rate measurements which were conducted at time of accident in working and residential locations, and the data from the route lists which contain information about behavior and migration during the period of interest. Aiming the reconstruction of individual doses of evacuees, we had conducted the wide scale public survey in 1988-1989. Data about more than 35000 individual behavior/migration histories were entered into the computer data bases. At the first stage of the investigation, these data was used for calculation of individual external gamma exposure doses with simplified deterministic models. Individual and collective doses of this cohort were assessed. The average effective dose due to external irradiation for this cohort was estimated to be 15 mSv, although individual values vary in an extremely wide range from 0.1 to 383 mSv. The collective dose of the whole evacuated population was found to be 1,300 person-Sv.

At the moment the revision of individual dosimetry of evacuated population is in progress. This effort, being carried out in close collaboration with GSF, Neuherberg touches basically two aspects of the dosimetric models: reassessment of the values of location factors for dwellings in the 30-km zone and the city of Pripjat, and application of the Mote Carlo method and stochastic models. These stochastic models allows us to take into consideration uncertainty and variability of parameters of the model. As a result, the dose distributions individually to every person would be produced. The details of this study are presented elsewhere [1].

3. Liquidators

3.1. Analytical dose reconstruction

Another cohort of the highest priority in the sense of post-Chernobyl follow up and dose reconstruction are the Chernobyl clean-up workers, so-called liquidators. Application of analytical method for reconstruction of doses to liquidators is even more complicate due to enhanced requirements to accuracy of description of the professional routes which is caused by extreme heterogeneity of the dose rate fields at the NPP site. Usage of the fuzzy sets approach allows us to assess individual doses received by personnel. At the moment, more than 1700 route lists of liquidators of 1986 were processed; results of this dose assessment had created the basis of Chernobyl clean-up workers dose registry.

Unfortunately, the above approach is applicable not to all of the Chernobyl liquidators. A large fraction of clean-up workers had performed their duties in the NPP being not familiar with topography of the site as well as with the aims and details of the work they performed. Therefore, lot of liquidators could not describe their movements and activities making, thus, preparation of the route list impossible. For this group, analytical dose reconstruction is not valid and, therefore, different methods of retrospective dosimetry should be employed.

3.2. ESR-dosimetry with teeth

One of the most promising methods of retrospective dosimetry is ESR spectroscopy of tooth enamel. The main principles and approaches of the ESR dosimetry were studied and developed in SCRM from the point of view of application of this method to the reconstruction of individual doses to the groups of exposed population, particularly liquidators. As a result of this investigation the semi-routine version of the ESR dosimetric technique was developed and adopted by the Ministry of Health of Ukraine as a tool for dosimetric support of the post-Chernobyl follow up. The technique allows for reliable reconstruction of doses in excess of 0.1 Gy with good capabilities for serial analyses. The quality o f the results produced by this technique was proven in series of intercomparisons and cross-validations, in particular in the First International Intercalibration of ESR dosimetry with teeth [2].

The ESR dosimetric technique is extensively used for practical dose reconstruction. So far, the teeth from more than 400 liquidators were collected and analyzed in SCRM. At the moment the special sample acquisition network and the central bank of bioprobes are being established in Ukraine. This network will be based on the centers of the compact residence of liquidators being driven from the center in SCRM. The central bank of bioprobes is called to register, store and retrieve the samples for further analysis and dose reconstruction. According to our approach, the bioprobes, available in course of the normal dental treatment of liquidators will be collected and transferred to SCRM; since the ESR-dosimetry with teeth is extremely labor intensive, the dose reconstruction will be conducted in due of time.

4. Workers of "Ukrytije"

It is well known that the most dose intensive activities in Chernobyl have been and are currently performed at the location of the destroyed reactor #4 of the Chernobyl power plant. Since the completion of the construction of the entombment over the ruins of the reactor (officially called "Object "Ukrytije") the special research group was performing investigations inside the reactor building. Due to concentration of the fuel masses and radioactive dust inside the entombment, dose fields in these locations were extremely heterogeneous achieving in some places very high levels of dose rate. Although, official dose records certify absence of the cases of overexposure, there are several evidences of exposure of workers to extremely high doses. Moreover, it was reported that some of workers had "shadow" dose records, supported by themselves during the working period, which demonstrate extremely high doses (tens of Gy) in the period of 1988-1994. A summary of these "legendary" data together with results of cytogenetic tests were published by Sevan'kaev et.al. in 1995 [3] confirming an extra high exposure of these workers.

Our task was to perform EPR dosimetry using teeth of workers who were occupationally exposed at "Ukrytije". There were 4 individuals who had donated their teeth to the investigation. Individual doses were determined using semi-routine EPR-dosimetric technique which was developed and implemented in SCRM. The teeth were processed mechanically and chemically in order to obtain samples of pure tooth enamel to be used in further measurements. Individual calibration using additive dose method was performed in order to account for variations in radiosensitivity of samples. In one case, the tooth was found to be demineralized and no radiation induced (RI) signal was found in the EPR spectrum. Additional irradiation up to the dose of 1.05 Gy had initiated no RI signals. This sample was considered to be unsuitable for retrospective dose reconstruction. In three other cases dose reconstruction was successful. Moreover, two individuals had donated two of teeth each with ca.1.5 year interval between analyses. Doses were determined over the lifetime period prior the moment of extraction. Teeth from the subject #1 were extracted in 1992 and 1994, from the subject #2 - in 1994 and 1995, and from the subject #3 in 1995. An analysis of the individual dose values has revealed that although the dose values are not as high as it was reported elsewhere, rather intense exposure of individuals who were involved into the activities inside the "Ukrytije". The data obtained for the sequential sampling from two persons give evidence that dose per annum was in order of 0.5 Gy/y (which is much higher than any accidental or occupational dose limits). It is desirable to perform a systematic comparison of doses, received by these individuals with their "private" dose records and results of biodosimetric (FISH) dose assessment.

5. Conclusions

The intense work is being performed in Ukraine now in order to provide retrospective dose assessments to the persons exposed due to Chernobyl. The cohorts with the highest doses are liquidators and the population evacuated from the 30-km zone. These cohorts are of the highest relevance from the point of view of biomedical follow up. An approach developed in Ukraine presumes the involvement of large populations into the dose reconstruction program. A large fraction of evacuees was covered with the survey allowing for reconstruction of individual doses to persons of different ages and gender; these data are open for use in biomedical research of this cohort. Reconstruction of doses to liquidators implies use of both analytical and instrumental methods. Being less time consuming, analytical dose estimation methods give a possibility to provide significant groups of liquidators with dose assessments. Due to high labor-intensity we are going to use ESR-spectroscopy of tooth enamel for selective verification of the results of analytical dosimetry and dosimetric support of the most critical participants of the epidemiological follow-up. The special efforts are undertaken in Ukraine in order to provide the ESR dosimetry with sufficient amount of material and to achieve sufficient coverage of the liquidator cohort with potential for instrumental dose assessment.

References.

[1] R.Meckbach and V.Chumak, Reconstruction of the external dose of evacuees from the contaminated areas based on simulation modeling. First International Conference of the European Commission, Belarus, Russian Federation and Ukraine on the Radiological Consequences of the Chernobyl Accident, proceedings, 18-22 March 1996, Minsk.
[2] V.Skvortsov et.al., International intercomparison of dose measurements using EPR spectrometry of tooth enamel. First International Conference of the European Commission, Belarus, Russian Federation and Ukraine on the Radiological Consequences of the Chernobyl Accident, proceedings, 18-22 March 1996, Minsk.
[3] A.Sevan'kaev et.al., High Exposures to radiation received by workers inside the Chernobyl sarcophagus. *Radiat. Prot. Dosim.* **59** (1995), 85-91.

Some Advances in the Instrumental Retrospective Dosimetry Techniques with Tooth Enamel and Quartz.

Sergei V. SHOLOM, Vadim V. CHUMAK, Larisa F. PASALSKAJA
and Juri V. PAVLENKO

*Scientific Center of Radiation Medicine AMS of Ukraine,
Melnikova str. 53, 252050 Kiev, Ukraine*

Abstract. Some aspects of retrospective dosimetry with tooth enamel and quartz have been considered. Firstly, the experimental and theoretical investigation had been carried out concerning influence of secondary electron equilibrium on the absorbed dose in enamel under the laboratory irradiation. The irradiation had been made with photons of energy 1.25 MeV, 662 and 100 keV. It is demonstrated that the influence of secondary electron equilibrium on the absorbed dose in enamel does not exceed few percent. Secondly, some of paramagnetic centers of enamel different from CO_2^- ones have been researched by using of the thermoactivation technique. The enamel for this experiment had been carefully purified from organic components and then irradiated following annealed to consecutively increasing temperature. It was established that at least four of EPR centers of enamel possess radiation sensitivity and could be used for dosimetry purposes. Thirsty, it was performed a thorough investigation of the influence of different stages in quartz separation and purification with respect to obtaining of samples for TL-dosimetry. The optimal procedure has been developed.

1. Introduction

One of the main problems that are founding in the retrospective dosimetry with teeth and quartz is the correct determination of low level accident doses. Solving of this problem is possible in two ways. It may be reached by refinement of steps of existing techniques or development the techniques based on the new principles. Concerning of first way the present work had deal with well-known EPR-dosimetry tooth enamel and TL-dosimetry quartz inclusion techniques. The refinements could be made either in preparation of samples (better purification of the enamel and quartz samples) or in the laboratory irradiation of samples (simpler conditions of additional irradiation with more accuracy). Concerning of second way the possibility of application of some new paramagnetic centers in enamel to the aims of retrospective dosimetry has been investigated.

2. Retrospective individual EPR-dosimetry technique with tooth enamel.

2.1. Additive laboratory irradiation of the tooth enamel samples in the EPR-dosimetry technique.

The determination of individual sensitivity of tooth enamel to gamma-radiation is one of the most important stages of the EPR-dosimetry technique with tooth enamel. This gives a possibility to convert from the EPR signal intensity to the dose absorbed in the enamel sample. The method of additive dose is used for this purpose and an enamel sample under investigation is irradiated with calibrated doses under the laboratory condition. It is needed to reproduce the conditions of teeth irradiation in vivo as adequately as possible. This is achieved by a choice of appropriate photon energy and geometry of irradiation. The influence of the first factor on the magnitude of the radiation EPR signal of enamel is not substantial for energy more than few hundreds keV; because of this, the Cs-137 or Co-60 sources of photons are most often used.

Regarding the irradiation geometry, Shimano et al. [1] had investigated influence of the phenomena of secondary electron equilibrium on the absorbed dose in enamel. The problem lies in the fact that the teeth in vivo are irradiated under the condition of secondary electron equilibrium since they are located behind the soft tissue (behind cheek). In the laboratory such equilibrium had been reached by the location of tooth between two plates of polymethyl methacrylate (PMMA) with effective atom number close to the one of the human tissue.

In the present work an experimental and theoretical investigation had been carried out concerning influence of secondary electron equilibrium on the absorbed dose in enamel under the laboratory irradiation of fine-grained tooth enamel samples which is commonly used in the retrospective dosimetry technique while determination of individual radiation sensitivity of enamel. The equilibrium of secondary electrons is achieved by using the PMMA plates with different thickness and different geometry. The irradiation had been made with photons of energy 1.25 MeV (Co-60), 662 keV (Cs-137) and X-ray with effective energy 100 keV. The Monte-Carlo method had been used for simulation of corresponding irradiation geometry. A contribution to total dose from the absorption of scattering photons had been taken into account. It is demonstrated that if enamel samples consist of the grains with a characteristic dimensions 1 mm or less, the influence of secondary electron equilibrium on the absorbed dose in enamel does not exceed few percent while irradiating these samples under the laboratory conditions in the EPR-dosimetry technique.

2.2. An investigation of radiation properties of thermoactivated EPR-centers of tooth enamel.

At present, tooth enamel retrospective dosimetry bases on the radiation properties of EPR-signal from the CO_2^- centers. The reconstruction dose technique based on these centers usually assumes measurement of EPR spectra at ambient temperature. In this case a special control takes place so that the temperature does not exceed 60-80 $^\circ$C in the process of the sample preparation for measurement of spectra (this is related to the lifetime of CO_2^- centers). At the same time it is known that the tooth enamel contains some number of paramagnetic centers differing from CO_2^-, which appears at the time of the enamel heating to the fixed temperature. Some of these centers have radiation sensitivity and could be used for dosimetry purposes.

In the present work the research of the radiation properties of the tooth enamel EPR signals have been performed by using of the thermoactivation technique. The enamel for investigations had been obtained as a result of the mixing of few tens of unirradiated teeth. The separation of enamel from dentine and its purification from contamination and carious parts have been conducted by a chemical method by putting of tooth parts into the alkali solution with simultaneous influence of the ultrasound and enhanced ($60\ ^{O}C$) temperature. The pure enamel has been crushed into 0.1-0.25 mm grains and weighed out to the 80 mg aliquots. The samples had been irradiated in the range from 0 to 200 Gy. Then the samples had been annealed to consecutively increasing temperature in the range from ambient temperature to $850^{O}C$. After each annealing, the recording of the EPR spectra of the enamel sample has been performed. It has been found as a result of the decomposition of acquired spectra that enamel contains at least four different radiation induced paramagnetic centers in the region of g-factor 2.0. Concerning 3 centers from them we did not find any mention in the special literature. The above-mentioned centers become apparent in the irradiated enamel samples after their annealing to $140^{O}C$ temperature or more. The temperature and dose characteristics these centers had been investigated.

3. The retrospective dosimetry of environmental objects with thermoluminescence of quartz.

3.1. Preparation of Pure Quartz Samples for TL-Dosimetry

Quartz is known to be the mineral, in which the TL-emission occurs within a characteristic temperature range, and in certain limits, the intensity of the emission appears to be proportional to the accumulated dose. For this reason, quartz-containing materials that are annealed at high temperature in course of manufacturing (e.g., bricks and ceramics), can be used as TL-dosimeters.

Currently, methods of TL-dosimetry of quartz including both high-temperature and more sensitive predose methods, have found widespread application. Most features of separation of quartz from quartz-containing materials have already been studied. The whole procedure usually includes the stages of crushing and sieving a demanded fraction of rough material, magnetic separation (in order to remove magnetic admixtures), and treatment of samples with a concentrated solution of HF. Sometimes, floatation is used instead of magnetic separation.

It should be noted, however, that there are many problems in each of the mentioned stages, and our ability to handle those problems will finally determine the correctness of the dose determination. For example, in order to crush bricks, it is preferable to use a press versus a mortar, since in the latter case some additional TL-peaks may arise on the curve due to mechanical generation of luminescence centers. The average size of grains in a sieved fraction and the method used for separation of quartz and removal of impurities are also of importance. Our experience allows us to conclude that it is not enough to perform a single act of magnetic separation before treatment of samples with HF, since the etching releases insoluble impurities fused into quartz grains. Therefore, additional purification is needed.

We have performed a thorough investigation of the influence of different stages in quartz separation with respect to the suitability of obtained samples for predose TL-dosimetry. The optimal procedure looks as follows: the core of a brick is cut off with a diamond saw and crushed up with a mechanic press. The fraction of 0.1 to 0.25 mm size is sieved out, poured with a solution of hydrochloric acid (1:6), and left over a period of a few hours (usually, over a night). This ensures the dissolution of carbonates, and, in part, feldspars. Then, using a rubber pestle, quartz particles are separated from slurry that is washed off by a large amount of water. Then, wet quartz samples are subjected to the treatment with concentrated solution of hydrofluoric acid for a few minutes, rinsed with distilled water, ethanol, and dried at $80^{\circ}C$. To remove some minerals, which are insoluble in acid solutions, the samples are placed into a sodium polytungstate solution of density 2.70-2.75 g cm^{-3}, where quartz grains float, whereas heavier impurities sink. After the floatation, quartz is washed with water, ethanol, and dried once again. Then, under a microscope (magnification to some tens), colored and cracked grains are removed. Grains of pure quartz are colorless and transparent. All procedures described above should be conducted in red light in order to avoid bleaching of the TL- peaks.

References

[1] T. Shimano et al., Human Tooth Dosimetry for Gamma-rays and Dental X-ray Using ESR, *Appl. Radiat. Isot.* **40**(1989) 1035-1038.

Biological Dosimetric Studies in the Chernobyl Radiation Accident, on Populations Living in the Contaminated Areas (Gomel Regions) and in Estonian Clean-Up Workers, Using FISH Technique

F. DARROUDI, A.T. NATARAJAN

MGC, Department of Radiation Genetics and Chemical Mutagenesis,
University of Leiden, Wassenaarseweg 72, NL-2333 AL Leiden, The Netherlands and
J.A. Cohen Institute,
InterUniversity Research Institute for Radiopathology and Radiation Protection,
Leiden, The Netherlands

Abstract. In order to perform retrospective estimations of radiation doses seven years after the nuclear accident in Chernobyl, the frequencies of chromosomal aberrations in the peripheral blood lymphocytes of individuals living in contaminated areas around Chernobyl and the Estonian clean-up workers were determined. The first study group composed of 45 individuals living in four areas (i.e. Rechitsa, Komsomolski, Choiniki and Zaspa) in the vicinity (80-125 km) of Chernobyl and 20 individuals living in Minsk (control group - 340 km from Chernobyl). The second study group (Estonian clean-up workers) composed of 26 individuals involved in cleaning up the Chernobyl for a different period of time (up to 7 months) and a matched control group consisting of 9 probands. Unstable aberrations (dicentrics and rings) were scored in Giemsa stained preparations and stable aberrations (translocations) were analyzed using chromosome specific DNA libraries and fluorescence in situ hybridization (FISH) technique. For both study groups the estimated average dose is between 0.1-0.4 Gy. Among the people living in the contaminated areas in the vicinity of Chernobyl, a higher frequency of numerical aberrations (i.e. trisomy, hyperdiploidy) was evident.

1. Introduction

On April 26, 1986, when the radiation accident in the Chernobyl nuclear power plant occurred, several thousands of subjects were exposed to high doses of beta and gamma irradiation and a very large number of populated areas were contaminated as a result of the nuclear fallout. Among subjects exposed, were personnel of the power plant, those living in the vicinity of Chernobyl and groups involved in rescue and clean-up operations [1,2]. Since reliable physical dose measurements are not available from Chernobyl victims, alternatively biological dosimetry can be used. Generally, chromosomal aberrations (in particular dicentrics) in peripheral blood lymphocytes are being used to estimate the absorbed dose immediately following the accident [3].

However, difficulties in dose estimation arise for the past exposure, due to a decline of cells containing such unstable chromosome aberrations [4-6]. Fluorescence in situ hybridization (FISH) technique employing chromosome specific DNA libraries to "paint" individual human chromosomes [7] has opened new prospectives for rapid and precise detection of stable chromosome aberrations such as translocations (which are generally assumed to remain fairly constant for a long period of time) in peripheral lymphocytes of irradiated individuals [6,8].

In the present study seven years after the accident, we have used conventional dicentrics analysis and FISH technique for retrospective biological dosimetry of past exposure. Different groups were selected, i.e. subjects living in the vicinity of Chernobyl and the Estonian clean-up workers.

2. Materials and methods

The first study group composed of 45 individuals living in four areas (Choiniki, Komsomolski, Rechitsa and Zaspa) and a control group from Minsk (Table 1). The second study group composed of 26 subjects divided in four groups on the basis of different exposure levels they might have received. Among these subjects one group working with the construction of the sarcophagus, second and third groups working in the vicinity or on the roof of the reactor for 4-7 and 0-3 months, respectively, and the fourth group working in the 10 to 30 km area of the prohibited zone or beyond this area and a matched control group from Estonia was selected (Table 2).

Whole blood from each individual was collected in heparinized tubes, 0.5 ml of blood is added to 5.0 ml culture medium of Ham's F10 containing 15 % foetal calf serum (heat inactivated at 56°C for 30 minutes), phytohaemagglutinin (PHA), L-glutamine (200 mM), heparin and antibiotics. Four cultures were set up for each donor, they were allowed to grow for 48 hours (2 cultures), 54 hours (1 culture) and 68 hours (1 culture) in an incubator at 37°C. Colcemid at a final concentration of 1.0 µg/ml was added to the cultures fixed at 48 and 54 hours after simulation for the last 2 hours, and cultures fixed at 68 hours treated with Colcemid (final concentration 0.3 µg/ml) for 14 hours (using this protocol, yield of mitotic index increased by a factor of 2-3 compared to 2 hours Colcemid treatment regimen. Cells that have divided more than once appear as tetraploid which can be easily distinguished from diploid cells). 5-Bromodeoxyuridine (at final concentration of 10 µM) was added to all cultures in order to analyze cell cycle progression and the scoring of chromosomal aberrations was confined to cells in the first mitotic division. A routine fixation protocol was employed following treatment with hypotonic solution (KCl 0.075M) and fixation in acetic-acid: methanol (1:3). Air-dried preparations were made under an infra-red lamp.

For detection of unstable chromosomal aberrations such as dicentric, ring and acentric fragment in the first mitotic division, slides were stained according to FPG-technique [9]. Between 100-1000 metaphases were scored for each individual.

Chromosome specific DNA libraries and fluorescence in situ hybridization were applied to detect stable aberrations (i.e. translocations). Bluescribe DNA libraries specific for chromosomes 1, 4, 10, 15, 17 and/or X were labeled with biotin-16-dUTP by nick translation following standard protocol (7, 10). Generally, a cocktail of DNA libraries specific for 4 chromosomes were employed, namely, 1, 4, 15, X for the first study group, and for Estonian clean-up workers namely 1, 4, 15, X and 1, 4, 17, X for two donors 1, 4, 10, X and for three donors, 1, 4, X and for one 4, 15, X, representing 19.9, 19.6, 21.1, 16.9 and 11.8 % of the human genome, respectively (for men, and for women plus 2.5 %).

The preparations were examined under a Zeiss or Leitz microscope equipped with DAPI and FITC epifluorescence optics. 100-1000 metaphases were analyzed for each individual. Two types of translocations namely reciprocal and terminal were observed. In addition to structural aberrations (i.e. translocations) by using DNA libraries for specific chromosomes presence of trisomy/hyperdiploidy could be easily visualized.

3. Statistical analysis

Assuming that the exchange events occur randomly in all the chromosomes in the genome, depending on the DNA content of the cocktail employed in the study, the frequencies of translocations for whole genome were estimated [8]. Average translocation frequency for the whole exposed group was calculated and the average dose was estimated from both, the in vitro dose-response curve [8] and from the in vivo, an estimated linear dose-response curve calculated from victims of the Goiania accident [11], those we estimated to have received doses less than 1 Gy. Exact methods, based on the poisson distribution, were used for significance tests and construction of confidence limits.

4. Results

The frequencies of chromosomal aberrations for the populations living in the vicinity of Chernobyl and Estonian clean-up workers (including matched control groups) are presented in Tables 1 and 2, respectively. Based on these tables, the frequency of dicentrics observed in the control group in Minsk was 0.15 % and in the contaminated areas in the range of 0 (in Choiniki, but frequency of ring chromosomes was found to be 0.16 %) and 0.3 % (in Rechitsa), and the average dicentric frequency in the total groups from four cities, were 0.18 per 100 cells. Frequency of translocations in control group (Minsk) was 0.6 and in the contaminated areas in the range of 1.1 (in Zaspa) and 1.8 (in Rechitsa) per 100 cells, and the average translocation frequency was 1.4 per 100 cells (Table 1). Compared to the control group, no significant increase was found in the frequencies of dicentric and translocation, though the latter is slightly higher than control but was found to be not significant (p > 0.05). In addition to translocations, using chromosome specific DNA libraries (i.e. 1, 4, 15 and X) numerical aberrations were also detected and it was found to be significantly high (p < 0.01) in Choiniki, Rechitsa and Komsomolski when compared to the control group (Table 1).

Table 1.   Average dicentric and translocation frequencies and retrospective dose estimates in populations living in the vicinity of Chernobyl

City	No. of subjects	Cells scored	Genomic (%)		Hyperdiploidy (%)	Estimated dose (Gy)
			Dicentric	Translocation		
Minsk (control)	20	3800	0.15		0	-
		2882		0.6	0	
Choiniki	9	1802	0		0	0.25
		1931		1.4	0.15	
Komsomolski	16	8830	0.12		0.28	0.25
		5070		1.5	0.45	
Rechitsa	7	5750	0.3		0.26	0.4
		1400		1.8	0.20	
Zaspa	13	2976	0.22		0	0.18
		2645		1.1	0	

For Estonian clean-up workers data are presented in Table 2. The average dicentric and translocation frequency was 0.28 and 1.2 per 100 cells. When compared to the control group (from Estonia), a significant increase ($p < 0.05$) was found in the dicentric frequency while the difference in translocation is not significant ($p = 0.19$, one sided test) (Table 2). However, when we pooled data for control from both groups, an increase in the frequency of dicentrics in clean-up workers from Estonia was not significantly different from control groups (Tables 1 and 2). When we compared these data with the historical control values, 1 per 1000 cells for dicentrics and 5 per 1000 cells for translocations both groups under study showed a significantly increased frequency of both dicentrics and translocations ($p < 0.01$).

Table 2.   Average dicentric and translocation frequencies and retrospective dose estimates in Estonian clean-up workers

Groups of subjects*	No. of subjects	Cells scored	Genomic (%)		Estimated dose (Gy)
			Dicentric	Translocation	
Control	9	2500	0		-
		1370		0.62	
a	5	1150	2.6		0.04
		1083		0.8	
b	5	1400	0.7		0.28
		979		1.7	
c	5	1200	1.7		0.23
		1146		1.5	
d	11	2800	4.3		0.13
		2880		1.1	

* a) at construction of sarcophagus, b) and c) at roof or vicinity of reactor, working for 4-7 and 0-3 months, respectively, d) at 10-30 km zone. A detailed report on this analysis is under publication [12].

Retrospective dose estimates were obtained on the basis of translocation frequencies using two different calibration curves: one using a linear dose response obtained from frequency of translocations in the victims of the Goiania accident, exposed to doses less than 1 Gy [11], and the other one, using the in vitro dose-response curve [8]. They yielded a dose estimate of 0.13-0.4 and 0.10-0.25 Gy, respectively.

5. Discussion

Provided that translocation frequencies, unlike unstable dicentrics, do not essentially change with post-exposure time, recently, the possibility of using translocation frequencies for estimating radiation doses of past exposure has been suggested [8]. In this study, the frequencies of translocations in atom bomb survivors seem to be related to the ones which were expected on the basis of in vitro dose response curve for the calculated doses during the accident. In the case of victims of the Goiania accident, using initial dicentric data and comparing with translocations data 6 years after accident [11], we have found that the translocation frequencies can be fitted as a function of estimated dose with the linear, quadratic and linear-quadratic dose response models for the total data set as well as for individuals with doses less than 1 Gy, but for doses above 1 Gy a correction factor is required to estimate accurately past exposure doses retrospectively [11].

The aim of the present study was to use FISH technique for biological dose estimation of the inhabitants of four cities in the vicinity of Chernobyl and Estonian clean-up workers almost 7 years after accident, which presumably received a rather high chronic and acute exposure, respectively. On the basis of the dose response curve established (in vivo) for the victims of Goiania accident, the dose estimates presented constitute an estimate of the average exposure, and it was found to be in the range of 0.18 up to 0.4 Gy, for which retrospective dose estimation is feasible at individual level. FISH technique can discern individuals exposed to high doses and groups with different levels of radiation exposures. However, in order to assess reliable statistical analysis, the required elements are: the number of cells analyzed and the proportion of genome labeled [12], the calibration curve used for back-calculation, back-ground level and persistence of "stable" translocations specially at doses above 1 Gy [11].

Acknowledgement:

This research was financially supported in part by grants from the National Institute of Public Health and Environmental Protection (The Netherlands).

References

[1]   UNSCEAR, United Nations Scientific Committee on the Effects of Atomic Radiation: Sources, Effects and Risks of Ionizing Radiation, United Nations, New York, 1988.

[2]   M. Tekkel et al., Estonian Chernobyl clean-up workers study, Epidemiology 6, abstract 153, 1995.

[3]   A.T. Ramalho et al., Dose assessment by cytogenetic analysis in the Goiania (Brazil) radiation accident, Radiation Protection Dosimetry 25 (1988) 97-100.

[4]   K.E. Buckton, Chromosome aberrations in patients with X-irradiation for ankylosing spondylitis. In: T. Ishihara and M.S. Sasaki (Eds.) Radiation-induced Chromosome Damage in Man. Liss, New York, 1983, pp. 491-511.

[5]   M. Bauchinger et al., Radiation induced chromosome aberrations analyzed by two-colour fluorescence in situ hybridization with composite whole chromosome-specific DNA probes and a pancentromeric DNA probe, International Journal of Radiation Biology 62 (1993) 673-678.

[6]   A.T. Natarajan et al., A cytogenetic follow-up study of the victims of a radiation accident in Goiania (Brazil), Mutation Research 247 (1991) 103-111.

[7]   D. Pinkel et al., Cytogenetic analysis using quantitative, high sensitivity, fluorescence hybridization, Proceedings of the National Academy of Sciences USA 83 (1986) 2934-2938.

[8]   J.N. Lucas et al., Rapid translocation frequency analysis decades after exposure to ionizing radiation, International Journal of Radiation Biology 62 (1992) 53-63.

[9]   P.E. Perry and S. Wolff, New Giemsa method for differential staining of sister-chromoatids, Nature (London) 251 (1974) 156-158.

[10]   A.T. Natarajan et al., Frequencies of X-ray-induced translocations in human peripheral lymphocytes as detected by in situ hybridization using chromosome-specific DNA libraries, International Journal of Radiation Biology 61 (1992) 199-203.

[11]   A.T. Natarajan et al., Biological dosimetric studies in Goiania Radiation Accident, International Atomic Energy Agency (in press).

[12]   F. Granath et al., Retrospective dose estimates in Estonian Chernobyl clean-up workers by means of FISH, Mutation Research (in press).

IV. OFF-SITE MANAGEMENT
OF FUTURE ACCIDENTS

Decision support systems for the off-site
management of future accidents

European Commission's Contribution to Improving Off-site Emergency Preparedness

G N Kelly
European Commission, DG XII/F/6,
Rue de la Loi 200, Brussels 1049, Belgium

Abstract.   Increasing attention is being given by the European Commission to off-site emergency preparedness as part of its broader contribution to improving nuclear safety in Eastern Europe. The main initiatives being taken or planned by the Commission in this area are summarised. Particular attention is given to two topics: firstly, the development of the RODOS (Real-time On-line DecisiOn Support) system for supporting off-site emergency management in the event of a nuclear accident; and, secondly, the work of an Inter-Service Group on nuclear Off-Site Emergency Preparedness (OSEP) in Eastern Europe that has recently been established within the Commission. The contribution that each is making to improving emergency preparedness, both in Eastern Europe and in Europe more widely, is described.

1.   Introduction

Over the past few years the Commission has taken a number of initiatives to improve nuclear safety in Eastern Europe[1]. While the main focus of these has been accident prevention and mitigation, increasing attention is now being given to off-site emergency preparedness as the third and final link in the nuclear safety chain. The main initiatives taken or planned by the Commission in this latter area are summarised in this paper. Particular attention is given to two topics: firstly, the development of the RODOS (Real-time On-line DecisiOn Support) system for supporting off-site emergency management in the event of a nuclear accident; and, secondly, the work of an Inter-Service Group on nuclear Off-Site Emergency Preparedness (OSEP) in Eastern Europe that has recently been established within the Commission. The contribution that each is making to improving emergency preparedness, both in Eastern Europe and in Europe more widely, is described.

2.   The RODOS Decision Support System

Following the Chernobyl accident, increased resources were allocated in many countries to improve systems to aid the off-site management of any future nuclear accident. Much has since been achieved but much yet remains to be done to ensure an integrated, coherent and consistent response to any accident that might in future affect Europe. The need for and

---

[1]   Unless otherwise indicated, the use of Eastern Europe in this paper refers to all countries in East and Central Europe and to the European Countries of the Former Soviet Union.

importance of a coherent and consistent response were amply demonstrated following the Chernobyl accident when differences in the countermeasures taken by national authorities contributed greatly to a loss of public confidence. The development of a Decision Support System (DSS) for off-site emergency management, that would be comprehensive and capable of finding broad application across Europe, was included as a major item in the Radiation Protection Research Action of the European Commission's 3rd Framework Programme. The following considerations were central to the inclusion of this item in the programme:

- to make better use of resources in the European Union (EU) for further improving off-site emergency management (eg, minimise unnecessary duplication, integrate best features of systems developed at national levels, etc)

- to benefit from the development of a comprehensive (ie, applicable at all distances, all times and to all important countermeasures) and fully integrated decision support system that was generally applicable across the EU (eg, seamless transition between different stages of an accident, greater continuity and consistency in decision support, etc)

- to provide greater transparency in the decision process as one input to improving public understanding and acceptance of off-site emergency actions

- to provide a common platform or framework for incorporating the best features of existing and future DSS

- to provide a basis for improved communication between countries of monitoring data, predictions of consequences, etc, in the event of any future accident, and

- the overriding consideration, to promote, through the development and use of the system, a more coherent and harmonised response to any future accident that may affect the EU.

It is evident that these considerations, set out above in the context of the EU, are equally if not more pertinent to Europe as a whole.

Development of the RODOS system began in late 1990. For institutional reasons participation in the project was initially restricted to EU institutes. Forschungszentrum Karlsruhe (FZK) has coordinated the project throughout and, inter alia, has overall responsibility for developing the system and integrating the software products of other contractors. About 10 EU institutes were initially involved in the project increasing since to almost 20. Means were subsequently found to broaden the scope of participation beyond the EU with two main benefits: firstly, greater and more diverse resources were made available to the project leading to a better final product and, secondly, the opportunity to develop a system that would be applicable across the whole of Europe as opposed to the narrower confines of the EU. The latter is particularly important in terms of achieving effective and timely response to any future nuclear emergency. About 10 institutes from Belarus, Russia and the Ukraine were formally integrated within the project in 1992 under the auspices of a collaborative programme on the consequences of the Chernobyl accident between the EC and the State Committees on Chernobyl Affairs in the respective countries. More recently, institutes from Poland (in 1993), Hungary, Romania and the Slovak Republic (in 1994) have joined the project under the auspices of the Commission's PECO programme (Pays d'Europe Centrale et Orientale - Scientific and Technical Cooperation with Central and East European Countries) and requests to participate have been received from others (Czech Republic and Slovenia). By 1995 almost 40 institutes were involved in the development of RODOS with about half from Eastern Europe. This wide and diverse participation in the development of the system augurs well both for its efficacy and future implementation.

2.1   Objectives of the system and potential users

The basic concept, design and software framework of the system were specified and agreed by participants at the outset of the project and its main conceptual features are indicated in Figure 1. Decision support can be provided at various levels ranging, in increasing sophistication, from the largely descriptive, to providing an evaluation of the benefits and disadvantages of different countermeasures' options, to ranking them according to the decisions makers' expressed preferences for different outcomes. Most decision support systems, developed to an operational state, are limited to providing analyses and predictions of the current and future radiological situation. Some extend to the simulation of countermeasures but are often limited in the range of countermeasures they address or in the completeness of the benefits and disadvantages that are considered. The few systems that have progressed to the evaluation and ranking of alternative countermeasures' options are limited in the range of countermeasures they address. RODOS is unique in that it will provide comprehensive support (ie, at all levels) for each potentially useful countermeasure at all times following an accident.

The system is being designed to fulfil a number of roles, the more important of which are:

*   full or partial integration into emergency arrangements at local, regional, national or supra-national levels (ie, subject to interfacing with radiological monitoring and meteorological networks and with the decision making process)

*   providing a more effective means for communication and exchange of monitoring data, prognoses of accident consequences, etc, between countries

*   a stand alone interactive training tool for use, inter alia, by those responsible for making decisions on off-site emergency management and their technical advisers at local, regional, national and supra-national levels

*   a more general interactive training and educational tool for radiation protection, nuclear safety and emergency planning personnel with a professional interest in and/or responsibility for off-site emergency management

*   contributing to improvements in existing decision support systems through the development and dissemination of improved stand alone modules

*   a research and development tool to explore the merits and limitations of new techniques or approaches prior to their integration into operational decision support systems

- providing greater transparency in the decision process as a contribution to better public understanding and acceptance of emergency actions

- a basis or framework for decision support systems for the management of non-nuclear emergencies with potential widespread off-site consequences.

Not all roles will be of interest or relevant to every potential user. Consequently, the system has been designed in a modular way so that it can be tailored to the user's particular needs.

The roles for which RODOS is being designed largely determine its potential users. These include those responsible at local, regional, national and supra-national levels for off-site emergency management and related training, for the operation of nuclear installations, for public information, or for communication and exchange of information (eg, in accord with bi-lateral or international agreements); the research and development community concerned with improving decision support for off-site emergency management; and developers of decision support systems for the off-site management of non-nuclear emergencies.

2.2   Status of the system and its future development

A first prototype (PRTY 1.0) of the RODOS system was completed in 1992 and has been installed in institutes in Belarus, Germany, Greece, Hungary, Russia, Poland, Romania, the Slovak Republic and the Ukraine; requests for the system have also been received from institutes in other EU and East European countries. A second prototype (PRTY 2.0) and a first pilot version (PV1) of the system were completed in the autumn of 1995, the end of the first phase of the project (ie, the Commission's 3rd R&D Framework Programme). The second prototype has greater functionality through the integration of software developed by the various EU and East European partners and expansion of the user interface. The first pilot version (PV1), with functionality limited to the early and intermediate stages of an accident, has been developed specifically for on-line testing in emergency centres where it will be interfaced with meteorological and radiation monitoring networks and the decision making process itself, albeit in a pre-operational mode. Summaries of the overall design of the system, its software/hardware environment, its current status and plans for its implementation in Belarus, Russia and Ukraine, and the technical content of its major modules can be found elsewhere in these Proceedings [1-7] and in references [8-11]. Current developments and progress with the system are reported in periodic issues of the RODOS Newsletter [12], copies of which can be obtained on request.

The system will be further developed with support from the Commission's 4th R&D Framework Programme (1995-98). Both EU and East European institutes will be involved in a fully integrated manner. Further developments will focus on extending the applicability of the system to encompass all stages of an accident (ie, to the late stage including the long term management of contaminated land and the subsequent return to "normality" after an accident) and making improvements in those areas where there is a demonstrable need. Two topics will receive particular attention in the latter context: firstly, the development of an integrated approach for the handling of uncertainties and their effective communication to decision makers (an important issue that has received insufficient attention in the past) and, secondly, the development of improved methods for assimilating and making better use of expert judgement, model predictions and monitoring data. Improvements will also be made in response to experience gained in the pre-operational use and testing of the system in several European countries, in particular its interface with meteorological and radiological monitoring networks and with the decision making process. Important feedback has already been obtained from using the prototype system in exercises with decision makers and this aspect will receive increasing attention in future. A fully operational and comprehensive version of RODOS, applicable throughout Europe, is scheduled for completion by mid-1999, the end of the second phase of the project.

With the completion of the pilot version and a commitment from the Commission to support the further development of RODOS, potential users are increasingly recognising the many benefits which the system offers. In particular, its potential role as part of a wider European network has become evident. The existence of such a network would promote a more effective and coherent response to any future emergency in Europe. Four factors will largely determine how far and how quickly the RODOS system (or elements of it) finds operational use as part of emergency arrangements within Europe: firstly, the results of pre-operational testing of the pilot version in several countries in 1996; secondly, the extent to which the technical objectives of the second phase of the project are achieved; thirdly, a commitment by countries in Europe to take advantage of these new developments, a matter which will be influenced by broader and largely non-technical considerations; and, fourthly, the extent to which assistance can be made available to accelerate the implementation process in Eastern Europe. The interest currently being shown in the system by many EU and East European countries augurs well for its future use. Subject to the successful pre-operational testing of the pilot version (applicable to the early and intermediate stages of an accident) and assistance to accelerate its implementation in Eastern Europe, the basis of a European network of RODOS centres could be in existence by 1997.

3. Inter-Service Group on Off-Site Emergency Preparedness (OSEP)

Considerable resources have been and continue to be allocated by the Commission, through its TACIS and PHARE programmes, to improve the safety of nuclear reactors in Eastern Europe. By far the majority of these resources have, quite properly, been directed towards accident prevention and "in-plant" mitigation should accidents occur. By comparison, few resources have so far been directed towards off-site emergency preparedness, the third and final link in the nuclear safety chain. Prior to 1995 informal arrangements were used within the Commission to coordinate the initiatives being taken by its various Services to improve off-site emergency preparedness in Eastern Europe. Such an approach was appropriate for the relatively restricted scale and nature of the initiatives then being taken by the Commission. More recently, however, the need to improve emergency preparedness in Eastern Europe has been recognised as a necessary part of a more balanced and integrated approach towards improving nuclear safety.

The OSEP Group was established in response to this need and in anticipation of an increase in the resources which the Commission might allocate for this purpose. In such circumstances, the informal arrangements of the past were no longer judged appropriate, in particular given the many Commission Services with an interest in, and/or formal involvement with, any assistance programme. The need for a more formal and structured approach within the Commission was, moreover, exemplified by the apparent lack of coordination in this area among donor countries, both within the EU and more widely. Assistance on off-site emergency preparedness has been and continues to be forthcoming from many countries and in many different guises but often in an uncoordinated way. In particular, there is evidence of duplication with consequential wastage (or at best under-utilisation) of limited resources.

The OSEP Group was established in mid-1995 with the following terms of reference:

- to develop a coherent and integrated programme within the Commission to improve off-site emergency preparedness in Eastern Europe

- to formulate priorities within this integrated programme

- to advise TACIS/PHARE (Nuclear Safety) in developing a well balanced programme for improving nuclear safety (ie, balanced in terms of accident prevention, on-site mitigation and off-site emergency preparedness)

- to advise the TACIS/PHARE programmes on how and where resources could be most effectively allocated to improve off-site emergency preparedness in Eastern Europe

- to be consulted and to advise on the scope and content of all projects supported by the Commission concerned with off-site nuclear emergency preparedness in Eastern Europe and to evaluate their progress

- to establish appropriate means to coordinate/integrate the Commission's programme with other bi- and multi-lateral initiatives being taken in this area, in particular to minimise duplication and achieve a better utilisation of resources

The main objectives of the OSEP Group are:

- to contribute to improvements in local, regional, and national off-site emergency preparedness arrangements in Eastern Europe

- to establish and/or improve arrangements for information exchange within Eastern Europe and with the EU in the event of any future nuclear emergency

The following Directorate Generals (DG) of the Commission are currently formally represented on the OSEP Group: ECHO (European Community Humanitarian Office), DG IA (External Political Relations - TACIS/PHARE nuclear safety programmes), DG III (Industry), DG XI (Environment, Nuclear Safety and Civil Protection), DG XII (Science, Research and Development), DG XIII (Telecommunications, Information Market and Exploitation of Research), DG XVII (Energy) and the Secretariat-General; arrangements exist to involve others as need or interest dictates. The secretariat and chairman of the Group are provided by ECHO given their mandate within the Commission for coordination of disaster preparedness activities outside the EU.

3.1   OSEP Initiatives

Following its creation the OSEP Group reviewed the Commission's existing and planned activities in this area as an input to establishing a coherent and integrated programme of assistance. Apart from the RODOS decision support system, being developed within the Commission's R&D programme (see Section 2), the only other major project was the implementation, with support from the TACIS programme, of the first stage of an early warning system for nuclear accidents in Belarus and Ukraine (GAMMA-1); the scope and content of this project are described elsewhere in these Proceedings [13]. Two immediate priorities were identified by the OSEP Group to enable it to meet its two main objectives set out above. The first (in response to the first of the Group's objective) was to commission an assessment of needs in each of the countries concerned; clearly these should largely determine the nature and content of any future assistance programme. The second (in response to the second of the Group's objectives) was to explore possible means for improving the exchange of radiological information across Europe as a whole in the event of any future accident. The scope and objectives of this assessment of needs are summarised below together with initiatives being taken to improve information exchange.

3.1.1 Needs Assessment

The need for assistance in the area of off-site emergency preparedness, as part of a balanced approach to improving nuclear safety, is beyond question. However, it was judged essential for these needs to be prioritised, both within and between countries, in order to develop a well considered and directed assistance programme within a constrained budget. Moreover, it was also essential to identify what assistance had been or was being provided through other bi- or multi-lateral assistance programmes, both to avoid duplication and to allocate future assistance more effectively.

The objectives of the needs assessment were:

- to determine the current status of off-site emergency preparedness in each country

- to identify the nature and content of past, current and foreseen bi- or multi-lateral assistance projects

- to evaluate current and/or foreseen arrangements

- to identify where assistance is needed to bring off-site emergency preparedness to an adequate level

- to establish priorities for assistance within and between countries

- to prepare a data base to facilitate the implementation and monitoring of any future EC assistance programme

The countries included initially in the needs assessment were Belarus, Bulgaria, Czech Republic, Estonia, Hungary, Kazakhstan, Latvia, Lithuania, Romania, Poland, Slovenia, Slovak Republic, Russia, and Ukraine. The topics to be covered by the needs assessment include the following:

- legislative bases

- organisational arrangements and responsibilities

- on-site arrangements (in so far as they affect off-site emergency preparedness)

- off-site monitoring and early warning

- communications (local to international)

- arrangements for informing the public

- arrangements and criteria for countermeasures

- impact forecasting and decision support capabilities

- emergency response services capabilities

- exercises and training

Consultants (ES-konsult, IGTS and ENCONET) were appointed by the Commission to carry out this "needs assessment" to be completed in January 1996. The findings of this assessment are being used by the Commission to develop its future assistance programme in this area.

3.1.2   International data exchange

In the event of a nuclear accident the rapid and reliable exchange of radiological information, both within and between potentially affected countries, is a pre-requisite for timely, effective and coherent emergency response. The timely and efficient exchange of information between countries is of particular importance within Europe given the widespread use and dispersal of nuclear installations and the potentially large number of countries which may be affected, directly or indirectly, by any future accident. The establishment and/or improvement of arrangements for information exchange within Eastern Europe and with the EU is the second of the two objectives of the OSEP group.

Following the Chernobyl accident major improvements were made in the exchange of information following a nuclear accident. In particular an international Convention was established by IAEA [14] and a Decision taken by the Council of Ministers of the European Communities [15]; the latter was implemented through the European Community Urgent Radiological Information Exchange (ECURIE) system [16]. In addition a large number of bi-lateral arrangements have been made between neighbouring countries. More recently a pilot project (EURDEP - European Union Radioactivity Data Exchange Plantform) [17] was initiated to investigate improved means for data exchange, in particular in a continuous and automatic manner; a number of topics are being addressed, including common data formats and the possible establishment of a "European network" based on existing monitoring stations on a 100 by 100 km grid. Notwithstanding the considerable progress that has been made since the Chernobyl accident, it is evident that much more could and needs to be done to make best use of the very large amounts of radiological monitoring and other data (eg, accident consequence prognoses, etc) that would now be generated in the event of any future accident.

The development of a system for the on-line exchange of radiological information between European countries would greatly improve the efficacy of off-site response to any future nuclear accident that may occur in Europe. In this context a contract was placed for Terms of Reference to be prepared for the development and testing of a prototype of such a system.

The following aspects were to be addressed:

- specification of the conceptual design of a European system for on-line information exchange

- specification of a prototype system

- specification of the radiological information to be exchanged

- specification of the functional requirements of the hardware, software, data exchange formats, etc

- development of a testing programme for the prototype including quality assurance aspects.

This study is currently being carried out and, subject to its outcome, a decision will be taken on whether to support the implementation of such a prototype system.

4. References

[1] Ehrhardt, J, Shershakov, V, Zheleznyak, M and Mikhalevitch, A, RODOS: Decision support system for off-site emergency management in Europe. IN First International Conference of the European Commission, Belarus, the Russian Federation and Ukraine on the Consequences of the Chernobyl Accident, Minsk, 18-22 March (1996) - this issue.

[2] Shershakov, V, Ehrhardt, J, Zheleznyak, M and Mikhalevitch, A, The implementation of RODOS in Belarus, Russia and Ukraine and future perspectives. Ibid.

[3] Schüle, O, Rafat, M and Kossykh, V, The software environments of RODOS. Ibid.

[4] Glushkova, V and Schichtel, T, Modelling of early countermeasures in RODOS. Ibid.

[5] Brown, J, et al, Modelling of agricultural countermeasures in RODOS. Ibid.

[6] Zheleznyak, M, et al, Modelling of hydrological pathways in RODOS. Ibid.

[7] Borzenko, V and French, S, Decision analytic methods in RODOS. Ibid.

[8] Ehrhardt, J. et al, RODOS, a real time on-line decision support system for nuclear emergency management in Europe. IN Proc. of Conf on International Aspects of Emergency Management and Environmental Technology (K H Drager, Ed), Oslo, June 18-21, pp188-194 (1995).

[9] Päsler-Sauer, J. et al. Meteorology and atmospheric dispersion, simulation of emergency actions and consequence assessment in RODOS. Ibid, pp 197-204 (1995).

[10] Brown, J. et al. The modelling of exposure pathways and relocation, decontamination and agricultural countermeasures in the European RODOS system. Ibid, pp 207-213 (1995).

[11] Raskob, W. et al. The modelling concept for the radioactive contamination of water bodies in RODOS, the decision support system for nuclear emergencies in Europe. IN Proc. of Int. Seminar on Freshwater and Estuarine Radioecology, Lisbon, 21-25 March (1994).

[12] RODOS Newsletter (J Ehrhardt, Ed), Forschungszentrum Karlsruhe GmbH, INR, Germany.

[13] Jousten, N and Jeanes, P, The GAMMA-1 project. IN First International Conference

of the European Commission, Belarus, the Russian Federation and Ukraine on the Consequences of the Chernobyl Accident, Minsk, 18-22 March (1996) - this issue.

[14]   IAEA, Convention on early notification of a nuclear accident. Vienna, (1986).

[15]   Council Decision, 87/600/EURATOM, on Community arrangements for the early notification of information in the event of a radiological emergency. OJ L371/76 of 30/12/67, (1987).

[16]   Vadé, S, Die Europäische Union und der Strahlenschutz der Bevölkerung bei einer radiologischen Notstandssituation, FS-94-74-I, Stand des Notfallschutzes in Deutschland und der Schweiz, ISSN 1013-4506, 385-390, (1994).

[17]   De Cort, M, Leeb, H, De Vries, G, Breitenbach, L and Weiss, W, International exchange of radiological information in the event of a nuclear accident - future perspectives. IN First International Conference of the European Commission, Belarus, the Russian Federation and Ukraine on the Consequences of the Chernobyl Accident, Minsk, 18-22 March (1996) - this issue.

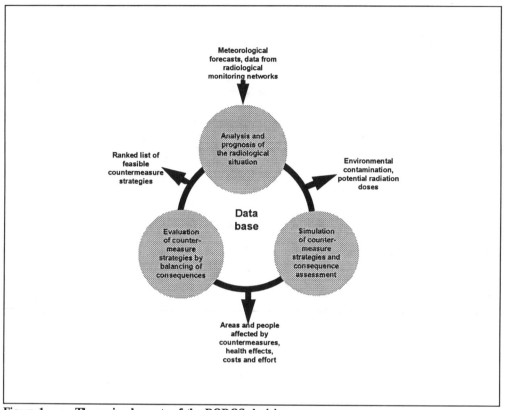

Figure 1   The main elements of the RODOS decision support system

RODOS: Decision Support System for Off-Site Emergency Management in Europe

Ehrhardt, J., Shershakov, V.[1], Zheleznyak, M.[2] , Mikhalevich, A.[3]
Forschungszentrum Karlsruhe, Institut für Neutronenphysik und Reaktortechnik,
Postfach 3640, D-76021 Karlsruhe
[1] Scientific Production Association TYPHOON, Emergency Centre,
Lenin St. 82, Obninsk, Kaluga Region, 249020, Russia
[2] Cybernetics Centre of the Ukrainian Academy of Sciences, Institute of Mathematical
Machines and Systems, Prospect Glushkova 42, Kiev, 252207, Ukraine
[3] Belarus Academy of Sciences, Institute of Power Engineering Problems,
Sosny, Minsk, 220109, Republic of Belarus

Abstract. The integrated and comprehensive real-time on-line decision support system, RODOS, for off-site emergency management of nuclear accidents is being developed under the auspices of the European Commission's Radiation Protection Research Action. A large number of both West and East European institutes are involved in the further development of the existing prototype versions to operational use with significant contributions coming from the partner institutes in the CIS Republics. This paper summarises the structure, the main functions and the status of the RODOS system.

1. Introduction

1.1. Background

Following the Chernobyl accident, increasing resources were allocated in many countries to the improvement of arrangements for off-site emergency response in the event of a nuclear accident. Within the European Commission's Radiation Protection Research Action, a major project was initiated in 1990 to develop a comprehensive Real-time On-line DecisiOn Support system, RODOS, for nuclear emergency management[1]. A large number of EC contractors are participating in this project and advantage is being taken of existing developments at national levels. Developments of a similar nature were being undertaken within the former Soviet Union and subsequently within individual Republics, taking account of the practical experience gained in responding to the Chernobyl accident. The working programme of the Joint Study Project 1 (JSP1) reflects the common efforts and ideas of the joint undertaking to develop a decision support system that would be broadly applicable and accepted in West and East Europe.

1.2. Objectives

The main objectives of the RODOS project are to provide the methodological basis, develop models and data bases and install the hardware and software framework of a system which offers comprehensive decision support from the very early stages of an accident up to many years after the release and from the vicinity of the site to far distant areas unperturbed by national boundaries. In this way it will be possible to achieve estimates, analyses, and prognoses of accident consequences, protective actions and countermeasures which are consistent throughout all accident phases and distance ranges. All relevant environmental data, including radiological and meteorological information and readings, are to be processed, by means of models and mathematical procedures, into understandable, interpretable pictures of the current and predicted future radiological situations. Simulation models for any kind of protective actions and countermeasures are designed not only to permit their extension in terms of time and space to be estimated, but, together with dose, health effects and economic models, also to allow their benefits and disadvantages to be quantified. Feasibility rules and subjective arguments of decision makers implemented in rule-based expert systems and other decision analytic methods will help to evaluate alternative countermeasure strategies and to provide a ranked order of countermeasure options together with an explanation for that order.

Within the collaborative arrangements between the institutes involved in JSP1, the following objectives were of particular importance:

- further development of a comprehensive decision support system for operational use generally applicable in the EU and CIS, using the RODOS system as a common platform
- improvement and validation of models and completion of data bases included in the RODOS system using monitoring and other data obtained during and after nuclear accidents in the CIS, such as those at Chernobyl and Tomsk
- implementation and adaptation of the RODOS system in each of the three CIS Republics, its link with meteorological and radiological monitoring networks and demonstration of its on-line operation

1.3. Project Overview

During the 3rd Framework Programme of the European Commission, 1990 - 1995, the RODOS project evolved as as an ambrella for four individual subprojects, each with its own contractors and coordinator:
- Co-ordination of atmospheric dispersion activities for the real-time decision support system under development at FZK, with RISØ National Laboratory as coordinator
- Development of a comprehensive decision support system for nuclear emergencies in Europe following an accidental release to atmosphere, with Forschungszentrum Karlsruhe as coordinator
- Evaluation and management of post-accident situations, with CEA/IPSN as coordinator
  The Joint Study Project 1 (JSP1) of the EC/CIS Collaborative Agreement for International Collaboration on the Consequences of the Chernobyl Accident

The contractual arrangements and the work performed within the first three contracts is described elsewhere[2,3,4]. This paper emphasises on the JSP1 contract and the institutes involved are as follows:

- Forschungszentrum Karlsruhe GmbH (FZK), D (EU coordinator)
- National Radiological Protection Board (NRPB), UK
- V. KEMA, NL
- Studiecentrum voor Kernenergie/Centre d'Etude de l'Energie Nucleaire (SCK/CEN), B
- SPA TYPHOON, Russia (CIS coordinator)
- Institute of Control Science Problems (ICSP), Russia
- Russian Institute of Agricultural Radiology (RIAR), Russia
- Institute of Mathematical Machines and Systems, Cybernetics Centre (IMMS CC), Ukraine
- Ukrainian Institute of Agricultural Radiology (UIAR), Ukraine
- Institute of Power Engeneering Problems (IPEP), Belarus
- Belorussian Institute of Agricultural Radiology (BIAR), Belarus
- Committee for Hydrometeorology (HYDROMET), Belarus

From the beginning of JSP1, the work programme was split into four subprojects and the tasks of each of them can be summarised as follows:

Development of decision support systems: Improvement and extension of the functions and capabilities of RODOS by incorporation of software available from and/or developed by the project partners and based on the experience with the operation of the actual RODOS prototype version in CIS and EU institutes (FZK; SPA TYPHOON, ICSP, IMMS CC, IPEP, CRKM, ONIL, HYDROMET).

Modelling of hydrological pathways: Development of a model chain for short- and long-term prognoses of the consequences of a radioactive contamination of the aquatic part of the environment (KEMA, FZK; IMMS CC, SPA TYPHOON).

Agricultural countermeasures: Modelling the efficiacy and cost of agricultural countermeasures by evaluating radiological and countermeasure data bases existing or under development in CIS institutes (NRPB; BIAR, RIAR, UIAR).

Data assimilation and interpretation: Development of a generic methodology for assessing the source term and the dose distributions during and shortly after an accidental release of radioactive material by evaluating model predictions and early radiological monitoring data and meteorological measurements (SCK/CEN, JRC ISPRA; SPA TYPHOON).

Close cooperation has been achieved with project JSP2 consequent upon the common interest in decision support systems for longer term countermeasures and with project ECP3 on hydrological modelling.

2. Main Functions and Characteristics of RODOS

2.1. Capacity for Generic Use in Europe

RODOS is designed as a comprehensive system incorporating models and data bases for assessing, presenting and evaluating the accident consequences in the near, intermediate and far distance ranges under due consideration of the mitigating effect of countermeasure actions. Its flexible coding allows it to cope with differing site and source term characteristics, differing amounts and quality of monitoring data, and differing national regulations and emergency plans. To facilitate its application over the whole of Europe, the software has been developed as a transportable package to run on workstations with UNIX operation system; in particular its software framework supports the integration of application software developed externally by many of the contractors[5]. The modular structure of RODOS allows an easy exchange of models and data, and thus facilitates the adaptation of the system to the local/regional and national conditions. Finally RODOS offers a variety of access tools to cope with the different capabilities, knowledge and aims of the future users.

2.2. Levels of Information Processing

If connected to on-line meteorological and radiological monitoring networks, the RODOS system provides decision support on various stages of information processing which conveniently can be categorised into four distinct levels. The functions performed at any given level include those specified together with those applying at all lower levels.

- Level 0: Acquisition and checking of radiological data and their presentation, directly or with minimal analysis, to decision makers, along with geographical and demographic information.
- Level 1: Analysis and prediction of the current and future radiological situation (i.e. the distribution over space and time in the absence of countermeasures) based upon monitoring data, meteorological data and models, incl. source term estimation.
- Level 2: Simulation of potential countermeasures (e.g. sheltering, evacuation , issue of iodine tablets, relocation, decontamination and food-bans), in particular, determination of their feasibility and quantification of their benefits and disadvantages.
- Level 3: Evaluation and ranking of alternative countermeasure strategies by balancing their respective benefits and disadvantages (e.g. costs, averted dose, stress reduction, social and political acceptability) taking account of societal preferences as perceived by decision makers.

Most decision support systems that have been developed to an operational state are limited to levels 0 or 1. A few extend to level 2 or even level 3 but, in general, are limited in the range of countermeasures they address or in the completeness of benefits and disadvantages that are considered.

2.3. Structure and Endpoints of RODOS

In recognition of the need for a unique and integrated real-time on-line decission support system that will provide consistent and comprehensive information from Level 0 to Level 3, the basic concept, content and design of RODOS were specified and agreed by participants at the outset of the project. The conceptual RODOS architecture is split into three distinct subsystems, which are denoted by ASY, CSY and ESY[1]. Each of the subsystems consists of a variety of modules developed for processing data and calculating endpoints belonging to the corresponding level of information processing. The modules are fed with data stored in four different data bases comprising real-time data with information coming from regional or national radiological and meteorological data networks, geographical data defining the environmental conditions, program data with results obtained and processed within the system, and facts and rules reflecting feasibility aspects and subjective arguments.

The content of the subsystems and the data bases will change with the application of RODOS in relation to a nuclear accident. The temporality of the decisions greatly influences both what information is available and how information is aggregated and integrated. At the different points in time different modules have to be chained together, at least one from each of the subsytems mentioned above, to produce the required output. For example, after the passage of the plume, meteorological forecasts are no longer necessary for the region considered, or after evacuation models for simulating sheltering or relocation in the same area are not needed. A Supervising Subsystem (SSY) under development will manipulate the components of RODOS in order to respond to user requests.

RODOS can be run in two modes of operation. In the *automatic mode* the system automatically presents all information which is relevant to decision making and quantifyable in accordance with the current state of knowledge in the real cycle time (e.g. 10 minutes in the early phase of emergency protection). For this purpose, all the data entered into the system in the preceding cycle (either on-line or by the user) are taken into account. Cyclic processing is carried out synchronous with the incoming monitoring data of automated radiological information networks. Interaction with the system is limited to a minimum of user input necessary to characterise the current situation and adapt models and data.

Either in parallel to the automatic mode or alone, RODOS can be operated in the *interactive mode*. In particular, in the later phases of an accident, when longer-term protective actions and countermeasures must be considered and no quick decisions are necessary, or for emergency planning, exercises and education under normal non-accident conditions, this mode is more important. Editors specially developed for the menu-driven user interaction allow specific modules to be called, different sequences of modules to be executed, input data and parameter values to be changed, and the output and representation of results to be varied.

The dialogue between RODOS and a user is performed via various user-interfaces tailored to the needs and qualification of the user. The access rights of different user groups determine the type of user-interface, which allows increasing access to models, data and system parameters in a hierarchical structure, with an easy understandable but very limited interface for training courses on emergency management on the top and the full spectrum of interaction tools for system developers familiar with the system ingredients and structures on the bottom.

The interconnection of all program modules, the input, transfer and exchange of data, the display of results, and the interactive and automatic modes of operation are all controlled by the specially designed operating system OSY. It has been developed following the Client-Server Model: each module requests a service from the system (typically this might be a request for data) and the system determines how this might be satisfied. If the data is already in the data base then it can be supplied directly. Alternatively, it may call another module which can calculate the required information[5].

3. Status and Future Planning

Since the end of 1995, the prototype version RODOS-PRTY-2.0 is available, which will be further developed in the next years with the first pre-operational version 3.1 ready by mid 1997 and the version 4.0 for full operational use by mid 1999. In the following the content of the current version will be briefly described together with selected key objectives for the 4th Framework Programme.

3.1. Diagnosis and Prognosis of the Radiological Situation

The analysing subsystem **ASY** can be operated with measured or historical meteorological data solely (diagnosis mode) or together with forecasted meteorological fields (prognosis mode). The meteorological data are converted in the meteorological preprocessor PAD into input data on the state of the atmospheric boundary layer for use by the subsequent meteorological model chain. It consists alternatively of the windfield models MCF or LINCOM and either the puff-model RIMPUFF or the simplified puff-model ATSTEP[6].

The mass consistent wind field model MCF allows a spatial wind vector field free of divergences to be set up over an area extending to a few tens of kilometres. The boundary layer and wind profile data are processed together with the site topography to get a wind vector field taking into consideration the influence of the underlying terrain over which material would be dispersed. The flow model LINCOM is a non-hydrostatic diagnostic model based on the solution of linearised continuity and momentum equations with a first order spectral turbulent diffusion closure. Both models can be operated in the prognosis mode with input from national forecast models to produce a wind field with higher resolution (e.g. 1 km), such as those from the German Weather Service with a resolution of 14 km or HIRLAM from Sweden[6].

Atmospheric dispersion and deposition as well as nuclide specific activity concentration and gamma radiation fields are calculated in RIMPUFF and ATSTEP. RIMPUFF is suitable for real-time simulation of puff and plume dispersion during time and space changing meteorology. Its modelling is consistent with the German-French atmospheric dispersion model now agreed by both countries for operational use in nuclear emergency situations. The ATSTEP code offers all features of RIMPUFF but calculates with a lower spatial and temporal resolution. It will mainly be applied as a fast atmospheric dispersion code for training and exercises with RODOS.

Main objective of the next project period will be the completion of the meteorological and atmospheric dispersion model chain for all distance ranges and its coupling to local synoptic stations and weather forecasts of the national weather services. Progress has already been made in this direction with support of SPA TYPHOON, who integrated in RODOS their long range atmosphric dispersion models together with software for accessing and evaluating weather forecasts for Europe from meteorological services (Washington, Moscow, Bracknell).

In plant data as well as off-site radiological measurement and monitoring data, such as air concentrations, ground contamination and gamma dose rates, allow comparisons between measurements and model predictions. With the help of data assimilation techniques presently under development, the model results and the observed data will be optimally used to achieve a consistent and realistic picture of the environmental contamination and to estimate the source term. The pilot version 3.1 of RODOS will contain such methods for near range atmospheric dispersion, by mid 1999 they will be extended to far distance calculations.

In connection with the development of data assimilation techniques, the quantification of the uncertainties in the predictions of the RODOS system are considered to be a key element of an advanced decision support system. Methodological investigations have already been started on how to assess and propagate uncertainty estimates throuh the various modules of the RODOS system. The further development of these techniques for operational use will be a main objective of the 4th Framework Programme, and the planning foresees the quantification of uncertainties in the near range models of RODOS by mid 1999.

3.2. Countermeasures and Consequences

The countermeasure and consequence subsystem **CSY** incorporates

- the module group ECOAMOR for calculating individual doses via all exposure pathways, in particular ingestion pathways,
- the module group EMERSIM for simulating sheltering, evacuation and distribution of stable iodine tablets,
- the module group FRODO for simulating relocation, decontamination and agricultural countermeasures,
- the module HEALTH for quantifying stochastic and deterministic health effects[6]
- the module ECONOM for estimating the economic costs of emergency actions, countermeasures and health effects[6].

ECOAMOR[7] is a system of program modules which has been developed on the basis of the dynamic radioecological model ECOSYS-87[8]. Inputs to ECOAMOR are the contamination of air and precipitation provided by ASY. Additionally, data on foodstuff production together with a large number of parameter values characterising the transfer processes in the radioecological scenario considered are required. Many of these parameters vary to a large extent over the different the regions in Europe; the modules are designed to facilitate the adaptation to these regional variations.

In its present version, ECOAMOR consideres 31 basic food products, 22 feedstuffs and 35 processed foodstuffs. The models describe the dynamics of the different radioecological transfer processes, such as the seasonality in the growing cycle of plants, the feeding practices of domestic animals and human dietary habits. In the dose modules of ECOAMOR, the ingestion doses are calculated from the activity concentrations in foodstuffs, age and possibly season dependent intake rates and age dependent dose factors. Doses due to short term inhalation from the passing plume as well as long-term inhalation of resuspended material are also calculated. In addition, doses resulting from external exposure pathways are determined, such as irradiation from the passing plume and from deposited material on ground surface and the skin. Dose reductions from nuclide migration into deeper soil layers and by the shielding of houses are considered, as well as the influence of variable deposition patterns at different urban environments.

The early emergency actions considered in EMERSIM[7,9] can be defined indirectly by dose intervention criteria or directly by graphical input of areas. Important endpoints are areas and number of people affected and individual doses with and without emergency actions. In the present version of EMERSIM, the assumption is made, that before and during evacuation the dose rate is constant and identical with the home location. In the next versions of EMERSIM, a more realistic modelling of exposure during evacuation will be possible by using the results of the evacuation simulation module EVSIM[9], which is capable of taking into account the most important factors that may influence the success and effectiveness of this measure, such as traffic network and conditions, availability of transportattion means, population distribution, weather conditions and time of accidental release. It will be completed by the optimisation module STOP[9], which is able to optimise routes for evacuation with respect to route length, dose saved, starting time and costs.

The relocation model in FRODO[7,10] uses criteria for the imposition and relaxation of permanent and temporary relocation in the form of dose levels. The endpoints evaluated relate to the areas of land interdicted, the time periods over which this occurs, the number of people relocated, the doses saved as a result of relocation, the doses received by those temporarily relocated following their return, and the doses received by individuals resettling in an area following the lifting of land interdiction after the permanent relocation of the original population.

The impact of decontamination on relocation can be evaluated for decontamination occurring either before or after relocation is implemented. The decontamintion of agricultural land is included in so far as its impact on the need for or reduction in food restrictions is evaluated. The other agricultural countermeasures considered in FRODO are: banning and disposal, food storage, food processing, supplementing animal feedstuffs with uncontaminated, lesser contaminated or different feedstuff, use of sorbents in animal feeds or boli, changes in crop variety and species grown, amelioration of land and change in land use.

The criteria for banning the consumption of food are defined in terms of the activity concentrations in foods. A database of information on the effectiveness of the agricultural countermeasures has been compiled. These data have come primarily from a database on the effectiveness of a range of agricultural countermeasures compiled for inclusion in RODOS under the JSP1. This database contains robust, representative data that can be applied to relatively large areas, potentially over long periods of time[10].

Important extensions of CSY in the next project period will be an improved treatment of the interaction between combined countermeasures, the inclusion of consequence and countermeasure models for natural and semi-natural environments and data assimilation in the models for deposition on soil and vegetation and foostuff and feedstuff contamination.

3.3. Hydrological Pathways

The evaluation of the radiological and environmental consequences of the Chernobyl accident demonstrated the significant contribution of contaminated water bodies. To complete the RODOS methodology and system, a hydrological model chain has been developed , which covers all the relevant processes such as the direct inflow into rivers, the migration and the run-off of radionuclides from watersheds, the transport of radionuclides in large river systems including exchange with sediments and the behaviour of radionuclides in lakes. The corresponding models RETRACE (run-off), RIVTOX and COASTOX (rivers) and LAKECO (lakes) have been coupled, implemented in RODOS and adapted to the Rhine river system[11]. Other river systems can be readily implemented in RODOS using the same model chain subject to gathering appropriate data.

3.4. Evaluation of Countermeasure Combinations

The evaluating subsystem **ESY** is being developed mainly to evaluate alternative countermeasure strategies under the aspects of feasibility in a given situation, public acceptance of the actions, socio-psychological and political implementations, and subjective arguments reflecting the judgements of the decision maker. These parameters can be taken into account in ESY using mathematical formulations as rules, weights, and preference functions. The application of these rules results in a ranked order of options together with those rules and preference functions which, above all, have led to this evaluation. This ranking order can be great help to a decision maker in taking a final decision. At present, both multi-attribute decision analysis techniques and expert systems are being studied as potential methodological tools in the evaluation of combinations of alternative actions. The ESY subsystem will become operational in the next project period; the sequence of calculations will be oriented at its internal structure:

First, a very simple expert system will be used to discard strategies which are incompatible with the principles of radiological protection, which do not five continuity of treatment, or which fail very coarse practicability rules. The remaining strategies will be passed to a multi-attribute value ranking module, which will identify the top 10 or 20 ranked strategies. The operator will be able to use interactive sensitivity analyses, such as that in the software packages HERESY[12] and M-CRIT[12], to confirm that these strategies are worth careful consideration. These strategies would then be passed to an expert system with a much finer and more sophisticated system or rules, each of which could be applied to each of the candidate strategies. The small number of strategies would allow a full set of explanations to be developed, which would give a critique of each of the strategies. Thus the output of RODOS will be a short list of strategies, each of which satisfies the constraints implied by intervention levels, practicability, etc. together with a detailed commentary on each strategy explaining its strength and weaknesses.

3.5. Customisation in Central and Eastern European Countries and CIS Republics

Under the auspices of the European Commission's R&D Programme, the basic hardware and software components of the RODOS system have been transferred to institutes in East European countries, namely, Belarus, Hungary, Poland, Romania, Russia, Slovak Republic and the Ukraine. Effective working arrangements between the project partners in the West and the East, in particular the institutes in the CIS Republics, have been established and a full integration in one coordinated working programme has been achieved[13].

Within 4 years of the start of the RODOS project, EU, Eastern European and CIS countries have agreed: the hardware and software concepts, the modelling and operating features, and the various levels of information processing and presentation. Moreover, the prospect of interconnected RODOS systems running in Western and Eastern European countries has been widely accepted as an important step forward to an improved emergency management in the case of any future nuclear accident.

Before the RODOS system can become fully operational in Europe, especially in the Central and Eastern European countries and the CIS Republics, a number of tasks need to be completed. These include

- customisation of data bases and models to local, regional and national conditions,
- establishing interfaces with national meteorological and radiological monitoring networks,
- integration of the system within national emergency management arrangements.

With support of the European Commission the process of bringing the RODOS system into operational use in the respective countries will be accelerated in the next months and years with attendant benefits for emergency preparedness both in the East and Europe more generally.

4. Benefits of RODOS

4.1. Potential Role for Improving Emergency Response in Europe

With the continuation of the R&D work by the institutes with a leading function already in the 3rd Framework Programme, existing skills, knowledge and co-operation will be carried forward into the Framework 4 programme to deliver a decision support system which provides benefits and functions unavailable elsewhere. These include:

- better use of resources allocated within the European Union to improve off-site emergency management, inter alia, minimising unnecessary duplication
- models, methods and data bases drawn from the best available at national and international levels
- comprehensive decision support will be provided (e.g. at all levels of information processing for each relevant countermeasure at all times and distances from a release)
- novel and enhanced technical features (e.g. assimilation of monitoring data and model predictions, integrated treatment of uncertainties)
- a seamless transition between all distance ranges and temporal phases of an accident offering continuity in providing public information and decision support
- a design for operational use at local, regional, national and supra-national levels and for training and exercises at these levels
- a modular design to facilitate long term development and adaptation to user requirements and local/regional conditions
- a stand-alone interactive training tool for use, inter alia, by those responsible for making decisions on off-site emergency management and their technical advisers at local, regional, national and supra-national levels
- a more general interactive training and educational tool for radiation protection, nuclear safety and emergency planning personnel with professional interest in or responsibility for off-site emergency management
- a software framework for developing decision support systems for the management of non-nuclear emergencies with potential off-site consequences.

Those institutes in Eastern and Western Europe, which already or in the near future will run the current version of RODOS, are directly or indirectly responsible for emergency management in their countries. In parallel to the ongoing R&D work during the period 1996-99, there are plans in at least some of these countries to integrate RODOS within national emergency response systems. In addition, the work on the realisation of a European wide network for the exchange of radiological information has started with the main partners of this proposal and the current RODOS users[14,15]. In this way, the RODOS system will facilitate communication and exchange of information and promote a more coherent and harmonised emergency response within Europe to any future nuclear accident.

4.2. Future Users

The roles for which RODOS is designed largely determine its potential users. These include those responsible at local, regional, national and supra-national levels for off-site emergency management and related training, for the operation of nuclear installations, for public information, or for communication and exchange of information (eg, in accord with bi-lateral or international agreements); the R&D community concerned with improving decision support for off-site emergency management; and developers of decision support systems for the off-site management of non-nuclear emergencies.

References

[1] Ehrhardt, J., Päsler-Sauer, J., Schüle, O., Benz, G., Rafat, M., Richter, J. "Development of RODOS, a comprehensive decision support system for nuclear emergencies in Europe-an overview", Radiation Protection Dosimetry, Vol. 50, Nos 2-4, pp 195-203 (1993)

[2] Final Report of contract FI3P-CT92-0044 „Co-ordination of atmospheric dispersion activities for the real-time decision support system under development at FZK", in: European Commission, Radiation Protection Research Action, Final Report 1992-95, EUR Report, to be published

[3] Final Report of contract FI3P-CT92-0036 „Development of a comprehensive decision support system for nuclear emergencies in Europe following an accidental release to atmosphere", in: European Commission, Radiation Protection Research Action, Final Report 1992-95, EUR Report, to be published

[4] Final Report of contract FI3P-CT92-0013b „Evaluation and management of post-accident situations", in: European Commission, Radiation Protection Research Action, Final Report 1992-95, EUR Report, to be published

[5] Schüle, O., Rafat, M., Kossykh, V. "The software environment of RODOS", First International Conference of the European Commission, Belarus, the Russian Federation and Ukraine on the Consequences of the Chernobyl Accident, Minsk, Belarus, 18-22 March 1996 (this issue)

[6] Päsler-Sauer,J., Schichtel, T., Mikkelsen, T., Thykier-Nielsen, S. "Meteorology and atmospheric dispersion, simulation of emergency actions and consequence assessment in RODOS", Proceedings for Oslo Conference on International Aspects of Emergency Management and Environmental Technology, 18-21 June 1995, Oslo, Norway, pp 197-204 (1995)

[7] Brown, J., Smith, K. R., Mansfield, P., Smith, J., Müller, H. "The modelling of exposure pathways and relocation, decontamination and agricultural countermeasures in the European RODOS system, Proceedings for Oslo Conference on International Aspects of Emergency Management and Environmental Technology, 18-21 June 1995, Oslo, Norway, pp 207-213 (1995)

[8] Müller, H., Pröhl, G., "ECOSYS-87: a dynamic model for assessing radiological consequences of nuclear accidents", Health Physics 64, pp 232-252 (1993)

[9] Glushkova, V., Schichtel, T. "Modelling of early countermeasures in RODOS", First International Conference of the European Commission, Belarus, the Russian Federation and Ukraine on the Consequences of the Chernobyl Accident, Minsk, Belarus, 18-22 March 1996 (this issue)

[10] Brown, J., Ivanov, Y. A., Perepeliatnikova, L., Priester, B. S. "Modelling of agricultural countermeasures in RODOS" First International Conference of the European Commission, Belarus, the Russian Federation and Ukraine on the Consequences of the Chernobyl Accident, Minsk, Belarus, 18-22 March 1996 (this issue)

[11] Zheleznyak, M., Heling, R. "Modelling of hydrological pathways in RODOS", First International Conference of the European Commission, Belarus, the Russian Federation and Ukraine on the Consequences of the Chernobyl Accident, Minsk, Belarus, 18-22 March 1996 (this issue)

[12] Borzenko, V., French, S. "Decision analytic methods in RODOS" First International Conference of the European Commission, Belarus, the Russian Federation and Ukraine on the Consequences of the Chernobyl Accident, Minsk, Belarus, 18-22 March 1996 (this issue)

[13] Shershakov, V., Ehrhardt, J., Mikhalevich, A., Zheleznyak, M. "The implementation of RODOS in Belarus, Russia and Ukraine, and future perspectives" First International Conference of the European Commission, Belarus, the Russian Federation and Ukraine on the Consequences of the Chernobyl Accident, Minsk, Belarus, 18-22 March 1996 (this issue)

[14] De Cort, M. "International exchange of radiological and meteorological information in the event of a nuclear accident-future possibilities" First International Conference of the European Commission, Belarus, the Russian Federation and Ukraine on the Consequences of the Chernobyl Accident, Minsk, Belarus, 18-22 March 1996 (this issue)

[15] Kelly, G. N. "EC contribution to improving off-site emergency planning in Eastern Europe" First International Conference of the European Commission, Belarus, the Russian Federation and Ukraine on the Consequences of the Chernobyl Accident, Minsk, Belarus, 18-22 March 1996 (this issue)

This work has been supported by the European Commission under contracts COSU-CT92-0087 and FI3P-CT92-0036 and the German Ministry of Environment under contract St. Sch. 1054/1

The Implementation of RODOS in Belarus, Russia and Ukraine, and Future Perspectives

Shershakov, V., Ehrhardt, J.[1], Zheleznyak, M.[2], Mikhalevich, A.[3]
Scientific Production Association TYPHOON, Emergency Centre,
Lenin St. 82, Obninsk, Kaluga Region, 249020, Russia
[1] Forschungszentrum Karlsruhe, Institut fur Neutronenphysik und Reaktortechnik,
Postfach 3640, D-76021 Karlsruhe
[2] Cybernetics Centre of the Ukrainian Academy of Sciences, Institute of Mathematical
Machines and Systems, Prospect Glushkova 42, Kiev, 252207, Ukraine
[3] Belarus Academy of Sciences, Institute of Power Engineering Problems,
Sosny, Minsk 220109, Republic of Belarus

Abstract. Broad agreement has been achieved between institutes and institutions in the European Union, Belarus, Russia and the Ukraine to cooperate in the development of the RODOS system, a decision support system for general application in Eastern and Western European countries. This is being coordinated within the Joint Study Project 1 (JSP1) of the EC/CIS Agreement for International Collaboration on the Consequences of the Chernobyl Accident. The ultimate goal of this joint venture is the integration of an operational RODOS system into the national emergency management arrangements. To provide a standard platform for the common R&D work within JSP1, the hardware and software components of RODOS have been implemented at SPA "Typhoon", Obninsk, IMMS CC, Kiev, and GLAVHYDROMET, Minsk. This paper summarises the activities of the CIS institutes in generating their RODOS teams, educating and training the personnel involved and to organise cooperation with other institutes for securing access to meteorological and radiological monitoring data, national geographical information and specific expertise necessary to adapt models and data of the RODOS system to local, regional and national conditions. The problems encountered during implementation, the status of the installation of data transmission lines for the remote operation of RODOS and for information exchange with RODOS teams in other countries and the current RODOS development activities are discussed.

1. Introduction

It is largely recognised that at present no alternative to nuclear energy exists and, this being so, not only the probability of potential harmful effects from nuclear installations need to be reduced, but also the preparedness for emergency response should be enhanced. By scale and severity, nuclear accidents have the potential of being comparable to natural catastrophes. They are no longer "national assets" and pooled efforts are required to ensure radiation protection of the general public and the environment in the case of any future emergency.

The proper selection and timely application of countermeasures can help to significantly reduce the consequences of an accident. For selecting effective countermeasures, a decision maker should take into account specific features of a given facility and the conditions under which the accident developed. The characteristics can change as a result of the application of radiation protection measures, which makes the task of a decision-maker even more difficult.

Considering all this, rational decisions are impossible to be made without a monitoring system operating on a constant basis and special information and decision supporting tools. Moreover, this being an international problem, a common monitoring system and a common computer technology for decision making information support is required.

The RODOS (Real - time On-line DecisiOn Support) system [1] being developed with support of the European Commission is meant to be a base for creating a unified programming environment for supporting collective decisions in the event of a nuclear accident. The RODOS system is being developed on the UNIX operation system using graphic workstations such as HP-9000/735 and DEC-ALPHA. The implementation of the RODOS components in the CIS Republics aimed at checking whether the RODOS system can be used as a common platform. The main efforts were concentrated on a series of problems related to emergency response:

- interaction between RODOS and national radiation monitoring systems with a view to collect and process on-line data ;
- organisation of a distributed system for collection and processing of on - line data on the basis of territorial, departmental and off-site systems of radiation monitoring ;
- using the RODOS software for analysing real accidents and their consequences (the Chernobyl accident, first of all);
- studies of the RODOS applicability and its acceptability by the potential RODOS users, i.e.national institutions that are responsible for the arrangements for off-site emergency response in the event of a nuclear accident;
- adaptation of the RODOS system in each of three CIS Republics to the needs of the specific RODOS users;
- preparing and providing data to international organisations under the Convention on Early Notification of nuclear accidents and other international agreements.

2. Directions in implementation of RODOS and its supporting infrastructure

2.1 Belarus

In Belarus the development of RODOS is carried out in the structure of the Ministry for Emergency Situations and Population's Protection from the Consequences of the Chernobyl NPP Catastrophe (Fig.1). A workstation HP-9000/735 has been installed in the Centre for Radiological and Environmental Monitoring (CREM) of the Commitee on Hydrometeorology which is in the structure of the Ministry for Emergency Situation.

The group of RODOS operators has been formed in CREM, the installation of software is being carried out as well as the link of RODOS with the systems of radiation monitoring and meteorological observations is being developed.

The preparedness of RODOS software (adaptation of program modules developed within the RODOS project, compilation and assimilation of databases, verification of models and programs, etc.) are being carried out at the Institute of Power Engineering Problems and the Institute of Radioecological Problems/Academy of Sciency of Belarus, CREM, Branch Research Laboratory of the Management Academy, etc.

2.2 Ukraine

In Ukraine the Cybernetics Center, Institute of Mathematical Machines and Systems (IMMS CC) has received since 1986 an important experience in developing and executing decision support systems for off-site emergency management after the Chernobyl accident. The

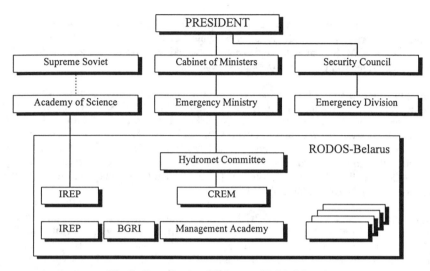

Fig.1   Organisational Scheme of RODOS-Belarus

Institute was responsible on the national level for the development of the decision support system for the Dnieper basin management after the Chernobyl accident. To that purpose, models and methods for treating releases of radioactive material into hydrological systems were developed and applied[2]. The Institute experience in developing efficient procedures for the problems of discrete and combinatorial optimisation was also used for developing decision making modules for different kinds of the decision support systems.

Since 1991 IMMS CC is involved in the R&D activities of RODOS in the fields of hydrological modelling, system development and optimisation of early emergency actions. The hardware and software components of RODOS PRTY 1.0 were installed at IMMS CC in spring 1994. The system is used both for the R&D work, including RODOS adaptation and customisation, and for the wide scale RODOS presentation for the Ukrainian emergency management institutions.

The following governmental institutions are reponsible for the off-site emergency management in Ukraine:

- State Emergency Commission, chaired by the Vice-Prime-Minister on Emergency Situations;
- Headquarters of the Civil Defense (HCD);
- Ministry of Environmental Protection and Nuclear Safety (MEPNS);
- Committee of Nuclear Power Production (GOSATOM);
- Administration of the Chernobyl Exclusive Zone (ACEZ) of the Ministry of Chernobyl Problems (Min Chernobyl).

All above-mentioned governmental institutions have expressed they will to implement the RODOS system for decision support at the upper level of data processing and monitoring systems. As a first step, the organisational conditons for the RODOS implementation were stablished at ACEZ. The joint programme of ACEZ and IMMS CC has been started in October 1995 to ensure the test-operation of RODOS in late 1996 and its operational use in late 1997 to support the off-site emergency response in the event of a nuclear accident in the Chernobyl Exclusive Zone. The system will be adapted for two kinds of the releases - from the Chernobyl NPP units still in operation and from the SHELTER (destroyed Unit 4).

The second Programme was signed in October 1995 between GOSATOM and IMMS

CC to develop in cooperation with some other Ukrainian institutions on the base of RODOS the decision support systems for the Emergency Centres of the NPPs and the National Crisis Centre of GOSATOM. The pioneer implementation would be done for the Zaporoshe NPP. The work programme comprises five work packages:

Coupling of RODOS to the UkrHydromet meteorological and radiological network: The on-line network will be installed at the RODOS centre in the IMMS CC and the Computer Center of the UkrHYDROMET to receive national level meteorological information. The regional and local level meteorological networks for the Zaporoshe NPP and Chernobyl NPP will be linked to RODOS via the national meteorological service branches and UkrHYDROMET communication net; a corresponding agreement has been recently signed by the IMMS CC and UkrHYDROMET. The developed software tools will be transferred to the RODOS systems at ACEZ and GOSATOM.

Development of modules to link RODOS with the Ukrainian radiological monitoring systems and the distributed data bank on post-Chernobyl radiological information: The RODOS system in the IMMS CC will be on-line connected via INTERNET and specific national channels with the MEPNS Data Processing Centres of the GAMMA system and the Zaporozhe GAMMA Regional Centre. Submodules will be developed for transferring the GAMMA radiological monitoring information to the RODOS data bases. The working package will include the development of on-line communication with the distributed data base of the MinChernobyl's "Inform-Chernobyl" system.

Extension of RoGIS and incorporation of country specific information: Several enterprises of the Ukrainian State Committee on Geodesy and Cartography and Defense Ministry are preparing digitised maps of the Ukrainian territory in different scales. Conversion routines will be developed to transfer the map data into the format of RoGIS, the geographical information system of RODOS. Further geographical data will be collected, such as on demography, topography, land use, agricultural production, building types, and economy, and data about territories with radioactive contamination after the Chernobyl accident. In a first application, data of the regions around the Zaporozhe NPP and the Chernobyl NPP will be provided, later regions of other Ukrainian NPPs and of the whole Ukrainian territory will be put into the RODOS databases.

Customisation of the RODOS modules: In the CIS Republics, decision making on off-site emergency actions and countermeasures is based on national radiation safety rules and standards different from those applied in the EU. Large differences exist in the economic systems. Another problem are the country specifics in radionuclide transfer through the environment and food chains and differences in the organisation of the agriculture production in the Ukraine and EU countries. Therefore the RODOS implementation in the Ukraine needs preparatory work to adopt the Ukrainian standards and economic specifics in its countermeasure, consequence and evaluation modules. In particular, when implementing RODOS at ACEZ, the specifics of the SHELTER as the potential release source and the environmental contamination in the Chernobyl zone have to be considered.

2.3 Russia

The Russian Federal Service on Hydrometeorology and Environmental Monitoring (Rosgydromet) is performing the development and implementation of RODOS in Russia.

In the Russian Federation, a series of regulations are in force such as for announcing a radiological emergency, real-time information transmission, special assistance to affected facilities, and population protection measures in the event of a radiation accident at a nuclear power plant. By these regulations, Rosgydromet is to play an active role in the emergency management on the national level.

In this context, Rosgydromet has established a special Emergency Centre for providing real-time and prognostic information about the radiological situation on the territory of Russia. The Centre has been installed in SPA "Typhoon", Obninsk, and its primary responsibility is to provide information and decision support for mitigating the consequences of a nuclear accident. Among others, data on atmospheric transport and dispersion of contamination in the environment are provided. The Centre works in close collaboration with organisations which are involved in the emergency management by Russian regulations.

Rosgydromet suggested that the Emergency Centre should be designated as a regional specialised meteorological centre (RSMC) of WMO with the responsibility to provide results of modelling radioactivity dispersion in the environment. The activities of the Emergency Centre as RSMC will be supported by the Russian Hydrometeorological Centre and the Institute of Experimental Meteorology of SPA "Typhoon" which have the resources required for this purpose: computer capabilities, telecommunication networks and qualified personnel.

3. Results: experience gained and problems arising on the way to implementation

The differences in the potentialities of the base organisations affected the implementation of the RODOS in the three countries. Yet, the active collaboration within JSP1 has enabled different aspects of the RODOS application to be analysed.

3.1. Organisation of the RODOS interconnection with national radiation monitoring systems

As part of the RODOS development, the Emergency Centre at SPA „Typhoon" has concentrated its efforts on
- collection of measurement data from the radiation monitoring network of Rosgydromet, managing the system of distributed databases on environmental contamination located in Rosgydromet divisions, providing data on the radiological situation on the territory of Russia and predicting its changes with migration and transformation of radioactivity;
- preparing prognostic data about dispersion of radioactivity in the environment in the case of a nuclear accident and estimation of the possible transboundary transport of radioactivity.

For accomplishing this, a tool has been developed as part of RODOS to process and present real-time meteorological and radiological data. The basic meteorological data (objective analysis data) and numerical predictions are transferred to the Centre in GRID (Grid code data in symbol form) and GRIB format (grid data in binary form) for the area extending from 25° N to 74° N and 45° W to 180° E. The data include surface pressure, geopotential, temperature, wind speed and direction at 1000, 925, 850 and 500 GPa. The predictions are made for 00, 12, 24 and 36 hours (00.00H and 12.00H).

At present, the Rosgydromet stationary network of radiological monitoring includes:
- 1355 meteostations located across the country and equipped with gamma-dose rate meters;
- 426 points of soil contamination monitoring based on measurements of atmospheric depositions;
- 32 points of atmospheric precipitation sampling to measure tritium concentration;
- 52 points of measurements of radioactive aerosols in the ambient air;
- 61 points at hydrological stations on the major rivers, lakes, reservoirs and seas where water samples are collected on a regular basis to be analysed in laboratories for Sr-90 and gamma-emitting radionuclides (in specific cases);
- 19 points of river water sampling for tritium analysis.

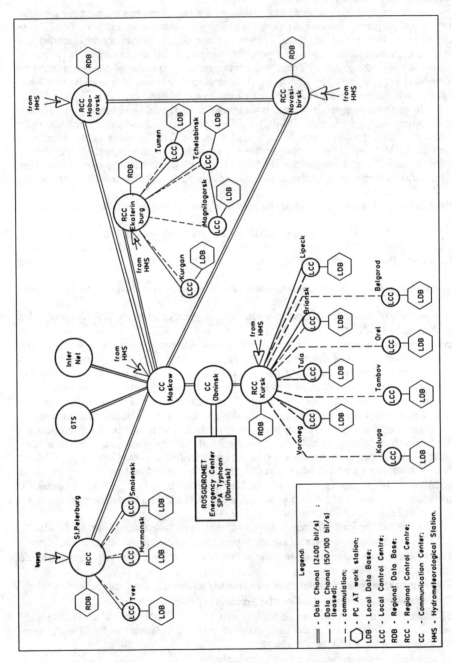

Fig. 2 The radiological network of Rosgidromet

The information is largely transmitted through the meteorological telecommunication system of Rosgydromet. The Rosgydromet system of collecting and processing real-time contamination data is being developed as a distributed system drawing on the regional and provincial divisions of Rosgydromet which are constantly in touch with the Emergency Centre (Fig.2). As basic software, the RECASS system [3] is used (Fig. 3).

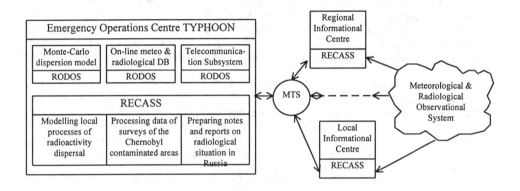

Fig. 3 RECASS/RODOS coupling to Rosgydromet

Since the RECASS system is oriented at PC IBM type computers with MS DOS operation system, one important task of the Emergency Centre is to make the RODOS and RECASS systems consistent in terms of both system management and data types.

As a result of this work, it has become possible to realise some of the RODOS objectives at the level of the Emergency Centre, namely:

- management of data transmission lines;
- management of the real-time meteorological and radiological data base;
- modelling processes of transboundary transport of contaminated air masses from the affected area (this function has enabled the efficiency of the entire system to be enhanced).

In the radiological monitoring network, information is transmitted in two modes: (1) routine passing and (2) request - answer mode.

In the first mode real time data are transmitted including measured daily dose rates in the 100 km zone around a nuclear facility, measured total beta-activity in the ambient air and from depositions, and monthly summary of average values for all types of measurements. If a threshold for any type of measurements is exceeded, an alarm message is sent. In the same mode all meteorological information is passed.

The "request-answer" mode is used to transmit data between information centres of different levels, both within Rosgydromet and between Rosgydromet and Rosenergoatom. In this mode, information can be accessed from the databases in the Centres including real-time data and more precise data of surveys around nuclear facilities.

In Belarus monitoring for radiation situation is mainly carried out in the network of synoptic stations of Hydromet (54 such stations are located on the territory of the Republic), which transmit the information daily to regional meteorological centres by telex and then from regional (district) centres to CREM over the allocated communication links. If necessary the information can be transmitted every three hours. At present in the framework of the European

Commission's "GAMMA-1" project, an automatic system of radiation monitoring is being created in the region adjacent to Ignalina NPP on the territory of Lithuania. It is supposed that as a first step RODOS will be adapted to this area for operational use. Afterwards the system will be continuously and consistently extended for application in other regions close to nuclear power plants surrounding the Republic of Belarus, such as: Chernobyl NPP, Rovno NPP and Smolensk NPP. The interaction with the RODOS systems operating in neighbouring countries is supposed to be developed in parallel.

The meteorological and radiological monitoring within Ukraine is provided by the State Committee on Hydrometeorology (UkrHYDROMET). The radiological monitoring around the NPP's will be provided also by the Automatic System of Radiation Situation Monitoring (ASKRO) that is under development by GOSATOM. The radiation monitoring system GAMMA-1 is being installed in Ukraine with support of the European Commission's TACIS programme.

The first stage of the GAMMA-1 includes the installation of the monitoring networks around Zaporoshe NPP and Rivno NPP. The information from these networks as also from the UkrHYDROMET will come to the Data Processing Centers in the MEPNS and Emergency Response Center in the HCD. The GAMMA monitoring network will be expanded later on other Ukrainian NPP's. GOSATOM has started a programme that includes the development of the Emergency Response Centres at each NPP and the National Crises Centre, based on the ASKRO network and UkrHYDROMET data.

In the future, real time data collection and processing will primarily be based on data processing in local centres, accompanied by an improvement of the request - answer mode to enable the Centres to interact on-line within the distributed databases using the client-server technique. The system should also be extended to allow for new types of data to be incorporated.

3.2. International data exchange

The RODOS concept is understood by the Russian participants as creating a distributed decision-making support system. To this end, tools have been developed to get access to the real-time meteorological and radiological data bases through Internet. Data can be accessed in the request-answer mode (as in the national system) and on-line. The developed tools can be used as basis for interrelating the RODOS local centres within the RODOS international network (Fig. 4) to exchange monitoring data and results of analyses and predictions of the radioactive contamination of the environment.

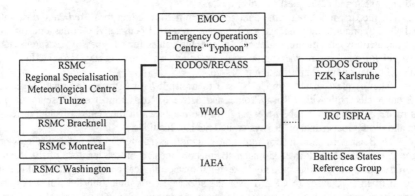

Fig. 4 Connection of RODOS local centres to international networks

Another important direction is organising links with specialised international systems which may be helpful for solving the RODOS tasks. With this aim in view, the Emergency Centre at SPA "Typhoon" is working out tools for interconnection with WMO regional meteorological centres assigned to provide prognostic data on transport of contaminaded air masses. The users can get this kind of information through national meteorological organisations. Similar activities are being developed by Hydromet of Belarus. This institution maintains contacts through regional centre of WMO in Moscow with international meteorological centres in Toulouse (France) and Bracknell (UK).

The experience suggests that creating an integrated network environment of the RODOS system with common data formats and data structures and common rules of access to distributed databases is the shortest way towards practical application of the RODOS system. Given this approach, three components of RODOS can be developed independently (Fig.4):

- a distributed system of data and expert knowledge based on a common network environment
- a decision - making support system combining tools to analyse available information and predict changes in the situation with allowance for countermeasures and protection measures
- a system of interfaces to interrelate the RODOS components between themselves and with national monitoring systems.

In the proposed arrangement, users may be involved at different levels - from data exchange to collective decision-making.

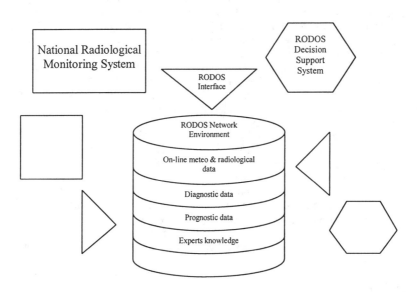

Fig. 5 Integrated network environment for RODOS

3.3. Practical application of RODOS for solving accident-related problems

During the last years, in Russia the RODOS/RECASS databases have been entered with a large body of data obtained in the surveys of the Russian territory contaminated after the Chernobyl and other accidents. The Centre prepares maps of contamination of the environmental media and provides them to users. The continuous inflow of real-time data from the 100 km monitoring zones around nuclear installations (such as gamma dose rate measurements) makes it possible to prepare information on the current radiological situation around nuclear installations.

The models incorporated in the system were actively used for analysing the radiological situation after the Chernobyl accident and, in particular, for reconstructing the dynamics of environmental contamination with short-lived and long-lived radionuclides. These data were used for estimating doses received by the population on the first day of the accident. The potential of the system was also tested during the accident at the Siberian chemical plant Tomsk-7[4]. The Centre was engaged in processing measurement data and modelling. The results presented in a short time included the fields of Nb-95 and Pu-239 environment contamination obtained from calculations with the RIMPUFF code, which were used to estimate the projected individual effective equivalent dose received by the population during the accident (Fig. 6).

In the future, the structure of the RODOS system, its software and databases will enable decision making support in accidents at nuclear installations and other types of disasters. One of the cases when the system was applied to such problems was the fire at the oil storage facilities near Grozny. The hydrological module of RODOS was used in Ukraine in summer 1995 to evaluate the consequences of the accidental release of the Kharkov municipal waste water to the rivers. As result of the heavy rainstorm at 29 June 1995 the pumps of the Dikanki Sewer Water Treatment Station that processed municipal waste water collected from the whole 2-million city of Kharkov was flooded by a 43-meters layer of water. During the following month, the waste water was directly released into the Kharkov's region small rivers - Udy River and its tributaries Lopan River and Kharkov River. The Udy River transported the contaminated water to the Siversky Donets River, that crosses the whole territory of the Donbass Region of Ukraine, which then transported the contaminated water to the Russian's Rostov Region. Under the impact of the heavy bioorganic pollution the dissolved oxygen concentrations fell down in the small Kharkov Region Rivers to practically zero. A heavy bacteriological contamination of the Siversky Donets was measured.

The hydrological module of RODOS was adapted to this emergency situation at IMMS CC within several days to simulate the chemical and bacteriological contamination of the Udy River - Siversky Donets River systems. The calculated predictions of the pollutants concentrations along the Siversky Donets River from the inflow of Udy River down stream till the Russian border were used for decision making by the State Emergency Commission. They were also used to evaluate the amount of water that should be pumped to the Siversky Donets through the channel from the Dnieper River to improve water quality parameters to the permissible levels. This countermeasure on the base of the RODOS' hydrological module calculations was successfully implemented by the Ukrainian State Committee on Water Resources.

4. Conclusion

The realisation of the projects described above will lead to the integration into the national emergency management arrangements of the operational version of the RODOS system in Russia, Belarus and Ukraine. The system will be adapted to the special conditions and coupled on-line to the local/regional/national meteorological and radiological monitoring

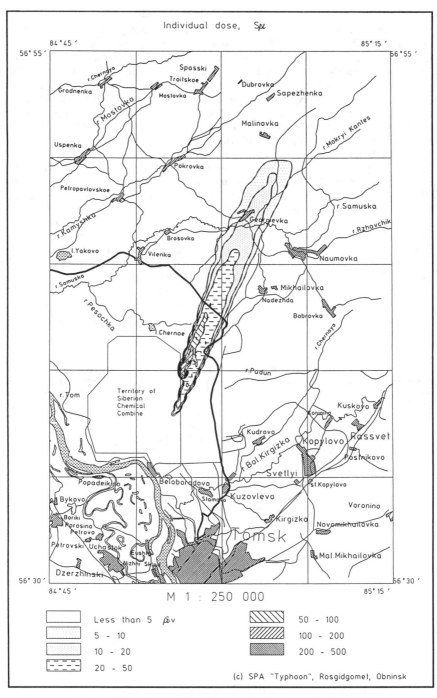

Fig. 6   Projected effective dose distribution around Tomsk-7

networks.

An improvement of the efficiency of the off-site emergency management arrangements in the Ukaraine, Belarus and Russia would be important for both sides, the CIS Republics and the neighbouring East and Central European Countries. The simulation of the accidental situation by national RODOS system coupled with the national meteorological and radiological monitoring networks and data bases would not only lead to a comprehensive and consistent evaluation of the present and future radiological situation and the actions to be taken for mitigating the off-site consequences on the national level, but it will also offer the possibility of early distributing the information on the emergency thoughout the whole of Europe via on-line connections between the European RODOS centres.

Experience that has been gained by the Ukranian, Belorussian and Russian specialists during the joint work with EU instituts and the training with the RODOS system has increased the scientific and technical level of those people, who have direct influence on nuclear safety in their countries. This will help to improve the national standards of the scientific work to be done on decision support for off-site emergency management.

Russian, Ukrainian and Belorussian scientists, engineers and programmers have gained experience on on-line support for decision making after the Chernobyl and other accidents. They will continue to collect comprehensive data bases on environmental contamination, efficiency of countermeasures and dose assessment after nuclear accidents. These data provide a unique base for validation studies with RODOS modules. Therefore the Russian, Belorusian and Ukrainian project partners will provide significant input for further development of the whole RODOS system to the benefit of all involved and the future emergency management in Europe more general.

References

[1] Ehrhardt, J., Shershakov, V., Zheleznyak, M., Mikhalevich, A. „RODOS: Decision support system for off-site emergency management in Europe", First International Conference of the European Commission, Belarus, the Russian Federation and Ukraine on the Consequences of the Chernobyl Accident, Minsk, Belarus, 18-22 March 1996 (this issue)
[2] Zheleznyak, M., Heling, R. "Modelling of hydrological pathways in RODOS", First International Conference of the European Commission, Belarus, the Russian Federation and Ukraine on the Consequences of the Chernobyl Accident, Minsk, Belarus, 18-22 March 1996 (this issue)
[3] Shershakov, V. M., et. al. Radioecological Analysis Support System (RECASS). Radiation Protection Dosimetry ,Vol.50, Nos 2-4, pp. 181-185 (1993)
[4] Shershakov, V. M., et.al. Analysis and prognosis of radiation exposure following the accident at the Siberian chemical combine Tomsk-7. Radiation Protection Dosimetry, V. 59, pp.93-126 (1995).

This work has been supported by the European Commission under contracts COSU-CT94-0087 and FI3P-CT92-0036.

The Software Environment of RODOS

Oliver Schüle, Mamad Rafat
Forschungszentrum Karlsruhe,
Institut für Neutronenphysik und Reaktortechnik,
P. O. Box 3640, D-76021 Karlsruhe, Germany

Valery Kossykh
Scientific Production Association TYPHOON, Emergency Centre,
82, Lenin Street, Obninsk, Kaluga Region, 249020, Russia

The Software Environment of RODOS provides tools for processing and managing a large variety of different types of information, including those which are categorised in terms of meteorology, radiology, economy, emergency actions and countermeasures, rules, preferences, facts, maps, statistics, catalogues, models and methods. The main tasks of the Operating Subsystem OSY, which is based on the Client-Server Model, are the control of system operation, data management, and the exchange of information among various modules as well as the interaction with users in distributed computer systems. The paper describes the software environment of RODOS, in particular, the individual modules of its Operating Subsystem OSY, its distributed database, the geographical information system RoGIS, the on-line connections to radiological and meteorological networks and the software environment for the integration of external programs into the RODOS system.

1. Introduction

The interconnection of all program modules, the input, transfer and exchange of data, the display of results and control of the interactive and automatic modes of operation of the system are all controlled by the Operating Subsystem OSY, which builds the central part of the Software Environment of RODOS. The main duties of OSY are the correct control of system operation, data management, and the exchange of information among various modules as well as the interaction with users in distributed computer systems. The flexibility of the whole system is defined by OSY and is independent of the development of program modules.

1.1. The Modular Design

The RODOS system is based on the Client-Server principle. It is built of modules, which are connected via a Communication Interface. Each of these modules can either be a

- Server, which provides special services to other modules, or a
- Client, which requests services from other modules,

or both. Well defined data structures allow the exchange of data between the client and the server.

This modular design is one of the key features of the RODOS system. It allows the easy extension of the system by adding new modules for special applications and the flexible control of the calculations. All program control, data management, input and output is done by the appropriate modules of the Operating Subsystem OSY. The task of the modules of the Analysing, Countermeasure and Evaluation Subsystems is just performing the model calculations for providing the required results.

1.2. Automatic and Interactive Mode

The dialogue between RODOS and a user can be organised in two different modes. In the so-called "automatic mode" the system automatically presents all information which is relevant to decision making and quantifiable in accordance with the current state of knowledge in the real cycle time (e.g., 10 minutes in the early phase of emergency protection). For this purpose, all the data entered into the system in the preceding cycle (either on-line or entered by the user) are taken into account in the current cycle. Interaction with the system is limited to a minimum amount of user input necessary to characterise the current situation and adapt models and data.

Either in parallel to the automatic mode or alone, RODOS can be operated in the "interactive mode". In this dialogue mode, the user of the system and RODOS communicate via a menu interface. Editors specially developed for this purpose allow specific modules to be called, different sequences of modules to be executed, input data and parameter values to be changed, and the output and representation of results to be varied. The Supervising Subsystem (control system), SSY, supports users by generating a suitable flowchart by which subsystems and modules can be called, which is based on the inherent logic of the spatial and temporal sequence of physical processes and protective actions and countermeasures.

2. Brief Description of the Modules of OSY

2.1. The Message Interface and Communication Server

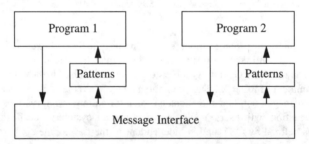

Figure 1: Message Interface of the RODOS System

The exchange of messages between the modules of the RODOS system is controlled by the Communication Interface. Each module can send messages to and receive messages from other modules. The messages contain fields which define the type, sender and recipient of the message. Three types of messages are considered by the Communication Interface:

- Requests are sent to other modules to ask for special services.
- Notification is sent back by the recipient if the request was successfully completed.
- Failures are sent back if some error occurred during the service.

On startup, each module sends a message to the Communication Interface, telling it the message patterns which should be sent to this module (c.f. figure 1).

The System Controller handles the program flow in the RODOS system. It uses information stored in the Database to decide which modules have to be called.

2.2. The Graphics System

The Graphics System must handle all graphics output from various modules of the system. A special graphics program for each module could not be the solution to this problem. It is better to have a universal graphics program, which can handle a large set of graphical output using a well defined data exchange format. As an additional feature, it should be possible to use parts of the Graphics System to create a graphical user interface for programs outside of the RODOS system. The main design aspects of the Graphics System are:

- Handling of graphics output from various external programs.
- Providing the main functionality of graphics programs, e.g zooming, scrolling, modification of graphics objects.
- Modular design to cope with different applications.
- Access to the functionality via a graphical user interface and a message interface.
- Providing functions to build graphical user interfaces for stand-alone programs.

The requirements for the Graphics System of RODOS lead to a modular design. It is divided into three parts:

Graphics Interface Toolbox is a set of functions, which allow the construction of graphics programs and user interfaces.

Graphics Server is a special graphics program designed for the needs of the RODOS system.

Graphics Manager is the interface between the Graphics Server and the RODOS system (mainly the Database Manager).

Each of the above parts uses the features of the previous parts. Graphics Server and Graphics Manager are independent programs, which communicate via a message interface. The user can select its configuration or create a new one using the above parts. Using this modular design, the Graphics System fits different requirements.

The **Graphics Interface Toolbox** contains all functions needed to create a graphics user interface. This user interface can handle graphics output as well as menus to control program execution.

The **Graphics Server** is a graphics program and user interface. It handles the graphics output, such as displaying results on geographical maps, histograms of function plots. A user interface gives the user the possibility of interacting with the Graphics Server (e.g. zoom the output, modify graphics objects).

The basic features of the Graphics Server are:

- A graphical user interface allows the user to control the Graphics Server.
- A message interface is used to parse messages from external programs.
- The picture is handled as a set of graphics objects, which are collected in layers.
- The user can zoom and scroll the picture. Objects can be selected.
- Basic drawing capabilities for the input and modification of graphics data are available.
- A well defined interface is used to send graphics data from different applications to the Graphics Server.

In a complex system – e.g. RODOS – more than one Graphics Server can be run. This allows users to work with the graphics data from the external programs in RODOS on different screens.

The **Graphics Manager** acts as an interface between external programs and the Graphics Server. The main task are:

* Transformation of graphics requests from external programs to commands for the Graphics Server.
* Handling of graphics data from several external programs.
* Control of several Graphics Servers in the system.
* Transformation of graphics data to the data interface of the Graphics Server.

There exist several instances of the Graphics Manager. They are customized to

* handle the communication with the Database Manager of RODOS or
* select graphics output directly from the shared memory of external programs.

Both programs – the Graphics Server and the Graphics Manager – can be connected to other programs via the Message Server of RODOS. These programs which use the capabilities of the RODOS message server can access the functionality of the Graphics System or the Graphics Manager by sending requests to these programs.

2.3. The Database Manager

A basic feature of the RODOS system is the centralized management of data by a Database Manager. It has to cope with different kinds of information, such as

* program parameters,
* geographical and statistical data,
* on-line measurement data,
* forecast weather data,
* result data from external programs.

The data have to be kept in some databases, sent to the programs on request and archived after calculations.

As the RODOS system will contain several different databases, only the Database Manager of RODOS is responsible for the exchange of data between the external programs and the different databases. A unique format for the transfer of data is used to facilitate the access to the data from the external programs.

3. The Databases of RODOS

Systems like RODOS have to manage, process and evaluate a large amount of data of different kinds and quantity, such as geographical, meteorological, radiological and economic data, messages, criteria, statistics, and expert knowledge (facts, rules, preferences). They may be stored in different data bases and computers with their own data structures and formats. In addition, the concept of developing RODOS distinguishes a stepwise progress with versions of improving functionality and for applications with differing complexity. Therefore, it is impossible to realise from the beginning a data bank for all applications and data-specific aspects.

This led to the concept of a distributed data base allowing for a decentralised data management and the parallel execution of multiple task operations. A corresponding Database Interface Manager program will transform the different data formats in the format of the RODOS operating system and will convert the system queries by means of the embedded SQL-interface, and thus increase the flexibility and efficiency of data access.

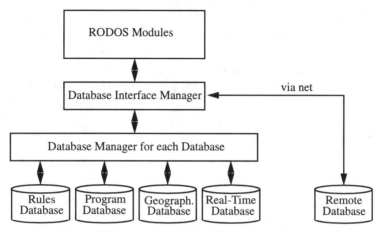

Figure 2: Structure of the RODOS Distributed Database

The data base of RODOS is designed as a distributed data base, which comprises special data bases for geographical information, real-time on-line monitoring data, program data and decision supporting rules (c.f. figure 2).

The program data base contains parameters and results of the application software implemented in RODOS. The real-time database will comprise all kinds of environmental monitoring data and measurements. The information in the rules database consists of expert judgments, facts, rules and preferences required for both evaluating alternative countermeasure combinations and controlling the user interaction and program flow in RODOS.

Each of the data bases of the RODOS system will be a stand-alone data base system, which has its own interface. A Database Interface Manager will give the programs of the RODOS system access to the data stored in these data bases with a unique interface format. The Interface Manager Program will convert the requests from the programs into a request to the appropriate data base. It will enable multiple clients to access multiple database servers. The Database Interface Manager will also facilitate access to external databases, such as the REM data bank of the ECURIE system maintained by JRC Ispra.

3.1. The Geographical Database

Geographical and statistical data are stored in the geographical database. These data are maintained by the geographical information system RoGIS, which is described in the next chapter.

4. The Geographical Information System RoGIS

The **Geographical Information System** RoGIS builds a system for

• handling various geographical and statistical information,

• storing environmental and radiological data,

• organising the access and interchange of data with other environmental data bases.

These features will make RoGIS to an interface between the external programs of the RODOS system and the geographical and statistical information stored as well as to external data bases.

RoGIS is designed as a stand alone program package, which includes all necessary tools for organising the data base and for handling various sets of data. Its structure allows an easy integration of different kinds of data structures. As part of the RoGIS system, an interface package will give external programs access to the data stored in RoGIS.

Another possible configuration of the RoGIS data base is the integration into the RODOS data base. In this case, the access to the data sets of RoGIS is controlled by the data base of RODOS. The close connection between the RODOS system and the RoGIS data base will help to install RoGIS at various sites. Main advantage of this will be the possibility of exchanging geographical and environmental data in an easy way, especially to allow radiological forecasts across boundaries.

Although there exist several socalled geographical information systems, with various applications, the RODOS developers have decided to create such a system of their own. This decision is a consequence of the main aim of the RODOS system, to be a transportable package running on various hardware platforms. The main advantage of RoGIS will be that it is adapted to the needs of RODOS. It will be available to other RODOS contractors with no license problems and no charge.

4.1. Objectives

The data sets of a Geographical Information System can be divided into three groups:

Map data: Landscape, buildings, streets, etc.

Statistical data: Population density, employment, land use, shelter factors, etc.

Environmental data: Wind field, rain areas, γ-measurements, etc.

The map data will mainly be used by the Graphical System as a background to display other information.

Statistical and environmental data are used by external programs as an input for their calculations. The output of these programs itself are mainly environmental data, so that this point has to be subdivided into

Monitoring data: Measurement data.

Modelling data: Output of the external programs.

Statistical data and monitoring data have to be updated. In the case of statistical data, this is done only from time to time. Monitoring data are real-time data, which can be supplied to the data base via an on-line connection by the measuring stations. The validation and preprocessing of the incoming data are tasks, which have to be treated by the geographical information system.

Due to the design of the distributed database in RODOS, the handling of on-line data and the results of external programs is done by other databases. Thus, RoGIS will mainly cope with statistical and geographical data.

4.2. Proposed Methods and Procedures

The term geographical information describes – as already mentioned – all information, which is linked to a geographical entity. It is therefore necessary to give a definition of these words, before the data structure can be described.

The division of an object into its physical appearance and the information available about this physical object is a normal way to describe an object. The information itself is divided into attributes.

4.2.1. Geographical Entity:

Looking at a map, one can see a lot of geometric objects. They can be houses, cities, streets or measurement points. Each of them is a geographical entity. Collections of such objects can build larger geographical entities (e.g. houses, streets and other objects build a city).

Speaking of geographical entities, one must also think of more abstract objects, like borders of countries or ethnic regions.

Definition: A **geographical entity** is an object which is used to describe the geographical structure of a region. It can either be a physical object or an abstract object with a well defined geometric dimension. Geographical entities build a hierarchical structure with various owner-member relations.

4.2.2. Attributes and Values:

The information about an object can be seen as a list of pairs consisting of attributes and values. Each of these pairs gives us a piece of information about this object, which is not divisible into smaller pieces.

Definition: Attribute characterizes a piece of information, which can be expressed by a single value.

Definition: Value is the magnitude of an attribute. It can be a number, a description or a key which is an abbreviation of a complex description or a link to another entry in the data base.

4.2.3. The Link between Attributes and Geographical Entities:

A piece of geographical information can be described by linking together the geographical entity with a list of attribute-value pairs.

The hierarchical structure of the objects allows a sparse use of attribute-value pairs. By defining the scope of the attributes – either local or global – the attribute is valid only for the given object or all objects which follow the given object in the hierarchy. Global attributes of objects preceding the given object in the hierarchy can be overwritten by defining a value for this attribute which is linked to the given object.

4.2.4. Geometric Objects:

The definition of geographical entity states, that each entity must have a well defined geometric dimension. This information is stored in geometric objects, which can be simple geometric primitives (as points, lines or area) or more complex ones (as grids).

To allow more complex structures for geometric objects, we have the following definitions:

Definition: A **primitive** is the basic geometric part of a geographical entity. It can be one of the following types:

Point: A geographical location, expressed by longitude and latitude.

Line: A set of points together with a relation between these points, which defines a simply connected graph.

Area: A closed path of lines.

Grid: A grid is defined by a rectangular region and the number of grid cells in both directions.

Definition: A **geometrical object** is the collection of primitives, which form the geometric information of an geographical entity.

4.2.5. The Hierarchical Organization of Geometric Objects:

Geographical entities are sometimes spread over a wide area. They also can be presented to the user in different ways, depending on the resolution of the map. To optimise the storage of this information, an hierarchical structure is introduced.

Definition: A **facet** is the collection of primitives, which belong to a well defined rectangular area.

Definition: A **level** is a collection of facets, containing the geometric information to be displayed at a given resolution.

Thus, we have on the top a collection of levels, collecting the geometric information for a given resolution. This geometric information itself is stored in primitives, which are collected in facets, each covering an area of the map.

4.2.6. The Link between Geometric Objects and Geographical Entities:

The link between a geographical entity and geometric object can be easily established by a common key. It is also possible to link a geometric object to more than one geographical entity (e.g. the border of a city can sometimes be also the border of a country).

4.3. The Informational Data Base InfDB:

All information, which is not geometrical will be stored in an informational database InfDB. This database will contain all records for the classes, objects, attributes, values and dictionaries.

4.4. The Geometrical Data Base GeoDB:

The geometrical database contains all information, needed to draw the objects in the informational database. It is structured in a hierarchical order to allow the storage of objects in different resolutions.

4.5. The Contents of the Geographical Database RoGIS

The geographical information system RoGIS is designed to hold various sets of information. Main feature of the record structure used in RoGIS is the ability to cope with different data structures. RoGIS is an open data base system, which means, that new hierarchical class structures to describe the information can be defined. Some of them are described here.

The Class **Administrative Data** and its subclasses hold statistical data for Administrative Objects (e.g. countries, cities). Geometrical objects linked to these Administrative Objects define the political and administrative border.

The Class **Nuclear Installation** and its subclasses are used to store information about different types of nuclear installations. Until now there is only one defined subclass for Nuclear Installation - Nuclear Power Plant (NPP). Geometrical object linked to these NPP objects are points which define the geographical coordinates for the object.

5. Remote Databases

The concept of distributed databases in the RODOS system allows the integration of remote databases, situated at different places. These remote databases can be either stand-alone databases (like ECURIE at ISPRA) or the databases of another RODOS system. An on-line connection to these databases is used to transfer the data.

As such remote databases can have different structures and contents, a tool has to be developed, which allows the easy exchange of data between the different database systems and handles the communication via the on-line network. The design of these tools uses the Client-Server approach to handle the data exchange and communication. The client part is responsible for the sending of the request, the server part handles the data exchange with the remote database. A well defined data format is used to send the requests and data via the on-line network.

In particular, the Client has to perform the following tasks:
- provide a user and program interface for the data access,
- create the message for the request and send it to the server,
- receive the data from the Server,
- provide the data to the user or the program.

The Server has to
- process the incoming requests for data,
- get the requested data from the remote database,
- send the data back to the Client via the on-line network.

Figure 3 shows the mechanism of processing requests to a database of a remote RODOS installation.

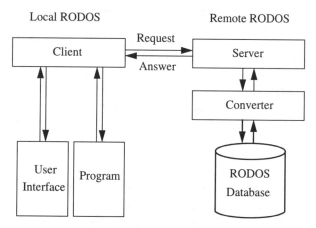

Figure 3: The Connection to Remote Databases

The Connection to the Databases at SPA TYPHOON was established as a first example of the remote database tools. This connection allows the exchange of weather forecasts (temperature, precipitation, wind).

A user interface was developed, which allows the user to define the data request. It allows the interactive input of the region, grid, type and other parameters, for the requested data. A second part of this user interface can display the requested data in form of fields on maps of different scale.

A server for processing the requests and sending the data back to the client via E-mail was developed. The data exchange is realised via the Internet data network. The request is sent using a mechanism of program interaction via Unix-Sockets, permitting a rather sophisticated interaction between the client and the server. E-mail was selected for the transfer of results as the most reliable tool available to pass large amounts of information.

The conducted trials with this software have shown good time characteristics, in particular, the request was sent from Karlsruhe (Germany) to Obninsk (Russia) in several seconds; the request was processed within one minute and the result (up to 50KB) was returned in about 3-5 minutes.

6. Integration of External Programs

The modular structure of the RODOS system and the Client-Server principle allows the integration of new modules in the system. Because the program control, data management, user input and graphical output is entirely handled by the Operating Subsystem OSY, the model developer can concentrate on the contents of his model. A further advantage of the use of RODOS to develop a new model is the possibility of testing it in connection with other – already verified – modules of the model chain.

Adding new modules to the RODOS system is done in several steps:

- Define the services which are provided by the new module (e.g. calculation of organ doses).
- Enter information needed for the program flow into the database (e.g. input data needed by the module, data produced by the module). This will allow the System Controller and the Supervising Subsystem to integrate the module into the program flow.
- Define the input and output data structures of the module.
- Enter the above definitions into the program database of RODOS. This is needed by the Database Manager for the exchange of data.
- Code the module, using a template for the message interface.
- Test the module in the RODOS system.

The integration of already existing stand-alone programs into the RODOS system is done in a similar way. Normally, such a program defines a whole model chain. It is therefore split into its modules, which are integrated into RODOS as described above. In particular, the following steps have to be done:

- Define the modules of the program and their interaction.
- For each module, perform the above steps for their integration.
- Enter the model chain into the database. This is done by defining the starting point and each calculation step based on the data flow of the model chain (e.g. start with meteorological data, calculate activity concentrations, calculate potential doses)

Acknowledgments

This work has been partially sponsored by the Commission of the European Communities, DG XII, Radiation Protection Research Action, under contracts COSU-CT92-0087 and FI3P-CT92-0036, and by the German Ministry of Environment, Nature Conservation and Reactor Safety under contract St.Sch. 1054/1.

Modelling of Early Countermeasures in RODOS

Glushkova, V. , Schichtel, T.[1] , Päsler-Sauer, J.[2]
Cybernetics Centre of the Ukrainian Academy of Sciences, Institute of Mathematical
Machines and Systems, Prospect Glushkova 42, Kiev, 252207, Ukraine
[1,2]Forschungszentrum Karlsruhe, Institut für Neutronenphysik und Reaktortechnik,
Postfach 3640, D-76021 Karlsruhe

Abstract.RODOS is a real-time on-line decision support system for emergency
management after a nuclear accident. One objective of RODOS is the simulation of
emergency actions (sheltering, evacuation, administration of iodine tablets) and the
assessment of the radiological and economic consequences without and with these
actions. In the first part of the paper the corresponding models and methods
integrated in the system are described. The action "evacuation of the population" is a
very complex process as many factors may influence the success and the
effectiveness of the measure. The second part of the paper describes the evacuation
simulation module EVSIM which accounts for most of these factors and which
offers an user interface to adapt the underlying data base to the emergency situation.
Decisions on the spatial and temporal implementation of evacuation are complicated
problems which have to take into account different criteria. The third part of the
paper describes the optimisation module STOP which has been developed for
optimising evacuation routes with respect to route length, minimal dose received,
optimal starting time and costs. The STOP module has been coupled to the EVSIM
module. The paper describes the methodology applied for the modules and
illustrates their interaction.

1. Introduction

RODOS is an integrated and comprehensive **R**eal-time **O**n-line **D**ecisi**O**n **S**upport system for
off-site emergency management of nuclear accidents. The system characteristics and the
current state of development are described in [1-3]. The simulation of the effect of emergency
actions on individual doses and the assessment of other radiological and economic
consequences without and with these actions is the main objective of the module group
EMERSIM, HEALTH and ECONOM in the Countermeasure Subsystem CSY of RODOS.
The emergency actions considered in EMERSIM are sheltering, evacuation and administration
of iodine tablets [2] . These actions are typically limited to areas up to of a few ten kilometres
from a nuclear power plant (NPP), and to time intervals from a few hours before the beginning
of the release to several hours after the cloud of released nuclides has left the near range. In a
given accident situation the areas with emergency actions may be defined by a series of dose
intervention levels and/or emergency zones. Whether, where and when the actions really can
be carried out is a question of the time left in comparison to the time needed for them, and of
the availability of technical and personnel support. This question has to be answered by the
decision maker and his technical advisors.

2. The Module Group EMERSIM, HEALTH and ECONOM

2.1. The EMERSIM Module

EMERSIM determines the areas with early emergency actions, simulates these actions, and quantifies individual organ doses with and without countermeasures. It allows the decision maker for choosing different temporal and spatial patterns of countermeasure combinations. All calculations of doses and consequences are carried out on the same coordinate grid as it is used in the RODOS subsystem ASY for calculating the concentration and radiation fields [2]. It is an orthogonal 41 x 41 cells grid. The typical cell size for near range problems is 1 x 1 km^2 or 2 x 2 km^2.

Modelling of emergency actions and dose calculation in EMERSIM. For the calculation of doses with protective actions an action scenario has to be defined. The definition of an action scenario includes the action areas and the action time intervals. The action areas are sets of grid cells marked with action specific tags. The action time intervals define when a certain action begins and ends. The action times are user input to EMERSIM. With given potential dose histories (a dose history is a time sequence of half hour dose segments; potential dose histories are calculated in ASY [2]), action areas and action time intervals the dose field calculations taking into account of emergency actions can be carried out (a dose field is an array of dose values for all grid cells). For this purpose in each grid cell without an action tag the potential dose segments are summed up. The cells with an action tag are treated in a different way: the potential dose segments are summed up only at times before the action has started. During or after the action the value of each dose segment is modified by a specific factor or function modelling the effect of actions on dose. Then all cell dose segments are added to get the cell dose.

The dose reduction during sheltering is simulated in EMERSIM using building type specific shielding factors for external cloud and ground gamma radiation. The shielding factors are defined as average values for each grid cell and are derived from the building types in that cell. In the present version of EMERSIM the simplifying assumption is made that before the start of and during the evacuation people get the potential dose of their home location. After the evacuation no additional dose is added (dose cut off). In the next versions of EMERSIM a more realistic modelling of exposure during the evacuation will be possible using the results of the evacuation simulation module EVSIM described below. The time dependence of the reduction factor resulting from administration of iodine tablets is modelled by an exponential function.

Modes of operation. EMERSIM can be run in an interactive and an automatic mode. In the automatic mode there is no user input of area and action defining variables. Fixed dose intervention levels are used for estimating emergency actions. In the resulting intervention areas a series of three basic actions is calculated: sheltering only, evacuation only and intake of stable iodine only.

In the interactive mode action areas can be defined indirectly starting from dose intervention levels which can be edited by the user, or directly by graphical input of these areas in the form of sectors and zones around the NPP or of arbitrary shape. The user can define several overlapping action areas, starting times and durations. Fig. 1 gives an example. The capital letters denote the action tags of the areas: **S**heltering, **E**vacuation, intake of stable **I**odine. The combination **SEI** means that **S**heltering, **E**vacuation and intake of stable **I**odine take place in the area tagged with it. 'No action´ denotes areas without emergency actions with normal living conditions. The starting times and durations of the actions can be chosen indepenently in different areas as follows: the sheltering time intervals in **S** and **SI** areas are equal; the sheltering time interval in areas with additional **E** tag (**SE** and **SEI** areas) may be completely different. The evacuation starting before the end of sheltering terminates sheltering. The intake of stable iodine is assumed to occur synchronously in all **I** tag areas. It is not carried out in **E** areas if evacuation starts before the time of iodine intake.

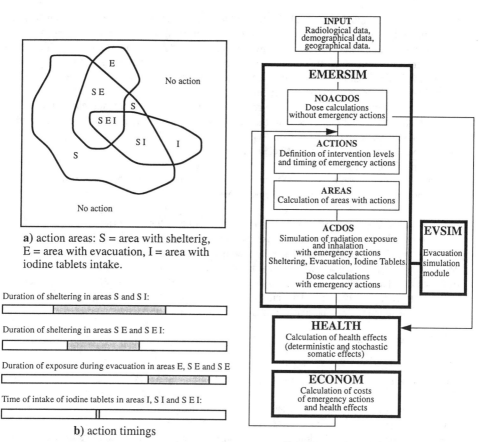

a) action areas: S = area with shelterig,
E = area with evacuation, I = area with
iodine tablets intake.

Duration of sheltering in areas S and S I:

Duration of sheltering in areas S E and S E I:

Duration of exposure during evacuation in areas E, S E and S E

Time of intake of iodine tablets in areas I, S I and S E I:

b) action timings

Fig.1: Example action areas and timings

Fig. 2: Program flow of EMERSIM, EVSIM
HEALTH and ECONOM

Dose results of EMERSIM.   Each dose is estimated for 7 different integration times: 24h, 7d, 14d, 30d, 0.5y, 1y and 50y (except the external cloud gamma dose). The doses are calculated separately for the three exposure pathways cloudshine, groundshine, inhalation, and as the sum of all three pathways. All doses are fields defined on each of the 41 x 41 cells grid. The organ doses calculated are: lung dose, bone marrow dose, thyroid dose, uterus dose and effective dose for adults and partially for children.

2.2. Health Effects and Economic Costs

The health effects module HEALTH estimates the numbers of people with deterministic and stochastic somatic health effects resulting from individual doses. The models used are the same as those implemented in the program package COSYMA [4]. Input to the health effects module are the organ dose fields calculated in the EMERSIM module and the population distribution. From these dose fields the following individual risk fields are calculated:
- Individual risk of deterministic health effects:
  - ◆ morbidity: lung function impairment, hypothyroidism, and mental retardation
  - ◆ mortality: pulmonary syndrome, hematopoietic syndrome, pre- and neo-natal death
- Risk of stochastic somatic health effects: cancer mortality

The number of people with different health effects are determined from the risk fields and the population distribution.

In the economic module ECONOM various off-site economic consequences of the accident are assessed in the form of monetary costs. The different kinds of costs are calculated in several models:
- model for costs of medical treatment of persons with health effects
- model for costs of evacuation and accommodation of people
- model for productivity losses to society due to illness or death of people

The models contain realistic economic parameters and data describing the economic structure of the areas affected by the accident. Input data to ECONOM are population data, the health effects data, regional economic data, and the numbers of persons affected by evacuation. The modelling of economic consequences is oriented at the methodology used in COSYMA [5]

2.3. Program Structure

The program structure and interaction of modules is shown in Fig. 2. The INPUT module provides geographic and demographic data and the potential dose segments stored in the RODOS data base. In the module NOACDOS fields of individual doses are calculated under the conditions of ´no action´. Two cases are distinguished: ´open air´ doses equal to potential individual organ doses and doses under ´normal living´ conditions calculated with location factors for urban areas. In the interactive mode the user can define his action scenarios in the module ACTIONS specifying dose intervention levels, a list of actions to be simulated and starting times and durations. The module AREAS determines the areas with grid cells in which the doses exceed the intervention levels. Once these areas are determined the user can modify them graphically on the screen. The dose calculations including actions are performed in the module ACDOS. The information about the movement of people during an evacuation provided by the module EVSIM will be input to ACDOS and allow a more realistic dose calculation. The modules HEALTH and ECONOM are called to estimate the health effects and the monetary costs with and without protective actions.

2.4. Presentation of Output

There is a variety of output presentations in RODOS: Emergency action areas, population, dose fields, individual risks and numbers of health effects are typically displayed as colour coded areas together with geographical information on the RODOS main window. In addition, local values can be shown on the screen by mouse clicking. Other forms of graphical dose presentation are dose frequency distributions in the population and the time dependence of dose accumulation. Output data not suited for graphical display are presented in tables: such data are collective doses, areas and numbers of people affected by protective actions, total numbers of health effects, costs, and important parameters of the simulated emergency actions like intervention levels, timing parameters, and logical flags. Both graphical output and table output can be printed.

3. The EVSIM Module

The task of the evacuation simulation module EVSIM developed at Forschungszentrum Karlsruhe is to estimate the time evolution of the spatial population distribution in the early countermeasure phase. Therefore, both the topology and the geometry of the traffic net have to be considered in the model. Because of the need of a very fast evacuation simulation the traffic net in EVSIM is modelled on a grid. The time evolution of the population distribution is modelled using constant timesteps. In the present version of EVSIM the length of a grid element is one kilometre and a timestep of a typical run is one minute of evacuation time. The model takes into consideration that the velocities on the road segments depend on the type of road. To improve the model efficiency, EVSIM uses the concept of representatives: individuals with the same or a very similar evacuation pattern are grouped. EVSIM simulates the movements of the group individuals by considering only one representative. The computed spatial population distributions for every timestep are output data from EVSIM which will be used by other RODOS modules (e.g. to estimate the doses of the individuals arriving at the emergency stations). Very important units of the EVSIM code are the EVSIM User Interface and the EVSIM Analysing module.

3.1. Traffic Net Data and User Interface

EVSIM reads the traffic net information from a data file. One part of this file contains a database of evacuation routes for the nuclear power plant considered. The traffic net information can be generated or modified by a text editor. The input module of EVSIM checks the information for consistency and fills the matrices in the computer memory with the traffic net and the information on the evacuation routes. The input module is able to recognise comment areas in the data file. Therefore it is possible to build new data files by the editor using one commented data file as an example.

The success of an evacuation strongly depends on a lot of factors which may change during daytime (e.g. actual number of people in settlements, availability of private cars, ...) and even during the evacuation (e.g. possible traffic blocks, availability of technical and personnel support). To handle all these actual conditions and to use the knowledge of experts about evacuation managment (e.g. by exercises and in the planning phase) EVSIM has its own User Interface. It includes a graphics package designed for the special needs of EVSIM. This graphics package is based on the graphics library used in RODOS [3]. The graphical information is organised in layers. This concept gives the user the opportunity to select and erase information from the graphics screen. So the user is able to specify the amount of data presented on the graphics screen. A HELP tool explains the functionality of EVSIM and offers example sessions to the untrained user.

3.2. EVSIM Scenarios and Simulation Control

Due to the modelling in EVSIM evacuation decisions have to be based on settlements (villages, cities, districts of cities, ...). The time schedule of the evacuation of the settlements in the evacuation zone is very important for the success in dose reduction. Therefore an evacuation scenario in EVSIM defines which settlements will be evacuated at which time. In EVSIM the user can either use predefined scenarios or define its own scenarios using the EVSIM User Interface. Furthermore it is possible to adapt the simulation data to the actual situation by means of this Interface (e.g. setting traffic blocks or the population distribution at the beginning of the evacuation)

The user can interrupt and resume the evacuation simulation by control buttons. He is able to make changes (e.g. set/unset traffic blocks) if the simulation is interrupted. EVSIM presents graphically roads by polygonal lines on the screen. The traffic net, the chosen scenario and the actual simulation status are depicted by special symbols. 'Information Windows´ of EVSIM give the symbol explanation. The user can specify the simulation time intervals between updates of the screen. So it is possible to follow the simulation in ´slow motion´. EVSIM detects when the evacuation is finished.

3.3. The EVSIM Analysing Module

Taking decisions on the most appropriate evacuation pattern and designing effective evacuation plans are complicated tasks. To support solving such problems EVSIM has its own Analysing Module. It is a module that evaluates the efficiency of the simulated evacuation. Endpoints are:
- Simulation data describing in summary the evacuation process simulated by EVSIM
- A graphical representation of the duration of the evacuation for each settlement.
- Driving time distributions
- Dose distributions computed during an EVSIM run

Furthermore it calculates so called "countermeasure indices", which are indicators for the efficiency of the evacuation decisions.

Simulation Data comprise the following data:
- Starting time and duration of the whole evacuation
- Number of evacuated persons
- Number of persons per private car
- Mean transport performance in persons per hour and in cars per hour
- Mean individual driving time and collective driving time in minutes
- Quality measure indices for the evacuation process

Duration of evacuation. To optimise the evacuation with respect to traffic jams and with respect to dose exposure it is extremely important to know the starting time and the end of the evacuations for each settlement. Therefore the starting time, the duration and the end of the evacuation of each of the settlements are graphically presented in one diagram. The connection of these data with the information about the movement of the cloud of radionuclides released by the NPP is necessary to optimise the evacuation with respect to dose exposure.

Driving Times and Countermeasure Indices. The Analysing Module calculates the driving time distributions for settlements and for the complete evacuated population using the simulation data. The distributions are presented as data files and as diagrams. The following diagrams are available:

- Absolute Frequency Distributions of Driving Times
- Normalised Frequency Distributions of Driving Times
- Cumulative Absolute Frequency Distributions of Driving Times
- Cumulative Normalised Frequency Distributions of Driving Times
- Countermeasure Quality Index 0 presents the inverse mean personal driving time in (hour)**-1 for every evacuated settlement and the whole evacuation zone
- Countermeasure Quality Index 1 presents the inverse collective driving time in (hour)**-1 for every evacuated settlement and the whole evacuation zone.

Individual Doses. Furthermore the user can select an EVSIM mode, in which dose field data from the EMERSIM module are used to compute individual doses of the people in the evacuation zone. In this case the Analysing Module of EVSIM builds a set of histograms for each combination of organ, exposure pathway and integration times (see chap. 2.1). A set of histograms consists of the following distributions and moments of distributions:

- Absolute Dose Frequency Distribution
- Normalised Dose Frequency Distribution
- Cumulative Absolute Dose Frequency Distribution
- Cumulative Normalised Dose Distribution
- Mean Dose presents the mean dose of the evacuated individuals got for every evacuated settlement and the whole evacuation zone.
- Collective Dose presents the sum dose of all evacuated individuals for every evacuated settlement and the whole evacuation zone.

If EVSIM is used in its standalone version this mode is very useful for optimising or testing evacuation plans. In this mode EVSIM is able to graphically present the potential dose histories calculated in a previous RODOS run. This enables the user to optimise the evacuation time schedule with respect to the individual dose (e.g. to compare the arrival time of the radioactive cloud at a settlement with the evacuation starting time and with the time when the evacuation is terminated.) Furthermore EVSIM presents the action areas from the EMERSIM module. This gives the decision maker the opportunity of defining the evacuation scenario consistent with the action area used in EMERSIM.

3.4. The EVSIM Data Manager

EVSIM simulations produce a huge amount of output. The data exchange with other RODOS modules is performed by shared memory segments (Shared Memory is a UNIX feature for interprocess communication). EVSIM stores huge data files on the fixed disk for its own purposes (e.g. data for the EVSIM Analysing Module). To avoid missing disk space the output data files of a new EVSIM simulation override the output files of older EVSIM simulations. For the optimisation and the development of evacuation plans it is necessary to compare the results of different simulations. Therefore EVSIM has a 'Data Manager Tool'. It allows to store the output data files in compressed archives and to expand such EVSIM archives. The data of EVSIM archives can be analysed by the Analysing Module of EVSIM.

4. The STOP Module

The optimisation module STOP (System of Transport Optimisation) has been developed at Cybernetics Centre in Kiev. For the decision maker it will be important to know the routes of minimal length, the routes on which people can receive the minimal doses, the starting time of evacuation (time when people finished sheltering and evacuation begins). STOP has been developed according to these needs. The methodology of solving problems in the STOP module is based on the theory of discrete optimisation and graph theory. STOP represents the road network of the region where the accident takes place as a graph. The nodes of this graph can be interpreted as intersections of the roads, the edges can be interpreted as road segments. Each edge of the graph has its weight. The weights can be interpreted differently according to the chosen optimisation criteria. For instance, if the time criteria is chosen, the weights will be the times, if doses are chosen the weights will be the doses (Fig. 3).

The STOP module can solve the following problems: (1) finding the "optimal" route between two nodes, (2) the travelling salesman problem and (3) finding the optimal starting time for evacuation. The module can optimise under different criteria such as route length, dose and time. In the future the dose criteria for the organs bone marrow, lungs, thyroid gland, uterus and effective dose will be used. First, the user chooses the problem he wants to solve and than selects the optimisation criteria. It is necessary to notice that the dose rates change in time. STOP receives the information about the dose rate in discrete time intervals as described above. The algorithms were chosen such that their calculation time is less than the time intervals, with which the dose rates change. This is essential for the travelling salesman problem which is NP-complete. (NP-complete i.e. NonPolynomial complete ; the computation time for an exact solution of NP-complete problems increases with N as exp(const.*N) .)

4.1. The problem of finding the "optimal" route between two nodes of the graph

Given the graph G=(V,A) with the weights of the edges w(e), with e member of A and two nodes s and t (Fig. 3). The problem is to find the path from s to t with the minimal sum of weights. Using the STOP module for solving this problem two nodes have to be set on the

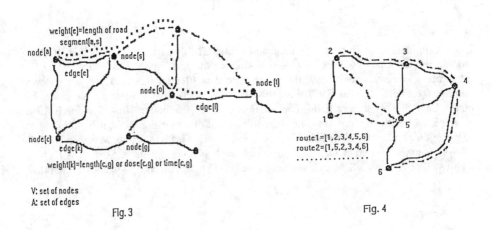

V: set of nodes
A: set of edges

Fig. 3

Fig. 4

graph and the optimisation criteria has to be chosen. The optimisation criteria tell the system how to interpret the weights of the graph. If the user wants to find the route with minimal length, the weights of the graph will be the length of the roads, which are given to STOP by EVSIM as input data.

In the case of finding the optimal route, which reduces the doses received by people during the evacuation, the weights of the graph will be the doses received by people riding on this road with constant velocity. When minimising the transport time through a contaminated zone, the weights will be interpreted as times. In STOP the transport velocity is considered constant for each road segment. For each road segment the velocity depends on the type of the road and its transport capacity. The module STOP not only solves the above mentioned problems. The user may set an upper dose value. STOP will calculate the transport velocity for each road on which the people will receive doses less than this upper value.

To solve the first problem the module uses the Dijkstra algorithm, known as most appropriate [6]. Because the complexity of this algorithm is only of order $n**2$ (where n is the number of graph nodes), the running time of the Dijkstra algorithm can be compared with the time for looking through the input data of the density graph. The STOP module uses two modifications of the Dijkstra algorithm, with and without metrics. In the case of the length and time criteria, where the axioms of metrics are fulfilled, the module applies the first algorithm and in the case of the dose criteria the module applies the second algorithm. We will not describe the problem, using the time criteria in detail, because from the mathematical point of view the time is equivalent to the length criteria in the case of the constant velocity of the transport.

4.2. The Travelling Salesman Problem

This problem appears when we have to visit some places and want to know the order of visiting, which gives us minimal times, length or doses. For example we have 6 places which have to be visited (Fig. 4). We know the length of each road which connects these places. The problem is to find the order of visiting which gives the minimal length of the route. The travelling salesman problem belongs to the NP-complete problems. That means, the time required solving the problem grows very fast, when the number of the nodes increases. Therefore the STOP module uses the local-optimisation methods, which are the analogies of gradient methods well-known in continuous optimisation. These methods permit to find the solution in acceptable time; they have been improved in the Cybernetic Centre, Kiev [7,8].

4.3. The problem of finding the optimal starting time for evacuation

If several cities or villages have to be evacuated it is a problem to know the time pattern of the evacuation process. If evacuation starts at the same time at all places, traffic jams on the roads will appear and the doses, received by the people, may grow. It is more convenient to begin the evacuation in different moments of time. The module can find the optimal starting time of the evacuation for each settlement. To solve this problem the algorithms of the local optimisation are also used.

4.4. Coupling of EVSIM and STOP and applications

The optimisation module STOP has been coupled to the evacuation simulation module EVSIM. The data flow between the two modules is realised by shared memory. The user may call the STOP-module to optimise the complete set of evacuation routes of an evacuation scenario or to find an optimised evacuation route for a specified settlement.

For all problems, which the STOP module can solve, the following input data will be sent to STOP automatically from the other parts of the RODOS system after calling the program: the length of the roads and their segments, the coordinates of the nodes, the population distribution and dose distribution. The user has to choose the number of the problem, criteria of optimisation from the list of criteria on the screen and the community of objects for which this problem will be solved.

Besides the RODOS version of the STOP module there exists an independent DOS version which consists of the following submodules:

- Interpreting the road net and graphics
- Calculating the weights of the graph in the case of dose optimisation
- Calculating the shortest path on the graph between two nodes for the length criteria
- Calculating the shortest path on the graph between two nodes for the dose criteria
- The local-optimisation methods.
- Graphics interface

This version is written in Pascal language. The module uses the network of the Chernobyl region and dose data of the Chernobyl accident.

The combined EVSIM and STOP modules will be used for an analysis of the Chernobyl accident scenario. Therefore work is in progress to create the corresponding input files. The application of the modules to the Chernobyl accident scenario will help to improve actual evacuation plans for nuclear power plants.

References

[1] Ehrhardt, J., Shershakov, V., Zheleznyak, M., Mikhalevich, A. RODOS: Decision Support System for Off-Site Emergency Managment in Europe. First International Conference of the European Commission, Belarus, the Russian Federation and Ukraine on the Consequences of the Chernobyl Accident, Minsk, Belarus, 18-22 March 1996 (this issue)

[2] Päsler-Sauer,J., Schichtel, T., Mikkelsen, T., Thykier-Nielsen, S., Meteorology and atmospheric dispersion simulation of emergency actions and consequence assessment in RODOS. Proceedings for Oslo Conference on International Aspects of Emergency Management and Environmental Technology, 18-21 June 1995, Oslo, Norway, (1995) 197-204.

[3] Schüle, O., Rafat, M., Kossykh, V., The software environment of RODOS. First International Conference of the European Commission, Belarus, the Russian Federation and Ukraine on the Consequences of the Chernobyl Accident, Minsk, Belarus, 18-22 March 1996 (this issue)

[4] COSYMA-A new programme package for accident consequence assessment. Report EUR13028 EN (Brussels: CEC)(1990).

[5] Faude, D., COSYMA - Modelling of economic consequences. KFZA report 4336 (Forschungszentrum Karlsruhe) (1992)

[6] Dijkstra I. B., A Note Two Problem in Connection with Graphs, *Num. Math.* **1** (1959) 269-271.

[7] Sergienko, I.V., Lebedeva, J.J., Roschin, V.A., The local methods in discrete optimisation (in Russian), 1980.

[8] Kaspshickaya, M., Glushkova, V., Some questions of the solving of the travelling salesman problem (in Russian), *Cybernetics* **5** (1985).

This work has been supported by the European Commission under contracts COSU-CT92-0087 and FI3P-CT92-0036 and the German Ministry of Environment under contract St. Sch. 1054/1 .

Modelling of Agricultural Countermeasures in RODOS

J Brown, B T Wilkins, A F Nisbet
National Radiological Protection Board, UK

Y A Ivanov, L V Perepelyatnikova
Ukrainian Institute of Agricultural Radiology, Ukraine

S V Fesenko, N I Sanzharova
Russian Institute of Agricultural Radiology and Agroecology, Russia

C N Bouzdalkin
Belarus Institute of Agricultural Radiology, Belarus

Abstract. Predictions of the effects of agricultural countermeasures taken to reduce doses are an important part of the decision making process following an accidental release of radioactive material into the environment. Models have been developed for this purpose within the EC decision support system, RODOS. This paper describes the methodology used and the development of databases on the practicability of agricultural countermeasures for use in such a system.

Within RODOS a wide range of potentially practicable countermeasures are considered and endpoints related to their imposition are calculated such that the economic and health impacts can be evaluated. The methodology utilises time-dependent information from a foodchain model, together with databases of empirical transfer parameters and effectiveness factors for the various countermeasures. Measures implemented in both the immediate aftermath of an accident and in the longer term are considered.

A database containing robust representative data that can be applied to relatively large areas is currently implemented in RODOS. The database utilises information from the Chernobyl accident compiled within an EC initiated collaborative project (JSP1) by institutes in the Ukraine, Russia, Belarus and the NRPB. Detailed data for four settlements in these countries have been compiled and compared with the robust database. The use of detailed databases at a local level in the three countries, where account is taken of factors such as soil type, is also discussed. The applicability of agricultural countermeasures implemented in the Ukraine, Russia and Belarus following the Chernobyl accident to agricultural systems in the UK has been evaluated and the appropriateness of the compiled databases for wider application is discussed.

1. Introduction

In the European Community's (EC) decision support system, RODOS[1], one requirement is a rapid and detailed prediction of the radiation exposure of the population after accidental releases of radionuclides to the atmosphere in 'real-time'. An assessment of the effect of countermeasures taken to reduce doses also forms an important input to a system of this type.

In RODOS, the calculation of doses and the consequences of agricultural countermeasures take place within two separate but closely linked modules, ECOAMOR (*ECOSYS ASY Mo*dules for *RODOS*) and FRODO (*Food, RelOcation and Decontamination Options*)[2]. These modules are linked with each other and with the input and output data interfaces. The full structure of RODOS is described elsewhere[1].

ECOAMOR, developed at GSF-Institut für Strahlenschutz, Germany comprises a foodchain transport module, based on the dynamic radioecological model ECOSYS-87[3], and a module for the estimation of doses. As input to FRODO, ECOAMOR provides activity concentrations in foods and animal feedstuffs as a function of location, nuclide and time, and doses from the relevant exposure pathways in the absence of any countermeasures. FRODO is a system of program modules which has been developed by the National Radiological Protection Board (NRPB) in the UK and comprises modules for controlling logic flow, relocation, decontamination and agricultural countermeasures[2].

This paper gives an overview of the modelling approach for agricultural countermeasures in FRODO and the databases used. Work carried out within Joint Study Project 1 (JSP1) under an EC initiated collaborative programme between a number of European research organisations and institutes in the Ukraine, Russia and Belarus is summarised and the input to the development of databases for FRODO is detailed. The applicability of agricultural countermeasures implemented in the Ukraine, Russia and Belarus since the Chernobyl accident to other situations is discussed.

2. The agricultural countermeasure module in RODOS, FRODO

The FRODO module includes models for the three countermeasure options relocation, decontamination and agricultural countermeasures. The options can be considered individually or the impact of each of the different options on the others can be evaluated to varying extents. The effect of decontamination of agricultural land on the need and duration of food restrictions is evaluated for a number of endpoints including the additional dose saving. Relocation is linked to agricultural countermeasures in so far as endpoints are calculated to provide information on the potential agricultural areas affected, the amount of food that could be produced within the relocated area and the proportion of food that would be subject to a ban. This enables the evaluation to be made of the use of this land and possible agricultural countermeasures that may be considered to make the area agriculturally productive.

Within FRODO, endpoints related to the imposition of countermeasures on food are evaluated. The agricultural countermeasures considered are: banning and disposal, food storage, food processing, supplementing animal feedstuffs with uncontaminated, lesser contaminated or different feedstuffs, use of sorbents in animal feeds or boli, changes in crop variety and species grown, amelioration of land and change in land use.

The approach taken to modelling agricultural countermeasure is outlined in Figure 1. The aim of the module is to determine whether there is need for intervention. If there is, the

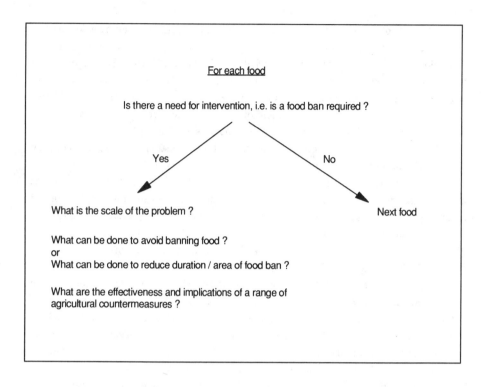

For each food

Is there a need for intervention, i.e. is a food ban required ?

Yes

No

What is the scale of the problem ?

Next food

What can be done to avoid banning food ?
or
What can be done to reduce duration / area of food ban ?

What are the effectiveness and implications of a range of
agricultural countermeasures ?

Figure 1. Approach to modelling agricultural countermeasures in FRODO

effectiveness of a number of countermeasure options is evaluated to determine if the need for
food restrictions can be avoided, or, if this can not be achieved, whether the duration of the
restrictions can be reduced.

The criteria for banning the consumption of food are defined in terms of the activity
concentrations in foods. As a default the EC maximum permitted levels in food are used
although the user of the system can change these criteria. The predicted activity concentrations
in foods are compared with the criteria as a function of time, nuclide and spatial grid point to
determine whether a ban is required. If a ban is not required for any of the foods then no
further measures are considered. If restrictions are required, a number of countermeasure
options are considered for each food. In the current version of RODOS, combinations of
countermeasures are not considered other than with a food ban. If restrictions on food are still
required following the implementation of a countermeasure option, the user of the system will
be informed of this requirement, together with the length of the restriction that would still be
required before activity concentrations fell to below the chosen criteria.

The aim of the module is to provide the information that underpins an assessment of possible courses of action for removing the need for or mitigating food bans. A brief description of the countermeasure options included is given below. Factors such as the timing of the implementation of an option or the duration of a given husbandry or farming practice can be changed by the user of the system so that a range of possible scenarios can be considered. A robust approach is taken in the modelling of the countermeasure options, some of which become very complex if the full flexibility of user choice is implemented.

The banning of foods is linked with disposal and the stopping of food production depending on the duration of the ban.

Food processing and the storage of food are closely linked. For fresh foods such as milk, processing into a form that can be stored is considered only if a ban would not be required on the processed product. Storage is considered with or without processing for all foods. However, a constraint is placed on the storage period such that these are practicable in terms of the 'shelf life' of processed or fresh foods. In practice, this means that storage is only considered when intervention is required only for short-lived radionuclides.

Changes in the dietary composition of grazing animals is considered. Factors that can be evaluated include the effect of administering clean feed for a chosen period at various times following deposition, changes in the proportion of contaminated feedstuffs in the diet and use of different feedstuffs. The effect of the administration of sorbents is modelled by reducing all activity concentrations in the animal by a factor for the period over which the sorbents are administered or, in the case of boli, for the period of efficacy in the gut.

Soil treatment such as the addition of fertilisers can be evaluated together with subsequent effects on the uptake of radionuclides by plants. This countermeasure is not considered until a minimum of 1 year after the accident in the current version of RODOS. Data for a range of techniques are utilised and are represented as a factor by which the activity concentrations in crops are reduced or enhanced. The effect of repeated applications can be considered.

The change of the crop variety or crop species grown is included as a countermeasure. The assumption is made that this option would only be considered if the existing crop could not be grown on the land over a chosen time and that by growing another crop the activity concentrations could be reduced to below the chosen criteria.

The change of land use from agricultural production to forestry can be assessed. The criteria for this option is that the land can not produce food at activity concentrations below the banning criteria for a chosen period of time.

Agrotechnical measures such as ploughing or soil removal are considered. The effectiveness of the techniques is assessed in terms of the reduction in activity concentrations found in food following their implementation. The effectiveness of these measures in reducing activity concentrations in crops is determined from a review of available data, primarily from the Ukraine, Russia, Belarus and western Europe following the Chernobyl accident[4,5,6], and the use of a dynamic foodchain model, FARMLAND[7]. A robust approach is taken such that a single reduction factor is used for all crops. In the current version of RODOS agrotechnical measures are not considered in conjunction with any other agricultural countermeasures.

A wide range of endpoints are calculated. These include the individual and collective doses received following the implementation of agricultural countermeasures and the doses saved by implementing the countermeasure. The extent and duration of food restrictions are also evaluated. Additional information on the impact of the agricultural countermeasures is also provided; this includes the numbers of animals affected and the required quantities of materials such as fertilisers and uncontaminated feedstuffs for animals. This information enables the economic cost of the countermeasures to be estimated.

4. Databases on the effectiveness of agricultural countermeasures

The agricultural countermeasures model in RODOS utilises time-dependent information on activity concentrations in food from the ECOSYS foodchain models together with databases of empirical transfer parameters which can be chosen by the user. Over the first one to two years following deposition it is important that the time-dependence of activity concentration in food and animal feedstuffs is taken into account in the evaluation of agricultural countermeasures; this is done using activity concentrations predicted by ECOSYS. At longer times the activity concentrations in food can be calculated using more simple empirical relationships either as part of a foodchain model or as simple multiplying factors.

4.1 Default agricultural database in RODOS

Within the EC initiated JSP1 project, a database of soil-plant transfer factors and agricultural countermeasure effectiveness for caesium and strontium has been developed for use within the EC RODOS system[4,5,6]. The database has been compiled through close collaboration between NRPB, the Ukrainian Institute of Agricultural Radiology, the Russian Institute of Agricultural Radiology and Agroecology and the Belarus Institute of Agricultural Radiology.

An early conclusion within this project was that for use in a decision support system, agricultural countermeasure data are required at two main levels, one to aid those with responsibility for advising persons who make broad, policy decisions, and the other to aid people at a more local level such as advisors to the farming community. A database has been compiled for RODOS on a scale appropriate for policy level decisions and is referred to as the Level 1 database. This contains robust, representative data that can be applied generally to relatively large areas, potentially over long periods of time. To achieve this, cautious values have been chosen where necessary so that the radiological impact would not be significantly underestimated.

The values in the Level 1 database are largely based on a compilation of data from the Ukraine, Russia and Belarus following the Chernobyl accident. Where data were available from the West these have been included either as supplementary data or to provide additional information. For Level 1 applications, the data on soil-plant transfer and countermeasure effectiveness showed a large degree of consensus between the three countries. This is largely to be expected because of the similar soil types, crops and agricultural practices found in these countries.

The enhanced transfer of radiocaesium from soil to plants in organic soils compared to mineral soils is well established[8]. For this reason, soils were broadly classified into mineral or organic for the purposes of the Level 1 database. Inspection of data from the three countries, together with published information[8], suggests that a similar approach is appropriate for radiostrontium. This is a simplified approach to soil classification but it is relatively easy to identify soils which fall into these categories. The database primarily contains data on mineral soils and includes data on organic soils where these are available. For Level 1 applications values for mineral soils have been chosen as a default because these soils are most typical for agricultural crops. However, information for both mineral and organic soils is provided so that a more specific assessment can be carried out if information on the broad soil-type is available. The default values for mineral soils were derived from data for a range of soil types within which there was considerable variability in characteristics such as pH.

Table 1 gives the recommended default values for the Level 1 database for strontium and caesium for some of the considered agricultural countermeasures.

4.2 Comparison of site-specific data and Level 1 default values

Data have been collected for four settlements; Rodina in Russia, Galuziya and Chapeaeva in the Ukraine and Kirovsky in Belarus[9]. The information compiled on each settlement has included soil type, contamination levels, crop species grown and animal stocks as well as the agricultural countermeasures that have been imposed and the resultant effectiveness factors. Generally, data have been compiled for both the collective farm and the associated private farms within these settlements.

Table 1: Effectiveness of some agricultural countermeasures for caesium and strontium in Level 1 database[4,5,6]

Countermeasure	Recommended value and range[a]	
	strontium	caesium
Sorbents: milk	not appropriate	5 (2-10)
meat		3 (2-10)
Ploughing meadows: discing (10 cm)	2 (1.3-3)	1.5 (1-2)
ploughing (20 cm)	3 (2-4.8)	3 (2-4)
ploughing with turnover of top layer (30-50 cm)	6 (5.9-6.7)	10 (7-16)
Lime application[b]	2 (1-2.7)	2 (1.5-4)
Organic fertiliser application[b]	not appropriate	2 (1-5)
Mineral fertiliser application[b]	1.3 (1-2)	2 (1-5)

Notes:
a. Expressed as a reduction in activity concentration in the food or feedstuff.
b. All crops

The comparison of the observed countermeasure effectiveness factors for caesium at the four settlements with the default Level 1 values and ranges given in references 4 and 5 showed that, in general, the differences were less than a factor of 3. Since the Level 1 database is intended for use at a broad, policy level, these differences are considered acceptable. Some categories of intervention, notably radical improvement, include a wide range of measures with very different effectiveness factors. Radical improvement of natural meadows was an important factor in the reduction of activity concentrations in milk but no firm recommendations were made for the Level 1 database because such measures would not be applicable to large areas of land. The overall conclusion was that modifications to the robust values in the Level 1 database would not be warranted, but more specific information would be required before decisions could be taken on a local level.

It should be noted that not all of the countermeasures considered in RODOS were implemented at the sites studied. For example, no changes to crop species were implemented.

4.3 Provision of advice at a local level

The user of the RODOS system has flexibility to use data for foodchain transfer, agricultural countermeasures and their effectiveness that are appropriate for the situation under consideration. RODOS could, therefore, be used on a local level, provided that appropriate detailed databases were available. In the longer term, practical, site-specific advice on countermeasure strategy would be sought from local agricultural advisers who might not have access to the RODOS system. Their requirements could be met by a system that dealt specifically with agricultural countermeasures. Such a system has been developed by RIRAE within the JSP1 project and this has been given the acronym, FORCON[10]. This system contains default databases for radiocaesium that are more detailed than those currently in RODOS. For example, five soil types are considered, classified on the basis of granularmetric composition. Alternatively, the user can encode site-specific information. The most recent version of FORCON can be run on the type of PC currently available in the newly independent states of the former Soviet Union; a detailed account of the system is available[10].

Provided that the same databases are used, assessment of countermeasure effectiveness in RODOS would be compatible with those made with FORCON. The two systems may therefore be regarded as complementary.

5. The use of data from the Ukraine, Russia and Belarus for wider application

A review of the applicability of the agricultural countermeasures taken or considered in the Ukraine, Russia and Belarus following Chernobyl to the UK has been carried out[11]. This review considered both radioecological effectiveness and ancillary factors such as costs, availability of resources, practicability and acceptability by farmers, the general public and the retail trade. The study considered 25 countermeasure options, many of which are included in FRODO. A few examples of the applicability of various countermeasures are given here.

Data from the three countries indicate that improvements in soil fertility have been effective in reducing radiocaesium uptake by plants. Experimental evidence suggest that such a measure would be ineffective on the high fertility soils prevalent in the Western Europe[8]. The application of fertilisers to poor quality soils, such as those found in semi-natural ecosystems, is not considered a practicable option: not only are these practical difficulties in administering the treatments in remote areas, but there is also the possibility of disturbance of the ecological balance.

The selection of crop species and variety also deserves comment. Values for soil-plant transfer in the Level 1 databases were distilled from a large amount of information and relate to climatic conditions and soil types in the Ukraine, Russia and Belarus. Given the variability in soil type and the crops grown, it would not be appropriate to apply these values broadly to western Europe. An investigation of radionuclide uptake by crops grown at a single site in the UK supports this view[12]. In this study it was concluded that a comprehensive datasets would be required before crop selection could be used with confidence as a countermeasure, since the variability between crops was similar to the inherent variability for the same crop grown in different years.

Some measures taken in the three countries considered would be generally applicable. Notable examples are the addition of sorbents to animal feeds or the administration of boli and the provision of clean feed to livestock through careful management of their feeding regime.

The overall conclusion was that few of the countermeasures considered would be widely applicable to any of the arable, grassland or semi-natural systems in the UK. This was due either to low radioecological effectiveness or low practicability. The conclusions of the study are likely to be applicable to other parts of Western Europe.

6. Conclusions

This paper has provided an outline of the agricultural countermeasures modelling within the FRODO module of the RODOS decision support system. The modelling approaches taken and assumptions made have been discussed and the endpoints calculated for use within the RODOS system identified. The paper presents the current status of the models that are incorporated within the modules of the RODOS system. It should be noted, however, that further development of some areas of these modules and links with other modules will be undertaken for inclusion in future versions of the system.

Databases for agricultural countermeasures effectiveness and soil-plant transfer have been compiled for the Ukraine, Russia and Belarus for use within RODOS under the EC initiated collaborative project, JSP1. The default values in the RODOS system are robust representative data that can be generally used and applied to large areas of land, potentially over long timescales. Detailed data have been compiled for four settlements in the three countries and these data compared to the robust values. Following this comparison modifications to the robust values in the Level 1 database were not warranted, although it was recognised that more specific information would be required before decisions could be taken on a local level. A PC based system for local use has been developed for use by local advisors to the farming community in Russia, Ukraine and Belarus which can utilise detailed default databases or, if available, site-specific information. This system is compatible with RODOS if equivalent databases are used.

The use of the databases compiled for the Ukraine, Russia and Belarus for wider application has been evaluated. Care is required in the use of these data. Some of the measures that were radioecologically effective in these countries would not be so in western Europe such as the use of fertilisers, which have been shown to be much less effective on the high fertility soils found in western Europe, while others would not be practicable. Further work on the applicability of countermeasures applied in the Ukraine, Russia and Belarus following Chernobyl to Western Europe is planned.

7. Acknowledgements

This work was partially funded by the European Commission.

8. References

[1] J. Ehrhardt, J. Päsler-Sauer et al, Development of RODOS, a comprehensive decision support system for nuclear emergencies in Europe - an overview. Rad. Prot. Dos, Nos 2-4, pp 195-203, 1993.

[2] J. Brown, K.R. Smith, P. Mansfield, J. Smith and H. Müller, The modelling of exposure pathways and relocation, decontamination and agricultural countermeasures in the European RODOS system. IN Proc. International Aspects of Emergency Management and Environmental Technology, Oslo, June 1995 (Ed. K H Drager, Oslo, 1995.

[3] H. Müller and G. Pröhl, ECOSYS-87: A dynamic model for assessing radiological consequences of nuclear accidents. Health Physics 64, pp 232-252, 1993.

[4] J. Brown, B.T. Wilkins et al, Compilation of data from the Ukraine, Russia and Belarus on the effectiveness of agricultural countermeasures for use in the RODOS system. Chilton, NRPB-M518, 1994.

[5] J. Brown, Y.A. Ivanov et al, Comparison of data from the Ukraine, Russia and Belarus on the effectiveness of agricultural countermeasures. Chilton, NRPB-M597, 1995.

[6] L.V. Perepelyatnikova, G.P. Perepelyatnikov et al, The behaviour of strontium-90 in agricultural systems: a compilation of data from the Ukraine, Russia and Belarus. NRPB Memorandum (in preparation).

[7] J. Brown and J..R. Simmonds, FARMLAND: A dynamic model for the transfer of radionuclides through terrestrial foodchains. Chilton, NRPB-R273, London HMSO, 1995.

[8] A.F. Nisbet, Effectiveness of soil-based countermeasures six months and one year after contamination of five diverse soil types with caesium-134 and strontium-90. Chilton, NRPB-M546, 1995.

[9] M. Paul, B.T. Wilkins et al, The development of countermeasure strategies at selected settlements in the areas affected by the Chernobyl accident. NRPB Memorandum (in preparation).

[10] S.V. Fesenko and B.P. Kulagin, Development of the FORCON decision support system for the provision of advice in agriculture at the local level (in preparation).

[11] A.F. Nisbet, Evaluation of the applicability of agricultural countermeasures for use in the UK. Chilton, NRPB-M551, 1995.

[12] N. Green, B.T. Wilkins et al, Transfer of radionuclides to vegetables and other crops in an area of land reclaimed from the sea: a compilation of data. Chilton, NRPB-M538, 1995.

Modelling of Hydrological Pathways in RODOS

Zheleznyak, M., Heling, R.[1], Raskob, W. [2]., Popov, A.[3], Borodin, R. [3], Gofman,D.,
Lyashenko,G., Marinets, A., Pokhil, A.[3], Shepeleva, T., Tkalich, P.
Cybernetics Centre of the Ukrainian Academy of Sciences, Institute of Mathematical
Machines and Systems, Prospect Glushkova 42, Kiev, 252207, Ukraine
[1] NV KEMA, P.O. Box 9035, Arnhem, Netherlands
[2]Forschungszentrum Karlsruhe GmbH, INR, P.O. Box 3640, D-76021, Karlsruhe,
/ D.T.I. Dr. Trippe Ing. G. m.b.H., Amalienstr. 63/65, 76133 Karlsruhe, Germany
[3] Scientific Production Association TYPHOON, Lenin str 82, Obninsk, Kaluga Region,
249020, Russia

Abstract. In 1992, a joint EC-CIS team of experts started to develop a
hydrological module for the decision support system RODOS. A model chain was
outlined covering the processes such as run-off of radionuclides from watersheds
following deposition from the atmosphere, transport of radionuclides in river
systems and the radionuclide behaviour in lakes and reservoirs. The output from
the hydrological transport chain is used to calculate the main exposure pathways
such as the doses derived from the consumption of drinking water, of fish, of
irrigated foodstuffs and the external irradiation. Test and validation studies of the
whole chain as well as for individual models were performed on the basis of
experimental data from the basins of Dnieper and Rhine. A user friendly graphical
interface was developed to operate the individual models inside the hydrological
module.

1. Introduction

Within its Radiation Protection Research Programme, the Commission of the European
Communities has embarked on a major project aiming at the development of an integrated
and comprehensive real-time on-line decision support system (RODOS) for nuclear
emergencies in Europe[1]. The Chernobyl accident demonstrated the importance of the
aquatic pathways in the radiological assessment of environmental consequences of an
accidental release of radionuclides from a nuclear installation. After the Chernobyl
accident, the CIS countries gained a lot of experience in supporting decision makers by
modeling the radionuclide contamination of large water systems [2,3].

In 1992, a joint EC-CIS team of experts started to develop a hydrological module for
RODOS in the frame of the Joint Study Project (JSP-1) [4,5]. A model chain was outlined
covering the processes such as run-off of radionuclides from watersheds following
deposition from the atmosphere (RETRACE-1 for small watersheds and RETRACE-2 for
large watersheds), transport of radionuclides in river systems (RIVTOX) and the
radionuclide behaviour in lakes and reservoirs (LAKECO and COASTOX). The near range
transport and dispersion of radionuclides following direct releases into the river are
described by the COASTOX model. The module H-DOSE uses subsequently the output
from the hydrological model chain to calculate the main exposure pathways such as the
doses from the consumption of drinking water, fish, irrigated foodstuffs, and from external

irradiation. Validation studies of the whole chain and of the individual models were performed on the basis of experimental data from the basins of the rivers Dnieper and Rhine. A user friendly graphical interface was developed to operate the individual models inside the hydrological module. This publication summarizes these collaborative activities.

2. Modeling of Radionuclide Transport via the Hydrological Pathways

The evaluation of the radiological consequences of accidental releases of radionuclides from various specific sites demonstrated a significant contribution from the contaminated waterbodies to the dose of the population. This was e.g. clearly shown for the Clinch River-Tennessee River basin (releases from Oak Ridge), for the Techa River-Ob River watershed (releases from "Mayak"), for the Dnieper river basin, and for the dose to the population in the vicinity of Scandinavian lakes (Chernobyl accident). The re-mobilisation of dry and wet deposited material by long term floods and heavy rain events, and the resuspension of sediments during storm events resulted in the migration of radionuclides and affected also uncontaminated agricultural areas together with drinking water supplies downstream from the source of the initial contamination. Additionally, the remobilisation of radionuclides stored in the bottom sediments of lakes and reservoirs caused a delayed transfer of activity to the aquatic environment.

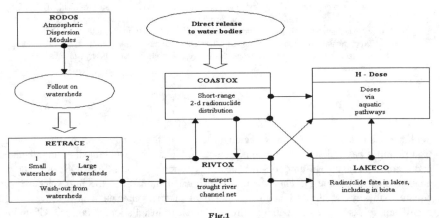

Fig.1

To facilitate and enhance the quality of emergency actions, the mathematical description of the processes involved is required. RODOS will therefore contain a chain of models, which cover all the relevant processes such as the direct inflow into rivers, the migration and the run-off of radionuclides from watersheds, the transport of radionuclides in large river systems including exchange with sediments and the behaviour of radionuclides in lakes and reservoirs. The hydrological model chain will be part of the analyzing subsystem of RODOS (ASY) to predict activity concentrations in waterbodies. These models will also operate in the consequence subsystem of RODOS - CSY, for identifying strategies of possible countermeasures. Starting points for the early phase can be a direct release into a river or lake and/or the predicted contamination of a land area following an atmospheric release of radionuclides. On later stages after the accident, the estimated deposition data will be corrected by monitoring data. As RODOS is designed to predict the short-term and long-term consequences of the accidental releases, the aquatic module contains models of different temporal and spatial scales (see Fig. 1).

2.1. Run-off models RETRACE-1 and RETRACE-2

The RETRACE code, simulating radionuclide transport by runoff from watersheds, is under development at SPA Typhoon, Obninsk, Russia [6]. RETRACE-1 describes the radionuclide wash-off at a local scale, i.e. small watersheds which sizes of less than 1000 km^2 and with a temporal resolution of hours, while RETRACE-2 covers the regional scale up to large watersheds (larger than 1000 km^2 and with a temporal resolution varying from days to years). Both models consist of a hydrological and a radionuclide transport submodel.

In RETRACE-1, the water dynamics of the soil surface following rain events is simulated on the basis of the two-dimensional kinematic wave equation including source/sink terms describing precipitation rate, infiltration rate, canopy interception rate, rate of losses in surface depression and evaporation rate. The two dimensional kinematic wave approach is also used for the description of the subsurface runoff.

The approach applied in the hydrological submodel of RETRACE-2 is situated between a lumped-parameter model and a distributed-parameter model. It operates with ordinary differential conceptual equations but for spatially distributed parts of the catchment.

The sediment concentration (from erosion) in the runoff water is calculated in both models on the base of empirical relationships. It is assumed, that the radionuclides in the upper soil layer with a thickness of 1 mm can contribute to the run-off process by water wash-off and by erosion processes. Additionally, it is assumed, that the concentration of the solved radionuclides in the surface and the subsurface water are in equilibrium (Kd approach). The transport equations of the radionuclides in the RETRACE are based on the conservation equation for the total activity of dissolved and sorbed components. RETRACE-2 uses ordinary differential equations for the radiological modules whereas RETRACE-1 is based on partial differential equations.

The required input data includes among others cartographic data (e.g. relief, soil, vegetation, rivers and lake location), parameters of the soil (e.g. infiltration capacity, moisture), weather (e.g. the probability of precipitation), vegetation (e.g. interception parameters of canopy for different seasons), and parameters of the radionuclide transport (e.g. distribution coefficient, transformation rates).

2.2. RIVTOX, a one dimensional river model

The one-dimensional model RIVTOX, developed at IMMS, Cybernetics Centre, Kiev[7,8], simulates the radionuclide transport in networks of river channels. Sources can be a direct release into a river or the runoff from a catchment. In the latter case, the output from RETRACE is used as the input of RIVTOX. The stream function, the transport of suspended sediments and the radionuclide dynamics are averaged over the cross-section of the river. A 'diffusion wave' model, derived from the one-dimensional Saint-Venant's equation, describes the water discharge. An advection-diffusion equation calculates the transport of the suspended sediments in the river channel. Its sink/source terms describe the rate of sedimentation and resuspension as a function of the difference between the actual and the equilibrium concentration of suspended matter with respect to the transport capacity of the flow. The latter is calculated on the base of semi-empirical relations. The dynamics of the upper contaminated river bed is driven by an equation for the erosion of the bottom layer.

The radionuclide transport submodel of RIVTOX describes the dynamics of the cross-sectionally averaged concentrations of activity in solution, in suspended sediments and in bottom depositions. The adsorption/desorption and diffusion transfer in the systems "solution - suspended sediments" and "solution - bottom deposition" is treated via the Kd approach assuming equilibrium. However, the exchange rates between solution and particles are taken into account too, for a more realistic simulation of the kinetics of the processes. It is assumed that the adsorption and desorption rates are not equal.

The most important input data are:

- parameters of the river channel network, e.g. length of branches and junction positions, dependence of the crossection on the water surface elevation, bottom roughness and typical scenarios of floods for the simulation of a direct release of radionuclides into the river.
- typical distribution of the grain size of suspended sediments and of bottom depositions.

2.3. COASTOX, a two-dimensional model calculating the lateral-longitudinal distribution of radionuclides in water bodies

The two-dimensional model COASTOX [7, 9] uses the depth averaged Navier Stokes equations to calculate the velocity field in rivers, lakes and reservoirs generated from the combined influence of discharge, wind and bottom friction. The steady state approximation without advection terms and the system of the unsteady shallow water equation are used. The same approach as in RIVTOX is applied to simulate the radionuclide exchange in the system: solution - suspended sediments - bottom depositions. The 2-D advection-diffusion equations and the equations of flow dynamics are solved numerically by using the finite difference methods. Necessary input to COASTOX is the geometrical data of the river/lake bed in a sufficient fine spatial resolution.

2.4. Lake model LAKECO

The box-type model LAKECO, developed by the KEMA, Arnhem, The Netherlands [10], is used for predicting the behaviour of radionuclides in lakes and reservoirs. It calculates the concentration of the activity in the water column, in sediments and in the biota dynamically. It is divided into an abiotic part, describing the change of the activity concentrations in the water/soil column by means of linear differential equations of first order and a biotic part predicting the transfer throughout the aquatic food chain.

The processes which are taken into account are: particle scavenging/sedimentation, molecular diffusion, enhanced migration of radionuclides in solution due to physical and biological mixing processes, particle reworking - also by physical and biological means - and the downward transfer of radionuclides in the seabed as a result of sedimentation. In sediments both the fractions of solved and dissolved radionuclides are modelled., A complex dynamic model, taking into account the position of the different species in the food web, has been developed to predict the transfer throughout the aquatic food chains. This dynamic uptake-model is based upon studies on mercury in fish [11].

Sensitivity analysis showed that the distribution coefficient water suspended matter, and the concentration factor water phytoplankton are the most sensitive parameters. Less sensitive were the reworking rate, and the biological half life of the aquatic organisms. To improve the predictive power and the flexibility of LAKECO, new submodels to assess these sensitive parameters were implemented. As a result, the modified model LAKECO-B has more environmental parameters, like the potassium concentration in the lake water, as input, but less model specific parameters. Thus, LAKECO-B has become an aquatic model where tuning is nearly impossible as environmental input parameters control the model.

2.5.H-DOSE - dose model for the hydrological chain

In order to provide the dose as endpoint, even in intermediate development stages of the aquatic module, a simple dose model was outlined at FZK Karlsruhe, Germany [12]. The model will be replaced in future by the dose model ECOAMOR presently implemented in RODOS [1]. However, the extension of the dose model as well as the data exchange of the aquatic model chain with RODOS has to be established in future. The computer code H-DOSE considers 4 different exposure pathways:

- Consumption of foodstuffs contaminated by irrigation (root vegetables, leafy vegetables, milk and milk products, meat and meat products)
- Consumption of contaminated drinking water
- Consumption of contaminated fish
- External radiation from the borderline of the river or lake.

As the present dose model is only of a preliminary state, rather simple approaches have been used mainly in accordance with the German Regulatory Guidelines [12]. Only the effective committed dose equivalent is assessed.

3. Software Framework of RODOS Hydrological Chain

A user friendly graphical interface was developed to operate the individual models inside the hydrological module. The interface provides the possibility to access easily all the information necessary to run the individual models as well as displaying the results in a way decision makers can handle them (e.g., Fig. 2). The interface was designed as stand alone program and allows:

- to integrate codes on the base of a RODOS-like technology with the possibility to allocate only as much shared memory as the program really uses for the simulation
- to input and edit data and parameters through a system of users-configured dialogs and input windows
- to run models separately or simultaneously with the possibility to exchange data between individual models via shared memory
- to manage the data base and to create predefined scenarios
- to present data base information and on-line results of the simulations in graphs and maps (e.g. contamination)
- to receive data from other RODOS modules (e.g. results of atmospheric dispersion)
- to support different modes with different user services: 2 automatic modes- "whole chain"(whole chain starting from data of the atmospheric dispersion till the dose model) ,"direct release"(RIVTOX, COASTOX, LAKECO, DOSE) and 2 manual modes -"decision maker"(with loading of predefined scenarios) and "scenario maker"(creating scenarios, data base updating)

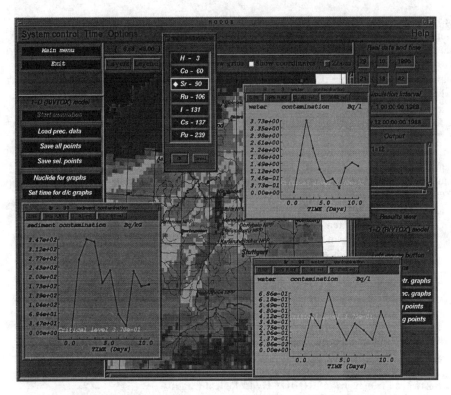

Fig. 2. Interface of RODOS Hydrological Model Chain. RIVTOX application for Rhine basin

New ideas realized in the RODOS hydrological interface are
- creation of predefined scenarios
- different automatic and manual modes for different categories of RODOS users
- a user configured system of input windows and dialogs
- new techniques of integration of external programs

At present, there are 5 models (RETRACE, RIVTOX, COASTOX, LAKECO and DOSE) integrated in the RODOS Hydrological module. For each of these models the interface provides the same basic set of the user interfaces and data base tools ,however great effort was made to consider all the specific needs of input and output of individual models. For example the operation of RETRACE in the interactive test and expert modes was realized by a specially developed interface - the RETRACE monitor. For the automatic and the "decision maker" mode however, RETRACE is under the complete control of the 'normal' tools of the hydrological interface.

4. Model Chain Validation Studies

To prove the reliability of the various aquatic models, validation and intercomparison studies were performed, among others, within the frame of the IAEA/CEC VAMP program and the

BIOMOVS II program, as well as within other special validation studies [13]. The knowledge gained herein has lead to further model improvements.

4.1. RETRACE - RIVTOX chain validation on the base of Ilya River case study

The RETRACE - RIVTOX chain validation study was performed for the catchment of Ilya River, a tributary of the Uzh River, flowing into the Kiev Reservoir. The watershed, situated mainly in the 30-km Chernobyl zone, has a size of about 20 km in longitudinal direction and about 15 km in lateral direction. The following data, measured in 1988 by the SPA Typhoon (Obninsk, Russia) and the Ukrainian Hydrometeorological Institute (Kiev, Ukraine) were used in the validation study:

- soil contamination
- meteorological data (daily precipitation);
- water discharge at the outlet;
- concentration of soluble and sorbed forms of ^{90}Sr and ^{137}Cs at the outlet;
- contamination of bottom sediments.

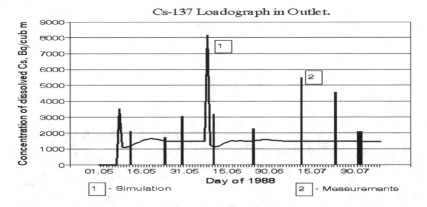

Fig. 3. Simulation by RETRACE of ^{137}Cs day averaged concentration in runoff from the Ilya River catchment

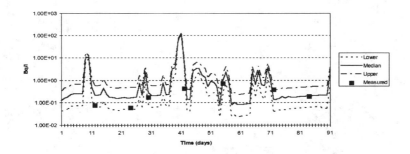

Fig. 4. Simulation by RIVTOX of ^{137}Cs concentration on suspended sediment at outflow from the Ilya River

The lateral inflow into the river net simulated by the RETRACE (Fig. 3) was used by the RIVTOX to calculate the transport in the river channels taking into account the interaction between radionuclides in dilution and bound on suspended sediments (Fig. 4). The incomplete set of the measurements at the river outlet did not cover the exact times of the short rainstorm events. Therefore, the measured and calculated concentrations should be compared rather for averaged than for peak values. The uncertainties for the water-sediments exchange parameters of RIVTOX were evaluated by using the Monte-Carlo technique. The measured data lay inside a 90% confidential interval of simulated ^{137}Cs concentration bound on sediments (Fig. 4).

RETRACE was tested within the BIOMOVS-II program with a scenario covering the run-off of cesium and strontium from small experimental plots located in the 30 kilometer Chernobyl zone. The agreement between the results of RETRACE and the experimental data was one of the best among the contributing models [6]

4.2. RIVTOX and COASTOX validation studies

The hydrological part of RIVTOX was tested with data of the Tvertsa river (Russia) and the Dniester rivers (Moldova-Ukraine) [8]. RIVTOX was successfully applied to simulate the fate of chemicals in the Rhine river which resulted of an accidental release at Sandoz, Basel, Switzerland [4]. In the frame of IAEA\CEC VAMP programme RIVTOX was validated on the scenario of the Clinch river -Tennessee river which were contaminated by radionuclide releases from Oak Ridge[13]. The VAMP scenario of the contamination of the Dnieper following the Chernobyl accident was used to calibrate ^{137}Cs and ^{90}Sr transfer parameters inside the RIVTOX code.

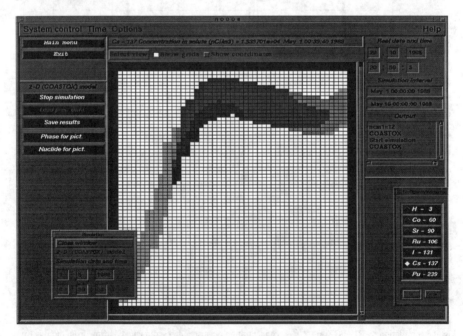

Fig. 5. Interface of COASTOX. Simulation of direct release into the Rhine River.

A special study was performed within the JSP-1 project to test RIVTOX on the basis of post-Chernobyl data of the Rhine basin. A reasonable good agreement was obtained with data measured in the two rivers Neckar and Mosel. The validation study for the combined RETRACE-RIVTOX chain on the basis of contamination data from the whole Rhine basin is still under way.

COASTOX (Fig. 5) was widely used by IMMS CC to simulate the radionuclide transport in the Kiev Reservoir and the Pripyat River floodplain close to the Chernobyl Nuclear Power Plant [3,4,7,9]. Measurements for a ^{90}Sr release from the floodplain after ice jams, January 1991 and February 1994, confirmed the results of predictions based on simulations with COASTOX [9]. Within JSP-1, COASTOX was tested and verified on the basis of measured ^{137}Cs and ^{90}Sr distributions during the 1987 spring flood in the Kiev reservoir. These results demonstrate the importance of application of different adsorption and desorption rates to describe the transfer of ^{137}Cs between solute and bottom deposition.

4.3. LAKECO validation

LAKECO was tested and validated within various international working groups. Within the IAEA/CEC VAMP-project the lake model was successfully applied to a wide range of lake ecosystems in Europe, different in terms of trophic status, climatology, deposition of radionuclides, and morphology. LAKECO participated in a blind test within BIOMOVS II, where a Cooling Pond Scenario was outlined. As tuning of the model was impossible, this study could be considered as a quality test. It showed that the original LAKECO model, with a relatively great number of parameters, most of them assessed on the basis of expert judgment, was not able to predict the concentration in the aquatic system with the required accuracy. The enhanced model LAKECO-B showed better results, which proved the increase of predictive power after the implementation of the new submodels. Fig. 6 shows in the left part the concentration in water, averaged over the entire cooling pond (Bq./l), and presents in the right part the activity in predatory fish (Bq/kg wet weight) for both model variants. Furthermore a fuel leaching submodel was added to the code, to govern the fact that in the vicinity of a reactor a high fraction of undissolvable particles can be expected.

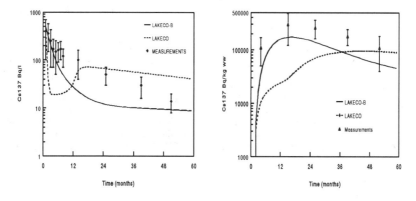

Fig . 6. LAKECO model results for Chernobyl NPP Cooling Pond

Conclusions

The hydrological model chain for the nuclear emergency decision support system RODOS was developed by the joint efforts of EC and CIS scientists in the frame of the JSP-1 project. Test and validation studies of the whole chain as well as of individual models were performed on the basis of post-Chernobyl data. The software framework for the hydrological model chain was developed and tested. The hydrological module will be part of the RODOS system starting from the PRT-version 2.0.

References

[1] J.Ehrhardt, V. Shershakov, M. Zheleznyak, A. Mikhalevich, RODOS: decision support system for off-site emergency management in Europe. -Proceedings of Minsk Conference. First International Conference of the European Commission, Belarus, the Russian Federation and Ukraine on the Consequences of the Chernobyl Accident, Minsk, Belarus, 18-22 March 1996 (this issue)

[2] V. Borzilev *et al.*, Forecasting of secondary radioactive contamination of the rivers in the 30-th kilometers zone of the Chernobyl NPP, *Meteorologica i Gidrologia*, No. 2 (1989), pp. 5-13, (in Russian)

[3] M. Zheleznyak *et al.*, Simulating the effectiveness of measures to reduce the transport of radionuclides in the Pripyat-Dnieper aquatic system - Proc. International Seminar on Intervention Levels and Countermeasures for Nuclear Accidents. Cadarache, France, 7-11 October 1991.-Commission of the European Communities, Radiation Protection-54, EUR 14469, 1992, p.336-362.

[4] M. Zheleznyak, P. Tkalich, G. Lyashenko, A. Marinets, Radionuclide aquatic dispersion model-first approach to integration into the EC decision support system on a basis of Post-Chernobyl experience. - *Radiation Protection Dosimetry*, N6 (1993), 37-43.

[5] W. Raskob, R. Heling, A. Popov , P. Tkalich, The modelling concept for the radioactive contamination of waterbodies in RODOS, the decision support system for nuclear emergencies in Europe, *The Science of the Total Environment,* (in press).

[6] A. Popov, R. Borodin, A. Pokhil. Description of a Physically Based Distributed RETRACE Model to Simulate Radionuclide Transport in Runoff Water, In: Hydrological Model Chain in RODOS, JCP-1 Final Report, Karlsruhe, 1995.

[7] M. Zheleznyak *et al.,* Mathematical modeling of radionuclide dispersion in the Pripyat-Dnieper aquatic system after the Chernobyl accident, *The Science of the Total Environment* **112** (1992), 89-114.

[8] P. Tkalich, M. Zheleznyak, G. Lyashenko, A. Marinets, RIVTOX - Computer Code to Simulate Radionuclides Transport in Rivers, In: A. Peters et al.(Eds.) , Computational Methods in Water Resources X, vol. 2, Kluwer Academic Publishers, Dordrecht, The Netherlands, 1994, pp. 1173-1180.

[9] M. Zheleznyak *et al.* Modeling of Radionuclides Transport in the Set of River Reservoirs, In: A. Peters et al.(Eds.), Computational Methods in Water Resources X, vol. 2, Kluwer Academic Publishers, Dordrecht, The Netherlands, 1994, pp. 1189 - 1196

[10] R. Heling, LAKECO - the model for predicting the behaviour of radionuclides in lakes and reservoirs, In: Hydrological Model Chain in RODOS, JCP-1 Final Report, Karlsruhe, 1995.

[11] M.B. De Vries, H. Pieters, Bioaccumulation in pike perch, data analysis on data of Lake IJsselmeer, Lake Ketelmeer, and Lake Markmeer. in: Accumulation of heavy metals in organics. Delft Hydraulics and National Institute of Fishery Investigations, 1989

[12] W. Raskob, Development of dose models for the hydrological chain, In: Hydrological Model Chain in RODOS, JCP-1 Final Report, Karlsruhe, 1995.

[13] M. Zheleznyak *et al.,* Modeling of radionuclide transfer in rivers and reservoirs: Validation study within the IAEA\CEC VAMP Program. International Symp. on Environmental Impact of Radioactive Releases, IAEA, Vienna, 8-12 May 1995, Extended Synopses IAEA-SM-339, IAEA, 1995, p.330-331

Decision Analytic Methods in RODOS

Vladimir BORZENKO
Institute of Control Sciences, Profsoyuznaya 65, Moscow, 117809 Russia
Simon FRENCH[1]
School of Computer Studies, University of Leeds, Leeds, LS2 9JT, UK

Abstract. In the event of a nuclear accident, RODOS seeks to provide decision support at all levels ranging from the largely descriptive to providing a detailed evaluation of the benefits and disadvantages of various countermeasure strategies and ranking them according to the societal preferences as perceived by the decision makers. To achieve this, it must draw upon several decision analytic methods and bring them together in a coherent manner so that the guidance offered to decision makers is consistent from one stage of an accident to the next. The methods used draw upon multi-attribute value and utility theories.
Keywords: Constraint satisfaction; decision support systems; expert systems; HERESY; M-Crit; multi-attribute value and utility theory.

1 Introduction

The lack of a uniform response to the Chernobyl accident, both in and beyond the former Soviet Union, has led to a number of projects supported by the Commission of the European Communities (CEC), under its Radiation Protection Programme. RODOS (Real-time On-line DecisiOn Support system) is one of these projects. It is designed to be a comprehensive decision support system (DSS) for off-site emergency management, which will provide support from the moment that an accident threatens through to long term countermeasures implemented months and years after an accident. A key feature of RODOS is that it seeks to provide decision support at all levels ranging from largely descriptive reports to a detailed evaluation of the benefits and disadvantages of various countermeasure strategies and their ranking according to the societal preferences as perceived by the decision makers: see Table 1. To provide such comprehensive decision support many design issues need to be addressed. In this paper, we describe the architecture of RODOS with specific reference to modules in the evaluation subsystems (ESY). For a more general introduction to the design of RODOS, we refer to papers earlier in this session, specifically [1], [2], and to [3].

Level 3 decision support (Table 1) requires complex modelling of preferences and values. The design of RODOS uses multi-attribute value and utility functions (MAV/UT) to provide this: see [4], [5] and [6] for discussions of these methods. Two modules, HERESY and M-Crit have been developed. These implement MAV/UT methods in subtly different

---

[1] Please address correspondence to Simon French

Table 1 Levels of decision support for off-site emergency management. Decision support can be provided at various levels, here categorised into four levels. The functions provided at any level include those at lower levels. RODOS is unique in that it will provide support at all levels, including Level 3 for all potentially useful countermeasures at all times following an accident.

Level 0: Acquisition and checking of radiological data and their presentation, directly or with minimal analysis, to decision makers, along with geographic and demographic information available in a geographic information system.

Level 1: Analysis and prediction of the current and future radiological situation (i.e. the distribution over space and time in the absence of countermeasures) based upon monitoring and meteorological data and models.

Level 2: Simulation of potential countermeasures (e.g. sheltering, evacuation, issue of iodine tablets, food bans, and relocation), in particular determination of their feasibility and quantification of their benefits and disadvantages.

Level 3: Evaluation and ranking of alternative countermeasure strategies in the face of uncertainty by balancing their respective benefits and disadvantages (e.g. costs, averted dose, stress reduction, social and political acceptability) taking account of societal preferences as perceived by decision makers.

ways. The design of RODOS also makes use of constraint management and expert system technologies within the ESY subsystems to help in problem structuring and in explaining to the decision makers the guidance provided by M-Crit and HERESY modules in formulating and evaluating countermeasure strategies.

The organisation of this paper is as follows. We begin by discussing a little further the decision support required during a nuclear accident. Section 3 illustrates multi-attribute modelling of consequences. Section 4 describes the ESY subsystem and the modules which form it, particularly the two MAV/UT modules: M-Crit and HERESY. Finally, the concluding section notes several issues relating to decision support, particularly matters related to the validation.

2 Decision support during a nuclear accident

RODOS is designed to support decision makers throughout all phases of a nuclear accident. Initially, RODOS will support the decision making of plant or site managers and local emergency management. Later, regional or national governments will become responsible for decision making, depending on how severe the accident is. Thus RODOS will support all decision makers and all decision making on countermeasures from initial evacuation, sheltering and issue of iodine tablets, through food bans to long term relocation.

The need to provide such comprehensive support introduces a number of issues that are seldom faced in the design of DSS's for other contexts.

- *Multiplicity of decision makers.* Many decision makers will be involved in the emergency management: plant managers, the emergency services, regional emergency planning officials, local, regional and national politicians. Each has differing levels of technical competence and differing information needs. Perhaps more importantly, they will have differing levels of authority to express value judgements. Plant managers will not be able to 'speak for the public', whereas politicians do have that authority.

- *Multi-criteria, public equity and risk.* There is no single criterion for choosing between countermeasure strategies. In addition to those directly relating to health risks arising from the radiation, there are issues related to psychological stress, public acceptability, and equity of risk sharing across the population. Equity issues are discussed in [8].
- *Many different levels of urgency.* Initially, decisions on countermeasures must be made in a matter of minutes; later, the timescales 'relax' with decision making able to take several hours or days.

In the early phases of an accident, the urgency and the probable lack of involvement of political decision makers means that issues related to 'intangibles' such as public acceptability and equity of treatment must be pre-programmed in some sense. Moreover, during this phase health issues related to the stochastic and non-stochastic effects of potential exposures will drive the decision making. In the later phases, decision making will have to address political and social imperatives in addition to pure health issues. This means that the support offered by RODOS must vary both in the range of criteria used in the analysis and in the software interface. Moreover, looking at Table 1, we may expect that full level 3 support may not be used in the first few hours of an accident.

3 Multi-attribute modelling of consequences

We shall use the term *attribute* to mean one of the dimensions along which we assess the consequences of a decision. Thus collective dose, individual dose and cost are attributes. A *criterion* or *objective* refers to an attribute and a direction of preference. Thus minimise collective dose, minimise individual dose and minimise cost are criteria. Some attributes are objective in that they correspond to physical measurements; others are more subjective requiring judgement in their definition, e.g. equity of treatment. It is conventional to organise the attributes involved in analysis into hierarchies. This offers many cognitive advantages and also helps structure decision analyses [6].

Figure 1 gives two attribute hierarchies that have been developed in case studies on relocation decisions after an accident. Note that some of the attributes are clearly objective, whereas others must inevitably reflect subjective judgements. Discussions of how these were defined and used to structure sensitivity analyses may be found in [4] and the

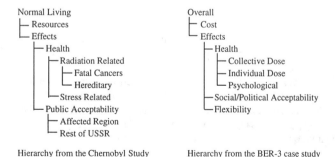

Hierarchy from the Chernobyl Study   Hierarchy from the BER-3 case study

Figure 1: Attribute hierarchies developed in two multi-criteria decision analyses on countermeasure strategies [4]

references cited therein. The use of event conditional attribute hierarchies to support decision making when there is a threat of an accident is discussed in [8].

4 The ESY subsystem

The conceptual architecture (not the physical software which includes, e.g., databases and user interfaces) of RODOS consists of three types of subsystem [1], [2], [3]:

- ASY (analysing subsystem) modules process incoming data and forecast the location and quantity of contamination including temporal variation.
- CSY (countermeasure subsystem) modules suggest possible countermeasures, check them for feasibility, and calculate their expected benefit in terms of a number of attributes (criteria).
- ESY (evaluation subsystem) modules rank countermeasure strategies according to their potential benefit and preference weights provided by the decision makers.

We focus on the ESY subsystems of modules, which support the evaluation of different countermeasure strategies. These implement the level 3 support offered by RODOS. Rules, weights and preference functions are encoded in the ESY and applied to a list of alternative countermeasures to provide a ranked short list to decision makers. Both the ASY and the CSY will use several models throughout the accident depending on the time, the location and the actual situation. However, the ESY may be based upon the same software module with different attribute trees and with the preference weights changing over time.

An ESY subsystem will operate in interactive mode using graphical interfaces to communicate with a variety of decision makers who may possess many qualitatively different skills and perspectives: e.g. scientists, medical personnel, engineers, emergency planners, government officials and senior politicians. It will present the countermeasures in a ranked short list together with those rules and preferences that determined the order of the list. Intuitive justifications for choices and underlying uncertainties inherent in the predictions will also be provided. The ESY will assist users in modifying rules, weights and preferences and other model parameters as well as exploring the consequences of each change. The importance of this exploration cannot be overemphasised. Any DSS helps decision makers not by *making* the decision itself, but by *enhancing the decision makers' understanding* of the problem, the issues before them and their value judgements. They are then better able to make the decision because of this greater understanding.

4.1 Architecture of an ESY subsystem

The ESY will[2] have the form of Figure 2. It comprises three further subsystems:

- A coarse expert system filter which rejects any strategies that are logically infeasible or do not satisfy some given constraints.
- A multi-attribute utility theory (MAV/UT) ranking module which takes the remaining list of strategies as input. It ranks the strategies for their relative effectiveness according to previously elicited utility attributes and preference weights from the decision makers. It may be necessary to revise and re-evaluate these preference weights in any given situation before a particular decision is taken.

---

[2] Prototypes of the ESY modules are written and are currently being evaluated; however, they are not fully implemented into the current realease of RODOS.

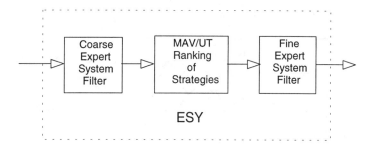

Figure 2   The conceptual structure of the ESY module

- A fine expert system filter which takes the top 10-15 strategies and produces a management summary report detailing the costs and benefits of each.

4.2   Constraint management coarse expert system

Although the number of protective measures which may be taken is limited, the number of potential countermeasure strategies is enormous. This arises because the region will be divided into a number of areas and a strategy specifies which measures should be applied in each of these areas. Moreover, each measure may have a time period associated with it, e.g. start and finish times for sheltering. Thus the number of strategies can grow combinatorially. However, very few of these conceptually possible strategies will be worthy of close examination. They may run break the guidance provided by intervention levels or may be infeasible in practical terms (e.g. requiring 100,000 people to be evacuated in 30 mins). Moreover, they may break simple principles describing public acceptability, e.g. protecting children less effectively than adults or evacuating lower risk areas in preference to higher ones. One would never evacuate part of a small village; the public would never accept or understand such an action. There must be continuity of treatment.

A very simple expert system will be used to discard strategies which are incompatible with such principles. This coarse expert system is being implemented using constraint management techniques [9], [10]. This technology is, in one sense, as old as combinatorial programming, for it does nothing other than identify objects that satisfy a set of constraints. But, in another sense, it is very new in drawing upon modern tree search and list manipulation algorithms implemented with artificial intelligence languages. Early experiments show that constraint satisfaction technology is well able to cope with the combinatorial problems we face here. In a simple example with a potential 17 billion possible countermeasure strategies were reduced to a list of about 500 which were worthy of further evaluation. The reduction took a matter of seconds on a Sun Sparcstation of comparable power to the HP workstation on which the full RODOS system runs.

The strategies satisfying the constraints imposed by the coarse expert system will be passed to a MAV/UT ranking module, which will identify the top ten or twenty ranking strategies. Two such modules have been written: M-Crit and HERESY.

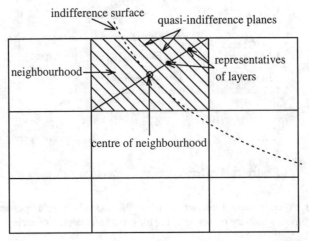

Figure 3 Splitting of the region of feasible solutions into neighbourhoods and the approximation of indifference planes by linear quasi-indifference planes within a neighbourhood

4.3 M-crit

M-Crit implements a piecewise linear approximation to the decision makers' indifference planes: an indifference plane is a surface or contour of points of equal value to the decision makers. The method is know by the acronym PLANT (piecewise linear approximation numbering technique) [11]. It assumes that the decision makers have a well formed set of preferences and that the problem is to model these and articulate them in the context of a particular problem. It approximates the decision makers' indifference surfaces by splitting the area of feasible solutions in the criteria space into a number of rectangular neighbourhoods; see Figure 3. Within each neighbourhood, the indifference surfaces are approximated linearly by eliciting the substitution coefficients from the decision makers. By 'chaining together' the approximating linear indifference (hyperplanes) the method can approximate complex preference structures to a reasonable degree of accuracy.

M-Crit is a window-based interactive implementation of the PLANT method. It allows the approximation to be constructed interactively and displays the approximation back to the decision maker visually for checking. While this visual reflection does allow the decision makers to consider whether their preferences are appropriate, no other consistency checking is built into the elicitation procedure. Nor are there any underlying preferential independence assumptions such as additive or utility independence which are often required in MAV/UT modelling [5], [6]. This makes the method appropriate for circumstances in which decision makers are sure of their value judgements.

4.4 HERESY

An alternative approach is taken by the HERESY module. This implements MAV/UT models based upon much stronger assumptions: e.g. preferential independence. These assumptions

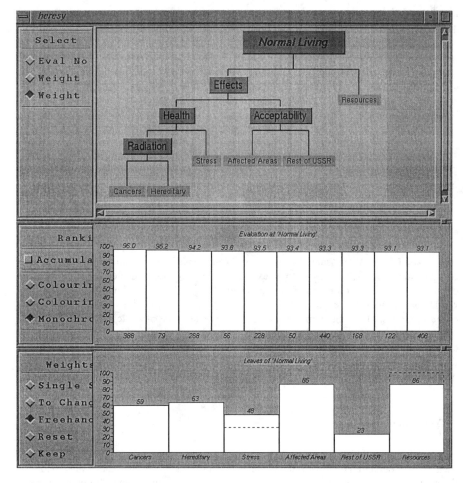

Figure 4 Screen dump of the HERESY module

restrict the form of value and utility functions considerably. The current implementation assumes sufficient independence to ensure additivity[3], but future implementations will also allow multiplicative and multi-linear forms. The advantage of introducing such restrictions is that the elicitation procedures can include more consistency checking and more detailed sensitivity analyses may be performed after the evaluation. The disadvantage is that the decision makers may be unable to articulate some of the value judgements which they might wish to express. Thus the strengths and weaknesses of M-Crit and HERESY complement each other.

HERESY's purpose is to identify the top few, say ten, ranking strategies and check the sensitivity of these to the choice of weights on different criteria. A screen dump is shown in Figure 4. The screen is divided into three areas. At the top the attribute hierarchy is shown.

---

[3] In an additive MAV model the overall value or score of an alternative is formed as a weighted sum of scores on individual attributes.

In the middle is a histogram showing the overall scores (values) of the top ten strategies. At the bottom is a histogram showing the current weights on (some of) the attributes. Not all the weights need be shown simultaneously. There may be cognitive advantages in concentrating attention on particular branches within the attribute hierarchy. All bars on the histogram are labelled appropriately. The user selects a weight with a mouse by clicking on the appropriate bar in the bottom histogram, and then increases or decreases the weight either by the keyboard or by pulling with a mouse. As the weight is changed, the middle histogram changes accordingly. When a change in the ranking of the top ten strategies occurs (or when one drops out of the top ten and another enters), the histogram rearranges itself. There is an audible beep and the user is informed of the change in a text window. Thus the user can identify the sensitivity of the ranking to the default weights in the model. The computational speed of the prototype confirms that the identification of the top 50 ranking strategies of about 10000 countermeasure strategies and associated sensitivity analysis can be performed almost instantly.

4.5 Fine expert system

After potential countermeasure strategies have been ranked using M-Crit or HERESY, the list of top ranking ones will be passed to an expert system with a sophisticated set of rules, each of which will be applied to each of the candidate strategies. The small number of strategies would allow a full set of explanations to be developed, which would give a critique of each of the strategies. Thus the output of RODOS will be a shortlist of strategies, each of which satisfies the constraints implied by intervention levels, practicability, etc., together with a detailed commentary on each strategy explaining its strengths and weakness. Klein [12] discusses a similar combination of expert system technology with MAV/UT ideas to provide decision makers with explanatory remarks on the ranking of strategies.

5 Discussion

The decision analytic issues involved in the design of RODOS are complex and we have only been able to touch upon a few of them here. A major omission is our lack of discussion of the relationship between uncertainty handling and preference modelling. Some discussion of this may be found in [7]; see also [13]. We have focused upon the prototype systems M-Crit and HERESY, which in their current implementations ignore uncertainty and risks. A later paper will report on the enhancement of their functionality to deal with these.

A more significant omission is the lack of discussion of validation and quality assurance. Validating a decision support system which one hopes and, indeed, plans should *never* be used brings a host of problems and issues. Moreover, the geographic and cultural spread of the many institutes involved in the software development itself raises many quality assurance issues. The latter issue of software quality assurance is being addressed very fully in the current fourth Framework Research and Development Programme of the CEC, during which quality assurance procedures are being applied. However, validation of the decision analytic methodologies is a more difficult matter. Essentially, we need to work very closely with a large number of decision makers to ensure that we are supporting them in the ways that they need.

Several exercises have been run and more are planned to explore and validate the use of RODOS in general and the ESY modules in particular. Our belief is that by working with decision makers as they make decisions – in albeit artificial circumstances – we shall discover where the design and implementation of RODOS is poor or deficient. Already it seems to be clear that the decision makers do not perceive a need for level 3 support in the early phase of an accident [14], whereas they do find level 3 MAU/VT support useful in the decision making on later countermeasures such as relocation [4]. However, it is also clear that whilst the team designing RODOS must learn the needs of decision makers, the decision makers themselves need to learn the potential of modern DSS's. Currently, they are unaware of what is possible. None the less, it does seem that for the present the methods and software described above will be more useful after the immediate emergency has passed. For further discussion of the contribution that these methods may make, see [4].

6 Acknowledgements

The development of the MAV/UT methods for RODOS described here was supported by grants from the Commission of the European Community (B17-0060-GB, F13P-CT92-0036, F13P-CT92-0136, Sub94-F15-028). The ideas in this paper were developed in close co-operation with R. Borodin, A. Despres, J.E. Ehrhardt, N. Papamichail, D.C. Ranyard, L. Simpson, J.Q. Smith, E. Trakhtengerts and D. Vanderpooten. N Papamichail has developed a prototype of the coarse expert system using the ILOG constraint satisfaction toolkit.. S. Young programmed the prototype of the HERESY module and R. Borodin programmed the M-Crit module.

References

[1] J. Ehrhardt, V. Shershakov, M. Zhelezniak, and A. Mikhalevitch, RODOS: Decision support for off-site emergency management in Europe, First International Conference of the European Commission, Belarus, Russian Federation, and Ukraine on the Radiological Consequences of the Chernobyl Accident.

[2] O. Schüle and V Kossykh, The software environment of RODOS, First International Conference of the European Commission, Belarus, Russian Federation, and Ukraine on the Radiological Consequences of the Chernobyl Accident.

[3] J. Ehrhardt, J Päsler-Sauer, O. Schüle, G. Benz, M. Rafat and J. Richter, Development of RODOS, a Comprehensive Decision Support System for Nuclear Emergencies in Europe - an Overview, *Radiation Protection Dosimetry*, **50** (1993), 195-.

[4] S. French, Multi-attribute decision analysis in the event of a nuclear accident, *Journal of Multi-Criteria Decision Analysis* **5** (1996) in press.

[5] S. French, Decision Theory: an Introduction to the Mathematics of Rationality, ISBN 0-85312-682-8. Ellis Horwood, Chichester, 1986.

[6] R.L. Keeney. Value Focused Thinking: a Path to Creative Decisionmaking. ISBN 0-674-93197-1, Harvard University Press, Cambridge, Mass., 1992.

[7] S. French, D. Ranyard and J.Q. Smith, Uncertainty in RODOS, RODOS(B)-RP(94)05, Research Report 95.10, School of Computer Studies, University of Leeds. Available by connecting to WWW at file://agora.leeds.ac.uk/scs/doc/reports/1995 or by anonymous ftp from agora.leeds.ac.uk of the file scs/doc/reports/1995/95_10.ps.Z. 1995.

[8] S.French, E. Halls and D.C. Ranyard, Equity and MCDA in the event of a Nuclear Accident. ". RODOS(B)-RP(95)03. Contributed paper at the XIIth International Conference on MCDM at Hagen, June 1995. Accepted for publication in the Proceedings.

[9] P. van Hentenryck, Constraint Satisfaction in Logic Programming. MIT Press, 1989.

[10] E. Tsang, Foundations of Constraint Satisfaction. Academic Press, New York, 1993.

[11] V. Borzenko, E. Trakhtengerts, R. Borodin and V.M. Shershakov, M-Crit: a Multi-Criterial Decision Support Subsystem for RODOS Prototype 2. Scientific Production Association "Typhoon", Emergency Center, Obninsk, Russia.

[12] D.A. Klein, Decision-Analytic Intelligent Systems: Automated Explanation and Knowledge Acquisition. New Jersey: Lawrence Erlbaum Associates, 1994.

[13] V.M. Shershakov and E. Trakhtengerts, Development of RODOS/RECASS System as a Distributed Decision Making Support in Emergency. Institut of Control Sciences, Moscow, Russia. 1995.

[14] M. Ahlbrecht, J.E. Ehrhardt and S. French, Designing the Evaluation Module in RODOS/RESY: Execution and Analysis of Elicitation Exercises with Emergency Management Teams. School of Computer Studies, University of Leeds, Leeds.

International Exchange of Radiological Information in the Event of a Nuclear Accident - Future Perspectives

Marc DE CORT
Environment Institute, CEC JRC-Ispra, Italy

Hermann LEEB
Bundesamt für Strahlenschutz, Institut für Strahlenhygiene,
Oberschleissheim, Germany

Gerhard DE VRIES, Lothar BREITENBACH
Environment Institute, CEC JRC-Ispra, Italy

Wolfgang WEISS
Bundesamt für Strahlenschutz, Institut für Atmosphärische Radioaktivität
Freiburg, Germany

Abstract.   Immediately after the Chernobyl accident most European countries established or enhanced their national radioactivity monitoring and information systems. The large transboundary effect of the radioactive release also triggered the need for bilateral and international agreements on the exchange of radiological information in case of a nuclear accident. Based on the experiences gained from existing bi- and multilateral data exchange the Commission of the European Communities has made provision for and is developing technical systems to exchange information of common interest.

Firstly the existing national systems and systems based on bilateral agreements are summarized. The objectives and technical realizations of the EC international information exchange systems ECURIE and EURDEP, are described. The experiences gained over the past few years and the concepts for the future, in which central and eastern European countries will be included, are discussed. The benefits that would result from improving the international exchange of radiological information in the event of a future nuclear accident are further being described.

1. Existing National and Bilateral Systems

After the Chernobyl accident monitoring networks and information systems of various kind have been established in many European countries. Furthermore there was the necessity to get quick information about the situation beyond national borders. Therefore within the framework of bilateral agreements between various European countries provision has been made and technical systems have been established to exchange data and information of mutual interest. In the following the situation in Europe is summarized [1, 2, 3, 6].

1.1. Objectives of the Assessment and Exchange of Radiological Data and Information

There are different objectives for the installation and permanent operation of technical systems for environmental monitoring and information exchange. The major objectives are:
- to monitor the normal radiological situation of the environment and to determine the permanent background levels;
- to provide early warning to competent authorities in cases of abnormal changes of the normal situation;
- to assess the dose and the risk to man, which might be caused by a radioactive contamination of the environment;
- to assure harmonized actions across national boundaries;
- to proof, that the accepted secondary intervention levels for food and feeding stuffs are enforced within the member states of the European Union;
- to inform national and international parliaments in due time;
- to inform the public fast and in an objective way;
- to improve the reliability of diagnostic and prognostic transport models.

Three different types of cross-boundary exchange of radiological data and information between European countries can be distinguished, e.g., the exchange between
- the local authorities on both sides of a national boundary;
- two or several countries at governmental level;
- two or several nations and multi- or supra-national organisations.

The objectives of the three types are different and so are the requirements for the exchange of data and information.

1.2. General Aspects for the Data Collection, -processing , and -exchange

There seems to be agreement that six different types of radiological data are required for the assessment of contamination of the environment, the dose to the population, and decision making, e.g., the
- gamma doserate;
- activity concentration of the air;
- deposition of radionuclides on the ground;
- specific activity of foodstuff;
- specific activity of feeding stuff;
- dose contributions via the relevant pathways.

Regional and national data centers are available in many European countries. Some of these centers have made provision or are planning to exchange radiological data and information as described in this paper. In the following chapters some of these systems are briefly described.

1.3. Present Status of the European Monitoring and Information Systems

The objectives defined above can be met in many ways. The existing monitoring systems in 31 European countries (Table 1) clearly show this by applying various strategies and technical systems. Automatic monitoring systems at fixed ground-based stations are combined with mobile units for in-situ measurement and with stationary and/or mobile units for sample collection and subsequent measurement in remote radiological laboratories.

Automatic monitoring networks play an important role in many national concepts for the surveillance of the radiological situation in the environment. Many of the networks operated in Europe, however, do not have technical facilities for real time in-situ measurements and on-line data transfer to a central computer facility. The areal density of the existing networks varies between 10 and 600 stations per 100,000 km^2 for gamma dose-rate and between <1 and 77 stations per 100,000 km^2 for aerosols. If we look at the present situation in Europe (Fig. 1) there seems to be no systematic difference between the systems in countries with and without nuclear power production. Other criteria such as the total population or the population density do not play a significant role for the design of the networks either.

Four types of networks can be distinguished:
- manual collection of data without using a computer;
- manual collection of data, evaluation of data in a central computer;
- online computer-controlled monitoring networks, where data are collected by and evaluated in a central computer;
- on-line monitoring and information systems which do not only collect and evaluate data but also support all competent authorities.

Most systems are used for internal tasks of the authorities, some are also designed to inform the public. Most European systems are computer-controlled monitoring networks.

1.4. Examples of Existing Systems

In the majority of the European countries on-line monitoring of gamma dose-rate with data transmission to a central computer is performed. Examples of existing national systems are:
- The French TELERAY system collects data of more than 165 gamma dose-rate probes and transmits them to a central data processing unit. In normal operation mode, every probe is called from the center once a day. The probes are able to transmit an alarm if predefined threshold-values are exceeded. The public is informed about the radiological situation via videotext (minitel) [7].
- The British RIMNET (Radioactive Incident Monitoring NETwotk) system is a complete national nuclear emergency response system. Its functions include storage, analysis, vizualisation and communication of large volumes of numerical and textual data around the UK [8].
- The German IMIS (Integrated Monitoring and Information System) [9] consists of three operational levels:
  - collection of radiological data from state-of-the-art monitoring networks and measurement laboratories;
  - presentation of measurements including transport and dose assessment models;
  - evaluation of data, management of the consequences of a given situation, legal enforcement of protective measures and information of the public.

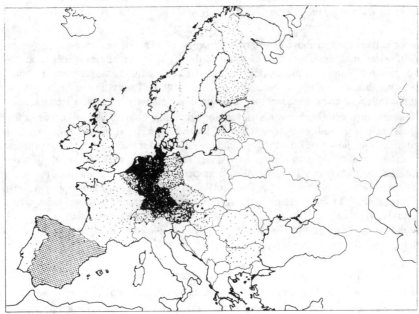

Euromap © IAR-BfS 71/37/20 [18.10.1995]

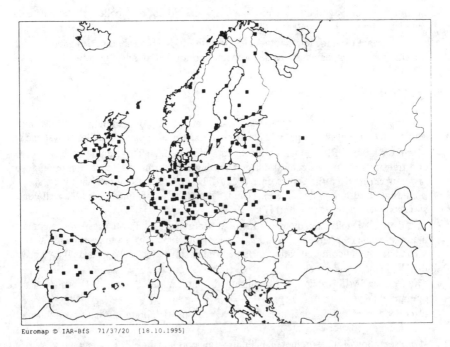

Euromap © IAR-BfS 71/37/20 [18.10.1995]

Figure 1: Overview of gamma dose-rate and aerosol monitoring networks in Europe
(Because of ignorance of the correct position of its stations, the French aerosol
monitoring network is indicated by a symbolic point)

Table 1: Overview of the European monitoring networks (September 1995). Planned monitoring stations or early warning systems are indicated in brackets.

Country	Area (km²)	Pop. (*1000)	NPP	gamma dose-rate	Air		Water	Early Warning	
					Aerosol/ Nuclide spec.	Iodine		gamma dose-rate	Air
Austria	83.853	8.000		336	8			x	
Belgium	30.518	10.000	7	183	7/3	7		x	x
Bulgaria	110.912	8.500	6						
Croatia	56.538	4.800	1		3				
Czech Rep.	78.864	10.300		(45)	9			x	
Denmark	43.093	5.190		11	3			x	x
Estonia	45.226	1.540			1				
Finland	338.145	5.060	4	30	7				
France	543.965	57.700	56	120	30/13		1	x	
Germany	356.854	81.200	21	2,200	53/40	11	40	x	x
Greece	131.957	10.300		1	1				
Hungary	93.032	10.300	4	50(+20)				x	
Ireland	69.895	3.500		12(+18)	8			x	x
Iceland	102.819	264			1				
Italy	301.277	57.200			6				
Latvia	64.600	2.600			8				
Lithuania	65.200	3.800	2						
Luxemburg	2.586	395		14	3		1	x	(x)
Netherlands	41.864	15.300	2	58	15	1			
Norway	323.878	4.300			8			x	
Poland	312.683	38.500			9				
Portugal	91.971	9.820		13	1			x	
Romania	237.500	22.800			8		1		
Russia	17.075.000	148.000	23		3				
Slovak Rep.	49.036	5.300	4	26				(x)	
Slovenia	20.256	2.000			8				
Spain	504.790	39.100	9	903	25	25		x	x
Sweden	449.964	8.730	12	35	8			x	
Switzerland	41.293	7.000	5	58	9			x	(x)
UK	244.100	58.100	31	92	8			x	
Ukraine	603.700	52.300	15		11				

Examples of bilateral agreements at governmental level are:

- Between Germany and France, Russia, and the Czech and the Slovak Republic bilateral agreements define the minimum requirements for a regular exchange of monitoring data obtained at a national level. Since 1992 for example daily averages of the gamma dose-rate obtained at representative stations of the nation-wide German IMIS system in major cities and of the German KFÜ systems in the vicinity of the NPP Cattenom and Fessenheim are provided once a day by fax to the French authorities [5]. The data are published daily together with the results of the French monitoring system in the widespread Minitel system. The problem of intercomparability of the data is solved by the continuous operation of a French (and a Swiss) gamma detector in the vicinity of a gamma detector of the German network near Freiburg. In an emergency situation the data exchange on a 2 hourly basis is aimed for. For this purpose a computer-based system for the automatic data exchange is currently being developed. It is expected to become operational by the end of 1995.
- There are similar activities in the Nordic countries Denmark, Finland, Iceland, Norway and Sweden which aim for the harmonization of the data exchange between these coun-

tries. The system includes gamma dose-rates from the fixed monitoring networks that exist.

- An alternative way is used for the exchange of processed data and information between Germany, Russia, the Czech and the Slovak Republic. Rather than exchanging monitoring data, documents are exchanged, which may consist of processed data such as maps with monitoring data, time series of data at particular sites, and free text messages. Routine exchange of documents of this kind has been started with the Russian Federation in January 1994. On a weekly basis documents with dose-rate data from the German IMIS and the corresponding IRIS (Integrated Radioactivity Information System) at the NPPs Smolensk and Novoworonesh are exchanged.

2. The EC International Information Exchange Systems

2.1. Introduction

Immediately after the accident at Chernobyl NPP, both the International Atomic Energy Agency (IAEA) and the Commission of the European Communities (CEC) set up a system to meet the requirements for early warning and exchange of information.

The IAEA system was established under the Early Notification Convention [10] (27 October 1986) and the Early Assistance Convention [11] (26 February 1987). The basis of the CEC system is the Council Decision of 14 December 1987 [12], which resulted in the European Community Urgent Radiological Information Exchange (ECURIE). By this system the EU Member States can exchange and submit to the Commission the information required by the Council Decision. Since the majority of the aims and needs for both systems overlap, it was decided between the IAEA and the CEC to harmonize as much as possible both systems, which resulted in the establishment of a common code (CIStructure) and coding/decoding software (CDS).

The current ECURIE system however is not tailored to the exchange of real-time monitoring data. Integration of national data exchange at a European level would speed up the availability of monitoring data at a large scale, essential for model calculations in view of accurate and timely predictions and indispensable to give actual and fast overviews of the contamination levels. The JRC detected these needs and - considering itself in the correct position for doing so having the necessary competences and taking part in the support activities REM (Radioactivity Environmental Monitoring) and ECURIE for CEC DG XI.C.1, Radioprotection, Luxembourg - decided to organize a workshop to investigate the technical aspects that would be needed for the creation of such a European-wide radiological information exchange system (EURDEP).

2.2. The ECURIE System

2.2.1 General Description of the Current System

The ECURIE system consists of a telex based communication network between the CEC and the Member States, through which radiological information can be exchanged.

At present, all ECURIE telex messages pass through the Commission Telex Centre, Brussels (See Figure 2). To receive priority treatment the messages are included in a IATA telex code, which allows for automatic recognition of ECURIE messages. The structure used for exchanging information is the Convention Information Structure (latest version of 6 June 1991) [13]. In order to avoid language translation problems it was developed as a short code of specific meaning. Each type of message that might be transmitted is represented as a block of code. Within each block the parts of a message are labeled as line numbers. The C.I.Structure does not only provide information on radiological measurements, but also on predicted values, site meteorological data and taken decisions (countermeasures).

Coding-decoding software (CDS) has been developed to reduce the time needed to code and decode the messages and to improve the reliability.

The part of the software for decoding messages was written by the IAEA whereas the encoding program was developed at JRC-Ispra. The encoding software is a menu driven program that helps the user to produce a valid C.I.Structure code, mostly by means of selection tables, and performs a validity check on the type of input. Additional help on line for every line number is available. In the present version of the CDS a hardcopy of the encoded message can be made to copy the message on a telex or the message can be sent away via an automatic telex device. Decoding software translates the encoded C.I.Structure message into a plain text message.

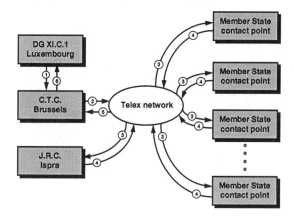

Figure 2: ECURIE message flow [15]

ECURIE provides four exercise levels with which the various aspects (communication, contact of responsible persons, exchange of data) are tested [14]. In Figure 2 the flow of a typical ECURIE message is shown: a notification message is generated by CEC DG XI/C/1 Radioprotection, Luxembourg (①), sent by internal electronic mail to the Commission Telex Centre (CTC) (②), converted to the telex network (③) and transmitted to the Member State contact points (④). The Member States answer by sending an appropriate response message via CTC (⑤) to DG XI/C/1 (⑥). In case of a level 3 exercise, this is followed by the transfer of two encoded messages containing data from each MS contact point to the CEC, who forwards it again to each contact point.

2.2.2. Experiences gained

Over the past three years, five exercises of level 3 have been held. Despite that in most cases the majority of the exercise objectives were met (transmission of the messages, contact of national radiological duty officers, coding and decoding of messages), important

inconveniences were detected [17]:
- the present communication system (telex) is too slow. Also the communication link between telex and the EC internal electronic mailing system (Figure 2: between ⑤ and ⑥) appeared to be not reliable enough for emergency situations;
- the joint development of the CDS by the IAEA and the JRC has lead to inconsistencies between the coding and decoding modules and a too high memory demand.

2.2.3. Future Developments

Because of above mentioned deficiencies the following three actions are foreseen [17]:
- improvement of the CDS software. The new CDS, called CoDecS, will be running under Windows 3.1, using a programming language that allows migration to a native version under Windows 95. The following requirements were defined:
  - automatic decoding/encoding of messages;
  - assistance when compiling the messages;
  - assistance in generating summary reports;
  - unattended and automatic daily exercises;
  - alarm signaling;
  - multi-language user interface;
  - automatic forwarding of messages.
- telex communications will remain available as backup but to improve respectively the integrity of the messages and the transmission speed, the X.400 protocol over Euro-ISDN will be used as default for the transmissions.
- creation of an on-line ECURIE database with WWW (World Wide Web) access.

2.3. The European Union Radioactivity Data Exchange Platform (EURDEP) System

2.3.1. General Description and Objectives

Resulting from the EC workshop on the 'Technical aspects of International Exchange of Radiological on-line monitoring data' (1-3 June 1994, Arona, Italy), agreements about the realization of a pilot project were reached. This project should include the following tasks:
- investigate the feasibility of using e-mail for the exchange of radiological data;
- setup a regular exchange of data between some European countries;
- use the information about the national monitoring networks submitted by the countries to define which radiological data could be exchanged now and in the future;
- define a common data-format for the exchange of the data and write software to convert the national format into the common format;
- establish a 'European network', based on existing monitoring stations on a 100 x 100 km grid.

2.3.2. Data Exchange Format

The various data items were based on the information available from the existing national formats, and are arranged in five main sections; geographical information about the sampling location, measurement information (primarily on gamma dose-rate and airborne concentration, but with the possibility to include additional sample types), additional sample characteristics, meteorological information at the sampling location and information about the sender.

The key-elements that influenced the data-exchange format (more detailed information can be found elsewhere [16,18]) are threefold:

- *flexibility*: the EURDEP data format has been defined such that it allows to include future radiological data as it becomes available. The structure of the format itself allows flexibility in the order of the data and the possibility to define static information only once as global data to reduce the size of the data;
- *compatibility*: the format was designed that it can be read and produced by software running on different platforms;
- *robustness* of the format: errors generated during the transmission should be detected and wrapping lines or adding headers or footers by e-mail software should have no impact on the successful interpretation conversion.

2.3.3. Experience gained from the Pilot Project

From the beginning of 1995 onwards consecutive participating institutes are being contacted by the JRC and invited to send their radiological information. These data were merged together at the JRC-Ispra and returned to the participating institutes. Until now (October 1995) successful bi-directional exchange of test-messages are regularly done with Germany, United Kingdom, Austria, Denmark, Norway, Sweden, Finland and Ireland. All communications are performed through SMTP (Simple Mail Transfer Protocol), except for

Table 2: Summary statistics of EURDEP exercise transmission delays (in hh:mm)

Date	D	DK	N	SF	S	A	UK	IRE	Avg	Min	Max	NP
06/02	1:16	0:50							1:03	0:50	1:16	2/2
20/02	-0:03[1]	1:05	?						0:31	-0:03[1]	1:05	3/3
06/03	?	-0:03[1]	-0:03[1]						-0:03[1]	-0:03[1]	-0:03[1]	3/3
20/03	0:58	1:03	0:31	0:01	?				0:38	0:01	1:03	5/5
03/04	0:00	?	0:49	0:01	?	0:08			0:15	0:00	0:49	6/6
18/04	3:00	3:04	?	2:02	?	3:22			2:52	2:02	3:22	6/6
02/05	2:06	7:48	?	1:26	?	2:08	?		3:22	1:26	7:48	7/7
15/05		2:05	1:02	2:04	?	1:28	?		1:40	1:02	2:05	6/7
29/05		0:04	0:01	1:20	?	0:02			0:22	0:01	1:20	5/7
12/06	0:00	0:00	0:02		?	0:00	?		0:01	0:00	0:02	6/7
26/06			0:00	?	?	-0:12[1]	?		-0:06[1]	-0:12[1]	2:00	6/7
10/07	0:00	0:00	0:05		?	0:02	?		0:02	0:02	0:05	5/7
04/09		-0:08[1]	-0:01[1]	-0:07[1]	?	-0:07[1]		?		-0:08[1]	-0:01[1]	6/8
11/09		0:14	0:05	0:52	?	0:07		?	0:20	0:05	0:52	6/8
25/09	0:55	4:21	3:40	5:01	?	0:50	?		2:57	0:50	5:01[2]	7/8
09/10	0:03	0:05	0:10	0:02	?	0:05	0:04		0:05	0:02	0:10	7/8
23/10	0:00	0:07	?	0:00	?	0:22	0:05	?	0:14	0:00	0:22	8/8

	D	DK	N	SF	S	A	UK	IRE
Avg	0:45	1:23	0:32	1:10		0:40		
Min	-0:03[1]	-0:03[1]	-0:03[1]	-0:07[1]		-0:12[1]		
Max	3:00	7:48	3:40	5:01		3:22		
NP	12/17	16/17	15/16	12/14	14/14	13/13	8/11	3/5

[1] **Negative delays are caused by inaccurate clocks. (the JRC mail server is often up to 15 minutes behind)**

[2] **The JRC IP-Server was down for about an hour.**

the UK where X.400 is used. The following experiences were gained:

- establishing a first contact using X.400 can several months due to the involvement of the different PTT's;
- the store and forward nature of e-mail makes that the delay in delivering mail varies largely. We also experienced several cases in which messages disappeared;
- mailboxes are not always frequently read and not checked by other persons in case of absence. Many e-mail communications sent to the contact points were reacted on very late;
- the maximum delays (see Table 2) of the transmissions clearly show that public e-mail cannot be used to exchange radiological data in case of an alert.

Table 2 lists the average, minimum and maximum travelling time and the number (NP) of messages by participant. The first information was not received from all contact points, because incompatible with the automated manner in which some send their data. It must also be said that the accuracy of the traveling time is in the order of several minutes due to the differences in clocks on the various computers, gateways and wrist watches!

2.3.4. Future Perspectives and Developments

Begin 1996 the development of an interface between the decision support system RODOS and EURDEP is planned.

A second workshop is foreseen for June '96 to discuss the results of the pilot project: the experiences gained will induce discussions on several modifications:

- e-mail has the advantage that it does not give security problems, but using it on top of Internet may not be reliable enough for the purposes of EURDEP. Two alternatives that will be proposed for a more reliable connection are the usage of X.400 and a protocol developed at the AR Institute of the German BfS, both on top of ISDN;
- review of the data format and porting of it to the EDIFACT (ISO 9735) standard.
- a more dense geographical coverage of the network than the initial 100 by 100 km;
- participation of central and eastern European countries;
- installation of the conversion programmes at the contact points;
- obtain an official status for the project.

3. Conclusions

The Chernobyl accident has triggered the development of regional on-line monitoring networks as well as the improvement of rapid national and international data transfer. However, the existing situation must be characterized as highly inhomogeneous. No accepted standards for data collection and information exchange are available yet. The national technical solutions differ in radiological as well as in data processing aspects. The data exchange systems on bilateral basis are also special solutions for the intended purpose.

On the other hand there is also the fast development of long range transport models and real-time decision support systems that require rapid access to international monitoring data in order to allow for accurate and timely predictions on a European scale. Therefore the development of EURDEP is necessary to bring together the essential elements for a European-wide emergency response system. The long term objectives of EURDEP are therefore:

- ensure the comparability of measured data;
- get an overview about the radiological situation in the whole of Europe;
- provide the CEC with all necessary data;
- provide national authorities with information from other European countries;
- interface with an on-line decision support system [4].

References

[1] W. Weiss, Exchange of Radiological Data and Information between European Countries, NEA-Workshop "Emergency Data Management", Zurich, 1995

[2] W. Weiss, Systeme zur Überwachung der radiologischen Lage- gegenwärtiger Stand im internationalen Vergleich; Ansätze zur Harmonisierung, FS-94-74-I, Stand des Notfallschutzes in Deutschland und der Schweiz, ISSN 1013 - 4506, 210 - 215, 1994

[3] S. Vadé, Die Europäische Union und der Strahlenschutz der Bevölkerung bei einer radiologischen Notstandssituation, FS-94-74-I, Stand des Notfallschutzes in Deutschland und der Schweiz, ISSN 1013 - 4506, 385-390, 1994

[4] J. Ehrhardt, J. Paesler-Sauer, O. Schuele, G. Benz, M. Raffat and J. Richter, Development of RODOS, a Comprehensive Decision Support System for Nuclear Emergencies in European Overview, Radiation Protection Dosimetry, Vol. 50, Nos 2-4, pp 195-203 (1993)

[5] J. Narrog and R. Obrecht, Transboundary Data Management for Plants near National Borders, NEA-Workshop "Emergency Data Management", Zurich, 1995

[6] W. Weiss, H. Leeb, F. Eberbach, Basic concepts and Objectives of Technical Systems for Computer-based National and International Exchange of Data and Information, EU workshop on the Technical Aspects of International Exchange of Radiological on-line Data, Ispra, June 1994

[7] G. Linden, "Teleray-Minitel, the French National Network for the Radiological Survey of the Territory", NEA-Workshop "Emergency Data Management", Zurich, 1995

[8] R. Jackson, "Treatment and Handling of Radiological Data for Nuclear Emergency Response: The United Kingdom RIMNET Arrangements", NEA-Workshop "Emergency Data Management", Zurich, 1995

[9] W. Weiss, H. Leeb, "IMIS - The German Integrated Radioactivity information and decision support System", Radiation Protection Dosimetry, Vol. 50, 1993

[10] IAEA - Convention on early notification of a nuclear accident, 27 October 1986, Vienna, Austria.

[11] IAEA - Convention on assistance in the case of a nuclear accident or radiological emergency, 26 February 1987, Vienna, Austria.

[12] Council Decision of 14 December 1987 on Community arrangements for the early exchange of information in the event of a radiological emergency, OJ No L 371/76 30 12 1987.

[13] CEC - Convention notification and information structure, DOC No 6019/1/91 EN, 6 June 1991.

[14] CEC - ECURIE level-3 exercise, 1991 - Position following the Meeting of the exercises Working Group 26-27 February, 1991, and its subsequent evolution. DOC No 6008/1/91 EN.

[15] M. De Cort, 'Regional Monitoring Networks and International Data Exchange', 'Environmental Monitoring, Off-site Emergency Response to Nuclear Accidents: Textbook based on training courses organized at SCK/CEN Mol, Belgium (in preparation)

[16] G. de Vries, M. De Cort, S. Giuliano and S. Gatti, Technical Note: 'Concept and development of an exchange format for European Radioactivity data' (in preparation)

[17] M. De Cort, L. Breitenbach and G. de Vries, The on-line European Community Urgent Radiological Information Exchange (ECURIE) information system, NEA-Workshop "Emergency Data Management", Zurich, 1995

[18] M. De Cort and G. de Vries, The EU Radiological Data Exchange Platform (EURDEP): recent developments, NEA-Workshop "Emergency Data Management", Zurich, 1995

IV. OFF-SITE MANAGEMENT
OF FUTURE ACCIDENTS

Decision support systems for the off-site

management of future accidents

Posters

THE GAMMA-1 PROJECT

Norbert Jousten
Principal Administrator,
DGI/E/4, Commission of the European Union

Phil Jeanes
Principal Consultant
PA Consulting Group, Cambridge, UK

Abstract. This paper describes the background and technical content of the GAMMA-1 Project. This project, funded by the CEU's TACIS Programme, is currently installing pilot radiation early warning systems near three nuclear power plants in Belarus and Ukraine. These pilot systems, operated in Belarus by the Committee for Hydrometeorology and in Ukraine by the Ministry for Environmental Protection and Nuclear Safety, are designed for maximum reliability and eventual expansion to provide nation-wide monitoring systems.

1.   Background

After the Chernobyl accident, the governments of Belarus and the Ukraine recognised the imperative of installing radiation early warning systems to detect radiation escapes that might arise from nuclear power plants. In 1992, the CEU's TACIS Programme was asked for assistance and financed a detailed study, performed by PA Consulting Group, to assess the feasibility of such systems being implemented using western technical and financial support. This was followed by a design study in 1993 which defined the first stage of a progressive implementation project to install on-line radiation monitors in a band from the Black Sea to the Baltic - *the Gamma Curtain*.

A competitive procurement exercise was launched at the beginning of 1994 and in October 1994 Hšrmann Systemtechnik was awarded the contract to implement the pilot project, GAMMA-1. The system is due to become operational in early 1996.

At an early stage the Committee for Hydrometeorology in Belarus (Glavhydromet) and the Ministry of Environmental Protection and Nuclear Safety of Ukraine (MEPNS) were identified as the appropriate organisations to support the installation and operate the systems. Thus the active co-operation of these organisations has been a critical factor in the progress made so far. Project direction is provided by an Executive Steering Committee which includes representatives of TACIS, the Beneficiaries as well as technical advisers.

PA Consulting Group has provided technical and project management support and advice to both TACIS and the Beneficiaries throughout this project.

2. Objectives of GAMMA-1

The primary objectives of the GAMMA-1 Radiation Early Warning System are to:

- Detect significant changes in the levels of radiation that might occur within the selected pilot areas to the highest possible levels of measurement reliability
- Alert the national authorities that a significant change has occurred and provide information to support counter-measures to protect populations and the environment
- Enable real time data exchange between local and national organisations
- Test and evaluate the key elements of the installed system.

The system has been implemented within the existing constraints of the communication technology and infrastructure available, as well as within financial and other resource limitations of both the Beneficiary states and the TACIS Programme's allocated funds.

3. Overall Technical Description

The overall system comprises of the following components:

- fully automatic stations performing continuous radiation monitoring, including fast response and water monitoring systems
- processing centres for collecting and analysing the monitoring data
- mobile response vehicles to focus response and provide maintenance facilities
- consistent communications to connect the components effectively.

Figure 1   The GAMMA-1 Pilot System

3.1   Monitoring Stations

Most monitoring stations consist of gamma probes (model IGS 421B) and precipitation detectors mounted on an antenna mast on a flat roof, usually of a public building. Two specialised sensors, an aerosol monitor and a water effluent monitor are also integrated into the system. Control units (model DLM 1440) are used to handle all measurements and controls and maintain communication with the local response centres.

The IGS421B gamma probes includes two low-dose and one high-dose Geiger-Mueller tube to ensure reliable readings. The intelligent probes calculate readings and provide diagnostic data ensuring that the response centres are aware of the equipment status at all times. These tasks can also be performed by a portable maintenance computer at the station. Some stations also have large public displays which show the current radiation levels.

3.2   Processing Centres

The primary tasks of the processing centres are to manage the data, provide analysis and alert the appropriate authorities if anomalies are detected. The six processing centres are equipped with dual Pentium 90 processors using the WINDOWS NT operating system. A modern client/server architecture separates communication, data storage and retrieval tasks from application software and ensures that the system's high specification for data communication and multi-tasking capabilities are met.

The processing centres are of two types:

- local response centres which are primarily for incident detection and management

- monitoring centres which are primarily for accumulating and analysing the monitored data.

The national monitoring centres also co-ordinate the national response to any radiation incident and maintains a national monitoring archive. They would also be responsible to transmit information to other organisations and international bodies in the event of an incident.

3.3.   Response Vehicles

The fixed subsystems are supplemented by special vehicles equipped with automatic gamma dose rate monitoring stations, called *Sentinels*. The vehicles will be used to provide maintenance for the fixed systems as well as additional monitoring coverage at critical times. The Sentinels have the same gamma probes and radio communications as the fixed gamma monitoring stations. They are powered by solar panels and operate automatically, communicating directly with the response centre to supplement the data from fixed stations.

4.   GAMMA-1 in Belarus

The pilot GAMMA-1 system in Belarus focuses on providing information on the radiation situation around Ignalina to local, regional and national authorities. A national centre is being established in Minsk, regional centre in Vitebsk and local centre in Braslav. The Monitoring Subsystem comprises the following items, all sited in Belarus within 30 km of the plant:

- 9 gamma dose-rate monitoring stations

- one automatic weather station 10 kms from Ignalina

- one local response centre located at Braslav, 30 kilometres from Ignalina.

Communications throughout the monitoring area is provided by radio using a frequency of 154 MHz. Information is transmitted to the regional centre at Vitebsk via radio at a similar frequency. A repeater station at Polotzk is used to boost the signal. From Vitebsk an existing leased line is used to pass information to the national centre in Minsk. The system is operated by Glavhydromet throughout.

5. GAMMA-1 in Ukraine

The GAMMA-1 system in Ukraine provides to monitoring subsystems around Rivne and Zaporizhya NPPs and three processing centres. Mobile facilities operate around both Rivne and Zaporizhya NPP's.

The extensive monitoring network around Rivne NPP is intended to establish the operational characteristics of a fully integrated high performance sensor system as a basis for future development. It comprises of the following:

- 27 gamma dose-rate monitoring stations located at distances from 2 to 30 kms
- one automatic alpha-beta aerosol monitor, located within 5 kms of the plant
- one automatic water effluent gamma monitoring station
- one automatic weather station
- a local response centre located in the city of Rivne, 100 kms from the NPP.

A polling station in Kuznetsovsk acts as a data concentrator before transmitting the measurements to the response centre in Rivne. This uses a frequency near 150 MHz.

The second subsystem around Zaporizhya NPP provides restricted monitoring facilities including the following:

- 11 gamma dose-rate monitoring stations within 15 kms of the plant
- one automatic weather station
- one local response centre located in the city of Zaporizhya, 80 kms from the NPP.

Communications uses similar frequencies to those in Belarus; the information is collected locally at Balki and transmitted, again by radio, to Zaporizhya response centre.

In addition to the local response centres, a national centre is installed in Kiev and connected to the local centres at Rivne and Zaporizhya via leased PSTN lines. All the processing centres are operated by MEPNS, with remote access facilities provided for the Civil Defence authorities at each location.

6. Future development

GAMMA-1 is intended to be the first stage of implementation to prove the key operational characteristics. A number of extensions to the system have always been foreseen and include:

- expanding the areas covered by the detection systems initially to include all NPPs, including Chernobyl
- improving the diagnostic capability of the monitoring subsystems
- strengthening the technical capabilities of the Beneficiary organisations
- linking to other international initiatives.

These are intended to take advantage of the strengths and local expertise wherever possible. Currently the details and prospects for future development are under discussion by the Executive Steering Committee.

7. Conclusions

The GAMMA-1 Project has demonstrated that major nuclear safety related projects can be implemented by drawing on the financial and technical resources of the European Union, represented by the TACIS Programme, with effective and enthusiastic co-operation from Beneficiary organisations. This is to the benefit of the peoples of Belarus and Ukraine as well as Europe in general.

REAL-TIME SOFTWARE FOR MULTI-ISOTOPIC SOURCE TERM ESTIMATION

Goloubenkov,A.[1], Borodin,R[1]., Sohier,A.[2]

[1] SPA TYPHOON, Lenin St. 82, Obninsk, Kaluga Region, 249020, Russia
[2] SCK*CEN, Radiation Protection Research Unit, Boeretang, 200, B-2400 ,Mol, Belgium

Abstract. Consideration is given to development of software for one of crucial components of the RODOS - assessment of the source rate (SR) from indirect measurements. Four components of the software are described in the paper. First component is a GRID system, which allow to prepare stochastic meteoro-logical and radioactivity fields using measured data. Second part is a model of atmospheric transport which can be adapted for emulation of practically any gamma dose/spectrum detectors. The third one is a method which allows space-time and quantitative discrepancies in measured and modelled data to be taken into account simultaneously. It bases on the preference scheme selected by an expert. Last component is a special optimisation method for calculation of multi-isotopic SR and its uncertainties. Results of a validation of the software using tracer experiments data and Chernobyl source estimation for main dose-forming isotopes are enclosed in the paper.

1. Introduction

The paper presents results of the joint Russian-Belgian research in the area Data Assimilation And Source Term Estimation of Joint Study Project 1 in 1993-1995. Consideration is given to development of software for one of crucial components of the RODOS[1] - a decision making support subsystem for an early stage of an accident. To calculate one of the most important criteria, population dose, data about dynamics of spatial fields of concentrations and radiation of dose - forming radioisotopes. Such assessments can not be made promptly without using a geophysical model for atmospheric transport and deposition of radioactivity (ATM). In turn, the reliability of model assessments depends on input parameters, primarily a source rate (SR) because it determines the extent of the accident. It is obvious that it is not always possible to measure directly the dynamics of the SR at the time of an accident. This makes important the task of reconstruction of these relationships from measurements of radioactivity around the source.

2. Problem statement

First of all, we suppose that SR may be presented as a piecewise-constant dynamic function with time step Δt. During Δt meteorological and radioactivity measurements are a constants also. These data are processed by Grid system, which prepares the meteorological fields needed for ATM and tests and deletes outliners from the radioactivity data set. The SR value is the ATM parameter. It is estimated with special method - ATOS, which allows to calculate the best degree of agreement of measured and calculated by ATM characteristics of radioactivity contamination of the atmosphere and surface ground. To evaluate the SR values for all time steps jointly and to take into account possible additional information we use multi-time&isotopic source solver - the optimisation method based on projective geometry. We use estimated SR in ATM calculations and then Grid system for presentation of the dose/concentration fields in the final stage of the algorithm. Using these fields and GIS for the vicinity of the source we can calculate Risk values for possible countermeasures strategies. The chart of this process is presented on Fig. 1.

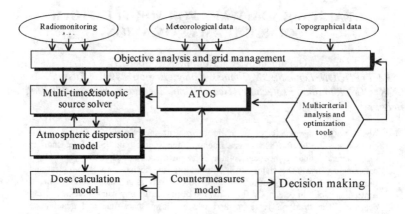

Figure 1. The chart of the interaction of the software components.

3. Grid system

The system of generation, management and visualisation (Grid) is the main tool for data exchange between software components. In addition to this function, it deals with the tasks of assessment of radiation situation in a populated point, presentation of results of modelling predictions and assimilation and evaluation of reliability of meteo/radioactivity measurements. Each of these problems has its specific features and it is impossible to elucidate all of them in one article. A more complete information about Grid can be found in [2]. In this paper we just underlay that Grid can calculate stochastic scalar/vector space-dynamic fields (including variance fields and semi-correlation functions).

4. Model for atmospheric transport and deposition of radioactivity (ATM)

Specific feature of the early stage of a radioactive accident is that measured gamma dose rate (GDR) and/or gamma-spectra are main data types that can be used for source term estimation (STE). In this particular case we need to simulate the detected GDR values with maximum accuracy. Calculation methods based on the semi-infinite cloud model may lead to large errors in the short range and are not suited in this situation. Because we use the RIMPUFF transport model [3] in our STE algorithm, it is natural to use a spherical reference system to describe the concentration field. On the other hand, as the STE method described below requires at least a ground-level data field, space integration process may be impossible in on-line mode. A compromise in this case is to use interpolation tables for normalised GDR for some energy intervals. The software includes the data bank containing radioisotopes spectra, decay chains and dose rate conversion coefficients. Therefore, ATM can emulate:
- instantaneous and time averaged concentrations of radioisotopes in the air;
- instantaneous and time averaged gamma-doses from a radioactive cloud;
- instantaneous and time averaged spectra of gamma shining from a radioactive cloud;
- time and/or space integrated concentrations of radioisotopes on the ground surface;
- time and/or space integrated gamma-spectra from the surface contamination,
and can be adapted for emulation of practically any measurement system.

5. Adequate transformation of observation space (ATOS)

The main technique of the STE is calculation of distance F between vectors of measured and model quantities (The Objective function). F is minimised via varying the ATM parameters. Note that in any case, F change can be superimposed on segment [0,1], with F=0 corresponding to a complete coincidence of model and measured quantities and F=1 - their complete noncoincidence. In this sense most of Objectives behave as a normalised Objective function of the Normalised Least Squares Method (NLSM). Note than main reason of failures in application of this method to the STE problem is uncertainties in meteorological parameters that can not be eliminated.. To improve the quality of results in this cases we use a method of Adequate Transformations of Observed Space (ATOS). Let us explain the main principles of the method.

In simple methods under simple meteorological conditions, minimising of F is achieved by including the effective wind direction in the list of parameters to be estimated. However, it is very difficult and time-consuming to seek the effective three-dimensional dynamic wind field under complex meteorological conditions. Therefore, we have decided to do otherwise: instead of varying the input model parameters, we modify the observed space (a ground level plane in the simplest case) using a series of affine transformations:
a) rotation around the source;
b) parallel transfer along the axis of the mass centre of measurements (MCM) - source;
c) proportional change of area around the mass centre of measurements;
d) ellipsoid change in the shape relative to the point of MCM and the axis MCM - source.

Without going into details, we note that these transformations are quite well defined in input model parameters.

By varying the degree of transformations, we obtain quite a small value of F and a good approximation of the source term.

However, two points should be emphasised. The value of transformation in the optimum with respect to F point determines the degree of uncertainty in the model prediction. A decrease in F, when it is already small enough, is explained by incidental reasons and does not correlate with the degree of reliability of the result.

With this in mind, we think that the optimisation problem should be stated as follows: *to find a rather small F at a rather small transformation.*

In this case, the quality of the solution must be evaluated by two criteria, i.e. two goals should be pursued at the same time. Note that the problem is made more difficult by the fact that the significance of criteria changes with the change in absolute estimates of variants by these criteria. For example, in the region of large values of F and small transformation it is essential to reduce the value of the objective function by, probably, more drastic transformation, while in the region of small values of F and large transformation it is more important to reduce the value of the latter.

In solving the problem of multicriteria optimisation, under these conditions the approximation approach [4] based on real preferences of a decision maker proved to be effective. The software for this approach (M-Crit) developed for application in decision making support system RODOS allows the structure of decision maker's preferences to be identified, fixed and saved for further use in interactive and automatic mode.

The proposed algorithm is implemented by multiple generation of random variants of transformation parameters, determining the rank of the variant by using a developed scheme of decision maker's preferences and selection of a variant corresponding to a minimum rank.

We have tested ATOS using tracer experiments data[5]. For simplifying we divided SR results on four groups 5(excellent), 4(good), 3(not good), 2(bad). Value TM in the Tab. 1 means the transformation magnitude and for F, TM and Rank we give average values.

Table 1. Summary results for different approaches to the source rate estimation.

Method used	Numb. of mark '2'	Numb. of mark '3'	Numb. of mark '4'	Numb. of mark '5'	F	TM	Rank
NLSM	14	10	13	40	0.67	0.00	17437
OP Transf.	4	4	18	51	0.29	0.76	19150
ATOS	4	3	9	61	0.40	0.28	11722

6. Multi-time&isotopic source solver

The necessity to take account of various information for the best possible determination of quantitative characteristics of the source term has led to development of a special optimisation method. It bases on application of the projection geometry methods to solving the restricted minimisation problem. It is worth noting that the algorithm not only estimates of the SR, but also gives its statistical evaluation, i.e. uncertainties related to optimal estimates can be evaluated.

The presented method was successfully used for reconstruction of the composition of the Chernobyl radionuclide fallout and external radiation absorbed dose of the population on the areas of Russia. All dose forming nuclides were subdivided in 6 groups and for each group "leader" was chosen (the dynamics of other nuclides reconstructed using the correlation relationship with the leader). These groups and leaders were:

1. ^{137}Cs - group ^{137}Cs, ^{136}Cs, ^{134}Cs, ^{125}Sb , 2. ^{131}I - group ^{131}I, ^{132}Te, ^{133}I,
3. ^{140}Ba- group ^{140}Ba, ^{140}La., 4. ^{95}Zr - group ^{95}Zr, ^{95}Nb,
5. ^{106}Ru- group ^{106}Ru, ^{103}Ru, 6. ^{144}Ce- group ^{144}Ce, ^{143}Ce, ^{141}Ce.

Results of STE and doses calculation can be found in [6]. The similar task was solved immediately after Tomsk accident. These materials were published in [7].

7. Conclusion

The main algorithms for assessment of the radiation situation and their validation and successful application for analyses in some radioactive accidents was described. Note that the presented software is a part of RECASS[2] - Russian national scale DMSS for cases of a radioactive accident.

References

[1] J. Ehrhardt et. al. Development of RODOS, a Comprehensive Decision Support System for Nuclear Emergencies in Europe - an Overview. Radiation Protection Dosimetry, Vol. 50, Nos 2-4(1993), p. 195-205.
[2] Shershakov V.M, Goloubenkov A.V, Vakulovsky S.M. Computer-information support of analysis of radiation environment in the territories polluted as a result of the Chernobyl accident. RADIATION&RISK, issue 3, 1993, ISSN 0131-3878, pp.40-61.
[3] Thykier-Nielsen S. and Mikkelsen T. (1995) RIMPUFF - User's guide/RIMDOS version. Available from Ris0 Library, Ris0 National Laboratory, DK-4000 Roskilde, Denmark.
[4] Borzenko V.I. (1989). Approximational Approach to Multicriteria Problems, Lecture Notes in Economics and Mathematical Systems, Springer-Verlag, v. 337.
[5] P. Thomas et al. Experimental determination of the atmospheric dispersion parameters at the Karlsruhe Nuclear Reset Centre. report KfK 3090,Sep.,1981, AND KfK 3456, Mar.,1983.
[6] Pitkevich V.A. et al. Reconstruction of composition of the Chernobyl radionuclide fallout in the territories of Russia. RADIATION & RISK, issue 3, 1993, ISSN 0131-3878, pp.62-93.
[7] V.M. Shershakov et al Analysis and prognosis of radiation exposure following the accident at the Siberian chemical combine Tomsk-7. Radiation Protection Dosimetry, V. 59, pp.93-126(1995).

Implementation of the Aquatic Radionuclide Transport Models RIVTOX and COASTOX into the RODOS System

Gofman, D., Lyashenko,G., Marinets, A., Mezhueva I., Shepeleva, T., Tkalich, P. , Zheleznyak, M.
Cybernetics Centre of the Ukrainian Academy of Sciences, Institute of Mathematical Machines and Systems, Prospect Glushkova 42, Kiev, 252207, Ukraine

Abstract. The one -dimensional model of radionuclide transport in a network of river channel RIVTOX and two-dimensional lateral-longitudinal model of radionuclide transport in rivers, reservoirs and shallow lakes COASTOX have been implemented into the hydrological model chain of the decision support system RODOS. The software framework is developed to operate the models and to support their coupling with the other parts of RODOS hydrological model chain. The validation studies were performed for RIVTOX and COASTOX on the base of the data sets from Ukrainian, German and United States rivers.

1. Introduction

Real-time on-line decision support system RODOS for nuclear emergencies in Europe includes the hydrological model chain [1,2]. The one -dimensional model of radionuclide transport in a network of river channel RIVTOX and two-dimensional lateral-longitudinal model of radionuclide transport in rivers, reservoirs and shallow lakes COASTOX, both developed to evaluate post-Chernobyl contamination of the Ukrainian rivers [3-6], have been implemented into the RODOS hydrological model chain. The main objectives of this models is to evaluate radionuclide concentration in water, bottom depositions and suspended sediments following radionuclide washing out into the rivers (RIVTOX) and direct releases of radioactivity to rivers, lakes and reservoirs (COASTOX). The simulated concentrations are used in the aquatic dose assessment module of hydrological chain [2]. The software framework was developed to operate RIVTOX and COASTOX and to support their coupling with other models of hydrological chain. The paper describes a brief these activities.

2. COASTOX model and its validation

COASTOX is two-dimensional model developed to simulate depth averaged concentrations of radionuclides in solute, suspended sediments and in bottom depositions of reservoirs, floodplains and coastal areas. The model is used to analyze radionuclide dispersion in water bodies with significant spatial variations of the concentrations (vicinity of the releases, transport above inhomogeniously contaminated bottom, etc.). The model describes currents (Fig.1), sediment transport, advection-diffusion pollutant transport, radionuclide - sediment interaction [5]. It also calculates the dynamics of the bottom deposition contamination and describes the rate of sedimentation and resuspension as a function of the difference between the actual and the equilibrium concentration of suspended matter depending on the transport capacity of the flow. The latter is calculated on the base of semi-empirical relations. For describing the adsorption/desorption and diffusion contamination transfer in the systems "solution - suspended sediments" and "solution - bottom deposition" the Kd approach has been used. For a more realistic simulation of the kinetics of the processes the exchange rates between solution and particles was taken into account. The adsorption and desorption rates assumes to be not equal. The finite-difference methods are used to solve the model's equations.

Fig. 1. Simulated flow pattern on the Pripyat River Floodplain at Chernobyl NPP

The validation studies were done for the COASTOX on the base of the experimental data obtained during flooding of Pripyat River Floodplain (Fig.1) [3,5] and during spring floods in the Dnieper reservoirs. The simulated ^{137}Cs distribution during 1987 spring flood in Kiev Reservoir (Fig.2) shows significant spatial variability in contamination field. The initial data for modeling are the measured by the Ukrainian Hydrometeorological Institute ^{137}Cs surface density of the bottom deposition and ten day averaged concentrations of ^{137}Cs in solute and on suspended sediments at the mouth of the Dnieper River and Pripyat River discharged into the Kiev Reservoir.

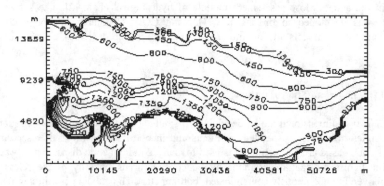

Fig. 2. Simulated ^{137}Cs (Bq/m^3) concentration in solute in the Kiev Reservoir, 30 April 1987.

The simulated results were averaged over the reservoir outflow and then were compared with experimental data. It was achieved the reasonable agreement with the averaged data of measurements at outflow from Kiev Reservoir (Fig. 2) - for April 1987 measured ^{137}Cs concentration equals to 751 (Bq/m^3) in solute and 130 (Bq/m^3) at suspended sediments.

3. RIVTOX model validation and implementation

The one-dimensional model RIVTOX was developed to simulate the radionuclide transport in networks of river channels [3, 4]. The model equations could be obtained by the averaging of COASTOX equations over the crossections of river channel.

RIVTOX simulate radionuclide transport in a river net following the direct release into the river (point source) or the wash-off of radionuclides from a watershed (lateral inflow). In the case of the wash-off the output from RETRACE is used as the input to RIVTOX.

Within JSP-1 Project RIVTOX was applied to simulate dispersion of pollutants in the Mosel River as result of radioactive fallout after the accident at Chernobyl NPP (Fig. 3).

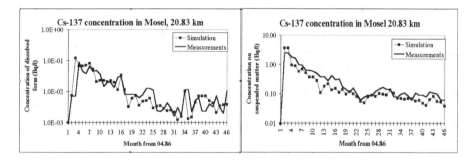

Fig. 3

The model tuning and calibration study has been done on the base of Dnieper River post-Chernobyl data within IAEA\CEC VAMP Programme [6]. RIVTOX was validated also within this Programme on the base of scenario of Clinch River -Tennessee River contamination due to radionuclide releases from Oak Ridge (Fig. 4)

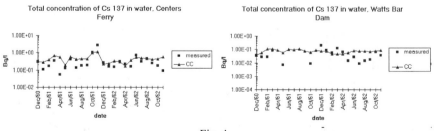

Fig. 4

RIVTOX has been used in Ukraine in summer 1995 to evaluate the consequences of the accidental continue release of the Kharkov municipal waste water to the rivers. It was customized and adopted in the IMMS as the part of the RODOS hydrological chain within several days to simulate chemical and bacteriological contamination of the Udy River - Siversky Donets River aquatic system. By the request of the State Emergency Commission some calculations have been provided to evaluate the amount of water that should be pumped to the Siversky Donets trough the channel from the Dnieper River to improve water quality parameters to the maximum permissible levels. This countermeasure on the base of RODOS' hydromodule calculations was successfully implemented by the Ukrainian State Committee on Water Resources.

4. User interfaces of RIVTOX and COASTOX.

User interfaces of RIVTOX and COASTOX models were realized in the frames of common RODOS Hydrological Module User Interface. Interface supports geographical object selection. It's possible to keep and select different river net systems for RIVTOX simulation and different 2-D objects for COASTOX simulation. Interface provides data base management functions for RIVTOX and COASTOX models. It is possible to input and to edit geographically depended and independent data. System of menus makes data base management easy and convenient.

An user can run RIVTOX and COASTOX separately or simultaneously in automatic and manual modes. It is possible to transfer data from RIVTOX to COASTOX and vice versa during the simulation. RIVTOX and COASTOX can use data exchange with other models of hydrological chain (RETRACE, H-DOSE and LAKECO). Results of simulation could be saved into the data base for later usage.

There are different graphical tools in the RIVTOX and COASTOX interfaces for simulation results and data base information representation. Data could be represented as maps of contamination and graphs.

References

[1] J.Ehrhardt, V.Shershakov, M. Zheleznyak, A.Mikhalevich, RODOS: decision support system for off-site emergency management in Europe. -Proceedings of Minsk Conference. First International Conference of the European Commission, Belarus, the Russian Federation and Ukraine on the Consequences of the Chernobyl Accident, Minsk, Belarus, 18-22 March 1996 (this issue)

[2] M. Zheleznyak, R. Heling, W. Raskob, A. Popov et al.,Modeling of Hydrological Pathways in RODOS. - Proceedings of Minsk Conference. First International Conference of the European Commission, Belarus, the Russian Federation and Ukraine on the Consequences of the Chernobyl Accident, Minsk, Belarus, 18-22 March 1996 (this issue)

[3] M. Zheleznyak, P.Tkalich, G.Lyashenko, A. Marinets, Radionuclide aquatic dispersion model-first approach to integration into the EC ision support system on a basis of Post-Chernobyl experience. - Radiation Protection Dosimetry, N6 (1993), 37-43.

[4] P. Tkalich, M. Zheleznyak, G. Lyashenko, A. Marinets, RIVTOX - Computer Code to Simulate Radionuclides Transport in Rivers, In: A. Peters et al.(Eds.) , Computational Methods in Water Resources X, vol. 2, Kluwer Academic Publishers, Dordrecht, The Netherlands, 1994, pp. 1173-1180.

[5] M. Zheleznyak et al. Modelling of Radionuclides Transport in the Set of River Reservoirs, In: A. Peters et al.(Eds.), Computational Methods in Water Resources X, vol. 2, Kluwer Academic Publishers, Dordrecht, The Netherlands, 1994, pp. 1189 - 1196

[6] M.Zheleznyak et al.,Modeling of radionuclide transfer in rivers and reservoirs: validation study whithin the IAEA\CEC VAMP Programme. International Symp. on Environmental Impact of Radioactive Releases, IAEA, Vienna, 8-12 May 1995, Extended Synopses IAEA-SM-339, IAEA, 1995, p.330-331

Assessment of agricultural countermeasures and the development of the decision support system "FORCON"

S.V. Fesenko, N.I. Sanzharova
Russian Institute of Radiology and Agroecology, Obninsk, Russia
L.V. Perepelyatnikova, B.S. Prister
Ukrainian Institute of Agricultural Radiology, Kiev, Ukraine
B.T. Wilkins
National Radiological Protection Board, Chilton, United Kingdom
C.N. Bouzdalkin
Belorussian Institute of Agricultural Radiology, Gomel, Belarus

Abstract. A large amount of data on agricultural countermeasures has been compiled and assessed within an EC-sponsored collaborative project. The results of the work are summarised, and the development of a predictive tool to aid decisions at the local level is briefly described.

1. Introduction

The major long-term exposure pathways in those areas of Russia, Ukraine and Belarus affected by the Chernobyl accident are external irradiation and the ingestion of contaminated foods. In the long-term, intervention in agricultural systems is a more practicable measure for the reduction of doses than the decontamination of settlements. Consequently, the development of optimal strategies for the application of countermeasures in agriculture was and continues to be a priority in these countries. In the 10 years since the accident, various measures have been implemented and a vast amount of data on their effectiveness has been generated, together with information on ancillary factors such as the required resources and costs. Within the EC-sponsored collaborative programme with Russia, Ukraine and Belarus, the underlying purpose of Joint Study Project 1 (JSP1) was to gather and assess data on agricultural countermeasures, with the intentions of improving both the countermeasure modelling in decision support systems such as RODOS[1] and the practical advice that could be provided in the affected areas. The development of databases for use in RODOS is discussed elsewhere[1]. The purpose of this paper is to describe briefly the evolution of countermeasure strategies at three settlements within the areas affected by the Chernobyl accident and the development and capabilities of a PC based system called FORCON, a predictive tool which is designed to aid decisions on agricultural countermeasures on a site specific basis. A detailed description of FORCON is already available[2]. Although both radiocaesium and radiostrontium have been considered within JSP1, this paper is confined to the former.

2. The evolution of countermeasures at selected settlements

The settlements considered here are "Rodina", Bryansk region, Russia, "Chapaeva", Rovno region Ukraine and "Kirovsky", Gomel region, Belarus, since these were the sites for which most comprehensive information was available. A detailed account of the agricultural countermeasure strategy at each of these settlements has been published[3], and only a few selected examples can be given here.

At all three settlements, the consumption of milk was an important contributor to the overall dose from ingestion, and so the reduction of activity concentrations in milk was a priority. In all cases, the general approach taken was to increase the proportion of "clean" feed in the animal's diet. The main dietary components were grass, hay and silage, all of which were produced on both managed pastures containing perennial grass and from unmanaged natural grassland. Transfer of ^{137}Cs from soil to plant can be about an order of magnitude greater for natural grass compared to perennial grass[3]. The overall approach was therefore to decrease the proportion of natural grass in the animal's diet, principally via radical and surface improvements of the unmanaged grassland. Radical measures can include the clearing of land, ploughing, seeding and fertilising; surface improvement is a cheaper option and includes levelling, improved drainage, fertilising and sowing additional grass. It should be noted however that not all natural grassland was amenable to improvement.

The effects of the measures taken can be distilled from the detailed data for each settlement[3]. Taking Chapaeva as an example, the areas of natural grassland that underwent radical improvement over a 3-year period were as follows: 1988, 30ha; 1989, 370ha; 1990, 120ha. As a result, the proportion of natural grass in the diet decreased from about 70% in 1987/88 to less than 40% in 1989, with a further decrease to about 20% in 1990; the estimated average activity of ^{137}Cs in the diet of dairy cattle in 1989 was 5 times lower than in the previous year, and decreased by a further factor of 2 in 1990. There was reasonable agreement between the changes in the estimated total activity in the diet and those in the resultant concentrations in milk, for which average values in 1989 were 3 times lower than in 1988, declining by a further factor of 2 the following year.

Experimental studies at NRPB have indicated that treatment of soil with ammonium ion enhances the uptake of radiocaesium by plants, while the addition of potassium to poor quality soils has a beneficial effect[4]. Using these data, temporal changes in activity concentrations in crops can be broadly rationalised in terms of changes in the composition of the fertiliser applied. As an example, concentrations in oats grown at Chapaeva decreased by a factor of about 2 in 1989 compared to the previous year, the result of an increase in the amount of potassium applied together with a reduction in the amount of ammonium. A further reduction in activity concentrations in the crops was observed in 1991 following another application of potassium fertiliser[3]. One important conclusion from this part of the study is that because of changes in fertilisation regimes, monitoring data should not be used indiscriminately to infer effective half times in crops.

3. The FORCON system

The FORCON system is intended for widespread use by agricultural specialists who are required to give advice at a local level or to individual farmers. The basis of the system is as follows:

(i)   the analysis of the existing radioecological situation and the identification of those agricultural products that require some form of intervention;

(ii) modelling of protective measures with regard to prevailing conditions, predicted effectiveness, available resources and costs;

(iii)   the identification of the most appropriate countermeasure strategy.

The need for intervention and the identification of countermeasure strategy can be based on either comparisons with Derived Intervention Levels (DILs), consideration of averted dose or economic and other ancillary factors.

The system contains default datasets for parameters such as soil:plant transfer factors and countermeasure effectiveness factors that have been derived from the assessment of data within JSP1, soils being divided into five categories, based on granulometric composition. Alternatively, the user can encode site-specific data.   The agricultural counte~ :easure options within the system are classified in the following groups: organisational (change in land use, use of "clean feeds", changes in animal husbandry, food processing), agro-technical (ploughing, liming, radical and surface improvements), agrochemical (application of mineral and organic fertilisers at various rates and in different combinations). There may however be practical constraints on some options at particular sites. For example, deep ploughing with turnover of the upper layer would in practice be confined to high fertility soils.

For each scenario chosen by the user, the following values are calculated: radionuclide concentrations in all of the foodstuffs produced; yields of each product;   quantity of resources necessary to implement the countermeasure(s); the associated costs of materials, transport, labour etc; the probability of the DIL being exceeded in each foodstuff after the countermeasures has been applied; averted dose; averted dose per unit of costs. For animal products, concentrations are presented separately for periods spent on pasture and indoors. In short, the user has a comprehensive dataset on which to base decisions, according to whether the radiological considerations are constrained by other factors such as finance or the availability of resources.

4.   An example of the use of FORCON

The FORCON system has been used to investigate the effects of various hypothetical countermeasure strategies at Rodina and Chapaeva.   The detailed data on land use, productivity, animal stocks, soil type and deposition for these settlements were used to estimate the collective and averted collective doses for a range of individual countermeasures. Averted dose was calculated from a baseline value for collective dose in the absence of any countermeasures and the corresponding value after a measure had been implemented. The exercise related to 1987, ie, the first year for which longer term measures would have been considered. It is important to note that cost estimates were derived from values for materials and labour specific to the affected areas of Russia and the Ukraine. Only a few examples of the results can be discussed here.

As noted earlier, the application of mineral fertilisers was an effective countermeasure in reducing the transfer of ^{137}Cs to crops: reductions in activity concentrations by factors of 1.5 - 2.5 were observed at both Rodina and Chapaeva[3]. Using FORCON, a reduction in collective dose of about 25% was predicted at Chapeava, compared to the predicted baseline value. However, at Rodina this countermeasure resulted in higher productivity, and as a result there was no significant change in collective dose. A similar finding in the highly-fertile areas of western Europe would be unlikely, since productivity is already high. However, this result does emphasise the need for care when deciding upon the criteria for countermeasure effectiveness.

The effects of radical and surface improvement of natural grassland were also investigated. At Rodina, the reductions in collective dose were around 45% for radical measures and around 35% for surface improvement. The corresponding values for Chapaeva were about 20% and 15%. As expected, the costs of radical improvement were greater than those for surface measures. However, because of the area of land that required treatment, the overall costs at both settlements was less than that for the application of fertiliser to the existing arable land. These results reinforce the practical decision to prioritise the improvement of natural grassland as a countermeasure.

The administration of sorbents in the diet of animals is a more recent development. At Rodina, the reduction in collective dose was around 40%, while the costs were comparable with those for radical improvement. At Chapaeva, the costs were less than those for radical improvement, but the reduction in collective dose was only about 12%.

A comparison of the actual approach adopted and the hypothetical optimum strategy has also been carried out, covering the period 1987-1992. In the longer term, some additional savings in collective dose could have been made. However, greater savings could have been achieved in the earlier years of 1987/8 had countermeasures such as improvements of natural grassland been imposed earlier.

5. Conclusions

The work within JSP1 has provided an opportunity to compile and assess a large amount of data on agricultural countermeasures from Russia, Ukraine and Belarus. If sufficient information is available, the evolution of countermeasure strategies can be discerned at specific sites. A comparison of hypothetical long-term countermeasure strategies has been carried out for two specific sites. Given the range of countermeasures that are, in principle, available to be used in agricultural systems, the results emphasise the need for a flexible predictive tool to provide practical advice at the local level in the event of a future accident.

6. Acknowledgements

We are indebted to the many colleagues in the participating institutes and elsewhere who made significant contributions to this work.

References

[1] J. Brown, B.T. Wilkins, A.F. Nisbet, Y.A. Ivanov, L.V. Perepelyatnikova, S.V. Fesenko, N.I. Sanzharova and C.N. Bouzdalkin, Modelling of Agricultural Countermeasures in RODOS. To be presented at the First International Conference of the European Commission, Belarus, Russian Federation and Ukraine on the Radiological Consequences of the Chernobyl Accident (Minsk, 18-22 March 1996).

[2] S.V. Fesenko and B.P. Kulagin, Development of the FORCON decision support system for the provision of advice in agricultural at the local level (in preparation).

[3] M. Paul, B.T. Wilkins, A.F. Nisbet, L.V. Perepelyatnikova, Y.A. Ivanov, S.V. Fesenko, N.I. Sanzharova and C.N. Bouzdalkin, The development of countermeasure strategies at selected settlements in the areas affected by the Chernobyl accident. Chilton, NRPB Memorandum (in preparation).

[4] A.F. Nisbet, Effectiveness of soil-based countermeasures six months and one year after contamination of five diverse soil types with caesium-134 and strontium-90. Chilton, NRPB-M546 (1995).

Aquatic Transfer Models to predict Wash-Off from Watersheds and the Migration of Radionuclides in Lakes for Implementation in RODOS

A.G. Popov

Emergency Centre of SPA TYPHOON, Leninstr. 82, Obninsk, Kaluga Region, Russia

R. Heling

KEMA Nederland B.V., P.O. Box 9035, 6800 ET Arnhem, The Netherlands

Abstract. The Chernobyl accident clearly demonstrated the necessity of the development of an integrated and comprehensive real-time on-line decision support system for nuclear emergencies. Aquatic exposure pathways are considered to be a significant part of a decision support system. In the decision support system RODOS a set of aquatic models are implemented to assess the dose due to the transfer of radionuclides throughout the aquatic environment. The runoff model RETRACE, developed by SPA Typhoon, Russia, models the land-to-water transfer, while the lake model LAKECO, developed by KEMA, the Netherlands, predicts the fate and behaviour of radionuclides in a lake ecosystem. RETRACE contains a hydrological and radiological submodel, while LAKECO has enhanced submodels for the water sediment interaction, and the transfer of radionuclides throughout the aquatic foodchain. To improve and prove the reliability of the models, several validation tests have been performed within the framework of the IAEA coordinated project VAMP, and within the BIOMOVS II program. Additionally the models were successfully tested by means of validation scenarios for the Rhine and the Ilya River. New empirically and physically based submodels improved the flexibility and the predictive power of the models. Within RODOS both models are integrated and coupled to dose and countermeasure models to assess in accidental circumstances the short and long term radiological consequences due to the aquatic exposure routes.

The Runoff model RETRACE

The RETRACE model is designed to operate as a part of a decision support system. It predicts the runoff in the early phase after a nuclear accident. The model uses a standard set of operative meteorological observations and cartographic data (relief, soil, vegetation and so on) stored in a database, available at this stage of an accident. RETRACE predicts the radionuclide transfer from 1 day to 2-3 months after the radioactive fallout. Two sets of submodels (RETRACE-1 and RETRACE-2) are aimed to model the migration of radionuclides from land to water both for relatively small (less than 1000 km^2) and large areas. The hydrological submodels consider the major hydrological processes, which govern radionuclide transport through a watershed: snowmelt, canopy interception, evapotranspiration, overland flow, interflow, and fast groundflow.

The physically based RETRACE-1 model is based on the well-known 2D-kinematic wave approach. The main principles of this model have been tested with the BIOMOVS II Scenario W [1].

The RETRACE-2 model was based on Yu.B.Vinogradov's approach [2] which was proved to be acceptable to consider hydrological processes on large territories (watershed scale). The hydrological submodel of RETRACE-2 has features of both the lumped-parameters and of the distributed-parameters models. The specially developed scenario on Ilya River data (30-km Chernobyl zone) was prepared to validate the RETRACE-2 model.

The radionuclide transport submodels are in balance with the main approaches of the hydrological submodels. All physio-chemical forms of radionuclides in soil-water systems are taken into account in both submodels. The well-known scheme [3] of radionuclide forms has been implemented. The balance of adsorbed and dissolved exchangeable radionuclide is described by using the distribution coefficient approach. The transport of radionuclides in the vertical direction is described by a relatively simple advective-diffusion model, enable to predict the total storage of a radionuclide in the top layer of the soil involved in the washing process.

Figure 1 demonstrates the comparison between measured and simulated radionuclide concentrations in the Ilya River system. The inset of figure 1 gives the location of the Ilya River.

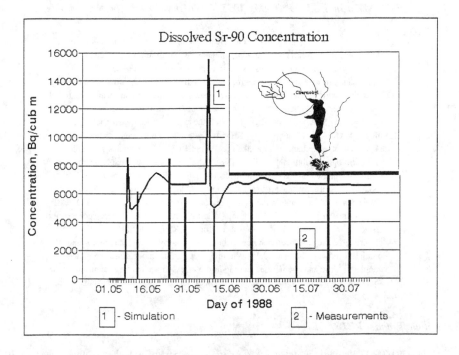

Figure 1.   The predicted ⁹⁰Sr levels compared with the measurements. Result RETRACE-2 model for Ilya River system. Inset: location of the Ilya River.

The lake ecosystem model LAKECO

LAKECO is based on the box model approach, in which complete mixing in boxes is assumed, and in which the radionuclide concentration after an accidental release can be predicted by a model based on a set of linear equations of the first order. The system of linear differential equations obtained by mass balances on all subcompartments is solved numerically. Figure 2 shows a general overview of the LAKECO model.

In the sediment layer two boxes, in which homogeneous concentrations are assumed, can be distinguished to describe the downward and upward transport of radionuclides. In the sediment layer both transport of adsorbed and dissolved radionuclides is modelled. The processes which are taken into account are: particle scavenging/sedimentation, molecular diffusion, enhanced migration of radionuclides in solution due to physical and biological mixing processes, particle reworking also due to physical and biological processes, burial, i.e. the downward transfer of radionuclides in the bottom sediment seabed as a result of the sedimentation process. Transports are the inflow of contaminated river water from the catchment and the outflow at the outlet of the lake [4].

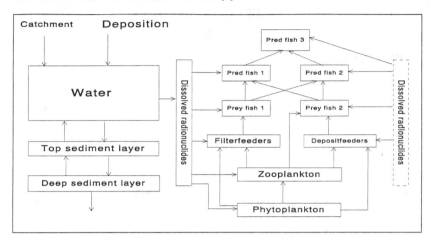

Figure 2. General outline of the lake ecosystem model LAKECO

LAKECO contains a dynamical biological uptake model to be applied on lake ecosystems with various environmental properties. This dynamic approach is necessary, since the concentration factor approach tends to overestimate the concentration in the first period after the initial contamination. The more complex and accurate dynamical approach takes the position in the foodchain into account i.e. the transfer of radionuclides from organism to organism due to the prey-predator relationship.

Tests with LAKECO on several lakes in Europe with a wide variation of environmental circumstances were a challenge to identify the predictive power of the model approach. Not only to obtain insight in the applicability of this model, but also to get more insight in all processes involved regarding ^{137}Cs as a tracer.

Finally validation exercises and a sensitivity analysis resulted in model improvements and modifications, these consisted out of empirically and physically based submodels derived from literature and moreover from studies performed within the framework of other EC-projects. The predictive power increased after the implementation of these submodels. On the basis of relatively easily measurable parameters like the chemical composition of the lake water, and the lake morphology, the model predicts the levels of radiocaesium in the lake ecosystem with a very high reliability. Other parameters, like the food extraction efficiency, were fixed on the basis of values of the VAMP validation [5], which gave a good agreement between the predictions and the measurements.

A good opportunity to test the increase of the predictive power of the model was the performance of a blind validation test by means of the Cooling Pond Scenario supplied by BIOMOVS II [6]. The improvement in the model predictions is presented in figure 3. The dashed line represents the results on the basis of the LAKECO model without the new submodels, the other line represents the results of the enhanced model. This indicates, that LAKECO-B has an accuracy and flexibility which makes the application on various types of lake ecosystems possible without special expert judgements or extended sets of data on biological and physical parameters.

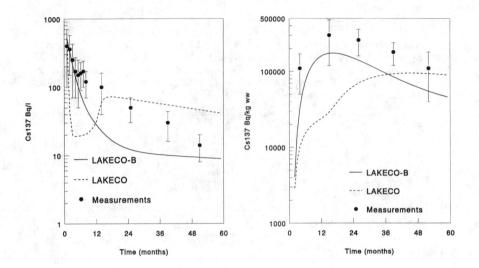

Figure 3.   Cooling Pond Scenario: model predictions with implemented submodels (LAKECO-B; line) versus results without submodels (LAKECO: dashed line). ^{137}Cs concentration in the water column of the Chernobyl Cooling Pond (left), ^{137}Cs in the top predator (right).

Conclusions

The results of the tests with the RETRACE model demonstrate the applicability of the RETRACE model to simulate radionuclide washout from watersheds. The approaches realized makes RETRACE a useful tool in decision support systems like RODOS.

The implementation of powerful submodels in LAKECO, after extensive validation tests, increased the predictive power of the lake ecosystem model. No longer expert knowledge is required to apply the model, since the input of the model has been changed from model specific parameters to environmental specific parameters. Blind tests on the Cooling Pond demonstrated the increase of the model accuracy by implementation of these empirically and physically based submodels. Therefore LAKECO can be regarded as a flexible tool to predict the fate and behaviour of radionuclides in lakes with a minimum of input parameters.

References

[1] BIOMOVS II. Wash-Off of ^{90}Sr and ^{137}Cs from Two Experimental Plots Established in the Vicinity of the Chernobyl Reactor. - Description of Scenario W , October 1993.
[2] Vinogradov Yu. B. Mathematical modelling of runoff formation. Critical review. Leningrad, Hydrometeorological Publishing House, 1988.
[3] Konoplev A.V., Bulgakov A.A., Popov V.E and Bobovnikova Ts. I. Behaviour of Long-lived Chernobyl Radionuclides in a Soil-Water System. - ANALYST, June, 1992, Vol.117, pp.1041-1047.
[4] Heling R. (1994) LAKECO, the ecological consequences of an accidental release of radionuclides on a lake ecosystem and its integration into the real-time-on-line decision support system for the off-site Emergency Management following a Nuclear Accident (RODOS). KEMA report 40352-NUC 93-5852.
[5] VAMP Aquatic Working Group. I. Modelling of Radiocaesium in Lakes. IAEA In press.
[6] BIOMOVS II. Assessment of the Consequences of the Radioactive Contamination of the Chernobyl NPP Cooling Pond. Description of the Scenario CP, stage 2. April 1994

European Commission

EUR 16544 — The radiological consequences of the Chernobyl accident

A. Karaoglou, G. Desmet, G. N. Kelly and H. G. Menzel

Luxembourg: Office for Official Publications of the European Communities

1996 — XXIV, 1192 pp. — 16 x 24 cm

ISBN 92-827-5248-8

Price (excluding VAT) in Luxembourg: ECU 92

The main aims of the conference are to present the major achievements of the 16 projects of the joint EC/CIS collaborative research programme (1992-95) on the consequences of the Chernobyl accident and to promote an objective evaluation of them by the international scientific community. The conference is taking place close to the 10th anniversary of the accident and we hope it will contribute to more objective communication of the health and environmental consequences of the Chernobyl accident and how these may be mitigated in future.

Five main objectives were assigned to the EC/CIS scientific collaborative programme:

(i)   improvement of the knowledge of the relationship between doses and radiation-induced health effects;

(ii)   updating of the arrangements for off-site emergency management response (short and medium term) in the event of a future nuclear accident;

(iii)   assisting the relevant CIS ministries to alleviate the consequences of the Chernobyl accident, in particular in the field of restoration of contaminated territories;

(iv)   elaboration of a scientific basis to define the content of Community assistance programmes;

(v)   updating of the local technical infrastructure and implementation of a large programme of exchange of scientists between both Communities.

The topics addressed during the conference mainly reflect the content of the joint collaborative programme:

1.   Environmental transfer and decontamination.

2.   Risk assessment and management.

3.   Health-related issues including dosimetry.